Ullmann's Polymers and Plastics

Excellence since 1914
Built from generations of expertise, for generations to come.

ULLMANN'S ENCYCLOPEDIA OF INDUSTRIAL CHEMISTRY

If you want to learn more about novel technologies in biochemistry or nanotechnology or discover unexpected new aspects of seemingly totally familiar processes in industrial chemistry – ULLMANN'S Encyclopedia of Industrial Chemistry is your first choice.

For over 100 years now, this reference provides top-notch information on the most diverse fields of industrial chemistry and chemical engineering.

When it comes to definite works on industrial chemistry, it has always been ULLMANN'S.

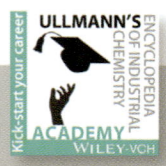

Welcome to the ULLMANN'S ACADEMY
Kick-start your career!

Key topics in industrial chemistry explained by the ULLMANN'S Encyclopedia experts – for teaching and learning, or for simply refreshing your knowledge.

What's new?

The Smart Article introduces new and enhanced article tools for chemistry content. It is now available within ULLMANN'S.

For further information on the features and functions available, go to wileyonlinelibrary.com/thesmartarticle

Only interested in a specific topic?

ULLMANN'S Energy
3 Volume Set • ISBN: 978-3-527-33370-7

ULLMANN'S Fine Chemicals
3 Volume Set • ISBN: 978-3-527-33477-3

ULLMANN'S Fibers
2 Volumes • ISBN: 978-3-527-31772-1

ULLMANN'S Modeling and Simulation
ISBN: 978-3-527-31605-2

ULLMANN'S Renewable Resources
ISBN: 978-3-527-33369-1

ULLMANN'S Agrochemicals
2 Volume Set • ISBN: 978-3-527-31604-5

ULLMANN'S Biotechnology and Biochemical Engineering
2 Volume Set • ISBN: 978-3-527-31603-8

ULLMANN'S Reaction Engineering
2 Volume Set • ISBN: 978-3-527-33371-4

ULLMANN'S Industrial Toxicology
2 Volume Set • ISBN: 978-3-527-31247-4

ULLMANN'S Chemical Engineering and Plant Design
2 Volume Set • ISBN: 978-3-527-31111-8

The information you need in the format you want.

DVD
- Released once a year.
- Fully networkable for up to 200 users.
- Time-limited access for up to 14 months (expires March 2016).
- 2015 Edition
  ISBN: 978-3-527-33754-5

Online
- Over 1,150 articles available online.
- Over 3,000 authors from over 30 countries have contributed.
- Offers flexible access 24/7 from your library, home, or on the road.
- Updated 4 times per year.
  ISBN: 978-3-527-30673-2

Print
- Available as a comprehensive 40 Volume-Set.
- The 7th Edition published in Aug 2011.
- ISBN: 978-3-527-32943-4

wileyonlinelibrary.com/ref/ullmanns

Visit our website to

Find sample chapters

Learn more about the history, the Editor-in-Chief and the Editorial Advisory Board

Discover the ULLMANN'S ACADEMY and more...

Chemistry that delivers... Continuous product innovation
Create Innovate Inspire

WILEY-VCH   **WILEY**

Ullmann's Polymers and Plastics

Products and Processes

Volume 1

WILEY-VCH
Verlag GmbH & Co. KGaA

Editor in Chief:

Dr. Barbara Elvers, Hamburg, Germany

All books published by **Wiley-VCH** are carefully produced. Nevertheless, authors, editors, and publisher do not warrant the information contained in these books, including this book, to be free of errors. Readers are advised to keep in mind that statements, data, illustrations, procedural details or other items may inadvertently be inaccurate.

Library of Congress Card No.:
applied for

British Library Cataloguing-in-Publication Data
A catalogue record for this book is available from the British Library.

Bibliographic information published by the Deutsche Nationalbibliothek
The Deutsche Nationalbibliothek lists this publication in the Deutsche Nationalbibliografie; detailed bibliographic data are available on the Internet at <http://dnb.d-nb.de>.

© 2016 Wiley-VCH Verlag GmbH & Co. KGaA, Boschstr. 12, 69469 Weinheim, Germany

All rights reserved (including those of translation into other languages). No part of this book may be reproduced in any form – by photoprinting, microfilm, or any other means – nor transmitted or translated into a machine language without written permission from the publishers. Registered names, trademarks, etc. used in this book, even when not specifically marked as such, are not to be considered unprotected by law.

Print ISBN: 978-3-527-33823-8
ePDF ISBN: 978-3-527-68595-0
ePub ISBN: 978-3-527-68596-7
Mobi ISBN: 978-3-527-68597-4

Cover Design Grafik-Design Schulz, Fußgönheim, Germany
Typesetting Thomson Digital, Noida, India
Printing and Binding Markono Print Media Pte Ltd, Singapore

Printed on acid-free paper

Preface

This handbook features selected articles from the 7th edition of *ULLMANN'S Encyclopedia of Industrial Chemistry*, including newly written articles that have not been published in a printed edition before. True to the tradition of the ULLMANN'S Encyclopedia, polymers and plastics are addressed from an industrial perspective, including production figures, quality standards and patent protection issues where appropriate. Safety and environmental aspects which are a key concern for modern process industries are likewise considered.

More content on related topics can be found in the complete edition of the ULLMANN'S Encyclopedia.

About ULLMANN'S

ULLMANN'S Encyclopedia is the world's largest reference in applied chemistry, industrial chemistry, and chemical engineering. In its current edition, the Encyclopedia contains more than 30,000 pages, 15,000 tables, 25,000 figures, and innumerable literature sources and cross-references, offering a wealth of comprehensive and well-structured information on all facets of industrial chemistry.

1,100 major articles cover the following main areas:

- Agrochemicals
- Analytical Techniques
- Biochemistry and Biotechnology
- Chemical Reactions
- Dyes and Pigments
- Energy
- Environmental Protection and Industrial Safety
- Fat, Oil, Food and Feed, Cosmetics
- Inorganic Chemicals
- Materials
- Metals and Alloys
- Organic Chemicals
- Pharmaceuticals
- Polymers and Plastics
- Processes and Process Engineering
- Renewable Resources
- Special Topics

First published in 1914 by Professor Fritz Ullmann in Berlin, the *Enzyklopädie der Technischen Chemie* (as the German title read) quickly became the standard reference work in industrial chemistry. Generations of chemists have since relied on ULLMANN'S as their prime reference source. Three further German editions followed in 1928–1932, 1951–1970, and in 1972–1984. From 1985 to 1996, the 5th edition of ULLMANN'S Encyclopedia of Industrial Chemistry was the first edition to be published in English rather than German language. So far, two more complete English editions have been published in print; the 6th edition of 40 volumes in 2002, and the 7th edition in 2011, again comprising 40 volumes. In addition, a number of smaller topic-oriented editions have been published.

Since 1997, *ULLMANN'S Encyclopedia of Industrial Chemistry* has also been available in electronic format, first in a CD-ROM edition and, since 2000, in an enhanced online edition. Both electronic editions feature powerful search and navigation functions as well as regular content updates.

Contents

Volume 1 ....
Symbols and Units IX
Conversion Factors XI
Abbreviations XIII
Country Codes XVIII
Periodic Table of Elements XIX

Part 1: Fundamentals 1
Plastics, General Survey, 1. Definition,
　Molecular Structure and Properties 3
Plastics, General Survey, 2. Production of
　Polymers and Plastics 149
Plastics, General Survey, 3. Supermolecular
　Structures 187
Plastics, General Survey, 4. Polymer
　Composites 205
Plastics, General Survey, 5. Plastics and
　Sustainability 223
Plastics, Analysis 231
Polymerization Processes, 1. Fundamentals 265
Polymerization Processes, 2. Modeling of
　Processes and Reactors 315
Plastics, Processing, 1. Processing of
　Thermoplastics 367
Plastics, Processing, 2. Processing of
　Thermosets 407
Plastics Processing, 3. Machining, Bonding,
　Surface Treatment 439
Plastics, Properties and Testing 471
Plastics, Additives 527
Plasticizers 581

Volume 2 ....
Part 2: Organic Polymers 601
Fluoropolymers, Organic 603
Polyacrylamides and Poly(Acrylic Acids) ... 659
Polyacrylates 675
Polyamides 697
Polyaspartates and Polysuccinimide 733
Polybutenes 747
Polycarbonates 763
Polyester Resins, Unsaturated 781
Polyesters 791
Polyethylene 817
Polyimides 859
Polymethacrylates 885

Polyoxyalkylenes 899
Polyoxymethylenes 911
Poly(Phenylene Oxides) 927
Polypropylene 937

Volume 3 ....
Polystyrene and Styrene Copolymers 981
Polyureas 1029
Polyurethanes 1051
Poly(Vinyl Chloride) 1111
Polyvinyl Compounds, Others 1141
Poly(Vinyl Esters) 1165
Poly(Vinyl Ethers) 1175
Poly(Vinylidene Chloride) 1181
Polymer Blends 1197
Polymers, Biodegradable 1231
Polymers, Electrically Conducting 1261
Polymers, High-Temperature 1281
Reinforced Plastics 1325
Specialty Plastics 1343
Thermoplastic Elastomers 1365

Volume 4 ....
Part 3: Films, Fibers, Foams 1405
Films 1407
Fibers, 4. Polyamide Fibers 1435
Fibers, 5. Polyester Fibers 1453
Fibers, 6. Polyurethane Fibers 1487
Fibers, 7. Polyolefin Fibers 1495
Fibers, 8. Polyacrylonitrile Fibers 1513
Fibers, 9. Polyvinyl Fibers 1529
Fibers, 10. Polytetrafluoroethylene Fibers 1539
High-Performance Fibers 1541
Foamed Plastics 1563

Part 4: Resins 1595
Alkyd Resins 1597
Amino Resins 1615
Epoxy Resins 1643
Phenolic Resins 1733
Resins, Synthetic 1751

Part 5: Inorganic Polymers 1775
Inorganic Polymers 1777

Author Index 1817
Subject Index 1823

Symbols and Units

Symbols and units agree with SI standards (for conversion factors see page XI). The following list gives the most important symbols used in the encyclopedia. Articles with many specific units and symbols have a similar list as front matter.

Symbol	Unit	Physical Quantity
a_B		activity of substance B
A_r		relative atomic mass (atomic weight)
A	m^2	area
c_B	mol/m^3, mol/L (M)	concentration of substance B
C	C/V	electric capacity
c_p, c_v	$J\,kg^{-1}\,K^{-1}$	specific heat capacity
d	cm, m	diameter
d		relative density (ϱ/ϱ_{water})
D	m^2/s	diffusion coefficient
D	Gy (=J/kg)	absorbed dose
e	C	elementary charge
E	J	energy
E	V/m	electric field strength
E	V	electromotive force
E_A	J	activation energy
f		activity coefficient
F	C/mol	Faraday constant
F	N	force
g	m/s^2	acceleration due to gravity
G	J	Gibbs free energy
h	m	height
$\hbar$	$W \cdot s^2$	Planck constant
H	J	enthalpy
I	A	electric current
I	cd	luminous intensity
k	(variable)	rate constant of a chemical reaction
k	J/K	Boltzmann constant
K	(variable)	equilibrium constant
l	m	length
m	g, kg, t	mass
M_r		relative molecular mass (molecular weight)
n_D^{20}		refractive index (sodium D-line, 20 °C)
n	mol	amount of substance
N_A	mol^{-1}	Avogadro constant ($6.023 \times 10^{23}\,mol^{-1}$)
P	Pa, bar[*]	pressure
Q	J	quantity of heat
r	m	radius
R	$J\,K^{-1}\,mol^{-1}$	gas constant
R	Ω	electric resistance
S	J/K	entropy
t	s, min, h, d, month, a	time
t	°C	temperature
T	K	absolute temperature
u	m/s	velocity
U	V	electric potential

Symbols and Units (Continued from p. IX)

Symbol	Unit	Physical Quantity
U	J	internal energy
V	m³, L, mL, μL	volume
w		mass fraction
W	J	work
x_B		mole fraction of substance B
Z		proton number, atomic number
α		cubic expansion coefficient
α	Wm⁻²K⁻¹	heat-transfer coefficient (heat-transfer number)
α		degree of dissociation of electrolyte
$[\alpha]$	10⁻² deg cm²g⁻¹	specific rotation
η	Pa·s	dynamic viscosity
θ	°C	temperature
$\varkappa$		c_p/c_v
λ	Wm⁻¹K⁻¹	thermal conductivity
λ	nm, m	wavelength
μ		chemical potential
ν	Hz, s⁻¹	frequency
ν	m²/s	kinematic viscosity (η/ϱ)
π	Pa	osmotic pressure
ϱ	g/cm³	density
σ	N/m	surface tension
τ	Pa (N/m²)	shear stress
φ		volume fraction
χ	Pa⁻¹ (m²/N)	compressibility

*The official unit of pressure is the pascal (Pa).

Conversion Factors

SI unit	Non-SI unit	From SI to non-SI multiply by
Mass		
kg	pound (avoirdupois)	2.205
kg	ton (long)	9.842×10^{-4}
kg	ton (short)	1.102×10^{-3}
Volume		
m^3	cubic inch	6.102×10^4
m^3	cubic foot	35.315
m^3	gallon (U.S., liquid)	2.642×10^2
m^3	gallon (Imperial)	2.200×10^2
Temperature		
°C	°F	°C $\times 1.8 + 32$
Force		
N	dyne	1.0×10^5
Energy, Work		
J	Btu (int.)	9.480×10^{-4}
J	cal (int.)	2.389×10^{-1}
J	eV	6.242×10^{18}
J	erg	1.0×10^7
J	kW·h	2.778×10^{-7}
J	kp·m	1.020×10^{-1}
Pressure		
MPa	at	10.20
MPa	atm	9.869
MPa	bar	10
kPa	mbar	10
kPa	mm Hg	7.502
kPa	psi	0.145
kPa	torr	7.502

Powers of Ten

E (exa)	10^{18}	d (deci)	10^{-1}
P (peta)	10^{15}	c (centi)	10^{-2}
T (tera)	10^{12}	m (milli)	10^{-3}
G (giga)	10^9	μ (micro)	10^{-6}
M (mega)	10^6	n (nano)	10^{-9}
k (kilo)	10^3	p (pico)	10^{-12}
h (hecto)	10^2	f (femto)	10^{-15}
da (deca)	10	a (atto)	10^{-18}

Abbreviations

The following is a list of the abbreviations used in the text. Common terms, the names of publications and institutions, and legal agreements are included along with their full identities. Other abbreviations will be defined wherever they first occur in an article. For further abbreviations, see page IX, Symbols and Units; page XVII, Frequently Cited Companies (Abbreviations), and page XVIII, Country Codes in patent references. The names of periodical publications are abbreviated exactly as done by Chemical Abstracts Service.

abs.	absolute	BGA	Bundesgesundheitsamt (Federal Republic of Germany)
a.c.	alternating current		
ACGIH	American Conference of Governmental Industrial Hygienists	BGB1.	Bundesgesetzblatt (Federal Republic of Germany)
ACS	American Chemical Society	BIOS	British Intelligence Objectives Subcommittee Report (see also FIAT)
ADI	acceptable daily intake		
ADN	accord européen relatif au transport international des marchandises dangereuses par voie de navigation interieure (European agreement concerning the international transportation of dangerous goods by inland waterways)	BOD	biological oxygen demand
		bp	boiling point
		B.P.	British Pharmacopeia
		BS	British Standard
		ca.	circa
		calcd.	calculated
ADNR	ADN par le Rhin (regulation concerning the transportation of dangerous goods on the Rhine and all national waterways of the countries concerned)	CAS	Chemical Abstracts Service
		cat.	catalyst, catalyzed
		CEN	Comité Européen de Normalisation
		cf.	compare
ADP	adenosine 5'-diphosphate	CFR	Code of Federal Regulations (United States)
ADR	accord européen relatif au transport international des marchandises dangereuses par route (European agreement concerning the international transportation of dangerous goods by road)	cfu	colony forming units
		Chap.	chapter
		ChemG	Chemikaliengesetz (Federal Republic of Germany)
AEC	Atomic Energy Commission (United States)	C.I.	Colour Index
		CIOS	Combined Intelligence Objectives Subcommitee Report (see also FIAT)
a.i.	active ingredient		
AIChE	American Institute of Chemical Engineers	CLP	Classification, Labelling and Packaging
		CNS	central nervous system
AIME	American Institute of Mining, Metallurgical, and Petroleum Engineers	Co.	Company
		COD	chemical oxygen demand
ANSI	American National Standards Institute	conc.	concentrated
AMP	adenosine 5'-monophosphate	const.	constant
APhA	American Pharmaceutical Association	Corp.	Corporation
API	American Petroleum Institute	crit.	critical
ASTM	American Society for Testing and Materials	CSA	Chemical Safety Assessment according to REACH
ATP	adenosine 5'-triphosphate	CSR	Chemical Safety Report according to REACH
BAM	Bundesanstalt für Materialprüfung (Federal Republic of Germany)	CTFA	The Cosmetic, Toiletry and Fragrance Association (United States)
BAT	Biologischer Arbeitsstofftoleranzwert (biological tolerance value for a working material, established by MAK Commission, see MAK)	DAB	Deutsches Arzneibuch, Deutscher Apotheker-Verlag, Stuttgart
		d.c.	direct current
		decomp.	decompose, decomposition
Beilstein	Beilstein's Handbook of Organic Chemistry, Springer, Berlin – Heidelberg – New York	DFG	Deutsche Forschungsgemeinschaft (German Science Foundation)
BET	Brunauer – Emmett – Teller	dil.	dilute, diluted

DIN	Deutsche Industrienorm (Federal Republic of Germany)		(regulation in the Federal Republic of Germany concerning the transportation of dangerous goods by rail)
DMF	dimethylformamide		
DNA	deoxyribonucleic acid	GGVS	Verordnung in der Bundesrepublik Deutschland über die Beförderung gefährlicher Güter auf der Straße (regulation in the Federal Republic of Germany concerning the transportation of dangerous goods by road)
DOE	Department of Energy (United States)		
DOT	Department of Transportation – Materials Transportation Bureau (United States)		
DTA	differential thermal analysis		
EC	effective concentration	GGVSee	Verordnung in der Bundesrepublik Deutschland über die Beförderung gefährlicher Güter mit Seeschiffen (regulation in the Federal Republic of Germany concerning the transportation of dangerous goods by sea-going vessels)
EC	European Community		
ed.	editor, edition, edited		
e.g.	for example		
emf	electromotive force		
EmS	Emergency Schedule		
EN	European Standard (European Community)		
EPA	Environmental Protection Agency (United States)	GHS	Globally Harmonised System of Chemicals (internationally agreed-upon system, created by the UN, designed to replace the various classification and labeling standards used in different countries by using consistent criteria for classification and labeling on a global level)
EPR	electron paramagnetic resonance		
Eq.	equation		
ESCA	electron spectroscopy for chemical analysis		
esp.	especially	GLC	gas-liquid chromatography
ESR	electron spin resonance	Gmelin	Gmelin's Handbook of Inorganic Chemistry, 8th ed., Springer, Berlin – Heidelberg – New York
Et	ethyl substituent ($-C_2H_5$)		
et al.	and others		
etc.	et cetera	GRAS	generally recognized as safe
EVO	Eisenbahnverkehrsordnung (Federal Republic of Germany)	Hal	halogen substituent ($-F$, $-Cl$, $-Br$, $-I$)
		Houben-Weyl	Methoden der organischen Chemie, 4th ed., Georg Thieme Verlag, Stuttgart
exp (...)	$e^{(\ldots)}$, mathematical exponent		
FAO	Food and Agriculture Organization (United Nations)		
		HPLC	high performance liquid chromatography
FDA	Food and Drug Administration (United States)		
		H statement	hazard statement in GHS
FD&C	Food, Drug and Cosmetic Act (United States)	IAEA	International Atomic Energy Agency
		IARC	International Agency for Research on Cancer, Lyon, France
FHSA	Federal Hazardous Substances Act (United States)		
		IATA-DGR	International Air Transport Association, Dangerous Goods Regulations
FIAT	Field Information Agency, Technical (United States reports on the chemical industry in Germany, 1945)		
		ICAO	International Civil Aviation Organization
Fig.	figure		
fp	freezing point	i.e.	that is
Friedländer	P. Friedländer, Fortschritte der Teerfarbenfabrikation und verwandter Industriezweige Vol. 1–25, Springer, Berlin 1888–1942	i.m.	intramuscular
		IMDG	International Maritime Dangerous Goods Code
		IMO	Inter-Governmental Maritime Consultive Organization (in the past: IMCO)
FT	Fourier transform		
(g)	gas, gaseous	Inst.	Institute
GC	gas chromatography	i.p.	intraperitoneal
GefStoffV	Gefahrstoffverordnung (regulations in the Federal Republic of Germany concerning hazardous substances)	IR	infrared
		ISO	International Organization for Standardization
		IUPAC	International Union of Pure and Applied Chemistry
GGVE	Verordnung in der Bundesrepublik Deutschland über die Beförderung gefährlicher Güter mit der Eisenbahn		
		i.v.	intravenous

Kirk-Othmer	Encyclopedia of Chemical Technology, 3rd ed., 1991–1998, 5th ed., 2004–2007, John Wiley & Sons, Hoboken	no.	number
		NOEL	no observed effect level
		NRC	Nuclear Regulatory Commission (United States)
(l)	liquid		
Landolt-Börnstein	Zahlenwerte u. Funktionen aus Physik, Chemie, Astronomie, Geophysik u. Technik, Springer, Heidelberg 1950–1980; Zahlenwerte und Funktionen aus Naturwissenschaften und Technik, Neue Serie, Springer, Heidelberg, since 1961	NRDC	National Research Development Corporation (United States)
		NSC	National Service Center (United States)
		NSF	National Science Foundation (United States)
		NTSB	National Transportation Safety Board (United States)
LC_{50}	lethal concentration for 50 % of the test animals	OECD	Organization for Economic Cooperation and Development
LCLo	lowest published lethal concentration	OSHA	Occupational Safety and Health Administration (United States)
LD_{50}	lethal dose for 50 % of the test animals		
LDLo	lowest published lethal dose	p., pp.	page, pages
ln	logarithm (base e)	Patty	G.D. Clayton, F.E. Clayton (eds.): Patty's Industrial Hygiene and Toxicology, 3rd ed., Wiley Interscience, New York
LNG	liquefied natural gas		
log	logarithm (base 10)		
LPG	liquefied petroleum gas		
M	mol/L	PB report	Publication Board Report (U.S. Department of Commerce, Scientific and Industrial Reports)
M	metal (in chemical formulas)		
MAK	Maximale Arbeitsplatzkonzentration (maximum concentration at the workplace in the Federal Republic of Germany); cf. Deutsche Forschungsgemeinschaft (ed.): Maximale Arbeitsplatzkonzentrationen (MAK) und Biologische Arbeitsstofftoleranzwerte (BAT), WILEY-VCH Verlag, Weinheim (published annually)		
		PEL	permitted exposure limit
		Ph	phenyl substituent (—C_6H_5)
		Ph. Eur.	European Pharmacopoeia, Council of Europe, Strasbourg
		phr	part per hundred rubber (resin)
		PNS	peripheral nervous system
		ppm	parts per million
		P statement	precautionary statement in GHS
max.	maximum	q.v.	which see (quod vide)
MCA	Manufacturing Chemists Association (United States)	REACH	Registration, Evaluation, Authorisation and Restriction of Chemicals (EU regulation addressing the production and use of chemical substances, and their potential impacts on both human health and the environment)
Me	methyl substituent (—CH_3)		
Methodicum Chimicum	Methodicum Chimicum, Georg Thieme Verlag, Stuttgart		
MFAG	Medical First Aid Guide for Use in Accidents Involving Dangerous Goods		
		ref.	refer, reference
MIK	maximale Immissionskonzentration (maximum immission concentration)	resp.	respectively
		R_f	retention factor (TLC)
min.	minimum	R.H.	relative humidity
mp	melting point	RID	réglement international concernant le transport des marchandises dangereuses par chemin de fer (international convention concerning the transportation of dangerous goods by rail)
MS	mass spectrum, mass spectrometry		
NAS	National Academy of Sciences (United States)		
NASA	National Aeronautics and Space Administration (United States)		
		RNA	ribonucleic acid
NBS	National Bureau of Standards (United States)	R phrase (R-Satz)	risk phrase according to ChemG and GefStoffV (Federal Republic of Germany)
NCTC	National Collection of Type Cultures (United States)		
		rpm	revolutions per minute
NIH	National Institutes of Health (United States)	RTECS	Registry of Toxic Effects of Chemical Substances, edited by the National Institute of Occupational Safety and Health (United States)
NIOSH	National Institute for Occupational Safety and Health (United States)		
NMR	nuclear magnetic resonance	(s)	solid

SAE	Society of Automotive Engineers (United States)		der Technischen Chemie, 4th ed., Verlag Chemie, Weinheim 1972–1984; 3rd ed., Urban und Schwarzenberg, München 1951–1970
SAICM	Strategic Approach on International Chemicals Management (international framework to foster the sound management of chemicals)	USAEC	United States Atomic Energy Commission
s.c.	subcutaneous	USAN	United States Adopted Names
SI	International System of Units	USD	United States Dispensatory
SIMS	secondary ion mass spectrometry	USDA	United States Department of Agriculture
S phrase (S-Satz)	safety phrase according to ChemG and GefStoffV (Federal Republic of Germany)	U.S.P. UV	United States Pharmacopeia ultraviolet
STEL	Short Term Exposure Limit (see TLV)	UVV	Unfallverhütungsvorschriften der Berufsgenossenschaft (workplace safety regulations in the Federal Republic of Germany)
STP	standard temperature and pressure (0°C, 101.325 kPa)		
T_g	glass transition temperature	VbF	Verordnung in der Bundesrepublik Deutschland über die Errichtung und den Betrieb von Anlagen zur Lagerung, Abfüllung und Beförderung brennbarer Flüssigkeiten (regulation in the Federal Republic of Germany concerning the construction and operation of plants for storage, filling, and transportation of flammable liquids; classification according to the flash point of liquids, in accordance with the classification in the United States)
TA Luft	Technische Anleitung zur Reinhaltung der Luft (clean air regulation in Federal Republic of Germany)		
TA Lärm	Technische Anleitung zum Schutz gegen Lärm (low noise regulation in Federal Republic of Germany)		
TDLo	lowest published toxic dose		
THF	tetrahydrofuran		
TLC	thin layer chromatography		
TLV	Threshold Limit Value (TWA and STEL); published annually by the American Conference of Governmental Industrial Hygienists (ACGIH), Cincinnati, Ohio		
		VDE	Verband Deutscher Elektroingenieure (Federal Republic of Germany)
		VDI	Verein Deutscher Ingenieure (Federal Republic of Germany)
TOD	total oxygen demand		
TRK	Technische Richtkonzentration (lowest technically feasible level)	vol	volume
		vol.	volume (of a series of books)
TSCA	Toxic Substances Control Act (United States)	vs.	versus
		WGK	Wassergefährdungsklasse (water hazard class)
TÜV	Technischer Überwachungsverein (Technical Control Board of the Federal Republic of Germany)	WHO	World Health Organization (United Nations)
TWA	Time Weighted Average	Winnacker-Küchler	Chemische Technologie, 4th ed., Carl Hanser Verlag, München, 1982-1986; Winnacker-Küchler, Chemische Technik: Prozesse und Produkte, Wiley-VCH, Weinheim, 2003–2006
UBA	Umweltbundesamt (Federal Environmental Agency)		
Ullmann	Ullmann's Encyclopedia of Industrial Chemistry, 6th ed., Wiley-VCH, Weinheim 2002; Ullmann's Encyclopedia of Industrial Chemistry, 5th ed., VCH Verlagsgesellschaft, Weinheim 1985–1996; Ullmanns Encyklopädie		
		wt	weight
		$	U.S. dollar, unless otherwise stated

Frequently Cited Companies (Abbreviations)

Air Products	Air Products and Chemicals	IFP	Institut Français du Pétrole
Akzo	Algemene Koninklijke Zout Organon	INCO	International Nickel Company
		3M	Minnesota Mining and Manufacturing Company
Alcoa	Aluminum Company of America	Mitsubishi Chemical	Mitsubishi Chemical Industries
Allied	Allied Corporation		
Amer. Cyanamid	American Cyanamid Company	Monsanto	Monsanto Company
		Nippon Shokubai	Nippon Shokubai Kagaku Kogyo
BASF	BASF Aktiengesellschaft		
Bayer	Bayer AG	PCUK	Pechiney Ugine Kuhlmann
BP	British Petroleum Company	PPG	Pittsburg Plate Glass Industries
Celanese	Celanese Corporation	Searle	G.D. Searle & Company
Daicel	Daicel Chemical Industries	SKF	Smith Kline & French Laboratories
Dainippon	Dainippon Ink and Chemicals Inc.	SNAM	Societá Nazionale Metandotti
Dow Chemical	The Dow Chemical Company	Sohio	Standard Oil of Ohio
		Stauffer	Stauffer Chemical Company
DSM	Dutch Staats Mijnen	Sumitomo	Sumitomo Chemical Company
Du Pont	E.I. du Pont de Nemours & Company	Toray	Toray Industries Inc.
Exxon	Exxon Corporation	UCB	Union Chimique Belge
FMC	Food Machinery & Chemical Corporation	Union Carbide	Union Carbide Corporation
GAF	General Aniline & Film Corporation	UOP	Universal Oil Products Company
W.R. Grace	W.R. Grace & Company	VEBA	Vereinigte Elektrizitäts- und Bergwerks-AG
Hoechst	Hoechst Aktiengesellschaft	Wacker	Wacker Chemie GmbH
IBM	International Business Machines Corporation		
ICI	Imperial Chemical Industries		

Country Codes

The following list contains a selection of standard country codes used in the patent references.

AT	Austria	IL	Israel
AU	Australia	IT	Italy
BE	Belgium	JP	Japan*
BG	Bulgaria	LU	Luxembourg
BR	Brazil	MA	Morocco
CA	Canada	NL	Netherlands*
CH	Switzerland	NO	Norway
CS	Czechoslovakia	NZ	New Zealand
DD	German Democratic Republic	PL	Poland
DE	Federal Republic of Germany (and Germany before 1949)*	PT	Portugal
		SE	Sweden
DK	Denmark	SU	Soviet Union
ES	Spain	US	United States of America
FI	Finland	YU	Yugoslavia
FR	France	ZA	South Africa
GB	United Kingdom	EP	European Patent Office*
GR	Greece	WO	World Intellectual Property Organization
HU	Hungary		
ID	Indonesia		

*For Europe, Federal Republic of Germany, Japan, and the Netherlands, the type of patent is specified: EP (patent), EP-A (application), DE (patent), DE-OS (Offenlegungsschrift), DE-AS (Auslegeschrift), JP (patent), JP-Kokai (Kokai tokkyo koho), NL (patent), and NL-A (application).

Periodic Table of Elements

element symbol, atomic number, and relative atomic mass (atomic weight)

- 1A "European" group designation and old IUPAC recommendation
- 1   group designation to 1986 IUPAC proposal
- IA "American" group designation, also used by the Chemical Abstracts Service until the end of 1986

[a] provisional IUPAC symbol

* radioactive element; mass of most important isotope given.

1A 1 IA	2A 2 IIA		3A 3 IIIB	4A 4 IVB	5A 5 VB	6A 6 VIB	7A 7 VIIB	8 8 VIII	8 9 VIII	8 10 VIII	1B 11 IB	2B 12 IIB	3B 13 IIIA	4B 14 IVA	5B 15 VA	6B 16 VIA	7B 17 VIA	0 18 VIIIA
1 **H** 1.0079																		2 **He** 4.0026
3 **Li** 6.941	4 **Be** 9.0122												5 **B** 10.811	6 **C** 12.011	7 **N** 14.007	8 **O** 15.999	9 **F** 18.998	10 **Ne** 20.180
11 **Na** 22.990	12 **Mg** 24.305												13 **Al** 26.982	14 **Si** 28.086	15 **P** 30.974	16 **S** 32.066	17 **Cl** 35.453	18 **Ar** 39.948
19 **K** 39.098	20 **Ca** 40.078		21 **Sc** 44.956	22 **Ti** 47.867	23 **V** 50.942	24 **Cr** 51.996	25 **Mn** 54.938	26 **Fe** 55.845	27 **Co** 58.933	28 **Ni** 58.693	29 **Cu** 63.546	30 **Zn** 65.409	31 **Ga** 69.723	32 **Ge** 72.61	33 **As** 74.922	34 **Se** 78.96	35 **Br** 79.904	36 **Kr** 83.80
37 **Rb** 85.468	38 **Sr** 87.62		39 **Y** 88.906	40 **Zr** 91.224	41 **Nb** 92.906	42 **Mo** 95.94	43 **Tc*** 98.906	44 **Ru** 101.07	45 **Rh** 102.91	46 **Pd** 106.42	47 **Ag** 107.87	48 **Cd** 112.41	49 **In** 114.82	50 **Sn** 118.71	51 **Sb** 121.76	52 **Te** 127.60	53 **I** 126.90	54 **Xe** 131.29
55 **Cs** 132.91	56 **Ba** 137.33			72 **Hf** 178.49	73 **Ta** 180.95	74 **W** 183.84	75 **Re** 186.21	76 **Os** 190.23	77 **Ir** 192.22	78 **Pt** 195.08	79 **Au** 196.97	80 **Hg** 200.59	81 **Tl** 204.38	82 **Pb** 207.2	83 **Bi** 208.98	84 **Po*** 208.98	85 **At*** 209.99	86 **Rn*** 222.02
87 **Fr*** 223.02	88 **Ra*** 226.03			104 **Rf*** 261.11	105 **Db***[a] 262.11	106 **Sg**[a]	107 **Bh**[a]	108 **Hs**[a]	109 **Mt**[a]	110 **Ds**[a]	111 **Rg**[a]	112 **Cn**[a]	113 **Uut**[a]	114 **Fl**[a]	115 **Uup**[a]	116 **Lv**[a]		118 **Uuo**[a]

57 **La** 138.91	58 **Ce** 140.12	59 **Pr** 140.91	60 **Nd** 144.24	61 **Pm*** 146.92	62 **Sm** 150.36	63 **Eu** 151.97	64 **Gd** 157.25	65 **Tb** 158.93	66 **Dy** 162.50	67 **Ho** 164.93	68 **Er** 167.26	69 **Tm** 168.93	70 **Yb** 173.04	71 **Lu** 174.97
89 **Ac*** 227.03	90 **Th*** 232.04	91 **Pa*** 231.04	92 **U*** 238.03	93 **Np*** 237.05	94 **Pu*** 244.06	95 **Am*** 243.06	96 **Cm*** 247.07	97 **Bk*** 247.07	98 **Cf*** 251.08	99 **Es*** 252.08	100 **Fm*** 257.10	101 **Md*** 258.10	102 **No*** 259.10	103 **Lr*** 260.11

Part 1

Fundamentals

Plastics, General Survey, 1. Definition, Molecular Structure and Properties

HANS-GEORG ELIAS, Michigan Molecular Institute, Midland, United States

ROLF MÜLHAUPT, Institute for Macromolecular Chemistry, Freiburg, Germany

1.	Introduction	5
1.1.	Polymers	5
1.1.1.	Fundamental Terms	5
1.1.2.	Nomenclature	6
1.2.	Plastics	8
1.2.1.	Fundamental Terms	8
1.2.2.	Designations	9
1.3.	History of Plastics	15
1.4.	Economic Importance	16
2.	Molecular Structure of Polymers	18
2.1.	Constitution	18
2.1.1.	Homopolymers	18
2.1.2.	Copolymers	19
2.1.3.	Branched Polymers	20
2.1.4.	Ordered Chain Assemblies	21
2.1.5.	Unordered Networks	22
2.2.	Molar Masses and Molar Mass Distributions	23
2.2.1.	Molar Mass Average	23
2.2.2.	Determination of Molar Mass	23
2.2.3.	Molar Mass Distributions	25
2.2.4.	Determination of Molar Mass Distributions	27
2.3.	Configurations	28
2.4.	Conformations	29
2.4.1.	Microconformations	29
2.4.2.	Conformations in Ideal Polymer Crystals	31
2.4.3.	Conformations in Polymer Solutions	32
2.4.4.	Unperturbed Coils	32
2.4.5.	Perturbed Coils	33
2.4.6.	Wormlike Chains	33
3.	Polymer Manufacture	34
3.1.	Raw Materials	34
3.1.1.	Wood	34
3.1.2.	Coal	35
3.1.3.	Natural Gas and Fracking	35
3.1.4.	Petroleum	35
3.1.5.	Other Renewable Resources and Biorefineries	35
3.2.	Polymer Syntheses: Overview	37
3.2.1.	Classifications	37
3.2.2.	Functionality	39
3.2.3.	Cyclopolymerizations	40
3.3.	Chain-Growth Polymerization	40
3.3.1.	Thermodynamics	40
3.3.2.	Free-Radical Polymerizations	41
3.3.3.	Anionic Polymerizations	47
3.3.4.	Cationic Polymerizations	48
3.3.5.	Ziegler–Natta, ROMP and ADMET Catalysis	48
3.3.6.	Copolymerizations	56
3.4.	Polycondensations and Polyadditions	57
3.4.1.	Bifunctional Polycondensations	57
3.4.2.	Multifunctional Polycondensations and Hyperbranched Polymers	60
3.4.3.	Polyaddition	60
3.5.	Polymer Reactions	61
3.5.1.	Polymer Transformations	61
3.5.2.	Block and Graft Formations	62
3.5.3.	Cross-Linking Reactions	62
3.5.4.	Degradation Reactions	62
4.	Plastics Manufacture	63
4.1.	Homogenization and Compounding	63
4.2.	Additives	63
4.2.1.	Overview	63
4.2.2.	Chemofunctional Additives	64
4.2.3.	Processing Aids	65
4.2.4.	Extending and Functional Fillers	65
4.2.5.	Plasticizers	68
4.2.6.	Colorants	68
4.2.7.	Blowing Agents	68
4.3.	Processing	69
5.	Supermolecular Structures	70
5.1.	Noncrystalline States	70

Ullmann's Polymers and Plastics: Products and Processes
© 2016 Wiley-VCH Verlag GmbH & Co. KGaA, Weinheim
ISBN: 978-3-527-33823-8 / DOI: 10.1002/14356007.a20_543.pub2

5.1.1.	Structure	70	7.3.3.	Melt Viscosities	98
5.1.2.	Orientation	70	**7.4.**	**Extensional Viscosities**	**100**
5.2.	**Polymer Crystals**	**71**	7.4.1.	Fundamentals	100
5.2.1.	Introduction	71	7.4.2.	Melts	100
5.2.2.	Crystal Structures	71	7.4.3.	Solutions	101
5.2.3.	Crystallinity	72	**8.**	**Mechanical Properties**	**101**
5.2.4.	Morphology	72	**8.1.**	**Introduction**	**101**
5.3.	**Mesophases and Liquid-Crystalline Polymers**	**76**	8.1.1.	Deformation of Polymers	101
5.3.1.	Introduction	76	8.1.2.	Tensile Tests	102
5.3.2.	Types of Liquid Crystals	76	8.1.3.	Moduli and Poisson Ratios	104
5.3.3.	Mesogens	77	**8.2.**	**Energy Elasticity**	**104**
5.3.4.	Lyotropic Liquid-Crystalline Polymers	79	8.2.1.	Theoretical Moduli	104
			8.2.2.	Real Moduli	105
5.3.5.	Thermotropic Liquid Crystal Polymers	81	8.2.3.	Temperature Dependence	105
			8.3.	**Entropy Elasticity**	**106**
5.3.6.	Block Copolymers	81	**8.4.**	**Viscoelasticity**	**107**
5.3.7.	Ionomers	82	8.4.1.	Fundamentals	107
5.4.	**Gels**	**83**	8.4.2.	Time−Temperature Superposition	108
5.5.	**Polymer Surfaces**	**84**			
5.5.1.	Structure	84	**8.5.**	**Dynamic Behavior**	**109**
5.5.2.	Interfacial Tension	84	8.5.1.	Fundamentals	109
5.5.3.	Adsorption	84	8.5.2.	Molecular Interpretations	110
6.	**Thermal Properties**	**85**	**8.6.**	**Fracture**	**111**
6.1.	**Molecular Motion**	**85**	8.6.1.	Overview	111
6.1.1.	Thermal Expansion	85	8.6.2.	Theoretical Fracture Strength	111
6.1.2.	Heat Capacity	85	8.6.3.	Real Fracture Strength	112
6.1.3.	Heat Conductivity	86	8.6.4.	Impact Resistance	114
6.2.	**Thermal Transitions and Relaxations**	**86**	8.6.5.	Stress Cracking	114
			8.6.6.	Fatigue	115
6.2.1.	Overview	86	**8.7.**	**Surface Mechanics**	**115**
6.2.2.	Crystallization	88	8.7.1.	Hardness	115
6.2.3.	Melting	89	8.7.2.	Friction	115
6.2.4.	Liquid Crystal Transitions	90	8.7.3.	Abrasion and Wear	116
6.2.5.	Glass Transitions	91	**9.**	**Electric Properties**	**116**
6.2.6.	Other Transitions and Relaxations	92	**9.1.**	**Dielectric Properties**	**116**
			9.1.1.	Relative Permittivity	116
6.2.7.	Technical Methods	92	9.1.2.	Dielectric Loss	117
6.3.	**Transport**	**92**	9.1.3.	Dielectric Strength and Tracking Resistance	118
6.3.1.	Self-Diffusion	92			
6.3.2.	Permeation	93	9.1.4.	Electrostatic Charging	119
7.	**Rheological Properties**	**95**	**9.2.**	**Electrical Conductivity**	**119**
7.1.	**Introduction**	**95**	**9.3.**	**Photoconductivity**	**121**
7.2.	**Shear Viscosities at Rest**	**95**	**10.**	**Optical Properties**	**121**
7.2.1.	Fundamentals	95	**11.**	**Polymer Composites**	**122**
7.2.2.	Molar Mass Dependence	96	**11.1.**	**Introduction**	**122**
7.2.3.	Concentrated Solutions	96	11.1.1.	Overview	122
7.3.	**Non-Newtonian Shear Viscosities**	**97**	11.1.2.	Mixture Rules	123
			11.2.	**Filled Polymers**	**126**
7.3.1.	Overview	97	11.2.1.	Microcomposites	126
7.3.2.	Flow Curves	98	11.2.2.	Molecular Composites	128

11.2.3.	Macrocomposites	128	12.2.	**Green Polymer Chemistry and Life Cycle Assessment.**	**139**
11.2.4.	Nanocomposites	129	12.3.	**Plastics Waste Recycling**	**140**
11.3.	**Homogeneous Blends**	**131**	12.3.1.	Mechanical Recycling	140
11.3.1.	Thermodynamics	131	12.3.2.	Feedstock Recycling	141
11.3.2.	Plastification	132	12.3.3.	Energy Recycling	141
11.3.3.	Rubber Blends	133	12.3.4.	Biodegradation and Bio-based Plastics	141
11.3.4.	Plastic Blends	134	12.3.5.	Plastics from Carbon Dioxide	142
11.4.	**Heterogeneous Blends**	**134**	12.4.	**Polymers for Sustainable Development**	**143**
11.4.1.	Compatible Polymers	134		References	144
11.4.2.	Blend Formation	134			
11.4.3.	Toughened Plastics	135			
11.4.4.	Thermoplastic Elastomers	136			
11.5.	**Expanded Plastics**	**136**			
12.	**Plastics and Sustainability**	**137**			
12.1.	**Plastics and Environment**	**137**			

1. Introduction

Plastics are commercially used materials that are based on polymers or prepolymers. The name plastics refers to their easy processability and shaping (Greek: *plastein* = to form, to shape). Plastics and polymers, also termed macromolecules (Greek: *macro* = large), are not synonyms. Polymers or prepolymers are raw materials for plastics; they become plastics only after processing. The same polymers may be used as plastics or as fibers, paints, rubbers, coatings, adhesives, thickeners, surfactants, and ion-exchange membranes. Properties of polymers are engineered by varying the molecular architecture or the formulation, combining them with different materials in multicomponent and multiphase systems. Typically, plastics contain additives that are needed to enhance their stability and to tune their properties. Unparalleled by any other material, polymers are exceptional regarding their attractive combination of facile processing with low mass and high versatility in terms of properties, applications, flexible choice of feedstocks, and recycling. The system integration of functional polymers represents the key to the development of advanced technologies with applications ranging from lightweight engineering to packaging, construction, aerospace and automotive industries, as well as biomedical engineering. Whereas most plastics are passive, advanced functional plastics are rendered interactive and capable of responding to environmental stimuli. Today, highly cost-, resource-, eco-, and energy-efficient polymers play a prominent role in sustainable development of advanced materials and technologies.

1.1. Polymers

1.1.1. Fundamental Terms [1–16]

The term polymer refers to macromolecules composed of many units (Greek: *poly* = many, *meros* = parts). As first proposed by HERMANN STAUDINGER in 1920, who was awarded the Nobel Price of Chemistry for his groundbreaking concept in 1953, macromolecules consist of many atoms, usually a thousand or more, thereby having high molar masses. Prior to STAUDINGER, the term polymer was not related to high molar mass. For example, benzene was originally called a polymer which was "polymerized" from three acetylene molecules. What is now called a polymer consists of molecules with hundreds and thousands of such units; it was therefore termed "high polymer" in the ancient literature. The term polymer carries with it the connotation of molecules composed of many equal "mers", such as the ethylene units in polyethylene, $R'(CH_2CH_2)_nR''$. The number n of monomer units in the polymer molecule is called the degree of polymerization X. There are, however, many polymer molecules (especially biopolymer molecules) with very different types of monomers per molecule, such as protein molecules $H(NH-CHR-CO)_nOH$ with up to 20 different R substituents. In accordance with STAUDINGER's view, a less constraining and more general term for polymer molecule is thus macromolecule. However, there is no sharp dividing line with respect to the number of units per molecule between macromolecules and low molar mass compounds. In principle, linear thermoplastic polymers are considered polymers when entanglement occurs, thus accounting for the viscoelastic properties typical for polymeric thermoplastics.

It was HERMANN STAUDINGER who recognized that polymers in industry and in nature are synthesized according to the same blueprint. Similar to pearls in a pearl necklace, monomer molecules are linked together by covalent bond formation. In the early days of polymer sciences and engineering, all polymers were biobased, because efficient synthetic polymerization processes were not at hand. Prominent examples of biopolymers include proteins, polynucleotides, cis-1,4-polyisoprene as natural rubber, polysaccharides (carbohydrates) with cellulose as the most abundant polymer and major component of biomass. Wood is a composite of two biopolymers, i.e., cellulose and lignin. As pointed out by STAUDINGER and others, most silicates are inorganic polymers derived by polymerization of silicic acid and its derivatives. Some of these naturally occurring polymers are used by man without further chemical transformation (e.g., cellulose for paper and cardboard). Yet, the chemical transformation of natural polymers with retention of their chain structures leads to solution and even melt processable semisynthetic materials, for example, cellulose acetates from cellulose. Chains of other natural polymers are cross-linked before commercial use. Examples are the hardening of casein (a protein) by formaldehyde to produce galalith (plastic) or the vulcanization of cis-1,4-polyisoprene (natural rubber) to afford an elastomer.

Today, most polymers are synthesized chemically from synthetic monomers. Examples include the preparations of polyethylene from ethylene, poly(vinyl chloride) from vinyl chloride, nylon 6 from ε-caprolactam, or nylon 66 from adipic acid and hexamethylenediamine. Whereas the majority of monomers are derived from oil and gas in petrochemistry, the progress in biotechnology and the quest for green economy are stimulating the production of bio-based synthetic monomers, such as butadiene, ethylene, propene, acrylic acid, glycol, and lactic acid in biorefineries using biomass as feedstock. Some industrial polymers result from the chemical conversion of other synthetic polymers, for example, poly(vinyl alcohol) from poly(vinyl acetate). In contrast to natural polymer syntheses and biotechnology, most large scale commercial synthetic polymers are produced in the absence of water either in bulk or gas phase. Moreover, compared to biopolymers, synthetic polymers, including the synthetic bio-based polymers derived from renewable feedstocks, are much easier to tailor with respect to their molar mass distribution, short- and long-chain branching and stereochemistry, thus meeting the demands of polymer melt processing and polymer applications. In contrast to protein synthesis, where polypeptide chains are produced with identical molar mass and comonomer sequence distribution, is rather problematic to produce thermoplastics with shear thinning resulting from long chain branching and multimodal molar mass distribution with enzyme-catalyzed reactions.

1.1.2. Nomenclature [1, 5, 17–19]

The nomenclature of individual polymers and plastics is as confusing as their classification according to properties. Various systems of nomenclature are used simultaneously. Abbreviations and acronyms abound, sometimes with different meanings for the same letter combinations and other times without explanation. In addition, many thousands of trade names are used worldwide for plastics, fibers, elastomers, and polymeric additives. Furthermore, a polymer from a certain company may come in many different grades depending on the processibility and application, sometimes up to 100 per polymer type. Some of these grades may even bear different trade names for various applications. The following nomenclature systems are commonly used for polymers.

Long-known Natural Polymers often have Trivial Names. Examples are cellulose, the polymeric sugar (-ose) of the plant cell; casein, the most important protein of milk and cheese (Latin: *caseus* = cheese); nucleic acids, the acids found in the cell nucleus; catalase, a catalyzing enzyme.

Synthetic Polymers are often Named after Their Monomers. Polymers of ethylene thus lead to polyethylene, styrene to polystyrene, vinyl chloride to poly(vinyl chloride), and lactic acid to a polylactic acid. This "polymonomer" nomenclature has the disadvantage that the constitution of monomeric units of the polymer molecules is not identical with the constitution of the monomers themselves. For example, the polymerization of ethylene, $CH_2=CH_2$, leads to $\sim(CH_2-CH_2)_n\sim$, a saturated compound and thus a polyalkane, not an unsaturated "ene" as the name polyethylene may suggest. The polymerization of lactams (cyclic amides) does not give macromolecules with intact lactam rings in the polymer chains but gives open-chain polyamides, etc. This polymonomer scheme is also ambiguous if a monomer can lead to more than one characteristic unit in a polymer. An example is acrolein, $CH_2=CH(CHO)$, which can polymerize via the ethylenic double bond to give $\sim[CH_2-CH(CHO)]_n\sim$, via the aldehyde group to $\sim[O-CH(CH=CH_2)]_n$, or via both to give six-membered rings in polymer chains.

For trade purposes, certain polymer names may denote not only homopolymers but also copolymers, contrary to what the "chemical" names imply. For example, the copolymers of ethylene with up to 10% butene-1, hexene-1, or octene-1 are known as linear low-density polyethylenes (LLDPEs). The commonly used chemical names of plastics thus often do not indicate the true chemical structure of the monomeric units of the polymers on which they are based.

Polymers are often Named after Characteristic Groups in Their Repeating Units. Polyamides are thus polymers with amide groups –NHCO– in their repeating units; for example $\sim[NHCO(CH_2)_5]_n\sim$ = polyamide 6 = nylon 6 = poly (ε-caprolactam). Other examples are polyesters with ester groups –COO– or polyurethanes with urethane groups –NH–CO–O– in the chains. A disadvantage is that this naming scheme is identical with that of organic chemistry where a polyisocyanate denotes a low molar mass compound with more than one isocyanate group per molecule [e.g., $C_6H_3(NCO)_3$]. A macromolecular polyisocyanate would thus be a polymer with many intact isocyanate groups per chain, for example, poly (vinyl isocyanate) $\sim[CH_2-CH(NCO)]_n\sim$. The polyisocyanates of polymer chemistry, on the other hand, possess polymerized isocyanate groups as, for example, in $\sim(NR-CO)_n\sim$. Such compounds are unfortunately also often called polyisocyanates.

IUPAC Names. IUPAC recommends the use of constitutive names, similar to those used in inorganic and organic chemistry. The nomenclature of low and high molar mass inorganic molecules follows the additivity principle; those of low molar mass organic molecules the substitution principle. The nomenclature of organic macromolecules is a hybrid of both principles: The smallest repeating units are thought of as biradicals according to the substitution principle; then their names are added according to the additivity principle, put in parentheses, and prefixed with "poly." Names of repeating units are written without spaces between words. The polymer $\sim[O-CH_2]_n\sim$ from formaldehyde, $H_2C=O$, is thus called poly(oxymethylene), abbreviated as POM. The polycondensation of ethylene glycol $HO-CH_2-CH_2-OH$ with terephthalic acid $HOOC-(p-C_6H_4)-COOH$ leads to a polymer $\sim[O-CH_2CH_2-O-OC(p-C_6H_4)CO]_n\sim$ with the systematic name poly(oxyethyleneoxyterephthaloyl). The trivial names of this polymer are poly (ethylene terephthalate) and poly(ethylene glycol terephthalate). It is also known as PET (an acronym) or PETP (an abbreviation) in the plastics literature, by the acronym PES in the fiber literature, and as PETE for recycling purposes.

With exception of POM, IUPAC names are rarely used in the plastics literature. They are however important for systematic searches in *Chemical Abstracts* and other literature services.

1.2. Plastics

1.2.1. Fundamental Terms [20–56]

Early plastics resembled natural resins. Natural resin refers mainly to oleoresins from tree sap but is also used for shellac, insect exudations, and mineral hydrocarbons (→ Resins, Natural). Early plastics were thus sometimes called synthetic resins. The word resin is today occasionally used for any organic chemical compound with medium to high molar mass that serves as a raw material for plastics (for a definition of the term resin according to current standards, see → Resins, Synthetic). Resin is not to be confused with rosin, which refers to mixtures of C_{20} fused-ring monocarboxylic acids, such as pine oil, tall oil, and kauri resin. Rosin is the main component of naval stores (→ Resins, Natural).

Plastics are usually divided into two groups according to their hardening processes. Those that yield solid materials by simple cooling of a polymer melt (a physical process) and soften while being heated are called *thermoplastics*. *Thermosets*, on the other hand, harden through chemical cross-linking reactions between polymer molecules; when heated, they do not soften but decompose chemically (→ Thermosets). The shaping of a thermoplastic is thus a reversible process: The same material can be melted and processed again. A thermoset cannot be remelted and reshaped; its formation is irreversible.

Thermoplastics are normally composed of fairly high molar mass molecules because many physical properties effectively become molar mass independent only above a certain molar mass enabling chain entanglement. Examples are melting temperatures and the moduli of elasticity. Other properties, however, increase with increasing molar mass and entanglement (e.g., melt viscosities).

Thermosets are usually generated from fairly low molar mass polymers, called oligomers (science) or prepolymers (industry). Oligomers containing functional endgroups are named telechelics, and telechelics containing polymerizable endgroups are termed macromonomers. High molar masses are unnecessary here because chemical reactions between prepolymer molecules lead to an interconnection of these molecules (cross-linking or advancement, respectively) and thus to a giant molecule with 100% conversion of the prepolymer. Prepolymers are thus thermosetting materials and become true thermosets only after the hardening reaction.

Plastics are usually divided into four groups: Commodity plastics (also called standard plastics or bulk plastics), engineering plastics (sometimes referred to as technical plastics or technoplastics), high-performance plastics, and functional plastics (or specialty plastics). A commodity, engineering, or high-performance plastic may have many different applications, whereas a functional plastic has one very specific application. Polyethylene, a commodity plastic, may according to its type or grade be used for containers, as packaging film, as agricultural mulch, etc. Poly(ethylene–co–vinyl alcohol) with a high content of vinyl alcohol units, on the other hand, is a functional plastic that is used only as an oxygen barrier resin. Other functional plastics are employed in optoelectronics, as resists, as piezoelectric materials, etc. Functional plastic is not synonymous with functional polymer or functionalized polymer, because the latter terms refer to polymers with functional chemical groups (i.e., groups with specific chemical reactivities).

Commodity plastics are manufactured in great amounts at low cost; hence, the terms bulk plastics or standard plastics. Engineering plastics possess improved mechanical properties and improved dimensional stability as compared to commodity plastics. Such improved properties may be higher moduli of elasticity, higher heat distortion temperature, smaller cold flows, higher impact strengths, etc. Engineering plastics are also often defined as those thermoplastics that maintain dimensional stability and most mechanical properties above 100 °C or below 0 °C. High-performance plastics, on the other hand, are engineering plastics with even more improved mechanical properties.

No sharp dividing lines exist between commodity plastics and engineering plastics on the one hand or engineering plastics and high-performance plastics on the other; nor are there generally agreed upon property levels beyond which polymers are designated as engineering or high-performance plastics. The three groups of plastics, in general, include the following:

Commodity Plastics. Poly(vinyl chloride), abbreviated as PVC, high density polyethylene (HDPE), low density polyethylene (LDPE), linear low density polyethylene (LLDPE), very low density polyethylene (VLDPE), isotactic polypropylene (PP), polystyrene (PS), polystyrene foam as expanded polystyrene (EPS) and poly(ethylene terephthalate) (PET).

Engineering Plastics. Polypropylene (PP), poly(ethylene terephthalate) (PET), poly(butylene terephthalate) (PBT), polyamides (aliphatic, amorphous, aromatic) (PA), polycarbonate (PC), polyoxymethylenes (POM), poly(methyl methacrylate) (PMMA), some modified polystyrenes such as styrene–acrylonitrile (SAN) and acrylonitrile–butadiene–styrene (ABS) copolymers, and high-impact polystyrenes (HIPS), as well as various blends such as poly(phenylene oxide)–polystyrene (PPO–PS), polyoxymethylene–polyurethane (POM–PUR), polypropylene–poly(ethylene-co-propene-co-diene) rubber (PP–EPDM), and polycarbonate–ABS (PC–ABS).

High-Performance Plastics. Liquid crystal-line polymers (LCPs), polyetheretherketone (PEEK), various polysulfones, polyimides, fluoropolymers, etc.

Some thermosets are also classified as engineering or high-performance plastics. In general, thermosets are, however, considered a separate group of plastics. They comprise alkyd (→ Alkyd Resins), phenolic (→ Phenolic Resins), and amino resins (→ Amino Resins) (melamine and urea resins), epoxides (epoxies) (→ Epoxy Resins), unsaturated polyesters (including so-called vinyl esters) (→ Polyester Resins, Unsaturated), polyurethanes (→ Polyurethanes), and allylics (→ Allyl Compounds).

1.2.2. Designations [57–61]

The plastics trade and technical literature indulges in trivial names, abbreviations, and acronyms. Acetal polymer is, for example, the name given to polymers with $-OCH_2-$ as the major repeating unit. Polymers of methacryl esters $\sim[CH_2-C(CH_3)(COOR)]_n\sim$ are known as methacrylics (or often only acrylics), but polyacrylonitrile $\sim[CH_2-CH(CN)]_n\sim$ is not considered an "acrylic" in the plastics literature (the textile literature calls it an acrylic fiber). Styrenics comprise homopolymers of styrene as well as copolymers of styrene with other monomers. Nylon, originally a protected trade name, is now a generic term for polyamides, especially aliphatic ones.

Abbreviations often have more than one meaning (Tables 1–4): PBT may, for example, denote poly(butylene terephthalate), but it is also used for poly(*p*-phenylenebenzbisthiazole). Exhaustive lists of commonly used abbreviations of polymers and additives as recommended by various organizations have been compiled [61].

Tables 1–4 give the recommendations for abbreviations of names of thermoplastics and thermosets by ASTM/ANSI (American Society for the Testing of Materials/American National Standards Institute [D 1600–86 a]), DIN (German Industrial Standards [7728]), ISO (International Standardization Organization [1043 – 1986 (ε)]), IUPAC (International Union of Pure and Applied Chemistry). "Other" means other abbreviations used in the field but not recommended by ASTM, DIN, ISO, or IUPAC. The recommended abbreviations are often neither identical with those of fibers of the same chemical structures nor with those of elastomers containing the same monomer units; some of these also deviate from those recommended for recyclable plastics. Several abbreviations have different meanings in either two or more of the ASTM, DIN, ISO and IUPAC systems and/or the technical literature.

The abbreviations are in part based on the poly(monomer) nomenclature, i.e., on the names of

Table 1. Polymers by chain polymerization (addition polymerization), copolymerization, polycondensation (condensation polymerization), and polyaddition which are characterized by abbreviations based on the poly(monomer) nomenclature

Polymers of	ASTM	DIN	ISO	IUPAC	Other
Acrylester+acrylonitrile+butadiene	ABA	A/B/A			
Acrylester+acrylonitrile+styrene	ASA	A/S/A	A/S/A		
Acrylester+ethylene	EEA[#]		E/EA		EAA
Acrylic acid	PAA				PAS[#]
Acrylonitrile	PAN	PAN		PAN	
Acrylonitrile+butadiene	PBAN				
Acrylonitrile+butadiene+methyl methacrylate+styrene					MABS
Acrylonitrile+butadiene+styrene	ABS	ABS	ABS	ABS	ABS
Acrylonitrile+ethylene-propene-diene+styrene		A/EPDM/S		AES	
Acrylonitrile+methyl methacrylate	AMMA	A/MMA	A/MMA		
Acrylonitrile+styrene	SAN	SAN	SAN	SAN	PSAN
Acrylonitrile+styrene+chlorinated polyethylene		A/PE-C/S			
Adipic acid+hexamethylenediamine	PA 6.6	PA 66	PA 6.6	PA 6.6	
Adipic acid+tetramethylene glycol					PTMA
Allyldiglycol carbonate	ADC				
p-Aminobenzoic acid					PAB
Aminotriazol					PAT
11-Aminoundecanoic acid		PA 11			
Azelaic anhydride					PAPA
Bisphenol A+phosgene	PC[#]	PC[#]	PC[#]	PC[#]	
Bitumen+ethylene				ECB	
Butadiene+methyl methacrylate+styrene		MBS			
Butadiene+styrene (thermoplastic)	SB	S/B	S/B	S/B	PASB
Butene-1	PB	PB	PB		BT
Butyl acrylate		PBA		PBA	
Butylene glycol+terephthalic acid	PBT	PBT	PBT		PTMT
Butyl methacrylate					PBMA
ε-Caprolactam	PA 6	PA 6	PA 6	PA 6	
Chlorotrifluoroethylene	PCTFE	PCTFE	PCTFE	PCTFE	
Chlorotrifluoroethylene+ethylene					ECTFE
Cresol+formaldehyde	CF	CF	CF	CF	
1,4-Cyclohexanedimethylol+terephthalic acid					PCDT
Diallyl chloroendate (=diallyl ester of 4,5,6,7,7-hexachlorobicyclo-[2,2,1]5-heptane-2,3-dicarbonic acid	PDAC				
Diallyl fumarate	PDAF				
Diallyl isophthalate	PDAIP				
Diallyl maleate	PDAM				
Diallyl phthalate	PDAP	PDAP	PDAP	PDAP	DAP
1,4-Dichlorobenzene+disodium sulfide	PPS	PPS			
2,6-Dimethylphenol+oxygen	POP				PPO™
Dodecanolactam (laurolactam)	PA 12	PA 12	PA 12	PA 12	
Ethyl acrylate					PEA
Ethyl acrylate+ethylene		E/EA			EEA[#]
Ethylene	PE	PE	PE	PE	PL
Cross-linked polymer		VPE			XLPE
High-density polymer	HDPE	PE-HD	PE-HD		
Low-density polymer	LDPE	PE-LD	PE-LD		
Low-density polymer, linear	LLDPE				
Medium density polymer	MDPE				
Ultra-high molar mass polymer	UHMW-PE				
Ethylene+methyl acrylate+vinyl chloride		VC/E/MA	VC/E/MA		
Ethylene+methyl methacrylate	EMA	E/MA			
Ethylene+propene		E/P	E/P		PEP
Ethylene+propene (+diene)		EPDM			
Ethylene+tetrafluoroethylene	ETFE	E/TFE			
Ethylene+vinyl acetate	EVA	E/VA E/VAC	EVA E/VAC		VAE
Ethylene+vinyl acetate+vinyl chloride		VC/E/VAC	VC/E/VAC		
Ethylene+vinyl chloride		VC/E	VC/E		
Ethylene glycol					PEG
Ethylene glycol+maleic anhydride	UP[#]				

Ethylene glycol+terephthalic acid (ester)	PET	PET	PETP	PETP	PETE
Fast crystallizing polymer					CPET
Oriented polymer					OPET
Ethylene oxide	PEO	PEOX	PEOX	PEO	
Formaldehyde (or trioxane)	POM#	POM#	POM#	POM#	
Formaldehyde+furan	FF				
Formaldehyde+melamine	MF	MF	MF	MF	
Formaldehyde+melamine+phenol		MPF	MPF		
Formaldehyde+phenol	PF	PF	PF	PF	
Formaldehyde+urea	UF	UF	UF	UF	
Furfural+Phenol	PFF				
Hexafluoropropene+tetrafluoroethylene	FEP	FEP	FEP		
Hexamethylenediamine+sebacic acid		PA 610			PA 6.10
p-Hydroxybenzoic acid	POB				PHB#
3-Hydroxybutyric acid					PHB#
2-Hydroxyethylmethacrylate					PHEMA
Isobutene	PIB	PIB	PIB	PIB	IM
Laurolactam (dodecanolactam)	PA 12	PA 12	PA 12	PA 12	
Linseed oil, epoxidized		ELO	ELO		
Maleic anhydride+ styrene	SMA	S/MA			
Methacrylimide		PMI			
Methyl acrylate					PMA
Methyl acrylate+vinyl chloride		VC/MA	VC/MA		
Methyl α-chloromethacrylate	PMCA				
Methyl methacrylate	PMMA	PMMA	PMMA	PMMA	
Methyl methacrylate+vinyl chloride			VC/MMA		
1,4-Methylpentene	PMP	PMP	PMP		TPX
α-Methylstyrene		PMS			PAMS
α-Methylstyrene+styrene	SMS	S/MS	S/MS		
Octyl acrylate+vinyl chloride		VC/OA	VC/OA		
Perfluoroalkoxyalkane	PFA	PFA			
Phosphoric acid					PPA#
Propene	PP	PP	PP	PP	
Oriented polymer					OPP
Oriented biaxially					BOPP
Propene+tetrafluoroethylene (alt.)					TFE-P
Propylene oxide	PPOX	PPOX	PPOX		
Soybean oil, epoxidized		ESO	ESO		
Styrene	PS	PS	PS	PS	
Expanded (foamed) polymer					EPS
High-impact polymer					HIPS
Impact resistant polymer	SRP	IPS			
Oriented polymer					OPS
Tetrafluoroethylene	PTFE	PTFE	PTFE	PTFE	
Tetrahydrofuran					PTHF
Polymer with hydroxy endgroups					PTMEG
Triallyl cyanurate	PTAC				
Trifluoroethylene				P 3 FE	
Trioxane (+comonomers)	POM#	POM#	POM#	POM#	
Vinyl acetate	PVAC	PVAC	PVAC	PVAC	PVA
Vinyl acetate+vinyl chloride	PVCA	VC/VAC	VC/VAC		
N-Vinylcarbazole	PVK	PVK	PVK		
Vinyl chloride	PVC	PVC	PVC	PVC	PCU, V
Polymerized in bulk					M-PVC
Polymerized in emulsion					E-PVC
Polymerized in suspension					S-PVC
Polymer as flexible film					FPVC
Polymer as oriented film					OPVC
Polymer as rigid film					RPVC
Vinyl chloride and vinylidene chloride		VC/VDC	VC/VDC		
Vinyl fluoride	PVF	PVF	PVF	PVF	
Vinylidene chloride	PVDC	PVDC	PVDC	PVDC	
Vinylidene fluoride	PVDF	PVDF	PVDF	PVDF	
Vinyl methyl ether					PVME
N-Vinylpyrrolidone	PVP	PVP	PVP		

Abbreviation with more than one meaning.

Table 2. Abbreviations for polymers named after a characteristic polymer group

Characteristic group	ASTM	DIN	ISO	IUPAC	Other
Amide	PA	PA	PA	PA	
Amide, aromatic	PARA				
Metal coated films	PA**				
Saran coated films	PA*				
Amide–imide	PAI	PAI			
Aryl amide	PARA				
Aryl sulfone					PAS#
Benzimidazole					PBI
Carbodiimide					PCD
Carbonate, aromatic	PC#	PC#	PC#	PC#	
Epoxide (epoxy)	EP	EP	EP	EP	
Glass-fiber reinforced					GEP
Epoxide–ester			EPE		
Ester, saturated		SP			
Ester, thermoplastic	TPES				
Metallized polymer film					MPE
Ester, unsaturated	UP#	UP#	UP#	UP#	
Glass fiber reinforced					FRP
Ester–alkyd	PAK				
Ester–ether (thermoplastic elastomer)	TEEE				
Ester–imide	PEI				
Ether–block-amide	PEBA	PEBA			
Etheretherketone	PEEK				
Ether–imide		PEI			
Ether sulfone	PES	PES			
	PESU				
Imide	PI	PI			
Isocyanurate		PIR			
Parabanic acid					PPA#
Phenylene ether	PPE	PPE			
Phenylene sulfone	PPSU	PPSU	PPSU		PSU
Silicone	SI	SI	SI		
Sulfone	PSUL	PSU			
Urethane	PUR	PUR	PUR	PUR	
Thermoplastic	TPUR				
Thermoset	TSUR				

*Saran coated polymer film.
**Metallized polymer film.
Abbreviation with more than one meaning.

the monomers used in the manufacture of polymers, sometimes, however, without a prefix "P" for "poly" (Table 1). The names of monomers for copolymers are given in alphabetical order without regard to their prevalence.

Other abbreviations are based on characteristic groups of polymers (Table 2) or indicate polymers synthesized by chemical transformation of base polymers (Table 3). Special abbreviations apply to blends, reinforced polymers, etc., (Table 4).

Special properties of polymers are symbolized by up to four additional letters. These letters are arranged after the symbol for the base polymer according to ISO and DIN; in the technical literature, they are, however, commonly placed in front of the base symbol, following ASTM. DIN recommends the following letters for special properties: C = chlorinated; D = density; E = expanded, expandable; F = flexible, liquid; H = high; I = impact resistant; L = linear, low; M = molar mass, average (mean), molecular; N = normal, novolac; P = plasticized; R = raised (enhanced), resol; U = ultra, plasticizer-free (unplasticized); V = very; W = mass (weight); X = cross-linked. The technical literature also uses BO = biaxially oriented and O = oriented (usually in one direction).

Companies have long used internal classification systems, and the military has issued plastics specifications. An industry-wide classification system for thermoplastic

Table 3. Abbreviations for polymers produced by chemical transformation of other polymers

Resulting polymer	ASTM	DIN	ISO	IUPAC	Other
Carboxymethyl cellulose	CMC	CMC	CMC	CMC	
Carboxymethyl hydroxyethyl cellulose					CMHEC
Casein, cross-linked with formaldehyde	CS	CS	CS	CS	
Cellulose, as Cellophan	C				
Ditto, coated with Saran	C*				
Cellulose acetate	CA	CA	CA	CA	
Cellulose acetobutyrate	CAB	CAB	CAB	CAB	
Cellulose acetopropionate	CAP	CAP	CAP	CAP	
Cellulose nitrate	CN	CN	CN	CN	NC
Cellulose plastics, unspecified	CE				
Cellulose propionate	CP	CP	CP	CP	
Cellulose triacetate	CTA	CTA			TA
Ethyl cellulose	EC	EC	EC	EC	
Hydroxyethyl cellulose					HEC
Hydroxypropyl cellulose					HPC
Hydroxypropyl methyl cellulose					HPMC
Methyl cellulose		MC			
Polyethylene, chlorinated	CPE	PE-C	PEC		
Chlorosulfonated	CSM				CSR
Poly(ethylene-co-vinyl alcohol)		E/VAL			EVAL, EVOH
Polypropylene, chlorinated		PPC			
Polyvinyl alcohol	PVAL	PVAL	PVAL	PVAL	PVOH
Polyvinyl butyral	PVB	PVB	PVB		
Polyvinyl chloride, chlorinated	CPVC	PVC-C	PVCC		PC#, PeCe
Polyvinyl formal	PVFM	PVFM	PVFM	PVFM	

* Saran coated polymer film.

materials introduced by ISO is based on an alphanumeric "line call-out," a special code (see below) [57, 58]. A material is characterized by several data blocks that indicate its composition and certain property data or ranges. These property data refer to essential criteria. Since different thermoplastics have different applications, property data are restricted to different "leading properties" for each type of thermoplastic.

The property data covered by this classification scheme are thus not comprehensive. Rather, various plastics types are characterized by one to three sets of leading criteria, selected from

- Chemical structure data, such as content of vinyl acetate (VAC) or acrylonitrile units (AN), isotacticity (IT), or density (D) as a measure of branching
- Molar mass data, such as intrinsic viscosity (IV) or Fikentscher's K value (FK)
- Bulk density (BD)
- Rheological data, such as melt flow rate (MFR) (melt flow index)

Table 4. Abbreviations for blends, reinforced polymers, etc

Resulting polymer	ASTM	DIN	ISO	IUPAC	Other
Elastomers, thermoplastic					
Containing ester and ether groups	TEEE				
Olefin-based	TEO				
Styrene-based	TES				
Epoxide, glass-fiber reinforced					GEP
Plastic, carbon-fiber reinforced		KEK			CFP
Glass-fiber reinforced					GFK
Man-made fiber reinforced					CFK
Metal fiber reinforced					MFK
Poly(acrylonitrile-co-styrene)+chlorinated polyethylene	ACS				
Sheet molding compound					SMC
with high glass fiber content					HMC

- Thermal data, such as Vicat temperature (VT) or torsional stiffness temperature (TST)
- Mechanical properties, such as modulus of elasticity (E), tensile stress at 100% strain (TS), shore hardness (SH), or impact strength (notched) (ISN)

The following leading criteria are used:

Polyethylene	D, MFR
Polypropylene	I, MFR
Polystyrene and acrylonitrile-styrene copolymers	VT, MFR
Styrene-butadiene, acrylonitrile-butadiene-styrene, and acrylonitrile-styrene-acrylic acid copolymers	VT, MFR, ISN
Ethylene-vinyl acetate copolymers	VAC, MFR
Poly(vinyl chloride)	IV, FK, BD
Poly(vinyl chloride), unplasticized	VT, E, ISN
Poly(vinyl chloride), plasticized	TS, SH, TST
Polyamides	IV, E
Polycarbonates	IV, MFR, E
Poly(methyl methacrylate)	IV, VT
Poly(ethylene terephthalate)	IV

The standard designation of a thermoplastic material consists of a description block, a standard number block, and a series of data blocks.

Designation Block. The designation block gives the type of material (e.g., thermoplastic, molding, etc.).

Standard Number Block. *Data block 1.* The standard number block consists of the number of the ISO (or DIN or other national) standard, followed by a hyphen and data block 1. This data block contains the abbreviation of the chemical name of the material [e.g., PE = polyethylene, PVC = poly(vinyl chloride)]. This may be followed by analytical data, such as the vinyl acetate content of ethylene–vinyl acetate copolymers (EVAC). However, these are not the exact analytical data but rather code numbers for the range (called "cell") in which this material can substitute for a similar one. Separated by a hyphen, supplemental information on this material may be given (e.g., H = homopolymer, P = plasticized, U = unplasticized, E = polymerized in emulsion).

Data block 2. may contain up to four letters that give qualitative information. The first letter denotes the intended application (e.g., B = blow molding, G = general purpose, P = paste resin, Y = textile yarn). The following one to three letters can code up to three essential additives or supplemental information, for example, A = antioxidant, D = powder (dry blend), L = light and weather stabilizer, S = slip agent.

Data Block 3. Quantitative information about the designated properties is contained in data block 3. The encoding of these data is different for each plastic material and each testing standard. For example, the code 20 – D 050 in data block 3 for a polyethylene tested according to DIN 16 776 means a material with a density of 0.918 g/cm^3 (cell 020) and a melt flow rate of 4.2 g/min (cell 050) measured at 190 °C under a load of 2.16 kg (D). On the other hand, a third data block 22 – 030 for a polyamide 12 means a polymer with an intrinsic viscosity of 210 cm^3/g (cell 22) and a modulus of elasticity of 280 MPa (cell 030). The definition of the cells can be found in special cell tables.

Data block 4. gives information about the type and content of fillers or reinforcing materials; these codes are independent of the types of polymers. The first letter gives the type of filler, for example, C = carbon or G = glass. The second letter indicates the shape of the filler (e.g., F = fiber, S = sphere). The letters are followed by a code for the mass content of the filler; 15 for the range 12.5–17.5 wt %, 60 for the range 57.5–62.5 wt %.

Data block 5. is reserved for specifications based on individual agreements between supplier and customer. It may code additional requirements, restrictions, or supplemental information.

A thermoplastic may thus carry the designation "Molding material DIN 7744-PC, XF, 55–045, GF 30," which indicates a polycarbonate (PC) with no special indication of its application (X) but with special burning characteristics (F), an intrinsic viscosity in a coded range of 55 (here for [η] = 56 cm^3/g), and a melt flow rate of 5.5 g/(10 min) (coded as 045), which contains 30% (30) of glass (G) fibers (F).

A designation system similar to the above ISO–DIN system has been recommended by ASTM [59, 60]. A typical ASTM designation, for example, would read "Molding material ASTM D-4000 PI 000 G 42 360"; this material is a polyimide (PI). Because no individual property table has been specified for polyimides, the cell table G of ASTM D-4000 has been used to characterize the properties of PI. This particular cell table identifies five different properties by designating cell limits to the five digits following the letter G. The first digit characterizes the minimum tensile strength, here "4", which according to cell table G means at least 85 MPa. The second digit gives the cell value for the flexural modulus (here "2" for 3 500 MPa). The third digit denotes the Izod impact strength (50 J/m → "3"), and the fourth digit the heat deflection temperature (300 °C → "6"). The fifth digit is undetermined; the "0" indicates an unspecified property.

1.3. History of Plastics [13–16, 62–67]

The first plastics were prepared long before their macromolecular nature was discovered. Decorative coatings based on polymers, such as egg white or blood proteins were used in the cave paintings of Altamira, Spain, as early as 15 000 B.C. Later painting methods utilized gelatin and polymers resulting from naturally drying vegetable oils. The use of these materials resulted from the eternal desire of artists for easy-to-handle materials that give an optimal artistic effect and have infinite stability.

These requirements were not restricted to surface coatings, which are two dimensionally applied plastics. Cow horns were used in the Middle Ages to prepare windows for lanterns and intarsia in wood. To this purpose, the horns had to be flattened by steaming, a very difficult process. The flattened horns also tended to curl up after a while. An early, easy-to-process substitute for natural horn was reported by the Bavarian monk WOLFGANG SEIDEL (1492–1562); it undoubtedly derived from far older recipes. This imitation horn was based on the protein casein, the white material from skim milk. Casein was extracted from skim milk with hot water, treated with warm lye, and shaped while being warmed; the desired shape was then fixed by immersion in cold water. Unknown inventors later discovered that the addition of inorganic fillers increased the mechanical stability of this early thermoplastic. The same material was used later for children's building blocks by the German aviation pioneer OTTO LILIENTHAL (1848–1896) and his brother GUSTAV. In 1885 a patent based on the same physical process was granted to the American EMERY EDWARD CHILDS. The properties of these plastics were improved markedly by the chemical reaction of casein with formaldehyde, as described in 1897 by the German inventors WILHELM KRISCHE and ADOLF SPITTELER. The resulting thermoset was called galalith (Greek: *gala* = milk, *lithos* = stone); it is still used today for haberdashery.

Another early thermoset used natural rubber as the raw material. In 1839, the American CHARLES GOODYEAR discovered the cross-linking (vulcanization) of natural rubber to an elastomer by sulfur under the action of "white lead" (basic lead carbonate) and heat. His brother NELSON used larger amounts of sulfur and in 1851 invented ebonite, a hard, black thermoset.

The first fully synthetic thermoset was invented in 1906 by the Belgian-born American chemist LEO H. BAEKELAND, who heated various phenols with formaldehyde under pressure and produced insoluble hard masses. These "bakelites" were recognized in 1909 as excellent electrical insulators and thus became one of the foundations of the modern electrical industry. BAEKELAND was not the first to prepare phenolic resins though; they were first observed by ADOLF VON BAEYER in 1872. BAEYER's substances were merely resinous materials, and BAEKELAND's "heat and pressure" process was necessary before commercially useful materials could be produced.

The first *semisynthetic thermoplastics* originated from cotton. Cotton fibers consist of cellulose; they have been used by man since prehistoric times. Because cotton is relatively easily grown, many attempts have been made to improve its textile properties. The Englishman JOHN MERCER discovered in 1844 that the treatment of cotton with aqueous solutions of caustic soda (sodium hydroxide) leads to fibers with increased strength, higher luster, and improved dyeability. The Frenchman L. FIGUIER demonstrated in 1846 that cellulose paper was strengthened in a manner similar to cotton when it was treated with sulfuric acid. In

1853 W. E. GAINE received an English patent for the same process, which delivered a parchment-like material. An English patent by THOMAS TAYLOR described the formation of very resistant materials from layers of paper sheets by the combined action of zinc chloride and pressure. These products were called "vulcanfiber" because they resembled the vulcanization products of natural rubber.

Mercerized cotton, artificial parchment, and vulcanized paper result from physical transformations of cellulose. If, however, cellulose materials are treated not by sulfuric acid alone but by a mixture of sulfuric and nitric acids, a chemical reaction to cellulose nitrate (nitrocellulose, gun cotton) occurs. This chance discovery by the German CHRISTIAN FRIEDRICH SCHÖNBEIN in 1846 paved the way for the invention of the first semisynthetic thermoplastic. In 1862 ALEXANDER PARKES found an easy method for processing cellulose nitrate into thermoplastic masses by the addition of castor oil, camphor, and dyes. The manufacture of the resulting "parkesine" was difficult, however, and production ceased in 1867. The use of alcoholic solutions of camphor by DANIEL W. SPILL was equally unsuccessful. In 1869, a patent was granted to the American JOHN WESLEY HYATT for the use of camphor without camphor oil or alcohol. The resulting "celluloid" is generally considered the pioneering thermoplastic [66].

Other early, fully synthetic thermoplastics have a far longer history, albeit not as industrial materials. Polymers of formaldehyde were found by JUSTUS VON LIEBIG in 1839; yet extensive scientific investigations by HERMANN STAUDINGER in the 1920s and 1930s and major industrial development work by Du Pont and Celanese were necessary before it became an engineering plastic in 1956.

Vinyl chloride was synthesized by H. V. REGNAULD in 1838, who also observed the formation of resins. Vinyl chloride was polymerized by KLATTE in Germany and OSTROMUISLENSKY in England in 1912, but not until 1931 was the first poly(vinyl chloride) produced commercially in Germany by I. G. Farben.

Styrene was discovered by a chemist called NEUMAN as reported in W. NICHOLSON's 1786 *Dictionary of Practical and Theoretical Chemistry*. The conversion of styrene to a solid mass was discovered by E. SIMON in 1839, who considered it a styrene oxide. A. W. VON HOFMANN and J. BLYTH showed in 1845 that the alleged styrene oxide was an isomer of styrene; they called it metastyrene. It was however only in 1920 that the polymeric nature of polystyrene was recognized by H. STAUDINGER; it was first produced commercially by I. G. Farben (Germany) in 1930.

A highly branched polyethylene is found in nature as the "mineral" elaterite. Lightly branched synthetic polyethylene was first obtained in 1933 by high-pressure polymerization of ethylene at ICI (England); its commercial production began in 1939. Linear polyethylenes were first synthesized in 1953 by low-pressure polymerization using catalysts based on transition metals (ZIEGLER), chromium oxide (Phillips Petroleum), and molybdenum oxide (Standard Oil of Indiana). For the first time, NATTA employed ZIEGLER's catalysts to polymerize α-olefins, such as propene, thus producing solid, crystalline stereoregular polymers. The commercial production of isotactic polypropylene (PP) began in Italy, Germany, and the United States in 1957. The use of these catalysts and the availability of inexpensive feedstocks from petroleum refining led to rapid growth of the polymer industry. Progress in heterogeneous polymerization catalysis enabled morphology control in olefin polymerization, thus producing spherical pellet-sized polyolefin particles without requiring pelletizing extrusion.

1.4. Economic Importance [68–73]

The rapid growth of plastics production is a result of three factors: (i) growth of world population with surging demand for cost-, resource-, eco-, and energy-efficient advanced materials, (ii) average increase in standard of living, and (iii) replacement of metal, wood, ceramics, and environmentally less friendly materials by plastics. The growth pattern may be economy driven [determined by the price of materials (substitutes); shortages (substitution), or governmental regulations; or technology driven (determined by technological requirements, internal knowhow, and accidental discoveries). At present, the expected declining

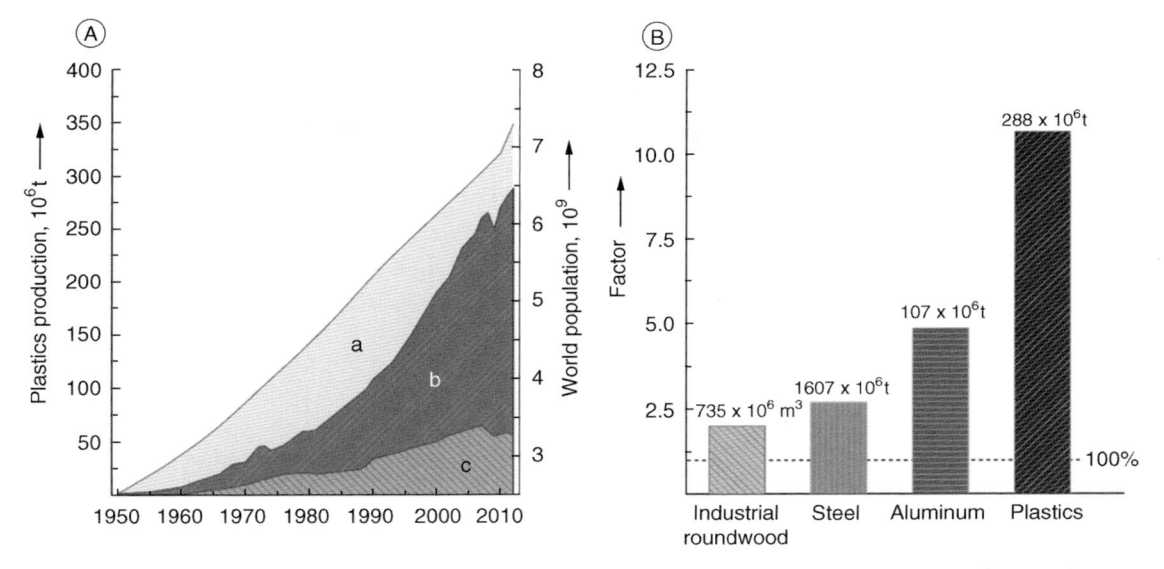

Figure 1. A) Growth of world plastics production and world population (data taken from [73, 74]); B) Production growth (1970–2012) of other materials
a) Population Growth; b) World plastic production; c) European production

supply of fossil resources, and public concerns relating to global warming and climate change, are stimulating the quest for polymers based on renewable resources. The world population grew from 2.532×10^9 in 1950 to 7.159×10^9 in 2014, and is expected to reach 9×10^9 in 2050. Because more people aspire the high living standards of the Western World, the rapid growing world population is accompanied by a surging demand for polymeric materials. Hence, the world plastics production (approximately equal to consumption) climbed from 0.6 kg per capita (1950) to ca. 42 kg per capita in 2014 (see Fig. 1). There are great differences in plastics consumption between various countries: Whereas plastic production is slowing down in Europe, it continues to grow rapidly in Asia and India.

According to a survey conducted by PlasticsEurope (the association of plastics manufacturers in Europe) in 2012, the European plastics industry producers, plastics converters, and plastics machinery with more than 62 000 companies accounted for approximately 1.4 million jobs in the European Union's 27 member states and had a combined turnover of around 3×10^9 Euro. By 2014, the world plastics production will pass the 300×10^6 t/a mark. In 2012, China was the leading plastics producer with 23.9% share of the world plastics production, followed by the rest of Asia, including Japan, (20.4%), Europe (20.4%), NAFTA (19.9%), Middle East/Africa (7.2%). In 2013, the world's plastic consumption was close to 300×10^6 t, of which commodity polymers, such as PE, PP, PVC, PS, and PET claim a major share of around 80% (Fig. 2). The growth of the plastics production during the first decade of the 21st century is unparalleled by any other class of materials including wood, aluminum and steel (Fig. 1). Close to half of the world's plastics production

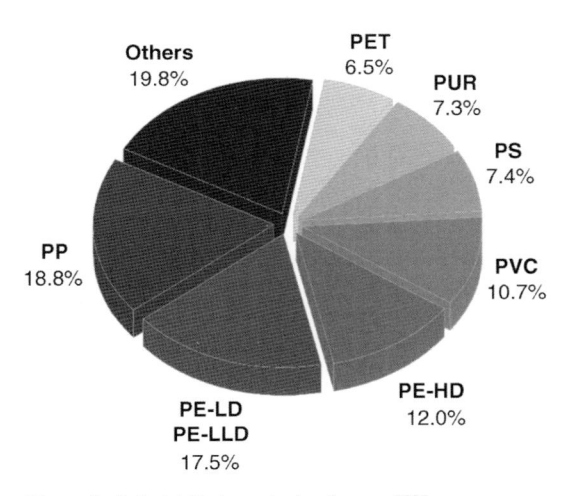

Figure 2. Industrially important polymers [73]

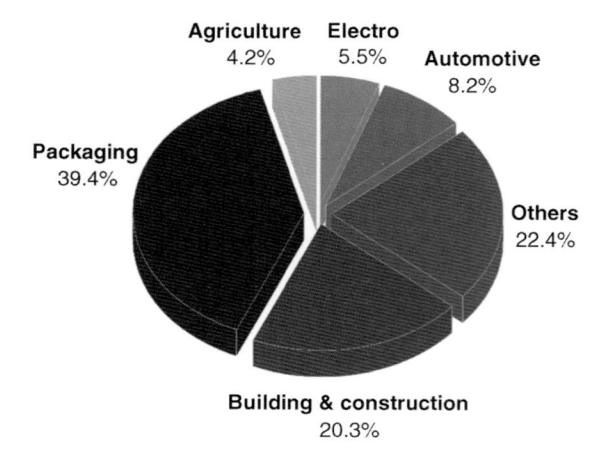

Figure 3. Application areas of plastics [73]

are polyolefins, such as PE and PP. Plastics serve the needs of a variety of markets (Fig. 3). In 2012, the largest plastic application area in Europe was packaging (39.4%), followed by building and construction (20.3%), automotive (8.2%), electrical and electronics (5.2%), agricultural applications (4.2%), and other diversified sectors (22.4%) including consumer and household products, furniture, sports, safety, as well as medical applications (see Fig. 3). Today, plastics wastes are recognized as very valuable source of raw materials and energy. Upon thermal treatments, commodity plastics are converted back into oil and gas, thus substituting fossil resources. In 2012, the plastics recycling and energy recovery reached 61.9% in Europe. The recycling of plastics and polymer products, ranging from lightweight engineering to membranes for water purification, offer great prospects for sustainable development, protecting resources, such as oil and water, for future generations, reducing carbon footprint, and securing the supply of energy and food.

2. Molecular Structure of Polymers [1–12]

2.1. Constitution

2.1.1. Homopolymers

A homopolymer is defined as a polymer derived from one type of monomer. The term homopolymer thus refers to the origin of the monomers in a polymer and not to the actual constitutional units. The polymerization of a monomer $CH_2=CHR$ leads to constitutional units $-CH_2CHR-$. These units may be connected to each other not only in head-to-tail positions but also in head-to-head and tail-to-tail arrangements:

$$-CH_2-CHR-CH_2-CHR- \quad -CH_2-CHR-CHR-CH_2-$$
$$\text{Head-to-tail} \qquad\qquad \text{Tail-to-tail}$$

$$-CHR-CH_2-CH_2-CHR-$$
$$\text{Head-to-head}$$

The proportion of head-to-head regioisomeric units increases with decreasing size of substituents R and with reduced resonance stabilization of the growing species. Free-radical polymerization of vinyl acetate [$CH_2=CH(OOCCH_3)$] leads to 1–2%, of vinyl fluoride ($CH_2=CHF$) to 6–10%, and of vinylidene fluoride ($CH_2=CF_2$) to 10–12% tail-to-tail connections. Head-to-head connection is detrimental to thermal stability, because the steric hindrance of the substituents reduces the bond strength of the C–C bonds.

Monomer molecules can also polymerize to a lesser or greater extent via "wrong" groups. The amount of such isomeric units depends on the constitution of the monomer and the mode of polymerization. Acrolein (**1**), for example, polymerizes free radically via the carbon–carbon double bond (**2**) and the aldehyde group (**3**); it forms branched (**4**) or cross-linked units and even intramolecular rings (**5**) from two units of (**4**):

Monomers may also isomerize during polymerization, especially cationic polymerization. 4,4-Dimethyl-pentene-1 (**6**) polymerizes at low temperature (≈ -130 °C) via the carbon–carbon double bond (**7**), but at higher temperature ($\approx$ 0 °C) leads to a "phantom polymer" (**8**) with three carbon atoms per monomeric unit:

$$\sim CH_2-CH\sim \quad \xrightarrow{-130\ °C} \quad CH_2=CH$$
$$||$$
$$CH_2-C(CH_3)_3 CH_2-C(CH_3)_3$$
$$\mathbf{7} \mathbf{6}$$

$$\downarrow 0\ °C$$

$$\sim CH_2-CH_2-CH\sim$$
$$|$$
$$C(CH_3)_3$$
$$\mathbf{8}$$

Irregular structures may also be created by chain-transfer reactions to polymer molecules (see Chap. 3); these reactions lead to branched polymers.

The groups at the ends of polymer chains are not shown, in part because they are often unknown and in part because their structure does not influence most of the polymer properties. End groups may be initiator or catalyst residues or groups that are generated by transfer reactions. These groups rarely affect mechanical properties; they may, however, negatively influence the thermal or photochemical stability of polymers.

2.1.2. Copolymers

Copolymers are generated from more than one type of monomer; they are called bipolymers, terpolymers, quaterpolymers, etc., according to the number of monomer types. The copolymerization of ethylene ($CH_2=CH_2$) and propene [$CH_2=CH(CH_3)$] thus leads to the bipolymer poly(ethylene-*co*-propene). Polymers of butadiene ($CH_2=CH-CH=CH_2$) with 1,4-units $-CH_2-CH=CH-CH_2-$ and 1,2-units $-CH_2-CH(CH=CH_2)-$ are not called copolymers because they are generated from only one monomer type. Rather, they are pseudocopolymers, such as the ones that result from partial saponification of poly(vinyl acetate), $\sim[CH_2-CH(OOCCH_3)]_n\sim$ to a pseudocopolymer with vinyl acetate units and vinyl alcohol units $-CH_2-CH(OH)-$. Thus, the term copolymer does not refer to the chemical structure of the resulting polymer, but to the monomers from which it derives.

The succession of monomeric units in copolymer chains is known as their sequence. Monomer units "a" and "b" alternate in *alternating copolymers* (Table 5), which are limiting cases of periodic copolymers $\sim(abb)_n\sim$, $\sim(aabb)_n\sim$, $\sim(abc)_n\sim$, etc.

The sequence of monomeric units in *statistical copolymers* is determined by the statistics of copolymerization (Markov statistics of zeroth, first, etc., order). *Random copolymers* are special cases of statistical copolymers: The sequence follows Bernoulli statistics (i.e., zeroth-order Markov statistics).

Graded copolymers (i.e., tapered copolymers) exhibit a compositional gradient along the chain: One chain end is enriched in "a" units, the other in "b" units. *Block copolymers* are extreme cases of such graded copolymers; they consist of blocks of homosequences that are joined via their ends. Linear multiblock copolymers with short blocks are called *segment(ed) copolymers*. In *graft copolymers*, "b" blocks are connected to "a" chains via center monomeric units.

Copolymers poly(A-co-B) of monomers A and B possess properties quite different from blends of homopolymers A and B.

Table 5. Types of copolymers from monomers A, B, and C (monomeric units are characterized by lower-case letters)

Name	Structure	Shorthand name
Copolymer without specified squende	$\sim(a/b)\sim$	poly(A-co-B)
Statistical copolymer (example)	$\sim$a-b-b-b-a-a-b-a-a-a-b$\sim$	poly(A-stat-B)
Random copolymer (with Bernoulli statistics of sequence)	ditto	poly(A-ran-B)
Alternating copolymer	$\sim$a-b-a-b-a-b-a-b-a-b$\sim$	poly(A-alt-B)
Periodic copolymer	$\sim$a-b-b-a-b-b-a-b-b-a-b$\sim$	poly(A-per-B-per-B)
	$\sim$a-b-c-a-b-c-a-b-c-a-b-c$\sim$	poly(A-per-B-per-C)
Diblock copolymer	$\sim$a . . . . .a-b . . . . .b$\sim$	poly(A)-block-poly(B)
Triblock copolymer	$\sim$a . . . . . . a-b . . . . . b-c . . . . . . c$\sim$	poly(A)-block-poly(B)-block-poly(C)
Graft copolymer	$\sim$a-a-a .... a-a-a .... a$\sim$ $||$ b_mb_n	poly(A)-graft-poly(B)

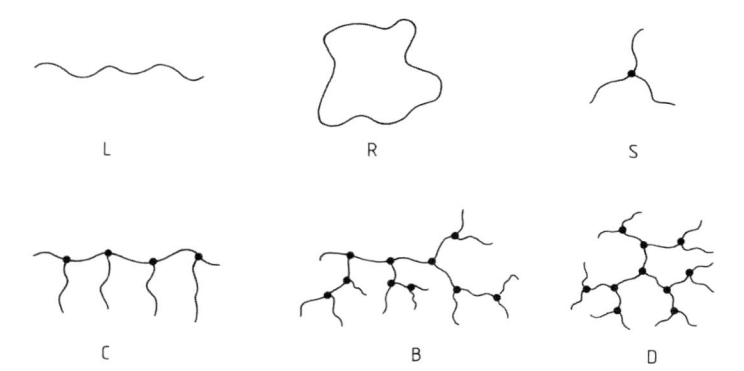

Figure 4. Schematic representation of linear and branched marcomolecules
L = Linear (unbranched) chain in almost extended state; R = ring molecule; S = starlike molecule with three subchains; C = comblike molecule with four branching points of functionality 3; B = randomly branched chain with subsequent branching; D = dendrimer with three-functional branch points and three generations of subchains

2.1.3. Branched Polymers

Open chains of the type $R-m_n-R$ exhibit the simplest structures. They possess n monomeric "m" units and two R end groups. They are also called linear chains (Fig. 4, L) because of their one-dimensional connectivity or their unbranched structures and because they were originally (and wrongly) assumed to be completely extended (stretched out). Spiropolymers and ladder polymers (see below) are also linear polymers because they are not "branched" in the macromolecular sense.

Cyclic polymers (ring polymers; Fig. 4, R) consist of linear polymer molecules that are joined via their ends. They do not have end groups and are not called macrocycles because that term is traditionally reserved for "big rings" of ca. 15–20 chain atoms.

Branched polymers contain branch points that connect three or more subchains (sometimes called subunits). The following types of branched polymers are distinguished:

1. *Star polymers* possess one branching point from which three or more subchains radiate (Fig. 4, S).
2. *Dendrimers* (dendritic polymers) are treelike star polymers in which the subchains are themselves starlike branched (Fig. 4, D). They are also called cascade polymers (because of the cascade-like sequence of branching points), isotropically branched polymers, or isobranched polymers (because each generation of newly added monomer molecules is "isotropically" arranged around the central core), or much more poetically, starburst polymers. Whereas ideal dendrimers require tedious multistep syntheses, a more industrially viable synthesis is to produce less perfect cascade macromolecules, which are known as hyperbranched polymers.
3. *Comb polymers* have "long" side chains, whose chemical structures usually differ from those of the main chains (Fig. 4, C). The term *graft copolymer* refers to comb polymers into which branches have been introduced by subsequent grafting onto a primary chain. Comb polymers whose side chains exhibit liquid crystalline (LC) behavior are called LC side-chain polymers.
4. *Randomly branched polymers* have a random distribution of branching points (Fig. 4, B); the subchains may be further branched (Christmas tree branching) or not. Short-chain and long-chain branching are distinguished. *Short-chain branches* are often generated by intramolecular transfer reactions, for example, the butyl side chain of polyethylene is formed during free-radical polymerization of ethylene:

$$\sim CH-CH_2-CH_2-CH_2-CH_3 \xrightarrow{+CH_2=CH_2} \sim CH-(CH_2)_3-CH_3$$
$$\quad\quad\quad\quad\quad\quad\quad\quad\quad\quad\quad\quad\quad |$$
$$\quad\quad\quad\quad\quad\quad\quad\quad\quad\quad\quad\quad CH_2$$
$$\quad\quad\quad\quad\quad\quad\quad\quad\quad\quad\quad\quad |$$
$$\quad\quad\quad\quad\quad\quad\quad\quad\quad\quad\quad\quad CH_2$$

Long-chain branches are usually generated by intermolecular transfer reactions. In the

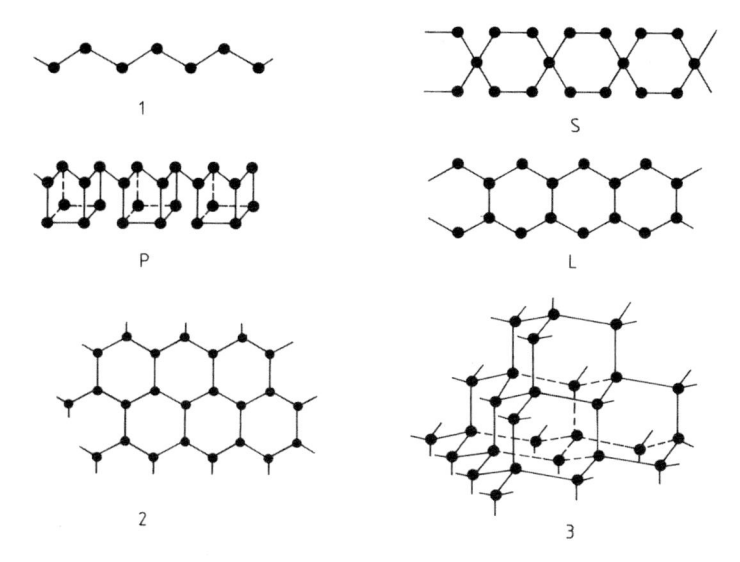

Figure 5. Schematic representation of some ordered chain assemblies
1 = Catena polymer in all-*trans* conformation; S = Spiro polymer; P = Pearl-stringlike polymer; L = Ladder polymer; 2 = Phyllo polymer; 3 = Tecto polymer

polymerization of vinyl acetate, side-chain radicals formed by transfer reactions may initiate the polymerization of additional vinyl acetate monomers.

$$\sim\!\!CH_2\!-\!\overset{\cdot}{C}H\!\!-\!\!OOC\!-\!CH_3 + CH_3\!-\!COO\!-\!\overset{\zeta}{C}H\;\longrightarrow$$

$$\sim\!\!CH_2\!-\!CH_2\!\!-\!\!OOC\!-\!CH_3 + {}^{\cdot}CH_2\!-\!COO\!-\!\overset{\zeta}{C}H$$

Such side-chain radicals are also formed by radical transfer from certain initiator radicals. Side chains introduced by the inherent structure of a (co)monomer are not called branches in polymer science. Neither linear poly(vinyl acetate) nor the low-crystallinity, low-density polymers formed by copolymerization of ethylene with small amounts of α-olefins (butene-1, hexene-1, octene-1) are considered branched; the latter are known as linear low-density polyethylenes. Although short-chain branching significantly reduces melting temperature and crystallinity of polyethylene, as reflected by reduced density, the incorporation of a few long chain branches greatly improves melt processing, such as shear thinning and melt strengthening in blow molding of polyolefin films.

2.1.4. Ordered Chain Assemblies

Two or more chains may also be joined at more than two points by chemical bonds in an ordered way. The chains are thought to be fully extended for classification purposes; a distinction is made among one-, two-, and three-dimensional assemblies or, according to IUPAC, among catena (Fig. 5, 1), phyllo (Fig. 5, 2) and tecto (Fig. 5, 3) compounds.

Linear polymers are catena polymers in this nomenclature because the fully extended chains extend in only one direction (Fig. 5, 1). Spiro polymers possess two chains with connecting atoms shared by both of them (Fig. 5, S).

Ladder polymers consist of two chains joined by bonds in regular intervals (Fig. 5, L); they are also called double-strand polymers. Such ladder polymers are generated by first forming a linear chain with pendant side groups that are subsequently polymerized. An example is the polymerization of 1,3-butadiene to 1,2-polybutadiene with subsequent cyclization to cyclopolybutadiene

$$CH_2\!=\!CH\!-\!CH\!=\!CH_2\;\longrightarrow\;\sim\!\!CH_2\!-\!\overset{}{C}H\!\!\sim\!\!\!-\!\!CH\!=\!CH_2\;\longrightarrow$$

Ladder polymers may also be synthesized by two-step polycondensation or polyaddition reactions; in general, however, only fairly short ladderlike units are formed in this way (see below). One-step ladder syntheses are very rare. Polymerization of the silicon analogue of cubane, $(C_6H_5)_8Si_8O_{12}$, does not lead to a ladder polymer of type L in Figure 5 but to a pearl-string-like polymer P (Fig. 5, P). Many silicates are ladder polymers or double-ladder polymers.

Phyllo polymers are also called layer or parquet polymers (Fig. 5, 2). Graphite and the silicates montmorillonite, bentonite, and mica are well-known examples. Cell walls of bacteria consist of bag-shaped polymers (i.e., layer polymers). Diamond and quartz are examples of tecto polymers (Fig. 5, 3). Two-dimensional carbon polymers, consisting of planar honeycomb-like sp^2-hybridzed carbons, are known as graphene which is equivalent to a single carbon layer of the graphite lattice.

2.1.5. Unordered Networks

Multifunctional oligomer and polymer molecules can be interconnected via chain groups or end groups to cross-linked polymers (networks). The cross-linking points must be at least trifunctional; that is, the cross-linkable groups must be at least bifunctional. The cross-linking bridges between chains can be short or long; the distribution of cross-linking points, at random or in clusters. Both chemical and physical bonds can be employed for cross-linking; chemically and physically cross-linked polymers are thus distinguished.

Chemical Networks. The structure of chemical networks depends on the relative amount of cross-linking sites and the state of the reactants during the cross-linking reaction. The network structure is to a first approximation independent of the chemical structure of the cross-linking sites. The relative amount of cross-linking sites determines the cross-link density. Light cross-linking does not change the mobility of chain segments between cross-links, and the resulting networks behave as elastomers if the chain segments are above the glass transition temperature. Networks with strong cross-linking or chain segments below the glass transition temperature are employed for thermosets.

Networks are fairly *homogeneous* with respect to the distribution of cross-links if they are prepared by polymerization in bulk or from homogeneous solutions and if the corresponding linear chains are soluble in their own monomers or in the applied solvent (Fig. 6,

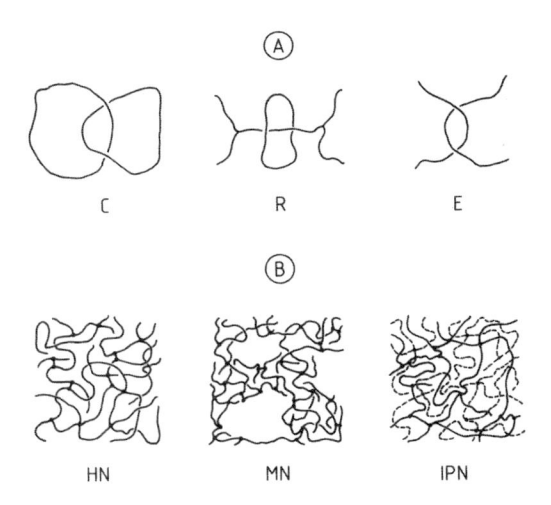

Figure 6. Schematic representation of cross-linked polymers
A) Physical cross-links; B) Chemical networks
C = catenane; E = entanglement; HN = homogeneous network; IPN = interpenetrating network; MN = macroreticular network; R = rotaxane
Reprinted with permission by Hüthig and Wepf Publ., Basel [5]

HN). Inhomogeneous networks are formed if phase separation occurs during polymerization; for example, if the low molar mass polymers generated early are insoluble in the applied solvent or the remaining monomer. The resulting *macroreticular networks* exhibit cavities (Fig. 6, MN).

In principle, two chemically different networks poly (a) and poly (b) can coexist independently of each other in interpenetrating networks (IPNs). In reality, these networks are not molecularly interdispersed; rather, each network forms interconnected domains with higher cross-link densities. Semi-interpenetrating networks consist of cross-linked poly (a) in uncross-linked poly (b).

Chemical cross-links are in most cases irreversible. These cross-linked polymers do not dissolve in solvents. With an increase in temperature, subchains decompose rather than revert to the original monomers. Chemically reversible cross-linked networks are being investigated because they would be of value in the recycling of thermosets.

Physical Networks. Physically cross-linked polymers, on the other hand, are in most cases reversibly cross-linked. Examples are microdomains in block copolymers, ion clusters in ionomers, or even crystallites in partly crystallized polymers. Physically cross-linked polymers often dissolve in suitable solvents.

Entanglements (Fig. 6, E) behave under strain as physical cross-links because the two chains cannot diffuse away rapidly enough; the chains disentangle, however, if sufficient time is allowed. Catenanes (Fig. 6, C) and rotaxanes (Fig. 6, R) have irreversible physical cross-links. Catenanes consist of two intertwined rings. Such structures are well known for deoxyribonucleic acids; they have also been observed for some synthetic polymers. Rotaxanes are rings on chains that are sealed on both ends by branched or cross-linked structures in such a way that the ring cannot slip out. Such structures are suspected to exist in some cross-linked polydimethylsiloxanes.

2.2. Molar Masses and Molar Mass Distributions

Molar masses of polymers can be varied over a wide range; they may extend from several hundreds to several millions, albeit not for every polymer type. The molar mass ranges used industrially depend on the limitations of the available polymerization procedures, the restrictions imposed by processing methods, and the desired properties of the plastics.

Presently available polymerization methods all employ statistical approaches to initiation and propagation of growing chains, as well as to termination and transfer, where applicable. The resulting polymer molecules thus show a distribution of molar masses; they are not molecularly uniform, and the experimentally determined molar masses represent averages. Synthetic copolymers are, in addition, nonhomogeneous with respect to the distribution of different types of monomer units; they are polydisperse with respect to both constitution and molar mass.

In contrast, biopolymers are synthesized via matrix polymerization. Many (if not all) biopolymers are monodisperse in their biological environments: Each macromolecule of a polymer has the same constitution and the same molar mass. Examples are nucleic acids and enzymes. Upon isolation from living cells and further processing, some degradation may occur. Industrially employed celluloses thus exhibit molar mass distributions.

Starburst dendrimers are synthetic polymers that are molecularly uniform. Their molecular homogeneity approaches that of isotopic distributions in atoms and molecules. It is caused by a self-limitation of molecular growth due to surface crowding after a certain number of growth generation steps [75].

2.2.1. Molar Mass Average

Many polymer properties depend on the type and range of molar mass distributions. In many cases, molar mass averages are determined instead, in part because it is experimentally easier and in part because a number is simpler to grasp than a function.

The averages employed in polymer science are usually arithmetic averages. The polymer molecules of size i (degree of polymerization X_i, molar mass M_i) are counted according to their statistical weight g_i, which may be their mole fraction $x_i \equiv n_i/\Sigma_i n_i$, their mass fraction $w_i \equiv m_i/\Sigma_i m_i = x_i M_i / M_\pi$, or their z-fraction defined as $Z_i \equiv z_i/\Sigma_i z_i = w_i M_i / M_w = x_i M_i^2/(M_\pi M_w)$. The number-, mass-, and z-average molar masses are defined as

$$\overline{M}_n \equiv \Sigma_i x_i M_i = \frac{\Sigma_i w_i}{\Sigma_i (w_i/M_i)}$$

$$\overline{M}_w \equiv \Sigma_i w_i M_i = \frac{\Sigma_i x_i M_i^2}{\Sigma_i x_i M_i}$$

$$\overline{M}_z \equiv \Sigma_i Z_i M_i = \frac{\Sigma_i w_i M_i^2}{\Sigma_i w_i M_i} = \frac{\Sigma_i x_i M_i^3}{\Sigma_i x_i M_i^2}$$

Each of these three average moments is essentially a single moment of the distribution of molar masses present. There are, however, molar mass averages that are composed of more than one moment, for example, molar masses from sedimentation and diffusion coefficients for certain types of molecular shapes and solvent interactions.

2.2.2. Determination of Molar Mass

Number-Average Molar Masses. Many methods exist for the determination of molar masses. Most important for the number-average molar mass is *osmometry*, which measures the osmotic pressure Π of polymer solutions at various polymer concentrations c against the solvent. Solution and solvent are separated by a semipermeable membrane that is permeable to the solvent but not to the polymer molecules.

Older osmometers determined osmotic pressure in true thermodynamic equilibrium, which was reached only after days and weeks because large volumes of solvent had to be moved through the membrane. Modern osmometers compensate for the increase in pressure difference between solution and solvent caused by the flow of solvent into the solution chamber via a servomechanism by a change of solvent pressure; equilibrium pressures are established after 10–30 min. Reduced osmotic pressures, Π/c, determined at various concentrations, are extrapolated to zero concentration to give the number-average molar mass according to the van't Hoff equation, $\overline{M}_n = RT\left[\lim_{c\to 0}(\Pi/c)\right]$. Number-average molar masses may be determined by osmometry in the range of 10^4 g/mol (lower limit of semipermeability of membranes) to ca. 10^6 g/mol (upper limit of sensitivity).

Mass-average molar masses are determined mainly via static *light scattering* of dust-free dilute solutions. Modern instruments measure the so-called Rayleigh ratio $R_o = K \cdot c \cdot \overline{M}_{w,app}$, at very low angles relative to the incident light beam, where K is an optical constant and $M_{w,\,app}$ an apparent mass-average molar mass that is calculated from R_o, c, and K. Values of $K \cdot c/R_o$ are measured at various polymer concentrations and extrapolated to zero concentration. The limiting value $\lim_{c\to 0} (K \cdot c/R_o)$ delivers the inverse mass-average molar mass, $1/\overline{M}_w$. Molar masses from several hundred to several million grams per mole can be determined by light scattering, the upper limit being given by experimental problems, such as multiple scattering. If measurements are performed over a wide range of scattering angles, important additional information about the radii of gyration of the molecules can be gained from the angular dependence of the Rayleigh ratio.

The *viscometric method* is fast and simple and therefore the most often applied method. From solvent viscosity η_o and viscosities η of dilute solutions of various concentrations c (measured at low shear rates), reduced viscosities (i.e., viscosity numbers) are calculated via $\eta_{red} = (\eta - \eta_o)/(\eta_o \cdot c)$. The reduced viscosities are extrapolated to zero concentration to give the *intrinsic viscosity* (limiting viscosity number) $[\eta]$ (DIN recommends the symbol J_o). This quantity has the physical unit of a specific volume because the concentration c is measured as mass per volume. The recommended units are milliliters per gram (cubic centimeters per gram), but data in 100 mL/g are still found in the literature; in the older literature, $[\eta]$ was given as Z_η in liters per gram.

Intrinsic viscosities measure the volume occupied by 1 g polymer at $c \to 0$, that is, the specific volumes (inverse densities) of isolated polymer molecules. Because molecular density varies with size for most molecule shapes, intrinsic viscosities can be used to measure molar masses for a homologous series of polymer molecules.

The viscosity increment η_i of unsolvated, compact spheres is, according to EINSTEIN:

$$\eta_i = (\eta - \eta_o)/\eta_o = (5/2)\phi = (5/2)(N_A V_H c/M)$$

where N_A is the Avogadro number, ϕ the volume fraction, $V_H = \rho_H/m$ the hydrodynamic volume, ρ_H the density, and m the mass of a sphere. Hard spheres from the same material have the same density. Because $m \sim M$ and $\eta_i/c \approx [\eta]$, intrinsic viscosities of spheres are independent of their molar masses (i.e., $[\eta] = K \cdot M^0$).

The hydrodynamic densities of other particle shapes, such as rods and coils change, however, with the molar mass

$$[\eta] = K \cdot M^a \tag{1}$$

The exponent $a = 2$ for rigid rods of infinite length. Theory and experiment indicate a value of $a = 1/2$ for random coils in the unperturbed state (see below); this is a limiting case for very flexible chains in theta solvents. A theta solvent is a solvent being at a temperature at which a polymer adopts the theta state, i.e., where coil molecules are in their unperturbed state with a Gaussian distribution of segments (no excluded volumes). A theta solvent is a thermodynamically bad solvent. Theory predicts $a = 0.764$ for nondraining coils of high molar mass in thermodynamically good solvents (Fig. 7). Experimentally, values of $0.5 < a < 0.764$ are often found for flexible chains (such as most

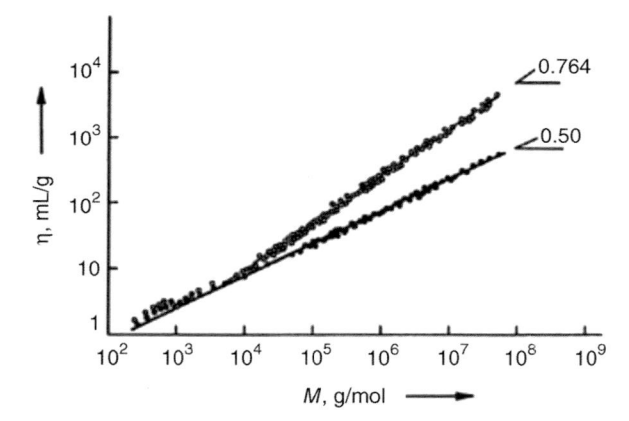

Figure 7. Molar mass dependence of intrinsic viscosities of flexible, coillike molecules: polystyrene in the theta solvent cyclohexane at 34.5°C (●) and in the thermodynamically good solvent benzene at 25°C (○) The slopes at high molar masses adopt the theoretical values of 0.764 (good solvent) and 0.500 (theta solvent)

thermoplastics in solution) because their molar masses are not high enough for the assumptions of the theory to be fulfilled.

Wormlike molecules may adopt values of $0 < a < 2$ because they resemble spheres and ellipsoids at low molar masses, rigid rods at medium ones, and random coils at very high molar masses. This behavior is found for stiff molecules such as the double helices of deoxyribonucleic acids in aqueous solutions or certain poly(α-amino acids) in helicogenic solvents (see Chap. 5).

Viscometry delivers a peculiar molar mass average, such as can be derived from Equation (1):

$$\overline{M}_\eta = ([\eta]/K)^{1/a} = \left(K^{-1} \cdot \Sigma_i w_i [\eta]_i\right)^{1/a} = \left(\Sigma_i w_i M_i^a\right)^{1/a}$$

Numerical values of viscosity-average molar masses $\overline{M}_\eta$ lie between number and mass averages for $0 < a < 1$ and become identical with mass averages if $a = 1$.

2.2.3. Molar Mass Distributions

The types and characteristic parameters of molar mass distributions of polymers are determined thermodynamically or kinetically by the synthesis conditions and, to a smaller extent, by the processing of plastics. The molar mass distributions are mathematically described by distribution functions, which may be discontinuous (discrete) or continuous and differential or integral (cumulative) (Fig. 8). In these functions, polymer molecules of size i (molar mass M_i, degree of polymerization X_i) are considered according to their statistical weights, which may be mole fractions x_i, mass fractions w_i, etc.

All molar mass distributions are discrete, because successive polymer molecules of a homologous series differ from each other by a degree of polymerization of 1 and the molar mass M_u of one monomeric unit. The discrete distribution functions can normally be replaced by continuous ones because so many different degrees of polymerization exist, and the difference in molar masses between molecules with M_i and M_{i+1} is small compared to the average molar mass.

The various types of distribution functions are usually named after their discoverers. The *Schulz–Zimm distribution* function derives from processes in which reactive chains add other monomer or polymer molecules until the chains are deactivated. Such deactivation processes may be chain termination reactions (chain polymerization, polyelimination) or the decrease of reactivity by lowering of temperature (polycondensation, polyaddition). The mass fraction of polymer molecules with degree of polymerization X_i is given by

$$w_i = \frac{(k/\overline{X}_n)^k \cdot X_i^k \cdot \exp(-kX_i/\overline{X}_n)}{\Gamma(k+1)} \quad (2)$$

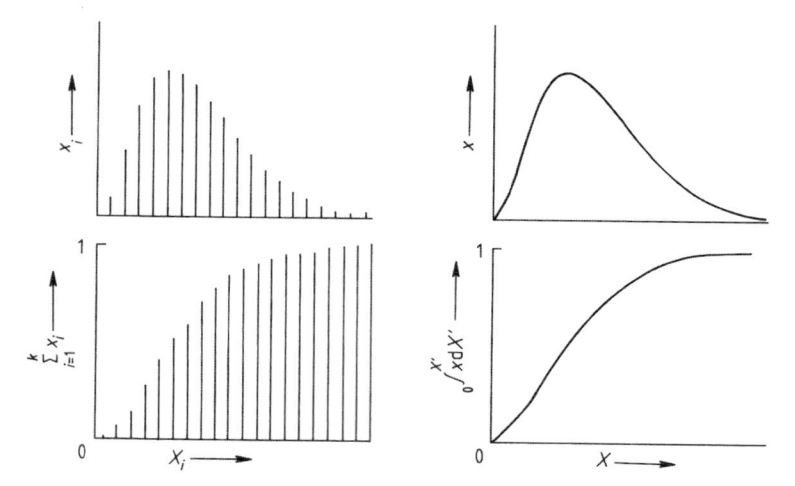

Figure 8. Distributions of mole fractions x of degrees of polymerization X
Left: discontinuous (discrete); right: continuous; top: differential; bottom: integral (cumulative)
Reprinted with permission by Hüthig and Wepf Publ., Basel [5]

where k is the degree of coupling of chains (e.g., $k = 2$ for recombination of two growing chains to one dead chain and $k = 1$ if two growing chains react by disproportionation to two dead chains) and γ ($k = 1$) the gamma function of ($k = 1$). This function is called *Schulz–Flory* or *Flory distribution* for the special case of $k = 1$ (polycondensation, radical chain polymerization with termination by disproportionation). Equation (2) reduces for $k = 1$ and high degrees of polymerization to

$$w_i = (1/\overline{X}_n)^2 \cdot X_i \cdot [1 - (1/\overline{X}_n)]^{x_i}$$

The various average degrees of polymerization are connected via

$$\overline{X}_n/k = \overline{X}_w/(k+1) = \overline{X}_z/(k+2)$$

Poisson distributions are formed if a constant number of chains starts to grow simultaneously and if monomer molecules are added to the chains at random and independent of previous additions. The mass fraction of molecules of degree of polymerization X_i is given by

$$w_i = \frac{X_i \cdot (\overline{X}_n - 1)^{X_i - 1} \cdot [\exp(1 - \overline{X}_n)]}{(X_i - 1)!(\overline{X}_n)}$$

and the interrelationship between the various averages by

$$\overline{X}_w/\overline{X}_n = 1 + (1/\overline{X}_n) - (1/\overline{X}_n)^2$$

The ratio of mass to number average approaches 1 at infinitely high molar masses. The Poisson distribution is thus a very narrow distribution, in contrast to the Schulz–Zimm distribution where the ratio $\overline{X}_w/\overline{X}_n$ equals 2 for $k = 1$ and 3/2 for $k = 2$.

The Schulz–Zimm distribution is a special case of the *Kubin distribution*, an empirical, generalized exponential distribution (GEX distribution) with three empirical, adjustable parameters γ, ε, and β:

$$w_i = \frac{\gamma \cdot \beta^{(\varepsilon+1)/\gamma} \cdot X_i^{\varepsilon} \cdot \exp(-\beta X_i^{\gamma})}{\Gamma[(\varepsilon + 1)/\gamma]}$$

This expression converts to the Schulz–Zimm distribution if $\gamma = 1$, $\varepsilon = k$, and $\beta = k\overline{X}_n$. The Kubin distribution also includes the *Tung distribution* ($\varepsilon = \gamma - 1$) and the logarithmic normal distribution ($\gamma = 0$; $\varepsilon = \infty$), both of which frequently describe molar mass distributions of polymers from Ziegler–Natta polymerizations. The Kubin distribution is a very adaptable distribution because it contains a experiments. Fractionations are inexpensive but time consuming.

The method of choice for the fast determination of molar mass distribution is *size exclusion chromatography* (SEC), also being referred to as gel permeation chromatography (GPC). Dilute polymer solutions are placed on the top of a column filled with a porous carrier. Low molar mass molecules can enter the pores, but high molar mass molecules cannot. Medium-sized molecules enter with difficulty and remain for shorter times than low molar mass molecules. Higher molar masses are thus eluted first, and elution curves are observed that give the concentration of eluted molecules as a function of the eluted volume V_e. The maximum of the elution curve is called the retention volume.

The retention volumes depend on the SEC system (carrier, solvent, temperature) and the polymer being investigated. The carrier may consist of rigid porous materials (e.g., porous glass beads) or swollen, cross-linked polymers (e.g., cross-linked polystyrenes or dextrans). In the latter case, the method is called gel permeation chromatography (synthetic polymers) or gel filtration (biopolymers).

Retention volumes are constant below and above certain molar masses of the polymers (Fig. 10). The volume V_o at lower molar masses gives the total volume available for the flow of solvent; the volume V_i at upper molar masses

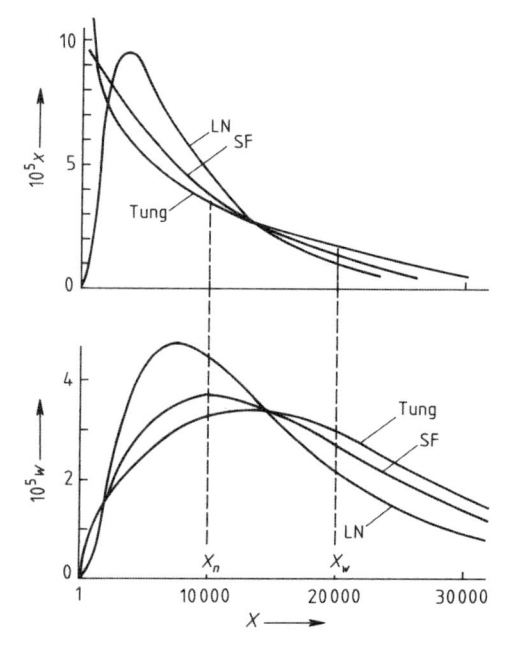

Figure 9. Continuous differential distributions of degrees of polymerization X for a) the logarithmic normal distribution (LN); b) the Schulz–Flory distribution (SF); and c) the Tung distribution (Tung), shown as distributions of mole fractions (top) and corresponding mass fractions (bottom) for a polymer with $\overline{X}_n = 10\,000$ and $\overline{X}_w = 20\,000$
The Poisson distribution is so narrow that it is practically identical with the vertical line for $\overline{X}_n = 10\,000$.
Reprinted with permission by Hüthig and Wepf Publ., Basel [5]

"stretched exponential" (i.e., a variable X with an exponent γ in the exponential term).

Some of these distribution functions are compared in Figure 9. Note that only the logarithmic normal distribution shows a maximum in the distribution of mole fractions and that none of the maxima in the distributions of the mass fractions correspond to simple molar mass averages.

2.2.4. Determination of Molar Mass Distributions

Molar mass distributions can be determined by preparative *fractionation* of polymers from solutions because the various species of a polymer-homologous series exhibit small differences in solubility. Fractionation occurs upon change of temperature or addition of a nonsolvent. The molar masses of the resulting fractions are determined in separate

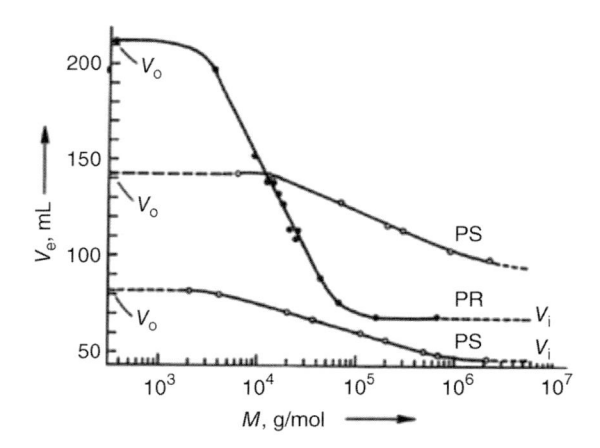

Figure 10. Molar mass dependence of elution volumes V_e
a) Coil-like linear polystyrenes PS with two different SEC columns from cross-linked polystyrenes; b) Various spheroidal proteins PR with cross-linked dextran
V_o = total volume; V_i = interstitial volume

indicates the interstitial volume between carrier particles. Polymer molecules can thus be separated only in the range $V_i < V_e < V_o$, where V_e is the elution volume.

Elution volumes depend to a first approximation on the logarithms of molar masses (Fig. 10). A function $V_e = K \cdot \ln M$ is thus often used to determine unknown molar masses with the help of a constant K derived from calibrations with narrow-distribution polystyrene standards. This procedure does not give absolute molar masses because polystyrenes and test polymers with the same molar masses possess different chromatographically effective volumes. A frequently used "universal" calibration method thus employs a function $V_e = f$ (log [η] · M) because [η] measures the specific hydrodynamic volume of the solute molecules and [η] · M has the physical unit of a molar volume.

2.3. Configurations

Polymers with symmetric repeating units, such as $-CH_2-CH_2-(-CH_2-CH_2-)$, $-CH_2-CF_2-$, and $-NH-CO(CH_2)_5-$ do not possess configurational isomers. Such isomers do exist, however, for polymer molecules with nonsymmetric repeating units, such as $-CH(CH_3)-CH_2-$ (polypropylene) and $-NH-CO-CH(CH_3)-$ (polyalanine). Polypropylene, for example, has two configurational repeating units **9** and **10** with two different monomeric units each:

$$
\begin{array}{cccc}
\mathrm{H} & \mathrm{CH_3} & \mathrm{H} & \mathrm{CH_3} \\
-\mathrm{C-CH_2-} & -\mathrm{C-CH_2-} & -\mathrm{CH_2-C-} & -\mathrm{CH_2-C-} \\
\mathrm{CH_3} & \mathrm{H} & \mathrm{CH_3} & \mathrm{H} \\
\mathbf{9a} & \mathbf{9b} & \mathbf{10a} & \mathbf{10b}
\end{array}
$$

The configurational units **9a** and **9b** are enantiomeric; they belong to the same constitutional unit $-CH(CH_3)-CH_2-$. The configurational units **9a** and **10a**, on the other hand, are based on two different constitutional units. Several configurational units may be joined to give steric repeating units; the three simplest steric repeating units for polypropylene are

$$
\begin{array}{cc}
\mathrm{H} & \mathrm{H} \quad \mathrm{CH_3} \\
-\mathrm{CH_2-} & -\mathrm{CH_2-}-\mathrm{CH_2-} \\
\mathrm{CH_3} & \mathrm{CH_3} \quad \mathrm{H} \\
\text{it} & \text{st}
\end{array}
$$

$$
\begin{array}{cccc}
\mathrm{H} & \mathrm{H} & \mathrm{CH_3} & \mathrm{CH_3} \\
-\mathrm{CH_2-}-\mathrm{CH_2-}-\mathrm{CH_2-}-\mathrm{CH_2-} \\
\mathrm{CH_3} & \mathrm{CH_3} & \mathrm{H} & \mathrm{H} \\
& & \text{ht}
\end{array}
$$

The repetition of these units leads to polymer chains that are called isotactic (it), syndiotactic (st), and heterotactic (ht). A heterotactic unit ht in a polymer chain consists of three monomeric units (the first or last three of ht), but the repetition of such ht-triads does not lead to a completely heterotactic polymer ht because it would consist of alternating heterotactic and syndiotactic triads. Whether **9a** or **10a** is used as the simplest configurational repeating unit is immaterial because infinitely long polypropylene chains of **9a** differ from those of **10a** only by the orientation of these units.

The term tacticity thus refers to the relative arrangements of configurational units in a chain. Relative configurations are classified by starting at one end of the polymer chain and considering the configuration around a central atom relative to the preceding one. This classification is different from that of the absolute configuration of organic chemistry where the configuration of each central atom is determined relative to the ligand with the lowest seniority.

The substituents R of isotactic polymer molecules are always "on the same side" if these molecules are shown in Fischer projections or in other stereo formulas with hypothetical cis conformations of the chains. In stereo formulas based on trans conformations, ligands are only on the same side for isotactic molecules consisting of two chain atoms per monomeric unit (Fig. 11), that is ~$(CHR-CH_2)$~.

Real polymer chains of the types ~CHR~, ~$CHR-CH_2$~, ~$X-CHR-CH_2$~, etc., contain configurational mistakes; they are neither 100% isotactic nor 100% syndiotactic. The tacticity of such chains is expressed by the

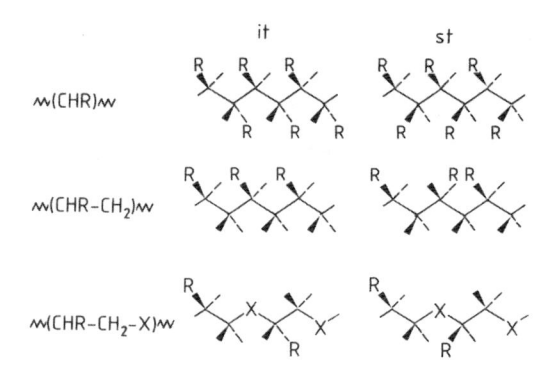

Figure 11. Isotactic (it) and syndiotactic (st) polymers with monomeric units –CHR–, –CHR–CH$_2$–, and –CHR–CH$_2$–X– in *trans* conformations
Reprinted with permission by Hüthig and Wepf Publ., Basel [5]

fraction X_J of their isotactic or syndiotactic diads, triads, etc. A diad consists of two monomeric units, a triad, of three; etc. Each monomeric unit of a polymer chain belongs to two tactic diads, three tactic triads, four tactic tetrads, and so on. Figure 11 thus shows five diads, four triads, three tetrads, etc., for ~CHR~. The sum of the mole fractions of all J-ads of given J must equal 1 ($x_i + x_s \equiv 1$, $x_{ii} + x_{is} + x_{si} + x_{ss} \equiv 1$, etc.), where i = isotactic diad, s = syndiotactic diad, ii = isotactic triad of two isotactic diads, etc. The mole fraction of isotactic diads is given by the mole fraction of isotactic triads plus 1/2 of the mole fraction of the sum of the two heterotactic triads, that is, $x_i = x_{ii} + (1/2)(x_{is} + x_{si})$, etc. The number-average sequence length of isotactic sequences is thus given by $(X_{\bar{i}})_n = 2 x_i/(x_{is} + x_{si})$.

The presence and/or fraction of tactic J-ads can be investigated by a number of experimental methods. 2D-NMR spectroscopy allows absolute determination of the types of J-ads and their amounts up to pentads, whereas conventional high-resolution NMR spectroscopy requires prior knowledge of the J-ad type by X-ray crystallography or other methods. Infrared spectroscopy detects only diads, whereas crystallinity, solubility, glass and melt temperatures, and chemical reactions may or may not indicate the presence of shorter or longer tactic sequences.

Highly isotactic polymers can be produced by Ziegler–Natta polymerization of α-olefins (e.g., it-polypropylene or it-polybutene-1). Special Ziegler–Natta catalysts lead to st-polypropylene and st-polystyrene; these syndiotactic polymers have no commercial use thus far. Highly syndiotactic polymers are also generated by most very low-temperature free-radical polymerizations of vinyl and acryl monomers. Conventional vinyl and acryl polymers are, however, synthesized at ambient temperature or above, and consist, at best, of slightly syndiotactic polymers; in general, they are considered atactic polymers.

Polymers with different geometric isomerism are also tactic. Depending on the configuration of the chain segments relative to the double bonds in the chain *cis*-tactic (ct) and *trans*-tactic (tt) structures are distinguished. The ct structure of 1,4-polyprenes[~ (CH$_2$–CR=CH–CH$_2$)$_n$ ~] from the polymerization of 1,3-dienes CH$_2$=CR–CH=CH$_2$ via both double bonds corresponds to an E isomer; the tt structure, to a Z isomer. If the polymerization proceeds via one double bond only, 1,2-polydienes are formed that may be isotactic, syndiotactic, atactic, etc.

cis-1,4 (E)   *trans*-1,4 (Z)   1,2 (it or st)

2.4. Conformations

2.4.1. Microconformations

Rotations of atoms or groups of atoms around single bonds create spatial arrangements called conformations (organic chemistry) or microconformations (macromolecular chemistry). The sequence of these microconformations determines the shape of the macromolecule, i.e., the macroconformation (in statistical mechanics, the configuration).

In principle, an infinite number of conformations are possible around each single bond. In practice, certain positions are energetically preferred; only these are called (micro)conformations. In two joined tetrahedrons, such as ethane (H$_3$C–CH$_3$), two extremes of energetically different positions are possible: The staggered

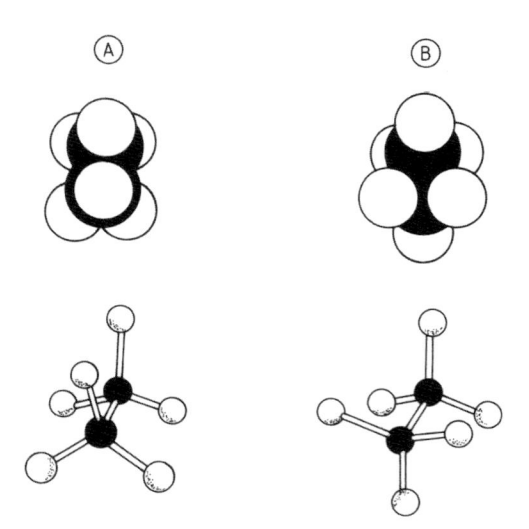

Figure 12. Microconformations of ethane
A) Eclipsed position (*cis* or synperiplanar); B) Staggered position (*trans* or antiperiplanar)
Reprinted with permission by Hüthig and Wepf Publ., Basel [5]

(one CH_3 group and two H atoms bonded to each carbon atom participating in the central C−C bond). In polymer chains, one of the bonded atoms is never equivalent because it is part of the chain; in polyethylene (∼CH_2−CH_2 ∼), the bonded atoms are two H atoms and the chain. There are two energetically different eclipsed positions (*cis* and *anti*) and two energetically different staggered ones (*trans* and *gauche*). Each *gauche* and each *anti* position can occur in two spatially different positions that are energetically equivalent (plus and minus) if two of the three bonded atoms are equivalent. These (micro)conformers have different names in macromolecular and in organic chemistry (Fig. 13).

All eclipsed positions are sterically hindered in polymer chains and therefore only *trans* and *gauche* positions must be considered. The conformational energy is the energy difference between the energies of the *trans* and *gauche* conformations. Activation energy (potential energy) is necessary to overcome the rotational barrier between *trans* and *gauche* conformations. This rotational barrier increases with decreasing length of the central bond and with increasing number and increasing size of bonded atoms. Its value is, for example, 12.1 kJ/mol for the C−C bonds in polyethylene chains ∼$(CH_2-CH_2)_n$∼ but only 2.1 kJ/mol for the CH_2−CO bonds in polyester chains with units ∼CH_2−CO−O−CH_2∼. Because these bond energies can easily be overcome by thermal energy, (hindered) rotations are possible around the chain bonds and the molecules may adopt many chain conformations (Section 2.4.3). Isolated aliphatic polyester

position corresponds to a minimum of energy and the eclipsed position to a maximum of energy if repulsive forces are prevalent (Fig. 12). On rotation by 360° around the C−C axis, three energetically equivalent eclipsed and three energetically equivalent staggered positions may be occupied as the three H atoms bound to the same C atom ("bonded atoms") are equivalent. Small molecules, such as ethane, can be considered as definite species in conformations with energy minima; they are called conformers, rotamers, or rotational isomers.

The number of types of conformers increases if the three bonded atoms are not equivalent, for example, in butane (CH_3−CH_2−CH_2−CH_3)

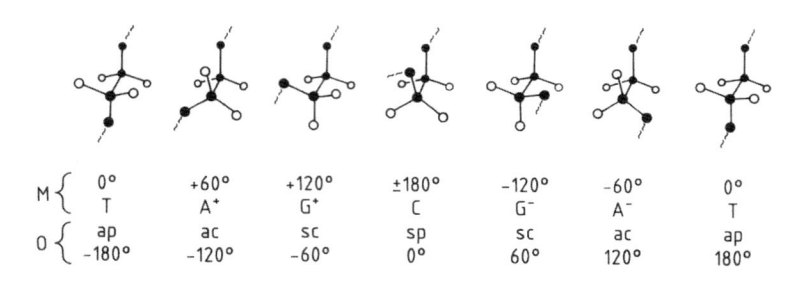

M	0° T	+60° A^+	+120° G^+	±180° C	−120° $G^−$	−60° $A^−$	0° T
O	ap −180°	ac −120°	sc −60°	sp 0°	sc 60°	ac 120°	ap 180°

Figure 13. Chain conformations and their names and symbols in macromolecular (M) and organic (O) chemistry
• Chain atoms; ○ substituents; T = *trans*, A = *anti*; G = *gauche*; C = *cis*; ap = antiperiplanar; ac = anticlinal; sc = synclinal; sp = synperiplanar

chains $\sim$OOC(CH$_2$)$_x\sim$ (with $x > 3$) are thus more flexible than carbon chains; such polymers are used as polymeric plasticizers.

Two different types of flexibility are distinguished. A chain molecule is said to be *statically flexible* if it possesses many accessible conformational minima. *Dynamic flexibilities* are characterized by low barriers between conformational minima.

2.4.2. Conformations in Ideal Polymer Crystals

A regular sequence of microconformations leads to regular macroconformations of polymer chains. The chains are linearly extended; the end-to-end distance is commonly called the contour length r_{cont} of the chain (this term originally referred to the contour of the chain along the individual chain bonds). Chains in all-*trans* microconformations are said to be fully extended (e. g., the chain in Fig. 5, 1).

Polyethylene crystallizes ideally in an all-*trans* conformation because the shortest distance between nonbonded hydrogen atoms (0.254 nm) is greater than the sum of the van der Waals radii of H atoms (0.24 nm). The size of substituents R in isotactic poly(α-olefins) [$\sim$(CH$_2$–CHR)$_n\sim$], however, forces the microconformation around each second chain bond to adopt a *gauche* position (see Fig. 13). All *gauche* conformations must be alike for steric reasons; conformational diads G$^+$G$^-$ and G$^-$G$^+$ are forbidden (Fig. 13). The chain thus adopts either a ... TG$^+$TG$^+$TG$^+$ ... or a TG$^-$TG$^-$TG$^-$ ... macroconformation, i.e., it becomes helical. The number of monomeric units per complete turn is determined mainly by the size of the immediate substituents. it-Polypropylene $\sim$[CH$_2$–CH(CH$_3$)]$_n\sim$ has three propene units per one turn (3$_1$ helix, Fig. 14 A), it-poly(4-methylpentene-1) (P4MP) $\sim$(CH$_2$–CH[CH$_2$CH(CH$_3$)$_2$])$_n\sim$ has seven units per two turns (7$_2$ helix = 3.5 helix, Fig. 14 B), and poly(3-methylbutene-1) (P3MB) has four units per one turn (4$_1$ helix, Fig. 14 C). The conformational angles are not necessarily the ideal ones of 0° for *trans* and 120° for *gauche* as found for it-polypropylene; they are rather −13°/110° for P4MP and −24°/96° for P3MB. Conformational positions with deviations up to ± 30° from the ideal conformational angles are still named after the ideal microconformations.

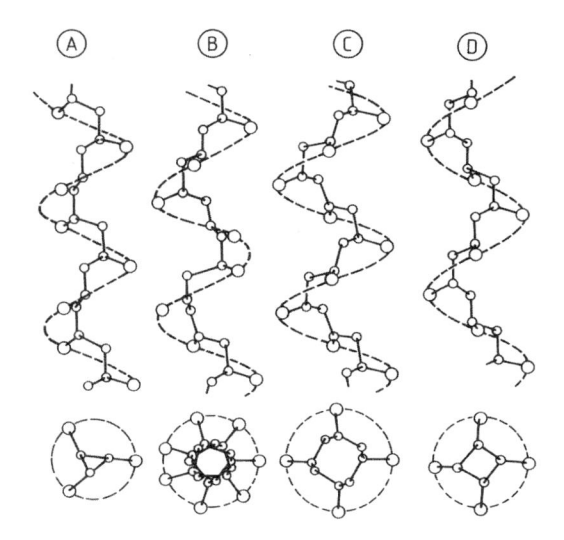

Figure 14. Helix types of poly(α-olefins) $\sim$CH$_2$–CHR$\sim$
A) 3$_1$ Helix; B) 7$_2$ Helix; C), D) Different types of 4$_1$ helices
Reprinted with permission by Societa Italiana di Fisica, Bologna, from [76]

The R substituents of syndiotactic vinyl polymers [$\sim$(CH$_2$–CHR)$_n\sim$] are farther apart than those of their isotactic counterparts (see Fig. 11). In general, *trans* conformations thus have the lowest energy in those st-polymers where only repulsive forces operate. Attractive forces, such as intramolecular hydrogen bonds between neighboring OH groups in poly(vinyl alcohol) ($\sim$[CH$_2$–CH(OH)]$_n\sim$, PVAL), lead to different conformational sequences: Isotactic PVAL exists in all-*trans* conformations, whereas syndiotactic PVAL forms helices.

Chain atoms with free electron pairs lead to *gauche* effects. The chains of crystallized polyoxymethylene [$\sim$(O–CH$_2$)$_n\sim$] exist in all-*gauche* conformation (9$_5$ helix). Polyoxyethylene [$\sim$(O–CH$_2$–CH$_2$)$_n\sim$] and polyglycine [$\sim$(NH–CO–CH$_2$)$_n\sim$], on the other hand, possess the conformational sequence TTG. The resulting 7$_2$ helices of polyglycine are stabilized by intramolecular hydrogen bonds between the first, fourth, seventh, etc., peptide bonds of a chain.

Polypeptides [$\sim$(NH–CO–CHR)$_n\sim$] are based on chiral monomeric units. An L-polymer may thus form two different helices: A right-handed and a left-handed one. These helices are diastereomers with different energy contents. In general, one "handedness" is preferred over the other in polymers with chiral base units. Poly(L-α-amino acids) generally form right-handed helices, whereas poly([S]-α-olefins) and most polysaccharides from D-sugars exist as left-handed ones.

2.4.3. Conformations in Polymer Solutions

Helices can survive melting and dissolution processes only if the helical structure is stabilized by intramolecular attractive forces. Examples are the hydrogen bonds in helices of poly(α-amino acids) or hydrogen bonds plus base stacking in double helices of deoxyribonucleic acids. These forces may be so strong that the molecules decompose rather than melt on heating. A delicate balance between intramolecular bonding and solvation of substituents is necessary to preserve helical structures in solution; examples are poly(α-amino acids) in helicogenic solvents.

Helices that are generated in crystals by packing of chains whose microconformations originate from repulsive forces do not survive the melting process intact. Each macromolecule can form many macroconformers that equilibrate rapidly. Only very short helical sequences in very low concentrations may thus exist in melts.

Two extreme cases must be considered for the dissolution of polymers. Only weak interactions (or none at all) exist between monomeric units of apolar polymers and apolar solvent molecules. Conformational changes are thus entropy driven; the sequence of microconformations is irregular, and the polymer molecule adopts the macroconformation of a coil.

In apolar enantiomeric polymers in apolar solvents, long conformational sequences are conserved, although fast conformational transformations from left- to right-handed helices and vice versa may occur in enantiomeric polymers. On average, few microconformations are converted into other ones. These helical sequences may be stabilized by association processes.

Polar solvents, on the other hand, interact strongly with polar polymers; these solvents cause strong changes in microconformations. Because the ligands around each chain bond can adopt various microconformations and these microconformations can change rapidly; a given macromolecule may exist in time in many macroconformations, similar to those of simple apolar polymers in apolar solvents:

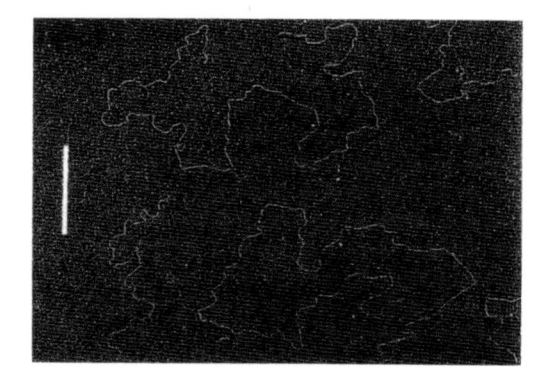

Figure 15. Electron micrograph of the double helix molecules of deoxyribonucleic acid chains showing two-dimensional random coils
Reprinted with permission by Academic Press, London, from [77]

A polymethylene $[\sim(CH_2)_n\sim]$ with degree of polymerization $n = 20\,001$ possesses 20 000 chain bonds, each of which can adopt three microconformations: *trans*, *gauche* (plus), and *gauche* (minus). According to statistics, such chain can exist in $3^n = 3^{20\,000} \approx 10^{9542}$ different macroconformations. Conversely, a collection of chains may have many macroconformations at any given time. None of these macroconformations adopts a simple geometric shape, not even instantaneously. Rather, the chains form rapidly changing coil structures that can be made visible by electron microscopy as two-dimensional projections of the three-dimensional coil shape for chains of sufficiently long chain diameter (Fig. 15).

2.4.4. Unperturbed Coils

The instantaneous shapes of coils cannot be determined by presently known methods. Their radius of gyration can be measured, however, by light scattering, small-angle X-ray scattering, and small-angle neutron scattering. The end-to-end distances of coiled linear chains can be modeled for various types of chains and calculated for individual chains via the rotational isomeric state method.

The simplest model is that of a freely jointed chain (no fixed bond angles) that assumes infinitely thin segments without interactions. The mean square end-to-end distance, r_∞^2, of such a coil is given by the number N of its unspecified segments with segment lengths b:

$$\langle r_{oo}^2 \rangle = N \cdot b^2$$

Real chains possess fixed bond angles τ between chain atoms and finite torsion (or dihedral) angles θ ($\theta = 0°$ for *trans* conformations). The mean square end-to-end distance for such a chain in the unperturbed state (zero net polymer–polymer and polymer–solvent interactions) is given by

$$\langle r_o^2 \rangle = N \cdot b^2 \cdot \left(\frac{1 - \cos\tau}{1 + \cos\tau}\right) \cdot \left(\frac{1 + \cos\theta}{1 - \cos\theta}\right)$$

$$\langle r_o^2 \rangle = N \cdot b^2 \cdot \left(\frac{1 - \cos\tau}{1 + \cos\tau}\right) \cdot \sigma^2$$

The factor σ is called the *hindrance parameter* or steric factor. Often the hindrance parameter and the bond angle are combined to give the *characteristic ratio* C_N:

$$C_N = \langle r_o^2 \rangle / (N^2 \cdot b^2)$$
$$= (1 - \cos\tau) \cdot (1 + \cos\tau)^{-1} \cdot \sigma^2$$

The characteristic ratio becomes approximately constant for chains with more than 100 atoms. A chain can also be characterized by its *Kuhn length* L_K, which can be calculated from its unperturbed mean square end-to-end distance, its degree of polymerization, and the effective length b_{eff} of monomeric units (i.e., length of units in chain direction projected on a plane; $r_{\text{cont}} = X \cdot b_{\text{eff}} = N_K \cdot L_K$)

$$\langle r_o^2 \rangle = N_K \cdot L_K^2$$

End-to-end distances can rarely be measured directly. They are however related to the experimentally accessible radii of gyration s via

$$\langle s_o^2 \rangle = \langle r_o^2 \rangle / 6$$

This relationship applies to all chains with random flight statistics (e.g., unperturbed chains and freely jointed chains). It is not valid for perturbed chains (Section 2.4.5).

Coil densities are very low (Fig. 15). The volume fraction of monomeric units is, for example, only 0.012 at the center of gyration for a polyethylene chain with a molar mass of 1.19×10^6 g/mol; 99.8% of the space of these polyethylene coils is occupied by solvent molecules (in very dilute solutions) or by units of other polymer chains (in melts).

A chain in a melt cannot distinguish between its own segments and segments of other chains. Because the interactions between the segments of various chains are also the same, such chains adopt their unperturbed dimensions, which has been confirmed by small-angle neutron scattering. Most of the coil volume is filled by segments of other chains to avoid space. Chains of sufficient length may thus become entangled, which manifests itself in properties, such as diffusion and melt viscosity.

2.4.5. Perturbed Coils

Coils adopt their unperturbed dimensions in melts or in certain solvents at certain temperatures (so-called theta solvents). In such theta solvents, polymer–solvent and polymer–polymer interactions cancel each other and the chain behaves as if it is infinitely thin. In thermodynamically good solvents, polymer–solvent interactions dominate and the coil is swollen. The volume requirements of such coils must thus lead to interpenetrations by other coils even at very low concentrations. Because segments are not infinitely thin, however, a part of the total space is excluded for segments of other chains (and also for other segments of the same chain).

2.4.6. Wormlike Chains

Real polymer chains are not totally flexible. Their finite thickness and the partially hindered rotation around chain bonds prevent them from adopting all possible positions in space. Such chains can be described by the model of the wormlike chain (Kratky–Porod model).

The characteristic parameter of this model is the *persistence length* a. This parameter is defined as the average of the projection of the end-to-end-distance of an infinitely long and infinitely thin chain in the direction of the first segment. It can be calculated from the radius of gyration and the conventional contour length via

$$\langle s_0^2 \rangle = a^2\{(y/3) - 1 + (2-y) - (2/y^2)[1 - \exp(-y)]\}$$

where $y = r_{cont}/a$. For flexible chains, it is related to the unperturbed radius of gyration via

$$\langle r_0^2 \rangle / 6 = \langle s_0^2 \rangle = a \cdot r_{cont}/3 = a \cdot N_K \cdot L_K/3$$

For infinitely stiff chains,

$$\langle s_0^2 \rangle = a^2 \cdot y^2/12 = (r_{cont})^2/12$$

which is an infinitely stiff chain, behaves like an infinitely thin rod. Wormlike chains thus describe the whole transition from rodlike molecules (small y) to random coils (large y). The model is strictly valid for infinitely thin chains, but the error produced by this assumption is negligible if the persistence length is much greater than the chain diameter.

3. Polymer Manufacture

3.1. Raw Materials [78]

The raw materials for most industrially used plastics are synthetic polymers and prepolymers; only a few are derived from naturally occurring polymers or monomers. The overwhelming majority of polymers for plastics are organic materials; very few are semiorganics (i.e., inorganic polymer chains with organic substituents). Raw materials for polymers include biomass, such as wood and fossil resources, such as coal, petroleum, natural gas, and gases supplied by fracking. In view of the increasing energy demand and the limited fossil resources, an important objective is to substitute fossil resources by renewable ones. Another important strategy aims at direct chemical fixation of carbon dioxide to produce monomers and polymers directly from carbon dioxide without endangering food supply.

3.1.1. Wood

Wood ($\rightarrow$ Wood) is a naturally occurring composite of oriented cellulose fibers in a continuous matrix of cross-linked lignin; it is plasticized by water and "foamed" by air (vacuoles). The water content varies from 40–60% for green wood to 10–20% for air-dried wood. Solids include cellulose (42%); hemicelluloses (28–38%); lignin (19–28%); and proteins, resins, and waxes (2–3%), depending on the type and age of the plant. Most of the wood is used directly as fuel or for construction purposes, but one-sixth is converted into wood pulp or cellulose pulp for the manufacture of paper, cardboard, or rayon. A very small amount of wood is filled with monomers that are subsequently polymerized to give polymer wood.

Repeating unit (cellobiose) of cellulose (R=H) and cellulose derivatives

Cellulose pulp is chemically transformed into cellulose derivatives ($\rightarrow$ Cellulose; $\rightarrow$ Cellulose Esters; $\rightarrow$ Cellulose Ethers):

- 2,5-Acetate with 2,5 acetate groups, R = OOC–CH$_3$, per glucose residue (fibers, cigarette filters)
- Cellulose triacetate with three acetate groups per glucose residue (fibers)
- Cellulose acetobutyrate with R=29–6 mol% acetyl and 17–48 mol% butyryl groups (a thermoplastic for tubes, packaging)
- Cellulose nitrate (R–NO$_2$; various degrees of substitution; gun cotton, films, lacquers, celluloid)
- Cellulose ethers with R–CH$_3$, C$_2$H$_5$, CH$_2$CH$_2$OR^1, or CH$_2$–CHOR^2–CH$_3$, where R^1 may be CH$_2$CH$_2$OR^1 or CH$_2$CH$_2$OH and R^2–CH$_2$–CHOR^2–CH$_3$ or CH$_2$–CHOH–CH$_3$ (thickeners, binders, suspension agents, injection molding materials, etc.)

Huge amounts of *lignin sulfonates* ($\rightarrow$ Lignin) are generated during wood pulp production, most of which are burned. Small amounts are used for street pavement, as binders for foundry sand, or as drilling agents, and even smaller amounts as raw materials for organic intermediates ($\rightarrow$ Lignin). Degradation products of lignin have been utilized for the synthesis of ion-exchange resins and other polymers. Lignocellulose of forestry and agricultural wastes represents an attractive source of glucose as a renewable feedstock for biorefineries without competing with food production. In biorefineries glucose is converted into a variety of monomers including lactic acid, butadiene, acrylic acid, and even ethylene and propene.

3.1.2. Coal

Coals ($\rightarrow$ Coal) are polymerized and partly cross-linked hydrocarbons with average compositions between $C_{75}H_{140}O_{56}N_2S$ (peat) and $C_{240}H_{90}O_4NS$ (anthracite). They were the primary raw materials for acetylene-based monomers during the first half of the 20th century but have since been replaced in most countries by oil and natural gas as monomer sources.

Coals deliver coke and coal tar. A number of aliphatic monomers are produced from coke via acetylene, such as chloroprene, various vinyl monomers, acrylonitrile, and hexamethylenediamine. Coal tar delivers aromatics, such as benzene, xylene, phenol, and phthalic anhydride, which lead subsequently to polystyrene, phenolic resins, and glyptal resins, to name a few.

3.1.3. Natural Gas and Fracking

Natural gas ($\rightarrow$ Natural Gas) is a gas with a high proportion of aliphatic hydrocarbons. European gas is rich in methane, whereas American and Saudi Arabian gases are relatively rich in higher hydrocarbons. Natural gas is processed to synthesis gas, ethylene, and acetylene. These gases are used to produce a variety of monomers for polymers. The increasing exploitation of fracking enables the production of hydrocarbons which offer great potential as cost-effective alternative to crude oil.

3.1.4. Petroleum

Petroleum (crude oil) ($\rightarrow$ Oil Refining) is the main feedstock for monomers. It consists of 95–98% hydrocarbons and 2–5% oxygen, nitrogen, and sulfur compounds. The hydrocarbons are largely aliphatic, partly naphthenic, and to a small extent aromatic, depending on the source. Petroleum distillation leads to saturated hydrocarbons (gas, naphtha, gasoline, kerosene, diesel fuel, heating oil, etc.). For monomer production, these hydrocarbons are subsequently cracked to mixtures of olefins that are fractionated by distillation. The resulting compounds are used as such or are converted further into the desired monomers.

The polymer industry is a major petrochemical customer: About 60% of the cracking and subsequent products of naphtha are used to produce polymers (ca. 50% for plastics–elastomers, ca. 10% for fibers).

3.1.5. Other Renewable Resources and Biorefineries [79–83]

Vegetable oils are a relatively large source of raw materials. They have high hydrocarbon content and consist of mixtures of fatty acid triglycerides and are usually subdivided into drying oils with high linolenic and linoleic acid content, semidrying oils with high linoleic and oleic acid content, and nondrying oils with high oleic acid content ($\rightarrow$ Fats and Fatty Oils). Some oils are used directly for paints; others are chemically converted into monomers.

Castor oil is converted into methyl ricinoleate, which is thermally cracked into methyl undecenate and heptanal. Methyl undecenate is the raw material for 11-aminoundecanoic acid, which is subsequently polycondensed to form polyamide 11. Alkali scission of castor oil or ricinoleic acid produces sebacic acid, a monomer used to prepare polyamide PA6,10. Moreover, castor is propoxylated and end-tipped with ethylene oxide in order to produce polyols, thus enabling the formation of polyurethanes containing up to 30 wt % renewable resources.

Soybean oil is converted by a series of chemical reactions into the methyl ester of the C_{10} amino acid $H_2N(CH_2)_9COOCH_3$, which is the monomer for PA 10. The plant *Crambe abyssinica* contains about 55% of erucic acid, which leads to the C_{13} amino acid and subsequently to PA 13 ($\sim[NH(CH_2)_{12}CO]_n\sim$).

Corn husks and other agricultural waste are rich in *pentosans*, which can be converted into tetrahydrofuran and further into polytetrahydrofuran

[HO(CH$_2$CH$_2$CH$_2$CH$_2$O)$_n$H], which serves as a soft block in certain polyurethanes. Moreover, epoxidized soybean oil is produced in large scale and employed as plasticizers for PVC.

Other Natural Oils. Although the production of polymers based upon soybean oil feedstocks may compete with food production, other nonfood oils, such as linseed and tung oil are attractive intermediates for plant-oil based polymers.

Bio-Based Monomers from Biorefineries. Progress made in biotechnology and in the production of biofuels is stimulating the development of new processes for exploiting biomass conversion in biorefineries as intermediates for bio-based plastics. In principle, there are three different strategies for biomass conversion (i) biomass-to-liquid conversion, (ii) biorefining exploiting glucose, bioethanol, glycerol, and plant oils as raw materials, and (iii) direct catalytic liquefaction and hydrothermal carbonization of biomass. Figure 16 gives an overview on the different strategies to produce bio-based monomers. Similar to the Fischer–Tropsch chemistry in coal-to-liquid conversion, biomass is converted by partial oxidation into syngas consisting of a mixture of carbon monoxide and hydrogen, which serves as feedstock for monomer production. This is a rather energy intensive process that is sensitive to catalyst poisoning by biomass components. A more energy-efficient process is the biotechnological conversion of glucose, preferably gained from nonfood sources, such as lignocellulose from agricultural and forestry wastes, into ethanol and a variety of other monomers, such as glycols and lactic acid. Bioethanol is readily converted into ethylene by splitting off water. In addition, butadiene and propene can be produced from bioethanol. In biodiesel production, vegetable oils are converted in their methyl ester by transesterification of oils with methanol. The byproduct glycerol can be converted into a variety of monomers including epichlorhydrin, acrylic acid, and polyols (see Fig. 16).

Whereas biomass-to-liquid conversion is energy intensive, the catalytic direct conversion of biomass into diesel represents another attractive green route to renewable oil. In this process, plastics wastes as well as biomass can be employed as carbon source. In the third approach, entire plants can be converted into coal by hydrothermal carbonization. This is an exothermic process that operates at much lower temperature than the Fischer–Tropsch process. This renewable coal obtained by hydrothermal biomass carbonization is an attractive feedstock

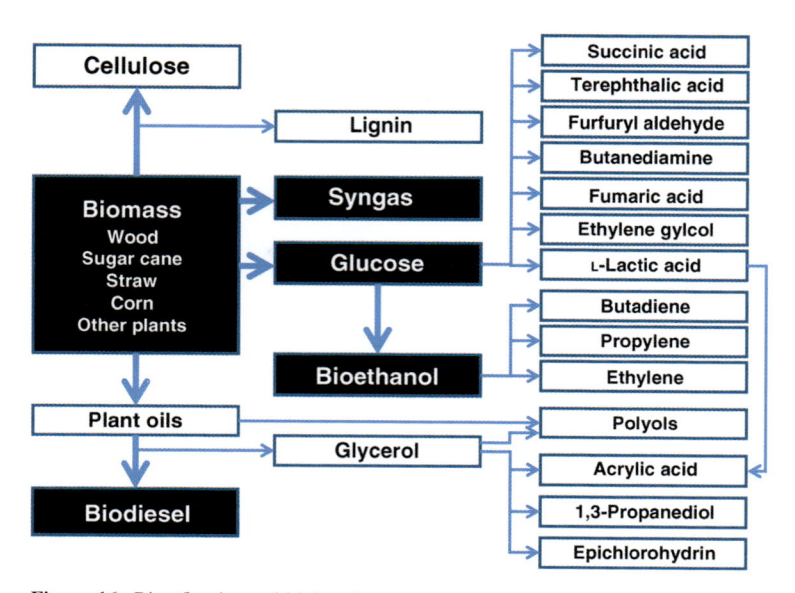

Figure 16. Biorefineries and biobased monomers

for catalytic liquefaction and production of renewable oil.

Carbon dioxide that is emitted by power plants, is the energy sink in combustion processes. Activation and chemical fixation of carbon dioxide requires energy, which can be supplied from renewable resources, thus enabling chemical energy storage of excess energy from wind and solar power plants. In contrast to traditional conversion of carbon dioxide into carbon monoxide and hydrogen by reduction with coal at very high temperatures, oil and gas are produced by electrochemical reduction of carbon dioxide in water. However, this process is not yet feasible for the production of monomers on industrial scale. Another more viable approach exploits the direct conversion of carbon dioxide with reactants, such as epoxides with a high energy content. Thus cyclic carbonates as well as linear aliphatic polycarbonates are available. Polyfunctional cyclic carbonates serve as intermediates for the formation of green polyurethanes without requiring phosgene or isocyanates.

3.2. Polymer Syntheses: Overview [84]

3.2.1. Classifications

Polymers may be synthesized from monomers by polymerization or from other polymers by chemical transformations. The nomenclature and classification of these reactions have grown and changed with the times; they are not very systematic and are sometimes confusing.

All reactions of small monomer molecules to macromolecules are called *polymerizations*. Polymerizations are classified according to the:

- Origin of polymers (bio and chemical polymer synthesis)
- Chemical structure of monomers (vinyl polymerizations, diene polymerizations, ring-opening polymerizations, etc.)
- Chemical structure of resulting polymers (linear, branching, cross-linking, isomerizing, cyclopolymerization, etc.)
- Relative composition of monomers and monomer units (addition, condensation, etc.)
- Formation of low molar mass byproducts (polycondensation)
- Type of polymerization initiation (thermal, catalytic, photochemical, enzymatic, etc.)
- Type of propagating species (free radical, anionic, cationic, polyinsertion, etc.)
- Type of mechanism (chain growth, step growth, living, controlled, catalytic, etc.)
- Reaction media (bulk, solution, emulsion, suspension, mesophase, etc.)
- State of matter (gaseous, homogeneous, heterogeneous, super critical, etc.)

The resulting four types of polymerization can be depicted schematically as

$P_n + P_m \rightarrow P_{n+m}$ $\quad$ $P_n + M \rightarrow P_{n+1}$
Polyaddition $\quad\quad\quad\quad$ Chain polymerization

$P_n + P_m \rightarrow P_{n+m} + L$ $\quad$ $P_n + M \rightarrow P_{n+1} + L$
Polycondensation $\quad\quad\quad$ Polyelimination

where M denotes monomer; P_n, P_m, P_{n+m}, P_{n+}, denote polymers and L leaving molecules (e.g., water in polyesterifications of diacids and diols).

The names of three of these four types have been used with different meanings by different researchers since the 1930s. The following discussion thus centers on the present nomenclature recommendations by IUPAC.

Chain-Growth Polymerization. Polymer chains grow by repeated addition of monomer molecules M to polymer molecules P_i, consisting of i monomeric units m, an initiator end group R, and an "active center ∗," without the formation of leaving molecules:

$R(m_n)m^* + M \rightarrow R(m_{n+1})m^*$

An example is the polymerization of vinyl monomers (CH_2=CHR) by chain carriers Y∗, which may be free radicals Y·, anions Y^-, or cations Y^+:

Y–CH_2–CHR* + CH_2=CHR → Y–CH_2–CHR–CH_2–CHR*

A *living polymerization* is a chain polymerization consisting only of fast initiating reactions (Y∗ + CH_2=CHR → Y–CH_2–CHR∗) and propagation reactions

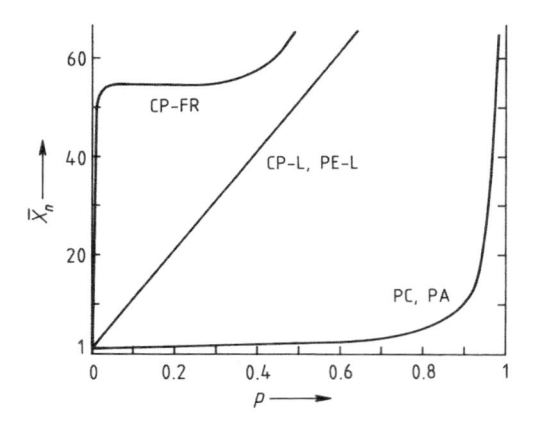

Figure 17. Change of number-average degree of polymerization $\bar{X}_n X^-$ with the extent of reaction of functional monomer groups p for living chain polymerizations CP–L and living polyeliminations PE–L (with initial monomer–initiator ratio 100 : 1 mol/mol), free-radical chain polymerizations CP–FR with gel effect (schematic for high initiator concentrations), and equilibrium polycondensations PC and polyadditions PA (numerically exact)

[Y–(CH$_2$–CHR)$_n$* + CH$_2$=CHR → Y–(CH$_2$–CHR)$_{n+1}$*] without termination of the growing chains and without side reactions. In such reactions, the degree of polymerization is directly proportional to the conversion u of monomer molecules or p of functional groups, respectively (Fig. 17, curve CP–L). An example is the polymerization of styrene CH$_2$=CH(C$_6$H$_5$) by the butyl anion C$_4$H$_9^-$ from butyllithium, LiC$_4$H$_9$.

The variation of the degree of polymerization with monomer conversion is quite different for *radical polymerizations*, such as the polymerization of styrene by benzoyloxy radicals C$_6$H$_5$COO˙ from the thermal decomposition of dibenzoyl peroxide, C$_6$H$_5$COO–OOCC$_6$H$_5$. Radicals are here formed successively and not "at once" as are the butyl anions in living polymerizations. Growing polymer chains are terminated by reaction with other radicals (recombination, disproportionation). Many side reactions may occur because of the high and fairly indiscriminate reactivity of radicals. Depending on reaction conditions, the degree of polymerization may remain constant (primarily at low conversions) or increase with monomer conversion (so-called gel effect, also referred to as autoacceleration or Trommsdorff effect) (Fig. 17, curve CP–FR). Whereas in living chain-growth polymerizations all chain ends are active during polymerization, in controlled polymerization reactions there is a rapid equilibrium between temporarily inactive ("dormant") chain ends and active chain ends. Similar to living polymerization, controlled chain-growth polymerizations afford narrow molar mass distribution, first order polymerization kinetics, and a degree of polymerization which is proportional to conversion.

Chain-growth polymerizations have generally been called *addition polymerizations* in the literature. Addition polymerization is not to be confused with polyaddition (see below).

Condensative Chain-Growth Polymerization is the name recommended by IUPAC for a polymerization by repeated addition of monomer molecules with elimination of low molar mass molecules L; this polymerization has thus also been called *polyelimination* [5]:

$$R(m_n)m^* + M \rightarrow R(m_{n+1})m^* + L \qquad (3)$$

Biological polymerizations to polysaccharides seem to proceed exclusively through polyeliminations. Cellulose of cotton, for example, is formed by polymerization of guanosine diphosphate D-glucose attached to a lipid matrix. The reaction proceeds via insertion of the glucose moiety into the chain with elimination of guanosine diphosphate.

Synthetic polymers are rarely formed by polyelimination. Examples are the polymerizations of α-amino acid N-carboxyanhydrides Leuchs anhydrides) and the "oxidative polymerization" of 2,6-dimethylphenol to polyphenylene oxide (PPO, PPE):

Polycondensation [85]. The step-growth polymerization, known as polycondensation, proceeds by reaction between molecules of all degrees of polymerization (not just between a polymer molecule and a monomer molecule as in chain polymerization and polyelimination) resulting in the formation of low molar mass molecules as byproducts. Many polymers with hetero atoms in the chains are generated by polycondensation. An example is the polycondensation of 11-aminoundecanoic acid to polyamide 11:

$$n\, H_2N(CH_2)_{10}COOH \rightarrow H[NH(CH_2)_{10}CO]_nOH + (n-1)H_2O \quad (4)$$

In these reactions, dimers ($n = 2$) are formed first, which then react to form tetramers ($n = 4$), etc. The reaction is thus statistical with respect to the size of the molecules to which the functional groups are attached. The degree of polymerization of the polymer is a number average $\overline{X}_n$ over all degrees of polymerizations of the polymer molecules.

All amino groups have the same chance to react if their reactivity is independent of molecule size. The reaction of amino groups attached to molecules with higher molar mass adds, however, considerably more to the increase of the degree of polymerization than the reaction of lower molar mass molecules: The degree of polymerization snowballs with increasing extent of reaction of functional groups (Fig. 17, PC). High conversion of functional groups and no side reactions are prime requirements for achieving high molar mass in step-growth polymerization. Molar mass is controlled either by adjusting the conversion of functional groups or by adding monofunctional chain terminating agents. The term polycondensation was used in the past as a synonym for condensation polymerization. The latter term included both the current concepts of polycondensation and polyelimination (condensative chain polymerization).

Polyaddition. This is a polymerization by reaction among molecules of all degrees of polymerization but without the formation of low molar mass byproducts. The number-average degree of polymerization of the resulting polymers varies with the extent of reaction of functional groups in the same way as it does in polycondensation. An example of polyaddition is the polymerization of diisocyanate with diol to form polyurethane:

$$n\, OCN-R-NCO + n\, HO-R'-OH$$
$$\rightarrow OCN(R-NH-CO-O-R'O)_nH \quad (5)$$

These polymerizations have also been called "polyadduct formations." Polyaddition is not to be confused with addition polymerization, the term used in the literature for chain-growth polymerization.

3.2.2. Functionality

The functionality of monomers is essential for controlling polymer constitution. (e.g., see Eq. 3). When difunctional monomers, such as divinylbenzenes $CH_2=CH-C_6H_4-CH=CH_2$, are added to the chain-growth polymerization of styrene, first branching and later cross-linking takes place. The second vinyl group of a divinylbenzene possesses, however, a different reactivity after polymerization of the first one. The two reactivities may differ so strongly in certain ionic polymerizations of divinylbenzenes that linear polymers, and not cross-linked ones, result (e.g., in the cationic polymerization of 1,4-divinylbenzene by acetyl perchlorate in dichloromethane).

The isocyanate group $-N=C=O$ of diisocyanates can react with hydroxy groups of diols to produce linear polyurethanes by step-growth polyaddition (Eq. 5). It may, however, undergo a chain-growth polymerization to form polyisocyanates via the polymerization of the N=C double bond

$$n\, R-N=C=O \rightarrow (NR-CO)_n$$

It should be noted that isocyanates react with urethane to afford branching via allophanate groups (Eq. 6)

$$\sim NH-CO-O\sim + O=C=N\sim$$
$$\longrightarrow \sim NH-CO-\underset{?}{N}-CO-O\sim \quad (6)$$

The functionality f of polymer molecules is given by the functionality f_o of the monomer molecules and the number X of mers if no intramolecular rings are formed between functional groups.

$$f = 2(f_o - 1) + (X - 2)(f_o - 2)$$
$$= 2 + X(f_o - 2)$$

A monomer with functionality $f_o = 3$ thus gives a dimer ($X = 2$) with a functionality $f = 4$ and a 100-mer ($X = 100$) with a functionality $f = 102$. Because the molecule needs to be bifunctional only for the formation of linear molecules, the "extra" functionalities lead to interconnections between various molecules (branching) and thus finally to cross-linked, "infinitely large" molecules.

3.2.3. Cyclopolymerizations

Polymerizations of multifunctional molecules may proceed not only intermolecularly to produce branched and cross-linked molecules but also intra/intermolecularly with the formation of rings in chains. These polymerizations are called cyclopolymerizations. They occur especially with monomers having carbon–carbon double bonds in the 1,5- or 1,6-position. An example is acrylic anhydride (**11**), $CH_2=CH-CO-O-OC-CH=CH_2$, which polymerizes free-radically even at high monomer concentrations to soluble polymers with 90–100% six-membered (**12**) and five-membered (**13**) ring structures, and only 0–10% "linear" monomeric units **14** (R symbolizes a polymer chain and ∗ an active chain end):

3.3. Chain-Growth Polymerization

Chain-growth polymerization and polyelimination (condensative chain polymerization) agree in their polymerization kinetics, molar mass distributions, etc., if similar polymerization mechanisms are compared. They differ in the formation of low molar mass byproducts and thus in the design of industrial polymerization reactors.

3.3.1. Thermodynamics

Polymerizations occur only if their molar standard Gibbs energies ΔG_p^o are negative. Because the Gibbs energy

$$\Delta G_p^o = \Delta H_p^o - T \cdot \Delta S_p^o = -RT \cdot \ln K \tag{7}$$

is determined by both the polymerization enthalpy ΔH_p^o and the polymerization entropy ΔS_p^o, four different cases are possible:

Case 1. Both enthalpy and entropy are negative; the entropy term $-T \cdot \Delta S_p^o$ becomes more positive with increasing temperature. At a certain temperature, the enthalpy and entropy terms balance each other; no polymerization to high molar mass molecules is possible above this *ceiling temperature*. Oligomers, however, may be formed at the ceiling temperature, because polymerization equilibria consist of a series of consecutive equilibria between monomer and growing chains of different degrees of polymerization.

Case 1 is the most common case. It is found for monomers with polymerizable double and triple bonds, such as C=C, C=O, C=S, or C=N. The polymerization entropies of such monomers are practically determined only by the loss of translational entropy; they are thus almost independent of the chemical structure of the monomers. The polymerization enthalpy of α-olefins $CH_2=CHR$ is influenced little by resonance stabilization and is practically independent of the size of the substituent R. Steric hindrance, however, strongly decreases the polymerization enthalpy of 1,1-disubstituted compounds $CH_2=CRR'$. Styrene and α-methylstyrene, have practically the same polymerization entropies (-105 J K^{-1} mol^{-1}) but very different polymerization enthalpies (-71 vs. -35 kJ/mol, respectively) for the polymerization of liquid monomers to condensed polymers. The ceiling temperatures in bulk are 303 °C (styrene) and 60 °C (α-methylstyrene), (i.e., the latter cannot be homopolymerized to high molar mass compounds above 60 °C).

σ-Bonds are opened and formed again in ring-opening polymerization of cyclic monomers. Differences of bond energies in monomers and polymers are thus practically zero, and the polymerization enthalpy is determined by delocalization and strain energies. Polymerization enthalpies are very negative for ring-opening polymerization of three-membered rings, increase to about zero for five- or six-membered rings, drop again to negative values, and return almost to zero for very large rings. The polymerization entropies of very small rings are very negative due to the release of rotational entropy upon polymerization. They become less negative with increasing ring size.

Case 2. The polymerization enthalpy is negative (or zero) and the polymerization entropy positive. Polymerization is possible at all temperatures in this rare case, the polymerization of cyclooctamethyltetrasiloxane being an example.

Case 3. The polymerization enthalpy is positive (or zero) and the polymerization entropy is negative. No polymerization is possible at any temperature. This case seems to apply to the polymerization of acetone.

Case 4. Both polymerization enthalpy and polymerization entropy are positive. A *floor temperature* exists below which no polymerization is possible. This rare case is fulfilled for the polymerization of tightly packed cyclic monomers, such as cyclooctasulfur or oxacycloheptane (oxepane) since the number of rotatory degrees of freedom increases strongly upon polymerization to open chains.

Polymerization enthalpies and entropies are influenced by the state of monomers and polymers (gaseous, liquid, crystalline, etc.), the physical interaction between reactants themselves or reactants and solvents, and pressure. Especially important for industrial polymerizations is the fact that most (but not all, see above) chain polymerizations are exothermic. The heat of polymerization can be considerable: Adiabatic polymerization of gaseous ethylene to crystalline polyethylene would result in a temperature increase of 1 800 K for complete monomer conversion. Polymerization reactors must be designed in such a way that the heat of polymerization can be removed rapidly because, otherwise, inhomogeneous charges may result or the reactor may even explode.

3.3.2. Free-Radical Polymerizations [86]

Free-radical polymerizations are initiated and propagated by free radicals. Such polymerizations are relatively insensitive to impurities in the reaction mixture and fairly easy to control. They are thus the methods of choice for the industrial production of plastics if the monomers can be subjected to free-radical polymerizations. This is true for ethylene and many ethylene derivatives: Most commodity plastics are thus obtained from free-radical polymerizations (Table 6).

Other industrial polymers formed by free-radical homopolymerizations include poly(vinyl acetate) (for coatings and adhesives), poly(acryl ester) (for adhesives), polyacrylamide (for thickeners), polyacrylonitrile (for fibers), polychloroprene (for elastomers), poly(acryl acid) (for thickeners), and poly(N-vinylpyrrolidone) (for various applications).

Table 6. Plastics by free-radical chain-growth polymerizations[*]

Monomers	Polymerization in						Monomeric units
	Gas phase	Bulk	Suspension	Emulsion	Solution	Precipitation	
Living polymerization							
p-Xylylene	+						$-CH_2-(p-C_6H_4)-CH_2-$
Irreversible polymerization to linear or slightly branched polymers							
Ethylene		+	+	(+)	(+)	(+)	$-CH_2-CH_2-$
Styrene		+	+	(+)	(+)		$-CH_2-CH(C_6H_5)-$
p-Methylstyrene			+	+			$-CH_2-CH(p-CH_3C_6H_4)-$
Vinyl acetate		(+)	(+)	+	(+)		$-CH_2-CH(OOCCH_3)-$
Vinyl chloride	+	(+)	(+)	+	(+)		$-CH_2-CHCl-$
Vinyl fluoride			+				$-CH_2-CHF-$
Vinylidene fluoride			+	+			$-CH_2-CF_2-$
Trifluorochloroethylene			+				$-CF_2-CFCl-$
Tetrafluoroethylene			+				$-CF_2-CF_2-$
Methyl methacrylate		+	+	+	+		$-CH_2-C(CH_3)(COOCH_3)-$
Cross-linking polymerization							
Diallyl phthalate		+					see Equation (7)

[*] + = major processes; (+) = minor processes.
Other industrial polymers by free radical homopolymerizations include poly(vinyl acetate) (for coatings and adhesives), poly(acryl ester) (for adhesives), polyacrylamide (for thickeners), polyacrylonitrile (for fibers), polychloroprene (for elastomers), poly(acrylic acid) (for thickeners), and poly(N-vinyl pyrrolidone) (for various applications).

Polymerizations may be carried out in the gas phase, bulk, suspension, emulsion, solution, or under precipitation. Most monomers are bifunctional; they lead to linear or slightly branched polymers and thus to thermoplastics. Thermosets obtained from free-radical homopolymerizations include bis- and trimethacrylates as well as diallyl and triallyl compounds that cross-link under polymerization conditions:

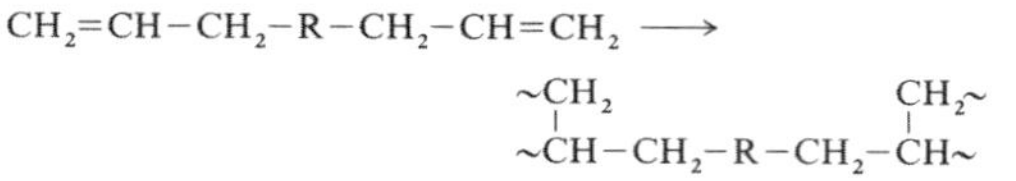

$$CH_2=CH-CH_2-R-CH_2-CH=CH_2 \longrightarrow$$

$$\sim\underset{|}{CH}-CH_2-R-CH_2-\underset{|}{CH}\sim$$
$$CH_2CH_2$$

where R may be $COO(CH_2)_2O(CH_2)_2OOC$ [diethylene glycol bis(allyl carbonate)], $o\text{-}C_6H_4(COO)_2$ (diallyl phthalate), or $m\text{-}C_6H_4(COO)_2$ (diallyl isophthalate). Other allyl compounds are used as cross-linking comonomers (diallyl fumarate, diallyl maleate, triallyl cyanurate).

Most free-radical polymerizations follow the kinetics of true chain reactions. In initiating (start) reactions (Eq. 9), radicals $R^\bullet$ are formed by the decomposition (Eq. 8) of initiator molecules I, that react with monomer molecules M. The resulting polymeric free radicals add further monomer molecules in the propagation reaction (Eq. 10) until the growing polymer radicals are terminated by recombination (Eq. 11) or disproportionation (Eq. 12) with other polymer radicals or by addition of initiator radicals (Eq. 13):

$$I \rightarrow 2\,R^\bullet \tag{8}$$

$$R^\bullet + M \rightarrow R-M^\bullet \tag{9}$$

$$R-M^\bullet + M \rightarrow R-M-M^\bullet, \text{etc.} \rightarrow R-M_{n-1}-M^\bullet \tag{10}$$

$$R-M_{n-1}-M^\bullet + {}^\bullet M-M_{m-1}-R \rightarrow R-M_{n+m}-R \tag{11}$$

$$R-M_{n-1}-M^\bullet + {}^\bullet M-M_{m-1}-R \rightarrow R-M_{n-2}-M$$
$$= M + X-M_m-R \tag{12}$$

$$R-M_{n-1}-M^\bullet + R^\bullet \rightarrow R-M_n-R \tag{13}$$

In Equation (12), X is an atom or a group that is transferred from the ultimate monomeric unit of one polymer radical to another polymer radical, for example, the H atom in the reaction

$$CH_2-C^\bullet HR + {}^\bullet CHR-CH_2 \rightarrow CH_2-CH_2R + CHR=CH_2$$

These reactions are kinetically classified as termination reactions and not as transfers because they lead to dead chains. The term "transfer" is rather reserved for chemical transfers that generate new propagating radicals, such as

$$CH_2-C^\bullet HR + AX \rightarrow CH_2-CH_2RX + A^\bullet \tag{14}$$

Initiation. Free-radical polymerizations can be initiated thermally by thermal initiators, by redox initiators, by photoinitiators, or electrolytically.

Thermal initiation is utilized in the polymerization of *p*-cyclophane to poly(*p*-xylylene):

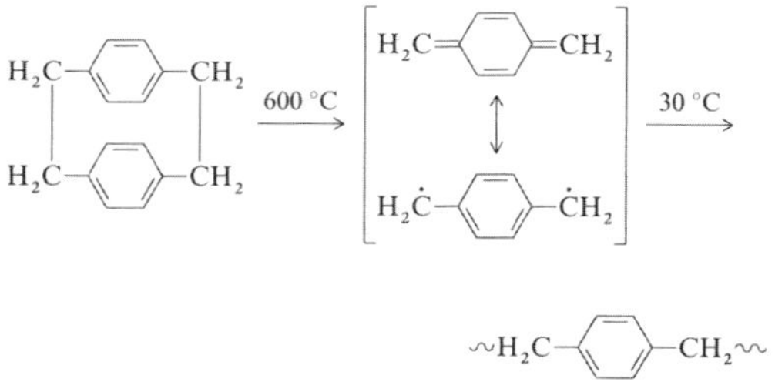

Propagation proceeds by addition of biradicals of various sizes i.e., by polyaddition and not by chain polymerization. The resulting polymer is cross-linked because radicals also attack methylene groups, which generates more than two radical sites per reactant and thus cross-linking sites. The reaction is utilized for the formation of thin coatings.

Styrene is the only monomer that is polymerized commercially by thermal initiation through monoradicals. The primary Diels–Alder adduct of two styrene molecules reacts with another styrene molecule to give two different starter radicals:

Only relatively small amounts of styrene are, however, polymerized commercially by this method. Most styrene polymerizations, like many other commercial free-radical polymerizations, are started by radicals from *thermal initiators*. These thermal initiators dissociate homolytically into two radicals at elevated temperature, usually 60–80 °C. For the bulk polymerization of styrene, high-temperature initiators, such as 1,2-dimethyl-1,2-diethyl-1,1-diphenylethane or vinylsilane triacetate are preferred. For styrene polymerization in suspension, dibenzoyl peroxide (BPO) ($C_6H_5CO-O-O-OCC_6H_5$) and *tert*-butyl perbenzoate [$C_6H_5CO-O-O-C(CH_3)_3$] are used.

Many other bulk polymerizations employ diisopropyl peroxydicarbonate (IPP), which decomposes according to

$$(CH_3)_2CHO-COO-OOC-OCH(CH_3)_2 \rightarrow 2(CH_3)_2CHO-COO^\bullet$$

Water-soluble initiators, such as dipotassium persulfate ($K_2S_2O_8$) and certain redox initiators are used in emulsion polymerizations. *Redox initiators* generate radicals by reaction of a reducing agent with an oxidizing agent (e.g., $Fe^{2+} + H_2O_2$). These systems need far less activation energy for the formation of radicals, and polymerization can thus be initiated at low temperature.

Photochemical initiations are utilized for the production of lithographic plates (→ Imaging Technology, 3. Imaging in Graphics Art) and for hardening lacquers but not in the manufacture of plastics. *Electrolytic polymerizations* find applications in the coating of metal sheets by plastics.

Only initiator radicals are formed at the early stages of free-radical polymerizations. These radicals are successively converted into monomer radicals by addition of initiator radicals to monomer molecules and into polymer radicals by further addition of monomers. At the same time, some polymer radicals are removed by termination reactions. Finally, a steady state is established in which as many radicals are formed as are removed by termination. This state of constant total radical concentration (ca. 10^{-8} mol/L) is in general observed within seconds at monomer conversions of 10^{-2}–10^{-4}%.

Propagation. Monomer molecules are usually added head-to-tail to growing polymer radicals:

$$CH_2-CHR^\bullet + CH_2=CHR \rightarrow CH_2-CHR-CH_2-CHR^\bullet$$

Tail-to-tail additions also occur (see Section 2.1).

Free-radical polymerizations lead in general to atactic or predominantly syndiotactic polymers. The syndiotacticity increases with decreasing polymerization temperature.

Termination. Termination reactions by recombination, disproportionation, and (at higher initiator concentrations) addition of initiator radicals, cannot be avoided. Because initiator radicals and thus also polymer radicals are formed successively and are deactivated at random by different termination reactions, distributions of molar masses are formed. The width of these distributions is given by the relative proportion of the different types of termination reactions.

Chain Transfer. The transfers of radicals to monomers, solvents, and initiators terminate polymer chains and generate others (Eq. 14). The ratios of rate constants k_{tr} of transfer to rate constants k_p of propagation are called transfer constants, $C_{tr} = k_{tr}/k_p$; at 60 °C they possess values between ca. 2×10^{-6} for the transfer of polystyryl radicals to benzene and 5 700 for the transfer of poly(vinyl acetate) radicals to carbon tetrabromide.

Chain transfers often lead to new radicals that have approximately the same reactivity as the disappearing ones. In this case, the rate of polymerization is not decreased by chain transfer, but the degree of polymerization is lowered. Certain transfer agents with high transfer constants (such as thiols) are thus often employed to regulate molar mass.

Inhibition and retardation occur if the new radical is more sluggish in adding monomers than the transferring polymer radical. In the case of *inhibition*, stable radicals are formed by chain transfer reactions. Typically, inhibitors, such as methylhydroquinone, are added. Moreover, retardation is observed when chain transfer to nonionic allyl monomers generates resonance-stabilized allyl radicals. The addition of monomer molecules to these radicals produces radicals that are more stabilized by resonance than their precursors and thus do not start a polymer chain. The polymerization of noncharged monoallyl monomers thus leads to oligomers with degrees of polymerization of ca. 10–20.

Controlled Radical Polymerization *[87, 88].* Reversible deactivation is the key feature of controlled radical polymerization. Unlike living polymerization, the free radicals at the chain end are rendered temporarily inactive ("dormant"). Because there is a rapid equilibrium between free radical and dormant sites, controlled radical polymerization exhibits all features typical for living polymerization, i.e., the degree of polymerization is proportional to conversion and the rate of polymerization follows first order kinetics. The reversible trapping of free radicals prevents undesirable chain termination by recombination and disproportionation.

$$P\text{-}X \xrightleftharpoons[k_c]{k_d} P\cdot \overset{k_p}{\longrightarrow} + X\cdot$$

By using controlled radical polymerization techniques, it is possible to synthesize block-, graft-, and star copolymers as well as polymers with narrow molar mass distribution and well-defined end group functionalities. In *nitroxide-mediated controlled radical polymerization* (NMRP), stable nitroxide free radicals, such as (2,2,6,6-tetramethylpiperidin-1-yl)oxyl (TEMPO) radicals are added to a free radical polymerization. Immediately after the nitroxide radical traps the free radial at the polymer chain end, thermal bond cleavage reinitiates chain growth by free radical polymerization. Because free radicals are temporarily trapped, the polymerization rate and monomer conversions are lower than in a living anionic polymerization.

In *atom transfer radical polymerization* (ATRP) organic halides react with redox metals, such as Cu^+, to produce organic free radicals via transfer of the halogen atom. This free radical at the chain end is temporarily terminated by transfer of the halogen, accompanied by changing of the oxidation state resulting from electron transfer of the redox metal. Although copper is preferred for ATRP, various other redox transition metals can be employed. The addition of ligands is required to prevent precipitation of the metal complex.

In *reversible addition chain tranfer polymerization* (RAFT) the polymer free radical

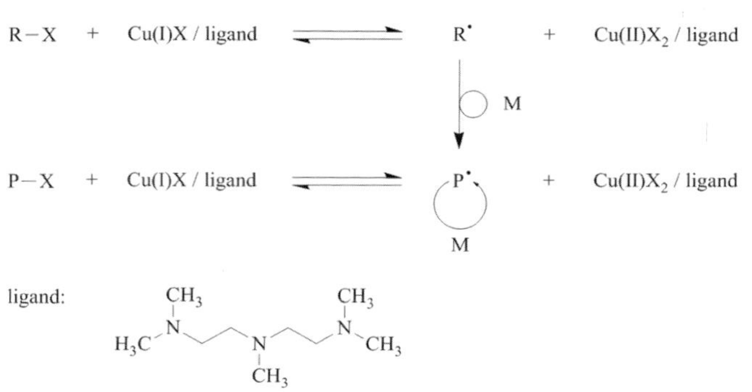

R—X + Cu(I)X / ligand ⇌ R• + Cu(II)X$_2$ / ligand

↓ M

P—X + Cu(I)X / ligand ⇌ P• + Cu(II)X$_2$ / ligand

M

ligand: H$_3$C—N(CH$_3$)—CH$_2$CH$_2$—N(CH$_3$)—CH$_2$CH$_2$—N(CH$_3$)—CH$_3$

adds to a thiocarbonyl group of dithioesters, dithiocarbamate or dithiocarbonates, forming a free radical that undergoes fragmentation to form a polymer radical and thiocarbonyl group.

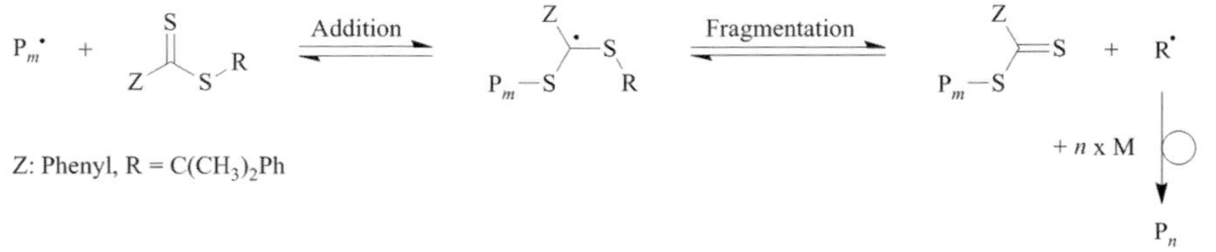

P$_m$• + (Z)C(=S)S—R →Addition→ Z—C(•)(S—P$_m$)(S—R) →Fragmentation→ Z—C(=S)—S—P$_m$ + R•

+ n x M ↓

P$_n$

Z: Phenyl, R = C(CH$_3$)$_2$Ph

Polymerization Kinetics of Free Radical Polymerization. Initiator radicals are generated by initiator decomposition (Eq. 15) with a rate of $2f \cdot k_d \cdot [I]$ where f is the radical yield (i.e., the fraction of radicals that start the polymerization). Initiator radicals disappear in the start reaction with a rate of $k_{st}[P][M]$. In the steady state, as recognized by BODENSTEIN, the polymer radical concentration is constant. Hence, the rate of the start reaction equals the rate of the termination reaction(s), thus $R_{st} = R_t$, and for terminations through deactivation by other polymer radicals,

$$d[P]/dt = R_{st} - R_t = 2f \cdot k_d \cdot [I] - k_t[P]^2 = 0 \quad (15)$$

Because monomer is practically consumed only by the propagation reaction with a rate of $R_p = -d[M]/dt = k_p[P][M]$, the propagation rate at diminishing initiator consumption is obtained from Equation (16)

$$R_p = -d[M]/dt = k_p(2f \cdot k_d/k_t)^{1/2} \cdot [I]_0^{1/2}[M] \quad (16)$$

The polymerization rate is directly proportional to the monomer concentration; bulk polymerization is thus always faster than solution polymerization. Because the kinetic chain length is equivalent to R_p/R_t, the degree of polymerization increases with increasing monomer concentration. Because ethylene has rather low reactivity, as reflected by its low k_p value, high pressure (i.e., high monomer concentration) and high temperature are required for achieving free radical ethylene polymerization. Due to inter- and intramolecular chain transfer reactions, the resulting polyethylene chains contain long- and short-chain alkyl branches. Whereas small amounts of long chain branches account for shear thinning during melt processing, the incorporation of short chain propyl- and butyl-branches impairs crystallization and lowers the polyethylene density. Therefore, high pressure free radical polymerization produces low density polyethylene (LDPE).

A "self-acceleration" of the polymerization and a concomitant increase of both the polymerization rate and degree of polymerization are observed at higher monomer conversion, especially in bulk and in concentrated solutions. This so-called gel-effect, also known as the Trommsdorff effect, is caused by a changing diffusion control of the termination by mutual reaction of two polymer radicals (i.e., a decrease of termination reactions), which in turn comes from an increasing entanglement of polymer chains. Because neither the rate of the initiator decomposition nor the rate of monomer addition is affected by increasing entanglements, and only the termination is slowed down, the Bodenstein steady-state principle does not apply anymore. Hence, both overall polymerization rates and molar masses simultaneously increase. This autoacceleration broadens molar mass distributions. Owing to the drastically reduced heat transfer at high viscosity, the autoacceleration can cause severe increase of temperature and even explosions, resulting from pyrolysis of polymers and gas evolution at temperatures above 400 °C. The addition of chain transfer agents, such as alkylthiols reduces molecular masses, thus preventing viscosity buildup and autoacceleration.

Cross-Linking Polymerization. Free-radical chain polymerizations are employed commercially in the manufacture of diallyl thermosets and in styrene−divinylbenzene copolymerizations for polymer beads. Primary macromolecules produced at very low monomer conversions are almost linear and carry pendant polymerizable groups. The likelihood of an attack at these groups increases with increasing mass average molar mass. Branched molecules are formed that convert to cross-linked polymers at the gel point. Monomer conversion at the gel point cannot at present be predicted theoretically for such free-radical polymerizations.

Process Engineering. Free-radical polymerizations may be performed in the gas phase, in bulk, solution, emulsion, suspension, or under precipitation (Table 6). Each method has advantages and disadvantages.

Bulk Polymerizations. Industrial bulk polymerizations are performed in *liquid monomers*, occasionally with the addition of 5–15 vol% solvent as a polymerization aid. The process leads to very pure products. However, much heat is generated per unit volume and, in addition, by the Trommsdorff effect. Local overheating may cause multimodal molar mass distributions, branching, polymer degradation, discoloring, or explosion. Bulk polymerizations are thus often terminated at 40–60% monomer conversion. The remaining monomer is distilled and recovered. Alternatively, polymerizations may be performed in two steps: First in large batch reactors, then in thin layers. Residual monomer is usually removed by steam.

Suspension polymerization is a "water-cooled" bulk polymerization. Water-insoluble monomers are suspended as small droplets of 0.001–1 cm diameter with the help of suspending agents, such as water-soluble polymers. The polymerization is initiated by oil-soluble initiators. The droplets are converted by polymerization into pearls; thus the process is also called *pearl polymerization*. The addition of protective colloids, such as water-soluble polymers, affords steric stabilization of the formed polymer particles and prevents their agglomeration. Suspension polymerization allows easy control of the reaction and delivers the polymer as easy-to-handle beads. A disadvantage is the cost of water removal and cleaning, along with the sometimes deleterious effects of incorporated suspension agents. The average size of spherical polymer particles decreases with increasing stirring speed. Prominent applications of styrene pearl polymerization include the formation of polystyrene pellets containing pentane as foaming agent and the production of cross-linked polystyrene beads that are sulfonated to produce polymeric cation exchange resins for water deionization.

Emulsion Polymerization. In emulsion polymerizations, water-insoluble monomers are solubilized in water by micelles of surfactant molecules (4–10 nm diameter); the monomers are also suspended in a few droplets (ca. 1 000 nm diameter). Water-soluble initiators decompose into radicals that travel through the water to the micelles where polymerization begins. Polymerization takes place in this micellar reaction compartment, referred to as polymer *latex*. Diffusion of monomer from the droplets to the latex replenishes the monomer that has been used up by reaction in the micelles. The average diameter of these latex particles is 500–5 000 nm. Each latex particle contains just one free radical, which is terminated when another initiator radical enters the latex particle; this remains dormant until a new initiator radical initiates the growth of a new chain. Because the reaction volume is very small in micelles (and in growing latex particles), termination reactions are rare, and polymerization proceeds in a quasi-living manner, leading to much higher molar masses than bulk polymerization. In terms of reaction control, emulsion polymerizations have advantages similar to suspension polymerizations. In addition, they offer higher polymerization rates by redox initiators and higher molar masses (which can be regulated by transfer agents). Emulsion polymerization is applied to produce homo- and copolymers based upon styrene, butadiene, vinyl chloride, vinyl acetate, acrylates as well as tetrafluoroethylene. The resulting latex can furthermore be used directly for producing water-borne paints, adhesives, and coatings forming corrosion resistant polymer films when water vaporizes and the polymers entangle.

Gas-phase polymerizations can be initiated photochemically. Each growing particle contains only one radical and the polymerization is thus quasi-living. The polymerization rate is determined by the rate of monomer adsorption in the precipitating particles and later by monomer diffusion to the occluded polymer radicals. The resulting polymers are very clean.

Solution polymerizations decrease polymerization rates by monomer dilution and allow easy removal of the heat of polymerization. Solvents are, however, costly to remove, and the process is used only if the resulting polymer solutions can be used directly or if monomers and polymers decompose in melts.

3.3.3. Anionic Polymerizations [89]

Anionic polymerizations are initiated by bases, such as alkali metals, alkoxides, amines, phosphines, Grignard compounds, and sodium naphthalene, very often in solution. The initiators dissociate "instantaneously" into the initiating species. These dissociations and the subsequent start reactions need little activation energy; anionic polymerizations thus often proceed with high speed even at −100 °C.

The initiating species and the growing macroions are rarely completely dissociated into anions and counterions. They are rather in equilibrium with various types of ion pairs (nondissociated species, contact ion pair, solvent-separated ion pair, free ions plus associates of these species). The type and amount of initiating species strongly affect the polymerization rates and the tacticity of polymers.

The initiating species are present in their effective concentrations at the beginning of polymerization; they are not formed one after another as in free-radical polymerization. Termination reactions are fairly rare if protons are absent. Hence, anionic polymerizations are often "living." Very low initiator concentrations are required for living polymerizations to high molar masses (see Section 3.2.1); however, high monomer–initiator ratios are difficult to control and may lead to low polymerization rates. In general, polymers from living polymerizations have very narrow molar mass distributions of the Poisson type if diffusion effects are avoided during the mixing of monomer and initiator solutions.

Contrary to traditional living polymerizations, a variety of controlled anionic polymerizations have been developed. Key feature is a rapid equilibrium between active and dormant endgroups. For example, in *group transfer polymerization (GTP)* there is a rapid equilibrium between inactive silyl ketene acetals and

Table 7. Industrial anionic homopolymerizations

Monomer	Repeating unit	Application
Butadiene	$\sim CH_2-CH=CH-CH_2\sim$	elastomer (1,4-*cis*)
Isoprene	$\sim CH_2-C(CH_3)=CH-CH_2\sim$	elastomer (1,4-*cis*)
Methyl cyanoacrylate	$\sim CH_2C(CN)(COOCH_3)\sim$	adhesive
Formaldehyde	$\sim O-CH_2\sim$	engineering plastics
Ethylene oxide	$\sim O-CH_2-CH_2\sim$	thickener
Glycolide	$\sim O-CO-CH_2\sim$	surgical threads
ε-Caprolactone	$\sim O-CO-(CH_2)_5\sim$	polymer plasticizer
ε-Caprolactam	$\sim NH-CO-(CH_2)_5\sim$	fiber, thermoplastic
Lauryllactam	$\sim NH-CO-(CH_2)_{11}\sim$	fiber, film
Hexamethylcyclotrisiloxane	$\sim O-Si(CH_3)_2\sim$	elastomer

enolate anions, catalyzed by nucleophilic catalysts ($[(CH_3)_3SiF_2]^-$, $[HF_2]^-$, CN^-, etc.):

$$(CH_3)_3SiO\diagdown \quad CH_3 \atop CH_3O\diagup C=C-CH_2-H \xrightarrow{+CH_3OOC-C(CH_3)=CH_2}$$

$$(CH_3)_3SiO\diagdown \quad CH_3 \quad\quad CH_3 \atop CH_3O\diagup C=C-CH_2-\underset{COOCH_3}{C}-CH_2-H$$

Very few anionic polymerizations are employed for the production of thermoplastics, most notably those of formaldehyde, ε-caprolactam (also used in reaction injection molding), and laurolactam (Table 7). Anionic polymerizations are, however, the method of choice for the syntheses of block copolymers, such as thermoplastic elastomers, e.g., polystyrene−block−polybutadiene−block− polystyrene (SBS).

3.3.4. Cationic Polymerizations [90]

Three groups of monomers can be polymerized cationically: (i) Olefin derivatives $CH_2=CHR$ with electron-rich double bonds, (ii) monomers with double bonds containing heteroatoms or heterogroups Z (e.g., $CH_2=Z$), and (iii) rings with heteroatoms. Initiators are Brønsted acids (perchloric acid, trichloroacetic acid, trifluoromethanesulfonic acid, etc.); Lewis acids ($AlCl_3$, $TiCl_4$, $SnCl_4$, etc.) with "coinitiators", such as water; and carbenium salts (acetyl perchlorate, tropylium hexachloroantimonate, etc.). The propagating macrocations are thermodynamically and kinetically unstable. They attempt to stabilize themselves by addition of nucleophilic species, which leads to very fast polymerization on one hand and to a host of transfer and termination reactions on the other. Very few monomers are thus polymerized cationically on an industrial scale (Table 8). Prominent examples include polyisobutene and poly(isobutene-co-isoprene) rubber.

3.3.5. Ziegler−Natta [91–95], ROMP [96] and ADMET [97] Catalysis

Ziegler−Natta polymerization catalysis involves a chain-growth polymerization on Ziegler catalysts, comprising a vast group of transition-metal compounds activated with main group metal alkyls, preferably aluminum alkyls. Today, in line with pioneering

Table 8. Industrial cationic polymerizations

Monomer		Application**
Name	Structure*	
Isobutene	$CH_2=C(CH_3)_2$	elastomers, adhesives, VI improvers
Vinyl ethers	$CH_2=CHOR$	adhesives, textile aids, plasticizers
Formaldehyde	$CH_2=O$	engineering plastic
Ethyleneimine	c-$(NHCH_2CH_2)$	paper additive, flocculant
Tetrahydrofuran	c-$[O(CH_2)_4]$	soft segment for polyurethanes or polyether–ester elastomers

* c = cyclo.
** VI = viscosity index.

advances at Phillips using activator-free silica supported Cr(III) catalysts, a variety of cationic transition metal alkyls containing weakly coordinating counterions are available that do not require the addition of main group metal activators. Because monomers are inserted into the transition metal alkyl bonds, this catalytic polymerization is named polyinsertion. In older literature, polyinsertions are also referred to as coordination polymerizations or anionic–coordination polymerizations, although coordination of a monomer to an initiator need not lead to a Ziegler–Natta type of polymerization catalysis. Whereas free-radical polymerization of ethylene requires temperatures above 150 °C and pressures well above 1 000 bar and produces branched low-density polyethylenes, catalytic ethylene polymerization takes place at ambient temperature and pressure below 10 bar, thus producing linear high density polyethylene. Short chain branches are introduced by copolymerization of ethylene with α-olefins. Stereospecific catalytic diene polymerization produces highly stereoregular poly(cis-1,4-isoprene) and poly(cis-1,4-butadiene) as synthetic alternative to natural rubber. Both KARL ZIEGLER and GIULIO NATTA were awarded the Nobel Prize for Chemistry in 1963, honoring their groundbreaking research, which paved the way for the industrial production of polyolefins. Half of the annual world production of polymers is based on this process.

Typical Ziegler catalysts for industrial polymerizations of olefins, such as ethylene, propene are shown in Table 9. Metallocenes and post-metallocene complexes have been developed since the early 1980s to produce tailor-made polyolefins in which the molecular architectures of polyolefins are tuned as a function of the molecular architectures of the catalytically active transition metal complexes. Traditionally, Ziegler catalysts are multi-site catalysts that contain different catalytically active sites and produce polymer chains with different molecular masses and molar-mass dependent comonomer sequence distribution. In contrast, modern single-site catalysts, which are based on metallocene and post-metallocene complexes, contain a single type of catalytically active site and give much more uniform polymers; molecular mass and molar-mass independent comonomer incorporation can be controlled easily.

Homogeneous catalysts are used in solution polymerization forming elastomers, such as ethylene–propene copolymers, which can contain a small amount of nonconjugated dienes, such as ethylidene norbornene as cross-linking sites. Slurry polymerizations, in which polymers precipitate during polymerization, require heterogeneous catalysts in order to prevent reactor fouling. Incorporating active transition metal alkyl sites on high surface area supports boosts catalyst activity and enables control of both polyolefin molecular architectures and polyolefin particle morphologies. In slurry polymerizations, most soluble catalysts produce small submicron dustlike polyolefin particles, thus causing severe reactor fouling and handling problems. Owing to fragmentation of supported catalysts during slurry polymerization (see Fig. 18), much larger micron-sized polyolefin particles are formed, which prevent reactor fouling. Moreover, spherical supported catalysts produce spherical pellet-sized polyolefin particles,

Table 9. Important industrial polymerizations with transition-metal catalysts

Monomer	Initiator	Application
Ethylene	$MgX_2/TiY_4/AlR_3$ (X = Cl, R, OR; Y = Cl, R, OR); (post)metallocene (Ti, Zr, Hf, Fe, Co), halfsandwich (Ti) and Cr(III) catalysts supported on silica	thermoplastic (HDPE, LLDPE)
Propene	$MgCl_2$/1,3-diether/$TiCl_4$/AlR_3; (post)metallocene (Ti, Zr, Hf) supported on silica	thermoplastic (PP)
Ethylene+propene+nonconjugated diene	Metallocene and halfsandwich catalysts, $VOCl_3$/R_2AlCl	elastomer (EPDM)
Butadiene	Co-, Ni-, Nd-, Sm-, Ln-catalysts, (post)metallocene and halfsandwich catalysts,	elastomer
Isoprene	Ti-, Co-, Ni-, Nd-, Sm-, Ln-catalysts, (post)metallocene and halfsandwich catalysts	elastomer
Butene-1	$MgCl_2$/1,3-diether/$TiCl_4$/AlR_3; (post)metallocene (Ti, Zr, Hf) supported on silica$TiCl_3$–Et_2AlCl	thermoplastic

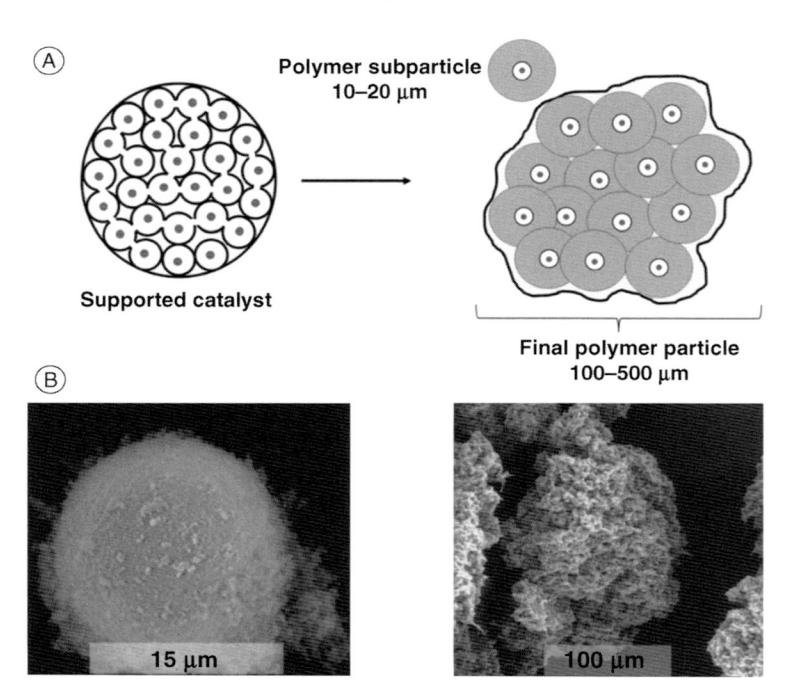

Figure 18. Morphology control in catalytic olefin polymerization by fragmentation of catalyst particles during polymerization A) Fragmentation of catalyst particle; B) Scanning Electron Microscope image of a catalyst particle (left) and a polyethylene particle (right)

thus eliminating the need for pelletizing extrusion.

The development of highly active and stereoselective supported catalysts was essential for enabling solvent-free gas phase and liquid pool olefin polymerization. The choice of the support materials and the selection of supporting strategies depend on the selected transition metal component. Because $MgCl_2$ has the same crystal structure as $TiCl_3$, it can substitute inactive bulk $TiCl_3$ that does not have vacant coordination sites. Hence, $MgCl_2$ supported titanium catalysts, activated with aluminum alkyls, afford orders of magnitude higher catalyst activities. In propene polymerization, the addition of small amounts of Lewis bases, such as 1,3-diethers promotes high stereospecificity, leading to the formation of highly isotactic polypropylene. Most (post)metallocene and chromium catalysts are preferably immobilized on high surface area silica supports.

Breakthroughs in catalyst and process technologies have greatly simplified olefin polymerization. With catalyst activities exceeding 1 000 kg polyolefin/g Ti, the titanium content in the polyolefin is below 1 ppm. Hence, catalyst residues can be left in the polymers, i.e., the catalyst does not need to be removed by extraction or by adsorber beds typical for early catalysts generations. Furthermore, highly energy- and resource-efficient polymerization in gas phase and bulk eliminates the need for organic solvents as polymerization media and for their recycling. Because of the excellent control of molecular mass and stereoregularity, polyolefin purification by solvent extraction of low molecular mass polyolefin waxes and atactic polypropylene is no longer required. Employing fossil and renewable feedstocks, such as oil, gas and even bioethanol in environmentally friendly and highly energy-efficient catalytic polymerization processes meets the demands of green chemistry. As hydrocarbon resins, polyolefins preserve oillike energy content. They are readily recycled by remolding and serve as valuable source of energy and "renewable oil", produced by thermal degradation of polyolefins.

The basic reaction mechanisms and the correlations between the architectures of catalytically active transition metal complexes and polyolefin architectures are quite well understood. As illustrated below for olefin polymerization on a cationic metallocene alkyl catalyst, the olefin forms a π-complex with the transition metal prior to insertion into the transition metal carbon bond. After repeated insertion β-hydride elimination terminates the polymer chain, thus forming vinyl-terminated polyethylene. The resulting transition metal hydride is the catalytically active intermediate, which adds to an olefin and starts the growth of a new polyethylene chain. Additionally, hydrogen can be added to control molecular mass by chain transfer that involves hydrogenation of the transition metal alkyls and formation of a transition metal hydride, which subsequently adds to the ethylene.

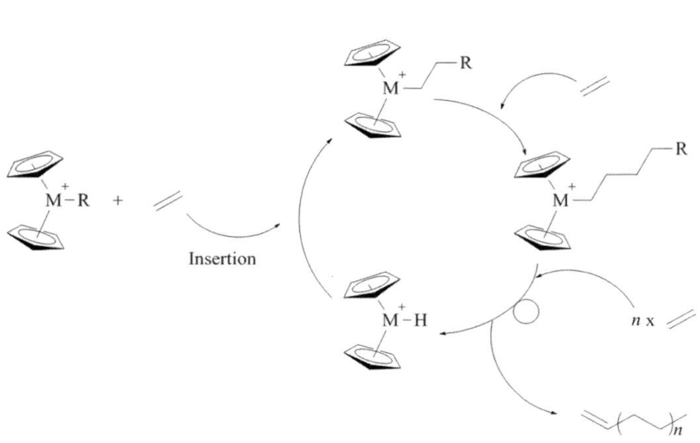

The regioselectivity in α-olefin and the polyolefin constitution is governed by the insertion type. Whereas 100% of 1,2- or 2,1-insertion exclusively afford head-to-chain polyolefins, lower regioselectivity accounts for head-to-head enchainment in the polyolefin backbone. Most α-olefins are polymerized by 1,2-insertion in highly regioselective catalytic polymerization. The insertion type is identified by end group analysis of poly(α-olefins), taking into account chain termination by β-hydride as well as β-methylide transfer.

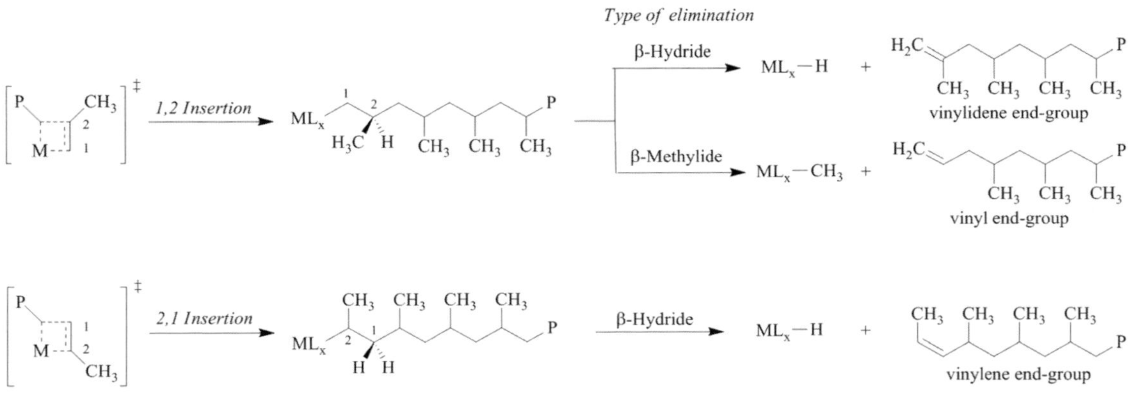

The stereospecificty of α-olefin polymerization is governed either by the configuration of transition metal complex (enantiomorphic site control) or by that of the asymmetric carbon atom of the monomeric unit at the chain end (chain end control). When a false insertion occurs, the resulting change of configuration is isolated in the case of enantiomorphic site control but is perpetuated in the case of chain end control (see below, m denotes meso, r denotes racemic). The analysis of polyolefin microstructures reveals that for most catalysts the chirality of the transition metal accounts for the control of the stereoregularity.

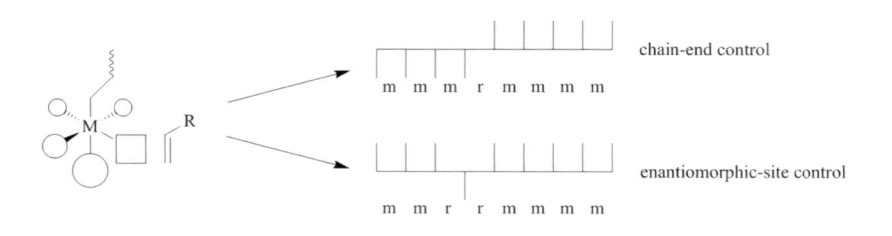

The ability of catalysts to control the stereochemistry of poly(α-olefins) and polydienes makes Ziegler–Natta polymerization catalysis industrially extremely useful. Propene, for example, polymerizes cationically to highly branched, atactic, low molar mass polymers that do not crystallize and have low glass transition temperatures. They can be used only as hotmelt adhesives and lubricants. Ziegler–Natta polymerization catalysis (Table 9) leads, however, to stereoregular polymers with high melting temperatures. Isotactic polypropylenes find extensive use as thermoplastics. Stereoblock polypropylenes can be engineered to produce thermoplastic elastomers containing highly flexible amorphous and rigid crystalline segments. The structure of single-site metallocene catalysts can be varied to produce virtually any kind of poly(α-olefin) microstructure industrially:

Similar to free radical polymerization, short and long chain branches are readily introduced by catalytic olefin polymerization. With increasing α-olefin comonomer incorporation, both crystallinity and density of polyethylene decrease. Whereas high density linear polyethylene (HDPE) is attractive for producing pipes, linear low density polyethylene (LLDPE) is highly flexible and has lower density, thus meeting the demands of packaging applications. A small amount of long chain branching promotes shear thinning and melt strengthening, which is beneficial to blow molding of films. At a high degree of branching, polyethylene is rendered amorphous and elastomeric. In principle, there are two synthetic strategies for branching of polyolefins: (i) Copolymerization of ethylene with various α-olefins, and (ii) branching homopolymerization of ethylene according to the "chain walking" mechanism. In chain walking, the catalytically active transition metal alkyl travels up and down the polymer chain via repeated β-hydride elimination followed by addition of the transition metal hydride to the resulting double bond. Whereas titanium catalysts form linear polyethylene chains, branching ethylene homopolymerization takes place on postmetallocene complexes, such as cationic alkyl Ni diimine. Alkyl Pd diimine complexes produce high molecular mass highly branched amorphous polyethylenes. Because Pd metal alkyl travels up and down the main chain and side chains, branches are formed.

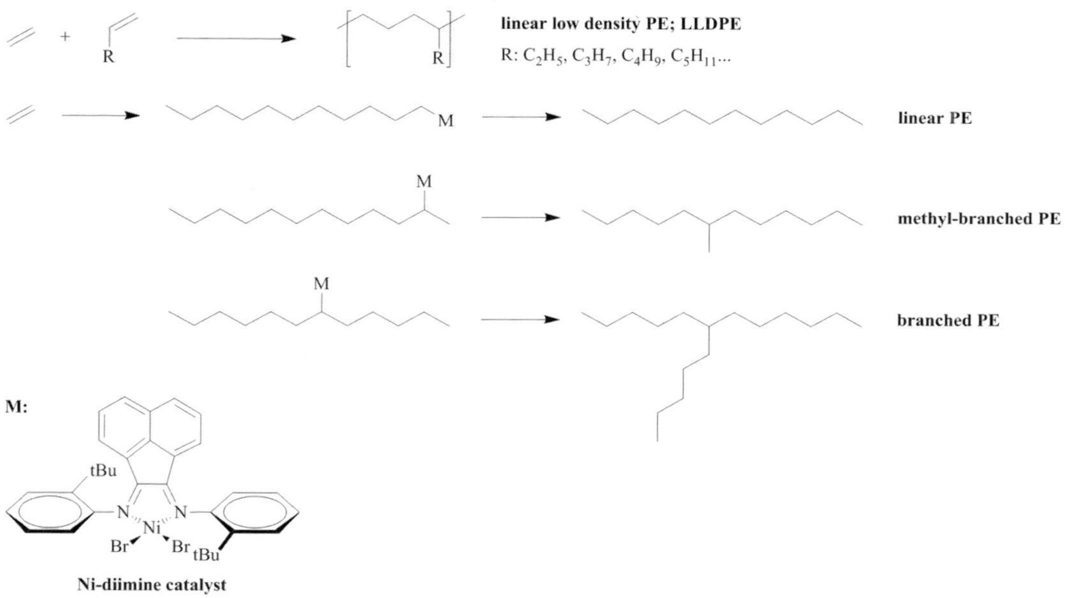

linear low density PE; LLDPE
R: C_2H_5, C_3H_7, C_4H_9, C_5H_{11}...

linear PE

methyl-branched PE

branched PE

M:

Ni-diimine catalyst

In the early days of catalyst development, the multisite nature of catalysts accounted for the formation of rather complex copolymer compositions in slurry polymerizations. Frequently, the α-olefin comonomer was predominantly incorporated in low molar mass waxlike fractions. More uniform copolymer compositions are obtained in high temperature solution polymerization operating at temperatures above 130 °C. In the past, very different catalyst systems were developed for producing HDPE, LLDPE, and EP(D)M rubbers. Since the 1980s, the development of single-site catalysts based upon metallocene, post-metallocene, and constrained geometry half-sandwich complexes has revolutionized the industrial synthesis of branched polyethylenes. As a function of the ligand framework, it is possible to control molecular mass independent incorporation of α-olefins into polyethylene backbone to produce LLDPE as well as very low density polyethylenes and elastomers by the same catalyst. Moreover, improved reactivity enables the efficient copolymerization of polyethylene vinyl endgroups, thus producing long chain branched polyethylene.

In a chain shuttling solution process, zinc alkyls transfer polymer chains between catalytically active sites producing exclusively linear polyethylene or ethylene−1-octene copolymers. As a consequence, block copolymers are obtained that contain alternating rigid HDPE segments and elastomeric poly(ethylene-co-1-octene) segments. Such block copolymers are attractive as thermoplastic elastomers.

thinning during melt processing. Moreover, small amounts of slightly branched ultrahigh molar mass poylolefins function as "tie molecules" bonding together polyethylene crystallites and significantly improve the fatigue resistance and durability of polyethylene pipes. Two strategies are exploited: (i) Ethylene polymerization in cascade reactors, and (ii) ethylene polymerization in a single reactor using multi-site catalysts, prepared by incorporating different single-site catalysts on the same support (see Fig. 19). In cascade reactors, ethylene is polymerized in the absence and the presence of hydrogen to control molar mass dstribution. The addition of α-olefin as comonomer selectively branches UHMWPE without affecting HDPE.

In multi-site catalysis, different catalytically active complexes that produce high and low molecular mass polyethylenes are incorporated on the same support, thus enabling simultaneous

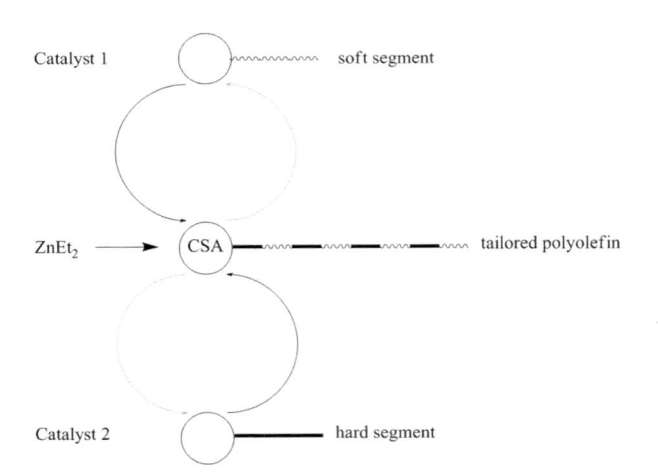

CSA: chain shuttling agent Et$_2$Zn-P

Although single-site catalysts afford uniform polyolefins, it is highly desirable to tailor polyolefins with broad molecular mass distributions and preferred incorporation of branches exclusively into the ultrahigh molar mass fraction. This greatly improves shear

production and blending of both polyolefins. Reactor blend formation is advantageous compared to blends prepared by conventional melt compounding, especially with respect to dispersion problems due to the high viscosity of ultrahigh molecular mass

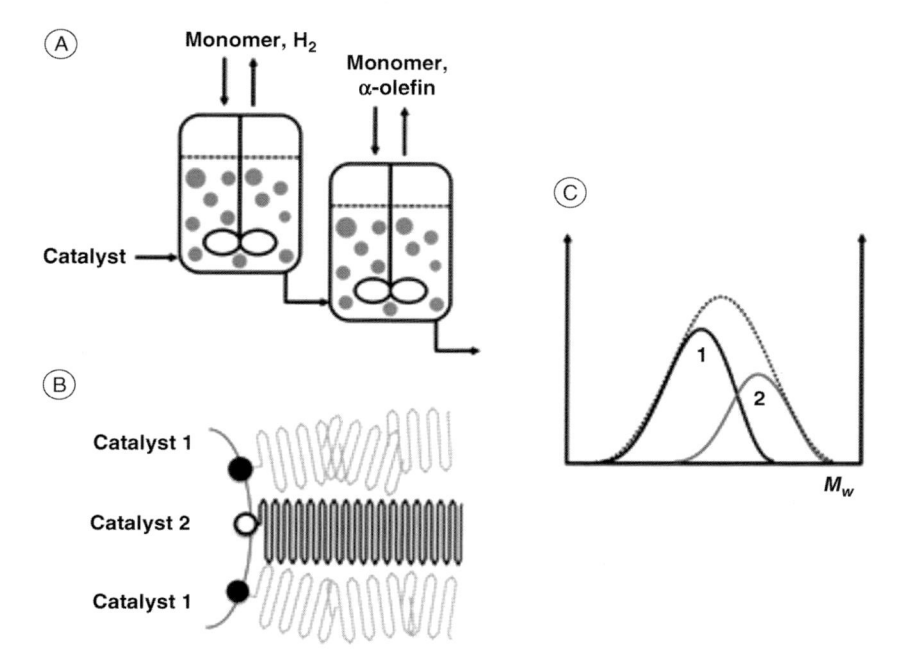

Figure 19. A) Reactor cascades; B) Multi-site catalysts producing reactor blends of HDPE with ultrahigh molecular mass polyethylene (UHMWPE); C) Polyolefin with bimodal molar mass distribution containing UHMWPE (2)

polyolefins and rubbers. The reactor blend formation of HDPE with UHMWPE (see Fig. 19) is the key to self-reinforcing polyethylene (see Section 11.2.4). The reactor granule technology, which combines liquid pool propene homopolymerization with subsequent propene−ethylene copolymerization in a gas phase reactor, incorporates EPM rubber phases directly into the matrix of the PP granules (Catalloy technology). Most conventional Ziegler and related catalyst systems polymerize cyclic olefin, such as cyclopentene via 1,2 polymerization, thus linking together cyclic hydrocarbons by covalent bonds to form rigid polymers. In contrast, transition carbene complexes enable ring-opening polymerization of cyclic olefins to produce flexible polyalkenamers. Catalytic intermediate of this ring-opening metathesis polymerizations (ROMP) are metallcylobutanes formed by cycloaddition of carbene complexes to cycloolefins.

Grubbs catalyst

1st generation

2nd generation

3rd generation

Metatheses reactions are exchange and disproportionation reactions of double bonds, mainly carbon–carbon double bonds in olefins and cycloolefins. A low molar mass example is the metathesis of pentene-2 to butene-2, pentene-2, and hexene-3 (in the ratio 1 : 2 : 1) by the catalyst system $WCl_6-C_2H_5AlCl_2-C_2H_5OH$. Stable carbene complexes are at hand for metathesis reaction and ROMP. Industrially, cyclooctene, norbornene, and dicyclopentadiene are polymerized by ROMP. Cyclooctene gives polyoctenamer $\sim CH=CH-(CH_2)_6\sim$; both the *cis* and the *trans* isomers are elastomers (→ Rubber, 6. Sythesis by Radical and Other Mechanisms). Norbornene polymerizes to a thermoplastic polymer that is plasticized with mineral oil to give an elastomer (→ Rubber, 6. Sythesis by Radical and Other Mechanisms). Dicyclopentadiene is polymerized by reaction injection molding (RIM) processes to produce rigid engineering materials.

In *acyclic diene metathesis polymerization* (ADMET), the polycondensation of a variety of nonconjugated linear dienes, such as 1,5-headiene, affords polyenes. Using functional nonconjugated diene monomers, ADMET enables precise control of the spacer length between two branches, whereas conventional copolymerization process affords broad distributions. Unlike industrial Ziegler catalysts, Ru carbene complexes are tolerant to polar groups, thus enabling the formation of polyolefins with polar groups attached as side chains.

3.3.6. Copolymerizations [98]

Copolymerizations are joint polymerizations of two or more monomers; they were also called interpolymerizations in the older literature. In the simplest case, both active chain ends $\sim a*$ and $\sim b*$ react irreversibly with the two monomers A and B:

$\sim a^* + A \rightarrow \sim a - a^*; R_{aA} = k_{aA}[a^*][A]$

$\sim a^* + B \rightarrow \sim a - b^*; R_{aB} = k_{aB}[a^*][B]$

$\sim b^* + A \rightarrow \sim b - a^*; R_{bA} = k_{bA}[b^*][A]$

$\sim b^* + B \rightarrow \sim b - b^*; R_{bB} = k_{bB}[b^*][B]$

Four rates R_{ij} and four rate constants k_{ij} are to be considered in this terminal model, which corresponds to Markov first-order statistics. At high molar masses, monomers are consumed only by the four propagation reactions. The relative monomer conversion is given by

$$\frac{-d[A]/dt}{-d[B]/dt} = \frac{R_{aA} + R_{bA}}{R_{bB} + R_{aB}} = \left(\frac{k_{bA} + k_{aA}([a^*]/[b^*])}{k_{bB} + k_{aB}([a^*]/[b^*])}\right) \cdot \frac{[A]}{[B]} = \frac{d[A]}{d[B]}$$

The ratios of rate constants of homopropagations and cross-propagations are called copolymerization parameters

$r_A \equiv k_{aA}/k_{aB}; \ r_B \equiv k_{bB}/k_{bA}$

Five different cases can be distinguished for each copolymerization parameter:

$r = 0$	The rate constant of homopropagation is zero; the active center adds only the other monomer.
$r<1$	The other monomer is added preferentially.
$r = 1$	Both monomers are added in the same amounts.
$r > 1$	The own monomer is added preferentially but not exclusively.
$r = \infty$	No copolymerization, only homopolymerization.

Depending on the relative numerical value of the two copolymerization parameters, five different types of copolymerization can be distinguished (Table 10).

In general, one of the monomers is consumed preferentially during copolymerization ($r_A \neq r_B$). This drift of copolymer composition can be avoided if the more reactive monomer is fed into the reactor according to its consumption. A conversion-independent polymer composition also results if copolymerization is performed under azeotropic conditions, which are defined by

$d[A]/d[B] \equiv [A]/[B]$

Table 10. Types of copolymerization of monomers A and B

Types	Copolymerization parameters				
	Azeotropic		Nonazeotropic		
	r_A	r_B	r_A	r_B	$r_A r_B$
Alternating	0	0	0	> 0	0
Statistical	< 1	< 1	< $1/r_B$	> 1	< 1
Ideal	1	1	$1/r_B$	$1/r_A$	1
Block forming	> 1	> 1	> $1/r_B$	< 1	> 1
Blend forming	∞	∞	∞	< ∞	∞

i.e., the relative rate of monomer consumption equals the ratio of monomer concentrations at any time: The monomer and the polymer compositions do not drift with the progress of polymerization.

This equation is a special case of the Lewis–Mayo equation, which describes the instantaneous composition of the copolymer as a function of the ratio of instantaneous monomer concentrations:

$$\frac{d[A]}{d[B]} = \frac{1 + r_A([A]/[B])}{1 + r_B([B]/[A])} = \frac{x_a}{x_b}$$

All types of copolymerization can in principle occur with the same monomer pair if various initiators are used. All copolymerizations of styrene and methyl methacrylate shown in Figure 20 are, for example, statistical. The free-radical-initiated copolymerization is, however, azeotropic, and the cationic and anionic ones are nonazeotropic. Initiation by $Et_3Al_2Cl_3$ and traces of oxygen leads to almost alternating copolymers, whereas cationic polymerization generates long styrene sequences and anionic polymerization produces long methyl methacrylate ones. The sequence length determines many polymer properties, most notably glass and melting temperatures.

Ionic copolymerizations very often lead to long block sequences, whereas free-radical copolymerizations tend in general to produce more random copolymers. Free-radical copolymerizations are thus the method of choice for industrial purposes if allowed by monomer structures (Table 11).

3.4. Polycondensations and Polyadditions [85]

3.4.1. Bifunctional Polycondensations

Polycondensations and polyadditions are generally divided into bifunctional reactions (two

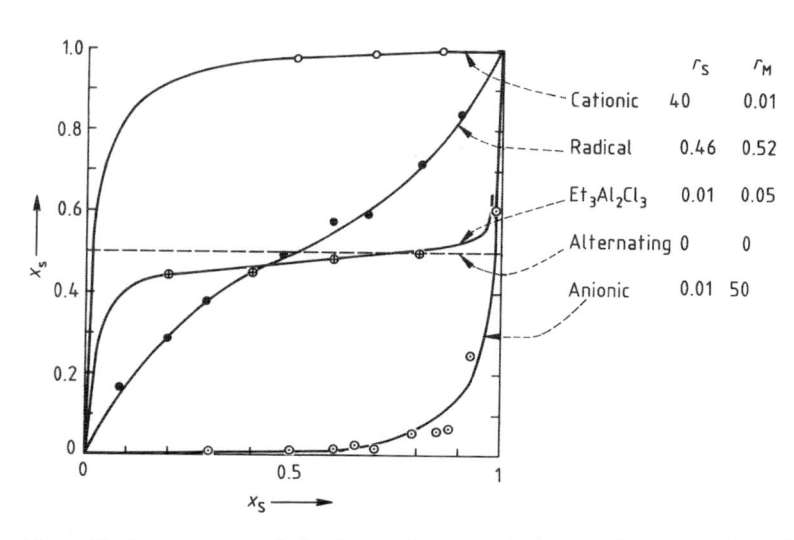

Figure 20. Instantaneous mole fraction x_s of styrene units in the polymer formed as a function of the instantaneous mole fraction x_s of styrene in the copolymerization of styrene with methyl methacrylate initiated by (○) cations; (●) free radicals; (⊕) $Et_3Al_2Cl_3$; (⊙) Anions
Solid lines: calculated with copolymerization parameters shown to the right; dotted line: theory for alternating copolymerizations.

	r_S	r_M
Cationic	40	0.01
Radical	0.46	0.52
$Et_3Al_2Cl_3$	0.01	0.05
Alternating	0	0
Anionic	0.01	50

Table 11. Industrial copolymerizations

Monomers	Polymerization	Application
Free radical, linear (or slightly branched)		
Ethylene + 10% vinyl acetate	bulk	shrink films
Ethylene + 10–35% vinyl acetate	bulk	thermoplastics
Ethylene + 35–40% vinyl acetate	under precipitation[a]	films
Ethylene + > 60% vinyl acetate	emulsion	elastomers
Ethylene + < 10% methacrylic acid	bulk	extrusion coating
Ethylene + trifluorochloroethylene		thermoplastics
Butadiene + styrene	emulsion	multipurpose elastomers
Butadiene + 37% acrylonitrile	emulsion	oil-resistant elastomers
Vinyl chloride + 3–20% vinyl acetate	solution[b]	Paints
Vinyl chloride + 15% vinyl acetate	solution[b]	Hi-fi records
Vinyl chloride + 3–10% propene	bulk	thermoplastics
Acryl esters + 5–15% acrylonitrile		oil-resistant elastomers
Acrylonitrile + 4% different monomers	under precipitation[c]	fibers with improved dyeability
Acrylonitrile + styrene		thermoplastics
Acrylonitrile + butadiene + styrene		thermoplastics
Tetrafluoroethylene + propene	emulsion	thermoplastics
Methacrylic acid + methacrylonitrile		hard foams (after cyclization)
Free radical, cross-linking		
Glycol methacrylate + 2–4% glycol dimethacrylate		contact lenses
Unsaturated polyesters + styrene or methyl methacrylate	bulk	glass-fiber-reinforced thermosets
Anionic		
Styrene + butadiene	solution	elastomers
Cationic		
Isobutene + 2% isoprene	solution	butyl rubber
Trioxane + ethylene oxide	solution	thermoplastics
Ethylene oxide + propylene oxide	solution	thickener, detergents
Ziegler–Natta polymerization		
Propylene oxide + nonconjugated dienes	solution	elastomers
Ethylene + propene + nonconjugated dienes		elastomers

[a] In *tert*-butanol.
[b] In acetone, 1,4-dioxane or hexane.
[c] In water.

functional groups per reactant) and multifunctional ones (three or more functional groups per monomer molecule). Bifunctional reactions lead to linear macromolecules; multifunctional ones, to branched and finally cross-linked polymers. Bifunctional reactions are further subdivided into AB reactions (e.g., that of $H_2N-R-COOH$) and AA/BB reactions (e.g., $H_2N-R-NH_2$ + $HOOC-R'-COOH$).

All reactants, from monomer to high polymer, may react to form macromolecules in bifunctional polycondensations and polyadditions. Only monomers are present at the beginning of the reaction; later, oligomers are present. If the reactivity of the functional groups is independent of molecule size, then the greater number of smaller molecules leads first to low molar mass polymers and only at high conversions to chemical compounds with high molar masses (see Fig. 17). In equilibrium, the number-average degree of polymerization of polymers from AB and stoichiometric AA−BB reactions is dictated by the extent of functional group reaction p:

$$\overline{X}_n = 1/(1-p)$$

At 50% conversion ($p = 0.5$), the number-average degree of polymerization of reactants is only 2 (this includes monomer molecules), and the number-average degree of polymerization of polymers ($2 < X_i < \infty$) is only 3. Industrial polycondensations rarely deliver number-average degrees of polymerization greater than 100–200 (i.e., $p = 0.99$–0.995 in equilibrium). The endgroups of these equilibrium polymers may condense further during processing by, e.g., extrusion or injection molding. The accompanying strong increase in viscosity is undesirable and these endgroups are therefore blocked against further polycondensation by "sealing" them with monofunctional reagents.

Equilibria are established not only by retro-reactions (reversal of polycondensations and polyadditions) but also by catalyzed exchange reactions between chain segments. In these trans reactions, number-average molar masses remain constant, but mass-average molar masses increase if the initial ratio of mass to number-average molar mass was smaller than dictated by equilibrium conditions. An example is the transesterification of poly(ethylene terephthalate):

reacting monomers. An example is the Schotten–Baumann reaction of diamines in water with diacid dichlorides in chloroform, for example. A polyamide film is formed at the interface of the two solutions. Functional groups of this polymer are buried in the interior of the film; they cannot react with each other. Exchange reactions and retroreactions are also absent. Only surface groups can react; the resulting irreversible reaction leads to high molar masses. Industrially, this reaction is utilized for one of the two industrial syntheses of polycarbonates, i.e., from the

$$\sim [O-CH_2-CH_2-O-OC-(p-C_6H_4)-CO]_n \quad [O-CH_2-CH_2-O-OC-(p-C_6H_4)-CO]_k \sim$$
$$\sim [OC-(p-C_6H_4)-CO-O-CH_2-CH_2-O]_j \quad + \quad [OC-(p-C_6H_4)-CO-O-CH_2-CH_2-O]_m \sim$$
$$\downarrow$$
$$\sim [O-CH_2-CH_2-O-OC-(p-C_6H_4)-CO]_n \text{———} [O-CH_2-CH_2-O-OC-(p-C_6H_4)-CO]_k \sim$$
$$\sim [OC-(p-C_6H_4)-CO-O-CH_2-CH_2-O]_j \text{———} [OC-(p-C_6H_4)-CO-O-CH_2-CH_2-O]_m \sim$$

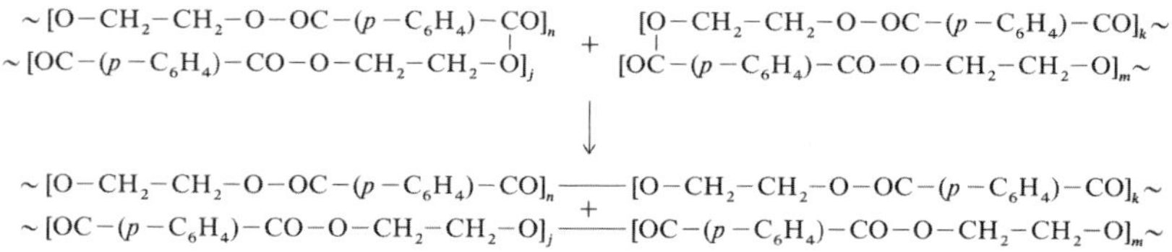

Higher degrees of polymerization at lower extents of reaction can be obtained by irreversible "activated reactions" and especially by heterogeneous reactions of fast reacting monomers. sodium salt of bisphenol A and phosgene (Table 12).

Most industrial polycondensations are of the AA–BB type (Table 12); they can be described

Table 12. Industrially important linear AA–BB polycondensations following Equation (17)

Polymer	Chemical structure				Remarks
	A	V	W	B	
Poly(ethylene terephthalate)	H	O(CH$_2$)$_i$O	CO–p-Ph–CO	OH	$i = 2$ (ethylene) or
	H	O(CH$_2$)$_i$O	CO–p-Ph–CO	OCH$_3$	4 (butene)
Unsaturated polyesters	H	O(CH$_2$)$_2$O	CO–CH=CH–CO	O$_{1/2}$	from maleic anhydride (isomerizes mainly to fumaric acid residues)
Polycarbonate	H	O–p-Ph–C(CH$_3$)$_2$–p-Ph–O	CO	OC$_6$H$_5$	
	Na	O–p-Ph–C(CH$_3$)$_2$–p-Ph–O	CO	Cl	
Polyarylates	Na	O–Ar–O	CO–Ar–CO	Cl	different Ar
Poly(phenylene sulfide)	Na	S	p-Ph	Cl	
Polysulfides	Na	S	R	Cl	different R
Polysulfones	K	O–p-Ph–O	SO$_2$–p-Ph–O–p-Ph–SO$_2$	Cl	different V and W units
Polyamides	H	NH(CH$_2$)$_i$NH	CO(CH$_2$)$_j$CO	OH	PA 66 ($i = 6; j = 4$) PA 610 ($i = 6; j = 8$) PA 46 ($i = 4; j = 4$)
Polyetheretherketone	K	O–p-Ph–CO–p-Ph–O	p-Ph–CO–p-Ph	F	

Ar = aromatic residue; p-Ph = para-substituted phenylene group; R = organic residue.

by

$$n A-V-A + n B-W-B \rightarrow A-(V-W)_n-B + (2n-1)AB \qquad (17)$$

where A— and B— are leaving groups, AB represents leaving molecules, —V— and —W— are monomeric units, and —V—W— are repeating units.

Industrial AB polycondensations are fairly rare. Examples are the formation of polyamide 11 (nylon 11) (Eq. 4), the polycondensation of carbodiimides

$$O=C=N-R-N=C=O \rightarrow R-N=C=N + CO_2$$

and the synthesis of certain polysulfones

$$C_6H_5O(p-C_6H_4)SO_2Cl \rightarrow$$
$$(p-C_6H_4)-O-(p-C_6H_4)-SO_2 + HCl$$

3.4.2. Multifunctional Polycondensations and Hyperbranched Polymers [99]

Multifunctional polycondensations are condensation reactions with the participation of monomers having higher functionality ($f_o > 2$). At low conversions, branched molecules are formed with increasingly higher number of functional end groups per molecule (i.e., increased functionality f):

$$f = 2(f_o - 1) + (X-2)(f_o - 2) = 2 + X(f_o - 2)$$

Only one such group is, however, needed to link two molecules; the probability of such a linkage thus increases dramatically with increasing degree of polymerization X. At a certain degree of conversion of functional groups, molecules are linked throughout the volume of the reactor. The viscosity increases strongly, and the mass-average molar mass of the cross-linked polymer approaches infinity at this "gel point." Not all reactants are linked at the gel point; some are still soluble up to high conversions. The number-average degrees of polymerization of reactants are thus fairly low at the gel point, usually in the range $10 < \overline{X}_n < 50$.

Knowledge of the gel point is technically very important. Polymers cannot, or can only with difficulty, be processed beyond the gel point and must thus be shaped before or during the cross-linking reaction. The gel point can be calculated from the functionality and relative amounts of reactive groups if intramolecular cyclizations are assumed absent. In practice, such cyclizations are often present and cross-linking occurs at higher extents of reaction than calculated. The theoretical calculations thus provide a safety margin.

Chemical cross-linking reactions are utilized for thermosets, certain leathers, and the preparation of elastomers from rubber. Thermosetting polycondensation reactions allow the hardening of phenol, amino, and alkyd resins. In the reaction of phenol(s) with formaldehyde, methylene, formal, and ether bridges are formed between the *ortho* and *para* positions of the phenols; the hardening of novolacs with hexamethylenetetramine (urotropin) also leads to imine structures ($\rightarrow$ Phenolic Resins). Amino plastics result from the reaction of urea or melamine with formaldehyde ($\rightarrow$ Amino Resins), and alkyd resins from multifunctional alcohols with bifunctional organic acids or their anhydrides ($\rightarrow$ Alkyd Resins).

Homopolymerization of AB_n monomers and copolymerization of AB with B_n monomers, respectively, affords hyperbranched polymers. Due to their low entanglement, hyperbranched polymers have reduced viscosity compared to linear polymer. Moreover, they have very high end-group functionality and can be tailored for applications in thermoset resin formulations and drug carriers. Amphiphilic hyperbranched polymers are effective dispersing agents for pigments and various other particles.

3.4.3. Polyaddition

No leaving molecules have to be removed during polyadditions, an important advantage for process engineering. Practically all industrially important polyadditions are thus performed with thermosetting resins. Polyurethanes are formed from diisocyanates (or triisocyanates) and polyols (see Eq. 5 for diols). Polyurethane formation may be so rapid that cross-linking and shaping can be performed in one step through reaction injection molding with short cycle times. Epoxides (epoxy resins) contain two or more epoxy groups per prepolymer molecule and are cured with either difunctional or multifunctional amines or with anhydrides of organic acids.

3.5. Polymer Reactions [100, 101]

3.5.1. Polymer Transformations

Several polymers are chemically transformed into industrially useful polymers by polymer-analogous reactions of their substituents (Table 13).

These reactions per group are chemically identical to their low molar mass analogues. They differ in the effect of side reactions, which do not lead to removable byproducts as in micromolecules but to "wrong" structures in the polymer chain. In most cases, such structures decrease the desired polymer properties. Polymer transformations that are prone to side reactions are thus generally avoided. This is one reason why only five types of polymer-analogous reactions are utilized commercially: Transesterification–saponification, chlorination–sulfochlorination, etherification, hydrogenation, and cyclization.

The reaction of poly(vinyl alcohol) with butyraldehyde leads to acetals. For statistical reasons, however, not all hydroxyl groups can be transformed:

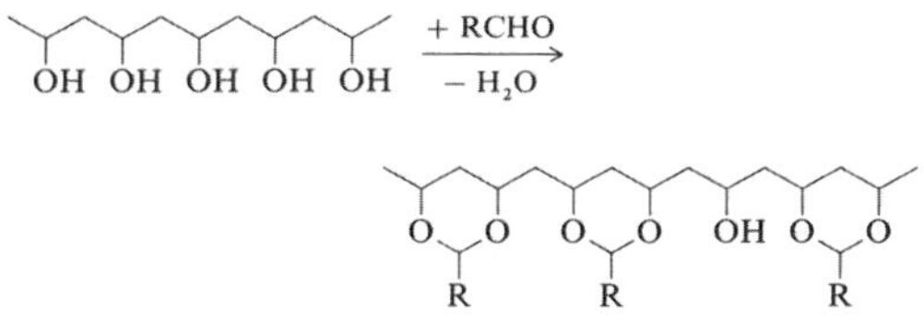

The resulting poly(vinyl butyral) is used as an intermediate layer in safety glass and for wash primers.

Heating of methacrylic acid–methacrylonitrile copolymers with ammonia produces polymethacrylimide structures that are hard foams:

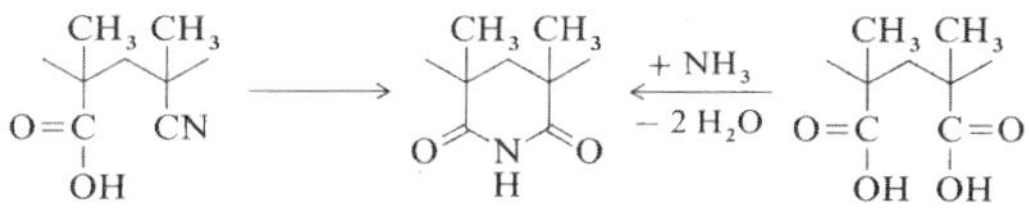

The thermal treatment and carbonization of polyacrylonitrile fibers represents the key to the industrial production of carbon fibers, which are highly attractive for reinforcement of thermosets and the formation of carbon-fiber reinforced polymers in lightweight engineering.

Table 13. Industrial polymer transformations (for ring formations see text)

Base units in primary polymer	Reagent	Conversion, %	Base units in final polymer	Application
$CH_2CH(OOCCH_3)$	ROH	98	CH_2-CHOH	thickener
$CH_2CH_2CH_2CH(OOCCH_3)$	CH_3OH	99	CH_2-CHOH/CH_2-CH_2	engineering plastics
Cellulose$-NHCOCH_3$	H_2O	?	Cellulose NH_2	paper additive
Cellulose OH	CH_3COOH	83–100	Cellulose $OOCCH_3$	fibers, films
Cellulose OH	HNO_3	67–97	Cellulose$-ONO_2$	plastics, fibers
Cellulose ONa	oxirane	53–87	Cellulose$-OCH_2CHRCH_3$	thickener
Cellulose OH	oxirane	100–400^a	Cellulose$-OCH_2CH(OR')CH_3$	thickener
CH_2-CH_2	Cl_2	25–40^b	CH_2-CHCl	elastomer
		>40^b	$CH_2-CHCl/CHCl-CHCl$	impact improver for PVC
CH_2-CHCl	Cl_2	64^b	$CH_2-CHCl/CHCl-CHCl$	adhesives, paints
CH_2-CH_2	Cl_2/SO_2	<42^b; <2^c		coatings
$CH_2CH=CHCH_2/CH_2-CHCN$	H_2	?	$(CH_2)_4/CH_2-CHCN$	elastomer
$CH_2-CH(C_6H_5)$	SO_3	low	$CH_2-CH(C_6H_4SO_3H)$	ion exchangers
$N=PCl_2$	RONa	?	$N=P(OR)_2$	elastomer

^{a}Further reaction of primary reaction products may lead to more than one monomeric unit of reagent per hydroxyl group of cellulose.
^{b}Chlorine content in %.
^{c}Ether groups per glucose unit.

3.5.2. Block and Graft Formations [102]

Reactive chain ends can either initiate the polymerization of other monomers or couple with other preformed macromolecules to give block polymers:

$$A_n^* + mB \rightarrow A_nB_m^*$$

$$A_n^* + B_m^* \rightarrow A_nB_m$$

Both methods are used industrially with a variety of strategies for the synthesis of diblock polymers A_nB_m as compatibilizing agents for polymer blends, triblock polymers $A_nB_mA_n$ as thermoplastic elastomers, and various multiblock polymers for different purposes, especially for thermoplastic elastomers. Anionic living polymerizations are used for the syntheses of poly(styrene−block-butadiene−block-styrene) and poly(styrene−block-isoprene−block-styrene). Polycondensations serve for the preparation of multi[{poly(butylene terephthalate)}−block-poly{(polytetrahydrofuran)−terephthalate}], and various polyesteramide and polyetheramide segmented block polymers.

Graft copolymers result from the grafting of monomers on chemically different polymer chains. Grafts of isobutene−isoprene on polyethylenes, vinyl chloride on poly(ethylene-stat-vinyl acetate), ethylene−propene on poly(vinyl chloride), or styrene−acrylonitrile on saturated acryl rubber also yield thermoplastic elastomers.

3.5.3. Cross-Linking Reactions

Cross-linking of preformed polymers is used for vulcanizing unsaturated rubbers with sulfur, vulcanizing saturated rubbers with peroxides, hardening unsaturated polyester chains with styrene or methyl methacrylate via copolymerization, or tanning collagen with certain reagents to leather.

3.5.4. Degradation Reactions

Degradation of polymers is undesirable for their application and often desirable for their disposal after use. The term degradation includes the decrease in degree of polymerization with preservation of the chemical structure of monomeric units (depolymerization), the unwanted chemical transformation of some monomeric units with preservation of the degree of polymerization (decomposition), and a combination of both. The terms degradation, depolymerization, and decomposition are sometimes used with the same meaning as in this article, but other times with broader or more restricted meanings; no accepted usage exists.

Depolymerizations involve chain scissions. They are retroreactions (i.e., the reverse of chain polymerizations, polyeliminations, polycondensations, or polyadditions). Chain scissions may be at random as in retropolycondensations and retropolyadditions or may be unzipping reactions as in retro-chain polymerizations and retropolyeliminations.

Unzipping can occur only with activated chains above their ceiling temperatures, that is, with living polymers, chain ends comprising certain functional groups, or radical ends generated by a prior chain scission. Unzipping at higher temperature results in gaseous monomers that lead to voids in molded parts. This can be prevented or reduced by shorter sequence lengths of the unzippable polymer blocks. This strategy is utilized, for example, in the formation of acetal copolymers by copolymerization of trioxane (gives unzippable oxymethylene sequences $\sim OCH_2 \sim$) with ethylene oxide (oxyethylene sequences $\sim OCH_2CH_2 \sim$ do not unzip).

Decomposition can be caused by heat, light, oxygen, water, or other environmental agents as well as mechanical forces such as extensional flow, ultrasound, or a combination of these. It may lead to undesirable changes of mechanical, electrical, or optical properties. Unstabilized poly(vinyl chloride) $\sim(CH_2CHCl)_n\sim$ discolors on decomposition because HCl is eliminated and colored polyene sequences $\sim(CH=CH)_i\sim$ are formed. The polymer ultimately becomes brittle.

Decomposition is the primary degradation process in the pyrolysis of polymers, which may be desirable (rocket fuels, waste disposal) or not (fires), and often produces toxic fumes or smoke at lower flame temperatures. The risk of pyrolysis is enhanced for polymers with branching points, electron-accepting groups, long methylene sequences, and all groups capable of forming five- or six-membered rings. Thermostability is enhanced by aromatic rings, ladder structures, fluorine as a substituent, and low hydrogen content. On heating above 400 °C, polyolefins degrade to form small molar mass hydrocarbons, which are equivalent to those obtained from fossil resources. This route to "renewable oil" from polyolefin wastes is attractive with respect to recycling and sustainability.

4. Plastics Manufacture [103]

4.1. Homogenization and Compounding

Polymerization reactors do not deliver polymers that can be used directly as raw materials for plastics. Polymer melts must be filtered; melts, powders, beads, pellets, and granulates must be degassed; and all particulates must be dried or conditioned to environmental surroundings. Beads from suspension polymerization are washed and polymer solutions concentrated.

Batchwise polymerizations often lead to slightly different products, which are mixed (blended) to guarantee customers polymer grades according to specifications (microhomogenization). Granulates and pellets are sometimes similarly blended by macrohomogenization.

Compounding is the mixing of polymers and additives. This process can be performed by adding single additives one at a time, by using additive systems, or by employing master batches. Additive systems are carefully adjusted mixtures of additives that are formulated to avoid mutually synergistic or antagonistic effects. Master batches are concentrates of additives in polymers; they facilitate the dosage of small amounts of additives.

The properties of compounds depend very much on the compounding process. Only heterogeneous compounds result when poly (vinyl chloride) is mixed with additives in regular mixers; these compounds cannot be processed directly to end-use products. High-performance mixers deliver free-flowing powders; these "dry blends" can be extruded and injection molded. Compounding is often carried out by specialized companies (compounders).

4.2. Additives [5, 104, 105]
(→ **Plastics, Additives**)

4.2.1. Overview

Polymers are rarely used directly as materials. They do not fulfill per se all technological requirements and become commercially useful only after they have been mixed with certain additives. These additives average ca. 23% of the total mass of plastics but may range for individual plastics from 0.01% (packaging films for food) to 90% (barium ferrite-filled ethylene−vinyl acetate copolymers for magnetizable sealing strips). Polymer additives are to be distinguished from *polymer auxiliaries*; the former are added to polymers after the polymerization process, whereas the latter are used for the manufacture of polymers (e.g., polymerization catalysts, emulsifiers, initiators). The free-radical initiators used for the hardening of unsaturated polyesters are sometimes considered additives (added to the solution of unsaturated polyester molecules in monomers such as styrene) and sometimes auxiliaries (promoting the thermosetting reaction).

Additives are usually subdivided according to their application into process additives and functional additives. Process additives aid the processing of plastics by either stabilizing the chemical composition of polymers (processing stabilizers) or facilitating the processing itself (processing aids). Functional additives stabilize the chemical composition against attacks by environmental agents (stabilizing additives) or improve certain end-use properties (modifiers). Additives may also be classified according to their primary mode of action (chemical or physical). Some additives act in more than one way:

A filler or colorant may also be a nucleating agent for crystallization; a pigment may enhance discoloration, etc.

4.2.2. Chemofunctional Additives

Polymers can be attacked by oxygen and ozone during processing and use. The attack generates radicals that cause chain reactions to produce even more radicals. Ultimately, changed chemical compositions, degradation, or cross-linking occur. Oxidation is reduced if the accessibility of the oxidizable groups is limited or if radical formation is prevented. Oxygen diffusion into the polymer is slowed by certain coatings that act as mechanical barriers or by additives that diffuse into surface layers and are preferentially oxidized there.

Antioxidants ($\rightarrow$ Antioxidants) prevent radical formation, at least during processing and for the targeted lifetime of the plastic. They are subdivided into deinitiators and chain terminators.

Deinitiators prevent the formation of radicals (i.e., preventive antioxidants) and are always used in combination with chain terminators (also called secondary antioxidants). They are further subdivided into peroxide deactivators, metal deactivators, and UV absorbers. Peroxide deactivators (tertiary amines, tertiary phosphines, sulfides) convert hydroperoxides into harmless compounds before they can form radicals. Metal deactivators are chelating agents that form inactive complexes with catalytically active metal species (mainly from Ziegler–Natta polymerizations).

Chain terminators, such as hindered phenols, amines, and anellated hydrocarbons, react with already formed radicals and terminate the kinetic chain (chain-breaking or primary antioxidants). A combination of deinitiators and chain terminators often leads to synergistic effects. Antagonistic effects are also known (e.g., with carbon black as filler).

Light-induced degradation reactions can be prevented by the reduction of light absorption or by the addition of *UV absorbers* and quenchers. Less light is absorbed by the plastic if the surfaces are reflective or if certain pigments are added (e.g., carbon black). Ultraviolet absorbers either convert the incident light into harmless infrared radiation or are transformed into other chemical compounds. Quenchers deactivate excited states and become excited themselves.

Heat stabilizers prevent chemical transformations of plastics at higher temperature; their main application is for the stabilization of poly(vinyl chloride) against HCl elimination during processing. PVC additives include HCl scavengers, such as metal stearates, metal oxides, and epoxidized soybean oil. PVC additives, such as dibutyltin dithiolate react with thermally labile allylic chloride and double bonds in the PVC backbone.

Flame retardants either prevent oxygen access to burning plastics by formation of nonburning gases or by "poisoning" radicals generated by the burning. Some plastics are self-extinguishing because either CO_2 (polycarbonates) or water vapor and a protecting carbon layer are formed (vulcanized fiber: a $ZnCl_2$-treated cellulose) during burning. Chlorine and bromine compounds generate radicals during burning; these radicals combine with radicals from the degradation of plastics and stop the kinetic chain. Phosphorus compounds are oxidized during the burning to nonvolatile phosphor oxides, which either form a protective layer or are converted by water into phosphorus acids that catalyze the elimination of water. Intumescent additives, such as ammonium polyphosphate, form a carbon form that serves as a thermal insulator.

The flammability of plastics is often characterized by the limiting oxygen index (LOI). The LOI value indicates the limiting value of the volume fraction of oxygen in an oxygen–nitrogen mixture that just allows the polymer to burn after ignition with a flame. Materials with LOI > 0.225 are called flame retardant; those with LOI > 0.27, self-extinguishing. Low LOI values (high flammabilities) are exhibited by polyoxymethylenes (0.14), polyolefins (0.17–0.18), saturated polyesters (0.20), and cellulose (0.20). High LOI values are exhibited by poly(vinyl chloride) (0.32), polybenzimidazole (0.48), and polytetrafluoroethylene (0.95). However, LOI values are not absolute measures of the flammability or combustibility of plastics because these properties also depend on flame temperatures, heat capacities, heat conductivities, melting temperatures, and melt viscosities. The hazard of burning plastics is also determined by smoke formation and the toxicity of the evolving gases. In view of growing environmental and health concerns, halogenated flame retardants are gradually substituted by halogen-free flame retardants, such as Al(OH)$_3$ hydrate, Mg(OH)$_2$, and ammonium polyphosphate.

4.2.3. Processing Aids

Processing aids facilitate the processing of plastics by enhancing either transport rates to the processing machines, flow behavior in these machines, achievement of final properties during processing, or removal of shaped articles from the machines or from each other.

Easy-flow grades of powdered polystyrenes, for example, often contain 3–4% mineral oil, which forms a low-viscosity film on the surface of the particles and thus reduces friction in polar polymers; amphiphilic compounds such as metal stearates or fatty acid amides are used for this purpose. Such *external lubricants* also reduce the friction between polymer particles and the walls of the processing machine and the friction of polymer melts at such walls; they also prevent the cleavage of particles to smaller flow units. External lubricants are always incompatible with polymers and are thus found predominantly at polymer surfaces. They are related to *release agents*, which facilitate the separation of shaped articles from the tools (molds), and *slip agents*, which prevent the sticking together of shaped articles. Slip agents are thus sometimes also called lubricants.

Internal lubricants improve the flow behavior and homogeneity of polymer melts; they also reduce the Barus effect (see Section 7.3.2) and the melt fracture. Internal lubricants probably act by desegregating larger units (aggregates), which were probably formed during polymerization and are still present shortly after the melting of the polymer to a macroscopically homogeneous material. Typical internal lubricants are amphipolar compounds, e.g., modified esters of long-chain fatty acids.

Nucleating agents promote the crystallization of crystallizable polymers by generating many nuclei for crystallites. They prevent the formation of larger spherulites and thus improve the mechanical properties of plastics.

4.2.4. Extending and Functional Fillers [106–108]

Fillers are solid inorganic or organic materials. Some fillers are added mainly to improve the economics of expensive polymers; they are *extenders*. Extenders are usually particulate materials of corpuscular nature, such as chalk and glass spheres (aspect ratio ca. 1). Functional fillers improve certain properties, such as stiffness, strength, scratch resistance, electrical conductivity, crystallization rate, etc. Reinforcing fillers possess aspect ratios higher than 1; they may be short fibers, platelets (e.g., kaolin, talc, mica) (→ Reinforced Plastics), or long fibers (continuous filaments). Active fillers are sometimes subdivided into property enhancers (aspect ratio < 100) and true reinforcing fillers (aspect ratio > 100). No sharp dividing line exists between fillers and reinforcing agents, nor can the term "reinforcement" be unambiguously defined (e.g., it may denote an increase in breaking or impact strength or a decrease of brittleness).

A variety of fillers are used for plastics (- Table 14). Glass-fiber-reinforced polymers often carry the abbreviation GRP (or FRP); those reinforced by carbon fibers, CRP. Syntactic plastics are polymers reinforced with hollow glass spheres. The amounts of fillers added vary widely: In general, industry standards are about 30 wt % for thermoplastics and 60 wt % for thermosets. Fillers act very differently in polymers. Some fillers form chemical bonds with polymers; an example is carbon black, which acts as a chemical cross-linker in elastomers. Other fillers can adsorb polymers on their surfaces, i.e., physical bonds are introduced between fillers and polymers. On impact, adsorbed chain segments may take up energy and slip from the surface, which increases impact strength. Still other fillers act as nucleating agents in crystallizable polymers. Fillers furthermore constitute impenetrable walls to polymer coils. They restrict the number of conformational positions of chain segments near the filler surface; chains become less

Table 14. Fillers for thermoplastics (T), thermosets (D), and elastomers (E). For definition of other acronyms see Tables [1–4]

Filler	Application in	Concentration in %	Improved property
Inorganic fillers			
Chalk	PE, PVC, PPS, PB, UP	< 33 in PVC	price, gloss
Potassium titanate	PA	40	dimensional stability
Heavy spar	PVC, PUR	< 25	density
Talc	PUR, UP, PVC, EP, PE, PS, PP		white pigment, impact strength, plasticizer uptake
Mica	PUR, UP	< 25	dimensional stability, stiffness, hardness
Kaolin	UP, vinyls	< 60	demolding
Glass spheres	T, D	< 40	modulus of elasticity, shrinkage, compressive strength, surface properties
Glass fibers	T, D	< 40	fracture strength, impact strength
Fumed silica	T, D	< 3	tear strength, viscosity (increase)
Quartz	PE, PMMA, EP	< 45	heat stability, fracture
Sand	EP, UP, PF	< 60	shrinkage (decrease)
Al, Zn, Cu, Ni, etc.	PA, POM, PP	< 100	conductivity (heat and electricity)
MgO	UP	< 70	stiffness, hardness
ZnO	PP, PUR, UP, EP	< 70	UV stability, heat conductivity
Organic fillers			
Carbon black	PVC, HDPE, PUR, PI, PE, E	< 60	UV stability, pigmentation, cross-linking
Graphite	EP, MF, PB, PI, PPS, UP, PMMA, PTFE	< 50	stiffness, creep
Wood flour	PF, MF, UF, UP	< 5	shrinkage (decrease), impact strength
Starch	PVAL, PE	< 7	biological degradation

flexible, and tensile strengths and moduli of elasticity increase.

Due to the significant progress made in nanotechnology, a large variety of nanofillers have emerged since the mid-1990s. As compared to conventional filled polymers that contain micron-sized filler particles nanocomposites have nine orders of magnitude larger number of nanometer-scaled particles at the same volume fraction (Fig. 21). As a consequence, most polymers are allocated in nanocomposites at the nanofiller interface. Converting bulk polymers into interfacial polymers account for novel property profiles. The high surface area lowers considerably the percolation threshold required for particle network formation. This improves the electrical conductivity at much lower filler content. However, in view of the nanoparticle interactions, dispersion of nanoparticles is more difficult than micron-sized particles. Therefore, special processing technology and nanofiller surface modification are required. As illustrated in Figure 22, the nanofiller family

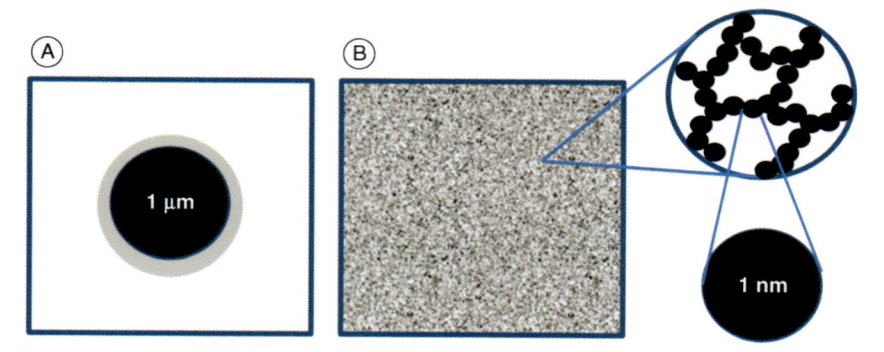

Figure 21. Different sized polymer composites containing the same volume fraction
A) One particle (1 μm) microfiller-sized fillers; B) 10^9 particles (1 nm) nanofiller with the same volume fraction like microfiller

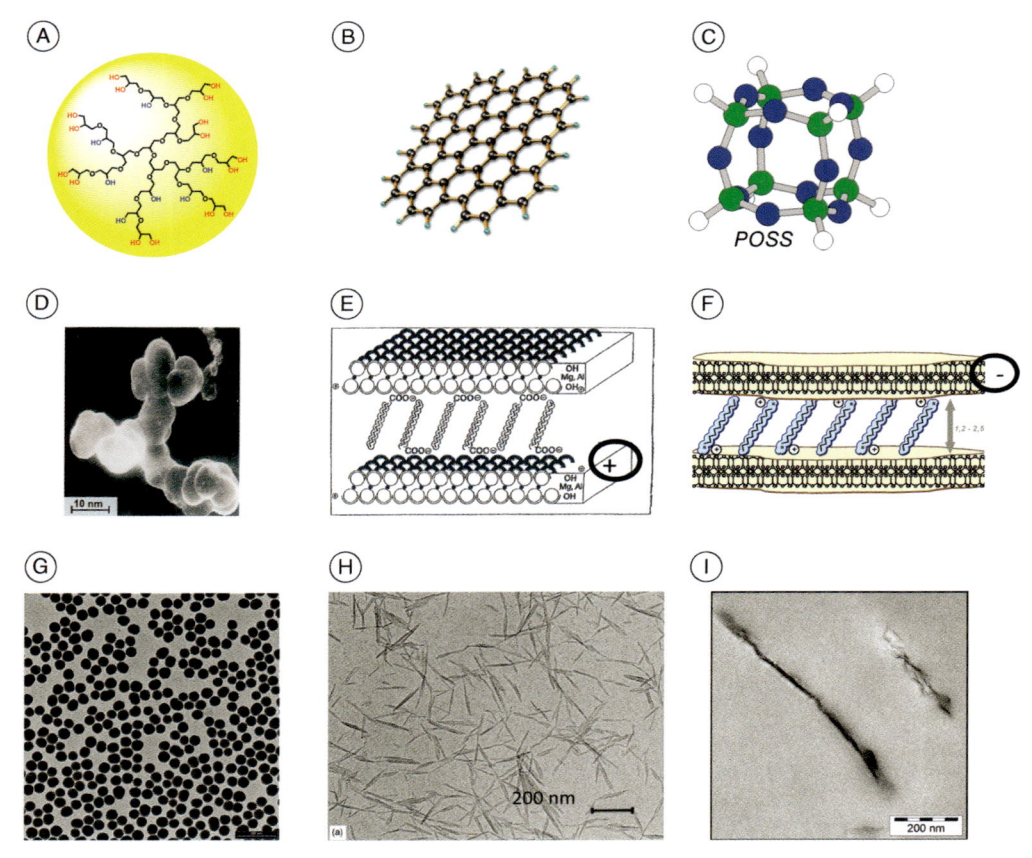

Figure 22. Different types of nanofillers
A) Nanomolecules; B) Molecular carbon; C) Molecular silica (POSS); D) Pyrogenic silica nanoparticles; E) Hydrotalcite; F) Organoclay; G) SiO$_2$ dispersion; H) Cellulose nanowhiskers; I) Graphene embedded in polypropylene

includes inorganic nanoparticles as well as organic−inorganic hybrid particles, among them dispersions of metal oxides and silicate nanoparticles with shapes varying from spheres to platelets and sheets. The smallest silica nanoparticles are polyhedral oligomeric silsesquioxanes (POSS), which are readily functionalized. The exfoliation of graphite affords graphene that consists of a honeycomb-like planar arrangement of sp^2 hybridized carbon atoms equivalent to a single layer of the graphite lattice. Other carbon allotropes, such as nanometer-scaled carbon particles ("conducting carbon black"), fullerenes, nano diamonds, and single- and multi-walled carbon nanotubes (CNT) are employed as nanofillers. Nanometer-sized carbon particles and especially carbon macromolecules render polymer nanocomposites electrically conductive without sacrificing optical clarity. Organophilic layered silicates ("organoclay") are obtained by cation exchange of the intergallery sodium cations of bentonite for alkyl ammonium. Alternatively, layered silicates with cationic layers, such as hydrotalcites are rendered organophilic by modification with long chain aliphatic carboxylates. As a function of the layered silicate modification and polymer processing conditions, either individual silicate layers are dispersed within the polymer matrix or the polymer is intercalated between silicates layers. The dispersion of sheetlike nanoparticles improves the barrier resistance of polymers. Going beyond wood flour, bio-based nanofillers (cellulose whiskers, nanocellulose) are derived from cellulose by etching of the amorphous regions.

4.2.5. Plasticizers (→ Plasticizers) [109]

Plasticizers are added to polymers to improve their flexibility, processibility, or foamability. About 500 different types of plasticizers are marketed. They are generally low molar mass liquids; polymer plasticizers are used in much smaller amounts. Of all plasticizers for thermoplastics, 80–85% are used for poly(vinyl chloride); the most important plasticizers are di-2-ethylhexyl phthalate ("dioctyl phthalate," DOP) and dinonyl phthalate. The main plasticizer for elastomers is mineral oil; tires contain up to 40%. Polymer plasticizers are mainly aliphatic polyesters and polyethers. The former are prepared by polycondensation; they thus possess fairly broad molar mass distributions and also contain monomeric and oligomeric molecules. Because their number-average molar mass is low (ca. 4 000 g/mol), they are called oligomer plasticizers. Plasticization can also be achieved by copolymerization of the parent monomer with certain other monomers; this effect is called internal plasticization, in analogy to external plasticization by added high and low molar mass plasticizers.

External plasticizers are subdivided into primary and secondary ones. Primary plasticizers interact directly with chains by way of solvation. Secondary plasticizers are merely extenders; they can be used only in combination with a primary plasticizer. A certain plasticizer may thus be a primary or a secondary one, depending on the chemical constitution of the polymer. Mineral oils are, for example, primary plasticizers for polydienes but extenders for poly(vinyl chloride).

4.2.6. Colorants

Colorants are subdivided into dyes (soluble in polymer matrix) and pigments (insoluble). Textile fibers are dyed mainly with dyes. Pigments are preferred for plastics because they have a higher light-fastness and are more stable against migration than dyes. Colorants for plastics are dominated by titanium dioxide (60–65%) and carbon black (20%); only 2% are dyes.

Pigment particles generally possess diameters of 0.3–0.8 μm, which allows the pigmentation of films and fibers of > 20 μm thickness. Very thin films and fibers are colored exclusively by organic pigments because these can be ground to much smaller diameters than inorganic ones. The hiding power of pigments increases with increasing difference between refractive indices of pigment and polymer.

Pigments need not have a special affinity for polymers. They must be wettable by the polymer melt, however, which can be achieved by treating them with surfactants. The aggregation of pigments is mainly the result of air inclusions; it can be removed by the application of vacuum. Pigments can be metered into plastics via master batches (in plastics), color concentrates (in plasticizers), or electric charging of the surface of polymer particles in granulate mixers. Up to 1 wt % of pigment can be mixed in by the last method.

4.2.7. Blowing Agents

Foamed plastics (plastic foams, cellular plastics, expanded plastics) are blends of polymers with gases (→ Foamed Plastics). They may be rigid (glass transition or melting temperature higher than use temperature) or flexible; their cell structure can be open or closed. The gases may be air, nitrogen, carbon dioxide, fluorinated hydrocarbons, etc.

Plastic foams can be produced by mechanical means (whipping, stirring), physical methods (shock volatilization of liquids, washing-out of solids), or chemical foaming either by internal foaming during the polymerization or by external foaming with chemical blowing agents. Chemical blowing agents are chemical compounds that decompose at elevated temperature with release of gases. The most widely used agent for natural rubber is N,N'-azobisisobutyronitrile $(CH_3)_2C(CN)-N=N-C(CN)(CH_3)_2$ (AIBN); for plasticized PVC, N,N'-dinitroso dimethyl terephthalamide $CH_3-N(NO)-CO-(p-C_6H_4)-CO-N(NO)-CH_3$ (NTA); and for other plastics, $1,1'$-azobisformamide $H_2N-CO-N=N-CO-NH_2$ (ABFA). Blowing agents are used in amounts of ca. 0.1% to eliminate sinks in injection molding; 0.2–0.8% for injection-molded structural foams; 0.3% for extended profiles; 1–15% for vinyl plastisol foaming; and 5–15% for compression-molded foam products. Because of the depletion of the ozone layer, some fluorine-containing blowing agents were banned.

4.3. Processing [110–113]

There are many different types of processes to convert monomers, prepolymers, or polymers into plastics. In general, four procedures can be distinguished for processing raw materials into shaped plastics:

1. From monomers to polymers by direct polymerization with simultaneous shaping
2. From monomers by oligomerization to prepolymers, followed by simultaneous polymerization and shaping
3. From monomers by polymerization to polymers, followed by separate shaping
4. From monomers by polymerization to polymers, shaping to semifinished product, and finally to the end product

In all four cases, end products must sometimes be after-treated (degated, polished, etc.).

Procedure 1 has only one process step and should be the most economical. It is widely used for wire coatings and casting. However, technological problems are often encountered for shaped articles because the process involves the conversion of a low-viscosity liquid into a high-viscosity body. Because very high polymerization rates are required for this process to achieve high cycle times (phase sequences) in molding (i.e., number of articles produced per unit time), very few monomers, types of polymerization, and shaping methods are suitable. The method is utilized industrially in the RIM (reaction injection molding) process, mainly for the polyaddition of diisocyanates with polyols. The anionic polymerization of lactams and the Diels–Alder polymerization of dicyclopentadienes and related compounds are also suitable for RIM. In resin transfer molding (RTM), thermoset resins and curing agents are mixed together, injected, and cured in the mold.

Procedure 2 is typical for thermosets. Oligomerizations are conducted to conversions shortly before the gel point. Additives are added, and the final processing step consists of a simultaneous cross-linking polymerization and shaping. Typical processing methods include molding, compression molding, cavity compression molding (hot pressing), injection–compression molding, and resin transfer molding.

Procedure 3 is generally chosen for thermoplastics. Polymers from the polymerization process are isolated before shaping and stored (e.g., as granulate) for a shorter or longer period of time. The stored polymers must be dried before further processing; otherwise, steam produced at higher processing temperatures may lead to voids and cavities in shaped articles. After compounding, the plastic raw material must be weighed out accurately for each cycle in discontinuous processing procedures. The filling factor (= raw material density/bulk density) must be known.

The choice of a processing technique is influenced technically by the rheological properties of the material and the form or shape of the desired article. The cost of processing machines and the cycle time are economically important. Commodity articles are produced by casting, centrifugal casting, hot pressing, compression molding, coating, spraying, roll milling, extrusion, injection molding, blow molding, cold forming, press forming, calendering, stretch forming, blowing, extrusion blowing, sintering, fluidized-bed sintering, flame spraying, or hot blast sprinkling, and specialty articles by molding.

Procedure 4 is the method of choice for difficult-to-manufacture articles. Procedures, such as welding, stamping, cutting, forging, sawing, boring, turning, milling (in the metalworking sense), sintering, and baking are employed for both thermoplastics and thermosets. The surfaces of plastics are sometimes treated further for technical (surface hardness, friction) or aesthetic reasons (gloss, color) by polishing, painting, metallizing, coating, etc.

Whereas conventional processing requires molds or machining, in additive processing, also referred to as additives processing and solid freeform fabrication, complex 3D objects are built by computer-assisted layer-by-layer assembly. Typically, 3D scanner or computer-assisted design are used to create virtual 3D objects, which are sliced and used to construct the real 3D objects layer by layer. Originally designed to serve the special needs of model making, tooling, and rapid prototyping, this mold-free fabrication is emerging as highly versatile processing technology for rapid manufacturing and digital 3D printing. Typical 3D printing technologies include stereolithography with laser-induced photopolymerization of acrylic monomers, selective laser sintering of metal, ceramic and polymer powders, fused deposition molding using 3D extrusion of thermoplastic filaments, and 3D printing by ink-jet printing of binders to bind together powder layers.

5. Supermolecular Structures (→ Plastics, Properties and Testing) [114–116]

Polymer properties are influenced not only by the chemical structure (constitution, molar mass, configuration, microconformation) but also by the physical structure of polymers. These structures may range from totally irregular arrangements of chain segments over shorter or longer parallelizations of chains, to voids and other defects in otherwise highly organized assemblies of polymer molecules. Two possible ideal structures exist in the solid state: Perfect crystals and totally amorphous polymers. Polymer molecules are perfectly ordered in ideal crystals. They convert at the thermodynamic melting temperature into melts, which ideally are totally disordered. Amorphous polymers can be viewed as frozen-in polymer melts. They are polymer glasses that convert to melts at the glass transition temperature.

5.1. Noncrystalline States

5.1.1. Structure

Isolated polymer coils possess approximately a Gaussian distribution of chain segments; their segment density decreases with increasing chain length. However, the macroscopic densities of polymer melts do not change with chain lengths if end group effects on small molar mass molecules are neglected. Coils must therefore overlap considerably in polymer melts. At the glass transition temperature, cooperative segmental movements freeze in, and the physical structure of the melt is conserved. Small-angle neutron scattering studies have shown that the radii of gyration of amorphous polymers are indeed essentially identical for their melts, glasses, and solutions in theta solvents.

The absence of long-range order in melts and amorphous polymers does not exclude the presence of short-range order in these states. Because of the persistence of polymer chains, a parallelization of short segments seems probable, as is found, for example, for alkanes according to X-ray investigation. This local order does not exceed 1 nm in each direction.

The packing of chain segments cannot be perfect. Amorphous polymers thus possess "free volumes", which are regions of approximately atomic diameters. The volume fraction of this free volume is ca. 2.5% at the glass transition temperature and is independent of polymer constitution. Polymer segments in melts can move more freely than in the glassy state; the densities of melts are thus higher than the densities of glasses at the same temperature.

5.1.2. Orientation

Polymer segments, polymer molecules, and crystalline domains may be oriented along the machine direction by drawing or other mechanical processes. The orientation of chain segments need not necessarily lead to crystallization, however. An example is injection-molded polystyrene, which shows optical birefringence due to the orientation of segments but no X-ray crystallinity.

The degree of orientation of segments or crystallites can be measured by wide-angle X-ray or small-angle light scattering, infrared dichroism, optical birefringence, polarized fluorescence, and ultrasound velocity. The orientation is usually characterized for each of the three directions a, b, and c by a *Hermans* orientation factor f_i, where β is the angle between the draw direction and the principal axis of the segments:

$$f_i = (1/2)(3\cos^2\beta - 1)$$

The orientation factor becomes 1 for a complete orientation of that axis in the draw direction ($\beta = 0°$), $-1/2$ for a complete orientation perpendicular to the draw direction ($\beta = 90°$), and 0 for a random orientation.

Since methods to determine orientation factors are often expensive, time consuming, or difficult to perform, the easy-to-calculate *draw ratio* (length after drawing/length before drawing) is often used to characterize orientation. The draw ratio is, however, not a good measure of the degree of orientation because it depends on the history of the specimen. Drawing may also lead to viscous flow of the polymer without any orientation of segments and crystallites.

5.2. Polymer Crystals [117, 118]

5.2.1. Introduction

Spheres and other geometrically simple entities can be arranged in crystal lattices irrespective of whether they are atoms, small molecules, large molecules such as enzyme, or latex particles. The centers of gravity of these entities occupy lattice sites that are regularly spaced in three dimensions. Lattices are called *superlattices* if the distances between lattice points significantly exceed atomic dimensions. The existence of three-dimensional order on the atomic level is called crystallinity.

Chain molecules can also be arranged in crystal lattices. The whole chain may be completely parallel to other chains as in the case of truly extended chain crystals; the microconformation of the chain can be all *trans* or helical (Fig. 23), and the chains may have a directional sense as in polyamide 6 or an inversion center as in polyamide 66 (Fig. 24). In other cases, only sections of the chains may be parallel to each other, either sections of a single chain as in folded chain crystals or those of neighboring chains as in fringed micelles. These ordered

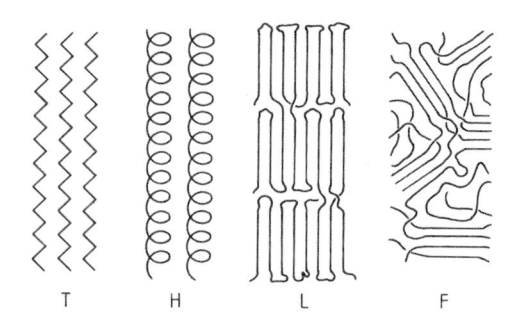

Figure 23. Schematic representation of crystalline polymers as extended chain crystals with chains in all-*trans* (T) or helical conformation (H), (L) lamellae of chains interconnected by tie molecules, and (F) fringed micelles. No bond angles are shown for H, L, and F.

segments form crystalline regions in otherwise less ordered (though not necessarily completely disordered) polymers.

Crystalline regions may be assembled into greater entities, which themselves may be more or less ordered. The types and degrees of order–disorder depend on external conditions such as temperature, pressure, cooling and heating rates, and presence of solvents. A variety of morphologies may thus exist for any crystallizable polymer [104], for example, spherical, platelet-like, fibrillar, and sheaflike structures of a polyamide 6 (Fig. 25).

5.2.2. Crystal Structures

The unit cell describes the smallest, regularly repeating structure of a parallelepiped that generates a crystal by parallel displacements in three directions. The parallelepipeds are characterized by three lattice constants (distances between lattice points in three directions) and three lattice angles (Table 15).

Lattice constants reflect the microconformations of chains in crystal lattices. The c-direction is normally assigned to the chain direction; a value of $c = 0.254$ nm thus gives the distance between repeating units $-CH_2-CH_2-$ in all-*trans* conformation (bond distance 0.154 nm, bond angle 112°). The c-values in carbon chains that are non-multiples of 0.254 indicate the presence of non-all-*trans* conformations (e.g., helical

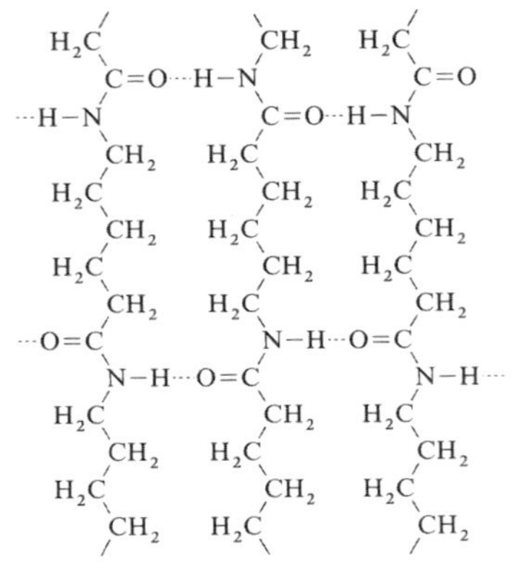

Figure 24. Pleated-sheet structures of polyamides PA 6 (anticlinal) and PA 66 (isoclinal)

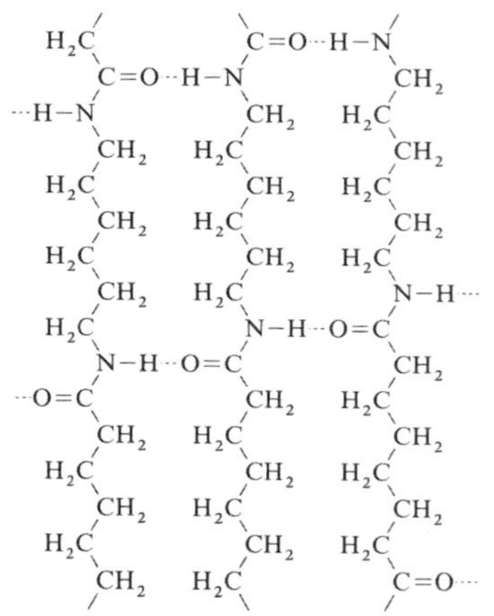

structures in it-polypropylene and it-polybutene-1). The a and b values reflect physical bonds (such as van der Waals or hydrogen bonds) that are longer than the chemical bonds in the c-direction. The anisotropy of bond lengths in the three spatial directions prevents the existence of cubic lattices for chain molecules. The other six lattice types are however found [hexagonal, tetragonal, trigonal, (ortho) rhombic, monoclinic, triclinic].

Most polymers form polymorphs (i.e., polymers of the same constitution and configuration can possess various energetically different crystal modifications). *Polymorphism* may be caused by different microconformations of polymer chains (e.g., polybutene-1) or by different packing of chains with identical chain conformations (e.g., it-polypropylene). Such differences can be generated by small changes in processing conditions (e.g., different crystallization temperatures) or by stretching.

Copolymers can furthermore show *isomorphism* if different monomeric units can replace each other without change of lattice structure. Isomorphism of chains is also possible if the two corresponding homopolymers have analogous crystal modifications, similar lattice constants, and the same helix types. An example is a mixture of the γ-modification of it-polypropylene and the modification I of it-polybutene-1 (Table 15). This phenomenon is sometimes also called allomerism. One company calls its crystalline copolymers from two or more olefinic monomers *polyallomers*, because the monomeric units are isomorphic.

5.2.3. Crystallinity

Perfect crystal lattices should give sharp X-ray reflections and sharp melting points. Polymers do not exhibit these features, even if they show a crystallike appearance in electron micrographs. In addition, X-ray patterns of polymers frequently exhibit continuous, diffuse scattering. These findings can be interpreted by either a one-phase model (voids in crystals) or a two-phase model (coexistence of completely crystalline and completely amorphous regions). Crystallinity is defined as the presence of three-dimensional order on the level of atomic dimensions.

In reality, segments are ordered to various degrees. Because most experimental methods are sensitive to different types and degrees of order, various "degrees of crystallinities" are found depending on the experiment. The degree of crystallinity is the fractional amount of crystallinity. Stretched poly(ethylene terephthalate) has, for example, a degree of crystallinity of 59% by IR spectroscopy, 20% by density determination, and 2% by X-ray diffraction.

5.2.4. Morphology

Earlier experimental findings were interpreted in terms of the two-phase model of partially crystalline polymers: The coexistence of

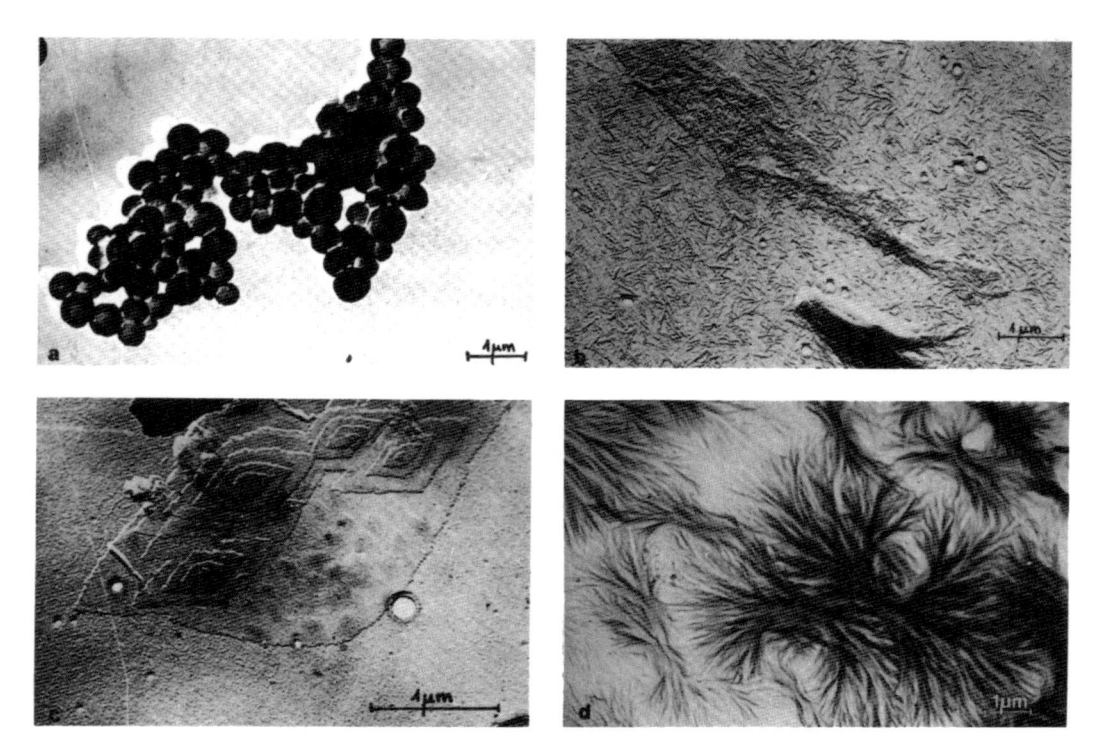

Figure 25. Electron micrographs of different morphologies of a polyamide 6
a) From a 260°C solution in glycerol quenched into 20°C glycerol; b) Same solution fast cooled (40 K/min); c) Same solution slowly cooled (1–2 K/min); d) Slow evaporation of formic acid solution at room temperature [119]

crystalline reflexes and amorphous halos in X-ray diagrams, the independence of X-ray short periodicities on molar masses, the disappearance of X-ray long periodicities with increasing molar masses, lower macroscopic densities than predicted by X-ray densities of unit cells, broad melting ranges, optical birefringence of oriented polymers, and heterogeneity of partially crystalline polymers against chemical reactions.

Table 15. Crystal structure of some polymers

Polymer			N_u	Lattice constants			Angles of unit cell			Helix type	Crystal system
				nm^a	nm^b	nm^c	α°	β°	γ°		
PE		I^b	2	0.742	0.495	0.254	90	90	90	2_1	R
		II$^{a\,b}$	2	0.809	0.479	0.253	90	90	107.9	2_1	M
PP	st		8	1.450	0.560	0.74	90	90	90	2_1	R
	it	α^b	12	0.665	2.09	0.495	90	99.6	90	3_1	M
	it	β^b	3	0.638	0.638	0.633	90	90	120	3_1	VI
	it	γ^b	12	0.647	2.140	6.50	89	100	99	3_1	T
PB	it	I^b	18	1.769	1.769	0.650	90	90	90	3_1	VI
		II^b	44	1.485	1.485	2.060	90	90	90	11_3	IV
		III^b	8	1.238	0.892	0.745	90	90	90	4_1	R

$^a N_u$ = Number of monomeric units per unit cell.
^{b} Crystal modification.
^{c} R = (ortho)rhombic; M = monoclinic; VI = hexagonal; T = triclinic; IV = tetragonal.

In 1957, however, three different groups found that rhombic platelets are obtained if very dilute solutions of polyethylene are cooled. Similar platelets of 5–20 nm thickness are now known from many other crystallizable polymers (e.g., see Fig. 25). Electron diffraction showed sharp reflections typical of single crystals and chain directions perpendicular to the surface of the platelets. Since the contour lengths of the chains are greater than the thicknesses of the platelets, chains must be folded back as shown in Figure 23, L.

Crystallization from polyethylene melts similarly shows stacked single crystals. These lamellae are 75–80% X-ray crystalline. After oxidation of the surface, a 100% crystalline material remains. The folds or the interface between lamellae must thus be "amorphous." Various models have been suggested for the surface layers: Loose folds with adjacent reentry of the chains, a switchboard-like structure, adsorbed chains not involved in chain folding, loose or entangled cilia (tangling chain ends), and crystal bridges composed of tie molecules (Fig. 26). Electron microscopy has shown that crystal bridges exist in crystallized mixtures of high and low molar mass polyethylenes. Such interlamellar connections are responsible for some remarkable properties of partially crystalline polymers (see below). Highly crystalline polymers result if tie molecules and surface layers are removed (e.g., by grinding); these fairly low molar mass materials are called microcrystalline polymers.

Crystallization of melts sometimes delivers spherical, polycrystalline entities (Fig. 27). These spherulites exhibit radial symmetry. They consist of lamellae with tangential chain axes (Fig. 28A); radial chain axes seem to be unknown (Fig. 28B). The formation of spherulites can be controlled by the crystallization temperature, the cooling rate, and the addition of nucleating agents.

Spherulites lead to opaque plastics if their diameters are greater than one-half the wavelength of incident light and their densities and

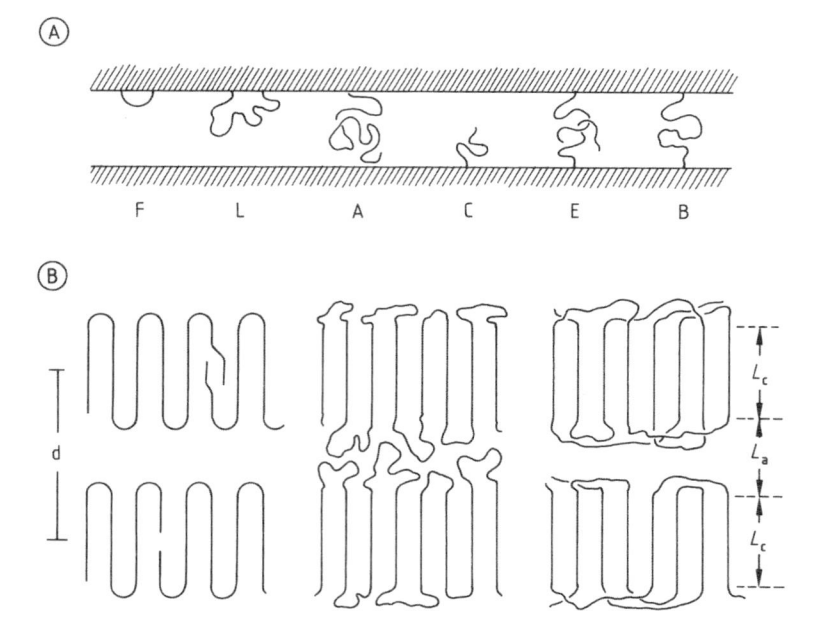

Figure 26. Models for the fine structure of lamellae of single crystals
A) Chains between two adjacent lamellae: F = sharp folds, L = loose loops, A = adsorbed chain, C = cilium, E = entangled cilium, B = crystal bridge or tie segment; B) Structures of lamellae: R=Regular lamellae with sharp folds, adjacent chain reentry, void from chain ends (below) and dislocation (above); L = lamellae with loose loops and adjacent reentry; S = switchboard model; d = crystallographic long period; L_c = thickness of lamella; L_a = thickness of "amorphous" surface layer ($d = L_c + L_a$)
Reprinted with permission by Hüthig and Wepf Publ., Basel [5]

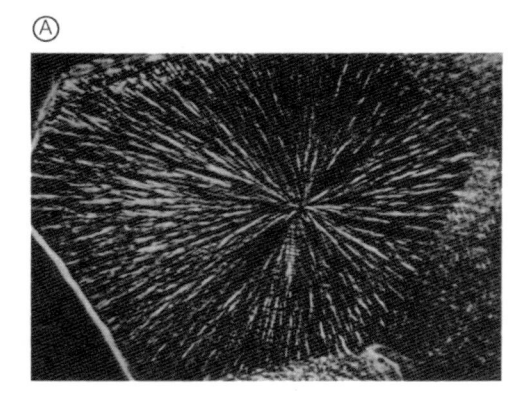

Figure 27. Spherulites of it-polypropylene under the phase contrast microscope (A) and the polarization microscope (B) [120]

refractive indices are different for crystalline and noncrystalline regions. The numbers and sizes of such spherulites are important for the fracture behavior of polymers.

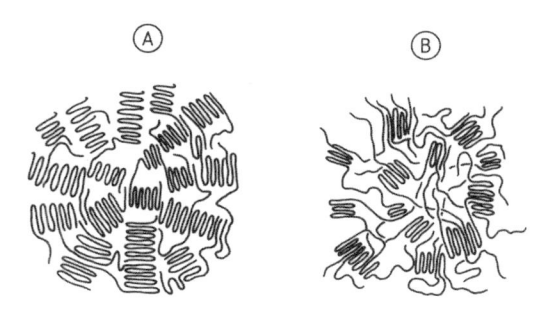

Figure 28. Schematic representation of spherulite structures
A) Tangential chain axes; B) Radial chain axes (hypothetical)
Reprinted with permission by Hüthig and Wepf Publ., Basel [5]

Spherulites develop in a viscous environment if crystallization nuclei are present and if equal crystallization rates prevail in all three directions. Different crystallization rates exist in strongly stirred dilute solutions. Polymer chains then try to orient themselves along the flow gradients and crystallize in an extended (nonfolded) conformation. The resulting fibrils are organized in bundles with chains parallel to the fibril axes. These bundles act as nuclei–substrate for the epitaxial growth of the remaining chains. Since the shear rate is strongly reduced between bundles, crystallization leads to lamellae with folded chains whose axes are parallel to the fibril axes (Fig. 29). The

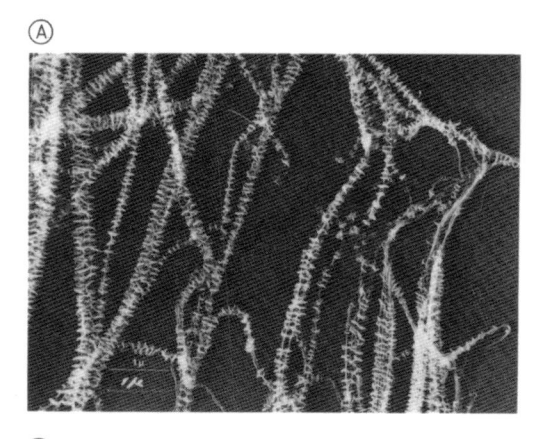

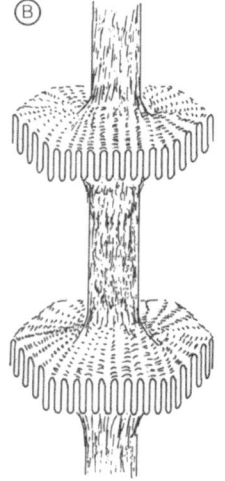

Figure 29. Shish kebab structures of linear polyethylene crystallized from rapidly stirred 5% solutions in toluene [76]
A) Electron micrograph; B) Schematic drawing of chain arrangements

resulting extended-chain structures are responsible for the remarkable properties of some ultraoriented polymers.

5.3. Mesophases and Liquid-Crystalline Polymers [121]

5.3.1. Introduction

Mesomorphic chemical compounds exhibit microscopic structures (mesophases) that are intermediate between crystals with long-range three-dimensional order and liquids without long-range order (Greek: *mesos* = middle, *morphe* = shape, form). Three types of mesophases are distinguished: Liquid crystals, plastic crystals, and condis crystals. Since liquid crystals were discovered first and are furthermore the most common of the three classes, the term mesophase is often used as a synonym for liquid crystal.

Liquid crystals (LCs) (→ Liquid Crystals for Display Application) possess a certain ordered structure like crystals, but they flow like liquids. *Thermotropic liquid crystals* exhibit this phenomenon in the pure state above a certain solid--mesophase transition temperature, where the solid may be a crystal or a glass. *Lyotropic liquid crystals* exhibit mesophase behavior in certain solvents above critical concentrations. Liquid crystalline behavior is present in both cases if the molecules contain "mesogens", that are short rodlike sections that comprise the whole molecule as in low molar mass LCs, or segments thereof as in polymer liquid crystals (LCPs). Mesogens are anisotropic; they are oriented with respect to their mesogen axes but not with respect to their positions.

Plastic crystals exhibit order of positions but disorder of orientations of molecules. They are found for certain spherical low molar mass molecules packed in cubic lattices (i.e., they are isotropic). The absence of strong attractive forces between molecules and the presence of slip planes lead to easy deformabilities of plastic crystals, sometimes even under their own weight. No polymers are known that form plastic crystals.

Condis crystals are *con*formationally *dis*ordered crystals containing several conformational isomers in more or less ideal crystalline positions with respect to position and orientation of segments (conformational isomorphs). An example is the hexagonal high-pressure phase of polyethylene extended-chain crystals. It is often difficult to determine whether polymer crystals are normal or condis crystals.

On cooling, thermotropic liquid crystals either crystallize or solidify to "mesomorphic glasses" with retention of their liquid crystal structures. These glasses do not carry special names, but since "liquid crystalline glass" and similar names are oxymorons, the terms "LC glass," "PC glass," and "CD glass" have been suggested for the three classes of mesomorphic materials. The LCPs are used mainly as LC glasses not as liquid crystals per se, except in processing.

Condensed matter can thus exist in several of the five possible classes of pure phases: Completely ordered crystal oC, mesomorphic glass mG, amorphous glass aG, ordered mesophase oM, and isotropic melt (liquid) iL. Seven types of biphasic materials exist: oC + mG, oC + aG, oC + oM, oC + iL, mG + aG, mG + iL, oM + iL, and three types of triphasic materials: oC + mG + aG, oC + mG + iL, oC + oM + iL. Because each class can be subdivided further, many physical structures (and thus properties) are possible for the same polymer.

Mesophasic polymers are characterized by domains of microscopic size (see below). "Microdomains" of similar size are also exhibited by certain block copolymers and, in much smaller sizes, by ionomers.

5.3.2. Types of Liquid Crystals

Liquid crystals and LC glasses are generated by rod- or disclike mesogens, which either comprise the whole molecule (as in low molar mass LCs) or sections of it (LCPs). Examples are

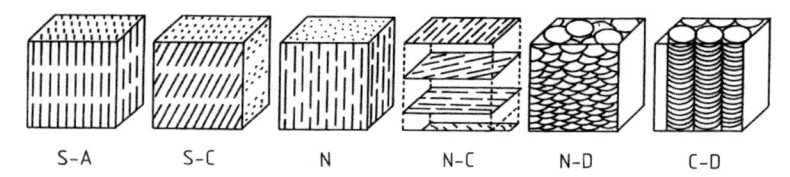

Figure 30. Schematic representation of mesogens in different types of mesophases
S = smectic (types A and C); N = nematic; N–C = nematic cholesteric; C–D = columnar discotic; N–D = nematic discotic
Reprinted with permission by Hüthig and Wepf Publ., Basel [5]

where X = O, OOC, COO, etc.; Y = COO, $p\text{-}C_6H_4$, CH=CH, N=N, etc.; R = CO $(CH_2)_n CH_3$, etc.; and phenylene residues may be replaced by 1,4-cyclohexane rings. The lengths of polymer mesogens are often comparable to those of repeating units; they seem to be identical with respect to persistence lengths.

Rodlike mesogens are also called calamitic (Greek: *calamos* = reed) and disclike ones discotic. Calamitic liquid crystals are further subdivided into smectic, nematic, and cholesteric mesophases. Smectic LCs show fan-shaped structures under the polarization microscope; they feel like soaps (Greek: *smegma* = soap). Nematic LCs form threadlike schlieren (Greek: *nema* = thread). Cholesteric LCs exhibit beautiful reflective colors, which were first discovered with cholesterol.

The birefringence of LCs originates from the anisotropy of mesogens, which are more or less aligned in microscopic domains with diameters in the micrometer range. These alignments lead to different refractive indices parallel and perpendicular to the direction of polarization of incident light. The domains themselves are arranged randomly. The presence of birefringence however is not proof for the existence of mesogens; it may also result from higher-melting crystallites in partially molten polymers or from crystallites in partially molten polymers or from crystallites in gels or concentrated solutions.

Liquid crystals are turbid because their domains are anisotropic structures whose dimensions are greater than the wavelength of incident light. At a certain "clearing temperature," these domains melt to a transparent (isotropic) liquid.

The soaplike character of *smectic LCs* is due to two-dimensional, layerlike arrangements of mesogens (Fig. 30). At present, twelve different smectic types are known: Four with the average direction of the long axes perpendicular to layer planes and seven with tilted axes. Examples are type S-A with mesogens perpendicular to the layer planes; type S-B, the same but with perfect hexagonal packing of mesogens; type S-C, with mesogen axes at an angle to the planes, etc.

Nematic mesophases (Fig. 30, N) are one-dimensionally ordered. The principal axes of mesogens are more or less parallel to each other; the centers of mass are however distributed at random.

Cholesteric mesophases (Fig. 30, N–C) are observed only with chiral mesogens. They consist of layers of nematically ordered mesogens. These layers have a sense of rotation relative to each other due to the presence of chirality centers in mesogens. Cholesteric mesophases are thus screwlike nematics.

Discotic mesophases may exist in many different structures. The disclike mesogens are randomly arranged in nematic-discotic types (Fig. 30, N–D) but stacked like coins in columnar-discotic ones (Fig. 30, C–D).

5.3.3. Mesogens

Mesogens of polymers can be contained in main chains (MC-LCPs) or in side chains (SC-LCPs). Thermotropic MC-LCP glasses are used as engineering plastics; lyotropic MC-LCP glasses, for fibers. Thermotropic SC-LCPs are not used commercially at present but may find use in optical recording devices.

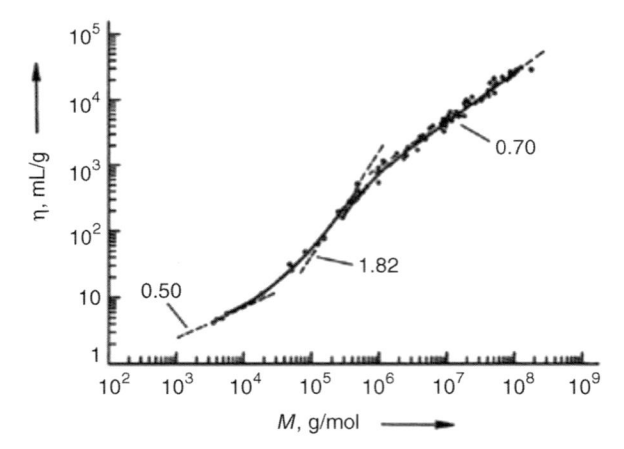

Figure 31. Molar mass dependence of intrinsic viscosities of wormlike chains
(○) Helices of poly(γ-benzyl-L-glutamate) (PLBG) in N,N-dimethylformamide at 25°C; (●) double helices of deoxyribonucleic acids in dilute aqueous NaCl solutions at 20°C. The two polymers are comparable because the helix diameters are similar (1.5 nm for PLBG, 2.0 nm for DNA). Numbers indicate exponents a in Equation (1)
Reprinted with permission by Hüthig and Wepf Publ., Basel [5]

Mesogens usually constitute the whole molecule in low molar mass LC molecules. In macromolecules, this is true only for dissolved helices up to certain molar masses. Such helicogenic macromolecules are rigid rods at small molar masses but become coillike at very high degrees of polymerization (cf. the behavior of short and very long garden hoses). This behavior is demonstrated by the molar mass dependence of their intrinsic viscosities, which are measures of the molecule volume per unit mass (Fig. 31). Rodlike molecules are characterized by their axial ratio (aspect ratio) Λ (length/diameter).

All other polymer chains form either flexible or semiflexible coils in dilute solution or in their isotropic melts (see examples in Fig. 32). Semiflexible (wormlike) chains SF allow rotations around some of their chain bonds. The rodlike character of their segments is however preserved in melts or concentrated solutions if the chain axes remain linear or if the segment linearity can be restored by a crankshaft motion. Conjugated bond structures are helpful but not necessarily required for mesogens since, for example, 1,4-phenylene rings may be replaced by 1,4-cyclohexylene units.

Linear chain axes are present in poly(p-phenylene) (PPP) or poly(p-phenylene benzoxazole) (PBO). Crankshaft motions around the chain axes are possible in the conjugated chains of poly(p-phenylene vinylene) (PPV), the partially conjugated chains of poly (p-hydroxybenzoic acid) (PHB), and even in the nonconjugated poly(p-phenylene alkylenes) as long as even numbers of methylene groups are present between phenylene rings as in poly(p-phenylene ethylene). Semiflexible molecules are characterized by their persistence lengths a or their Kuhn lengths $L_K = 2\,a$.

Rodlike molecules and semiflexible molecules with long persistence lengths form many physical bonds between parallel chains. The total energy of such assemblies is so high that individual intramolecular chemical bonds rather than all of the physical bonds are broken upon heating: These molecules decompose instead of producing thermotropic mesophases. They may however form lyotropic mesophases in suitable solvents. An example is poly (p-phenylene bisbenzoxazole).

The "stiffness" of the chain segments can be removed or reduced by several means (Fig. 33). Bulky substituents prevent the parallelization of chains; they "frustrate" crystallization. Nonlinear chain elements (such as *ortho* and *meta* substituents) work in the same way. Flexible chain segments reduce the persistence lengths of the chains.

Figure 32. Examples of rodlike (R), wormlike or semiflexible (SF), and flexible (F) molecules and segments. See text for further explanation.
Reprinted with permission by Hüthig and Wepf Publ., Basel [5]

5.3.4. Lyotropic Liquid-Crystalline Polymers

Molecule axes of low molar mass rodlike molecules are disordered in isotropic melts, whereas mesogen axes are more or less parallel to each other in nematic and smectic mesophases. Very short rods resemble spheroids. They can arrange themselves in various stable ways, but the parallel ordering need not be much more stable than other arrangements. Therefore, a critical axial ratio exists above which the simple geometric anisotropy is sufficient to stabilize a mesophase.

This critical axial ratio has been calculated with the lattice theory of polymer solutions as $\Lambda_{crit} \approx 6.42$. A geometric stabilization by

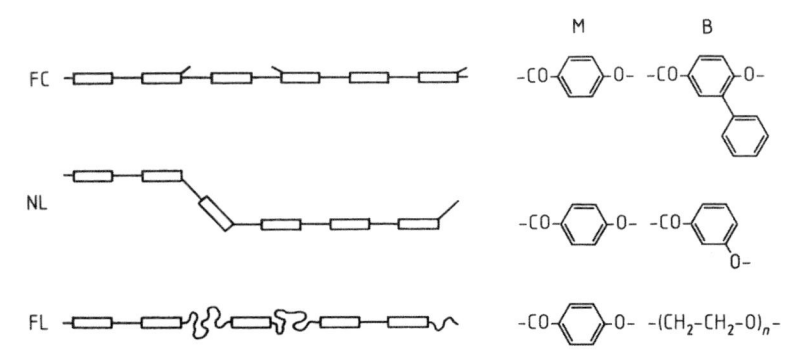

Figure 33. Flexibilization of rigid chain segments by frustrated crystallization (FC), nonlinear chains (NL), or flexible chain elements (FL); M = mesogen-forming monomeric units; B = stiffness-breaking units
Reprinted with permission by Hüthig and Wepf Publ., Basel [5]

repulsion is insufficient for $\Lambda < \Lambda_{\text{crit}}$ and the mesogens must exert additional orientation-dependent attractive forces.

In solution, repulsive forces between mesogens can act only at sufficiently high mesogen concentrations. At a critical volume fraction, Φ_p^* phase separation occurs into a polymer-rich mesophase and a dilute isotropic phase. The dependence of Φ_p^* on Λ can be described to a first approximation by [122]

$$\Phi_p^* \approx 8(1 - 2L^{-1})/L \qquad (18)$$

The axial ratio Λ is the molar mass-dependent true axial ratio of rodlike molecules (i.e., helices) and the molar mass-independent Kuhn length of semiflexible molecules. Equation (18) is remarkably well fulfilled for both helical and semiflexible molecules (Fig. 34). An exception is the tobacco mosaic virus, probably because of additional charge effects that are not considered by the theory.

Lyotropic mesophases are formed by the helices of tobacco mosaic virus ($c > 2\%$) and deoxyribonucleic acids ($c > 6\%$) in dilute aqueous salt solutions as well as by poly (α-amino acids) in helicogenic solvents, e.g., poly(γ-methyl L-glutamate) in dichloromethane–ethyl acetate (12 : 5) ($c > 15\%$).

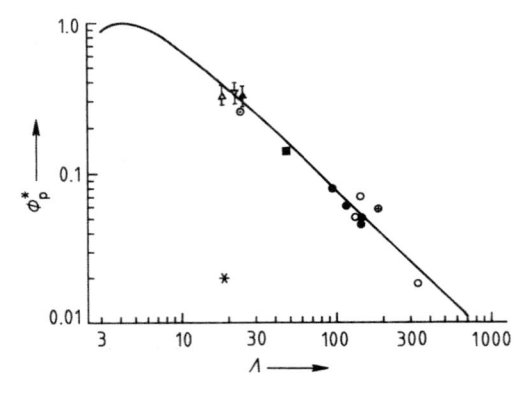

Figure 34. Critical volume fractions ϕ_p^* for phase separations of isotropic solution → nematic mesophase as function of Λ [i.e., the axial ratio of helical (rigid) molecules (circles) and the Kuhn length of semiflexible molecules (triangles, square)]
*Tobacco mosaic virus; solid line: prediction of Equation (18)
Reprinted with permission by Hüthig and Wepf Publ., Basel [5]

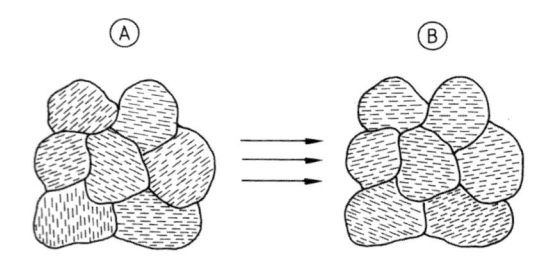

Figure 35. Alignment of mesogen axes upon shearing
A) Mesophase domains in liquid crystalline state at rest;
B) During shearing or as LC glass with "frozen in sheared state"

Hydroxypropyl cellulose (HPC) forms a cholesteric mesophase at room temperature in water at $w_{\text{HPC}} > 0.41$.

On shearing, mesophase domains become oriented with respect to the mesogen axes (Fig. 35). This alignment becomes frozen-in on rapid quenching of thermotropic LCs or by precipitation of lyotropic ones into baths (half-lifes of orientation between a few seconds and a few hundredths seconds). The resulting LC glasses possess good orientations of mesogen axes in the shear direction (e.g., fiber direction), high tensile strengths, and low extensibilities.

A number of LC fibers are produced commercially. Poly(p-phenylene terephthalamide) (PPB−T) was originally spun from concentrated sulfuric acid solutions, whereas chlorinated N-methyl pyrrolidine is now used:

∿NH—⟨⟩—NH−CO—⟨⟩—CO∿   Kevlar; PPB−T

A similar copolyamide is also on the market

−NH—⟨⟩—NH−CO—⟨⟩—CO−
−NH—⟨⟩—O—⟨⟩—NH−CO—⟨⟩—CO−   } Technora

Even better LC fibers are produced from poly (p-phenylene-2,6-benzoxazoles) and poly(p-phenylene-2,6-benzthiazoles) (PBT), which are spun from poly(phosphoric acid) solutions, at present on a bench scale. Both *"cis"* (shown) and *"trans"* (X and N exchanged at one ring) structures are known:

PBO (X = O); PBT (X = S)

5.3.5. Thermotropic Liquid Crystal Polymers

Non-Newtonian shear viscosities of liquid crystal polymers drop drastically with increasing shear rates due to the orientation of mesogen axes upon shearing. Less energy is thus needed for processing thermotropic LCPs than isotropic thermoplastics. The high orientation times permit a freezing-in of the mesogen orientation in injection molding, for example. Because of the higher than usual moduli of elasticity and tensile strengths, such LCPs are also called *self-reinforcing polymers*.

The first self-reinforcing polymer was the copolyester X7 G with 60 mol% *p*-hydroxybenzoyl and 40 mol% terephthaloyl glycol units. It is no longer produced because the less expensive glass-fiber-reinforced saturated polyesters possess similar properties. Xydar and Vectra are, however, industrially produced:

Polymeric side-chain LCs can serve for the thermooptical storage of information. A guided laser beam increases the temperature locally, whereupon phase transformations occur. The change of order can then be frozen-in on cooling below the glass transition temperature. The resolution is ca. 0.3 µm.

5.3.6. Block Copolymers [101]

Block copolymers consist of two and more blocks of constitutionally or configurationally different monomeric units (e.g., A_n-B_m, $A_n-B_m-A_n$). In most cases these blocks are thermodynamically or kinetically immiscible because their homopolymers A_n, B_m, etc., are not miscible. The demixing of block copolymers cannot proceed, however, to macroscopic phase separations because the blocks are chemically coupled. Similar blocks of different block copolymer molecules can thus only aggregate and form domains in a matrix of the other blocks. This phenomenon is called microphase separation.

Three different applications of block copolymers are known: Polymeric detergents, thermoplastic elastomers, and compatibilizers for blends. Polymeric detergents consist of hydrophilic and hydrophobic blocks. Examples are the multiblock copolymers from watersoluble ethylene oxide blocks $[-(OCH_2CH_2)_n-]$ and water-insoluble propene oxide blocks $-[OCH_2CH(CH_3)]_n-$. The hydrophobic blocks associate in water, which produces high viscosities.

A similar action is exerted by *compatibilizers* for nonmiscible blends of A_n and B_m polymers. These polymers are A_p-B_q diblock copolymers: The A_p block resides in the A_n phases and the B_q block in the B_m ones. Compatibilizers thus anchor the two types of phases (Fig. 36).

Thermoplastic elastomers [123] consist of tri- or multiblock copolymers. Triblock copolymers $S_n-B_m-S_n$ possess a soft center polybutadiene block—B_m—(glass transition temperature below use temperature) and two hard end polystyrene blocks S_n—(glass transition temperature above use temperature). Below certain *m/n* ratios, styrene blocks form spherical microdomains in a continuous matrix of

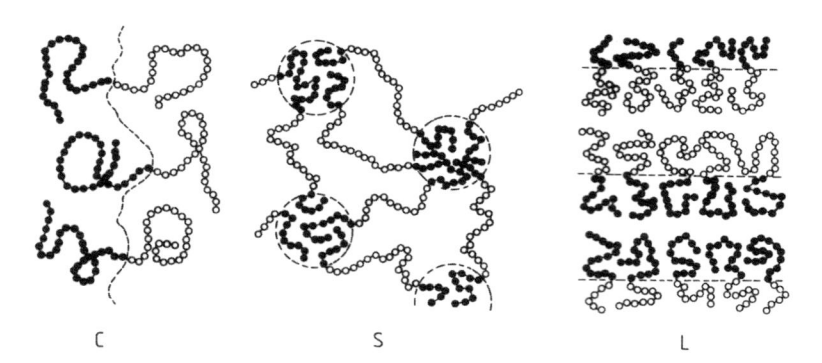

Figure 36. Arrangement of A_n blocks (with A units ●) and B_m blocks (with B units ○) in diblock (C, L) and triblock copolymers (S)
C: compatibilizers at an interface - - -; S: spherical A_n domains in continuous matrix B_m; L: lamellae of A_n and B_m

butadiene blocks. These spherical domains thus act as cross-linking units in an elastomer (Fig. 36, S). The hard domains "melt" above their glass transition temperatures, and the block copolymers can then be processed like thermoplastics.

The morphology of *diblock copolymers* is governed mainly by the relative spatial requirements of the blocks. Both blocks, if independent, would form unperturbed random coils. If the unperturbed volumes of both blocks are of equal size, all A_n blocks would line up in one layer and all B_m blocks in another because the blocks are (1) coupled to each other and (2) incompatible. The A_n layer faces another A_n layer, where the A_n blocks are coupled to B_m blocks, etc. (Fig. 36, L). The diblock polymers thus form lamellae with a thickness of two coil diameters. Similar structures result for triblock polymers A_r–B_m–A_r, where $r = n/2$.

If the volume of the A_n blocks is much lower than the volume of the B_m blocks, then the A_n blocks can no longer be packed into layers without violating the demand for tightest packing or deviating from the shape of unperturbed coils. Both possibilities are energetically unfavorable, and the smaller A_n blocks thus cluster together and form spherical domains in a continuous matrix of the larger B_m blocks. If the A_n blocks are somewhat larger (but not big enough to form lamellae), then cylinders would result (Fig. 37).

A great number of thermoplastic elastomers of both the triblock and the multiblock type is known [70]. Their chemical structures range from (styrene)$_n$–(butadiene)$_m$–(styrene)$_n$ triblock copolymers to polyether–urethane multiblock polymers, ethylene–propene copolymer–polyolefin blends, polyether–esters, and graft copolymers of butyl rubber on poly-ethylenes.

5.3.7. Ionomers

Ionomers are copolymers from primarily hydrophobic monomers with small amounts of ionic comonomers. The ionic groups may be in the main chain or in substituents. Four different types of ionomers are produced commercially.

$-CH_2-CH_2-\ +\ < 10\ mol\%\ -CH_2-C(CH_3)-$ $\qquad\qquad\qquad\qquad\qquad\qquad\quad\ \ \|$ $\qquad\qquad\qquad\qquad\qquad\qquad\ \ COOH$	Surlyn
$-CH_2-CH_2-\ +\ < 3.5-20\ mol\%\ -CH_2-CH-$ $\qquad\qquad\qquad\qquad\qquad\qquad\qquad\qquad\ \|$ $\qquad\qquad\qquad\qquad\qquad\qquad\qquad\ COOH$	EEA Copolymer
$-CF_2-CF_2-\ +\ ?\ mol\%\ -CF_2-CF-$ $\qquad\qquad\qquad\qquad\qquad\qquad\quad\ \|$ $\qquad\qquad\qquad\qquad\qquad\ O-[CF_2-CF(CF_3)-O]_n-(CF_2)_2SO_3H$	Nafion
$-CH_2-CH_2-\ +$ $-CH_2-CH(CH_3)-$ + (norbornene)$-C(CH_3)SO_3H$	Thionic

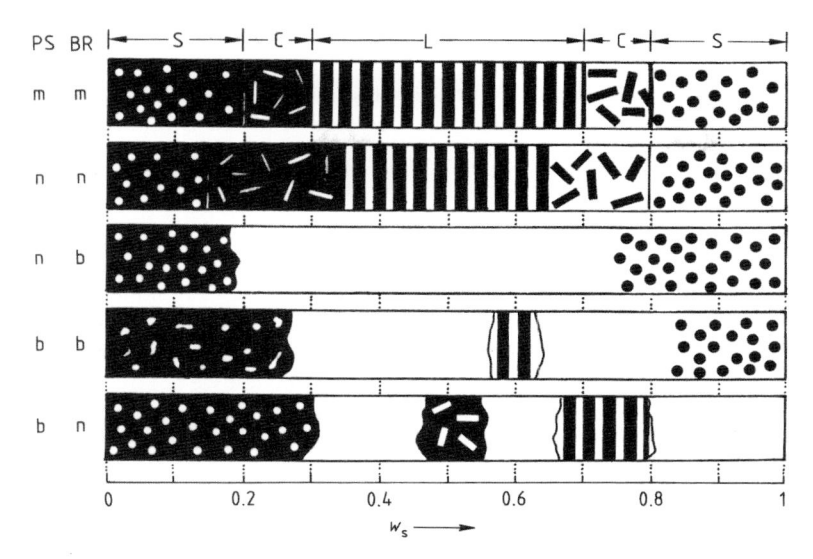

Figure 37. Morphology of styrene–butadiene diblock copolymers as a function of the mass fraction of the styrene units. White: polystyrene blocks (PS); black: polybutadiene blocks (BR). The block length distribution may be (m) molecularly homogeneous (theory), (n) narrow, or (b) broad
S = Spherical domains, C = cylinders (rods), L = lamellae (layers). Upper row: Predictions of Meier theory; rows 2–5: schematic after experimental results of many authors
Reprinted with permission by Hüthig and Wepf Publ., Basel [5]

Surlyn, EEA, and Nafion are synthesized by direct copolymerization of the corresponding monomers. Thionic results from the postpolymerization sulfonation (< 5%) of the diene units (usually 5-ethylidene-2-norbornene) of the primary ethylene–propene–diene (EPDM) rubber. Both EEA and Nafion are used as acids, whereas Surlyn and Thionic are obtained by aftertreatment that leads to sodium or zinc salts.

The introduction of ionic groups into hydrophobic polymers leads to ion associations and subsequently to microphase separations. The ion association is controlled by the coordination number of the ions; that is, sodium ions with valence 1 and coordination number 6 are as good cross-linkers as zinc ions with valence 2. Cross-linking by ions does not occur via ion pairs but via clusters and domains of many ions. These ionic bonds dissociate at higher temperature, and the ionomers can then be processed like thermoplastics. At lower temperature, they behave either as thermoplastic elastomers (T_G of polymer segments lower than use temperature) or as "reversible" thermosets (T_G of segments higher than use temperature).

5.4. Gels

Chemically lightly cross-linked polymers swell upon addition of solvent to form gels. The maximum degree of swelling results from the attempt of segments located between cross-linking sites to attain their energetically most favored coil dimensions which in turn is counteracted by the elastic retraction due to cross-links. Physically cross-linked polymers behave in the same way if the physical cross-links (crystallites, ion clusters, etc.) survive the dissolution process.

The latter phenomenon is utilized in poly(vinyl chloride) pastes. The strong dipole–dipole interactions between C–Cl bonds leads to associations of chain segments which at room temperature resist dissolution by plasticizers such as phthalates, adipates, sebacates, or citrates. Heating of PVC with these plasticizers and subsequent cooling leads to gelation by formation of a lightly physically cross-linked network. Plasticized PVC thus behaves like an elastomeric material. The physical cross-links dissociate at higher temperature, and the resulting PVC

5.5. Polymer Surfaces [124]

5.5.1. Structure

Outer layers of solid or liquid polymers do not exhibit, in contrast to air, water, metal surfaces, etc., the same average composition as the interiors of these polymers. Groups or segments with the lowest Gibbs interfacial energy will reside preferentially at the surface or interface. Surface structures may also be altered by chemical attack during processing or on prolonged use (e.g., by oxidation) or by physical processes such as transcrystallization.

This behavior can be studied by a wide variety of new spectroscopic methods, for example, the concentration of immediate surface groups by Fourier transform infrared spectroscopy (FT-IR); the composition of the upper 0.3–0.5 nm layer by ion scattering spectroscopy (ISS, LEIS); of the upper 1–10 nm by photoelectron spectroscopy [UPS, PE (S)], electron spectroscopy for chemical analysis (ESCA, XPES, XPS, IEE), and Auger spectroscopy (AES).

These methods have shown that poly-(2-vinylpyridine) and poly(4-vinylpyridine) have much more hydrophobic surfaces than their chemical structure indicates: CH_2- and CH- groups face the surface; pyridine residues (C_5H_4N), the interior. Imide–carbonyl residues of polyimides, on the other hand, are preferentially found at the surface and aromatic rings in the interior. The surface is enriched by siloxane residues in polycarbonate–polydimethylsiloxane block copolymers but by styrene units in polystyrene–poly(ethylene oxide) block copolymers. Because the surface energy depends on the contact (air, water, metals, etc.) and also on kinetic effects (thermal history, solvents, etc.), one and the same polymer may possess various surface compositions and thus also surface properties.

5.5.2. Interfacial Tension

Interfacial tension is the force that acts at the interface between two phases; it is called surface tension for the interface condensed phase–gas phase.

Surface tensions γ_{lv} of liquid polymers against air decrease with the two-third power of the molar mass according to

$$\gamma_{lv} = \gamma_{lv}^{\infty} + K_e \times M^{-2/3}$$

They do not vary markedly with temperature. Typical values (at 150 °C) range from 13.6 mN/m (polydimethylsiloxane), 22.1 mN/m (it-polypropylene), 28.1 mN/m (polyethylene), and 33.0 mN/m poly(ethylene oxide). At present, no correlation of these surface tensions with chemical constitution is possible because the surface structures of liquid polymers are unknown.

Interfacial tensions between two liquid polymers are always lower than surface tensions. They are low for two apolar polymers but higher if one polymer or both of them are polar. For example, the interfacial tension of polyethylene–it-polypropylene is only 1.1 mN/m, but that of polyethylene–poly(ethylene oxide) is 9.5 mN/m and poly(ethylene oxide)–polydimethylsiloxane 9.8 mN/m.

The interfacial tensions between solid polymer and air are generally unknown. They do not differ markedly, however, from the critical surface tensions determined by the *Zisman method*. In this method, the liquid–air surface tensions of various liquids are plotted against the cosine of contact angles θ of these liquids against air and extrapolated to cos θ → 0. The resulting "critical" surface tensions of polymers are always lower than the surface tension of water: Polytetrafluoroethylene, 18.5; polyethylene, 33; poly(vinyl alcohol), 37; polyamide 66, 46; and urea–formaldehyde resins, 61 mN/m. Fluorinated polymers have especially low critical surface tensions. They are not wetted by water (72 mN/m) or oils and fats (20–30 mN/m) and are therefore used as surface coatings to prevent sticking.

5.5.3. Adsorption

Macromolecules possess many adsorbable groups and segments. The type of adsorption depends on the adsorption energy per segment (group), the concentration of adsorbable macromolecules, and the duration of the experiment.

"Adsorption equilibria" are established in minutes to hours at smooth surfaces but may take days to attain at rough surfaces or powders. The adsorption time increases with the concentration and molar mass of the polymer.

Polymer coils tend to overlap even in fairly dilute solutions: The adsorbed polymer layers are almost always multilayers, except for the adsorption of oligomers ($X < 10$–100) or from very dilute solutions (volume fractions $< 10^{-4}$–10^{-3}). At higher concentrations or molar masses, adsorbed amounts and structures of polymer layers are mainly kinetically controlled. Reorganization may occur with time; for example, the adsorption of polar polymers at polar surfaces can change from an initial loose loop structure to a more compact, flat covering. The layer of poly(ethylene oxide) adsorbed on chromium surfaces is only ca. 2 nm thick and so tightly packed that the refractive index of the adsorbed layer is identical with that of the crystalline polymer. The thickness of the surface layer of polystyrene of $M = 176\,000$ g/mol adsorbed from a 5 mg/mL solution in the thermodynamically bad solvent cyclohexane is however about 27 nm (i.e., identical with the end-to-end distance in the unperturbed state). The physical structures of adsorbed polymers obviously play an important, yet widely unknown, role in many technological applications of polymers.

6. Thermal Properties [116]

6.1. Molecular Motion

6.1.1. Thermal Expansion

Isotropic bodies expand upon heating equally in all three spatial directions because of the increasing thermal motions of atoms, groups, and molecules. The expansion is characterized by the cubic expansion coefficient $\beta = V^{-1}(\partial V/\partial T)_p$, which is usually converted by $\beta = 3\alpha$ into the linear expansion coefficient $\alpha = L^{-1}(\partial L/\partial T)_p$. Such isotropic materials are, for example, diamond ($\alpha = 1.06 \times 10^{-6}$ K^{-1}), iron (12×10^{-6} K^{-1}), water (70×10^{-6} K^{-1}), and carbon disulfide (380×10^{-6} K^{-1}) (all data at 25°C). All of these materials exhibit the same types of bonds in the three directions: All covalent bonds between carbon atoms in diamond, all metallic bonds between iron atoms, all hydrogen bonds between water molecules, and all dispersion forces between CS_2 molecules.

Polymer chains are, however, anisotropic: The intramolecular bonds along the chain are chemical (almost always covalent); the intermolecular bonds perpendicular to the chain, physical (dispersion forces, dipole–dipole interactions). On thermal expansion of polymer crystals, chains contract because of the increasing amplitude of the lateral motions. The thermal expansion coefficient in the chain direction is thus zero to negative, whereas the overall expansion coefficient is positive. The linear thermal expansion coefficients of polymers are thus averages over the three spatial directions; they lie between those of metals and liquids. Typical values are $\alpha = 60 \times 10^{-6}$ K^{-1} (polyamide 6) and 80×10^{-6} K^{-1} [poly(vinyl chloride)].

Significant problems may thus arise due to different expansion coefficients for polymer −metal composites upon thermal stress. Another problem is the low dimensional stability of polymers on temperature change. This problem may be aggravated by a concomitant change of the water content of polymers or by recrystallization phenomena, both of which can lead to warping.

6.1.2. Heat Capacity

The molar heat capacity can be $3R$ per atom according to the law of equal distribution of energy. In reality, degrees of freedom are always frozen in and the molar heat capacity is lowered. Empirically, a value of ca. $1R$ has been found for solid polymers at room temperature. Poly(2,6-dimethylphenylene oxide) [$(C_8H_8O)_n$] at 25°C has a specific heat capacity of 1.22 J K^{-1} g^{-1} and a molar heat capacity (per monomeric unit) of 146.4 J K^{-1} mol^{-1}. The molar heat capacity (per mole of atoms) is thus 146.4 J K^{-1} mol^{-1}/17 = 8.61 J K^{-1} mol^{-1} (i.e., approximately $1R = 8.314$ J K^{-1} mol^{-1}).

Below the glass transition temperature T_G, heat capacities are not influenced by the degree of crystallinity of the polymer. At T_G, a stepwise increase is observed. The heat capacity passes through a maximum a few degrees below the macroscopic melting temperature; that is, the true melting temperature is given by the upper end of the melting range where the largest and most perfect crystals melt.

6.1.3. Heat Conductivity

Conventional polymers are electrical insulators. Heat is thus not transported by electrons but by elastic waves (phonons in the corpuscular model). The free path length of phonons is defined as the distance at which the intensity of elastic waves has decreased to 1/e. This free path length is about 0.7 nm for glasses, amorphous polymers, and liquids; it is practically independent of temperature. The slight decrease of heat conductivities (thermal conductivities) of amorphous plastics and elastomers below their glass transition temperatures must thus be caused by the decrease of heat capacities with decreasing temperature. For crystalline polymers, a strong decrease of heat capacities is observed at their melting points because packing densities decline drastically at these temperatures.

6.2. Thermal Transitions and Relaxations

6.2.1. Overview

Thermal transitions and relaxations are characterized by large changes of physical properties at the corresponding temperatures. In a true thermal transition, chemical compounds are in equilibrium on both sides of the transition temperature. An example is the melting transition.

Thermal relaxations, on the other hand, are kinetic effects. They depend on the frequency of the experimental method and thus on the time scale. Typical thermal relaxations are caused by the onset of translations and rotations of charges, dipoles, and chemical groups (i.e., by atomic motions).

Some experimental methods work at frequencies that such relaxations appear to be thermal transitions. The best-known example is the glass transition temperature at which hard, glassy polymers convert to soft, rubbery materials and vice versa. In many cases, a thermal effect cannot be unambiguously classified as either transition or relaxation.

Thermal transitions and relaxations can be detected and determined by many different experimental methods. The most commonly applied methods for the determination of *thermal transitions* are differential thermoanalysis (measures temperature differences between specimen and standard on heating or cooling with constant rate), differential scanning calorimetry (does the same for enthalpy differences), thermomechanical analysis (deformation of specimen under load), dynamical–mechanical analysis (either free or forced vibration of specimen), and torsional braid analysis (specimen on vibrating support). Figure 38 shows a typical thermogram.

Many methods are available for the study of molecular motions and thus *thermal relaxations*. These methods work with frequencies ν that correspond to correlation times t_c of 10^{-12} s $< 1/\nu < 10^6$ s (11.5 d). Typical methods include quasi-elastic neutron scattering ($10^{-12} < t_c/\text{s} < 10^{-8}$), NMR spin-lattice relaxation ($10^{-12} < t_c/\text{s} < 10^{-5}$), dielectric relaxation ($10^{-10} < t_c/\text{s} < 10^{-5}$), and photon correlation spectroscopy ($10^{-4} < t_c/\text{s} < 10^2$). Thermal relaxations furthermore manifest themselves in sudden changes of mechanical properties, such as rebound elasticity ($t_c/\text{s} \approx 10^{-5}$), penetrometry ($t_c/\text{s} \approx 10^2$), mechanical loss ($10^3 < t_c/\text{s} < 10^7$), and thermal expansion ($t_c/\text{s} \approx 10^4$). Slow methods (high t_c) are called "static" methods; fast ones, "dynamic." Transition–relaxation phenomena are also detected by several empirical, standardized methods that measure the resistance of specimens against flow under various loads (Vicat temperature, heat distortion temperature, Martens temperature, etc.).

Various characteristic signals are observed at a fixed temperature for a given frequency (see insert in Fig. 39). They often cannot be correlated with molecular processes and are commonly indicated with descending temperature by letters in the sequence of the Greek alphabet, starting with the melting temperature

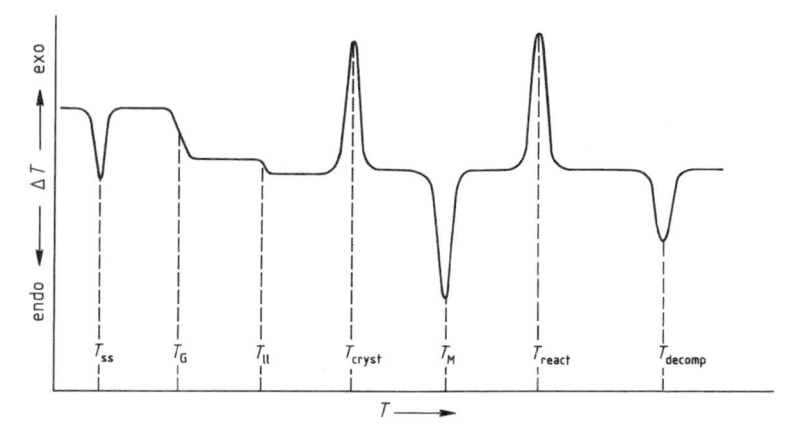

Figure 38. Idealized thermogram of a partially crystalline polymer with solid–solid transition T_{ss}, glass transition temperature T_G, liquid–liquid transition T_{ll}, maximum crystallization temperature T_{cryst}, melting temperature T_M, maximum temperature T_{react} of a chemical transformation, and maximum temperature T_{decomp} of chemical degradation
Reprinted with permission by Hüthig and Wepf Publ., Basel [5]

(crystalline polymers, subscript c) or glass transition temperature (amorphous polymers, subscript a).

The various methods work at different frequencies and thus give different relaxation temperatures for the same molecular process. The frequency dependence of relaxation temperatures can be described by the Eyring equation for rate processes:

$$\nu = (k_B T/2\pi h) \exp(-\Delta H^{\neq}/RT) \exp(\Delta S^{\neq}/R) \qquad (19)$$

where k_B is the Boltzmann constant, h the Planck constant, $\Delta H^{\neq}$ the activation enthalpy,

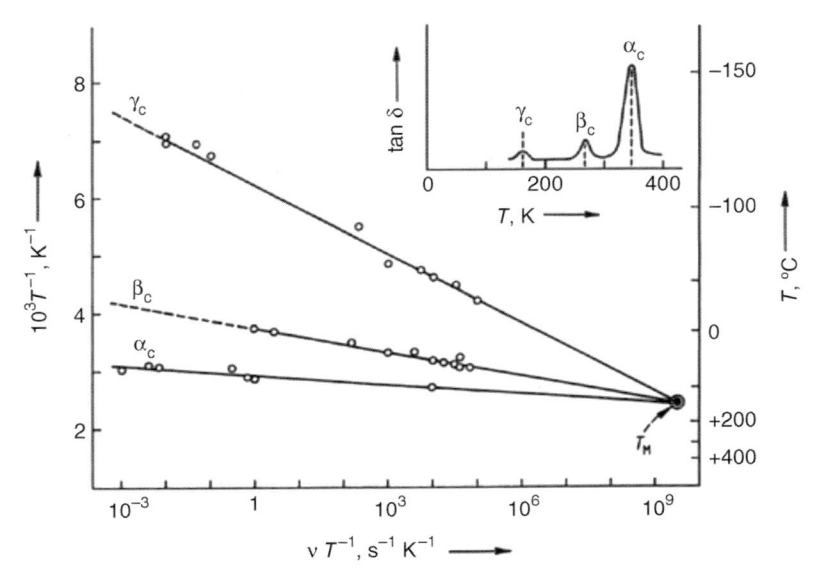

Figure 39. Dependence of inverse relaxation temperature $1/T$ on the logarithm of reduced frequency ν/T for various relaxation processes of a low-density polyethylene
Insert shows the mechanical relaxation spectrum at $\nu = 1\,000$ Hz.
Reprinted with permission by Hüthig and Wepf Publ., Basel [5]

and $\Delta S^{\neq}$ the activation entropy. Transformation of Equation (19) leads to

$$\frac{1}{T} = \frac{R}{\Delta H^{\neq}} \cdot \left[\left(\frac{\Delta S^{\neq}}{R} + \ln \frac{k_B}{2\pi h}\right) - \ln \frac{\nu}{T}\right]$$

In a $(1/T) = f[\ln(\nu/T)]$ plot, lines for various processes intersect at the melting temperature ($T_M = 131$ °C in Fig. 39). Such common intersects seem to be general for nonhelical polymers.

6.2.2. Crystallization

The crystallization of coillike polymers from dilute solutions leads to platelets in which folded polymer chains are arranged with their stems perpendicular to the fold surface (Fig. 26). From concentrated solutions and melts at rest, lamellae are organized into spherulites (Fig. 28), which may be transformed to row structures by shearing or drawing (Fig. 40B).

Crystallization can be subdivided into two elementary processes: Primary nucleation and crystal growth (secondary nucleation). Both processes determine the crystallization rates, which depend strongly on both temperature and polymer structure. At 30 K below the melt temperature, the linear crystallization rate may, for example, range between 5000 μm/min for polyethylene and 0.01 μm/min for poly(vinyl chloride). Symmetrically structured polymers usually crystallize rapidly; polymers with bulky groups or low tacticities, only slowly. Quenching of poly(ethylene terephthalate) melts leads, for example, to amorphous polymers, whereas quenching of polyethylene melts never gives amorphous polymers, even if liquid nitrogen is used.

Primary nucleation may be homogeneous (spontaneous, sporadic, thermal) or heterogeneous (simultaneous, athermal). Homogeneous nuclei are formed from segments of the crystallizing polymer molecules; they are very rare. Heterogeneous nuclei result from extraneous materials such as additives, dust particles, container walls, or specially added nucleation agents. Such nuclei must have minimum sizes of 2–10 nm. Their concentrations can range from ca. 1 nucleus per cubic centimeter (polyoxyethylene) to ca. 10^{12} nuclei per cubic centimeter (polyethylene).

Above the melting temperature, fragments of crystallites may survive for certain time periods. These fragments act as athermal primary nucleation agents on subsequent cooling. They are responsible for the "memory effect" (i.e., the

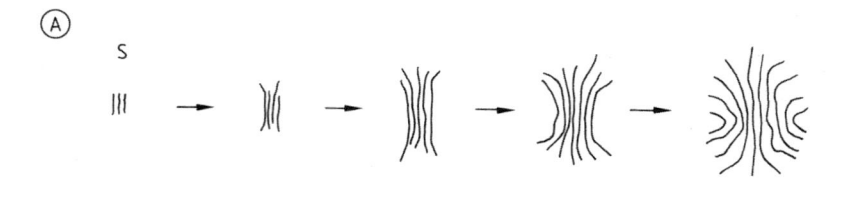

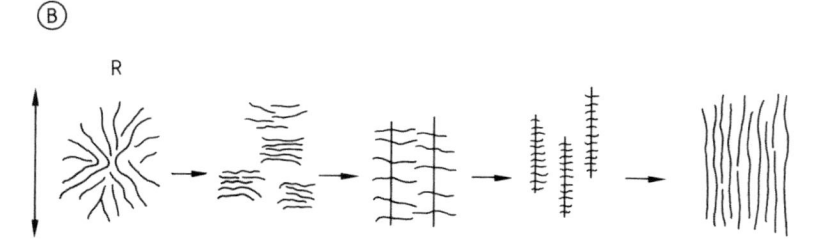

Figure 40. A) Formation of spherulites from lamellae on crystallization of melts at rest; B) Transformation of spherulites into row structures by shearing or drawing in ↕ direction.
Reprinted with permission by Hüthig and Wepf Publ., Basel [5]

reappearance of spherulites at the same locations after and before the melting), which occurs because of low diffusion rates at very high melt viscosities.

Growth of primary nuclei occurs by a *secondary nucleation process*. The growth rate is low just below the melting temperature because secondary nuclei are formed and dissolved rapidly. About 50 K below the glass transition temperature, on the other hand, motions of molecule segments are practically zero and the crystal growth rate is therefore low as well. Crystallization rates must thus exhibit a maximum between the melting and glass transition temperatures; the maximum crystallization rates are usually at (0.80–0.87) T_M (in kelvin).

In addition, the entire crystallization process can be subdivided into a primary and a secondary phase. The *primary phase* comprise the conversion of the total volume to a solid. At the end of this phase, the volume may be filled (e.g., with spherulites), but not all polymer segments between spherulites or between the lamellae of the spherulites may have crystallized. The polymer has not attained its maximal crystallinity (the crystallizability). Crystallization may thus continue during the *second phase*. In this after-crystallization, lamellae may thicken, lattices become more perfect, etc.

Primary crystallization can be characterized by the Avrami equation:

$$\phi/\phi_\infty = 1 - \exp(-z \cdot t^n)$$

where ϕ = fraction of crystallized volume, ϕ_∞ = fraction of maximal attainable crystallinity for a given entity (e.g., spherulite), and z and n are constants that depend on both the nature of the nucleation process (homogeneous, heterogeneous) and the type of growing entity (rod, disk, sphere, sheaflet, etc.). The exponents n range between 1 and 7; empirically, they may assume fractional values.

6.2.3. Melting

Melting is defined as the thermal transition of a crystal to an isotropic melt. The melting temperature T_M (fusion temperature) is defined as the temperature at which crystallites are in equilibrium with the melt. Melting starts at the corners and edges of crystal surfaces; in contrast to crystallization, no nuclei are needed.

Segments of about 60–100 chain atoms participate in the melting process. During heating, segments are redistributed continuously between crystalline and noncrystalline regions; a melting range exists and no sharp melting point is observed. The melting temperature is defined as the upper end of the melting range because the biggest and most perfect crystals melt there. Published melting temperatures often refer to the maxima of $\Delta T = f(T)$ curves, however. Observed melting temperatures are in general lower than the thermodynamic melting temperatures of perfect crystals but may be occasionally higher because of overheating effects.

Melting temperatures increase with molar mass and become practically constant at molar masses of ca. 50 000–150 000 g/mol. The melt can be considered as a dilute solution of end groups in monomeric units, and the reduction of the melting temperature with decreasing molar mass (i.e., increasing concentration of end groups) can be described by the thermodynamic law for the lowering of freezing temperatures:

$$\frac{1}{T_M} = \frac{1}{T_M^0} + \left(\frac{2R}{\Delta H_{M,u}^m}\right) \cdot \frac{1}{\overline{X}_n}$$

where $\Delta H_{M,u}^m$ is the molar melt enthalpy per monomeric unit, $\overline{X}_n$ the number-average degree of polymerization, and T_M and T_M^0 are the thermodynamic melting temperatures at finite and infinite molar masses. Similar depression of the melting point is caused by addition of low molar mass solvents and amorphous polymers or by statistical copolymerization; in all these cases, the term $2/\overline{X}_n$ must be replaced by

$$(^*V_u^m/^*V_1^m)[(1 - \phi_2) - \chi(1 - \phi_2)^2] \text{ (solvent)}$$

$$(^*V_u^m/^*V_A^m) - \chi(1 - \phi_2)^2] \text{ (amorphous polymer)}$$

$$[-\ln x_u - \chi(1 - \phi_2)^2] \text{ (statistical copolymer)}$$

where $^*V^m$ is the partial molar volume of solvent 1, amorphous polymer A, or monomeric units u of copolymer; x_u the mole fraction of units u; ϕ_2 the volume fraction of crystallizable polymer 2; and χ an interaction parameter.

Melting temperatures $T_M = \Delta H_M^m / \Delta S_M^m$ are determined by the changes in molar melting enthalpies ΔH_M^m and molar melting entropies ΔS_M^m. The *melting entropy* results from conformational changes and volume changes upon melting. The melting entropy theoretically adds $R \cdot \ln 3 = 9.12$ J K^{-1} mol^{-1} for the formation of three conformers with equal energy and 7.41 J K^{-1}mol^{-1} for one *trans* and two *gauche* conformers in the case of polymethylene $\sim(CH_2)_n\sim$. The entropy change due to the volume change should add another 10.9 J K^{-1}mol^{-1} so that the theoretical melt entropy of polymethylene should be ca. 18.3–20.0 J K^{-1}mol^{-1}. Experimentally, only 9.9 J K^{-1}mol^{-1} is observed, which points toward either the existence of local order in melts or a high segment mobility below the melting temperature. The latter was found experimentally by broad-line NMR for *cis*-1,4-polyisoprene ($\Delta S_M^m = 4.8$ J K^{-1} mol^{-1}).

Melting enthalpies are usually between 1 and 5 kJ per mole of chain atom. Low values are to be expected for polymers with high chain mobilities below the melting temperature (*cis*-1,4-polyisoprene, aliphatic polyesters and polyethers). High values are found for polymers with strong interactions between chains and tight packing of chains in crystals (polyoxymethylene, it-polystyrene). Some of these strong interactions may survive the melting process; for example, most of the hydrogen bonds of polyamides are still detected by IR spectroscopy above the melting temperature.

Thus the primary factors for high melting temperatures are not intermolecular interactions (e.g., cohesive energies) but reduced flexibilities of chains. Low melting temperatures are found for polymers with low rotational barriers (ester, oxygen, sulfide groups in chains), high melting temperatures for tightly packed helices [polyoxymethylene, it-poly(3-methylbutene)] and for ladder and ladderlike polymers [poly(*p*-phenylene), polybenzimidazole, etc.]. Such factors are responsible for the variation of melting temperatures with the number of methylene units in aliphatic polymer chains of the type $\sim X-(CH_2)_n\sim$ (Fig. 41).

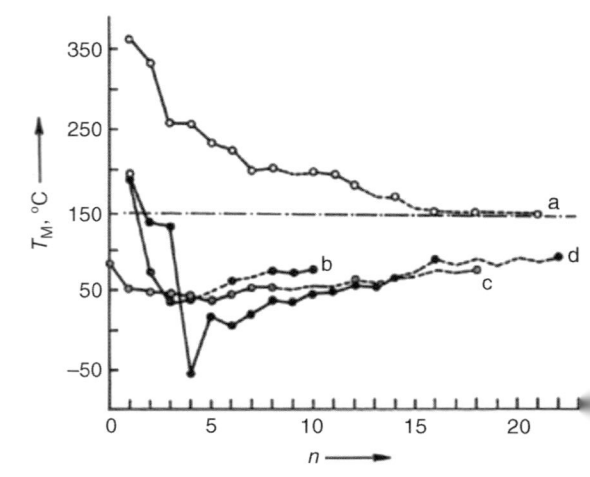

Figure 41. Melting temperatures of (○, ⊙, ⊕) aliphatic polymers $\sim[X-(CH_2)_n]\sim$ and (●) isotactic poly(α-olefins) $\sim\{CH_2-CH[(CH_2)_nH]\}\sim$ as function of the length n of methylene sequences
a) X = NHCO (polyamides); b) X = O (polyoxides);
c) X = COO(CH$_2$)$_3$OOC (aliphatic polyesters of trimethylene glycol)
– ·—·– theoretical melting temperature of polyethylene

6.2.4. Liquid Crystal Transitions

Thermal transitions of thermotropic LC polymers from their crystals to smectic (T_{cs}) or nematic phases (T_{cn}), from smectic to nematic mesophases (T_{sn}), and from nematic phases to isotropic melts (T_{ni}, clearing temperature) are thermodynamic first-order transitions. They exhibit steplike changes in volume, enthalpy, and entropy, just like the melting of three-dimensional crystals to isotropic melts. Thermodynamically stable phases can exist only between melting and clearing temperatures (T_{cs}, $T_{cn} < T_{ni}$); they are called *enantiotropic phases*.

Mesophases form dispersions in supercooled isotropic melts if clearing temperatures are lower than melting temperatures (T_{cs}, $T_{cn} > T_{ni}$). Such phases are thermodynamically unstable compared to the crystalline; they are called *monotropic*.

Mesophases may be supercooled below $T_M = T_{cs}$ to smectic liquids sL* and below T_{sn} to nematic liquids nL* if crystallization can be suppressed. At even lower temperatures

T_{gn} and T_{gs}, these supercooled liquids may yield anisotropic glasses nG and sG, respectively. Some of these transition temperatures cannot be measured directly, but their existence can be deduced from extrapolations of transition temperatures of copolymers to 100% of the pure mesogenic compound (virtual transition temperatures).

The transition temperatures T_{trans} (i.e., T_{cs}, T_{gs}, T_{sn}, T_{cn}, T_{gn}, T_{ni}) depend on the degree of polymerization X in the same way the melting temperature does: $1/T_{trans} = f(1/X)$. The transition enthalpy ΔH_{ni} is always lower than the transition enthalpies ΔH_{gs} and ΔH_{sn} because the n → i transition is from order to disorder, whereas the g → s and s → n transitions are from order to less order. Transition entropies of mesophases are lower than melt entropies; they usually have values of ca. 0.5–1.5 J K^{-1}mol^{-1}, with $\Delta S_{sn} < \Delta S_{ni}$.

6.2.5. Glass Transitions

Glass transitions are phenomenologically characterized by a change from a "hard," noncrystalline, glasslike material to a rubbery to highly viscous "melt." The viscosities at glass transitions are ca. 10^{12} Pa · s, independent of chemical structures. Glass transitions were thus thought to be "isoviscous" phenomena. Today, glass transitions are considered to occur at that physical state where all materials exhibit the same "free volume."

Various free-volume fractions are discussed in the literature. The empirical Boyer–Simha rule relates a free-volume fraction f_{exp} to the cubic expansion coefficients β of liquid (L) and amorphous, glasslike (G) polymers and their glass transition temperatures T_G (see Table 16):

$$f_{exp} \approx (\beta_L - \beta_G) \cdot T_G \approx 0.11 \pm 0.02 \quad (20)$$

Table 16. Glass transition temperatures and free-volume fractions of polymers (for explanation of symbols, see text)

Polymer	T_G, °C	f_{exp}	f_{WLF}	f_{fluc}
Polyethylene	−80	0.098	0.025	
Polyisobutene	−73	0.079	0.026	0.0017
Poly(butyl methacrylate)	20	0.13	0.026	0.0010
Poly(vinyl acetate)	27	0.128	0.028	0.0023
Polystyrene	100	0.133	0.025	0.0035
Poly(methyl methacrylate)	105	0.118	0.025	0.0015

The Williams–Landel–Ferry (WLF) approach relates a free-volume fraction f_{WLF} to the probability of segment movements. Empirically, values of $f_{WLF} \approx 0.025 \pm 0.01$ were found; these values can be calculated, for example, from $K = f_{WLF}/(\beta_L - \beta_G)$ and $K' = \log e/f_{WLF}$ of the semi-empirical WLF equation (Eq. 21) where $-\log a_t = \Delta(\log t)$ is a shift factor

$$T = T_G + \frac{K \cdot \log a_t}{K' - \log a_t} = T_G + \frac{51.6 \cdot \log a_t}{17.4 - \log a_t} \quad (21)$$

The Williams–Landel–Ferry equation applies to all relaxation processes; its use is restricted to temperatures in the range $T_G < T < (T_G + 100$ K$)$.

The WLF equation allows calculation of the static glass transition from the various dynamic glass transition temperatures if the deformation times (inverse effective frequencies) of the methods are known. The glass transition temperatures of poly(methyl methacrylate) (PMMA) are given as 105 °C (thermal expansion, "static"), 120 °C (penetrometry), and 160 °C (rebound elasticity). The same polymer may thus exhibit very different mechanical properties if subjected to different stresses; at 140 °C, PMMA behaves as either a glass (rebound elasticity) or an elastomer (penetrometry).

The glass transition temperature indicates the onset of cooperative movements of chain segments of 25–50 chain atoms, which can be deduced from the ratio of molar activation energies and melt energies. These cooperative movements very probably involve *trans–gauche* transitions that proceed cooperatively along greater distances since only small changes of chain axes are involved according to deuterium NMR. The participation of segments of 25–50 chain atoms is also indicated by cross-linking experiments: As long as the average segment length N_{seg} between two cross-linking points is less than 25–50 chain atoms, no change of glass transition temperatures is observed. At $N_{seg} < 25$–50, T_G increases with the inverse molar mass of the segments.

Since both glass transition and melt temperatures depend on segmental motions, close relationship between these two temperatures can be expected. The empirical Beaman–Boyer rule states that $T_G \approx (2/3)T_M$, which holds reasonably well for many polymers except for chains such as polyethylene and polyoxyethylene for which $T_G/T_m \approx 1/2$.

A vast literature exists about effects of constitution on T_G. Cyclic macromolecules possess no end groups, and thus no free-volume effects from these. Small rings are furthermore strained (less possible microconformations). The greater the molar mass, the more microconformations can be adopted, the greater is the chain flexibility and the lower is T_G. The same is true for segment flexibilities of star-branched polymers and long side chains in comblike molecules (side-chain "crystallization").

Linear relationships are found between the logarithms of glass transition temperatures T_G and the logarithms of cross-sectional areas A of carbon, carbon–oxygen, or carbon–nitrogen chains (another measure of segment flexibilities). The three lines intersect at $A = 0.17$ nm^2 and $T_G = 141$ K, which should be the lowest glass transition temperature possible. The lowest experimentally found glass transition temperature (150 K) is that of polydimethylsiloxane, $\sim[O-Si(CH_3)_2]_n\sim$.

Glass transitions can be decreased (or increased) by copolymerization with suitable monomers (internal plasticization) and by addition of external plasticizers (see Section 4.2.5).

6.2.6. Other Transitions and Relaxations

Experimentally, a number of other transition–relaxation temperatures are observed, mostly of unknown origin. Amorphous polymers exhibit weak "liquid–liquid" transitions at ca. $T_{ll} \approx 1.2\ T_G$. Below the critical molar mass for entanglements, transition temperatures T_{ll} equal flow temperatures T_F at which polymers start to flow under their own weight. At higher molar masses, $T_F > T_{ll}$.

Another transition temperature $T_U \approx 1.2\ T_M$ seems to exist for crystalline polymers. This transition has been interpreted as the dissolution of smectic structures.

Few β-relaxations have been correlated with molecular phenomena. An example is the frequency-dependent boat–chair transition of cyclohexane rings, which occurs at e.g., -125 °C (10^{-4} Hz) and $+80$ °C (10^5 Hz).

6.2.7. Technical Methods

The technical testing on thermal transitions and relaxations of plastics is usually performed with simple methods under standardized conditions and always under load. *Martens numbers* measure temperatures at which the specimen has experienced a certain bend, *Vicat softening temperatures* give the temperatures for a certain penetration of a rod into the plastics, and the *heat distortion temperatures* (heat deflection temperatures) indicate the temperatures for a certain bending with a three-point method. The temperatures from these three methods do not only depend on transitions or relaxations but also on the elasticity of the specimen; Vicat and heat distortion temperatures are in addition affected by the surface hardness. The resulting softening temperatures are neither identical with glass transition nor with melting temperatures; they are often also not a good measure of the continuous service temperature of a plastic.

6.3. Transport

6.3.1. Self-Diffusion

Brownian movements cause molecules and their segments to interchange positions in fluid phases. If all entities are of the same type, such interchanges lead to a "self-diffusion", which involves no net transport of polymers. Self-diffusions can be measured by pulsed field-gradient spin-echo NMR (segments) and radioactive tracers (molecules).

Coiled molecules behave in their melts as unperturbed coils with low coil densities, which are filled with segments of other coils. Self-diffusion of a segment must therefore occur by interchanging position with a segment of another molecule.

Self-diffusion coefficients decrease with the squares of molar masses (Fig. 42). According to *reptation theory*, this functionality is caused by temporary (but fairly long-lived) entanglements of polymer chains. Such entanglements cause the sudden change of molar mass dependences

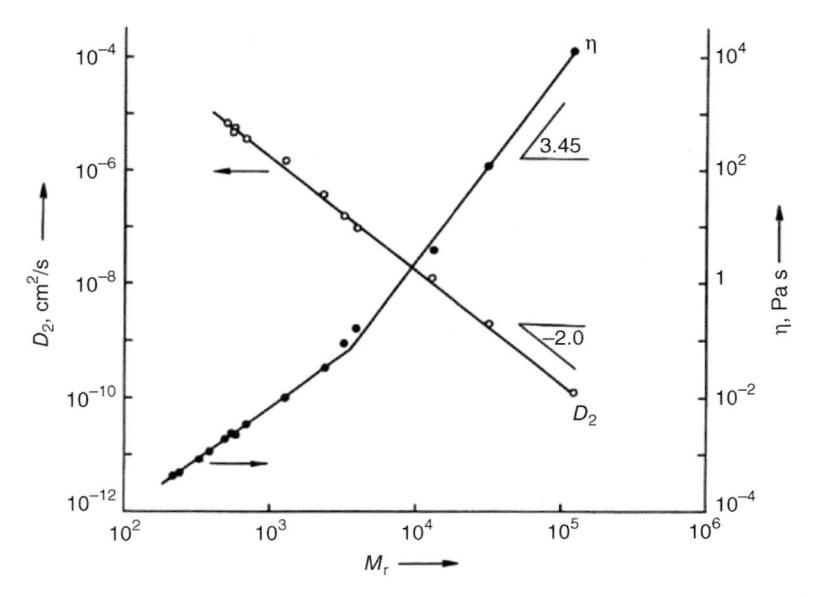

Figure 42. Segmental self-diffusion coefficients D_2 (○) and Newtonian melt viscosities η of alkanes and narrow-distribution polyethylenes (●) as function of the relative molecular mass M_r at 175°C
Reprinted with permission by Hüthig and Wepf Publ., Basel [5]

of melt viscosities above a critical molar mass M_c (Fig. 42).

The resulting topological restraints make the polymer chain move through the maze of other segments like a reptile through brush. According to the Doi–Edwards theory, the test chain reptates in a tube of ca. 5 nm diameter, which is formed by other segments (Fig. 43). The theory predicts the experimentally found function $D_2 = f(M^{-2})$ albeit only for $M > M_c$.

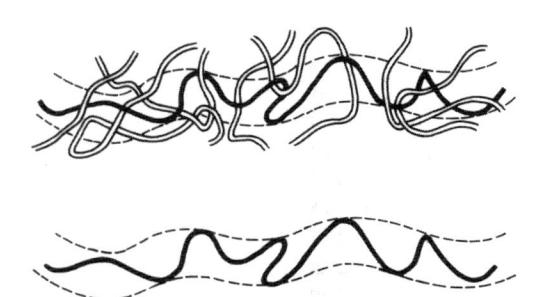

Figure 43. Reptation of a test chain (black) through the segments of a matrix (white). The "walls" of the tube are indicated by dotted lines.
Reprinted with permission by Hüthig and Wepf Publ., Basel [5]

Experimentally, the same function is also observed for $M < M_c$, probably because end groups cause successively greater free-volume fractions, which are assumed by the reptation theory to be molar mass independent.

The same dependence of self-diffusion coefficients of test chains on the squares of their molar masses is theoretically predicted and experimentally found for test chains in higher molar mass matrices of the same constitution and configuration ($M_{test} \ll M_{matrix}$); the molar mass of the matrix exerts no influence. If however $M_{test} \gg M_{matrix}$, then a function $D_{test} = f[(M_{test})^{-1/2} \cdot (M_{matrix})^{-1}]$ is predicted and observed.

6.3.2. Permeation

The transport of extraneous material through polymers is called permeation. The resulting net flow of mass is caused by differences in chemical potentials, that is, concentration differences (at constant temperature) or thermal gradients (at constant concentration). Permeation may be desirable as in the dyeing of textiles or the controlled transdermal delivery of pharmaceuticals, or undesirable as in the loss of carbon dioxide from plastic bottles for carbonated soft drinks or in migration of plasticizers.

Permeation of chemical compounds through amorphous polymers below their glass transition temperatures or through crystalline polymers below their melting temperatures can occur either by flow through pores or by molecular transport. Pores have diameters much greater than the diameters of permeating substances (diameter of spheres, cross section of coillike chains); the interactions of permeants and pore walls are negligible. Molecular transport, on the other hand, depends on such interactions between permeant and matrix (i.e., on the solubility of the former in the latter). Both types of transport can be distinguished by their temperature dependence: Permeation coefficients of gases decrease with temperature at pore membranes; they increase with temperature at solubility membranes.

The counteraction of the two types of permeation can be utilized in lamination. Oxygen permeates through pores in aluminum films but by molecular transport through polyethylene films. At 1 bar, permeation rates decrease from 5×10^{-5} cm^3/s for 0.025-mm-thick aluminum films to 5×10^{-13} cm^3/s for the same films laminated with 0.025-mm-thick PE films.

The permeation coefficient P is given by the product of the diffusion coefficient D and the solubility coefficient S of the permeant in the matrix and, in the steady state, by the expression to the right of the second equality sign:

$$P = D \cdot S = \frac{Q \cdot L}{A \cdot \Delta p \cdot t} \tag{22}$$

where Q = permeated amount, L = thickness of film (membrane, etc.), A = area, Δp = pressure difference, and t = time. The literature often uses different "practical units" for the various quantities of Equation (22), for example, Q in cm^3, L in mm, A in m^2, Δp in atm, and t in 24 h; P is then given in (cm$^3 \cdot$ mm)/(24 h m^2 atm). If like units are used, then the unit of P is of course length$^2 \cdot$ time$^{-1} \cdot$ pressure^{-1} and the unit of S is pressure^{-1}. Permeation coefficients P^* of liquids are usually measured without a pressure differential; their unit is cm^2/s and the unit of S is 1.

Table 17. Permeation coefficients of gases (P) and water vapor (P^*) through polymers at 30 °C [a]: $P = 1 \times 10^{-14}$ cm^2 s^{-1} Pa^{-1} corresponds to $P^* = 1 \times 10^{-9}$ cm^2/s at normal pressure ($p = 1 \times 10^5$ Pa)

Polymer	$\frac{10^{14} P}{\text{cm}^2\,\text{s}^{-1}\,\text{Pa}^{-1}}$		$\frac{10^9 P^*}{\text{cm}^2\,\text{s}^{-1}}$
	O$_2$	CO$_2$	H$_2$O
Polydimethylsiloxane	25 000	85 000	40
cis-1,4-Polyisoprene	2 000	10 000	0.3
Butyl rubber	100	500	0.1
Polystyrene			
regular	200	1 000	1
biaxially oriented	0.1	100	0.5
Poly(ethylene terephthalate)			
regular	4	20	0.2
biaxially oriented	0.2	1	0.2
Poly(vinylidene chloride)	0.05	0.15	0.02
Cellulose	0.03	0.1	10
Polyacrylonitrile, required for bottles	0.002	0.02	0.02
Cola	1	0.5	0.14
Beer	0.05	0.5	0.14

[a] Literature values vary widely because polymers are often not identical with respect to crystallinity, orientation, water absorption, etc.

Permeation coefficients of gases in polymers vary widely; for example, oxygen permeates 10 million times faster through polydimethylsiloxane than through polyacrylonitrile (Table 17). Gases in general, have lower permeation coefficients in thermoplastics than in elastomers because segmental movements of the former are frozen-in below the glass transition temperature. Bulky substituents, orientation of polymer segments, crystalline domains, and added fillers all increase the pathway for a gas molecule through a polymer matrix; these tortuosity factors decrease permeation coefficients.

The *permeation of nondissolving liquids* through a polymer is proportional to t^n. The exponent n depends on the ratio of the relaxation time of the polymer–solvent system to the diffusion time of the solvent (i.e., on the Deborah number DB).

Three different regimes are normally considered. In the regime denoted as *Case I*, the mobility of the permeant is much smaller than the relaxation of the polymer segments ($DB < 0.1$). The movement of the permeant causes "instantaneous" conformational changes of the polymer segments. Both permeant and polymer behave as viscous liquids: The system can be described by Fickian diffusion laws (i.e., $n = 1/2$).

In the *Case II* regime, the mobility of the permeant is much higher than the relaxation of the polymer segments ($DB > 10$). The physical structure of the polymer does not change during the permeation; the polymer appears to the permeant as an elastic body. Case II is characterized by a sharp demarcation line between the glassy inner polymer core and the swollen zone advancing with constant speed. The permeating amount is directly proportional to time (i.e., $n=1$).

Relaxation and permeation become comparable for $0.1 < DB < 10$. This *third regime* is usually called anomalous diffusion or viscoelastic diffusion ($1/2 < n < 1$).

7. Rheological Properties [125, 126]

7.1. Introduction

Materials exhibit two limiting types of behavior against deformation. Typical liquids such as water flow under their own weight and are irreversibly deformed (viscous behavior). Typical solids such as iron resist deformation; they return from small deformations to their former states after removal of loads (elastic behavior). Polymers commonly combine both types of behavior: They are viscoelastic materials. Their melts exhibit viscous behavior at small deformations and elastic properties at larger ones. Polymer solids respond elastically at small deformations but begin to flow at larger ones.

Polymer melts and solutions have extremely high viscosities which may, in addition, be dependent on deformation rates and duration of the experiment. Air has, for example, a viscosity of 10^{-5} Pa · s; water, 10^{-3} Pa · s; and glycerol, 1 Pa · s, whereas polymer melts exhibit viscosities of ca. 10^2–10^8 Pa · s.

Three types of viscosity are usually distinguished:

1. Shear viscosity describes the rate of shear flow as function of the applied stress.
2. Extensional viscosity measures the rate of extensional (elongational) flow as function of tensile stress.
3. Bulk viscosity relates the rate of deformation of volume to the applied hydrostatic pressure.

Shear viscosities are the most often studied rheological properties of polymers; they are of great importance for the processing of plastics by extrusion or injection molding, for example. Much less is known about *extensional viscosities* (important for blow forming and fiber spinning) and practically nothing about the *bulk viscosities* of polymers.

7.2. Shear Viscosities at Rest

7.2.1. Fundamentals

Nine different shear stresses may be assigned to a three-dimensional body: One parallel to each of the three spatial directions (σ_{11}, σ_{22}, σ_{33}) and six perpendicular to these (σ_{12}, σ_{13}, etc.). A body is by definition sheared in the 2–1 direction. The ratio of shearing force K to contact area is called the shear(ing) stress $\sigma_{21} = K/A$; it produces a shear strain γ. The ratio $G = \sigma_{21}/\gamma$ is the shear modulus. Between layers of distance y moving parallel to each other with different rates v, a shear gradient $\dot\gamma = dv/dy$ thus exists. The ratio of shear stress to shear rate is the (dynamic) viscosity (Newton's law);

$$\eta = \sigma_{21}/\dot\gamma$$

its inverse $1/\eta$ is called fluidity.

Viscosities are independent of shear rates for Newtonian liquids ($\eta = \eta_o$); these viscosities are also called zero-shear viscosities, viscosities at rest, or stationary viscosities. For Newtonian liquids, the shear modulus G_o is independent of the extent of deformation. η_o and G_o are true material constants, whereas η and G depend on shear rates and sometimes also on shearing time.

Shear stresses and shear gradients (and thus viscosities) can be measured with a variety of instruments that usually belong to one of three groups: Capillary, rotatory, and cone–plate viscometers. A number of industrially used instruments provide viscometric indicators (but neither shear stress, shear gradient, nor viscosity); this group includes Höppler viscometers, Cochius tubes, Ford beakers, and instruments that measure melt flow indices or Mooney values (Mooney viscosities).

Thermoplastics are usually characterized by their *melt flow indices* $MFI_{T/F}$. The melt flow index measures the mass of polymer extruded in 10 min by a standard load F from a standard plastometer at temperature T. It is a measure of the (usually non-Newtonian) fluidity. The higher the melt flow indices, the lower are the molar masses of polymers of the same constitution.

The Mooney "viscosity" is really a measure of elasticity; it is used mainly for elastomers but also for polymer melts. A polymer is deformed in a standardized cone–plate viscometer at constant rotational speed and constant temperature T; after t minutes, the force to recover is read.

7.2.2. Molar Mass Dependence

Newtonian viscosities exhibit two different regimes for their (mass-average) molar mass dependencies $\eta_o = K \cdot M^a$: A weaker dependency ($a = 1/2–1$) for low molar masses and a stronger one ($a \approx 3.4$) at higher molar masses (Fig. 42). The transition from one to the other is thought to be caused by the onset of molecular entanglements that cause the molecules to behave as physically cross-linked networks.

The number of entanglements can be assumed to be constant at low shear rates. The elasticity of such a network is described by its shear modulus G (unit of pressure). Because shear viscosities have the unit pressure · time, $\eta_o = G_o \cdot t$. The reptation model identifies the time t as the time required for a chain to leave the tube. This time is proportional to the third power of the number of segments per molecule (i.e., the viscosity should be proportional to the third power of the molar mass). The deviation between the theoretically predicted $\alpha = 3$ and the experimental value of $\alpha = 3.4$ is thought to be due to "breathing" of the tube. Breathing pushes nonentangled chain loops back into the surrounding matrix; chain ends cause additional relaxations and the tube length decreases.

7.2.3. Concentrated Solutions

Newtonian viscosities of concentrated solutions increase with both solute concentrations c and molar masses M or, since $[\eta] = K \cdot M^a$, with intrinsic viscosity $[\eta]$ as well. The concentration c measures the mass of polymer per unit volume of solution; the intrinsic viscosity, the volume of polymer molecules per unit mass of polymer. The product $c \cdot [\eta]$ is thus a measure of the volume fraction of polymer molecules that would be occupied by isolated polymer coils. At higher concentrations, coils start to overlap and the total occupied volume is smaller than the one demanded by isolated coils (i.e. $c \cdot [\eta] > 1$).

At low values of $c \cdot [\eta]$, η_i (the "specific viscosity") and $c \cdot [\eta]$ are proportional (Fig. 44), where $\eta_i = (\eta/\eta_1) - 1$, η is the viscosity of the polymer solution at rest, and η_1 the viscosity of the solvent. At higher $c \cdot [\eta]$ values, $\eta_i \sim (c \cdot [\eta])^q$ or, because $[\eta] \sim M^a$, $\eta_i \sim c^q \cdot M^{aq}$. Because at very high concentrations $\eta_i \approx \eta/\eta_1$ and η approaches the Newtonian melt viscosity $\eta_o \sim M^{3.4}$ for high molar masses, $a \cdot q = 3.4$ is obtained. In theta solvents and melts, coils are unperturbed ($a = 1/2$) and $q = 6.8$ (Fig. 44). In good solvents, $a \to 0.764$ and $q = 4.55$. These relationships are

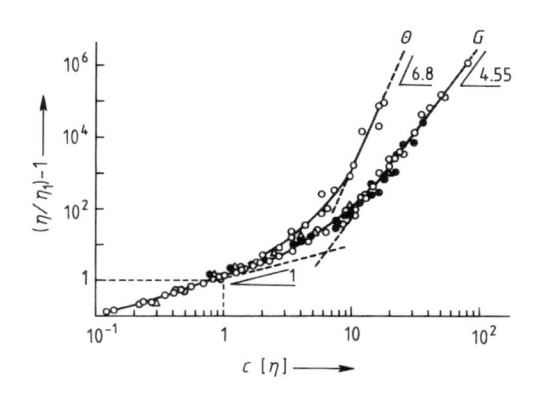

Figure 44. Relative viscosity increment ("specific viscosity") as function of $c \cdot [\eta]$ for (○) polystyrenes in *trans*-decalin (theta solvent) or toluene (good solvent G) at 25 °C; (●) *cis*-1,4-polyisoprene in toluene at 34°C; and (△) hyaluronates in water at 25°C
Reprinted with permission by Hüthig and Wepf Publ., Basel [5]

independent of the chemical nature of polymers and solvents.

7.3. Non-Newtonian Shear Viscosities

7.3.1. Overview

The shear stress σ_{21} is directly proportional to the shear gradient γ for Newtonian liquids; these relationships and thus the viscosities η are moreover, independent of the duration of the experiments. Non-Newtonian viscosities (apparent viscosities), on the other hand, vary with shear rate and sometimes even with time.

Various dependencies of (apparent) viscosity on shear rate are found for time-independent non-Newtonian liquids (Fig. 45). *Plastic bodies* (Bingham bodies) exhibit a yield value y, i.e., shear stresses have finite values σ_o at $\dot{\gamma} \to 0$. Above σ_o, such bodies may behave in a Newtonian (ideal Bingham bodies) or non-Newtonian (pseudoplastic Bingham bodies) manner. This behavior seems to be due to the disappearance of aggregates. An example for a Bingham body is tomato ketchup.

In *dilatant liquids*, the shear stress increase is more than linearly proportional to shear rate; viscosities increase with shear rate (shear thickening). This behavior occurs frequently in polymer dispersions.

A decrease of apparent viscosity with shear rate is most common for polymer melts. Because such behavior resembles that of pseudoplastic Bingham bodies, it is called *pseudoplasticity* in the English literature ("structural viscosity" in German), although such "pseudoplastic" materials do not possess a yield value. Pseudoplastic behavior eases polymer processing from melts and reduces the energy required (e.g., lower pressures can be applied in injection molding). It is usually accompanied by an orientation of chain segments that is most pronounced in the processing of liquid crystalline polymers with rigid mesogens and the ultradrawing of flexible polymers.

Dilatant and pseudoplastic liquids are characterized by "instantaneous" adoption of shear rates on application of shear stresses (i.e., by time-independent apparent viscosities).

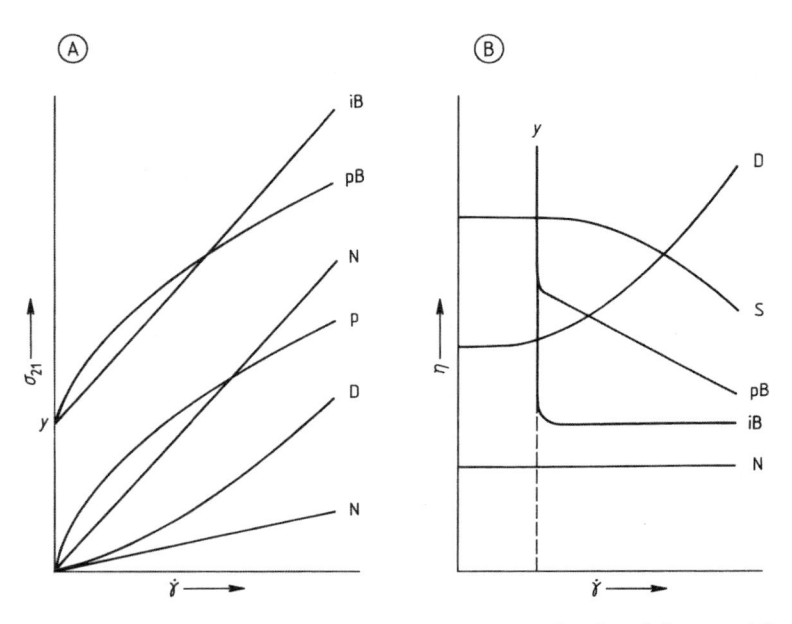

Figure 45. A) Shear stress σ_{21} and B) shear viscosity η as a function of shear rate $\dot{\gamma}$ for Newtonian (N), dilatant (D), and pseudoplastic (P) liquids and for ideal (iB) and pseudoplastic (pB) Bingham bodies (y = yield)
Reprinted with permission by Hüthig and Wepf Publ., Basel [5]

Thixotropic materials exhibit a decrease of apparent viscosity with time at constant shear rate; examples are dispersions of bentonite and other platelet-like silicates. *Rheopectic* or *antithixotropic* materials show an increase of apparent viscosity with time at constant shear rate. The flow behavior may be further complicated by wall effects. Certain dispersions and gels exude liquid on application of a shear stress. The liquid acts as an external lubricant and a pluglike flow results (e.g., tooth pastes). An additional complication may result from the onset of turbulence, which usually occurs at much lower Reynolds numbers in non-Newtonian than in Newtonian liquids.

7.3.2. Flow Curves

A plot of $\log \dot\gamma = f(\sigma_{21})$ or vice versa is usually called a flow curve. A generalized flow curve may contain an initial Newtonian region, followed by pseudoplasticity and a second Newtonian regime. Because experiments are difficult to conduct at high shear rates, many rheologists doubt the existence of true second Newtonian regions. Dilatancy may or may not be present. Finally, turbulence sets in and melt fracture occurs (Fig. 46).

Several empirical laws have been suggested for the description of flow curves, all of which usually apply only to limited ranges of shear rates. Examples include

$\dot\gamma = a \cdot (\sigma_{21})^m$ Ostwald – de Waele
$\dot\gamma = b \cdot \sinh(\sigma_{21}/d)$ Prandtl – Eyring
$\dot\gamma = f \cdot \sigma_{21} + g \cdot (\sigma_{21})^3$ Rabinowitsch – Weissenberg

where a, b, d, f, g, and m are empirical constants. The exponent m is known as the "flow exponent"; it takes a value of 1 (Newtonian liquid) and < 1 (pseudoplasticity). No constitutive equations are known that cover the entire range of rheological phenomena.

7.3.3. Melt Viscosities

Polymers below the critical molar masses M_c for entanglements show extended Newtonian ranges. For $M > M_c$, non-Newtonian behavior appears at lower shear rates as molar mass increases (lower melt flow index) (Fig. 47). At high shear rates, a constant exponent q is approached for the function $\eta = f(\dot\gamma^q)$. The non-Newtonian viscosities are no longer proportional to the mass-average molar masses; the broader the molar mass distribution, the more the number average seems to be the correct corresponding quantity.

Shear thinning, i.e., the decrease of apparent viscosity with shear rate, is very important for plastics processing. Viscosities describe a frictional behavior: The higher the viscosity, the higher is the internal friction of the melt and the greater is the proportion of energy provided that is converted into heat. Strong non-Newtonian behavior thus saves energy. Polymer melts are therefore processed at the highest possible shear rates. The upper range is given by the processing method (calendering, extrusion, etc.), and the polymer properties (thermal degradation, melt fracture, etc.).

Surface roughness of barrel walls, diameter changes, etc., create additional rate components in extrusion that are dampened by the viscosity of the liquid. These disturbances become stronger with increasing flow rate and can no longer be dampened at high shear rate. Finally, turbulence sets in. In entangled polymer melts, additional elastic vibrations occur due to the presence of physical cross-links. The resulting elastic turbulences lead to rough surfaces of the

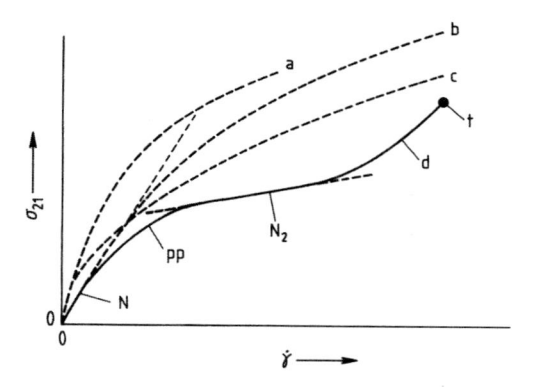

Figure 46. Generalized flow curve with first Newtonian region (N), pseudoplasticity (pp), second Newtonian region (N$_2$), dilatancy (d), and onset of turbulence (t)
- - - Schematic representations of various viscosity "laws": a) Prandtl–Eyring;   b) Rabinowitsch–Weissenberg; c) Ostwald–de Waele
Reprinted with permission by Hüthig and Wepf Publ., Basel [5]

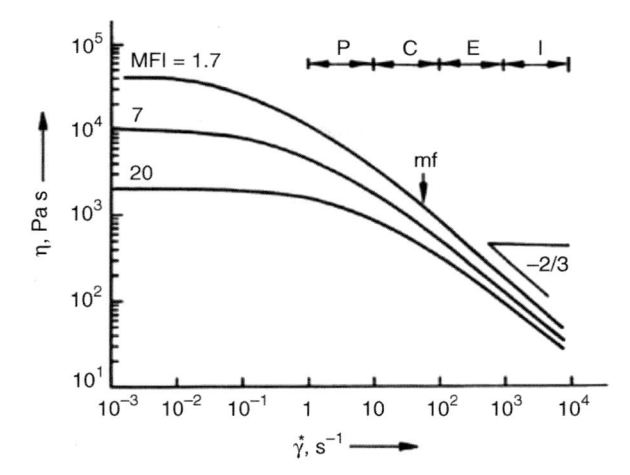

Figure 47. Dependence of melt viscosities of polyethylenes with different melt flow indices (MFI, inversely proportional to molar mass) on shear rates (mf = onset of melt fracture for the highest molar mass polyethylene) The ranges of the various processing regimes overlap: P = compression molding, C = calendering, E = extrusion, I=injection molding. Shear rates refer to those at orifices and are much lower in the mold (tool, die).
Reprinted with permission by BASF AG, Ludwigshafen [127]

extrudate, which are subsequently frozen-in upon exit from the orifice. The polymer surface appears "fractured"; "melt fracture" thus does not refer to a breakage of the extrudate strand.

Molecular coils are deformed if their melt is pressed through an orifice (Fig. 48). For $M > M_c$, segments can no longer slip from each other at high stresses and short times because of entanglements; a "normal" stress builds up perpendicular to the stress direction. At the die exit, this stress is relieved and the coil returns to the thermodynamically more favorable shape of an unperturbed coil. The melt expands perpendicular to the flow direction. This phenomenon is known as Barus effect or memory effect (melts), the Weissenberg effect (solutions), parison swell (extrusion), swelling (blow molding), etc. It is especially pronounced

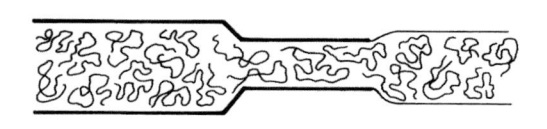

Figure 48. Parison swell upon extrusion (Barus effect)
Reprinted with permission by Hüthig and Wepf Publ., Basel [5]

for high molar mass tails in polymers with $M > M_c$ because of the 3.4 power dependence of η on $\overline{M}_w$.

Negative Barus effects are known for solutions of rodlike molecules or LCPs with rodlike mesogens (diameter of strand smaller than diameter of orifice). If such molecules crystallize after exiting from the die, the strand contracts perpendicular to the extrusion direction and the strand diameter becomes smaller than the diameter of the orifice.

The effect of normal stress can be determined by the Bagley equation. A force $F_f = \pi R^2 p$ is exerted on a liquid during the flow through a capillary with radius R and length L under a pressure p. It is counteracted by a frictional force $F_r = 2\pi R L\, \sigma_{21}$. In steady state, $F_f = F_r$; thus $p = 2\sigma_{21}(L/R) = 2\eta \cdot \dot{\gamma}(L/R)$ because $\sigma_{21} = \eta \cdot \dot{\gamma}$. In the Bagley diagram, pressure p is accordingly plotted at constant shear rate against die geometry L/R (Fig. 49). For non-Newtonian liquids, a relationship

$$p = p_o + K \cdot (L/R)$$

is found. The intercept p_o at $L/R \to 0$ is identified with the pressure loss caused by the elastically stored energy of the flowing melt and the formation of a steady-state flow

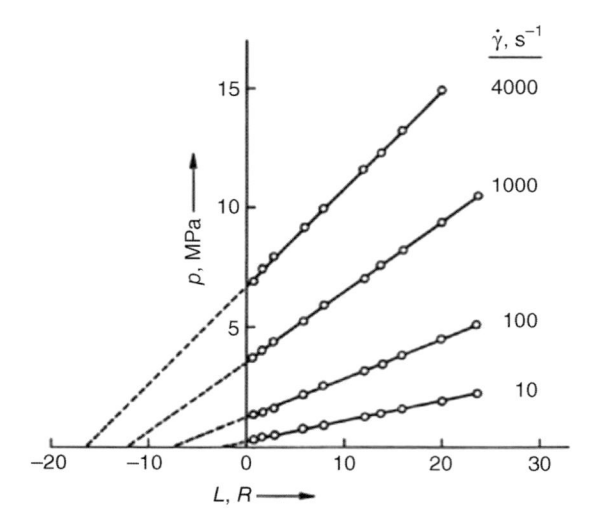

Figure 49. Bagley diagram for a high-impact polystyrene at 189°C and different shear rates
The intercept at $L/R \to 0$ is the pressure correction; the intercept at $p \to 0$, the Bagley correction a) $\dot{\gamma} = 4000$ s^{-1}; b) $\dot{\gamma} = 1000$ s^{-1}; c) $\dot{\gamma} = 100$ s^{-1}; d) $\dot{\gamma} = 10$ s^{-1}
Reprinted with permission by BASF AG, Ludwigshafen [127]

profile at both ends of the capillary (die). The higher L/R is, the higher must be the applied pressure. Die lengths should thus be as short as possible.

7.4. Extensional Viscosities

7.4.1. Fundamentals

Polymer liquids can be elongated considerably without being broken. This extensibility allows fiber spinning from melts and solutions, blow molding of hollow bodies, vacuum forming of parts, etc.

The extensional viscosity η_e (elongational viscosity) is given by the ratio of tensile stress σ_{11} in draw direction 11 to the elongational rate $\dot{\varepsilon}$

$$\eta_e = \sigma_{11}/\dot{\varepsilon} \qquad (23)$$

The type of deformation must always be indicated for extensional viscosities, contrary to shear viscosities. The three principal deformation rates can be defined in such a way that $\dot{\varepsilon}_{11} \geq \dot{\varepsilon}_{22} \geq \dot{\varepsilon}_{33}$. The ratio $m = \dot{\varepsilon}_{22}/\dot{\varepsilon}_{11}$ characterizes the special type of elongational flow: Theory gives $m = -1/2$ for uniaxial elongation, $m = 1$ for equal-biaxial, and $m = 0$ for planar (pure shear). Uniaxial elongational viscosities are important for fiber spinning; they are the only ones used to characterize fluids. Biaxial elongational viscosities play an important role in blow and vacuum forming; very little is known about them.

The elongation in Equation (23) is the true strain $\varepsilon' = \ln (L/L_o)$ (Hencky strain) and not the nominal strain $\varepsilon = (L-L_0)/L_0$ (Cauchy strain, engineering strain). The rate of elongation is thus $\dot{\varepsilon} = d\varepsilon/dt = d \ln L/dt = L^{-1} (dL/dt)$.

Extensional viscosities are very difficult to measure. Elastomers and melts of entangled polymer coils can be stretched between rotating rollers. Extensional viscosities of solutions can be determined if two liquid jets streaming toward each other are redirected by rollers or siphoned off.

7.4.2. Melts

Extensional and shear viscosities depend very differently on deformation rates. At low rates, extensional viscosities are independent of the rate of extension (Fig. 50). The uniaxial extensional viscosity at rest (historically called "Trouton viscosity") is three times the Newton viscosity $[(\eta_e)_0 = 3 (\eta_s)_0]$, whereas the biaxial

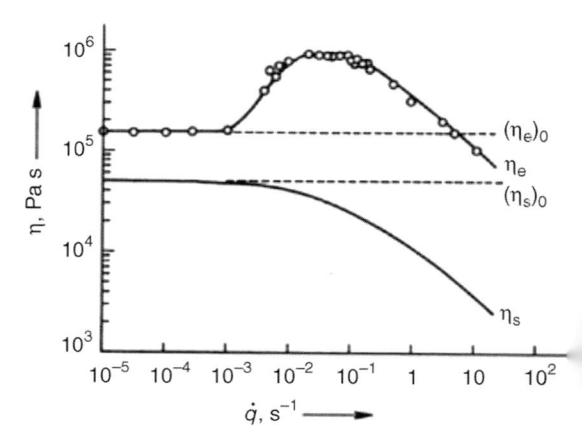

Figure 50. Shear viscosity η_s as function of shear rate $\dot{q} = \dot{\gamma}$, and extensional viscosity η_e as function of the uniaxial extensional rate $\dot{q} = \dot{\varepsilon}$ of a polyethylene at 150 °C
Reprinted with permission by Steinkopff Verlag, Darmstadt [128]

extensional viscosity at rest is six times its shear counterpart.

Above a critical deformation rate, shear thinning is observed for the shearing and extension of melts of linear coil molecules. At the same rate, melts of branched coil molecules become dilatant on extension. The apparent extensional viscosities then pass through a maximum and decrease with further increase of extension rates. The maximum of the extensional viscosity increases with broader molar mass distribution and increased long chain branching, which indicates the strong influence of entanglements on extensional viscosities.

7.4.3. Solutions

Rodlike molecules and segments are increasingly oriented in the extensional direction with increasing elongational rates. The molecular axes are no longer distributed at random, and the solution becomes anisotropic and thus birefringent. A limiting value is asymptotically reached if all molecular axes are completely aligned in the flow direction.

Flexible molecules in solution are only slightly deformed and oriented at comparable extension rates because elastic (entropic) forces cause the chains to return to the thermodynamically favored coil shape. At high critical extension rates, these retraction forces can be overcome and the chain axes become oriented in the flow direction. Only an incremental increase of extension rates is needed to orient the chains completely. Molecules are stressed more and more above the critical extension rates until they finally break. This fracture occurs primarily in the middle of the molecule so that degradation products possess 1/2, 1/4, 1/8, etc., of the initial molar mass. Such degradation occurs under very low deformation rates for rigid macromolecules, for example, during pipetting of dilute solutions of high molar mass deoxyribonucleic acids.

These degradations by extensional flow are not caused by turbulence because they happen at lower Reynolds numbers than those of pure solvents. Chain degradations by turbulence do occur on shearing of very dilute solutions of flexible coil molecules (e.g., 10^{-4} g/mL aqueous solutions of polyoxyethylenes). The degradation reduces the frictional resistance of liquids up to 75%. This "Toms effect" by added small amounts of such polymers eases the flow of crude oil through pipelines and increases the distance and height to which water can be directed at fires.

8. Mechanical Properties [116, 129, 130]

8.1. Introduction

8.1.1. Deformation of Polymers

Mechanical properties of a polymer include the deformation of bulk polymers or their surfaces, the resistance to such deformation, and the fracture under static or dynamic loads. Deformations may be reversible or irreversible; they can be caused by drawing, shearing, compression, bending, and torsion, as well as by combinations of these.

Reversible deformations are due to the presence of elasticity. Irreversible deformations are also called inelastic; they are further subdivided into deformation by viscous flow, plasticity, phase transformations, craze formation, cracking, viscoelasticity, creep, etc. An inelastic deformation of metals by viscoelasticity is known as anelasticity; in polymer physics, anelasticity denotes a reversible elasticity with retardation, which does not lead to energy dissipation. Deformation of the upper layers of a polymer body is characterized by its "hardness," which influences friction and abrasion.

The term elasticity may refer to either energy or entropy elasticity. These elasticities differ in their molecular mechanisms and the resulting phenomenological behavior. In *energy-elastic bodies* (steel, plastics, and elastomers at low strains), torsion and bond angles are changed and bond distances are enlarged on deformation, whereas the macroconformations of chains, for example, remain basically the same. A deformation of *entropy-elastic bodies* (elastomers at high strains), on the other hand, leads to entropically unfavorable positions of chain segments which, however, cannot slip irreversibly from each other because of their cross-linking ("rubber elasticity"). The deformation thus changes the macroconformation (molecular picture), decreases entropy (thermodynamics), and creates normal stresses (mechanics).

These molecular changes are reflected in the properties of energy- and entropy-elastic materials:

	Energy-elastic	Entropy-elastic
Elastic moduli	large	small
Reversible deformation	small	large
Temperature change on deformation	cooling	warming
Length change on heating	expansion	contraction

The deformation behavior of plastics depends on the molar mass of their constituting polymers and on the testing temperature: That is, whether the molar mass is greater than the one needed for the establishment of entanglements between chains ($M > M_{ent}$) and whether the testing temperature is lower than the glass transition temperature of the polymers ($T < T_G$). All mechanical deformations recover for $M > M_{ent}$ and $T < T_G$ because chain entanglements cannot reorganize below the glass transition temperature (memory effect). Yielded plastics and crazed plastics recover on (not too long) heating above the glass transition temperature. These plastics thus do not show true plastic flow.

8.1.2. Tensile Tests

Mechanical properties of polymers are most commonly evaluated by tensile tests in which a specimen is drawn with constant speed. The tensile stress σ_{11} is recorded as a function of time t, draw ratio (strain ratio) $\lambda = L/L_o$, or tensile strain (elongation) $\varepsilon = (L-L_o)/L_o = \lambda - 1$. If a specimen is extended to $L = 2.5\, L_o$ of the original length L_o, then it is said to have been drawn by 150%.

The tensile stress of *elastomers* increases continuously with increasing tensile strain until the polymer finally ruptures at σ_R and ε_R (Fig. 51). Typical *thermoplastics* (and most fibers) follow Hooke's law

$$\sigma_{11} = (F/A_o) \cdot \varepsilon = E \cdot \varepsilon$$

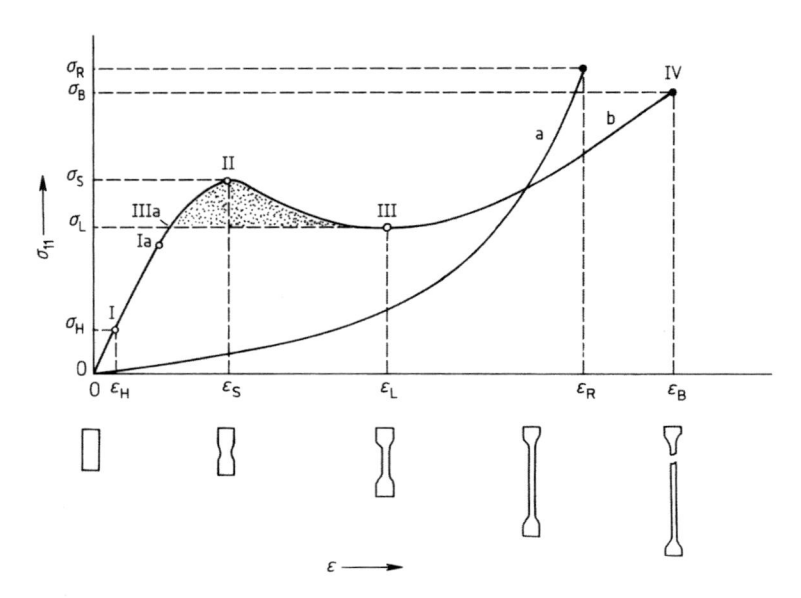

Figure 51. Stress–strain curves of an elastomer (a) and a partially crystalline thermoplastic (b) (schematic) (for further explanation, see text)
The necking effect shown below is characteristic for conventional thermoplastics; it is not found for elastomers and hard-elastic thermoplastics.
Reprinted with permission by Hüthig and Wepf Publ., Basel [5]

for small strains (up to point I), where A_o is the original cross section of the specimen, F is the force, and E is the tensile modulus (Young's modulus). Point I is thus called the proportionality limit or elastic limit; it is defined for a remaining strain of 0.1% after removal of the stress.

The maximum of the stress–strain curve is called the upper yield (point II), and the subsequent minimum is the lower yield (point III). The ratio of upper yield stress σ_S to tensile modulus E is practically constant for all polymers; the numerical value of $\sigma_S/E \approx 0.025$ suggests that van der Waals bonds are broken and molecule segments begin to move more freely.

Brittle polymers break at the upper yield. Tough polymers continue to extend and the stress either remains constant (see below) or decreases. The latter phenomenon is called stress softening. It is typical for polymers with neck formation (telescope effect) and is nominal since it disappears if the stress is given relative to the actual cross section instead of the initial one.

The region between points IIIa–II–III is known as the ductile region; its area describes the absorbed energy and thus the toughness of the specimen. The subsequent increase of stress with strain is called stress hardening. At point IV, the ultimate strength (tensile strength, tenacity at break) σ_B and ultimate elongation (elongation at break) ε_B occur.

Stress–strain curves may differ considerably for plastics. Polymers with high Young's moduli (steep initial slopes) are called hard polymers; those with low moduli, soft; this hardness should not be confused with surface hardness. Typical hard polymers are phenolics [phenol–formaldehyde (PF)], polyacetals [polyoxymethylene (POM)], polycarbonates (PC), and poly(ethylene terephthalates) (Fig. 52). Polymers are further characterized by their stress–strain behavior between upper yield and failure. Polymers without yield cannot absorb energy and thus break easily; they are hard-brittle polymers (e.g., PF). Polycarbonates, on the other hand, show an extended ductile region and a fairly high fracture strain; they are called hard-tough. Polyethylene is similar with respect to ductile behavior and strain hardening; the modulus is however much lower, and PE is considered soft-tough.

The stress–strain behavior described above is typical for tensile experiments. Tensile stresses

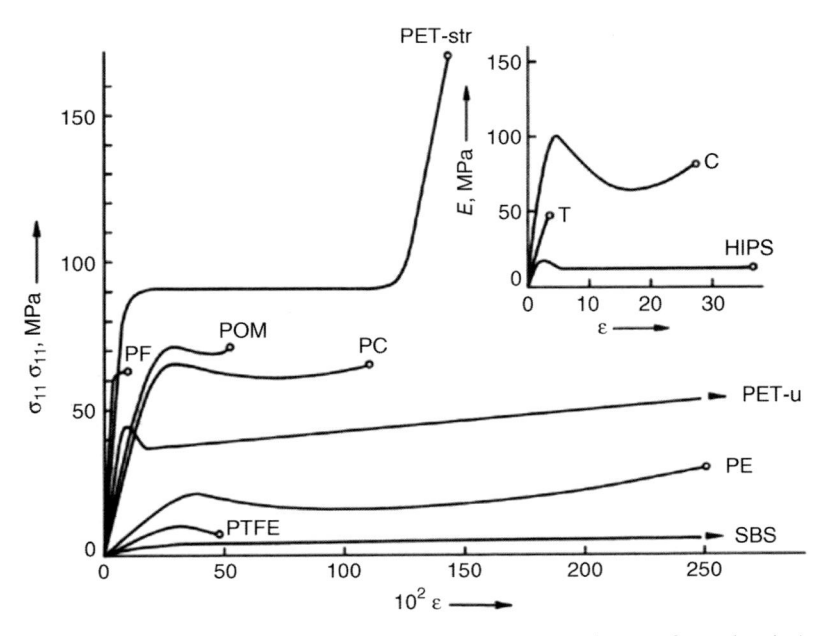

Figure 52. Stress–strain curves of polymers at room temperature (see text for explanation)

lead to strong deformations in neck zones and cause microscopic voids at which fracture originates. Atactic polystyrene (PS) is such a hard-brittle polymer under tension T. No voids can be formed, however, under compression C, and PS appears as a hard-tough material (Fig. 52, insert). Rubber modification of PS leads to high-impact polystyrene (HIPS), which behaves quite differently from PS under tension (see Section 8.6.4).

No yield value is found if drawn polymers are further subjected to tensile tests. In biaxially stretched poly(ethylene terephthalate) (PET-str) some chain segments are already oriented, whereas other remain in their original positions. Biaxially stretched films are thus under stress, which is utilized in shrink films. Such films are used for the packaging of goods. On heating semicrystalline polymers above the glass transition temperature and below the melting temperature, chain segments become more mobile. Molecules attempt to attain their unperturbed dimensions and the films shrink, covering the goods tightly.

8.1.3. Moduli and Poisson Ratios

Elasticities can be described by three elastic moduli: Tensile modulus or Young's modulus E, shear modulus G, and bulk modulus or compressive modulus K. Their values are inversely proportional to the corresponding compliances for static deformations (but not for dynamic ones):

Moduli	Compliances
$E = \sigma_{11}/\varepsilon$	$D = 1/E$
$G = \sigma_{21}/\gamma$	$J = 1/G$
$K = p/(-\Delta V/V_o)$	$B = 1/K$

The three simple moduli are related to each other for small deformations of simple isotropic bodies:

$$E = 2G \cdot (1 + \mu) = 3K \cdot (1 - 2\mu) \qquad (24)$$

where $\mu = (\Delta d/d_o)/(\Delta L/L_o)$ is the Poisson ratio (Poisson number), d is the diameter, and L is the length of the specimen. Poisson ratios can only adopt values $0 < \mu < 1/2$ for isotropic bodies but

Table 18. Poisson ratios and elastic constants of various materials

Material	μ	E, GPa	G, GPa	K, GPa
Water	0.50	≈0	≈0	≈2
Gelatin (80% water)	0.50	0.002		
Natural rubber	0.495	0.0009	0.0003	≈2
Polyethylene, LD	0.49	0.20	0.070	3.3
Polystyrene	0.38	3.4	1.2	5.0
Granite	0.30	30	12	25
Steel	0.28	211	80	160
Glass	0.23	60	25	37
Quartz	0.07	101	47	39
Aluminum oxide Fibers	0	2000	1000	667

may assume $\mu > 1/2$ for anisotropic ones. Equation (24) is invalid for anisotropic bodies and viscoelastic materials; $E/3 < G < E/2$ and $0 < K < E/3$, are however, still valid for these materials.

Common polymers behave with respect to μ, E, G, and K more like liquids than like metals (Table 18). Ultradrawn and self-reinforcing polymers may, however, exhibit tensile moduli that exceed those of steel (see below).

8.2. Energy Elasticity

8.2.1. Theoretical Moduli

The tensile moduli of common polymers are far lower than the theoretical moduli deduced from their chemical and physical structures (- Table 19). Such theoretical moduli can be calculated from bond lengths, valence angles, and force constants for the deformation of these quantities. The theoretical moduli agree well

Table 19. Modulus of elasticity

Polymer	$E_\parallel$, GPa			$E_\perp$, GPa
	Theory	Lattice	Tensile	Lattice
Polyethylene	340	325	<1	3.4
Polypropylene, it	50	42	<3	2.9
Polyoxymethylene				
orthorhombic	220	189	<2	7.8
trigonal	48	54	2	
Poly(p-phenylene terephthalamide)	182	200	132	10
Poly(4-methylpentene)	6.7		1	2.9

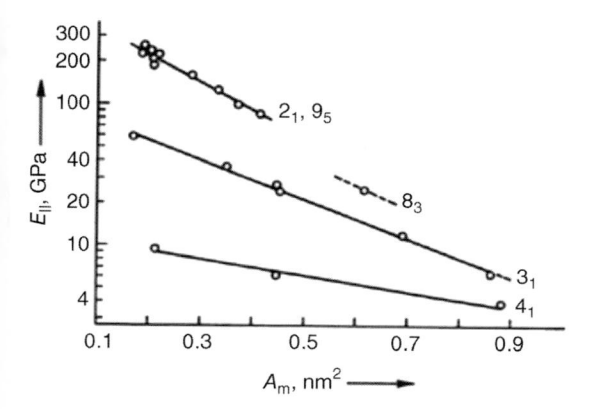

Figure 53. Longitudinal lattice moduli $E_\parallel$ as function of cross-sectional area A_m of polymer chains in all-*trans* (2_1) or helical conformations (9_5, 8_3, 3_1, 4_1) of their main chains Reprinted with permission by Hüthig and Wepf Publ., Basel [5]

with the microscopic lattice constants, which are determined experimentally by X-ray diffraction (change of Bragg reflexes), Raman spectroscopy, or inelastic neutron scattering under load. The longitudinal theoretical moduli $E_\parallel$ are far greater than the transverse moduli $E_\perp$; because the former are controlled by covalent bonds and the latter by van der Waals forces.

The larger the cross-section of the single polymer chain, the larger is the force distributed to fewer chains per unit area and the smaller are the moduli (Fig. 53). Chains in the all-*trans* conformation always exhibit higher theoretical moduli than do helical chains since the elongation of the former changes bond angles, whereas the extension of the latter involves only lower-energy torsion angles.

Polyethylene has the highest theoretical modulus of all one-dimensional chains ($E_\parallel = 340$ GPa), about one-third of that of diamond ($E_\parallel = 1\,160$ GPa) with its "naked" carbon chains. Polypropylene has the highest theoretical modulus of helical chains ($E_\parallel = 50$ GPa).

Moduli approaching these theoretical values have been realized by ultradrawing of mats of polyethylene single crystals, giving polymers with $E_\parallel = 240$ GPa. Industrially, high-modulus polyethylene fibers are manufactured by gel spinning of ultra-high molecular polyethylene ($E_\parallel = 97$ GPa 90 $\stackrel{\triangle}{=}$ N/tex, 1 tex = 1 g per 1 000 m).

8.2.2. Real Moduli

The much lower tensile moduli of conventionally processed polymers result from their disordered physical structures. In amorphous polymers, chain segments are oriented at random. Even in partially crystallized polymers, amorphous layers exist and chain axes (stems of lamellae) are distributed at random. The moduli of flexible polymers can be increased somewhat by processing under external force fields, for example, by partial orientation of chain segments during fiber spinning (Table 20, column L). Extrusion of solid polymers is a particularly effective method: The longitudinal modulus of polyoxymethylene increased to 24 GPa from 2 GPa on hydrostatic extrusion.

Rodlike mesogens of liquid semiflexible polymers align in mesophase domains. The domains orient themselves in shear fields; the orientations can be frozen-in to LCP glasses. Such self-orienting polymers possess much higher moduli in the longitudinal direction than conventionally processed flexible polymers (Table 19). Their transverse moduli are also higher because of intermolecular dipole–dipole interactions.

8.2.3. Temperature Dependence

Young's moduli change characteristically with temperature for the various classes of polymers

Table 20. Moduli and fracture strengths of conventional polymers as isotropic molding masses (I) or in draw direction of fibers (L) and of thermotropic (TT) and lyotropic (LT) glasses longitudinal (L) and transverse (T) to draw direction compared to isotropic polymers (I)

Polymer	E, GPa			σ_B, MPa		
	L	T	I	L	T	I
PE-LD	?	?	0.15	?	?	23
PA 6.6	13	?	2.5	1 000	?	74
PET	19	?	0.13	1 400	?	54
TT X 7 G[a]	54	1.4	2.2	151	10	63
TT Vectra[b]	11	2.6	5.0	144	54	97
LT Kevlar[c]	138	7	?	2 800	?	?
LT PPBT[d]	120	17	62	1 500	680	700

[a] X 7 G = poly(*p*-hydroxybenzoate-*co*-ethylene terephthalate).
[b] Vectra = poly(*p*-hydroxybenzoate-*co*-2-hydroxy-6-naphthalate).
[c] Kevlar 49 = poly(*p*-phenylene terephthalamide).
[d] PPBT = 30% poly(*p*-phenylene benzbisthiazole) in poly(2,5-benzimidazole).

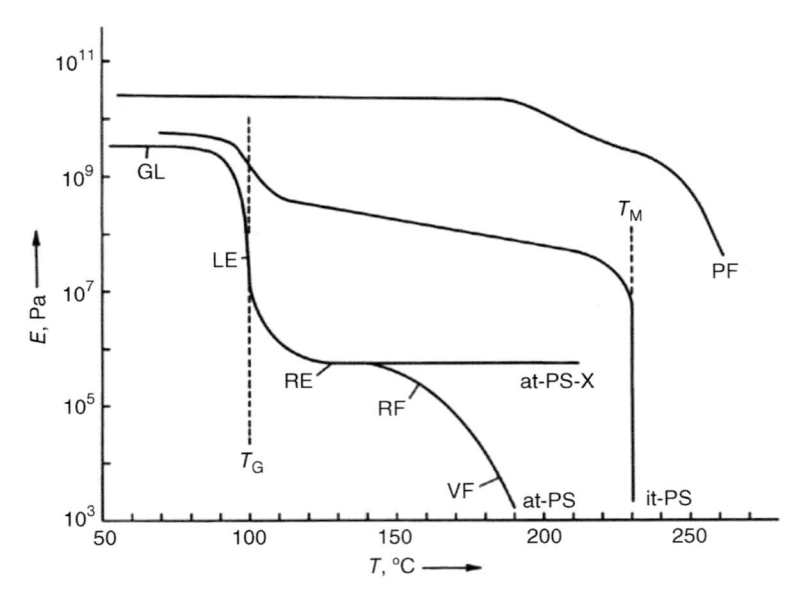

Figure 54. Temperature dependence of Young's moduli for conventional polymers
at-PS = amorphous (atactic) polystyrene ($M > M_c$); at-PS-X = its slightly cross-linked product; it-PS = partially crystalline (isotactic) polystyrene; PF = hardened phenol–formaldehyde resin; GL = glasslike; LE = leatherlike; RE = rubber (entangled); RF = viscoelastic; VF = viscous flow
Reprinted with permission by Hüthig and Wepf Publ., Basel [5]

(Fig. 54). Five characteristic ranges can be distinguished: Glassy (GL), leatherlike (LE), rubbery (RE), viscoelastic (RF), and viscous (VF).

Amorphous Polymers. Moduli (0.1–1 GPa) are practically independent of temperature below the glass transition temperature T_G but drop to 10^5–10^6 Pa at $T = T_G$; the polymer appears leathery around T_G and rubbery above the glass transition temperature. The rubbery region is maintained for entangled (high molar mass) polymers but is nonexistent for low molar masses ($M < M_c$). Moduli decrease on further temperature increase until the polymers behave like viscous liquids ($E \approx 10^3$ Pa).

Elastomers. At use temperatures, these polymers are above their T_G and thus behave like cross-linked thermoplastics above the T_G of the latter. The decrease in modulus at T_G is limited by (and a measure of) the degree of cross-linking. At higher temperatures, elastomers decompose and the moduli decrease drastically.

Crystalline polymers have only a weak leathery region around T_G because their amorphicity is slight and their crystallites act as physical cross-linkers. The modulus at the beginning of the rubbery plateau is a measure of the degree of crystallinity. The moduli decrease slowly during the rubbery "plateau" because more and more crystallites are molten. The remaining crystallites practically disappear at the melt temperature: The moduli drop drastically and the polymers behave like viscous liquids.

Thermosets. Thermosets are highly cross-linked polymers. They have high Young's moduli below T_G, usually one decade higher than thermoplastics. A very weak glass transition is accompanied by a somewhat leathery behavior. The subsequent rubbery region is not very marked because the cross-linking density is high.

8.3. Entropy Elasticity

Elastomers show pronounced entropy elasticities. They exhibit simultaneously some characteristics of solids, liquids, and gases. Like solids, they display Hookean behavior at not too high deformations (i.e., they show no permanent deformation after removal of the load). Moduli and expansion coefficients, on the other hand, resemble those of liquids. Like compressed gases, stresses increase with increasing temperature for elastomers at $T > T_G$.

The elastic behavior can be modeled with various theories. In the simplest case, dislocations of network junctions are assumed to be affine to the macroscopic deformation of the network. The tensile stress σ_{11} varies with the elongation $\lambda = L/L_o$ according to

$$\sigma_{11} = RT \cdot [M_c] \cdot (V_o/V)^{-1/3}(\lambda - \lambda^{-2})$$

It depends on the volumes before (V_o) and after (V) deformation and on the molar concentration $[M_c]$ of network junctions. True Young's moduli can be obtained from the limiting value of $\sigma_{11}/(\lambda - \lambda^{-2})$ at $\lambda \to 1$, for example, for volume-constant deformations ($V_o/V = 1$):

$$RT \cdot [M_c] = lim_{\lambda \to 1}[\sigma_{11}/(\lambda - \lambda^{-2})]$$
$$= \sigma_{11}/[3(\lambda - 1)] = \sigma_{11}/(3\varepsilon) = E_o \quad (25)$$

The Young's modulus is thus directly proportional to the molar concentration of network junctions (cross-linking sites), independent of the nature of the latter. On shearing, elastomers behave like Hookean bodies because

$$RT \cdot [M_c] = \sigma_{21}/\gamma = G$$

They are, however, non-Hookean for elongations because $E_o \ne \sigma_{11}/\varepsilon$ (Eq. 25).

8.4. Viscoelasticity ($\to$ Plastics, Properties and Testing)

8.4.1. Fundamentals

Most polymers do not revert "instantaneously" to their initial states after the removal of loads; they are neither ideal energy elastic (Section 8.2) nor ideal entropy elastic (Section 8.3). These processes take certain times, (i.e., time-independent elastic and time-dependent viscous properties work together to produce a viscoelastic behavior). If stress, strain, and strain rate can be combined linearly, then the process is said to be linear elastic. In addition, some polymers may be irreversibly deformed.

The two ideal cases of response to deformations can be well described by mechanical models: A spring for a Hookean body (instantaneous response) and a dashpot for a Newtonian liquid (linear time dependence of response) (Fig. 55). The Maxwell element combines spring and dashpot in a series; the Voigt–Kelvin element, in a parallel manner.

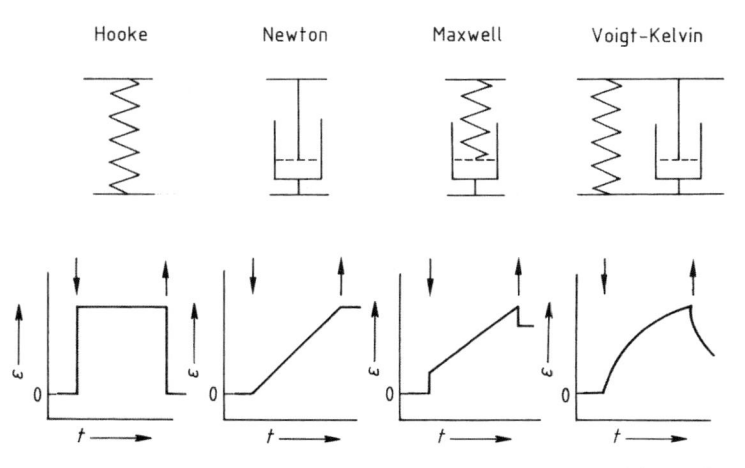

Figure 55. Time dependence of deformations according to various models. Loads are added at ↓ and removed at ↑ Reprinted with permission by Hüthig and Wepf Publ., Basel [5]

Table 21. Simple mechanical models for the deformation of polymers

Model	Function	Behavior* Initial	Final
Newtonian liquid	$\sigma = \eta \cdot \dot{\gamma}$	L	L
Voigt–Kelvin element	$\sigma = G \cdot \gamma + \eta \cdot \dot{\gamma}$	L	S
Maxwell element	$\sigma + (\eta/G) \cdot \dot{\sigma} = \eta \cdot \dot{\gamma}$	S	L
Hookean body	$\sigma = G \cdot \gamma$	S	S
Jeffrey's body	$\sigma + (\eta/G) \cdot \dot{\sigma} = G \cdot \gamma + \eta \cdot \dot{\gamma}$	S	S

* S = solid-like behavior; L = liquid-like.

The Maxwell element describes a *relaxation* (i.e., the decrease of stress at constant deformation; Table 21). Linear combination of elastic deformation rates $d\gamma_e/dt = (1/G)(d\sigma/dt)$ and viscous deformation rates $d\gamma_\eta/dt = \sigma/\eta$ yields, after integration of the resulting expression and indexing the time for this particular behavior for $d\gamma/dt = 0$,

$$\sigma = \sigma_o \cdot \exp(-G \cdot t_e/\eta) = \sigma_o \cdot \exp(-t_e/\tau)$$

The relaxation time $\tau = \eta/G$ indicates the time after which the stress has fallen to the e−1th fraction of its initial value. The ratio τ/t_e of the relaxation time τ to the time scale t_e of the experiment is called the Deborah number ($DB = \tau/t_e$); it is 0 for liquids, ∞ for ideal-elastic solids, and approximately 1 for polymers near their glass transition temperatures.

Retardation is defined as the increase of deformation with time at constant stress. It is characterized by a "creep" of the material. Since this phenomenon was first observed on seemingly solid polymers at room temperature, it is also called "cold flow." In principle, retardation phenomena can be described by a Maxwell element. Because of mathematical difficulties in the solution of the equations, a special model is prefered (Voigt–Kelvin element) (Fig. 55), which yields for the deformation γ_r at constant stress σ_o after indexation for retardations r

$$\gamma_r = (\sigma_o/G_r) \cdot [1 - \exp(-G_r \cdot t_r/\eta)] \qquad (26)$$

where G_r is the retardation modulus, also often called the relaxation modulus. The retardation time t_r indicates the time at which the deformation has reached $(1 - 1/e) = 0.632$ of the final deformation σ_o/G_r. Retardation times and relaxation times are of the same magnitude, but not equal because they rely on different models.

The term "viscoelasticity" is sometimes used to describe the reversible deformation according to Equation (26). However, it is often applied to the total deformation, which is composed of the contributions by Equation (26), a Hookean body with $\gamma_e = \sigma_o/G_o$, and a Newtonian liquid [$\gamma_\eta = (\sigma_o/\eta_o) \cdot t$]:

$$\gamma_{tot} = \{(1 - G_o) + (t/\eta_o) + (1/G_r) \cdot [1 - \exp(-t/\tau)]\} \cdot \sigma_o$$

$$= \gamma_e + \gamma_\eta + \gamma_r$$

The three deformation terms γ_e (elastic), γ_η (viscous), and γ_r (viscoelastic) are often not explicitly evaluated. The time-dependent viscous and viscoelastic parts are rather combined into a new parameter $\gamma_c = \gamma \cdot t^n$ (Findlay law). The resulting function for the creep curve allows an extrapolation to long-time behavior from short-time experiments:

$$\gamma_{tot} = \gamma_e + \gamma_c = \gamma_e + \gamma \cdot t^n$$

8.4.2. Time–Temperature Superposition

Deformations, shear moduli, and shear compliances are time and temperature dependent (Fig. 56). The moduli vary about one decade for six decades in time at constant temperature, and up to one decade for each 10 K at constant frequency. Since no single experimental method can cover the 15–20 decades of frequency that are required for good characterization of a polymer, time–temperature data from various techniques are usually combined with the help of the Boltzmann superposition principle.

This principle states that the deformation (or recovery) caused by an additional load (or removal thereof) is independent of previous loads or their removal. The $G = f(t, T)$ curves can be combined if (1) the relaxation time spectrum is temperature independent and (2) the thermal activation is the same over the entire time and temperature range (no transition or relaxation temperatures). A reference temperature is chosen close to the static glass transition temperature (115 vs. 105 °C in

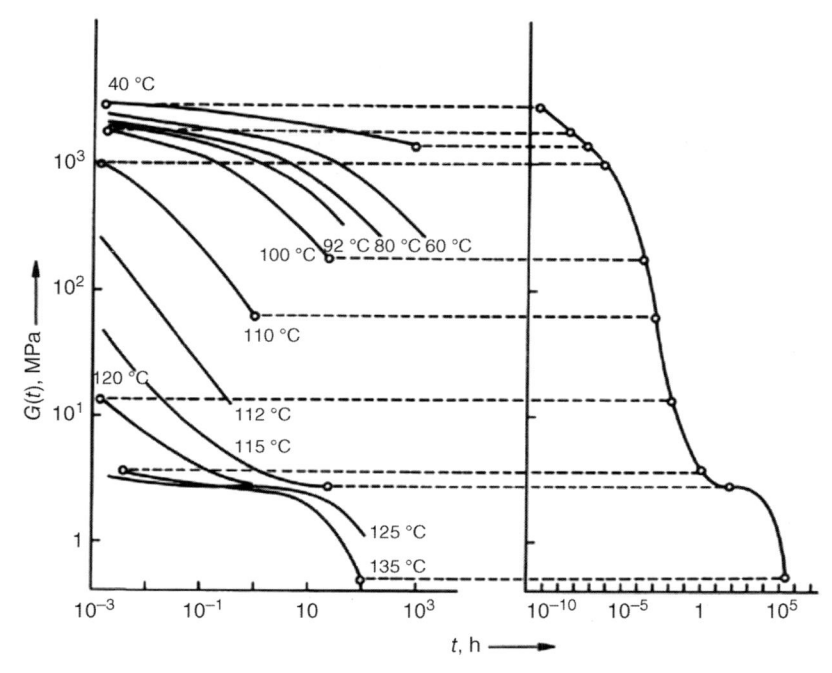

Figure 56. Time dependence of the shear modulus from measurements of the stress relaxation of a poly(methyl methacrylate) ($\overline{M}_n$ = 3 600 000 g/mol) at various temperatures (left) and the resulting time–temperature superpositions for a reference temperature of 115°C (right) (circles represent equivalent positions)
Reprinted with permission by Hüthig and Wepf Publ., Basel [5]

Fig. 56) and the G values are shifted horizontally with the help of a shift factor from the WLF equation (Eq. 21).

8.5. Dynamic Behavior

8.5.1. Fundamentals

Dynamic–mechanical methods expose the specimen to periodic stresses. The polymer either is put under torsion once and then oscillates freely (torsion pendulum) or is subjected continuously to forced oscillations (e.g., Rheovibron). In addition, ultrasound, dielectric, and NMR methods can be used to study the dynamic properties of polymers.

In the simplest case, the applied stress is sinusoidal with a frequency ω ($\sigma_t = \sigma_o \cdot \sin \omega \cdot t$). The deformation of ideal-elastic bodies follows the stress instantaneously ($\gamma_t = \gamma_o \cdot \sin \omega \cdot t$) but that of viscoelastic polymers experiences a delay ($\gamma_t = \gamma_o \cdot \sin(\omega \cdot t - \vartheta)$). The stress vector is assumed to be a sum of two components: One component is in phase with the deformation ($\sigma' = \sigma_o \cdot \cos \theta$); the other is not ($\sigma'' = \sigma_o \cdot \sin \theta$).

Each of these two components possesses a modulus. The *real modulus (shear storage modulus)* G' measures the stiffness and shape stability of the specimen:

$$G' = \sigma'/\gamma_o = (\sigma_o/\gamma_o) \cdot \cos \theta = G^* \cdot \cos \theta$$

whereas the *imaginary modulus (shear loss modulus)* G'' describes the loss of usable mechanical energy by dissipation into heat:

$$G'' = \sigma''/\gamma_o = G^* \cdot \sin \theta$$

The same quantities can also be derived if complex variables are introduced

$$G^* = G' + i \cdot G'' = [(G')^2 + (G'')^2]^{-1/2}$$

The loss factor δ is the ratio of imaginary to real modulus. It is the same for shear and

Young's moduli but not for compression moduli

$$\Delta = \tan\theta = G''/G' = E''/E' < K''/K'$$

8.5.2. Molecular Interpretations

The *shear storage moduli* of low molar mass polymer melts with narrow molar mass distributions increase continually with increasing frequency (Fig. 57). At high normalized frequencies $\alpha_T\omega$, all storage moduli asymptotically approach a limiting line, regardless of molar mass. This part of the relaxation spectrum thus originates from the mobility of chain segments; it is called the transition range.

High molar mass polymers show a corresponding frequency dependence of *loss moduli* at very low frequencies (called end range), followed by a plateau at higher frequencies (plateau modulus G_N^o), and finally the transition range. The end ranges of these spectra are molar mass dependent; this behavior must come from long-range conformational changes. Since the transition range characterizes viscous behavior, and the end range viscoelastic behavior, the plateau range must reflect rubbery behavior (see Fig. 54 for the temperature dependence).

The *rubbery behavior* of polymer melts can be described by the theories of entropy elasticity according to which the shear modulus of chemically cross-linked polymers depends on the molar concentration of network junctions. The plateau modulus G_N^o of melts thus indicates the concentration of temporary junctions (entanglements). The molar mass M_e of segments between such junctions of polymers with volume fractions ϕ_p in solution ($\phi_p = 1$ for melts) and polymer melt densities ρ_p is given by

$$M_e = RT \cdot r_p \cdot \phi_p / G_N^o$$

These dynamic entanglement molar masses M_e are a factor 2.0 ± 0.2 lower than the corresponding molar masses M_c from rest viscosities (Table 22).

The plateau is not well developed or may even be absent for polymers with broad molar mass distributions; a complicated dependence of shear compliances on higher molar mass averages has been predicted by reptation theory.

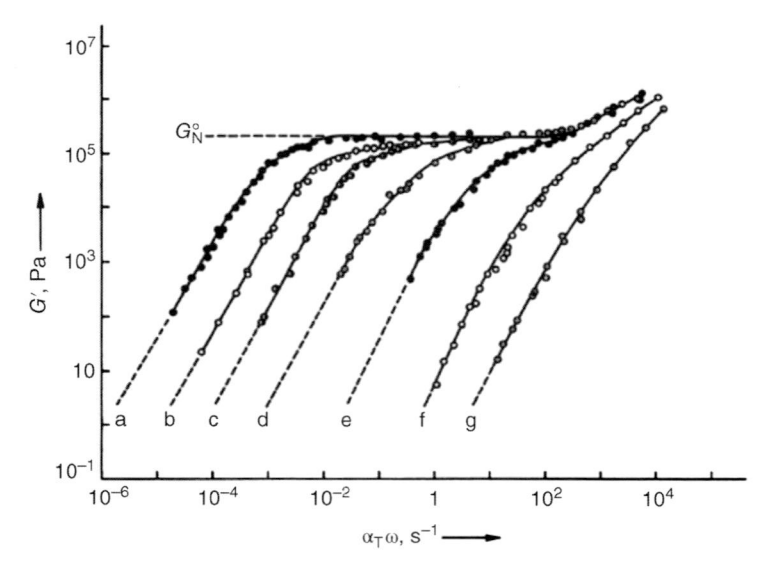

Figure 57. Frequency dependence of the shear storage modulus G' for narrow-distribution polystyrene melts of different molar masses M. A shift factor α_T was used to convert to 160°C data measured at various temperatures
a) $M = 581\,000$ g/mol; b) $M = 351\,000$ g/mol; c) $M = 215\,000$ g/mol; d) $M = 113\,000$ g/mol; e) $M = 46\,900$ g/mol; f) $M = 14\,800$ g/mol; g) $M = 8\,900$ g/mol
Reprinted with permission by Hüthig and Wepf Publ., Basel [5] after data of [131]

Table 22. Critical molar masses for entanglement from shear moduli (M_e) and rest viscosities (M_c)

Polymer	T, °C	M_c, g/mol	M_e, g/mol	M_c/M_e
Polyethylene	190	3 800	1 790	2.1
Polypropylene, it	190	7 000		
Polyisobutene	25	15 200	8 800	1.7
Polydimethylsiloxane	25	24 500	10 500	2.4
Poly(vinyl acetate), at	57	24 500	12 000	2.0
Poly(α-methylstyrene), at	100	28 000	13 500	2.1
Polystyrene, at	190	35 000	18 100	1.9

8.6. Fracture [129]

8.6.1. Overview

Polymers break very differently depending on their chemical and physical structure; environment (humidity, solvents, temperature); and the type, duration, and frequency of deformation. Some polymers break immediately; others are unchanged even after months. The fracture surface can be smooth or splintery; the elongation at fracture, less than 1% or greater than 1 000%.

Two fracture modes can be distinguished, brittle and tough (ductile). Brittle polymers fracture perpendicularly to the stress direction, tough polymers longitudinally (Fig. 58). A polymer is defined as brittle if its elongation at break is less than 20%.

Brittle fractures are rare for ideal solids since many bonds must be severed simultaneously. Real polymers however contain many small imperfect regions that act as "nuclei" for the formation of microcracks. Brittle polymers usually possess "natural" microvoids, which may also appear in drawn amorphous polymers or through separation of crystal lamellae in hard-tough polymers.

Tough failures (ductile fractures) are caused by viscous flow ("plastic flow"). This process may involve the slipping of chain segments past each other (amorphous polymers) or the movement of crystalline domains (partly crystalline polymers). Polymer chains may also de-entangle at long times and under small stresses. The same processes and additional ones may occur on failure of composites (Fig. 58).

Polymers are subjected to very different stress conditions in typical applications; they thus experience different failure modes. Test methods try to simulate complex real-life situations by standardized procedures. They include long-term experiments such as static deformations under constant load by tension, compression, or bending; short-term methods such as tensile tests under various speeds or impact tests with unnotched or notched specimens; dynamic testing with variation of the number of loadings–unloadings, impacts, vibrations, etc.

8.6.2. Theoretical Fracture Strength

The fracture of brittle polymers generates free radicals. Since the probabilities of such homolyses depend on bond strengths, which also determine tensile moduli, relationships must exist between the theoretical moduli and the theoretical fracture strengths of polymers.

Bonds are severed if atoms are separated from each other by certain distances L_b greater

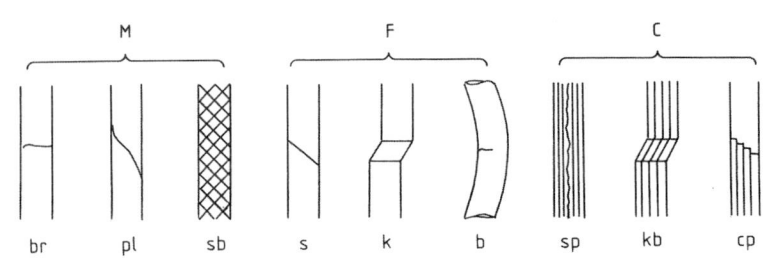

Figure 58. Failure modes of polymers (matrix M), fibers (F), and fiber-reinforced polymer composites (C) by brittle failure (br), plastic flow (pl), shear band formation (sb), shearing (s), kink formation (k), bending (b), longitudinal splicing (sp), formation of kink bands (kb), and step formation by compression (cp)
Reprinted with permission by Hüthig and Wepf Publ., Basel [5]

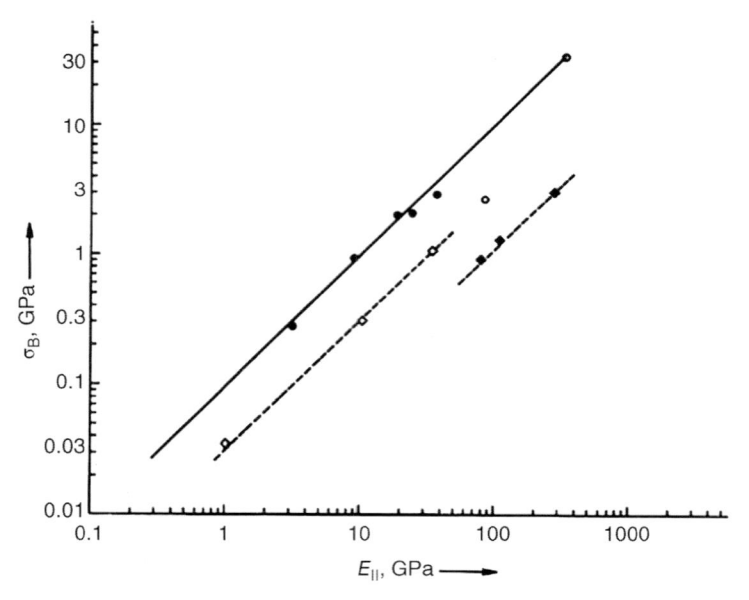

Figure 59. Tensile strength at break as function of tensile modulus: (●) Experimentally ultradrawn ultra-high-modulus polyethylenes; (○) Industrially manufactured ultradrawn polyethylene (Dyneema); (⊙) Theory for perfectly aligned polyethylene; (◆) Heterogeneous molecular composites of ABPBI fibers or PBT fibers or films in ABPBI matrix; (◆) Homogeneous molecular composites of PBT fibers or PBT fibers for films in ABPBI. Solid line corresponds to $\sigma_B = 0.095 \cdot E_\|$

than their equilibrium distances L_o. The necessary theoretical strength $\sigma_\|^o$ is given by [5].

$$\sigma_\|^o = \frac{E_\|^o \cdot (L_b - L_o)}{\pi L_o} = K \cdot E_\|^o \qquad (27)$$

Polymer main-chain bonds break at approximately the same relative distance ($L_b \approx 1.3\, L_o$) because bond lengths and strengths are not too different for bonds such as C–C, C–O, and C–N. Thus, $K \approx 0.095$ and the theoretical fracture strength $\sigma_\|^o$ should be ca. one-tenth of the theoretical tensile modulus $E_\|^o$ in the chain direction, regardless of the chemical nature of the polymer. Polyethylene with a theoretical modulus of 340 GPa (Table 19) should thus have a theoretical fracture strength of ca. 32 GPa [i.e., much higher than the theoretical strength of steel (ca. 20 GPa)]. Industrially manufactured ultradrawn polyethylene fibers have higher experimental fracture strengths than steel (2.9 GPa = 2.7 N/tex vs. 2.5 GPa).

The theoretical fracture strength–tensile modulus relationship of Equation (27) has been realized for certain ultradrawn poly-ethylenes in which both the predicted proportionality constant $K = 0.095$ and the first power of E were found (Fig. 59). The type of same relationship is also observed for molecular composites of poly (p-phenylene-2,6-bisbenzthiazole) (PPBT) with rodlike mesogens in coillike polybenzimide (ABPBI), albeit with a lower proportionality constant. In other ultradrawing experiments, a power dependence $\sigma = K' \cdot E^n$ was found, however.

8.6.3. Real Fracture Strength

The lower than theoretical fracture strengths of most polymers (see Tables 19 and 20) are caused by many factors. Theoretical fracture strengths relate to infinitely long, completely aligned, immobile polymer chains. End groups and chain folds act as disturbances: Fracture strengths of conventional polymers increase with increasing molar mass and become practically constant above a "critical" molar mass.

At the latter molar mass range, chain segments of amorphous polymers are distributed at random. A brittle fracture across such polymers will create two new surfaces with a total surface energy of 2 γ_{lv}. The theoretical fracture strength for brittle, energy-elastic bodies

$$\sigma_a^o = (E \cdot \gamma_{lv}/L_o)^{1/2} \qquad (28)$$

now depends on the product of modulus and surface energy (Ingles theory). The Ingles theory works well for silicate glasses.

The experimentally found fracture strengths σ_b of plastics are however much lower than the strengths predicted by Equation (28). The ratios σ_a^o/σ_{exp} of molded (unoriented) plastics decrease with increasing elongation ε_b at break (Fig. 60), whereas those of drawn fibers (partially oriented chain segments) increase.

The reason for the lower than expected fracture strengths of amorphous polymers is the presence of microvoids, which act on drawing as nuclei for cracks. According to the Griffith theory, a crack can grow only if the energy required for the fracture of chemical bonds is just surpassed by the stored elastic energy. This theory predicts a dependence of the fracture strength on crack length L:

$$\sigma_B = [(2E \cdot \gamma_{lv})/(\pi \cdot L)]^{1/2} \qquad (29)$$

This functionality is indeed observed for artificially introduced long cracks. The predicted fracture strengths are however much lower than those found by experiment. Furthermore, deviations from Equation (29) occur at small crack lengths because the fracture behavior of plastics is not dominated by the cleavage of chemical bonds but by other types of energy absorption (crazing; shear flow).

On drawing, stresses are imposed on microvoids. The polymer reaches its upper yield stress at sufficiently high stress concentrations at the tip of such a void and relieves the stress by stress softening (Fig. 51). The induced cooperative movements of chain segments cause long-range changes of macroconformations. In partially crystalline polymers, these changes can occur

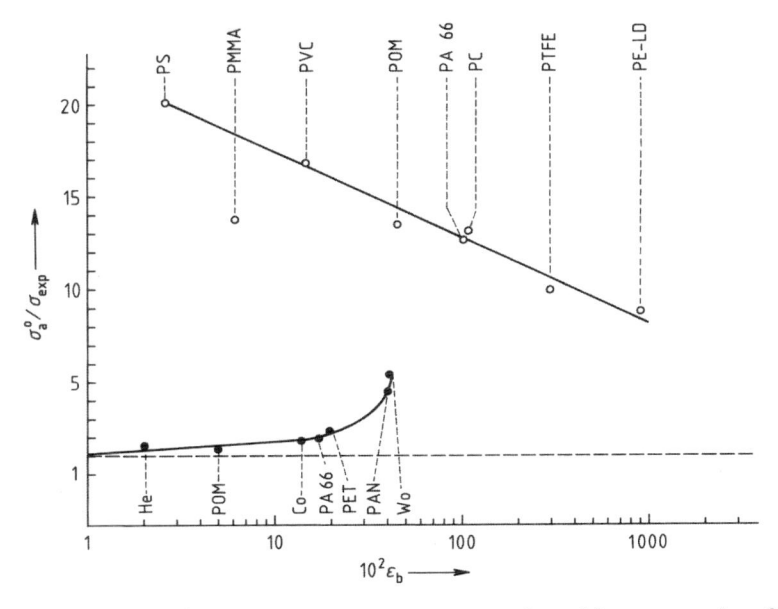

Figure 60. Ratio σ_a^o/σ_{exp} of theoretical [Eq. (28)] and experimental fracture strengths as function of elongation at break ε_b for (○) nonoriented (molded plastics) and (●) oriented (fibers) polymers
He = hemp, Co = cotton, Wo = wool (for other abbreviations see Tables 1–4)

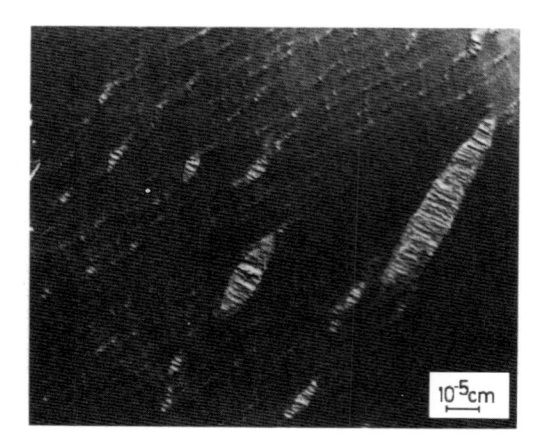

Figure 61. Crazes in a polystyrene drawn to 25% [132]

only in amorphous domains; spherulitic polymers break accordingly either between spherulites or in the radial spherulite direction.

The cooperative movements of segments lead to either shear or normal stress yielding. On shearing, the whole specimen yields either homogeneously or heterogeneously (localized). In the latter case, shear bands are formed at angles of 38–45° to the stress direction (Fig. 58). Chain segments are arranged at angles between shear bands and stress directions.

All polymers with upper yield values form crazes upon stress softening (Fig. 61), regardless of whether they are amorphous, crystalline, linear, or cross-linked. Crazes can be up to 100 μm long and 10 μm wide; their long axes are parallel to the stress direction. They are not voids since their interior is filled with amorphous microfibrils of 0.6–30 nm diameter; these microfibrils are oriented in the stress direction (i.e., perpendicular to the craze long axes). On further deformation, microvoids are formed.

The formation of crazes is the primary mechanism for the dissipation of stress energy. It is utilized in the rubber reinforcement of polystyrene. Rubber-modified polypropylene, on the other hand, deforms mainly by shear flow.

8.6.4. Impact Resistance

Impact strength is the resistance of a material to impact. It is one of the many quantities used to characterize the strength of a material under (the usually complex) use conditions; all test methods are thus standardized. Most test methods measure the energy required to break a notched or unnotched specimen (Izod, Charpy, high-speed tensile). Impact speeds range from $10^{-5}–10^{-1}$ m/s in conventional tensile tests to 20–240 m/s for high-speed tensiles; elongation speeds are usually from 10^{-3} to 10^{4} s^{-1}.

Impact strengths depend on experimental conditions. The smaller the radius of the notch, the higher is the stress concentration at the tip and the lower is the impact strength. At very low temperature, all polymers are brittle. The mobility of chain segments increases with increasing temperature, allowing stresses to be relieved by shear-band or craze formation. Impact strengths increase with temperature, especially near the glass transition temperature. Polymers with additional transition temperatures below the glass transition temperatures are for the same reason almost always more impact resistant than polymers without such transitions. Nonentangled polymers exhibit very low impact strengths because no crazes can be formed. The impact behavior of polymers can be improved considerably by modification with rubber (see Section 11.4.3).

8.6.5. Stress Cracking

Stress cracking (stress corrosion, stress crazing) is the formation of crazes under the physical action of chemicals, especially surfactants. Stress corrosion starts at polymer surfaces and proceeds into the interior until the polymer finally cracks. The appearance and the extent of stress cracking depend on the polymer–reagent interaction and the magnitude of the stress.

Effects are weak in nonwetting liquids but strong in polymer–liquid systems with solubility parameters of polymers and liquids matching each other and even more dramatic under tension in the presence of surfactants. Stress cracking decreases with increasing molar mass of the polymer since entanglements allow stresses to relax elastically. Cross-linked polymers are less prone to stress cracking for the same reason. Stress cracking is also reduced if polymer plasticizers are present in plastics because these additives increase the mobility of chain segments and thus the ability to relieve stresses. The same action is responsible for the fact that no stress corrosions are observed above glass transition temperatures.

8.6.6. Fatigue

Materials may be damaged not only "instantaneously" (i.e., on impact) but also by static or periodic loads after certain times or number of loadings. This fatigue is characterized by the *fatigue limit* (endurance) at which the plastics are not damaged even after infinite time and the *fatigue strength*, which indicates the load at which damage sets in after a certain time.

Plastics may be subjected to static loads for certain times t, after which their fracture strengths σ_B are measured by tensile tests. The logarithms of strength of amorphous polymers usually decrease linearly with logarithms of time due to viscous flow (Fig. 62). Partially crystalline polymers show a bend in these lines after certain times, which indicates a change from tough fracture (short times) to brittle fracture (long times), probably caused by recrystallization phenomena.

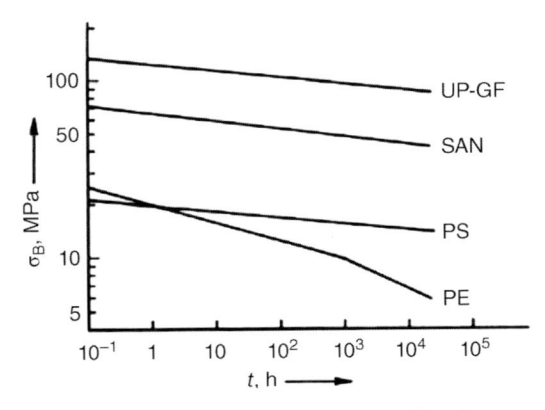

Figure 62. Time dependence of tensile fracture strengths after static loading during time t. UP-GF = glass-fiber-reinforced unsaturated polyester; SAN = styrene–acrylonitrile copolymer (impact polystyrene); PS = polystyrene; PE = polyethylene
Reprinted with permission by BASF AG, Ludwigshafen [127]

8.7. Surface Mechanics

8.7.1. Hardness

The hardness of a material is its resistance to penetration by another body. Hardness is a very complex quantity; it depends on Young's modulus, yield stress, and stress hardening. A general definition of hardness, applicable to all materials, does not exist; neither does a universally applicable testing method. The various technical test methods thus emphasize one or another factor that contributes to the hardness of a specific class of materials.

The hardness of *hard plastics* is normally characterized by various Rockwell (ISO, ASTM) or ball indentation hardnesses (DIN, ISO). These methods measure the indentation of a polymer by a steel sphere under load and thus the compression set and the recoverable deformation. The plastic deformations of polymers increase with time (creep), whereas those of metals are time independent. Because of the short duration of the hardness test polymers exhibit relatively high Rockwell hardnesses.

The hardness of *soft plastics* is characterized by their durometer (ASTM) or various Shore hardnesses (ISO, DIN). These methods measure the resistance to penetration by a truncated cone (static methods). Hardness properties of metals and hard plastics are evaluated by another Shore hardness that uses the rebound of a small steel sphere (dynamic method).

All methods measure the hardness of surfaces, not of the interior of the specimen. The surface may, for example, be plasticized by the humidity of the air. Crystallizable polymers may have lower surface hardnesses than interior hardnesses (if the plastic had been injected into a cold mold) or the reverse may be true (if transcrystallization occurred).

8.7.2. Friction

Friction, the resistance against the relative movement of two bodies contacting each other is measured by the friction coefficient $\mu = R/L$ (i.e., the ratio of friction R to total load L). Friction depends in a complex and not understood way

Table 23. Friction coefficients of various sliding bodies

Plastic	Friction coefficient of		
	Plastic on plastic	Plastic on steel	Steel on plastic
Poly(methyl methacrylate)	0.8	0.5	0.45
Polystyrene	0.5	0.3	0.35
Polyethylene, high density	0.1	0.15	0.20
Polyethylene, low density	0.3		0.80
Polytetrafluoroethylene	0.04	0.04	0.10

Table 24. Abrasion coefficients K for moving plastics against resting materials

Resting material	$10^{10}K$, MPa^{-1}	Moving polymer	$10^{10}K$, MPa^{-1}
Polycarbonate A	200 000	polyamide 66	11 000
Polyamide 66	250	polycarbonate A	9 800
Polyamide 66	220	polyamide 66	510
Polyamide 66	10	polyacetal	12
Polyacetal	11	polyamide 66	15
Steel		polyamide 66	8 600
Steel		polyamide 66 with 30% glass fibers	1

on both the surface roughness of the specimen and its mechanical properties.

The rolling of *hard bodies on soft materials* is determined almost exclusively by the deformation of the soft base (i.e., its viscoelastic properties). Elastomers thus have fairly high friction coefficients of $0.5 < \mu < 3.0$, depending on the contacting body and its type of movement (rolling, sliding).

The sliding of *hard bodies on other hard bodies* occurs on the tops of the microscopic surfaces: The true contact area is much smaller than the geometric one. The applied load thus acts on very small effective areas. Local stresses are high and the tops are leveled. Large adhesion forces exist between chemical groups of the resulting effective contact areas of both bodies, which must be overcome by breaking the bonds or by shearing one of the materials. The adhesive friction $R = A_w \cdot \sigma_b$ is given by the effective contact surface A_w and the shear strength σ_b. Soft materials possess high effective surfaces (large A_w) and are easily sheared (small σ_b), whereas the opposite is true for hard materials. Plastics, metals, and ceramics therefore often exhibit very similar friction coefficients (Table 23).

8.7.3. Abrasion and Wear

Abrasion is the loss of material from surfaces by friction. It is thus affected by both friction properties and hardnesses of the specimens. The abrasion coefficient K is given by the applied force F, the linear speed v of the contacting body, the total time t, and the volume loss ΔV of the abraded material: $K = \Delta V/(F \cdot v \cdot t)$.

Abrasion coefficients vary widely with polymer type and state (resting, mobile) (Table 24).

The best resistance against abrasion is shown by polyureas, followed by polyamides and polyacetals. It can be enhanced greatly by addition of certain fillers (e.g., short fibers).

9. Electric Properties [116, 126, 133, 134]

Matter is subdivided according to its specific electrical conductivity σ into insulators ($\sigma = 10^{-14}$–10^{-22} S/cm), semiconductors (10^2–10^{-9} S/cm), conductors ($> 10^3$ S/cm), and superconductors ($\approx 10^{20}$ S/cm). Most plastics are insulators (Section 9.1), but certain polymers are intrinsic semiconductors and some may even be conductors after doping (Section 9.2).

9.1. Dielectric Properties

9.1.1. Relative Permittivity

Groups within a molecule and entire molecules of an insulator are polarized by applied electric fields. Polarization is usually measured by the ratio of capacitances of a condensor in vacuo and in the specimen (i.e., the relative

Table 25. Electrical properties of plastics

Polymer	Δw^a		ρ_v^c	ρ_s^d	S^f		
	%	ε_r^b	$\Omega \cdot cm$	Ω	$\tan \delta^e$	$kV^{-1} mm^{-1}$	U^g/V
PTFE	0	2.15	10^{18}		0.0001	40	> 600
PE	0.05	2.3	10^{17}	10^{13}	0.0007	70	600
PS	0.1	2.5	10^{18}	10^{15}	0.0002	140	500
SAN		3			0.0070	100	
ABS		3.2	10^{15}	10^{13}	0.02	15	600
PVC	$< 1.8^h$	$< 3.7^h$	10^{15}	10^{13}	0.015	< 50	< 600
PA 6							
Dry	0	3.7	10^{15}		0.03	< 150	600
Conditioned	9.5	7	10^{12}		0.3	80	600
CA	4.7	5.8	10^{13}		0.03	35	
UP		3.4	10^{13}	10^{12}	0.01	50	500
PUR		4	10^{13}		0.05	30	
PF		8	10^{14}		0.05	12	
With inorganic fillers			10^{10}		< 0.5	14	< 150
Without inorganic fillers			10^{9}		< 0.5	10	125
UF		6	10^{11}		0.1	10	

[a] Δw = water absorption (at 50% relative humidity).
[b] ε_r = relative permittivity ("dielectric constant").
[c] ρ_v = volume resistivity (inverse specific electrical conductivity).
[d] ρ_s = surface resistivity (inverse surface conductivity).
[e] $\tan \delta$ = dissipation factor (loss tangent, at 1 MHz).
[f] S = dielectric strength.
[g] U = tracking resistance (method KC).
[h] Depends on impurities from polymerization (e.g., emulsifier residues).

permittivity ε_r of the specimen; formerly called the dielectric constant). Relative permittivities are low for apolar polymers [polytetrafluoroethylene (PTFE); PE], higher for polymers with polarizable groups (PS, PC), and still higher for polar materials [dry polyamide (PA)] (Table 25).

The relative permittivity thus increases with increasing water content of plastics ($\varepsilon_r = 81$ for water). It also increases with increased segmental mobility ($\varepsilon_r = 13.0$ for cis-1,4-polyisoprene); rubber-modified plastics have higher relative permittivities than conventional plastics. Expanded plastics and elastomers are composites with air ($\varepsilon_r = 1.00058$) and subsequently have low relative permittivities of ca. 1–2. Fillers increase the relative permittivities to values up to 170 (filled thermoplastics) and up to 18 000 (filled elastomers).

9.1.2. Dielectric Loss

Dipoles try to follow the direction of an electric field if an alternating current is applied. The required adjustment times correspond to the orientation times of groups and molecules. The faster the alternation, the longer the orientation lags behind the field and the greater is the electrical energy consumed. Available output power is decreased because electric power is lost by conversion into thermal energy.

The ratio of power loss N_v to total power output N_b is called the dielectric dissipation factor or loss tangent $\tan \delta$, which can also be expressed as the ratio of imaginary relative permittivity (ε'') to real relative permittivity (ε'):

$$N_v/N_b = \tan \delta = \sin \delta / \cos \delta = \varepsilon'' / \varepsilon'$$

Sometimes power factors $\sin \delta$ are given instead of loss tangents $\tan \delta$.

Polymers with high loss factors $\varepsilon \cdot \tan \delta$ can be heated and thus welded by high-frequency fields; PVC is an example. Polymers with low loss factors [PE, PS, polyisobutene (PIB)], on the other hand, are excellent insulators for high-frequency conductors.

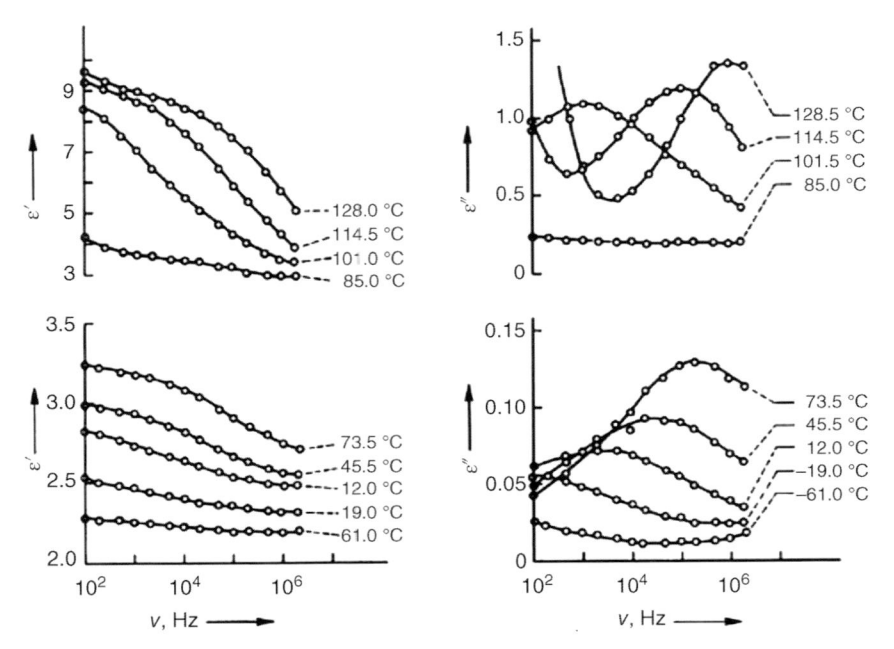

Figure 63. Frequency dependence of real (ε') and imaginary relative permittivities (ε'') of a poly(vinyl chloride) at various temperatures
Upper: β-dispersion; lower: γ-dispersion.
Reprinted with permission by Hüthig and Wepf Publ., Basel [5]

Real and imaginary relative permittivities depend on the frequencies ν of the alternating current (Fig. 63). The function $\varepsilon' = f(\nu)$ corresponds to a dispersion and the function $\varepsilon'' = f(\nu)$ to an absorption of energy. Transitions and relaxations consume energy and therefore inflection points (ε'-curves) and maxima (ε''-curves) are found at appropriate temperatures and frequencies.

9.1.3. Dielectric Strength and Tracking Resistance

The imaginary part of the relative permittivity is caused by the dissociation of polar groups that may be either inherent to the polymer or introduced by extraneous impurities. These polar groups must be of ionic nature since the electrical conductivities of conventional plastics are strongly temperature dependent (electronic conductivities are far less temperature dependent).

Heat is caused to develop by the imaginary part of the relative permittivity. The low thermal conductivities of plastics do not allow this heat to dissipate, and the temperature increases. Ionic conductivities are thus increased until a breakdown (arcthrough) finally occurs. The resistance against such a breakdown is measured by the electric strengths S of a plastic ("dielectric strength") (Table 25).

A breakdown can also occur through tracking on the surface of a plastic. The tracking resistance is difficult to measure because surface resistivities are 2–3 decades lower than volume resistivities (Table 25). The tracking resistance is thus measured by standardized methods, such as the maximal voltage that does not cause tracking if 50 drops of an aqueous 0.1 wt% NH_4Cl solution are applied between two platinum electrodes that are under an alternating current and 4 mm apart on the specimen surface. A polymer has a good tracking resistance if it forms volatile products and no carbon upon degradation (volatile monomer by depolymerization of PMMA, volatile oligomers by degradation of PE or PA). Poly(N-vinyl carbazole) does not form volatile products and thus has a poor tracking resistance although it is a good insulator.

9.1.4. Electrostatic Charging

Static electricity originates from an excess or a deficiency of electrons on isolated or ungrounded surfaces. It can be created by rubbing two surfaces against each other (triboelectric charging) or by contact of a surface with ionized air. Matter is charged electrostatically if specific conductivities are lower than ca. 10^{-8} S/cm and relative humidities lower than ca. 70%. All conventional plastics can thus be electrostatically charged (Table 25). Charge densities may vary between, e. g., 8.2 C/g for polychlorotrifluoroethylene (PCTFE) and -13.9 C/g for phenolic resins, and charges may vary between 3 000 V/cm (POM vs. PA 6) and -1 700 V/cm (ABS vs. PA 6).

Static charging can be reduced by incorporation of conducting fillers into the plastics, such as carbon black or metal powders (internal antistatics). External antistatics reduce surface resistivities by increasing the polarity of the surface via application of humidity-absorbing additives or by reducing friction through lubricants or coating with PTFE.

9.2. Electrical Conductivity
(→ Polymers, Electrically Conducting) [126, 133–137]

Electrical conductivities in metals are caused by $N/V = 10^{21}$–10^{22} (quasi) free electrons per cubic centimeter with electric charges $e \equiv 1.6 \times 10^{-19}$ C and mobilities $\mu = 10$–10^6 cm^2 V^{-1} s^{-1}. According to

$$\sigma = (N/V) \cdot \mu \cdot e$$

specific electrical conductivities of metals are therefore between ca. 630 000 (Ag) and 10 400 (Hg) S/cm. Graphite exhibits $\sigma = 10^4$ S/cm in plane direction and 1 S/cm perpendicular to it. Semiconductors possess carrier mobilities similar to metals but at far lower carrier concentrations.

Charge transfer occurs fairly easily in metals and semimetals because atoms are tightly packed. Chain atoms of polymer molecules are also close in the chain direction because of covalent bonds. Only van der Waals and/or dipole forces act between chains, however, resulting in large intermolecular atomic distances and very difficult charge transfers. All electrons are furthermore localized in covalent polymer chains. Conventional polymers are thus insulators.

Polymers with conjugated chains [e.g., *trans*-polyacetylene $\sim(CH=CH)_n\sim$] are, for these reasons, merely semiconductors and insulators even if the chains are planar. These low conductivities can be increased substantially, however, if the polymers are doped with substances such as I_2, AsF_5, BF_3, etc.: Doping with AsF_5 increases the specific electrical conductivity of *trans*-polyacetylene from 10^{-9} to 1 200 S/cm and of poly(p-phenylene) from 10^{-15} to 500 S/cm. Such doped polymers can be processed like thermoplastics to any shape desired, which together with their light weight makes them attractive for many applications.

The action of these dopants is quite different from those of small amounts of dopants in inorganic semiconductors. The doping of inorganic semiconductors such as GaP, InSb, or Ge generates quasi-free electrons (n-carriers) or defect electrons (p-carriers), whereas the doping of suitable organic polymers leads to oxidation (p-doping) or reduction (n-doping) reactions. Sizable effects in organics are thus achieved only if large amounts of dopants are used, often up to 1 : 1 molar ratios of dopants to repeating units.

Doped polymers exhibit neither Curie paramagnetism (localized charge carriers) nor Pauli paramagnetism (electrons delocalized over the entire system). Thus the electrical conductivity of such systems is assumed to be due to *solitons* or *polarons*.

Double and single bonds alternate in *trans*-polyacetylene. Since these bonds are exchangeable, two low-energy states A and B with equal energy must exist. A soliton is a kind of topological kink that separates the A state from the B state with opposite bond alternation. Polarons are similar kinks between aromatic and quinoid structures:

Soliton

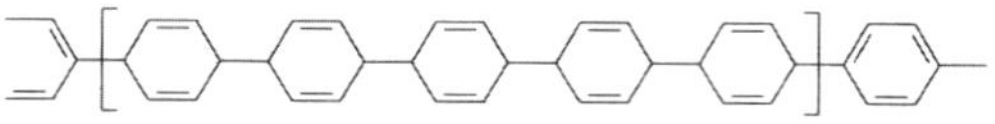

Polaron

The kink is small (about 14 chain atoms in *trans*-polyacetylene) and thus very mobile. Two types of solitons and polarons exist. Neutral solitons and polarons possess radicals that are produced as defects during isomerization of the polymers. Charged solitons and polarons are created by doping, either as carbonium ions or as carbanions.

Solitons can move only along the chain; there is no tunnel effect between chains. Because of this anisotropic behavior, doped polymers (and conducting low molar mass organic molecules) are called low-dimensional conductors, synthetic metals, or organic metals.

The most important condition for the existence of electrical conductivity in organic polymers seems to be the ability to form overlapping orbitals. The planar structure of *trans*-polyacetylene promotes the overlapping of its π and p orbitals. In poly(*p*-phenylene sulfide) (PPS) $\sim[S-(p-C_6H_4)]_n\sim$, p and d orbitals of sulfur atoms probably overlap with the π systems of phenylene groups; PPS–AsF$_5$ shows an electrical conductivity of 10 S/cm, although the chain is not planar and the phenylene residues are arranged at angles of 45 °C to the planar zigzag chain of the sulfur atoms.

Electrically conducting polymers are presently used in small batteries, for example, a complex of poly(2-vinylpyridine)/I$_2$ as cathode in Li/I$_2$ batteries for pacemakers ($\sigma = 10^{-3}$ S/cm). Other possible applications of electrically conducting polymers are in solar cells, electrolysis membranes, microwave shielding, and integrated circuits.

One prime example of an electrically conducting polymer is poly-3,4-ethylendioxythiophene (PEDOT), which consists of 2,5-coupled 3,4-ethylendioxythiophene repeat units and is made by oxidative coupling. In its oxidized form, PEDOT is an almost transparent, light-blue conductor, and, depending on the application, several counterions can be employed. The most common form of PEDOT is the commercially available blend with polystyrene sulfonate (PSS) in aqueous suspension (PEDOT:PSS, Baytron P). High conductivities up to 1 000 S/cm allow to apply this material as thin film for electrode coatings in solar cells or sensor devices to replace the traditionally used indium tin oxide.

Although high electrical conductivities of conjugated polymers obtained by chemical doping are important for e.g., electrode applications, conjugated absorber materials in organic photovoltaics are required to exhibit semiconducting behavior. An appropriate design of conjugated polymers for organic photovoltaics is of critical importance, as several parameters, such as absorption, energy levels, charge carrier mobility, molar mass, or solubility have to be considered simultaneously and are mutually dependent. Consequently, numerous new conjugated polymer structures have been prepared by transition metal-catalyzed cross coupling since the 1990s, whereby the most successful materials have moved from the class of polyphenylene–vinylenes (PPV) and poly(3-alkylthiophene)s (P3ATs) to donor–acceptor polymers with alternating electron-rich and electron-deficient repeat units. The photovoltaic performance of the best donor–acceptor copolymers are now approaching 10%, which is considered an important threshold value to commercialization. Two successful copolymers DPPTT and PTB7 are shown below:

9.3. Photoconductivity

Light generates radical ions and thus photoconductivities in certain systems. This effect is used in xerography to generate pictures of objects (e.g., copies of documents). The early photoconducting material As_2Se_3 has since been replaced by poly(*N*-vinylcarbazole), which absorbs UV light and forms an exciton, which is ionized by an electric field. The polymer is nonconducting in visible light; upon sensibilization by certain electron donors, charge-transfer complexes are formed, however. Another photoconducting system consists of polycarbonate A and triphenylamine.

10. Optical Properties [116]

Many optical properties of plastics depend on their *refractive index n*, for example, reflection, gloss, transparency, and hiding power. Refractive indices are in turn determined by the polarizabilities Q according to the Lorenz–Lorentz relationship

$$(n^2 - 1)/(n^2 - 2) = (4/3)\pi Q = (4/3)\pi (N/V) \cdot \alpha$$

where N/V is the number concentration of molecules with polarization α. The polarization is a function of the dipole moments of all groups in a molecule (i.e., the mobility and the number of electrons per molecule). Contributions to the refractive index are thus much higher for carbon atoms than for hydrogen atoms. Because the contributions of the latter can be neglected and because carbon atoms dominate the structures of polymers, all polymers possess approximately the same refractive index of 1.5 (e.g., PMMA 1.492, PP-it 1.53, PS 1.59). Deviations from this rule exist for strongly polarizable polymers (PTFE 1.37), polymers with bulky conjugated substituents (PVK 1.69), or polymers with a high content of noncarbon atoms (PDMS 1.40). According to the molecular structure of all known polymers, their refractive indices should be between ca. 1.33 and 1.73.

Molecules are more tightly packed in crystalline polymers than in amorphous ones. Refractive indices thus increase with increasing crystallinity. Because crystalline polymers are always anisotropic, different polarizabilities and refractive indices are exhibited in the chain direction and perpendicular to it.

A part of the light falling on a homogeneous, transparent body is reflected. *Reflectivity* is defined as the ratio of the intensities of reflected and incident light $R = I_r/I_o$, which according to Fresnel's law depends on the angles of incidence α and refraction β:

$$R = \frac{I_r}{I_0} = \frac{1}{2}\left[\frac{\sin^2(\alpha-\beta)}{\sin^2(\alpha+\beta)} + \frac{\tan^2(\alpha-\beta)}{\tan^2(\alpha+\beta)}\right] \quad (30)$$

The reflectivity is low at small incident angles ($R = 0.040$ for $n = 1.5$ at $10°$) and rises sharply at higher ones ($R = 0.388$ for $n = 1.5$ at $80°$).

Gloss is the ratio of the reflection of the specimen to that of a standard, for example, a body with $n_D = 1.567$ in the paint industry. The maximum theoretical gloss R_o is given by Equation (30). It increases with increasing refractive index of the specimen and with increasing angle of incidence (Fig. 64). The maximum theoretical gloss is almost never achieved because of surface roughness.

The maximum *transparencies* $\tau_i = 1 - R_o$ can be calculated from $R_o = (n-1)^2/(n+1)^2$ by Fresnel's law for $\alpha \to 0$ and $\beta \to 0$. They can have values between 98.0% ($n = 1.33$) and 92.8% ($n = 1.73$) for polymers, (i.e., between 2 and 7.2% of the incident light is maximally reflected at the polymer–air interface). These ideal transparencies are rarely achieved for polymers because a small portion of the light is always absorbed and/or scattered. Poly(methyl methacrylate), one of the most transparent plastics, never exceeds a transparency of ca. 93% between wavelengths of 430 and 1 100 nm (theory 96.1%). Transparencies decrease above wavelengths of ca. 1 150 nm and (rapidly) below 380 nm. All polymers except halogenated polyethylenes absorb infrared radiation.

A distinction is usually made between transparent and translucent polymers. *Transparent* polymers have transparencies > 90%; they look clear even at greater thicknesses. *Translucent* polymers ($\tau_i < 0.90$) appear clear only as thin films. They are also called contact clear because films look clear in contact with packaged goods but turbid when viewed alone.

An additional loss of clarity is caused by light scattering. Electromagnetic waves lose part of their energy by scattering in inhomogeneous systems. The loss of contrast by forward scattering is called *haze*. The combined loss by forward and backward scattering makes a specimen milky. A body appears opaque if local fluctuations of refractive indices or orientations of anisotropic volume elements are present. The different volume elements must also be larger than the wavelength of incident light. The clarity of a material can thus be increased considerably if the size of the different volume elements (e.g., microdomains) is decreased. Diminished differences in refractive indices improve the clarity to only a small extent. Lamellar structures are optically less heterogeneous than spherulites of approximately the same diameter. Under certain conditions, clear polyethylene films can be produced by quenching and orientation, although crystalline lamellae with dimensions greater than the wavelength of light are present in these films.

11. Polymer Composites (→ Composite Materials) [138]

11.1. Introduction

11.1.1. Overview

The term composite has various meanings. In general, it denotes a complex material in which

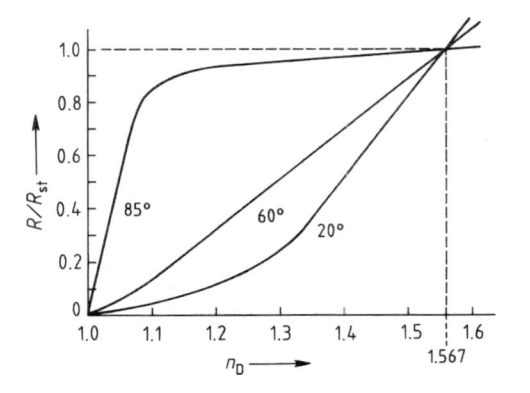

Figure 64. Maximum theoretical gloss as a function of the refractive index of specimen for various angles of incidence α and a standard with $n_D = 1.567$
Reprinted with permission by Hüthig and Wepf Publ., Basel [5]

two or more distinct substances combine to produce some properties not present in any individual component. In biomaterials, it means any two joined materials (e.g., polymer-coated titanium parts). In engineering, composites are more narrowly defined as physical admixtures of various materials that are present as distinct phases. The term "phase" is used here in the descriptive sense, not in the thermodynamic one. The engineering term "composite" thus includes only heterogeneous composites. Heterocomposites can be further subdivided into nanocomposites (dispersed phase has at least one dimension smaller than 100 nm), microcomposites (dispersed phase in the micrometer to millimeter range), and macrocomposites. Polymer composites are defined as composites in which at least one component is of a polymeric nature. In accordance with the general definition of a composite and in contrast to the engineering use of the word, the term "polymer composite" is often used to cover not only heterogeneous mixtures of a polymer and another material (minerals, fibers, other plastics, elastomers, etc.), but also homogeneous (single-phase) materials of two polymers (homogeneous polymer blends). Such polymer composites are a subgroup of "multicomponent polymer systems", which also include copolymers (two or more types of monomeric units chemically bound together).

Polymer composites occur naturally (wood, bamboo, lobster shells, bone, muscle tissue, etc.) or are manufactured synthetically. Synthetic polymer composites can also be subdivided into single-phase (homogeneous) and multiphase (heterogeneous) composites. Further subdivisions may be according to the second component's size (microcomposites, macrocomposites), chemical nature (air, low molar mass plasticizer, elastomer, plastic, mineral), geometry (particulate, fiber, platelet, fabric), and macroconformation (if polymeric: coil, rod), etc. (see Fig. 65). Most of these subclasses have different technical names depending on whether the glass transition temperature of the continuous phase is above (plastics) or below (rubbers) the use temperature. The same name is often given to various subtypes.

A mixture of two (or more) different polymers is often called a *blend* when both polymers are either above (rubber) or below (plastics) their glass transition temperatures. Blends are distinguished from composites of two polymers such as polymer–fiber-reinforced or rubber-toughened plastics. The term blend is sometimes restricted to mean only "incompatible polymer mixtures", whereas compatible polymer mixtures are called polymer alloys. The latter term is, however, occasionally more narrowly used for mixtures of two crystallizable polymers.

The term *miscibility* refers to admixtures on the molecular level (nanometer range), that is true thermodynamic solubility. It is sometimes used more loosely for any polymer mixture that exhibits only one glass transition temperature. *Compatibility* denotes the ability of admixtures to be blended to heterogeneous microcomposites that do not separate into macroscopic phases.

11.1.2. Mixture Rules

The composition dependence of properties Q of composites can often be described by mixture rules. The *generalized simple mixing law* relates Q to the fractions $f_A = 1 - f_B$ of components A and B with properties Q_A and Q_B:

$$Q^n = Q_A^n \cdot f_A + Q_B^n \cdot f_B \qquad (31)$$

The generalized simple mixing law is usually applicable to single-phase composites without specific interactions of its components and to two-phase systems with regularly dispersed discrete phases of "infinite" dimensions. It includes three special cases that are known by different names in different fields (Table 26) and are illustrated in Figure 66.

The *rule of mixtures* ($n = 1$) applies, for example, to various moduli if both components behave elastically. An example is Young's modulus parallel to the fiber direction for epoxide resins reinforced with long glass fibers (see below). The same composites follow the *inverse rule of mixtures* ($n = -1$) if the modulus is measured perpendicular to the fiber direction and no slippage occurs. The *logarithmic mixture rule*

$$\log Q = f_A \cdot \log Q_A + f_B \cdot \log Q_B \qquad (32)$$

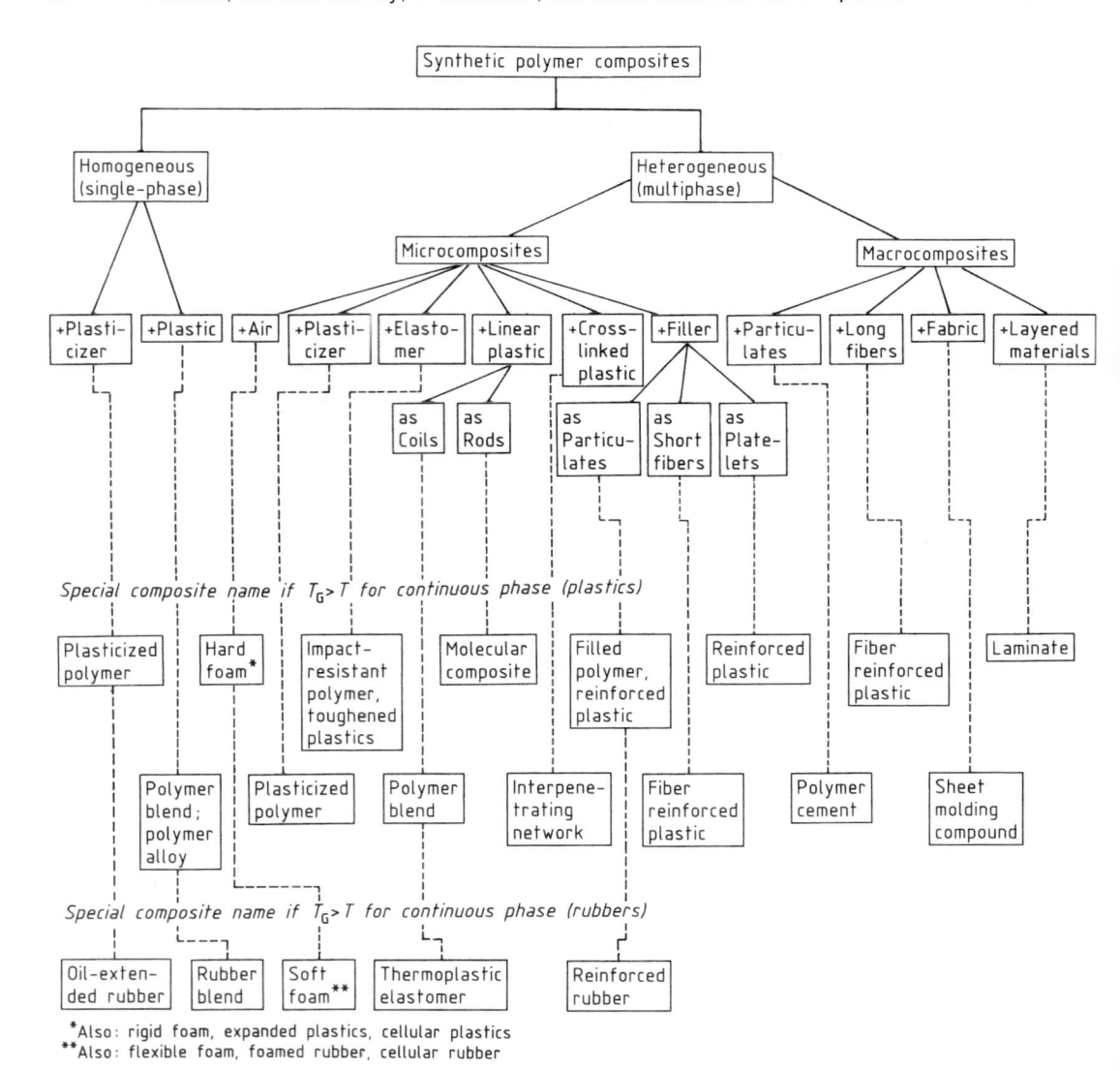

Figure 65. Classification of polymer composites of a base polymer (matrix polymer) with other materials (plasticizer, plastic, air, long fiber, layered materials). Some of these materials have their molecules in different shapes (coils, rods) or have various outer shapes (particulates, short fibers, platelets). Adopted from [1]

Table 26. Names of simple mixing rules

Field	Property exponent in generalized mixing law		
	$n = 1$	$n = 0$	$n = -1$
Mathematics	arithmetic mean	harmonic mean	geometric mean
Chemical engineering	rule of mixtures	logarithmic mixture rule	inverse rule of mixtures
Mechanical engineering	Voigt model		Reuss model
Electrical engineering	parallel		series
Materials science	upper bound		lower bound

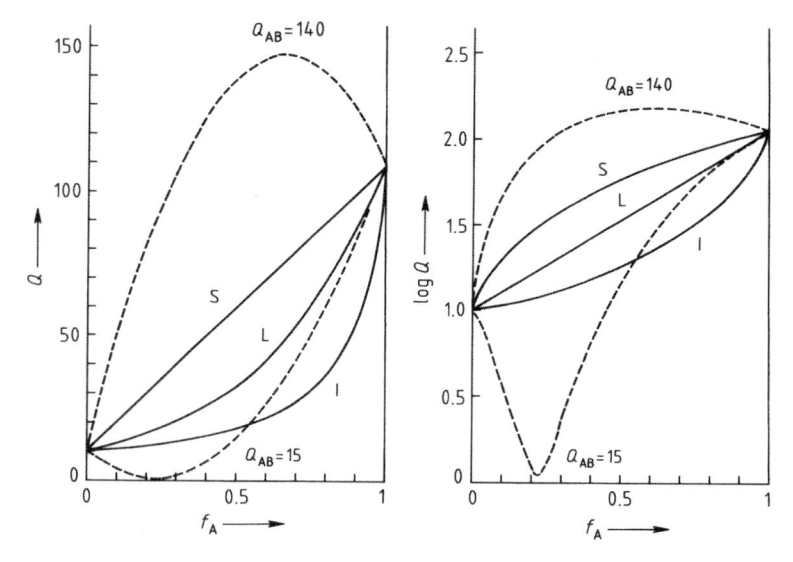

Figure 66. Mixing laws for composites with $Q_A = 110$ and $Q_B = 10$: left: $Q = f(f_A)$; right: $\log Q = f(f_A)$
Simple mixing laws with interaction parameter
$Q_{AB} = 0$ (S = simple mixing law; L = logarithmic mixing law; I = inverse mixing law
- - - Composites with interaction parameters $Q_{AB} = 140$ and 15, as indicated

is obtained from Equation (31) for $n = 0$ after some mathematical manipulation. The only known case concerns the permeation of oxygen through ABS resins. The proper fraction f_i in Equations (31) and (32) is always the volume fraction. If mass fractions are used instead, deviations from the curves shown in Figure 66 occur, which may be confused with either deviations from additivity or specific interactions.

True deviations from the upper bound S and lower bound I (Fig. 66) may be caused by irregularities in phase distributions (discrete phases) or phase continuities (continuous phases), phase alignments (anisotropic phases), bonding (adhesion), or nonuniform stresses. The resulting functions always lie between the upper and the lower bound if specific interactions are absent. Many empirical, semiempirical, and theoretical expressions have been proposed for such composites. A widely used function is the Halpin–Tsai equation

$$Q = Q_B \cdot \frac{1 + K_1 K_2 f_A}{1 - K_2 f_A} ; K_2 = \frac{Q_A - Q_B}{Q_A + K_1 Q_B}$$

where K_1 is an adjustable constant. The Halpin–Tsai equation gives the lower bound for $K_1 \to 0$ and the upper bound for $K_1 \to \infty$. Particulate fillers often have $K_1 \approx 0$, and fiber-reinforced plastics $K_1 \approx 2\,L/d$, where $2\,L/d$ is the aspect ratio of the fiber.

Specific interactions between components can be considered by an additional interaction term. This term is differently defined, for example:

$$Q = Q_A f_A + Q_B f_B + K_i f_A f_B;$$
$$K_i = 2(2Q_{AB} - Q_A - Q_B)$$

where the property term Q_{AB} is taken at $f_A = f_B = 1/2$. The value Q_{AB} may be an interaction parameter between components A and B in homogeneous mixtures or the property of the interphase between A and B phases in heterogeneous mixtures. The upper bound for absent interactions or interphases is retrieved for the condition $Q_{AB} = (Q_A + Q_B)/2$. The functions for $Q_{AB} \neq (Q_A + Q_B)/2$ may lie fully or partially outside the range between upper and lower bounds (Fig. 66); that is synergistic effects (above the upper bound) or antagonistic effects (below the lower bound) may be present. They may also show maxima and minima, depending on the relative magnitudes of Q_{AB}, Q_A, and Q_B.

Maxima or minima that exceed the highest property value (either Q_A or Q_B) are sometimes considered the true criteria for synergistic and antagonistic effects, respectively.

11.2. Filled Polymers

11.2.1. Microcomposites

The addition of fillers (low thermal expansion) to polymers (high thermal expansion) reduces the proportion of the latter and thus the shrinkage of plastics on processing (Table 27). The melt flow is also lower. Heat deflection temperatures (HDT) of amorphous polymers are either the same as those of unfilled plastics or no more than ca. 10 K higher (Fig. 67), group a), probably because of the increased viscosity of the filled plastic. The HDTs of amorphous thermoplastics are almost identical with glass transition temperatures.

The HDTs of crystalline polymers depend on a number of factors. Unfilled or particulate-filled crystalline polymers show HDTs that are often considerably lower than their melting temperature. Upon fiber-filling, HDTs of crystalline polymers increase strongly; they may reach values near the melting temperatures. This behavior is thought to be due to additional polymer crystallization near, or polymer adsorption on, the fiber surface, epitaxial

Table 27. Effect of fillers on some properties of amorphous (A) and crystalline (C) thermoplastics [(↑) weak increase, ↑ increase, ↑↑ strong increase, (↓) weak decrease, ↓ decrease, ↓↓ strong decrease]

Property	Extenders		Reinforcing agents	
	A	C	A	C
Shrinkage	↓	↓	↓	↓
Heat deflection temperature		(↑↓)		↑
Melt flow	↓	↓	↓	↓
Young's modulus	(↑)	(↑)	↑	↑
Flexural modulus	↑	↑	↑	↑↑
Fracture strength	↓	↓	↑	↑
Brittleness	↑	↑	↑↓*	↑↓*

*Tough plastics become more brittle, and brittle plastics become tough.

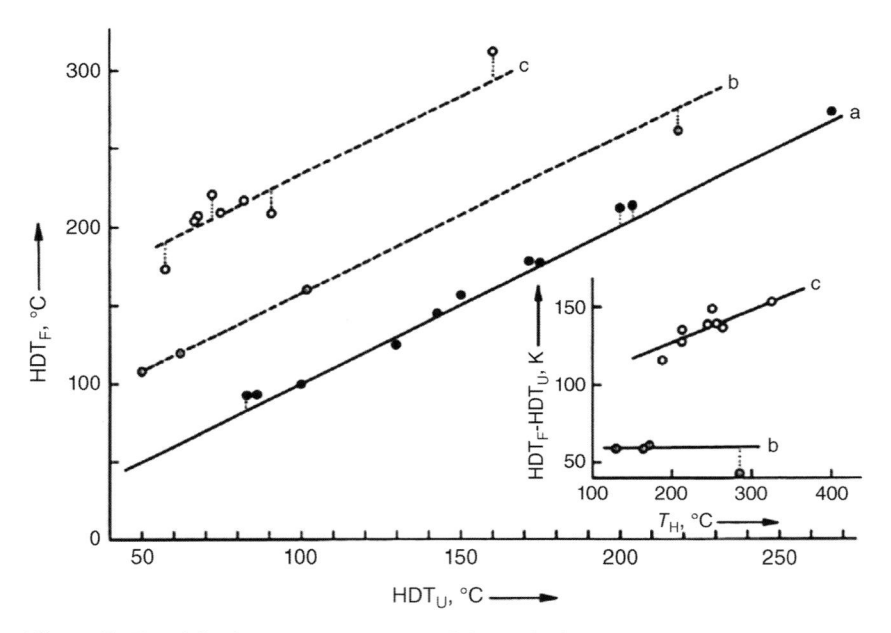

Figure 67. Heat deflection temperatures HDT_F of thermoplastics filled with 30 wt % short glass fibers as a function of heat deflection temperatures HDT_u of the corresponding unfilled polymers, both at 1.85 N/mm² (ISO/R 75 A)
(●) Amorphous polymers; (⊙, ○) crystalline polymers; solid line: $HDT_F = HDT_u$; broken lines: empirical
a) Amorphous polymers; b) Apolar crystalline polymers; c) Polar crystalline polymers
Insert: Difference $HDT_F - HDT_u$ as function of melting temperatures T_M of unfilled crystalline polymers

layers have been found to extend from the filler up to 150 nm into the matrix. The surface layers overlap; the resulting physical network reduces creep.

Two groups of HDTs of glass-fiber-filled crystalline polymers can be distinguished (Fig. 67). The HDTs of group b are about 50–60 K above those of their unfilled counterparts. This group comprises apolar polymers such as polyethylene, polypropylene, and poly(phenylene sulfide); polyoxymethylene also belongs to group b because its oxygen groups are buried deep inside compact helical structures, which are thus apolar on their surfaces. The difference in HDTs of filled and unfilled group b polymers does not vary with melting temperature. Group c includes polar crystalline polymers (polyesters, polyamides, etc.). For this group, the difference in HDTs of filled and unfilled polymers increases with the melting temperature of the unfilled polymer.

Young's moduli of thermoplastics increase with filler content (Fig. 68). The increase is small for spheroidal fillers (glass spheres, chalk), stronger for platelets (kaolin, talc), and strongest for fibers (glass). Short fibers are usually added in amounts of ca. 30 wt %, which corresponds to ca. 17 vol % for glass fibers ($\rho = 2.55$ g/cm^3) in an average polymer ($\rho \approx 1.2$ g/cm^3). Higher fiber content often leads to packing problems since the maximum volume fraction of three-dimensional randomly packed rods with $L/d = 20$ is already ca. 25% (glass fibers typically have diameters $d = 10$–20 µm and lengths of ca. 0.2 mm after injection molding). Particulate fillers can be added in far greater amounts without sacrificing properties (see Table 14) because maximum packing densities of spheres are, for example, 0.745 for the hexagonally most dense and 0.637 for the randomly most dense packing.

Young's moduli E of filled materials can be described by a modified mixing law

$$E = E_M \phi_M + f E_F \phi_F = E_M + (f E_F - E_M) \cdot \phi_F \quad (33)$$

where f is an adjustable parameter. The data in Figure 68 can be described with $f = 0.39 \pm 0.03$ (chalk), 0.66 ± 0.04 (kaolin), and 1.61 (talc). The effect of 30 wt % of short glass fiber was tested in 14 thermoplastics, and f was found to be 0.57 ± 0.04. These numbers increase with the specific area of the fillers, which are given as 0.3–2.2 (ground limestone), 0.5–1 (glass fibers), 6–22 (kaolin), and 6–17 m^2/g (talc) [139]. Composites obviously increase in stiffness as the contact between filler and polymer segments increases. Good adhesion is not necessary, since E is measured for strains approaching zero.

The adjustable parameter f of Equation (33) adopts a physical meaning for short fiber reinforced plastics. In such composites, forces are transferred to the hard fiber from the soft matrix if (i) the Young's modulus of the fiber is greater than the Young's modulus of the matrix ($E_F > E_M$), (ii) the fiber length L_F exceeds a certain critical value $L_{F,\,crit}$, and (iii) the shear strength in the matrix is smaller than the shear strength between fiber and matrix (the fiber would be otherwise pulled out of the matrix unless the fiber breaks). In this case, $f = 1 - [L_{F,crit}/(2\, L_F)]$.

Fracture strengths are, however, measured for maximum strains. The failure mode of filled thermoplastics is matrix dominated. Electron micrographs show that at low fiber contents, planar cracks spread through the matrices and

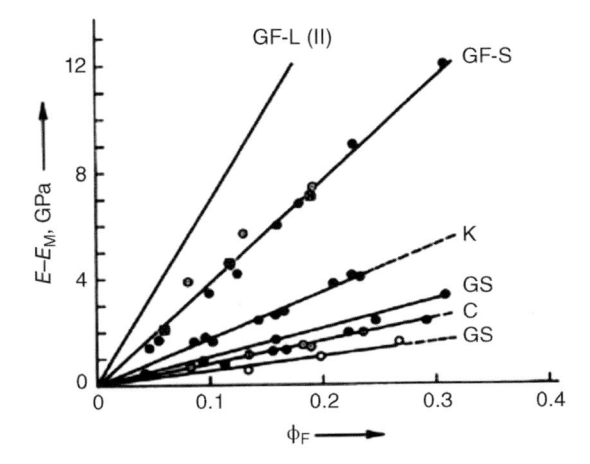

Figure 68. Influence of the volume fraction ϕ_F of fillers on the difference in Young's moduli of filled thermoplastics (E) and their polymer matrices (E_M)
GS = glass spheres; C = chalk; K = kaolin; GF = glass fibers; (S = short; L = long)
○ = polyethylene; (⊙) = polypropylene; ⊗ = polyoxymethylene; ⊕ = poly(butylene terephthalate); ● = polyamide 6
The solid line for GF-L (∥) corresponds to the simple mixing law. From data of [140]

debonded fibers are pulled out of either surface; the matrix must thus adhere well to the filler.

The tensile strength σ_B at break is given by the rule of mixtures ($\sigma_B = \sigma_F \phi_F + \sigma'_{M'} \phi_M$) where $\sigma'_{M'}$ is the tensile stress of the matrix upon application of a load σ_B to the composite. This relationship applies if (i) the tensile strength σ_B is greater than the contribution $\sigma_M \phi_M$ of the matrix component to the total tensile strength σ_B (that is $\sigma_B > \sigma_M \phi_M$), and (ii) the volume fraction ϕ_F of the fiber exceeds a certain minimum value ($\phi_F > \phi_{F, min}$).

Particulate fillers usually strongly reduce the fracture strengths of composites of amorphous thermoplastics, whereas those of crystalline polymers are reduced less or not at all. *Short fibers* generally increase the fracture strengths by factors of 1.5–2 for amorphous and crystalline polymers alike. Rigid PVC is an exception because its fracture strength is reduced on fiber-filling. The moduli and strengths can be further enhanced if coupling agents are used, which improves the bonding between fiber and matrix (e.g., silanes for glass-fiber-filled polymers).

11.2.2. Molecular Composites

The problem of insufficient bonding between fiber and matrix can be dramatically reduced if so-called molecular composites are used. These materials are dispersions of semiflexible (rodlike) polymer molecules with mesogenic units in chemically similar polymer matrices, e.g., poly(*p*-phenylenebenzobisthiazoles) in polybenzimides (see Fig. 59). The mesogens are not truly molecularly dissolved; rather, they form microdomains in the matrix. The segment axes are randomly distributed in these domains if the molecular composites are formed from polymer solutions with total polymer concentrations c below the critical isotrope phase–mesophase concentration $c_{i/n}$. Microdomains show mesophase behavior if films or fiber are formed at $c > c_i/_n$. Such microcomposites have excellent moduli and fracture strengths (Fig. 59).

11.2.3. Macrocomposites

Polymer macrocomposites are heterogeneous composites of polymers and "macro"-sized additives. Examples are polymer concrete, composition board, long fiber-reinforced polymers, and plywood.

Polymer concrete is composed of polymer [usually polyester or poly(methyl methacrylate)], aggregate (sand, ground limestone, etc.), and other additives (dyes, etc.); it does not contain cement. *Polymer–concrete hybrids* include polymer–cement hybrids (cement is partially replaced by polymers) and polymer-impregnated concrete (conventional concrete saturated with a monomer that is polymerized in place). Polymer concretes have higher moduli and higher tensile and compressive strengths than ordinary concretes. They are not vulnerable to damage by freeze-thaw cycles.

Composition boards are panel products. They include particleboard, fiberboard, hard board from flakes or shearings, etc. The fillers are mainly wood products; the resins, urea–formaldehyde (interior applications), and polyurethanes or phenolic resins (exterior).

Long-fiber-filled polymer macrocomposites range in structure from polymers filled with "long" fibers 6–12 mm in length to fiber strands and mats filled with polymers. The filler content can often reach 65 wt%. Upon application of tensile stress, local stress concentrations are transferred by shear forces onto the plastic–fiber interface and distributed over the much greater fiber surface area. The fibers must therefore bond well to the plastic (use of coupling agents) and must have a certain length to avoid slipping.

Young's moduli of long glass-fiber-reinforced epoxide resins follow the simple mixing law if the tensile stress is in the fiber direction (Fig. 69). The inverse mixing law is not observed as well for stresses perpendicular to the fiber direction since some fibers bend under stress.

The simple mixing law corresponds to $f = 1$ in Equation (33). For long fibers, f has the meaning of an orientation factor. It becomes 1/6 for fibers distributed three-dimensionally at random, 1/2 for 90° cross-plies measured in either fiber direction, and 3/8 for a uniform, planar distribution of fibers.

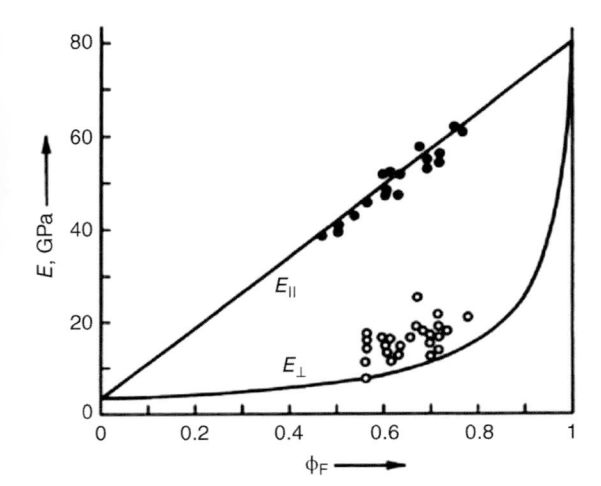

Figure 69. Young's moduli of epoxy–glass-fiber composites parallel (∥) and perpendicular (⊥) to the fiber direction

This group of macrocomposites also includes prepregs (fiber mats soaked with unsaturated polyester resins) and the composites generated by filament winding (wound filaments saturated with epoxies or unsaturated polyesters), both of which are subsequently polymerized.

11.2.4. Nanocomposites [108, 141, 142]

Polymeric nanocomposites are defined as multiphase materials in which at least one dimension of the dispersed or continuous phase is smaller than 100 nm. This includes organic–inorganic hybrid materials as well as "all-polymer" nanocomposites in which matrix and reinforcing phase consist of the same polymer. In contrast to micro- and macrocomposites, most polymers are allocated due to the presence of nanostructures at interfaces that affect the chain mobility and entanglement (see Fig. 20). Hence, the properties of interfacial polymers substantially differ from those of bulk polymers, especially regarding their thermal, mechanical, electrical, optical, and biological properties. For instance, polymeric nanocomposites can be optically transparent, exhibit improved stiffness, high surface gloss, high scratch resistance, and even electrical conductivity. Three different strategies lead to polymeric nanocomposites: (i) Dispersion of functional nanofillers in a polymer matrix, (ii) formation of organic–inorganic hybrid polymers, and (iii) oriented polymer crystallization.

Nanofiller Dispersion [108, 141–144]. Since the 1990s, significant progress in nanotechnology has made a large variety of nanometer-scaled inorganic and organic fillers as well as nanofibers commercially available for nanocomposite preparation (see Section 4.2.4) and Fig. 21). When the size of inorganic fillers is reduced, a transition from the inorganic solids to the molecules takes place. Molecular nanofillers include nanosilica, such as POSS; 2D carbon polymers, such as graphene; and multifunctional hyperbranched polymers with nanometer dimensions.

Nanometer-scaled silica spheres, prepared by a flame process, are known as pyrogenic silica of around 5 nm diameter, which is widely applied as thixotropic agents in coatings, adhesives, and sealants. Due to the shear-sensitive formation of hydrogen bridges between silanol groups at the silica particle surface, silica percolation networks are destroyed upon shearing, thus accounting for shear thinning. This thixotropic effect, caused by less than 0.5 wt% silica nanoparticles, prevents sedimentation of fillers in formulated thermoset systems. Moreover, functionalized silica fillers play an important role in improving scratch resistance in coatings. Nano silica together with disulfide-functional silane coupling agents are used for the preparation of silica–rubber nanocomposites.

For many years, carbon black nanoparticles have been commercially employed to improve electrical conductivity of plastics and coatings. This is important for antistatics and electromagnetic interference shielding. Compared to micron-sized carbon black, the percolation threshold of nanocarbon black is much lower, which drastically reduces the amount of carbon black required for achieving electrical conductivity. Highly anisotropic carbon nanofillers, such as carbon nanotubes and graphene, further reduce the percolation threshold. Moreover, both carbon nanotubes and ultrathin but micron-sized graphene enable combining electrical conductivity with improved matrix reinforcement. Alignment of graphene in a polymer matrix improves barrier resistance to permeation of gas and liquids ("labyrinth effect") together with improved electrical conductivity and matrix reinforcement. Another approach to improved barrier resistance combined with halogen-free flame retardency and matrix reinforcement employs organophilic layered silicates. When layered silicates are made organophilic by cation exchange of sodium cations for alkyl ammonium cations (see Fig. 21), they can be intercalated and exfoliated during polymer melt compounding. The family of organophilic nanofillers is rapidly expanding.

Inorganic—Organic Hybrids [108, 145, 146]. Polymer hybrids combine the properties typical for ceramics with facile processing typical for polymers. Sol–gel chemistry is employed to equip inorganic particles and resins with reactive groups that enable efficient covalent coupling with various polymers. For example, functionalized alkoxysilanes are cocondensed with tetraethoxysilane and other metal alkoxides, such as zirconates and aluminates to produce reactive liquid oligomers or reactive nanoparticles. They are mixed and cured together with thermoset resins to form nanostructured polymer–ceramic hybrids. Sol–gel coatings are well known for their extraordinary scratch resistance. Moreover, sol–gel chemistry is used to functionalize inorganic nanoparticles by incorporating initiator and transfer agents, which enables covalent attachment of polymer chains by "grafting-from" and "grafting-to" processes. Polymer grafting of nanoparticles greatly facilitates nanoparticle dispersion and prevents undesirable agglomeration.

Self-Reinforcing Plastics and All-Polymer Nanocomposites [147, 148]. Going well beyond the scope of liquid crystalline polymers and conventional nanocomposites, the self-reinforcement of polymers does not require mesogens as comonomers and eliminates the special handling and safety precautions related to the use of potentially hazardous alien nanoparticles and nanofibers. Moreover, such all-polymer nanocomposites, in which both the matrix and the reinforcing phase consist of the same polymer, have much lower mass and are much easier to recycle than conventional polymer composites. This is highly advantageous for applications in sustainable lightweight engineering. Key concept for achieving self-reinforcement is the oriented polymer crystallization during processing. Whereas one-dimensional crystallization affords in situ aligned polymer nanofibers, the two-dimensional crystallization produces nanosheets and all-polymer multilayer composites. Because this nanostructure formation of polymers does not require nanofibers and nanoparticles, nanometer-scaled particles will not be emitted. In contrast to many other nanocomposites, all-polymer nanocomposites are environmentally benign. In one strategy, oriented films are laminated or coextruded to produce multilayer all-polymer nanocomposites. In another strategy, multilayer all-polymer nanocomposites are produced by pressure- and flow-induced crystallization near the melting temperature of thermoplastics. Extrusion of polyethylene containing ultrahigh molecular mass polyethylene (UHMWPE) close to the PE melting temperature can afford shish-kebab-like aligned UHMWPE nanofibers. However, because UHMWPE has very high viscosity, the incorporation of UHMWPE in PE reactor blends has been restricted to a few percent. Exploiting advanced multisite ethylene polymerization catalysis, it became possible to incorporate much larger amounts of UHMWPE into reactor blends without impairing processing. Because no entanglement with HDPE takes place, the nanometer-scaled UHMWPE forms a low viscosity phase separated melt and elongates during processing. Because the lower HDPE fraction crystallizes much faster, it crystallizes onto UHMWPE nanofibers to produce shish-kebab-like UHMWPE fibers in conventional melt processing without narrowing the processing window. The development of self-reinforcing plastics and all-polymer nanocomposites, which do not require the addition of (nano)fillers and fibers, holds great promise for the sustainable development of lightweight engineering plastics.

11.3. Homogeneous Blends [149, 150]

Mixtures of polymers with other polymers are called polymer blends (→ Polymer Blends). Such blends may be composed of two thermoplastics (plastic blends), two elastomers (rubber blends), a plastic filled with an elastomer as the dispersed phase (rubber-modified plastics), an elastomer with a plastic as the dispersed phase (polymer-filled elastomer), or a plastic filled with a polymer melt or a low molar mass liquid (plasticized polymers). Whether or not these blends are truly miscible on a molecular level depends on the thermodynamics of the systems.

11.3.1. Thermodynamics

If two components 1 and 2 are mixed, entropy S, enthalpy H, and volume V change. In many cases, volume changes can be neglected and the Gibbs energy of mixing is given by the changes in enthalpy and entropy: $\Delta G_{mix} = \Delta H_{mix} = T \cdot \Delta S_{mix}$. These changes can be calculated by the Flory–Huggins theory, which assumes that the mixture can be modeled by a three-dimensional lattice on which the monomeric units (or solvent molecules) can be placed. Each monomeric unit experiences the same force field; the theory is thus a mean-field theory. The assumption of a constant force field is fairly well fulfilled for polymer blends and concentrated polymer solutions but not for dilute solutions.

The entropy and enthalpy of mixing are calculated separately. Entropies are calculated for ideal solutions (zero mixing enthalpy); all environment-dependent entropy changes are thus zero (translational, vibrational, and inner rotational entropies do not change). Units can however be arranged relative to each other in many ways; however, there is a combinatorial entropy that is also called "configurational entropy" in statistical mechanics.

The mixing of two components is modeled as a quasi-chemical reaction. The mixing enthalpy is thus determined by the change in interaction energies and the number and type of nearest neighbors. Enthalpy and entropy terms are added to give the molar Gibbs energy of mixing:

$$(\Delta G_{mix})^m / RT = \chi \cdot \phi_1 \phi_2 + \phi_1 \cdot \ln \phi_1 + (X_1/X_2) \cdot \phi_2 \cdot \ln \phi_2$$

which, on differentiation with respect to the molar amounts of the components, yields the chemical potentials, for example, of component 1

$$\Delta \mu_1 = RT \{ \chi \cdot \phi_2^2 + \ln(1 - \phi_2) + [1 - (X_1/X_2)] \cdot \phi_2^2 \}$$

where χ is the parameter for the interaction between component 1 and component 2 (usually with numerical values between 0 and 2), also called the Flory–Huggins parameter. All three terms in braces must be considered if a polymer 2 is mixed with a solvent 1, for example, a plasticizer ($X_1 \ll X_2$). If both components are polymers however, then $X_1 \approx X_2$ and the entire entropy contribution comes from the relatively small logarithmic term. Thus little combinatorial entropy is gained on mixing two polymers, and the miscibility is determined mainly by the mixing enthalpy, that is, by the term $\chi \phi_2^2$.

Phase separation occurs if the curve $(\Delta G_{mix})^m = f(\phi_2)$ (at $T = $ const.) has a shape that can be touched at two points by a tangent. The compositions at these points are ϕ_2' and ϕ_2'' (with $\phi_2' < \phi_2''$). Two phases exist for the unstable range $\phi_2' < \phi_2 < \phi_2''$, which is separated from the stable ranges $\phi_2 < \phi_2'$ and $\phi_2'' > \phi_2$ by the so-called binodal. The unstable range itself is subdivided by the spinodal into two metastable ranges and one unstable one. Spinodals are characterized by inflection points in the Gibbs energy–composition functions and extremal values in the chemical potential–composition functions. Maximum, minimum, and inflection point become identical at the critical point, which is defined by a zero value of the second derivative of the chemical potential

$$\partial^2 \Delta \mu_1 / \partial \phi_2^2 = RT \cdot [2\chi - (1 - \phi_2)^{-2}] = 0 \qquad (34)$$

This critical point is characterized by a critical volume fraction $\phi_{2,crit} = 1/(1 + X_2^{1/2})$ and a critical interaction parameter $\chi_{crit} \approx (1/2) + (1/X_2)^{1/2}$.

The temperature dependence of χ is approximately

$$\chi \approx A + (B/T) \qquad (35)$$

where A and B are system-dependent constants; B is positive for endothermal mixtures [i.e., χ increases with increasing temperature and a homogeneous solution exists above an "upper critical solution temperature" (UCST)]. Other systems are homogeneous only below a "lower critical solution temperature" (LCST). The demixing is correspondingly either enthalpy (UCST) or entropy (LCST) induced.

The terms UCST and LCST do not indicate the absolute position of demixing temperatures since some systems show LCST > UCST (Fig. 70). Depending on the system, either an hourglass-type diagram or a closed miscibility gap is shown.

11.3.2. Plastification

Plastification of a plastic is the flexibilization of the material either by added plasticizers (external plastification) or by incorporated flexibilizing comonomer units (internal plastification).

The molecular action of plasticizers consists of a flexibilization of chain segments, which can have various causes. *Polar plasticizers* increase the proportion of *gauche* conformations in polar chains, thus decreasing rotational barriers. *Primary plasticizers* can dissolve helical structures and crystalline regions if they are thermodynamically good solvents. Furthermore both *primary and secondary plasticizers* are diluents; their addition increases the distance between polymer segments and thus decreases the activation energy needed for cooperative movement of segments. Solvation does not per se increase chain flexibility, however, because solvent shells act as substituents and increase the rotational barriers.

The molecular effect of an increase in chain flexibility by plasticization is measured as a decrease of glass transition temperatures (Fig. 71). Small molecules are better plasticizers than bigger ones of similar chemical

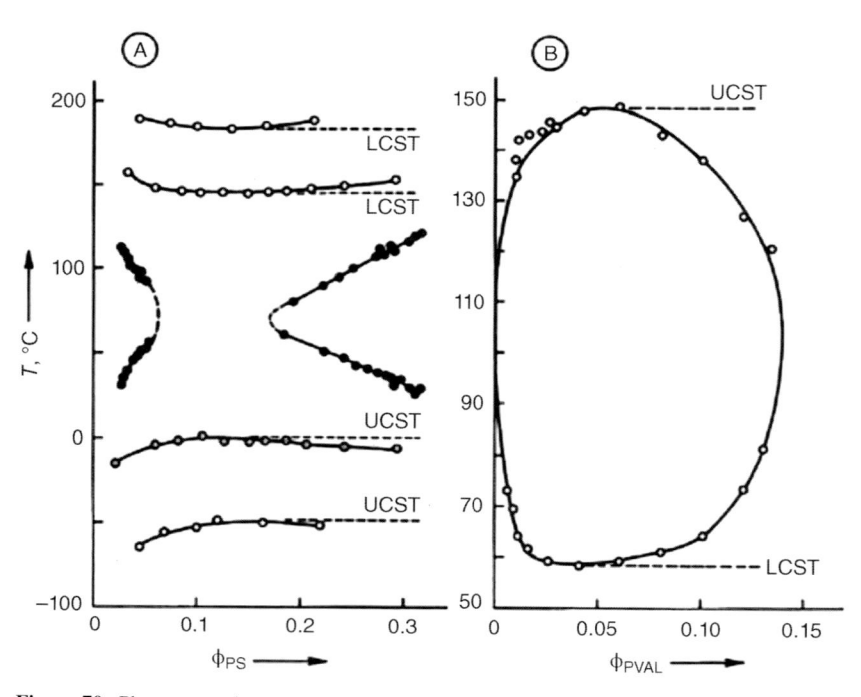

Figure 70. Phase separation temperature as function of polymer–solvent composition
A) Polystyrenes with various molar masses in acetone; B) Poly(vinyl alcohol-*co*-vinyl acetate) (93 : 7) in water
Reprinted with permission by Hüthig and Wepf Publ., Basel [5]

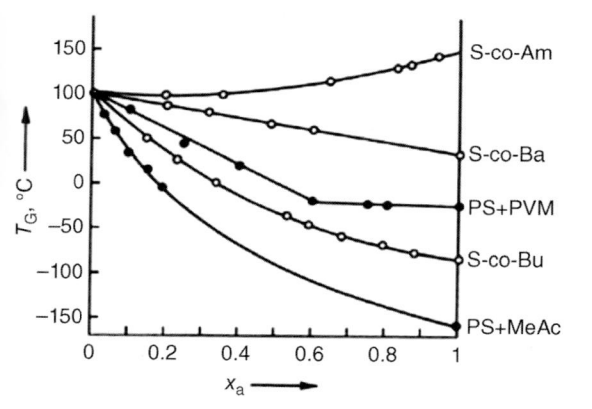

Figure 71. Glass transition temperatures of (●) polystyrene plasticized with methyl acetate (MeAc) or poly(vinyl methyl ether) (PVM) and (○) copolymers of styrene with butadiene (Bu), butyl acrylate (Ba), or acryl amide (Am)

constitution. The plasticizer efficiency decreases with increasing thermodynamic "goodness" of the plasticizer for the polymer (greater interaction, i.e., bigger solvation). Good plasticizers with respect to lowering glass transition temperatures are thus small molecules that act as theta solvents. These plasticizers are however bad for technological applications because they bleed (transport of plasticizer to the surface of the plastic), migrate (transport of plasticizer into another contacting material), or are extractable (e.g., transport into packaged foods). These transport phenomena can be reduced if polymer plasticizers are used. Increasing molar masses lead to decreasing thermodynamic compatibility, however (see below). Technical plasticizers are therefore in most cases a compromise between thermodynamic plasticizer efficiency and kinetically hindered plasticizer transport.

Plastification can also be followed by the changes of certain mechanical properties such as lowering of fracture strength, increase of elongation, and decrease of tensile moduli. Such methods sometimes show "antiplasticizations" at low plasticizer concentrations (Fig. 72) (e.g., increases of moduli). Glass transition temperatures are not increased, however. The effect thus cannot be the result of increased segment mobility. It may be due to a healing of microvoids in amorphous polymers such as polystyrene or to additional crystallization of slightly crystalline polymers, such as poly(vinyl chloride).

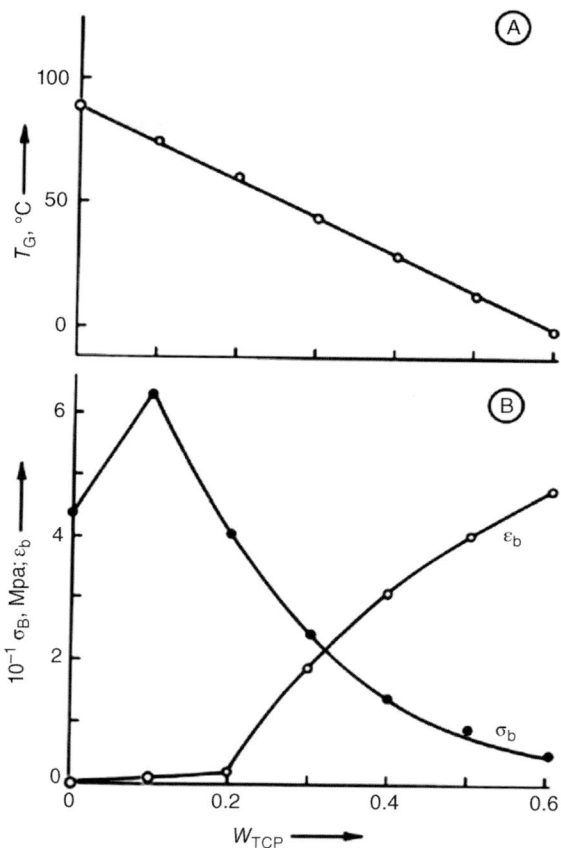

Figure 72. A) Glass transition temperature T_G and B) fracture strength σ_B and elongations ε_B at break of a poly(vinyl chloride) plasticized with tricresyl phosphate (TCP) Reprinted with permission by Hüthig and Wepf Publ., Basel [151]

11.3.3. Rubber Blends [149, 150, 152, 153]

About 75% of all elastomers are employed as blends. Heterogeneous blends of natural (rubber–styrene)–butadiene rubber or (cis-butadiene rubber–styrene)–butadiene rubber are used mainly for tire treads since they reduce abrasion. Homogeneous blends are more rare. They possess only one glass transition temperature which, for 50 : 50 blends, is an average of the glass transition temperatures of the two parent polymers. Heterogeneous blends exhibit two glass transition temperatures, both practically identical to those of the parent polymers. Homogeneous blends are usually formed if the difference in solubility parameters is smaller than 0.7 $cal^{1/2}$ $cm^{-3/2}$ (1.44 $J^{1/2}$ $cm^{-3/2}$).

11.3.4. Plastic Blends

Plastic blends are usually immiscible because the combinatorial entropy of mixing is too small (high molar masses) and the enthalpy changes on mixing are often positive. Miscibility is observed for three types of systems:

1. Chemically similar polymers, for example, polystyrene–poly(o-chlorostyrene), showing both LCST and UCST
2. Systems having specific interactions between different components, for example, polystyrene–poly(vinyl methyl ether); these systems show only LCSTs
3. Systems consisting of oligomers, for example, oligo(ethylene oxide)–oligo(propylene oxide); these systems possess only UCSTs

Homogeneous polymer blends show only single glass transition temperatures; the composition dependence of these T_G's can be described by the simple rule of mixtures. Two-phase systems always show two glass transition temperatures if the phase diameters are greater than ca. 3 nm. Transparency, on the other hand, is no indication of a single-phase blend since opacity can be observed only if the refractive indices of the two phases are sufficiently different and if the phase diameters are greater than ca. one-half the wavelength of incident light. The presence of two phases in clear specimens can often be detected with electron microscopy by special staining techniques.

Modified PPO, a blend of poly(2,6-dimethylphenylene oxide) and polystyrene, is the industrially leading *homogeneous blend*. Blends of poly(vinyl chloride) with poly(methyl methacrylate), poly(ethylene-co-vinyl acetate) or chlorinated polyethylene are also said to be one-phase systems.

11.4. Heterogeneous Blends [149, 150, 152, 153]

11.4.1. Compatible Polymers

Two polymers may be thermodynamically immiscible but still mechanically compatible. These compatible blends are two- or multiphase systems that show multiple glass transition temperatures, which do not change with composition. The components do not separate, however, under mechanical stress during the expected life of the product. Such mechanical compatibilities can be improved or introduced by the addition of diblock polymers. One block of these diblock polymers is compatible with one phase type of the heterogeneous blend and the other one with the other type (see Section 5.3.6). The two blocks need not necessarily be chemically identical with the two parent polymers, that is, a compatibility of poly (A) and poly (B) can be achieved not only by addition of poly (A)–*block*-poly (B) but also by poly (A)–*block*-poly (C) if the C block is either miscible with poly (B) or can form mixed crystals with poly (B).

11.4.2. Blend Formation

Heterogeneous polymer blends can be produced by mixing together two polymers (as melts, lattices, or in solution) or by in situ polymerization of a monomer in the presence of a dissolved polymer.

The energy taken up during *melt mixing* is used for flow processes and for the generation of surfaces of new microdomains. After some time, a steady state is established and the domain size becomes constant. No macroscopic demixing occurs because of the low diffusion coefficients resulting from the high viscosities.

Latex blending consists of the mixing of aqueous dispersions of two polymers. Far lower temperatures and lower shear fields can be employed compared to melt blending. The good mixing of the latex particles remains after coagulation. The domain size is, however, restricted to the size of the latex particles themselves; it is not altered by subsequent melting of the coagulate.

Solution blending involves the mixing of two polymer solutions. Miscible polymers can be blended to domains of molecular size. Solutions of immiscible polymers demix, however, at very low concentrations, sometimes under fractionation with respect to molar masses. Domains grow further on solvent removal by distillation or freeze drying.

In situ polymerization involves solutions or gels of polymers in monomers, which are subsequently polymerized. The in situ polymerization of styrene in a styrenic polydiene solution, which leads to rubber-toughened polystyrene (HIPS) is most important industrially. The polymerization of a cross-linkable monomer in a gel of a cross-linked rubber in the very same monomer results in interpenetrating networks.

Various blending processes produce blends with very different properties (Fig. 73). The high notched impact strengths of the polymerization-blended materials (line P) are caused by a strong anchoring of phases due to the formation of graft copolymers of styrene on polybutadiene and cross-linking within the rubber domains, both caused by free-radical initiators. Such processes are less prevalent during melt blending (radical formation by high shearing at elevated temperature) (line M) and latex blending (low shearing at ambient temperature) (line L). In situ polymerization is thus the method of choice for unsaturated rubber (easy cross-linking and grafting by chain transfer) and monomer–rubber pairs with favorable Q, e-values for copolymerizations. In all other cases, blends are formed by melt mixing.

11.4.3. Toughened Plastics

Rubber-modified plastics (toughened plastics, high-impact plastics) consist of rubber domains dispersed in plastic matrices. The domain size varies with the blending process. Typical values are ca. 0.1 µm for melt-blended poly(vinyl chloride)–acryl rubber and 1 µm for polystyrene–polybutadiene. The domains are often multiphased; small plastic domains are imbedded in the rubber domains (Fig. 74). The morphologies depend strongly on the blending process. During stirred in situ polymerizations, phase inversions often occur if the amount of newly formed plastic approaches that of the incipient rubber. During phase inversion, already present plastic particles may be embedded in the newly formed rubber domains.

Rubber-toughened plastics are valued because of their improved impact strength. On impact, very many crazes are formed near the equators of the rubber particles. The crazes propagate until they encounter an obstacle (e.g., rubber particle or shear band) or until

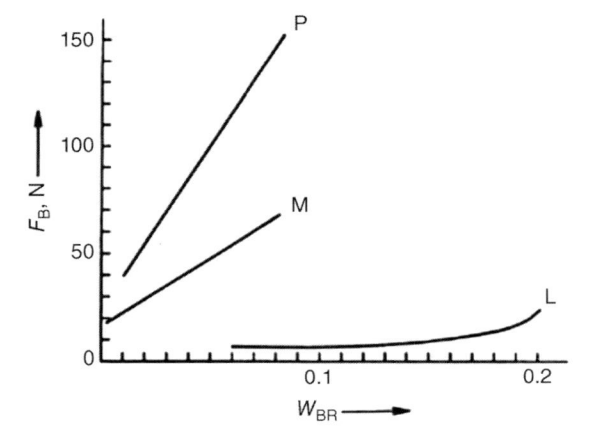

Figure 73. Impact strength with notch F_B as function of the mass concentration w_{BR} of *cis*-1,4-polybutadiene in its blends with polystyrene. P = in situ polymerization of styrene; M = melt blending; L = latex blending
Reprinted with permission by Plenum, New York [154]

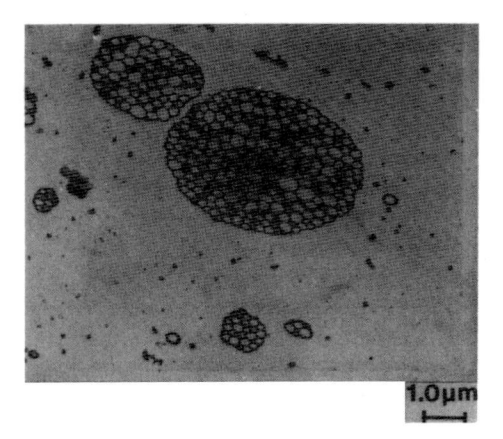

Figure 74. Rubber domains in high-impact polystyrene by in situ free-radical polymerization of styrenic polybutadiene solutions
Reprinted with permission by Society of Plastics Engineers [155]

the stress concentration at the tip of the craze becomes very low. Many small crazes result, and the stress is evenly distributed if the rubber phase is cross-linked and binds well to the thermoplast phase [e.g., by in situ formed graft copolymers (HIPS)]. In contrast, stress peaks concentrate at a few defect points in normal thermoplasts.

11.4.4. Thermoplastic Elastomers
(→ **Thermoplastic Elastomers**)

Thermoplastic elastomers (elastoplastics, thermoplastics, plastomers) are processed like thermoplastics but applied like elastomers. Their unique properties follow from their molecular structures; their chains consist of "soft" and "hard" segments in block, graft, or segmented copolymers composed of monomeric units A and B. The A and B segments are mutually incompatible and form locally separated regions. With a well-designed molecular architecture, domains of the "hard" A segments (transition temperature > service temperature) act as physical cross-links in the continuous matrix of the "soft" B segments (transition temperature < service temperature) (Section 5.3.6). Such transition temperatures may be glass transition temperatures in amorphous polymers or melting temperatures in partially crystalline polymers. The A segments attain mobility above these transition temperatures and the elastoplasts become processible.

Thermoplastic elastomers comprise linear triblock polymers of the polystyrene−*block*-polydiene−*block*-polystyrene type; radial, star, or "teleblock" polymers of the same monomeric units; urethane segment or block copolymers with polyester or polyether soft segments; polyesteramides and polyesteretheramides; graft copolymers of butyl rubber on polyethylene, vinyl chloride on poly(ethylene-co-vinyl acetate), styrene−acrylonitrile on saturated acryl rubbers, and various ionomers.

Thermoplastic elastomers can also be produced by "dynamic vulcanization." The mastication of blends of conventional rubbers and crystalline poly(α-olefins) leads to chain scissions. The resulting macroradicals cross-link the rubber domains.

The mechanical properties of thermoplastic elastomers are determined mainly by their morphologies. In styrene−butadiene−styrene triblock polymers, for example, morphologies are governed by the spatial requirements of the various blocks (Section 5.3.6). With a low content of hard styrene segments, small spherical polystyrene microdomains are formed in the soft polybutadiene matrix. These microdomains act as physical cross-linkers. The distances between the domains are large, and the polymer behaves as a weakly linked elastomer with correspondingly high extension (Fig. 75). The domains are larger and their distances are shorter at 28% styrene units and the polymer strengthens (stiffens). The stiffening becomes stronger for rodlike styrene domains (39% S). Lamellar morphologies (53% S) show an "unruly" behavior initially because of reorientation of lamellae. At even higher styrene content, the polymer behaves as a plasticized tough thermoplast (65% S) or almost like polystyrene itself (80% S).

11.5. Expanded Plastics
(→ **Foamed Plastics**) [156, 157]

Expanded plastics (foams, foamed plastics, cellular plastics) are blends of plastics with air. They are subdivided according to their rigidity, their cell structure, and the nature of the parent plastics. Rigid expanded plastics are used mainly for thermal insulation, and flexible foams for damping and cushioning materials.

The rigidity ("hardness") of expanded plastics follows the properties of the parent polymers. Phenolic and urea resins thus yield brittle-rigid plastics; polystyrene and hard poly(vinyl chloride), tough-rigid; and polyethylenes, polyurethanes, and plasticized PVC, semirigid to flexible foams. The elastic moduli of expanded plastics decrease approximately proportional to their polymer content (simple mixing law). Because the stiffness of an article increases with the third power of the wall thickness, the gases in expanded plastics work as enhancers that reduce material costs and weights.

Tensile strengths of foamed plastics also follow the simple mixing law. Because the tensile strengths of gases are diminishingly

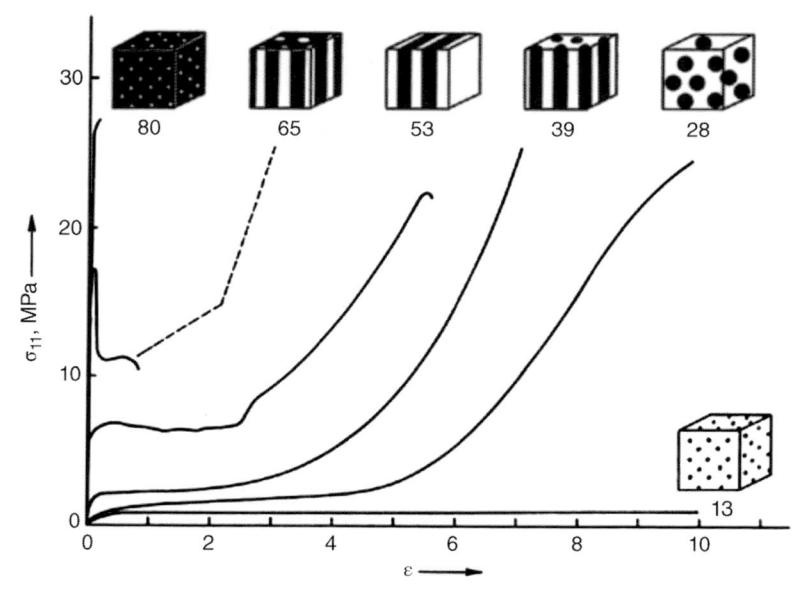

Figure 75. Stress–strain curves and morphologies of styrene–butadiene–styrene triblock polymers with various styrene contents
Reprinted with permission by Hüthig and Wepf Publ., Basel [5]

small, tensile strengths σ_B of foamed plastics are determined by the volume fraction ϕ_p and strengths $\sigma_{B,p}$ of the plastics themselves [i.e., $\sigma_B \approx \sigma_{B,p} \cdot \phi_p = \sigma_{B,p} \cdot (\rho/\rho_p)$] (Fig. 76). Compression strengths follow a similar simple mixing law.

The cell structure can be open, closed, or mixed. Open-celled foams are always air filled, regardless of the blowing or expanding gases used for foam manufacture. The trapped gases in closed-cell structures can be exchanged with the surrounding air only by slow diffusion through the polymer matrix. Because thermal conductivities of low-density foams ($\rho_p < 0.3$) obey, to a first approximation, a logarithmic mixing law, they are considerably affected by the thermal conductivities of trapped gases [(λ = 93 J m^{-1} h^{-1} K^{-1} (nitrogen), 56 J m^{-1} h^{-1} K^{-1} (CO$_2$), 26.4 J m^{-1} h^{-1} K^{-1} (CCl$_3$F)]. The gas exchange is considerably reduced in integral skin foams (dense skin, low-density core) and in syntactic foams (plastics filled with hollow spheres; spheres may be made of glass, ceramics, or plastics and contain either gases or a vacuum).

12. Plastics and Sustainability
(→ Plastics, Recycling)

12.1. Plastics and Environment [158–162]

During the first decade of the 21st century, more plastics were produced than during the entire 20th century. The plastics production rapidly increases and follows the growth of the world population (see Fig. 1). Originally introduced as rather poor imitation of natural materials, such as silk, natural rubber, and ivory, modern plastics have emerged as advanced materials with tailored property profiles, which is unparalleled in nature. Synthetic polymers are engineered to meet the demands of diversified technologies ranging from food packaging to textiles, shelter, communication, mobility, and health care. Plastics play an important role in daily life. Without plastics, the high quality of modern life with secure supply of food, water, and energy would not be feasible. Highly cost-efficient and versatile polymeric materials have made hightech

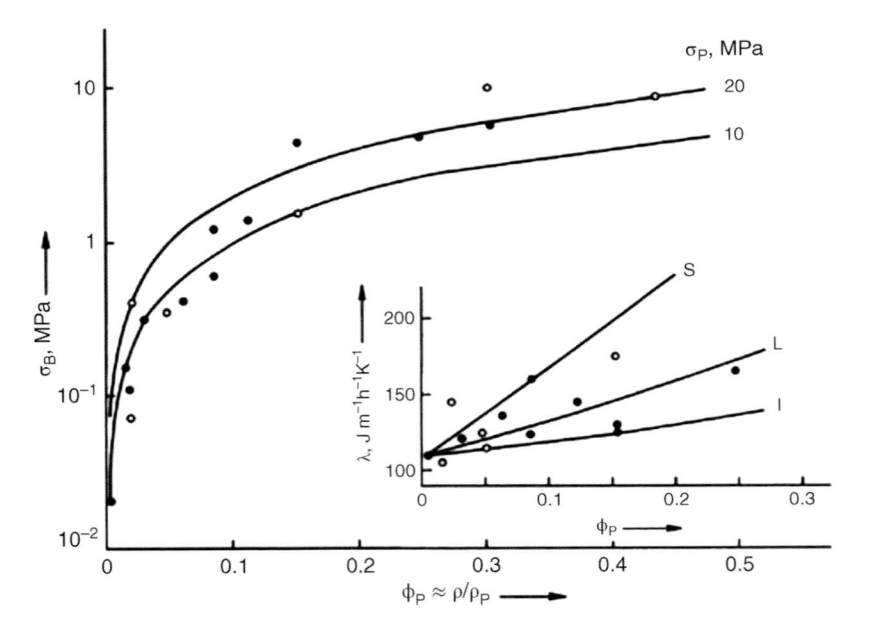

Figure 76. Tensile strengths of expanded plastics as a function of the volume fraction of the rigid (●) or flexible (○) parent polymers P
Solid lines indicate a) Simple mixing law for $\sigma_p = 20$ MPa; b) Simple mixing law for $\sigma_p = 10$ MPa;
Insert: Thermal conductivities (for $\lambda_p = 700$ and $\lambda_{air} = 110$ J (m^{-1} h^{-1} K^{-1}) of rigid (●) and flexible (○) plastic foams
S: Simple mixing law; L: Logarithmic mixing law; I: Inverse mixing law

affordable for people of all income groups. At present, plastics production consumes around 5% of the world's fossil resources, which is equivalent to about 60% of the petrochemical production. At present around 90% of the world's fossil resources are burned in energy production accompanied by severe emission of carbon dioxide, which contributes to global warming and climate change. The emission of green house gases, i.e., carbon dioxide and methane, is also being referred to as carbon footprint. More than 30% of durable plastics are used in packaging, such as bags, wraps, and bottles, all of which possess short product life of a few months. According to the statistics of the Environmental Protection Agency (EPA), around 251 ×10^6 t of municipal trash composed of paper (27.4%), food wastes (14.5%), yard trimmings (13.5%), plastics (12.7%), metals (8.9%), rubber, leather, and textiles (8.7%), wood (6.3%), and glass (4.6%) were created in the United States in 2012. Archaeological studies of landfills have revealed that the biodegradation processes of synthetic polymers and biopolymers are very slow in landfills, especially in the absence of light, air, and water. Some buried news papers could be read after residing for 25 years in a landfill. Clearly, recycling and incineration are the only possible ways for disposing of plastics. Because most plastics have an oillike, high energy content, plastic wastes are valuable raw materials that need to be reused. While in 2012 the plastics recycling and energy recovery reached 61.9% in Europe, it was around 9% in the United States (for comparison: paper recycling reaches 50% in the United States). When plastic recycling fails, the consequences of littering are clearly visible in nature. Driven by marine and wind currents, nontoxic plastics debris together with hazardous chemical sludge accumulate in the ocean, in particular in a region known as the northern pacific garbage vortex. Addressing high sustainability is an important issue in plastics development. As stated in 1987 by the Brundtland Commission of the UN General Assembly, sustainability means meeting the needs of the present without compromising the ability of future generations to meet their own needs. High resource-, eco-, and energy efficiency of plastics together with their low carbon footprint and high energy content, typical for the large-scale commodity plastics, holds great promise for achieving high sustainability. Efficient plastics recycling, exploiting renewable feedstocks, and especially tailoring advanced polymeric materials for sustainable development are important tasks in the plastics industry.

12.2. Green Polymer Chemistry [163, 164] and Life Cycle Assessment [165]

According to the definition by ANASTAS, important elements of green chemistry include high resource effectiveness by maximizing the content of raw materials in the product; clean and lean production processes; catalytic reactions at ambient temperature and pressure; preventing wastes and reducing greenhouse gas emissions; high safety standards; no use of auxiliary substances, such as organic solvents, blocking groups etc.; low energy demand of production and processing; no health and environmental hazards by substituting toxic chemicals; using renewable resources; low carbon footprint,; and controlled product life cycles with effective waste recycling. This definition of "green" is not at all synonymous with biomaterials and biotechnology.

An important tool for assessing the environmental impact of materials is the life cycle analysis, which is also referred to as ecobalance, cradle-to-grave analysis, and life cycle assessment (LCA). LCA addresses the environmental aspects of all stages of a product's life time from feedstock recovery to monomer and polymer manufacturing, processing, distribution, use, maintenance, disposal, and recycling. Taking into account all relevant materials and energy inputs and output together with environmental aspects, the following LCA ranking of plastics was determined in 2010:

1. Polypropylene (PP),
2. High density polyethylene (HDPE),
3. Low density polyethylene (LLDPE).

The LCA ranking of bio-based polylactic acid (rank 9) was similar to that of PVC (rank 7). According to LCA, converting petrochemical plastics into green plastics by using renewable feedstocks lowers the LCA ranking. The success of polyolefins, such as PP, HDPE, and LLDPE is clearly associated with significant progress made in polyolefin production since the 1950s. The polyethylene life cycle is displayed in Figure 77. Produced in solvent-free catalytic polymerization processes with low energy demand using a highly active stereospecific catalysts, polyolefins are readily tailored to meet the demands of diversified applications ranging from food packaging to baby diapers, carpets, and automotive bumpers. Modern reactor granule technology produces pellet-sized polyolefin particles, which eliminates pelletizing extrusion. The heat of polymerization can be used as power generator. No water and no purification are required. This is in contrast to biotechnology processes and to pulping. The latter has very high water demand and produces byproducts, such as organic compounds and salts. Compared with paper, LLDPE packaging has much lower mass and is easy to process by blow molding. After completing their product life, polyolefins are readily recycled by mechanical, feedstock, and energy recycling. As hydrocarbon resins, they have oillike energy content. On heating above 300 °C, they are converted into oil and gas, which are valuable sources for feedstocks and energy. The raw materials of polyolefins are highly flexible. Virtually any carbon source, ranging from oil, gas, and coal to biomass and bioethanol, can be employed in polyolefin production.

From the LCA it is apparent that the substitution of plastics packaging by paper and other materials would substantially increase mass, energy demand, and greenhouse gas emission. Contrary to the public opinion, increased use of

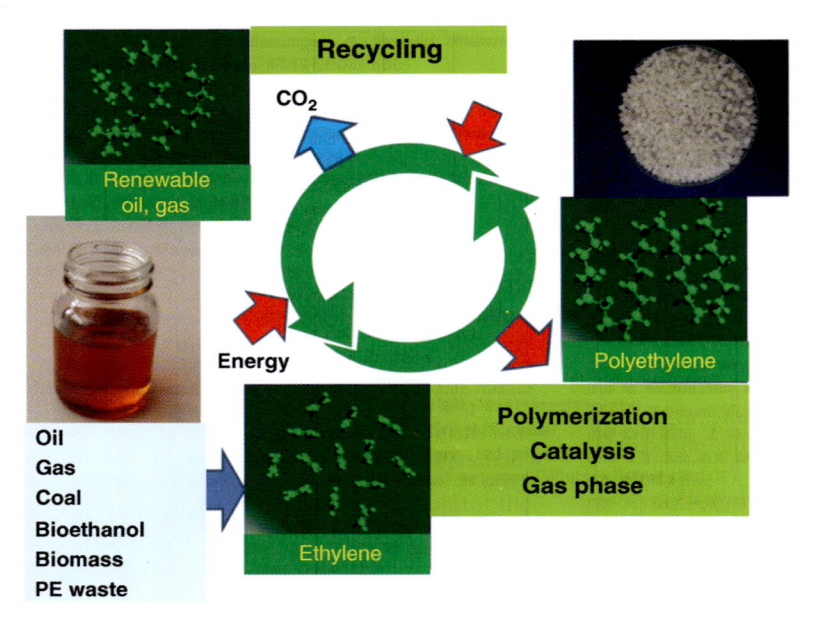

Figure 77. Life cycle of polyethylene

plastics reduces the greenhouse gas emission, whereas the substitution of plastics by other materials would have a stringent detrimental environmental impact and drastically enhance greenhouse gas emissions. At the end of their service life, plastics wastes are too valuable to be thrown away.

12.3. Plastics Waste Recycling

12.3.1. Mechanical Recycling [166, 167]

For identification and for facilitating their recycling, plastic products bear special codes consisting of label with three broken arrows, arranged in a triangle, and a number, sometimes also combined with letters and/or bar code. The numbers and letter combinations indicate the type of plastics. The letter combinations are not always identical with the acronyms and abbreviations recommended by ISO, ASTM, DIN, IUPAC, etc., for the same polymers. The following number and letter codes are used:

- 1 = PET [poly(ethylene terephthalate)]
- 2 = HDPE (high-density polyethylene)
- 3 = PVC (polyvinylchloride)
- 4 = LDPE (low-density polyethylene)
- 5 = PP (polypropylene)
- 6 = PS (polystyrene)
- 7 = for all other plastics, including multi-layered materials

Different waste management systems have been established to collect and sort wastes from different applications, such as packaging, automotive, electronics, and construction. A large variety of waste-pretreatment and sorting technologies are in place including shredding, magnetic separation, flotation, and air separation assisted by plastics identification by spectroscopy, such as near infrared sensing.

Especially when plastics wastes are clean and composed of as single type of plastics, mechanical recycling is attractive. In thermoplastics mechanical recycling, wastes are sorted, purified, and extruded to produce granulates. For example, mechanical recycling converts PET bottle waste into new PET bottles or PET textile fleeces for jackets. Due to polymer ageing by degradation and autoxidation during product life and recycling, most recycled plastics are somewhat inferior with respect to the corresponding virgin plastics. Therefore, most recycled plastics are blended together with virgin polymers in mechanical recycling. Because most polymers are highly immiscible, mechanical recycling of mixed plastics wastes produces blends, which require the addition of compatibilizers. Because cross-linked plastics do not melt, they can be recycled by grinding. Recycled thermoset resin powders or ground rubber are used as organic fillers for thermoset or rubber formulations.

12.3.2. Feedstock Recycling [167, 168]

During pyrolysis plastics are broken down at elevated temperatures in the absence of air to produce liquid and gaseous fragments. Thermolysis is of particular interest for recycling highly polluted complex plastic mixtures. In the case of PVC, HCl is formed and removed by washing the thermolysis gases with water. Hydrocarbons, such as polyolefins, are readily thermolyzed to produce oil and gas in essentially quantitative yields. Plastics, such as polyoxymethylene and poly(methyl methacrylate) are thermally depolymerized to recover the corresponding monomers. On thermolysis, carbohydrates decompose to form a considerable amount of water and ashes. Whereas pyrolysis requires rather high temperatures, typically above 500 °C, catalytic liquefaction converts biomass and plastics wastes into diesel fuel at temperature well below 500 °C. In this process, water is catalytically converted into hydrogen, which improves the yield of low molecular mass hydrocarbons by hydrogenation. In sharp contrast, hydrogenation of plastics wastes and distillation residues in oil refinery require much higher hydrogen pressure and temperature. The presence of hydrogen increases the amount of liquid hydrocarbon. In the case of biomass and biopolymers, anaerobic digestion can be employed to convert biowastes into methane or bioethanol. Lignocellulose from forestry and agricultural wastes is an attractive source of sugar that serves as raw material for biorefineries without competing with food production.

When small amounts of oxygen are present, gasification by incomplete combustion converts plastics into a mixture of carbon monoxide and hydrogen (syngas), which can be used as feedstock for producing a variety of chemicals and monomers in refineries. Similar to the Fischer–Tropsch process for coal-to-liquid conversion, this process can also be applied to biomass (biomass-to-liquid conversion) and plastics wastes. Moreover, syngas is of interest as reducing gas for metal oxide ores in steel mills, thus substituting oil and gas. Polycondensates, such as polyesters and polyurethanes are readily cleaved by hydrolysis and glycolysis, thus recovering monomers useful in polycondensation reactions.

12.3.3. Energy Recycling [166–168]

Taking into account the high energy content and hydrocarbon nature of commodity plastics, in many countries incineration is the preferred method for municipal waste recycling, which does not need extensive sorting and purification of the plastics wastes. In fact, the plastics wastes represent an attractive source of energy for the combustion of municipal wastes, which contain components with much less energy content. In contrast to inorganic wastes and biomass, commodity plastics do not form large amounts of ashes but completely decompose to produce gas and heat. For many years, public fears concern the formation of dioxin by waste incineration. Today, the incineration off-gases are carefully purified to remove gaseous and dustlike particulate pollutants. This includes the catalytic conversion of dioxin and furan traces. Incineration is the preferred process for disposing of hazardous wastes containing microorganisms, toxins, or chemicals.

12.3.4. Biodegradation and Bio-based Plastics [79–83, 164, 169]
(→ Polymers, Biodegradable)

In the early 1900s, all plastics were bio-based because carbohydrates and proteins were the only industrially available feedstocks for plastics production. Because abundant biopolymers, such as cellulose, are infusible and insoluble, extensive chemical modifications are required to enable melt and solution processing of cellulosics. Within very short time during the first half of the 20th century, the emerging highly versatile and low-cost synthetic plastics, based on fossil feedstocks and petrochemistry, have outperformed biopolymers and conventional bio-based plastics. In spite of the ongoing commercial significance of natural rubber, claiming 40% of the rubber market, and the successful development of wood plastics compounds and natural fiber reinforced composites, the total biopolymer share of the annual world's plastics production is still below 5%. However, there is a renaissance of the use of renewable and natural resources as well as the development of biopolymers. In view of the dwindling fossil resources, which are rapidly being depleted in energy production, the shift from fossil to renewable resources is thought to safeguard plastics production against an expected future oil crisis. Moreover, the growing consumers' concerns regarding global warming and the fear that a collapse of the biosphere is imminent, stimulate the surging demand for renewable and bio-based products. Because most natural polymers still fail to match the performance of synthetic polymers, synthetic plastics are rendered "green" and bio-based by using renewable raw materials for the production of monomers. For example, bioethanol is converted into ethylene and propene, which are polymerized in highly efficient gas phase processes to afford "green" polyolefins, which have properties equivalent to petrochemical polyolefins. In contrast to bacterial poly-L-lactic acid (PLA), which requires separation of PLA from cell proteins and water recovery, the solvent-free ring-opening polymerization of bio-based dilactide affords much better control of polymer properties, such as crystallization rate and shear thinning.

In life cycle assessment, most bio-based products have lower performance with respect to the corresponding petrochemical products due to the additional process steps. Moreover, in many industrialized countries the biomass supply is insufficient to substitute oil in energy and plastics production. PLA, polyolefins, or paper are not completely biodegradable in landfill, especially when air and water are absent. In contrast to the public opinion, biodegradation and bioerosion, which are highly dependent on the climate, do not instantaneously degrade biopolymers to produce carbon dioxide and water. Instead, biodegradation makes polymers brittle. They disintegrate to form tiny particles, invisible for human eyes, which are carried away by wind or rain and serve as breeding ground for spores and bacteria. Because the label "biodegradable" stimulates consumers to carelessly dump wastes and ignore recycling efforts, biodegradable plastics are contra productive to recycling. Moreover, the high energy content of polymers is wasted during biodegradation in landfill. Hence, biodegradable polymers do not seem to be the universal solution to the plastics waste problem. Yet, they offer prospects for special applications like bioresorbable implants, surgical sutures, drug release, as well as agricultural films and packaging. Today, it is well recognized that the surging biomass demand in bioenergy production, accompanied by extensive farming of energy crops, endangers both the biosphere and the food supply of the rapidly growing world population. Many industrialized countries do not have enough biomass to serve the need of both energy and plastic production.

12.3.5. Plastics from Carbon Dioxide [164, 170] (→ Carbon Dioxide 12.4.10. Polymers from Carbon Dioxide)

In order to prevent a conflict with food production, several successful attempts have been made to exploit the direct chemical fixation of carbon dioxide in plastics production. Because carbon dioxide is the energy sink in the combustion process, it must be activated in order to be useful in polymer production. In copolymerization this is achieved by reacting carbon dioxide with energy-rich highly strained

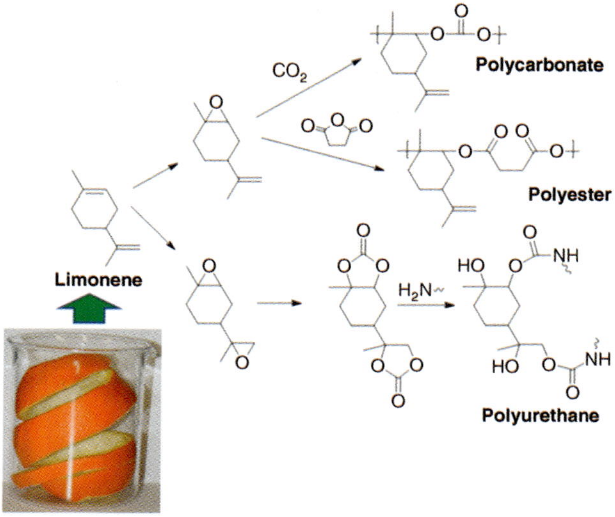

Figure 78. Bio-based polyesters, polycarbonates, and nonisocyanate polyurethanes from orange peels
Reprinted with permission by Wiley

oxirane and oxetane rings, thus producing polyfunctional cyclic carbonates and linear polycarbonates. Dihydroxy-terminated poly(propylene ether carbonates) are of special interest as polyols in polyurethane synthesis. Polyfunctional cyclic carbonates, prepared by reacting epoxy resins with carbon dioxide, are attractive intermediates for the isocyanate- and phosgene-free production of hydroxyl-functional polyurethanes (NIPU for nonisocyanate polyurethane). Biowastes, such as orange peels can be used as a source of limonene, which upon oxidation and carbonation affords mono- and difunctional cyclic carbonates as intermediates for bio-based polyesters and NIPU (Fig. 78). Electrochemical processes are developed to exploit excess renewable energy supplied by wind mills and solar power plants to convert carbon dioxide into methane, which is useful as chemical storage of renewable energy, as fuel, and as intermediate for the production of chemicals and plastics.

12.4. Polymers for Sustainable Development [164]

In many ways plastics serve the needs of sustainable development. Produced in highly energy- and resource-efficient polymerization processes, plastics have low carbon footprint and preserve high oillike energy content. After completing their product life, they are mechanically recycled or serve as valuable source of energy and feedstocks. This attractive balance of low energy demand during production and processing combined with low carbon footprint and high energy content is typical for all large-volume commodity and most engineering plastics and rubbers, amounting to more than 90% of the world's plastics production. Among them, hydrocarbon plastics, such as polypropylene and polyethylene, produced by solvent-free catalytic polymerizations processes, meet the demands of green chemistry and achieve top rankings in life cycle assessment. Moreover, plastics have a low mass and are corrosion resistant. In lightweight engineering and packaging, polymeric materials contribute to considerable weight savings, thus reducing fuel consumption and dioxide emission in transportation. Moreover, unparalleled by other materials, plastics are highly versatile with respect to tailoring property profiles, processing, and applications. Owing to their low mass together with thermal and electrical insulation, plastics contribute to high energy efficiency and secure the supply of food, water, and energy.

Although plastics consume around 5% of fossil fuels, they help to preserve close to 20% oil by thermal insulation and mass reduction in transportation. As advanced functional materials and integrated into systems, polymers bring high resource efficiency to a variety of applications ranging from lightweight engineering to health care. For example, as filters and membranes, functional plastics enable water and air purification, water desalination, and recovery of precious metals from wastes. This unique combination of high cost-, eco-, resource-, and energy efficiency with high versatility typical for most plastics holds great promise for sustainable development.

References

1. H.-G. Elias: Technologie, *Makromoleküle*, vol. **2**, Hüthig and Wepf, Basel 1992.
2. F.W. Billmeyer, Jr.: *Textbook of Polymer Science*, 3rd ed., Wiley, New York 1984.
3. F. Rodriguez: *Principles of Polymer Systems*, 3rd ed., Hemisphere Publ., New York 1989.
4. G. Champetier, R. Buvet, J. Néel, P. Sigwalt, (eds.): *Chimie Macromoléculaire*, 2 vols., Hermann, Paris 1970–1972.
5. H.-G. Elias: Grundlagen, *Makromoleküle*, 5th ed., vol. 1, Hüthig and Wepf, Basel 1989.
6. L. Mandelkern: *An Introduction to Macromolecules*, 2nd ed., Springer, New York 1972, 1982.
7. J.M.G. Cowie: *Polymers: Chemistry and Physics of Modern Materials*, Intext Educational Publ., New York 1974; *Polymers: Chemistry and Physics of Modern Materials*, 2nd ed., Routledge, Chapman and Hall, New York 1991.
8. F.A. Bovey, F.H. Winslow (eds.): *Macromolecules*, Academic Press, New York 1979.
9. R.B. Seymour, C.E. Carraher, Jr.: *Polymer Chemistry*, 2nd ed., Dekker, New York 1989.
10. A. Rudin: *The Elements of Polymer Science and Engineering*, Academic Press, New York 1982.
11. P.C. Hiemenz: *Polymer Chemistry*, Dekker, New York 1984.
12. P. Munk: *Introduction to Macromolecular Science*, Wiley, New York 1989.
13. H. Morawetz (eds.): *Polymers: The Origins and Growth of a Science*, John Wiley & Sons, New York 1985.
14. Y. Furukawa (ed.): *Inventing Polymer Science, Staudinger, Carothers, and the Emergence of Macromolecular Chemistry*, University of Pennsylvania Press, Philadelphia 1998.
15. H. Ringsdorf, *Angew. Chem. Int. Ed.* **43** (2004) 1064.
16. R. Mülhaupt, *Angew. Chem. Int. Ed.* **43** (2004) 1054.
17. International Union of Pure and Applied Chemistry: *Compendium of Macromolecular Nomenclature (Purple Book)*, Blackwell Sci. Publ., Oxford 1991.
18. R.C. Hiorns, R.J. Boucher. R. Duhlev, K.-H. Hellwich, P. Hodge, A.D. Jenkins, R.G. Jones, J. Kahovec, G. Moad, C. K. Ober, D.W. Smith, R.F.T. Stepto, J.-P. Vairon, J. Vohlídal, *Pure Appl. Chem.* **84** (2012) 2167.
19. J. Brandrup, H.E. Immergut, E.A. Grulke (eds.): *Polymer Handbook*, John Wiley & Sons, New York 1999.
20. J. Schultz: *Polymer Materials Science*, Prentice-Hall, Englewood Cliffs, N.J., 1974.
21. J.A. Brydson: *Plastics Materials*, 3rd ed., Butterworth, London 1975.
22. S. Rosen: *Fundamental Principles of Polymeric Materials for Practicing Engineers*, Cahners Books, Boston 1973.
23. J.H. DuBois, F.W. John: *Plastics*, 5th ed., Van Nostrand-Reinhold, New York 1976.
24. W.E. Driver: *Plastics Chemistry and Technology*, Van Nostrand-Reinhold, Cincinnati 1974.
25. *Houben-Weyl*, 5th ed., vol. 20; *Science of Synthesis*, vol. 1, Georg Thieme, Stuttgart 2001, p. 57; *Science of Synthesis*, vol. 2, Georg Thieme, Stuttgart 2002, p. 291; *Science of Synthesis*, vol. 2, Georg Thieme, Stuttgart 2002, p. 906.
26. G.W. Becker et al. (eds.): *Kunststoff-Handbuch*, 2nd ed., Hanser, München 1983 ff. (many vols.).
27. H. Mark, C. Overberger, G. Menges, N.M. Bikales (eds.): *Encyclopedia of Polymer Science and Engineering*, 2nd ed., Wiley, New York 1985–1990.
28. W.A. Kargin (ed.): *Enciklopedia Polimerov*, 3 vols., Sovietskaya Enciklopedia Publ., Moscow 1972.
29. G. Allen, J.C. Bevington (eds.): *Comprehensive Polymer Science*, Pergamon, Oxford 1989.
30. *Chemiefasern auf dem Weltmarkt*, 7th ed., Deutsche Rhodiaceta, Freiburg/Br., 1969, Supplement 1976 (trade names of fibers).
31. The International Plastics Selector, *Commercial Names and Sources*, Cordura Publ., San Diego 1978.
32. Fachinformationszentrum Chemie: *Index of Polymer Trade Names (Parat)*, VCH, Weinheim 1987.
33. J. Brandrup, E.H. Immergut (eds.): *Polymer Handbook*, 3rd ed., Wiley, New York 1989.
34. O. Griffin Lewis: *Physical Constants of Linear Homopolymers*, Springer, Berlin 1968.
35. R.E. Schramm, A.F. Clark, R.P. Reeds (eds.): *A Compilation and Evaluation of Mechanical, Thermal, and Electrical Properties of Selected Polymers*, U.S. National Bureau of Standards, Washington, D.C., 1973.
36. P.A. Schweitzer: *Corrosion Resistance Tables (Metals, Plastics, Nonmetallics, Rubbers)*, Dekker, New York 1976.
37. The International Plastics Selector, San Diego 1977.
38. H. Saechtling: *Kunststoff-Taschenbuch*, 24th ed., Hanser, Munich 1989; *International Plastics Handbook*, 2nd ed., Hanser, München 1989.
39. *Modern Plastics Encyclopedia*, McGraw-Hill, New York (annually in October as issue 10 A).
40. *Encyclopedie Francaise des Matieres Plastiques*, Les Publicateurs Techniques Association, Paris, annually.
41. J.E. Williams (ed.): *Computer-Readable Databases*, American Library Association, Chicago 1985.
42. German Plastics Institute, Plastics Databank POLYMAT, Darmstadt; Fachinformationszentrum Chemie, Berlin (6000 polymers from 80 manufacturers).
43. Plastics Databank CAMPUS (= Computer Aided Material Preselection by Uniform Standards), (5000 thermoplastics with 50 most important properties each), issued for their company products by 17 European companies.
44. E.R. Yescombe: *Plastics and Rubbers: World Sources of Information*, 2nd ed., Appl. Sci. Publ., Barking 1976.
45. G.J. Patterson: *Plastics Book List*, Technomic Publ., Westport 1975.
46. P. Eyerer: *Informationsführer Kunststoffe*, VDI-Verlag, Düsseldorf 1976.

47 J. Schrade: *Kunststoffe (Hochpolymere): Bibliographie aus dem deutschen Sprachgebiet*, 1st ed., Schweiz. Aluminium AG., Zürich 1976; 2nd ed., Zürich 1980.
48 O.A. Battista, *The Polymer Index*, McGraw Hill, New York 1976.
49 S.M. Kaback: "Literature of Polymers," *Encycl. Polym. Sci. Technol.* **8** (1968) 273; J.T. Lee, *Encycl. Polym. Sci. Eng.* **9** (1987) 62.
50 W.J. Roff, J.R. Scott: *Handbook of Common Polymers*, Butterworths, London 1971.
51 M. Ash, I. Ash: *Encyclopedia of Plastics, Polymers and Resins*, Chem. Publ. Co., New York 1982–1988.
52 M.S.M. Alger: *Polymer Science Dictionary*, Elsevier Applied Sci. Publ., Barking 1989.
53 American Society for Testing Materials, ASTM D-1600–86 a; D-1418–67; D-4020–81.
54 German Industrial Standards DIN 7723; 7728 (1988); 16 913; 55 950; 60 001.
55 European Textile Characterization Law.
56 International Organization for Standardization, ISO 1043–1978, ISO 1629–1980.
57 Guide for the Harmonisation of Designations of Thermoplastic Materials, ISO/TC 61/SC 9 N 435, April 1981; ASTM D 4000.
58 K. Wiebusch, *Kunststoffe-German Plastics* **72** (1982) 22, 167.
59 F.O. Swanson, *Plastics Engng.* (1983) Jan. 31.
60 R.H. Wehrenberg, II, *Mat. Engng.* (1984) Feb. 48.
61 H. Elias et al., *Polymer News* **9** (1983) 101; **10** (1985) 169.
62 J.H. DuBois: *Plastics History–U.S.A.*, Cahners, Boston 1971.
63 J.K. Craver, R.W. Tess (eds.): Applied Polymer Science, *Am. Chem. Soc.*, Washington, D.C., 1975.
64 F.M. McMillan: *The Chain Straigtheners: Fruitful Innovation. The Discovery of Linear and Stereoregular Polymers*, MacMillan, London 1981.
65 R.B. Seymour (ed.): History of Polymer Science and Technology, Dekker, New York 1982 (Reprint of articles in *J. Macromol. Sci.-Chem.* **A 17** (1982) 1065–1460.
66 R. Friedel: *Pioneer Plastic. The Making and Selling of Celluloid*, University of Wisconsin Press, Madison 1983.
67 W. Glenz (ed.): *Kunststoffe–Ein Werkstoff macht Karriere*, Hanser, München 1985.
68 United Nations, Statistical Yearbooks. United Nations, http://esa.un.org/unpd/wpp/Excel-Data/population.htm accessed March 7, 2015
69 Calculated from data in *Kunststoffe* **73** (1983) no. 10.
70 H.-G. Elias, F. Vohwinkel: *New Commercial Polymers*, Gordon and Breach, New York 1986; *Neue polymere Werkstoffe*, Hanser, München 1983.
71 H.-G. Elias, *Grosse Moleküle*, Springer, Berlin 1985; *Mega Molecules*, Springer, Berlin 1987; *Megamolekuly*, Khimiya, Leningrad 1990.
72 United States, Statistical Abstracts (annually).
73 Plastics Europe, http://www.plasticseurope.org/Document/plastics-the-facts-2013 (accessed April 26, 2014).
74 United Nations: *World population to 2300*, United Nations, New York 2004, http://www.un.org/esa/population/publications (accessed August 5, 2014).
75 D. Tomalia, A.M. Naylor, W.A. Goddard III, *Angew. Chem.* **102** (1990) 119; *Angew. Chem. Intern. Ed.* **29** (1990) 138.
76 G. Natta, P. Corradini, *Nuovo Cimento, Suppl.* **15** (1960) 11.
77 D. Lang, H. Bujard, B. Wolff, D. Russell, *J. Mol. Biol.* **23** (1967) 163.
78 N. Karak (ed.): *Fundamentals of Polymers: Raw Materials to Finish Products*, PHI Learning Private Limited, New Dehli 2010.
79 V. Mittal (ed.): *Renewable Polymers*, Scrivener Publishing and John Wiley & Sons, Hoboken, Salem 2012.
80 M.N. Belgacem, A. Gandini (eds.): *Monomers, Polymers and Composites from Renewable Resources*, Elsevier, Oxford 2008.
81 A. Gandini, *Macromolecules* **41** (2008) 9491.
82 B. Rieger, A. Künkel, G.W. Coates, R. Reichardt, E. Dinjus, T. A. Zevaco: "Synthetic Biodegradable Polymers" in *Advanced Polymer Science*, vol 245, Springer, Berlin 2012.
83 A. Demirbas (ed.): *Biorefineries*, Springer, Berlin 2010.
84 D.M. Mandal (ed.): *Fundamentals of Polymerization*, World Scientific Publishing Company, Singapore 2013.
85 H.R. Kricheldorf (ed.): *Polycondensation*, Springer, Berlin 2014.
86 K. Matyjaszewski, T.P. Davis (eds.): *Handbook of Radical Polymerization*, John Wiley & Sons, New York 2003.
87 N.V. Tsarevsky, B.S. Sumerlin (eds.): *Fundamentals of Controlled/Living Radical Polymerization*, RCS Polymer Chemistry Series, RCS Publishing, London 2013.
88 K. Matyjaszewski, B.S. Sumerlin, N.V. Tsarevsky (eds.): *Progress in Controlled Radical Polymerization: Mechanisms and Techniques*, American Chemical Society, Washington 2012.
89 N. Hadjichristidis, M. Pitsikalis, S. Pispas, H. Iatrou: "Polymers with Complex Architecture by Living Anionic Polymerization," *Chem. Rev.* **101** (2001) 3747–3792.
90 S. Aoshima, S. Kanaoka: "A Renaissance in Living Cationic Polymerization," *Chem. Rev.* **109** (2009) 5245–5287.
91 H. Brintzinger, D. Fischer, R. Mülhaupt, B. Rieger, R.M. Waymouth, *Angew. Chem. Int. Ed.* **34** (1995) 1143–1170.
92 V.C. Gibson, S.K. Spitzmesser, *Chem. Rev.* **103** (2003) 283–315.
93 R. Mülhaupt, *Macromol. Chem. Phys.* **204** (2003) 289.
94 J.R. Severn, J.C. Chadwick (eds.): *Tailor-Made Polymers: Via Immobilization of Alpha-Olefin Polymerization Catalysts*, Wiley-VCH, Weinheim 2008.
95 J.L. White, D.C. Choi (eds.): *Polyolefins: Processing, Structure Development, and Properties*, Hanser Publ., Munich 2004.
96 M. Buchmeiser, *Chem. Rev.* **100** (2000) 1565.
97 T.M. Baughman, K.B. Wagener, *Adv. Polym. Sci.* **176** (2005) 1–42.
98 C. Hagiopol (eds): *Copolymerization: Toward a Systematic Approach*, Kluwer Academic/Plenum Publishers, New York 1999.
99 D. Yan, C. Gao, H. Frey (eds.): *Hyperbranched Polymers*, Wiley-VCH, Weinheim 2010.
100 J.J. Meister (ed.): *Polymer Modification, Principles, Techniques, and Applications*, Marcel Dekker, New York 2000.
101 P. Theato, H.-A. Klok (eds.): *Functional Polymers by Post-Polymerization Modification*, Wiley-VCH, Weinheim 2013.
102 N. Hadjichristidis, S. Pispas, G.A. Floudas (eds.): *Block Copolymers: Synthetic Strategies, Physical Properties, and Applications*, John Wiley & Sons, Hoboken 2003.
103 E.S. Guerra, E.V. Lima: *Handbook of Polymer Synthesis, Characterization and Processing*, John Wiley & Sons, Hoboken 2013.
104 R. Gächter, H. Müller (eds.): *Plastics Additives Handbook* 3rd ed., Hanser, München 1989.
105 H. Zweifel, R.D. Maier, M. Schiller (eds.): *Plastics Additives Handbook*, 6th ed., Hanser Publ., Munich 2008.
106 H.S. Katz, J.V. Milevski (eds.): *Handbook of Fillers for Plastics*, Van Nostrand Reinhold, New York 1987.
107 M. Xanthos (ed.): *Functional Fillers for Plastics*, Wiley-VCH, Weinheim 2010.

108 R.K. Gupta, E.B. Kennel, K.-J. Kim (eds.): *Polymer Nanocomposites Handbook*, CRC Press, Boca Raton 2010.
109 A.D. Godwin, W. Arendt, P. Daniels (eds.): *New Plasticizer Technology*, Wiley-VCH, Weinheim 2014.
110 W. Michaeli (ed.): *Plastics Technology*, Hanser Publ., Munich 2000.
111 T.A. Oswald (ed.): *Polymer Processing Fundamentals*, Hanser Publ., Munich 1998.
112 Z. Tadmor, C.C. Gogos (eds.): *Principles of Polymer Processing*, John Wiley & Sons, Hoboken 2006.
113 M. Reyne (ed.): *Plastic Forming Processes*, Wiley-VCH, Weinheim 2008.
114 R.B. Brown (ed.): *Handbook of Polymer Testing*, Marcel Dekker, New York 1999.
115 W. Grellmann, S. Seidler (eds.): *Polymer Testing*, Hanser Publ. Munich 2013.
116 D.W. Van Krevelen, K. Te Nijenhuis (eds.): *Properties of Polymers*, Elsevier, Amsterdam 2009.
117 A.E. Woodward, *Atlas of Polymer Morphology*, Hanser, Munich 1988.
118 L. Mandelkern (ed.): *Crystallization of Polymers*, Cambridge University Press, Cambridge 2004.
119 C. Ruscher, E. Schulz, private communication.
120 R.J. Samuels, private communication.
121 A.M. Donald, A.H. Windle, S. Hanna (eds.): *Liquid Crystalline Polymers*, Cambridge University Press, Cambridge 2009.
122 P.J. Flory, *Adv. Polym. Sci.* **59** (1984) 1.
123 J.G. Drobny (eds): *Handbook of Thermoplastic Elastomers*, Elsevier, Amsterdam 2007.
124 J. Martin (ed.): *The Concise Encyclopedia of the Properties of Materials Surfaces and Interfaces*, Elsevier Science, Amsterdam 2008.
125 Y.G. Yanovsky (ed.): *Polymer Rheology: Theory and Practice*, Springer, Berlin 2014.
126 E. Riande, R. Diaz-Celleja (eds.): *Electrical Properties of Polymers*, Marcel Dekker, New York 2004.
127 *Kunststoff-Physik im Gespräch*, 8th ed., BASF, Ludwigshafen 1988.
128 H.M. Laun, H. Münstedt, *Rheol. Acta* **17** (1978) 415.
129 G.H. Michler, F.J. Balta-Calleja (eds.): *Nano- and Micromechanics of Polymers*, Hanser Publ., Munich 2012.
130 R.F. Landel, L.E. Nielsen (eds.): *Mechanical Properties of Polymers and Composites*, CRC Press, Boca Raton 1993.
131 S. Onogi, T. Matsuda, K. Kitagawa, *Macromolecules* **3** (1970) 109.
132 S. Wellinghoff, E. Baer, private communication.
133 T.A. Skotheim, R.L. Elsenbaumer, J.R. Reynolds (eds.): *Handbook of Conducting Polymers*, New York 1998.
134 T.A. Skotheim, J. Reynolds (eds.): *Conjugated Polymers: Theory, Synthesis, Properties and Characterization*, CRC Press, Boca Raton 2006.
135 J.-F. Morin, M. Leclerc (eds.): *Design and Synthesis of Conjugated Polymers*, Wiley-VCH, Weinheim 2010.
136 C. Brabec, V. Dyakonov, U. Scherf (eds.): *Organic Photovoltaics*, Wiley-VCH, Weinheim 2008.
137 G. Hadziioannou, G.G. Maliaras (eds.): *Simconducting Polymers*, Wiley-VCH, Weinheim 2006.
138 S. Thomas, K. Joseph, S.K. Malhotra, K. Goda, M.S. Sreekala (eds.): *Polymer Composites*, Wiley-VCH, Weinheim 2013.
139 T.H. Ferrigno, in S. Katz, J.V. Milewski (eds.): *Handbook of Fillers and Reinforcements for Plastics*, Van Nonstrand, New York 1978, p. 11.
140 G.E. Ehrenstein, R. Wurmb, *Angew. Makromol. Chem.* **60/61** (1977) 157.
141 M. Moniruzzaman, K.J. Winey, *Macromolecules* **39** (2006) 5194.
142 V. Mittal (ed.): *Polymer Nanotube Nanocomposites: Synthesis, Properties, and Applications*, John Wiley & Sons, Hoboken 2010.
143 G. Carotenuto, L. Nicolais (eds.): *Graphene-Polymer Composites*, John Wiley & Sons, Hoboken 2012.
144 S. Ray, M. Okamoto: "Polymer/Layered Silicate Nanocomposites: A Review from Preparation to Processing," *Prog. Polym. Sci.* **28** (2003) 1539.
145 C.J. Brinker, G.W. Scherer (eds.): *Sol-Gel Science*, Academic Press, San Diego 1990.
146 S. Sakka: *Handbook of Sol-Gel Science and Technology*, vol 3, Springer, Berlin 2005.
147 A. Kmetty, T. Barany, J.J. Karger-Kocis, *Progr. Polym. Sci.* **35** (2010) 1288.
148 M. Stürzel, Y. Thomann, M. Enders, R. Mülhaupt, *Macromolecules* **47**(2014), 4979. DOI: 10.1021/ma500769g.
149 L.A. Utracki, C. Wilkie (eds.): *Polymer Blends Handbook*, Springer, Berlin 2014.
150 A. Boudenne, L. Ibos, Y. Candau, S. Thomas (eds.): *Multiphase Polymer Systems*, John Wiley & Sons, Hoboken 2001.
151 *Makromoleküle*, 4th ed., Hüthig and Wepf, Basel 1981.
152 R.A. Pearson: "Introduction to the Toughening of Plastics," *ACS Symp. Ser.* **759** (2000) 1.
153 A.A. Collyer (ed.): *Rubber-Toughened Engineering Plastics*, Chapman & Hall, London 1994.
154 J.A. Manson, L.H. Sperling: *Polymer Blends and Composites*, Plenum, New York 1976.
155 S.L. Aggarwal, R.L. Livigni, *Polym. Eng. Sci.* **17** (1977) 498.
156 D. Klempner, H.C. Frisch (eds.): *Handbook of Polymeric Foams and Foam Technology*, Hanser Publ., Munich 2004.
157 N. Mills (ed.): *Polymer Foams Handbook*, Elsevier, Amsterdam 2007.
158 R.A. Meyers (ed.): *Encyclopedia of Sustainability Science and Technology*, Springer, Berlin 2012.
159 P. Eyrer, M. Weller, C. Hübner (eds.): *The Handbook of Environemtal Chemistry, Polymers–Opportunities and Risks II*, 12 (2010).
160 G. Payne, P. Smith: "Renewable and Sustainable Polymers," *ACS Symp. Ser.*, American Chemical Society, Washington 2010.
161 M. Tolinski (ed.): *Plastics and Sustainability: Towards a Peaceful Coexistence between Bio-based and Fossil Fuel-based Plastics*, Scrivener Publishing, Salem 2012.
162 A. Azapagic, A. Emsley, I. Hamerton (eds.): *Polymers: The Environment and Sustainable Development*, John Wiley & Sons, Chichester 2003.
163 P.T. Anastas, T. Horvath (eds.): *Green Chemistry for a Sustainable Future*, John Wiley & Sons, Hoboken 2015.
164 R. Mülhaupt: "Green Polymer Chemistry and Bio-based Plastics: Dreams and Reality," *Macromol. Chem. Phys.* **214** (2013) 159.
165 M.D. Tabone, J.J. Cregg, E.J. Beckman, A.E. Landis, *Environ. Sci. Technol.* **44** (2010) 8264.
166 S. Thomas, M. Sebastian, A. George, Y. Weimin (eds.): *Recycling and Reuse of Materials and Their Products (Advances in Materials Science)*, Apple Academic Press, Point Pleasant 2013.
167 S. M. Al-Salem, P. Lettieri, J. Baeyens, *Waste Manage.* **29** (2009) 2625.

168 J. Scheirs, W. Kaminsky (eds.): *Feedstock Recycling and Pyrolysis of Waste Plastics: Converting Waste Plastics Into Diesel and Other Fuels (Wiley Series in Polymer Science)*, John Wiley & Sons, Chichester 2006.
169 R.P. Wool (ed.): *Bio-based Polymers and Composites*, Elsevier Academic Press, New York 2005.
170 H. Blattmann, M. Fleischer, M. Bähr, R. Mülhaupt, *Macromol. Rapid Commun.* **35** (2014) 1238.

Further Reading

P.C. Painter, M.M. Coleman (eds.): *Essentials of Polymer Science and Engineering*, DEStech Publication, Lancaster 2009.

E.S. Guerra, E.V. Lima (eds.): *Handbook of Polymer Synthesis, Characterization and Processing*, John Wiley & Sons, Hoboken 2013.

H.G. Elias (ed.): *Macromolecules: Industrial Polymers and Synthesis*, Wiley-VCH, Weinheim 2007.

H.G. Elias (ed.): *Macromolecules*, vols. 1-4, Wiley-VCH, Weinheim 2008.

K. Matyjaszewski, M. Möller (eds.): *Polymer Science: A Comprehensive Reference*, Elsevier, Amsterdam 2012.

G.H. Michler, F.J. Balta-Calleja (eds.): *Nano- and Micromechanics of Polymers*, Hanser Publ., Munich 2012.

N. Rudolph, T. Osswald (eds.): *Polymer Rheology: Fundamentals and Applications*, Hanser Publ. Munich 2014.

Plastics, General Survey, 2. Production of Polymers and Plastics

Hans-Georg Elias, Michigan Molecular Institute, Midland, United States

Rolf Mülhaupt, Institute for Macromolecular Chemistry, Freiburg, Germany

1.	Polymer Manufacture	149
1.1.	Raw Materials	149
1.1.1.	Wood	149
1.1.2.	Coal	150
1.1.3.	Natural Gas and Fracking	150
1.1.4.	Petroleum	150
1.1.5.	Other Renewable Resources and Biorefineries	151
1.2.	Polymer Syntheses: Overview	152
1.2.1.	Classifications	152
1.2.2.	Functionality	155
1.2.3.	Cyclopolymerizations	155
1.3.	Chain-Growth Polymerization	156
1.3.1.	Thermodynamics	156
1.3.2.	Free-Radical Polymerizations	157
1.3.3.	Anionic Polymerizations	163
1.3.4.	Cationic Polymerizations	163
1.3.5.	Ziegler–Natta, ROMP and ADMET Catalysis	164
1.3.6.	Copolymerizations	172
1.4.	Polycondensations and Polyadditions	173
1.4.1.	Bifunctional Polycondensations	173
1.4.2.	Multifunctional Polycondensations and Hyperbranched Polymers	175
1.4.3.	Polyaddition	176
1.5.	Polymer Reactions	176
1.5.1.	Polymer Transformations	176
1.5.2.	Block and Graft Formations	177
1.5.3.	Cross-Linking Reactions	178
1.5.4.	Degradation Reactions	178
2.	Plastics Manufacture	178
2.1.	Homogenization and Compounding	178
2.2.	Additives	179
2.2.1.	Overview	179
2.2.2.	Chemofunctional Additives	179
2.2.3.	Processing Aids	180
2.2.4.	Extending and Functional Fillers	181
2.2.5.	Plasticizers	183
2.2.6.	Colorants	184
2.2.7.	Blowing Agents	184
2.3.	Processing	184
	References	185

1. Polymer Manufacture

1.1. Raw Materials [1]

The raw materials for most industrially used plastics are synthetic polymers and prepolymers; only a few are derived from naturally occurring polymers or monomers. The overwhelming majority of polymers for plastics are organic materials; very few are semiorganics (i.e., inorganic polymer chains with organic substituents). Raw materials for polymers include biomass, such as wood and fossil resources, such as coal, petroleum, natural gas, and gases supplied by fracking. In view of the increasing energy demand and the limited fossil resources, an important objective is to substitute fossil resources by renewable ones. Another important strategy aims at direct chemical fixation of carbon dioxide to produce monomers and polymers directly from carbon dioxide without endangering food supply.

1.1.1. Wood

Wood (→ Wood) is a naturally occurring composite of oriented cellulose fibers in a continuous matrix of cross-linked lignin; it is plasticized by water and "foamed" by air (vacuoles). The water content varies from 40–60% for green wood to 10–20% for air-dried wood. Solids include cellulose (42%); hemicelluloses (28–38%); lignin (19–28%); and proteins, resins, and waxes (2–3%), depending on the type and age of the plant. Most of the wood is used directly as fuel or for construction purposes, but one-sixth is converted into wood pulp or cellulose pulp for the manufacture of paper, cardboard, or rayon. A very small amount of wood is filled with monomers that are subsequently polymerized to give polymer wood.

Repeating unit (cellobiose) of cellulose (R=H) and cellulose derivatives

Cellulose pulp is chemically transformed into cellulose derivatives (→ Cellulose; → Cellulose Esters; → Cellulose Ethers):

- 2,5-Acetate with 2,5 acetate groups, R = OOC−CH_3, per glucose residue (fibers, cigarette filters)
- Cellulose triacetate with three acetate groups per glucose residue (fibers)
- Cellulose acetobutyrate with R=29–6 mol% acetyl and 17–48 mol% butyryl groups (a thermoplastic for tubes, packaging)
- Cellulose nitrate (R−NO_2; various degrees of substitution; gun cotton, films, lacquers, celluloid)
- Cellulose ethers with R−CH_3, C_2H_5, $CH_2CH_2OR^1$, or CH_2−$CHOR^2$−CH_3, where R^1 may be $CH_2CH_2OR^1$ or CH_2CH_2OH and R^2−CH_2−$CHOR^2$−CH_3 or CH_2−CHOH−CH_3 (thickeners, binders, suspension agents, injection molding materials, etc.)

Huge amounts of *lignin sulfonates* (→ Lignin) are generated during wood pulp production, most of which are burned. Small amounts are used for street pavement, as binders for foundry sand, or as drilling agents, and even smaller amounts as raw materials for organic intermediates (→ Lignin). Degradation products of lignin have been utilized for the synthesis of ion-exchange resins and other polymers. Lignocellulose of forestry and agricultural wastes represents an attractive source of glucose as a renewable feedstock for biorefineries without competing with food production. In biorefineries glucose is converted into a variety of monomers including lactic acid, butadiene, acrylic acid, and even ethylene and propene.

1.1.2. Coal

Coals (→ Coal) are polymerized and partly cross-linked hydrocarbons with average compositions between $C_{75}H_{140}O_{56}N_2S$ (peat) and $C_{240}H_{90}O_4NS$ (anthracite). They were the primary raw materials for acetylene-based monomers during the first half of the 20th century but have since been replaced in most countries by oil and natural gas as monomer sources.

Coals deliver coke and coal tar. A number of aliphatic monomers are produced from coke via acetylene, such as chloroprene, various vinyl monomers, acrylonitrile, and hexamethylenediamine. Coal tar delivers aromatics, such as benzene, xylene, phenol, and phthalic anhydride, which lead subsequently to polystyrene, phenolic resins, and glyptal resins, to name a few.

1.1.3. Natural Gas and Fracking

Natural gas (→ Natural Gas) is a gas with a high proportion of aliphatic hydrocarbons. European gas is rich in methane, whereas American and Saudi Arabian gases are relatively rich in higher hydrocarbons. Natural gas is processed to synthesis gas, ethylene, and acetylene. These gases are used to produce a variety of monomers for polymers. The increasing exploitation of fracking enables the production of hydrocarbons which offer great potential as cost-effective alternative to crude oil.

1.1.4. Petroleum

Petroleum (crude oil) (→ Oil Refining) is the main feedstock for monomers. It consists of 95–98% hydrocarbons and 2–5% oxygen, nitrogen, and sulfur compounds. The hydrocarbons are largely aliphatic, partly naphthenic, and to a small extent aromatic, depending on the source. Petroleum distillation leads to saturated hydrocarbons (gas, naphtha, gasoline, kerosene, diesel fuel, heating oil, etc.). For monomer production, these hydrocarbons are subsequently cracked to mixtures of olefins that are fractionated by distillation. The resulting compounds are used as such or are converted further into the desired monomers.

The polymer industry is a major petrochemical customer: About 60% of the cracking and subsequent products of naphtha are used to produce polymers (ca. 50% for plastics–elastomers, ca. 10% for fibers).

1.1.5. Other Renewable Resources and Biorefineries [2–6]

Vegetable oils are a relatively large source of raw materials. They have high hydrocarbon content and consist of mixtures of fatty acid triglycerides and are usually subdivided into drying oils with high linolenic and linoleic acid content, semidrying oils with high linoleic and oleic acid content, and nondrying oils with high oleic acid content (→ Fats and Fatty Oils). Some oils are used directly for paints; others are chemically converted into monomers.

Castor oil is converted into methyl ricinoleate, which is thermally cracked into methyl undecenate and heptanal. Methyl undecenate is the raw material for 11-aminoundecanoic acid, which is subsequently polycondensed to form polyamide 11. Alkali scission of castor oil or ricinoleic acid produces sebacic acid, a monomer used to prepare polyamide PA6,10. Moreover, castor is propoxylated and end-tipped with ethylene oxide in order to produce polyols, thus enabling the formation of polyurethanes containing up to 30 wt % renewable resources.

Soybean oil is converted by a series of chemical reactions into the methyl ester of the C_{10} amino acid $H_2N(CH_2)_9COOCH_3$, which is the monomer for PA 10. The plant *Crambe abyssinica* contains about 55% of erucic acid, which leads to the C_{13} amino acid and subsequently to PA 13 ($\sim[NH(CH_2)_{12}CO]_n\sim$).

Corn husks and other agricultural waste are rich in *pentosans*, which can be converted into tetrahydrofuran and further into polytetrahydrofuran $[HO(CH_2CH_2CH_2CH_2O)_nH]$, which serves as a soft block in certain polyurethanes. Moreover, epoxidized soybean oil is produced in large scale and employed as plasticizers for PVC.

Other Natural Oils. Although the production of polymers based upon soybean oil feedstocks may compete with food production, other nonfood oils, such as linseed and tung oil are attractive intermediates for plant-oil based polymers.

Bio-Based Monomers from Biorefineries. Progress made in biotechnology and in the production of biofuels is stimulating the development of new processes for exploiting biomass conversion in biorefineries as intermediates for bio-based plastics. In principle, there are three different strategies for biomass conversion (i) biomass-to-liquid conversion, (ii) biorefining exploiting glucose, bioethanol, glycerol, and plant oils as raw materials, and (iii) direct catalytic liquefaction and hydrothermal carbonization of biomass. Figure 1 gives an overview on the different strategies to produce bio-based monomers. Similar to the Fischer–Tropsch chemistry in coal-to-liquid conversion, biomass is converted by partial oxidation into syngas consisting of a mixture of carbon monoxide and hydrogen, which serves as feedstock for monomer production. This is a rather energy intensive process that is sensitive to catalyst poisoning by biomass components. A more energy-efficient process is the biotechnological conversion of glucose, preferably gained from nonfood sources, such as lignocellulose from agricultural and forestry wastes, into ethanol and a variety of other monomers, such as glycols and lactic acid. Bioethanol is readily converted into ethylene by splitting off water. In addition, butadiene and propene can be produced from bioethanol. In biodiesel production, vegetable oils are converted in their methyl ester by transesterification of oils with methanol. The byproduct glycerol can be converted into a variety of monomers including epichlorhydrin, acrylic acid, and polyols (see Fig. 1).

Whereas biomass-to-liquid conversion is energy intensive, the catalytic direct conversion

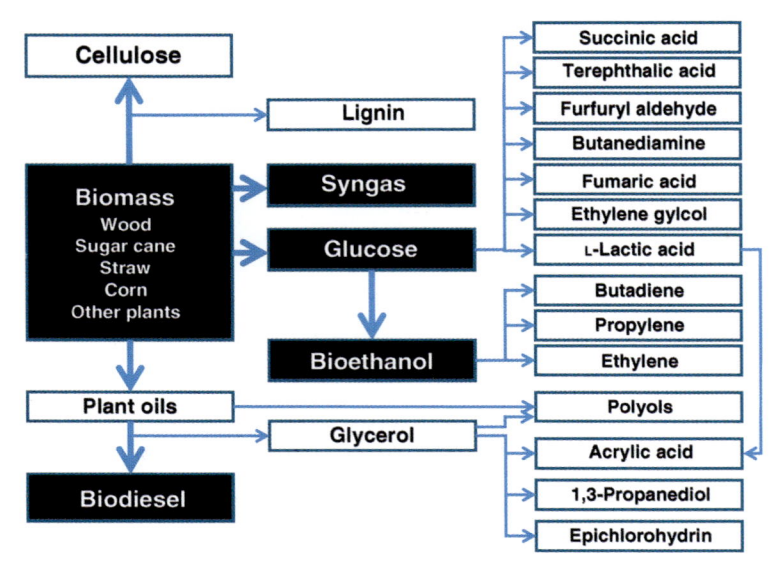

Figure 1. Biorefineries and biobased monomers

of biomass into diesel represents another attractive green route to renewable oil. In this process, plastics wastes as well as biomass can be employed as carbon source. In the third approach, entire plants can be converted into coal by hydrothermal carbonization. This is an exothermic process that operates at much lower temperature than the Fischer–Tropsch process. This renewable coal obtained by hydrothermal biomass carbonization is an attractive feedstock for catalytic liquefaction and production of renewable oil.

Carbon dioxide that is emitted by power plants, is the energy sink in combustion processes. Activation and chemical fixation of carbon dioxide requires energy, which can be supplied from renewable resources, thus enabling chemical energy storage of excess energy from wind and solar power plants. In contrast to traditional conversion of carbon dioxide into carbon monoxide and hydrogen by reduction with coal at very high temperatures, oil and gas are produced by electrochemical reduction of carbon dioxide in water. However, this process is not yet feasible for the production of monomers on industrial scale. Another more viable approach exploits the direct conversion of carbon dioxide with reactants, such as epoxides with a high energy content. Thus cyclic carbonates as well as linear aliphatic polycarbonates are available. Polyfunctional cyclic carbonates serve as intermediates for the formation of green polyurethanes without requiring phosgene or isocyanates.

1.2. Polymer Syntheses: Overview [7]

1.2.1. Classifications

Polymers may be synthesized from monomers by polymerization or from other polymers by chemical transformations. The nomenclature and classification of these reactions have grown and changed with the times; they are not very systematic and are sometimes confusing.

All reactions of small monomer molecules to macromolecules are called *polymerizations*. Polymerizations are classified according to the:

- Origin of polymers (bio and chemical polymer synthesis)
- Chemical structure of monomers (vinyl polymerizations, diene polymerizations, ring-opening polymerizations, etc.)

- Chemical structure of resulting polymers (linear, branching, cross-linking, isomerizing, cyclopolymerization, etc.)
- Relative composition of monomers and monomer units (addition, condensation, etc.)
- Formation of low molar mass byproducts (polycondensation)
- Type of polymerization initiation (thermal, catalytic, photochemical, enzymatic, etc.)
- Type of propagating species (free radical, anionic, cationic, polyinsertion, etc.)
- Type of mechanism (chain growth, step growth, living, controlled, catalytic, etc.)
- Reaction media (bulk, solution, emulsion, suspension, mesophase, etc.)
- State of matter (gaseous, homogeneous, heterogeneous, super critical, etc.)

The resulting four types of polymerization can be depicted schematically as

$P_n + P_m \rightarrow P_{n+m}$ $P_n + M \rightarrow P_{n+1}$
Polyaddition   Chain polymerization

$P_n + P_m \rightarrow P_{n+m} + L$ $P_n + M \rightarrow P_{n+1} + L$
Polycondensation   Polyelimination

where M denotes monomer; P_n, P_m, P_{n+m}, P_{n+}, denote polymers and L leaving molecules (e.g., water in polyesterifications of diacids and diols).

The names of three of these four types have been used with different meanings by different researchers since the 1930s. The following discussion thus centers on the present nomenclature recommendations by IUPAC.

Chain-Growth Polymerization. Polymer chains grow by repeated addition of monomer molecules M to polymer molecules P_i, consisting of i monomeric units m, an initiator end group R, and an "active center $*$," without the formation of leaving molecules:

$R(m_n)m^* + M \rightarrow R(m_{n+1})m^*$

An example is the polymerization of vinyl monomers ($CH_2=CHR$) by chain carriers $Y*$, which may be free radicals $Y^{\cdot}$, anions Y^-, or cations Y^+:

$Y-CH_2-CHR^* + CH_2 = CHR \rightarrow Y-CH_2-CHR-CH_2-CHR^*$

A *living polymerization* is a chain polymerization consisting only of fast initiating reactions ($Y* + CH_2=CHR \rightarrow Y-CH_2-CHR*$) and propagation reactions [$Y-(CH_2-CHR)_n* + CH_2=CHR \rightarrow Y-(CH_2-CHR)_{n+1}*$] without termination of the growing chains and without side reactions. In such reactions, the degree of polymerization is directly proportional to the conversion u of monomer molecules or p of functional groups, respectively (Fig. 2, curve CP–L). An example is the polymerization of styrene $CH_2=CH(C_6H_5)$ by the butyl anion $C_4H_9^-$ from butyllithium, LiC_4H_9.

The variation of the degree of polymerization with monomer conversion is quite different for *radical polymerizations*, such as the polymerization of styrene by benzoyloxy radicals $C_6H_5COO^*$ from the thermal decomposition of dibenzoyl peroxide, $C_6H_5COO-OOCC_6H_5$. Radicals are here formed successively and not "at once" as are the butyl anions in living

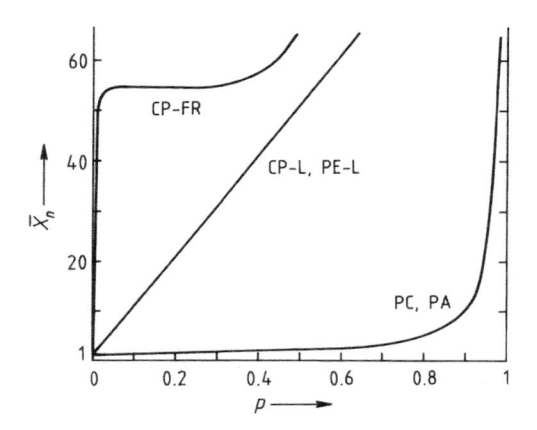

Figure 2. Change of number-average degree of polymerization $\overline{X}_n X^-$ with the extent of reaction of functional monomer groups p for living chain polymerizations CP–L and living polyeliminations PE–L (with initial monomer–initiator ratio 100 : 1 mol/mol), free-radical chain polymerizations CP–FR with gel effect (schematic for high initiator concentrations), and equilibrium polycondensations PC and polyadditions PA (numerically exact)

polymerizations. Growing polymer chains are terminated by reaction with other radicals (recombination, disproportionation). Many side reactions may occur because of the high and fairly indiscriminate reactivity of radicals. Depending on reaction conditions, the degree of polymerization may remain constant (primarily at low conversions) or increase with monomer conversion (so-called gel effect, also referred to as autoacceleration or Trommsdorff effect) (Fig. 2, curve CP–FR). Whereas in living chain-growth polymerizations all chain ends are active during polymerization, in controlled polymerization reactions there is a rapid equilibrium between temporarily inactive ("dormant") chain ends and active chain ends. Similar to living polymerization, controlled chain-growth polymerizations afford narrow molar mass distribution, first order polymerization kinetics, and a degree of polymerization which is proportional to conversion.

Chain-growth polymerizations have generally been called *addition polymerizations* in the literature. Addition polymerization is not to be confused with polyaddition (see below).

Condensative Chain-Growth Polymerization is the name recommended by IUPAC for a polymerization by repeated addition of monomer molecules with elimination of low molar mass molecules L; this polymerization has thus also been called *polyelimination* [8]:

$$R(m_n)m^* + M \rightarrow R(m_{n+1})m^* + L \quad (1)$$

Biological polymerizations to polysaccharides seem to proceed exclusively through polyeliminations. Cellulose of cotton, for example, is formed by polymerization of guanosine diphosphate D-glucose attached to a lipid matrix. The reaction proceeds via insertion of the glucose moiety into the chain with elimination of guanosine diphosphate.

Synthetic polymers are rarely formed by polyelimination. Examples are the polymerizations of α-amino acid *N*-carboxyanhydrides Leuchs anhydrides) and the "oxidative polymerization" of 2,6-dimethylphenol to polyphenylene oxide (PPO, PPE):

[Reaction scheme: 2,6-dimethylphenol + 1/2 O$_2$ → polyphenylene oxide + H$_2$O]

Polycondensation [9]. The step-growth polymerization, known as polycondensation, proceeds by reaction between molecules of all degrees of polymerization (not just between a polymer molecule and a monomer molecule as in chain polymerization and polyelimination) resulting in the formation of low molar mass molecules as byproducts. Many polymers with hetero atoms in the chains are generated by polycondensation. An example is the polycondensation of 11-aminoundecanoic acid to polyamide 11:

$$n\, H_2N(CH_2)_{10}COOH \rightarrow H[NH(CH_2)_{10}CO]_nOH + (n-1)H_2O \quad (2)$$

In these reactions, dimers ($n = 2$) are formed first, which then react to form tetramers ($n = 4$), etc. The reaction is thus statistical with respect to the size of the molecules to which the functional groups are attached. The degree of polymerization of the polymer is a number average $\overline{X}_n$ over all degrees of polymerizations of the polymer molecules.

All amino groups have the same chance to react if their reactivity is independent of molecule size. The reaction of amino groups attached to molecules with higher molar mass adds, however, considerably more to the increase of the degree of polymerization than the reaction of lower molar mass molecules: The degree of polymerization snowballs with increasing extent of reaction of functional groups (Fig. 2, PC). High conversion of functional groups and no side reactions are prime requirements for achieving high molar mass in step-growth polymerization. Molar mass is controlled either by adjusting the conversion of functional groups or by adding monofunctional chain terminating agents. The term polycondensation was used in the past as a synonym for condensation polymerization. The latter term included both the current concepts of polycondensation and polyelimination (condensative chain polymerization).

Polyaddition. This is a polymerization by reaction among molecules of all degrees of polymerization but without the formation of low molar mass byproducts. The number-average degree of polymerization of the resulting polymers varies with the extent of reaction of functional groups in the same way as it does in polycondensation. An example of polyaddition is the polymerization of diisocyanate with diol to form polyurethane:

$$n \, OCN-R-NCO + n \, HO-R'-OH$$
$$\rightarrow OCN(R-NH-CO-O-R'O)_n H \qquad (3)$$

These polymerizations have also been called "polyadduct formations." Polyaddition is not to be confused with addition polymerization, the term used in the literature for chain-growth polymerization.

1.2.2. Functionality

The functionality of monomers is essential for controlling polymer constitution. (e.g., see Eq. 1). When difunctional monomers, such as divinylbenzenes $CH_2=CH-C_6H_4-CH=CH_2$, are added to the chain-growth polymerization of styrene, first branching and later cross-linking takes place. The second vinyl group of a divinylbenzene possesses, however, a different reactivity after polymerization of the first one. The two reactivities may differ so strongly in certain ionic polymerizations of divinylbenzenes that linear polymers, and not cross-linked ones, result (e.g., in the cationic polymerization of 1,4-divinylbenzene by acetyl perchlorate in dichloromethane).

The isocyanate group $-N=C=O$ of diisocyanates can react with hydroxy groups of diols to produce linear polyurethanes by step-growth polyaddition (Eq. 3). It may, however, undergo a chain-growth polymerization to form polyisocyanates via the polymerization of the N=C double bond

$$n \, R-N=C=O \rightarrow (NR-CO)_n$$

It should be noted that isocyanates react with urethane to afford branching via allophanate groups (Eq. 4)

$$\sim NH-CO-O\sim + O=C=N\sim$$
$$\longrightarrow \sim NH-CO-\underset{\mid}{N}-CO-O\sim \qquad (4)$$

The functionality f of polymer molecules is given by the functionality f_o of the monomer molecules and the number X of mers if no intramolecular rings are formed between functional groups.

$$f = 2(f_o - 1) + (X - 2)(f_o - 2)$$
$$= 2 + X(f_o - 2)$$

A monomer with functionality $f_o = 3$ thus gives a dimer ($X = 2$) with a functionality $f = 4$ and a 100-mer ($X = 100$) with a functionality $f = 102$. Because the molecule needs to be bifunctional only for the formation of linear molecules, the "extra" functionalities lead to interconnections between various molecules (branching) and thus finally to cross-linked, "infinitely large" molecules.

1.2.3. Cyclopolymerizations

Polymerizations of multifunctional molecules may proceed not only intermolecularly to produce branched and cross-linked molecules but also intra/intermolecularly with the formation of rings in chains. These polymerizations are called cyclopolymerizations. They occur especially with monomers having carbon–carbon double bonds in the 1,5- or 1,6-position. An example is acrylic anhydride (**1**), $CH_2=CH-CO-O-OC-CH=CH_2$, which polymerizes free-radically even at high monomer concentrations to soluble polymers with 90–100% six-membered (**2**) and five-membered (**3**) ring structures, and only 0–10% "linear" monomeric units **4** (R symbolizes a polymer chain and * an active chain end):

1.3. Chain-Growth Polymerization

Chain-growth polymerization and polyelimination (condensative chain polymerization) agree in their polymerization kinetics, molar mass distributions, etc., if similar polymerization mechanisms are compared. They differ in the formation of low molar mass byproducts and thus in the design of industrial polymerization reactors.

1.3.1. Thermodynamics

Polymerizations occur only if their molar standard Gibbs energies ΔG_p^o are negative. Because the Gibbs energy

$$\Delta G_p^o = \Delta H_p^o - T \cdot \Delta S_p^o = -RT \cdot \ln K \quad (5)$$

is determined by both the polymerization enthalpy ΔH_p^o and the polymerization entropy ΔS_p^o, four different cases are possible:

Case 1. Both enthalpy and entropy are negative; the entropy term $-T \cdot \Delta S_p^o$ becomes more positive with increasing temperature. At a certain temperature, the enthalpy and entropy terms balance each other; no polymerization to high molar mass molecules is possible above this *ceiling temperature*. Oligomers, however, may be formed at the ceiling temperature, because polymerization equilibria consist of a series of consecutive equilibria between monomer and growing chains of different degrees of polymerization.

Case 1 is the most common case. It is found for monomers with polymerizable double and triple bonds, such as C=C, C=O, C=S, or C=N. The polymerization entropies of such monomers are practically determined only by the loss of translational entropy; they are thus almost independent of the chemical structure of the monomers. The polymerization enthalpy of α-olefins CH_2=CHR is influenced little by resonance stabilization and is practically independent of the size of the substituent R. Steric hindrance, however, strongly decreases the polymerization enthalpy of 1,1-disubstituted compounds CH_2=CRR′. Styrene and α-methylstyrene, have practically the same polymerization entropies (-105 J K^{-1} mol^{-1}) but very different polymerization enthalpies (-71 vs. -35 kJ/mol, respectively) for the polymerization of liquid monomers to condensed polymers. The ceiling temperatures in bulk are 303°C (styrene) and 60°C (α-methylstyrene), (i.e., the latter cannot be homopolymerized to high molar mass compounds above 60°C).

σ-Bonds are opened and formed again in ring-opening polymerization of cyclic monomers. Differences of bond energies in monomers and polymers are thus practically zero, and the polymerization enthalpy is determined by delocalization and strain energies. Polymerization enthalpies are very negative for ring-opening polymerization of three-membered rings, increase to about zero for five- or six-membered rings, drop again to negative values, and return almost to zero for very large rings. The polymerization entropies of very small rings are very negative due to the release of rotational entropy upon polymerization. They become less negative with increasing ring size.

Case 2. The polymerization enthalpy is negative (or zero) and the polymerization entropy positive. Polymerization is possible at all temperatures in this rare case, the polymerization of cyclooctamethyltetrasiloxane being an example.

Case 3. The polymerization enthalpy is positive (or zero) and the polymerization entropy is negative. No polymerization is possible at any temperature. This case seems to apply to the polymerization of acetone.

Case 4. Both polymerization enthalpy and polymerization entropy are positive. A *floor temperature* exists below which no polymerization is possible. This rare case is fulfilled for the polymerization of tightly packed cyclic monomers, such as cyclooctasulfur or oxacycloheptane (oxepane) since the number of rotatory degrees of freedom increases strongly upon polymerization to open chains.

Polymerization enthalpies and entropies are influenced by the state of monomers and polymers (gaseous, liquid, crystalline, etc.), the physical interaction between reactants themselves or reactants and solvents, and pressure. Especially important for industrial polymerizations is the fact that most (but not all, see above) chain polymerizations are exothermic. The heat of polymerization can be considerable: Adiabatic polymerization of gaseous ethylene to crystalline polyethylene would result in a temperature increase of 1 800 K for complete monomer conversion. Polymerization reactors must be designed in such a way that the heat of polymerization can be removed rapidly because, otherwise, inhomogeneous charges may result or the reactor may even explode.

1.3.2. Free-Radical Polymerizations [10]

Free-radical polymerizations are initiated and propagated by free radicals. Such polymerizations are relatively insensitive to impurities in the reaction mixture and fairly easy to control. They are thus the methods of choice for the industrial production of plastics if the monomers can be subjected to free-radical polymerizations. This is true for ethylene and many ethylene derivatives: Most commodity plastics are thus obtained from free-radical polymerizations (Table 1).

Other industrial polymers formed by free-radical homopolymerizations include poly(vinyl acetate) (for coatings and adhesives), poly(acryl ester) (for adhesives), polyacrylamide (for thickeners), polyacrylonitrile (for fibers), polychloroprene (for elastomers), poly(acryl acid) (for thickeners), and poly(N-vinylpyrrolidone) (for various applications).

Polymerizations may be carried out in the gas phase, bulk, suspension, emulsion, solution, or under precipitation. Most monomers are bifunctional; they lead to linear or slightly branched polymers and thus to thermoplastics. Thermosets obtained from free-radical homopolymerizations include bis- and trimethacrylates as well as diallyl and triallyl compounds that cross-link under polymerization conditions:

$$CH_2=CH-CH_2-R-CH_2-CH=CH_2 \longrightarrow$$

$$\sim CH_2 \qquad CH_2\sim$$
$$\sim CH-CH_2-R-CH_2-CH\sim$$

where R may be $COO(CH_2)_2O(CH_2)_2OOC$ [diethylene glycol bis(allyl carbonate], o-$C_6H_4(COO)_2$ (diallyl phthalate), or m-$C_6H_4(COO)_2$ (diallyl isophthalate). Other allyl compounds are used as cross-linking comonomers (diallyl fumarate, diallyl maleate, triallyl cyanurate).

Most free-radical polymerizations follow the kinetics of true chain reactions. In initiating (start) reactions (Eq. 7), radicals $R^\bullet$ are formed by the decomposition (Eq. 6) of initiator

Table 1. Plastics by free-radical chain-growth polymerizations[*]

Monomers	Polymerization in						Monomeric units
	Gas phase	Bulk	Suspension	Emulsion	Solution	Precipitation	
Living polymerization							
p-Xylylene		+					$-CH_2-(p-C_6H_4)-CH_2-$
Irreversible polymerization to linear or slightly branched polymers							
Ethylene		+	+	(+)	(+)	(+)	$-CH_2-CH_2-$
Styrene		+	+	(+)	(+)		$-CH_2-CH(C_6H_5)-$
p-Methylstyrene			+	+			$-CH_2-CH(p-CH_3C_6H_4)-$
Vinyl acetate		(+)	(+)	+	(+)		$-CH_2-CH(OOCCH_3)-$
Vinyl chloride	+		(+)	(+)	+	(+)	$-CH_2-CHCl-$
Vinyl fluoride			+				$-CH_2-CHF-$
Vinylidene fluoride			+	+			$-CH_2-CF_2-$
Trifluorochloroethylene			+				$-CF_2-CFCl-$
Tetrafluoroethylene			+				$-CF_2-CF_2-$
Methyl methacrylate		+	+	+	+		$-CH_2-C(CH_3)(COOCH_3)-$
Cross-linking polymerization							
Diallyl phthalate			+				see Equation (5)

[*] + = major processes; (+) = minor processes.
Other industrial polymers by free radical homopolymerizations include poly(vinyl acetate) (for coatings and adhesives), poly(acryl ester) (for adhesives), polyacrylamide (for thickeners), polyacrylonitrile (for fibers), polychloroprene (for elastomers), poly(acrylic acid) (for thickeners), and poly(N-vinyl pyrrolidone) (for various applications).

molecules I, that react with monomer molecules M. The resulting polymeric free radicals add further monomer molecules in the propagation reaction (Eq. 8) until the growing polymer radicals are terminated by recombination (Eq. 9) or disproportionation (Eq. 10) with other polymer radicals or by addition of initiator radicals (Eq. 11):

$$I \rightarrow 2\,R^\bullet \tag{6}$$

$$R^\bullet + M \rightarrow R - M^\bullet \tag{7}$$

$$R - M^\bullet + M \rightarrow R - M - M^\bullet, \text{etc.} \rightarrow R - M_{n-1} - M^\bullet \tag{8}$$

$$R - M_{n-1} - M^\bullet + {}^\bullet M - M_{m-1} - R \rightarrow R - M_{n+m} - R \tag{9}$$

$$\begin{aligned} R - M_{n-1} - M^\bullet + {}^\bullet M - M_{m-1} - R \rightarrow R - M_{n-2} - M \\ = M + X - M_m - R \end{aligned} \tag{10}$$

$$R - M_{n-1} - M^\bullet + R^\bullet \rightarrow R - M_n - R \tag{11}$$

In Equation (10), X is an atom or a group that is transferred from the ultimate monomeric unit of one polymer radical to another polymer radical, for example, the H atom in the reaction

$$CH_2 - C^\bullet HR + {}^\bullet CHR - CH_2 \rightarrow CH_2 - CH_2R + CHR = CH_2$$

These reactions are kinetically classified as termination reactions and not as transfers because they lead to dead chains. The term "transfer" is rather reserved for chemical transfers that generate new propagating radicals, such as

$$CH_2 - C^\bullet HR + AX \rightarrow CH_2 - CH_2\,RX + A^\bullet \tag{12}$$

Initiation. Free-radical polymerizations can be initiated thermally by thermal initiators, by redox initiators, by photoinitiators, or electrolytically.

Thermal initiation is utilized in the polymerization of *p*-cyclophane to poly(*p*-xylylene):

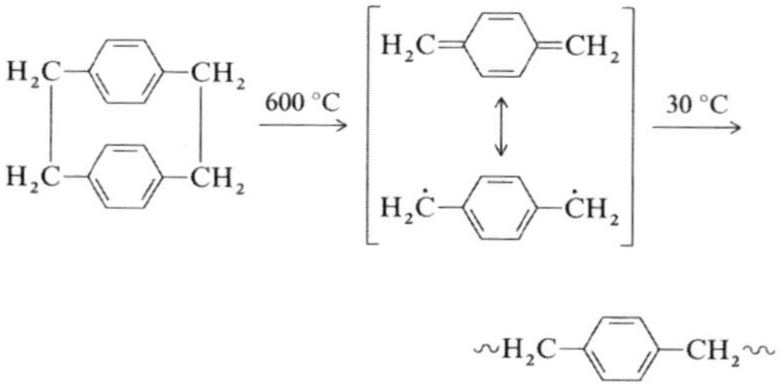

Propagation proceeds by addition of biradicals of various sizes i.e., by polyaddition and not by chain polymerization. The resulting polymer is cross-linked because radicals also attack methylene groups, which generates more than two radical sites per reactant and thus cross-linking sites. The reaction is utilized for the formation of thin coatings.

Styrene is the only monomer that is polymerized commercially by thermal initiation through monoradicals. The primary Diels–Alder adduct of two styrene molecules reacts with another styrene molecule to give two different starter radicals:

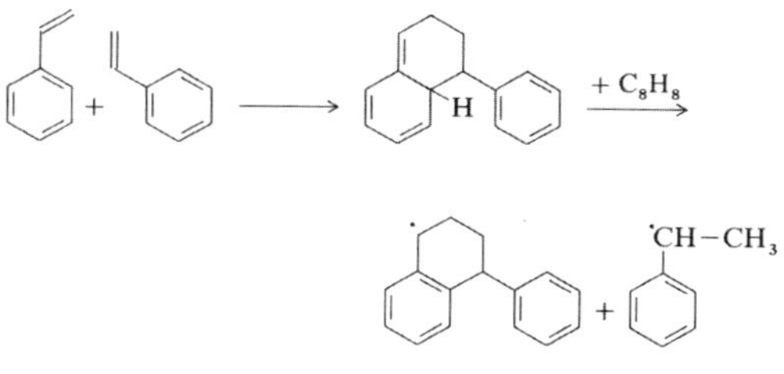

Only relatively small amounts of styrene are, however, polymerized commercially by this method. Most styrene polymerizations, like many other commercial free-radical polymerizations, are started by radicals from *thermal initiators*. These thermal initiators dissociate homolytically into two radicals at elevated temperature, usually 60–80°C. For the bulk polymerization of styrene, high-temperature initiators, such as 1,2-dimethyl-1,2-diethyl-1,1-diphenyl-ethane or vinylsilane triacetate are preferred. For styrene polymerization in suspension, dibenzoyl peroxide (BPO) ($C_6H_5CO-O-O-OCC_6H_5$) and *tert*-butyl perbenzoate [$C_6H_5CO-O-O-C(CH_3)_3$] are used.

Many other bulk polymerizations employ diisopropyl peroxydicarbonate (IPP), which decomposes according to

$(CH_3)_2CHO - COO - OOC - OCH(CH_3)_2 \rightarrow 2(CH_3)_2CHO - COO^\bullet$

Water-soluble initiators, such as dipotassium persulfate ($K_2S_2O_8$) and certain redox initiators are used in emulsion polymerizations. *Redox initiators* generate radicals by reaction of a reducing agent with an oxidizing agent (e.g., $Fe^{2+} + H_2O_2$). These systems need far less activation energy for the formation of radicals, and polymerization can thus be initiated at low temperature.

Photochemical initiations are utilized for the production of lithographic plates ($\rightarrow$ Imaging Technology, 3. Imaging in Graphics Art) and for hardening lacquers but not in the manufacture of plastics. *Electrolytic polymerizations* find applications in the coating of metal sheets by plastics.

Only initiator radicals are formed at the early stages of free-radical polymerizations. These radicals are successively converted into monomer radicals by addition of initiator radicals to monomer molecules and into polymer radicals by further addition of monomers. At the same time, some polymer radicals are removed by termination reactions. Finally, a steady state is established in which as many radicals are formed as are removed by termination. This state of constant total radical concentration (ca. 10^{-8} mol/L) is in general observed within seconds at monomer conversions of $10^{-2}-10^{-4}\%$.

Propagation. Monomer molecules are usually added head-to-tail to growing polymer radicals:

$CH_2 - CHR^\bullet + CH_2 = CHR \rightarrow CH_2 - CHR - CH_2 - CHR^\bullet$

Tail-to-tail additions also occur (see $\rightarrow$ Plastics, General Survey).

Free-radical polymerizations lead in general to atactic or predominantly syndiotactic polymers. The syndiotacticity increases with decreasing polymerization temperature.

Termination. Termination reactions by recombination, disproportionation, and (at higher initiator concentrations) addition of initiator radicals, cannot be avoided. Because initiator radicals and thus also polymer radicals are formed successively and are deactivated at random by different termination reactions, distributions of molar masses are formed. The width of these distributions is given by the relative proportion of the different types of termination reactions.

Chain Transfer. The transfers of radicals to monomers, solvents, and initiators terminate polymer chains and generate others (Eq. 12). The ratios of rate constants k_{tr} of transfer to rate constants k_p of propagation are called transfer constants, $C_{tr} = k_{tr}/k_p$; at 60°C they possess values between ca. 2×10^{-6} for the transfer of polystyryl radicals to benzene and 5 700 for the transfer of poly(vinyl acetate) radicals to carbon tetrabromide.

Chain transfers often lead to new radicals that have approximately the same reactivity as the disappearing ones. In this case, the rate of polymerization is not decreased by chain transfer, but the degree of polymerization is lowered. Certain transfer agents with high transfer constants (such as thiols) are thus often employed to regulate molar mass.

Inhibition and retardation occur if the new radical is more sluggish in adding monomers than the transferring polymer radical. In the case of *inhibition*, stable radicals are formed by chain transfer reactions. Typically, inhibitors, such as methylhydroquinone, are added. Moreover, retardation is observed when chain transfer to nonionic allyl monomers generates resonance-stabilized allyl radicals. The addition of monomer molecules to these radicals produces radicals that are more stabilized by resonance than their precursors and thus do not start a polymer chain. The polymerization of noncharged monoallyl monomers thus leads to oligomers with degrees of polymerization of ca. 10–20.

Controlled Radical Polymerization [11, 12]. Reversible deactivation is the key feature of controlled radical polymerization. Unlike living polymerization, the free radicals at the chain end are rendered temporarily inactive ("dormant"). Because there is a rapid equilibrium between free radical and dormant sites, controlled radical polymerization exhibits all features typical for living polymerization, i.e., the degree of polymerization is proportional to conversion and the rate of polymerization follows first order kinetics. The reversible trapping of free radicals prevents undesirable chain termination by recombination and disproportionation.

chain growth by free radical polymerization. Because free radicals are temporarily trapped, the polymerization rate and monomer conversions are lower than in a living anionic polymerization.

By using controlled radical polymerization techniques, it is possible to synthesize block-, graft-, and star copolymers as well as polymers with narrow molar mass distribution and well-defined end group functionalities. In *nitroxide-mediated controlled radical polymerization* (NMRP), stable nitroxide free radicals, such as (2,2,6,6-tetramethylpiperidin-1-yl)oxyl (TEMPO) radicals are added to a free radical polymerization. Immediately after the nitroxide radical traps the free radial at the polymer chain end, thermal bond cleavage reinitiates

In *atom transfer radical polymerization* (ATRP) organic halides react with redox metals, such as Cu^+, to produce organic free radicals via transfer of the halogen atom. This free radical at the chain end is temporarily terminated by transfer of the halogen, accompanied by changing of the oxidation state resulting from electron transfer of the redox metal. Although copper is preferred for ATRP, various other redox transition metals can be employed. The addition of ligands is required to prevent precipitation of the metal complex.

In *reversible addition chain tranfer polymerization* (RAFT) the polymer free radical adds to a thiocarbonyl group of dithioesters, dithiocarbamate or dithiocarbonates, forming a free radical that undergoes fragmentation to form a polymer radical and thiocarbonyl group.

low k_p value, high pressure (i.e., high monomer concentration) and high temperature are required for achieving free radical ethylene polymerization. Due to inter- and intramolecular chain transfer reactions, the resulting polyethylene chains contain long- and

$$P_m^{\bullet} + \underset{Z}{\overset{S}{\underset{\|}{C}}}\text{-}S\text{-}R \xrightarrow{\text{Addition}} \underset{P_m-S}{\overset{Z}{\underset{\|}{C}}}\text{-}S\text{-}R \xrightarrow{\text{Fragmentation}} \underset{P_m-S}{\overset{Z}{\underset{\|}{C}}}\text{=}S + R^{\bullet}$$

Z: Phenyl, R = C(CH$_3$)$_2$Ph

$+ n \times M \to P_n$

chain alkyl branches. Whereas small amounts of long chain branches account for shear thinning during melt processing, the incorporation of short chain propyl- and butyl-branches impairs crystallization and lowers the polyethylene density. Therefore, high pressure free radical polymerization produces low density polyethylene (LDPE).

Polymerization Kinetics of Free Radical Polymerization. Initiator radicals are generated by initiator decomposition (Eq. 13) with a rate of $2f \cdot k_d \cdot [I]$ where f is the radical yield (i.e., the fraction of radicals that start the polymerization). Initiator radicals disappear in the start reaction with a rate of $k_{st}[P][M]$. In the steady state, as recognized by BODENSTEIN, the polymer radical concentration is constant. Hence, the rate of the start reaction equals the rate of the termination reaction(s), thus $R_{st} = R_t$, and for terminations through deactivation by other polymer radicals,

$$d[P]/dt = R_{st} - R_t = 2f \cdot k_d \cdot [I] - k_t[P]^2 = 0 \quad (13)$$

Because monomer is practically consumed only by the propagation reaction with a rate of $R_p = -d[M]/dt = k_p[P][M]$, the propagation rate at diminishing initiator consumption is obtained from Equation (14)

$$R_p = -d[M]/dt = k_p(2f \cdot k_d/k_t)^{1/2} \cdot [I]_o^{1/2}[M] \quad (14)$$

The polymerization rate is directly proportional to the monomer concentration; bulk polymerization is thus always faster than solution polymerization. Because the kinetic chain length is equivalent to R_p/R_t, the degree of polymerization increases with increasing monomer concentration. Because ethylene has rather low reactivity, as reflected by its

A "self-acceleration" of the polymerization and a concomitant increase of both the polymerization rate and degree of polymerization are observed at higher monomer conversion, especially in bulk and in concentrated solutions. This so-called gel-effect, also known as the Trommsdorff effect, is caused by a changing diffusion control of the termination by mutual reaction of two polymer radicals (i.e., a decrease of termination reactions), which in turn comes from an increasing entanglement of polymer chains. Because neither the rate of the initiator decomposition nor the rate of monomer addition is affected by increasing entanglements, and only the termination is slowed down, the Bodenstein steady-state principle does not apply anymore. Hence, both overall polymerization rates and molar masses simultaneously increase. This autoacceleration broadens molar mass distributions. Owing to the drastically reduced heat transfer at high viscosity, the autoacceleration can cause severe increase of temperature and even explosions, resulting from pyrolysis of polymers and gas evolution at temperatures above 400°C. The addition of chain transfer agents, such as alkylthiols reduces molecular masses, thus preventing viscosity buildup and autoacceleration.

Cross-Linking Polymerization. Free-radical chain polymerizations are employed commercially in the manufacture of diallyl thermosets and in styrene−divinylbenzene copolymerizations for polymer beads. Primary macromolecules produced at very low monomer conversions are almost linear and carry pendant polymerizable groups. The likelihood of an attack at these groups increases with increasing mass average molar mass. Branched molecules are formed that convert to cross-linked polymers at the gel point. Monomer conversion at the gel point cannot at present be predicted theoretically for such free-radical polymerizations.

Process Engineering. Free-radical polymerizations may be performed in the gas phase, in bulk, solution, emulsion, suspension, or under precipitation (Table 1). Each method has advantages and disadvantages.

Bulk Polymerizations. Industrial bulk polymerizations are performed in *liquid monomers*, occasionally with the addition of 5–15 vol% solvent as a polymerization aid. The process leads to very pure products. However, much heat is generated per unit volume and, in addition, by the Trommsdorff effect. Local overheating may cause multimodal molar mass distributions, branching, polymer degradation, discoloring, or explosion. Bulk polymerizations are thus often terminated at 40–60% monomer conversion. The remaining monomer is distilled and recovered. Alternatively, polymerizations may be performed in two steps: First in large batch reactors, then in thin layers. Residual monomer is usually removed by steam.

Suspension polymerization is a "water-cooled" bulk polymerization. Water-insoluble monomers are suspended as small droplets of 0.001–1 cm diameter with the help of suspending agents, such as water-soluble polymers. The polymerization is initiated by oil-soluble initiators. The droplets are converted by polymerization into pearls; thus the process is also called *pearl polymerization*. The addition of protective colloids, such as water-soluble polymers, affords steric stabilization of the formed polymer particles and prevents their agglomeration. Suspension polymerization allows easy control of the reaction and delivers the polymer as easy-to-handle beads. A disadvantage is the cost of water removal and cleaning, along with the sometimes deleterious effects of incorporated suspension agents. The average size of spherical polymer particles decreases with increasing stirring speed. Prominent applications of styrene pearl polymerization include the formation of polystyrene pellets containing pentane as foaming agent and the production of cross-linked polystyrene beads that are sulfonated to produce polymeric cation exchange resins for water deionization.

Emulsion Polymerization. In emulsion polymerizations, water-insoluble monomers are solubilized in water by micelles of surfactant molecules (4–10 nm diameter); the monomers are also suspended in a few droplets (ca. 1 000 nm diameter). Water-soluble initiators decompose into radicals that travel through the water to the micelles where polymerization begins. Polymerization takes place in this micellar reaction compartment, referred to as polymer *latex*. Diffusion of monomer from the droplets to the latex replenishes the monomer that has been used up by reaction in the micelles. The average diameter of these latex particles is 500–5 000 nm. Each latex particle contains just one free radical, which is terminated when another initiator radical enters the latex particle; this remains dormant until a new initiator radical initiates the growth of a new chain. Because the reaction volume is very small in micelles (and in growing latex particles), termination reactions are rare, and polymerization proceeds in a quasi-living manner, leading to much higher molar masses than bulk polymerization. In terms of reaction control, emulsion polymerizations have advantages similar to suspension polymerizations. In addition, they offer higher polymerization rates by redox initiators and higher molar masses (which can be regulated by transfer agents). Emulsion polymerization is applied to produce homo- and copolymers based upon styrene, butadiene, vinyl chloride, vinyl acetate, acrylates as well as tetrafluoroethylene. The resulting latex can furthermore be used directly for producing water-borne paints, adhesives, and coatings forming corrosion resistant polymer films when water vaporizes and the polymers entangle.

Gas-phase polymerizations can be initiated photochemically. Each growing particle contains only one radical and the polymerization is thus quasi-living. The polymerization rate is determined by the rate of monomer adsorption in the precipitating particles and later by monomer diffusion to the occluded polymer radicals. The resulting polymers are very clean.

Solution polymerizations decrease polymerization rates by monomer dilution and allow easy removal of the heat of polymerization. Solvents are, however, costly to remove, and the process is used only if the resulting polymer solutions can be used directly or if monomers and polymers decompose in melts.

1.3.3. Anionic Polymerizations [13]

Anionic polymerizations are initiated by bases, such as alkali metals, alkoxides, amines, phosphines, Grignard compounds, and sodium naphthalene, very often in solution. The initiators dissociate "instantaneously" into the initiating species. These dissociations and the subsequent start reactions need little activation energy; anionic polymerizations thus often proceed with high speed even at -100 °C.

The initiating species and the growing macroions are rarely completely dissociated into anions and counterions. They are rather in equilibrium with various types of ion pairs (non-dissociated species, contact ion pair, solvent-separated ion pair, free ions plus associates of these species). The type and amount of initiating species strongly affect the polymerization rates and the tacticity of polymers.

The initiating species are present in their effective concentrations at the beginning of polymerization; they are not formed one after another as in free-radical polymerization. Termination reactions are fairly rare if protons are absent. Hence, anionic polymerizations are often "living." Very low initiator concentrations are required for living polymerizations to high molar masses (see Section 1.2.1); however, high monomer–initiator ratios are difficult to control and may lead to low polymerization rates. In general, polymers from living polymerizations have very narrow molar mass distributions of the Poisson type if diffusion effects are avoided during the mixing of monomer and initiator solutions.

Contrary to traditional living polymerizations, a variety of controlled anionic polymerizations have been developed. Key feature is a rapid equilibrium between active and dormant endgroups. For example, in *group transfer polymerization (GTP)* there is a rapid equilibrium between inactive silyl ketene acetals and enolate anions, catalyzed by nucleophilic catalysts ($[(CH_3)_3SiF_2]^-$, $[HF_2]^-$, CN^-, etc.):

$$\underset{CH_3O}{\overset{(CH_3)_3SiO}{\diagdown}}C=\underset{CH_3}{\overset{}{C}}-CH_2-H \xrightarrow{+CH_3OOC-C(CH_3)=CH_2}$$

$$\underset{CH_3O}{\overset{(CH_3)_3SiO}{\diagdown}}C=\underset{CH_3}{\overset{}{C}}-CH_2-\underset{COOCH_3}{\overset{CH_3}{C}}-CH_2-H$$

Very few anionic polymerizations are employed for the production of thermoplastics, most notably those of formaldehyde, ε-caprolactam (also used in reaction injection molding), and laurolactam (Table 2). Anionic polymerizations are, however, the method of choice for the syntheses of block copolymers, such as thermoplastic elastomers, e.g., polystyrene –block–polybutadiene– block–polystyrene (SBS).

1.3.4. Cationic Polymerizations [14]

Three groups of monomers can be polymerized cationically: (i) Olefin derivatives $CH_2=CHR$ with electron-rich double bonds, (ii) monomers with double bonds containing heteroatoms or heterogroups Z (e.g., $CH_2=Z$), and (iii) rings with heteroatoms. Initiators are Brønsted acids (perchloric acid, trichloroacetic acid, trifluoromethanesulfonic acid, etc.); Lewis acids ($AlCl_3$, $TiCl_4$, $SnCl_4$, etc.) with "coinitiators", such as water; and carbenium salts (acetyl perchlorate, tropylium hexachloroantimonate, etc.). The propagating macrocations are thermodynamically and kinetically unstable. They attempt to stabilize themselves by addition

Table 2. Industrial anionic homopolymerizations

Monomer	Repeating unit	Application
Butadiene	~CH$_2$−CH=CH−CH$_2$~	elastomer (1,4-cis)
Isoprene	~CH$_2$−C(CH$_3$)=CH−CH$_2$~	elastomer (1,4-cis)
Methyl cyanoacrylate	~CH$_2$C(CN)(COOCH$_3$)~	adhesive
Formaldehyde	~O−CH$_2$~	engineering plastics
Ethylene oxide	~O−CH$_2$−CH$_2$~	thickener
Glycolide	~O−CO−CH$_2$~	surgical threads
ε-Caprolactone	~O−CO−(CH$_2$)$_5$~	polymer plasticizer
ε-Caprolactam	~NH−CO−(CH$_2$)$_5$~	fiber, thermoplastic
Lauryllactam	~NH−CO−(CH$_2$)$_{11}$~	fiber, film
Hexamethylcyclotrisiloxane	~O−Si(CH$_3$)$_2$~	elastomer

of nucleophilic species, which leads to very fast polymerization on one hand and to a host of transfer and termination reactions on the other. Very few monomers are thus polymerized cationically on an industrial scale (Table 3). Prominent examples include polyisobutene and poly(isobutene-co-isoprene) rubber.

1.3.5. Ziegler−Natta [15–19], ROMP [20] and ADMET [21] Catalysis

Ziegler−Natta polymerization catalysis involves a chain-growth polymerization on Ziegler catalysts, comprising a vast group of transition-metal compounds activated with main group metal alkyls, preferably aluminum alkyls. Today, in line with pioneering advances at Phillips using activator-free silica supported Cr(III) catalysts, a variety of cationic transition metal alkyls containing weakly coordinating counterions are available that do not require the addition of main group metal activators. Because monomers are inserted into the transition metal alkyl bonds, this catalytic polymerization is named polyinsertion. In older literature, polyinsertions are also referred to as coordination polymerizations or anionic −coordination polymerizations, although coordination of a monomer to an initiator need not lead to a Ziegler−Natta type of polymerization catalysis. Whereas free-radical polymerization of ethylene requires temperatures above 150°C and pressures well above 1 000 bar and produces branched low-density polyethylenes, catalytic ethylene polymerization takes place at ambient temperature and pressure below 10 bar, thus producing linear high density polyethylene. Short chain branches are introduced by copolymerization of ethylene with α-olefins. Stereospecific catalytic diene polymerization produces highly stereoregular poly(cis-1,4-isoprene) and poly(cis-1,4-butadiene) as synthetic alternative to natural rubber. Both KARL ZIEGLER and GIULIO NATTA were awarded the Nobel Prize for Chemistry in 1963, honoring their groundbreaking research, which paved the way for the

Table 3. Industrial cationic polymerizations

Monomer		Application[**]
Name	Structure[*]	
Isobutene	CH$_2$=C(CH$_3$)$_2$	elastomers, adhesives, VI improvers
Vinyl ethers	CH$_2$=CHOR	adhesives, textile aids, plasticizers
Formaldehyde	CH$_2$=O	engineering plastic
Ethyleneimine	c-(NHCH$_2$CH$_2$)	paper additive, flocculant
Tetrahydrofuran	c-[O(CH$_2$)$_4$]	soft segment for polyurethanes or polyether–ester elastomers

[*] c = cyclo.
[**] VI = viscosity index.

Table 4. Important industrial polymerizations with transition-metal catalysts

Monomer	Initiator	Application
Ethylene	$MgX_2/TiY_4/AlR_3$ (X = Cl, R, OR; Y = Cl, R, OR); (post)metallocene (Ti, Zr, Hf, Fe, Co), halfsandwich (Ti) and Cr(III) catalysts supported on silica	thermoplastic (HDPE, LLDPE)
Propene	$MgCl_2/1,3$-diether/$TiCl_4/AlR_3$; (post)metallocene (Ti, Zr, Hf) supported on silica	thermoplastic (PP)
Ethylene+propene+nonconjugated diene	Metallocene and halfsandwich catalysts, $VOCl_3/R_2AlCl$	elastomer (EPDM)
Butadiene	Co-, Ni-, Nd-, Sm-, Ln-catalysts, (post)metallocene and halfsandwich catalysts,	elastomer
Isoprene	Ti-, Co-, Ni-, Nd-, Sm-, Ln-catalysts, (post)metallocene and halfsandwich catalysts	elastomer
Butene-1	$MgCl_2/1,3$-diether/$TiCl_4/AlR_3$; (post)metallocene (Ti, Zr, Hf) supported on silica$TiCl_3$–Et_2AlCl	thermoplastic

industrial production of polyolefins. Half of the annual world production of polymers is based on this process.

Typical Ziegler catalysts for industrial polymerizations of olefins, such as ethylene, propene are shown in Table 4. Metallocenes and post-metallocene complexes have been developed since the early 1980s to produce tailor-made polyolefins in which the molecular architectures of polyolefins are tuned as a function of the molecular architectures of the catalytically active transition metal complexes. Traditionally, Ziegler catalysts are multi-site catalysts that contain different catalytically active sites and produce polymer chains with different molecular masses and molar-mass dependent comonomer sequence distribution. In contrast, modern single-site catalysts, which are based on metallocene and post-metallocene complexes, contain a single type of catalytically active site and give much more uniform polymers; molecular mass and molar-mass independent comonomer incorporation can be controlled easily.

Homogeneous catalysts are used in solution polymerization forming elastomers, such as ethylene−propene copolymers, which can contain a small amount of nonconjugated dienes, such as ethylidene norbornene as cross-linking sites. Slurry polymerizations, in which polymers precipitate during polymerization, require heterogeneous catalysts in order to prevent reactor fouling. Incorporating active transition metal alkyl sites on high surface area supports boosts catalyst activity and enables control of both polyolefin molecular architectures and polyolefin particle morphologies. In slurry polymerizations, most soluble catalysts produce small submicron dustlike polyolefin particles, thus causing severe reactor fouling and handling problems. Owing to fragmentation of supported catalysts during slurry polymerization (see Fig. 3), much larger micron-sized polyolefin particles are formed, which prevent reactor fouling. Moreover, spherical supported catalysts produce spherical pellet-sized polyolefin particles, thus eliminating the need for pelletizing extrusion.

The development of highly active and stereoselective supported catalysts was essential for enabling solvent-free gas phase and liquid pool olefin polymerization. The choice of the support materials and the selection of supporting strategies depend on the selected transition metal component. Because $MgCl_2$ has the same crystal structure as $TiCl_3$, it can substitute inactive bulk $TiCl_3$ that does not have vacant coordination sites. Hence, $MgCl_2$ supported titanium catalysts, activated with aluminum alkyls, afford orders of magnitude higher catalyst activities. In propene polymerization, the addition of small amounts of Lewis bases, such as 1,3-diethers promotes high stereospecifity, leading to the formation of highly isotactic polypropylene. Most (post)metallocene and chromium catalysts are preferably immobilized on high surface area silica supports.

Breakthroughs in catalyst and process technologies have greatly simplified olefin

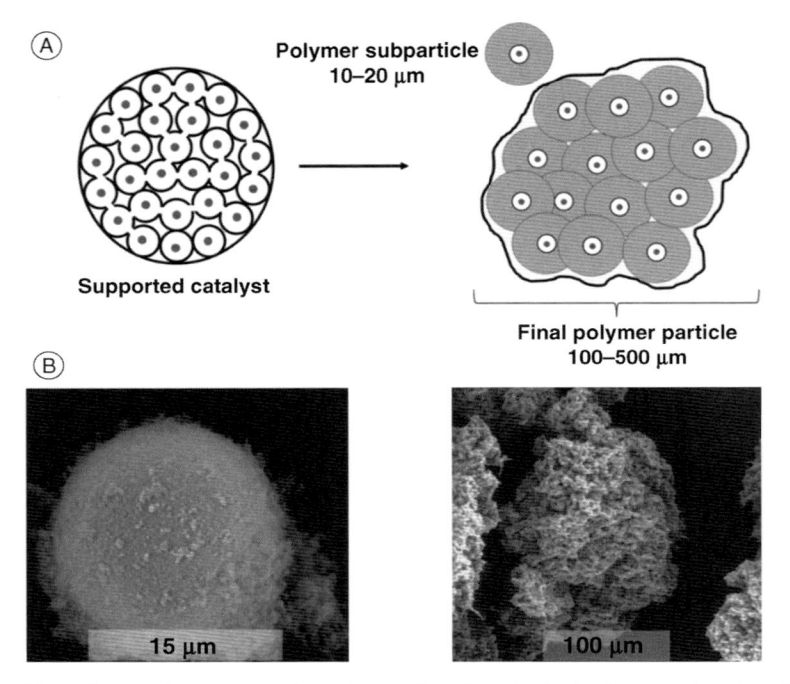

Figure 3. Morphology control in catalytic olefin polymerization by fragmentation of catalyst particles during polymerization A) Fragmentation of catalyst particle; B) Scanning Electron Microscope image of a catalyst particle (left) and a polyethylene particle (right)

polymerization. With catalyst activities exceeding 1 000 kg polyolefin/g Ti, the titanium content in the polyolefin is below 1 ppm. Hence, catalyst residues can be left in the polymers, i.e., the catalyst does not need to be removed by extraction or by adsorber beds typical for early catalysts generations. Furthermore, highly energy- and resource-efficient polymerization in gas phase and bulk eliminates the need for organic solvents as polymerization media and for their recycling. Because of the excellent control of molecular mass and stereoregularity, polyolefin purification by solvent extraction of low molecular mass polyolefin waxes and atactic polypropylene is no longer required. Employing fossil and renewable feedstocks, such as oil, gas and even bioethanol in environmentally friendly and highly energy-efficient catalytic polymerization processes meets the demands of green chemistry. As hydrocarbon resins, polyolefins preserve oillike energy content.

They are readily recycled by remolding and serve as valuable source of energy and "renewable oil", produced by thermal degradation of polyolefins.

The basic reaction mechanisms and the correlations between the architectures of catalytically active transition metal complexes and polyolefin architectures are quite well understood. As illustrated below for olefin polymerization on a cationic metallocene alkyl catalyst, the olefin forms a π-complex with the transition metal prior to insertion into the transition metal carbon bond. After repeated insertion β-hydride elimination terminates the polymer chain, thus forming vinyl-terminated polyethylene. The resulting transition metal hydride is the catalytically active intermediate, which adds to an olefin and starts the growth of a new polyethylene chain. Additionally, hydrogen can be added to control molecular mass by chain transfer that involves hydrogenation of the transition metal alkyls and formation of a transition metal hydride, which subsequently adds to the ethylene.

The stereospecificty of α-olefin polymerization is governed either by the configuration

The regioselectivity in α-olefin and the polyolefin constitution is governed by the insertion type. Whereas 100% of 1,2- or 2,1-insertion exclusively afford head-to-chain polyolefins, lower regioselectivity accounts for head-to-head enchainment in the polyolefin backbone. Most α-olefins are polymerized by 1,2-insertion in highly regioselective catalytic polymerization. The insertion type is identified by end group analysis of poly(α-olefins), taking into account chain termination by β-hydride as well as β-methylide transfer.

of transition metal complex (enantiomorphic site control) or by that of the asymmetric carbon atom of the monomeric unit at the chain end (chain end control). When a false insertion occurs, the resulting change of configuration is isolated in the case of enantiomorphic site control but is perpetuated in the case of chain end control (see below, m denotes meso, r denotes racemic). The analysis of polyolefin microstructures reveals that for most catalysts the chirality of the transition metal accounts for the control of the stereoregularity.

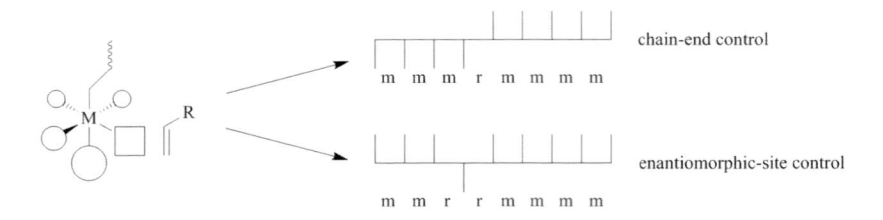

The ability of catalysts to control the stereochemistry of poly(α-olefins) and polydienes makes Ziegler–Natta polymerization catalysis industrially extremely useful. Propene, for example, polymerizes cationically to highly branched, atactic, low molar mass polymers that do not crystallize and have low glass transition temperatures. They can be used only as hotmelt adhesives and lubricants. Ziegler–Natta polymerization catalysis (Table 4) leads, however, to stereoregular polymers with high melting temperatures. Isotactic polypropylenes find extensive use as thermoplastics. Stereoblock polypropylenes can be engineered to produce thermoplastic elastomers containing highly flexible amorphous and rigid crystalline segments. The structure of single-site metallocene catalysts can be varied to produce virtually any kind of poly(α-olefin) microstructure industrially:

Similar to free radical polymerization, short and long chain branches are readily introduced by catalytic olefin polymerization. With increasing α-olefin comonomer incorporation, both crystallinity and density of polyethylene decrease. Whereas high density linear polyethylene (HDPE) is attractive for producing pipes, linear low density polyethylene (LLDPE) is highly flexible and has lower density, thus meeting the demands of packaging applications. A small amount of long chain branching promotes shear thinning and melt strengthening, which is beneficial to blow molding of films. At a high degree of branching, polyethylene is rendered amorphous and elastomeric. In principle, there are two synthetic strategies for branching of polyolefins: (i) Copolymerization of ethylene with various α-olefins, and (ii) branching homopolymerization of ethylene according to the "chain walking" mechanism. In chain walking, the catalytically active transition metal alkyl travels up and down the polymer chain via repeated β-hydride elimination followed by addition of the transition metal hydride to the resulting double bond. Whereas titanium catalysts form linear polyethylene chains, branching ethylene homopolymerization takes place on postmetallocene complexes, such as cationic alkyl Ni diimine. Alkyl Pd diimine complexes produce high molecular mass highly branched amorphous polyethylenes. Because Pd metal alkyl travels up and down the main chain and side chains, branches are formed.

developed for producing HDPE, LLDPE, and EP(D)M rubbers. Since the 1980s, the development of single-site catalysts based upon metallocene, post-metallocene, and constrained geometry half-sandwich complexes has revolutionized the industrial synthesis of branched polyethylenes. As a function of the ligand framework, it is possible to control molecular mass independent incorporation of α-olefins into polyethylene backbone to produce LLDPE as well as very low density polyethylenes and elastomers by the same catalyst. Moreover, improved reactivity enables the efficient copolymerization of polyethylene vinyl endgroups, thus producing long chain branched polyethylene.

linear low density PE; LLDPE
R: C_2H_5, C_3H_7, C_4H_9, C_5H_{11}...

linear PE

methyl-branched PE

branched PE

M: Ni-diimine catalyst

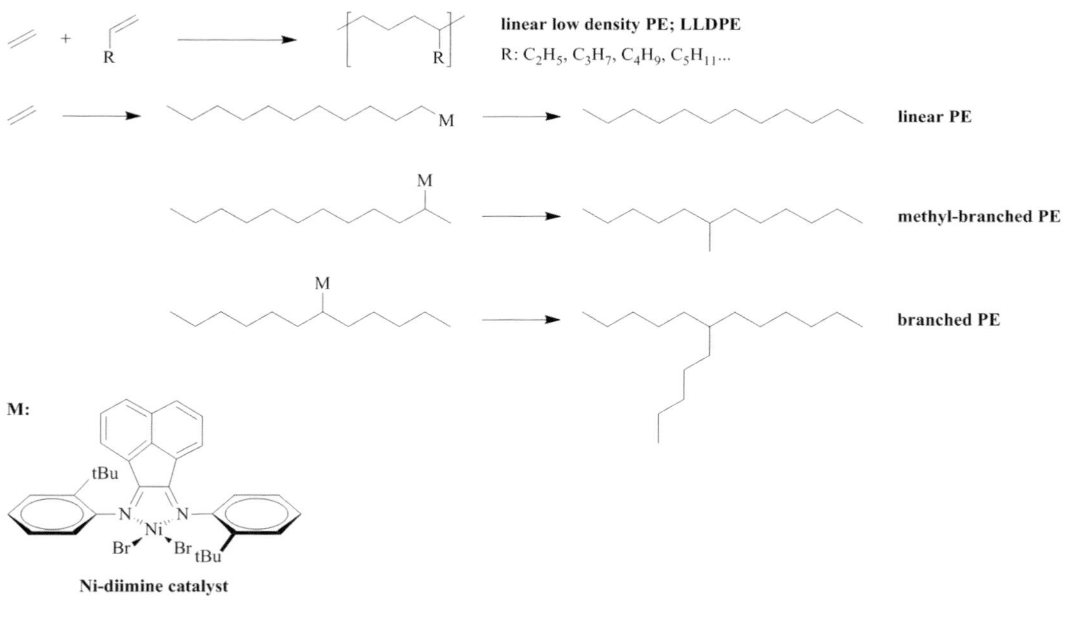

In the early days of catalyst development, the multisite nature of catalysts accounted for the formation of rather complex copolymer compositions in slurry polymerizations. Frequently, the α-olefin comonomer was predominantly incorporated in low molar mass waxlike fractions. More uniform copolymer compositions are obtained in high temperature solution polymerization operating at temperatures above 130°C. In the past, very different catalyst systems were

In a chain shuttling solution process, zinc alkyls transfer polymer chains between catalytically active sites producing exclusively linear polyethylene or ethylene–1-octene copolymers. As a consequence, block copolymers are obtained that contain alternating rigid HDPE segments and elastomeric poly(ethylene-co-1-octene) segments. Such block copolymers are attractive as thermoplastic elastomers.

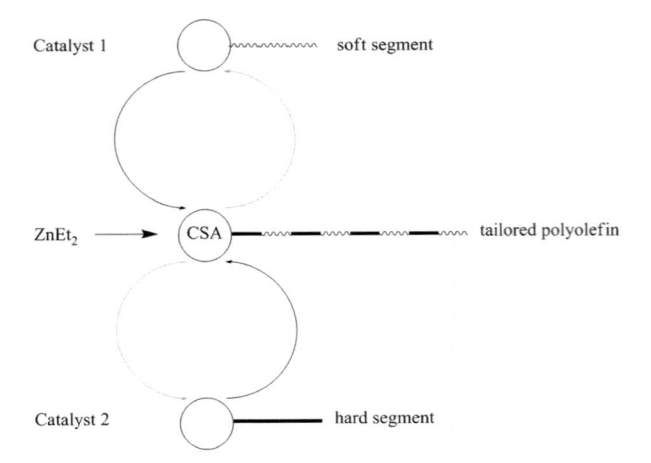

Although single-site catalysts afford uniform polyolefins, it is highly desirable to tailor polyolefins with broad molecular mass distributions and preferred incorporation of branches exclusively into the ultrahigh molar mass fraction. This greatly improves shear thinning during melt processing. Moreover, small amounts of slightly branched ultrahigh molar mass poylolefins function as "tie molecules" bonding together polyethylene crystallites and significantly improve the fatigue resistance and durability of polyethylene pipes. Two strategies are exploited: (i) Ethylene polymerization in cascade reactors, and (ii) ethylene polymerization in a single reactor using multi-site catalysts, prepared by incorporating different single-site catalysts on the same support (see Fig. 4). In cascade reactors, ethylene is polymerized in the

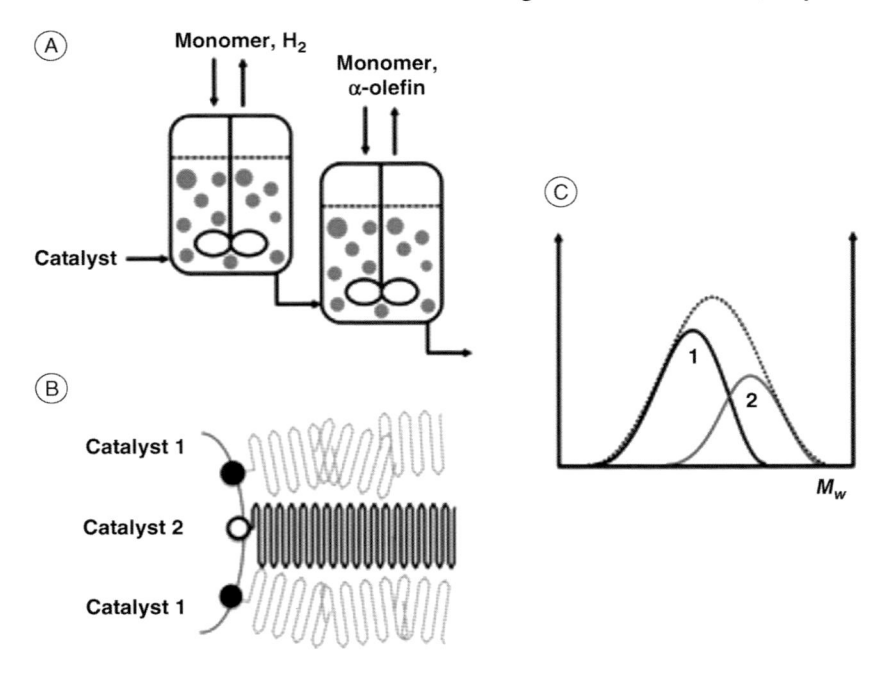

Figure 4. A) Reactor cascades; B) Multi-site catalysts producing reactor blends of HDPE with ultrahigh molecular mass polyethylene (UHMWPE); C) Polyolefin with bimodal molar mass distribution containing UHMWPE (2)

absence and the presence of hydrogen to control molar mass dstribution. The addition of α-olefin as comonomer selectively branches UHMWPE without affecting HDPE.

In multi-site catalysis, different catalytically active complexes that produce high and low molecular mass polyethylenes are incorporated on the same support, thus enabling simultaneous production and blending of both polyolefins. Reactor blend formation is advantageous compared to blends prepared by conventional melt compounding, especially with respect to dispersion problems due to the high viscosity of ultra-high molecular mass polyolefins and rubbers. The reactor blend formation of HDPE with UHMWPE (see Fig. 4) is the key to self-reinforcing polyethylene (see Plastics, General Survey, 4. Polymer Composites). The reactor granule technology, which combines liquid pool propene homopolymerization with subsequent propene–ethylene copolymerization in a gas phase reactor, incorporates EPM rubber phases directly into the matrix of the PP granules (Catalloy technology). Most conventional Ziegler and related catalyst systems polymerize cyclic olefin, such as cyclopentene via 1,2 polymerization, thus linking together cyclic hydrocarbons by covalent bonds to form rigid polymers. In contrast, transition carbene complexes enable ring-opening polymerization of cyclic olefins to produce flexible polyalkenamers. Catalytic intermediate of this ring-opening metathesis polymerizations (ROMP) are metallcylobutanes formed by cycloaddition of carbene complexes to cycloolefins.

Metatheses reactions are exchange and disproportionation reactions of double bonds, mainly carbon–carbon double bonds in olefins and cycloolefins. A low molar mass example is the metathesis of pentene-2 to butene-2, pentene-2, and hexene-3 (in the ratio 1 : 2 : 1) by the catalyst system $WCl_6-C_2H_5AlCl_2-C_2H_5OH$. Stable carbene complexes are at hand for metathesis reaction and ROMP. Industrially, cyclooctene, norbornene, and dicyclopentadiene are polymerized by ROMP. Cyclooctene gives polyoctenamer $\sim CH=CH-(CH_2)_6\sim$; both the *cis* and the *trans* isomers are elastomers (→ Rubber, 6. Sythesis by Radical and Other Mechanisms). Norbornene polymerizes to a thermoplastic polymer that is plasticized with mineral oil to give an elastomer (→ Rubber, 6. Sythesis by Radical and Other Mechanisms). Dicyclopentadiene is polymerized by reaction injection molding (RIM) processes to produce rigid engineering materials.

In *acyclic diene metathesis polymerization* (ADMET), the polycondensation of a variety of nonconjugated linear dienes, such as 1,5-headiene, affords polyenes. Using functional nonconjugated diene monomers, ADMET enables precise control of the spacer length between two branches, whereas conventional copolymerization process affords broad distributions. Unlike industrial Ziegler catalysts, Ru carbene complexes are tolerant to polar groups, thus enabling the formation of polyolefins with polar groups attached as side chains.

1.3.6. Copolymerizations [22]

Copolymerizations are joint polymerizations of two or more monomers; they were also called interpolymerizations in the older literature. In the simplest case, both active chain ends ~a∗ and ~b∗ react irreversibly with the two monomers A and B:

$$\sim a^* + A \to \sim a - a^*; R_{aA} = k_{aA}[a^*][A]$$

$$\sim a^* + B \to \sim a - b^*; R_{aB} = k_{aB}[a^*][B]$$

$$\sim b^* + A \to \sim b - a^*; R_{bA} = k_{bA}[b^*][A]$$

$$\sim b^* + B \to \sim b - b^*; R_{bB} = k_{bB}[b^*][B]$$

Four rates R_{ij} and four rate constants k_{ij} are to be considered in this terminal model, which corresponds to Markov first-order statistics. At high molar masses, monomers are consumed only by the four propagation reactions. The relative monomer conversion is given by

$$\frac{-d[A]/dt}{-d[B]/dt} = \frac{R_{aA} + R_{bA}}{R_{bB} + R_{aB}} = \left(\frac{k_{bA} + k_{aA}([a^*]/[b^*])}{k_{bB} + k_{aB}([a^*]/[b^*])}\right) \cdot \frac{[A]}{[B]} = \frac{d[A]}{d[B]}$$

The ratios of rate constants of homopropagations and cross-propagations are called copolymerization parameters

$$r_A \equiv k_{aA}/k_{aB}; \; r_B \equiv k_{bB}/k_{bA}$$

Five different cases can be distinguished for each copolymerization parameter:

$r = 0$	The rate constant of homopropagation is zero; the active center adds only the other monomer.
$r < 1$	The other monomer is added preferentially.
$r = 1$	Both monomers are added in the same amounts.
$r > 1$	The own monomer is added preferentially but not exclusively.
$r = \infty$	No copolymerization, only homopolymerization.

Table 5. Types of copolymerization of monomers A and B

Types	Copolymerization parameters				
	Azeotropic		Nonazeotropic		
	r_A	r_B	r_A	r_B	$r_A r_B$
Alternating	0	0	0	> 0	0
Statistical	< 1	< 1	$< 1/r_B$	> 1	< 1
Ideal	1	1	$1/r_B$	$1/r_A$	1
Block forming	> 1	> 1	$> 1/r_B$	< 1	> 1
Blend forming	∞	∞	∞	$< \infty$	∞

Depending on the relative numerical value of the two copolymerization parameters, five different types of copolymerization can be distinguished (Table 5).

In general, one of the monomers is consumed preferentially during copolymerization ($r_A \neq r_B$). This drift of copolymer composition can be avoided if the more reactive monomer is fed into the reactor according to its consumption. A conversion-independent polymer composition also results if copolymerization is performed under azeotropic conditions, which are defined by

$$d[A]/d[B] \equiv [A]/[B]$$

i.e., the relative rate of monomer consumption equals the ratio of monomer concentrations at any time: The monomer and the polymer compositions do not drift with the progress of polymerization.

This equation is a special case of the Lewis–Mayo equation, which describes the instantaneous composition of the copolymer as a function of the ratio of instantaneous monomer concentrations:

$$\frac{d[A]}{d[B]} = \frac{1 + r_A([A]/[B])}{1 + r_B([B]/[A])} = \frac{x_a}{x_b}$$

All types of copolymerization can in principle occur with the same monomer pair if various initiators are used. All copolymerizations of styrene and methyl methacrylate shown in Figure 5 are, for example, statistical. The free-radical-initiated copolymerization is, however, azeotropic, and the cationic and anionic ones are nonazeotropic. Initiation by $Et_3Al_2Cl_3$ and traces of oxygen leads to almost alternating

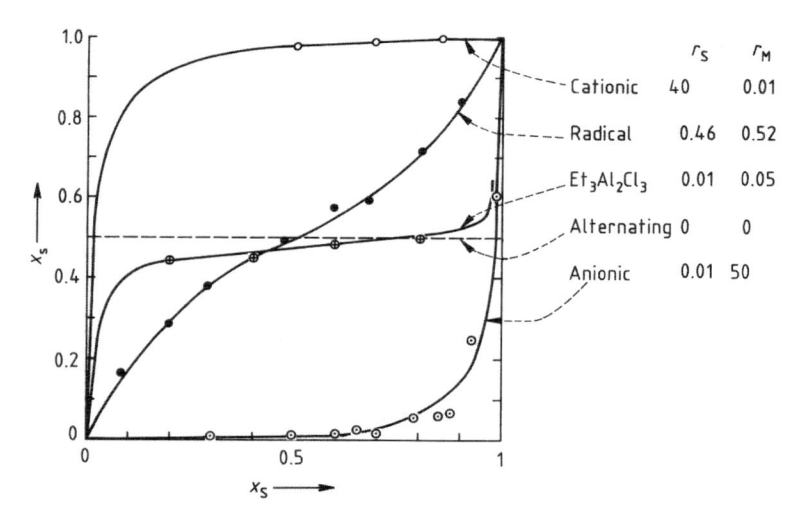

Figure 5. Instantaneous mole fraction x_s of styrene units in the polymer formed as a function of the instantaneous mole fraction x_s of styrene in the copolymerization of styrene with methyl methacrylate initiated by (○) cations; (●) free radicals; (⊕) $Et_3Al_2Cl_3$; (◎) Anions
Solid lines: calculated with copolymerization parameters shown to the right; dotted line: theory for alternating copolymerizations.

copolymers, whereas cationic polymerization generates long styrene sequences and anionic polymerization produces long methyl methacrylate ones. The sequence length determines many polymer properties, most notably glass and melting temperatures.

Ionic copolymerizations very often lead to long block sequences, whereas free-radical copolymerizations tend in general to produce more random copolymers. Free-radical copolymerizations are thus the method of choice for industrial purposes if allowed by monomer structures (Table 6).

1.4. Polycondensations and Polyadditions [9]

1.4.1. Bifunctional Polycondensations

Polycondensations and polyadditions are generally divided into bifunctional reactions (two functional groups per reactant) and multifunctional ones (three or more functional groups per monomer molecule). Bifunctional reactions lead to linear macromolecules; multifunctional ones, to branched and finally cross-linked polymers. Bifunctional reactions are further subdivided into AB reactions (e.g., that of $H_2N-R-COOH$) and AA/BB reactions (e.g., $H_2N-R-NH_2$ + $HOOC-R'-COOH$).

All reactants, from monomer to high polymer, may react to form macromolecules in bifunctional polycondensations and polyadditions. Only monomers are present at the beginning of the reaction; later, oligomers are present. If the reactivity of the functional groups is independent of molecule size, then the greater number of smaller molecules leads first to low molar mass polymers and only at high conversions to chemical compounds with high molar masses (see Fig. 2). In equilibrium, the number-average degree of polymerization of polymers from AB and stoichiometric AA−BB reactions is dictated by the extent of functional group reaction p:

$$\bar{X}_n = 1/(1-p)$$

At 50% conversion ($p = 0.5$), the number-average degree of polymerization of reactants is only 2 (this includes monomer molecules), and the number-average degree of polymerization of polymers ($2 < X_i < \infty$) is only 3.

Table 6. Industrial copolymerizations

Monomers	Polymerization	Application
Free radical, linear (or slightly branched)		
Ethylene + 10% vinyl acetate	bulk	shrink films
Ethylene + 10–35% vinyl acetate	bulk	thermoplastics
Ethylene + 35–40% vinyl acetate	under precipitation[a]	films
Ethylene + > 60% vinyl acetate	emulsion	elastomers
Ethylene + < 10% methacrylic acid	bulk	extrusion coating
Ethylene + trifluorochloroethylene		thermoplastics
Butadiene + styrene	emulsion	multipurpose elastomers
Butadiene + 37% acrylonitrile	emulsion	oil-resistant elastomers
Vinyl chloride + 3–20% vinyl acetate	solution[b]	Paints
Vinyl chloride + 15% vinyl acetate	solution[b]	Hi-fi records
Vinyl chloride + 3–10% propene	bulk	thermoplastics
Acryl esters + 5–15% acrylonitrile		oil-resistant elastomers
Acrylonitrile + 4% different monomers	under precipitation[c]	fibers with improved dyeability
Acrylonitrile + styrene		thermoplastics
Acrylonitrile + butadiene + styrene		thermoplastics
Tetrafluoroethylene + propene	emulsion	thermoplastics
Methacrylic acid + methacrylonitrile		hard foams (after cyclization)
Free radical, cross-linking		
Glycol methacrylate + 2–4% glycol dimethacrylate		contact lenses
Unsaturated polyesters + styrene or methyl methacrylate	bulk	glass-fiber-reinforced thermosets
Anionic		
Styrene + butadiene	solution	elastomers
Cationic		
Isobutene + 2% isoprene	solution	butyl rubber
Trioxane + ethylene oxide	solution	thermoplastics
Ethylene oxide + propylene oxide	solution	thickener, detergents
Ziegler–Natta polymerization		
Propylene oxide + nonconjugated dienes	solution	elastomers
Ethylene + propene + nonconjugated dienes		elastomers

[a] In *tert*-butanol.
[b] In acetone, 1,4-dioxane or hexane.
[c] In water.

Industrial polycondensations rarely deliver number-average degrees of polymerization greater than 100–200 (i.e., $p = 0.99$–0.995 in equilibrium). The endgroups of these equilibrium polymers may condense further during processing by, e.g., extrusion or injection molding. The accompanying strong increase in viscosity is undesirable and these endgroups are therefore blocked against further polycondensation by "sealing" them with monofunctional reagents.

Equilibria are established not only by retroreactions (reversal of polycondensations and polyadditions) but also by catalyzed exchange reactions between chain segments. In these trans reactions, number-average molar masses remain constant, but mass-average molar masses increase if the initial ratio of mass to number-average molar mass was smaller than dictated by equilibrium conditions. An example is the transesterification of poly(ethylene terephthalate):

$$\sim[O-CH_2-CH_2-O-OC-(p-C_6H_4)-CO]_n \\ \sim[OC-(p-C_6H_4)-CO-O-CH_2-CH_2-O]_j \quad + \quad [O-CH_2-CH_2-O-OC-(p-C_6H_4)-CO]_k\sim \\ [OC-(p-C_6H_4)-CO-O-CH_2-CH_2-O]_m\sim$$

$$\downarrow$$

$$\sim[O-CH_2-CH_2-O-OC-(p-C_6H_4)-CO]_n - [O-CH_2-CH_2-O-OC-(p-C_6H_4)-CO]_k\sim \\ \sim[OC-(p-C_6H_4)-CO-O-CH_2-CH_2-O]_j + [OC-(p-C_6H_4)-CO-O-CH_2-CH_2-O]_m\sim$$

Table 7. Industrially important linear AA–BB polycondensations following Equation (15)

Polymer	Chemical structure				Remarks
	A	V	W	B	
Poly(ethylene terephthalate)	H	$O(CH_2)_iO$	$CO-p$-Ph$-CO$	OH	$i = 2$ (ethylene) or
	H	$O(CH_2)_iO$	$CO-p$-Ph$-CO$	OCH_3	4 (butene)
Unsaturated polyesters	H	$O(CH_2)_2O$	$CO-CH=CH-CO$	$O_{1/2}$	from maleic anhydride (isomerizes mainly to fumaric acid residues)
Polycarbonate	H	$O-p$-Ph$-C(CH_3)_2-p$-Ph$-O$	CO	OC_6H_5	
	Na	$O-p$-Ph$-C(CH_3)_2-p$-Ph$-O$	CO	Cl	
Polyarylates	Na	$O-$Ar$-O$	$CO-$Ar$-CO$	Cl	different Ar
Poly(phenylene sulfide)	Na	S	p-Ph	Cl	
Polysulfides	Na	S	R	Cl	different R
Polysulfones	K	$O-p$-Ph$-O$	SO_2-p-Ph$-O-p$-Ph$-SO_2$	Cl	different V and W units
Polyamides	H	$NH(CH_2)_iNH$	$CO(CH_2)_jCO$	OH	PA 66 ($i = 6; j = 4$) PA 610 ($i = 6; j = 8$) PA 46 ($i = 4; j = 4$)
Polyetheretherketone	K	$O-p$-Ph$-CO-p$-Ph$-O$	p-Ph$-CO-p$-Ph	F	

Ar = aromatic residue; p-Ph = *para*-substituted phenylene group; R = organic residue.

Higher degrees of polymerization at lower extents of reaction can be obtained by irreversible "activated reactions" and especially by heterogeneous reactions of fast reacting monomers. An example is the Schotten–Baumann reaction of diamines in water with diacid dichlorides in chloroform, for example. A polyamide film is formed at the interface of the two solutions. Functional groups of this polymer are buried in the interior of the film; they cannot react with each other. Exchange reactions and retroreactions are also absent. Only surface groups can react; the resulting irreversible reaction leads to high molar masses. Industrially, this reaction is utilized for one of the two industrial syntheses of polycarbonates, i.e., from the sodium salt of bisphenol A and phosgene (Table 7).

Most industrial polycondensations are of the AA–BB type (Table 7); they can be described by

$$n\,A-V-A + n\,B-W-B \rightarrow A-(V-W)_n-B + (2n-1)AB \quad (15)$$

where A– and B– are leaving groups, AB represents leaving molecules, –V– and –W– are monomeric units, and –V–W– are repeating units.

Industrial AB polycondensations are fairly rare. Examples are the formation of polyamide 11 (nylon 11) (Eq. 2), the polycondensation of carbodiimides

$$O=C=N-R-N=C=O \rightarrow R-N=C=N + CO_2$$

and the synthesis of certain polysulfones

$$C_6H_5O(p-C_6H_4)SO_2Cl \rightarrow$$
$$(p-C_6H_4)-O-(p-C_6H_4)-SO_2 + HCl$$

1.4.2. Multifunctional Polycondensations and Hyperbranched Polymers [23]

Multifunctional polycondensations are condensation reactions with the participation of monomers having higher functionality ($f_o > 2$). At low conversions, branched molecules are formed with increasingly higher number of functional end groups per molecule (i.e., increased functionality f):

$$f = 2(f_o - 1) + (X - 2)(f_o - 2) = 2 + X(f_o - 2)$$

Only one such group is, however, needed to link two molecules; the probability of such a linkage thus increases dramatically with increasing degree of polymerization X. At a certain degree of conversion of functional groups, molecules are linked throughout the volume of the reactor. The viscosity increases strongly, and the mass-average molar mass of the cross-linked polymer approaches infinity at this "gel point." Not all reactants are linked at the gel point; some are still soluble up to high conversions. The number-average degrees of polymerization of reactants are thus fairly low at the gel point, usually in the range $10 < \overline{X}_n < 50$.

Knowledge of the gel point is technically very important. Polymers cannot, or can only with difficulty, be processed beyond the gel point and must thus be shaped before or during the cross-linking reaction. The gel point can be calculated from the functionality and relative amounts of reactive groups if intramolecular cyclizations are assumed absent. In practice, such cyclizations are often present and cross-linking occurs at higher extents of reaction than calculated. The theoretical calculations thus provide a safety margin.

Chemical cross-linking reactions are utilized for thermosets, certain leathers, and the preparation of elastomers from rubber. Thermosetting polycondensation reactions allow the hardening of phenol, amino, and alkyd resins. In the reaction of phenol(s) with formaldehyde, methylene, formal, and ether bridges are formed between the *ortho* and *para* positions of the phenols; the hardening of novolacs with hexamethylenetetramine (urotropin) also leads to imine structures (→ Phenolic Resins). Amino plastics result from the reaction of urea or melamine with formaldehyde (→ Amino Resins), and alkyd resins from multifunctional alcohols with bifunctional organic acids or their anhydrides (→ Alkyd Resins).

Homopolymerization of AB_n monomers and copolymerization of AB with B_n monomers, respectively, affords hyperbranched polymers. Due to their low entanglement, hyperbranched polymers have reduced viscosity compared to linear polymer. Moreover, they have very high end-group functionality and can be tailored for applications in thermoset resin formulations and drug carriers. Amphiphilic hyperbranched polymers are effective dispersing agents for pigments and various other particles.

1.4.3. Polyaddition

No leaving molecules have to be removed during polyadditions, an important advantage for process engineering. Practically all industrially important polyadditions are thus performed with thermosetting resins. Polyurethanes are formed from diisocyanates (or triisocyanates) and polyols (see Eq. 3 for diols). Polyurethane formation may be so rapid that cross-linking and shaping can be performed in one step through reaction injection molding with short cycle times. Epoxides (epoxy resins) contain two or more epoxy groups per prepolymer molecule and are cured with either difunctional or multifunctional amines or with anhydrides of organic acids.

1.5. Polymer Reactions [24, 25]

1.5.1. Polymer Transformations

Several polymers are chemically transformed into industrially useful polymers by polymer-analogous reactions of their substituents (Table 8).

These reactions per group are chemically identical to their low molar mass analogues. They differ in the effect of side reactions, which do not lead to removable byproducts as in micromolecules but to "wrong" structures in the polymer chain. In most cases, such structures decrease the desired polymer properties. Polymer transformations that are prone to side reactions are thus generally avoided. This is one reason why only five types of polymer-analogous reactions are utilized commercially: Transesterification–saponification, chlorination
–sulfochlorination, etherification, hydrogenation, and cyclization.

The reaction of poly(vinyl alcohol) with butyraldehyde leads to acetals. For statistical

Table 8. Industrial polymer transformations (for ring formations see text)

Base units in primary polymer	Reagent	Conversion, %	Base units in final polymer	Application
$CH_2CH(OOCCH_3)$	ROH	98	CH_2–CHOH	thickener
$CH_2CH_2CH_2CH(OOCCH_3)$	CH_3OH	99	CH_2–$CHOH/CH_2$–CH_2	engineering plastics
Cellulose–$NHCOCH_3$	H_2O	?	Cellulose NH_2	paper additive
Cellulose OH	CH_3COOH	83–100	Cellulose $OOCCH_3$	fibers, films
Cellulose OH	HNO_3	67–97	Cellulose–ONO_2	plastics, fibers
Cellulose ONa	oxirane	53–87	Cellulose–OCH_2CHRCH_3	thickener
Cellulose OH	oxirane	100–400[a]	Cellulose–$OCH_2CH(OR')CH_3$	thickener
CH_2–CH_2	Cl_2	25–40[b]	CH_2–CHCl	elastomer
		>40[b]	CH_2–CHCl/CHCl–CHCl	impact improver for PVC
CH_2–CHCl	Cl_2	64[b]	CH_2–CHCl/CHCl–CHCl	adhesives, paints
CH_2–CH_2	Cl_2/SO_2	<42[b]; <2[c]		coatings
CH_2CH=$CHCH_2/CH_2$–CHCN	H_2	?	$(CH_2)_4/CH_2$–CHCN	elastomer
CH_2–$CH(C_6H_5)$	SO_3	low	CH_2–$CH(C_6H_4SO_3H)$	ion exchangers
N=PCl_2	RONa	?	N=$P(OR)_2$	elastomer

[a] Further reaction of primary reaction products may lead to more than one monomeric unit of reagent per hydroxyl group of cellulose.
[b] Chlorine content in %.
[c] Ether groups per glucose unit.

reasons, however, not all hydroxyl groups can be transformed:

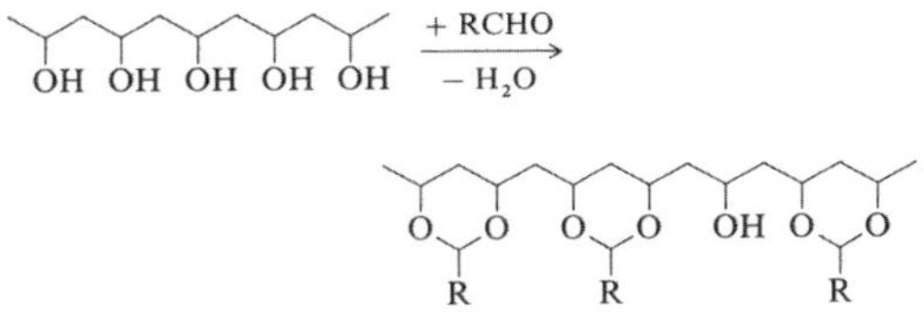

The resulting poly(vinyl butyral) is used as an intermediate layer in safety glass and for wash primers.

Heating of methacrylic acid–methacrylonitrile copolymers with ammonia produces polymethacrylimide structures that are hard foams:

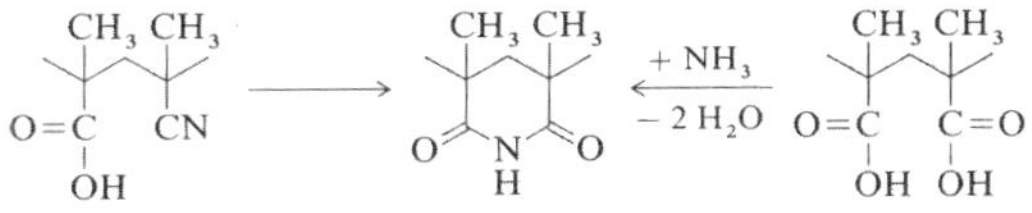

The thermal treatment and carbonization of polyacrylonitrile fibers represents the key to the industrial production of carbon fibers, which are highly attractive for reinforcement of thermosets and the formation of carbon-fiber reinforced polymers in lightweight engineering.

1.5.2. Block and Graft Formations [26]

Reactive chain ends can either initiate the polymerization of other monomers or couple with other preformed macromolecules to give block polymers:

$$A_n^* + mB \rightarrow A_n B_m^*$$

$$A_n^* + B_m^* \rightarrow A_n B_m$$

Both methods are used industrially with a variety of strategies for the synthesis of diblock polymers $A_n B_m$ as compatibilizing agents for polymer blends, triblock polymers $A_n B_m A_n$ as thermoplastic elastomers, and various multiblock polymers for different purposes, especially for thermoplastic elastomers. Anionic living polymerizations are used for the syntheses of poly(styrene–block-butadiene–block-styrene) and poly(styrene–block-isoprene–block-styrene). Polycondensations serve for the preparation of multi[{poly(butylene terephthalate)}–block-poly{(polytetrahydrofuran)–terephthalate}], and various polyesteramide and polyetheramide segmented block polymers.

Graft copolymers result from the grafting of monomers on chemically different polymer chains. Grafts of isobutene–isoprene on polyethylenes, vinyl chloride on poly (ethylene-stat-vinyl acetate), ethylene–propene on poly(vinyl chloride), or styrene--acrylonitrile on saturated acryl rubber also yield thermoplastic elastomers.

1.5.3. Cross-Linking Reactions

Cross-linking of preformed polymers is used for vulcanizing unsaturated rubbers with sulfur, vulcanizing saturated rubbers with peroxides, hardening unsaturated polyester chains with styrene or methyl methacrylate via copolymerization, or tanning collagen with certain reagents to leather.

1.5.4. Degradation Reactions

Degradation of polymers is undesirable for their application and often desirable for their disposal after use. The term degradation includes the decrease in degree of polymerization with preservation of the chemical structure of monomeric units (depolymerization), the unwanted chemical transformation of some monomeric units with preservation of the degree of polymerization (decomposition), and a combination of both. The terms degradation, depolymerization, and decomposition are sometimes used with the same meaning as in this article, but other times with broader or more restricted meanings; no accepted usage exists.

Depolymerizations involve chain scissions. They are retroreactions (i.e., the reverse of chain polymerizations, polyeliminations, polycondensations, or polyadditions). Chain scissions may be at random as in retropolycondensations and retropolyadditions or may be unzipping reactions as in retro-chain polymerizations and retropolyeliminations.

Unzipping can occur only with activated chains above their ceiling temperatures, that is, with living polymers, chain ends comprising certain functional groups, or radical ends generated by a prior chain scission. Unzipping at higher temperature results in gaseous monomers that lead to voids in molded parts. This can be prevented or reduced by shorter sequence lengths of the unzippable polymer blocks. This strategy is utilized, for example, in the formation of acetal copolymers by copolymerization of trioxane (gives unzippable oxymethylene sequences $\sim$OCH$_2\sim$) with ethylene oxide (oxyethylene sequences $\sim$OCH$_2$CH$_2\sim$ do not unzip).

Decomposition can be caused by heat, light, oxygen, water, or other environmental agents as well as mechanical forces such as extensional flow, ultrasound, or a combination of these. It may lead to undesirable changes of mechanical, electrical, or optical properties. Unstabilized poly(vinyl chloride) $\sim$(CH$_2$CHCl)$_n\sim$ discolors on decomposition because HCl is eliminated and colored polyene sequences $\sim$(CH=CH)$_i\sim$ are formed. The polymer ultimately becomes brittle.

Decomposition is the primary degradation process in the pyrolysis of polymers, which may be desirable (rocket fuels, waste disposal) or not (fires), and often produces toxic fumes or smoke at lower flame temperatures. The risk of pyrolysis is enhanced for polymers with branching points, electron-accepting groups, long methylene sequences, and all groups capable of forming five- or six-membered rings. Thermostability is enhanced by aromatic rings, ladder structures, fluorine as a substituent, and low hydrogen content. On heating above 400°C, polyolefins degrade to form small molar mass hydrocarbons, which are equivalent to those obtained from fossil resources. This route to "renewable oil" from polyolefin wastes is attractive with respect to recycling and sustainability.

2. Plastics Manufacture [27]

2.1. Homogenization and Compounding

Polymerization reactors do not deliver polymers that can be used directly as raw materials for plastics. Polymer melts must be filtered; melts, powders, beads, pellets, and granulates must be degassed; and all particulates must be dried or conditioned to environmental surroundings. Beads from suspension polymerization are washed and polymer solutions concentrated.

Batchwise polymerizations often lead to slightly different products, which are mixed (blended) to guarantee customers polymer grades according to specifications (microhomogenization). Granulates and pellets are sometimes similarly blended by macrohomogenization.

Compounding is the mixing of polymers and additives. This process can be performed by adding single additives one at a time, by using additive systems, or by employing master batches. Additive systems are carefully adjusted mixtures of additives that are formulated to avoid mutually synergistic or antagonistic effects. Master batches are concentrates of additives in polymers; they facilitate the dosage of small amounts of additives.

The properties of compounds depend very much on the compounding process. Only heterogeneous compounds result when poly(vinyl chloride) is mixed with additives in regular mixers; these compounds cannot be processed directly to end-use products. High-performance mixers deliver free-flowing powders; these "dry blends" can be extruded and injection molded. Compounding is often carried out by specialized companies (compounders).

2.2. Additives [8, 28, 29] (→ **Plastics, Additives**)

2.2.1. Overview

Polymers are rarely used directly as materials. They do not fulfill per se all technological requirements and become commercially useful only after they have been mixed with certain additives. These additives average ca. 23% of the total mass of plastics but may range for individual plastics from 0.01% (packaging films for food) to 90% (barium ferrite-filled ethylene–vinyl acetate copolymers for magnetizable sealing strips). Polymer additives are to be distinguished from *polymer auxiliaries*; the former are added to polymers after the polymerization process, whereas the latter are used for the manufacture of polymers (e.g., polymerization catalysts, emulsifiers, initiators). The free-radical initiators used for the hardening of unsaturated polyesters are sometimes considered additives (added to the solution of unsaturated polyester molecules in monomers such as styrene) and sometimes auxiliaries (promoting the thermosetting reaction).

Additives are usually subdivided according to their application into process additives and functional additives. Process additives aid the processing of plastics by either stabilizing the chemical composition of polymers (processing stabilizers) or facilitating the processing itself (processing aids). Functional additives stabilize the chemical composition against attacks by environmental agents (stabilizing additives) or improve certain end-use properties (modifiers). Additives may also be classified according to their primary mode of action (chemical or physical). Some additives act in more than one way: A filler or colorant may also be a nucleating agent for crystallization; a pigment may enhance discoloration, etc.

2.2.2. Chemofunctional Additives

Polymers can be attacked by oxygen and ozone during processing and use. The attack generates radicals that cause chain reactions to produce even more radicals. Ultimately, changed chemical compositions, degradation, or cross-linking occur. Oxidation is reduced if the accessibility of the oxidizable groups is limited or if radical formation is prevented. Oxygen diffusion into the polymer is slowed by certain coatings that act as mechanical barriers or by additives that diffuse into surface layers and are preferentially oxidized there.

Antioxidants (→ Antioxidants) prevent radical formation, at least during processing and for the targeted lifetime of the plastic. They are subdivided into deinitiators and chain terminators.

Deinitiators prevent the formation of radicals (i.e., preventive antioxidants) and are always used in combination with chain terminators (also called secondary antioxidants). They are further subdivided into peroxide deactivators, metal deactivators, and UV absorbers. Peroxide deactivators (tertiary amines, tertiary phosphines, sulfides) convert hydroperoxides into harmless compounds before they can form radicals. Metal deactivators are chelating agents that form inactive complexes with catalytically active metal species (mainly from Ziegler–Natta polymerizations).

Chain terminators, such as hindered phenols, amines, and anellated hydrocarbons, react with already formed radicals and terminate the kinetic chain (chain-breaking or primary antioxidants). A combination of deinitiators and chain terminators often leads to synergistic effects. Antagonistic effects are also known (e.g., with carbon black as filler).

Light-induced degradation reactions can be prevented by the reduction of light absorption or by the addition of *UV absorbers* and quenchers. Less light is absorbed by the plastic if the surfaces are reflective or if certain pigments are added (e.g., carbon black). Ultraviolet absorbers either convert the incident light into harmless infrared radiation or are transformed into other chemical compounds. Quenchers deactivate excited states and become excited themselves.

Heat stabilizers prevent chemical transformations of plastics at higher temperature; their main application is for the stabilization of poly(vinyl chloride) against HCl elimination during processing. PVC additives include HCl scavengers, such as metal stearates, metal oxides, and epoxidized soybean oil. PVC additives, such as dibutyltin dithiolate react with thermally labile allylic chloride and double bonds in the PVC backbone.

Flame retardants either prevent oxygen access to burning plastics by formation of nonburning gases or by "poisoning" radicals generated by the burning. Some plastics are self-extinguishing because either CO_2 (polycarbonates) or water vapor and a protecting carbon layer are formed (vulcanized fiber: a $ZnCl_2$-treated cellulose) during burning. Chlorine and bromine compounds generate radicals during burning; these radicals combine with radicals from the degradation of plastics and stop the kinetic chain. Phosphorus compounds are oxidized during the burning to nonvolatile phosphor oxides, which either form a protective layer or are converted by water into phosphorus acids that catalyze the elimination of water. Intumescent additives, such as ammonium polyphosphate, form a carbon form that serves as a thermal insulator.

The flammability of plastics is often characterized by the limiting oxygen index (LOI). The LOI value indicates the limiting value of the volume fraction of oxygen in an oxygen–nitrogen mixture that just allows the polymer to burn after ignition with a flame. Materials with LOI > 0.225 are called flame retardant; those with LOI > 0.27, self-extinguishing. Low LOI values (high flammabilities) are exhibited by polyoxymethylenes (0.14), polyolefins (0.17–0.18), saturated polyesters (0.20), and cellulose (0.20). High LOI values are exhibited by poly(vinyl chloride) (0.32), polybenzimidazole (0.48), and polytetrafluoroethylene (0.95). However, LOI values are not absolute measures of the flammability or combustibility of plastics because these properties also depend on flame temperatures, heat capacities, heat conductivities, melting temperatures, and melt viscosities. The hazard of burning plastics is also determined by smoke formation and the toxicity of the evolving gases. In view of growing environmental and health concerns, halogenated flame retardants are gradually substituted by halogen-free flame retardants, such as $Al(OH)_3$ hydrate, $Mg(OH)_2$, and ammonium polyphosphate.

2.2.3. Processing Aids

Processing aids facilitate the processing of plastics by enhancing either transport rates to the processing machines, flow behavior in these machines, achievement of final properties during processing, or removal of shaped articles from the machines or from each other.

Easy-flow grades of powdered polystyrenes, for example, often contain 3–4% mineral oil, which forms a low-viscosity film on the surface of the particles and thus reduces friction in polar polymers; amphiphilic compounds such as metal stearates or fatty acid amides are used for this purpose. Such *external lubricants* also reduce the friction between polymer particles and the walls of the processing machine and the friction of polymer melts at such walls; they also prevent the cleavage of particles to smaller flow units. External lubricants are always incompatible with polymers and are thus found predominantly at polymer surfaces. They are related to *release agents*, which facilitate the separation of shaped articles from the tools (molds), and *slip agents*, which prevent the sticking together of shaped articles. Slip agents are thus sometimes also called lubricants.

Internal lubricants improve the flow behavior and homogeneity of polymer melts; they also reduce the Barus effect and the melt fracture. Internal lubricants probably act by desegregating larger units (aggregates), which were probably formed during polymerization and are still present shortly after the melting of the polymer to a macroscopically homogeneous material. Typical internal lubricants are amphipolar compounds, e.g., modified esters of long-chain fatty acids.

Nucleating agents promote the crystallization of crystallizable polymers by generating many nuclei for crystallites. They prevent the formation of larger spherulites and thus improve the mechanical properties of plastics.

2.2.4. Extending and Functional Fillers [30–32]

Fillers are solid inorganic or organic materials. Some fillers are added mainly to improve the economics of expensive polymers; they are *extenders*. Extenders are usually particulate materials of corpuscular nature, such as chalk and glass spheres (aspect ratio ca. 1). Functional fillers improve certain properties, such as stiffness, strength, scratch resistance, electrical conductivity, crystallization rate, etc. Reinforcing fillers possess aspect ratios higher than 1; they may be short fibers, platelets (e.g., kaolin, talc, mica) (→ Reinforced Plastics), or long fibers (continuous filaments). Active fillers are sometimes subdivided into property enhancers (aspect ratio < 100) and true reinforcing fillers (aspect ratio > 100). No sharp dividing line exists between fillers and reinforcing agents, nor can the term "reinforcement" be unambiguously defined (e.g., it may denote an increase in breaking or impact strength or a decrease of brittleness).

A variety of fillers are used for plastics (- Table 9). Glass-fiber-reinforced polymers often carry the abbreviation GRP (or FRP); those reinforced by carbon fibers, CRP. Syntactic plastics are polymers reinforced with hollow glass spheres. The amounts of fillers added vary widely: In general, industry standards are about 30 wt % for thermoplastics and 60 wt % for thermosets. Fillers act very differently in polymers. Some fillers form chemical bonds with polymers; an example is carbon black, which acts as a chemical cross-linker in elastomers. Other fillers can adsorb polymers on their surfaces, i.e., physical bonds are

Table 9. Fillers for thermoplastics (T), thermosets (D), and elastomers (E). For definition of other acronyms see Tables [1–4]

Filler	Application in	Concentration in %	Improved property
Inorganic fillers			
Chalk	PE, PVC, PPS, PB, UP	< 33 in PVC	price, gloss
Potassium titanate	PA	40	dimensional stability
Heavy spar	PVC, PUR	< 25	density
Talc	PUR, UP, PVC, EP, PE, PS, PP		white pigment, impact strength, plasticizer uptake
Mica	PUR, UP	< 25	dimensional stability, stiffness, hardness
Kaolin	UP, vinyls	< 60	demolding
Glass spheres	T, D	< 40	modulus of elasticity, shrinkage, compressive strength, surface properties
Glass fibers	T, D	< 40	fracture strength, impact strength
Fumed silica	T, D	< 3	tear strength, viscosity (increase)
Quartz	PE, PMMA, EP	< 45	heat stability, fracture
Sand	EP, UP, PF	< 60	shrinkage (decrease)
Al, Zn, Cu, Ni, etc.	PA, POM, PP	< 100	conductivity (heat and electricity)
MgO	UP	< 70	stiffness, hardness
ZnO	PP, PUR, UP, EP	< 70	UV stability, heat conductivity
Organic fillers			
Carbon black	PVC, HDPE, PUR, PI, PE, E	< 60	UV stability, pigmentation, cross-linking
Graphite	EP, MF, PB, PI, PPS, UP, PMMA, PTFE	< 50	stiffness, creep
Wood flour	PF, MF, UF, UP	< 5	shrinkage (decrease), impact strength
Starch	PVAL, PE	< 7	biological degradation

introduced between fillers and polymers. On impact, adsorbed chain segments may take up energy and slip from the surface, which increases impact strength. Still other fillers act as nucleating agents in crystallizable polymers. Fillers furthermore constitute impenetrable walls to polymer coils. They restrict the number of conformational positions of chain segments near the filler surface; chains become less flexible, and tensile strengths and moduli of elasticity increase.

Due to the significant progress made in nanotechnology, a large variety of nanofillers have emerged since the mid-1990s. As compared to conventional filled polymers that contain micron-sized filler particles nanocomposites have nine orders of magnitude larger number of nanometer-scaled particles at the same volume fraction (Fig. 6). As a consequence, most polymers are allocated in nanocomposites at the nanofiller interface. Converting bulk polymers into interfacial polymers account for novel property profiles. The high surface area lowers considerably the percolation threshold required for particle network formation. This improves the electrical conductivity at much lower filler content. However, in view of the nanoparticle interactions, dispersion of nanoparticles is more difficult than micron-sized particles. Therefore, special processing technology and nanofiller surface modification are required. As illustrated in Figure 7, the nanofiller family includes inorganic nanoparticles as well as organic–inorganic hybrid particles, among them dispersions of metal oxides and silicate nanoparticles with shapes varying from spheres to platelets and sheets. The smallest silica nanoparticles are polyhedral oligomeric silsesquioxanes (POSS), which are readily functionalized. The exfoliation of graphite affords graphene that consists of a honeycomb-like planar arrangement of sp^2 hybridized carbon atoms equivalent to a single layer of the graphite lattice. Other carbon allotropes, such as nanometer-scaled carbon particles ("conducting carbon black"), fullerenes, nano diamonds, and single- and multi-walled carbon nanotubes (CNT) are employed as nanofillers. Nanometer-sized carbon particles and especially carbon macromolecules render polymer nanocomposites electrically conductive without sacrificing optical clarity. Organophilic layered silicates ("organoclay") are obtained by cation exchange of the intergallery sodium cations of bentonite for alkyl ammonium. Alternatively, layered silicates with cationic layers, such as hydrotalcites are rendered organophilic by modification with long chain aliphatic carboxylates. As a function of the layered silicate modification and polymer processing conditions, either individual silicate layers are dispersed within the polymer matrix or the polymer is intercalated between silicates layers. The dispersion of sheetlike nanoparticles improves the barrier resistance of polymers. Going beyond wood flour, bio-based nanofillers (cellulose whiskers, nanocellulose) are derived from cellulose by etching of the amorphous regions.

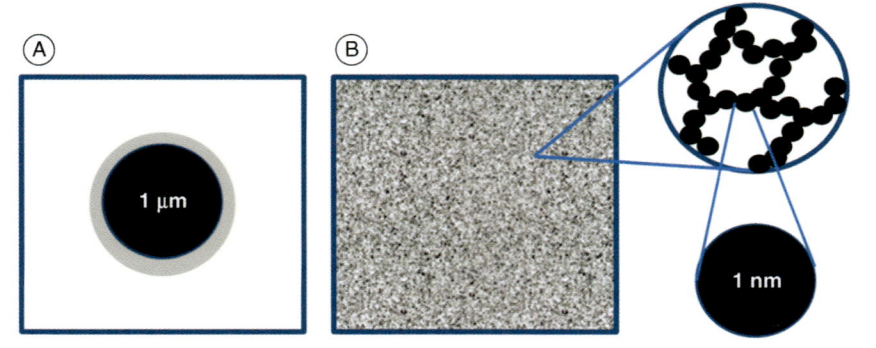

Figure 6. Different sized polymer composites containing the same volume fraction
A) One particle (1 μm) microfiller-sized fillers; B) 10^9 particles (1 nm) nanofiller with the same volume fraction like microfiller

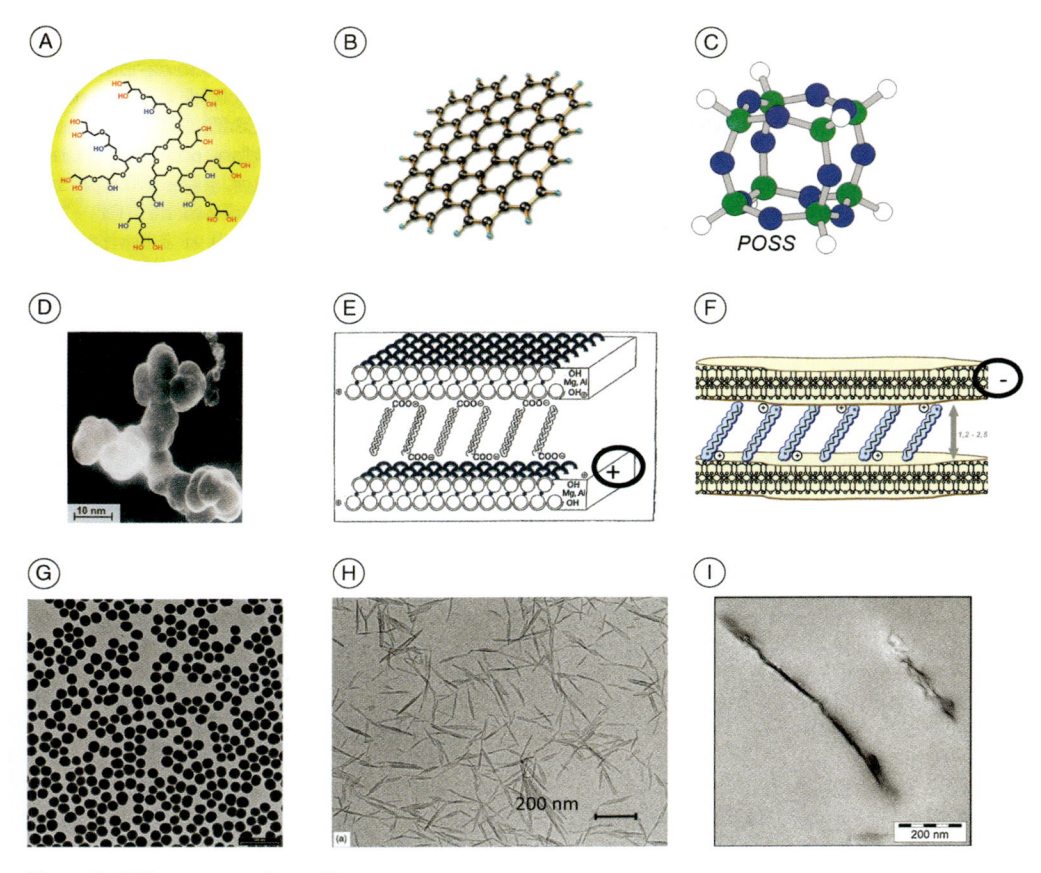

Figure 7. Different types of nanofillers
A) Nanomolecules; B) Molecular carbon; C) Molecular silica (POSS); D) Pyrogenic silica nanoparticles; E) Hydrotalcite; F) Organoclay; G) SiO$_2$ dispersion; H) Cellulose nanowhiskers; I) Graphene embedded in polypropylene

2.2.5. Plasticizers (→ Plasticizers) [33]

Plasticizers are added to polymers to improve their flexibility, processibility, or foamability. About 500 different types of plasticizers are marketed. They are generally low molar mass liquids; polymer plasticizers are used in much smaller amounts. Of all plasticizers for thermoplastics, 80–85% are used for poly(vinyl chloride); the most important plasticizers are di-2-ethylhexyl phthalate ("dioctyl phthalate," DOP) and dinonyl phthalate. The main plasticizer for elastomers is mineral oil; tires contain up to 40%. Polymer plasticizers are mainly aliphatic polyesters and polyethers. The former are prepared by polycondensation; they thus possess fairly broad molar mass distributions and also contain monomeric and oligomeric molecules. Because their number-average molar mass is low (ca. 4 000 g/mol), they are called oligomer plasticizers. Plasticization can also be achieved by copolymerization of the parent monomer with certain other monomers; this effect is called internal plasticization, in analogy to external plasticization by added high and low molar mass plasticizers.

External plasticizers are subdivided into primary and secondary ones. Primary plasticizers interact directly with chains by way of solvation. Secondary plasticizers are merely extenders; they can be used only in combination with a primary plasticizer. A certain plasticizer may thus be a primary or a secondary one, depending on the chemical constitution of the polymer. Mineral oils are, for example, primary plasticizers for polydienes but extenders for poly(vinyl chloride).

2.2.6. Colorants

Colorants are subdivided into dyes (soluble in polymer matrix) and pigments (insoluble). Textile fibers are dyed mainly with dyes. Pigments are preferred for plastics because they have a higher light-fastness and are more stable against migration than dyes. Colorants for plastics are dominated by titanium dioxide (60–65%) and carbon black (20%); only 2% are dyes.

Pigment particles generally possess diameters of 0.3–0.8 µm, which allows the pigmentation of films and fibers of > 20 µm thickness. Very thin films and fibers are colored exclusively by organic pigments because these can be ground to much smaller diameters than inorganic ones. The hiding power of pigments increases with increasing difference between refractive indices of pigment and polymer.

Pigments need not have a special affinity for polymers. They must be wettable by the polymer melt, however, which can be achieved by treating them with surfactants. The aggregation of pigments is mainly the result of air inclusions; it can be removed by the application of vacuum. Pigments can be metered into plastics via master batches (in plastics), color concentrates (in plasticizers), or electric charging of the surface of polymer particles in granulate mixers. Up to 1 wt % of pigment can be mixed in by the last method.

2.2.7. Blowing Agents

Foamed plastics (plastic foams, cellular plastics, expanded plastics) are blends of polymers with gases (→ Foamed Plastics). They may be rigid (glass transition or melting temperature higher than use temperature) or flexible; their cell structure can be open or closed. The gases may be air, nitrogen, carbon dioxide, fluorinated hydrocarbons, etc.

Plastic foams can be produced by mechanical means (whipping, stirring), physical methods (shock volatilization of liquids, washing-out of solids), or chemical foaming either by internal foaming during the polymerization or by external foaming with chemical blowing agents. Chemical blowing agents are chemical compounds that decompose at elevated temperature with release of gases. The most widely used agent for natural rubber is N,N'-azobisisobutyronitrile $(CH_3)_2C(CN)-N=N-C(CN)(CH_3)_2$ (AIBN); for plasticized PVC, N,N'-dinitroso dimethyl terephthalamide $CH_3-N(NO)-CO-(p-C_6H_4)-CO-N(NO)-CH_3$ (NTA); and for other plastics, 1,1′-azobisformamide $H_2N-CO-N=N-CO-NH_2$ (ABFA). Blowing agents are used in amounts of ca. 0.1% to eliminate sinks in injection molding; 0.2–0.8% for injection-molded structural foams; 0.3% for extended profiles; 1–15% for vinyl plastisol foaming; and 5–15% for compression-molded foam products. Because of the depletion of the ozone layer, some fluorine-containing blowing agents were banned.

2.3. Processing [34–37]

There are many different types of processes to convert monomers, prepolymers, or polymers into plastics. In general, four procedures can be distinguished for processing raw materials into shaped plastics:

1. From monomers to polymers by direct polymerization with simultaneous shaping
2. From monomers by oligomerization to prepolymers, followed by simultaneous polymerization and shaping
3. From monomers by polymerization to polymers, followed by separate shaping
4. From monomers by polymerization to polymers, shaping to semifinished product, and finally to the end product

In all four cases, end products must sometimes be after-treated (degated, polished, etc.).

Procedure 1 has only one process step and should be the most economical. It is widely used for wire coatings and casting. However, technological problems are often encountered for shaped articles because the process involves the conversion of a low-viscosity liquid into a high-viscosity body. Because very high polymerization rates are required for this process to achieve high cycle times (phase sequences) in molding (i.e., number of articles produced per unit time), very few monomers, types of polymerization, and shaping methods are suitable. The method is utilized industrially in the RIM (reaction injection molding) process, mainly for the polyaddition of diisocyanates with polyols. The anionic polymerization of lactams and the Diels–Alder polymerization of dicyclopentadienes and related compounds are also suitable for RIM. In resin transfer molding (RTM), thermoset resins and curing agents are mixed together, injected, and cured in the mold.

Procedure 2 is typical for thermosets. Oligomerizations are conducted to conversions shortly before the gel point. Additives are added, and the final processing step consists of a simultaneous cross-linking polymerization and shaping. Typical processing methods include molding, compression molding, cavity compression molding (hot pressing), injection–compression molding, and resin transfer molding.

Procedure 3 is generally chosen for thermoplastics. Polymers from the polymerization process are isolated before shaping and stored (e.g., as granulate) for a shorter or longer period of time. The stored polymers must be dried before further processing; otherwise, steam produced at higher processing temperatures may lead to voids and cavities in shaped articles. After compounding, the plastic raw material must be weighed out accurately for each cycle in discontinuous processing procedures. The filling factor (= raw material density/bulk density) must be known.

The choice of a processing technique is influenced technically by the rheological properties of the material and the form or shape of the desired article. The cost of processing machines and the cycle time are economically important. Commodity articles are produced by casting, centrifugal casting, hot pressing, compression molding, coating, spraying, roll milling, extrusion, injection molding, blow molding, cold forming, press forming, calendering, stretch forming, blowing, extrusion blowing, sintering, fluidized-bed sintering, flame spraying, or hot blast sprinkling, and specialty articles by molding.

Procedure 4 is the method of choice for difficult-to-manufacture articles. Procedures, such as welding, stamping, cutting, forging, sawing, boring, turning, milling (in the metalworking sense), sintering, and baking are employed for both thermoplastics and thermosets. The surfaces of plastics are sometimes treated further for technical (surface hardness, friction) or aesthetic reasons (gloss, color) by polishing, painting, metallizing, coating, etc.

Whereas conventional processing requires molds or machining, in additive processing, also referred to as additives processing and solid freeform fabrication, complex 3D objects are built by computer-assisted layer-by-layer assembly. Typically, 3D scanner or computer-assisted design are used to create virtual 3D objects, which are sliced and used to construct the real 3D objects layer by layer. Originally designed to serve the special needs of model making, tooling, and rapid prototyping, this mold-free fabrication is emerging as highly versatile processing technology for rapid manufacturing and digital 3D printing. Typical 3D printing technologies include stereolithography with laser-induced photopolymerization of acrylic monomers, selective laser sintering of metal, ceramic and polymer powders, fused deposition molding using 3D extrusion of thermoplastic filaments, and 3D printing by ink-jet printing of binders to bind together powder layers.

References

1 N. Karak (ed.): *Fundamentals of Polymers: Raw Materials to Finish Products*, PHI Learning Private Limited, New Dehli 2010.
2 V. Mittal (ed.): *Renewable Polymers*, Scrivener Publishing and John Wiley & Sons, Hoboken, Salem 2012.
3 M.N. Belgacem, A. Gandini (eds.): *Monomers, Polymers and Composites from Renewable Resources*, Elsevier, Oxford 2008.
4 A. Gandini, *Macromolecules* **41** (2008) 9491.
5 B. Rieger, A. Künkel, G.W. Coates, R. Reichardt, E. Dinjus, T.A. Zevaco: "Synthetic Biodegradable Polymers" in *Advanced Polymer Science*, vol 245, Springer, Berlin 2012.
6 A. Demirbas (ed.): *Biorefineries*, Springer, Berlin 2010.
7 D.M. Mandal (ed.): *Fundamentals of Polymerization*, World Scientific Publishing Company, Singapore 2013.
8 H.-G. Elias: Grundlagen, *Makromoleküle*, 5th ed., vol. 1, Hüthig and Wepf, Basel 1989.
9 H.R. Kricheldorf (ed.): *Polycondensation*, Springer, Berlin 2014.
10 K. Matyjaszewski, T.P. Davis (eds.): *Handbook of Radical Polymerization*, John Wiley & Sons, New York 2003.
11 N.V. Tsarevsky, B.S. Sumerlin (eds.): *Fundamentals of Controlled/Living Radical Polymerization*, RCS Polymer Chemistry Series, RCS Publishing, London 2013.

12. K. Matyjaszewski, B.S. Sumerlin, N.V. Tsarevsky (eds.): *Progress in Controlled Radical Polymerization: Mechanisms and Techniques*, American Chemical Society, Washington 2012.
13. N. Hadjichristidis, M. Pitsikalis, S. Pispas, H. Iatrou: "Polymers with Complex Architecture by Living Anionic Polymerization," *Chem. Rev.* **101** (2001) 3747–3792.
14. S. Aoshima, S. Kanaoka: "A Renaissance in Living Cationic Polymerization," *Chem. Rev.* **109** (2009) 5245–5287.
15. H. Brintzinger, D. Fischer, R. Mülhaupt, B. Rieger, R.M. Waymouth, *Angew. Chem. Int. Ed.* **34** (1995) 1143–1170.
16. V.C. Gibson, S.K. Spitzmesser, *Chem. Rev.* **103** (2003) 283–315.
17. R. Mülhaupt, *Macromol. Chem. Phys.* **204** (2003) 289.
18. J.R. Severn, J.C. Chadwick (eds.): *Tailor-Made Polymers: Via Immobilization of Alpha-Olefin Polymerization Catalysts*, Wiley-VCH, Weinheim 2008.
19. J.L. White, D.C. Choi (eds.): *Polyolefins: Processing, Structure Development, and Properties*, Hanser Publ., Munich 2004.
20. M. Buchmeiser, *Chem. Rev.* **100** (2000) 1565.
21. T.M. Baughman, K.B. Wagener, *Adv. Polym. Sci.* **176** (2005) 1–42.
22. C. Hagiopol (eds): *Copolymerization: Toward a Systematic Approach*, Kluwer Academic/Plenum Publishers, New York 1999.
23. D. Yan, C. Gao, H. Frey (eds.): *Hyperbranched Polymers*, Wiley-VCH, Weinheim 2010.
24. J.J. Meister (ed.): *Polymer Modification, Principles, Techniques, and Applications*, Marcel Dekker, New York 2000.
25. P. Theato, H.-A. Klok (eds.): *Functional Polymers by Post-Polymerization Modification*, Wiley-VCH, Weinheim 2013.
26. N. Hadjichristidis, S. Pispas, G.A. Floudas (eds.): *Block Copolymers: Synthetic Strategies, Physical Properties, and Applications*, John Wiley & Sons, Hoboken 2003.
27. E.S. Guerra, E.V. Lima: *Handbook of Polymer Synthesis, Characterization and Processing*, John Wiley & Sons, Hoboken 2013.
28. R. Gächter, H. Müller (eds.): *Plastics Additives Handbook* 3rd ed., Hanser, München 1989.
29. H. Zweifel, R.D. Maier, M. Schiller (eds.): *Plastics Additives Handbook*, 6th ed., Hanser Publ., Munich 2008.
30. H.S. Katz, J.V. Milevski (eds.): *Handbook of Fillers for Plastics*, Van Nostrand Reinhold, New York 1987.
31. M. Xanthos (ed.): *Functional Fillers for Plastics*, Wiley-VCH, Weinheim 2010.
32. R.K. Gupta, E.B. Kennel, K.-J. Kim (eds.): *Polymer Nanocomposites Handbook*, CRC Press, Boca Raton 2010.
33. A.D. Godwin, W. Arendt, P. Daniels (eds.): *New Plasticizer Technology*, Wiley-VCH, Weinheim 2014.
34. W. Michaeli (ed.): *Plastics Technology*, Hanser Publ., Munich 2000.
35. T.A. Oswald (ed.): *Polymer Processing Fundamentals*, Hanser Publ., Munich 1998.
36. Z. Tadmor, C.C. Gogos (eds.): *Principles of Polymer Processing*, John Wiley & Sons, Hoboken 2006.
37. M. Reyne (ed.): *Plastic Forming Processes*, Wiley-VCH, Weinheim 2008.

Further Reading

P.C. Painter, M.M. Coleman (eds.): *Essentials of Polymer Science and Engineering*, DEStech Publication, Lancaster 2009.

E.S. Guerra, E.V. Lima (eds.): *Handbook of Polymer Synthesis, Characterization and Processing*, John Wiley & Sons, Hoboken 2013.

H.G. Elias (ed.): *Macromolecules: Industrial Polymers and Synthesis*, Wiley-VCH, Weinheim 2007.

H.G. Elias (ed.): *Macromolecules*, vols. 1-4, Wiley-VCH, Weinheim 2008.

K. Matyjaszewski, M. Möller (eds.): *Polymer Science: A Comprehensive Reference*, Elsevier, Amsterdam 2012.

G.H. Michler, F.J. Balta-Calleja (eds.): *Nano- and Micromechanics of Polymers*, Hanser Publ., Munich 2012.

N. Rudolph, T. Osswald (eds.): *Polymer Rheology: Fundamentals and Applications*, Hanser Publ. Munich 2014.

Plastics, General Survey, 3. Supermolecular Structures

HANS-GEORG ELIAS, Michigan Molecular Institute, Midland, United States

ROLF MÜLHAUPT, Institute for Macromolecular Chemistry, Freiburg, Germany

1.	Supermolecular Structures	187
1.1.	Noncrystalline States	187
1.1.1.	Structure	187
1.1.2.	Orientation	188
1.2.	Polymer Crystals	188
1.2.1.	Introduction	188
1.2.2.	Crystal Structures	189
1.2.3.	Crystallinity	190
1.2.4.	Morphology	191
1.3.	Mesophases and Liquid-Crystalline Polymers	191
1.3.1.	Introduction	191
1.3.2.	Types of Liquid Crystals	194
1.3.3.	Mesogens	195
1.3.4.	Lyotropic Liquid-Crystalline Polymers	196
1.3.5.	Thermotropic Liquid Crystal Polymers	198
1.3.6.	Block Copolymers	198
1.3.7.	Ionomers	199
1.4.	Gels	201
1.5.	Polymer Surfaces	201
1.5.1.	Structure	201
1.5.2.	Interfacial Tension	201
1.5.3.	Adsorption	202
	References	202

1. Supermolecular Structures (→ Plastics, Properties and Testing) [1–3]

Polymer properties are influenced not only by the chemical structure (constitution, molar mass, configuration, microconformation) but also by the physical structure of polymers. These structures may range from totally irregular arrangements of chain segments over shorter or longer parallelizations of chains, to voids and other defects in otherwise highly organized assemblies of polymer molecules. Two possible ideal structures exist in the solid state: Perfect crystals and totally amorphous polymers. Polymer molecules are perfectly ordered in ideal crystals. They convert at the thermodynamic melting temperature into melts, which ideally are totally disordered. Amorphous polymers can be viewed as frozen-in polymer melts. They are polymer glasses that convert to melts at the glass transition temperature.

1.1. Noncrystalline States

1.1.1. Structure

Isolated polymer coils possess approximately a Gaussian distribution of chain segments; their segment density decreases with increasing chain length. However, the macroscopic densities of polymer melts do not change with chain lengths if end group effects on small molar mass molecules are neglected. Coils must therefore overlap considerably in polymer melts. At the glass transition temperature, cooperative segmental movements freeze in, and the physical structure of the melt is conserved. Small-angle neutron scattering studies have shown that the radii of gyration of amorphous polymers are indeed essentially identical for their melts, glasses, and solutions in theta solvents.

The absence of long-range order in melts and amorphous polymers does not exclude the presence of short-range order in these states.

Because of the persistence of polymer chains, a parallelization of short segments seems probable, as is found, for example, for alkanes according to X-ray investigation. This local order does not exceed 1 nm in each direction.

The packing of chain segments cannot be perfect. Amorphous polymers thus possess "free volumes", which are regions of approximately atomic diameters. The volume fraction of this free volume is ca. 2.5% at the glass transition temperature and is independent of polymer constitution. Polymer segments in melts can move more freely than in the glassy state; the densities of melts are thus higher than the densities of glasses at the same temperature.

1.1.2. Orientation

Polymer segments, polymer molecules, and crystalline domains may be oriented along the machine direction by drawing or other mechanical processes. The orientation of chain segments need not necessarily lead to crystallization, however. An example is injection-molded polystyrene, which shows optical birefringence due to the orientation of segments but no X-ray crystallinity.

The degree of orientation of segments or crystallites can be measured by wide-angle X-ray or small-angle light scattering, infrared dichroism, optical birefringence, polarized fluorescence, and ultrasound velocity. The orientation is usually characterized for each of the three directions a, b, and c by a *Hermans orientation factor* f_i, where β is the angle between the draw direction and the principal axis of the segments:

$$f_i = (1/2)(3\cos^2\beta - 1)$$

The orientation factor becomes 1 for a complete orientation of that axis in the draw direction ($\beta = 0°$), $-1/2$ for a complete orientation perpendicular to the draw direction ($\beta = 90°$), and 0 for a random orientation.

Since methods to determine orientation factors are often expensive, time consuming, or difficult to perform, the easy-to-calculate *draw ratio* (length after drawing/length before drawing) is often used to characterize orientation. The draw ratio is, however, not a good measure of the degree of orientation because it depends on the history of the specimen. Drawing may also lead to viscous flow of the polymer without any orientation of segments and crystallites.

1.2. Polymer Crystals [4, 5]

1.2.1. Introduction

Spheres and other geometrically simple entities can be arranged in crystal lattices irrespective of whether they are atoms, small molecules, large molecules such as enzyme, or latex particles. The centers of gravity of these entities occupy lattice sites that are regularly spaced in three dimensions. Lattices are called *superlattices* if the distances between lattice points significantly exceed atomic dimensions. The existence of three-dimensional order on the atomic level is called crystallinity.

Chain molecules can also be arranged in crystal lattices. The whole chain may be completely parallel to other chains as in the case of truly extended chain crystals; the microconformation of the chain can be all *trans* or helical (Fig. 1), and the chains may have a directional sense as in polyamide 6 or an inversion center as in polyamide 66 (Fig. 2). In other cases, only sections of the chains may be parallel to each other, either sections of a single chain as in folded chain crystals or those of neighboring chains as in fringed micelles. These ordered segments form crystalline regions in otherwise less ordered (though not necessarily completely disordered) polymers.

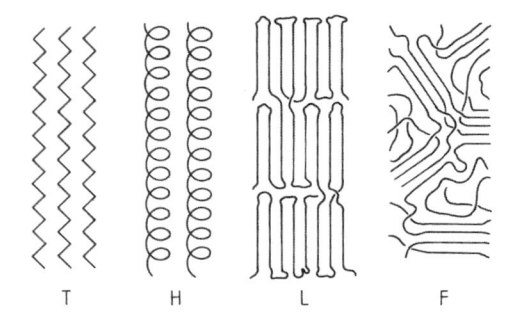

Figure 1. Schematic representation of crystalline polymers as extended chain crystals with chains in all-*trans* (T) or helical conformation (H), (L) lamellae of chains interconnected by tie molecules, and (F) fringed micelles. No bond angles are shown for H, L, and F.

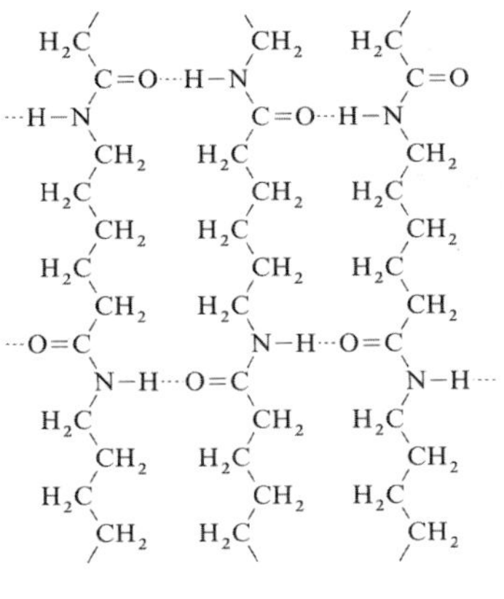

Figure 2. Pleated-sheet structures of polyamides PA 6 (anticlinal) and PA 66 (isoclinal)

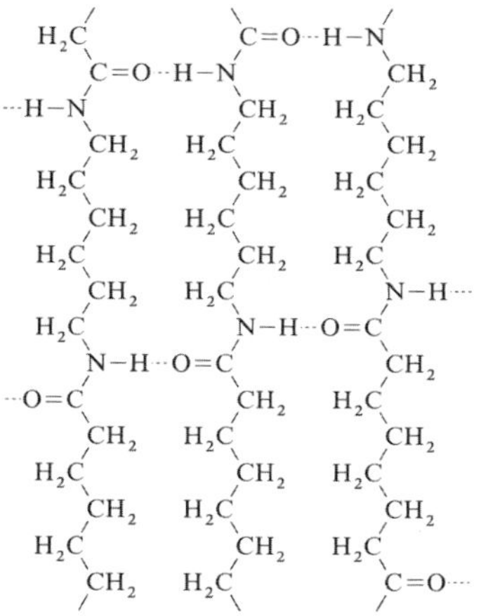

Crystalline regions may be assembled into greater entities, which themselves may be more or less ordered. The types and degrees of order−disorder depend on external conditions such as temperature, pressure, cooling and heating rates, and presence of solvents. A variety of morphologies may thus exist for any crystallizable polymer [6], for example, spherical, platelet-like, fibrillar, and sheaflike structures of a polyamide 6 (Fig. 3).

1.2.2. Crystal Structures

The unit cell describes the smallest, regularly repeating structure of a parallelepiped that generates a crystal by parallel displacements in three directions. The parallelepipeds are characterized by three lattice constants (distances between lattice points in three directions) and three lattice angles (Table 1).

Lattice constants reflect the microconformations of chains in crystal lattices. The c-direction is normally assigned to the chain direction; a value of $c = 0.254$ nm thus gives the distance between repeating units $-CH_2-CH_2-$ in all-*trans* conformation (bond distance 0.154 nm, bond angle 112°). The c-values in carbon chains that are nonmultiples of 0.254 indicate the presence of non-all-*trans* conformations (e.g., helical structures in it-polypropylene and it-polybutene-1). The a and b values reflect physical bonds (such as van der Waals or hydrogen bonds) that are longer than the chemical bonds in the c-direction. The anisotropy of bond lengths in the three spatial directions prevents the existence of cubic lattices for chain molecules. The other six lattice types are however found [hexagonal, tetragonal, trigonal, (ortho) rhombic, monoclinic, triclinic].

Most polymers form polymorphs (i.e., polymers of the same constitution and configuration can possess various energetically different crystal modifications). *Polymorphism* may be caused by different microconformations of polymer chains (e.g., polybutene-1) or by different packing of chains with identical chain conformations (e.g., it-polypropylene). Such differences can be generated by small changes in processing conditions (e.g., different crystallization temperatures) or by stretching.

Copolymers can furthermore show *isomorphism* if different monomeric units can replace each other without change of lattice structure. Isomorphism of chains is also possible if the two corresponding homopolymers have analogous crystal modifications, similar lattice constants, and the same helix types. An example is a mixture of the γ-modification of it-polypropylene and the modification I of it-polybutene-1 (Table 1). This phenomenon is sometimes also called allomerism. One company calls its crystalline copolymers from two or more olefinic monomers *polyallomers*, because the monomeric units are isomorphic.

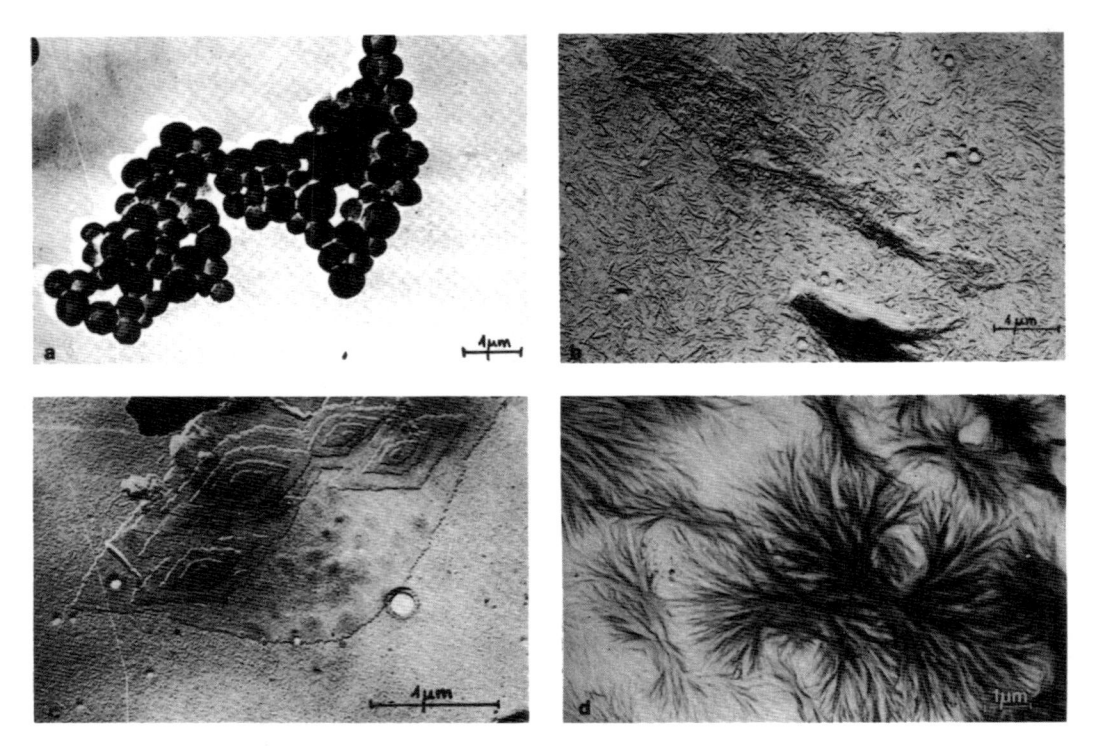

Figure 3. Electron micrographs of different morphologies of a polyamide 6
a) From a 260°C solution in glycerol quenched into 20°C glycerol; b) Same solution fast cooled (40 K/min); c) Same solution slowly cooled (1–2 K/min); d) Slow evaporation of formic acid solution at room temperature [7]

1.2.3. Crystallinity

Perfect crystal lattices should give sharp X-ray reflections and sharp melting points. Polymers do not exhibit these features, even if they show a crystallike appearance in electron micrographs. In addition, X-ray patterns of polymers frequently exhibit continuous, diffuse scattering. These findings can be interpreted by either a one-phase model (voids in crystals) or a two-phase model (coexistence of completely crystalline and completely amorphous regions).

Table 1. Crystal structure of some polymers

Polymer				Lattice constants			Angles of unit cell			Helix type	Crystal system
			N_u	nm^a	nm^b	nm^c	$\alpha°$	$\beta°$	$\gamma°$		
PE		I^b	2	0.742	0.495	0.254	90	90	90	2_1	R
		II$^{a\,b}$	2	0.809	0.479	0.253	90	90	107.9	2_1	M
PP	st		8	1.450	0.560	0.74	90	90	90	2_1	R
	it	α^b	12	0.665	2.09	0.495	90	99.6	90	3_1	M
	it	β^b	3	0.638	0.638	0.633	90	90	120	3_1	VI
	it	γ^b	12	0.647	2.140	6.50	89	100	99	3_1	T
PB	it	I^b	18	1.769	1.769	0.650	90	90	90	3_1	VI
		II^b	44	1.485	1.485	2.060	90	90	90	11_3	IV
		III^b	8	1.238	0.892	0.745	90	90	90	4_1	R

$^a N_u$ = Number of monomeric units per unit cell.
^{b}Crystal modification.
^{c}R = (ortho)rhombic; M = monoclinic; VI = hexagonal; T = triclinic; IV = tetragonal.

Crystallinity is defined as the presence of three-dimensional order on the level of atomic dimensions.

In reality, segments are ordered to various degrees. Because most experimental methods are sensitive to different types and degrees of order, various "degrees of crystallinities" are found depending on the experiment. The degree of crystallinity is the fractional amount of crystallinity. Stretched poly(ethylene terephthalate) has, for example, a degree of crystallinity of 59% by IR spectroscopy, 20% by density determination, and 2% by X-ray diffraction.

1.2.4. Morphology

Earlier experimental findings were interpreted in terms of the two-phase model of partially crystalline polymers: The coexistence of crystalline reflexes and amorphous halos in X-ray diagrams, the independence of X-ray short periodicities on molar masses, the disappearance of X-ray long periodicities with increasing molar masses, lower macroscopic densities than predicted by X-ray densities of unit cells, broad melting ranges, optical birefringence of oriented polymers, and heterogeneity of partially crystalline polymers against chemical reactions.

In 1957, however, three different groups found that rhombic platelets are obtained if very dilute solutions of polyethylene are cooled. Similar platelets of 5–20 nm thickness are now known from many other crystallizable polymers (e.g., see Fig. 3). Electron diffraction showed sharp reflections typical of single crystals and chain directions perpendicular to the surface of the platelets. Since the contour lengths of the chains are greater than the thicknesses of the platelets, chains must be folded back as shown in Figure 1, L.

Crystallization from polyethylene melts similarly shows stacked single crystals. These lamellae are 75–80% X-ray crystalline. After oxidation of the surface, a 100% crystalline material remains. The folds or the interface between lamellae must thus be "amorphous." Various models have been suggested for the surface layers: Loose folds with adjacent reentry of the chains, a switchboard-like structure, adsorbed chains not involved in chain folding, loose or entangled cilia (tangling chain ends), and crystal bridges composed of tie molecules (Fig. 4).

Electron microscopy has shown that crystal bridges exist in crystallized mixtures of high and low molar mass polyethylenes. Such interlamellar connections are responsible for some remarkable properties of partially crystalline polymers (see below). Highly crystalline polymers result if tie molecules and surface layers are removed (e.g., by grinding); these fairly low molar mass materials are called microcrystalline polymers.

Crystallization of melts sometimes delivers spherical, polycrystalline entities (Fig. 5). These spherulites exhibit radial symmetry. They consist of lamellae with tangential chain axes (Fig. 6A); radial chain axes seem to be unknown (Fig. 6B). The formation of spherulites can be controlled by the crystallization temperature, the cooling rate, and the addition of nucleating agents.

Spherulites lead to opaque plastics if their diameters are greater than one-half the wavelength of incident light and their densities and refractive indices are different for crystalline and noncrystalline regions. The numbers and sizes of such spherulites are important for the fracture behavior of polymers.

Spherulites develop in a viscous environment if crystallization nuclei are present and if equal crystallization rates prevail in all three directions. Different crystallization rates exist in strongly stirred dilute solutions. Polymer chains then try to orient themselves along the flow gradients and crystallize in an extended (nonfolded) conformation. The resulting fibrils are organized in bundles with chains parallel to the fibril axes. These bundles act as nuclei–substrate for the epitaxial growth of the remaining chains. Since the shear rate is strongly reduced between bundles, crystallization leads to lamellae with folded chains whose axes are parallel to the fibril axes (Fig. 7). The resulting extended-chain structures are responsible for the remarkable properties of some ultraoriented polymers.

1.3. Mesophases and Liquid-Crystalline Polymers [8]

1.3.1. Introduction

Mesomorphic chemical compounds exhibit microscopic structures (mesophases) that are

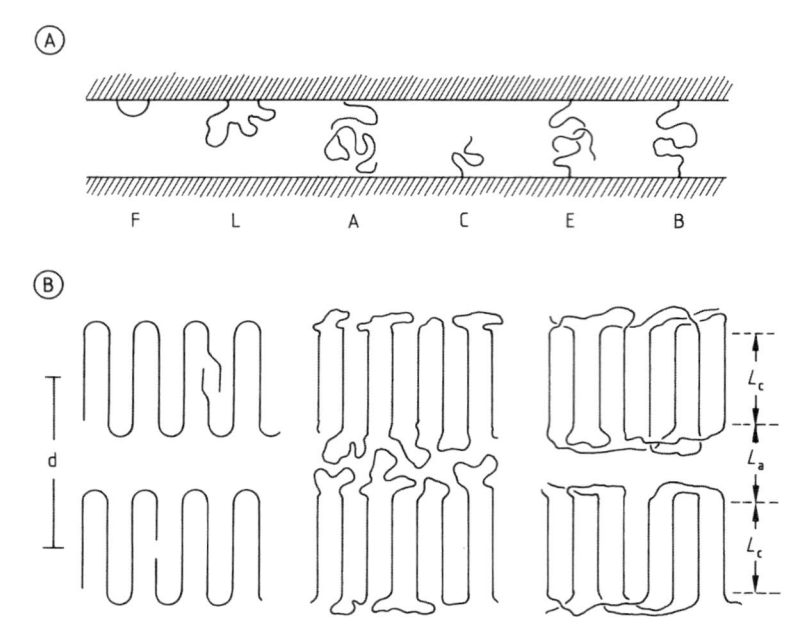

Figure 4. Models for the fine structure of lamellae of single crystals
A) Chains between two adjacent lamellae: F = sharp folds, L = loose loops, A = adsorbed chain, C = cilium, E = entangled cilium, B = crystal bridge or tie segment; B) Structures of lamellae: R=Regular lamellae with sharp folds, adjacent chain reentry, void from chain ends (below) and dislocation (above); L = lamellae with loose loops and adjacent reentry; S = switchboard model; d = crystallographic long period; L_c = thickness of lamella; L_a = thickness of "amorphous" surface layer ($d = L_c + L_a$)
Reprinted with permission by Hüthig and Wepf Publ., Basel [9]

intermediate between crystals with long-range three-dimensional order and liquids without long-range order (Greek: *mesos* = middle, *morphe* = shape, form). Three types of mesophases are distinguished: Liquid crystals, plastic crystals, and condis crystals. Since liquid crystals were discovered first and are furthermore the most common of the three classes, the term mesophase is often used as a synonym for liquid crystal.

Liquid crystals (LCs) (→ Liquid Crystals for Display Application) possess a certain ordered structure like crystals, but they flow like liquids. *Thermotropic liquid crystals* exhibit this phenomenon in the pure state above a certain solid–mesophase transition temperature, where the solid may be a crystal or a glass. *Lyotropic liquid crystals* exhibit mesophase behavior in certain solvents above critical concentrations. Liquid crystalline behavior is present in both cases if the molecules contain "mesogens", that are short rodlike sections that comprise the whole molecule as in low molar mass LCs, or segments thereof as in polymer liquid crystals (LCPs). Mesogens are anisotropic; they are oriented with respect to their mesogen axes but not with respect to their positions.

Plastic crystals exhibit order of positions but disorder of orientations of molecules. They are found for certain spherical low molar mass molecules packed in cubic lattices (i.e., they are isotropic). The absence of strong attractive forces between molecules and the presence of slip planes lead to easy deformabilities of plastic crystals, sometimes even under their own weight. No polymers are known that form plastic crystals.

Condis crystals are *con*formationally *dis*ordered crystals containing several conformational isomers in more or less ideal

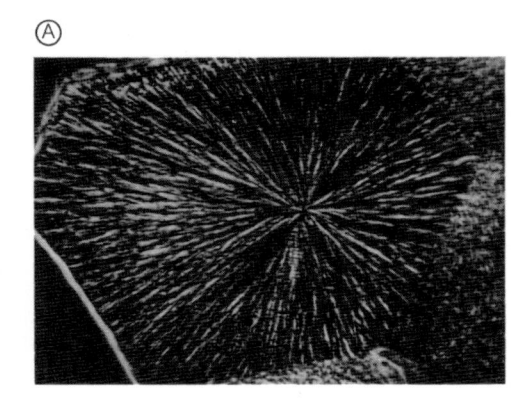

Figure 5. Spherulites of it-polypropylene under the phase contrast microscope (A) and the polarization microscope (B) [10]

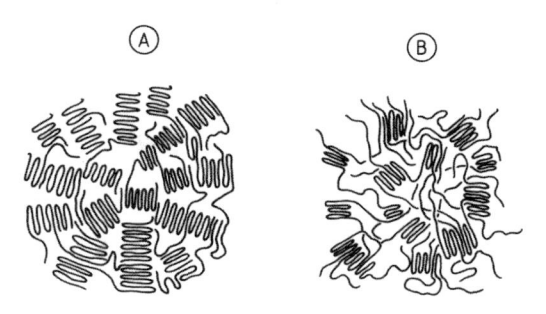

Figure 6. Schematic representation of spherulite structures A) Tangential chain axes; B) Radial chain axes (hypothetical) Reprinted with permission by Hüthig and Wepf Publ., Basel [9]

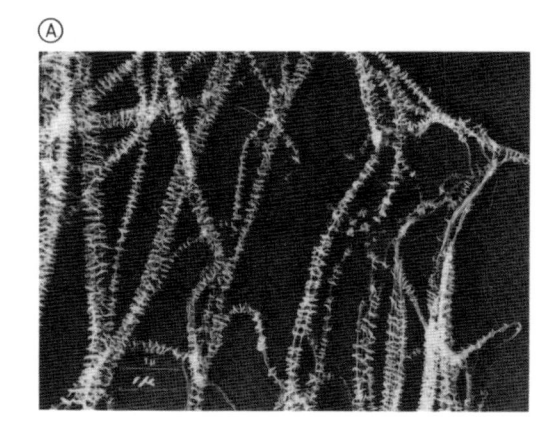

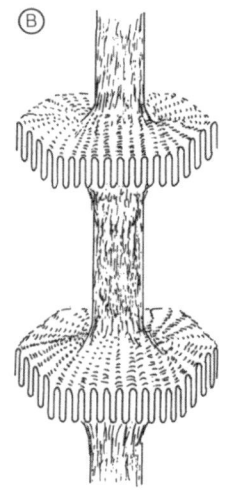

Figure 7. Shish kebab structures of linear polyethylene crystallized from rapidly stirred 5% solutions in toluene [11] A) Electron micrograph; B) Schematic drawing of chain arrangements

crystalline positions with respect to position and orientation of segments (conformational isomorphs). An example is the hexagonal high-pressure phase of polyethylene extended-chain crystals. It is often difficult to determine whether polymer crystals are normal or condis crystals.

On cooling, thermotropic liquid crystals either crystallize or solidify to "mesomorphic glasses" with retention of their liquid crystal structures. These glasses do not carry special names, but since "liquid crystalline glass" and similar names are oxymorons, the terms "LC glass," "PC glass," and "CD glass" have been suggested for the three classes of mesomorphic materials. The LCPs are used mainly as LC glasses not as liquid crystals per se, except in processing.

Condensed matter can thus exist in several of the five possible classes of pure phases:

Completely ordered crystal oC, mesomorphic glass mG, amorphous glass aG, ordered mesophase oM, and isotropic melt (liquid) iL. Seven types of biphasic materials exist: oC + mG, oC + aG, oC + oM, oC + iL, mG + aG, mG + iL, oM + iL, and three types of triphasic materials: oC + mG + aG, oC + mG + iL, oC + oM + iL. Because each class can be subdivided further, many physical structures (and thus properties) are possible for the same polymer.

Mesophasic polymers are characterized by domains of microscopic size (see below). "Microdomains" of similar size are also exhibited by certain block copolymers and, in much smaller sizes, by ionomers.

1.3.2. Types of Liquid Crystals

Liquid crystals and LC glasses are generated by rod- or disclike mesogens, which either comprise the whole molecule (as in low molar mass LCs) or sections of it (LCPs). Examples are

Rodlike

$-X-\bigcirc-Y-\bigcirc-X-$

Disclike

where X = O, OOC, COO, etc.; Y = COO, p-C_6H_4, CH=CH, N=N, etc.; R = CO$(CH_2)_n CH_3$, etc.; and phenylene residues may be replaced by 1,4-cyclohexane rings. The lengths of polymer mesogens are often comparable to those of repeating units; they seem to be identical with respect to persistence lengths.

Rodlike mesogens are also called calamitic (Greek: *calamos* = reed) and disclike ones discotic. Calamitic liquid crystals are further subdivided into smectic, nematic, and cholesteric mesophases. Smectic LCs show fan-shaped structures under the polarization microscope; they feel like soaps (Greek: *smegma* = soap). Nematic LCs form threadlike schlieren (Greek: *nema* = thread). Cholesteric LCs exhibit beautiful reflective colors, which were first discovered with cholesterol.

The birefringence of LCs originates from the anisotropy of mesogens, which are more or less aligned in microscopic domains with diameters in the micrometer range. These alignments lead to different refractive indices parallel and perpendicular to the direction of polarization of incident light. The domains themselves are arranged randomly. The presence of birefringence however is not proof for the existence of mesogens; it may also result from higher-melting crystallites in partially molten polymers or from crystallites in partially molten polymers or from crystallites in gels or concentrated solutions.

Liquid crystals are turbid because their domains are anisotropic structures whose dimensions are greater than the wavelength of incident light. At a certain "clearing temperature," these domains melt to a transparent (isotropic) liquid.

The soaplike character of *smectic LCs* is due to two-dimensional, layerlike arrangements of mesogens (Fig. 8). At present, twelve different smectic types are known: Four with the average direction of the long axes perpendicular to layer planes and seven with tilted axes. Examples are type S-A with mesogens perpendicular to the layer planes; type S-B, the same but with perfect hexagonal packing of mesogens; type S-C, with mesogen axes at an angle to the planes, etc.

Nematic mesophases (Fig. 8, N) are one-dimensionally ordered. The principal axes of

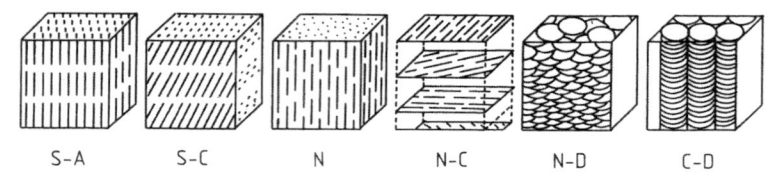

Figure 8. Schematic representation of mesogens in different types of mesophases
S = smectic (types A and C); N = nematic; N–C = nematic cholesteric; C–D = columnar discotic; N–D = nematic discotic
Reprinted with permission by Hüthig and Wepf Publ., Basel [9]

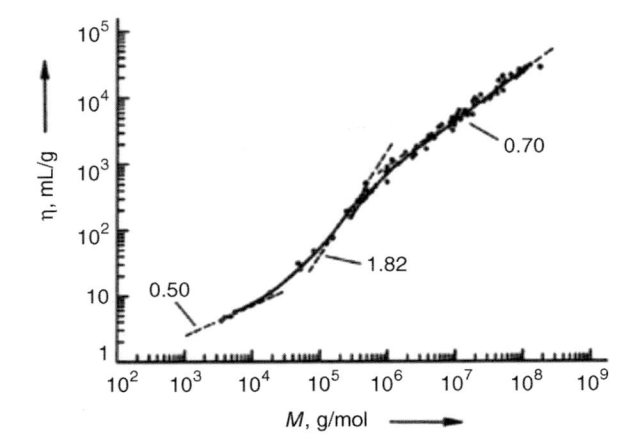

Figure 9. Molar mass dependence of intrinsic viscosities of wormlike chains
(○) Helices of poly(γ-benzyl-L-glutamate) (PLBG) in N,N-dimethylformamide at 25°C; (●) double helices of deoxyribonucleic acids in dilute aqueous NaCl solutions at 20°C. The two polymers are comparable because the helix diameters are similar (1.5 nm for PLBG, 2.0 nm for DNA). Numbers indicate exponents a in Equation (1)
Reprinted with permission by Hüthig and Wepf Publ., Basel [9]

mesogens are more or less parallel to each other; the centers of mass are however distributed at random.

Cholesteric mesophases (Fig. 8, N–C) are observed only with chiral mesogens. They consist of layers of nematically ordered mesogens. These layers have a sense of rotation relative to each other due to the presence of chirality centers in mesogens. Cholesteric mesophases are thus screwlike nematics.

Discotic mesophases may exist in many different structures. The disclike mesogens are randomly arranged in nematic-discotic types (Fig. 8, N–D) but stacked like coins in columnar-discotic ones (Fig. 8, C–D).

1.3.3. Mesogens

Mesogens of polymers can be contained in main chains (MC-LCPs) or in side chains (SC-LCPs). Thermotropic MC-LCP glasses are used as engineering plastics; lyotropic MC-LCP glasses, for fibers. Thermotropic SC-LCPs are not used commercially at present but may find use in optical recording devices.

Mesogens usually constitute the whole molecule in low molar mass LC molecules. In macromolecules, this is true only for dissolved helices up to certain molar masses. Such helicogenic macromolecules are rigid rods at small molar masses but become coillike at very high degrees of polymerization (cf. the behavior of short and very long garden hoses). This behavior is demonstrated by the molar mass dependence of their intrinsic viscosities, which are measures of the molecule volume per unit mass (Fig. 9). Rodlike molecules are characterized by their axial ratio (aspect ratio) Λ (length/ diameter).

All other polymer chains form either flexible or semiflexible coils in dilute solution or in their isotropic melts (see examples in Fig. 10). Semiflexible (wormlike) chains SF allow rotations around some of their chain bonds. The rodlike character of their segments is however preserved in melts or concentrated solutions if the chain axes remain linear or if the segment linearity can be restored by a crankshaft motion. Conjugated bond structures are helpful but not necessarily required for mesogens since, for example, 1,4-phenylene rings may be replaced by 1,4-cyclohexylene units.

Linear chain axes are present in poly(p-phenylene) (PPP) or poly(p-phenylene benzoxazole) (PBO). Crankshaft motions around the chain axes are possible in the conjugated chains of poly(p-phenylene vinylene) (PPV), the partially conjugated chains of poly(p-hydroxybenzoic

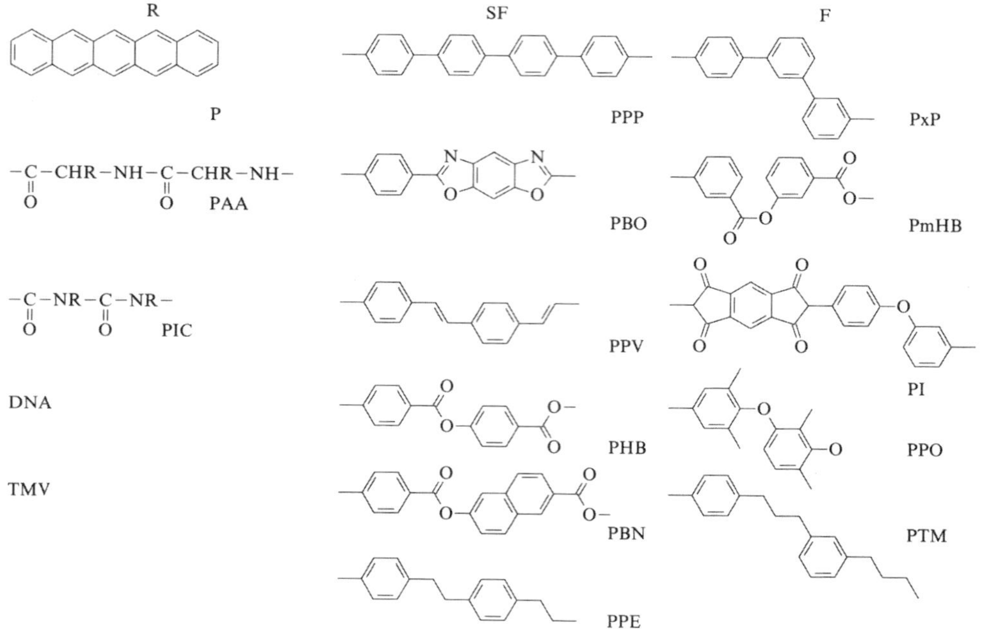

Figure 10. Examples of rodlike (R), wormlike or semiflexible (SF), and flexible (F) molecules and segments. See text for further explanation.
Reprinted with permission by Hüthig and Wepf Publ., Basel [9]

acid) (PHB), and even in the nonconjugated poly(p-phenylene alkylenes) as long as even numbers of methylene groups are present between phenylene rings as in poly(p-phenylene ethylene). Semiflexible molecules are characterized by their persistence lengths a or their Kuhn lengths $L_K = 2\ a$.

Rodlike molecules and semiflexible molecules with long persistence lengths form many physical bonds between parallel chains. The total energy of such assemblies is so high that individual intramolecular chemical bonds rather than all of the physical bonds are broken upon heating: These molecules decompose instead of producing thermotropic mesophases. They may however form lyotropic mesophases in suitable solvents. An example is poly (p-phenylene bisbenzoxazole).

The "stiffness" of the chain segments can be removed or reduced by several means (Fig. 11). Bulky substituents prevent the parallelization of chains; they "frustrate" crystallization. Nonlinear chain elements (such as *ortho* and *meta* substituents) work in the same way. Flexible chain segments reduce the persistence lengths of the chains.

1.3.4. Lyotropic Liquid-Crystalline Polymers

Molecule axes of low molar mass rodlike molecules are disordered in isotropic melts, whereas mesogen axes are more or less parallel to each other in nematic and smectic mesophases. Very short rods resemble spheroids. They can arrange themselves in various stable ways, but the parallel ordering need not be much more stable than other arrangements. Therefore, a critical axial ratio exists above which the simple geometric anisotropy is sufficient to stabilize a mesophase.

This critical axial ratio has been calculated with the lattice theory of polymer solutions as $\Lambda_{crit} \approx 6.42$. A geometric stabilization by repulsion is insufficient for $\Lambda < \Lambda_{crit}$ and the mesogens must exert additional orientation-dependent attractive forces.

In solution, repulsive forces between mesogens can act only at sufficiently high mesogen

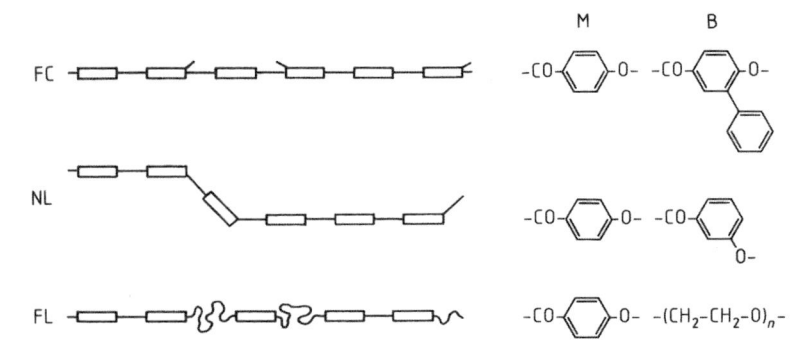

Figure 11. Flexibilization of rigid chain segments by frustrated crystallization (FC), nonlinear chains (NL), or flexible chain elements (FL); M = mesogen-forming monomeric units; B = stiffness-breaking units
Reprinted with permission by Hüthig and Wepf Publ., Basel [9]

concentrations. At a critical volume fraction, Φ_p^* phase separation occurs into a polymer-rich mesophase and a dilute isotropic phase. The dependence of Φ_p^* on Λ can be described to a first approximation by [12]

$$\Phi_p^* \approx 8(1 - 2L^{-1})/L \tag{1}$$

The axial ratio Λ is the molar mass-dependent true axial ratio of rodlike molecules (i.e., helices) and the molar mass-independent Kuhn length of semiflexible molecules. Equation 1 is remarkably well fulfilled for both helical and semiflexible molecules (Fig. 12). An exception

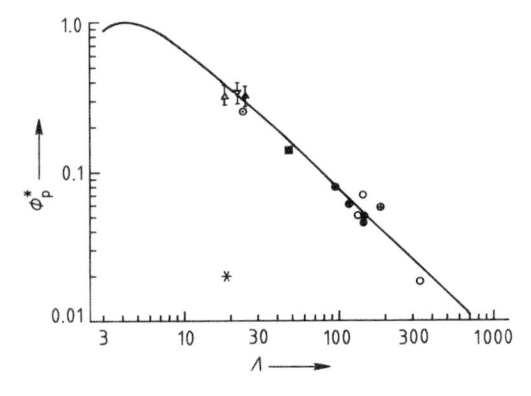

Figure 12. Critical volume fractions ϕ_p^* for phase separations of isotropic solution → nematic mesophase as function of Λ [i.e., the axial ratio of helical (rigid) molecules (circles) and the Kuhn length of semiflexible molecules (triangles, square)]
∗ Tobacco mosaic virus; solid line: prediction of Equation (1)
Reprinted with permission by Hüthig and Wepf Publ., Basel [9]

is the tobacco mosaic virus, probably because of additional charge effects that are not considered by the theory.

Lyotropic mesophases are formed by the helices of tobacco mosaic virus ($c > 2\%$) and deoxyribonucleic acids ($c > 6\%$) in dilute aqueous salt solutions as well as by poly(α-amino acids) in helicogenic solvents, e.g., poly(γ-methyl L-glutamate) in dichloromethane–ethyl acetate (12 : 5) ($c > 15\%$). Hydroxypropyl cellulose (HPC) forms a cholesteric mesophase at room temperature in water at $w_{HPC} > 0.41$.

On shearing, mesophase domains become oriented with respect to the mesogen axes (Fig. 13). This alignment becomes frozen-in on rapid quenching of thermotropic LCs or by precipitation of lyotropic ones into baths (half-lives of orientation between a few seconds and a few hundredths seconds). The resulting

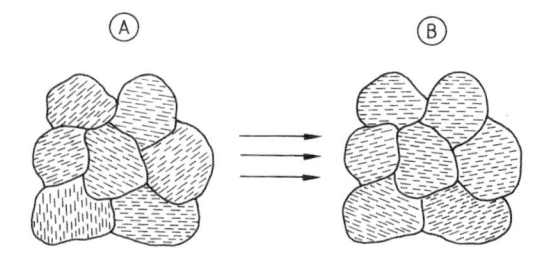

Figure 13. Alignment of mesogen axes upon shearing
A) Mesophase domains in liquid crystalline state at rest; B) During shearing or as LC glass with "frozen in sheared state"

LC glasses possess good orientations of mesogen axes in the shear direction (e.g., fiber direction), high tensile strengths, and low extensibilities.

A number of LC fibers are produced commercially. Poly(p-phenylene terephthalamide) (PPB–T) was originally spun from concentrated sulfuric acid solutions, whereas chlorinated N-methyl pyrrolidine is now used:

~NH–⟨⟩–NH–CO–⟨⟩–CO~   Kevlar; PPB–T

A similar copolyamide is also on the market

–NH–⟨⟩–NH–CO–⟨⟩–CO– / –NH–⟨⟩–O–⟨⟩–NH–CO–⟨⟩–CO– } Technora

Even better LC fibers are produced from poly(p-phenylene-2,6-benzoxazoles) and poly(p-phenylene-2,6-benzthiazoles) (PBT), which are spun from poly(phosphoric acid) solutions, at present on a bench scale. Both "cis" (shown) and "trans" (X and N exchanged at one ring) structures are known:

PBO (X = O); PBT (X = S)

1.3.5. Thermotropic Liquid Crystal Polymers

Non-Newtonian shear viscosities of liquid crystal polymers drop drastically with increasing shear rates due to the orientation of mesogen axes upon shearing. Less energy is thus needed for processing thermotropic LCPs than isotropic thermoplastics. The high orientation times permit a freezing-in of the mesogen orientation in injection molding, for example. Because of the higher than usual moduli of elasticity and tensile strengths, such LCPs are also called *self-reinforcing polymers*.

The first self-reinforcing polymer was the copolyester X7 G with 60 mol% p-hydroxybenzoyl and 40 mol% terephthaloyl glycol units. It is no longer produced because the less expensive glass-fiber-reinforced saturated polyesters possess similar properties. Xydar and Vectra are, however, industrially produced:

–O–⟨⟩–CO– + –O–CH$_2$–CH$_2$–O– + –OC–⟨⟩–CO– } X7G

–O–⟨⟩–CO– + –O–⟨⟩–⟨⟩–O– + –OC–⟨⟩–CO– } Xydar

–O–⟨⟩–CO– + –O–⟨⟩⟨⟩–CO–   Vectra

Polymeric side-chain LCs can serve for the thermooptical storage of information. A guided laser beam increases the temperature locally, whereupon phase transformations occur. The change of order can then be frozen-in on cooling below the glass transition temperature. The resolution is ca. 0.3 µm.

1.3.6. Block Copolymers [13]

Block copolymers consist of two and more blocks of constitutionally or configurationally different monomeric units (e.g., A_n–B_m, A_n–B_m–A_n). In most cases these blocks are thermodynamically or kinetically immiscible because their homopolymers A_n, B_m, etc., are not miscible. The demixing of block copolymers cannot proceed, however, to macroscopic phase separations because the blocks are chemically coupled. Similar blocks of different block copolymer molecules can thus only aggregate and form domains in a matrix of the other blocks. This phenomenon is called microphase separation.

Three different applications of block copolymers are known: Polymeric detergents, thermoplastic elastomers, and compatibilizers for

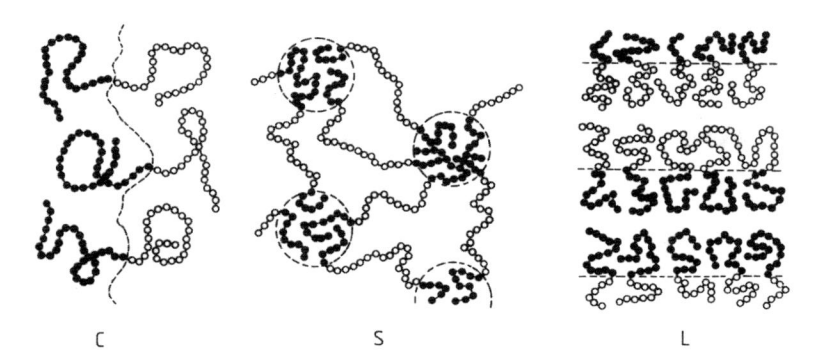

Figure 14. Arrangement of A_n blocks (with A units ●) and B_m blocks (with B units ○) in diblock (C, L) and triblock copolymers (S)
C: compatibilizers at an interface - - -; S: spherical A_n domains in continuous matrix B_m; L: lamellae of A_n and B_m

blends. Polymeric detergents consist of hydrophilic and hydrophobic blocks. Examples are the multiblock copolymers from watersoluble ethylene oxide blocks $[-(OCH_2CH_2)_n-]$ and water-insoluble propene oxide blocks $-[OCH_2CH(CH_3)]_n-$. The hydrophobic blocks associate in water, which produces high viscosities.

A similar action is exerted by *compatibilizers* for nonmiscible blends of A_n and B_m polymers. These polymers are A_p-B_q diblock copolymers: The A_p block resides in the A_n phases and the B_q block in the B_m ones. Compatibilizers thus anchor the two types of phases (Fig. 14).

Thermoplastic elastomers [14] consist of tri- or multiblock copolymers. Triblock copolymers $S_n-B_m-S_n$ possess a soft center polybutadiene block $-B_m-$ (glass transition temperature below use temperature) and two hard end polystyrene blocks S_n- (glass transition temperature above use temperature). Below certain m/n ratios, styrene blocks form spherical microdomains in a continuous matrix of butadiene blocks. These spherical domains thus act as cross-linking units in an elastomer (Fig. 14, S). The hard domains "melt" above their glass transition temperatures, and the block copolymers can then be processed like thermoplastics.

The morphology of *diblock copolymers* is governed mainly by the relative spatial requirements of the blocks. Both blocks, if independent, would form unperturbed random coils. If the unperturbed volumes of both blocks are of equal size, all A_n blocks would line up in one layer and all B_m blocks in another because the blocks are (1) coupled to each other and (2) incompatible. The A_n layer faces another A_n layer, where the A_n blocks are coupled to B_m blocks, etc. (Fig. 14, L). The diblock polymers thus form lamellae with a thickness of two coil diameters. Similar structures result for triblock polymers $A_r-B_m-A_r$, where $r = n/2$.

If the volume of the A_n blocks is much lower than the volume of the B_m blocks, then the A_n blocks can no longer be packed into layers without violating the demand for tightest packing or deviating from the shape of unperturbed coils. Both possibilities are energetically unfavorable, and the smaller A_n blocks thus cluster together and form spherical domains in a continuous matrix of the larger B_m blocks. If the A_n blocks are somewhat larger (but not big enough to form lamellae), then cylinders would result (Fig. 15).

A great number of thermoplastic elastomers of both the triblock and the multiblock type is known [15]. Their chemical structures range from (styrene)$_n$–(butadiene)$_m$–(styrene)$_n$ triblock copolymers to polyether–urethane multiblock polymers, ethylene–propene copolymer–polyolefin blends, polyether–esters, and graft copolymers of butyl rubber on polyethylenes.

1.3.7. Ionomers

Ionomers are copolymers from primarily hydrophobic monomers with small amounts

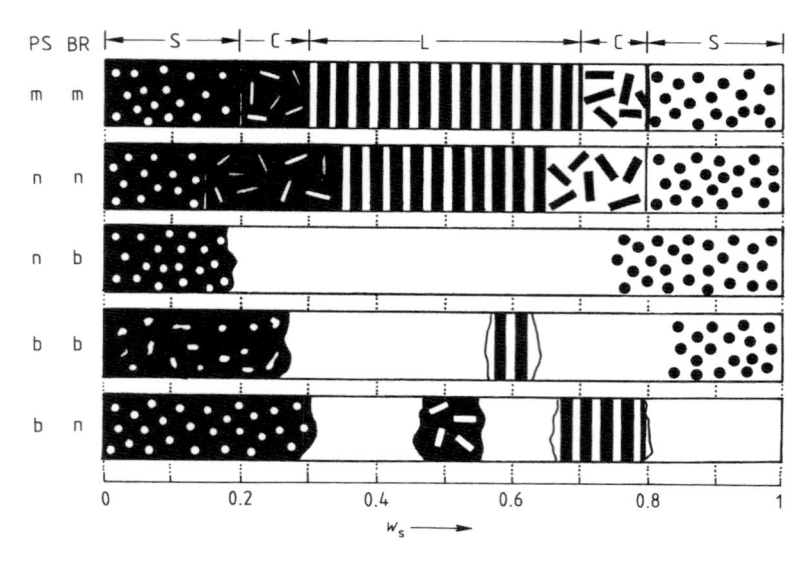

Figure 15. Morphology of styrene–butadiene diblock copolymers as a function of the mass fraction of the styrene units. White: polystyrene blocks (PS); black: polybutadiene blocks (BR). The block length distribution may be (m) molecularly homogeneous (theory), (n) narrow, or (b) broad
S = Spherical domains, C = cylinders (rods), L = lamellae (layers). Upper row: Predictions of Meier theory; rows 2–5: schematic after experimental results of many authors
Reprinted with permission by Hüthig and Wepf Publ., Basel [9]

of ionic comonomers. The ionic groups may be in the main chain or in substituents. Four different types of ionomers are produced commercially.

rubber. Both EEA and Nafion are used as acids, whereas Surlyn and Thionic are obtained by aftertreatment that leads to sodium or zinc salts.

$-CH_2-CH_2-$ + < 10 mol% $-CH_2-C(CH_3)-$
                                       |
                                      COOH   Surlyn

$-CH_2-CH_2-$ + < 3.5–20 mol% $-CH_2-CH-$
                                         |
                                        COOH   EEA Copolymer

$-CF_2-CF_2-$ + ? mol% $-CF_2-CF-$
                                 |
                                 $O-[CF_2-CF(CF_3)-O]_n-(CF_2)_2SO_3H$ Nafion

$-CH_2-CH_2-$ +
$-CH_2-CH(CH_3)-$ $C(CH_3)SO_3H$ Thionic

Surlyn, EEA, and Nafion are synthesized by direct copolymerization of the corresponding monomers. Thionic results from the postpolymerization sulfonation (< 5%) of the diene units (usually 5-ethylidene-2-norbornene) of the primary ethylene–propene–diene (EPDM)

The introduction of ionic groups into hydrophobic polymers leads to ion associations and subsequently to microphase separations. The ion association is controlled by the coordination number of the ions; that is, sodium ions with valence 1 and coordination number 6 are as good cross-linkers as zinc ions with valence 2. Cross-linking by ions does not occur via ion pairs but via clusters and domains of many ions. These ionic bonds dissociate at higher temperature, and the ionomers can then be processed like thermoplastics. At lower temperature, they behave either as thermoplastic elastomers (T_G of polymer segments lower than use temperature) or as "reversible" thermosets (T_G of segments higher than use temperature).

1.4. Gels

Chemically lightly cross-linked polymers swell upon addition of solvent to form gels. The maximum degree of swelling results from the attempt of segments located between cross-linking sites to attain their energetically most favored coil dimensions which in turn is counteracted by the elastic retraction due to cross-links. Physically cross-linked polymers behave in the same way if the physical cross-links (crystallites, ion clusters, etc.) survive the dissolution process.

The latter phenomenon is utilized in poly(vinyl chloride) pastes. The strong dipole–dipole interactions between C−Cl bonds leads to associations of chain segments which at room temperature resist dissolution by plasticizers such as phthalates, adipates, sebacates, or citrates. Heating of PVC with these plasticizers and subsequent cooling leads to gelation by formation of a lightly physically cross-linked network. Plasticized PVC thus behaves like an elastomeric material. The physical cross-links dissociate at higher temperature, and the resulting PVC paste can be processed with extruders or roll mills.

1.5. Polymer Surfaces [16]

1.5.1. Structure

Outer layers of solid or liquid polymers do not exhibit, in contrast to air, water, metal surfaces, etc., the same average composition as the interiors of these polymers. Groups or segments with the lowest Gibbs interfacial energy will reside preferentially at the surface or interface. Surface structures may also be altered by chemical attack during processing or on prolonged use (e.g., by oxidation) or by physical processes such as transcrystallization.

This behavior can be studied by a wide variety of new spectroscopic methods, for example, the concentration of immediate surface groups by Fourier transform infrared spectroscopy (FT-IR); the composition of the upper 0.3–0.5 nm layer by ion scattering spectroscopy (ISS, LEIS); of the upper 1–10 nm by photoelectron spectroscopy [UPS, PE (S)], electron spectroscopy for chemical analysis (ESCA, XPES, XPS, IEE), and Auger spectroscopy (AES).

These methods have shown that poly-(2-vinylpyridine) and poly(4-vinylpyridine) have much more hydrophobic surfaces than their chemical structure indicates: CH_2- and CH-groups face the surface; pyridine residues (C_5H_4N), the interior. Imide–carbonyl residues of polyimides, on the other hand, are preferentially found at the surface and aromatic rings in the interior. The surface is enriched by siloxane residues in polycarbonate–polydimethylsiloxane block copolymers but by styrene units in polystyrene–poly(ethylene oxide) block copolymers. Because the surface energy depends on the contact (air, water, metals, etc.) and also on kinetic effects (thermal history, solvents, etc.), one and the same polymer may possess various surface compositions and thus also surface properties.

1.5.2. Interfacial Tension

Interfacial tension is the force that acts at the interface between two phases; it is called surface tension for the interface condensed phase–gas phase.

Surface tensions γ_{lv} of liquid polymers against air decrease with the two-third power of the molar mass according to

$$\gamma_{lv} = \gamma_{lv}^{\infty} + K_e \times M^{-2/3}$$

They do not vary markedly with temperature. Typical values (at 150 °C) range from 13.6 mN/m (polydimethylsiloxane), 22.1 mN/m (it-polypropylene), 28.1 mN/m (polyethylene), and 33.0 mN/m poly(ethylene oxide). At present, no correlation of these surface tensions with chemical constitution is possible because the surface structures of liquid polymers are unknown.

Interfacial tensions between two liquid polymers are always lower than surface tensions. They are low for two apolar polymers but higher if one polymer or both of them are polar. For example, the interfacial tension of polyethylene–it-polypropylene is only 1.1 mN/m, but that of polyethylene–poly(ethylene oxide) is 9.5 mN/m and poly(ethylene oxide)–polydimethylsiloxane 9.8 mN/m.

The interfacial tensions between solid polymer and air are generally unknown. They do not differ markedly, however, from the critical surface tensions determined by the *Zisman method*. In this method, the liquid–air surface tensions of various liquids are plotted against the cosine of contact angles $\cos \theta$ of these liquids against air and extrapolated to $\cos \theta \rightarrow 0$. The resulting "critical" surface tensions of polymers are always lower than the surface tension of water: Polytetrafluoroethylene, 18.5; polyethylene, 33; poly(vinyl alcohol), 37; polyamide 66, 46; and urea–formaldehyde resins, 61 mN/m. Fluorinated polymers have especially low critical surface tensions. They are not wetted by water (72 mN/m) or oils and fats (20–30 mN/m) and are therefore used as surface coatings to prevent sticking.

1.5.3. Adsorption

Macromolecules possess many adsorbable groups and segments. The type of adsorption depends on the adsorption energy per segment (group), the concentration of adsorbable macromolecules, and the duration of the experiment.

"Adsorption equilibria" are established in minutes to hours at smooth surfaces but may take days to attain at rough surfaces or powders. The adsorption time increases with the concentration and molar mass of the polymer.

Polymer coils tend to overlap even in fairly dilute solutions: The adsorbed polymer layers are almost always multilayers, except for the adsorption of oligomers ($X < 10–100$) or from very dilute solutions (volume fractions $< 10^{-4}$–10^{-3}). At higher concentrations or molar masses, adsorbed amounts and structures of polymer layers are mainly kinetically controlled. Reorganization may occur with time; for example, the adsorption of polar polymers at polar surfaces can change from an initial loose loop structure to a more compact, flat covering. The layer of poly(ethylene oxide) adsorbed on chromium surfaces is only ca. 2 nm thick and so tightly packed that the refractive index of the adsorbed layer is identical with that of the crystalline polymer. The thickness of the surface layer of polystyrene of $M = 176\,000$ g/mol adsorbed from a 5 mg/mL solution in the thermodynamically bad solvent cyclohexane is however about 27 nm (i.e., identical with the end-to-end distance in the unperturbed state). The physical structures of adsorbed polymers obviously play an important, yet widely unknown, role in many technological applications of polymers.

References

1 R.B. Brown (ed.): *Handbook of Polymer Testing*, Marcel Dekker, New York 1999.
2 W. Grellmann, S. Seidler (eds.): *Polymer Testing*, Hanser Publ. Munich 2013.
3 D.W. Van Krevelen, K. Te Nijenhuis (eds.): *Properties of Polymers*, Elsevier, Amsterdam 2009.
4 A.E. Woodward, *Atlas of Polymer Morphology*, Hanser, Munich 1988.
5 L. Mandelkern (ed.): *Crystallization of Polymers*, Cambridge University Press, Cambridge 2004.
6 R. Gächter, H. Müller (eds.): *Plastics Additives Handbook* 3rd ed., Hanser, München 1989.
7 C. Ruscher, E. Schulz, private communication.
8 A.M. Donald, A.H. Windle, S. Hanna (eds.): *Liquid Crystalline Polymers*, Cambridge University Press, Cambridge 2009.
9 H.-G. Elias: Grundlagen, *Makromoleküle*, 5th ed., vol. 1, Hüthig and Wepf, Basel 1989.
10 R.J. Samuels, private communication.
11 G. Natta, P. Corradini, *Nuovo Cimento, Suppl.* **15** (1960) 11.
12 P.J. Flory, *Adv. Polym. Sci.* **59** (1984) 1.
13 P. Theato, H.-A. Klok (eds.): *Functional Polymers by Post-Polymerization Modification*, Wiley-VCH, Weinheim 2013.
14 J.G. Drobny (eds): *Handbook of Thermoplastic Elastomers*, Elsevier, Amsterdam 2007.
15 H.-G. Elias, F. Vohwinkel: *New Commercial Polymers*, Gordon and Breach, New York 1986;Neue polymere Werkstoffe, Hanser, Munchen 1983.
16 J. Martin (ed.): *The Concise Encyclopedia of the Properties of Materials Surfaces and Interfaces*, Elsevier Science, Amsterdam 2008.

Further Reading

P.C. Painter, M.M. Coleman (eds.): *Essentials of Polymer Science and Engineering*, DEStech Publication, Lancaster 2009.
E.S. Guerra, E.V. Lima (eds.): *Handbook of Polymer Synthesis, Characterization and Processing*, John Wiley & Sons, Hoboken 2013.

H.G. Elias (ed.): *Macromolecules: Industrial Polymers and Synthesis*, Wiley-VCH, Weinheim 2007.

H.G. Elias (ed.): *Macromolecules*, vols. 1-4, Wiley-VCH, Weinheim 2008.

K. Matyjaszewski, M. Möller (eds.): *Polymer Science: A Comprehensive Reference*, Elsevier, Amsterdam 2012.

G.H. Michler, F.J. Balta-Calleja (eds.): *Nano- and Micromechanics of Polymers*, Hanser Publ., Munich 2012.

N. Rudolph, T. Osswald (eds.): *Polymer Rheology: Fundamentals and Applications*, Hanser Publ. Munich 2014.

Plastics, General Survey, 4. Polymer Composites

HANS-GEORG ELIAS, Michigan Molecular Institute, Midland, United States

ROLF MÜLHAUPT, Institute for Macromolecular Chemistry, Freiburg, Germany

1.	Overview.	205
2.	Mixture Rules	207
3.	Filled Polymers	208
3.1.	Microcomposites	208
3.2.	Molecular Composites	210
3.3.	Macrocomposites	211
3.4.	Nanocomposites	211
4.	Homogeneous Blends	213
4.1.	Thermodynamics	213
4.2.	Plastification	214
4.3.	Rubber Blends	216
4.4.	Plastic Blends	216
5.	Heterogeneous Blends	217
5.1.	Compatible Polymers	217
5.2.	Blend Formation	217
5.3.	Toughened Plastics	218
5.4.	Thermoplastic Elastomers	218
6.	Expanded Plastics	219
	References	220

1. Overview

The term composite has various meanings. In general, it denotes a complex material in which two or more distinct substances combine to produce some properties not present in any individual component. In biomaterials, it means any two joined materials (e.g., polymer-coated titanium parts). In engineering, composites are more narrowly defined as physical admixtures of various materials that are present as distinct phases. The term "phase" is used here in the descriptive sense, not in the thermodynamic one. The engineering term "composite" thus includes only heterogeneous composites. Heterocomposites can be further subdivided into nanocomposites (dispersed phase has at least one dimension smaller than 100 nm), microcomposites (dispersed phase in the micrometer to millimeter range), and macrocomposites. Polymer composites are defined as composites in which at least one component is of a polymeric nature. In accordance with the general definition of a composite and in contrast to the engineering use of the word, the term "polymer composite" is often used to cover not only heterogeneous mixtures of a polymer and another material (minerals, fibers, other plastics, elastomers, etc.), but also homogeneous (single-phase) materials of two polymers (homogeneous polymer blends). Such polymer composites are a subgroup of "multicomponent polymer systems", which also include copolymers (two or more types of monomeric units chemically bound together).

Polymer composites occur naturally (wood, bamboo, lobster shells, bone, muscle tissue, etc.) or are manufactured synthetically. Synthetic polymer composites can also be subdivided into single-phase (homogeneous) and multiphase (heterogeneous) composites. Further subdivisions may be according to the second component's size (microcomposites, macrocomposites), chemical nature (air, low molar mass plasticizer, elastomer, plastic, mineral), geometry (particulate, fiber, platelet, fabric), and macroconformation (if polymeric: coil, rod), etc. (see Fig. 1). Most of these subclasses have different technical names

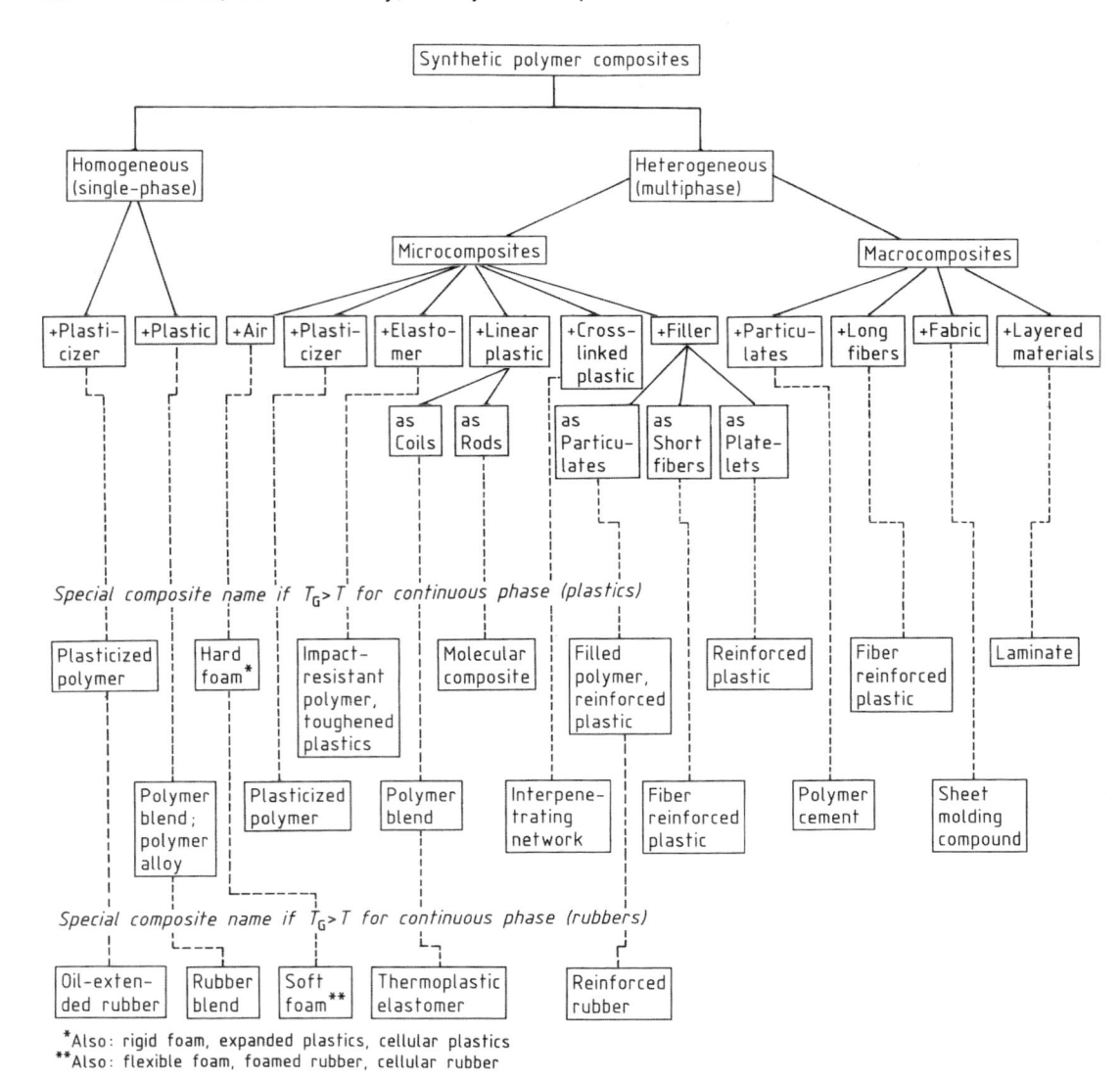

Figure 1. Classification of polymer composites of a base polymer (matrix polymer) with other materials (plasticizer, plastic, air, long fiber, layered materials). Some of these materials have their molecules in different shapes (coils, rods) or have various outer shapes (particulates, short fibers, platelets). Adopted from [1]

depending on whether the glass transition temperature of the continuous phase is above (plastics) or below (rubbers) the use temperature. The same name is often given to various subtypes.

A mixture of two (or more) different polymers is often called a *blend* when both polymers are either above (rubber) or below (plastics) their glass transition temperatures.

Blends are distinguished from composites of two polymers such as polymer–fiber-reinforced or rubber-toughened plastics. The term blend is sometimes restricted to mean only "incompatible polymer mixtures", whereas compatible polymer mixtures are called polymer alloys. The latter term is, however, occasionally more narrowly used for mixtures of two crystallizable polymers.

Table 1. Names of simple mixing rules

Field	Property exponent in generalized mixing law		
	$n = 1$	$n = 0$	$n = -1$
Mathematics	arithmetic mean	harmonic mean	geometric mean
Chemical engineering	rule of mixtures	logarithmic mixture rule	inverse rule of mixtures
Mechanical engineering	Voigt model		Reuss model
Electrical engineering	parallel		series
Materials science	upper bound		lower bound

The term *miscibility* refers to admixtures on the molecular level (nanometer range), that is true thermodynamic solubility. It is sometimes used more loosely for any polymer mixture that exhibits only one glass transition temperature. *Compatibility* denotes the ability of admixtures to be blended to heterogeneous microcomposites that do not separate into macroscopic phases.

2. Mixture Rules

The composition dependence of properties Q of composites can often be described by mixture rules. The *generalized simple mixing law* relates Q to the fractions $f_A = 1 - f_B$ of components A and B with properties Q_A and Q_B:

$$Q^n = Q_A^n \cdot f_A + Q_B^n \cdot f_B \quad (1)$$

The generalized simple mixing law is usually applicable to single-phase composites without specific interactions of its components and to two-phase systems with regularly dispersed discrete phases of "infinite" dimensions. It includes three special cases that are known by different names in different fields (Table 1) and are illustrated in Figure 2.

The *rule of mixtures* ($n = 1$) applies, for example, to various moduli if both components behave elastically. An example is Young's modulus parallel to the fiber direction for

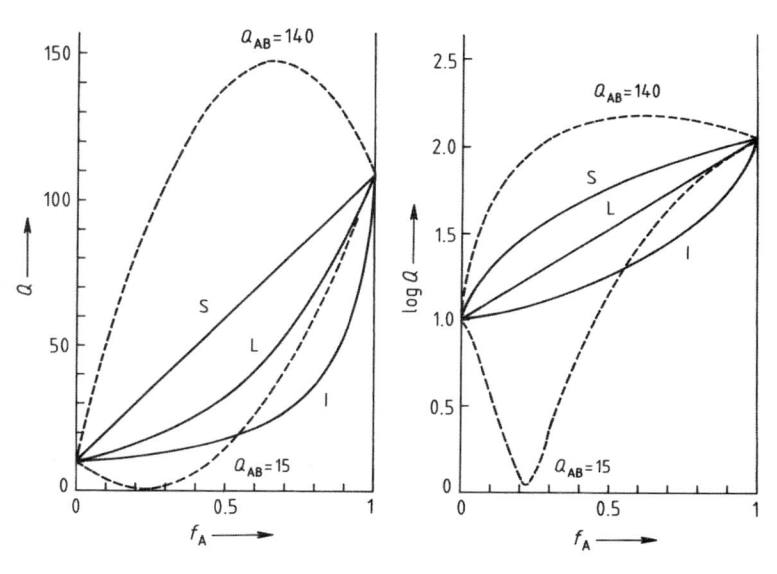

Figure 2. Mixing laws for composites with $Q_A = 110$ and $Q_B = 10$: left: $Q = f(f_A)$; right: log $Q = f(f_A)$
Simple mixing laws with interaction parameter
$Q_{AB} = 0$ (S = simple mixing law; L = logarithmic mixing law; I = inverse mixing law)
- - - Composites with interaction parameters $Q_{AB} = 140$ and 15, as indicated

epoxide resins reinforced with long glass fibers (see below). The same composites follow the *inverse rule of mixtures* ($n = -1$) if the modulus is measured perpendicular to the fiber direction and no slippage occurs. The *logarithmic mixture rule*

$$\log Q = f_A \cdot \log Q_A + f_B \cdot \log Q_B \tag{2}$$

is obtained from Equation 1 for $n = 0$ after some mathematical manipulation. The only known case concerns the permeation of oxygen through ABS resins. The proper fraction f_i in Equations 1 and 2 is always the volume fraction. If mass fractions are used instead, deviations from the curves shown in Figure 2 occur, which may be confused with either deviations from additivity or specific interactions.

True deviations from the upper bound S and lower bound I (Fig. 2) may be caused by irregularities in phase distributions (discrete phases) or phase continuities (continuous phases), phase alignments (anisotropic phases), bonding (adhesion), or nonuniform stresses. The resulting functions always lie between the upper and the lower bound if specific interactions are absent. Many empirical, semiempirical, and theoretical expressions have been proposed for such composites. A widely used function is the Halpin–Tsai equation

$$Q = Q_B \cdot \frac{1 + K_1 K_2 f_A}{1 - K_2 f_A}; K_2 = \frac{Q_A - Q_B}{Q_A + K_1 Q_B}$$

where K_1 is an adjustable constant. The Halpin–Tsai equation gives the lower bound for $K_1 \to 0$ and the upper bound for $K_1 \to \infty$. Particulate fillers often have $K_1 \approx 0$, and fiber-reinforced plastics $K_1 \approx 2\,L/d$, where $2\,L/d$ is the aspect ratio of the fiber.

Specific interactions between components can be considered by an additional interaction term. This term is differently defined, for example:

$$Q = Q_A f_A + Q_B f_B + K_i f_A f_B;$$
$$K_i = 2(2Q_{AB} - Q_A - Q_B)$$

where the property term Q_{AB} is taken at $f_A = f_B = 1/2$. The value Q_{AB} may be an interaction parameter between components A and B in homogeneous mixtures or the property of the interphase between A and B phases in heterogeneous mixtures. The upper bound for absent interactions or interphases is retrieved for the condition $Q_{AB} = (Q_A + Q_B)/2$. The functions for $Q_{AB} \neq (Q_A + Q_B)/2$ may lie fully or partially outside the range between upper and lower bounds (Fig. 2); that is synergistic effects (above the upper bound) or antagonistic effects (below the lower bound) may be present. They may also show maxima and minima, depending on the relative magnitudes of Q_{AB}, Q_A, and Q_B. Maxima or minima that exceed the highest property value (either Q_A or Q_B) are sometimes considered the true criteria for synergistic and antagonistic effects, respectively.

3. Filled Polymers

3.1. Microcomposites

The addition of fillers (low thermal expansion) to polymers (high thermal expansion) reduces the proportion of the latter and thus the shrinkage of plastics on processing (Table 2). The melt flow is also lower. Heat deflection temperatures (HDT) of amorphous polymers are either the same as those of unfilled plastics or no more than ca. 10 K higher (Fig. 3, group a), probably because of the increased viscosity of the filled plastic. The HDTs of amorphous thermoplastics are almost identical with glass transition temperatures.

The HDTs of crystalline polymers depend on a number of factors. Unfilled or particulate-filled crystalline polymers show HDTs that are

Table 2. Effect of fillers on some properties of amorphous (A) and crystalline (C) thermoplastics [(↑) weak increase, ↑ increase, ↑↑ strong increase, (↓) weak decrease, ↓ decrease, ↓↓ strong decrease]

Property	Extenders		Reinforcing agents	
	A	C	A	C
Shrinkage	↓	↓	↓	↓
Heat deflection temperature		(↑↓)		↑
Melt flow	↓	↓	↓	↓
Young's modulus	(↑)	(↑)	↑	↑
Flexural modulus	↑	↑	↑	↑↑
Fracture strength	↓	↓	↑	↑
Brittleness	↑	↑	↑↓*	↑↓*

*Tough plastics become more brittle, and brittle plastics become tough.

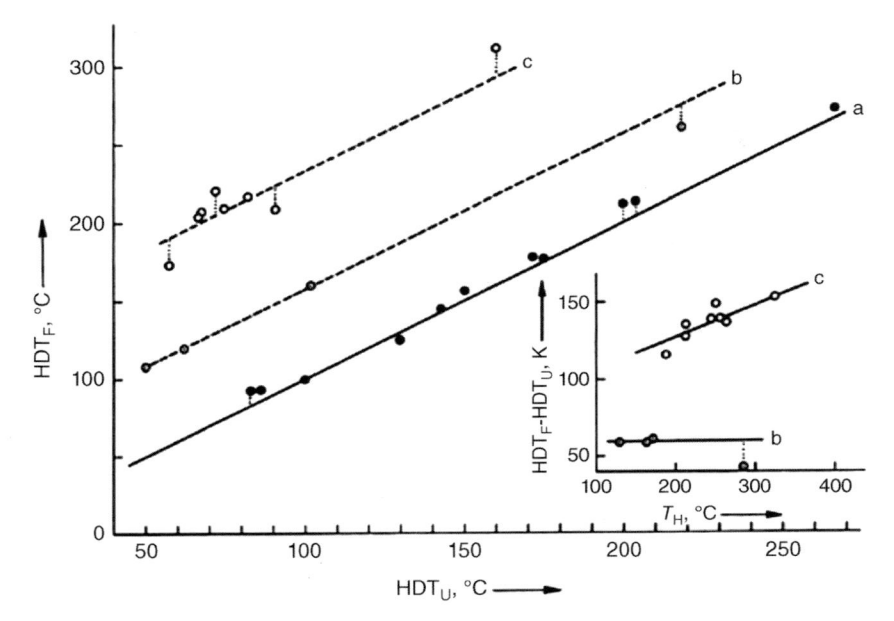

Figure 3. Heat deflection temperatures HDT_F of thermoplastics filled with 30 wt% short glass fibers as a function of heat deflection temperatures HDT_u of the corresponding unfilled polymers, both at 1.85 N/mm² (ISO/R 75 A)
(●) Amorphous polymers; (⊙, ○) crystalline polymers; solid line: $HDT_F = HDT_u$; broken lines: empirical
a) Amorphous polymers; b) Apolar crystalline polymers; c) Polar crystalline polymers
Insert: Difference $HDT_F - HDT_u$ as function of melting temperatures T_M of unfilled crystalline polymers

often considerably lower than their melting temperature. Upon fiber-filling, HDTs of crystalline polymers increase strongly; they may reach values near the melting temperatures. This behavior is thought to be due to additional polymer crystallization near, or polymer adsorption on, the fiber surface, epitaxial layers have been found to extend from the filler up to 150 nm into the matrix. The surface layers overlap; the resulting physical network reduces creep.

Two groups of HDTs of glass-fiber-filled crystalline polymers can be distinguished (Fig. 3). The HDTs of group b are about 50–60 K above those of their unfilled counterparts. This group comprises apolar polymers such as polyethylene, polypropylene, and poly(phenylene sulfide); polyoxymethylene also belongs to group b because its oxygen groups are buried deep inside compact helical structures, which are thus apolar on their surfaces. The difference in HDTs of filled and unfilled group b polymers does not vary with melting temperature. Group c includes polar crystalline polymers (polyesters, polyamides, etc.). For this group, the difference in HDTs of filled and unfilled polymers increases with the melting temperature of the unfilled polymer.

Young's moduli of thermoplastics increase with filler content (Fig. 4). The increase is small for spheroidal fillers (glass spheres, chalk), stronger for platelets (kaolin, talc), and strongest for fibers (glass). Short fibers are usually added in amounts of ca. 30 wt%, which corresponds to ca. 17 vol% for glass fibers ($\rho = 2.55$ g/cm³) in an average polymer ($\rho \approx 1.2$ g/cm³). Higher fiber content often leads to packing problems since the maximum volume fraction of three-dimensional randomly packed rods with $L/d = 20$ is already ca. 25% (glass fibers typically have diameters $d = 10$–20 µm and lengths of ca. 0.2 mm after injection molding). Particulate fillers can be added in far greater amounts without sacrificing properties because maximum packing densities of spheres are, for example, 0.745 for the hexagonally most dense and 0.637 for the randomly most dense packing.

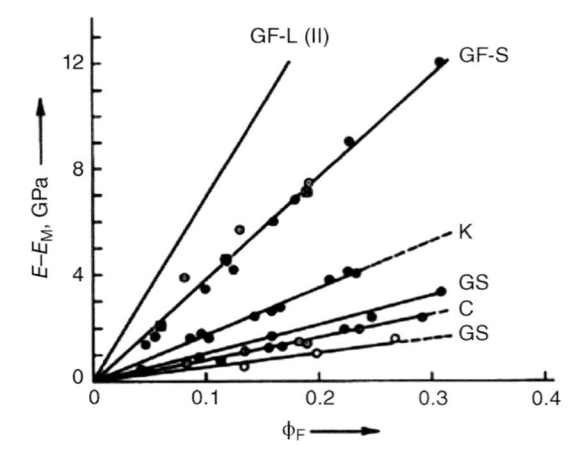

Figure 4. Influence of the volume fraction ϕ_F of fillers on the difference in Young's moduli of filled thermoplastics (E) and their polymer matrices (E_M)
GS = glass spheres; C = chalk; K = kaolin; GF = glass fibers; (S = short; L = long)
○ = polyethylene; (◉) = polypropylene; ⊗ = polyoxymethylene; ⊕ = poly(butylene terephthalate); ● = polyamide 6
The solid line for GF-L (∥) corresponds to the simple mixing law. From data of [2]

Young's moduli E of filled materials can be described by a modified mixing law

$$E = E_M \phi_M + f E_F \phi_F = E_M + (f E_F - E_M) \cdot \phi_F \quad (3)$$

where f is an adjustable parameter. The data in Figure 4 can be described with $f = 0.39 \pm 0.03$ (chalk), 0.66 ± 0.04 (kaolin), and 1.61 (talc). The effect of 30 wt% of short glass fiber was tested in 14 thermoplastics, and f was found to be 0.57 ± 0.04. These numbers increase with the specific area of the fillers, which are given as 0.3–2.2 (ground limestone), 0.5–1 (glass fibers), 6–22 (kaolin), and 6–17 m²/g (talc) [3]. Composites obviously increase in stiffness as the contact between filler and polymer segments increases. Good adhesion is not necessary, since E is measured for strains approaching zero.

The adjustable parameter f of Equation 3 adopts a physical meaning for short fiber reinforced plastics. In such composites, forces are transferred to the hard fiber from the soft matrix if (i) the Young's modulus of the fiber is greater than the Young's modulus of the matrix ($E_F > E_M$), (ii) the fiber length L_F exceeds a certain critical value $L_{F,crit}$, and (iii) the shear strength in the matrix is smaller than the shear strength between fiber and matrix (the fiber would be otherwise pulled out of the matrix unless the fiber breaks). In this case, $f = 1 - [L_{F,crit}/(2 L_F)]$.

Fracture strengths are, however, measured for maximum strains. The failure mode of filled thermoplastics is matrix dominated. Electron micrographs show that at low fiber contents, planar cracks spread through the matrices and debonded fibers are pulled out of either surface; the matrix must thus adhere well to the filler.

The tensile strength σ_B at break is given by the rule of mixtures ($\sigma_B = \sigma_F \phi_F + \sigma'_{M'} \phi_M$) where $\sigma'_{M'}$ is the tensile stress of the matrix upon application of a load σ_B to the composite. This relationship applies if (i) the tensile strength σ_B is greater than the contribution $\sigma_M \phi_M$ of the matrix component to the total tensile strength σ_B (that is $\sigma_B > \sigma_M \phi_M$), and (ii) the volume fraction ϕ_F of the fiber exceeds a certain minimum value ($\phi_F > \phi_{F,min}$).

Particulate fillers usually strongly reduce the fracture strengths of composites of amorphous thermoplastics, whereas those of crystalline polymers are reduced less or not at all. *Short fibers* generally increase the fracture strengths by factors of 1.5–2 for amorphous and crystalline polymers alike. Rigid PVC is an exception because its fracture strength is reduced on fiber-filling. The moduli and strengths can be further enhanced if coupling agents are used, which improves the bonding between fiber and matrix (e.g., silanes for glass-fiber-filled polymers).

3.2. Molecular Composites

The problem of insufficient bonding between fiber and matrix can be dramatically reduced if so-called molecular composites are used. These materials are dispersions of semiflexible (rod-like) polymer molecules with mesogenic units in chemically similar polymer matrices, e.g., poly(p-phenylenebenzobisthiazoles) in polybenzimides (→ Plastics, General Survey, 1. Definition, Molecular Structure and Properties, Section 5.6.2 Theoretical Fracture Strength). The mesogens are not truly molecularly dissolved; rather, they form microdomains in the matrix. The segment axes are randomly distributed in these domains if the molecular composites are formed from polymer solutions with total polymer concentrations c below the critical isotrope phase–mesophase concentration $c_{i/n}$. Microdomains show mesophase behavior if films or fiber are formed at $c>c_i/_n$. Such microcomposites have excellent moduli and fracture strengths (Fig. 2).

3.3. Macrocomposites

Polymer macrocomposites are heterogeneous composites of polymers and "macro"-sized additives. Examples are polymer concrete, composition board, long fiber-reinforced polymers, and plywood.

Polymer concrete is composed of polymer [usually polyester or poly(methyl methacrylate)], aggregate (sand, ground limestone, etc.), and other additives (dyes, etc.); it does not contain cement. *Polymer–concrete hybrids* include polymer–cement hybrids (cement is partially replaced by polymers) and polymer-impregnated concrete (conventional concrete saturated with a monomer that is polymerized in place). Polymer concretes have higher moduli and higher tensile and compressive strengths than ordinary concretes. They are not vulnerable to damage by freeze-thaw cycles.

Composition boards are panel products. They include particleboard, fiberboard, hard board from flakes or shearings, etc. The fillers are mainly wood products; the resins, urea–formaldehyde (interior applications), and polyurethanes or phenolic resins (exterior).

Long-fiber-filled polymer macrocomposites range in structure from polymers filled with "long" fibers 6–12 mm in length to fiber strands and mats filled with polymers. The filler content can often reach 65 wt%. Upon application of tensile stress, local stress concentrations are transferred by shear forces onto the plastic–fiber interface and distributed over the much greater fiber surface area. The fibers must therefore bond well to the plastic (use of coupling agents) and must have a certain length to avoid slipping.

Young's moduli of long glass-fiber-reinforced epoxide resins follow the simple mixing law if the tensile stress is in the fiber direction (Fig. 5). The

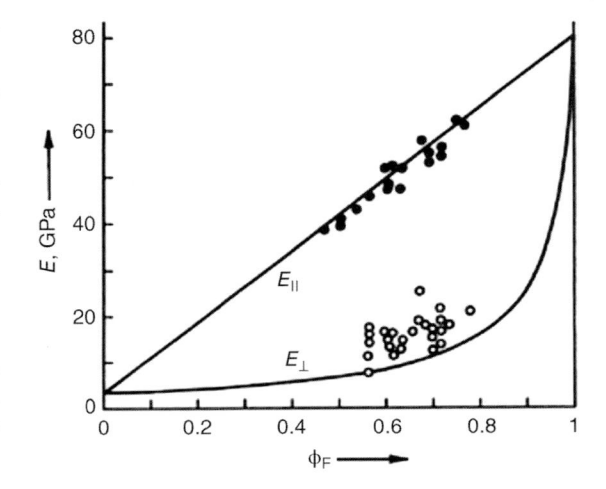

Figure 5. Young's moduli of epoxy–glass-fiber composites parallel ($\parallel$) and perpendicular ($\perp$) to the fiber direction

inverse mixing law is not observed as well for stresses perpendicular to the fiber direction since some fibers bend under stress.

The simple mixing law corresponds to $f = 1$ in Equation (3). For long fibers, f has the meaning of an orientation factor. It becomes 1/6 for fibers distributed three-dimensionally at random, 1/2 for 90° cross-plies measured in either fiber direction, and 3/8 for a uniform, planar distribution of fibers.

This group of macrocomposites also includes prepregs (fiber mats soaked with unsaturated polyester resins) and the composites generated by filament winding (wound filaments saturated with epoxies or unsaturated polyesters), both of which are subsequently polymerized.

3.4. Nanocomposites [4–6]

Polymeric nanocomposites are defined as multiphase materials in which at least one dimension of the dispersed or continuous phase is smaller than 100 nm. This includes organic–inorganic hybrid materials as well as "all-polymer" nanocomposites in which matrix and reinforcing phase consist of the same polymer. In contrast to micro- and macrocomposites, most polymers are allocated due to the presence of nanostructures at interfaces that affect the chain mobility and entanglement. Hence, the properties of interfacial polymers substantially differ from those of bulk polymers, especially regarding their thermal, mechanical, electrical, optical, and biological properties. For instance, polymeric nanocomposites can be optically transparent, exhibit improved stiffness, high surface gloss, high scratch resistance, and even electrical conductivity. Three different strategies lead to polymeric nanocomposites: (i) Dispersion of functional nanofillers in a polymer matrix, (ii) formation of organic–inorganic hybrid polymers, and (iii) oriented polymer crystallization.

Nanofiller Dispersion [4–8]. Since the 1990s, significant progress in nanotechnology has made a large variety of nanometer-scaled inorganic and organic fillers as well as nanofibers commercially available for nanocomposite preparation (see Section 2.2.4 Extending and Functional Fillers, in Plastics, General Survey, 2. Production of Polymers and Plastics). When the size of inorganic fillers is reduced, a transition from the inorganic solids to the molecules takes place. Molecular nanofillers include nanosilica, such as POSS (polyhedral oligomeric silsesquioxane); 2D carbon polymers, such as graphene; and multifunctional hyperbranched polymers with nanometer dimensions.

Nanometer-scaled silica spheres, prepared by a flame process, are known as pyrogenic silica of around 5 nm diameter, which is widely applied as thixotropic agents in coatings, adhesives, and sealants. Due to the shear-sensitive formation of hydrogen bridges between silanol groups at the silica particle surface, silica percolation networks are destroyed upon shearing, thus accounting for shear thinning. This thixotropic effect, caused by less than 0.5 wt% silica nanoparticles, prevents sedimentation of fillers in formulated thermoset systems. Moreover, functionalized silica fillers play an important role in improving scratch resistance in coatings. Nanosilica together with disulfide-functional silane coupling agents are used for the preparation of silica–rubber nanocomposites.

For many years, carbon black nanoparticles have been commercially employed to improve electrical conductivity of plastics and coatings. This is important for antistatics and electromagnetic interference shielding. Compared to micron-sized carbon black, the percolation threshold of nanocarbon black is much lower, which drastically reduces the amount of carbon black required for achieving electrical conductivity. Highly anisotropic carbon nanofillers, such as carbon nanotubes and graphene, further reduce the percolation threshold. Moreover, both carbon nanotubes and ultrathin but micron-sized graphene enable combining electrical conductivity with improved matrix reinforcement. Alignment of graphene in a polymer matrix improves barrier resistance to permeation of gas and liquids ("labyrinth effect") together with improved electrical conductivity and matrix reinforcement. Another approach to improved barrier resistance combined with halogen-free flame retardancy and matrix reinforcement employs organophilic layered silicates. When layered silicates are made organophilic by cation exchange of sodium cations for alkyl ammonium cations, they can be intercalated and exfoliated during polymer melt compounding. The family of organophilic nanofillers is rapidly expanding.

Inorganic–Organic Hybrids [4, 9, 10]. Polymer hybrids combine the properties typical for ceramics with facile processing typical for polymers. Sol–gel chemistry is employed to equip inorganic particles and resins with reactive groups that enable efficient covalent coupling with various polymers. For example, functionalized alkoxysilanes are cocondensed with tetraethoxysilane and other metal alkoxides, such as zirconates and aluminates to produce reactive liquid oligomers or reactive nanoparticles. They are mixed and cured together with thermoset resins to form nanostructured polymer–ceramic hybrids. Sol–gel coatings are well known for their extraordinary scratch resistance. Moreover, sol–gel chemistry is used to functionalize inorganic nanoparticles by incorporating initiator and transfer agents, which enables covalent attachment of polymer chains by "grafting-from" and "grafting-to" processes. Polymer grafting of nanoparticles greatly facilitates nanoparticle dispersion and prevents undesirable agglomeration.

Self-Reinforcing Plastics and All-Polymer Nanocomposites [11, 12]. Going well beyond the scope of liquid crystalline polymers and conventional nanocomposites, the self-reinforcement of polymers does not require mesogens as comonomers and eliminates the special handling and safety precautions related to the use of potentially hazardous alien nanoparticles and nanofibers. Moreover, such all-polymer nanocomposites, in which both the matrix and the reinforcing phase consist of the same polymer, have much lower mass and are much easier to recycle than conventional polymer composites. This is highly advantageous for applications in sustainable lightweight engineering. Key concept for achieving self-reinforcement is the oriented polymer crystallization during processing. Whereas one-dimensional crystallization affords in situ aligned polymer nanofibers, the two-dimensional crystallization produces nanosheets and all-polymer multilayer composites. Because this nanostructure formation of polymers does not require nanofibers and nanoparticles, nanometer-scaled particles will not be emitted. In contrast to many other nanocomposites, all-polymer nanocomposites are environmentally benign. In one strategy, oriented films are laminated or coextruded to produce multilayer all-polymer nanocomposites. In another strategy, multilayer all-polymer nanocomposites are produced by pressure- and flow-induced crystallization near the melting temperature of thermoplastics. Extrusion of polyethylene containing ultrahigh molecular mass polyethylene (UHMWPE) close to the PE melting temperature can afford shish-kebab-like aligned UHMWPE nanofibers. However, because UHMWPE has very high viscosity, the incorporation of UHMWPE in PE reactor blends has been restricted to a few percent. Exploiting advanced multisite ethylene polymerization catalysis, it became possible to incorporate much larger amounts of UHMWPE into reactor blends without impairing processing. Because no entanglement with HDPE takes place, the nanometer-scaled UHMWPE forms a low viscosity phase separated melt and elongates during processing. Because the lower HDPE fraction crystallizes much faster, it crystallizes onto UHMWPE nanofibers to produce shish-kebab-like UHMWPE fibers in conventional melt processing without narrowing the processing window. The development of self-reinforcing plastics and all-polymer nanocomposites, which do not require the addition of (nano)fillers and fibers, holds great promise for the sustainable development of lightweight engineering plastics.

4. Homogeneous Blends [13, 14]

Mixtures of polymers with other polymers are called polymer blends ($\rightarrow$ Polymer Blends). Such blends may be composed of two thermoplastics (plastic blends), two elastomers (rubber blends), a plastic filled with an elastomer as the dispersed phase (rubber-modified plastics), an elastomer with a plastic as the dispersed phase (polymer-filled elastomer), or a plastic filled with a polymer melt or a low molar mass liquid (plasticized polymers). Whether or not these blends are truly miscible on a molecular level depends on the thermodynamics of the systems.

4.1. Thermodynamics

If two components 1 and 2 are mixed, entropy S, enthalpy H, and volume V change. In many cases, volume changes can be neglected and the Gibbs energy of mixing is given by the changes in enthalpy and entropy: $\Delta G_{mix} = \Delta H_{mix} - T \cdot \Delta S_{mix}$. These changes can be calculated by the Flory–Huggins theory, which assumes that the mixture can be modeled by a three-dimensional lattice on which the monomeric units (or solvent molecules) can be placed. Each monomeric unit experiences the same force field; the theory is thus a mean-field theory. The assumption of a constant force field is fairly well fulfilled for polymer blends and concentrated polymer solutions but not for dilute solutions.

The entropy and enthalpy of mixing are calculated separately. Entropies are calculated for ideal solutions (zero mixing enthalpy); all environment-dependent entropy changes are thus zero (translational, vibrational, and inner rotational entropies do not change). Units can however be arranged relative to each other in many ways; however, there is a combinatorial entropy that is also called "configurational entropy" in statistical mechanics.

The mixing of two components is modeled as a quasi-chemical reaction. The mixing enthalpy is thus determined by the change in interaction energies and the number and type of nearest neighbors. Enthalpy and entropy terms are added to give the molar Gibbs energy of mixing:

$$(\Delta G_{\text{mix}})^{\text{m}}/RT = \chi \cdot \phi_1\phi_2 + \phi_1 \cdot \ln\phi_1 + (X_1/X_2) \cdot \phi_2 \cdot \ln\phi_2$$

which, on differentiation with respect to the molar amounts of the components, yields the chemical potentials, for example, of component 1

$$\Delta\mu_1 = RT\{\chi \cdot \phi_2^2 + \ln(1 - \phi_2) + [1 - (X_1/X_2)] \cdot \phi_2^2\}$$

where χ is the parameter for the interaction between component 1 and component 2 (usually with numerical values between 0 and 2), also called the Flory–Huggins parameter. All three terms in brackets must be considered if a polymer 2 is mixed with a solvent 1, for example, a plasticizer ($X_1 \ll X_2$). If both components are polymers however, then $X_1 \approx X_2$ and the entire entropy contribution comes from the relatively small logarithmic term. Thus little combinatorial entropy is gained on mixing two polymers, and the miscibility is determined mainly by the mixing enthalpy, that is, by the term $\chi \phi_2^2$.

Phase separation occurs if the curve $(\Delta G_{\text{mix}})^{\text{m}} = f(\phi_2)$ (at $T = $ const.) has a shape that can be touched at two points by a tangent. The compositions at these points are ϕ_2' and ϕ_2'' (with $\phi_2' < \phi_2''$). Two phases exist for the unstable range $\phi_2' < \phi_2 < \phi_2''$, which is separated from the stable ranges $\phi_2 < \phi_2'$ and $\phi_2'' > \phi_2$ by the so-called binodal. The unstable range itself is subdivided by the spinodal into two metastable ranges and one unstable one. Spinodals are characterized by inflection points in the Gibbs energy–composition functions and extremal values in the chemical potential–composition functions. Maximum, minimum, and inflection point become identical at the critical point, which is defined by a zero value of the second derivative of the chemical potential

$$\partial^2 \Delta\mu_1/\partial\phi_2^2 = RT \cdot [2\chi - (1 - \phi_2)^{-2}] = 0 \quad (4)$$

This critical point is characterized by a critical volume fraction $\phi_{2,\text{crit}} = 1/(1 + X_2^{1/2})$ and a critical interaction parameter $\chi_{\text{crit}} \approx (1/2) + (1/X_2)^{1/2}$.

The temperature dependence of χ is approximately

$$\chi \approx A + (B/T) \quad (5)$$

where A and B are system-dependent constants; B is positive for endothermal mixtures [i.e., χ increases with increasing temperature and a homogeneous solution exists above an "upper critical solution temperature" (UCST)]. Other systems are homogeneous only below a "lower critical solution temperature" (LCST). The demixing is correspondingly either enthalpy (UCST) or entropy (LCST) induced.

The terms UCST and LCST do not indicate the absolute position of demixing temperatures since some systems show LCST > UCST (Fig. 6). Depending on the system, either an hourglass-type diagram or a closed miscibility gap is shown.

4.2. Plastification

Plastification of a plastic is the flexibilization of the material either by added plasticizers (external plastification) or by incorporated flexibilizing comonomer units (internal plastification).

The molecular action of plasticizers consists of a flexibilization of chain segments, which can have various causes. *Polar plasticizers* increase the proportion of *gauche* conformations in polar chains, thus decreasing rotational barriers. *Primary plasticizers* can dissolve helical structures and crystalline regions if they are thermodynamically good solvents. Furthermore both *primary and secondary plasticizers* are diluents; their addition increases the distance between polymer segments and thus decreases the activation energy needed for cooperative movement of segments. Solvation does not per se increase chain flexibility, however, because solvent shells act as substituents and increase the rotational barriers.

The molecular effect of an increase in chain flexibility by plasticization is measured as a decrease of glass transition temperatures (Fig. 7). Small molecules are better plasticizers than bigger ones of similar chemical constitution. The plasticizer efficiency decreases with increasing thermodynamic "goodness" of the

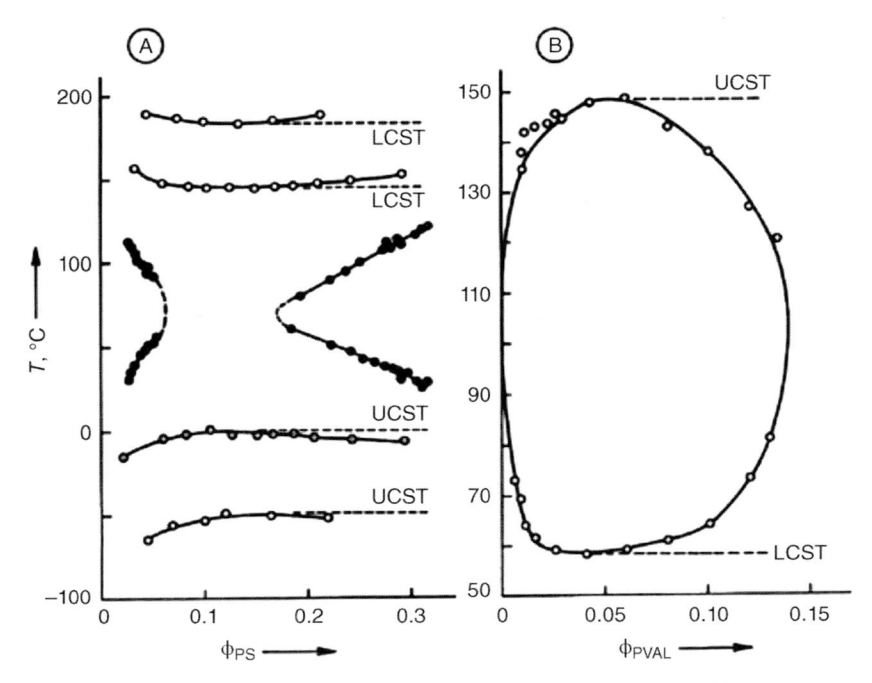

Figure 6. Phase separation temperature as function of polymer–solvent composition
A) Polystyrenes with various molar masses in acetone; B) Poly(vinyl alcohol-co-vinyl acetate) (93 : 7) in water
Reprinted with permission by Hüthig and Wepf Publ., Basel [15]

plasticizer for the polymer (greater interaction, i.e., bigger solvation). Good plasticizers with respect to lowering glass transition temperatures are thus small molecules that act as theta solvents. These plasticizers are however bad

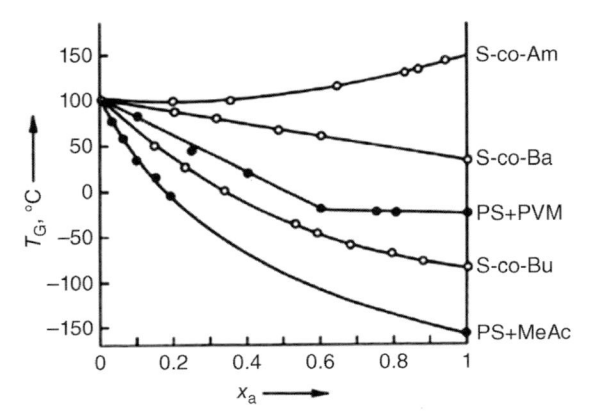

Figure 7. Glass transition temperatures of (●) polystyrene plasticized with methyl acetate (MeAc) or poly(vinyl methyl ether) (PVM) and (○) copolymers of styrene with butadiene (Bu), butyl acrylate (Ba), or acryl amide (Am)

for technological applications because they bleed (transport of plasticizer to the surface of the plastic), migrate (transport of plasticizer into another contacting material), or are extractable (e.g., transport into packaged foods). These transport phenomena can be reduced if polymer plasticizers are used. Increasing molar masses lead to decreasing thermodynamic compatibility, however (see below). Technical plasticizers are therefore in most cases a compromise between thermodynamic plasticizer efficiency and kinetically hindered plasticizer transport.

Plastification can also be followed by the changes of certain mechanical properties such as lowering of fracture strength, increase of elongation, and decrease of tensile moduli. Such methods sometimes show "antiplasticizations" at low plasticizer concentrations (Fig. 8) (e.g., increases of moduli). Glass transition temperatures are not increased, however. The effect thus cannot be the result of increased segment mobility. It may be due to a healing of microvoids in amorphous polymers such as polystyrene or of the two parent polymers. Heterogeneous blends exhibit two glass transition temperatures, both practically identical to those of the parent polymers. Homogeneous blends are usually formed if the difference in solubility parameters is smaller than 0.7 cal$^{1/2}$ cm$^{-3/2}$ (1.44 J$^{1/2}$ cm$^{-3/2}$).

4.4. Plastic Blends

Plastic blends are usually immiscible because the combinatorial entropy of mixing is too small (high molar masses) and the enthalpy changes on mixing are often positive. Miscibility is observed for three types of systems:

1. Chemically similar polymers, for example, polystyrene–poly(o-chlorostyrene), showing both LCST and UCST
2. Systems having specific interactions between different components, for example, polystyrene–poly(vinyl methyl ether); these systems show only LCSTs
3. Systems consisting of oligomers, for example, oligo(ethylene oxide)–oligo(propylene oxide); these systems possess only UCSTs

Homogeneous polymer blends show only single glass transition temperatures; the composition dependence of these T_G's can be described by the simple rule of mixtures. Two-phase systems always show two glass transition temperatures if the phase diameters are greater than ca. 3 nm. Transparency, on the other hand, is no indication of a single-phase blend since opacity can be observed only if the refractive indices of the two phases are sufficiently different and if the phase diameters are greater than ca. one-half the wavelength of incident light. The presence of two phases in clear specimens can often be detected with electron microscopy by special staining techniques.

Modified PPO, a blend of poly(2,6-dimethylphenylene oxide) and polystyrene, is the industrially leading *homogeneous blend*. Blends of poly(vinyl chloride) with poly(methyl methacrylate), poly(ethylene-co-vinyl acetate) or chlorinated polyethylene are also said to be one-phase systems.

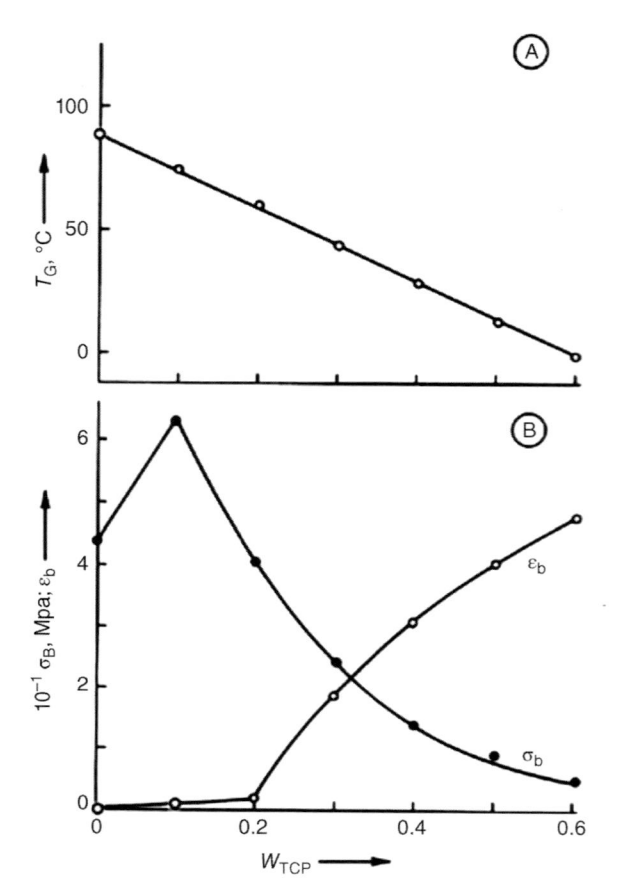

Figure 8. A) Glass transition temperature T_G and B) fracture strength σ_B and elongations ε_B at break of a poly(vinyl chloride) plasticized with tricresyl phosphate (TCP) Reprinted with permission by Hüthig and Wepf Publ., Basel [16]

to additional crystallization of slightly crystalline polymers, such as poly(vinyl chloride).

4.3. Rubber Blends [13, 14, 17, 18]

About 75% of all elastomers are employed as blends. Heterogeneous blends of natural (rubber–styrene)–butadiene rubber or (*cis*-butadiene rubber–styrene)–butadiene rubber are used mainly for tire treads since they reduce abrasion. Homogeneous blends are more rare. They possess only one glass transition temperature which, for 50 : 50 blends, is an average of the glass transition temperatures

5. Heterogeneous Blends [13, 14, 17, 18]

5.1. Compatible Polymers

Two polymers may be thermodynamically immiscible but still mechanically compatible. These compatible blends are two- or multiphase systems that show multiple glass transition temperatures, which do not change with composition. The components do not separate, however, under mechanical stress during the expected life of the product. Such mechanical compatibilities can be improved or introduced by the addition of diblock polymers. One block of these diblock polymers is compatible with one phase type of the heterogeneous blend and the other one with the other type (→ Plastics, General Survey, 3. Supermolecular Structures). The two blocks need not necessarily be chemically identical with the two parent polymers, that is, a compatibility of poly(A) and poly(B) can be achieved not only by addition of poly(A)–*block*-poly(B) but also by poly(A)–*block*-poly(C) if the C block is either miscible with poly(B) or can form mixed crystals with poly(B).

5.2. Blend Formation

Heterogeneous polymer blends can be produced by mixing together two polymers (as melts, lattices, or in solution) or by in situ polymerization of a monomer in the presence of a dissolved polymer.

The energy taken up during *melt mixing* is used for flow processes and for the generation of surfaces of new microdomains. After some time, a steady state is established and the domain size becomes constant. No macroscopic demixing occurs because of the low diffusion coefficients resulting from the high viscosities.

Latex blending consists of the mixing of aqueous dispersions of two polymers. Far lower temperatures and lower shear fields can be employed compared to melt blending. The good mixing of the latex particles remains after coagulation. The domain size is, however, restricted to the size of the latex particles themselves; it is not altered by subsequent melting of the coagulate.

Solution blending involves the mixing of two polymer solutions. Miscible polymers can be blended to domains of molecular size. Solutions of immiscible polymers demix, however, at very low concentrations, sometimes under fractionation with respect to molar masses. Domains grow further on solvent removal by distillation or freeze drying.

In situ polymerization involves solutions or gels of polymers in monomers, which are subsequently polymerized. The in situ polymerization of styrene in a styrenic polydiene solution, which leads to rubber-toughened polystyrene (HIPS) is most important industrially. The polymerization of a cross-linkable monomer in a gel of a cross-linked rubber in the very same monomer results in interpenetrating networks.

Various blending processes produce blends with very different properties (Fig. 9). The high notched impact strengths of the polymerization-blended materials (line P) are caused by a strong anchoring of phases due to the formation of graft copolymers of styrene on polybutadiene and cross-linking within the rubber domains, both caused by free-radical initiators. Such processes are less prevalent during melt blending (radical formation by high shearing at elevated temperature) (line M) and latex blending (low

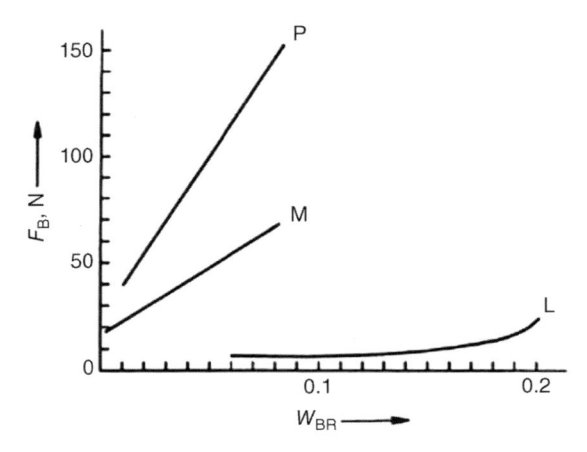

Figure 9. Impact strength with notch F_B as function of the mass concentration w_{BR} of *cis*-1,4-polybutadiene in its blends with polystyrene. P = in situ polymerization of styrene; M = melt blending; L = latex blending
Reprinted with permission by Plenum, New York [19]

shearing at ambient temperature) (line L). In situ polymerization is thus the method of choice for unsaturated rubber (easy cross-linking and grafting by chain transfer) and monomer–rubber pairs with favorable Q, e-values for copolymerizations. In all other cases, blends are formed by melt mixing.

5.3. Toughened Plastics

Rubber-modified plastics (toughened plastics, high-impact plastics) consist of rubber domains dispersed in plastic matrices. The domain size varies with the blending process. Typical values are ca. 0.1 μm for melt-blended poly(vinyl chloride)–acryl rubber and 1 μm for polystyrene–polybutadiene. The domains are often multiphased; small plastic domains are imbedded in the rubber domains (Fig. 10). The morphologies depend strongly on the blending process. During stirred in situ polymerizations, phase inversions often occur if the amount of newly formed plastic approaches that of the incipient rubber. During phase inversion, already present plastic particles may be embedded in the newly formed rubber domains.

Rubber-toughened plastics are valued because of their improved impact strength. On impact, very many crazes are formed near the equators of the rubber particles. The crazes propagate until they encounter an obstacle (e.g., rubber particle or shear band) or until the stress concentration at the tip of the craze becomes very low. Many small crazes result, and the stress is evenly distributed if the rubber phase is cross-linked and binds well to the thermoplast phase [e.g., by in situ formed graft copolymers (HIPS)]. In contrast, stress peaks concentrate at a few defect points in normal thermoplasts.

5.4. Thermoplastic Elastomers (→ Thermoplastic Elastomers)

Thermoplastic elastomers (elastoplastics, thermoplastics, plastomers) are processed like thermoplastics but applied like elastomers. Their unique properties follow from their molecular structures; their chains consist of "soft" and "hard" segments in block, graft, or segmented copolymers composed of monomeric units A and B. The A and B segments are mutually incompatible and form locally separated regions. With a well-designed molecular architecture, domains of the "hard" A segments (transition temperature > service temperature) act as physical cross-links in the continuous matrix of the "soft" B segments (transition temperature < service temperature) (→ Plastics, General Survey, 3. Supermolecular Structures Section 3.6). Such transition temperatures may be glass transition temperatures in amorphous polymers or melting temperatures in partially crystalline polymers. The A segments attain mobility above these transition temperatures and the elastoplasts become processable.

Thermoplastic elastomers comprise linear triblock polymers of the polystyrene–*block*-polydiene–*block*-polystyrene type; radial, star, or "teleblock" polymers of the same monomeric units; urethane segment or block copolymers with polyester or polyether soft segments; polyesteramides and polyesteretheramides; graft copolymers of butyl rubber on polyethylene, vinyl chloride on poly(ethylene-co-vinyl acetate), styrene–acrylonitrile on saturated acryl rubbers, and various ionomers.

Thermoplastic elastomers can also be produced by "dynamic vulcanization." The

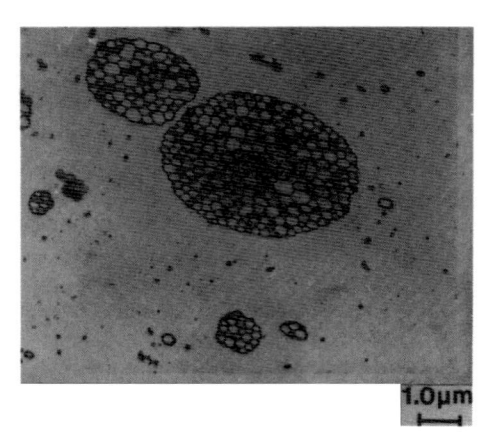

Figure 10. Rubber domains in high-impact polystyrene by in situ free-radical polymerization of styrenic polybutadiene solutions
Reprinted with permission by Society of Plastics Engineers [20]

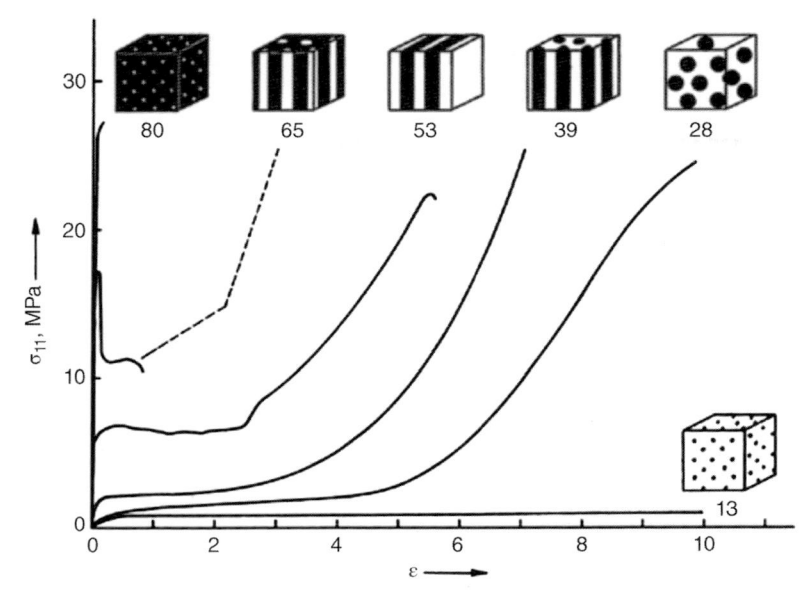

Figure 11. Stress–strain curves and morphologies of styrene–butadiene–styrene triblock polymers with various styrene contents
Reprinted with permission by Hüthig and Wepf Publ., Basel [15]

mastication of blends of conventional rubbers and crystalline poly(α-olefins) leads to chain scissions. The resulting macroradicals crosslink the rubber domains.

The mechanical properties of thermoplastic elastomers are determined mainly by their morphologies. In styrene–butadiene–styrene triblock polymers, for example, morphologies are governed by the spatial requirements of the various blocks (→ Plastics, General Survey, 3. Supermolecular Structures, Section 3.6). With a low content of hard styrene segments, small spherical polystyrene microdomains are formed in the soft polybutadiene matrix. These microdomains act as physical cross-linkers. The distances between the domains are large, and the polymer behaves as a weakly linked elastomer with correspondingly high extension (Fig. 11). The domains are larger and their distances are shorter at 28% styrene units and the polymer strengthens (stiffens). The stiffening becomes stronger for rodlike styrene domains (39% S). Lamellar morphologies (53% S) show an "unruly" behavior initially because of reorientation of lamellae. At even higher styrene content, the polymer behaves as a plasticized tough thermoplast (65% S) or almost like polystyrene itself (80% S).

6. Expanded Plastics
(→ Foamed Plastics) [21, 22]

Expanded plastics (foams, foamed plastics, cellular plastics) are blends of plastics with air. They are subdivided according to their rigidity, their cell structure, and the nature of the parent plastics. Rigid expanded plastics are used mainly for thermal insulation, and flexible foams for damping and cushioning materials.

The rigidity ("hardness") of expanded plastics follows the properties of the parent polymers. Phenolic and urea resins thus yield brittle-rigid plastics; polystyrene and hard poly(vinyl chloride), tough-rigid; and polyethylenes, polyurethanes, and plasticized PVC, semirigid to flexible foams. The elastic moduli of expanded plastics decrease approximately proportional to their polymer content (simple mixing law). Because the stiffness of

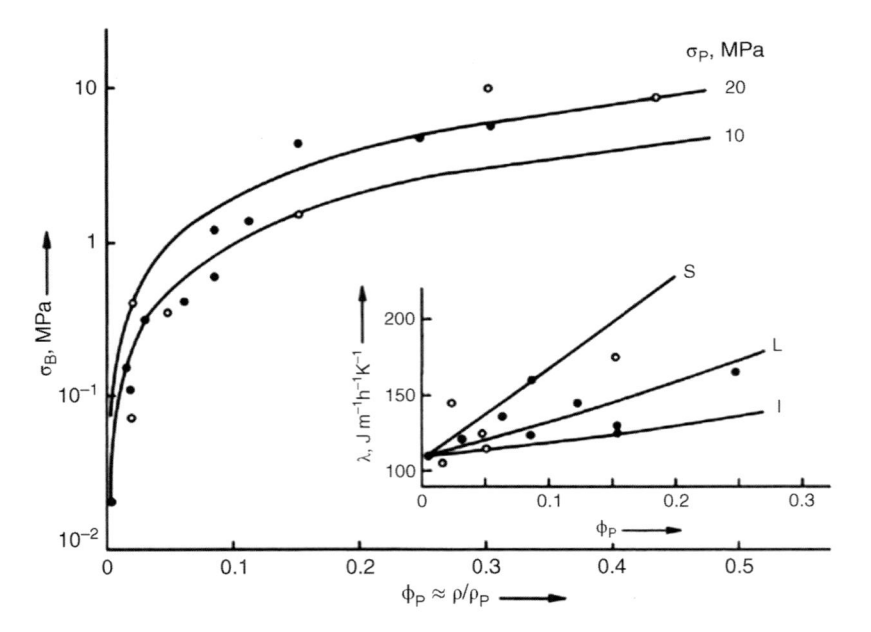

Figure 12. Tensile strengths of expanded plastics as a function of the volume fraction of the rigid (●) or flexible (○) parent polymers P
Solid lines indicate a) Simple mixing law for $\sigma_p = 20$ MPa; b) Simple mixing law for $\sigma_p = 10$ MPa; Insert: Thermal conductivities (for $\lambda_p = 700$ and $\lambda_{air} = 110$ J (m^{-1} h^{-1} K^{-1}) of rigid (●) and flexible (○) plastic foams
S: Simple mixing law; L: Logarithmic mixing law; I: Inverse mixing law

an article increases with the third power of the wall thickness, the gases in expanded plastics work as enhancers that reduce material costs and weights.

Tensile strengths of foamed plastics also follow the simple mixing law. Because the tensile strengths of gases are diminishingly small, tensile strengths σ_B of foamed plastics are determined by the volume fraction ϕ_p and strengths $\sigma_{B,p}$ of the plastics themselves [i.e., $\sigma_B \approx \sigma_{B,p} \cdot \phi_p = \sigma_{B,p} \cdot (\rho/\rho_p)$] (Fig. 12). Compression strengths follow a similar simple mixing law.

The cell structure can be open, closed, or mixed. Open-celled foams are always air filled, regardless of the blowing or expanding gases used for foam manufacture. The trapped gases in closed-cell structures can be exchanged with the surrounding air only by slow diffusion through the polymer matrix. Because thermal conductivities of low-density foams ($\rho_p < 0.3$) obey, to a first approximation, a logarithmic mixing law, they are considerably affected by the thermal conductivities of trapped gases [($\lambda = 93$ J m^{-1} h^{-1} K^{-1} (nitrogen), 56 J m^{-1} h^{-1} K^{-1} (CO_2), 26.4 J m^{-1} h^{-1} K^{-1} (CCl_3F)]. The gas exchange is considerably reduced in integral skin foams (dense skin, low-density core) and in syntactic foams (plastics filled with hollow spheres; spheres may be made of glass, ceramics, or plastics and contain either gases or a vacuum).

References

1 H.-G. Elias: Technologie, *Makromoleküle*, vol. **2**, Hüthig and Wepf, Basel 1992.
2 G.E. Ehrenstein, R. Wurmb, *Angew. Makromol. Chem.* **60/61** (1977) 157.
3 T.H. Ferrigno, in S. Katz, J.V. Milewski (eds.): *Handbook of Fillers and Reinforcements for Plastics*, Van Nonstrand, New York 1978, p. 11.
4 R.K. Gupta, E.B. Kennel, K.-J. Kim (eds.): *Polymer Nanocomposites Handbook*, CRC Press, Boca Raton 2010.
5 M. Moniruzzaman, K.J. Winey, *Macromolecules* **39** (2006) 5194.
6 V. Mittal (ed.): *Polymer Nanotube Nanocomposites: Synthesis, Properties, and Applications*, John Wiley & Sons, Hoboken 2010.
7 G. Carotenuto, L. Nicolais (eds.): *Graphene-Polymer Composites*, John Wiley & Sons, Hoboken 2012.

8. S. Ray, M. Okamoto: "Polymer/Layered Silicate Nanocomposites: A Review from Preparation to Processing," *Prog. Polym. Sci.* **28** (2003) 1539.
9. C.J. Brinker, G.W. Scherer (eds.): *Sol-Gel Science*, Academic Press, San Diego 1990.
10. S. Sakka: *Handbook of Sol-Gel Science and Technology*, vol 3, Springer, Berlin 2005.
11. A. Kmetty, T. Barany, J.J. Karger-Kocis, *Progr. Polym. Sci.* **35** (2010) 1288.
12. M. Stürzel, Y. Thomann, M. Enders, R. Mülhaupt, *Macromolecules* **47**(2014), 4979. DOI: 10.1021/ma500769g.
13. L.A. Utracki, C. Wilkie (eds.): *Polymer Blends Handbook*, Springer, Berlin 2014.
14. A. Boudenne, L. Ibos, Y. Candau, S. Thomas (eds.): *Multiphase Polymer Systems*, John Wiley & Sons, Hoboken 2001.
15. H.-G. Elias: Grundlagen, *Makromoleküle*, 5th ed., vol. 1, Hüthig and Wepf, Basel 1989.
16. *Makromoleküle*, 4th ed., Hüthig and Wepf, Basel 1981.
17. R.A. Pearson: "Introduction to the Toughening of Plastics," *ACS Symp. Ser.* **759** (2000) 1.
18. A.A. Collyer (ed.): *Rubber-Toughened Engineering Plastics*, Chapman & Hall, London 1994.
19. J.A. Manson, L.H. Sperling: *Polymer Blends and Composites*, Plenum, New York 1976.
20. S.L. Aggarwal, R.L. Livigni, *Polym. Eng. Sci.* **17** (1977) 498.
21. D. Klempner, H.C. Frisch (eds.): *Handbook of Polymeric Foams and Foam Technology*, Hanser Publ., Munich 2004.
22. N. Mills (ed.): *Polymer Foams Handbook*, Elsevier, Amsterdam 2007.

Further Reading

P.C. Painter, M.M. Coleman (eds.): *Essentials of Polymer Science and Engineering*, DEStech Publication, Lancaster 2009.

E.S. Guerra, E.V. Lima (eds.): *Handbook of Polymer Synthesis, Characterization and Processing*, John Wiley & Sons, Hoboken 2013.

H.G. Elias (ed.): *Macromolecules: Industrial Polymers and Synthesis*, Wiley-VCH, Weinheim 2007.

H.G. Elias (ed.): *Macromolecules*, vols. 1-4, Wiley-VCH, Weinheim 2008.

K. Matyjaszewski, M. Möller (eds.): *Polymer Science: A Comprehensive Reference*, Elsevier, Amsterdam 2012.

G.H. Michler, F.J. Balta-Calleja (eds.): *Nano- and Micromechanics of Polymers*, Hanser Publ., Munich 2012.

N. Rudolph, T. Osswald (eds.): *Polymer Rheology: Fundamentals and Applications*, Hanser Publ. Munich 2014.

Plastics, General Survey, 5. Plastics and Sustainability

HANS-GEORG ELIAS, Michigan Molecular Institute, Midland, United States

ROLF MÜLHAUPT, Institute for Macromolecular Chemistry, Freiburg, Germany

1.	Plastics and Environment......	223	3.4.	Biodegradation and Bio-based
2.	Green Polymer Chemistry and Life			Plastics 226
	Cycle Assessment.............	224	3.5.	Plastics from Carbon Dioxide..... 227
3.	Plastics Waste Recycling	225	4.	Polymers for Sustainable
3.1.	Mechanical Recycling	225		Development 228
3.2.	Feedstock Recycling...........	226		References................... 229
3.3.	Energy Recycling..............	226		

1. Plastics and Environment [1–5]

During the first decade of the 21st century, more plastics were produced than during the entire 20th century. The plastics production rapidly increases and follows the growth of the world population. Originally introduced as rather poor imitation of natural materials, such as silk, natural rubber, and ivory, modern plastics have emerged as advanced materials with tailored property profiles, which is unparalleled in nature. Synthetic polymers are engineered to meet the demands of diversified technologies ranging from food packaging to textiles, shelter, communication, mobility, and health care. Plastics play an important role in daily life. Without plastics, the high quality of modern life with secure supply of food, water, and energy would not be feasible. Highly cost-efficient and versatile polymeric materials have made hightech affordable for people of all income groups. At present, plastics production consumes around 5% of the world's fossil resources, which is equivalent to about 60% of the petrochemical production. At present around 90% of the world's fossil resources are burned in energy production accompanied by severe emission of carbon dioxide, which contributes to global warming and climate change. The emission of greenhouse gases, i.e., carbon dioxide and methane, is also being referred to as carbon footprint. More than 30% of durable plastics are used in packaging, such as bags, wraps, and bottles, all of which possess short product life of a few months. According to the statistics of the U.S. Environmental Protection Agency (EPA), around 251×10^6 t of municipal trash composed of paper (27.4%), food wastes (14.5%), yard trimmings (13.5%), plastics (12.7%), metals (8.9%), rubber, leather, and textiles (8.7%), wood (6.3%), and glass (4.6%) were created in the United States in 2012. Archaeological studies of landfills have revealed that the biodegradation processes of synthetic polymers and biopolymers are very slow in landfills, especially in the absence of light, air, and water. Some buried newspapers could be read after residing for 25 years in a landfill. Clearly, recycling and incineration are the only possible ways for disposing of plastics. Because most plastics have an oillike, high energy content, plastic wastes are valuable raw materials that need to be reused. While in 2012 the plastics recycling and energy recovery reached 61.9% in Europe, it was around 9% in the United States (for comparison: paper recycling reaches 50% in the United States). When plastic recycling fails, the consequences of littering are clearly visible in nature. Driven by marine and wind currents, nontoxic plastics debris together with hazardous chemical sludge accumulate in the ocean, in particular in a region known as the northern pacific garbage vortex. Addressing high sustainability is an important issue in plastics development. As stated in 1987 by the Brundtland Commission of the UN General Assembly, sustainability means meeting the needs of the present without compromising the ability of future generations to meet their own needs. High resource-, eco-, and energy efficiency of plastics together with their low carbon footprint and high energy content, typical for the large-scale commodity plastics, holds great promise for achieving high sustainability. Efficient plastics recycling, exploiting renewable feedstocks, and especially tailoring advanced polymeric materials for sustainable development are important tasks in the plastics industry.

2. Green Polymer Chemistry [6, 7] and Life Cycle Assessment [8]

According to the definition by ANASTAS, important elements of green chemistry include high resource effectiveness by maximizing the content of raw materials in the product; clean and lean production processes; catalytic reactions at ambient temperature and pressure; preventing wastes and reducing greenhouse gas emissions; high safety standards; no use of auxiliary substances, such as organic solvents, blocking groups etc.; low energy demand of production and processing; no health and environmental hazards by substituting toxic chemicals; using renewable resources; low carbon footprint; and controlled product life cycles with effective waste recycling. This definition of "green" is not at all synonymous with biomaterials and biotechnology.

An important tool for assessing the environmental impact of materials is the life cycle analysis, which is also referred to as ecobalance, cradle-to-grave analysis, and life cycle assessment (LCA). LCA addresses the environmental aspects of all stages of a product's life time from feedstock recovery to monomer and polymer manufacturing, processing, distribution, use, maintenance, disposal, and recycling. Taking into account all relevant materials and energy inputs and output together with environmental aspects, the following LCA ranking of plastics was determined in 2010, yielding the best marks for polyolefins:

1. Polypropylene (PP),
2. High-density polyethylene (HDPE),
3. Low-density polyethylene (LLDPE).

The LCA ranking of bio-based polylactic acid (rank 9) was similar to that of PVC (rank 7). According to LCA, converting petrochemical plastics into green plastics by using renewable feedstocks lowers the LCA ranking. The success of polyolefins, such as PP, HDPE, and LLDPE is clearly associated with significant progress made in polyolefin production since the 1950s. The polyethylene life cycle is displayed in Figure 1. Produced in solvent-free catalytic polymerization processes with low energy demand using a highly active stereospecific catalysts, polyolefins are readily tailored to meet the demands of diversified applications ranging from food packaging to baby diapers, carpets, and automotive bumpers. Modern reactor granule technology produces pellet-sized polyolefin particles, which eliminates pelletizing extrusion. The heat of polymerization can be used as power generator. No water and no purification are required. This is in contrast to biotechnology processes and to pulping. The latter has very high water demand and produces byproducts, such as organic compounds and salts. Compared with paper, LLDPE packaging has much lower mass and is easy to process by blow molding. After completing their product life, polyolefins are readily recycled by mechanical, feedstock, and energy recycling. As hydrocarbon resins, they have oillike energy content. On heating above 300°C, they are converted into oil and gas, which are valuable sources for feedstocks and energy. The raw materials of polyolefins are highly flexible. Virtually any carbon source, ranging from oil, gas, and coal to biomass and bioethanol, can be employed in polyolefin production.

From the LCA it is apparent that the substitution of plastics packaging by paper and other materials would substantially increase mass, energy demand, and greenhouse gas emission.

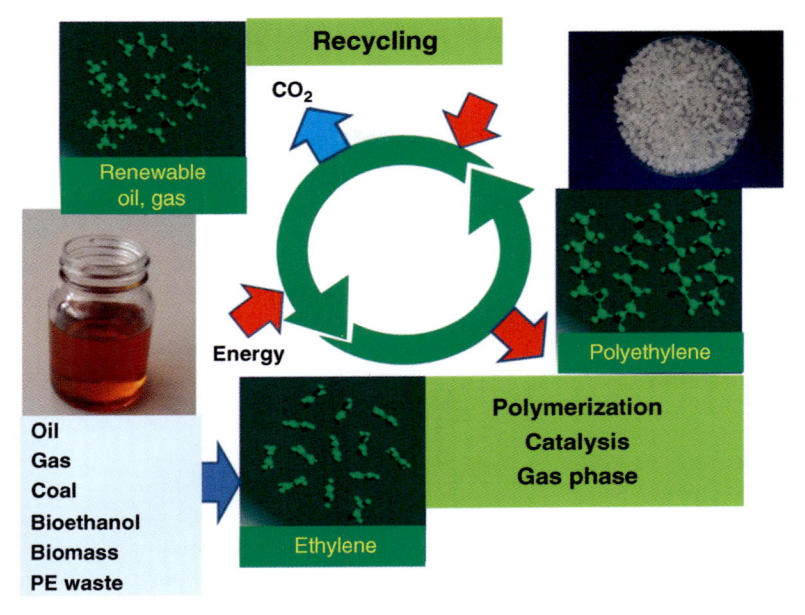

Figure 1. Life cycle of polyethylene

Contrary to the public opinion, increased use of plastics reduces the greenhouse gas emission, whereas the substitution of plastics by other materials would have a stringent detrimental environmental impact and drastically enhance greenhouse gas emissions. At the end of their service life, plastics wastes are too valuable to be thrown away.

3. Plastics Waste Recycling

3.1. Mechanical Recycling [9, 10]

For identification and for facilitating their recycling, plastic products bear special codes consisting of label with three broken arrows, arranged in a triangle, and a number, sometimes also combined with letters and/or bar code. The numbers and letter combinations indicate the type of plastics. The letter combinations are not always identical with the acronyms and abbreviations recommended by ISO, ASTM, DIN, IUPAC, etc., for the same polymers. The following number and letter codes are used:

- 1 = PET [poly(ethylene terephthalate)]
- 2 = HDPE (high-density polyethylene)
- 3 = PVC [poly(vinyl chloride)]
- 4 = LDPE (low-density polyethylene)
- 5 = PP (polypropylene)
- 6 = PS (polystyrene)
- 7 = for all other plastics, including multi-layered materials

Different waste management systems have been established to collect and sort wastes from different applications, such as packaging, automotive, electronics, and construction. A large variety of waste-pretreatment and sorting technologies are in place including shredding, magnetic separation, flotation, and air separation assisted by plastics identification by spectroscopy, such as near infrared sensing.

Especially when plastics wastes are clean and composed of as single type of plastics, mechanical recycling is attractive. In thermoplastics mechanical recycling, wastes are sorted, purified, and extruded to produce granulates. For example, mechanical recycling converts PET bottle waste into new PET bottles or PET textile fleeces for jackets. Due to polymer ageing by degradation and autoxidation during product life and recycling, most recycled plastics are somewhat inferior with respect to the corresponding virgin plastics. Therefore, most recycled plastics are blended together with virgin polymers in mechanical recycling. Because most polymers are highly immiscible, mechanical recycling of mixed plastics wastes produces blends, which require the addition of compatibilizers. Because cross-linked plastics do not melt, they can be recycled by grinding. Recycled thermoset resin powders or ground rubber are used as organic fillers for thermoset or rubber formulations.

3.2. Feedstock Recycling [10, 11]

During pyrolysis plastics are broken down at elevated temperatures in the absence of air to produce liquid and gaseous fragments. Thermolysis is of particular interest for recycling highly polluted complex plastic mixtures. In the case of PVC, HCl is formed and removed by washing the thermolysis gases with water. Hydrocarbons, such as polyolefins, are readily thermolyzed to produce oil and gas in essentially quantitative yields. Plastics, such as polyoxymethylene and poly(methyl methacrylate) are thermally depolymerized to recover the corresponding monomers. On thermolysis, carbohydrates decompose to form a considerable amount of water and ashes. Whereas pyrolysis requires rather high temperatures, typically above 500°C, catalytic liquefaction converts biomass and plastics wastes into diesel fuel at temperature well below 500°C. In this process, water is catalytically converted into hydrogen, which improves the yield of low molecular mass hydrocarbons by hydrogenation. In sharp contrast, hydrogenation of plastics wastes and distillation residues in oil refinery require much higher hydrogen pressure and temperature. The presence of hydrogen increases the amount of liquid hydrocarbon. In the case of biomass and biopolymers, anaerobic digestion can be employed to convert biowastes into methane or bioethanol. Lignocellulose from forestry and agricultural wastes is an attractive source of sugar that serves as raw material for biorefineries without competing with food production.

When small amounts of oxygen are present, gasification by incomplete combustion converts plastics into a mixture of carbon monoxide and hydrogen (syngas), which can be used as feedstock for producing a variety of chemicals and monomers in refineries. Similar to the Fischer–Tropsch process for coal-to-liquid conversion, this process can also be applied to biomass (biomass-to-liquid conversion) and plastics wastes. Moreover, syngas is of interest as reducing gas for metal oxide ores in steel mills, thus substituting oil and gas. Polycondensates, such as polyesters and polyurethanes are readily cleaved by hydrolysis and glycolysis, thus recovering monomers useful in polycondensation reactions.

3.3. Energy Recycling [9–11]

Taking into account the high energy content and hydrocarbon nature of commodity plastics, in many countries incineration is the preferred method for municipal waste recycling, which does not need extensive sorting and purification of the plastics wastes. In fact, the plastics wastes represent an attractive source of energy for the combustion of municipal wastes, which contain components with much less energy content. In contrast to inorganic wastes and biomass, commodity plastics do not form large amounts of ashes but completely decompose to produce gas and heat. For many years, public fears concern the formation of dioxin by waste incineration. Today, the incineration off-gases are carefully purified to remove gaseous and dustlike particulate pollutants. This includes the catalytic conversion of dioxin and furan traces. Incineration is the preferred process for disposing of hazardous wastes containing microorganisms, toxins, or chemicals.

3.4. Biodegradation and Bio-based Plastics [7, 12–16, 17] (→ Polymers, Biodegradable)

In the early 1900s, all plastics were bio-based because carbohydrates and proteins were the only industrially available feedstocks for plastics production. Because abundant biopolymers, such as cellulose, are infusible and insoluble, extensive chemical modifications are required to enable melt and solution processing of cellulosics. Within very short time during the first half of the 20th century, the emerging highly versatile and low-cost synthetic plastics, based on fossil feedstocks and petrochemistry, have outperformed biopolymers and conventional bio-based plastics. In spite of the ongoing commercial significance of natural rubber, claiming 40% of the rubber market, and the successful development of wood plastics compounds and natural fiber reinforced composites, the total biopolymer share of the annual world's plastics production is still below 5%. However, there is a renaissance of the use of renewable and natural resources as well as the development of biopolymers. In view of the dwindling fossil resources, which are rapidly being depleted in energy production, the shift from fossil to renewable resources is thought to safeguard plastics production against an expected future oil crisis. Moreover, the growing consumers' concerns regarding global warming and the fear that a collapse of the biosphere is imminent, stimulate the surging demand for renewable and bio-based products. Because most natural polymers still fail to match the performance of synthetic polymers, synthetic plastics are rendered "green" and bio-based by using renewable raw materials for the production of monomers. For example, bioethanol is converted into ethylene and propene, which are polymerized in highly efficient gas phase processes to afford "green" polyolefins, which have properties equivalent to petrochemical polyolefins. In contrast to bacterial poly-L-lactic acid (PLA), which requires separation of PLA from cell proteins and water recovery, the solvent-free ring-opening polymerization of bio-based dilactide affords much better control of polymer properties, such as crystallization rate and shear thinning.

In life cycle assessment, most bio-based products have lower performance with respect to the corresponding petrochemical products due to the additional process steps. Moreover, in many industrialized countries the biomass supply is insufficient to substitute oil in energy and plastics production. PLA, polyolefins, or paper are not completely biodegradable in landfill, especially when air and water are absent. In contrast to the public opinion, biodegradation and bioerosion, which are highly dependent on the climate, do not instantaneously degrade biopolymers to produce carbon dioxide and water. Instead, biodegradation makes polymers brittle. They disintegrate to form tiny particles, invisible for human eyes, which are carried away by wind or rain and serve as breeding ground for spores and bacteria. Because the label "biodegradable" stimulates consumers to carelessly dump wastes and ignore recycling efforts, biodegradable plastics are contra productive to recycling. Moreover, the high energy content of polymers is wasted during biodegradation in landfill. Hence, biodegradable polymers do not seem to be the universal solution to the plastics waste problem. Yet, they offer prospects for special applications like bioresorbable implants, surgical sutures, drug release, as well as agricultural films and packaging. Today, it is well recognized that the surging biomass demand in bioenergy production, accompanied by extensive farming of energy crops, endangers both the biosphere and the food supply of the rapidly growing world population. Many industrialized countries do not have enough biomass to serve the need of both energy and plastic production.

3.5. Plastics from Carbon Dioxide [7, 18] (→ Carbon Dioxide 12.4.10. Polymers from Carbon Dioxide)

In order to prevent a conflict with food production, several successful attempts have been made to exploit the direct chemical fixation of carbon dioxide in plastics production. Because carbon dioxide is the energy sink in the combustion process, it must be activated in order to be useful in polymer production. In copolymerization this is achieved by reacting

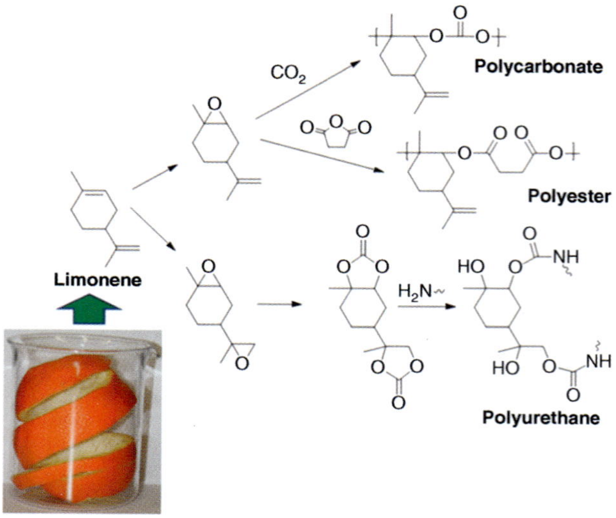

Figure 2. Bio-based polyesters, polycarbonates, and nonisocyanate polyurethanes from orange peels
Reprinted with permission by Wiley

carbon dioxide with energy-rich highly strained oxirane and oxetane rings, thus producing polyfunctional cyclic carbonates and linear polycarbonates. Dihydroxy-terminated poly(propylene ether carbonates) are of special interest as polyols in polyurethane synthesis. Polyfunctional cyclic carbonates, prepared by reacting epoxy resins with carbon dioxide, are attractive intermediates for the isocyanate- and phosgene-free production of hydroxyl-functional polyurethanes (NIPU for nonisocyanate polyurethane). Biowastes, such as orange peels can be used as a source of limonene, which upon oxidation and carbonation affords mono- and difunctional cyclic carbonates as intermediates for bio-based polyesters and NIPU (Fig. 2). Electrochemical processes are developed to exploit excess renewable energy supplied by wind mills and solar power plants to convert carbon dioxide into methane, which is useful as chemical storage of renewable energy, as fuel, and as intermediate for the production of chemicals and plastics.

4. Polymers for Sustainable Development [7]

In many ways plastics serve the needs of sustainable development. Produced in highly energy- and resource-efficient polymerization processes, plastics have low carbon footprint and preserve high oillike energy content. After completing their product life, they are mechanically recycled or serve as valuable source of energy and feedstocks. This attractive balance of low energy demand during production and processing combined with low carbon footprint and high energy content is typical for all large-volume commodity and most engineering plastics and rubbers, amounting to more than 90% of the world's plastics production. Among them, hydrocarbon plastics, such as polypropylene and polyethylene, produced by solvent-free catalytic polymerizations processes, meet the demands of green chemistry and achieve top rankings in life cycle assessment. Moreover, plastics have a low mass and are corrosion resistant. In lightweight engineering and packaging, polymeric materials contribute to considerable weight savings, thus reducing fuel consumption and dioxide emission in transportation. Moreover, unparalleled by other materials, plastics are highly versatile with respect to tailoring property profiles, processing, and applications. Owing to their low mass together with thermal and electrical insulation, plastics contribute to high energy efficiency and secure the supply of food, water, and energy. Although plastics consume around 5%

of fossil fuels, they help to preserve close to 20% oil by thermal insulation and mass reduction in transportation. As advanced functional materials and integrated into systems, polymers bring high resource efficiency to a variety of applications ranging from lightweight engineering to health care. For example, as filters and membranes, functional plastics enable water and air purification, water desalination, and recovery of precious metals from wastes. This unique combination of high cost-, eco-, resource-, and energy efficiency with high versatility typical for most plastics holds great promise for sustainable development.

References

1. R.A. Meyers (ed.): *Encyclopedia of Sustainability Science and Technology*, Springer, Berlin 2012.
2. P. Eyrer, M. Weller, C. Hübner (eds.): *The Handbook of Environemtal Chemistry, Polymers–Opportunities and Risks II*, 12 (2010).
3. G. Payne, P. Smith: "Renewable and Sustainable Polymers," *ACS Symp. Ser.*, American Chemical Society, Washington 2010.
4. M. Tolinski (ed.): *Plastics and Sustainability: Towards a Peaceful Coexistence between Bio-based and Fossil Fuel-based Plastics*, Scrivener Publishing, Salem 2012.
5. A. Azapagic, A. Emsley, I. Hamerton (eds.): *Polymers: The Environment and Sustainable Development*, John Wiley & Sons, Chichester 2003.
6. P.T. Anastas, T. Horvath (eds.): *Green Chemistry for a Sustainable Future*, John Wiley & Sons, Hoboken 2015.
7. R. Mülhaupt: "Green Polymer Chemistry and Bio-based Plastics: Dreams and Reality," *Macromol. Chem. Phys.* **214** (2013) 159.
8. M.D. Tabone, J.J. Cregg, E.J. Beckman, A.E. Landis, *Environ. Sci. Technol.* **44** (2010) 8264.
9. S. Thomas, M. Sebastian, A. George, Y. Weimin (eds.): *Recycling and Reuse of Materials and Their Products (Advances in Materials Science)*, Apple Academic Press, Point Pleasant 2013.
10. S. M. Al-Salem, P. Lettieri, J. Baeyens, *Waste Manage.* **29** (2009) 2625.
11. J. Scheirs, W. Kaminsky (eds.): *Feedstock Recycling and Pyrolysis of Waste Plastics: Converting Waste Plastics Into Diesel and Other Fuels (Wiley Series in Polymer Science)*, John Wiley & Sons, Chichester 2006.
12. V. Mittal (ed.): *Renewable Polymers*, Scrivener Publishing and John Wiley & Sons, Hoboken, Salem 2012.
13. M.N. Belgacem, A. Gandini (eds.): *Monomers, Polymers and Composites from Renewable Resources*, Elsevier, Oxford 2008.
14. A. Gandini, *Macromolecules* **41** (2008) 9491.
15. B. Rieger, A. Künkel, G.W. Coates, R. Reichardt, E. Dinjus, T. A. Zevaco: "Synthetic Biodegradable Polymers" in *Advanced Polymer Science*, vol 245, Springer, Berlin 2012.
16. A. Demirbas (ed.): *Biorefineries*, Springer, Berlin 2010.
17. R.P. Wool (ed.): *Bio-based Polymers and Composites*, Elsevier Academic Press, New York 2005.
18. H. Blattmann, M. Fleischer, M. Bähr, R. Mülhaupt, *Macromol. Rapid Commun.* **35** (2014) 1238.

Further Reading

P.C. Painter, M.M. Coleman (eds.): *Essentials of Polymer Science and Engineering*, DEStech Publication, Lancaster 2009.

E.S. Guerra, E.V. Lima (eds.): *Handbook of Polymer Synthesis, Characterization and Processing*, John Wiley & Sons, Hoboken 2013.

H.G. Elias (ed.): *Macromolecules: Industrial Polymers and Synthesis*, Wiley-VCH, Weinheim 2007.

H.G. Elias (ed.): *Macromolecules*, vols. 1-4, Wiley-VCH, Weinheim 2008.

K. Matyjaszewski, M. Möller (eds.): *Polymer Science: A Comprehensive Reference*, Elsevier, Amsterdam 2012.

G.H. Michler, F.J. Balta-Calleja (eds.): *Nano- and Micromechanics of Polymers*, Hanser Publ., Munich 2012.

N. Rudolph, T. Osswald (eds.): *Polymer Rheology: Fundamentals and Applications*, Hanser Publ. Munich 2014.

Plastics, Analysis

KARL-FRIEDRICH ELGERT, Institut für Textilchemie, Denkendorf, Federal Republic of Germany

1.	Introduction	231
2.	Polymer Isolation	232
3.	Preliminary Tests for Polymer Identification	234
4.	Chemical Analysis	234
4.1.	Heteroelements and Functional Groups	234
4.2.	Separation and Identification Schemes	238
4.3.	Spectroscopic Methods	238
5.	Molecular Mass Determination	242
5.1.	Absolute Methods	244
5.1.1.	Membrane Osmometry	244
5.1.2.	Lowering of Vapor Pressure	245
5.1.3.	Ultracentrifugation	245
5.1.4.	Light Scattering	247
5.2.	Relative Methods	249
5.3.	Determination of Molecular Mass Distribution	250
6.	Determination of Sequential Structure	255
6.1.	NMR Spectroscopy	255
6.2.	Pyrolysis Gas Chromatography	258
7.	Determination of Chemical Heterogeneity	258
	References	262

1. Introduction

The polymeric chains of most industrially and economically important plastics are built up from only a small number of elements: mainly C, H, O, N, Cl, F, and Si. Nevertheless, plastics analysis constitutes an extremely complicated and wide-ranging field that requires the application of many chemical and physical analytical methods.

The subject of plastics analysis is treated here in terms specifically of the analysis of polymer chains. Most plastics – that is to say, materials formulated on the basis of synthetic polymers – contain significant amounts of additives such as pigments, plasticizers, stabilizers, and fillers, but analysis of these secondary components constitutes a separate field of study. In certain cases the desirable properties of a particular plastic arise only during the course of subsequent thermal or mechanical treatment associated with shaping or molding. There has also been increasing interest in the determination of polymer degradation products and monomer residues.

The fundamental problem in plastics analysis is the heterogeneity of macromolecular materials with respect to molecular mass and the numerous isomeric possibilities open to a polymeric chain.

For example, vinyl chloride is a gas under normal conditions, but its lower oligomers are liquids, and poly(vinyl chloride) is a solid. Thus the physical, physicochemical, and applications characteristics of a particular batch of poly(vinyl chloride), such as solubility, crystallinity, strength, and elasticity, are a function not only of the structure of the polymeric chain, but also of the degree of polymerization.

Isomerism. The empirical formula obtained by elemental analysis (see Section 4) is generally of little help in the precise identification of a plastic or a polymer. This is due to the constitutional, configurational, and conformational isomerism characteristic of a polymer chain, which is reflected in a structure that can be described at primary through quaternary levels. Thus, an empirical formula such as $(C_4H_8)_n$ is equally valid for achiral polyisobutene and its positional isomer poly(1-butene), which contains a series of pseudoasymmetric carbon atoms (C^*).

$$\left[-CH_2-\underset{\underset{CH_3}{|}}{\overset{\overset{CH_3}{|}}{C}}-\right]_n \quad \left[-CH_2-\underset{\underset{CH_3}{|}}{\overset{\overset{\overset{*}{}}{CH-}}{CH_2}}\right]_n$$

Poly(isobutene) Poly(1-butene)

Polymolecularity in a polymer is another form of constitutional isomerism, which results from the stochastic character of the polymerization reaction but can also arise from decomposition, aging, or cross-linking reactions that occur during processing or utilization. The determination of polymolecularity is discussed in Chapter 5.

Another source of constitutional isomerism may be the way in which monomer units are linked, although head-to-tail linkage usually predominates. An example is poly(1-butene):

$$\underset{\text{head-to-tail}}{-CH_2-\underset{\underset{CH_3}{|}}{\overset{\overset{CH_2}{|}}{CH}}-CH_2-\underset{\underset{CH_3}{|}}{\overset{\overset{CH_2}{|}}{CH}}-} \qquad \underset{\text{head-to-head}}{-CH_2-\underset{\underset{CH_3}{|}}{\overset{\overset{CH_2}{|}}{CH}}-\underset{\underset{CH_3}{|}}{\overset{\overset{CH_2}{|}}{CH}}-CH_2-}$$

Apart from the nature of polymer end groups and the extent of branching, the sequence of structural units within the polymer chain is also of major importance. Analysis of sequential structure is treated in Section 4.3 and Chapter 6.

Plastics that contain pseudoasymmetric carbon atoms in their polymer chains, such as poly (1-butene), can also exhibit *configurational isomerism*. The sequence of such pseudoasymmetric C atoms has a major effect on properties such as melting point, glass transition temperature, and crystallinity. There is also isomerism of the *cis – trans* type to be considered in the case of unsaturated polydienes, whose structural units include C=C double bonds. The number of possibilities for isomerism – and thus the analytical challenge – increases dramatically on moving from homopolymers to copolymers. Analysis of configurational isomers is the subject of Section 6.2.

As a result of the wide diversity in types of plastic materials and the range of potential problems, it is impossible to describe a single general scheme for polymer analysis. Both the number of questions that might be raised regarding the nature of a completely unknown polymer and the catalog of applicable analytical methods are very extensive. However, it is often possible to monitor product quality by determining a single structural property by means of a standardized analytical method. The scope and complexity of an analytical investigation is thus a direct function of the nature of the question.

The most superficial description of an unknown plastic must include some characterization of the *primary structure*: chemical composition, molecular mass and molecular mass distribution, and the sequential arrangement of subunits. This leads to the following series of analyses for identifying an unknown plastic:

1. Determination of solubility (Chap. 2)
2. Isolation of macromolecular components (Chap. 2)
3. Preliminary tests for polymer identification (Chap. 3)
4. Chemical analysis (Chap. 4)
5. Molecular mass determination (Chap. 5)
6. Sequence analysis (Chap. 6)
7. In the case of a copolymer, analysis of chemical heterogeneity (Chap. 7)

For a discussion of structure at the secondary through quaternary levels (e.g., conformation, orientation, and crystallinity), see → Plastics, Properties and Testing.

Literature. Plastics analysis is treated in detail in various standard works [1–26]. There also exists an extensive specialized literature [27–63] devoted to specific problems (e.g., analysis of silicones [51], or fiber-reinforced plastics [52], which is further supplemented by tabular works [53] and data from handbooks [54–57]. The review literature should be consulted for more recent developments in plastics analysis, especially with respect to highly specialized questions [58], [59]. For the analysis of selected additives, see [60–63].

2. Polymer Isolation

Solubility is the basis for preliminary separation of a plastic into polymer components and additives. Insoluble fillers and pigments can be removed from a polymer solution by filtration or centrifugation. A plastic that is received in the form of a solution, emulsion, or suspension must first be dried by evaporation of low molecular mass solvents; nonvolatile additives remain in the residue.

A polymeric material that is to be extracted or brought into solution for analytical investigation should be first converted into the most finely divided state possible. Prior freezing of plastic or elastic samples in liquid nitrogen often facilitates size reduction.

Solubility. Polymers and solvents are miscible only if the corresponding free energy of mixing is negative [64], [65]:

$$\Delta G_{cm^3} = \frac{RT}{V_s}\left(\Phi_s \ln\Phi_s + \frac{\Phi_p}{m}\ln\Phi_p + \mu\Phi_s\Phi_p\right) \quad (1)$$

G_{cm^3} Gibbs free energy of mixing per cm³ solution
V_s Molar volume of solvent
Φ Volume fraction (s = solvent; p = polymer)
m Ratio of molar volumes of polymer and solvent
μ Constant

The interaction parameter μ is a function of temperature that takes into account the enthalpy of mixing. Miscibility can be anticipated for any polymer – solvent system in which $\mu<0.5$. In the case of an amorphous polymer it is possible to ignore, to a first approximation, the change in entropy, which is the justification for focusing here strictly on the enthalpy of mixing. In the expression:

$$\Delta H_{cm^3} = \Phi_s\Phi_p\left[(E_s/V_s)^{1/2} - (E_p/V_p)^{1/2}\right]^2 \quad (2)$$

ΔH_{cm^3} Enthalpy of mixing for 1 cm³ solution
Φ Volume fraction
E Molar cohesion energy
V Molar volume

the term E/V is referred to as the cohesion energy density, from which can be derived the solubility parameter $\delta = \sqrt{E/V}$. The following list gives δ values for various polymers and solvents.

Polymers	
Polyethylene	7.9
Polyisoprene	7.9
Polyisobutylene	8.05
Styrene rubber	8.60
Polybutadiene	8.6
Polystyrene	8.7
Neoprene	9.2
Poly(dimethyl siloxane)	9.4
Nitrile rubber	9.5
Poly(vinyl chloride)	9.7
Poly(methyl methacrylate)	10.2
Cellulose acetate	10.9
Nylon 66	14.5
Poly(vinyl alcohol)	23.4
Solvents	
Heptane	7.43
Diethyl ether	7.70
Cyclohexane	8.25
Carbon tetrachloride	8.62
Benzene	9.21
Chloroform	9.40
Chlorobenzene	9.70
Acetonitrile	9.75
Acetone	9.89
Dichloromethane	10.04
Carbon disulfide	10.10
Dioxane	10.15
Methanol	14.48
Water	23.41

Solubility can be expected when polymer and solvent have comparable solubility parameters. Such δ values provide useful orientation in a search for a suitable solvent, although many exceptions are known, attributable to such factors as molecular mass, degree of branching and cross-linking, configuration, and crystallinity.

Precipitation. A considerable degree of polymer purification can often be achieved by precipitation with a precipitating agent, a process that may need to be repeated several times. However, some oligomers may remain in solution, as a result of which precipitation may cause a change in the molecular mass distribution.

If a plastic and a solvent have different δ values the plastic will be insoluble in that solvent. Therefore, addition of such a solvent to a polymer already dissolved in another solvent should lead to precipitation. Methanol is a poor solvent for most polymers but miscible with many other organic solvents, so it is frequently used as a precipitating agent. Generally, the lower the molecular mass of a plastic, and the more atactic or amorphous it is, the higher its solubility. Raising the temperature also leads to an increase in solubility. Appropriate solvents and precipitating agents have been established empirically for a wide range of plastics; for a tabular summary see [53].

If precipitation proves unsuccessful for the removal of additives it is usually necessary to resort to preparative chromatography [66–68].

Extraction. Plastics can be extracted with organic solvents either in a Soxhlet apparatus or in a high-pressure extractor with supercritical solvents such as CO_2. Extraction is the method of choice for pretreatment of an insoluble plastic. A prerequisite to success is finding an extractant that will dissolve the additives at the temperature of operation but act as a nonsolvent for polymeric components. If necessary, several extractants can be utilized in sequence, although inorganic fillers or pigments that are insoluble in organic solvents will always remain in the sample after extraction (see Table 1).

3. Preliminary Tests for Polymer Identification

Once a set of macromolecular components has been isolated, characterization can begin on the basis of physical methods. Powerful and rapid spectroscopic procedures are now available for establishing chemical structure, although their successful utilization presupposes access to the required instrumentation and an appropriate data base. In the absence of such information it may still be possible to narrow the range of structural possibilities in relatively simple ways.

Analysis is most straightforward for soluble polymers. A material consisting of a mixture of various polymers must first be separated into its macromolecular components, preferably by preparative chromatography [66–71] or precipitation. Such a separation may prove problematic in the case of a copolymer, since separability then becomes a function of heterogeneity with respect to both molecular mass and chemical composition of the polymer molecules (see Chap. 7).

Density. Density measurements are conducted either with a pycnometer or on the basis of buoyancy. A rough determination of density can be made by introducing an unknown sample into a series of nonsolvents of known density. By using Figure 1, the nature of the material can then be established with a reasonably high degree of certainty. Crystalline polymers are generally denser than amorphous materials. Density can also be affected by the presence of fillers. Density measurements on plastic materials that contain entrapped gases (e.g., foams) must be preceded by melting or precipitation.

Combustion Characteristics. The combustibility of a plastic can be investigated in the flame of a Bunsen burner. Apart from distinguishing between flammable and inflammable materials, it is also important to observe whether a substance burns only while it is in the flame or if burning persists outside the flame as well. Table 2 provides an overview of flammability characteristics for various plastics.

Thermal Degradation. Pyrolysis behavior and the characteristics of a pyrolyzate give further information on the nature of an unknown plastic. The sample is heated to redness in a pyrolyzer (Fig. 2). Volatile products are either condensed or collected in a cotton wool filter. Approximately 100 mg of sample is sufficient for this analysis.

The nature and odor of a pyrolyzate can be characteristic for a particular polymer. Highly volatile pyrolyzates are produced by polystyrene, polyisobutene, polybutadiene, and polyacrylates. Oily, resinous, or waxy products arise from polyethylene, polypropylene, and poly(1-butene) as well as from copolymers of styrene with butadiene and acrylonitrile. Black carbonaceous residues result from the pyrolysis of cellulose and its derivatives, as well as from poly(vinyl chloride) and resins.

Formation of ammonia suggests a polyamide or a resin based on urea, melamine, aniline, or casein. Pyrolyzate with a pungent odor is released by poly(vinyl chloride), polyacrylates, and poly(vinyl formates).

The combination of pyrolysis with hydrogenation of the resulting fragments, followed by gas chromatographic separation and identification, can provide considerable information on the sequential structure of a polymer (see Section 6.3).

4. Chemical Analysis

4.1. Heteroelements and Functional Groups

Heteroelements. In conjunction with quantitative analysis for the principal elements C and H, the potassium fusion test for heteroelements remains a very important analytical technique [3] (see Table 3). Roughly 100 mg of a finely divided plastic is fused in a test tube with a pea-sized portion of potassium, heated to a red glow,

Table 1. Solubility of common plastics and polymers [14]*

Plastic/polymer	Soluble in	Insoluble in
Alkyd resin	chlorinated hydrocarbons, lower alcohols, esters	hydrocarbons
Amine – formaldehyde resin, cured	benzylamine (60 °C), ammonia	
Methyl cellulose	water, dilute sodium hydroxide, dichloromethane, methanol	acetone, ethanol
Cellulose esters	ketones, esters	aliphatic hydrocarbons, water
Cellulose nitrate	lower alcohols, acetic esters, ketones	ether, benzene, chlorinated hydrocarbons
Chlorinated rubber	esters, ketones, carbon tetrachloride, tetrahydrofuran	aliphatic hydrocarbons
Chloroprene rubber	toluene, chlorinated hydrocarbons	alcohols
Polychlorotrifluoroethylene	hot fluorinated solvents	all common solvents
Poly(vinyl chloride)	dimethyl formamide, tetrahydrofuran, cyclohexanone	alcohols, butyl acetate, dioxane, hydrocarbons
Chlorinated poly(vinyl chloride)	dichloromethane, cyclohexane, benzene, tetrachloroethylene	
Poly(vinylidene chloride)	tetrahydrofuran, ketones, butyl acetate, dimethylformamide (hot), chlorobenzene	alcohols, hydrocarbons
Acrylonitrile – butadiene – styrene copolymer	dichloromethane	alcohols, water, aliphatic hydrocarbons
Styrene – butadiene copolymer	ethyl acetate, benzene, dichloromethane	alcohols, water
Rubber	chlorinated and aromatic hydrocarbons	oxygen-containing solvents
Polytetrafluoroethylene	fluorocarbon oil (hot)	all solvents
Poly(vinyl fluoride)	above 110 °C: cyclohexanone, propylene carbonate, dimethyl sulfoxide, dimethylformamide	
Poly(vinylidene fluoride)	dimethyl sulfoxide, dioxane	
Phenolic resin	alcohol, ketones	chlorinated hydrocarbons, aliphatic hydrocarbons
Polyacrylamide	water	alcohols, esters, hydrocarbons
Polyacrylonitrile	dimethylformamide, butyrolactone nitrophenol, dimethyl sulfoxide, mineral acids	alcohols, esters, ketones, formic acid, hydrocarbons
Polyacrylates	aromatic hydrocarbons, esters, chlorinated hydrocarbons, acetone, tetrahydrofuran	aliphatic hydrocarbons
Polymethacrylates	aromatic hydrocarbons, dioxane, chlorinated hydrocarbons, esters, ketones	ether, alcohols, aliphatic hydrocarbons
Polyamides	phenols, formic acid, conc. mineral acids	alcohols, esters, hydrocarbons
Polybutadiene	aromatic hydrocarbons, cyclohexane, dibutyl ether	alcohols, esters
Polycarbonates	chlorinated hydrocarbons, dioxane, cyclohexanone	alcohols, aliphatic hydrocarbons, water
Polyesters, unsaturated, uncured	ketones, styrene, acrylic esters	aliphatic hydrocarbons
Poly(ethylene terephthalate)	cresol, conc. sulfuric acid, chlorophenol	
Polyethylene	dichloroethylene, tetralin, hot hydrocarbons	polar solvents, alcohols, esters
Poly(ethylene glycol)	chlorinated hydrocarbons, alcohols, water	aliphatic hydrocarbons
Polyformaldehyde	hot solvents, phenols, benzyl alcohol, dimethylformamide	alcohols, ketones, esters, hydrocarbons
Polyisoprene	benzene	alcohols, ketones, esters, hydrocarbons
Polypropylene	at high temperature: aromatic and chlorinated hydrocarbons, tetralin	alcohols, esters, cyclohexanone
Polystyrene	aromatic and chlorinated hyrocarbons, pyridine, ethyl acetate, methyl ethyl ketone, dioxane, tetralin	alcohols, water, aliphatic hydrocarbons
Polyurethane	tetrahydrofuran, pyridine, dimethylformamide, formic acid, dimethyl sulfoxide	ether, alcohols, benzene, water, hydrogen chloride (6 N)
Poly(vinyl acetal)	esters, ketones, tetrahydrofuran	methanol, aliphatic hydrocarbons
Poly(vinyl formal)	dichloroethane, dioxane, glacial acetic acid, phenols	aliphatic hydrocarbons
Poly(vinyl acetate)	aromatic and chlorinated hydrocarbons, acetone, methanol, esters	aliphatic hydrocarbons
Poly(vinyl alcohol)	formamide, water	ether, alcohols, esters, ketones, aliphatic and aromatic hydrocarbons

*For comprehensive table, see [14].

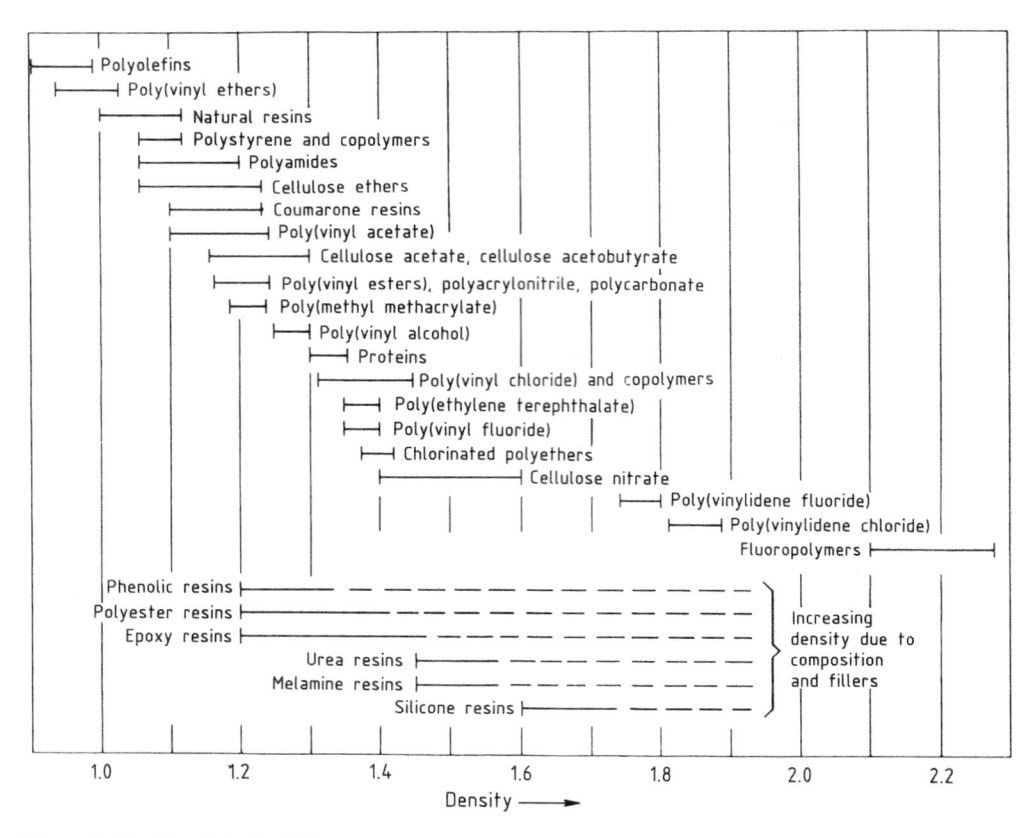

Figure 1. Density of plastics [62]

and then quenched in 10 cm^3 of water. The resulting solution is filtered and divided into six equal portions for the identification of Cl, F, N, S, P, and Si.

Chlorine. The filtrate is acidified with HNO_3 and treated with $AgNO_3$ solution; a white precipitate soluble in ammonia indicates chlorine.

Table 2. Flammability of plastics and polymers [14]

Nonflammable	Self-extinguishing	Flammable
Polychlorotrifluoroethylene	poly(vinyl chloride)	polyolefins
Polytetrafluoroethylene	vinyl chloride copolymers	polystyrene
Chlorinated poly(vinyl chloride)	poly(vinyl fluoride)	copolymers of styrene
Urea resins	polyvinylcarbazole	poly(vinyl ethers)
Melamine resins	polyvinylpyrrolidone	poly(vinyl esters)
Silicone resins	polyacrylonitrile	poly(vinyl alcohol)
	filled phenolic, epoxy, and polyester resins	polyacrylates
		polyamides
		polycarbonates
		poly(ethylene terephthalate)
		polyurethanes
		cellulose
		alkyd, coumarone, phenolic, epoxy resins

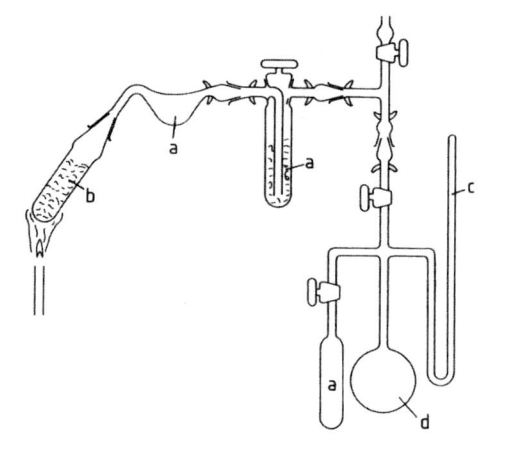

Figure 2. Pyrolysis of plastics by vacuum distillation [3]
a) Condensate traps; b) Sample; c) Manometer; d) IR sample cell

Fluorine. After acidification with HCl the solution is treated with $CaCl_2$ solution; a white precipitate of CaF_2 is characteristic of fluorine.

Nitrogen. The solution is treated with a spatulaful of $FeSO_4$, boiled, cooled, and a few drops of $FeCl_3$ solution are added. A precipitate of Prussian blue upon acidification with HCl indicates nitrogen.

Sulfur. The cold solution is treated with a freshly prepared solution of sodium prussiate. A deep violet color represents a positive test for sulfur.

Phosphorus. Acidification of the solution with HNO_3 and addition of a few drops of ammonium molybdate solution results in a yellow precipitate in the presence of phosphorus.

Silicon. The polymer sample is boiled with a mixture of sulfuric and nitric acids and then evaporated. The residue is pyrolyzed to SiO_2, which is fumed with a mixture of sulfuric and hydrofluoric acids. Appearance of a white precipitate on a moist piece of black paper is evidence of silicon.

Functional Groups. Chemical tests for heteroelements are supplemented by tests for specific functional groups. In addition to establishing the presence of C=C double bonds by Br_2 addition, determination of the saponification number is useful. For this purpose a known amount of plastic is dissolved in alcohol, benzene, or toluene, and refluxed for 2 – 3 h with a known excess of 1 N KOH. The excess alkali is then determined by back-titration with 1 N HCl. The saponification number is defined as the number of milligrams of KOH consumed per gram of polymer. Many polymers containing no heteroelements other than oxygen can be distinguished on the basis of

Table 3. Chemical analysis of common polymers [14]

Elements (other than C and H)	Polymers
None	polymerized hydrocarbons (e.g., polyethylene, polypropylene, polybutadienes, polystyrene, poly(methyl styrene), poly(vinyl benzene), butadiene – styrene copolymer)
O	Carbohydrates (e.g., celluloses, starches), phenol – formaldehyde resins, phenol – ether resins, phenol – furfural resins, phthalic esters and related compounds, alkyds, poly(vinyl esters), poly(allyl esters), polyacrylic and related esters, polycarbonates, aliphatic polyesters, epoxy resins, aldehyde resins, poly(vinyl alcohol)
N (O)	cellulose nitrate, polyacrylonitrile and copolymers, polyacrylamide, nylons and other polyamides, poly(ethylene imines), urea – formaldehyde resins, melamine – formaldehyde resins, polyurethane, polyester – urethane
Cl (O)	Chlorinated polyhydrocarbons, poly(vinyl chloride), poly(vinylidene chloride), poly(vinyl chloride acetate), polychloroacrylates, chlorinated alkyds
F (O)	polytetrafluoroethylene, poly(vinyl fluoride)
F, Cl	Polychlorotrifluoroethylene
S (O)	alkyl polysulfides, polysulfones, thiophenol condensates
N, S	thiourea – formaldehyde resins
Cl, S	thioplastics, vulcanized neoprenes
N, S, P	casein, protein
Si	silicones
Ti	poly(titanic esters)
B	poly(boric esters)
N, P, Cl	poly(phosphonitrilic halides)

Table 4. Saponification number of plastics, resins, and polymers [6]

< 20	100 – 200	> 200
Polyolefins	natural resins	polyacrylates
Methyl cellulose	alkyd resins	polymethacrylate
Polystyrene		poly(vinyl acetate)
Poly(vinyl alcohol)		cellulose esters
Phenol – formaldehyde resins		poly(butylene terephthalate)
Natural and synthetic rubber		polycarbonates
Poly(vinyl chloride)		
Chlorinated poly(vinyl chloride)		
Polyamides		
Natural waxes		

their saponification numbers, as indicated in Table 4.

A specific color reaction between a plastic and dichloroacetic acid sometimes serves as a rapid and simple means of identification [3].

4.2. Separation and Identification Schemes

A simple identification scheme based on polymer pyrolysis is illustrated in Table 5.

Figure 3 shows a more elaborate scheme that uses a set of solubility tests, and combines pyrolysis with both thin-layer chromatography and chemical identification. Specific tests applied to the pyrolyzate result in identification of a particular plastic. In the case of a copolymer, various isomeric possibilities with respect to the principal chain require that chemical analysis be supplemented by physical methods of structure determination. Purely chemical methods may also fail to distinguish clearly between a copolymer and a blend of two homopolymers.

If detailed and unambiguous structural analysis is required spectroscopic methods should be used. Not only do they often allow nondestructive testing of the sample, they also give considerably more information than elemental analysis, particularly with regard to structural subunits and their sequence within the polymer chain (see Sections 4.3 and 6.2).

4.3. Spectroscopic Methods

Infrared spectroscopy (IR) [72–77] is now a very important tool in the routine analysis of plastics, both for establishing macromolecular composition and for identifying additives.

The IR portion of the electromagnetic spectrum can be divided into three regions:

Table 5. Pyrolysis of polymers [14]

Observations	Tests	Conclusions
Smell of burnt paper	vapors burn with characteristic flame – yellow with little smoke	cellulose
+ smell of acetic acid		cellulose acetate
+ smell of butyrates		cellulose butyrate
Smell of acetic acid	fumes; acidic to indicator paper	acetate polymers
Cyanide		polyacrylonitrile
Smell of burning vegetation	test vapor with damp universal indicator paper; alkaline vapors	polyamides (nylon), polyurethanes, protein-type resins
Smell of formaldehyde		urea – formaldehyde
+ amine-like smell		melamine – formaldehyde
+ phenolic smell		phenol – formaldehyde
Sweetish odor		methacrylate and related compounds
Acrid odor	allow vapors to impinge on a glass rod moistened with silver nitrate solution; cloudy precipitate	poly(vinyl chloride) and its copolymers; natural and synthetic chlorinated rubbers, chloroprene and chlorinated polyolefins
+ heavy black ash	allow vapors to impinge on a glass rod moistened with silver nitrate solution; cloudy precipitate	probably poly(vinylidene chloride)
Sulfurous	allow vapors to impinge on a glass rod moistened with barium chloride solution; cloudy precipitate	sulfur-containing rubbers, etc.

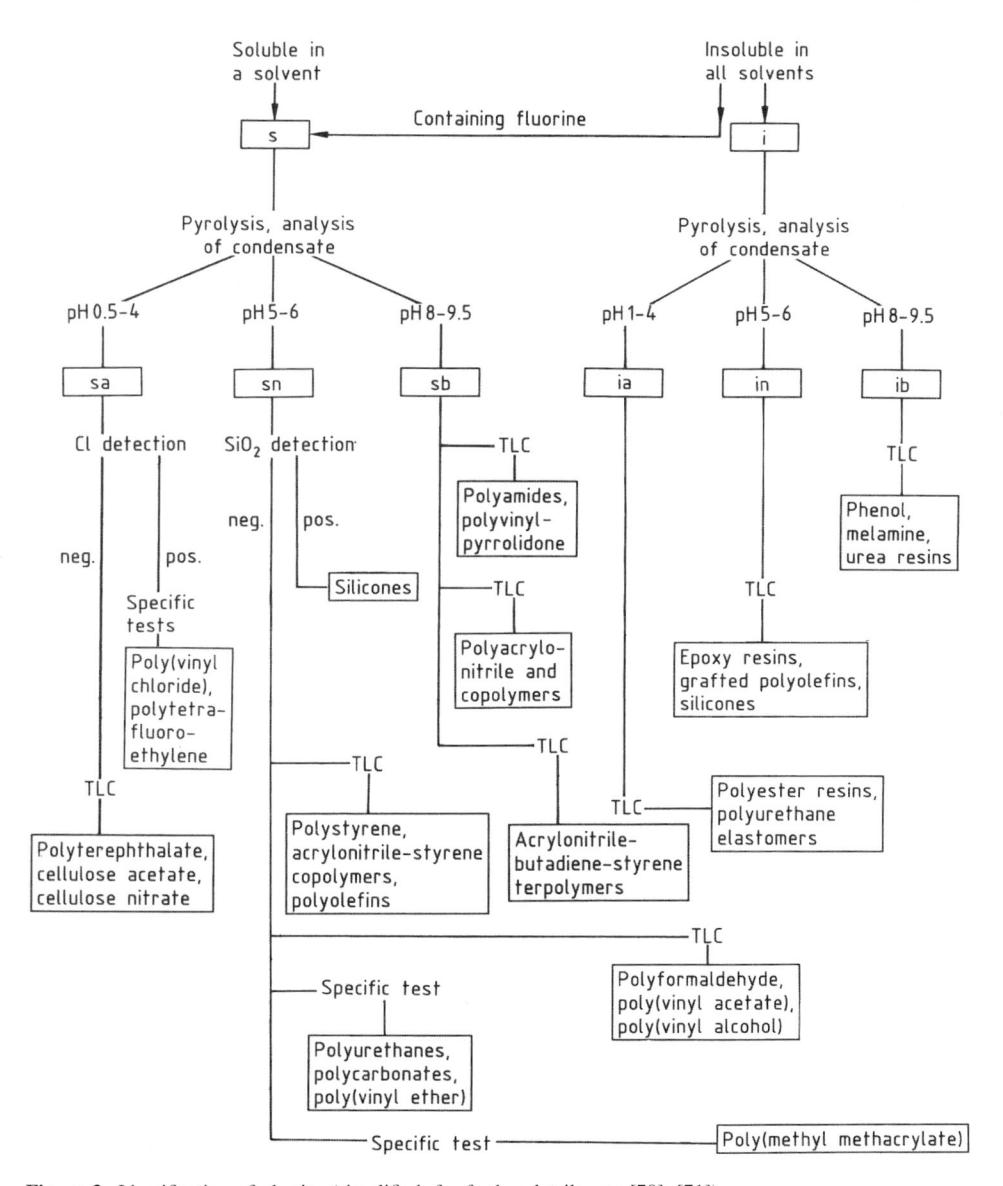

Figure 3. Identification of plastics (simplified; for further details, see [70], [71])
Solubility test with the solvents: water, tetrahydrofuran, dimethylformamide, xylene, formic acid, nitrobenzene
s = soluble; i = insoluble; a = acid; n = neutral; b = basic

Near IR (NIR)	0.76 – 2.5 μm, 13 200 – 4000 cm^{-1}	
Mid IR (MIR)	2.5 – 50 μm, 4000 – 200 cm^{-1}	
Far IR (FIR)	50 – 1000 μm, 200 – 10 cm^{-1}	

Plastics analysis can be performed in all three regions, but MIR is preferred for routine work. Each structural element in a polymer chain has associated with it several absorption lines at various positions in the spectrum, corresponding to the number and type of IR-active vibrational modes (Table 6). For a macromolecule with a degree of polymerization P and a structural unit that consists of N atoms, $3NP - 5 \cong 3PN$ vibrations can be anticipated. However, the number of vibrations actually observed is considerably

Table 6. Positions of characteristic IR bands [3]

Polymer	Position of band, μm
Homopolymers	
cis-1,4-Polybutadiene	3.31 – 6.06 – 7.12 – 7.64 – 8.07 – 9.82 – 12.8. . . .12.9 – 13.5 – 14.5
trans-1,4-Polybutadiene	7.48 – 8.09 – 9.49 – 10.3. . .10.4 (double band) – 12.93
1,2-Polybutadiene	5.47 – 6.10 – 7.32 – 8.10 – 10.07 – 11.0 – 12.3. . . .12.4
Copolymers	
Butadiene – styrene	6.12 – 6.7 – 10.8 – 10.35 – 10.99 – 13.2 – 14.3
Butadiene – methyl methacrylate	5.8 – 7.25 – 8.38 – 8.75 – 10.3 – 10.95 – 13.1
Butadiene – acrylonitrile	4.48 – 6.1 – 10.3 – 10.9

smaller, so that, contrary to what might be expected, polymers are not completely opaque to infrared radiation. By using group theory together with symmetry considerations, it can be shown that $3N - 4$ fundamental vibrations are expected for a linear polymer chain.

In practice, direct assignment of the large number of observable vibrations that still remain is not always feasible. The coupling of molecular vibrations is generally so weak that a spectrum of an unknown polymer can usually be identified most easily by comparing it with spec-tra of low molecular mass model compounds, or with spectra of polymers of known structure. In those cases in which inter- or intramolecular coupling is evident, however, the IR spectrum can provide important additional structural information (Figs. 4 and 5) and the sequence of structural units (Table 7) [3].

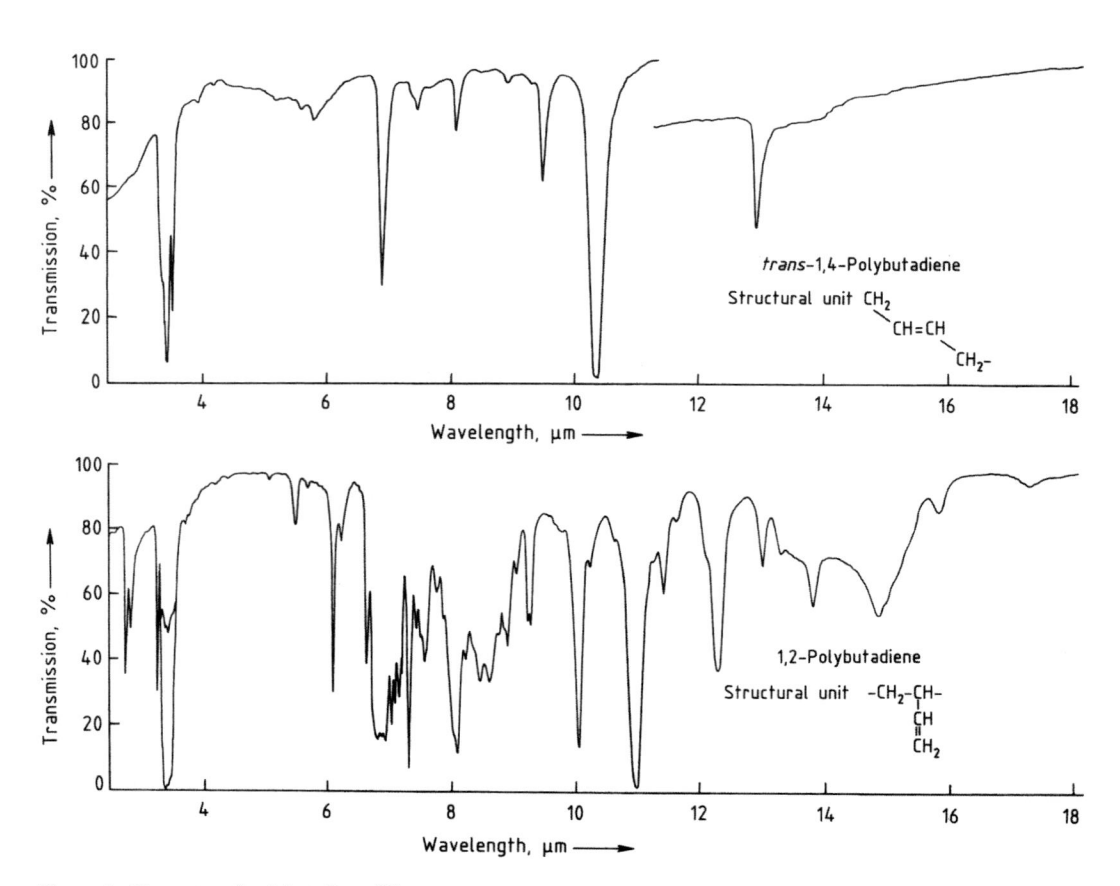

Figure 4. IR spectra of polybutadiene [3]

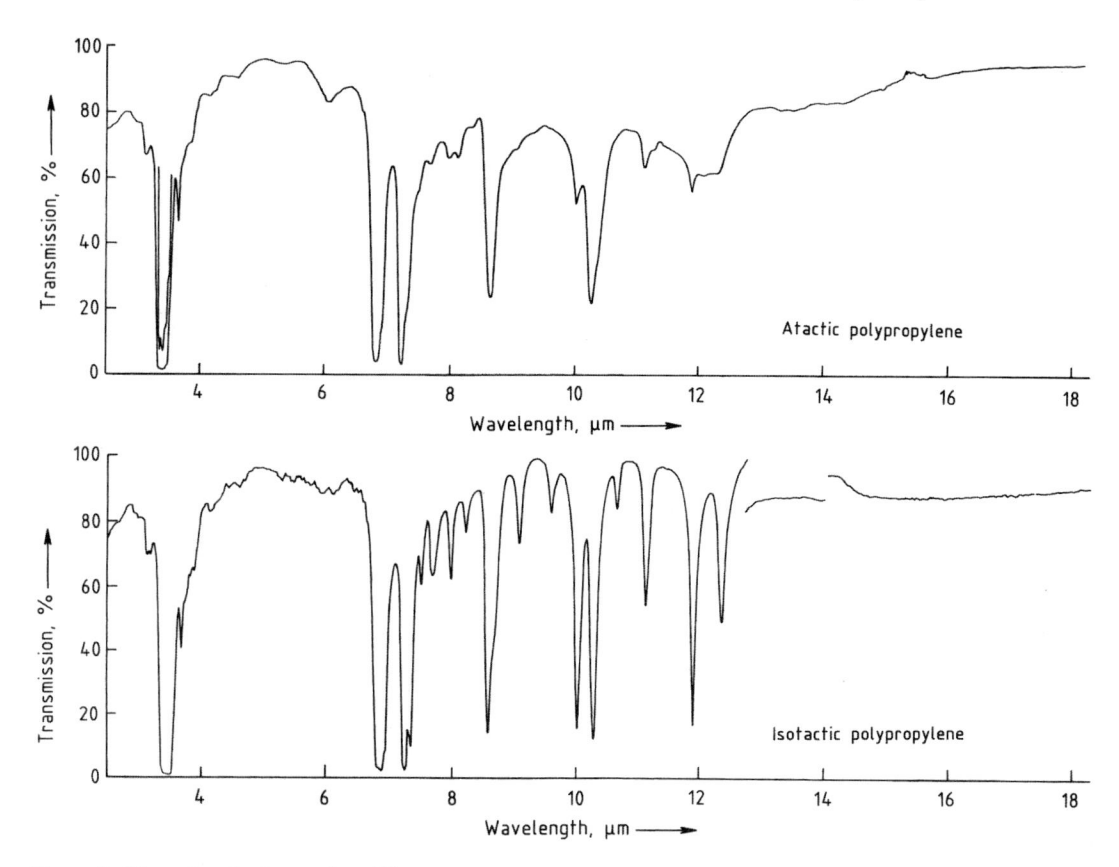

Figure 5. IR spectra of polypropylene [3]

Table 7. Chain structure of ethylene – propylene copolymers from IR spectra [3]

Wavelength, μm	Structure	Sequence of monomer units*
12.25	$-\underset{\underset{CH_3}{\vert}}{CH}-CH_2-\underset{\underset{CH_3}{\vert}}{CH}-CH_2-$	polypropylene units in ht polymerization
13.3	$-CH_2-\underset{\underset{CH_3}{\vert}}{CH}-CH_2-CH_2-\underset{\underset{CH_3}{\vert}}{CH}-CH_2-$	polypropylene units in tt polymerization equivalent to ethylene unit between two propylene units in hh polymerization
13.6	$-\underset{\underset{CH_3}{\vert}}{CH}-CH_2-CH_2-CH_2-\underset{\underset{CH_3}{\vert}}{CH}-CH_2-$	ethylene unit between two propylene units in ht polymerization
13.7	$-CH_2-\underset{\underset{CH_3}{\vert}}{CH}-CH_2-CH_2-CH_2-CH_2-\underset{\underset{CH_3}{\vert}}{CH}-CH_2-$	ethylene unit between two propylene units in tt polymerization two ethylene units between two propylene units in hh polymerization
13.83	$-\underset{\underset{CH_3}{\vert}}{CH}-CH_2-CH_2-CH_2-CH_2-CH_2-\underset{\underset{CH_3}{\vert}}{CH}-CH_2-$	two ethylene units between two propylene units in ht polymerization

*hh = head-to-head; tt = tail-to-tail; ht = head-to-tail.

Extensive spectral catalogs are now available containing tables of group frequencies [3]. Nevertheless, one should never rely too heavily on an analysis based solely on spectral comparisons. Confirmatory tests should be conducted as well. Elemental analysis (Section 4.1) and other physical methods such as NMR spectroscopy (Section 6.2) are especially recommended.

Quantitative Spectroscopy. IR spectra of polymers are preferably recorded with highly automated Fourier transform instruments (FT-IR).

Spectra of soluble plastics can be recorded in largely IR-transparent solvents such as carbon disulfide, carbon tetrachloride, and dichloromethane. However, the preferred method is to examine a polymer film, prepared either as a thin section or cast directly as a film. A film can also be prepared from a solution evaporated on a sodium chloride plate. A film thickness of ca. 10 μm is usually adequate for a reliable measurement.

An insoluble plastic can also be analyzed in the form of a fine powder. In this case the sample is first mixed with a powdered alkali metal halide (potassium bromide) and then compressed into a pellet.

The ATR (attenuated total reflection) method [77] allows IR analysis to be carried out on the surface of a plastic sample. The method is based on examining IR radiation that has been reflected from a plastic surface in contact with an IR-transparent substance of lower refractive index. Selective absorption results in a spectrum analogous to a transmission spectrum. Multiple total reflection techniques increase the sensitivity of the analysis [3].

Other Spectroscopic Methods. Comprehensive reviews are provided in [9] and [33]. Methods for detecting traces of metals in plastics include flame photometry [1] and atomic absorption spectroscopy, applicable in both the visible and ultraviolet regions of the spectrum.

Elemental analysis and the spectroscopic methods so far discussed provide results that correspond to averages over an entire plastic sample. In the case of plastics containing inorganic pigments (titanium white, cadmium yellow) or fillers (glass fibers, carbon black) information on the distribution of the additives is often required. Appropriate analytical methods here include X-ray fluorescence spectroscopy and especially the electron beam microprobe. The latter method employs an electron beam to probe systematically the nature of a plastic surface. Adjusting the analyzer so that it corresponds to X-ray lines of a particular element (e.g., silicon in the glass beads of a filler) gives a picture of the local concentration distribution of that particular element (Fig. 6). The resolution of an electron microprobe is ca. 1 μm. Elements of atomic number > 5 can be detected with a sensitivity up to 100 ppm (ca. 10^{-14} g/cm^3).

5. Molecular Mass Determination

Introduction. Molecular masses for industrially important polymers extend to values as high as 10^6.

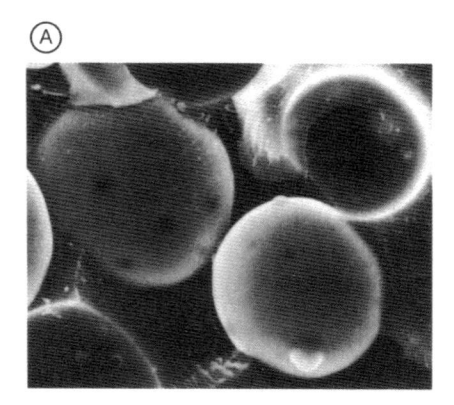

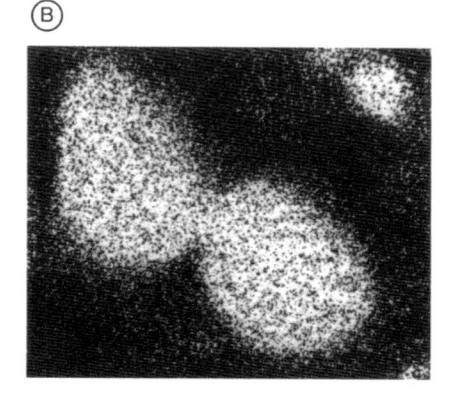

Figure 6. Scanning electron micrograph of the fracture plane of a phenol – formaldehyde resin filled with glass beads (A) and identification of silicon by X-ray fluorescence (B) Magnification × 300

Methods of molecular mass determination can be divided into two categories: absolute and relative [78–104]. Evaluation based on an absolute method involves in addition to experimental data (e.g., osmotic pressure, intensity of dispersed light) only such universal factors as the gas constant. The most important absolute methods are membrane osmosis, vapor pressure osmosis, combined sedimentation and diffusion, sedimentation equilibrium, and light scattering [78], [79], [83], [87].

Membrane osmosis is based on the equilibrium established by a membrane separating a polymer solution from pure solvent. Molecular mass determination by combined sedimentation and diffusion requires analysis of two types of material transport within a polymer solution: (1) sedimentation under the influence of a powerful centrifugal field, and (2) free diffusion. Alternatively, establishment of a stationary state in an analytical cell under the influence of a weak centrifugal field and consideration of the resulting equilibrium between sedimentation and diffusion also permits calculation of an absolute molecular mass. The method of light scattering requires the analysis of the intensity of Tyndall light as a function of the angle of observation.

Relative methods also involve polymer properties that are dependent upon molecular mass, but ones that do not lend themselves to direct conversion into unique molecular mass values. Essential to this approach is a calibration curve, prepared by using an absolute method. The most important relative methods utilize intrinsic viscosity (Staudinger index) or gel permeation chromatography (GPC). The experimental results in both cases depend not only on molecular mass but also on the conformation of polymer molecules in solution.

Since polymerization reactions are stochastic processes, most polymer samples are polymolecular. Standard separation methods do not allow the separation of macromolecules differing in degree of polymerization by only a single monomeric unit. For this reason, all absolute methods for molecular mass determination give characteristic mean values. Investigation of the molecular heterogeneity of a polymer must therefore include two discrete steps: (1) separation of the polymer components, and (2) determination of molecular mass for the separated components. Both ultracentrifugation and gel chromatography are effective for separating homopolymers of differing molecular mass, so these methods are widely used in the determination of molecular mass distribution.

Average Molecular Masses. Each molecule of solute in a solution makes an equal contribution to the osmotic pressure regardless of molecular mass. Osmotic analysis of a polymer thus gives the number-average molecular mass $\overline{M}_n$. If n_i is the number of moles of molecules with molecular mass M_i, the corresponding mass concentration will be $n_i = c_i/M_i$. It therefore follows that:

$$\overline{M}_n = \frac{\sum_i n_i M_i}{\sum_i n_i} = \frac{\sum_i c_i}{\sum_i c_i/M_i} \qquad (3)$$

Dispersion of light depends on the molecular mass, and therefore gives the weight-average molecular mass $\overline{M}_w$:

$$\overline{M}_w = \frac{\sum_i n_i M_i^2}{\sum_i n_i M_i} = \frac{\sum_i c_i M_i}{\sum_i c_i} \qquad (4)$$

In a molecularly uniform polymer $\overline{M}_w = \overline{M}_n$, whereas for a polymolecular sample $\overline{M}_w > \overline{M}_n$. The discrepancy becomes greater the broader the mass distribution. The molecular non-uniformity U

$$U = (\overline{M}_w/\overline{M}_n) - 1 \qquad (5)$$

is a measure of the molecular heterogeneity so long as the samples compared have similar molecular mass distributions. In some cases, samples with differing molecular mass distributions may actually have identical values of $\overline{M}_w$ and $\overline{M}_n$, a consequence of the fact that Equations (3) and (4) are based on summations.

Not all analytical methods lead to simple mean values such as $\overline{M}_w$ and $\overline{M}_n$. A more general definition of an average $\overline{M}$ is

$$\overline{M}_\beta = \frac{\sum_i c_i M_i^\beta}{\sum_i c_i M_i^{\beta-1}} \qquad (6)$$

For $\overline{M}_n$, $\beta = 0$, whereas $\beta = 1$ for $\overline{M}_w$. Depending upon the experimental method, the mean that is determined may have $\beta > 1$ or an exponent β with a fractional value. For example, the method that combines sedimentation velocity with diffusion leads to a mean somewhat smaller than $\overline{M}_w$, and the same is true for the intrinsic viscosity $[\eta]$.

5.1. Absolute Methods

5.1.1. Membrane Osmometry

The osmotic pressure Π of a solution is a function of the chemical potential μ_1 and partial molar volume V_1 of the solvent:

$$\Pi = -\Delta\mu_1/V_1 \tag{7}$$

The molecular mass is derived from the Vant'Hoff equation, which applies to ideal solutions

$$\Pi = \frac{RTc}{\overline{M}_n} \tag{8}$$

The assumption of ideality for a polymer solution is valid only in exceptional cases. Nonideality of a solution results in an osmotic pressure and, therefore, an apparent molecular mass that depends upon the concentration of the solute, a consequence of the differing spatial demands of solvent and polymer molecules. However, the additional enthalpy and entropy terms that determine μ_1 may cancel out at the so-called Θ temperature, resulting in a pseudoideal solution. The Θ temperature is of major significance in the determination of molecular masses and molecular mass distributions of polymers. Such Θ systems are in fact known for a large number of polymers [53]. At the Θ temperature, plotting Π/c versus c gives a line parallel to the c axis. The advantage of acquiring data at the Θ temperature is that this permits measurement of the molar mass distribution at a single concentration.

An example is given in Figure 7 for the case of polystyrene dissolved in cyclohexane. The molecular-mass-independent Θ temperature in this case is 34 °C. Increasing the temperature leads to an additional term for the osmotic pressure, reflecting the effect of the chemical potential. Lowering the temperature makes this term negative. The nonideal concentration dependence observed at temperatures other than the Θ temperature can be represented by a power series in c:

$$\frac{\Pi}{c} = \frac{RT}{\overline{M}_n} + A_2 c + \ldots \tag{9}$$

As shown in Figure 7, the molecular mass $\overline{M}_n$ can be determined by extrapolation to $c \to 0$.

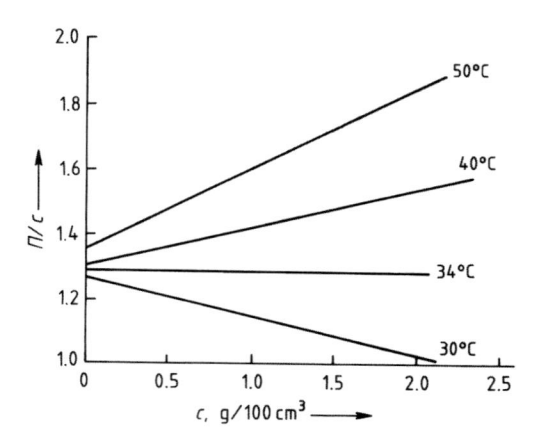

Figure 7. The dependence of osmotic pressure on temperature [21]

Experimental Technique. In an osmometer, such as that illustrated schematically in Figure 8, a polymer solution and a pure solvent are separated by a semipermeable membrane, ideally one that permits passage only of the solvent. Each of the chambers is surmounted by a capillary. The difference in chemical potential between the two chambers creates an osmotic pressure Π, which can be quantified in terms of a difference in height Δh of the liquid in the two capillaries. Equilibration of the osmotic pressure may require from several hours to several days. Automatic osmometers eliminate the need for waiting until equilibrium has been achieved.

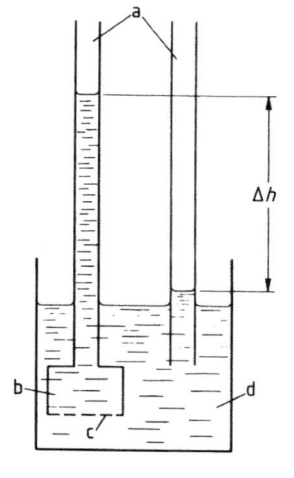

Figure 8. Schematic of a membrane osmometer [21]
a) Capillaries; b) Polymer solution; c) Membrane; d) Pure solvent

Instead, a servo-activated device applies an external pressure continuously to the solution chamber, and the magnitude of the pressure required to maintain equivalence is taken as a quantitative measure of the osmotic pressure Π. Another type of osmometer utilizes chambers with fixed volumes, permitting direct evaluation of Π by means of anelectronic pressure sensor.

Since the osmotic pressure decreases with increasing molecular mass, the scope of membrane osmometry is limited by the minimum detectable pressure, corresponding to a molecular mass of at most 10^6. The reliability of an osmotic determination is largely a function of the permeability of the membrane. If molecules of low molecular mass (e.g., oligomers, foreign solvent introduced during precipitation) can enter the solvent chamber then the apparent value for the molecular mass of the polymer will lie above $\overline{M}_n$. The extent of such error is a function of membrane permeability and the proportion of foreign molecules.

Osmosis is not limited to molecular mass determination on homopolymers. It is also the method of choice for copolymers regardless of the sequence of structural units or the extent of chemical heterogeneity, because osmotic pressure is a function only of the number of molecules present, irrespective of their chemical structures.

5.1.2. Lowering of Vapor Pressure

Vapor Pressure Osmometry. According to Raoult's law, the ratio of the vapor pressure of a pure solvent p_1^0 to that of a solution p_1 is equal to the mole fraction of the solvent x_1:

$$p_1/p_1^0 = x_1 = 1 - x_2 \tag{10}$$

x_2 mole fraction of solute and the relative decrease in vapor pressure is:

$$\Delta p_1/p_1^0 = x_2 = \frac{n_2}{n_1} \tag{11}$$

n_1 number of moles of solvent
n_2 number of moles of solute

In a vapor-pressure osmometer, solvent distills, a consequence of the difference in vapor pressure exerted by solvent and solution, onto droplets of solution. The heat of condensation is measured, which for an ideal system is proportional to the number of moles of dissolved polymer (i.e., with a constant sample size, inversely proportional to the molecular mass). The range of applicability of vapor-pressure osmometry is thus a function of the precision and sensitivity with which temperature differences can be measured. Determinations of this type are possible up to $\overline{M}_n = 10^4$. The results must be extrapolated to infinite dilution, as in the case of membrane osmometry, and for the same reasons.

Freezing Point Depression, Boiling Point Elevation. The relative decrease in vapor pressure of a polymer solution can also be determined by measuring the depression of freezing point ΔT_f or elevation of boiling point ΔT_b. The corresponding cryoscopic and ebullioscopic methods are based on the relationship that exists between vapor pressure and temperature, as given by the Clausius – Clapeyron equation:

$$dp/dT = \Delta H/\Delta V \cdot T \tag{12}$$

ΔH heat of vaporization
ΔV volume change on evaporation

The ebullioscopic method is the more important in practice. Molecular masses are derived from the equation:

$$\frac{\Delta T_b}{c} = \frac{RT^2}{\overline{M}_n} \cdot \frac{\overline{M}_1}{\Delta H_s} \tag{13}$$

$\overline{V}_1$ specific volume of solvent
H_s specific heat of vaporization of solvent

The term $RT^2 \times \overline{V}_1/\Delta H_s$ is the ebullioscopic constant K for a particular solvent.

5.1.3. Ultracentrifugation

Combined Sedimentation and Diffusion. Sedimentation rates are measured by subjecting a polymer solution to a high centrifugal acceleration in an ultracentrifuge, depicted schematically in Figure 9. The centrifugal acceleration is expressed as $\omega^2 r$, where ω is angular velocity and r is the distance from the axis of rotation [83–85]. Centrifugal force causes dissolved molecules to be sedimented at the bottom of the sample cell as a function of molecular mass and conformation in solution. Directed material transport is opposed by forces of friction. With F as the molar coefficient of friction, $\overline{V}_2$ the partial specific volume of the solute, and ϱ_1 the

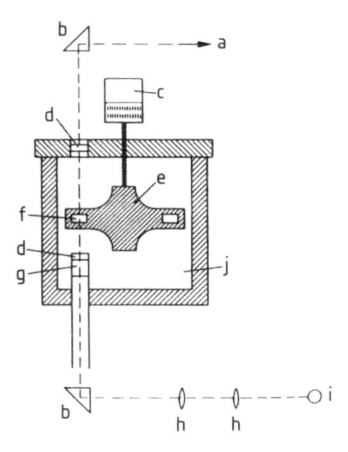

Figure 9. Schematic of an ultracentrifuge [21] a) Camera; b) Prism; c) Motor; d) Window; e) Rotor; f) Sample cell; g) Filter; h) Condenser; i) Light source; j) Vacuum chamber

density of the solvent, the following equation of motion is applicable:

$$M(1 - \tilde{V}_2 \varrho_1) = F \frac{t}{\omega^2 r} = F \cdot s \tag{14}$$

The sedimentation constant s thus represents a rate of sedimentation through a uniform field. At infinite dilution, Einstein's law is applicable, providing a relationship between the coeffficient of friction F and the diffusion constant D:

$$F = RT/D \tag{15}$$

Equations (14) and (15) can be combined to give the Svedberg equation:

$$M = \frac{s_0}{D_0} \cdot \frac{RT}{(1 - \tilde{V}_2 \varrho_1)} \tag{16}$$

By combining the results obtained in separate sedimentation and diffusion experiments it is possible to eliminate the effect of the coefficient of friction on both transport processes. In this way, determination of molecular mass via the Svedberg equation becomes an absolute method. The procedure requires optimization of the rotational speed of the centrifuge as well as the choice of a suitable solvent [53]. For sedimentation measurements, the density of the polymer must be greater than that of the solvent. The transport velocity for the polymer in a centrifuge cell is measured by detecting either the UV spectrum or the refractive index of the polymer.

As in all experiments involving molecular mass determination, concentration dependencies must also be considered. If measurements are not made at the Θ temperature, then a series of s and D at various concentrations must be measured. The corresponding concentration dependencies can be expressed as:

$$\frac{1}{s} = \frac{1}{s_0}(1 + k_s c) \tag{17}$$

$$D = D_0(1 + k_D c) \tag{18}$$

The resulting molecular mass is a mean that arises from averaging the two measured variables; its value is close to $\overline{M}_w$. The method is applicable over the molecular mass range $10^4 - 10^7$ [83].

Sedimentation Equilibrium. If the rotational velocity of a centrifuge is so adjusted that a mass-transport equilibrium is established between sedimentation and diffusion, then during the time interval dt the same number of molecules will migrate through a cross-section of the cell in the direction of the cell floor (due to sedimentation) as toward the axis of rotation (due to diffusion). This equilibrium follows the relationship:

$$\frac{\partial c}{c} = \frac{M(1 - \tilde{V}_2 \varrho_1)\omega^2 r \, dr}{RT} \tag{19}$$

Integration between r_1 and r_2 leads to an equation for determining the molecular mass:

$$M = \frac{2RT \ln(c_1/c_2)}{(1 - \tilde{V}_2 \varrho_1)\omega^2(r_2^2 - r_1^2)} \tag{20}$$

Equation 20 is applicable only to an ideal system and requires extrapolation to $c \to 0$ for a real system.

The nature of the resulting mean is dependent upon the method used for interpreting the experiment. In the simplest case the z-mean, corresponding to $\beta = 2$ in Equation (6) is obtained. The range of applicability is extremely large, extending from monomers to molecular masses of 10^6. Analyses involving high molecular mass require long periods of time for the establishment of equilibrium. The time factor can be reduced by utilizing cells with low fill heights together with a multicell rotor. Solvents are selected on the same basis as in sedimentation transport.

5.1.4. Light Scattering

Introduction. The electric field associated with a light wave causes the electron shells of a polymer molecule to vibrate with a frequency corresponding to that of the monochromatic excitation beam. Excited molecules then emit scattered light in all directions at a wavelength λ_0 identical to that of the primary irradiation (the Tyndall effect) [86–88].

The intensity I of a light beam passing through a nonabsorbing polymer solution is reduced relative to the initial intensity I_0 by an amount equivalent to the light that is scattered. The overall scattering intensity attributable to N molecules within a unit volume (1 cm^3) is expressed by the turbidity number τ:

$$\tau = \frac{I}{I_0} = \frac{128}{3}\pi^5 N \lambda_0^4 \alpha^2 \qquad (21)$$

The polarizability α of a substance is a function of its refractive index n:

$$\alpha = \frac{1}{N} \frac{n^2 - 1}{4\pi} \qquad (22)$$

Combining Equations (21) and (22) leads to:

$$\tau = \frac{I}{I_0} \frac{32}{3}\pi^3 \frac{(n^2 - 1)^2}{N \lambda_0^4} \qquad (23)$$

The above equations apply strictly only to gases but can also be used for dilute solutions of polymers. For an ideal solution, the scattering intensity is a function of the difference between the refractive index of the solvent n_1 and that of the dissolved polymer n, introduced as the refractive index increment $\partial n/\partial c$ [53]. Replacing the number of macromolecules by the mass concentration c and the molecular mass $\overline{M}_w$ leads to the expression:

$$\frac{\tau}{c} = \overline{M}_w \frac{32}{3}\pi^3 \frac{n_1^2}{N_A \lambda_0^4}(\partial n/\partial c) \qquad (24)$$

N_A is the Avogadro number from which the weight-average molecular mass $\overline{M}_w$ of an unknown polymer can be calculated. Equation (24) contains no term for scattering due to the solvent, so this must be subtracted from the overall scattering τ of the polymer solution. The value of τ is determined in practice by measuring the intensity of laterally scattered radiation as a function of the angle Θ. All measurements and their interpretation are influenced by the dimensions and shapes of the dissolved macromolecules relative to λ_0.

Macromolecules Smaller than $\lambda/20$**.** If the solute molecules responsible for scattering are small relative to the wavelength of the source they can be treated as point sources of radiation. This is the case for flexible chain molecules up to the dimension $\lambda/20$. If the irradiating light is polarized in a plane perpendicular to the plane of observation, the resulting intensity I_θ of light scattered through an angle Θ is independent of the angle of observation Θ. In this case the relative scattering intensity R_θ per unit volume and distance r for any angle of observation Θ is a function of the turbidity number τ according to:

$$R_\Theta = \frac{r^2 I_\Theta}{I_0} = \frac{3}{8\pi}\tau \qquad (25)$$

where I_0 is the intensity of the incident light. Introducing the optical constant K,

$$K = 4\pi^2 \frac{n_1^2}{N_A \lambda_0^4}(\partial n/\partial c) \qquad (26)$$

which is characteristic of a particular system and wavelength leads to an equation applicable to a dilute ideal solution. This defines the molecular mass as a function of the amount of light scattered through an angle Θ, an expression which is also appropriate for pseudoideal systems (see Section 5.2.1):

$$K \frac{c}{R_\Theta} = \frac{1}{\overline{M}_w} \qquad (27)$$

Analogous to the situation with osmotic pressure, the concentration dependence of the reduced scattering intensity for real polymer solutions is taken into account by the Debye relationship:

$$K \frac{c}{R_\Theta} = \frac{1}{\overline{M}_w} + 2A_2 c + \ldots \qquad (28)$$

where A_2 is the second virial coefficient of the osmotic pressure. In this case $\overline{M}_w$ is established by measurement of a concentration series and then extrapolating to infinite dilution.

Macromolecules larger than $\lambda/20$ can no longer be regarded as point oscillators because of interference with the laterally scattered light. Path differences and, therefore, the amount of interference quenching of the scattered radiation increase with increasing angle of observation and with increasing size of the macromolecules. However, at an angle of 0° – where direct measurement is of course impractical – no interference occurs, so Equation (27) remains valid even for large molecules. The angular dependence of scattering intensity for large macromolecules is a function not only of molecular size but also of shape and molecular heterogeneity. These factors are accounted for in the equation for $\overline{M}_w$ by the scattering function P_Θ:

$$K\frac{c}{R_\Theta} = \frac{1}{P_\Theta \overline{M}_w} + \frac{2A_2 c}{P_\Theta} + \dots \quad (29)$$

For statistically entangled, uniform macromolecules with a mean distance of $\sqrt{h^2}$, the following expression is applicable to P_Θ:

$$P_\Theta^{-1} = 1 + (8\pi^2/9\lambda_0^2) \cdot h^{-2} \cdot \sin^2(\Theta/2) \quad (30)$$

Scattering functions have been computed for a wide range of molecular shapes [86], [87]. Their utility extends beyond providing a correction factor for use in determining $\overline{M}_w$. Quantitative evaluation of the angular dependence of the scattering function also provides additional information on macroscopic conformation (size and shape) of the dissolved macromolecules. It follows from Equations (29) and (30) that:

$$K\frac{c}{R_\Theta} = \frac{1}{\overline{M}_w} + D \cdot \sin^2(\Theta/2) + 2A_2 c + \dots \quad (31)$$

where

$$D = \frac{1}{\overline{M}_w} \cdot \frac{8\pi^2}{9\lambda_0^2} \cdot h^{-2} \quad (32)$$

The determination of the molecular mass of a large macromolecule by means of Equation (31) involves extrapolation to $\Theta \to 0$ and $c \to 0$ with the aid of a Zimm diagram, as shown by the example in Figure 10. Here the expression Kc/R_Θ is plotted simultaneously against the concentration c and the angular function $\sin^2(\Theta/2)$. This facilitates both of the required extrapolations, which are linear and intersect the ordinate at $1/\overline{M}_w$. In a similar way one can also evaluate the concentration dependence of the virial coefficient A_2 and the angular dependence of the molecular dimension $\sqrt{h^2}$. The method of light scattering is applicable up to a molecular mass of 10^8.

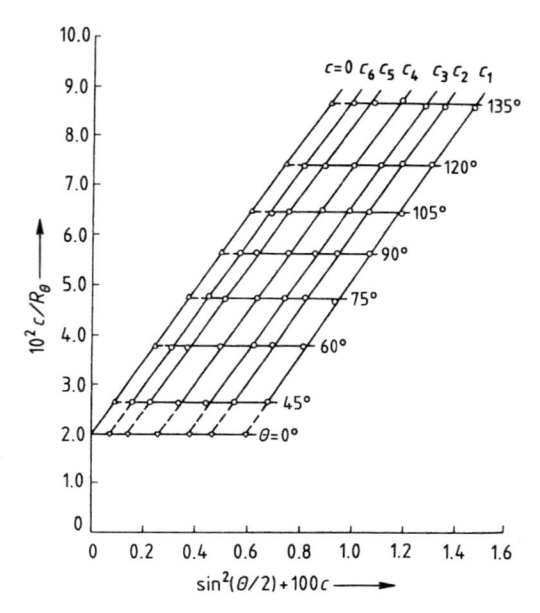

Figure 10. Zimm diagram of polytetrahydrofuran ($\overline{M}_w = 2.26 \times 10^6$) in 2-propanol ($T = 317$ K) [87]

Molecular Masses of Copolymers. Equation (28) for $\overline{M}_w$ assumes a macromolecular system with a uniform index of refraction; in other words, it is strictly valid only for homopolymers. A copolymer, on the other hand, consists of structural units that differ chemically, and thus would be expected to differ with respect to refractive index as well. Each component makes its own characteristic contribution to the overall scattering through the term $(\partial n/\partial c)^2$.

The refractive index increment for a copolymer is a function of the mass fractions m_A and m_B of the individual components and the refractive index increments of the corresponding homopolymers:

$$(\partial n/\partial c)_{AB} = m_A (\partial n/\partial c)_A + m_B (\partial n/\partial c)_B \quad (33)$$

The measured scattering intensity for a copolymer therefore depends not only on molecular mass but also on chemical composition.

The molecular mass of a copolymer that is heterogeneous both chemically and in degree of polymerization can be determined by conducting light-scattering experiments in at least three solvents [88]. Considerable experimental effort is required, but it is possible to establish in this way a mean mass $\overline{M}_w$ and thereby the non-uniformity U.

The equations used in studying homopolymers can also be applied to chemically heterogeneous copolymers provided structural units in the latter are isorefractive. This is the case to a good approximation, for example, with ethylene – propylene copolymers.

Experimental Technique. Figure 11 illustrates the design of a light-scattering apparatus. The measurement requires total absence of dust from the solutions and the cuvettes. Solutions can be purified by filtration or ultracentrifugation. The refractive index increment is determined by means of a differential refractometer. In order to achieve a high scattering intensity it is important that the refractive indices of solvent and solute differ as much as possible.

5.2. Relative Methods

Intrinsic Viscosity. Even a low concentration of macromolecules leads to a significant increase in the viscosity of a solvent. The increase observed for a given amount of a particular polymer depends not only upon molecular mass but also upon the dimensions of the molecules in solution. Macroconformation for a statistically entangled macromolecule is a function of both solvent and temperature. Use of intrinsic viscosity for characterizing a molecular mass requires that the experiment be conducted at constant temperature [53], [89–91]. The intrinsic viscosity [η] (Staudinger index) is defined as:

$$[\eta] = \lim_{\substack{c \to 0 \\ G \to 0}} \left[\frac{\eta_{sp}}{c} \right] \quad (34)$$

This represents a limiting value for the reduced specific viscosity η_{sp} for a concentration $c \to 0$ and a shear rate $G \to 0$. Extrapolation to $c \to 0$ can be accomplished graphically by plotting η_{sp}/c against η_{sp} or c. In most cases a linear relationship results. The reduced viscosity is determined by measuring the flow times t for the solution and t_1 for the solvent, leading to:

$$\eta_{sp} = \frac{t\varrho - t_1\varrho_1}{t_1\varrho_1} \quad (35)$$

ϱ, ϱ_1 Density of solution and solvent, respectively

At low concentration $\varrho_1 = \varrho$, so this can be simplified to:

$$\eta_{sp} = \frac{t - t_1}{t_1} \quad (36)$$

Extrapolation to $G \to 0$ can be dispensed with for the low molecular mass range and by employing an Ubbelohde viscometer with a low shear rate. Large macromolecules tend to become partially oriented upon passage through a capillary as a consequence of a reduction in frictional resistance by solvent shearing. This structural viscosity is taken into account by measuring the viscosity of a solution as a function of the applied pressure p. An Ostwald viscometer is convenient for the purpose. Extrapolation to $G \to 0$ can be accomplished graphically by plotting η_{sp} against p. Linear extrapolation is possible here as well. Automatic viscometers have also been developed. These devices take into account the role played by shear rate as well as measuring the flow time.

The relationship between intrinsic viscosity [η] and molecular mass for a linear, statistically entangled thread-like chain molecule is given by the Kuhn expression:

$$[\eta] = KM^\alpha \quad (37)$$

A log – log plot of [η] versus $\overline{M}_{w,n}$ gives a straight line with slope α. The value of the exponent depends upon the flexibility of the entanglement. It approaches 1 in a thermodynamically favorable solvent. The minimum value of 0.5 is obtained with Θ systems (cf. Section 5.2.1). Representative values of α lie in the range 0.6 – 0.9. The intrinsic viscosity [η] leads to a mean of the form:

$$M_\eta = \left[\frac{\sum_i c_i M_i^\alpha}{\sum_i c_i} \right]^{1/\alpha} \quad (38)$$

This is identical to $\overline{M}_w$ when $\alpha = 1$, but it becomes increasingly smaller as α decreases.

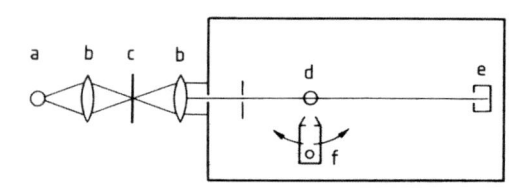

Figure 11. Schematic of a light-scattering apparatus
a) Light source; b) Collimator; c) Filter and polarizer; d) Sample cell; e) Light trap; f) Photomultiplier

For this reason it is advantageous to calibrate the intrinsic viscosity [η] by means of an absolute method such as light scattering. Figure 12 illustrates [η] – M calibration curves for various temperatures and solvents in the case of polystyrene. For additional systems see [53].

Relative molecular mass determination on the basis of viscometry is of major technical importance. Absolute methods are preferable for reasons of time and expense only with polymers for which an expression corresponding to Equation (35) has yet to be established. Relative methods provide reliable results whenever there is reasonable agreement between the molecular mass distributions of the reference and the sample.

5.3. Determination of Molecular Mass Distribution

Introduction. The nonuniformity U is incapable of providing an unambiguous characterization of polymolecularity. Samples with identical values of $\overline{M}_w$ and $\overline{M}_n$ may differ in their molecular mass distributions. The polymolecularity of a polymer sample can be established fully only by fractionation. In a batch separation procedure the total mass of polymer molecules in the system is represented in terms of an integral distribution over a discrete set of molecular masses between M_0 and M_i:

$$J_w(M) = \sum_{M_0}^{M_i} w_i \Delta M \quad (39)$$

Equation (39) describes a step function which progresses in steps of the monomer mass dM. A corresponding experimental measurement assumes prior molecular mass evaluation of the various fractions with molecular mass M_i. However, changes in physical characteristics for similar macromolecules differing only in size are so small that isolation of a fraction containing only molecules of a single degree of polymerization is essentially impossible. In a continuous separation method the step function of Equation (39) is replaced by a continuous integral distribution function:

$$J_w(M) = \int_{M_0}^{M_i} w(M) dM \quad (40)$$

with the normalization:

$$J_w(M) = \int_0^{\infty} w(M) dM = 1 \quad (41)$$

The first derivative of this expression is the differential molecular mass distribution, which provides a mass fraction for macromolecules in the increment M+dM:

$$w(M) = \frac{dJ_w(M)}{dM} \quad (42)$$

Experimental determination of a molecular mass distribution thus always entails use of a method of separation combined with molecular mass analysis of the individual fractions.

Polymers can be separated by taking advantage of macromolecular properties that depend exclusively on molecular mass, including solubility (precipitation fractionation) and sedimentation rate. Chromatographic procedures have also assumed considerable practical importance.

Precipitation Fractionation. The solubility of a polymer depends not only on the nature of the structural units present, but also on molecular mass. This characteristic of polymer solutions serves as the basis for a simple and effective approach to establishing a molecular mass distribution. The process begins with a ca. 1 % solution of a polymer, to which is slowly added dropwise with stirring a precipitating agent, usually methanol. Turbidity is observed after the addition of a certain amount of the precipitating agent. The mixture is then warmed slightly to restore complete dissolution. Subsequent lowering of the temperature again leads to precipitation. After some period of time the

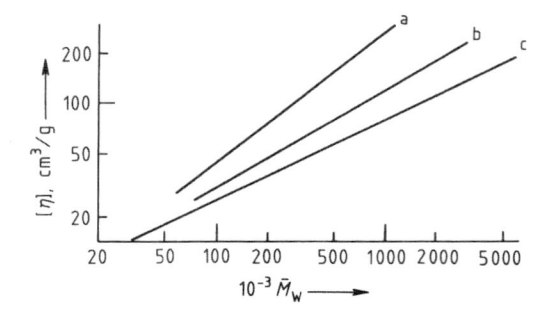

Figure 12. Logarithmic plot of viscosity versus molecular mass of polystyrene a) In benzene (T = 290 K); b) In methyl ethyl ketone (T = 290 K); c) In cyclohexane (T = 305 K)

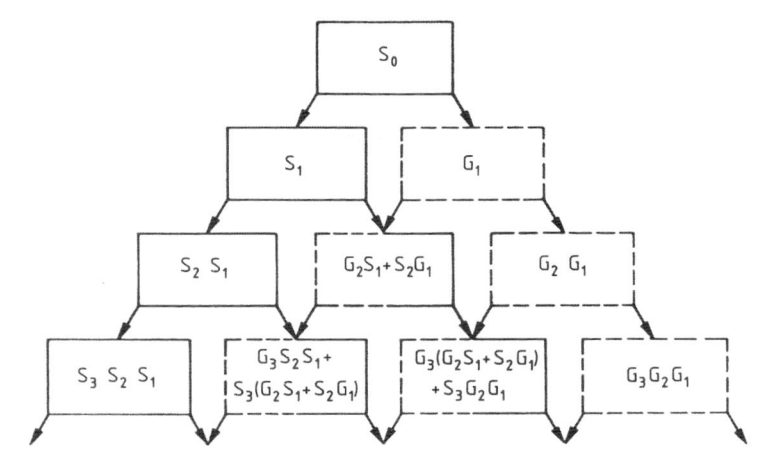

Figure 13. Schematic fractionation of polymers by dissolution and precipitation [9]
S = Solution; G = Gel

system separates into a polymer-rich gel phase (G_1) and a polymer-poor sol phase (S_1). In most cases the polymer is the more dense component, so the gel phase collects at the bottom of the precipitating vessel. Evaporation of this phase gives the polymer fraction with the highest molecular mass. The sol S_1 is then treated with additional precipitating agent leading to a second stage of separation, etc. This procedure allows the isolation of up to 10 fractions from a single polymer sample. Fractions do not consist exclusively of molecules with uniform molecular masses, however, but instead contain components with both higher and lower masses as well. Even repetition of the process will not alter this situation [92], [93].

In order to improve the fractionation the gel phase G_1 can be treated again with precipitating agent, producing another sol phase that can be combined with S_1 subfractionation procedure. The triangular fractionation method illustrated schematically in Figure 13 represents another effective approach to fractionation.

Once a set of fractions has been isolated the corresponding masses must be determined, along with mass fractions relative to the total sample. Molecular masses are determined by one of the methods described in Sections 5.2 and 5.3. The results permit one to construct a molecular mass distribution consistent with Equation (39) (Fig. 14).

Precipitation fractionation is a very simple procedure. Its chief limitation lies in the small

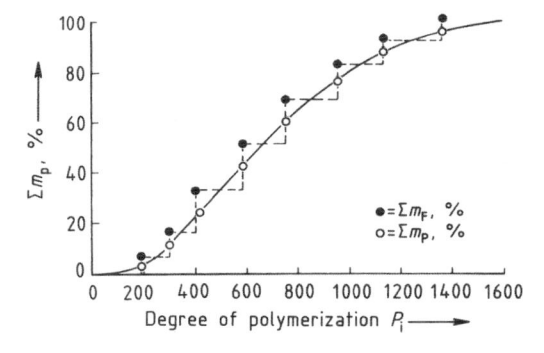

Analysis of typical fractionation data

Fraction	Degree of polymerization P_i	wt% of fraction m_F	Σm_F, %	Σm_p, %
VIII	190	6.4	6.4	3.2
VII	300	9.0	15.4	10.9
VI	420	17.2	32.6	24.0
V	585	17.6	50.2	41.4
IV	755	18.5	68.7	59.4
III	955	13.5	82.2	75.5
II	1140	9.7	92.0	87.0
I	1370	8.1	100.0	95.0

Figure 14. Molecular mass distribution by dissolution fractionation [21]
m_F = mass fraction of a fraction; m_p = mass fraction with degree of polymerization P_i

number of fractions generated, which might, for example, fail to reveal a bimodal character of the molecular mass distribution. The time required for separating the gel phases can also be very great, especially since the sol phase increases in volume at each stage in the process, requiring the use of large amounts of solvent. Nevertheless, precipitation has the advantage compared to preparative chromatography that it represents an economical source of gram-quantity fractions, which can in turn serve as starting points for analyzing a polymer by other methods (e.g., configurational analysis with NMR; see Section 6.2).

Gel permeation chromatography (GPC) is a form of liquid column chromatography in which carrier solvent is passed at a constant rate through a column consisting of cross-linked polymer beads previously allowed to swell in the same solvent. The total volume V_t of the column includes the volume occupied by both mobile (V_0) and stationary (V_x) phases. Only a portion of V_x is actually utilized in the separation process: the pore volume V_p [94–103].

Figure 15 shows a simplified model of the separation process. Separation occurs on the basis of the hydrodynamic volume of a given macromolecule, which is in turn a function of molecular mass. It is this volume that governs the extent to which the macromolecule can permeate into the pore structure of a suitably dimensioned microporous gel.

The sample that is to be separated is introduced onto the top of the column as a solution in the mobile phase. Polymer molecules then distribute themselves between the mobile phase V_0 and the pore volume of the stationary phase V_p on the basis of a diffusion-driven equilibrium defined by their hydrodynamic volumes and the pore diameter of the gel. As solvent flows through the column, large molecules are unable to penetrate into the gel matrix due to their diameters, so these are eluted first within the elution volume V_0. By contrast, molecules with diameters smaller than the pore openings diffuse into the interiors of the pores. The smaller the macromolecule the more deeply it is able to penetrate into the stationary phase, and the more its progress is retarded relative to solvent flow outside the pores. In the absence of special interactions between the polymer and the surface of the stationary phase, the observed order of elution and the corresponding elution volume V_e will be a function only of molecular size. Small macromolecules capable of passing freely between V_p and V_0 are eluted last. The difference between the elution volumes for totally excluded molecules and the smallest macromolecules present corresponds to V_p.

The *gel chromatogram* is generated by measuring the polymer concentration in the eluate as a function of elution volume. The polymer concentration can be measured by monitoring the refractive index and/or ultraviolet absorption of the eluate. Larger columns can be used for

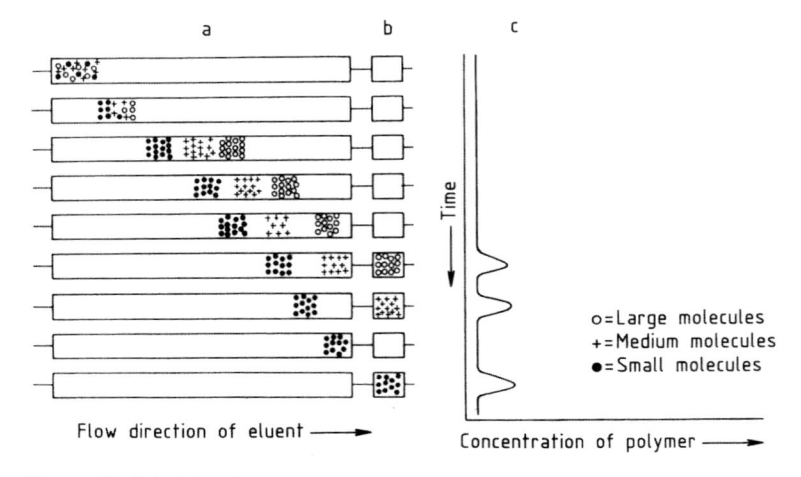

Figure 15. Principle of polymer fractionation by size exclusion chromatography (SEC) [31] a) Column; b) Detector; c) Detector signal

preparative GPC, providing an effective alternative to precipitation fractionation.

Even for a completely homogeneous fraction, the quality of the column packing and effects of axial diffusion during the separation process cause peak broadening. An elution curve can be calibrated by chromatography of a set of uniform polymers with known masses determined by an absolute method, as illustrated in Figure 16. The linear portion of the elution curve is described by the function:

$$\log M = \log M_a - AV_e \qquad (43)$$

where M_a is the molecular mass above which separation is impossible and A is a constant specific to a particular system and apparatus. This method is applicable within the molecular mass range $10^2 - 10^6$.

The relationship shown in Figure 16 between elution volume V_e and molecular mass M is strictly valid only for a particular homologous series of polymers, and it must be established with the aid of calibration samples. If the intrinsic viscosity $[\eta]$ is used to take into account hydrodynamic volumes of polymers that differ chemically by plotting $[\eta]M$ vs. V_e, the result is a universal elution curve applicable to multiple polymer systems. Within its range of linearity GPC can be used as a relative method for the determination of molecular mass [98].

Evaluation. Gel chromatograms often consist of several broad, unsymmetrical, and overlapping peaks, as illustrated by the example of polypropylene adipate in Figure 17. If the relationship between detector signal and polymer concentration is known, mass fractions $w(M)$ can be computed from the corresponding surface areas in the chromatogram of a series of oligomers. Equation (39) is then used to construct a step distribution function for the various molecular masses. It is not possible in the high molecular mass range to isolate fractions with uniform molecular mass. The typical gel chromatogram must instead be regarded as a superposition of uniform fractions characterized by differing molecular masses. The problem in interpretation is therefore to distinguish the contribution to peak broadening that can be attributed directly to molecular mass distribution in the sample from peak-broadening effects inherent in the system.

The chromatogram of a single fraction can be described as a Gaussian curve of the form:

$$f(V) = A(h/\sqrt{\pi})\exp[-h^2(V - V_{\max})^2] \qquad (44)$$

where $V_{\max}$ is the elution volume at the maximum of the curve, h represents the breadth of the Gaussian curve, and A is the area under the chromatogram. For a system of i fractions each with area A_i:

$$f(V) = \sum_i A_i(h_i/\sqrt{\pi})\exp[-h_i^2(V - V_{\max,i})^2] \qquad (45)$$

If the number of fractions is very large, or if the chromatogram cannot be resolved into uniform fractions in the high molecular mass range, the summation A_i can be replaced by the continuous function $w(Y)$. For h=constant it follows that:

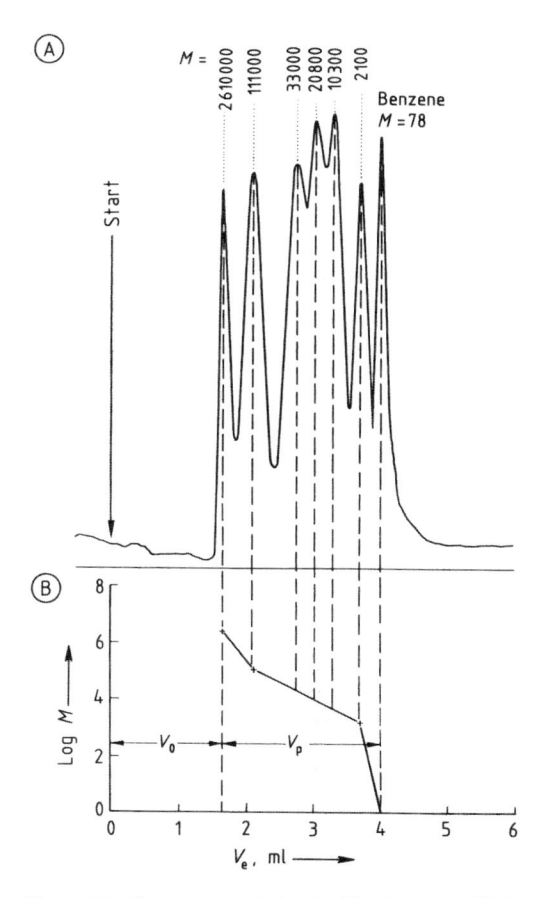

Figure 16. Chromatogram (A) and calibration curve (B) for SEC [96]
V_0 = Volume of eluent in column; V_p = pore volume; V_e = volume of eluate

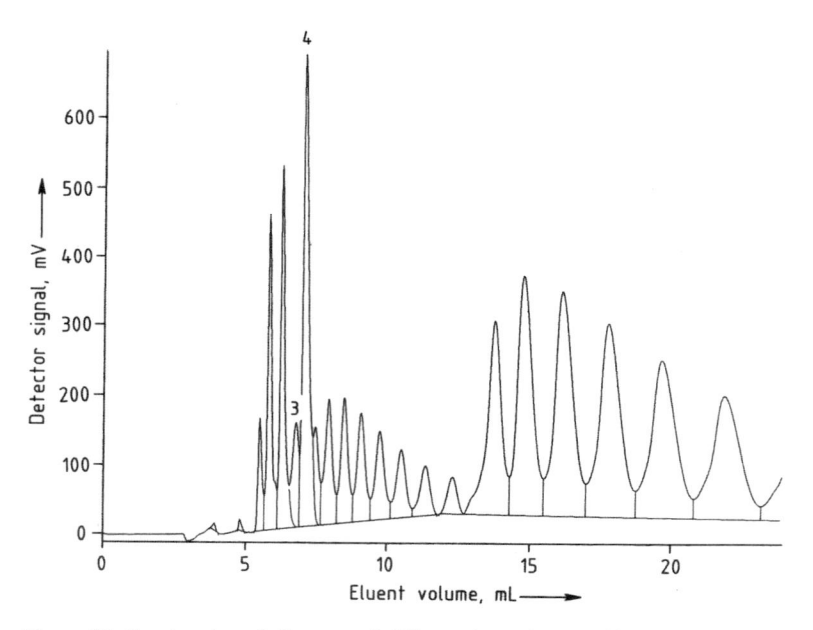

Figure 17. Fractionation of oligomers of different size and composition [101]
Chromatographic separation for coupling with NMR detection (see Fig. 29). Sample: polypropylene adipate; Eluent: D_2O/methanol-d_4; Column: Chromegabond 5 μm

$$f(V) = \int w(Y)(h/\sqrt{\pi})\exp[-h^2(V+Y)^2] \quad (46)$$

where $w(Y)$ represents a chromatogram from which diffusion broadening has been removed. This can in turn be transformed into the molecular mass distribution by using a calibrated elution curve; various methods are available for this [99].

A second and very effective approach for determining a molecular mass distribution entails greater experimental effort: extrapolation to zero of the data from a gel chromatogram of a polymer in terms of both concentration and elution velocity [100].

GPC is a very flexible analytical method as a result of the availability of a wide variety of column packings and solvents. Calibration curves permit coverage of a molecular mass range extending from oligomers up to 10^6. GPC coupled with GC has become the most important method for the identification and molecular mass determination of oligomers.

Ultracentrifugation. The sedimentation constant

$$s_0 = \left(\frac{\dot{r}}{\omega^2 r}\right)_{c \to 0, M, T, S} \quad (47)$$

where $\dot{r} = dr/dt$ describes the rate of sedimentation (extrapolated to $c \to 0$) in a uniform field for a macromolecule of molecular mass M in the solvent S at temperature T. The dependency upon molecular mass is expressed by:

$$s_0 = kM^\alpha \quad (48)$$

A log–log plot of the experimental results gives a straight line. The observed value of the exponent is a function of the thermodynamic quality of the solvent. The smallest values are associated with sedimentation systems in which $\alpha = 0.5$. Numerous relationships based on Equation (48), calibrated by using absolute methods, have been tabulated [53]. For this reason the determination of molecular mass by sedimentation is also important as a relative method (for the absolute method, see Section 5.2.3).

Recording the polymer concentration during sedimentation as a function of time and location gives a set of curves in which the information relevant to molecular mass distribution is initially masked by other parameters influencing the separation process. Methods for determining molecular mass distribution by ultracentrifugation are described in detail in [78], [79], [83–85].

Ultracentrifugation is capable of producing forces up to 400 000 g. The lower limit for establishing a molecular mass distribution is $M = 10^4$. The unmatched flexibility of ultracentrifugation with respect to relative and absolute mass determinations as well as sample separation comes at the expense of high investment and operating costs, factors which must be taken into account in any comparison with GPC.

6. Determination of Sequential Structure

Introduction. The properties and applications of plastics are determined largely by the primary structures of their mainly linear macromolecules, which are formed by the sequential coupling of a single monomer or several chemically distinct species (see Chap. 7). One important type of problem in plastics analysis derives from the numerous types of isomerism, a situation that becomes increasingly complex in the case of co- and multipolymers [1], [104–109]. The sequence of two structural characteristics A and B can be block-like, alternating, or random. Structural units comprising *configurational sequences* of a single monomer contain pseudoasymmetric carbon atoms. Such polymers can be classed as isotactic (meso, m), syndiotactic (racemic, r), or atactic (random distribution of m and r linkages). Table 8 provides a summary of possible configurational sequences for discrete groups of three (triads) and five (pentads) structural units. The number of possible sequences N increases dramatically with the sequence length n:

Sequence length n:	2	3	4	5	6
Sequences $N(n)$:	2	3	6	10	20

If addition of a new monomer unit occurs independently of the structure of the growing chain, then the probability P for a particular sequence can be shown to be $P_r = 1 - P_m$. One speaks in this case of Bernoulli statistics [107].

Configurational sequences also arise in the 1,4 addition of 1,3-dienes such as butadiene. The resulting polymer chain contains one C=C double bond per structural unit, which may have either the *cis* or *trans* configuration. The term *constitutional sequences* refers to the situation in which identical monomer units are capable of bonding in different ways (see Table 8). *Compositional sequences* are characteristic of copolymers into which two different comonomers have been incorporated [108].

Table 8. Tactical sequences in poly-α-olefins [3]

The monomer unit $-CH_2-CHR-$ is symbolized by ┼┼.

	Sequence	Symbol	Probability
Triads	mm		P_m^2
	mr		$2P_m(1-P_m)$
	rr		$(1-P_m)^2$
Pentads	mmmm		P_m^4
	mmmr		$2P_m^3(1-P_m)$
	rmmr		$P_m^2(1-P_m)^2$
	mmrm		$2P_m^3(1-P_m)$
	mmrr		$2P_m^2(1-P_m)^2$
	rmrm		$2P_m^2(1-P_m)^2$
	rmrr		$2P_m(1-P_m)^3$
	mrrm		$P_m^2(1-P_m)^2$
	rrrm		$2P_m(1-P_m)^3$
	rrrr		$(1-P_m)^4$

6.1. NMR Spectroscopy

The most useful technique for the analysis of sequences in polymer chains is NMR spectroscopy. It often facilitates quantitative evaluation of even the smallest structural details of a polymer. For a comprehensive treatment of the fundamentals of NMR, see [109–119], [121]; general principles are treated in [106]. Nuclei that are important in the analysis of polymers by NMR are listed in Table 9; they suffice for characterizing all the important types of polymers.

Nuclear magnetic resonance spectroscopy of polymers exploits the fact that the resonance frequencies of the investigated nuclei depend strongly on the constitutional, configurational, and conformational structure of the polymer. Each structural feature plays a role, so that each

Table 9. NMR properties of nuclei important in the analysis of plastics [115]

Nucleus	Natural abundance	NMR frequency in 1 T magnetic field, MHz	Sensitivity relative to proton	Magnetogyric moment	Spin
^{1}H	99.9844	42.577	1.000	2.79270	1/2
^{2}H (D)	0.0156	6.536	0.00964	0.85738	1
^{13}C	1.108	10.705	0.0159	0.70216	1/2
^{14}N	99.635	3.076	0.00101	0.40357	1
^{15}N	0.365	4.315	0.00104	− 0.28304	1/2
^{17}O	0.037	5.772	0.0291	− 1.8930	5/2
^{19}F	100	40.055	0.834	2.6273	1/2
^{29}Si	4.70	8.460	0.0785	− 0.55477	1/2
^{31}P	100	17.235	0.0664	1.1305	1/2
^{35}Cl	75.4	4.172	0.00471	0.82089	3/2
^{37}Cl	24.6	3.472	0.00272	0.68329	3/2

of the sequences listed in Table 8 is associated with specific resonance signals. Primary structures have been assigned for a large number of polymers in solution on the basis of ^{1}H and ^{13}C NMR spectroscopy [3], [104], [113], and NMR spectroscopy of heteroelements is gaining in importance [115].

Poly(Vinyl Chloride). The ^{13}C NMR spectrum of poly(vinyl chloride) is an example of a simple atactic polymer.

In Figure 18 configurative sequences corresponding to those listed in Table 8 can be identified. Configurational differentiation is possible for sequences of up to seven successive monomer units.

Polybutadiene. The ^{13}C NMR spectrum of polybutadiene is particularly instructive with respect to the effectiveness of the NMR method. Butadiene molecules can be linked in either a 1,4 or a 1,2 configuration

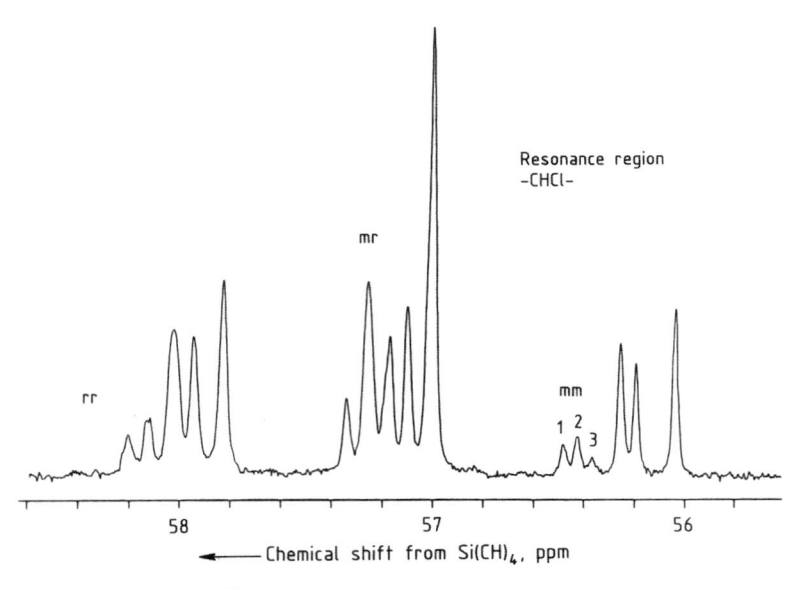

Figure 18. 67.88 MHz ^{13}C NMR spectrum of poly(vinyl chloride)
The signals 1 – 3 indicate configurational sequences (see Table 8) of up to 7 monomer units, rmmmmr, mmmmmr, and mmmmmm, respectively. Solvent: tetrahydrofuran-d$_8$; Temperature: 373 K; Broadband decoupling

cis-1,4-block

trans-1,4-block

1,2-block

Figure 19 A shows a ^{13}C NMR spectrum of a mixture of *cis*-1,4- and *trans*-1,4-polybutadiene. Separate resonance signals are observed in the C=C double bond region for each of the possible structures. The transition from *cis* to *trans* and vice versa during isomerization of the polymers can also be observed by means of NMR spectroscopy (Fig. 19 B and C).

Depending on the polymerization conditions, polybutadiene can contain both 1,4 and 1,2 linkages. In the NMR spectrum (Fig. 20) it is possible to identify all eight of the possible triad sequences for combinations of *cis* (c), *trans* (t), and 1,2 (v) linkage between monomer units.

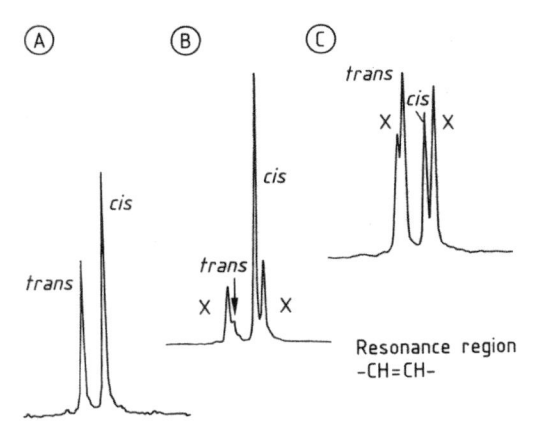

Figure 19. 67.88 MHz ^{13}C NMR spectrum of 1,4-polybutadiene
x = transition from *cis* to *trans* sequences; Solvent: deuterochloroform; Temperature: 300 K; All samples with random sequence distribution A) Mixture of *cis*- and *trans*-1,4-polybutadiene; B) *cis*- with 20 % *trans*-1,4-polybutadiene sequences; C) 1,4-Polybutadiene with *cis* and *trans* sequences (1: 1)

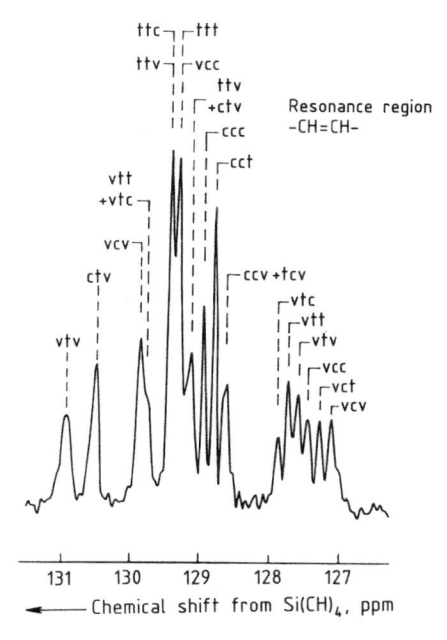

Figure 20. ^{13}C NMR spectrum of polybutadiene with 1,4- and 1,2-sequences
Conditions and sequence distribution as in Figure 19

Evaluation of the spectrum reveals that monomer addition in this sample was independent of the structure of the previously formed chain.

The 1,2 linkage of two butadiene units corresponds formally to an α-olefin, and it is in this way that the ^{13}C NMR spectrum in Figure 21 must be interpreted with respect to the configurative sequences listed in Table 8. The results are consistent with a Bernoulli distribution of the sequences. The observed distribution depends upon the polymerization conditions.

2D NMR Spectroscopy. Two-dimensional NMR spectroscopy allows the investigation of the nature of magnetic interactions between adjacent nuclei. Verifying the presence of a bond between particular nuclei significantly simplifies the interpretation of an NMR spectrum. Figure 22 illustrates a typical coupling experiment [111], [114] conducted on a fluorine- containing polymer [119].

3D NMR spectroscopy is a tomographic procedure for solid polymers. In addition to providing primary structural information on insoluble polymers it can also be used to test the quality of finished products made of plastic.

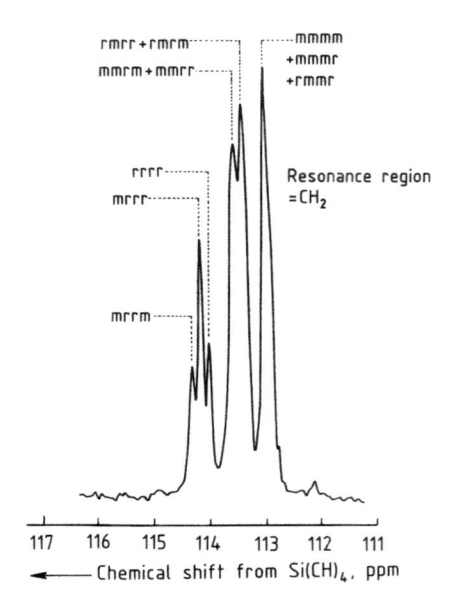

Figure 21. ^{13}C NMR spectrum of 1,2-polybutadiene at 67.88 MHz
Conditions and sequence distribution as in Figure 19

$$-CH_2-CH_2-CH_2-CH_2-$$
$$\downarrow$$
$$-CH_2-CH_2\cdot \quad \cdot CH_2-CH_2-$$
$$\downarrow$$
$$-CH=CH_2 \quad CH_3-CH_2-$$
$$\downarrow \text{Hydrogenation}$$
$$-CH_2-CH_3 \quad CH_3-CH_2-$$

Scheme 1. Pyrolysis of polyethylene

chain. Under constant experimental conditions, each polymer gives a characteristic pyrogram, permitting structural assignment by comparison with standard pyrograms from structurally defined polymers. In favorable cases it is sometimes possible to identify stereoisomeric fragments [126] (Fig. 25).

Use of isotopically labeled polymers permits the application of pyrolysis chromatography in kinetic investigations (depolymerization, sequences in copolymers).

From a methodological standpoint 3D NMR bridges the gap between structural analysis and quality control. Figure 23 shows the structure of a polybutadiene molding [120].

6.2. Pyrolysis Gas Chromatography

The pyrolytic methods described in Chapter 3 can be combined with gas chromatography [122–125] and mass spectrometry [126] to provide an extremely sensitive analysis of the chain structure of a polymer. Pyrolysis conditions are chosen such that the chain fragments released contain up to 30 monomer units, requiring temperatures of 500 – 800°C. The nature and number of observed fragments depends not only on the empirically determined experimental conditions but also on the composition and structure of the polymer (structural units, branching, tacticity, and sequence in the case of copolymers).

The pyrolysis of polyethylene proceeds as shown in Scheme 1. Simultaneous hydrogenation of the fragments often greatly simplifies the resulting pyrogram. The pyrogram of polyethylene consists exclusively of *n*-alkanes (Fig. 24). The presence of isoalkanes in a pyrogram is evidence for branching of the polymer

7. Determination of Chemical Heterogeneity

Methods that combine the online detection of chemical structure and molecular mass distribution are of growing importance.

Molecular Heterogeneity of Copolymers. Copolymerization of the monomers A and B can be described in terms of four different growth reactions:

$$\sim\!\!\sim A \xrightarrow[K_{AA}]{+A} \sim\!\!\sim AA \qquad \sim\!\!\sim B \xrightarrow[K_{BB}]{+B} \sim\!\!\sim BB$$
$$\sim\!\!\sim A \xrightarrow[K_{AB}]{+B} \sim\!\!\sim AB \qquad \sim\!\!\sim B \xrightarrow[K_{BA}]{+A} \sim\!\!\sim BA$$

$K_{AA}, K_{AB}, K_{BB}, K_{BA}$ Rate constants

Monomers typically differ in their reactivity towards a growing chain. The reactivity relationship can be expressed in terms of two copolymerization parameters:

$$r_A = \frac{K_{AA}}{K_{AB}}; \; r_B = \frac{K_{BB}}{K_{BA}} \qquad (49)$$

It follows for a radical copolymerization that the structure of emerging copolymer will vary as

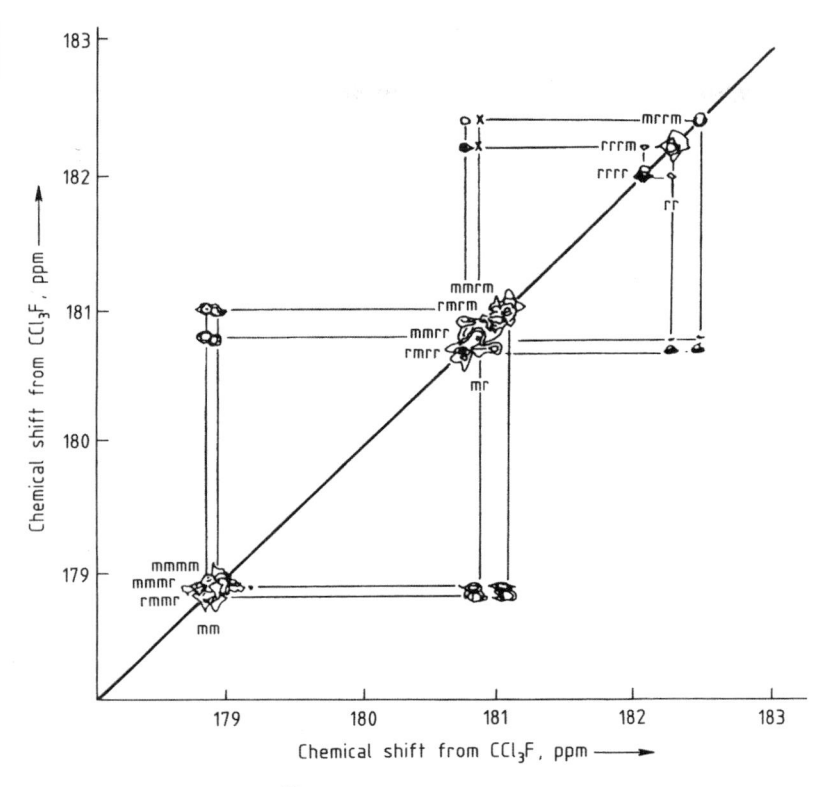

Figure 22. Two-dimensional ^{19}F NMR spectrum of poly(vinyl fluoride) [119]
Contour map of two-dimensional (refers to two different observation frequencies) J-correlated ^{19}F NMR spectrum. Sequences of five monomer units are assigned unambiguously

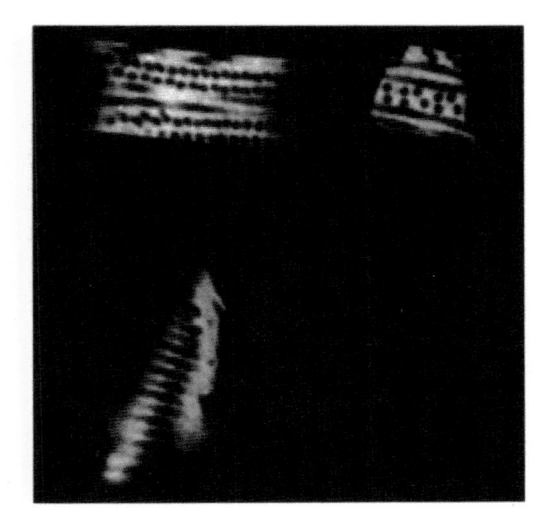

Figure 23. NMR micrograph of filled synthetic rubber [120] Three-dimensional (refers to space coordinates) ^{1}H NMR spectrum of a sample of reinforced tire rubber. Position of inhomogeneously distributed cord can be located without degradation of sample

a function of the extent of conversion:

$$\frac{m_A}{m_B} \sim \frac{A}{B} \cdot \frac{r_A A + B}{r_B B + A} \tag{50}$$

A, B mole fractions of monomers A and B in initial monomer mixture

m_A, m_B mole fractions of monomers A and B in copolymer

Generally, the macromolecules of a copolymer are heterogeneous with respect to molecular mass, chemical composition, and sequential structure. The differential distribution for a typical copolymer with respect to molecular mass and chemical composition is illustrated in Figure 26. Macromolecules with a particular molecular mass M_1 vary in their composition m_B. By the same token, molecules of a particular composition m_B exhibit a molecular mass distribution. Such a distribution function for a copolymer cannot be established by a single method of separation, since the

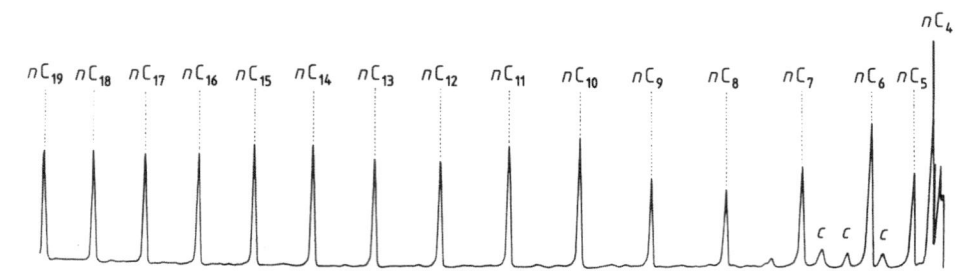

Figure 24. Pyrogram of polyethylene in a hydrogen atmosphere [124]
n = Linear olefins; c = Cyclic olefins

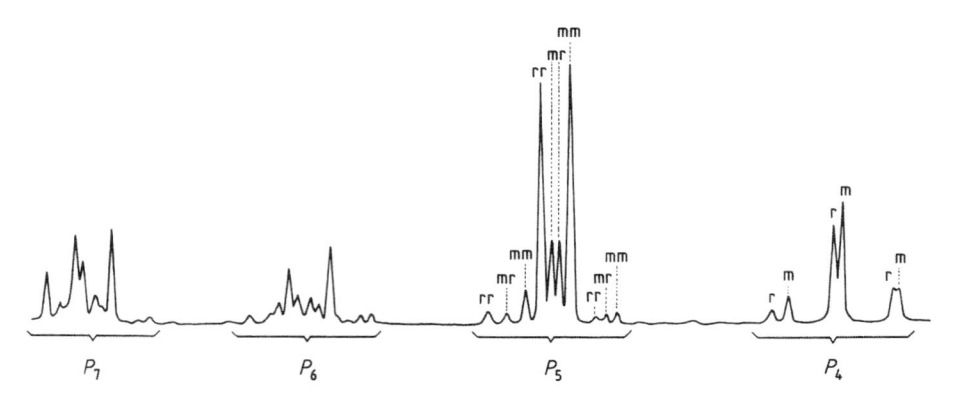

Figure 25. Pyrogram of isotactic polypropylene in a hydrogen atmosphere [124]
$P_4 - P_7$ = Degree of polymerization. For assignment of configurational sequences, see Table 8

experimental parameter that distinguishes one molecule from another is no longer determined only by molecular mass, but also by chemical composition and sequential structure. As a result, several structure-specific separation techniques must be used.

This condition can be met to a first approximation by the application of turbidity titration to eluate fractions obtained by gel permeation chromatography (see Gol Permeation Chromatography (GPC)). A copolymer sample is first separated by gel chromatography to give a series

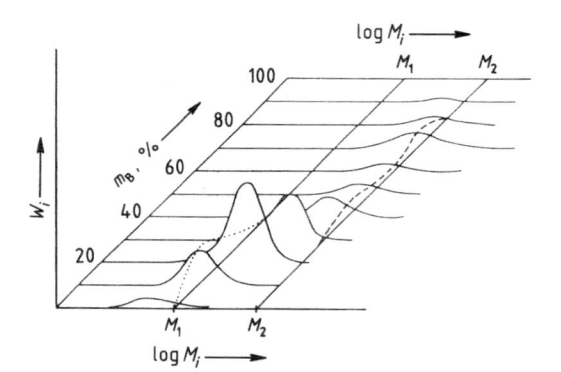

Figure 26. Heterogeneity of copolymer [9]
W_i = Frequency of the macromolecules with molecular mass M_i and chemical composition m_B

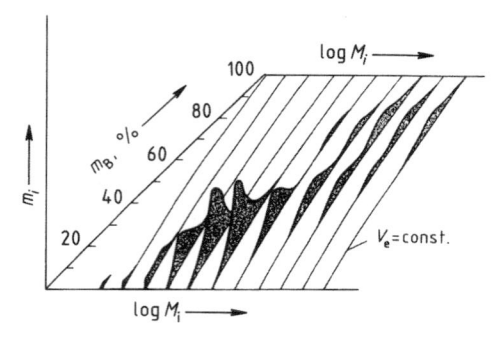

Figure 27. Distribution function of copolymer [9]
Methods: SEC and turbidity titration
m_i = Fraction of copolymer AB with molecular mass M_i and composition m_B; V_e = GPC elution volume

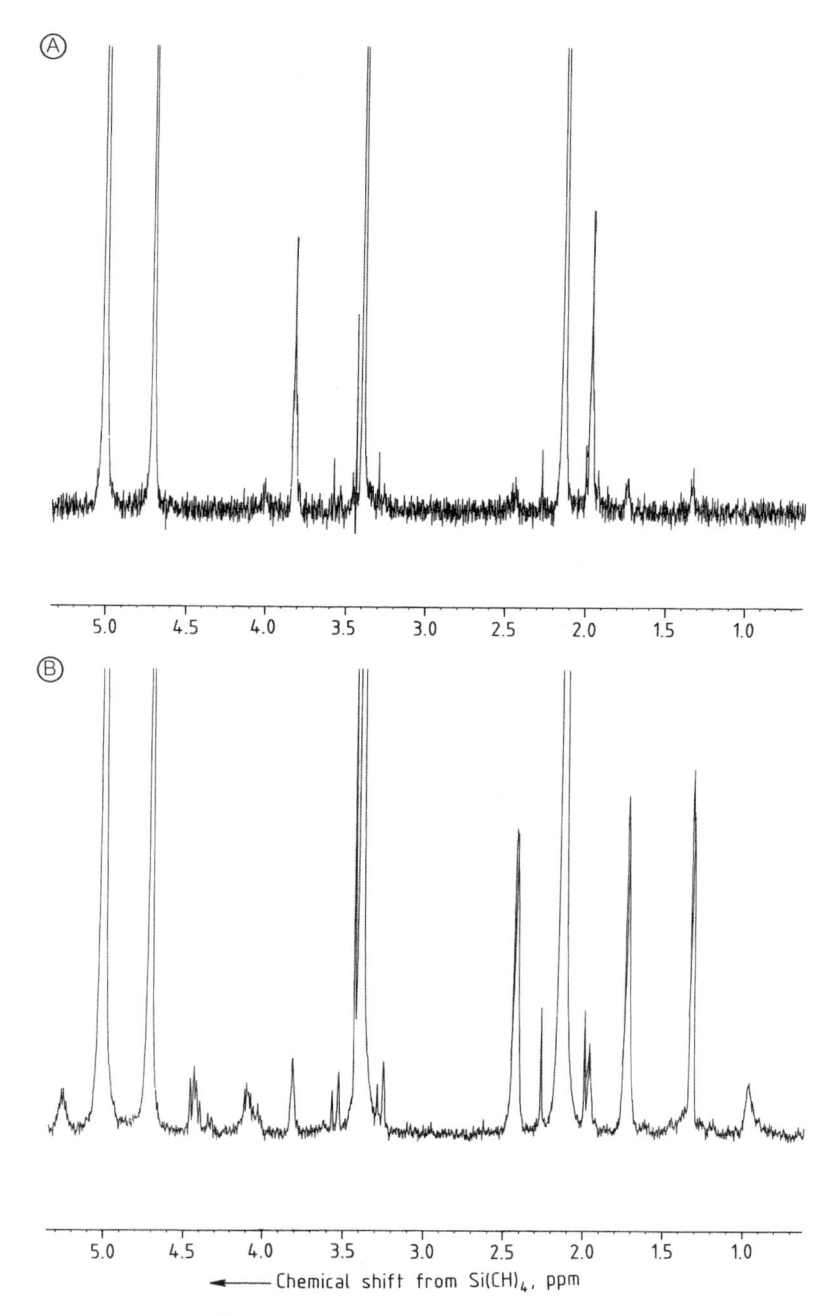

Figure 28. 500 MHz ^{1}H NMR spectrum of fractions from LC-NMR [127]
Online NMR spectra of the fractions 3 (A) and 4 (B) from Figure 17. The chemical structure of different components can be identified by analysis of the NMR spectrum.

of fractions containing molecules of comparable hydrodynamic volume, but not necessarily with the same molecular mass. Under certain conditions subsequent turbidity titration – essentially a precipitation fractionation monitored on the basis of light scattered laterally at an angle of 90°C (see Section 5.2.4) – makes it possible to determine the chemical composition of the polymer. Here both the amount of precipitant and the rate of precipitant addition are characteristic of the composition of the copolymer, whereas these factors are largely insensitive to molecular mass. The result is a differential distribution function for the copolymer, as shown in Figure 27 [9].

Structural Heterogeneity of Oligomers. Coupling of gradient HPLC and NMR spectroscopy opens up new ways for the online analysis of polymerization reactions. This method is applicable to polymers containing heteroelements such as ^{19}F, ^{31}P etc. The minimal amount of substance for detection of a fraction is ca. 30 μg. The spectra in Figure 28 demonstrate the considerable potential of the method [127].

References

1. J. L. Koenig: *Spectroscopy of Polymers*, American Chemical Society, Washington 1991.
2. C. D. Craver (ed.): "Polymer Characterisation," *Advances in Chemistry Series*, vol. 203, American Chemical Society, Washington 1983.
3. D. O. Hummel, F. Scholl: *Atlas of Polymer and Plastic Analysis*, vols. 1–3, VCH, Weinheim 1991.
4. D. McIntyre (ed.): *Characterisation of Macromolecular Structure*, National Academy of Sciences, Washington 1968.
5. J. V. Dawkins (ed.): *Developments in Polymer Characterisation*, Appl. Science Publishers, London 1982.
6. G. M. Kline: "Analysis of Monomers and Polymer Materials," *Analytical Chemistry of Polymers*, part 1, Interscience Publ., New York 1959.
7. G. M. Kline: "Analysis of Molecular Structure," *Analytical Chemistry of Polymers*, part 2, Interscience Publ., New York 1960.
8. G. M. Kline: Identification Procedures and Chemical Analysis, *Analytical Chemistry of Polymers*, part 3, Interscience Publ., New York 1962.
9. M. Hoffmann, H. Krömer, R. Kuhn: *Polymeranalytik*, vols. 1 and 2, Thieme Verlag, Stuttgart 1976.
10. R. Nitsche, K. A. Wolf: *Kunststoffe, Struktur, physikalisches Verhalten und Prüfung*, Springer Verlag, Berlin 1961.
11. H. Batzer (ed.): *Polymere Werkstoffe*, vols. 1–3, Thieme Verlag, Stuttgart 1985.
12. A. Cooper (ed.): *Determination of Molecular Weight*, John Wiley and Sons, New York 1989.
13. R. A. Komoroski: *NMR Spectroscopy of Synthetic Polymers in Bulk*, VCH, Weinheim Germany 1986.
14. L. S. Bark, N. S. Allen: *Analysis of Polymer Systems*, Applied Science Publishers, London 1983.
15. F. Rodriguez: *Principles of Polymer Systems*, Hemishere Publishing Corp., New York 1989.
16. L. Mandelkern: *An Introduction to Macromolecules*, Springer-Verlag, New York 1983.
17. S. L. Rosen: *Fundamental Principles of Polymeric Material*, Wiley-Interscience, New York 1982.
18. D. McIntyre: *Characterisation of Macromolecular Structure*, Nat. Acad. Sciences, Publ. 1573, Washington 1968.
19. J. L. Koenig: *Chemical Microstructure of Polymer Chains*, Wiley-Interscience, New York 1980.
20. J. F. Rabek: *Experimental Methods in Polymer Chemistry*, Wiley-Interscience, New York 1980.
21. B. H. Vollmert, *Polymer Chemistry*, Springer-Verlag, New York 1973.
22. P. J. Flory: *Principles of Polymer Chemistry*, Cornell University Press, Ithaca 1953.
23. P. E. M. Allen, C. R. Patrick: *Kinetics and Mechanisms of Polymerisation Reactions*, Wiley-Interscience, New York 1974.
24. C. Tanford: *Physical Chemistry of Macromolecules*, Wiley-Interscience, New York 1961.
25. G. M. Bartenev, J. V. Zelenev: *Physik der Hochpolymere*, VEB Deutscher Verlag für Grundstoffindustrie, Leipzig 1979.
26. B. Wunderlich: *Macromolecular Physics*, vols. 1 and 2, Academic Press, New York 1973.
27. C. D. Craver in T. Provder (ed.): *Polymer Characterisation: Physical Property, Spectroscopic and Chromatographic Methods*, American Chemical Society, Washington 1990.
28. P. J. Flory: *Statistical Mechanics of Chain Molecules*, Interscience Publ., Wiley 1969.
29. F. A. Bovey, F. H. Winslow: *Macromolecules*, Academic Press, New York 1979.
30. A. J. Hopfinger: *Conformational Properties of Macromolecules*, Academic Press, New York 1973.
31. F. Scholl, R. Vieweg, D. Braun: *Kunststoffhandbuch*, vol. 1, Hanser Verlag, München 1975.
32. G. Bandel, R. Houwink: *Chemie und Technologie der Kunststoffe*, vol. 1, Akademie Verlagsges., Leipzig 1954.
33. B. Carroll: *Physical Methods in Polymer Chemistry*, vol. 1, Marcel Dekker, New York 1969.
34. N. W. Johnston, H. J. Harwood, *Macromolecules* **3** (1970) 20.
35. H. Domininghaus: *Die Kunststoffe und ihre Eigenschaften*, VDI Verlag, Düsseldorf 1976.
36. G. C. Ives, J. A. Mead, M. M. Riley: *Handbook of Plastic Test Methods*, Iliffe Books, London 1971.
37. D. McIntire: *Characterisation of Macromolecular Structure*, Nat. Acad. Sci., Washington, D.C. 1968.
38. P. W. Allen: *Techniques of Polymer Characterisation*, Butterworth, London 1959.
39. M. Stratmann: *Erkennen und Identifizierung der Faserstoffe*, Spohr Verlag, Stuttgart 1973.
40. G. J. Domsch, *Kunststoffe* **61** (1971) 669.
41. E. A. Turi (ed.): *Thermal Analysis in Polymer Characterisation*, Heyden, Philadelphia 1981.
42. H. G. Kilian in A. Weiss (ed.): *Progress in Colloid & Polymer Science*, vols. 1–86, Steinkopf Verlag, Darmstadt 1991.
43. A. Abe, H. Benoit in H. J. Cantow (ed.): *Progress in Polymer Science*, vols. 1–101, Springer Verlag, New York 1991.
44. S. R. Palit, B. M. Mandal, *J. Macromol. Sci. Rev. Macromol. Chem.* **2** (1968) 225.
45. J. Mitchell, Jen Chiu, *Anal. Chem.* **43** (1971) 267.
46. J. M. Zeigler, F. W. G. Fearon: *Silicon-based Polymer Science: A Comprehensive Resource*, American Chemical Society, Washington 1989.
47. R. A. Dickie, S. S. Labana, R. S. Bauer: *Cross-Linked Polymers: Chemistry, Properties and Applications*, American Chemical Society, Washington 1989.
48. T. R. Crompton: *The Analysis of Plastics*, Pergamon Press, Oxford 1984.
49. W. C. Wake, M. J. Loadman in B. K. Tidd (ed.): *Analysis of Rubber and Rubber-like Polymers*, Elsevier, Amsterdam 1983.
50. E. Schröder, G. Müller, K.-F. Arndt: *Polymer Characterisation*, Oxford University Press, New York 1989.

51. A. L. Smith: *Analysis of Silicones*, Wiley-Interscience, New York 1974.
52. R. Taprogge: *Faserverstärkte Hochleistungsverbundstoffe*, Vogel Verlag, Würzburg 1974.
53. J. Brandrup, E. H. Immergut: *Polymer Handbook*, 3rd. ed., Wiley-Interscience, New York 1984.
54. J. Urbanski, W. Czerwinski, F. Majewska: *Handbook of Analysis of Synthetic Polymers and Plastics*, John Wiley and Sons, New York 1977.
55. H. Mark, N. G. Gaylord, N. M. Bikales: *Encyclopedia of Polymer Science and Technology*, vols. 1 – 16, Interscience Publ., New York 1964.
56. E. Müller (ed.): "Makromolekulare Stoffe," *Methoden der Organischen Chemie*, vol. 14, part 1/2, Georg Thieme Verlag, Stuttgart 1961.
57. R. Vieweg, D. Braun: *Kunststoffhandbuch*, vols. 1 –9, Hanser Verlag, München 1963.
58. American Chemical Society: *Chemical Abstracts and CAS Online*, Washington.
59. DKI, Deutsches Kunststoff Institut: *DKI Datenbank*, Darmstadt.
60. T. R. Crompton: *Chemical Analysis of Additives in Plastics*, Pergamon Press, Oxford 1971.
61. H. Wandel, H. Tengler, H. Ostramow: *Die Analyse von Weichmachern*, Springer-Verlag, Berlin 1967.
62. A. Krause, A. Lange, M. Ezrin: *Plastics Analysis Guide*, Hanser Verlag, München 1983.
63. J. Mitchell, Jr. (ed.): *Applied Polymer Analysis and Characterisation*, Hanser Verlag, München 1987.
64. H. Morawetz: *Macromolecules in Solution*, Interscience, New York 1965.
65. A. F. M. Barton (ed.): *CRC Handbook of Solubility Parameters and other Cohesion Parameters*, CRC Press Inc., Ohio 1983.
66. P. A. Bristow: *Liquid Chromatography in Practise*, hept, Wilmslow 1976.
67. A. Zlatkis, R. E. Kaiser, *J. Chromatogr. Libr.* **9** (1976).
68. J. Cazes: *Liquid Chromatography of Polymers and Related Materials*, Marcel Dekker, New York 1977.
69. A. Krause, A. Lange: *Kunststoff-Bestimmungsmöglichkeiten*, Hanser Verlag, München 1970.
70. D. Braun: *Simple Methods for Identification of Plastics*, Macmillan, New York 1982.
71. D. Braun, J. C. Jung, *Asbest Kunstst.* **23** (1970) 651.
72. W. Brügel: *Einführung in die Ultrarotspektroskopie*, 4th ed., Steinkopf Verlag, Darmstadt 1969.
73. S. Krimm, *Fortschr. Hochpolym. Forsch.* **2** (1960) 51.
74. A. Elliott: *Infrared Spectra and Structure of Organic Long Chain Polymers*, Arnold Verlag, London 1969.
75. R. Zbinden: *Infrared Spectroscopy of High Polymers*, Academic Press, New York 1964.
76. G. Schnell, *Ber. Bunsenges. Phys. Chem.* **70** (1966) 297.
77. N. J. Harrick: *Internal Reflectance Spectrometry*, Interscience Publ., New York 1967.
78. P. E. Slade Jr. *Polymer Molecular Weights*, vols. **1**, 2, Marcel Dekker, New York 1975.
79. L. H. Peebles Jr. *Molecular Weight Distributions in Polymers*, Interscience Publ., New York 1971.
80. J. S. Mattson, H. B. Mark, H. C. MacDonald Jr. *Computers in Polymer Science*, Marcel Dekker, New York 1977.
81. T. Provder (ed.): *Computer Applications in Applied Polymer Science II*, American Chemical Society, Washington 1989.
82. G. C. Lowry (ed.): *Markov Chains and Monte Carlo Calculations in Polymer Science*, Marcel Dekker, New York 1970.
83. H. Fujita: *Mathematical Theory of Sedimentation Analysis*, Academic Press, New York 1962.
84. J. W. Williams (ed.): *Ultracentrifugal Analysis*, Academic Press, New York 1963.
85. H. K. Schachmann: *Ultracentrifugation in Biochemistry*, Academic Press, New York 1959.
86. H. A. Stuart: *Die Physik der Hochpolymeren*, vol. 1, Springer-Verlag, Berlin 1952.
87. M. B. Huglin: *Light Scattering from Polymer Solutions*, Academic Press, London 1972.
88. H. Benoit, *Ber. Bunsenges. Phys. Chem.* **70** (1966) 286.
89. J. Schurz: *Viskositätsmessungen an Hochpolymeren*, Kohlhammer Verlag, Stuttgart 1972.
90. G. Meyerhoff, *Fortschr. Hochpolym. Forsch.* **7** (1970) 1.
91. G. V. Schulz, H. A. Stuart: *Physik der Hochpolymeren*, vol. 2, Springer-Verlag, Berlin 1953.
92. B. Vollmert: *Polymer Chemistry*, Springer-Verlag, Berlin 1973.
93. M. J. R. Cantow: *Polymer Fractionation*, Academic Press, New York 1969.
94. K. H. Altgelt, L. Segal: *Gel Permeation Chromatography*, Marcel Dekker, New York 1971.
95. H. Determann: *Gel-Chromatographie*, Springer-Verlag, Berlin 1967.
96. H. Engelhardt: *Hochdruck-Flüssigkeits-Chromatographie*, Springer-Verlag, Berlin 1975.
97. T. Provder (ed.): *Detection and Analysis in Size Exclusion Chromatography*, American Chemical Society, Washington 1987.
98. T. Provder (ed.): "Size Exclusion Chromatography," *ACS Symp. Ser.* **254** (1984).
99. H. L. Tung: *Fractionation of Synthetic Polymers*, Marcel Dekker, New York 1977.
100. J. Cazes (ed.): "Liquid Chromatography of Polymers," *Chromatographic Science*, vol. 19, Marcel Dekker, New York 1981.
101. W. Heitz, *Angew. Chem.* **82** (1970) 675.
102. K. C. Berger, G. V. Schulz, *Makromol. Chem.* **136** (1970) 221.
103. H.-J. Cantow, *Makromol. Chem.* **30** (1959) 169.
104. F. A. Bovey: *Polymer Conformation and Configuration*, Academic Press, New York 1969.
105. H. Günther: *NMR-Spektroskopie*, Thieme Verlag, Stuttgart 1973.
106. T. C. Farrar, E. D. Becker: *Pulse and Fourier Transform NMR*, Academic Press, New York 1971.
107. F. A. Bovey: *High Resolution NMR of Macromolecules*, Academic Press, New York 1972.
108. E. Klesper, G. Sielaff, D. O. Hummel: *Polymer Spectroscopy*, Verlag Chemie, Weinheim, Germany 1974.
109. J. Schaefer, G. C. Levy: *Carbon-13 NMR Spectroscopy*, Wiley-Interscience, New York 1979.
110. Q. T. Pham, R. Petiaud, L. M. France: *Proton and Carbon NMR Spectra of Polymers*, John Wiley and Sons, Chichester 1984.
111. W. R. Croasmun, R. M. K. Carlson: *Two-Dimensional NMR Spectroscopy*, VCH Publishers, New York 1987.
112. W. M. Pasika: *Carbon-13 in Polymer Science*, American Chemical Society, Washinton 1979.
113. A. E. Tonelli: *NMR-Spectroscopy and Polymer Microstructure*, VCH Publishers, New York 1989.
114. G. E. Martin, A. S. Zektzer: *Two-Dimensional NMR Methods for Establishing Molecular Connectivity*, VCH Publishers, New York 1988.
115. R. K. Harris, B. E. Mann: *NMR and the Periodic Table*, Academic Press, London 1978.
116. G. C. Levy (ed.): *Topics in Carbon-13-NMR Spectroscopy*, vols. 1 – 4, Wiley-Interscience, New York 1974.
117. J. B. Strothers: *Carbon-13 NMR Spectroscopy*, Academic Press, New York 1972.

118 R. R. Ernst, G. Bodenhausen, A. Wokau: *Principles of Nuclear Magnetic Resonance in One and Two Dimensions*, Clarendon Press, Oxford 1991.
119 M. Bruch, F. A. Bovey, *Macromolecules* **17** (1984) 2547.
120 D. Groß, V. Lehmann, W. E. Hull: *NMR Microscopy, Applications in Biology, Biomedicine and Material Science*, Bruker Analytische Meßtechnik, Rheinstetten 1991.
121 R. K. Harris (ed.): *Nuclear Magnetic Resonance (A Specialist Periodical Report)*, vols. 1–18, The Chemical Society, London 1972.
122 M. P. Stevens: *Characterisation and Analysis of Polymers by Gas Chromatography*, Marcel Dekker, New York 1969.
123 O. Hummel: *Polymer Spectroscopy*, Verlag Chemie, Weinheim, Germany 1974.
124 M. Seeger, H.-J. Cantow, *Makromol. Chem.* **176** (1975) 2059.
125 S. A. Liebmann, E. J. Levy: "Pyrolysis and GC in Polymer Analysis," *Chromatographic Science Series*, vol. 29, Marcel Dekker, New York 1985.
126 A. L. Yergey, C. G. Edmonds, I. A. S. Lewis: *Liquid Chromatography/Mass Spectrometry*, Plenum Press, New York 1990.
127 M. Spraul, M. Hofmann, H. Glauner: *LC-NMR – New Life for an Old Technique*, Bruker Report 2/1990, Bruker Analytische Meßtechnik, Rheinstetten 1990.

Further Reading

M. Belenkiæi, A. Tunik: *Handbook on Plastic Analysis in Engineering*, Backbone Publ., Fair Lawn, NJ 2006.

M. Bolgar, J. Hubball, J. Groeger: *Handbook for the Chemical Analysis of Plastic and Polymer Additives*, CRC Press, Boca Raton, FL 2008.

V. Shah: *Handbook of Plastics Testing and Failure Analysis*, 3rd ed., Wiley-Interscience, Hoboken, NJ 2007.

M. B. Wong: *Plastic Analysis and Design of Steel Structures*, Elsevier/Butterworth-Heinemann, Amsterdam 2009.

M. Yu;J. Li;G. Ma: *Structural Plasticity. Limit, Shakedown and Dynamic Plastic Analyses of Structures*.Springer, Heidelberg 2009.

Polymerization Processes, 1. Fundamentals

HIDETAKA TOBITA, University of Fukui, Fukui, Japan

1.	Introduction	265
2.	Polymerization Mechanisms and Kinetics	266
3.	Kinetics of Step Polymerization	267
4.	Basic Concepts of Molecular Mass Distribution (MMD)	268
5.	Chain Polymerization	271
6.	Free-Radical Polymerization	272
6.1.	Elementary Reactions	272
6.1.1.	Initiation	273
6.1.2.	Propagation	274
6.1.3.	Termination	275
6.1.4.	Chain Transfer to Small Molecules	277
6.2.	Kinetics of Free-Radical Polymerization	277
6.2.1.	Polymerization Rate	278
6.2.2.	Molecular Mass Distribution	279
6.3.	Effect of Temperature	283
7.	Copolymerization	283
7.1.	Copolymer Composition	284
7.2.	Kinetics of Free-Radical Copolymerization	287
8.	Living Polymerization	288
8.1.	Ideal Living Polymerization	289
8.2.	Reversible-Deactivation Radical Polymerization	290
8.2.1.	Polymerization Rate	291
8.2.2.	Molecular Mass Distribution	293
9.	Polymers with Branches and Crosslinks	298
9.1.	Crosslinked Polymers	298
9.2.	Branched Polymers	303
9.3.	Note on Nonlinear Polymerization	308
	List of Symbols	309
	General References	312
	Specific References	312

1. Introduction

Plastics have shaped our modern life, and various types of synthetic polymers are found everywhere in our daily life. Polymers are macromolecules, usually possessing molecular mass larger than 10 000, built up by a large number of small molecules called monomers. There is no significant meaning in the number "10 000", but the term oligomer is often used for polymers with molecular mass smaller than several thousand. Chemical reactions to combine monomers into polymers are called polymerization. The extraordinary versatility of synthetic polymers results from the complexities of polymer structures. The most fundamental polymer structure is determined during the polymerization processes, and therefore, the understanding of the processes is crucially important to produce high quality polymer products.

In the polymerization processes, monomer molecules that move around freely are joined together to form polymers. In general, the entropy change of polymerization ΔS is negative. In order for the free energy change, $\Delta G = \Delta H - T\Delta S$, to be negative, the enthalpy change ΔH must be negative, meaning that the polymerization processes are exothermic. In addition, the viscosity of the reaction system becomes very high as the polymerization proceeds, due to the entanglements of polymer chains, which tends to result in the incomplete mixing and drastic decrease in the heat transfer coefficient during polymerization. Therefore, failure in temperature control may lead to runaway reactions. The control of polymerization processes is a challenging field for the process engineers; however, at the same time, it is highly rewarding because the polymer properties are usually very sensitive to the processes used and good operation may turn commodity polymers into high quality specialty polymers.

In this article, some of fundamental elements to understand the polymerization reactions and their mathematical representations are discussed, focusing attention mainly on the polymerization rate and molecular mass distribution (MMD).

2. Polymerization Mechanisms and Kinetics

Polymerization mechanisms are conveniently divided into two categories, step and chain polymerization. Polymerizations in which the polymer chains grow step-wise by reactions that connect any two molecular species are step polymerizations. Polymerizations in which a polymer chain grows only by reaction of monomers with a reactive chain end of the growing chain are chain polymerizations. It is important to note that this is a classification of reaction mechanisms, not of the structure of the repeating unit, since the same polymer could be synthesized either by step or chain polymerization. Officially, IUPAC recommends classifying polymerization reactions into four categories, chain polymerization, condensation chain polymerization, polycondensation, and polyaddition. However, it is still convenient to use the terms, step and chain polymerization to start learning polymerization processes, and these terms are used in this article. The relationships of various definitions are summarized in Table 1 [1]. The terms, step-growth and chain-growth polymerizations have long been used. Although the meanings are the same, for brevity and the modern preference [2], step polymerization and chain polymerization are used in this article.

Generally, polymer physical properties can differ significantly depending on the polymerization mechanism, and this is often due to the difference in molecular mass, i.e., polymers synthesized by chain polymerization often have higher molecular masses. In step polymerization, the size of polymer molecules increases at a relatively slow rate over a much longer period of time. With step polymerization, the reactions that link monomers, oligomers, and polymers involve the same reaction mechanism, and any two molecular species (monomer, oligomer, or polymer) can be coupled. The growth of a polymer chain proceeds slowly from monomer to dimer, trimer, tetramer, and so on, until full-sized polymer molecules are formed at high monomer conversions. Polymer chains continue to grow from both ends throughout the polymerization and, therefore, both chain lifetimes and polymerization times are usually of the order of hours.

Chain polymerization is a chain reaction of the active center, radical or ion, and addition of monomer to the active center is very fast. Once an active center is created, a polymer chain grows rapidly, and when the growing chain is deactivated by a termination reaction, the polymer chain is dead and no longer takes part as a reactant in the polymerization solely producing linear polymer molecules. Except for living polymerization, in which chain transfer and termination reactions are eliminated, polymer molecules generally grow to full size in a timescale that is much smaller than the time required for high conversion of monomer to polymer. In the case of chain polymerization, unreacted monomer is clearly distinct from polymer, and normally, monomers are not included in

Table 1. Classification of polymerization reactions

	Terms used in this article	Chain polymerization	Step polymerization
	Growth mechanism	monomers reacting with active polymer chains	molecules of all sizes reacting together
	Reaction type	chain reaction	usually non-chain reaction
Stoichiometry	With low-molar-mass byproduct	condensative chain polymerization	polycondensation
	Without low-molar-mass byproduct	chain polymerization	polyaddition

the polymer molecular mass distribution. On the other hand, in step polymerization there is no clear distinction between polymers and monomers, and monomers are included in the polymer molecular mass distribution.

3. Kinetics of Step Polymerization

In step polymerization, there is generally only one type of chemical reaction that links molecules of all sizes. Some of the typical chemical reactions are esterification, amidation, the formation of urethanes, and aromatic substitution. The growth reaction in step polymerization can be represented by the general reaction:

$$m \text{ mer} + n \text{ mer} \rightarrow (m+n) \text{mer} \qquad (1)$$

The kinetic study of such reactions would be extremely difficult if the rate constant for the coupling reaction depended on the size of both species. Fortunately, various kinetic studies have shown that the rate constant is effectively independent of chain length except perhaps for oligomers. This is often referred to as the concept of equal reactivity of functional groups.

Consider the example of step polymerization shown below.

$$n\text{A} - \text{A} + n\text{B} - \text{B} \rightarrow [\text{A} - \text{A} - \text{B} - \text{B}]_n \qquad (2)$$

In the case of the polyesterification of a diol and a diacid, A may be a hydroxyl group and B may be a carboxyl group, although the low molecular mass condensation byproduct is not shown. As shown later, an almost exact equivalence in the number of functional groups is necessary to obtain polymers with high molecular weight. In the case of exact stoichiometric ratio of the two types of functional groups, i.e., [A] = [B], the polymerization rate or the rate of disappearance of functional groups is given by

$$-\frac{1}{V}\frac{d(V[\text{A}])}{dt} = k[\text{A}][\text{B}] = k[\text{A}]^2 \qquad (3)$$

except for self-catalyzed polymerization, in which case the rate is third order in monomer. The self-catalyzed polymerization may not be a useful reaction from the practical point of view of productivity. Neglecting the volume (V) change during polymerization, integration of Equation (3) gives

$$\frac{1}{1-p} = 1 + k[\text{A}]_0 t \qquad (4)$$

where $[\text{A}]_0$ is the initial (at $t = 0$) concentration of A groups, and p is the conversion of functional groups, which is defined as

$$p = (N_{\text{A}_0} - N_\text{A})/N_{\text{A}_0} \qquad (5)$$

where N_{A_0} and N_A are the total number of moles of A groups at $t = 0$ and at any later time t, respectively, i.e., $N_{\text{A}_0} = V_0[\text{A}]_0$ and $N_\text{A} = V[\text{A}]$.

Consider the number-average chain length of the reaction mixture $\bar{P}_n$, which is simply given by the following ratio:

$$\bar{P}_n = \frac{(\text{total number of monomeric units})}{(\text{total number of molecules})} \qquad (6)$$

Note that in the case of step polymerization, monomer molecules are also counted in both denominator and numerator.

For the polymerization of A-A and B-B type monomers, if $[\text{A}]_0 = [\text{B}]_0$ holds, the total number of monomeric units in the reaction system, which is the numerator of Equation (6) is $N_{\text{A}_0}/2 + N_{\text{B}_0}/2 = N_{\text{A}_0}$. On the other hand, because the total number of molecules decreases by one each time a bond is formed, the total number of molecules is given by $N_{\text{A}_0} - (N_{\text{A}_0} - N_\text{A}) = N_\text{A}$, and therefore:

$$\bar{P}_n = \frac{N_{\text{A}_0}}{N_\text{A}} = \frac{1}{1-p} \qquad (7)$$

Equation (7) shows that very high conversions are necessary to obtain large chain lengths. For example, $\bar{P}_n$ is only 20 even when the conversion is as large as $p = 0.95$, and $p = 0.99$ is required to obtain $\bar{P}_n = 100$.

Combining equations (4) and (7) gives:

$$\bar{P}_n = 1 + k[\text{A}]_0 t \qquad (8)$$

which shows that $\bar{P}_n$ increases linearly with time. In terms of conversion of functional groups p, $\bar{P}_n$ increases significantly only at the final stage of polymerization as shown in

Equation (7). The assumption with respect to time that $\bar{P}_n$ also increases rapidly in the final stage, is not so as shown in Equation (8). Because the melt viscosity increases with 3.4 power of molecular mass [3], the viscosity does increase tremendously at the final stage even with respect to time, but the $\bar{P}_n$-value does not. In fact, the increase in viscosity makes it difficult to remove condensation byproduct, which may lead to a slower increase in $\bar{P}_n$ with respect to time during the final stage of polymerization.

Equation (7) assumes an exact stoichiometric ratio of unity. If a slight excess of one bifunctional monomer is used, all chain ends will eventually consist of the group present in excess. Assuming $N_{A_0} < N_{B_0}$, and defining the stoichiometric imbalance α by $\alpha = N_{A_0}/N_{B_0}$. The total number of monomer molecules initially present is given by $(N_{A_0} + N_{B_0})/2 = N_{A_0}(1 + 1/\alpha)/2$. Now consider the situation at conversion p. Note that p is usually defined with respect to the deficient group, so that p is defined for A groups in this example. Because the total number of molecules decreases by one each time a bond is formed as in the exact stoichiometric case, the total number of molecules is given by $(N_{A_0} + N_{B_0}/2 - N_{A_0}p) = [N_{A_0}(1-p) + N_{B_0}(1-\alpha p)]/2$. Therefore, Equation (6) leads to:

$$\bar{P}_n = \frac{1 + 1/\alpha}{1 - p + (1 - \alpha p)/\alpha} = \frac{1 + \alpha}{1 + \alpha - 2\alpha p} \qquad (9)$$

When $\alpha = 1$, Equation (9) reduces to Equation (7). As conversion p approaches unity, $\bar{P}_n$ goes to $(1 + \alpha)/(1 - \alpha)$. Thus if $\alpha = 0.99$, the maximum number-average chain length attainable is only 199. This example illustrates the importance of precise control of the stoichiometric ratio in order to obtain a desired chain length.

In general, in order to produce high molecular mass polymers by step polymerization, the system must satisfy the following requirements:

1. Very accurate control of the stoichiometric ratio of functional groups
2. Absence of side reactions
3. Availability of high-purity monomers
4. Reasonably high polymerization rate
5. Little tendency towards cyclization reactions

4. Basic Concepts of Molecular Mass Distribution (MMD)

The representative MMDs used in polymer science are the distribution on a number and a mass basis. A convenient way to understand the concept of the MMD, or equivalently, the chain length distribution, is to consider the fraction of polymer with chain length r, as the probability of finding r-mer in the polymer mixture. The random sampling technique [5] is a convenient method to elucidate the difference of the number and mass basis. Figure 1 shows the concept of selection on a number and a mass basis. Suppose there are an infinite number of polymer molecules. The number fraction distribution is obtained when polymer molecules are sampled by selecting chain ends randomly, as shown in

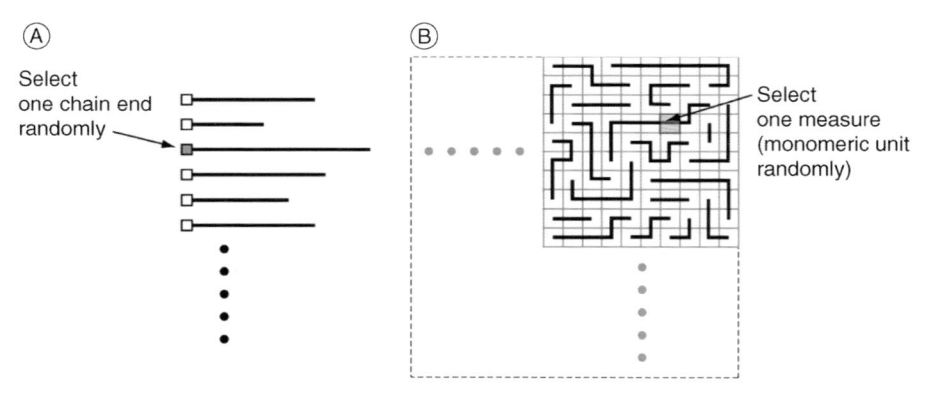

Figure 1. Concept of the chain length distribution on a number basis (A) and on a mass basis (B), from the point of view of the random sampling technique

Figure 1A. The number-average chain length is the expected chain length of a polymer molecule selected. On the other hand, suppose that all polymer molecules are embedded into 2D space, so that each monomeric unit occupies one measure (Fig. 1B). When one measure, or equivalently, one monomeric unit is chosen randomly, the polymer molecule that involves this particular unit is selected on a mass basis.

Consider the chain length distribution formed in an ideal step polymerization, in which the probability of connecting the next unit is given by p for all units. The number fraction distribution can be obtained by selecting chain ends randomly, as shown in Figure 2A. The polymer molecule having r units must have $(r-1)$ linkages, and at the same time, the final unit must not be connected. Therefore, the probability of finding polymer with length r is given by:

$$N(r) = p^{r-1}(1-p) \tag{10}$$

Obviously, the mass fraction distribution $M(r)$ can be calculated from the number fraction distribution, as follows:

$$M(r) = \frac{rN(r)}{\sum_{r=1}^{\infty} rN(r)} = rp^{r-1}(1-p)^2 \tag{11}$$

On the basis of the random sampling technique, the above equation can be obtained with the following reasoning. Suppose a polymer molecule has been chosen by selecting one monomeric unit randomly, as shown in Figure 2B. This polymer molecule contains $(r-1)$ linkages, and both chain ends must not be connected. In addition, the same polymer molecule could have selected by selecting any monomeric units in the chain, and there are r selection possibilities. Therefore, the probability of finding polymer with length r on a mass basis, which is equal to the mass fraction distribution, is given by:

$$M(r) = rp^{r-1}(1-p)^2 \tag{12}$$

The chain length distribution represented by Equation (10) on a number basis and Equation (12) on a mass basis is called the most probable distribution [4]. The most probable distribution is formed when the probability of connecting next unit is a constant, and is the most fundamental distribution of synthetic polymers. When the number-average chain length is large enough, i.e., $p \to 1$, the distribution can be represented by the following continuous functions, by using the relationships, $p^{r-1} \cong \exp[-(1-p)r]$ and Equation (7).

$$N(r) = \frac{1}{\bar{P}_n} \exp\left(-\frac{r}{\bar{P}_n}\right) \tag{13}$$

$$M(r) = \frac{r}{(\bar{P}_n)^2} \exp\left(-\frac{r}{\bar{P}_n}\right) \tag{14}$$

The mass average chain length is the expected chain length when one polymer molecule is selected on a mass basis. The averages can always be obtained from the distribution functions, however, for the most probable distribution, the mass-average chain length can be obtained by using Figure 2, just arithmetically as follows. In Figure 2B, the connection statistics are completely the same as that for Figure 2A, when approaching the randomly selected unit from the right, which shows the number-based selection. Therefore, the

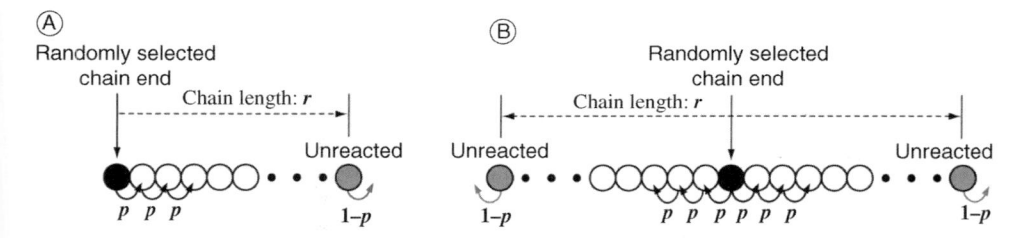

Figure 2. Number (A) and mass (B) fraction distribution of polymers formed in step polymerization

expected chain length from the right direction starting from the randomly selected unit is $\bar{P}_n$. Similarly, approaching from the left direction from the randomly selected unit, the connection statistics are again the same as for Figure 2A, and the expectation is $\bar{P}_n$. Remembering that the randomly selected unit is counted twice, the mass-average chain length of the most probable distribution is given by:

$$\bar{P}_M = 2\bar{P}_n - 1 = \frac{1+p}{1-p} \quad (15)$$

On the basis of the random sampling technique, it is obvious that the polydispersity index $\bar{P}_M/\bar{P}_n$ of the most probable distribution is practically equal to 2, when the chains are long enough.

Figure 3A shows the most probable distribution profiles on a number basis $N(r)$, and on a mass basis $M(r)$. There are a greater number of smaller chains and $N(r)$ is simply an exponentially decreasing function. On the other hand, a peak appears on a mass basis $M(r)$. The peak chain length conforms to the number average chain length $\bar{P}_n$, which is 100 in the case shown in Figure 3.

The most probable distribution is most frequently seen distribution in synthetic polymers. The most probable distribution is obtained if the probability of connecting the next unit is the same for all units in a chain. Therefore, this type of distribution is obtained if the reactivity of all functional groups is equal and independent of the chain length. In step polymerization, the most probable distribution is obtained not only for the cases with complete equilibrium, but also or the cases with irreversible polymerization in a batch reactor starting from monomer.

In free-radical polymerization, the polymer radicals follow the most probable distribution, as shown in Section 6.2.2. Another example includes the degradation of polymer chains. Starting from any kind of distribution, random degradation leads to the formation of the most probable distribution [6–8]. Irreversible depolymerization by chain-end scission (unzipping) also leads to the formation the most probable distribution [9]. Because of the random nature of synthetic polymer formation, it can be found everywhere in polymer science, except for the living polymers.

The chain length distribution profiles are often represented by using a logarithm of chain length as an independent variable. In the measurement of MMDs using size exclusion chromatography (SEC, also known as gel permeation chromatography, GPC), the MMD is commonly given by the mass fraction distribution whose independent variable is the logarithm of molecular mass. Now consider how to convert independent variable to the logarithm of it. Note that the MMD or its equivalent chain length distribution is the probability density distribution, and it is always the area that represents the probability. The mass fraction of polymers whose chain length is between a and b is given by:

$$\int_a^b M(r)dr = \int_{\log_{10}a}^{\log_{10}b} M(\log_{10}r)d(\log_{10}r) \quad (16)$$

This is a simple problem on the change of variables. By using the relationship, $\log_{10}r = \ln r/\ln 10$, one obtains $dr = r\ln 10 d(\log_{10}r)$. By

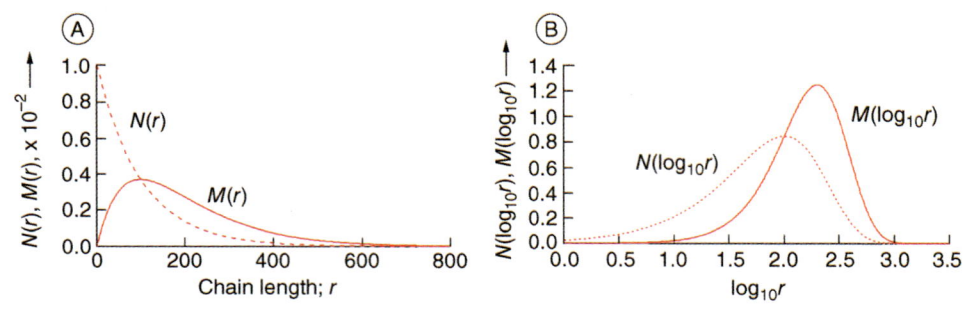

Figure 3. Most probable distribution profiles plotted as a function of chain length r (A), and as a function of the logarithm of r (B), for $\bar{P}_n = 100$

substituting into Equation (16),

$$\int_{\log_{10}a}^{\log_{10}b} M(r)r\ln 10\, d(\log_{10}r) = \int_{\log_{10}a}^{\log_{10}b} M(\log_{10}r)\, d(\log_{10}r) \quad (17)$$

Equation (17) shows the transformation rule, represented by:

$$M(\log_{10}r) = M(r)r\ln 10 \quad (18)$$

Figure 3B shows the most probable distribution, plotted as a function of $\log_{10}r$. Both number and mass fraction distribution possess a peak. The peak chain length is $\overline{P}_n$ for $N(\log_{10}r)$, and $\overline{P}_w$ for $M(\log_{10}r)$.

All 4 curves in Figure 3 show the same most probable distribution, and the physical meaning is completely the same, although appearances of the profiles are significantly different. When discussing the MMDs, one needs to clarify (1) whether the distribution is on the number basis or on the mass basis, and (2) whether the independent variable is a normal scale or a logarithmic scale.

The chain length distributions can be converted to the MMDs in a straightforward manner, simply bearing in mind that it is the area that represents the probability. Assuming the molecular mass of monomeric unit is m, the molecular mass is given by $M_M = mr$. When the independent variable is a normal scale, the x-axis is enlarged m times ($M_M = mr$), while the y of the polymer-axis is made smaller by m times, i.e., $N(M_M) = N(r)/m$ and $M(M_M) = M(r)/m$. On the other hand, when the independent variable is changed to $\log_{10}M_M$ from $\log_{10}r$, only the reading of the x-axis is shifted by $\log_{10}m$, i.e., $\log_{10}M_M = \log_{10}r + \log_{10}m$, while the value of y-axis is the same, $N(\log_{10}M_M) = N(\log_{10}r)$ and $M(\log_{10}M_M) = W(\log_{10}r)$.

5. Chain Polymerization

Chain polymerization is initiated by a reactive species, R_{in}^*, produced from an initiator or catalyst I.

$$I \xrightarrow{k_d} nR_{in}^* \quad (19)$$

Depending on the type of active center, chain polymerization can be divided into free-radical, anionic, and cationic polymerization. The reactive species R_{in}^* adds to a monomer to form a new active center, and monomer molecules are added to the active center successively. This process is called the propagation reaction:

$$R_{in}^* \xrightarrow{+M} P_1^* \xrightarrow{+M} P_2^* \xrightarrow{+M} P_3^* \ldots \quad (20)$$

where M represents a monomer molecule, and P_r^* is an active polymer molecule with chain length r. In general, the propagation reaction is represented by:

$$P_r^* + M \xrightarrow{k_p} P_{r+1}^* \quad (21)$$

In chain polymerization, only molecules with an active center can propagate, so that polymer molecules once formed may be considered dead polymer for linear chain polymerization. Dead polymer molecules do not take part as reactants thereafter. The active center is always on the chain end when linear chains are being produced exclusively. Polymer chain growth is terminated at some point by termination. Bimolecular termination between active centers occurs only in free-radical polymerization.

Carbon–carbon double bonds and the carbon–oxygen double bond in aldehydes and ketones are the two main types of functional groups that undergo chain polymerization. The polymerization of the carbon–carbon double bond is much more important, as most commercial monomers with carbon–carbon double bonds readily undergo free-radical polymerization (an important exception is propene). The carbonyl bond is not generally susceptible to polymerization by radical initiators due to its highly polarized structure. Another reason is that most of the carbonyl monomers (except formaldehyde) possess very low ceiling temperatures [10], the temperature above which active polymer chains depolymerize rather than grow.

Most of the commercial vinyl monomers ($CH_2=CHX$ and $CH_2=CXY$, and monomers in which fluorine is substituted for hydrogen) can be polymerized with free radicals. Whether a vinyl monomer can be polymerized by anionic or cationic mechanisms strongly depends on the type of monomer. Monomers with electron-donating groups attached to the doubly bonded carbon atoms form stable carbenium ions and polymerize best with cationic initiators. Conversely, monomers with electron-withdrawing substituents form stable carbanions and require anionic initiators. It should be noted that ions of low stability might be expected to react with carbon–carbon double bonds; however, in many cases they cannot be formed or else are easily consumed by side reactions.

6. Free-Radical Polymerization

As one of the most representative examples of chain polymerization, the kinetics of free-radical polymerization is discussed in detail. Controlled/living radical polymerization, whose IUPAC name is "reversible-deactivation radical polymerization" [11], also belongs to free-radical polymerization; however, it possesses a living nature and will be discussed in Chapter 8.

6.1. Elementary Reactions

Generally, free-radical polymerization consists of four types of elementary reaction; initiation, propagation, termination, and chain transfer reactions.

1. Initiation reactions, which continuously generate radicals during polymerization.

$$I \xrightarrow{k_d} n R_{in}^{\bullet} \quad (22)$$

$$R_{in}^{\bullet} + M \xrightarrow{k_i} R_1^{\bullet} \quad (23)$$

The stoichiometric coefficient n is two for the thermal decomposition of initiators. A free-radical $R_{in}^{\bullet}$ derived from the initiator is called a primary or initiator radical.

2. Propagation reactions, which are responsible for the growth of polymer chains by addition of monomer to a radical center.

$$R_r^{\bullet} + M \xrightarrow{k_p} R_{r+1}^{\bullet} \quad (24)$$

where $R_r^{\bullet}$ is a polymer radical (or macroradical) of chain length r.

3. Bimolecular termination reactions between two radical centers, which give a net consumption of radicals. These consist of disproportionation (Eq. 25) and combination (Eq. 26).

$$R_r^{\bullet} + R_s^{\bullet} \xrightarrow{k_{td}} P_r + P_s \quad (25)$$

$$R_r^{\bullet} + R_s^{\bullet} \xrightarrow{k_{tc}} P_{r+s} \quad (26)$$

where P_r is a polymer molecule of chain length r and does not have a radical center.

4. Chain transfer to small molecules that causes the cessation of the growth of polymer radicals while generating small transfer radicals simultaneously. Chain-transfer reactions do not give a net consumption of radicals, and if the transfer radicals are as reactive as polymer radical (or more reactive) these reactions should not affect the polymerization rate or monomer consumption rate when the bimolecular termination reactions are chemically controlled. Chain-transfer reactions to small molecules reduce the size of polymer radicals and therefore would increase bimolecular termination rates when these reactions are diffusion controlled.

$$R_r^{\bullet} + X \xrightarrow{k_{fX}} P_r + X^{\bullet} \quad (27)$$

$$X^{\bullet} + M \xrightarrow{k'_p} R_1^{\bullet} \quad (28)$$

X may be monomer, a solvent molecule, or a chain-transfer agent. When X is a polymer molecule, polymer molecules with long-chain branches are formed. Long-branch formation is discussed separately in Section 9.2.

The sequence of elementary reactions in Equations (22–28) result in total radical concentrations of the order 10^{-9}–10^{-5} mol/L for most commercial polymerizations. Since polymer molecules with high molecular masses are produced from the very start of polymerization, the reacting solution can be quite viscous over most of the monomer conversion range. The high viscosities not only cause problems in mixing and heat removal, but also can affect reaction rates (reactions such as bimolecular termination of polymer radicals).

Free-radical polymerization is the most commonly used method for the synthesis of polymers from vinyl and divinyl monomers. Some typical monomers which readily undergo free-radical polymerization are ethylene, styrene, vinyl chloride, vinylidene chloride, acrylonitrile, vinyl acetate, methyl methacrylate, methyl acrylate, acrylamide, etc. Of all chain polymerization processes, it is the most widely studied and best understood.

6.1.1. Initiation

Free radicals may be generated in a monomer in a number of ways. Some of representative methods to generate free radicals are as follows.

1. Thermal decomposition of initiators.
    Chemical initiators, such as azo and peroxide compounds, are added to the monomer in low concentrations (usually < 1 wt% based on monomer). When heated, the initiator decomposes, generating radicals which act as active centers for monomer addition. For example, organic peroxides (ROOR′) decompose thermally by O–O bond cleavage to produce two initiator radicals as follows.

$$ROOR' \rightarrow RO^{\bullet} + R'O^{\bullet} \quad (29)$$

2. Redox initiation.
    This is a method to generate radicals by an oxidation–reduction reaction. The advantage of using this method is that it is possible to generate radicals at lower temperature, normally 0-50°C.

3. Irradiation.
    Irradiation with UV, high-energy electrons, and γ-rays can initiate polymerization with or without the presence of initiators. Radiation initiation has been used widely for polymer modification (chain scission, long-chain branching, cross-linking, and grafting). These radiation processes are characterized by a zero activation energy for radical generation and as a consequence a low activation energy for polymerization. Therefore, they are effective at both low and high temperatures.

4. Thermal polymerization.
    It is known that some types of monomers, such as styrene (and some substituted styrenes such as p-methylstyrene) and methyl methacrylate generate free radicals at elevated temperatures in the absence of a free-radical initiating system.

The most often used method of free-radical generation is to use the thermal decomposition of initiators, and the kinetics of initiation is discussed for such systems, represented by:

$$I \xrightarrow{k_d} 2R^{\bullet}_{in} \quad (30)$$

where k_d is a thermal decomposition rate constant with units of inverse time, most often s^{-1}. This is a first order reaction, and the change in the number of moles of initiator N_I for a batch reactor is given by:

$$dN_I/dt = -k_d N_I \quad (31)$$

For isothermal decomposition (isothermal polymerization) Equation (31) can be integrated analytically to obtain:

$$N_I = N_{I_0} \exp(-k_d t) \quad (32)$$

where N_{I_0} is the number of moles of initiator at time $t = 0$.

The half-life of an initiator is given by:

$$t_{1/2} = -\ln(0.5)/k_d = 0.693/k_d \quad (33)$$

Equation (33) shows that the knowledge of k_d for an initiator permits calculation of the initiator half-life $t_{1/2}$. In reverse, it is possible to determine k_d from $t_{1/2}$, and they are equivalent. Since k_d has an Arrhenius temperature dependence, k_d and $t_{1/2}$ both depend on temperature. Activation energies for peroxides and azo initiators are ca. 120 kJ/mol, so the decomposition rate is highly temperature dependent, and the useful temperature range is quite small (decomposition rate is either too fast or too slow outside the useful temperature range, which normally spans about 30°C).

To complete the initiation step, the initiator radicals ($R^{\bullet}_{in}$) must add to the double bond of a monomer molecule to generate a polymer radical of unit chain length $R_1^{\bullet}$. In most polymerizations, this step (Eq. 23) is much faster than the rate of initiator decomposition (Eq. 22). The homolysis of the initiator is the rate-determining step in the initiation sequence, and the initiation rate R_I is given by:

$$R_I = 2k_d f[I] \qquad (34)$$

where f is the initiator efficiency. The coefficient 2 comes from the two initiator radicals which are generated from a single initiator decomposition. The initiator efficiency is defined as the fraction of radicals produced by initiator decomposition that initiate polymer radicals. Some primary radicals may react with themselves or with other molecules to form stable species that do not form polymer radicals. The initiator efficiency usually has values in the range 0.2–1.0 at low monomer conversions, where polymer concentrations are low.

The major cause of low initiator efficiency is recombination of the radical pairs before they diffuse apart, which is called the cage effect. When an initiator decomposes, the primary radicals $R_{in}^{\bullet}$ are nearest neighbors for about $10^{-10} - 10^{-9}$ s. During this interval they are surrounded by a "cage" of solvent and monomer molecules through which they must diffuse to escape from the cage. Since reactions between radicals are extremely fast, there is reasonable probability that reaction between primary radicals occurs. Direct recombination may simply regenerate the original initiator molecules, but other reactions can also occur that consume initiator radicals without forming polymer chains. In particular, since azo initiators decompose with the elimination of a nitrogen molecule, recombination of the primary radicals results in the formation of a stable molecule that cannot generate radicals, and thus there may be a significant decrease in initiator efficiency. Efficiency decreases with increasing viscosity of the reaction medium. Thus f decreases during the course of polymerization and may approach zero at very high polymer concentrations where the diffusion coefficient of primary radicals in the "cage" is very small [12, 13], which may result in stoppage of bulk polymerization at below 100% monomer conversion.

When selecting an initiator type, in general the magnitude of the decomposition rate constant, water and oil solubility, stability of initiator fragments on chain ends, and other factors should be considered. Another important point is the activity of the initiator radical center towards the abstraction of atoms (e.g., hydrogen atoms) from the polymer backbone. This can lead to chain scission, long-chain branching, and possibly crosslinking.

6.1.2. Propagation

The propagation reaction (Eq. 24) controls both the rate of growth and the structure of the polymer chain. Monomers which undergo free-radical polymerization are commonly monosubstituted or 1,1-disubstituted ethylenes, $CH_2=CHX$ or $CH_2=CXY$. With 1,1-disubstituted ethylenes both substituents should not be large, since propagation would be sterically hindered. 1,2-Disubstituted ethylenes are normally considered very difficult to polymerize since the approach of the propagating radicals to a monomer is sterically hindered. 1,2-Disubstituted ethylenes can, however, often be incorporated into copolymers.

Due to steric and resonance effects, vinyl monomers predominantly undergo head-to-tail addition. In certain cases when the substituents are small and do not have large resonance stabilizing effects, head-to-head propagation may occur. For example, approximately 16% head-to-head placement has been reported for poly(vinyl fluoride) [14].

In free-radical polymerization, chain microstructure is largely independent of initiation mechanism and initiator type. Polymers produced by free-radical polymerization are largely atactic. The slight preference for syndiotactic over isotactic placement is caused by steric and/or electrical repulsion between substituents in the chain, although at high temperatures their effects are progressively diminished, due to thermal motion. For example, the fraction of syndiotactic diads of poly(vinyl chloride) changes from 0.67 to 0.51 as the synthesis temperature increases from $-78°C$ to $120°C$ [15]. For methyl methacrylate, it is 0.86 at $-40°C$ and 0.64 at $250°C$ [16, 17].

For most chain polymerization conditions, either free-radical or ionic, the propagation reactions can be considered irreversible. However, it is important to bear in mind that the rate of depropagation becomes significant at elevated temperature [18, 19]. The ceiling temperature T_c, which is the temperature above which active polymer chains depolymerize rather than grow, is reached when the propagation and depropagation rates are equal. Note that the ceiling temperature T_c is not a singular value but is a function of monomer concentration. At any temperature, a concentration of monomer exists at which polymerization/depolymerization is at equilibrium. The existence of this equilibrium concentration prevents monomer conversion reaching 100%. Normally, this equilibrium monomer concentration is very low. A notable exception is α-methylstyrene whose T_c for 100% monomer concentration is 61°C, and the equilibrium monomer concentration at 25°C is 2.2 mol/L [20].

In a strict sense, the propagation rate constant k_p changes with chain length. The first propagation step could be much faster than the long chain propagation reactions. However, experimental results have shown that k_p is independent of chain length for chain lengths larger than about 5 for many monomer types [21]. When producing polymers with high molecular masses, the chain-length dependence could be neglected. The propagation rate constant k_p is relatively insensitive to the viscosity of the system, except at very high polymer concentrations [13, 22].

On the basis of the elementary reaction shown in Equation (24), the propagation rate R_p is represented by:

$$R_p = k_p[M][R^\bullet] \tag{35}$$

where k_p is the propagation rate constant, [M] is the monomer concentration, and [R$^\bullet$] is the total polymer radical concentration, i.e., $[R^\bullet] = \sum_{r=1}^{\infty}[R_r^\bullet]$. When the chain lengths of produced polymers are long enough, monomer consumption other than propagation is negligibly small, and Equation (35) is directly equal to the polymerization rate for usual free-radical polymerizations (long-chain approximation).

For a batch polymerization, the time development of monomer conversion, x can calculated from:

$$\frac{dx}{dt} = k_p(1-x)[R^\bullet] \tag{36}$$

The average number of monomer units added to a single radical in a second, $\bar{n}_M$ is given by:

$$\bar{n}_M = R_p/[R^\bullet] = k_p[M] \tag{37}$$

Usually, $k_p[M]$ is quite large, and one would understand that a single polymer chain can be formed within a couple of seconds in free-radical polymerization. The average time needed to add a monomer unit to a radical center can be estimated by $1/k_p[M]$.

6.1.3. Termination

Bimolecular termination in free-radical polymerization is either by disproportionation (Eq. 25) or by combination (Eq. 26). In the disproportionation termination, the terminal double bonds could be formed. However, these double bonds are often both substituted and usually inactive toward free-radical polymerization. Even when they are active, the reaction of the terminal double bonds can be neglected until the polymerization proceeds to a high conversion region where the terminal double bond fraction among the residual double bonds is noticeable.

Termination by combination and disproportionation can occur simultaneously, and the relative importance of these two modes of termination depends most importantly on the monomer type, and secondary on the polymerization temperature. For methyl methacrylate, combination and disproportionation are both important at low temperature, with disproportionation becoming the dominant mode at high temperatures [23].

Distinction of disproportionation and combination is important when discussing the MMD, but for the description of polymerization rate, only the sum of these reaction rates is needed, and the overall termination rate constant k_t can be used.

$$k_t = k_{td} + k_{tc} \tag{38}$$

There are two types of definition for the termination rate constant. One school of thought, such as the compilation of *Polymer Handbook* and also in this article, employs the expression for the termination reaction rate R_t, which is the radical consumption rate as follows.

$$R_t = k_t[R^\bullet]^2 \tag{39}$$

Another school of thought uses to add coefficient 2:

$$R_t = 2k_t[R^\bullet]^2 \tag{40}$$

Equation (40) is based on the argument that two radicals are consumed by a single termination reaction. On the other hand, when the combination of possible radical pairs is considered, the coefficient 2 disappears. In a termination reaction, two radicals are chosen as a pair. Within a system consisting N radicals, the possible combination is $N(N-1)/2 \cong N^2/2$. When the fate of each radical is considered faithfully, $R_t = 2k_t[R^\bullet]^2/2 = k_t[R^\bullet]^2$.

Each conversion for termination rate constant can be employed as long as used consistently, but these two cannot be mixed. The termination rate constant given by Equation (39) is exactly twice as large as k_t in Equation (40). Careful attention should be paid when deciding which conversion of termination rate constant is used when reading published papers. As mentioned later, Equation (39) is preferable, in order to avoid mistakes in building up the population balance equations.

Because bimolecular termination reactions are intrinsically very fast, these reactions are likely to be diffusion controlled when they involve radical centers on polymeric reactants. The autoacceleration of polymerization rate that is a consequence of diffusion-controlled termination is usually called the gel effect or Trommsdorff–Norrish effect. Figure 4 illustrates the autoacceleration in rate for the polymerization of methyl methacrylate [24]. The interpretation proposed was that the increase in rate is a consequence of a decrease in the rate of termination, due to the large increase in viscosity of the reacting medium, thus giving an increase in radical concentration.

Originally, upon examination of the autoacceleration of the conversion–time curve, it was thought that the termination reaction only became diffusion controlled at some monomer conversions greater than zero and that this occurred when the polymer chains were

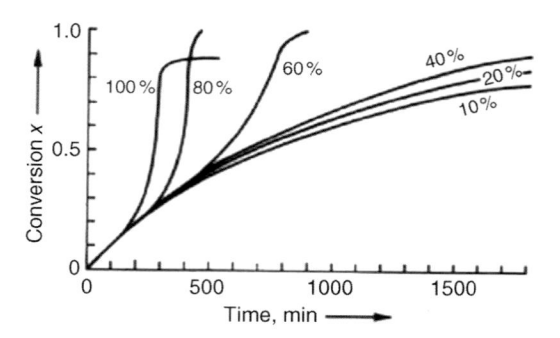

Figure 4. Conversion development for the polymerization of methyl methacrylate in benzene initiated by benzoyl peroxide at 50°C. Different curves are for various initial monomer concentrations in solvent [24]

sufficiently entangled (with a sufficient number of physical entanglement points). It is now recognized that the bimolecular termination reactions may be diffusion controlled even at zero monomer conversion (zero polymer concentration) [25, 26]. This interpretation would be reasonable, because the bimolecular termination constants for polymeric systems are 10^6–10^8 in Lmol^{-1}s^{-1}, while those for small molecules are 10^9–10^{10} in solution. When bimolecular termination is diffusion controlled, the chain-length dependent, bimolecular termination is dominated by interactions between short chains and long chains [27, 28].

Strictly, the termination rate constant should depend on the size of the polymeric radical reactants, the concentration and molecular mass distribution of the accumulated polymer, solvent type, temperature, etc. [29] Considering the complexity of the mechanisms of diffusion-controlled termination reactions, a practical way is to use an empirical approach for reactor calculations [30–32]. For example, k_t may be approximated by:

$$k_t = k_t^0 exp[-(A_1 x + A_2 x^2 + A_3 x^3)] \tag{41}$$

where x is the monomer conversion, k_t^0 is the termination rate constant at zero monomer conversion ($x = 0$), and A_1, A_2, A_3 are adjustable parameters. The adjustable parameters A_1, A_2, A_3 are usually estimated by fitting the isothermal conversion–time curve, for a given set of polymerization conditions.

6.1.4. Chain Transfer to Small Molecules

During free-radical polymerization, chain transfer to small molecules X may occur. The small molecule may be an initiator, a monomer, a chain-transfer agent, a solvent, an inhibitor, or an impurity. In general, these chain-transfer reactions can be represented by Equations (27) and (28).

When k'_p in Equation (28) is approximately zero (i.e., X• is a stable radical) X is called an inhibitor. If k'_p is smaller than the propagation rate constant k_p, X is called a retarder. For an added agent to be a chain-transfer agent, k'_p must be approximately equal to k_p (or $k_p < k'_p$, if chain length is large enough). Therefore, the chain-transfer agent reduces the molecular mass but does not affect rates of polymerization.

Monomer is always required in polymerization, and chain transfer to monomer cannot be totally eliminated. The monomer transfer constant $C_m (=k_{fm}/k_p)$ is normally rather small, i.e., $C_m = 10^{-5}–10^{-4}$, but without other chain stoppage mechanism, it may dominate the dead polymer formation. The value of C_m places an upper limit on the average polymer molecular mass that can be obtained with a given monomer, i.e., $\bar{P}_{n,Max} = 1/C_m$.

Consider a batch polymerization with an added chain-transfer agent, CTA. Neglecting the volume change during batch polymerization, the consumption rate of CTA is given by:

$$\frac{d[CTA]}{dt} = k_{fCTA}[CTA][R^\bullet] \quad (42)$$

From Equations (36) and (42), one obtains:

$$\frac{d[CTA]}{dx} = C_{fCTA}\frac{[CTA]}{(1-x)} \quad (43)$$

where C_{fCTA} ($=k_{fCTA}/k_p$) is the chain-transfer constant. Equation (43) can be solved to obtain:

$$\frac{[CTA]}{[CTA]_0} = (1-x)^{C_{fCTA}} \quad (44)$$

where $[CTA]_0$ is the initial CTA concentration.

Figure 5 shows the calculated results. It is possible to think that CTA with a larger chain transfer constant is effective in reducing the polymer chain length. However, as shown in Figure 5, with a larger C_{fCTA}-values, the CTA is used up in the earlier part of polymerization, and the chain length is uncontrolled in the later stage of polymerization. When $C_{fCTA} = 1$, the ratio [CTA]/[M] is kept constant during a batch polymerization, and the average chain length is kept constant if the chain transfer reaction dominates the dead polymer formation.

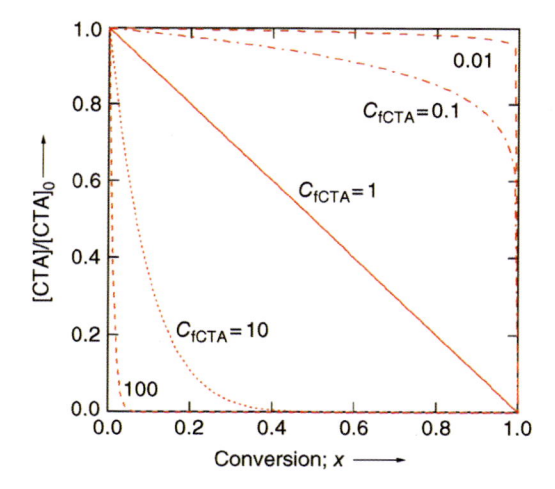

Figure 5. CTA concentration change during batch polymerization

6.2. Kinetics of Free-Radical Polymerization

Bearing in mind the elementary reactions discussed in Section 6.1, consider a free-radical polymerization consisting of the following elementary reactions.

Initiation

$$I \xrightarrow{k_d} 2R^\bullet_{in} \quad (45)$$

$$R^\bullet_{in} + M \xrightarrow{k_d} R^\bullet_1 \quad (46)$$

Propagation

$$R^\bullet_r + M \xrightarrow{k_p} R^\bullet_{r+1} \quad (47)$$

Chain transfer to monomer

$$R^\bullet_r + M \xrightarrow{k_{fm}} P_r + R^\bullet_1 \quad (48)$$

Chain transfer to small molecule, X

$$R_r^\bullet + X \xrightarrow{k_{fX}} P_r + X^\bullet \qquad (49)$$

$$X^\bullet + M \xrightarrow{k'_p} R_1^\bullet \qquad (50)$$

Termination by disproportionation

$$R_r^\bullet + R_s^\bullet \xrightarrow{k_{td}} P_r + P_s \qquad (51)$$

Termination by combination

$$R_r^\bullet + R_s^\bullet \xrightarrow{k_{tc}} P_{r+s} \qquad (52)$$

Complexities in dealing with free-radical polymerization come from the fact that the elementary reactions (45)–(52) occur simultaneously.

The polymer radical with chain length unity, $R_1^\bullet$ is formed in elementary reactions (46), (48), and (50). The chemical structure of these three types of radical is different; however, the effect of chain ends on the chemical reactivity of a radical center diminishes well before the chain length becomes large enough to be called a "polymer". One does not have to distinguish these species, except for the production of oligomers.

When dealing with the kinetics of free-radical polymerization, the following assumptions are usually made, without any explicit statement. If any other assumption is employed, it is usually stated clearly.

1. All rate constants are independent of chain length. Note, however, the kinetic rate constants may change with time.
2. Chain lengths are sufficiently large that the total rate of monomer consumption may be equated to the rate of monomer consumption by the propagation reactions alone. This is often called the long-chain approximation (LCA).
3. Radicals generated in chain-transfer reactions propagate with monomer rapidly and thus do not affect the polymerization rate.
4. The stationary-state hypothesis (SSH) is valid for radical reactions. The rates of radical generation and consumption are balanced very quickly.

6.2.1. Polymerization Rate

With the long-chain approximation, the polymerization rate is already represented by Equation (35). However, the total radical concentration [R$^\bullet$] is usually too small to measure directly. In addition, the recipe of polymerization has to be decided before starting experiments, and therefore, [R$^\bullet$] needs to be represented by the concentrations of the basic ingredients.

Looking at the elementary reactions (45)–(52) carefully, it is shown that the change in the number of radicals occurs only in the initiation and termination steps. Therefore, the balance equation for the total radical concentration is given by:

$$\frac{d[R^\bullet]}{dt} = R_I - R_t = 2k_d f[I] - k_t[R^\bullet]^2 \qquad (53)$$

Note that $k_t = k_{td} + k_{tc}$, and the definition of k_t is Equation (39), not Equation (40).

By using the stationary-state hypothesis (SSH) for the total radical concentration, d[R$^\bullet$]/dt=0,

$$[R^\bullet] = \sqrt{\frac{2k_d f[I]}{k_t}} \qquad (54)$$

Note that the SSH, d[R$^\bullet$]/dt = 0 does not mean that the radical concentration is constant throughout the polymerization. As shown in Equation (54), [R$^\bullet$] changes with [I], k_d, f, and/or k_t during polymerization. The SSH means that the rates of radical generation and consumption are balanced much faster than the time required for the changes in the other concentrations and parameters.

By substituting Equation (54) into Equation (35), one obtains:

$$R_p = \left(\frac{k_p}{k_t^{0.5}}\right)(2k_d f[I])^{0.5}[M] \qquad (55)$$

Equation (54) can be obtained also by using the population balance equations for radicals having different chain lengths. For the radicals with chain length unity,

$$\frac{d[R_1^\bullet]}{dt} = R_I + k_{fm}[M]\sum_{r=1}^{\infty}[R_r^\bullet] + k'_p[X^\bullet][M] - k_p[R_1^\bullet][M]$$
$$- k_{fm}[R_1^\bullet][M] - k_{fX}[R_1^\bullet][X] - (k_{tc}+k_{td})[R_1^\bullet][R^\bullet] \quad (56)$$

Note if the conversion for termination rate constant represented by Equation (40) instead of Equation (39) is used, the coefficient 2 is needed in the final term. In this term, only one molecule of $R_1^\bullet$ is consumed in a single termination reaction, and it is difficult to logically explain why the coefficient 2 is needed. Equation (39) may be better to use in building the population balance equations to avoid mistakes.

For the radicals with chain length r (≥ 2),

$$\frac{d[R_r^\bullet]}{dt} = k_p[R_{r-1}^\bullet][M] - k_p[R_r^\bullet][M] - k_{fm}[R_r^\bullet][M] - k_{fX}[R_r^\bullet][X]$$
$$- (k_{tc}+k_{td})[R_r^\bullet][R^\bullet] \quad (57)$$

Invoking the SSH for each radical and summing up for all chain lengths,

$$\frac{d[R^\bullet]}{dt} = \sum_{r=1}^{\infty}\frac{d[R_r^\bullet]}{dt} = R_I + k'_p[X^\bullet][M] - k_{fX}[R^\bullet][X] - (k_{tc}+k_{td})[R^\bullet]^2 = 0 \quad (58)$$

Using the SSH also for $X^\bullet$, $k'_p[X^\bullet][M] = k_{fX}[R^\bullet][X]$, and therefore:

$$R_I = (k_{tc}+k_{td})[R^\bullet]^2 \quad \text{or} \quad [R^\bullet] = \sqrt{R_I/k_t} \quad (59)$$

which agrees with Equation (54).

6.2.2. Molecular Mass Distribution

Invoking the SSH for Equations (56) and (57), one obtains:

$$[R_1^\bullet] = \frac{R_I + k_{fm}[M][R^\bullet] + k'_p[X^\bullet][M]}{k_p[M] + k_{fm}[M] + k_{fX}[X] + (k_{tc}+k_{td})[R^\bullet]}$$
$$= \frac{(k_{tc}+k_{td})[R^\bullet] + k_{fm}[M] + k_{fX}[X]}{k_p[M] + k_{fm}[M] + k_{fX}[X] + (k_{tc}+k_{td})[R^\bullet]}[R^\bullet] \quad (60)$$

$$[R_r^\bullet] = \frac{k_p[M][R_{r-1}^\bullet]}{k_p[M] + k_{fm}[M] + k_{fX}[X] + (k_{tc}+k_{td})[R^\bullet]} \quad (r \geq 2) \quad (61)$$

Note that the relationships, $R_I = (k_{tc}+k_{td})[R^\bullet]^2$ and $k'_p[X^\bullet][M] = k_{fX}[R^\bullet][X]$ are used in deriving the final form of Equation (60).

Equations (60) and (61) hold at each instant. As discussed using Equation (37), a single polymer molecule in free-radical polymerization is formed within a couple of seconds, while it usually takes several hours to obtain high conversion of monomer to polymer. Therefore, the concentrations and kinetic rate constants involved in Equations (60) and (61) can be considered constant during the time required for the formation of a single polymer chain, which shows that $[R_r^\bullet]$ is represented by a geometric progression.

Let us introduce the following dimensionless groups:

$$\tau = \frac{R_{td} + R_f}{R_p} = \frac{k_{td}[R^\bullet] + k_{fm}[M] + k_{fX}[X]}{k_p[M]} \quad (62)$$

$$\beta = \frac{R_{tc}}{R_p} = \frac{k_{tc}[R^\bullet]}{k_p[M]} \quad (63)$$

where

$R_p = k_p[R^\bullet][M]$; propagation rate
$R_{td} = k_{td}[R^\bullet]^2$; rate of termination by disproportionation
$R_{tc} = k_{tc}[R^\bullet]^2$; rate of termination by combination
$R_f = k_{fm}[R^\bullet][M] + k_{fX}[R^\bullet][X]$; rate of chain transfer.

The values of τ and β can be considered constant during the formation of a single chain. Note that in order to produce polymers with sufficient chain lengths, both τ and β must be much smaller than unity, i.e., $\tau \ll 1$ and $\beta \ll 1$.

By using τ and β, Equations (60) and (61) reduce to:

$$[R_1^\bullet] = \frac{\tau + \beta}{1 + \tau + \beta}[R^\bullet] \quad (64)$$

$$[R_r^\bullet] = \frac{1}{1 + \tau + \beta}[R_{r-1}^\bullet] \quad (r \geq 2) \quad (65)$$

Equation (65) shows the probability of connecting the next unit, $p = 1/(1 + \tau + \beta)$ is constant for chains formed within a very small time interval. As clarified in Chapter 4, the most probable distribution is obtained when the probability of connecting the next unit is the same for all units in the sequence of the chain, and therefore, the number and mass fraction distribution of polymer radicals are given by the following most probable distribution.

$$N^{\bullet}(r) = p^{r-1}(1-p) \qquad (66)$$

$$M^{\bullet}(r) = rp^{r-1}(1-p)^2 \qquad (67)$$

Note that $p = 1/(1 + \tau + \beta)$ and $(1 - p) = (\tau + \beta)/(1 + \tau + \beta)$. Referring to Equations (7) and (15), the number- and mass-average chain length of polymer radicals are given respectively by:

$$\overline{P}_n^{\bullet} = \frac{1}{1-p} = \frac{1+\tau+\beta}{\tau+\beta} \cong \frac{1}{\tau+\beta} \qquad (68)$$

$$\overline{P}_M^{\bullet} = \frac{1+p}{1-p} = \frac{2+\tau+\beta}{\tau+\beta} \cong \frac{2}{\tau+\beta} \qquad (69)$$

If one uses the continuous functions given by Equations (13) and (14), the distributions are given by:

$$N^{\bullet}(r) = (\tau + \beta)\exp[-(\tau + \beta)r] \qquad (70)$$

$$M^{\bullet}(r) = r(\tau + \beta)^2 \exp[-(\tau + \beta)r] \qquad (71)$$

Now, consider the instantaneous mass fraction distribution of dead polymers, i.e., the mass fraction distribution of dead polymers formed within a small time interval. On the basis of the random sampling technique, represented by Figure 1, the mass fraction distribution can be obtained by randomly selecting one monomeric unit in the chain. Figure 6 shows the random selection of a polymer radical, just before the dead polymer formation. Because this polymer radical is selected on a mass basis, the chain length distribution is given by $M^{\bullet}(r)$. With probability $\tau/(\tau + \beta)$, the dead polymer chain is formed by disproportionation termination or chain-transfer reaction. In this case, the formed dead polymer length is the same as the original polymer radical, i.e., $M_\tau(r) = M^{\bullet}(r)$. On the other hand, with probability $\beta/(\tau + \beta)$, the dead polymer chain is formed by combination termination. In this case, the chain length of the dead polymer chain is longer than the original polymer radical.

Figure 7 shows the schematic representation of the combination termination. Neglecting the chain length dependence of bimolecular termination reaction, the partner is selected by choosing a radical center randomly, i.e., by choosing a chain end randomly. This is the selection process based on a number, as was shown in Figure 1. Therefore, given the chain is formed by combination termination, the probability to obtain dead polymer chain having length r is given by:

$$\begin{aligned} M_\beta(r) &= \sum_{s=1}^{r-1} M^{\bullet}(s) N^{\bullet}(r-s) \\ &\cong \int_0^r M^{\bullet}(s) N^{\bullet}(r-s) ds = (\tau+\beta)^3 \exp[-(\tau+\beta)r] \int_0^r s\, ds \\ &= \frac{r^2}{2}(\tau+\beta)^3 \exp[-(\tau+\beta)r] \end{aligned} \qquad (72)$$

Therefore, the instantaneous mass fraction distribution is given by:

$$\begin{aligned} M_{\text{inst.}}(r) &= \frac{\tau}{\tau+\beta} M_\tau(r) + \frac{\beta}{\tau+\beta} M_\beta(r) \\ &= (\tau+\beta)\left\{\tau + \frac{\beta}{2}(\tau+\beta)r\right\} r \exp[-(\tau+\beta)r] \end{aligned} \qquad (73)$$

The mass-average chain length of the dead polymer formed instantaneously, $\overline{P}_{M,\text{inst.}}$ can be calculated from the mass fraction distribution,

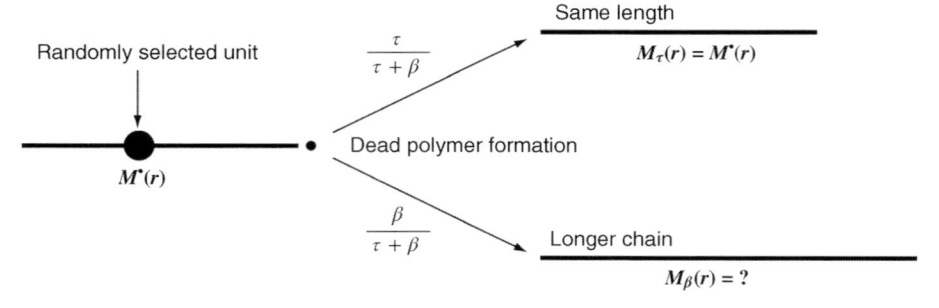

Figure 6. Process of dead polymer chain formation

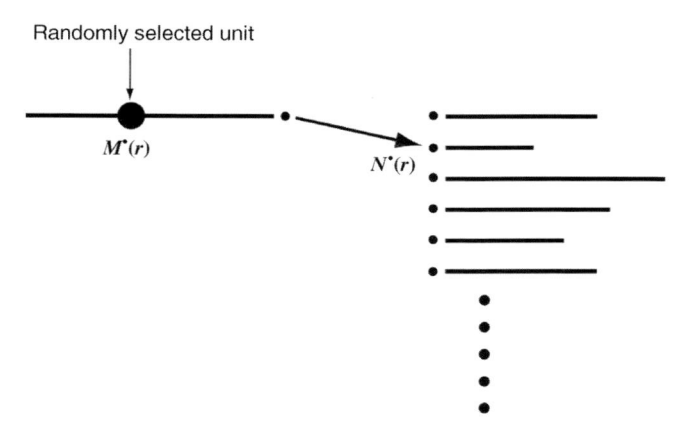

Figure 7. Schematic representation of the process of combination termination

but on the basis of the random sampling technique, it can be obtained as follows.

As was shown in Figure 6, the expected chain length of the randomly selected polymer radical is $\overline{P}_M^\bullet$, because it is selected on a mass basis. Only with combination termination, whose probability is $\beta/(\tau+\beta)$, does the chain length increase. As shown in Figure 7, the expected chain length of the partner is $\overline{P}_n^\bullet$. Therefore, the mass-average chain length of the dead polymer chains, $\overline{P}_{M,\text{inst.}}$ is given by:

$$\overline{P}_{w,\text{inst.}} = \overline{P}_w^\bullet + \frac{\beta}{\tau+\beta}\overline{P}_n^\bullet = \frac{2}{\tau+\beta} + \frac{\beta}{(\tau+\beta)^2} = \frac{2\tau+3\beta}{(\tau+\beta)^2} \quad (74)$$

Note that the number and mass average chain lengths of radicals are given by Equations (68) and (69), respectively.

Next consider the number-average chain length of the dead polymer chain formed instantaneously, $\overline{P}_{n,\text{inst.}}$. Figure 8 shows a schematic representation of a random sampling of a dead polymer chain on a number basis, i.e., one dead polymer chain end is selected randomly. Note that the number of dead polymer chains becomes smaller than that for polymer radicals because of the combination termination. Just before the dead polymer formation, the number ratio of the polymer radicals causing the τ-process and β-process is $\tau:\beta$. When one dead polymer chain end is selected randomly, the expected chain length of the radical part is $\overline{P}_n^\bullet$. Only with combination termination (β-process), whose probability is $(\beta/2)/(\tau+\beta/2)$ as shown in Figure 8, does the chain length increase. The expected chain length of the coupled part is $\overline{P}_n^\bullet$, because the active radical center is chosen randomly. Therefore, the expected chain length when one dead polymer chain end is selected randomly, which is the number-average chain length of the dead polymer chains, $\overline{P}_{n,\text{inst.}}$ is given by:

$$\overline{P}_{n,\text{inst.}} = \overline{P}_n^\bullet + \frac{\beta/2}{\tau+\beta/2}\overline{P}_n^\bullet = \frac{\tau+\beta}{\tau+\beta/2}\overline{P}_n^\bullet = \frac{1}{\tau+\beta/2} \quad (75)$$

The polydispersity index (PDI) for polymer produced instantaneously is given by:

$$\text{PDI}_{\text{inst.}} = \frac{\overline{P}_{M,\text{inst.}}}{\overline{P}_{n,\text{inst.}}} = \frac{(2\tau+3\beta)(\tau+\beta/2)}{(\tau+\beta)^2} \quad (76)$$

If $\beta = 0$, i.e., termination by combination does not occur, the polydispersity index takes on the maximum value, PDI = 2. On the other hand, if $\tau = 0$, i.e., chain termination is solely by bimolecular termination through combination, PDI takes on the minimum value, 1.5. Note that the combination termination makes MMD narrower. This is because although the combination termination makes the number-average chain length twice as large as that of polymer radicals $\overline{P}_{n,\text{inst.}} = 2\overline{P}_n^\bullet$, it makes the mass-average only 1.5 times larger than for polymer radicals $\overline{P}_{M,\text{inst.}} = \overline{P}_M^\bullet + \overline{P}_n^\bullet = 1.5\overline{P}_M^\bullet$, as was shown in Figure 7.

Figure 9 shows some examples of the calculated instantaneous mass fraction distribution profiles, with $\tau+\beta = 0.002$. Because the

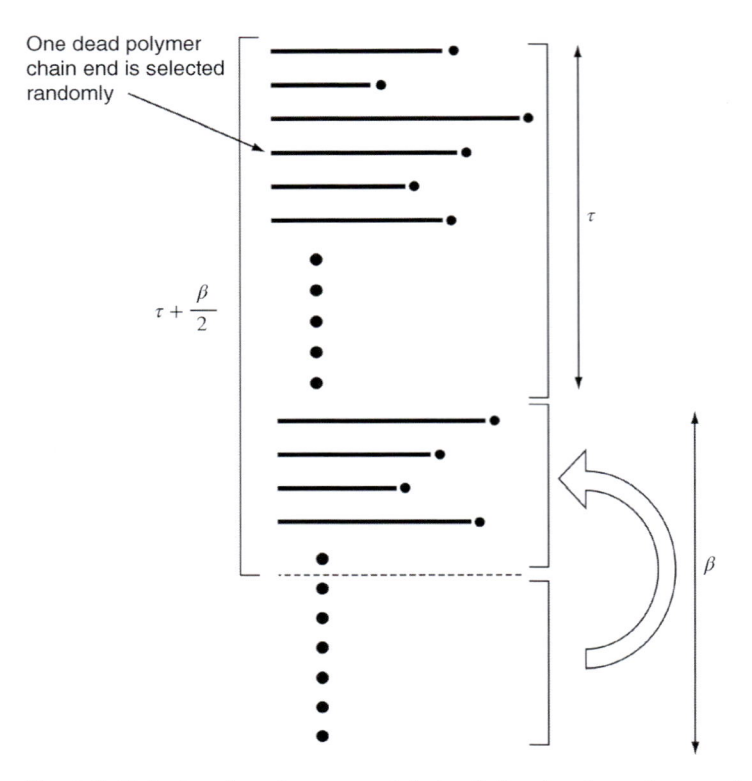

Figure 8. Derivation of number-average chain length, based on the random sampling technique

mass fraction distribution of the polymer radicals is given by Equation (71), the distributions of polymer radicals are the same for all three cases. The difference of chain length distribution of dead polymer molecules is caused by the contribution of dead polymer formation by combination termination, whose probability is represented by the ratio, $\beta/(\tau+\beta)$. One would

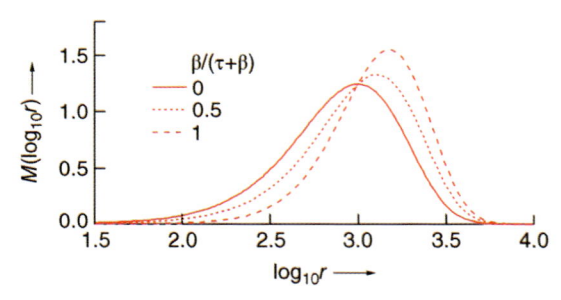

Figure 9. Instantaneous mass fraction distribution. The chain length distributions of polymer radicals are the same for all three cases, with $\tau+\beta=0.002$, while the formation mechanism of dead polymer chains is different

notice that the distribution becomes narrower as the contribution of combination termination becomes significant.

The chain length distribution derived here gives the instantaneous distribution, which is formed within a very small time interval. In linear free-radical polymerization the polymer molecules once formed are inert and do not react further. In general, since the concentrations of monomer, initiator, and chain-transfer agent change with time, the chain length distribution of the accumulated polymer is broader than the instantaneous distribution. The accumulated mass fraction distribution $M_{\text{accum.}}(r)$ and its mass-average $\overline{P}_{M,\text{accum.}}$ for batch polymerization are given by:

$$M_{\text{accum.}}(r) = \frac{1}{x}\int_0^x M_{\text{inst.}}(r)\mathrm{d}x$$

$$= \frac{1}{x}\int_0^x (\tau+\beta)\left\{\tau+\frac{\beta}{2}(\tau+\beta)r\right\}r\exp[-(\tau+\beta)r]\mathrm{d}x$$

(77)

$$\overline{P}_{M,\text{accum.}} = \frac{1}{x}\int_0^x \overline{P}_{M,\text{inst.}}\,dx = \frac{1}{x}\int_0^x \frac{2\tau + 3\beta}{(\tau+\beta)^2}\,dx \quad (78)$$

where x is the monomer conversion to polymer. The conversion is based on mass, and therefore, simple integration by conversion leads to the accumulated properties. Note that τ and β change with time, i.e., with conversion x.

For the calculation of number-average, the definition given by Equation (6) should be revisited. The number of polymer chains within a conversion interval dx can be evaluated by $dx/\overline{P}_{n,\text{inst.}}$, and the number-average chain length of the accumulated polymers is given by:

$$\overline{P}_{n,\text{accum.}} = \frac{x}{\int_0^x \dfrac{dx}{\overline{P}_{n,\text{inst.}}}} = \frac{x}{\int_0^x \left(\tau + \dfrac{\beta}{2}\right) dx} \quad (79)$$

The accumulated molecular mass properties can be calculated for semi-batch and continuous reactors in a similar manner. In an ideal continuous stirred tank reactor (CSTR), the concentration of every component is kept constant, and the chain length distribution produced is the instantaneous distribution, which is the narrowest. On the other hand, however, because the polymerization in a CSTR proceeds with a high polymer concentration, the polymer transfer reaction may not be neglected, which could make the distribution broader.

6.3. Effect of Temperature

In free-radical polymerization initiated by the thermal decomposition of an initiator, the polymerization rate is given by Equation (55), and the effect of temperature can be estimated by the change in the ratio of three rate constants, $k_p (k_d/k_t)^{0.5}$. Since each kinetic rate constant is considered to follow the Arrhenius equation, the activation energy of polymerization E_R is given by:

$$E_R = E_p + \frac{E_d}{2} - \frac{E_t}{2} \quad (80)$$

where E_p, E_d, E_t are the activation energies for propagation, initiator decomposition, and bimolecular termination. Typical values for E_p, E_d, E_t are 30, 120, and 15 kJ/mol, respectively. Therefore, E_R is 82.5 kJ/mol and this is largely due to the high activation energy for initiator decomposition. The large activation energy for polymerization means that the rate of polymerization increases strongly with increasing temperature. With redox initiation, E_d is ca. 40 kJ/mol and therefore E_R is considerably smaller at 42.5 kJ/mol. With radiation initiation, E_d is close to zero and E_R is 22.5 kJ/mol, and the effect of temperature is not very significant.

Now consider the effect of temperature on the molecular mass of polymer obtained. For a simple example, consider the case where $\tau \ll \beta$, that is, termination by combination produces most of the polymer chains. Equation (74) gives:

$$\overline{P}_{M,\text{inst.}} \propto \frac{1}{\beta} = \frac{k_p[M]}{k_{tc}(2k_d f[I]/k_{tc})^{0.5}} \propto \frac{k_p}{(k_{tc}k_d)^{0.5}} \quad (81)$$

The activation energy for average chain lengths E_L is given by:

$$E_L = E_p - \frac{E_d}{2} - \frac{E_t}{2} \quad (82)$$

Using the same activation energies used the above, E_L is -37.5 kJ/mol. The average chain length decreases with increasing temperature when initiators are used and bimolecular termination controls the polymer chain lengths. With radiation initiation, E_L is 22.5 kJ/mol and molecular mass increases moderately with temperature.

When chain transfer reactions dominate in the production of polymer chains, E_L is given by:

$$E_L = E_p - E_f \quad (83)$$

where E_f is the activation energy for the chain-transfer reaction. E_f depends on the type of chain transfer agent, but often $E_p - E_f < 0$, and therefore, the chain length may decrease with increasing temperature when chain transfer to small molecules controls the dead polymer formation.

7. Copolymerization

Copolymerization permits the synthesis of an almost unlimited number of polymer types and is therefore often used to obtain a better balance of properties for commercial applications. Copolymers may be synthesized by chain and step polymerization. In step polymerization, different monomers with the same type of functional group generally show only minor differences in reactivity. As a result, most copolymers prepared by step polymerization contain essentially random placements of repeat units, with the composition of the copolymers essentially the same as those of the original monomer mixture.

In contrast, strong selective effects often occur in chain copolymerizations, and the composition of the copolymer formed may differ greatly from the composition of the original monomer mixture. This section deals exclusively with chain copolymerization.

7.1. Copolymer Composition

The composition of copolymers cannot be determined from a knowledge of the homopolymerization rates of each monomer. In 1944, the instantaneous copolymer composition equation was proposed independently by several researchers by assuming that the chemical activity of a propagating chain depends solely on the terminal monomer unit on which the active center is located [33–36]. This model is called the terminal model, and the copolymer chain can be considered as a first-order Markov chain. For binary systems, the following four propagation reactions are possible.

$$P^*_{m,n,1} + M_1 \xrightarrow{k_{11}} P^*_{m+1,n,1} \quad (84)$$

$$P^*_{m,n,1} + M_2 \xrightarrow{k_{12}} P^*_{m,n+1,2} \quad (85)$$

$$P^*_{m,n,2} + M_1 \xrightarrow{k_{21}} P^*_{m+1,n,1} \quad (86)$$

$$P^*_{m,n,2} + M_2 \xrightarrow{k_{22}} P^*_{m,n+1,2} \quad (87)$$

where $P^*_{m,n,1}$ is a live copolymer chain with m units of monomer 1 (M_1) and n units of monomer 2 (M_2) bound in the polymer chain and with the active center located on terminal monomer unit 1.

The reactivity of the propagating species may be affected by the penultimate monomer unit. In such cases, the model is referred to as the penultimate model or a second-order Markov chain, and propagation consists of eight reactions. Although the penultimate effects could be found in the sequence of units in chain, the copolymer composition is usually represented, at least approximately, by the terminal model. Higher order Markov chain could be used, but the number of propagation rate constants increases exponentially. For the nth order Markovian model consisting of m types of monomer, the necessary number of propagation rate constants is m^{n+1}.

It is customarily assumed that the propagation rate constants are independent of chain length and that the chains are sufficiently large (long-chain approximation, LCA). The LCA includes the approximation that monomer consumed in reactions other than propagation is negligible and that the SSH is valid for each type of active center; i.e., the rate of formation of any type of active center is equal to its rate of consumption.

Consider a binary copolymerization whose propagation reactions follow the terminal model. Our present concern is to develop the instantaneous copolymer composition equation, so that the active species does not have to be distinguished by the size. The total concentration of active center of type 1 is given by:

$$[P^*_1] = \sum_{m=1}^{\infty} \sum_{n=1}^{\infty} [P^*_{m,n,1}] \quad (88)$$

The active center of type 1 is created solely by Equation (86). Equation (84) does not lead to an increase in the amount of P^*_1, because the number of active centers is not changed by the reaction. Similarly, the active center of type 1 is consumed solely by Equation (85), and therefore, the population balance of P^*_1 is given by:

$$\frac{d[P^*_1]}{dt} = k_{21}[P^*_2][M_1] - k_{12}[P^*_1][M_2] \quad (89)$$

Invoking the SSH provides:

$$k_{21}[P^*_2][M_1] = k_{12}[P^*_1][M_2] \quad (90)$$

The instantaneous copolymer composition is equal to the ratio of the consumption rates of M_1 and M_2. The specific polymerization rates of monomers 1 and 2 are given by:

$$-\frac{d[M_1]}{dt} = k_{11}[P_1^*][M_1] + k_{21}[P_2^*][M_1] \quad (91)$$

$$-\frac{d[M_2]}{dt} = k_{12}[P_1^*][M_2] + k_{22}[P_2^*][M_2] \quad (92)$$

Let F_1 and F_2 be the mole fractions of monomers 1 and 2 in the copolymer produced instantaneously.

$$\frac{F_1}{F_2} = \frac{-d(V[M_1])}{-d(V[M_2])} = \frac{k_{11}[P_1^*][M_1] + k_{21}[P_2^*][M_1]}{k_{12}[P_1^*][M_2] + k_{22}[P_2^*][M_2]} \quad (93)$$

By using Equation (90) and eliminating concentrations of active centers, one obtains:

$$\frac{F_1}{F_2} = \frac{[M_1](r_1[M_1] + [M_2])}{[M_2]([M_1] + r_2[M_2])} \quad (94)$$

where r_1 and r_2 are the reactivity ratios defined respectively by:

$$r_1 = \frac{k_{11}}{k_{12}} \quad (95)$$

$$r_2 = \frac{k_{22}}{k_{21}} \quad (96)$$

By using the mole fractions of unreacted monomer, $f_1 = [M_1]/([M_1]+[M_2])$ and $f_2 = 1-f_1$, Equation (94) can also represented by:

$$F_1 = \frac{r_1 f_1^2 + f_1 f_2}{r_1 f_1^2 + 2 f_1 f_2 + r_2 f_2^2} \quad (97)$$

Equation (94), or equivalently, Equation (97) represents the composition of monomeric units bound into polymer chains in a small time interval, which is called the instantaneous copolymer composition. In free-radical polymerization, because each polymer chain is formed instantaneously, the instantaneous copolymer composition equation shows the composition of each polymer molecule. On the other hand, in the case of living polymerization, this is an imaginary composition of monomer molecules bound into different polymer chains instantaneously when such monomer molecules are collected.

Figure 10 shows the calculated results of the copolymer composition for various reactivity ratio pairs. Except for the azeotropic point, the copolymer composition is different from the monomer composition, which results in a compositional drift during copolymerization. In free-radical polymerization, the compositional drift occurs among polymer chains born at different times. For living polymers, it occurs along a polymer chain. The compositional drift during batch polymerization will be discussed later.

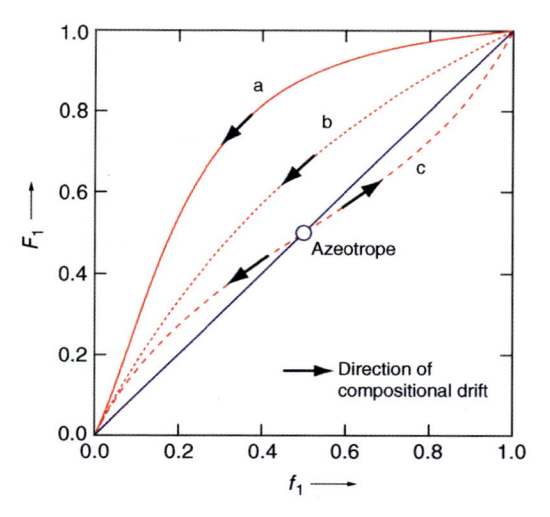

Figure 10. Instantaneous copolymer composition for various reactivity ratio pairs (a) $r_1 = 10$, $r_2 = 0.5$, (b) $r_1 = 2$, $r_2 = 0.5$, and (c) $r_1 = 0.5$, $r_2 = 0.5$.

Now consider free-radical polymerization, in which dead polymer chains are formed instantaneously and the instantaneous copolymer composition represents the composition of each polymer molecule. In this case, the instantaneous copolymer composition equation introduced here is strictly valid only for infinitely long chains. For finite chains, there exists statistical variations of composition, as well as the chain length distribution. For longer chains, say chain length $r > 20$, Equation (97) is practically equal to the average composition of the instantaneous polymers, $F_{1,av}$. For the oligomers, the effect of initiation and chain stoppage mechanism cannot be neglected, and it may not appropriate to use the copolymer composition equation [37–39].

For copolymer chains produced instantaneously, in general, there is a bivariate distribution of composition and chain length. Except for oligomers, Stockmayer's bivariate distribution of chain length and composition is a convenient equation to apply for a binary copolymerization with the terminal model [40]. The Stockmayer bivariate distribution, $M(r,F_1)$ consists of the product of mass-based chain length distribution $M(r)$ and the composition distribution $Comp(F_1|r)$, which is represented by the conditional probability given the chains with length r.

$$M(r,F_1) = M(r)Comp(F_1|r) \quad (98)$$

The composition distribution $Comp(F_1|r)$ is given by the following Gaussian distribution, whose average is $F_{1,av}$ and the variance is inversely proportional to chain length r, as follows:

$$Comp(F_1|r) = \frac{1}{\sqrt{2\pi\sigma^2}}\exp\left[-\frac{(F_1-F_{1,av})^2}{2\sigma^2}\right] \quad (99)$$

$$\sigma^2 = \frac{F_{1,av} F_{2,av} K}{r} \quad (100)$$

$$K = \sqrt{1 + 4F_{1,av} F_{2,av}(r_1 r_2 - 1)} \quad (101)$$

where the average composition, $F_{1,av}$ can be determined from the copolymer composition equation, Equation (97). Extension of the bivariate distribution to the multicomponent polymerization can be found elsewhere [38, 39, 41].

Figure 11 shows the effect of chain length on the copolymer composition distribution. For chains with $r = 50$, the composition distribution is quite large and polymers with ca. $F_1 = 0.3$ to 0.7 exist in the instantaneous distribution. The distribution becomes narrower as the chain length increases. Figure 12 shows an example

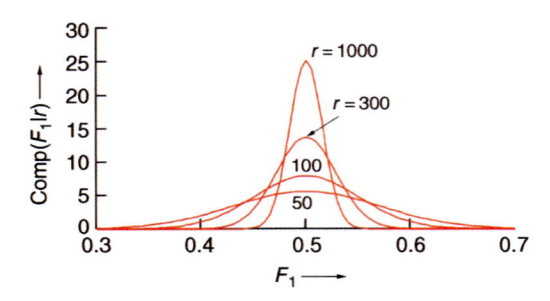

Figure 11. Instantaneous copolymer composition distribution calculated with $r_1 r_2 = 1$ and $F_{1,av} = 0.5$

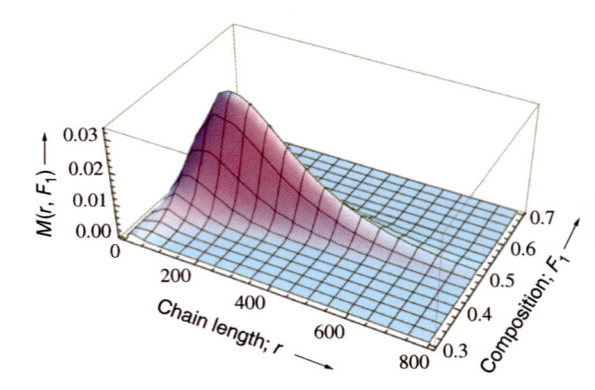

Figure 12. Bivariate distribution of chain length and composition, with $\bar{P}_{n,inst.} = 100$, $r_1 r_2 = 1$ and $F_{1,av} = 0.5$

of the bivariate distribution in a 3D form. This is the instantaneous distribution profile, and the distribution of accumulated polymers is obtained by the integration of instantaneous distribution. The chain length distribution formed in copolymerization, $M(r)$ will be discussed in the next section.

In a batch copolymerization, the compositional drift can be calculated as follows. Let N_M be the number of moles of total unreacted monomer. When a small amount of monomer is polymerized, the decrease in monomer 1 must be equal to the increase in M_1 units in polymer, and therefore:

$$\Delta(N_M f_1) = F_1 \Delta N_M \quad (102)$$

From Equation (102), one obtains the following differential equation:

$$\frac{df}{dN_M} = \frac{F_1 - f_1}{N_M} \quad (103)$$

By using total monomer conversion, $x = (N_M^0 - N_M)/N_M^0$, where N_M^0 is the initial moles of total amount of monomer,

$$\frac{df_1}{dx} = \frac{f_1 - F_1}{(1-x)} \quad (104)$$

One can calculate the monomer composition by solving Equation (104) with the initial condition, $f_1 = f_1^0$ and $F_1 = F_1^0$ at $x = 0$. Obviously, the copolymer composition equation is needed, i.e., Equation (97) for the terminal model with a binary system, to solve Equation (104). The accumulated copolymer composition, $\overline{F}_1$ can be calculated from:

$$\overline{F}_1 = \frac{f_1^0 - (1-x)f_1}{x} \quad (105)$$

Note that $\overline{F}_1 = F_1^0$ at $x = 0$.

If the terminal model is applicable to a binary system, Equation (104) can be integrated analytically to obtain the following equation [42, 43]:

$$x = 1 - \left(\frac{f_1}{f_1^0}\right)^a \left(\frac{f_2}{f_2^0}\right)^b \left(\frac{f_1^0 - d}{f_1 - d}\right)^c \quad (106)$$

where

$a = r_2/(1 - r_2)$,
$b = r_1/(1 - r_1)$,
$c = (1 - r_1 r_2)/\{(1 - r_1)(1 - r_2)\}$,
$d = (1 - r_2)/(2 - r_1 - r_2)$.

On the other hand, however, because f_1 is not given explicitly by x in Equation (106), it may be straightforward to solve Equation (104) numerically, rather than using Equation (106).

One method to avoid compositional drift is to use an ideally micromixed continuous stirred tank reactor. In this case, Stockmayer's bivariate distribution of chain length and composition will be obtained directly. Another method to avoid compositional drift is to use semi-batch operation in which monomers are fed to maintain a constant ratio of monomer concentrations in the reactor [44, 45].

7.2. Kinetics of Free-Radical Copolymerization

Consider the kinetics of free-radical copolymerization of monomers M_1 and M_2, assuming the terminal model is applicable. Important elementary reactions are:

Initiation

$$I \xrightarrow{k_d} 2R_{in}^{\bullet}$$

$$R_{in}^{\bullet} + M_1 \xrightarrow{k_1} R_{1,0,1}^{\bullet}$$

$$R_{in}^{\bullet} + M_2 \xrightarrow{k_2} R_{0,1,2}^{\bullet}$$

Propagation

$$R_{m,n,1}^{\bullet} + M_1 \xrightarrow{k_{11}} R_{m+1,n,1}^{\bullet}$$

$$R_{m,n,1}^{\bullet} + M_2 \xrightarrow{k_{12}} R_{m,n+1,2}^{\bullet}$$

$$R_{m,n,2}^{\bullet} + M_1 \xrightarrow{k_{21}} R_{m+1,n,1}^{\bullet}$$

$$R_{m,n,2}^{\bullet} + M_2 \xrightarrow{k_{22}} R_{m,n+1,2}^{\bullet}$$

Chain transfer to monomer

$$R_{m,n,1}^{\bullet} + M_1 \xrightarrow{k_{fm11}} P_{m,n} + R_{1,0,1}^{\bullet}$$

$$R_{m,n,1}^{\bullet} + M_2 \xrightarrow{k_{fm12}} P_{m,n} + R_{0,1,2}^{\bullet}$$

$$R_{m,n,2}^{\bullet} + M_1 \xrightarrow{k_{fm21}} P_{m,n} + R_{1,0,1}^{\bullet}$$

$$R_{m,n,2}^{\bullet} + M_2 \xrightarrow{k_{fm22}} P_{m,n} + R_{0,1,2}^{\bullet}$$

Chain transfer to small molecule, X

$$R_{m,n,1}^{\bullet} + X \xrightarrow{k_{fX1}} P_{m,n} + X^{\bullet}$$

$$R_{m,n,2}^{\bullet} + X \xrightarrow{k_{fX2}} P_{m,n} + X^{\bullet}$$

$$X^{\bullet} + M_1 \xrightarrow{k'_1} R_{1,0,1}^{\bullet}$$

$$X^{\bullet} + M_2 \xrightarrow{k'_2} R_{0,1,2}^{\bullet}$$

Termination by disproportionation

$$R_{m,n,1}^{\bullet} + R_{r,s,1}^{\bullet} \xrightarrow{k_{td,11}} P_{m,n} + P_{r,s}$$

$$R_{m,n,1}^{\bullet} + R_{r,s,2}^{\bullet} \xrightarrow{k_{td,12}} P_{m,n} + P_{r,s}$$

$$R_{m,n,2}^{\bullet} + R_{r,s,2}^{\bullet} \xrightarrow{k_{td,22}} P_{m,n} + P_{r,s}$$

Termination by combination

$$R_{m,n,1}^{\bullet} + R_{r,s,1}^{\bullet} \xrightarrow{k_{tc,11}} P_{m+r,n+s}$$

$$R_{m,n,1}^{\bullet} + R_{r,s,2}^{\bullet} \xrightarrow{k_{tc,12}} P_{m+r,n+s}$$

$$R_{m,n,2}^{\bullet} + R_{r,s,2}^{\bullet} \xrightarrow{k_{tc,22}} P_{m+r,n+s}$$

where $R_{m,n,1}^{\bullet}$ is a polymer radical with m units of monomer 1 (M_1) and n units of monomer 2 (M_2) bound in the polymer chain with active center located on monomer unit 1. $P_{m,n}$ is a dead polymer molecule with m units of monomer 1 and n units of monomer 2. The polymerization rate, R_p is given by:

$$R_p = k_{11}[R_1^\bullet][M_1] + k_{12}[R_1^\bullet][M_2] + k_{21}[R_2^\bullet][M_1] + k_{22}[R_2^\bullet][M_2] \quad (107)$$

where $[R_1^\bullet]$ is the total concentration of polymer radical of type 1 explicitly defined by Equation (88). By using the mole fractions, $f_1 = [M_1]/([M_1]+[M_2])$ and $\phi_1^\bullet = [R_1^\bullet]/\left[[R_1^\bullet]+[R_2^\bullet]\right]$, Equation (107) can be written as:

$$R_p = (k_{11}\phi_1^\bullet f_1 + k_{12}\phi_1^\bullet f_2 + k_{21}\phi_2^\bullet f_1 + k_{22}\phi_2^\bullet f_2)[R^\bullet][M] \quad (108)$$

where $[M]$ and $[R^\bullet]$ are the total concentrations of monomer and radical, respectively, explicitly defined by $[M] = [M_1] + [M_2]$ and $[R^\bullet] = [R_1^\bullet] + [R_2^\bullet]$.

Therefore, by defining the pseudo-kinetic rate constant for propagation, $k_p = k_{11}\phi_1^\bullet f_1 + k_{12}\phi_1^\bullet f_2 + k_{21}\phi_2^\bullet f_1 + k_{22}\phi_1^\bullet f_2$, the explicit expression of the polymerization rate, $R_p = k_p[R^\bullet][M]$ becomes the same as that for homopolymerization. In general for the multicomponent polymerization consisting of N components, the pseudo-kinetic rate constants can be defined as follows:

Propagation

$$k_p = \sum_{i=1}^{N}\sum_{j=1}^{N} k_{ij}\phi_i^\bullet f_j \quad (109)$$

Chain transfer to monomer

$$k_{fm} = \sum_{i=1}^{N}\sum_{j=1}^{N} k_{fm,ij}\phi_i^\bullet f_j \quad (110)$$

Chain transfer to small molecule, X

$$k_{fX} = \sum_{i=1}^{N} k_{fX,i}\phi_i^\bullet \quad (111)$$

Termination by disproportionation

$$k_{td} = \sum_{i=1}^{N}\sum_{j=1}^{N} k_{td,ij}\phi_i^\bullet \phi_j^\bullet \quad (112)$$

Termination by combination

$$k_{tc} = \sum_{i=1}^{N}\sum_{j=1}^{N} k_{tc,ij}\phi_i^\bullet \phi_j^\bullet \quad (113)$$

By application of the above pseudo-kinetic rate constants, a multicomponent polymerization reduces to a homopolymerization, and therefore Equations (73–79) are all applicable for copolymerization. With appropriate definitions for pseudo-kinetic rate constants, this method may also be usefully applied when copolymerization kinetics follow the penultimate model [46]. Note that for copolymerization, τ and β in Equations (73–79) change with the progress of polymerization not only by the change in the concentrations but by the change in the pseudo-kinetic rate constants.

8. Living Polymerization

Living polymerization is strictly defined as the chain polymerization from which chain termination and irreversible chain transfer are absent [11]. Practically, however, polymers are considered as living if their end groups retain the propensity of growth for at least as long a period as needed for the completion of an intended synthesis, or any other desired task [47]. Living polymerization was first reported for ionic polymerization, more specifically in the anionic polymerization [48]. Unlike radical polymerization, bimolecular termination between active centers does not occur in ionic polymerization. Termination of an active center on a polymer chain occurs by reaction with the counterion, solvent, monomer, or other species. If the termination and chain-transfer reactions are suppressed effectively, living polymerization proceeds. Often, the initiation reactions are very fast, and the initiator is consumed in the early stages of polymerization before the polymer chains have grown much beyond oligomeric size. Living polymerization was made possible also in cationic polymerization [49, 50]. In cationic living polymerization, monomers are not added one by one to an active center, but reversible deactivation occurs and monomer molecules are added to an active center during intermittent active period. In free-radical polymerization, bimolecular termination reactions between polymer radicals are inevitable, and therefore, it cannot be living polymerization in a strict sense, but by utilizing the reversible deactivation, living polymers in a practical sense can be synthesized [51–53]. Living polymerization has been made possible for all three types of active centers in chain polymerization, i.e., anion, cation, and radical.

Conventionally, polycondensation proceeds with step polymerization. However, by preventing the reaction between monomer and by making monomer react only with the propagating polymer chain end, it has become possible to conduct living polycondensation [54, 55]. Ideally, monomers are added to an activated polymer chain end one by one in this case.

8.1. Ideal Living Polymerization

In this section, the kinetic characteristics of the ideal living polymerization where the initiation reactions are instantaneous and monomers are added one by one to an active center without reversible deactivation processes are considered. The effect of reversible deactivation will be discussed in the next section.

Suppose active centers are created instantaneously with concentration $[P^*]$. The polymerization rate is represented by:

$$R_p = k_p[P^*][M] \qquad (114)$$

In the absence of termination, $[P^*]$ may be kept constant, and $k_p[P^*]$ could be regarded as a constant. In this case, Equation (114) can be integrated to give:

$$\int_0^x \frac{dx}{1-x} = k_p[P^*] \int_0^t dt \qquad (115)$$

or

$$\ln \frac{1}{1-x} = k_p[P^*]t \qquad (116)$$

where x is the monomer conversion to polymer, $x = ([M]_0 - [M])/[M]_0$.

Note, however, Equation (116) applies for the systems with a constant concentration of the active center $[P^*]$ during polymerization, and it does not necessarily mean that the reaction system is living. Balanced rates of initiation and termination may also lead to a constant $[P^*]$. In addition, with the reversible deactivation, which will be discussed in the next section, $[P^*]$ may change during polymerization, even when the system can be considered living polymerization.

The number-average chain length, $\bar{P}_n$ is the ratio between the total number of monomeric units incorporated into polymer and the total number of polymer chains, as represented by Equation (6). Therefore, $\bar{P}_n$ is given by:

$$\bar{P}_n = \frac{x[M]_0}{[P]} \qquad (117)$$

where $[P]$ represents the concentration of all polymer molecules, i.e., $[P] = Q_0 = \sum_{r=1}^{\infty}[P_r]$, where all polymer chains possess active centers in the present simplified model. With reversible deactivation, most of the polymer chains are dormant, but the total number of active and dormant polymer chains is kept constant for the most part of polymerization. The linear increase of $\bar{P}_n$ with respect to conversion, x is one of the most important characteristics of living polymerization.

Now consider the chain length distribution of the ideal living polymerization, by using the model shown in Figure 13. Assuming that

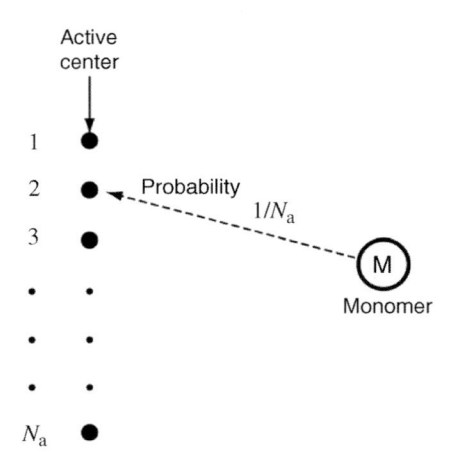

Figure 13. Model to derive the chain length distribution in ideal living polymerization. A monomer molecule is added to an active center selected randomly, with probability $1/N_a$, where N_a is the total number of growing chains having an active center

initiation is instantaneous, the total number of growing chains having an active center N_a is constant throughout the polymerization. Now a monomer molecule adds to a polymer chain as shown in Figure 13. The probability that a monomer adds to a particular polymer chain (2 in the figure) is $1/N_a$, and that it does not add to the polymer chain (2) is $1-1/N_a$. Therefore, when the total of N_M monomer units have been consumed and bound into polymer chains the probability that a randomly selected polymer chain possesses r monomer units is given by a binomial distribution:

$$N(r) = {}_{N_M}C_r \left(\frac{1}{N_a}\right)^r \left(1 - \frac{1}{N_a}\right)^{N_M - r} \quad (118)$$

where ${}_nC_k$ is the binomial coefficient, ${}_nC_k = \frac{n!}{k!(n-k)!}$.

Because $1/N_a \ll 1$ and N_M is large, the binomial distribution reduces to the Poisson distribution:

$$N(r) = \frac{e^{-\eta}\eta^r}{r!} \quad (119)$$

where $\eta = N_M/N_a$, which is equal to the number-average chain length, $\eta = \bar{P}_n$.

The mass fraction distribution $M(r)$ and the mass-average chain length $\bar{P}_M$ can be calculated, as follows:

$$M(r) = \frac{rN(r)}{\sum_{r=1}^{\infty} rN(r)} = \frac{e^{-\eta}\eta^{r-1}}{(r-1)!} \quad (120)$$

$$\bar{P}_M = \sum_{r=1}^{\infty} rM(r) = \bar{P}_n + 1 \quad (121)$$

On the basis of the random sampling technique, Equation (121) can be rationalized as follows. The mass-average chain length is the expected chain length when one monomeric unit in polymer is selected randomly, as was schematically shown in Figure 1. Now suppose that the unit shown by an arrow in Figure 14 is selected randomly. The expected chain length other than the randomly selected units is $\bar{n}_1 + \bar{n}_2$, as shown in Figure 14. Except for this randomly selected unit, we have a total of $N_M - 1$ monomer molecules initially. The

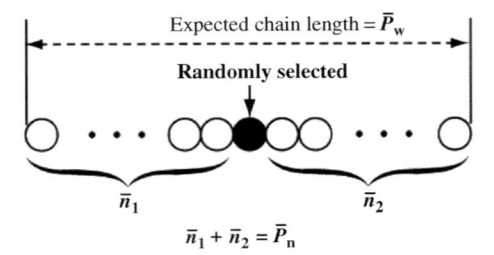

Figure 14. Weight-average chain length of the ideal living polymerization, derived on the basis of the random sampling technique

expected length, $\bar{n}_1 + \bar{n}_2$ is equal to the number-average chain length when $N_M - 1$ monomer molecules are distributed randomly, through the process shown in Figure 13. Obviously, $N_M \gg 1$, and $N_M - 1 \cong N_M$. Therefore, the expected length of a single chain distributed as shown in Figure 13 must be equal to $\bar{P}_n$, and $\bar{n}_1 + \bar{n}_2 = \bar{P}_n$. By including the randomly selected unit, the equation, $\bar{P}_M = \bar{P}_n + 1$ holds.

From Equation (121), the polydispersity index is given by:

$$\text{PDI} = \frac{\bar{P}_M}{\bar{P}_n} = 1 + \frac{1}{\bar{P}_n} \quad (122)$$

With sufficiently large average chain length, the PDI approaches unity.

On the other hand, in order for the Poisson distribution to be valid, the rate of initiation must be much faster than that of propagation. Slow initiation may lead to significant broadening of the MMD. In anionic polymerization, active chain ends with various states whose propagation rate constant is different greatly could exist [56]. If the exchange rate between these states is slow, the MMD could become much broader than the Poisson distribution.

8.2. Reversible-Deactivation Radical Polymerization

In free-radical polymerization, the bimolecular termination reactions are inevitable. Therefore, the living polymerization in a strict sense in which the chain termination reactions are totally absent is impossible. However, if a large percentage of polymer chains are dormant and can

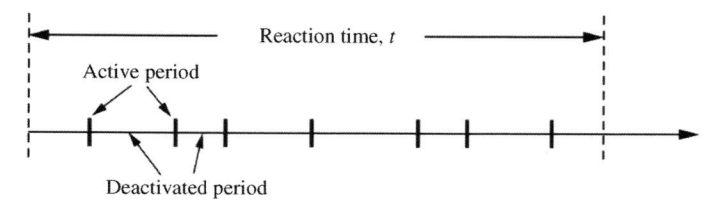

Figure 15. Schematic representation of the formation history of a polymer chain in reversible-deactivation radical polymerization (RDRP)

potentially grow further, such free-radical polymerization systems can be regarded as pseudo-living polymerization. By introducing the reversible-deactivation process in free-radical polymerization, polymers having a narrow polydispersity can be obtained, as long as the number of terminated chains is small compared with the potentially active chains.

This type of radical polymerization has been referred to as, "controlled", "controlled/living", or "living" radical polymerization. In the present article, the IUPAC name [11], reversible-deactivation radical polymerization, RDRP is used.

Because the lifetime of a generated radical in free-radical polymerization is short, normally less than a few seconds, a basic strategy to keep the chain potentially active is to distribute very short active periods throughout the whole reaction time. Figure 15 shows a schematic representation of the pseudo-living polymer formation during reversible-deactivation radical polymerization. The thickness of each vertical line represents the time length of an active period, which is typically $10^{-4} - 10^{-2}$ s, and therefore, the thickness should be much thinner than the figure shows. On the other hand, it may take hours for the whole polymerization time. From the point of view of the formation history of a chain, most of the time is spent as a deactivated form. Only when the number of added monomeric units in chain during a single active period is always unity, the formed polymers would have the Poisson distribution.

8.2.1. Polymerization Rate

Depending on the reversible-deactivation processes, various RDRPs have been proposed. RDRPs include stable-radical-mediated polymerization (SRMP), atom-transfer radical polymerization (ATRP), reversible-addition-fragmentation chain-transfer polymerization (RAFT), and degenerative-transfer radical polymerization (DTRP). The reversible-deactivation reactions are shown in Figure 16. Other elementary reactions involved are the same as the conventional free-radical polymerization, as shown in Equations (45)–(52). The deactivation in SRMP involves reversible coupling with stable (persistent) free radical, X in Figure 16. In ATRP, the deactivation of radicals involves atom transfer or reversible group transfer catalyzed often by transition-metal complexes. Both SRMP and ATRP are based on the principle in which active radicals are protected by reversible capping with a trapping agent, before termination or chain transfer occurs.

RAFT and DTRP are based on the reversible chain-transfer reactions. The kinetic chain growth process is schematically shown in Figure 17, neglecting the irreversible chain transfer reactions given by Equation (48)–(50). There are lots of dormant chains XP's, and when a radical is generated, it forms a single

SRMP $P_i X \underset{k_2}{\overset{k_1}{\rightleftarrows}} R_i^{\bullet} + X$
        RGS   Trap

ATRP $P_i X + Y \underset{k_2}{\overset{k_1'}{\rightleftarrows}} R_i^{\bullet} + XY$
        RGS   Trap

RAFT $R_i^{\bullet} + XP_j \underset{k_1}{\overset{k_2}{\rightleftarrows}} P_i X P_j \underset{k_2}{\overset{k_1}{\rightleftarrows}} P_i X + R_j^{\bullet}$
         Trap   RGS   Trap

DTRP $R_i^{\bullet} + XP_j \underset{k_{ex}}{\overset{k_{ex}}{\rightleftarrows}} P_i X + R_j^{\bullet}$

Figure 16. Reversible deactivation reaction scheme in each type of reversible-deactivation radical polymerization (RDRP). In the figure, $P_i X$ or XP_i is the dormant polymer with chain length i. $R_i^{\bullet}$ is the active polymer radical with chain length i

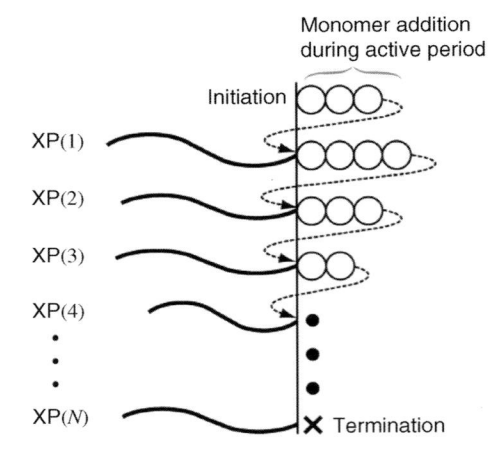

Figure 17. Schematic representation of the chain formation in RAFT and DT polymerization

kinetic chain, but this kinetic chain is separated into many small chain fractions as shown in Figure 17. The pseudo-livingness comes from the fact that an active radical is relayed to a large number of chains, N in Figure 17, before finally being stopped by bimolecular termination. If the value of N is large enough, sufficient percentage of chains are still potentially active and good livingness is kept. From the point of view of a single chain, the growth process is the intermittent chain growth, as was shown in Figure 15. This is why the theoretical MMD function is essentially the same for all RDRPs. Both RAFT and DTRP are based on the degenerative chain transfer. In RAFT it occurs by a two-step addition-fragmentation mechanism, while it is a single step in DTRP.

In order to formulate the kinetics of various types RDRPs in a unified manner, the radical generating species is represented by RGS, and the trapping agent to deactivate the active radical is represented by Trap, as shown in Figure 16. In the case of DTRP, XP (or equivalently, PX) acts as both RGS and Trap, depending on the reversible deactivation reaction type.

Similarly with Equations (35) and (114), the polymerization rate is represented by:

$$R_p = k_p[M][R^\bullet] \quad (123)$$

Unlike with Equation (114), the concentration of the active species $[R^\bullet]$ may change during polymerization. In the case of conventional free-radical polymerization, the radical concentration is determined by the balance of initiation and termination, as was shown in Equation (53). On the other hand, $[R^\bullet]$ is determined by the reversible deactivation reactions. (DTRP is a special case, as will be discussed later.)

For the calculation of polymerization rate and resulting conversion development, one can conveniently use Equation (123), by simultaneously solving differential population balance equations for various species [57]. On the other hand, however, Equation (123) does not involve concentrations of characteristic components of reversible deactivation reactions shown in Figure 16, essential to preserve the pseudo-livingness. Equation (123) may not be suitable for the prediction and control of the RDRPs, based on the reaction mechanism. The polymerization rate expression unique to the RDRPs can be derived as follows.

In order for the pseudo-living condition to be valid, the deactivation rate R_{deact} must be much larger than the bimolecular termination of active radicals R_t, i.e., $R_{\text{deact}} \gg R_t$. If not, a large amount of dead polymer chains are formed. Similarly, if the initiation reaction R_I is involved, the activation reaction in RDRP R_{act} must be much larger than R_I, i.e., $R_{\text{act}} \gg R_I$. The activation and deactivation rates are given by:

$$R_{\text{act}} = k_1[\text{RGS}] \quad (124)$$

$$R_{\text{deact}} = k_2[R^\bullet][\text{Trap}] \quad (125)$$

In the case of ATRP, k_1 is the pseudo-kinetic rate constant given by $k_1 = k'_1[Y]$. For DTRP, both R_{act} and R_{deact} are given by $k_{\text{ex}}[XP][R^\bullet]$.

Because $R_{\text{deact}} \gg R_t$, the average time interval for an active radical $\bar{t}_{R^\bullet}$ is represented as follows. The average frequency of a radical to be deactivated is given by $R_{\text{deact}}/[R^\bullet]$ s^{-1}. Therefore, the average time interval for the radical deactivation is:

$$\bar{t}_{R^\bullet} = \frac{[R^\bullet]}{R_{\text{deact}}} = \frac{1}{k_2[\text{Trap}]} \quad (126)$$

The average number of monomeric units added to a single radical was given by Equation (37). The average number of monomeric units added to an active radical during a single active period is given by:

$$\bar{n}_M \bar{t}_{R^\bullet} = \frac{k_p[M]}{k_2[\text{Trap}]} \quad (127)$$

Every time a radical is generated, $\bar{n}_M \bar{t}_{R^\bullet}$ of monomeric units are added. Therefore, the polymerization rate is given by multiplying the radical generation rate and $\bar{n}_M \bar{t}_{R^\bullet}$. Because $R_{act} \gg R_I$, the radical generation rate is given by R_{act}. Therefore, the polymerization rate, R_p is given by:

$$R_p = R_{act} \bar{n}_M \bar{t}_{R^\bullet} = k_p[M] K \frac{[\text{RGS}]}{[\text{Trap}]} \quad (128)$$

where $K = k_1/k_2$.

Equation (128) shows the characteristic polymerization rate expression that can be applied to all of RDRPs. The living cationic polymerization that involves reversible-deactivation reactions can also be represented by this equation. For a given reaction system, a large [RGS] leads to larger polymerization rate, while a large [Trap] leads to smaller polymerization rate. Obviously, in order for the pseudo-living polymerization to be valid, the conditions $R_{deact} \gg R_t$ and $R_{act} \gg R_I$ must be satisfied.

For DTRP, both R_{act} and R_{deact} are given by $k_{ex}[XP][R^\bullet]$, and $R_{act}\bar{n}_M\bar{t}_{R^\bullet}$ reduces to $k_p[M][R^\bullet]$, resulting in Equation (123). The radical concentration is controlled by the balance of R_t and R_I, but one needs to bear in mind that the conditions $R_{deact} \gg R_t$ and $R_{act} \gg R_I$ must be valid.

In all types of RDRPs, bimolecular termination is inevitable, and longer polymerization time tends to increase the number of dead polymer chains.

Unique characteristics of RDRP conducted in a very small reaction locus, such as for the miniemulsion polymerization in which the polymerization rate might be different from the corresponding bulk polymerization, can be elucidated on the basis of Equation (128), (→ Polymerization Processes, 2. Modeling of Processes and Reactors).

8.2.2. Molecular Mass Distribution

Consider the MMD formed in RDRPs, referring to the chain formation history in which the chain growth is intermittent, as shown in Figure 15. The time length of each active period is very small in order to suppress termination reactions effectively, and the concentration changes of various components during a single active period can be neglected. Note that the duration of a single active period must be much smaller than the lifetime of a radical in the conventional free-radical polymerization. After all, RDRP is a free-radical polymerization in which the lifetime of a radical in the conventional free-radical polymerization is divided into small pieces of active periods. Therefore, the probability of connection to the next unit during a single active period would be a constant, leading to the most probable distribution. The MMD of RDRP is formed by adding a large number of short chain parts that follow the most probable distribution.

Consider SRMP as an example. The average time length of a deactivated time $\bar{t}_D$ can be obtained as follows. The frequency of activation for a particular dormant chain PX is given by $k_1[PX]/[PX]$ s^{-1}, and therefore:

$$\bar{t}_D = 1/k_1 \quad \text{(for SRMP)} \quad (129)$$

Equation (129) applies also for ATRP, by defining $k_1 = k'_1[Y]$. The explicit expression of $\bar{t}_D$ for each type of RDRP is summarized in Table 2. The coefficient 2 for RAFT comes from the fact that another dormant period is repeated with probability 1/2 [58].

The probability of connecting next unit during the active period for SRMP is given by:

$$p = \frac{R_p}{R_p + R_{deact}} = \frac{k_p[M]}{k_p[M] + k_2[X]} \quad \text{(for SRMP)} \quad (130)$$

The p-values for other types of RDRPs are also shown in Table 2. The coefficient 1/2 for the reactivation rate in RAFT comes from the fact that another active period is repeated with probability 1/2 [58].

As shown in Table 2, both the average time length of a deactivated time $\bar{t}_D$ and the probability of growth during a single active period p change with the progress of polymerization. In order to highlight the important characteristics of RDRP polymers, however, let us consider a simplified case in which both $\bar{t}_D$ and p are kept constant during polymerization [59].

Table 2. Average time of deactivated period $\bar{t}_D$, and probability of connecting next unit p during the active period, for various types of RDRPs

	SRMP	ATRP	RAFT	DTRP
$\bar{t}_D$	$1/k_1$	$1/k_1 = 1/(k'_1[Y])$	$2(\bar{t}_{XP} + \bar{t}_{PXP}) = 2\left[\dfrac{1}{k_2[R^\bullet]} + \dfrac{1}{k_1}\right]$	$1/(k_{ex}[R^\bullet])$
p	$\dfrac{k_p[M]}{k_p[M] + k_2[X]}$	$\dfrac{k_p[M]}{k_p[M] + k_2[XY]}$	$\dfrac{k_p[M]}{k_p[M] + k_2[XP]/2}$	$\dfrac{k_p[M]}{k_p[M] + k_2[XP]}$

First, the chain length distribution during a single active period is considered. Figure 18 shows a schematic representation of chain growth during a single active period. Because the probability of connection p is assumed constant, the distribution is most probable. However, in the present case, the possibility of event in which no monomeric units are added must be accounted for. Similarly with the derivation of the most probable distribution, as shown in Figure 2, the number fraction distribution during a single active period, $N_{SA}(r)$ is given by:

$$N_{SA}(r) = p^r(1-p) \tag{131}$$

Obviously, the number- and mass-average chain length can be obtained from the distribution function; however, to show the versatility, the random sampling technique [5] is used here. The number-average chain length is the expected chain length when one chain end is chosen randomly. Suppose one has selected a chain that has just become active, as shown in Figure 18. The probability that a unit is connected to this active chain end is p, and if the first unit is connected, the chain length increases by one. Therefore, the expected length up to this first unit is $p \times 1$. The probability that the second unit is connected on the growing chain is p^2, and the expected number of units connected up to the second unit is $p \times 1 + p^2 \times 1$. As a consequence, the total expected length, which is equal to the number-average chain length during a single active period, $\bar{P}_{n,SA}$ is given by:

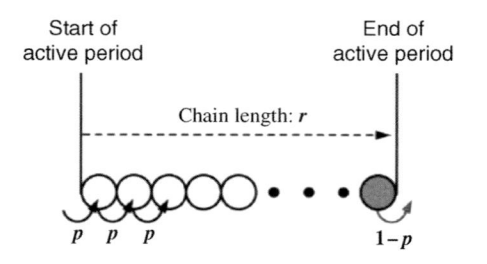

$$\bar{P}_{n,SA} = \sum_{r=1}^{\infty} p^r = \frac{p}{1-p} \tag{132}$$

The mass-average chain length is the expected length when one unit is selected randomly. Suppose a unit shown in Figure 19 has been selected randomly. Looking from the initially selected unit toward the right direction, the connection statistics is completely the same as shown in Figure 18, and therefore, the expected chain length is $\bar{P}_{n,SA}$. The same discussion applies also for the left direction. Remembering to include the initially selected unit, the mass-average chain length $\bar{P}_{M,A}$ is given by:

$$\bar{P}_{M,SA} = 2\bar{P}_{n,SA} + 1 = \frac{1+p}{1-p} \tag{133}$$

Note that the mass-average chain length is the same as the usual most probable distribution given by Equation (15), although the p-value for the usual RDRP is not very close to unity. The polydispersity index is given by:

$$\bar{P}_{M,SA}/\bar{P}_{n,SA} = \frac{1+p}{p} \tag{134}$$

In this case, the polydispersity index decreases to approach 2 when p is increased, while the usual most probable distribution leads to PDI $= 1 + p$ which goes from 1 to 2.

Let the reaction time be t_R, then the average number of active periods, $\bar{n}_A$ is given by:

$$\bar{n}_A = t_R/\bar{t}_D \tag{135}$$

Note that $\bar{t}_D \gg \bar{t}_R$ in RDRP.

Figure 18. Schematic representation of chain growth during a single active period in RDRP

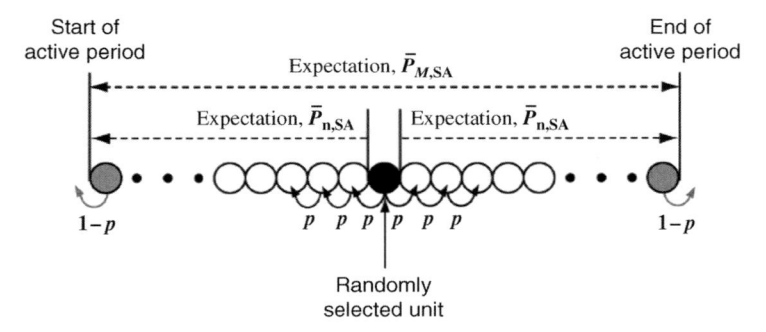

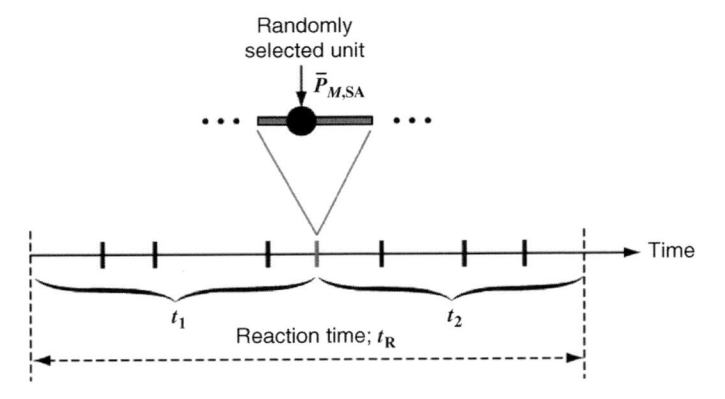

Figure 19. Derivation of the mass-average chain length during a single active period based on the random sampling technique

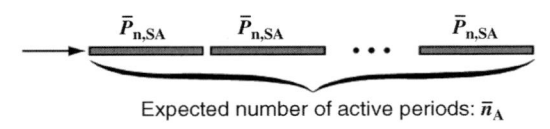

Figure 20. Derivation of the number-average chain length

When a chain end is chosen randomly, as shown in Figure 20, the expected number of active periods is $\bar{n}_A$, and the expected length formed during a single active period is $\bar{P}_{n,SA}$. Therefore, the expected length of a polymer chain when one chain end is selected randomly, which is equal to the number-average chain length, $\bar{P}_n$ is given by:

$$\bar{P}_n = \bar{n}_A \bar{P}_{n,SA} \tag{136}$$

The mass-average chain length is obtained by considering the expected chain length when one unit is selected randomly. Suppose a unit formed during a single active period at $t = t_1$ is selected, as shown in Figure 21. The expected length of this part of a polymer chain is $\bar{P}_{M,SA}$ because the selection is on a mass basis. All active periods are considered instantaneous, and therefore, the total reaction time except for the formation of this particular part of chain is still t_R, namely, $t_1 + t_2 = t_R$. The average number of active periods during the time interval t_R is $\bar{n}_A$. All other segments in this polymer chain are connected through their chain ends, and therefore, they are considered selected on the number basis. The mass-average chain length, $\bar{P}_M$ is, therefore, given by:

$$\bar{P}_M = \bar{P}_{M,SA} + \bar{n}_A \bar{P}_{n,SA} \tag{137}$$

Figure 21. Derivation of the mass-average chain length

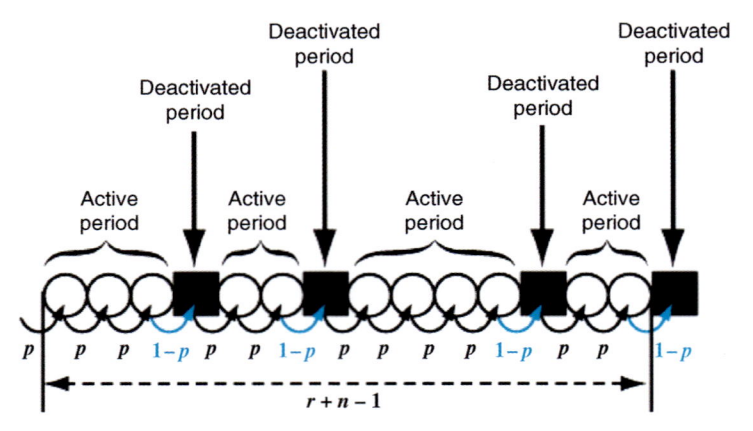

Figure 22. Schematic representation of a sequence of monomer units and deactivated periods

The polydispersity index is given by:

$$\bar{P}_M/\bar{P}_n = 1 + \frac{\bar{P}_{M,SA}}{\bar{P}_n} \tag{138a}$$

$$= 1 + \frac{1+p}{\bar{n}_A p} \tag{138b}$$

In the Poisson distribution, the PDI is $1 + 1/\bar{P}_n$ as shown in Equation (122), and therefore, the distribution given by Equation (138a) is broader than the Poisson distribution because $\bar{P}_{M,SA} > 1$.

It is usually stated that to obtain a good control in RDRP, the number of added monomeric units during a single active period must be kept small enough. Obviously, the condition, $R_{\text{deact}} \gg R_t$ is a requisite to have a good livingness, and a smaller p-value would be preferable. For a given $\bar{P}_n$-value, a smaller $\bar{P}_{M,SA}$ leads to a smaller PDI as shown in Equation (138a). The magnitude of $\bar{n}_A$ that is in the denominator of Equation (138b) has a significant effect, and by increasing the number of times of intermittent chain growth, the PDI can be made smaller. On the other hand, however, Equation (138b) also shows that if the $\bar{n}_A$-value is kept constant, a larger p-value leads to a smaller PDI. With a large p-value, the polymerization rate can also be increased. A good balance of $\bar{n}_A$ and p should be considered for commercial production.

The full chain length distribution function for the present simplified model, with constant values of $\bar{t}_D$ and p is as follows. First, consider the probability that a polymer chain with length r is formed through n times of active periods $N(r|n)$, as schematically shown in Figure 22. The probability of obtaining such a sequence of chain is $p^r(1-p)^n$. The total number of arrangements can be obtained by considering the total number of ways to place $(n-1)$ deactivated periods (depicted in square in Fig. 22) within r monomeric units. This can be done by choosing $(n-1)$ sites for the deactivated periods from the total of $(r+n-1)$ sites. Therefore, $N(r|n)$ is given by:

$$N(r|n) = {}_{r+n-1}C_{n-1} p^r (1-p)^n \tag{139}$$

Equation (139) approaches the Schulz–Zimm distribution [60, 61] (gamma distribution), sometimes used as a polymer MMD function, at $p \to 1$. In the present reaction system, however, p is usually less than 0.9, and the Schulz–Zimm distribution cannot be used instead of Equation (139).

Looking from a dormant chain, activation is a random process, similarly as in the classical examples of radioactive disintegrations, or incoming calls at a telephone exchange [62]. Therefore, the number of active periods within a given reaction time is given by the Poisson distribution. The probability that the chain experiences n times of active periods $P(n)$ is given by:

$$P(n) = \frac{(\bar{n}_A)^n e^{-\bar{n}_A}}{n!} \tag{140}$$

The derivation of Equation (140) for the present situation can be found in the appendix of [59].

The number fraction distribution, $N(r)$ is therefore given by [59]:

$$N(r) = \sum_{n=0}^{\infty} P(n)N(r|n) = (1-p)p^r \bar{n}_A e^{-\bar{n}_A} F(1+r, 2; (1-p)\bar{n}_A) \quad (141)$$

where $F(a, b; x)$ is the confluent hypergeometric function (Kummer's function of the first kind), represented by:

$$F(a, b; x) = 1 + \frac{ax}{b} + \frac{a(a+1)}{b(b+1)} \cdot \frac{x^2}{2!} + \ldots \quad (142)$$

The mass fraction distribution is given by:

$$M(r) = \frac{rN(r)}{\bar{P}_n} = (1-p)^2 p^{r-1} r e^{-\bar{n}_A} F(1+r, 2; (1-p)\bar{n}_A) \quad (143)$$

The fundamental distribution function for the living polymers formed in RDRP is a hypergeometric function that results from the superposition of two types of distribution, the most probable and the Poisson distribution. Another method to derive the present fundamental distribution for RDRP can be found in [63].

Figure 23 shows the calculated mass fraction distribution, $M(\log_{10} r)$. The comparison with the Poisson distribution is made at the same number-average chain lengths, $\bar{P}_n = 10$ and 100. The distribution is always broader than the Poisson distribution. At the same number-average chain length, the distribution becomes broader by increasing the p-value, i.e., by making the number of monomer addition during a single active period larger for a given $\bar{P}_n$.

Figure 24 shows the mass fraction distribution of the ideal RDRP polymers when the average number of active period is $\bar{n}_A = 10$. If the $\bar{n}_A$-value is the same, larger p-value makes the chain longer, leading to narrower distribution, as discussed earlier. Assuming the time length of the inactive period $\bar{t}_D$ is the same for both cases, these two profiles show the MMDs at the same reaction time. Larger p-value may lead to higher productivity with a sufficiently narrow distribution.

To calculate the MMDs for more realistic cases, including termination and irreversible

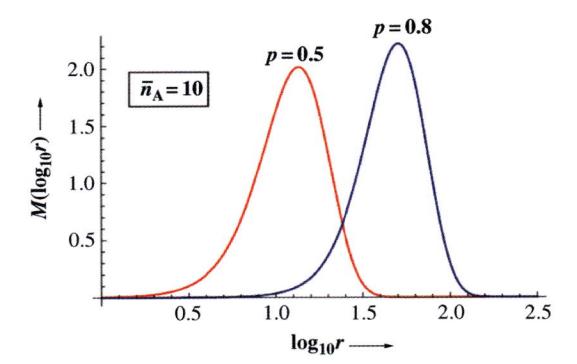

Figure 24. Mass fraction distribution of the ideal RDRP polymers when the average number of active period is set to be $\bar{n}_A = 10$

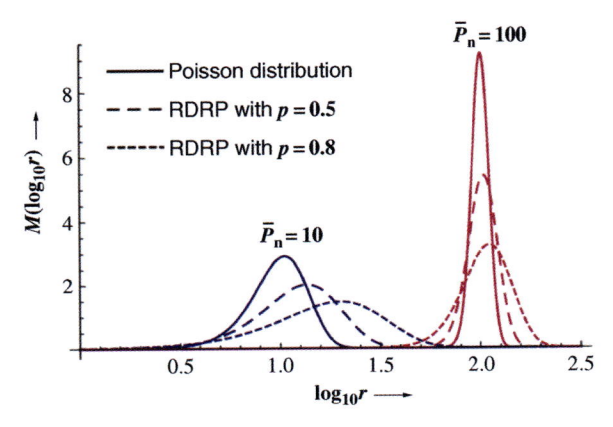

Figure 23. Mass fraction distribution for the ideal RDRP polymers and ideal living polymers (Poisson distribution), at the same number-average chain lengths, $\bar{P}_n = 10$ and 100

chain transfer reactions, deterministic approaches to solve an infinite number of differential population balance equations approximately [64, 65], or a Monte Carlo simulation method [66–68] can be applied.

9. Polymers with Branches and Crosslinks

We have, until now, considered the cases where produced polymer molecules are linear. While linear polymers are important, especially to understand the fundamentals of polymer chain formation, introduction of branches and/or crosslinks endows polymers with further versatility. However, nonlinear polymer formation is a kind of advanced topic, and in this section, discussions are made only for the most important characteristics of nonlinear polymers, assuming idealized conditions.

9.1. Crosslinked Polymers

An important characteristic of the introduction of crosslinks to linear polymer systems is the formation of a gel molecule. A gel molecule is considered to be a crosslinked polymer with infinite molecular mass, and as a consequence, a gel molecule is insoluble in any solvent under conditions where polymer degradation does not occur.

Gel is a molecule that spans the whole reaction system, and therefore, if one chooses one monomeric unit in polymer randomly, as was shown in Figure 1B, the probability of choosing a gel molecule must be larger than zero. On the other hand, because the number of polymer molecules in the system could be infinite, the probability of selecting a gel molecule on a number basis could be zero. Therefore, the onset of gelation can be recognized as a point when the mass-average degree of polymerization (chain length) $\bar{P}_M$ goes to infinity [4].

Let us consider the mass-average chain length of the whole system when crosslinks are introduced randomly to a linear polymer system, as shown in Figure 25. The linear polymer chains that constitute a crosslinked polymer system are called the primary chains. Here, we have primary chains whose mass-average chain length is $\bar{P}_{Mp}$. The probability for a unit (monomeric unit) to have a crosslink point is called the crosslinking density, represented by ρ in this article. Note that a crosslink point is defined as a unit to have a tri-branch point [4], and therefore, two crosslink points are formed by a single crosslinkage. In the random crosslinking considered here, the crosslinking density ρ is the same for all units.

Suppose one monomeric unit shown in Figure 26 is selected randomly from the crosslinked polymer system. Because the selection is made on a mass basis, the expected chain length of the primary chain so selected is the mass-average chain length of the primary chains, $\bar{P}_{Mp}$. This primary chain is considered belonging to the 0th generation, as shown in Figure 26. The expected number of crosslink points on the 0th generation chain is $\bar{P}_{Mp}\rho$, which must be equal to the expected number of primary chains belonging to the 1st generation. The probability of possessing a crosslink point is the same for all units, and therefore, if one chooses a primary

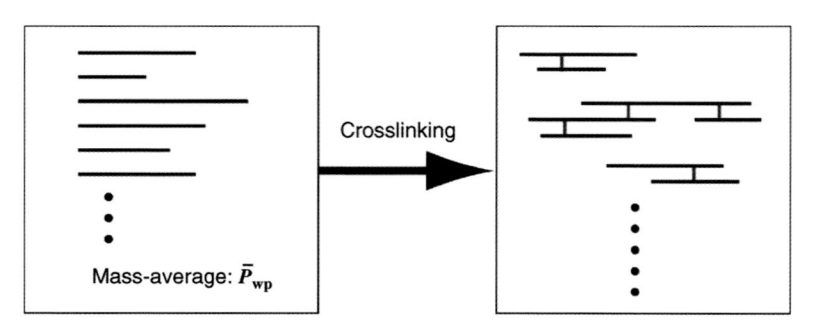

Figure 25. Schematic representation of random crosslinking of linear polymer chains

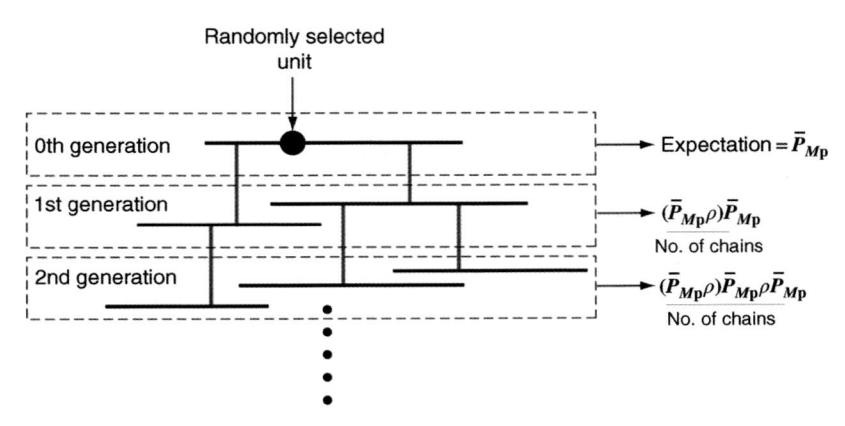

Figure 26. Derivation of the mass-average chain length $\bar{P}_M$ for a random crosslink system

chain by selecting a crosslink point randomly, the expected primary chain length so selected is the mass-average chain length, $\bar{P}_{Mp}$. The primary chains belonging to the 1st generation are selected by choosing crosslink points randomly, and the total expected monomeric units (chain length) belonging to the first generation is given by $(\bar{P}_{Mp}\rho) \times \bar{P}_{Mp}$.

The total expected number of crosslink points on the 1st generation chains connected to the 2nd generation is $(\bar{P}_{Mp}\rho)(\bar{P}_{Mp} - 1)\rho$. Note that $(\bar{P}_{Mp} - 1)$ is used here because one unit has already been used to connect with the 0th generation chain. Assuming long primary chains, $(\bar{P}_{Mp} - 1) \cong \bar{P}_{Mp}$. The number of primary chains belonging to the second generation is given by $(\bar{P}_{Mp}\rho)\bar{P}_{Mp}\rho$. The primary chains belonging to the second generation must be selected on a mass basis, and the expected size of each primary chain is again $\bar{P}_{Mp}$. The total expected chain length belonging to the 2nd generation is $(\bar{P}_{Mp}\rho)^2 \bar{P}_{Mp}$.

Extending the present discussion to further generations, the expected chain length of the whole polymer molecule when one unit is selected randomly, which is the mass-average chain length of the whole reaction mixture, $\bar{P}_M$ is given by:

$$\bar{P}_M = \bar{P}_{Mp}\{1 + (\bar{P}_{Mp}\rho) + (\bar{P}_{Mp}\rho)^2 + (\bar{P}_{Mp}\rho)^3 + \cdots\}$$
$$= \bar{P}_{Mp} \sum_{i=0}^{\infty} (\rho \bar{P}_{Mp})^i = \frac{\bar{P}_{Mp}}{1 - \rho \bar{P}_{Mp}} \quad (144)$$

Equation (144) shows that $\bar{P}_M$ goes to infinity when the crosslinking density ρ reaches:

$$\rho_{gp} = 1/\bar{P}_{Mp} \quad (145)$$

which is the gel point. Note that as long as the mass-average chain length of primary chains, $\bar{P}_{Mp}$ is kept constant, the gel point, ρ_{gp} is the same irrespective of the chain length distribution of the primary chains.

The number-average chain length, $\bar{P}_n$ is simply the ratio of the total number of monomeric units in polymer and the number of polymer molecules, as represented by Equation (6). Because the total number of monomeric units in the system does not change with the crosslinking process, $\bar{P}_n$ can be obtained simply by considering how the crosslinking process changes the number of polymer molecules. By introducing a single crosslinkage, the number of polymer molecules decreases by one, and the following equation holds:

$$\frac{1}{\bar{P}_n} = \frac{1}{\bar{P}_{np}} - \frac{\rho}{2} \quad (146)$$

where $\bar{P}_{np}$ is the number-average chain length of the primary chains, and $\rho/2$ represents the number of crosslinkages. Remember that one crosslinkage consists of two crosslink points by definition [4]. Therefore, the number-average chain length of the crosslinked polymer system,

$\bar{P}_n$ is given by:

$$\bar{P}_n = \frac{\bar{P}_{np}}{1 - \rho\bar{P}_{np}/2} \tag{147}$$

With Equation (145), the number-average chain length at the gel point, $\bar{P}_{n,gp}$ is given by:

$$\bar{P}_{n,gp} = \frac{\bar{P}_{np}}{1 - \bar{P}_{np}/2\bar{P}_{Mp}} \tag{148}$$

Assuming the primary chain length distribution is the most probable, $\bar{P}_{Mp}/\bar{P}_{np} = 2$, and therefore, $\bar{P}_{n,gp} = (4/3)\bar{P}_{np}$. The number-average chain length is only 4/3 times as large as that of the primary chains, which is rather small, even though $\bar{P}_M$ goes to infinity.

Next, consider the mass fraction distribution of the random crosslink system. Let $M_p(r)$ be the mass chain-length distribution of the primary chains. Because the most probable distribution is the most fundamental polymer distribution, consider the case where $M_p(r)$ follows the most probable distribution, represented by Equation (14), that is:

$$M_p(r) = \frac{r}{(\bar{P}_{np})^2} \exp\left(-\frac{r}{\bar{P}_{np}}\right) \tag{149}$$

Consider the fractional mass-based chain length distribution of the polymer molecules having k crosslinkages, $M_k(r)$ defined by:

$$M(r) = \sum_{k=0}^{\infty} M_k(r) \tag{150}$$

To obtain a polymer molecule without any crosslinkage ($k = 0$) by randomly choosing a unit, the selected primary chain must not possess any crosslink point. The fractional mass-based distribution, $M_0(r)$ is given by:

$$M_0(r) = (1 - \rho)^r M_p(r) \cong \exp(-\rho r) M_p(r) \quad (\text{for } \rho \ll 1) \tag{151}$$

For the primary chains following the most probable distribution:

$$M_0(r) = \frac{r}{(\bar{P}_{np})^2} \exp\left[-\left(\frac{1 + \rho\bar{P}_{np}}{\bar{P}_{np}}\right)r\right] \tag{152}$$

Next, consider $M_1(r)$. When one unit is selected randomly, as shown in Figure 26, a polymer molecule with chain length r having one crosslinkage is obtained by connecting two primary chains with length s and $r - s$. The 0th generation chain must possess one crosslink point and one $(s - 1)$ uncrosslinked point. Note that any one of the units can possess a crosslink point, there are s ways of choosing a crosslink point on this chain. On the other hand, the 1st generation chain must not possess any further crosslink point. Therefore, $M_1(r)$ is given by:

$$M_1(r) = \sum_{s=1}^{r-1}\left[M_p(s)s\rho(1-\rho)^{s-1}\right]\left[M_p(r-s)(1-\rho)^{r-s-1}\right] \tag{153}$$

$$\cong \rho(1-\rho)^{r-2} \int_0^r s M_p(s) M_p(r-s) ds$$

For the case with $M_p(r)$ given by Equation (149):

$$M_1(r) = \frac{\rho(1-\rho)^{r-2}}{(\bar{P}_{np})^4} \exp\left(-\frac{r}{\bar{P}_{np}}\right) \int_0^r s^2(r-s) ds \tag{154}$$

$$\cong \frac{r}{(\bar{P}_{np})^2}\left(\frac{\rho\bar{P}_{np}}{2!3!}\right)\left(\frac{r}{\bar{P}_{np}}\right)^3 \exp\left[-\left(\frac{1+\rho\bar{P}_{np}}{\bar{P}_{np}}\right)r\right]$$

By continuing the present reasoning, it is straightforward to obtain the fractional mass-based distribution having k crosslinkages, $M_k(r)$ is given by [5]:

$$M_k(r) = \frac{r}{(\bar{P}_{np})^2}\left(\frac{(\rho\bar{P}_{np})^k}{(k+1)!(2k+1)!}\right)\left(\frac{r}{\bar{P}_{np}}\right)^{3k} \exp\left[-\left(\frac{1+\rho\bar{P}_{np}}{\bar{P}_{np}}\right)r\right] \tag{155}$$

The whole distribution consisting of all values of k is given by:

$$M(r) = \sum_{k=0}^{\infty} M_k(r)$$

$$= {}_0F_2\left(;\frac{3}{2},2;\frac{\rho r^3}{4(\bar{P}_{np})^2}\right)\frac{r}{(\bar{P}_{np})^2}\exp\left[-\left(\frac{1+\rho\bar{P}_{np}}{\bar{P}_{np}}\right)r\right] \tag{156}$$

In Equation (152), ${}_pF_q[a_1, \cdots, a_p; b_1, \cdots, b_q; z]$ represents a generalized hypergeometric function [69] defined by:

$${}_pF_q[a_1, \cdots, a_p; b_1, \cdots, b_q; z] = \sum_{k=0}^{\infty} \frac{(a_1)_k \cdots (a_p)_k}{k!(b_1)_k \cdots (b_q)_k} z^k \tag{157}$$

where $(a)_k$ is the Pochhammer symbol defined by:

$$(a)_k \equiv \frac{\Gamma(a+k)}{\Gamma(a)} \tag{158}$$

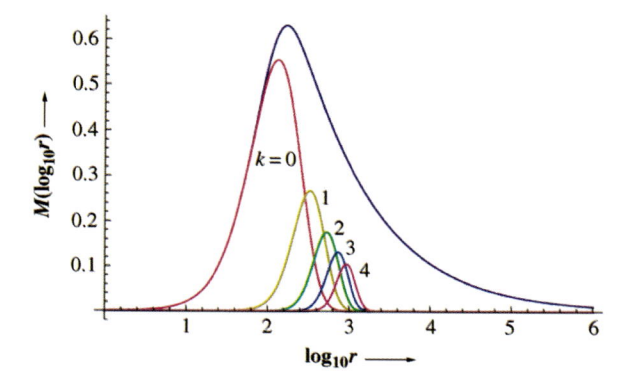

Figure 27. Calculated mass fraction distribution right at the gel point, with $\overline{P}_{np} = 100$ and $\rho = 5 \times 10^{-3}$. The fractional chain length distributions containing k crosslinkages are also shown

For the actual calculation of the hypergeometric function, one can simply use a build-in function of the calculation software.

Figure 27 shows the calculated whole mass fraction distribution, Equation (156) and the fractional distribution, Equation (155) right at the gel point. Even at the gel point, the mass fraction of linear chains ($k = 0$) is the largest within the fractional MMDs.

The mass fraction of polymers containing k crosslinkages is given by:

$$M_k = \int_0^\infty M_k(r)dr = \frac{(3k+1)!(\rho \overline{P}_{np})^k}{(k+1)!(2k+1)!}\left(\frac{1}{1+\rho \overline{P}_{np}}\right)^{3k+2} \quad (159)$$

When the fractional MMD given by Equation (155) is normalized to make the total area of each $M_k(r)$ function unity, $M_{k,\text{norm}}(r)$ is given by the following Schulz–Zimm distribution [60, 61], or equivalently, the gamma distribution:

$$M_{k,\text{norm}}(r) = \frac{M_k(r)}{M_k} = \frac{\sigma^\sigma}{\overline{P}_{n,k}\Gamma(\sigma)}\left(\frac{r}{\overline{P}_{n,k}}\right)^\sigma \exp\left(-\frac{\sigma r}{\overline{P}_{n,k}}\right) \quad (160)$$

where $\sigma = 3k + 1$, and $\overline{P}_{n,k}$ is the number-average chain length of the polymers having k crosslinkages, given by:

$$\overline{P}_{n,k} = \frac{(3k+1)\overline{P}_{np}}{1+\rho \overline{P}_{np}} \quad (161)$$

For the Schulz–Zimm distribution, σ is represented by the number- and mass-average as follows:

$$\sigma = \frac{\overline{P}_{n,k}}{\overline{P}_{M,k} - \overline{P}_{n,k}} \quad (162)$$

From Equation (162), the mass-average chain length of the polymers having k crosslinkages, given by:

$$\overline{P}_{M,k} = \frac{(3k+2)\overline{P}_{np}}{1+\rho \overline{P}_{np}} \quad (163)$$

The PDI of the polymers having k crosslinkages is given by:

$$\overline{P}_{M,k}/\overline{P}_{n,k} = \frac{3k+2}{3k+1} \quad (164)$$

The PDI decreases with increasing the number of crosslinkages k, and when $k = 3$, the PDI has already become so small as 1.1.

Even with the random crosslinking, the crosslinking density of each polymer molecule is different. Each small dot in Figure 28 shows the chain length and crosslinking density of each polymer molecule generated by using a Monte Carlo simulation method [5]. The dashed line shows the crosslinking density of the whole reaction system, $\rho = 4 \times 10^{-3}$. As clearly shown in Figure 28, most of cross-linked polymer molecules are distributed around the asymptotic crosslinking density with $r \to \infty$, i.e., $\lim_{r \to \infty} \overline{\rho}(r)$, which is considered a representative value of the crosslinked polymers, rather than ρ.

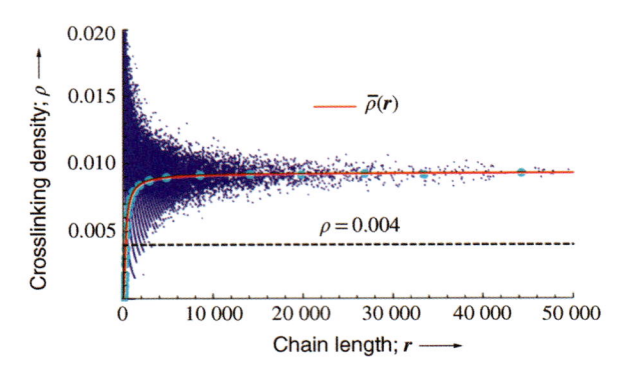

Figure 28. Relationship between crosslinking density and chain length of each polymer molecule generated by a Monte Carlo simulation (small dot), their average crosslinking density within the chain length intervals (circular symbols), with $\overline{P}_{np} = 100$ and $\rho = 4 \times 10^{-3}$. The solid curve, $\overline{\rho}(r)$ is calculated from Equation (165)

The average crosslinking density of polymers having chain length r, $\overline{\rho}(r)$ can be obtained from $M_k(r)$-function, as follows:

$$\overline{\rho}(r) = \frac{\sum_{k=0}^{\infty} \frac{2k}{r} M_k(r)}{\sum_{k=0}^{\infty} M_k(r)} = \left(\frac{\rho r^2}{6(\overline{P}_{np})^2}\right) \frac{{}_0F_2\left(;\frac{5}{2},3;\frac{\rho r^3}{4(\overline{P}_{np})^2}\right)}{{}_0F_2\left(;\frac{3}{2},2;\frac{\rho r^3}{4(\overline{P}_{np})^2}\right)} \quad (165)$$

The solid curve in Figure 28 shows the calculated result of Equation (165). The value of $\lim_{r \to \infty} \overline{\rho}(r)$ can be determined from this equation, at least, numerically.

The average crosslinking density of polymers having k crosslinkages, is given by:

$$\overline{\rho}(k) = \frac{\int_0^{\infty} \left[\frac{2k}{r} M_k(r)\right] dr}{\int_0^{\infty} M_k(r) dr} = \frac{2k(1 + \rho \overline{P}_{np})}{(3k+1)\overline{P}_{np}} \quad (166)$$

As was shown in [70], $\lim_{r \to \infty} \overline{\rho}(r) \leq \lim_{k \to \infty} \overline{\rho}(k)$, and the equality holds only right at the gel point (gp), i.e.;

$$\lim_{r \to \infty} \overline{\rho}(r)|gp = \lim_{k \to \infty} \overline{\rho}(k)|gp = \frac{2(1 + \rho \overline{P}_{np})}{3\overline{P}_{np}} \quad (167)$$

The crosslinking density at the gel point is $\rho = 1/\overline{P}_{Mp}$, and therefore, the following equation holds:

$$\lim_{r \to \infty} \overline{\rho}(r)|gp = 2\rho_{gp} \quad (168)$$

Note that $\overline{P}_{Mp}/\overline{P}_{np} = 2$ for the most probable distribution. At the gel point, the crosslinking density of a gel molecule just formed is twice as large as the crosslinking density of the whole system, $\rho_{gp} = 1/\overline{P}_{Mp}$. Equation (168) is valid, irrespective of the primary chain length distribution [4].

More details on random crosslinking, including the post-gelation behavior can be found in [4–6, 70–75]. Although the random crosslinking is an idealized reaction system, the obtained results are important for understanding the fundamental characteristics of crosslinked polymer systems, at least qualitatively. In the present classical theory, the ring formation is neglected, and it is called the ring-free model. In a real system, rings would be formed. However, one needs to understand that the ring-free model is a natural consequence as long as the usual chemical kinetics is valid, in which the reaction rate is represented by the product of concentrations of functional groups. Suppose an active site is looking for a partner to react. The concentration of functional groups that can react with this active site on its own polymer molecule is zero for all sol-polymer molecules, because the mass fraction of any single sol-polymer molecule is zero. The ring formation is prohibited within sol-polymer molecules, as long as the conventional chemical kinetics is valid. On the other hand, a single gel molecule possesses a nonzero mass fraction, the intramolecular reaction or the ring formation within the gel molecule is not prohibited and is accounted for even in the ring-free model. In fact, the distribution functions shown in Equations (155) and (156) apply to the chain length distribution of sol molecules in the post-gelation period.

An example in which rings are formed in the context of the ring-free model can be found in the cases of emulsion polymerization [76–79]. In emulsion polymerization, because the number of molecules involved in a single reaction locus is limited, each polymer molecule possesses a nonzero mass fraction, and the usual chemical kinetics naturally includes cycle formation.

9.2. Branched Polymers

There are two distinct types of branching; short-chain branching and long-chain branching. The short-chain branching is formed via backbiting reaction in free-radical polymerization, in which the active radical on the chain end abstracts a hydrogen atom intramolecularly. The backbiting is a chain transfer reaction to its own chain, and a small branch is left after the transferred radical grows. In this case, the chains consisting of only several carbons in length are the branches. High pressure free-radical polymerization of ethylene is a representative reaction to produce short-chain branches.

On the other hand, long-chain branches are formed by intermolecular reactions among polymer molecules. In free-radical polymerization, long-chain branches are formed through chain transfer reactions to other polymers and the terminal double-bond polymerization. In this section, important characteristics of long-chain branched polymers are highlighted.

Figure 29 shows an example of random branched polymer structure, formed from the primary chains having the most probable distribution. *Random branching* means that the probability of possessing a branch point is the same for all units in polymer, which provides a kind of standard structure of branched polymers. In addition, the most probable distribution is one of the most fundamental distributions in synthetic polymers, and therefore, this figure could be considered a typical structure of highly branched polymer. From the figure, a large number of relatively small branches are shown, although their lengths are much longer than the branches in the short-chain branching. In the

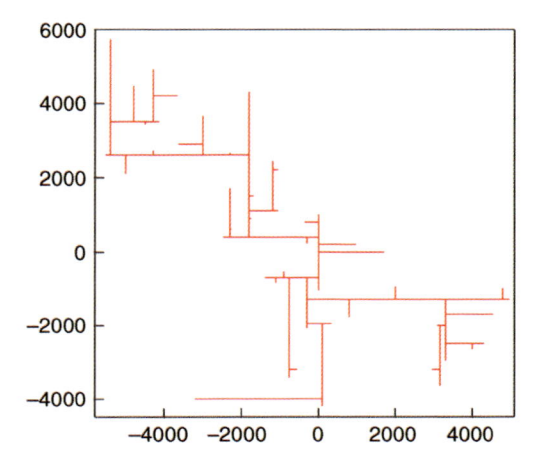

Figure 29. Representative example of a random branch polymer, in which the probability of possessing a branch point is the same for all units in chain and the primary chains follow the most probable distribution. In this particular example, the number-average chain length of primary chains is $\bar{P}_{np} = 1000$, and the chain length of this particular branched polymer is $r = 54\,000$, consisting of 46 primary chains [5]

random branched polymers, the probability of any chain end being a branch point is the same for all chain ends. In the most probable distribution, smaller chains are larger in number, as was shown in Figure 3A, and these smaller chains tend to become the branch chains. A longer chain possesses a larger number of units in it, and therefore, the probability of having a branch point is larger for longer chains and longer chains tend to become the backbone chains. Note that only one of two chain ends of a primary chain can be connected to a backbone chain in the present discussion. If both chain ends are connected, they are crosslinks, not the branches, as discussed later.

The probability that a unit possesses a branch point is defined as the branching density, ρ_b. The ρ_b-value is the same for all units in the random branched polymers. On the other hand, ρ_b is not a constant in nonrandom branching. For example, in the polymer transfer reactions in free-radical polymerization, the primary chains formed at earlier stages of polymerization are subjected to branching reactions for a longer period of time, the expected branching density is larger for the primary chains born earlier [5, 80].

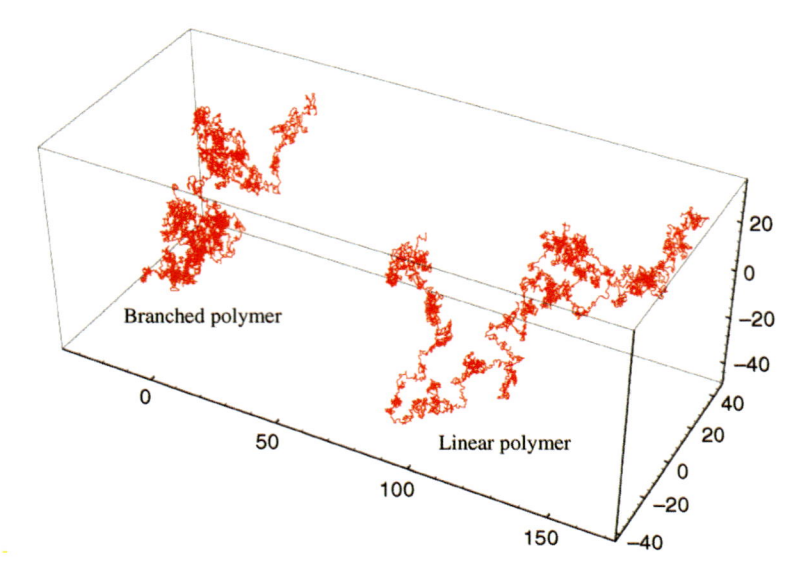

Figure 30. Examples of 3D structure of a random branch polymer and a linear polymer under unperturbed conditions. The chain lengths of both polymers are the same, $r = 6294$. The branch polymer consists of 8 primary chains [5]

Polymers containing branches/crosslinks possess smaller 3D size than linear polymers. Figure 30 shows an example of 3D structure of a random branch polymer under perturbed conditions, as well as that for a linear polymer having the same chain length $r = 6294$. The 3D size is clearly smaller for the branched polymers.

The radius of gyration $\sqrt{\langle s^2 \rangle}$ of the random branch polymer whose primary chains follow the most probable distribution is given by [81]:

$$\sqrt{\langle s^2 \rangle} = \sqrt{\langle s^2 \rangle_{\text{linear}}} \left\{ \left(1 + \frac{\overline{m}}{7}\right)^{0.5} + \frac{4\overline{m}}{9\pi} \right\}^{-0.25} \quad (169)$$

where $\overline{m}$ is the average number of branch points within the polymer molecules having the same chain length r, and $\sqrt{\langle s^2 \rangle_{\text{linear}}}$ is the radius of gyration for the linear polymers with chain length r. When using Equation (169), $\overline{m}$ is calculated from $\overline{m} = r\overline{\rho}_b(r)$. Similarly with the random crosslinking system shown in Figure 28, the branching density of each polymer molecules is distributed around the asymptotic value, $\lim_{r \to \infty} \overline{\rho}_b(r)$, not the average branching density of the whole system [82], as will be graphically shown in Figure 35. The value of $\lim_{r \to \infty} \overline{\rho}_b(r)$ is much more important than the average branching density of the whole reaction system. Further note that $\overline{m} = r \lim_{r \to \infty} \overline{\rho}_b(r)$ could be a good approximation for $\overline{m} = r\overline{\rho}_b(r)$ when using Equation (169) [83].

The number-average chain length, $\overline{P}_n$ can be determined as follows, because the number of polymer molecules decreases by one due to the formation of a single branch point:

$$\frac{1}{\overline{P}_n} = \frac{1}{\overline{P}_{np}} - \rho_b \quad \text{or} \quad \overline{P}_n = \frac{\overline{P}_{np}}{1 - \rho_b \overline{P}_{np}} \quad (170)$$

Next, consider the mass-average chain length $\overline{P}_M$. Suppose a monomeric unit shown in Figure 31 is chosen randomly. Before considering the total expected number of monomeric units connected into this polymer molecule $\overline{P}_M$, let us consider the branched chain part in the downward direction, as shown in Figure 31.

The expected chain length of the 0th generation primary chain is obviously $\overline{P}_{Mp}$, because it is selected on a mass basis, as was shown in Figure 1B. The expected number of branch points on the 0th generation chain is $\overline{P}_{Mp}\rho_b$, which is equal to the number of primary chains belonging to the 1st generation. Because only the chain end can be connected to the branch

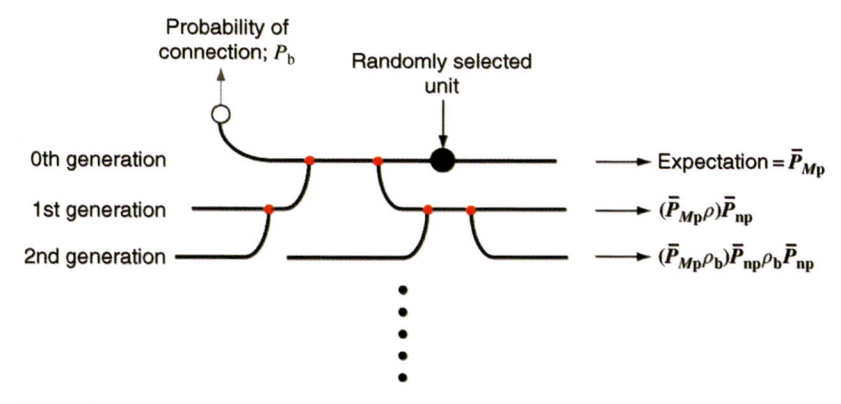

Figure 31. Schematic representation of a randomly selected polymer molecule toward the downward direction

point, the branch chains must be selected on a number basis, and therefore, the expected chain length of a branch chain is $\bar{P}_{np}$. The total chain length belonging to the 1st generation is given by $(\bar{P}_{Mp}\rho_b)\bar{P}_{np}$. Similarly, the total chain length belonging to the 2nd generation is given by $(\bar{P}_{Mp}\rho_b)\bar{P}_{np}\rho_b\bar{P}_{np}$ as shown in Figure 31. By continuing the present processes repeatedly, the total expected mass in the downward direction, $\bar{P}_{M,A}$ is given by:

$$\bar{P}_{M,A} = \bar{P}_{Mp} + \bar{P}_{Mp}(\rho_b\bar{P}_{np}) + \bar{P}_{Mp}(\rho_b\bar{P}_{np})^2 + \cdots$$
$$= \bar{P}_{Mp}\sum_{i=0}^{\infty}(\rho_b\bar{P}_{np})^i = \frac{\bar{P}_{Mp}}{1-\rho_b\bar{P}_{np}} \quad (171)$$

Up to the present stage, the chains connected upward direction has not been considered. The chain end of the 0th generation primary chain is connected with probability P_b, which is given by:

$$P_b = \frac{(\text{total number of branch points})}{(\text{total number of primary chains})} = \frac{\rho}{1/\bar{P}_{np}} = \rho_b\bar{P}_{np} \quad (172)$$

Note that $P_b < 1$, i.e., $\rho_b\bar{P}_{np} < 1$. Therefore, $\bar{P}_{M,A}$ given by Equation (171) never goes to infinity.

Figure 32 shows the connection of a head unit of the 0th generation primary chain. Suppose the head unit is connected. The connected primary chain is selected on a mass basis, because any of the monomeric units in the system can be connected, and therefore, the expected chain length is $\bar{P}_{Mp}$. Except for the branched unit of the connected chain, there are $\bar{P}_{Mp} - 1$ units that can have a branch point. Therefore, the expected number of branch points on the connected primary chain is $(\bar{P}_{Mp} - 1)\rho_b \cong \bar{P}_{Mp}\rho_b$. Looking at Figure 31 again, one would notice that the expected chain length toward the downward direction is $\bar{P}_{M,A}$, excluding the branch unit connected from the 0th generation chain. By continuing the connection through the head unit, the total expected mass of a polymer molecule by choosing one unit randomly, which is the mass-average chain length

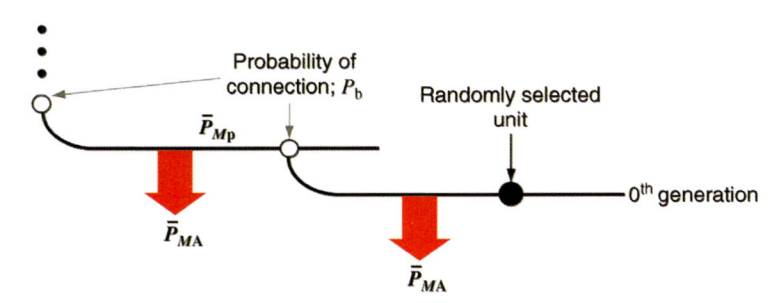

Figure 32. Schematic representation of the connection of a head unit of the randomly selected primary chain

of the whole system, $\overline{P}_M$ is given by:

$$\overline{P}_M = \overline{P}_{M,A} + P_b\overline{P}_{M,A} + (P_b)^2\overline{P}_{M,A} + \cdots = \overline{P}_{M,A}\sum_{i=0}^{\infty}(\rho_b\overline{P}_{np})^i$$

$$= \frac{\overline{P}_{Mp}}{(1-\rho_b\overline{P}_{np})^2} \quad (173)$$

The PDI of a random branched polymer system is given by:

$$\overline{P}_M/\overline{P}_n = \frac{(\overline{P}_{Mp}/\overline{P}_{np})}{1-\rho_b\overline{P}_{np}} \quad (174)$$

Equation (174) shows that the PDI of random branch polymers is always larger than the PDI of the primary chains.

Note that Equations (170), (173), and (174) are valid for the random branched polymers, irrespective of the primary polymer distribution.

Because $p_b = \rho_b\overline{P}_{np} < 1$, the mass-average chain length never goes to infinity, which means that gelation can never occur. In general, gelation never occurs only through branches [84], and crosslinks are required for gelation.

Figure 33 shows some examples of H-shaped (crosslinking) and T-shaped (branching) chain connection. The H-shaped chain connections, i. e., the crosslinkages are required for gelation to occur. In the case of free-radical polymerization that involves chain transfer to polymer, combination termination is required to cause gelation. Conversely, if the reaction system causes gelation, there should exist some reaction that forms the H-shaped chain connections.

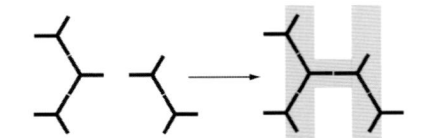

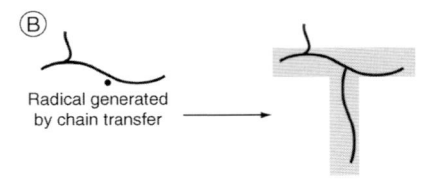

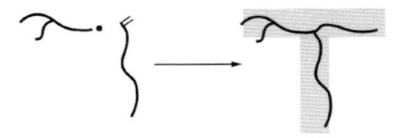

As was done for the random crosslink system, consider the mass fraction distribution formed through random branching of the primary chains having the most probable distribution. By using the similar random sampling method introduced in Section 9.1, the fractional mass-based chain length distribution of the polymers possessing k branch points is given by [5, 82]:

$$M_k(r) = \left(\frac{1-\rho_b\overline{P}_{np}}{\overline{P}_{np}}\right)\left(\frac{(\rho_b\overline{P}_{np})^k}{k!(k+1)!}\right)\left(\frac{r}{\overline{P}_{np}}\right)^{2k+1}\exp\left[-\left(\frac{1+\rho_b\overline{P}_{np}}{\overline{P}_{np}}\right)r\right] \quad (175)$$

The whole distribution can be obtained by summing up for all k's:

$$M(r) = \sum_{k=0}^{\infty}M_k(r)$$

$$= \left(\frac{1-\rho_b\overline{P}_{np}}{\overline{P}_{np}}\right)\left(\frac{1}{\sqrt{\rho_b\overline{P}_{np}}}\right)I_1\left(2r\sqrt{\frac{\rho_b}{\overline{P}_{np}}}\right)\exp\left[-\left(\frac{1+\rho_b\overline{P}_{np}}{\overline{P}_{np}}\right)r\right] \quad (176)$$

where I_1 is the modified Bessel function of the first kind and of the first order.

It is interesting to note that the above equations can also be obtained by utilizing the Stockmayer bivariate distribution for the instantaneous copolymers, Equations (98–101) [85, 86].

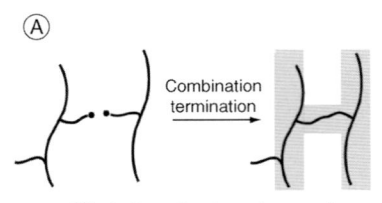

Figure 33. Examples of (A) H-shaped (crosslinking) and (B) T-shaped (branching) chain connections

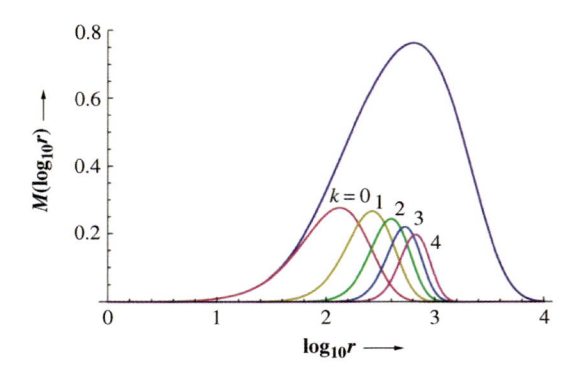

Figure 34. Calculated mass fraction distribution of random branch polymers whose primary chains follow the most probable distribution, with $\overline{P}_{np}=100$ and $\rho_b = 5 \times 10^{-3}$. The fractional chain length distributions containing k branches are also shown

Figure 34 shows the calculated mass fraction distribution of the random branched polymer system with $\overline{P}_{np} = 100$ and $\rho_b = 5 \times 10^{-3}$, by using Equations (175) and (176). Compared with a random crosslinking system shown in Figure 27, the high molecular mass tail decreases quickly, i.e., the persistence of the high molecular mass tail is a characteristic of gelling system.

The mass fraction of polymers containing k branches, M_k is given by:

$$M_k = \int_0^\infty M_k(r) dr$$
$$= \frac{(2k+1)!(\rho_b \overline{P}_{np})^k (1-\rho_b \overline{P}_{np})}{k!(k+1)!} \left(\frac{1}{1+\rho_b \overline{P}_{np}}\right)^{2k+2} \quad (177)$$

Therefore, the normalization of the distribution $M_k(r)$ to make the total area unity leads again to the Schulz–Zimm distribution, as follows:

$$M_{k,\text{norm}}(r) = \frac{M_k(r)}{M_k} = \frac{\sigma^\sigma}{\overline{P}_{n,k}\Gamma(\sigma)} \left(\frac{r}{\overline{P}_{n,k}}\right)^\sigma \exp\left(-\frac{\sigma r}{\overline{P}_{n,k}}\right) \quad (178)$$

where $\sigma = 2k + 1$, and $\overline{P}_{n,k}$ is the number-average chain length of the polymers having k branch points, which is given by:

$$\overline{P}_{n,k} = \frac{(2k+1)\overline{P}_{np}}{1+\rho_b \overline{P}_{np}} \quad (179)$$

For the Schulz–Zimm distribution, σ is represented by Equation (162), and therefore, the mass-average chain length of the polymers having k brach points, $\overline{P}_{M,k}$ is given by:

$$\overline{P}_{M,k} = \frac{(2k+2)\overline{P}_{np}}{1+\rho_b \overline{P}_{np}} \quad (180)$$

The polydispersity index, PDI of the polymers having k branch points is given by:

$$\overline{P}_{M,k}/\overline{P}_{n,k} = \frac{2k+2}{2k+1} \quad (181)$$

Similarly with the random crosslinking system, the PDI decreases with increasing the number of branch points k, but the PDI is slightly larger than that for random crosslinking for a given k. For example, when $k = 3$, the PDI for random branch is about 1.14 while it is 1.1 for random crosslinking.

Similarly with the random crosslinking system, the branching density of each polymer molecule is different. Each small dot in Figure 35 shows the chain length and branching density of each polymer molecule generated by

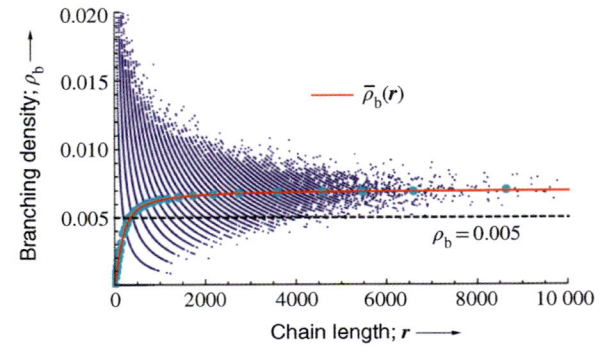

Figure 35. Relationship between branching density and chain length of each polymer molecule generated by a Monte Carlo simulation (small dot), their average crosslinking density within chain length intervals (circular symbols), with $\overline{P}_{np} = 100$ and $\rho_b = 5 \times 10^{-3}$. The solid curve, $\overline{\rho}_b(r)$ is calculated from Equation (182)

using a Monte Carlo simulation method [5]. The dashed line shows the branching density of the whole system, $\rho_b = 5 \times 10^{-3}$. Similarly with the random crosslinking case, as was shown in Figure 28, most of branched polymer molecules are distributed around the asymptotic branching density with $r \to \infty$, i.e., $\lim_{r \to \infty} \bar{\rho}_b(r)$, which is considered a representative value of the branched polymers, rather than ρ_b.

The $\bar{\rho}_b(r)$ function can be obtained from $M_k(r)$, as follows:

$$\bar{\rho}_b(r) = \frac{\sum_{k=0}^{\infty} \frac{k}{r} M_k(r)}{\sum_{k=0}^{\infty} M_k(r)} = \frac{\sqrt{\frac{\rho_b}{\bar{P}_{np}}} I_2\left(2r\sqrt{\frac{\rho_b}{\bar{P}_{np}}}\right)}{I_1\left(2r\sqrt{\frac{\rho_b}{\bar{P}_{np}}}\right)} \quad (182)$$

where I_2 is the modified Bessel function of the first kind and of the second order. The solid curve in Figure 35 shows the calculated result of Equation (182). The value of $\lim_{r \to \infty} \bar{\rho}(r)$ can be determined from Equation (182), at least, numerically.

The average branching density of polymers having k branch points, is given by:

$$\bar{\rho}_b(k) = \frac{\int_0^\infty \left[\frac{k}{r} M_k(r)\right] dr}{\int_0^\infty M_k(r) dr} = \frac{k(1+\rho\bar{P}_{np})}{(2k+1)\bar{P}_{np}} \quad (183)$$

9.3. Note on Nonlinear Polymerization

While randomly branched or crosslinked systems are important to understand the fundamental characteristics of branched or crosslinked polymer systems, real systems usually show some degrees of non-randomness in their structure and system-specific consideration is required. The number-average chain length, on the other hand, is always obtainable based on the stoichiometric investigation, because it is simply the ratio between the total number of monomeric units and the total number of polymer molecules.

Assuming that the effects of the size and/or structure dependent kinetics are negligible, the random sampling technique introduced in this article makes it possible to formulate the mass-average chain length in a matrix formula for nonrandom branching and/or crosslinking, generally represented by [87–90]:

$$\bar{P}_{\text{non-randomness}} = \bar{P}_{\text{non-randomness},0} + \mathbf{mF}(\mathbf{I} - \mathbf{M})^{-1}\mathbf{s} \quad (184)$$

where $\bar{P}_{\text{non-randomness},0}$ is the chain length of the initially selected chain or unit on a mass basis, $\mathbf{m}$ represents the mass fraction vector, $\mathbf{F}$ is the matrix showing the connection statistics from the 0th to the first generation chain or unit, $\mathbf{I}$ is the identity matrix, $\mathbf{M}$ is the transition matrix representing the connection statistics among chains or units, and $\mathbf{s}$ is a vector representing the expected size of chain or unit of various species. Gelation is predicted to occur when the largest eigenvalue of the transition matrix, $\mathbf{M}$ reaches unity. One would find the similarities of Equation (184) with Equation (144) and (173).

For the calculation of average properties, the method of moments has been applied to various complex reaction systems [91–94]. The ith moment of the polymer distribution is defined by:

$$Q_i = \sum_{r=1}^{\infty} r^i [P_r] \quad (185)$$

For example, Q_0 describes the number of polymer molecules, and Q_1 means the total mass of polymers. The differential equations describing the time development of Q_i can be obtained by taking the population balance equations, similarly as was shown in Equations (56) and (57) for polymer radicals. Once the moments of the distributions are determined, the number- and mass-average chain length are given by:

$$\bar{P}_n = \frac{Q_1}{Q_0} \quad (186)$$

$$\bar{P}_M = \frac{Q_2}{Q_1} \quad (187)$$

The method of moments is a convenient mathematical technique to determine the averages. Because any order of moments can be obtained, one may think that it would be possible to obtain the full chain length distribution, based on the calculated moments. A numerical method to calculate the distribution from its moments by approximating the distribution as a sum of Laguerre polynomials [95, 96], or the generalized Laguerre polynomials [97]. However, the method is limited to the distribution close to the most probable distribution, and accurate determination of the full distribution is disturbed by the amplification of oscillations in the sum of polynomials as the number of terms increases [98, 99]. In general, it is better to think that the method of moments is limited to obtain the averages, not the full distribution.

For the problems where an infinite set of differential equations describing the population balance equations can be written down, the approximate numerical solution for the full distribution may be obtained by using a discrete Galarkin method [100].

On the other hand, there are problems where the population balance equations cannot be set up. A notable example is a random scission of branched polymers [101]. All 4 types of branched polymer shown in Figure 36 possess 12 monomeric units with 4 branches. Suppose we want to have a linear polymer with 3 monomeric units by cutting one bonding. For the polymer (A), it is impossible to obtain a linear polymer with $r = 3$. On the other hand, there is one possible scission point in the polymer (B). The number of possible scission points changes with the detailed branched structure, and it is impossible to set up the population balance equations fully describing the time development during the scission reactions.

On the other hand, the random sampling technique can be applied also for such cases, as long as the size and structure dependence of branching/crosslinking/scission reaction can be neglected. With the matrix formula shown in Equation (184), the mass-average chain length can be obtained. The full chain length distribution can be obtained by application the random sampling technique to the Monte Carlo simulation method [5, 102–106]. With the Monte Carlo simulation, one can observe the structure of each polymer molecule directly on the computer screen as was shown in Figures 29 and 30, and one can determine very detailed structural information. On the basis of the detailed structural information, it is possible to determine the viscoelastic properties of branched polymers [107]. The Monte Carlo simulation technique promises to develop the computer-aided polymer design system that enables one to estimate the synthesis-structure-property relationships.

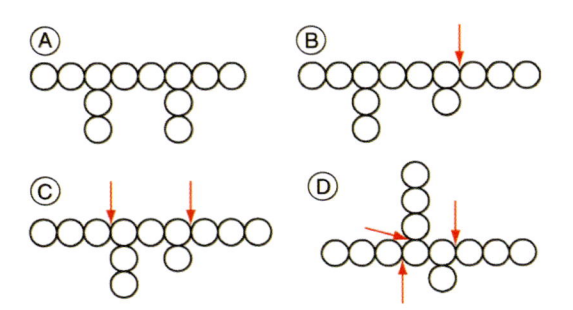

Figure 36. Possible scission point to obtain linear polymer with $r = 3$ by cutting one bonding

List of Symbols

A	functional group of reactant that reacts with B in step polymerization
A_1, A_2, A_3	adjustable parameter used for the empirical conversion dependence of k_t
ATRP	atom-transfer radical polymerization
B	functional group of reactant that reacts with A in step polymerization
C_{fCTA}	chain transfer constant of chain-transfer agent, CTA
C_{fm}	chain transfer constant of monomer
$Comp(F_1\|r)$	composition distribution of polymer molecules with chain length r
CSTR	continuous stirred tank reactor
CTA	chain-transfer agent
E_d	activation energy for initiator decomposition
E_f	activation energy for chain-transfer reaction
E_L	activation energy for the average chain length
E_p	activation energy for propagation
E_R	activation energy for polymerization
E_t	activation energy for termination
f	initiator efficiency

f_1	mole fraction of monomer 1 in copolymerization	k_p	propagation rate constant
f_1^0	initial mole fraction of monomer 1 in copolymerization	k_p'	propagation rate constant for transfer radical
F_1	mole fraction of monomer 1 bound in polymer produced instantaneously in copolymerization	k_t	rate constant for bimolecular termination ($=k_{tc}+k_{td}$)
		k_t^0	termination rate constant at zero monomer conversion ($x=0$)
$\overline{F}_1$	mole fraction of monomer 1 bound in accumulated copolymer	k_{tc}	rate constant for bimolecular termination by combination
F_1^0	instantaneous copolymer composition of polymers formed at $x=0$	k_{td}	rate constant for bimolecular termination by disproportionation
$F_{1,av}$	average composition of polymer formed instantaneously in Stockmayer's bivariate distribution	K	equilibrium constant ($=k_1/k_2$)
		LCA	long-chain approximation
		m	molecular mass of monomeric unit
$F(a,b;x)$	confluent hypergeometric function	$\overline{m}$	average number of branch points within the polymer molecules having the same chain length
$_pF_q$	generalized hypergeometric function	M	monomer
GPC	gel permeation chromatography	M_1	monomer of type 1 in copolymerization
I	initiator		
I_1	modified Bessel function of the first kind of the first order	M_k	mass fraction of polymer molecules containing k crosslinkages or k branch points
I_2	modified Bessel function of the first kind of the second order	$M(r)$	mass chain length distribution (mass-fraction of polymer molecules of length r)
k	reaction rate constant	$M^\bullet(r)$	mass chain length distribution of polymer radicals
k	number of crosslinkages or branch points in a polymer molecule	$M_{accum.}(r)$	mass chain length distribution of accumulated polymer
k_1	activation rate constant in RDRP	$M_{inst.}(r)$	"instantaneous" mass chain length distribution
k_2	deactivation rate constant in RDRP		
k_d	initiator decomposition rate constant	$M_k(r)$	fractional mass chain length distribution of polymer molecules having k crosslinkages or k branch points
k_{ex}	exchange reaction rate constant in degenerative-transfer radical polymerization	$M_{k,norm}(r)$	normalized mass chain length distribution of polymer molecules having k crosslinkages or k branch points
k_i	rate constant for monomer adding to an initiator radical		
k_{ij}	propagation rate constant for monomer of type i adding to a double bond on a monomer unit of type j (copolymerization)	$M_p(r)$	mass chain length distribution of primary chains
		$M_\beta(r)$	"instantaneous" mass chain length distribution formed by combination termination (β-mode)
k_{fCTA}	rate constant for chain transfer to CTA		
k_{fm}	rate constant for chain transfer to monomer	$M_\tau(r)$	"instantaneous" mass chain length distribution formed by other than combination termination (τ-mode)
k_{fX}	rate constant for chain transfer to small species, X		

MMD	molecular mass distribution	$\bar{P}_{n,k}$	number-average chain length of polymer molecules having k crosslinkages or k branch points
$\bar{n}_A$	average number of active periods		
$\bar{n}_M$	average number of monomeric units bound to an active center in a second	$\bar{P}_{np}$	number-average chain length of primary chains
N_a	total number of growing chains having an active center in an ideal living polymerization	$\bar{P}_{n,SA}$	number-average chain length added during a single active period
N_A	number of moles of A-functional groups	$\bar{P}_{n,gp}$	number-average chain length at the gel point
N_{A_0}	initial number of moles of A-functional groups	$P^*_{m,n,1}$	live copolymer chain with m units of monomer 1 and n units of monomer 2 bound in polymer chain and with active center located on terminal unit 1
N_I	number of moles of initiator in the reactor		
N_{I_0}	initial number of moles of initiator in the reactor		
N_M	number of monomer molecules	$\bar{P}_M$	mass-average chain length (degree of polymerization)
$N(r)$	number chain length distribution (number-fraction of polymer molecules of length r)	$\bar{P}_{M,A}$	expected mass-average chain length toward the downward direction in Figure 31
$N^\bullet(r)$	number chain length distribution of polymer radicals	$\bar{P}^\bullet_M$	mass-average chain length of polymer radicals
$N_{SA}(r)$	number chain length distribution formed during a single active period	$\bar{P}_{M,accum.}$	mass-average chain length of accumulated polymer
p	conversion of functional group in step polymerization, or the probability of connection to the next unit	$\bar{P}_{M,inst.}$	mass-average chain length of dead polymer chains formed instantaneously
		$\bar{P}_{M,k}$	mass-average chain length of polymer molecules having k crosslinkages or k branch points
[P]	concentration of all polymer molecules		
[P*]	concentration of active center	$\bar{P}_{Mp}$	mass-average chain length of primary chains
[P*₁]	concentration of active center of type 1 (copolymerization)	$\bar{P}_{M,SA}$	mass-average chain length during a single active period in RDRP
P_b	probability that a chain end is connected to a backbone chain (branched polymer)		
		PXP	adduct radical formed in RAFT
PDI	polydispersity index ($=\bar{P}_M/\bar{P}_n$)	Q_i	i-th moment of polymer distribution
P_r	polymer with chain length r		
P^*_r	active polymer molecule with chain length r	r	polymer chain length (degree of polymerization)
$P(n)$	probability that the chain experiences n times of active periods	r_1, r_2	reactivity ratios
		[R•]	total radical concentration
$\bar{P}_n$	number-average chain length (degree of polymerization)	R_{act}	rate of activation in RDRP
		RAFT	reversible-addition-fragmentation chain-transfer
$\bar{P}^\bullet_n$	number-average chain length of polymer radicals		
		RDRP	reversible-deactivation radical polymerization
$\bar{P}_{n,accum.}$	number-average chain length of accumulated polymer		
		R_{deact}	rate of deactivation in RDRP
$\bar{P}_{n,inst.}$	number-average chain length of dead polymer chains formed instantaneously	R_f	rate of chain transfer
		RGS	radical generating species in RDRP

R_I	initiation rate
R_{in}^*	initiator fragment with an active center
$R_{in}^{\bullet}$	initiator radical
R_p	propagation rate or polymerization rate
$R_r^{\bullet}$	polymer radical of chain length r
R_t	termination rate ($=R_{tc} + R_{td}$)
R_{tc}	rate of termination by combination
R_{td}	rate of termination by disproportionation
$\sqrt{\langle s^2 \rangle}$	radius of gyration
SEC	size exclusion chromatography
SRMP	stable-radical-mediated polymerization
SSH	stationary-state hypothesis
t	time
$t_{1/2}$	half-life
$\bar{t}_D$	average time length of a deactivated time
t_R	reaction time used for the development of MMD in RDRP
$\bar{t}_{R^{\bullet}}$	average lifetime of a radical
T	temperature
T_c	ceiling temperature
Trap	trapping agent in RDRP
V	volume of reacting mixture
x	monomer conversion
X	general description of small molecule that can act as a chain-transfer agent
X	trapping agent in SRMP
α	stoichiometric imbalance ($= N_{A_0}/N_{B_0}$)
β	kinetic dimensionless parameter ($= R_{tc}/R_p$)
Γ	gamma function
ΔG	free energy change
ΔH	enthalpy change
ΔS	entropy change
η	number average for the Poisson distribution
ρ	crosslinking density
ρ_b	branching density
$\bar{\rho}_b(k)$	average branching density of polymer molecule having k branch points
$\bar{\rho}_b(r)$	average branching density of polymer molecule having chain length r
ρ_{gp}	crosslinking density at the gel point
$\bar{\rho}(k)$	average crosslinking density of polymer molecule having k crosslinkages
$\bar{\rho}(r)$	average crosslinking density of polymer molecule having chain length r
σ	standard deviation
σ	parameter representing the narrowness of the Schulz–Zimm distribution
τ	kinetic dimensionless parameter [$=(R_{td} + R_f)/R_p$]
$\phi_1^{\bullet}$	mole fraction of polymer radicals of type 1

General References

G. Odian: *Principles of Polymerization*, 4th ed., Wiley-Interscience, New York 2004.

J.M. Asua (ed.): *Polymer Reaction Engineering*, Blackwell, Oxford 2007.

A.H.E. Muller, K. Matyjaszewski (eds): *Controlled and Living Polymerizations*, Wiley-VCH, Weinheim 2009.

R.J. Young, P.A. Lovell: *Introduction to Polymers*, 3rd ed., CRC Press, Boca Raton 2011.

W.F. Su: *Principles of Polymer Design and Synthesis*, Springer, Heidelberg 2013.

B.M. Mandal: *Fundamentals of Polymerization*, World Scientific, Singapore 2013.

W.F. Reed, A.M. Alb (eds.): *Monitoring Polymerization Reactions*, John Wiley & Sons, Hoboken 2014.

Specific References

1 I. Mita, R.F.T. Stepto, U.W. Suter, *Pure Appl. Chem.* **66** (1994) 2483.
2 R.J. Young, P.A. Lovell: *Introduction to Polymers*, 3rd ed., CRC Press, Boca Raton 2011, p. 15.
3 W. Graessley: *Adv. Polym. Sci.* **16** (1974) 1.
4 P.J. Flory: *Principles of Polymer Chemistry*, Cornell University Press, Ithaca, New York 1953.
5 H. Tobita, *Macromol. Theory Simul.* **5** (1996) 1167.
6 O. Saito, *J. Phys. Soc. Jpn.* **13** (1958) 198.
7 A. Charlesby: *Atomic Radiation of Polymers*, Pergamon Press, Oxford 1960.
8 H. Tobita, *Macromol. React. Eng.* **4** (2010) 333.
9 H. Tobita, *Macromol. Theory Simul.* **16** (2007) 399.
10 O. Vogl: "Aldehyde Polymers", *Encyclopedia of Polymer Science and Engineering*, vol. 1, Wiley-Interscience, New York 1985, p. 623.
11 A.D. Jenkins, R.G. Jones, G. Moad, *Pure Appl. Chem.* **82** (2010) 483.
12 G.T. Russell, D.H. Napper, R.G. Gilbert, *Macromolecules* **21** (1988) 2141.
13 S. Zhu, A.E. Hamielec, *Macromolecules* **22** (1989) 3098.

14. C.W. Wilson III, E.R. Santee, Jr., *J. Polym. Sci., Part C* **8** (1965) 97.
15. G. Talamini, G. Vidotto, *Makromol. Chem.* **100** (1967) 48.
16. T.G. Fox, H.W. Schnecko, *Polymer* **3** (1962) 575.
17. T. Otsu, B. Yamada, M. Imoto, *J. Macoromol. Sci. Chem.* **1** (1966) 61.
18. F.S. Dainton, K.J. Ivin, *Nature* **162** (1948) 705.
19. H. Sawada: *Thermodynamics of Polymerization*, Marcel Dekker, New York 1976.
20. H.W. McCormick, *J. Polym. Sci.* **25** (1957) 488.
21. J.P.A. Heuts, G.T. Russell, *Eur. Polym. J.* **42** (2006) 3.
22. M.J. Ballard, R.G. Gilbert, D.H. Napper, P.J. Pomery, P.W. O'Sullivan, J.H. O'Donnell, *Macromolecules* **19** (1986) 1303.
23. M. Stickler, D. Panke, A.E. Hamielec, *J. Polym. Sci., Polym. Chem. Ed.* **22** (1984) 2243.
24. G.V. Schulz, G. Haborth, *Makromol. Chem.* **1** (1948) 106.
25. I. Mita, K. Horie, *J. Macromol. Sci., Rev. Macromol. Chem. Phys. C* **27** (1987) 91.
26. K.F. O'Driscoll: "Kinetics of Bimolecular Termination", *Comprehensive Polymer Science*, vol. 3, Pergamon Press, London 1989, p. 161.
27. G.T. Russell, R.G. Gilbert, D.H. Napper, *Macromolecules* **25** (1992) 2459.
28. I.A. Maxwell, G.T. Russell, *Macromol. Theory Simul.* **2** (1993) 95.
29. S. Beuermann, M. Buback, *Prog. Polym. Sci.* **27** (2002) 191.
30. A. Husain, A.E. Hamielec, *J. Appl. Polym. Sci.* **22** (1978) 1207.
31. G.Z.A. Wu, L.A. Denton, R.L. Laurence, *Polym. Eng. Sci.* **22** (1982) 1.
32. C.J. Kim, A.E. Hamielec, *Polymer* **25** (1984) 845.
33. F.R. Mayo, F.M. Lewis, *J. Am. Chem. Soc.* **66** (1944) 1594.
34. T. Alfrey, Jr., G. Goldfinger, *J. Chem. Phys.* **12** (1944) 205.
35. F.T. Wall, *J. Am. Chem. Soc.* **66** (1944) 2050.
36. I. Sakurada: *Kyojugo Hanno*, Society of Polymer Chemistry, Tokyo 1944, p. 35.
37. T. Fueno, J. Furukawa, *J. Polym. Sci.* **A2** (1964) 3681.
38. F.P. Price: "Copolymer Composition and Tacticity in Markov Chains and Monte Carlo Calculations", in G.G. Lowry (ed.): *Polymer Science*, Marcel Dekker, New York 1970, p. 187.
39. H. Tobita, *Polymer* **39** (1998) 2367.
40. W.H. Stockmayer, *J. Chem. Phys.* **13** (1945) 199.
41. H. Tobita, *Macromol. Theory Simul.* **12** (2003) 463, 470.
42. I.H. Spinner, B.C.-Y. Lu, W.F. Graydon, *J. Am. Chem. Soc.* **77** (1955) 2198.
43. V.E. Meyer, G.G. Lowry, *J. Polym. Sci., Part A* **3** (1965) 2843.
44. A.E. Hamielec, J.F. MacGregor: "Modelling Copolymerization-Control of Composition, Chain Microstructure, Molecular Weight Distribution, Long Chain Branching and Cross-linking", in K.-H. Reichert, W. Geiseler (eds.): *Polymer Reaction Engineering*, Hanser Publishers, New York 1983, p. 21.
45. A.E. Hamielec, J.F. MacGregor, A. Penlidis, *Makromol. Chem., Macromol. Symp.* **10/11** (1987) 521.
46. H. Tobita, A.E. Hamietec, *Polymer* **32** (1991) 2641.
47. M. Szwarc, M.V. Beylen: *Ionic Polymerization and Living Polymers*, Chapman & Hall, London 1993, p. 12.
48. M. Szwarc, M. Levy, R. Milkovich, *J. Am. Chem. Soc.* **78** (1956) 2656.
49. K. Matyjaszewski (ed.): *Cationic Polymerization: Mechanism, Synthesis, and Applications*, Marcel Dekker, New York 1996.
50. P. De, R. Faust: "Carbocationic Polymerization", in A.H.E. Muller, K. Matyjaszewski (eds.): *Controlled and Living Polymerizations*, Wiley-VCH, Weinheim 2009, p. 57.
51. A. Goto, T. Fukuda, *Prog. Polym. Sci.* **29** (2004) 329.
52. W.A. Braunecker, K. Matyjaszewski, *Prog. Polym. Sci.* **32** (2007) 93.
53. K. Matyjaszewski: "Radical Polymerization", in A.H.E. Muller, K. Matyjaszewski (eds): *Controlled and Living Polymerizations*, Wiley-VCH, Weinheim 2009, p. 103.
54. A. Yokoyama, T. Yokozawa, *Macromolecules* **40** (2007) 4093.
55. T. Yokozawa, A. Yokoyama, *Prog. Polym. Sci.* **32** (2007) 147.
56. D. Baskaran, A.H.E. Muller: "Anionic Vinyl Polymerization", in A.H.E. Muller, K. Matyjaszewski (eds.): *Controlled and Living Polymerizations*, Wiley-VCH, Weinheim 2009, p. 1.
57. H. Tobita, *Macromol. React. Eng.* **4** (2010) 643.
58. H. Tobita, *Macromol. React. Eng.* **2** (2008) 371.
59. H. Tobita, *Macromol.Theory Simul.* **15** (2006) 12.
60. G.V. Schulz, *Z. Phys. Chem. (Leipzig)* **B43** (1939) 25.
61. B.H. Zimm, *J. Chem. Phys.* **16** (1948) 1099.
62. W. Feller: *An Introduction to Probability Theory and Its Applications*, vol. 1, John Wiley & Sons, New York 1950.
63. A.H.E. Muller, D. Yan, G. Litvinenko, R. Zhuang, H. Dong, *Macromolecules* **28** (1995) 7335.
64. P. Vana, T.P. Davis, C. Barner-Kowollik, *Macromol. Theory Simul.* **11** (2002) 823.
65. H. Chaffey-Miller, B. Busch, T.P. Davis, M.H. Stenzel, C. Barner-Kowollik, *Macromol. Theory Simul.* **14** (2005) 143.
66. J. He, H. Zhang, J. Chen, Y. Yang, *Macromolecules* **30** (1997) 8010.
67. H. Tobita, *Macromol. Theory Simul.* **15** (2006) 23.
68. H. Tobita, F. Yanase, *Macromol. Theory Simul.* **16** (2007) 476.
69. R.L. Graham, D.E. Knuth, O. Parashnik: *Concrete Mathematics: A Foundation for Computer Science*, 2nd ed., Addison-Wesley, Reading 1994, p. 204.
70. H. Tobita, *J. Polym. Sci., Polym. Phys. Ed.* **33** (1995) 1191.
71. O. Saito: "Statistical Theory of Crosslinking", in M. Dole (ed.): *The Radiation of Macromolecules*, Academic Press, New York 1972, p. 223.
72. T. Kimura, *J. Phys. Soc., Japan* **17** (1962) 1884.
73. T. Kimura, *J. Phys. Soc., Japan* **19** (1964) 777.
74. H. Tobita, *J. Polym. Sci., Polym. Phys. Ed.* **36** (1998) 2423.
75. H. Tobita, *J. Polym. Sci., Polym. Phys. Ed.* **38** (2000) 2333.
76. H. Tobita, K. Yamamoto, *Macromolecules* **27** (1994) 3389.
77. H. Tobita, *Acta Polymer.* **46** (1995) 185.
78. H. Tobita, Y. Uemura, *J. Polym. Sci., Polym. Phys. Ed.* **34** (1996) 1403.
79. H. Tobita, Y. Yoshihara, *J. Polym. Sci., Polym. Phys. Ed.* **34** (1996) 1415.
80. H. Tobita, *Polym. React. Eng.* **1** (1993) 357.
81. B.H. Zimm, W.H. Stockmayer, *J. Chem. Phys.* **17** (1949) 1301.
82. H. Tobita, *Macromol. Theory Simul.* **5** (1996) 129.
83. H. Tobita, K. Hatanaka, *J. Polym. Sci., Polym. Phys. Ed.* **33** (1995) 841.
84. H. Tobita, *J. Polym. Sci., Polym. Phys. Ed.* **36** (1998) 2015.
85. J.B.P. Soares, A.E. Hamielec, *Macromol. Theory Simul.* **5** (1996) 547.
86. J.B.P. Soares, *Macromol.React. Eng.* **8** (2013) 235.
87. H. Tobita, *J. Polym. Sci., Polym. Phys. Ed.* **36** (1998) 2423.
88. H. Tobita, *Macromol.Theory Simul.* **7** (1998) 675.
89. H. Tobita, *Macromol.Theory Simul.* **12** (2003) 24.
90. H. Tobita, *Macromol.Theory Simul.* **23** (2014) 477.
91. W.H. Ray, *Polym. Rev. Macromol. Chem.* **8** (1972) 1.

92 H. Tobita, A.E. Hamielec, *Macromolecules* **22** (1989) 3098.
93 S. Zhu, A.E. Hamielec, *Macromolecules* **26** (1993) 3131.
94 J.C. Hernandez-Ortiz, E. Vivaldo-Lima, M.A. Dube, A. Penlidis, *Macromol. Theory Simul.* **23** (2014) 147.
95 C.H. Bamford, H. Tompa, *J. Polym. Sci.* **10** (1953) 345.
96 C.H. Bamford, H. Tompa, *Trans. Faraday Soc.* **50** (1954) 1097.
97 K.W. Min, *J. Appl. Polym. Sci.* **22** (1978) 589.
98 O. Saito, K. Nagasubramanian, W.W. Graessley, *J. Polym. Sci.*, Part 2-A **7** (1969) 1937.
99 H. Tobita, K. Ito, *Polym. React. Eng.* **1** (1993) 407.
100 M. Wulkow, *Macromol. Theory Simul.* **5** (1996) 393.
101 H. Tobita, *Macromolecules* **29** (1996) 3000, 3010.
102 H. Tobita, *Macromolecules* **26** (1993) 836.
103 H. Tobita, *J. Polym. Sci., Polym. Phys.* **31** (1993) 1363.
104 H. Tobita, *J. Polym. Sci., Polym. Phys.* **39**, (2001) 391.
105 H. Tobita, *Macromol. React. Eng.* **7** (2013) 181.
106 H. Tobita, *Macromol. Theory Simul.* **23** (2014) 182.
107 D.J. Read, D. Auhl, C. Das, J. den Doelder, M. Kapnistos, I. Vittorias, T.C.B. McLeish, *Science* **333** (2011) 1871.

Polymerization Processes, 2. Modeling of Processes and Reactors

HIDETAKA TOBITA, University of Fukui, Fukui, Japan

ARCHIE E. HAMIELEC, Institute for Polymer Production Technology, McMaster University, Canada

1.	Introduction	315	6.	Precipitation and Dispersion Polymerization	337
2.	Fundamental Effects of Reactor Types	316	6.1.	Polymerization without Solvent	337
2.1.	Reactor Types and Their Models	316	6.2.	Polymerization with Solvent	338
2.2.	General Effects on the Molecular Mass Distribution (MMD)	319	7.	Suspension Polymerization	339
3.	Processes and Reactor Modeling for Step Polymerization	320	7.1.	Process Description	339
			7.2.	Polymerization Kinetics	341
3.1.	Types of Reactors and Reactor Modeling	320	8.	Emulsion Polymerization	342
			8.1.	Process Description	342
3.2.	Specific Processes	323	8.2.	Polymerization Kinetics	344
3.2.1.	Polyamides	323	8.3.	Molecular Mass Distribution	350
3.2.2.	Polyesters	324	8.3.1.	Linear Polymerization	350
4.	Processes and Reactor Modeling for Chain Polymerization	326	8.3.2.	Nonlinear Polymerization	353
			8.4.	Effect of Small Reaction Loci in Reversible-Deactivation Radical Polymerization	355
4.1.	Introduction to Polymerization Techniques	326			
4.2.	Fundamentals of Material Balance Equations	327	8.4.1.	Stable Radical-Mediated Polymerization (SRMP) and Atom-Transfer Radical Polymerization (ATRP)	356
4.3.	Effect of Reactor Types on Copolymer Composition Distribution	328			
			8.4.2.	Reversible Addition–Fragmentation Chain Transfer (RAFT) Polymerization	358
4.4.	Effect of Reactor Types on Nonlinear Polymer Formation	329			
5.	Bulk and Solution Polymerization	334		List of Symbols	359
5.1.	Removal of Solvent and Residual Monomer	335		General References	362
				Specific References	363
5.2.	Systems with Polymer–Polymer Demixing	336			

1. Introduction

The polymer reactor model is accepted as a valuable tool whose use contributes significantly to all aspects of process technology for polymer manufacture. This includes process design, optimization, state estimation, and control. Through process design, polymers with a unique and desirable combination of properties can often be obtained. Process parameters such as residence time distribution (RTD) are usually not considered by polymer synthesis chemists, although RTD can influence chemical composition distribution (CCD), molecular mass distribution (MMD), long-chain branching and gel/sol ratios. In the early days of the polymer industry, the chemists played the major role in product and process development and scale-up. This has changed, with the process engineers now playing a significant role in all phases of commercialization of new and improved polymer products. Their broad experience with process fundamentals and computer modeling are essential to obtain high-quality products, safely and economically.

Dynamic reactor models can be used in a variety of ways. Stability and control of polymer reactors should be considered at the design stage and control problems minimized then, rather than take corrective action after the plant is built. Complex interactions which are involved in polymerization (highly nonlinear temperature and concentration effects) preclude optimal design based on experimentation alone because the cost would be prohibitive. Models can be used to identify potential sources of product variability and strategies to minimize their effects. They can also be used to store information on process technology in a concise and readily retrievable and modifiable form.

Process models are often used to train chemists, chemical engineers, and plant operators and give them a feel for the dynamics of the polymerization process. Model-based simulation that requires a large amount of calculation is now feasible with the help of modern high speed computers. Still simpler models are preferable in order to comprehend important characteristics of the polymerization processes.

The most expensive aspect of model development is the experimental estimation of model parameters; highly instrumented bench-, pilot-, and plant-scale reactors are required. Statistically designed experiments should be performed to permit efficient parameter estimation and model development. Modeling is an iterative process and the very act of developing a deterministic and/or probabilistic model permits a greater understanding of the relevant microscopic processes that occur during polymerization or polymer modification. As additional (plant, pilot-plant, and bench-scale) data become available, model structure and parameters can be updated.

The applications to various reactor operations discussed in this articles are based on the fundamentals of the polymerization models presented in (→ Polymerization Processes, 1. Fundamentals).

2. Fundamental Effects of Reactor Types

2.1. Reactor Types and Their Models

All chemical reactors, whether industrial or experimental in scale, fall into one of three broad categories: batch, semibatch (or semicontinuous), and continuous. The readers who are not familiar with these names should refer to a chemical reaction engineering textbook, such as [1]. Molecular mass and copolymer composition distributions, long-chain branching and cross-linking are affected significantly by the reactor types used, and important characteristics of polymerization reactions and reactors must be properly accounted for.

Figure 1 shows three representative ideal reactors, characterizing them on the basis of the RTD and the temporal and spatial course of the chemical reaction.

The batch reactor is a closed system, and no material enters or leaves during polymerization. In a basic modeling, it is assumed that there are no spatial variations in temperature and concentrations. Mixing at the molecular level is often called "micromixing". All fluid elements have the same reaction time in a batch reactor, which is assumed also for the plug–flow reactor (PFR). In a basic modeling, it is assumed that there are no variations in temperature and concentrations in the radial direction of a PFR, with no fluid elements overtaking or mixing with any other elements ahead or behind. The design equation for an ideal PFR is essentially the same as for a batch reactor; the reaction time in a batch reactor is changed to the length in the flow direction.

$$\text{Batch reactor}: \quad -\frac{dN_M}{dt} = R_p V \quad (1)$$

$$\text{PFR}: \quad -\frac{dF_M}{dV} = R_p \quad (2)$$

Where N_M is the number of moles of monomer in the batch reactor, F_M is the molar flow rate of monomer in a PFR, R_p is the polymerization rate, and V is the reactor volume. Note that V in a batch reactor is the whole reaction volume, while V in Equation (2) is a variable indicating

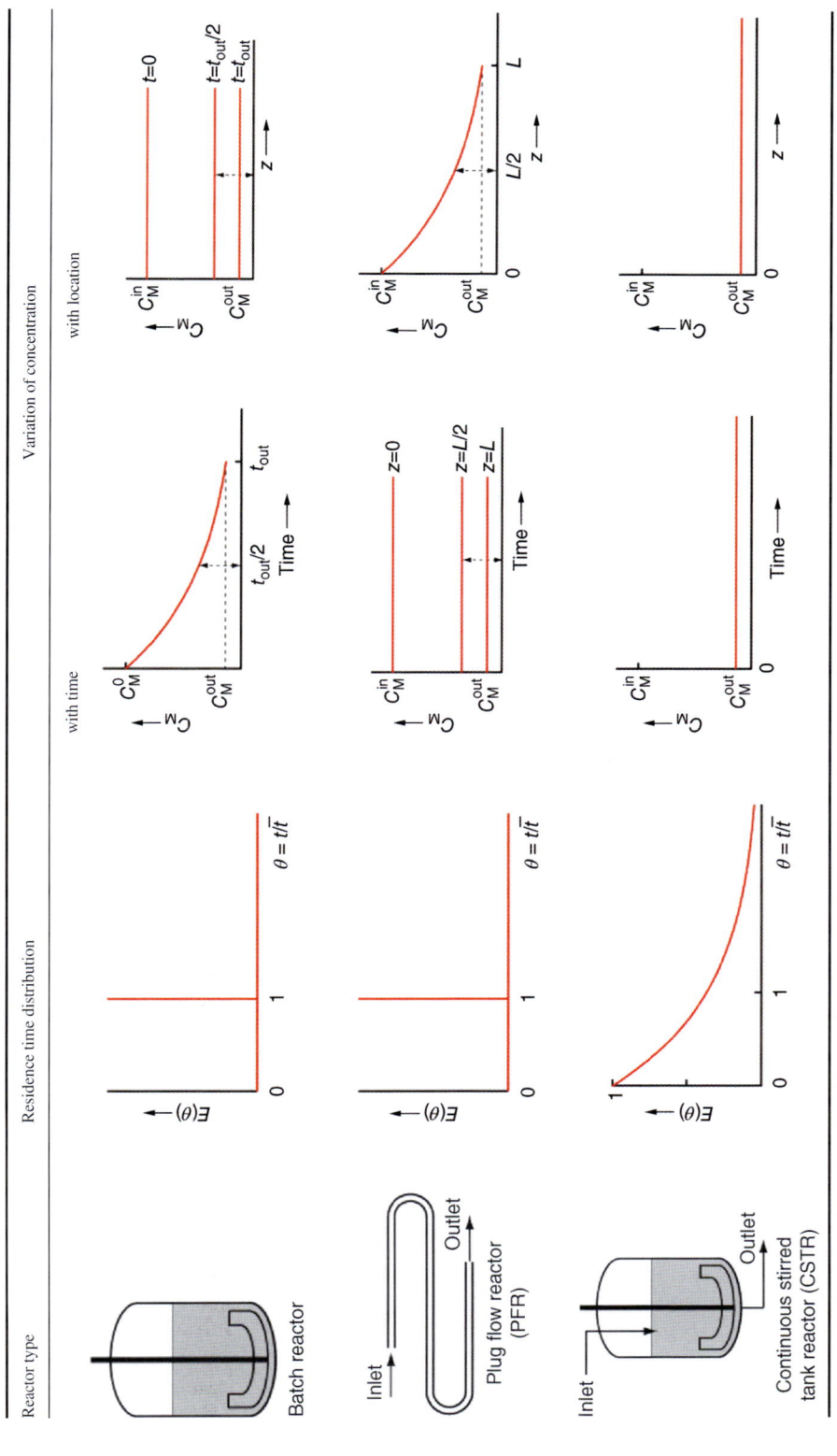

Figure 1. Schematic illustration of the course of reaction in various types of reactors. $E(\theta)$ = residence time distribution function; C_M = monomer concentration; t = time; $\bar{t}$ = mean residence time; z = spatial coordinate; L = reactor length

the reactor volume up to a given location of the PFR.

With a continuous stirred tank reactor (CSTR), the micromixing is normally assumed, and when it is necessary to emphasize this, it is called the homogeneous continuous stirred tank reactor (HCSTR). At steady state, the concentrations of all components in an HCSTR are kept constant, and the polymerization proceeds at a constant rate. The design equation for an HCSTR is:

$$\text{HCSTR}: \quad \frac{F_M^{in} - F_M^{out}}{V} = R_p \quad (3)$$

Although the concentrations are kept constant in an HCSTR, the residence time of the fluid elements have a large variation, whose residence time distribution is given by [1]:

$$E(\theta) = \exp(-\theta) \quad (4)$$

where $\theta = t/\bar{t}$ and $\bar{t}$ is the mean residence time.

Compare the reactor performance of a PFR and an HCSTR in productivity. Because the monomer conversion x in a continuous reactor is given by $x = (F_M^{in} - F_M)/F_M^{in}$, the required reactor volume to make the final conversion x_{out} is given by:

$$\text{PFR}: \quad V = F_M^{in} \int_0^{x_{out}} \frac{dx}{R_p} \quad (5)$$

$$\text{HCSTR}: \quad V = \frac{F_M^{in} x_{out}}{R_p} \quad (6)$$

Normally, the polymerization rate R_p decreases with the decrease in monomer concentration, the inverse of R_p is an increasing function of x, as shown in Figure 2. The area under the curve represents the required reactor volume of a PFR, while that for an HCSTR is given by the rectangular area. Obviously, $V_{PFR} < V_{CSTR}$, and a PFR is better in its performance. On the other hand, when the autoacceleration due to the gel effect in free-radical polymerization is significant, the $1/R_p$ curve may show a complex behavior. If the $1/R_p$ curve possesses a single minimum at x_{min} between $x = 0$ and x_{out}, the best choice is a train of two reactors, with the first

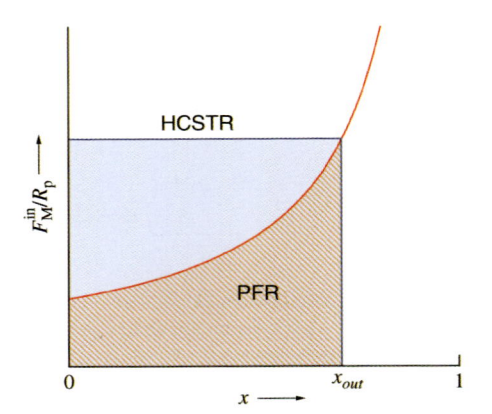

Figure 2. Necessary reactor volume to obtain polymer with conversion x_{out}

reactor being an HCSTR operated up to x_{min} and the second being a PFR.

Even when a single PFR is the best choice, the polymers may stick to the wall and it may be difficult to obtain a stable flow in a tube. In such cases, a train of HCSTRs would be the choice to approximately accomplish a plug flow. Assuming a series of N HCSTRs having the same volume, the residence time distribution function is given by [1]:

$$E(\theta) = \frac{N(N\theta)^{N-1}}{(N-1)!} \exp(-N\theta) \quad (7)$$

where $\theta = t/\bar{t}$ and the mean residence time $\bar{t} = V/v$ is calculated for the total volume of N CSTRs.

Figure 3 illustrates the calculated RTDs. By increasing the number of tanks N, the RTD

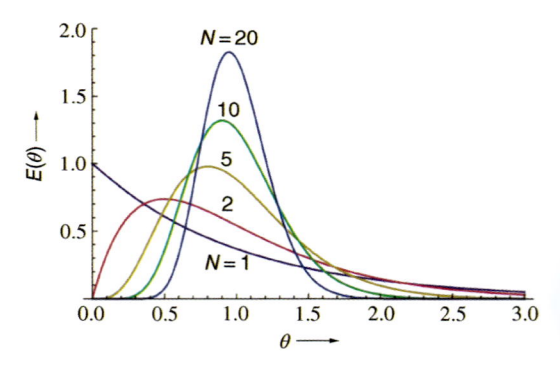

Figure 3. Residence-time distribution of the tanks-in-series model N = number of tanks; $\theta = 1$ for all fluid elements of a PFR

becomes narrower to approach that for a PFR. In fact, the tanks-in-series model is used widely as the one-parameter model to represent nonideal flow in various types of actual continuous reactors. Another notable model to represent the nonideal flow is the dispersion model [1].

In a CSTR, the micromixing condition is often not achieved in practice, however, and for this reason an alternative type of model is introduced, the segregated continuous stirred tank reactor (SCSTR) [1]. Here the fluid phase is regarded as subdivided into many small isolated compartments. Each compartment contains a large number of molecules, which are permanently confined within the limits of that compartment; therefore, the individual compartments function as miniature batch reactors with different residence times in the flow reactor. The compartments themselves are taken to be ideally mixed, with total segregation of molecules in different compartments, what is called "macromixing". Thus, the sum of all the compartments in an SCSTR have the same residence time distribution as the contents of the HCSTR, represented by Equation (4). A macroscopic mean taken over all the compartments in the effluent stream and in the reactor itself would show concentrations and temperature that are constant both spatially and with respect to time. On the other hand, a probe capable of microscopic sampling of individual compartments would reveal concentrations that varied in a statistical manner from one compartment to another. Given the high viscosity of a typical polymerizing solution it is likely that some solution polymerizations may occur in segregated systems [2]. On the other hand, however, when a CSTR is mentioned, normally, an HCSTR is envisaged unless otherwise noted about the segregation.

2.2. General Effects on the Molecular Mass Distribution (MMD)

To clarify general effects of various reactor types on the formed MMD, the linear polymerization in a homogeneous media is considered. In step polymerization and also in living polymerization, a longer reaction time, i.e., a larger residence time leads to the production of longer chains. Therefore, the variance of the residence time distribution directly influences the breadth of the MMD. In these polymerization systems, the MMD is the narrowest in a batch or a PFR. For an ideal step polymerization, the most probable distribution, whose polydispersity index (PDI) $\overline{M}_w/\overline{M}_n$ is 2, is formed (→ Polymerization Processes, 1. Fundamentals). For an ideal living polymerization, polymers having the Poisson distribution are expected (→ Polymerization Processes, 1. Fundamentals).

On the other hand, much broader MMDs are obtained for these polymerization systems when using a CSTR [3, 4], because of a broader RTD leading to a wide variation of growth time. An HCSTR and an SCSTR are compared. In an HCSTR, all chains grow in a constant environment. On the other hand, in an SCSTR, the compartments with larger residence time will possess lower monomer concentrations, leading to smaller growth rate. In addition, in the compartments with smaller residence time, the monomer concentration level is higher than in the corresponding HCSTR, leading to faster chain growth. Both high- and low-molecular-mass tails are reduced compared with the corresponding HCSTR, leading to a narrower MMD in an SCSTR [3, 4].

In step polymerization, a high monomer conversion is required to attain a high degree of polymerization (chain length). A CSTR could be used for low conversion region, however, the horizontal reactor shown later in Figure 5 (Section 3.1) whose RTD is close to a PFR, is often used for the final stage of polymerization.

In a conventional free-radical polymerization, each chain is formed in a very small time interval, and it is convenient to consider the instantaneous MMD (→ Polymerization Processes: 1. Fundamentals). The instantaneous MMD is dominated by the concentrations of the ingredients at that instant, and the residence time after the formation of dead polymer chain is not relevant. In an HCSTR at steady state, the concentration in the reactor is kept constant, as shown in Figure 1. The instantaneous MMD, which is the narrowest for a given free-radical polymerization system producing only linear polymers, is formed in an HCSTR.

On the other hand, the use of a batch or a PFR leads to a broad distribution for a conventional free-radical polymerization in which time varying instantaneous MMDs are superimposed, resulting in an accumulated MMD (→ Polymerization Processes, 1. Fundamentals). The MMD is broader in a batch or a PFR for the free-radical polymerization.

A subtle note when using an HCSTR in free-radical polymerization is that polymers may exist in a high concentration throughout the polymerization in an HCSTR, which promotes the reaction with polymers, such as chain transfer to polymer. Even for a system where the polymer transfer reaction is not significant in a batch reactor, it may have significant effects in a HCSTR. With polymer transfer reactions, the MMD becomes broader (→ Polymerization Processes, 1. Fundamentals).

Another note on the conventional free-radical polymerization is the effect of segregation on the MMD. With an SCSTR, compartments with various monomer concentrations coexist in a reactor, which makes the MMD broader. Because each compartment can be considered as an independent batch reactor, the MMD becomes broader than the batch/PFR.

In step and living polymerization, the breadth of the MMD is batch (PFR)<SCSTR<HCSTR. On the other hand for the conventional free-radical polymerization, the breadth is HCSTR<batch (PFR)<SCSTR.

3. Processes and Reactor Modeling for Step Polymerization

3.1. Types of Reactors and Reactor Modeling

In step polymerization, high molecular mass polymers are usually not produced until the final stage of reactions (→ Polymerization Processes, 1. Fundamentals), so that thermal control and mixing of the reaction mixture do not present serious problems in the earlier stages. However, since the final stage of polymerization is very important for the production of polymers with high molecular mass, handling of very high viscosities and temperatures, as well as a high interfacial area to remove small molecules, are

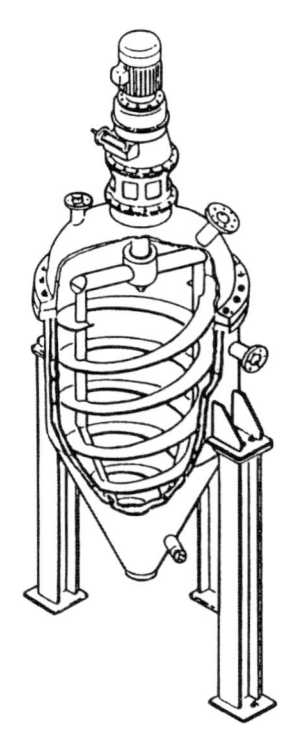

Figure 4. Vertical cone ribbon blade reactor (Mitsubishi Heavy Industries)

required. Various polymerization processes and reactor types, both for batch and continuous production, have been proposed. Examples of reactors for high viscosities are shown in Figures 4 and 5. Careful selection of the polymerization reactor is important to produce high-quality polymers [5, 6]. The horizontal reactor shown in Figure 5 is normally used for continuous production processes. Desirable characteristics of this type of reactor are (1) the axial flow must be close to piston flow, (2) the mixing blades must have self-cleaning performance with no dead zone, and (3) the gas-liquid interfacial area must be large enough. The item (3) is important particularly to remove condensation byproduct effectively. The rotation direction of the blades shown in Figure 5c is the opposite of that of 5a and b, which leads to a larger gas–liquid interfacial area.

The batch reactor is the most versatile reactor type and is used extensively for specialty polymers at low production volumes.

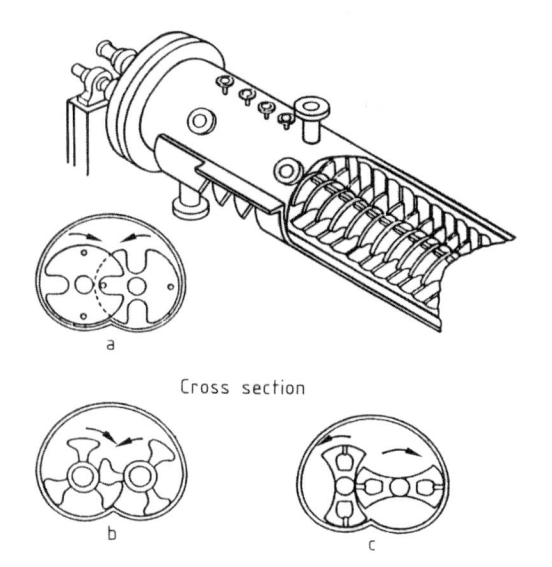

Figure 5. Horizontal high-viscosity reactor

formaldehyde (→ Amino Resins). In polycondensation reactions, it may be necessary to remove condensation products to attain sufficient conversion. When the volume of the reaction mass decreases continuously with time, such reactors can be considered as semibatch reactors. For example, in the production of poly(ethylene terephthalate), PET, because methanol or ethylene glycol evaporates during polymerization, the batchwise production of PET is considered a semibatch operation (→ Fibers, 5. Polyester Fibers). Polycondensation reactions often require high temperature, and bulk polymerization is preferred in order to suppress undesirable side reactions.

Some examples of step polymerizations carried out in such reactors are nylon 6 (→ Polyamides); Fibers, → Fibers, 4. Polyamide Fibers), phenol–formaldehyde (→ Phenolic Resins), urea–formaldehyde, and melamine-

High-capacity plants often use continuous processes. The first approximation for a continuous process is a model that consists of PFRs and CSTRs in various combinations, as shown in Figure 6, although various nonideal effects such as flow pattern in the reactor, mass- and heat-transfer limitations, and residence time distribution must be considered for a detailed analysis and design of commercial reactors [1, 7–10].

The MMD of linear polymers produced by step polymerization in a batch or a PFR

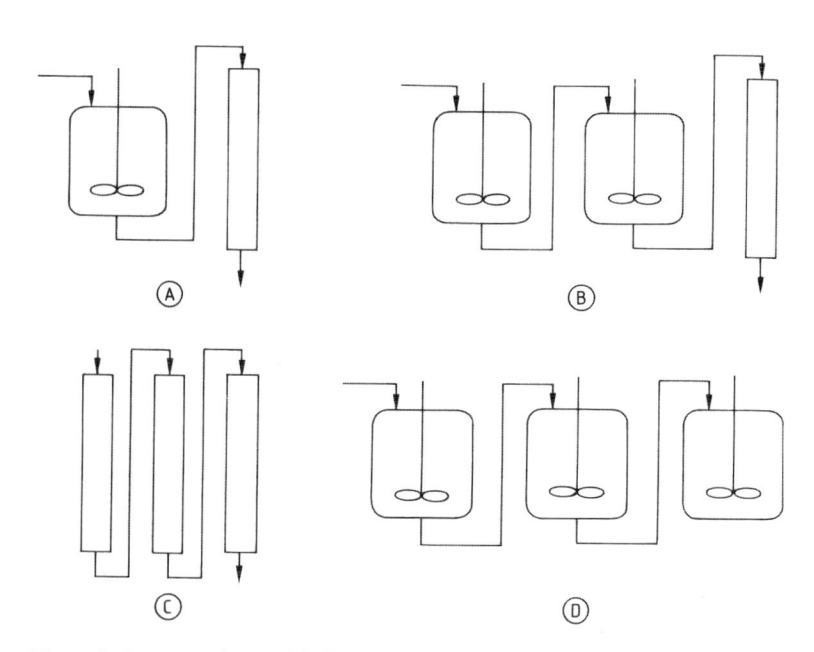

Figure 6. Representative models for continuous polymerization processes
A) CSTR + PFR; B) CSTR + CSTR + PFR; C) PFRs; D) CSTRs

basically follows the most probable distribution (→ Polymerization Processes, 1. Fundamentals). The MMD may be controlled by varying its reaction path if the reaction system is in a nonequilibrium state. Assuming irreversible step polymerization without interchange reactions, the effect of reactor types was discussed in Section 2.2. An important feature of step polymerization is that the variance of MMD is smallest in a batch reactor or PFR and is very large in an HCSTR, which is quite contrary to that for free-radical polymerization. This result may be disappointing, since it is, in principle, impossible to produce polymers whose PDI $\overline{M}_w/\overline{M}_n$ is smaller than 2 in step polymerization at sufficiently high conversions. The PDI of polymers produced in a batch reactor is given by $1 + p$, where p is the conversion of the functional group (→ Polymerization Processes, 1. Fundamentals). However, in commercial polymeric materials, polymers with narrower distributions are not always superior to those with broader distributions, since various levels of properties are required at the same time. The use of a cascade of CSTRs and/or PFRs with recycle loops may be one method to obtain a MMD with a PDI larger than two. In practice, however, these methods may have shortcomings because they need a long start-up period and problems may occur with the stability of the reaction system. A method in which additional monomers are fed intermediately to a batch reactor or a PFR was proposed [11], and it was shown that the polydispersity index can be easily controlled over a wide range with values greater than 2.

In a batch reactor, the reverse reactions and the interchange reactions (redistribution reactions) do not change the MMD from the most probable distribution [12–14]. However, these reactions do change the MMD of polymers produced in CSTR, PFR with a recycle loop, and intermediate monomer feed method. Some consideration of these reactions in a CSTR is given in [15]. Qualitatively, these effects lower the polydispersity and make the MMD approach the most probable distribution. This result seems reasonable, because any MMD approaches the most probable distribution with a PDI of two when polymer chains are severed randomly [16–18].

Other than the common reactor types discussed above, other special types of reactor system may be applied. For example, in the polymerization of urethanes, the reaction rates are so high that reaction takes place even when the monomers are being mixed and pumped into molds. In situ polymerization to form the desired articles directly from monomeric liquids is known as reaction injection molding (RIM) [19–21]. A schematic of the RIM process is shown in Figure 7. The RIM processing of polyesters, epoxy resins, polyamides, and dicyclopentadienes has also been introduced, although polyurethanes are the most common product via RIM process. In the RIM process, the reaction is almost complete by the time the material fills the mold, and therefore the mixing and flow equations must be solved simultaneously with those for chemical reactions in a rational model for these complex situations.

In the finishing stage of nylon 6, nylon 66, and PET polymerizations, higher molecular-mass polymers may be obtained by solid-state polymerization in which polymerization occurs by heating chips or flakes of a material below

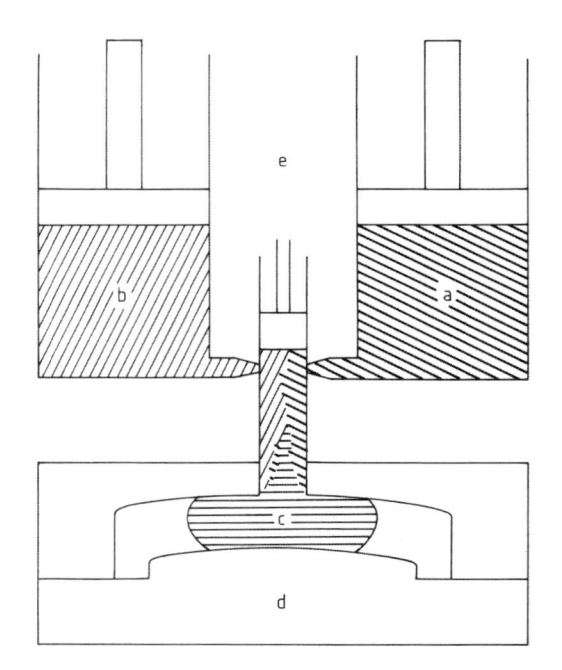

Figure 7. Schematic drawing of the RIM process a) Monomer A; b) Monomer B; c) Polymerizing mixture; d) Mold; e) Mixer

its melting point in a stream of hot gases in a fluidized bed or in a drier operated under vacuum [22–26]. The monomer, condensation products, and various byproducts diffuse out, and further reaction takes place inside the solid. The progress of these types of reaction is affected significantly by the diffusion of the condensation products and the morphology of the solid.

Although step polymerization has a very long history, a systematic kinetic treatment like that available for free-radical polymerization does not exist because of limitations due to system-specific side reactions and the scarcity of reliable kinetic data. However, this synthetic route is becoming more important due to the development of engineering plastics synthesized by step polymerization such as aramides, PPS, PEK, and PES.

3.2. Specific Processes

3.2.1. Polyamides (→ Polyamides), (→ Fibers, 4. Polyamide Fibers)

Polyamides are manufactured by two basic routes. One of these is synthesis from cyclic monomers such as lactams. Polymerization of these substances requires ring opening and subsequent chain growth. Another class of synthetic polyamides is formed from diamines and diacids.

The most common types of polyamides are nylon 6 and nylon 66. The term nylon is often used for synthetic aliphatic polyamides. One number indicates that the product was prepared from a single monomer and represents the number of carbon atoms in the repeating unit. Two numbers refer to the number of carbon atoms in the diamine and that in the diacid, respectively.

Nylon 6 is typically produced by the hydrolytic polymerization of ε-caprolactam [27–33], although polymers with higher molecular mass can be produced by ionic polymerization [27, 28, 34, 35]. The major reactions in the hydrolytic polymerization are:

1. Ring opening of ε-caprolactam by water (Eq. 8), which produces aminocaproic acid (ACA)
2. Polycondensation of ACA (Eq. 9)
3. Acid-catalyzed polyaddition (ring-opening polymerization) by nucleophilic attack of the amine nitrogen on the lactam (Eq. 10)

$$\text{caprolactam} + H_2O \rightleftharpoons H_2N(CH_2)_5COOH \quad \text{Aminocaproic acid} \tag{8}$$

$$NH_2\text{\textapprox}COOH + NH_2\text{\textapprox}COOH \rightleftharpoons NH_2\text{\textapprox\textapprox}COOH + H_2O \tag{9}$$

$$\text{caprolactam} + NH_2\text{\textapprox} \rightleftharpoons NH_2(CH_2)_5CONH\text{\textapprox} \tag{10}$$

Step polymerization of the amino acid (Eq. 9) accounts for only a few percent of total polymerization of ε-caprolactam. However, step polymerization is important because it usually determines the final degree of polymerization at equilibrium. The MMD is essentially the most probable distribution, except for the presence of monomer and cyclic oligomers. As low molecular mass substances lower the polymer quality, they are usually removed by leaching or vacuum treatment of the polymer melt. The formation of cyclic oligomers is an important side reaction [36, 37].

Various types of polymerization reactors have been proposed both for batch and continuous processes. A commonly used industrial reactor for a continuous process is a tubular reactor such as the conventional VK column (Vereinfacht Kontinuierliches Rohr) [27, 29, 33], which consists of a vertical tube operating at atmospheric pressure. The feed enters the top of the column and is heated to approximately 220–270°C. The simplest model for this type of reactor is a PFR. However, according to impulse response experiments, the flow is approximately laminar rather than plug flow [29, 38], and the reactor could be modeled as a CSTR or a tanks-in-series with recycle streams followed by a tubular reactor when a large quantity of water is used, since a significant convection current and mixing is provided by the evaporating water [39, 40].

Nylon 66 is manufactured by polycondensation of hexamethylenediamine and adipic acid [39], usually in a multistage process. First, nylon salt (hexamethylenediammonium adipate) is prepared from stoichiometric quantities of hexamethylenediamine and adipic acid in water. The salt can easily be separated by precipitation with methanol.

$$H_2N(CH_2)_6NH_2 + HOOC(CH_2)_4COOH$$

$$\longrightarrow \begin{bmatrix} ^-O_2C(CH_2)_4CO_2^- \\ H_3\overset{+}{N}(CH_2)_6\overset{+}{N}H_3 \end{bmatrix} \quad (11)$$

The use of nylon salt guarantees the presence of equimolar amounts of $-NH_2$ and $-COOH$ groups. Close control of diamine diacid balance is important to control the final polymer molecular mass and reactive end groups.

Nylon 66 is fairly unstable at high temperatures in the presence of oxygen. Not only degradation but also cross-linking may occur. Because of this instability, polymerization was carried out solely in batch processes. However, complete elimination of oxygen has made it possible to carry out continuous polymerization. An example of a continuous melt polymerization process is shown in Figure 8. The aqueous nylon salt solution is heated to above 200°C at > 17 bar in an oxygen-free atmosphere. Thereafter, the pressure is reduced to atmospheric and vapor is separated from polymer to promote polymerization to the desired high molecular mass. It is also possible to polymerize molten hexamethylenediamine and adipic acid directly [31, 39]. Polymerization can also be completed in the solid state.

Some examples of the kinetic simulation studies on the synthesis of nylon 66 can be found in [41–44].

3.2.2. Polyesters (→ Polyesters)

The production of high molecular mass polyesters differs somewhat from that of polyamides. In the case of nylons, the chemical equilibrium favors the polyamide under polymerization conditions. With polyester formation, however, the equilibrium is much less favorable. In order to drive the reaction in the forward direction, the condensation product must be removed continuously, usually by application of high vacuum. For polyester reactors, a high vacuum, a high temperature, and a high interfacial area with sufficient surface renewal are required, especially at high conversions.

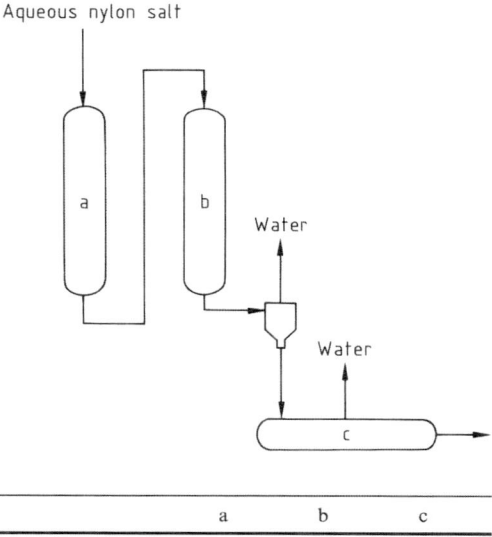

	a	b	c
Hold-up time, min	15–30	15–30	60
Temperature, °C	200–235	270–290	270–290
Pressure, 10^5 Pa	17–24	17–24	atmospheric

Figure 8. Continuous melt polymerization of nylon 66

Both saturated and unsaturated polyesters are produced. Among the saturated polyesters, PET is produced in the largest quantity, and is used for production of fibers, films, molding plastics, and beverage containers. In this section, the engineering aspects of PET formation are illustrated as an example of a polyester production process.

There are two major routes to synthesize PET industrially, although the objective in each case is to obtain an intermediate product, i.e., bis (hydroxyethyl) terephthalate (BHET). Two major routes to synthesize BHET are ester interchange of dimethyl terephthalate (DMT) and direct esterification of terephthalic acid. Figure 9 shows an example of the PET

Figure 9. Continuous polymerization process of PET via ester interchange route

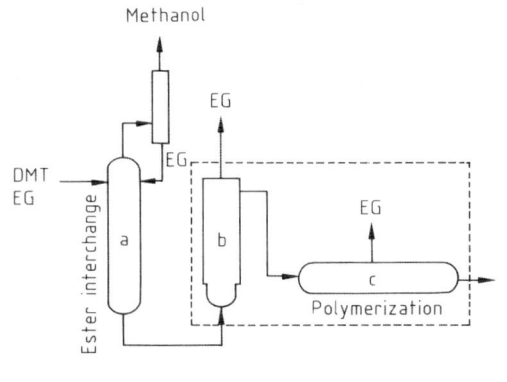

	a	b	c
Hold-up time, h	4–6	2–3	2–3
Temperature, °C	150–210	265–285	265–285
Pressure, 10^2 Pa	atmospheric	13–133	≪ 6

production process via the ester interchange route. The ester interchange reaction

$$CH_3O-\overset{O}{\underset{\|}{C}}-\underset{}{\bigcirc}-\overset{O}{\underset{\|}{C}}-OCH_3 + 2\,HOCH_2CH_2OH \rightleftharpoons$$

$$HOCH_2CH_2O-\overset{O}{\underset{\|}{C}}-\underset{}{\bigcirc}-\overset{O}{\underset{\|}{C}}-OCH_2CH_2OH + 2\,CH_3OH \quad (12)$$

is operated in the temperature range 150–210 °C at atmospheric pressure. The use of a catalyst is common [45, 46]. The methanol and ethylene glycol (EG) emerging from the reactor are passed through a rectifying column and EG is fed back to the reactor. It is very difficult to force the ester interchange reaction to completion, and therefore after a particular conversion (usually 90–95%), the reaction mixture is passed on to the polycondensation stage. The reaction mixture consists of oligomers of various types. Oligomers with degrees of polymerization as high as three may be formed [45, 47, 48]. Several examples of reactor models for both batch and continuous processes can be found in [44–52]. An optimization study showed that the ester interchange reactor should be operated initially at high temperature to obtain high conversion; the temperature should be lowered to reduce side reactions [46, 50].

In the polycondensation stage, the reaction temperature is raised to 265–285 °C to keep the reaction mixture molten and polymerization fast.

$$n\,HOCH_2CH_2O-\overset{O}{\underset{\|}{C}}-\underset{}{\bigcirc}-\overset{O}{\underset{\|}{C}}-OCH_2CH_2OH \rightleftharpoons$$

$$H\!\left[OCH_2CH_2OCO-\overset{O}{\underset{\|}{C}}-\underset{}{\bigcirc}-\overset{O}{\underset{\|}{C}}\right]_{\!n}\!OCH_2CH_2OH \quad (13)$$

$$+\,(n-1)\,HOCH_2CH_2OH$$

For PET production, a dual catalyst system in which one component is specially active for ester interchange and the other for polymerization is often used [53]. The production of high molecular mass polymer requires the complete removal of ethylene glycol due to the unfavorable equilibrium, and therefore a vacuum is applied. Especially in the final stage of the polycondensation reaction, a very high vacuum is required since the reaction system becomes highly viscous. Consideration of the limitations of mass and heat transfer is very important. Various types of reactors such as rotating disc contactors, wiped film reactors, partially filled screw extruders have been developed as finishers for the polycondensation reaction [45, 46, 51]. Details of fluid mechanics, mixing, and mass and heat transfer characteristics are required for a rational analysis and design of such high-viscosity reactors, as was shown in Figure 5. In addition to polycondensation reactions, various side reactions must also be considered since a very high temperature is used.

Melt polycondensation of PET is not generally carried out beyond a particular extent of polymerization, since the degradation reactions dominate the process and the product quality may suffer from various undesirable byproducts. To attain higher molecular masses, the products may be subjected to solid-state polymerization [22–26]. For beverage and food containers, the production processes may prefer to stop melt polymerization at lower conversion, and solid-state polymerization is applied.

Direct esterification of terephthalic acid (TPA) and ethylene glycol was generally not preferred earlier because of the difficulties in the purification of TPA due to its low solubility and high melting point. However, with improvements in technology, the direct esterification method has been gaining in importance. The process is claimed to give polyesters with superior quality due to their low content of carboxyl end groups and diglycol linkages [53]. In the modeling of this process, aside from the difficulties caused by the various reactions and mass balances involved, it is necessary to take account of the heterogeneity of the reactions due to the low solubility of TPA in EG. Simulation and control of the direct esterification reactors is reported in [46, 51, 54–56]. Figure 10 shows a flow diagram of a continuous process for PET production by direct esterification.

Additional information for the industrial polycondensation processes can be found in [57].

4. Processes and Reactor Modeling for Chain Polymerization

4.1. Introduction to Polymerization Techniques

Four major categories of the industrial polymerization methods used for chain polymerizations are bulk polymerization, solution polymerization, suspension polymerization, and emulsion polymerization. The characteristics of these techniques are summarized in Table 1.

If the final product is a solid polymer, bulk polymerization might be the best method one should aim for, because a highly pure polymer is produced. Major step polymerization processes employ bulk polymerization technique, mainly to prevent side reactions at a high polymerization temperature. Other techniques provide alternatives to overcome the major difficulties of bulk polymerization: mixing and heat removal, except when the product's final form is any one of solution, suspension, or emulsion.

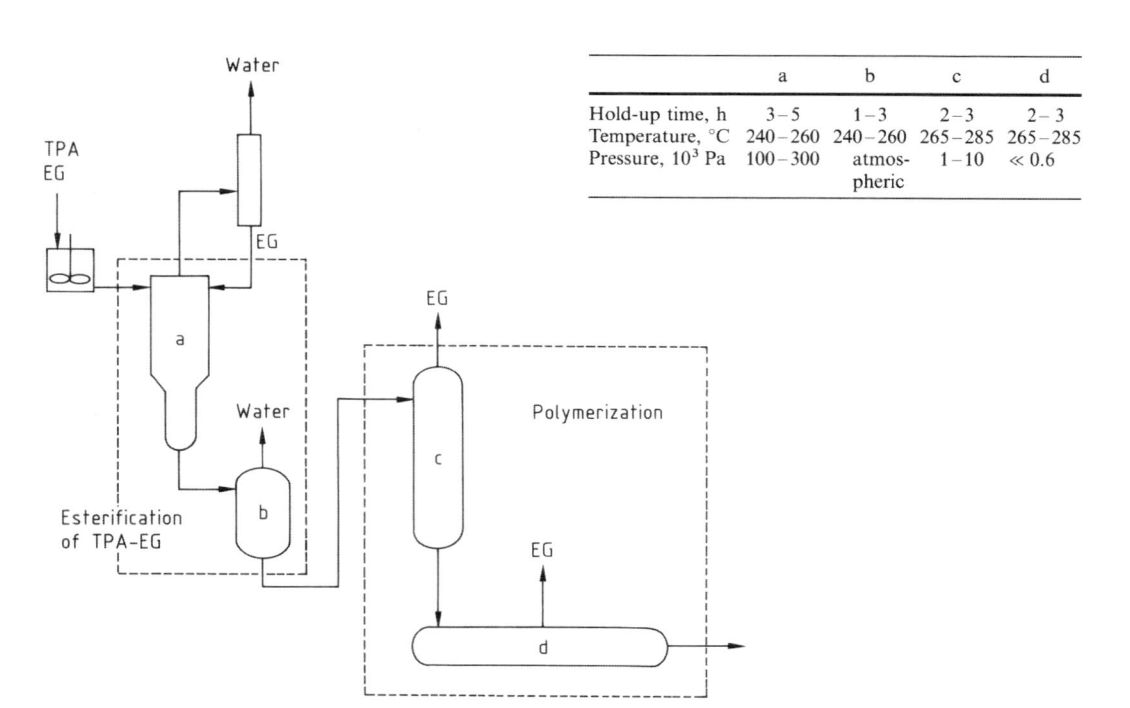

	a	b	c	d
Hold-up time, h	3–5	1–3	2–3	2–3
Temperature, °C	240–260	240–260	265–285	265–285
Pressure, 10^3 Pa	100–300	atmospheric	1–10	≪ 0.6

Figure 10. Continuous polymerization process of PET via direct esterification route

Table 1. Characteristics of major industrial polymerization methods

Method	Advantage	Disadvantage
Bulk polymerization	Minimum contamination of product	High viscosity (mixing, heat removal, pumping, etc.)
	High reactor efficiency	Adhesion to reactor wall and mixing apparatus (fouling)
Solution polymerization	Lower viscosity than bulk	Lower reactor efficiency
	Smaller fouling problems	Separation cost of polymer
	End-usage as a solution form	Inflammability and toxicity of the solvent used
		Side reaction, such as chain transfer
Suspension polymerization	Low viscosity	Lower reactor efficiency
	Good heat transfer	No large-scale continuous processes have been developed
	Higher purity than the emulsion polymers	Waste water treatment
		Adhesion and aggregation (fouling)
Emulsion polymerization	Low viscosity	Lower reactor efficiency
	Good heat transfer	High cost for separation of polymers
	High polymerization rate with large molar mass products in conventional free-radical polymerization	Impurity (emulsifier)
		Adhesion and aggregation (fouling)
	End-usage as a latex form	

4.2. Fundamentals of Material Balance Equations

The following material balance equations apply for multicomponent polymerization, accommodate operation of a well-stirred reactor (no spatial variations in temperature and concentrations), and may be used to simulate different comonomer systems under a variety of operating conditions. Bulk (suspension) and solution polymerizations are considered; extensions required for multiphase systems (emulsion, inverse emulsion, suspension, dispersion, and gas-phase processes) may be found elsewhere [58].

Many bulk, solution, and suspension polymerization systems are characterized by the fact that all of the reactions proceed in a single phase with no spatial variations in temperature and concentration. A model for a reactor carrying out such polymerizations would consist of a set of material balances giving the rates of accumulation, inflow, outflow and a reaction source (sink) term for the various monomers, initiators, chain-transfer agents, and polymer in the reactor. These balance equations are given herein in general form.

Monomer Balances

$$dN_{Mi}/dt = F^{in}_{Mi} - (N_{Mi}/V)v_{out} - R_{pi}V \qquad (14)$$

where

N_{Mi} = number of moles of monomer i in the reactor
F^{in}_{Mi} = molar flow rate of monomer i into the reactor
V = reaction volume in the reactor
v_{out} = total volumetric flow rate of all species out of the reactor
R_{pi} = net rate of disappearance of monomer i by reaction

Reaction Volume. Since the density of a polymer is usually greater than that of its monomer, the reaction volume V decreases with conversion for isothermal polymerization in a batch reactor. This shrinkage must also be taken into account in semibatch and continuous operations. Neglecting volume change on mixing polymer and monomers (thermodynamic data are most often not available and deviation from ideality is often not great) the change in reaction volume may be calculated by using (assuming V is the equilibrium volume):

$$dV/dt = v_{in} - v_{out} - \text{shrinkage rate}$$

$$= v_{s,in} + \sum_{i=1}^{n} F^{in}_{Mi} M_{mi}/\rho_{m_i} \qquad (15)$$

$$- \sum_{i=1}^{n} R_{pi} M_{mi} \left(\frac{1}{\rho_{mi}} - \frac{1}{\rho_p} \right) V - v_{out}$$

where n is the number of monomer types, $v_{s,in}$ is the volumetric flow rate of inert solvent into the reactor, M_{mi} is the molecular mass of monomer i, ρ_{mi} is the density of monomer i, ρ_p is the density of polymer produced instantaneously.

Polymer Balances. With a batch reactor, where there is no inflow and outflow of polymer from the reactor, the total amount of polymer formed and its composition can be obtained directly from the monomer balances. However, with semibatch and continuous operation, additional balances are required and these follow:

$$dP_i/dt = F_{Pi}^{in} - (P_i/V)v_{out} + R_{pi}V \qquad (16)$$

where P_i is the number of moles of monomer i chemically bound in the polymer "in the reactor", F_{Pi}^{in} is the molar flow rate of monomer i bound in the polymer flowing into the reactor.

Additional Ingredient Balances. In order to calculate R_{p_i} for free-radical systems, the total polymer radical concentration and, therefore, initiator concentration are required. In addition, balances for the chain-transfer agent (for molecular mass calculations) and for the solvent in solution polymerizations are required. These balances follow:

$$dN_{Ii}/dt = F_{Ii}^{in} - (N_{Ii}/V)v_{out} - R_{Ii}V \qquad (17)$$

$$dV_s/dt = v_{s,in} - (V_s/V)v_{out} \qquad (18)$$

$$dN_{CTA}/dt = F_{CTA}^{in} - (N_{CTA}/V)v_{out} - R_{CTA}V \qquad (19)$$

where N_{Ii} is the number of moles of initiator of type i in the reactor, F_{Ii}^{in} is the molar flow rate of initiator i into the reactor, R_{Ii} is the consumption rate of initiator i by reaction, V_s is the volume of inert solvent in the reactor, N_{CTA} is the number of moles of chain-transfer agent (CTA) in the reactor, $F_{CTA,\,in}$ is the molar flow rate of CTA into the reactor, and R_{CTA} is the consumption rate of CTA by reaction.

It is convenient to sum the monomer balance equation (Eq. 14) over n, the number of monomer types to give

$$dN_M/dt = F_M^{in} - (N_M/V)v_{out} - R_pV \qquad (20)$$

where N_M is the total number of moles of monomer in the reactor, F_M^{in} is the total molar flow rate of monomer to the reactor, R_p is the total molar consumption rate of monomer by reaction.

The equations introduced in this section are solved to determine the time developments of various components, by incorporating with the rate expression introduced in → Polymerization Processes, 1. Fundamentals. For the multicomponent polymerization, the use of pseudo-kinetic rate constants allows to use the same kinetic rate expression as for homopolymerization. At steady state in an HCSTR, all time derivatives are set to be zero.

4.3. Effect of Reactor Types on Copolymer Composition Distribution

The copolymer composition distribution of polymers formed instantaneously follows the Gaussian distribution [59], which gives the narrowest composition distribution. For the conventional free-radical polymerization, this distribution represents that polymer chains formed instantaneously. On the other hand, for living polymers, the distribution describes the distribution of monomeric units incorporated into various polymer chains instantaneously. The instantaneous copolymer composition can be obtained in an HCSTR at steady state, because the concentrations of all components are kept constant. Note that for the conventional free-radical polymerization, both distributions of molecular mass and composition are the narrowest in an HCSTR.

In a batch or a PFR, a compositional drift occurs, whose trajectory can be calculated by using the equations presented in → Polymerization Processes, 1. Fundamentals. For the conventional free-radical polymerization, the compositional drift occurs among polymer chains formed at different conversion levels, while the drift occurs along a single chain in living polymerization. Obviously, the

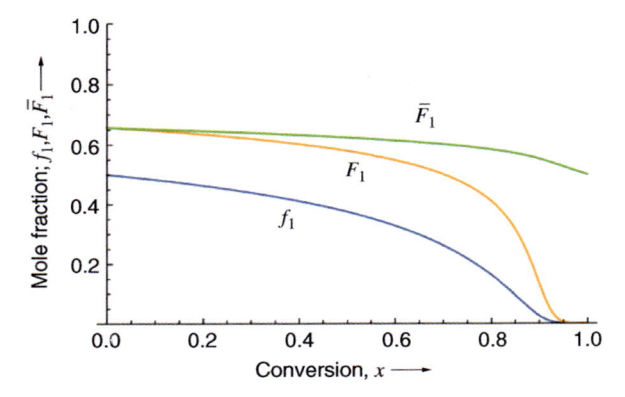

Figure 11. Calculated compositional drift during batch copolymerization of methyl methacrylate (M_1) and acrylonitrile (M_2) reactivity ratio $r_1=1.2$, $r_2=0.15$; initial composition monomer $f_1^0 = 0.5$

composition distribution becomes broader than for an HCSTR.

Figure 11 shows an example of compositional drift during batch free-radical copolymerization. The drift of monomer composition f_1, instantaneous copolymer composition F_1, and the accumulated copolymer composition $\overline{F}_1$ are shown. Even when the accumulated copolymer composition $\overline{F}_1$, which is usually measured in experiment, does not change significantly during polymerization, the composition of individual polymer chain (instantaneous composition) may differ significantly, as shown in the figure. Other than using an HCSTR, the compositional drift can be suppressed by using a semibatch operation in which monomers are fed to maintain a constant ratio of monomer concentrations in the reactor [58].

In an SCSTR, the compositional drift occurs within the compartments whose conversion levels are different significantly because of the broad residence time distribution, leading to a broader copolymer composition distribution than that for the batch/PFR [60].

4.4. Effect of Reactor Types on Nonlinear Polymer Formation

Branching/cross-linking reactions in chain polymerization involves reaction with polymer. In a batch polymerization, the polymer concentration increases during polymerization, starting from 0 to the final conversion level. On the other hand, in an HCSTR the polymer concentration is kept constant, and the polymerization proceeds under a high polymer concentration level. Therefore, the branching/cross-linking density is higher for an HCSTR than in a batch reactor at the same conversion level.

The effects of reactor types are considered in detail by taking free-radical polymerization that involves chain transfer to polymer, as an example. Figure 12 shows the process of chain transfer to polymer. Some time after the formation of the primary polymer chain **A**, which already possesses two branch chains, a primary polymer radical **B** attacks **A**, and an internal radical on the backbone chain is formed. At the same time, this internal radical grows to add monomeric units to form a long-branch chain

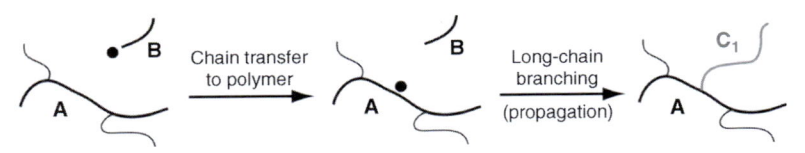

Figure 12. Schematic illustration of the process of chain transfer to polymer, leading to form polymer with long-chain branches

C_1. In free-radical polymerization, each primary polymer molecule is formed within a very small interval, and it is reasonable to consider that all the processes depicted in Figure 12 occur instantaneously.

As any of the monomeric units in polymer can react with a polymer radical, the polymer transfer reaction rate is represented by:

$$R_{fp} = k_{fp}[R^\bullet][M]_0 x \qquad (21)$$

where k_{fp} is the rate constant for the polymer transfer reaction, $[R^\bullet]$ is the total polymer radical concentration, $[M]_0$ is the initial (feed) monomer concentration, and x is the monomer conversion to polymer.

The polymerization rate is given by ($\rightarrow$ Polymerization Processes, 1. Fundamentals):

$$R_p = k_p[R^\bullet][M]_0(1-x) \qquad (22)$$

The branching density, ρ_b is a probability that a monomeric unit in polymer possesses a branch point. In an HCSTR, both the rates of monomer addition (Eq. 22) and branch point formation (Eq. 21) do not change with time, and therefore, the average branching density is given by:

$$\bar{\rho}_{b,HCSTR} = \frac{R_{fp}}{R_p} = C_{fp}\frac{x}{(1-x)} \qquad (23)$$

where C_{fp} is the polymer transfer constant ($= k_{fp}/k_p$).

On the other hand, this rate ratio changes with time in a batch reactor. Therefore, the average branching density for a batch reactor is given by:

$$\frac{d(x\bar{\rho}_{b,BR})}{dx} = C_{fp}\frac{x}{(1-x)} \qquad (24)$$

which is integrated to give [61]:

$$\bar{\rho}_{b,BR} = C_{fp}\left[\frac{1}{x}\ln\left(\frac{1}{1-x}\right) - 1\right] \qquad (25)$$

Figure 13 compares the average branching density, which shows that the use of an HCSTR leads to form larger average branching density than using a batch reactor at the same monomer conversion, x.

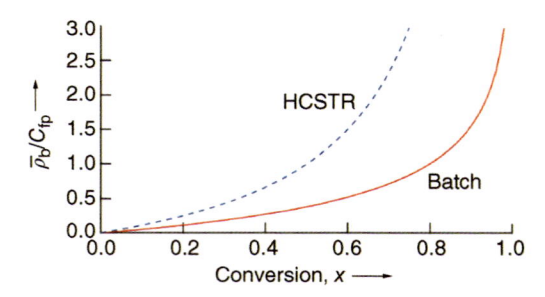

Figure 13. Calculated average branching density for an HCSTR and a batch reactor

The primary polymer chain length distribution in an HCSTR follows the instantaneous distribution, which is the narrowest. On the other hand, branch formation leads to broader distribution, as discussed in $\rightarrow$ Polymerization Processes, 1. Fundamentals. Normally, branch formation dominates the breadth of the MMD, except for extremely low branching density cases, and a larger average branching density in an HCSTR leads to a larger MM with a broader MMD, compared with a batch reactor at the same conversion level.

As already discussed $\rightarrow$ Polymerization Processes, 1. Fundamentals, the present reaction system cannot cause gelation without combination termination. Consider the special case where the contribution of combination termination in forming primary chains is negligible, i.e., $R_{tc} \ll R_{td} + R_{fX} + R_{fp}$, where R_{tc} is the rate of termination by combination ($=k_{tc}[R^\bullet]^2$), R_{td} is the rate of termination by disproportionation ($=k_{td}[R^\bullet]^2$), R_{fX} is the rate of chain transfer to small molecule X ($=k_{fX}[R^\bullet][X]$), and R_{fp} is the rate of chain transfer to polymer. In this case, the probability that the head of a primary polymer chain is connected to a backbone chain, P_b is given by:

$$P_b = \frac{R_{fp}}{R_{td} + R_{fX} + R_{fp}} \qquad (26)$$

P_b is kept constant for an HCSTR, but it changes with the birth time of the primary polymer molecule in a batch reactor.

In an HCSTR, the head unit of a newly formed primary chain is connected to a backbone chain with probability P_b. Because a larger polymer molecule has a better chance of being attacked by a polymer radical in the polymer transfer reaction, the probability to connect a newly formed primary chain is proportional to its molecular mass. This type of primary chain connection statistics leads to form an interesting MMD, in which the high MM tail follows the power-low distribution [62], with the relationship, $M(r) \sim r^{-1/P_b}$ [63], where r is the chain length, and $M(r)$ represents the mass fraction distribution. The same primary chain connection statistics operates also in emulsion polymerization during Interval 2, for which the polymer monomer/polymer concentration is kept constant [64, 65].

In a gelling system, the power-low distribution is formed only right at the gel point, with the relationship $W(M) \sim M^{-1.5}$ in the mean-field theory [66, 67], and it is impossible to actually obtain such polymer mixture because the polymerization cannot be stopped right at the gel point. In addition, the power exponent cannot be changed. In the present branched polymer system, the exponent can be controlled by changing the P_b value.

A strange characteristic of the power-law distribution is that the higher order moments always go to infinity, as represented by:

$$Q_i = \sum_{r=1}^{\infty} r^i [P_r] \propto \int_0^{\infty} r^{i-1} M(r) dr \sim \int_0^{\infty} r^{-(\alpha+1-i)} dr \quad (27)$$

where Q_i is the ith moment of polymer distribution, r is the chain length (degree of polymerization), and α is the exponent of the power-law distribution, which is $\alpha = 1/P_b$ in the present case. Note that by using the moments, the number-average $(\bar{P}_n)$ and mass-average $(\bar{P}_M)$ chain lengths are represented by $\bar{P}_n = Q_1/Q_0$ and $\bar{P}_M = Q_2/Q_1$, respectively.

Equation (27) shows Q_i stays finite for $\alpha > i$, i.e., for $P_b < 1/i$. In other words, Q_i goes to infinity for $P_b \geq 1/i$, meaning that the first moment (number-average) always reaches a steady state, while the second moment (mass-average) goes to infinity for $P_b \geq 1/2$, and the third moment (z average) goes to infinity for $P_b \geq 1/3$. This phenomenon should not be confused with gelation, because it takes an infinite time for the second and higher moments to go to infinity [68]. Gelation is a critical phenomenon, and it involves a critical change in the mass-average MM during polymerization. The present type of strange behavior does not occur in an SCSTR, and the polydispersity index, $\bar{P}_M/\bar{P}_n$ becomes much smaller than HCSTR [69].

Another interesting effect of reactor type is the distribution of branch points among primary chains. Referring to the branching process shown in Figure 12, consider the expected branching density of the primary chain born at $x = \xi$ in a batch polymerization. Suppose the primary chain **A** was born at $x = \xi$. The primary chains formed earlier are subjected to branching reaction for a longer period of time, and the branching density is expected to be larger for these chains rather than the chains formed lately. Let $\rho_b(\xi, x)$ be the expected branching density of the primary chains born at $x = \xi$, when the conversion at the present time is x ($> \xi$). The balance equation for the number of branch points is given by:

$$\rho_b(\xi, x + \Delta x) - \rho_b(\xi, x) = k_{fp}\{1 - \rho_b(\xi, x)\}[R^{\cdot}]\Delta t \cong k_{fp}[R^{\cdot}]\Delta t \quad (28)$$

The final approximation is reasonable, because the branching density for this kind of reaction system is normally much smaller than unity. By using the conversion–time relationship:

$$\frac{dx}{dt} = k_p(1-x)[R^{\cdot}] \quad (29)$$

One obtains the following differential equation:

$$\frac{\partial \rho_b(\xi, x)}{\partial x} = \frac{C_{fp}}{1-x} \quad (30)$$

Equation (30) can be integrated to obtain:

$$\rho_b(\xi, \psi) = C_{fp} \int_{\xi}^{\psi} \frac{dx}{1-x} = C_{fp} \ln\left(\frac{1-\xi}{1-\psi}\right) \quad (31)$$

where ψ is the conversion at the present time.

Figure 14 shows the calculated branching density distribution at the time when the conversion at the present time is $\psi = 0.8$. The primary chains formed at $x = 0$ possess the largest expected branching density because

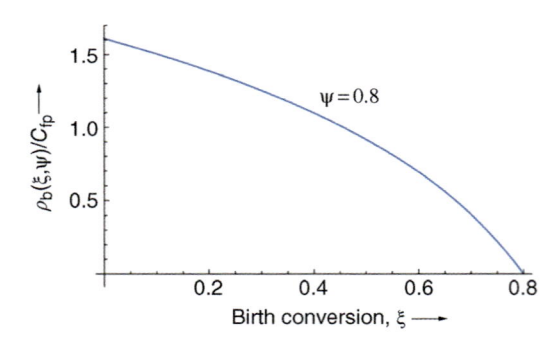

Figure 14. Calculated branching density distribution when the conversion at the present time is $\psi = 0.8$ in a batch polymerization

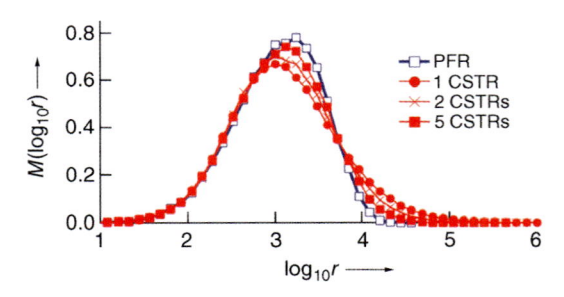

Figure 16. Calculated mass-fraction distribution [71]

they are subjected to branching reaction for the longest period of time in a batch reactor. On the other hand, the primary chains born at $x = 0.8$ do not possess any branching point. The chains formed in later stages of polymerization tend to become branched chains rather than backbone chains.

In the case of HCSTR, the branching reaction rate is kept constant. Therefore, the expected branching density increases linearly with its residence time [70]. The branching density distribution functions for various types of reactors can also be found therein.

Figure 15 shows an example of the calculated branching density distribution, represented as the mass-based probability density distribution $M(\rho_b)$ [71]. In the figure, the area under the curve within a certain branching density interval shows the mass fraction of primary chains possessing such expected branching density values. A PFR is equivalent to a batch reactor.

In the figure, the average branching density $\bar{\rho}_b$ is set to be the same for all reactor types. For a PFR (batch reactor), there exists a maximum branching density, represented by the y-intercept in Figure 14. For an HCSTR, the branching density distribution is very broad, because of a broad residence time distribution, given by Equation (4) and Figure 3. As the number of tanks increases, the branching density distribution approaches to that for a PFR (batch).

By application of the random sampling technique discussed in → Polymerization Processes, 1. Fundamentals to a Monte Carlo simulation method, it is possible to reconstruct the whole branched polymer molecules, on the basis of the branching density distribution function [71–73]. Figure 16 shows comparison of the full mass-fraction distribution [71]. In the calculation, not only the long-chain branching but random scission reactions are also accounted for, bearing low-density polyethylene (LDPE) synthesis in mind. However, the contribution of scission used for the present calculation is rather minor, so the major effect

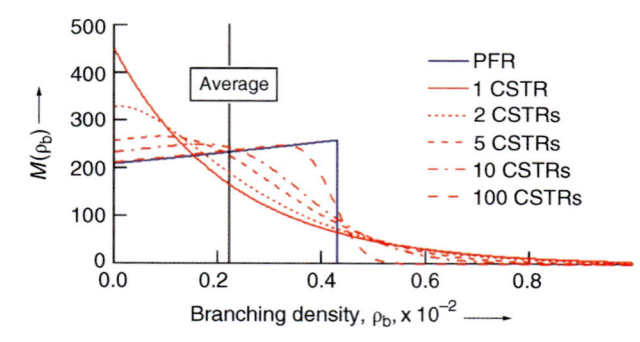

Figure 15. Calculated branching density distribution, represented as the mass-based probability density distribution, for a PFR, an HCSTR, and a tanks-in-series model [71] $\bar{\rho}_b = 2.22 \times 10^{-3}$; average branching density

is through the long-chain formation. Both the average branching density $(\bar{\rho}_b = 2.22 \times 10^{-3})$ and the average scission density $(\bar{\rho}_s = 1.11 \times 10^{-4})$ are set to be the same for all calculation conditions. Even when the average branching density is the same, the MMD is broader for an HCSTR than a batch reactor (PFR). As the number of tanks increases in a tanks-in-series model, the MMD approaches to that for a batch reactor.

In the Monte Carlo simulation, each polymer molecule formed in a reactor can be investigated directly on the computer screen, and therefore, very detailed structural information can be obtained. Figure 17 shows an example of branched polymer structure determined in the Monte Carlo simulation, in which the sphere represented by small dots has a radius that is equal to the radius of gyration of this particular structure. Figure 18 gives the mean-square radii of gyration as a function of chain length for the polymer molecules formed.

Each small dot in Figure 18 shows the mean-square radius of gyration (y axis) and its chain length (x axis) of the polymer molecule formed in a batch/PFR (A) and in an HCSTR (B). In the y axis of Figure 18, U is the number of monomeric units in a random walk segment, and L is the segment length. Here, the radius of gyration

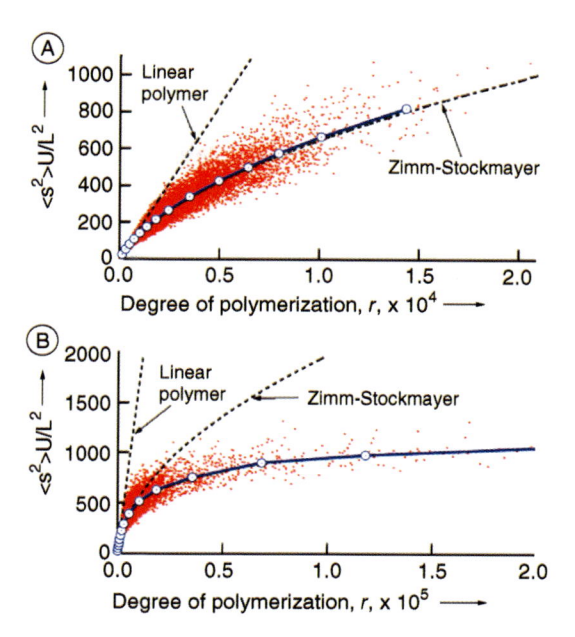

Figure 18. Mean-square radii of gyration as a function of chain length for the polymer molecules formed [74]
A) Batch/PFR; B) HCSTR

was determined by conducting the random walk 100 times, although exact mean-square radius of gyration can be determined by calculating the Wiener index [75], based on the structural information obtained in the Monte Carlo simulation [76]. In the figure, the mean-square radii of gyration for the linear polymers, and those for random branched polymer molecules represented by the Zimm–Stockmayer equation ([77]) are also shown by the broken lines. In the case of batch/PFR, the radii of gyration are close to those for the random branch polymers, although the branched structure is not random, as was shown by the branching density distribution, Figures 14 and 15. As the polymerization proceeds, the curve moves to upward direction and the radii of gyration become slightly larger than the random branched polymers [78, 79].

On the other hand, in an HCSTR, the radii of gyration of large polymers become much smaller than the random branched polymers, as shown in Figure 18B. This is because the primary chains whose residence times are large have large branching density values, as shown in Figure 15. In a highly branched, large-sized polymer molecule, it is expected there are

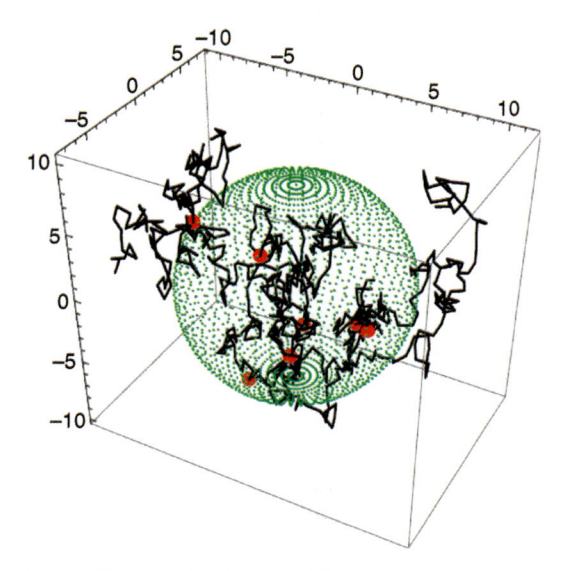

Figure 17. Example of branched polymer structure determined by Monte Carlo simulation [74]

regions in which many branch points are concentrated, leading to a compact dimension of a polymer. A tanks-in-series model has clarified that as the number of tanks increases, the branched structure becomes closer to that for a PFR [71]. The residence time distribution has a significant effect on the branched polymer structure.

On the basis of the detailed structural information obtained through the MC simulation is now being utilized to estimate the physical properties, such as the viscoelastic properties [80]. The computer-aided polymer design system for complex polymer systems are being developed.

5. Bulk and Solution Polymerization

Bulk polymerization might be the best method to produce pure, solid polymers. To minimize contamination, poly(methyl methacrylate) used for fiber optics, in which high transparency is required, is produced via bulk polymerization. The major operating problem of bulk polymerization is the extremely high viscosity, making it difficult to accomplish good mixing, as well as efficient heat removal. The difference of bulk and solution polymerization is simply the degree of high viscosity when the polymer is miscible to monomer and solvent. In the initial stages of polymerization, monomer acts as the solvent for polymer, and stirred-tank reactor can be used to obtain good mixing. In the later stages of polymerization, however, plug–flow-type reactors with high mechanical strength are often employed.

The standard polymerization kinetics discussed in → Polymerization Processes, 1. Fundamentals are basically valid for bulk and solution polymerization. In the viscous polymer solution, the chain-length-dependent rate constants, resulting from diffusion-controlled reactions are the major problem in the theoretical calculations. Although the chain-length-dependent rate constants are being clarified [81, 82], more research must be done before these topics can be considered standard engineering practice.

One of the most successful classical examples in the modeling work in bulk polymerization is the thermal polymerization of styrene at temperatures > 100°C [83–90]. In this reaction system, the Mayo mechanism for thermal initiation of radicals is valid [91], and the initiation reaction does not become diffusion controlled at monomer conversions as high as 97 % [92]. The polystyrene chains are linear. The Trommsdorff–Norrish effect, although significantly affecting the polymerization rate, has at most a minor effect on molecular-mass development because most of the polymer chains are produced by chain transfer to the Diels–Alder intermediate [91].

Because the ratio of cooling surface area to volume of reacting mixture decreases as the reactor volume increases, the heat removal is one of the major technical problems. When jacket cooling is no longer sufficient to maintain isothermal polymerization, additional modes of heat transfer must be used, as shown in Figure 19. Internal cooling coils

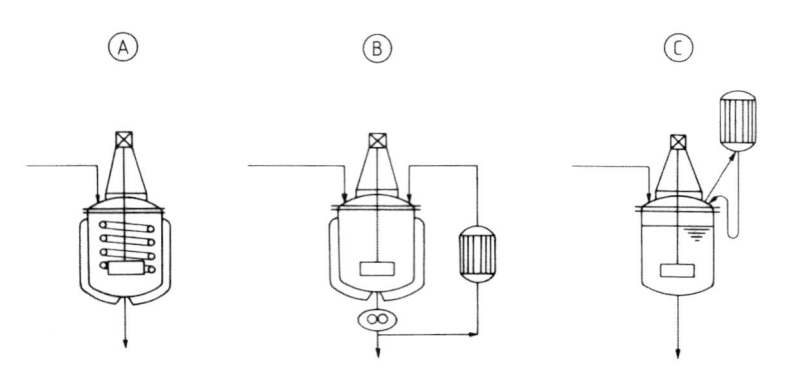

Figure 19. Alternative methods of heat removal for polymerization conducted in a stirred reactor
A) Internal cooling coils; B) External cooler; C) Reflux cooling

are often not practical because they tend to interfere with stirring. External tubular coolers can, in principle, provide a very large heat-transfer area, but they may have very large pumping requirements in the case of highly viscous solutions. Reflux cooling removes the heat of polymerization by evaporation of solvent and/or monomer; the condensed vapor is recycled to the reacting mass. Condensers may be as large as necessary, with the limiting factor usually the amount of vapor that can be treated without causing intense foaming or spattering of the polymer solution. Remixing of the condensed liquid with the more viscous reacting mass may also be difficult.

Another industrially important class of bulk polymerization can be found in photoinitiated cross-linking polymerization [93], applied for coating and printing.

The addition of a solvent in which both monomer and polymer are miscible lowers the viscosity of the reacting mass, thereby improving its flow and heat-transfer characteristics. As a result, and depending upon the nature and concentration of the solvent, the Trommsdorff–Norrish effect can be reduced significantly. Use of solvent can be especially beneficial when evaporative cooling is used. In choosing a solvent, it is important to take into account the possibility of chain transfer to solvent with a concomitant reduction in polymer molecular mass.

5.1. Removal of Solvent and Residual Monomer

Removal of solvent and residual (unreacted) monomer from the highly viscous polymer solution requires very high surface areas to permit rapid devolatilization at the moderate temperatures required to minimize polymerization and degradation of chains during devolatilization to reduce off-spec polymer. Figure 20 illustrates several designs of devolatilizers. One of these, the vacuum degasser, incorporating a type of spray device (A), operates adiabatically, which means that the heat for evaporation is supplied by the solution itself. In contrast, a degasser in series with a tubular heat exchanger causes some of the monomer/solvent to evaporate during passage through the heat exchanger. This evaporation accelerates the flow of product, thereby increasing the heat-transfer rate (B). Twin-screw extruders, as were shown in Figure 5, with one or more vapor outlets that can be connected to a vacuum source are also utilized for the removal of monomer and solvent (C). Intermeshing and self-cleaning screws provide continuous renewal of the evaporating surface. The thin-film evaporators with roatating wiper blades can also be applied (D). More details on devolatilization techniques and their modeling can be found in [94, 95].

Solution polymerization is frequently used in laboratory experiments, but not so often in industry because of the cumbersome solvent

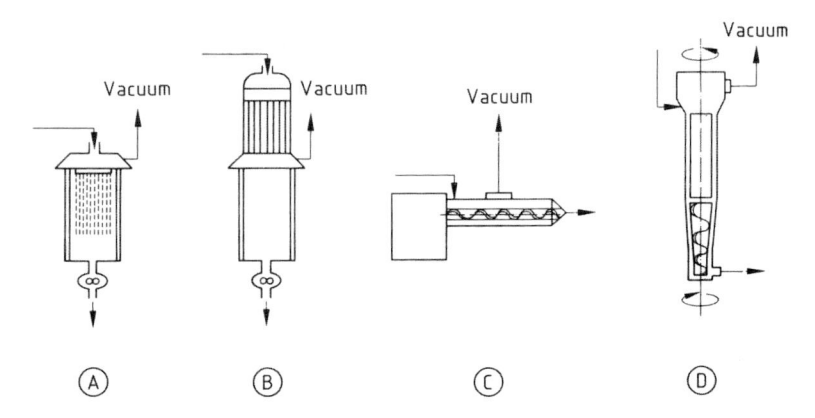

Figure 20. Options for removing residual monomer and solvent
A) Falling strand devolatilizer; B) Tubular evaporator; C) Devolatilizing extruder; D) Thin-film evaporator

recovery process required. On the other hand, to produce well-defined, highly functional polymers through the controlled/living radical polymerization, or reversible-deactivation radical polymerization according to the IUPAC recommendation [96], solution polymerization could be applied.

Recovered monomer/solvent can be recycled to the reactor. However, special attention must be paid to prevent the buildup of impurities (e.g., inhibitors, chain-transfer agents) over time, by using appropriate purge streams.

5.2. Systems with Polymer–Polymer Demixing

Polymer–polymer demixing is important in the production of thermoplastics whose application characteristics are enhanced by the presence of dispersed domains containing an elastomer. Generally, the thermodynamics are unfavorable for complete miscibility between different types of polymers. For a system polymer A–polymer B–solvent, the following generalizations are usually applicable [97]: dilution with solvent and, in most cases, an increase in temperature increases the compatibility; increasing the molecular masses of the polymers has the opposite effect. Except in the case of high dilution, incompatibility in solution is the rule for pairs of polymers, even when the solvent is a good solvent for both polymer types.

Systems such as (polymer A)–(polymer B)–block or graft copolymer AB can be regarded as a polymeric oil-in-oil emulsion in which the copolymer functions as an emulsifying agent. Such systems arise, for example, in the manufacture of high-impact polystyrene (HIPS), or in the preparation of ABS (acrylonitrile–butadiene–styrene) [98]. In the simplest case, ca. 5–10% polybutadiene is dissolved in monomeric styrene to give a homogeneous solution suitable for polymerization. The polystyrene synthesized is incompatible with the polybutadiene present, causing phase separation even at very low monomer conversion and producing a polybutadiene–styrene continuous phase and a polystyrene–styrene disperse phase. Simultaneously, graft polymerization produces polystyrene branches on polybutadiene backbone. The graft copolymer serves as an emulsifier, accumulating at the interface and stabilizing the oil-in-oil emulsion.

With increasing conversion, the volume fraction of the polystyrene phase (which is initially small) increases considerably due to formation of additional polystyrene that absorbs styrene monomer. Finally, often at a phase–volume ratio of about unity, a phase reversal occurs, with the rubber phase now the disperse phase, and the polystyrene phase the continuous phase. Agitation is essential for the completion of phase reversal, since rapid approach to equilibrium with respect to transfer of monomer/polymer must be attained in a highly viscous medium. A common result of inadequate agitation is the interpenetration of two continuous phases.

Phase reversal can lead to a complex morphology; e.g., polystyrene–styrene droplets can be occluded within the rubber particles of the disperse phase, which is often referred to as salami morphology, as shown in Figure 21 [99]. This is actually desirable as it increases the volume fraction of the disperse rubber phase. A reinforcing effect is observed, with the resulting HIPS having higher impact strength for a given mass of rubber in comparison with materials prepared by emulsion grafting, in which occluded polystyrene is not formed [100, 101].

Examples of the modeling and simulation of the bulk HIPS process can be found in [102, 103].

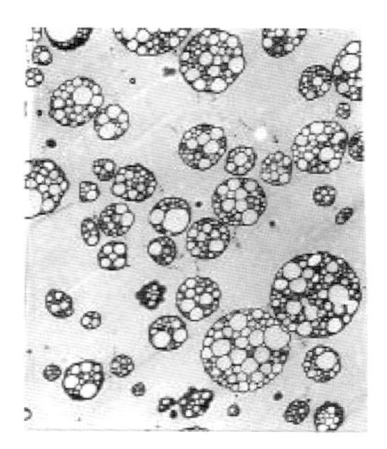

Figure 21. Salami structure of HIPS [99]

6. Precipitation and Dispersion Polymerization

The term precipitation polymerization refers to processes in which the initial ingredients of a recipe are soluble, giving a homogeneous solution, but the synthesized polymer precipitates during the course of polymerization. According to the IUPAC recommendation [104], "polymerization in which monomer(s), initiator(s) and colloid stabilizer(s) are dissolved in a solvent and this continuous phase that is a nonsolvent for the formed polymer beyond a critical molecular mass" is called the precipitation polymerization. The precipitation polymerization that leads to form polymer particles, normally of colloidal dimensions is called dispersion polymerization [104]. Precipitation and dispersion polymerizations were occasionally distinguished by the use of dispersion stabilizer. However, according to the IUPAC recommendation, the precipitation polymerization leading to form colloidal particles can be called a dispersion polymerization. Therefore, the preparation of poly(N-isopropylacrylamide) microgels in aqueous medium without using stabilizer, leading to uniform particles [105], can also be called a dispersion polymerization.

6.1. Polymerization without Solvent

A notable example of precipitation polymerizations is a bulk polymerization of vinyl chloride in which polymer is insoluble in its monomer. To ensure convenient handling, the disperse phase must be finely divided, which is achieved by effective agitation. Table 2 shows representative industrial production processes of poly(vinyl chloride). About 80% of worldwide production involves suspension polymerization, in which poly(vinyl chloride) precipitates in the monomer droplets. Bulk polymerization accounts for less than 10%. The final product of bulk and suspension polymerization is made up of primary particles and its agglomerates. The nucleation, growth and aggregation are responsible for the formation internal morphology and associated properties such as porosity, pore size distribution and specific surface area. For details, see also → Poly(Vinyl Chloride).

Polymers derived from halo or pseudohalo-substituted ethylenes such as vinyl bromide, vinylidene chloride, trifluoroethylene, and acrylonitrile are insoluble in their monomer. Polyethylene also falls into this category, at least when produced under moderately high pressure.

The conversion–time curves for the bulk polymerization of such monomer–polymer systems display rate increases with increasing conversion. This phenomenon has been especially thoroughly investigated for the bulk polymerization of vinyl chloride. Its cause may be interpreted as follows: even at very low conversion, poly(vinyl chloride) precipitates, forming a monomer-swollen polymer-rich phase. Initiator is partitioned between the phases, and thus radical generation and polymerization occur in both phases. The bimolecular termination rate is diffusion controlled in the polymer-rich phase and hence the concentration of radicals is higher in this phase. Even though the monomer concentration is lower in the polymer-rich phase, the specific polymerization rate R_p is higher. In addition, the volume of the polymer-rich phase grows at the expense of the monomer-rich phase with increasing monomer conversion. This explains the autoacceleration in rate as monomer conversion increases. Bulk polymerization of vinyl chloride may be a good example to study fundamental aspects of precipitation polymerization, and its mathematical model development can be found in [106–109].

Table 2. Industrial polymerization techniques of poly(vinyl chloride)

Techniques	Initiator	Stabilizer	Particle size, μm	Processing/application
Bulk polymerization	oil soluble	no	100–150	extrusion, injection, calendering
Suspension polymerization	oil soluble	polymeric dispersant	100–150	extrusion, injection, calendering
Emulsion polymerization	water soluble	emulsifier	0.1–2	paste, latex
Microsuspension polymerization	oil soluble	emulsifier	0.1–2	paste, latex
Solution polymerization	oil soluble	no	solution	paint, coating

Precipitation polymerization also encompasses what has been called "gas-phase" polymerization, a process in which polymer particles form within a monomer vapor. Notable commercial application includes gas-phase reactors for low-pressure polymerization of ethylene and propylene. Such polymerization does not actually occur in the gas phase, however, because the catalyst resides either within or on the surface of existing polymer particles, and a significant amount of monomer is dissolved in the polymer. The actual site of polymerization is thus within the polymer particle, to which a continuous supply of monomer flows from the gaseous phase.

Figure 22 shows a classical example called the Unipol process [110] developed by Union Carbide in which polyethylene powder is produced using ethylene as a fluidizing gas in a fluidized-bed reactor. Linear polyethylene, called high-density polyethylene (HDPE), is produced. Copolymerization of ethylene with propene, 1-butene, and 1-hexene gives products with a controlled amount of short-chain branching and lower polymer density, called linear low-density polyethylene (LLDPE). HDPE and LLDPE are produced via coordination polymerization [111], which belongs to the ionic polymerization, not free-radical polymerization. Gas-phase precipitation polymerization and liquid-phase precipitation polymerization mentioned in the next section are the major production methods for HDPE and LLDPE. Note that low-density polyethylene (LDPE) is produced in the high-pressure free-radical polymerization process.

6.2. Polymerization with Solvent

Solvents that can dissolve monomers but precipitates polymers are used for precipitation polymerization. A notable example is the so-called slurry process for manufacture of high-density polyethylene (HDPE), isotactic polypropylene, and their copolymers. Similarly with the gas-phase processes, the polymer is formed around heterogeneous catalyst. The slurry processes use an autoclave-type reactor with agitator or a loop reactor. A loop reactor is a continuous tube or pipe, which connects the outlet of a circulation pump to its inlet, with a cooling jacket outside. The continuous production process using a loop reactor is a variation of a CSTR, and it offers a large heat removal capacity and prevents adhesion through a high-speed circulation of slurry. Two or more reactors can be operated in series to produce polyolefins with more complex microstructures [112, 113].

If the polymer is soluble in its monomer, precipitation polymerization requires the addition of a precipitant that is miscible with the monomer. Polymerization then begins in a solvent – nonsolvent mixture and it ends in a phase consisting only of pure precipitant once conversion of the monomer is complete. Thus, the solubility relationships change during the course of the polymerization, and polymerization may initially occur in a homogeneous manner prior to the onset of precipitation, induced by the enrichment in precipitant that accompanies the consumption of monomer. In precipitation polymerization of this type, it is especially important to ensure that the polymer precipitates in finely divided form. Among other things this ensures that the particles do not overheat during polymerization. The stabilizers can be added to prevent agglomeration.

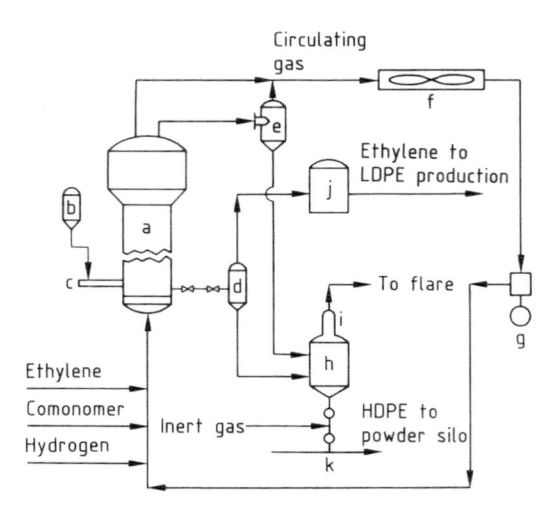

Figure 22. Schematic diagram of the Union Carbide gas-phase process for manufacturing HDPE [110]
a) Fluidized-bed reactor; b) Catalyst transfer tanks; c) Catalyst feeders; d) Product discharge tanks; e) Multiclone dust separator; f) Air coolers; g) Compressor; h) Product degassing tank; i) Filter; j) Ethylene tank; k) Pneumatic transport system

When the stabilizer (dispersing agent) works effectively in precipitation polymerization, nice uniform spherical particles of micron-sizes can be obtained, which belongs to the dispersion polymerization [114–116]. In this polymerization, the monomers, the initiator, and the stabilizer are dissolved in a solvent. Polymerization starts in a homogeneous phase and polymer precipitates to form unstable nuclei. Nucleation ends when the number of stable polymer particles reaches to the point in which all new nuclei are captured by the existing stable particles. Shorter nucleation period is preferable to obtain uniform particles. The polymerization locus changes from the continuous phase to the dispersed phase.

The key to controlling particle size is the selection of stabilizer type. Among the most effective stabilizers are the so-called amphipathic molecules: block and graft copolymers made up of two polymeric components, only one of which is insoluble in the continuous, diluent-containing phase. Graft copolymers of this type often form during the polymerization as a result of grafting on the dissolved polymer, but it is not absolutely essential that the insoluble portion of the polymeric stabilizer be identical to or soluble in the disperse phase. In many cases its insolubility in the diluent is sufficient to ensure adequate adsorption on the particle surface. These amphipathic dispersing agents act as steric stabilizers [117].

Figure 23A is a schematic representation of the adsorption of di-block, multiple-block, and graft copolymers on the surface of growing polymer particles. Soluble and insoluble portions of the dispersant molecules must be kept carefully in balance. If the insoluble part is too small, or if it interacts too weakly with polymer particles, then adequate adsorption will occur only when the dispersant concentration in the continuous phase is very high. If the soluble portion is too large, the dispersant will be present largely as aggregates or micelles with little tendency to dissociate and be adsorbed on the interface. Figure 23B depicts the equilibrium situation. Finally, it is important to note that multifunctional amphipathic molecules like those shown in Figure 23C can also function as weak flocculating agents.

By selecting appropriate reaction condition, monodisperse polymer particles can be prepared

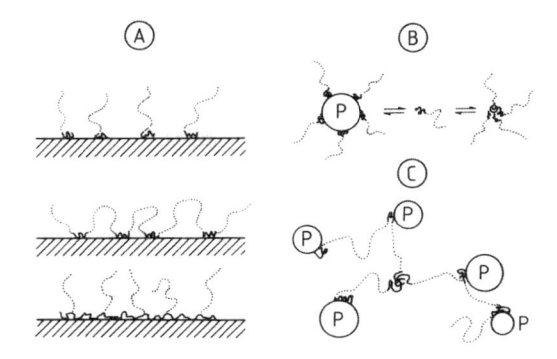

Figure 23. Steric stabilization of precipitating polymer/monomer particles with the aid of amphipathic block and graft copolymers [118]
A) Schematic representation of adsorption of amphipathic molecules on the polymer particles; — insoluble groups, ---- soluble groups; B) Equilibrium established in the course of a precipitation polymerization P = (growing) polymer particle; C) Schematic representation of flocculation due to multifunctional amphipathic molecules

in dispersion polymerization, as shown in Figure 24 [119]. Dispersion polymerization connects nicely the particle size between emulsion polymerization and suspension polymerization. Figure 25 shows approximate particle sizes that can be prepared from various heterogeneous polymerization techniques.

Precipitation/dispersion polymerization can be conducted not only for conventional free-radical polymerization but with reversible-deactivation radical polymerization (RDRP) (or controlled/living radical polymerization) [120, 121]. A problem in conducting RDRP in dispersion polymerization is a longer nucleation period because of slow chain growth in RDRP, which makes the particle size distribution broader. Newer topics in precipitation polymerization include the use of CO_2 as the solvent [122, 123].

7. Suspension Polymerization

7.1. Process Description

The term "suspension polymerization" includes a series of processes, in which polymer is formed in monomer droplets dispersed in a continuous phase that is nonsolvent for both

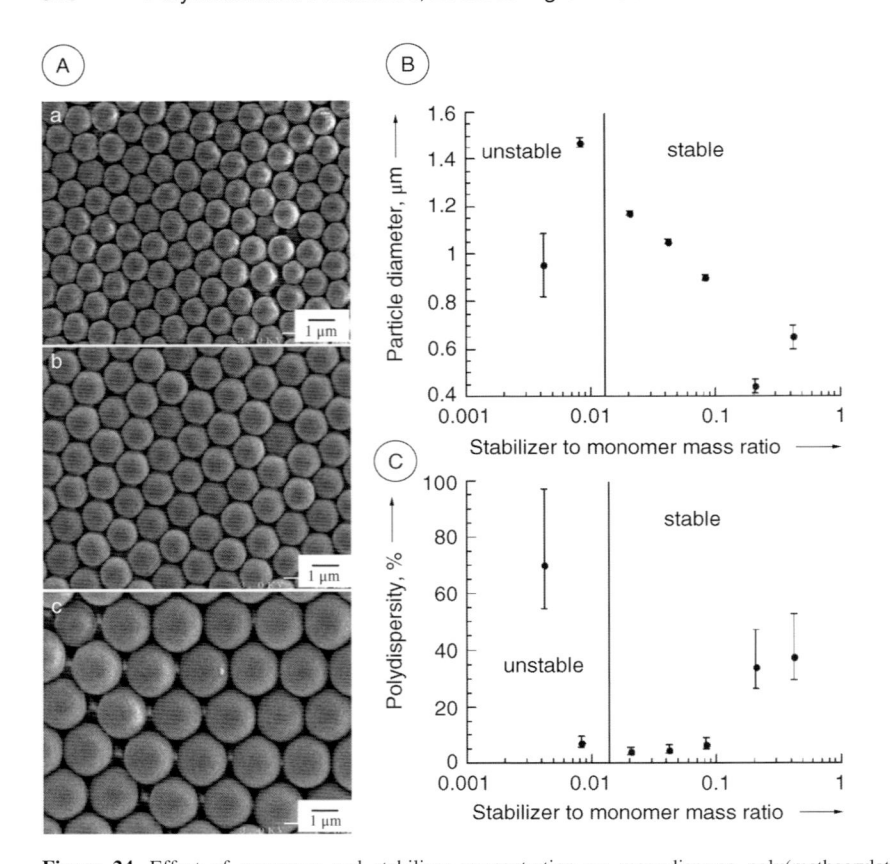

Figure 24. Effect of monomer and stabilizer concentration on monodisperse poly(methacrylate particles prepared by dispersion polymerization in hexane [119]
A) Scanning electron micrographs, a) 15 wt%, b) 26.3 wt%, c) 47.2 wt% monomer concentration; B) Particle diameter as a function of concentration of stabilizer; C) Polydispersity as a function of concentration of stabilizer

monomer and the formed polymer. In suspension polymerization, the initiator is located in the monomer phase. The monomer droplets have diameters usually exceeding 10 µm, which can be considered as miniature bulk polymerization reactors. Although the viscosity inside the droplets increases during polymerization, the effective viscosity of the suspension remains

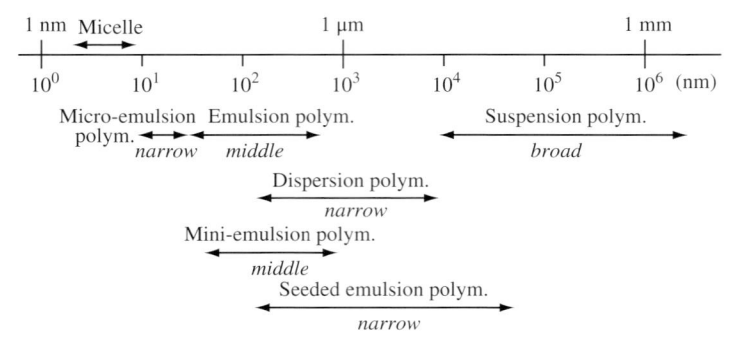

Figure 25. Approximate particle size ranges and distribution width (*narrow*, *broad*, *middle*) that can be prepared from various heterogeneous polymerization techniques

low and effective agitation is possible. Large specific surface area of small droplets and low viscosity of suspension make it possible to have a good heat removal from the reactor. Suspension polymerization is suitable for the preparation of cross-linked polymer particles, such as ion exchange resins and superabsorbent polymers. Cross-linked sodium polyacrylate used for disposable diapers can be produced by the inverse-suspension polymerization in which a hydrophobic organic suspending medium is used.

One of the major disadvantages of heterogeneous polymerization is lower reactor efficiency. The volume ratio monomer/continuous phase in suspension polymerization is usually between 25 : 75 and 50 : 50. Incidentally, the most space-efficient way to pack uniform spheres is in a pyramid shape and the volume fraction of spheres is $\pi/\sqrt{18} \cong 0.74$, which has been known as the Kepler conjecture for more than 400 years and a formal proof was announced completed in 2014 [124].

For the dispersants, water-soluble polymers or insoluble, usually inorganic, particles, the so-called Pickering stabilizers [125] are used. Their function is first to assist in the formation of the initial monomer droplets and then to stabilize the resulting polymer particle suspension. The inorganic stabilizers can be removed easily after polymerization.

In a large-scale commercial suspension polymerization, stirred batch reactors are commonly employed, and huge reactors larger than 100 m^3 are used for the production of poly (vinyl chloride). The continuous suspension polymerization processes are being developed [126, 127]. In the continuous processes, narrower RTD is preferable to prevent coexistence of polymer particles whose monomer conversion is not high enough.

The monomer droplets may experience both breakage and coalescence phenomena. The droplet breakage mainly occurs in the regions of high shear stress, i.e., near the agitator blades, and would be significant at low monomer conversions. After reaching so-called "sticky-stage", the coalescence becomes significant, and the particle size increases. At higher monomer conversions, the particles are sufficiently hard and the coalescence ceases. The particle size distribution remains constant from this point, called the identity point, except for some minor shrinkage due to density difference between monomer and polymer. Increasing the stabilizer concentration (i.e., reducing the size of particles) displaces the identity point towards lower conversion.

The control of the particle size distribution is important in many industrial applications. To control particle sizes, various models and equations were proposed for specific polymerization systems [128], and these empirical approaches are still useful for the industrial applications. To follow the dynamic evolution of the particle size distribution theoretically, the population balance approaches could be used [129–131].

7.2. Polymerization Kinetics

The monomer droplets are large enough to contain a large number of radicals. Assuming that the polymer radical concentration $[R^\bullet] = 1 \times 10^{-7}$ mol/L, a monomer droplet with diameter $d = 100$ µm possesses 3×10^7 polymer radicals. With so many radicals in a polymerization locus, bulk/solution polymerization kinetics is valid. When the polymerization locus becomes so small that radicals are isolated in different particles, one needs to use the emulsion polymerization kinetics that is different from the bulk/solution polymerization kinetics. Each droplet/particle can be considered as a miniature bulk polymerization reactor in suspension polymerization. The statistical variation of the conversion levels among particles is small enough, because each droplet/particle contains a large enough number of monomeric units in it. Assuming a particle with $d = 100$ µm, density 1 g/cm^3, and the molar mass of a monomeric unit 100 g/mol, there are 3×10^{15} monomeric units in it. Therefore, even when droplet/particle breakage and coalescence occur, the concentrations of the various components in the reaction loci are not changed. However, if the effect of residence time distribution cannot be neglected in a continuous production process, one needs to account for the difference in the monomer conversion levels among particles.

The Trommsdorff–Norrish and glass effects must be taken into account, as in the case of bulk polymerization. The process is sometimes referred to as a water-cooled bulk polymerization. To ensure that conversion is as complete as possible, it is common to employ mixtures of initiator types with different half lives, and to allow the polymerization temperature to increase in the final stages of conversion.

When the polymer formed is soluble in the monomer, nonporous spherical "beads" are formed, hence the term "suspension bead polymerization" is sometimes used. If, however, the polymer precipitates during polymerization, the resulting polymer particles are composed of many smaller primary particles. They are opaque, usually possess an irregular surface, and may have substantial internal porosity. This type of polymerization has been called "suspension powder polymerization". Suspension powder polymerization is the most important commercial process for the manufacture of poly(vinyl chloride). Porous powders with a rough surface are advantageous for rapid incorporation of plasticizers, as well as for a fast removal of residual monomer. The polymer particle morphology development during suspension polymerization of vinyl chloride [132, 133] is an important issue to produce high quality product.

Some of the factors that might be different from bulk/solution polymerizations are: (1) the reflux from condenser may have difficulty in returning to the particles [134], which implies that semibatch operation requires careful attention, and (2) because of the existence of continuous phase, some amount of the components may coexist in the continuous phase. This may change the apparent reactivity ratios when one of the monomers is more soluble in the continuous phase than the others [128].

8. Emulsion Polymerization

8.1. Process Description

Emulsion polymerization involves emulsification of monomers in dispersed phase, normally forming an oil-in-water emulsion, although inverse emulsion polymerization is also possible. The initiator is normally located in water (continuous) phase, but an oil-soluble initiator may also be used for special cases. At the end of emulsion polymerization, a milky fluid called "polymer latex", which is a colloidal dispersion of polymer particles whose diameter is about several tens to a few hundred nanometers, is obtained. In industry, the product latexes contain 40–60 wt% polymer solids, although much smaller solid content latex might be prepared in the laboratory scale experiments. Assuming a 50 wt% latex with the particle diameter $d = 100$ nm and the density of polymer 1 g/cm^3, there are about $N = 1 \times 10^{18}$ polymer particles per liter of latex.

In a batch ab initio emulsion polymerization, the polymerization may be conveniently divided into three intervals, as shown in Figure 26. At the start of polymerization, the monomer droplets whose size is about $d = 1$–10 μm and number is about $N = 10^{12}$–10^{14} per liter of latex

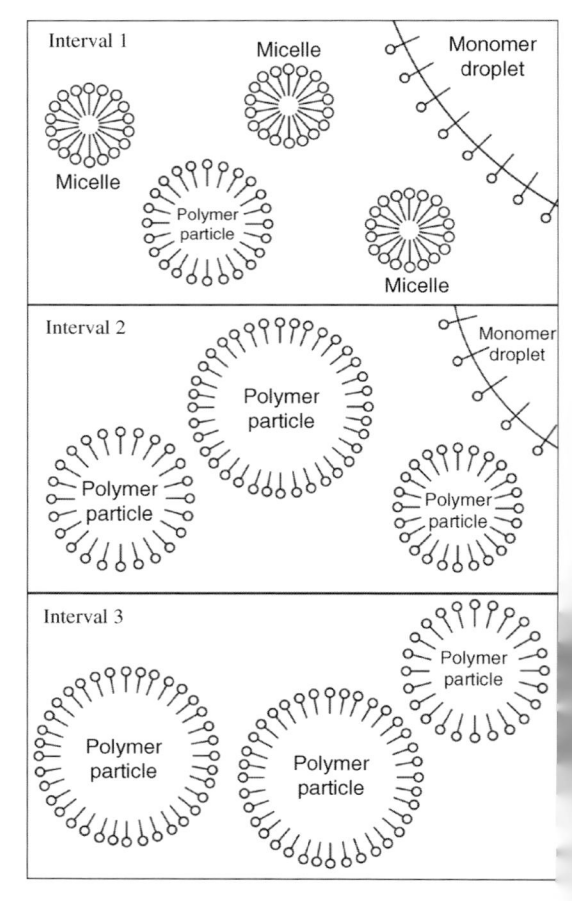

Figure 26. Schematic representation of ab initio emulsion polymerization

are formed, with much larger number of monomer-swollen micelles if the surfactant concentration is higher than the critical micelle concentration (CMC), which is the case for usual emulsion polymerization. The micelles are usually $d = 5$–10 nm in size, and $N = 10^{19}$–10^{21} in number.

Interval 1 is the nucleation period of polymer particles. As one can guess easily from the size relationship, (micelles) $\ll$ (final polymer particles) $\ll$ (monomer droplets), the monomer droplets are not the main loci of polymerization. The initiator decomposes in the water phase to generate primary radicals, which propagate with monomer dissolved in water to form oligomer radicals to obtain sufficient hydrophobicity. The mass transfer coefficient around a spherical particle is given by $2D_w/d$, where D_w is the diffusion coefficient of the oligomer radicals in water phase. Therefore, the total entry frequency of an oligomer radical is given by $(2D_w/d) \times (\pi d^2) \times N = 2\pi D_w N d$. For the $(N \times d)$ value, there is a relationship, (monomer droplets) $\ll$ (polymer particles) $\ll$ (micelles). Therefore, the oligomer radicals will have a strong tendency to enter monomer-swollen micelles, and they propagate rapidly with solubilized monomer to form a polymer particle. This is a modern interpretation of the micellar (or heterogeneous) nucleation theory [135–137]. As new polymer particles are formed and grow, the particles absorb monomer from the monomer droplets and the monomer-swollen micelles. At the same time, the polymer particles adsorb surfactant to obtain colloidal stability. Eventually, the micelles are gone and the nucleation of polymer particles ceases. This is the end of Interval 1. Only about one of every 10^2–10^3 micelles are converted to the polymer particles.

The polymer particles formed at the beginning of Interval 1 are expected to be larger than those formed at the end of Interval 1. Therefore, smaller duration of Interval 1 is preferable to obtain a narrower particle size distribution.

At the end of Interval 1, as illustrated in Figure 27, there is a rapid drop in the free surfactant concentration (which throughout Interval 1 is equal to the CMC due to equilibrium with micelles) and the surface tension (head-space gas/latex) rises rapidly from its previously stationary value. Because the

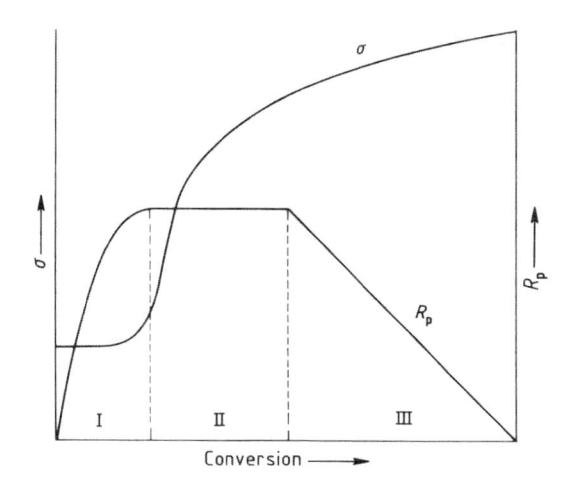

Figure 27. Overall rate of reaction R_p and surface tension σ as a function of conversion during the three intervals of emulsion polymerization (schematic)

fractional coverage of the surface of polymer particles falls after the end of Interval 1, problems with particle stability and coagulation may occur.

As for the polymer particle nucleation, one may heard of the homogeneous nucleation theory [138–140], which assumes that the polymer particles are formed when the oligomer radicals have grown to the critical chain length for precipitation. The precipitated polymer chain is stabilized by the emulsifier and monomer diffuses into this new organic phase, which allows a fast growth of the polymer radical. The homogeneous nucleation would be important when the initial surfactant concentration is below the CMC. In the surfactant-free emulsion polymerization, the ionic initiator fragment (e.g., $SO_4^{\bullet-}$ from persulfate) provides colloidal stability. Nucleation ceases when the number of polymer particles is high enough to capture all oligomer radicals. Regardless of the particle nucleation mechanism, the newly formed polymer particles are very small, and grow rapidly with increasing surface area that needs to be covered by the emulsifier molecules. If the diffusion of emulsifier is not fast enough, the embryo polymer particles may coagulate. The process combining nucleation with subsequent coagulation is referred to as the coagulative nucleation [141].

Interval 2 is known as the polymer particle growth stage, during which the number of particles remains constant (in the absence of coagulation), as does the monomer concentration $[M]_p$ in the latex particles as a result of monomer diffusion from the reservoir of monomer droplets. A constant $[M]_p$ during Interval 2 can be rationalized, at least qualitatively, as follows. Because of the extremely large interfacial areas (polymer particle/water and monomer droplet/water) and associated very rapid mass transfer of monomer, there is an equilibrium with respect to monomer transfer from monomer droplets to polymer particles (the chemical potential of monomer is the same in all three phases, monomer droplet/water/polymer particle). The interfacial energy per unit volume for the small polymer particles contributes significantly to the free energy and is accounted for as follows [142, 143]:

$$\frac{2V_m\sigma}{r_0 RT} = -\left\{\ln(1-\phi_p) + \phi_p + \chi\phi_p^2\right\} \quad (32)$$

where V_m is the molar volume of monomer, σ is the interfacial tension (latex particle/aqueous phase), r_0 is the radius of unswollen polymer particle, R is the gas constant, T is the absolute temperature, ϕ_p is the volume fraction of polymer in the latex particle, and χ is the Flory–Huggins polymer–solvent interaction parameter.

For a given set of values, V_m, r_0, T, and χ, one can determine ϕ_p. The saturated monomer concentration in the polymer particle can be calculated from:

$$[M]_p = \frac{1-\phi_p}{V_m} \quad (33)$$

Because σ and r_0 increase simultaneously during Interval 2, $[M]_p$ remains relatively constant [142–144], as long as the monomer droplets are present. Practically, it is more convenient to actually measure the saturated amount of monomer in the polymer particles experimentally. Anyway, the kinetic behavior during Interval 2 could be modeled as a nano-sized semibatch reactor for which monomer is fed from the monomer droplets.

Interval 3, known as the depletion or monomer finishing stage, begins with the disappearance of all monomer droplets. The only reservoir of monomer for the polymerization in the latex particles is the aqueous phase. This is hardly sufficient and in Interval 3 the monomer concentration $[M]_p$ falls with time and conversion. For a hydrophobic monomer, the monomer conversion at which Interval 2 ends, x_{c2} is equal to the mass fraction of polymer during Interval 2 that is approximately constant during Interval 2. Therefore, x_{c2} can be estimated by measuring the saturated monomer concentration experimentally.

For more detailed information and realistic problems on emulsion polymerization, one can refer to the review articles and books [145–150].

8.2. Polymerization Kinetics

Emulsion polymerization follows unique polymerization kinetics, and both polymerization rate and molecular mass of the formed polymers can be increased simultaneously, which is difficult to achieve in bulk/solution/suspension polymerization.

Unique characteristics of polymerization kinetics result from the smallness of the polymerization locus, which is the polymer particle, quite often smaller than $d = 100$ nm. In a submicron particle, even a single molecule occupies a sufficient concentration. Table 3 shows the calculated single molecule concentration in a particle with diameter d, i.e., $[\text{Single}]_p = 6/(\pi N_A d^3)$, where N_A is the Avogadro constant. In bulk/solution polymerization, the polymer radical concentration is usually about $[R^\bullet]_{bulk} = \sqrt{R_I/k_t} = \cdot 10^{-8} - 10^{-6}$ mol/L, where R_I is the initiation rate and k_t is the termination rate constant. At least one polymer radical is required to provoke polymerization in

Table 3. Concentration of a single molecule inside a particle with diameter d

Particle diameter d, nm	Concentration of a single molecule, $[\text{Single}]_p$, mol/L
1000 (=1 μm)	3.18×10^{-9}
200	3.97×10^{-7}
150	9.43×10^{-7}
100	3.18×10^{-6}
50	2.55×10^{-5}
25	2.04×10^{-4}

a polymer particle. For the particles with $d < 100$ nm, the radical concentration is significantly larger than in a bulk polymerization. On the other hand, the contribution of a single molecule is negligibly small for $d = 1$ µm, and normally, the bulk polymerization kinetics apply.

In emulsion polymerization, the average time interval between radical entry to a polymer particle $\bar{t}_{entry}$, is normally larger than 10 seconds. During this time interval, a single polymer radical can grow without termination (although it is likely to cause chain transfer reaction in this time interval). When the second radical comes in, the average time required for this radical to cause bimolecular termination is given by $\bar{t}_{term} = 1/(k_t[R^\bullet])$. Assuming $k_t = 1 \times 10^7$ L mol^{-1} s^{-1}, $\bar{t}_{term.} = 8.48 \times 10^{-4}$ s for $d = 25$ nm, $\bar{t}_{term.} = 3.93 \times 10^{-3}$ s for $d = 50$ nm, and $\bar{t}_{term.} = 3.14 \times 10^{-2}$ s for $d = 100$ nm. Therefore, the second incoming radical causes bimolecular termination almost instantaneously ($\bar{t}_{term.} \ll \bar{t}_{entry}$). Figure 28 shows such behavior in a polymer particle schematically. For a half of the time, it possesses one radical $n = 1$, and $n = 0$ for the rest of one half. Statistically, this situation applies to all particles, and the average number of radicals per particle is $\bar{n} = 0.5$. This is so-called zero–one behavior. It is possible to retain a very high average radical concentration, $[R^\bullet]_p = \bar{n} \times [\text{Single}]_p$ by isolating each radical into different polymer particles.

By hypothetically assuming that both a polymer particle and a monomer droplet are the isolated microreactors, the increase rate of local conversion, x, is simply given by that for homogeneous batch polymerization, as follows:

$$\frac{dx}{dt} = k_p(1-x)[R^\bullet] \tag{34}$$

Because $[R^\bullet]$ is much larger for the polymer particles, the polymer fraction wants to grow much faster in the polymer particles. Therefore, the polymer particles will absorb monomer from the water phase, and the monomer in water phase is supplied from the monomer droplets.

On the other hand, however, in order for the polymer particles to be the main loci of polymerization, the total monomer consumption rate in the polymer particles must be much larger than that in the monomer droplets. Consider the competition of monomer consumption between the polymer particles and the monomer droplets, which is represented by the following ratio, ς.

$$\varsigma = \frac{[R^\bullet]_p[M]_p V_p}{[R^\bullet]_d[M]_d V_d} \tag{35}$$

where the subscripts p and d represent the polymer particle and the monomer droplet, respectively. Note that V represents the total volume, not each particle/droplet.

Suppose that we have polymer particles with diameter 50 nm at conversion 20%. Assuming polymer exists only in the polymer particles and the volume fraction of monomer $\phi_m = 0.5$, then $[M]_p/[M]_d = 0.5$ and $V_p/V_d = 20 \times 2/(100-20 \times 2) = 0.67$, neglecting the density

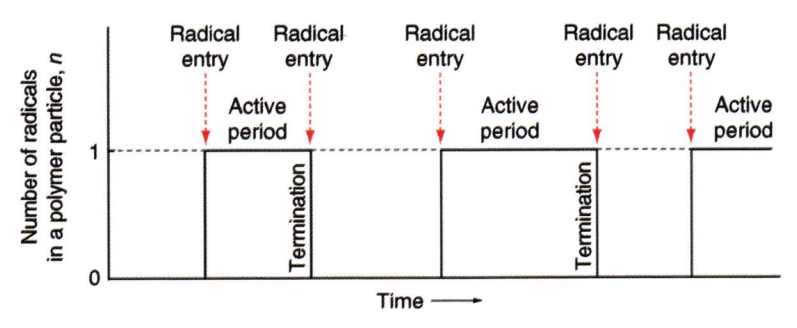

Figure 28. Schematic representation of the time sequence of the number of radicals in a small polymer particle in which bimolecular termination is instantaneous, i.e., so-called zero-one behavior

difference of monomer and polymer. The monomer droplets are large, so the maximum value of $[R^\bullet]_d$ attainable would be the radical concentration in the corresponding bulk polymerization, $[R^\bullet]_{bulk}$. For $\bar{n} = 0.5$ and $[R^\bullet]_{bulk} = 1 \times 10^{-7}$ mol/L, $[R^\bullet]_p/[R^\bullet]_d$ is calculated to be $(0.5)(2.55 \times 10^{-5})/(1 \times 10^{-7})=128$. Therefore, the ç value is $(128)(0.5)(0.67) = 43 \gg 1$. In addition to a larger total radical entry frequency over monomer droplets, as estimated in the previous section, the faster polymerization rate in smaller particles leads to make the polymer particles the dominant loci of polymerization. The emulsion polymerization rate can be represented by the polymerization rate in the polymer particles.

Another explanation as to why the polymer particles win the competition might be given on the basis of a large number of polymer particle combined with the zero-one behavior. At the beginning of Section 8.1, we have estimated that the number of polymer particles is ca. $N_p = 1 \times 10^{18}$ per liter of latex. With the ideal zero-one behavior, one half of polymer particles contain a radical. Therefore, the radical concentration in the latex, meaning within the volume that involves the water phase, is as large as $0.5 \times 10^{18}/N_A = 8.3 \times 10^{-7}$ mol/L, which is already higher than that for usual bulk polymerization.

The main loci of polymerization is the polymer particles, and the emulsion polymerization rate, R_p in mol s^{-1} (L-water)$^{-1}$ is given by:

$$R_p = k_p[M]_p \frac{\bar{n}}{N_A} N_T \quad (36)$$

where k_p is the propagation rate constant, N_T is the number of polymer particles per unit volume of water phase. Normally, the radicals are generated in water phase, and it is convenient to take the water-phase volume as a base unit. Obviously, R_p in mol s^{-1} (L-latex)$^{-1}$ can be obtained simply by changing N_T to N_p.

Equation (36) shows that the polymerization rate can be increased by making the number of particles, N_T larger. Given the radical generation rate in the water phase is kept constant, a larger number of polymer particles makes the average time interval between radical entry larger, leading to larger kinetic chain length. Therefore, both the polymerization rate and molecular mass of polymer can be made larger simultaneously.

By larger polymerization rates in emulsion polymerization it may also be understood that the polymer radicals are segregated into different polymer particles, resulting in the suppression of bimolecular termination reactions. There is nothing wrong with this statement in terms of the radical lifetime, however, notably, the total initiation rate and termination rate within the whole reactor is still balanced, i.e., the total termination rate in the whole reactor is not decreased in emulsion polymerization.

If a chain-transfer reaction occurs during the active period, a small radical formed by chain transfer reaction may exit from the particle. With the exit of small radicals, the active period becomes shorter, while the radical entry frequency may not increase significantly. In such a case, the average number of radical becomes smaller than 0.5 even when the zero–one behavior dominates. Because the chain transfer reaction tend to increase the radical desorption frequency, the addition of chain-transfer agent (CTA) may decrease the polymerization rate by reducing the $\bar{n}$ value [151]. This chain transfer effect in reducing the polymerization rate is a unique characteristic of emulsion polymerization.

On the other hand, as the polymer particle grows, it takes longer time to cause the bimolecular termination reaction. If the third radical comes into the polymer particle before bimolecular termination, the $\bar{n}$ value increases.

Now, let us consider how the polymer radicals are distributed in polymer particles. N_n is considered to be the number of particles containing n radicals per unit volume of water phase. The population balance equation for N_n is given by:

$$\frac{dN_n}{dt} = \left(\frac{\rho_A}{N_T}\right)N_{n-1} + k_f(n+1)N_{n+1} + \frac{k_t}{v_p N_A}_{n+2}C_2 N_{n+2}$$
$$- \left(\frac{\rho_A}{N_T}\right)N_n - k_f n N_n - \frac{k_t}{v_p N_A} {}_n C_2 N_n \quad (37)$$

where

$\left(\frac{\rho_A}{N_T}\right)N_{n-1}$: Generation of N_n through the entry of a radical into the

$k_f(n+1)N_{n+1}$: Generation of N_n through the exit of a radical from the particle with $n+1$ radicals. Here, k_f is the desorption rate constant of a radical.

$\frac{k_t}{v_p N_A} {}_{n+2}C_2 N_{n+2}$: Generation of N_n through the bimolecular termination in the polymer particle containing $n+2$ radicals. ${}_{n+2}C_2$ shows the number of possible combination of radicals to cause bimolecular termination reaction. The number of termination events per unit volume is given by $\frac{k_t}{2}\frac{(n+2)(n+1)}{(v_p N_A)^2}$, where v_p is the particle volume. Note that k_t in which the termination rate is represented by $R_t = k_t[R^\bullet]^2$ is used ($\rightarrow$ Polymerization Processes, 1. Fundamentals). The number of termination events per particle is given by

$$\frac{k_t}{2}\frac{(n+2)(n+1)}{(v_p N_A)^2} \times (v_p N_A)$$
$$= \frac{k_t}{2}\frac{(n+2)(n+1)}{v_p N_A}.$$

particle with $n-1$ radicals. Here, ρ_A is the total radical entry rate from the aqueous phase to the polymer particles.

The 4th to 6th terms in Equation (37) represent consumption of N_n. Note that the bimolecular termination rate constant, k_t might be significantly smaller than that for the bulk polymerization at conversion $x = 0$, because a high mass fraction of polymer exists in the polymer particle from the beginning.

By application of the stationary state hypothesis ($dN_n/dt = 0$), Equation (37) was solved to give the following solutions for $\bar{n}$ and the number fraction of particles containing n radicals, $\Phi_p(n)$ [152, 153].

$$\bar{n} = \frac{a}{4}\frac{I_m(a)}{I_{m-1}(a)} \tag{38}$$

$$\Phi_p(n) = \frac{a^n 2^{(m-1-3n)/2} I_{m-1+n}(2^{-0.5}a)}{n! I_{m-1}(a)} \tag{39}$$

where $I_m(x)$ is the modified Bessel function of the first kind, $a = (8\alpha)^{0.5}$, $\alpha = 2\rho_A v_p N_A/(k_t N_T)$, and $m = 2k_f v_p N_A/k_t$. The dimensionless variables α and m are the ratios of the radical entry rate ρ_A/N_T and the radical exit rate k_f, with respect to the termination rate of a single pair of radicals divided by 2 and $k_t/(2v_p N_A)$, respectively. (The problem of the division by 2 is caused by the difference in the definition of the termination rate constant, as discussed ($\rightarrow$ Polymerization Processes, 1. Fundamentals). Therefore, the α value represents the significance of radical entry, and m shows that of radical exit. The validity of the stationary state hypothesis was confirmed for $m = 0$ by comparing the numerical solution of Equation (37) [154].

Figure 29 shows the calculated results from Equations (38) and (39), for the case of $m = 0$. The comparison was made with the Monte Carlo simulation results that agree perfectly with the calculated results [155]. In Figure 29A, ξ represents the magnitude of bimolecular termination, $\xi = k_t/(k_p[M]_p v_p N_A)$, and faster termination results in $\bar{n} = 0.5$, as expected. When the k_t value is large enough to retain the zero–one behavior, the magnitude of k_t does not affect the $\bar{n}$ value. On the other hand, smaller k_t value (or to be more precise, smaller ξ value) leads to make the $\bar{n}$ value larger. Figure 29B shows that the particles with up to $n = 6$ exist when $\bar{n} \cong 2$.

With the radical desorption ($m > 0$), a radical once exits from the particle and may enter another particle. The ρ_A value in a steady state is given by:

$$\rho_A = k_A[R_w^\bullet]N_T = \rho_w + k_f \bar{n} N_p - k_{tw}[R_w^\bullet]^2 \tag{40}$$

or equivalently,

$$\alpha = \alpha_w + m\bar{n} - Y\alpha^2 \tag{41}$$

where ρ_w is the initiation rate in the water phase, $[R_w^\bullet]$ is the radical concentration in the water phase, k_{tw} is the bimolecular termination rate constant in the water phase, $\alpha_w = 2\rho_w v_p N_A/(k_t N_T)$ and $Y = \frac{k_{tw}k_t}{2k_A^2 v_p N_A N_T}$. The k_{tw} value is expected to be much larger than k_t in the particle, because of low viscosity and oligomeric chain lengths in the water phase. The radical concentration in water phase, $[R_w^\bullet]$ is

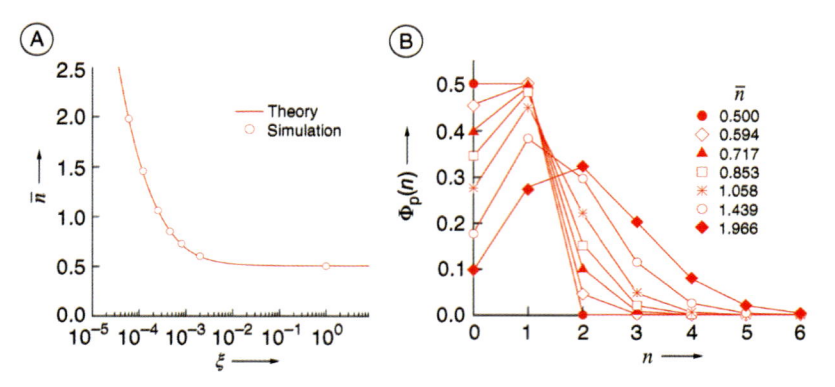

Figure 29. Radical distribution in polymer particles, calculated results [155]
A) Average number of radicals per particle $\bar{n}$; B) number-fraction distribution of the particles containing n radicals
Solid lines: calculation; Symbols: Monte Carlo simulation results

extremely small especially for hydrophobic monomers, and the final term in Equation (41) could possibly be neglected, i.e., $Y \cong 0$.

It is much more convenient if ρ_w is separated from ρ_A in the experimental design, because it is straightforward to control ρ_w, not ρ_A. To do this, Equations (38) and (41) must be solved simultaneously. Because $\bar{n}$ is an implicit function of α_w, a recursive calculation is required. It is not very difficult to do this in the present age of efficient computers. However, for the cases with $Y = 0$, the following approximate equation [147, 156] that is an explicit function of α_w and m is very convenient to use.

$$\bar{n} = \frac{1}{2}\left[\left\{\left(\alpha_w + \frac{\alpha_w}{m}\right)^2 + 2\left(\alpha_w + \frac{\alpha_w}{m}\right)\right\}^{1/2} - \left(\alpha_w + \frac{\alpha_w}{m}\right)\right] + \left(\alpha_w + \frac{\alpha_w}{m}\right)^{1/2} - \frac{1}{2} \quad (42)$$

Figure 30 shows the calculated results for the exact solution of Equations (38) and (41) and the approximation using Equation (42). The maximum deviation of Equation (42) from the exact solution is smaller than 5%. The fully converged lines in the large α_w region is where the pseudo-bulk kinetics are valid, and $\bar{n} \propto \alpha_w^{0.5}$. It seems reasonable that the unique characteristics of emulsion polymerization kinetics are lost for $\bar{n} > 2$. Larger m values mean that the exit of radicals is significant. For hydrophobic monomer such as styrene, the m value is rather small, and a wide region of $\bar{n} = 0.5$ exists. On the other hand, for monomers with higher solubilities in water, such as vinyl acetate and vinyl chloride, the m value is large and close to the linear line with slope 1/2 in Figure 30 prevails. The experimental data of the relationship between $\bar{n}$ and α_w for representative monomers,

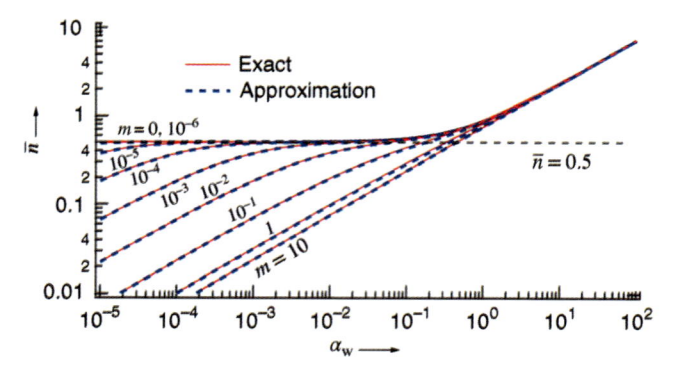

Figure 30. Average number of radicals per particle $\bar{n}$ as predicted from Equations 38 and 41 (Exact), and the approximation from Equation (42) for $Y = 0$

such as styrene, methyl methacrylate, and vinyl chloride, agree reasonably well with the theory presented herein [147].

With the progress of polymerization, the particle volume v_p increases. In addition, k_t may decrease significantly because of the gel effect, especially during Interval 3 where the polymer fraction in the polymer particle increases. Both effects make the α_w value larger, leading to a larger $\bar{n}$ value, as shown in Figure 30. Figure 31 shows an example of the calculated results for the styrene emulsion polymerization [157].

To calculate the polymerization rate by using Equation (36), the number of polymer particles N_T needs to be determined. It is practical to determine the N_T value experimentally. On the other hand, the micellar nucleation theory gives [136]:

$$N_p \propto (\rho_w/\mu)^{0.4}(a_s[S]_0)^{0.6} \propto [I]_0^{0.4}[S]_0^{0.6} \tag{43}$$

where μ is the volumetric growth rate per particle, a_s is the surface area occupied by a unit amount of surfactant, $[S]_0$ is the charged surfactant concentration, and $[I]_0$ is the initial initiator concentration. For a hydrophobic monomer, such as styrene, this agrees with Equation (43).

When the radical desorption is taken into account, Equation (43) can be modified as follows [158–160].

$$N_p \propto (\rho_w/\mu)^{1-z}(a_s[S]_0)^z \propto [I]_0^{1-z}[S]_0^z \tag{44}$$

where $0.6 < z < 1.0$. The z value increases with the water solubility of monomer, starting from $z = 0.6$ for styrene to $z = 1.0$ for vinyl acetate.

Miniemulsion Polymerization. The polymerization rate expression given by Equation (36) can be applied also for miniemulsion polymerization [161–163], in which all of polymerization can be considered as occurring within the preexisting monomer droplets without the formation of new particles. Initial monomer droplets are small enough (typically, <500 nm), and each monomer droplet could be regarded as an independent microreactor, at least, approximately.

The particle size below which the miniemulsion polymerization rate becomes noticeably larger than the corresponding bulk polymerization can be determined as follows. The effective radical concentration in a particle is given by $\bar{n}[\text{Single}]_p$. If this concentration becomes larger than the corresponding bulk polymerization, the miniemulsion polymerization would win the competition. Therefore, the threshold particle diameter below which the minimusion polymerization rate becomes larger than the corresponding bulk polymerization, d_c is given by:

$$[R^\bullet]_{\text{bulk}} = \bar{n}[\text{Single}]_p = \frac{6\bar{n}}{\pi N_A d_c^3} \tag{45}$$

or

$$d_c = \left(\frac{6\bar{n}}{\pi N_A [R^\bullet]_{\text{bulk}}}\right)^{1/3} \tag{46}$$

Figure 32 shows the Monte Carlo simulation results of the miniemulsion polymerization assuming a uniform particle size distribution without the radical exit. In the comparison, the radical generation rate per initial amount of monomer is set to be the same for all calculations, including the bulk polymerization. For the bulk polymerization, $[R^\bullet]_{\text{bulk}} = \sqrt{R_I/k_t} = \sqrt{1 \times 10^{-7}/1 \times 10^7} = 1 \times 10^{-7}$ mol/L in the present calculation, and therefore, the

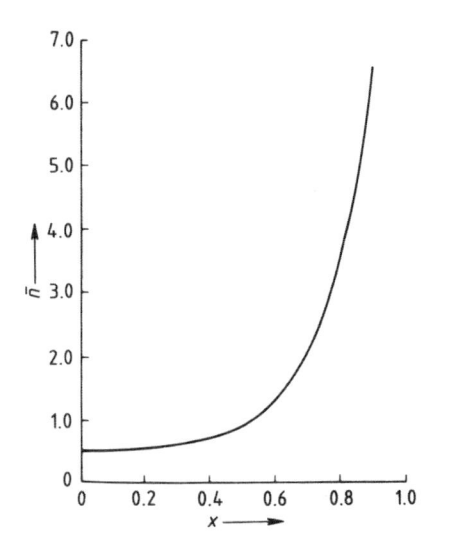

Figure 31. Theoretical course of the average number of radicals $\bar{n}$ per polymer particle with increasing conversion [157]

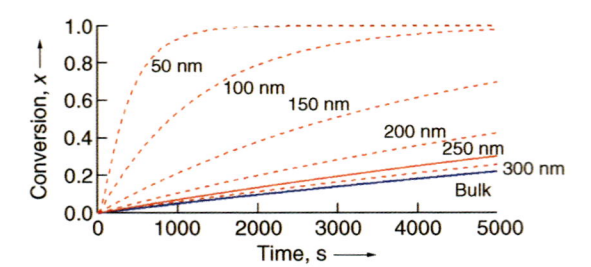

Figure 32. Calculated conversion development for miniemulsion polymerization in which the radical generation rate per unit oil phase is $R_I = 1 \times 10^{-7}$ mol L^{-1} s^{-1} [164] Rate constants: $k_p = 500$ L mol^{-1} s^{-1}, $k_t = 1 \times 10^7$ L mol^{-1} s^{-1}; initial monomer concentration [M]$_0$ = 8 mol/L

threshold diameter is calculated to be $d_c = 252$ nm. As shown in Figure 32, a significant polymerization rate increase is observed for $d < 250$ nm.

8.3. Molecular Mass Distribution

8.3.1. Linear Polymerization

For MMD theory in homogeneous free-radical polymerization, see → Polymerization Processes, 1. Fundamentals. The standard kinetics of conventional free-radical polymerization in homogeneous media leads to form the most probable distribution for the polymer radicals, the mass fraction distribution of which is given by:

$$M^{\bullet}(r) = r(\tau + \beta)^2 \exp[-(\tau + \beta)r] \quad (47)$$

$$\tau = \frac{R_{td} + R_f}{R_p} \quad (48)$$

$$\beta = \frac{R_{tc}}{R_p} \quad (49)$$

where r is the chain length, R_p is the propagation rate, R_{td} is the rate of termination by disproportionation, R_{tc} is the rate of termination by combination, and R_f is the rate of chain transfer reaction.

Chain-Transfer Dominated Systems. When the chain-transfer reaction dominates for the dead polymer chain formation, i.e., for the cases with $R_f \gg R_{tc} + R_{td}$, the instantaneous mass fraction distribution of the dead polymers is simply given by the following most probable distribution:

$$M_{inst.}(r) = r\tau_f^2 \exp(-\tau_f r) \quad (50)$$

where $\tau_f = R_f/R_p = (k_{fm}[M]_p + k_{fCTA}[CTA]_p/)k_p[M]_p$, and k_{fm} and k_{fCTA} are the rate the constants for chain transfer reaction to monomer and to the chain transfer agent, CTA, respectively.

Notably, Equation (50) is valid, irrespective of the magnitude of $\bar{n}$. The number- and mass average chain lengths are given respectively by (→ Polymerization Processes, 1. Fundamentals):

$$\bar{P}_{n,inst.} = 1/\tau_f \quad (51)$$

$$\bar{P}_{M,inst.} = 2/\tau_f \quad (52)$$

This situation is exactly the same as that for homogeneous free-radical polymerization. In general, the concentrations [M]$_p$ and [CTA]$_p$ changes with the progress of polymerization, although [M]$_p$ may not change significantly during Interval 2 in emulsion polymerization. The accumulated chain length distribution can be obtained by summing up all instantaneous dead polymer chain length distribution. For a batch polymerization, the accumulated mass fraction distribution is given by:

$$M_{accum.}(r) = \frac{1}{x}\int_0^x M_{inst.}(r)dx \quad (53)$$

MMD of the Zero–One Systems. For the zero–one system, because the second incoming radical will cause bimolecular termination instantaneously, $(R_{tc} + R_{td})/R_p$ is modified to $1/(k_p[M]_p \bar{t}_{entry})$, by referring to the radical entry rate to a polymer particle $\rho_A/N_T = 1/\bar{t}_{entry}$. Notably, $k_p[M]_p$ represents the number of monomeric units added to a radical center in 1 s (→ Polymerization Processes, 1. Fundamentals). Therefore, for the zero-one system in emulsion polymerization, Equation (47) can be used by changing the definition of $\tau+\beta$ to:

$$\tau + \beta = \tau_f + \frac{1}{k_p[M]_p \bar{t}_{entry}} \quad (54)$$

If the dead polymer chain is formed by the chain-transfer reaction, the chain length is the same as that of the polymer radical just before chain stoppage, which is given by Equation (47). When a dead chain is formed by combination termination, an oligomeric radical chain is added to a polymer radical. Because the contribution of the oligimeric chain part is small, the dead polymer chain length is essentially the same as that of the polymer radical just before chain stoppage. Finally, if a dead chain is formed by disproportionation, a dead polymer chain with the same chain length as the polymer radical and an oligomeric dead polymer chain are formed. In summary, if one neglects the oligomeric dead chain formed by disproportionation termination whose mass fraction is negligibly small, the instantaneous dead polymer chain length distribution is essentially the same as that given by Equation (47), with $\tau + \beta$ represented by Equation (54). The effect of oligomeric chains formed by disproportionation termination will be discussed again for the cases where most of the dead polymer chains are formed by termination reactions.

In emulsion polymerization, $\bar{t}_{entry}$ is usually larger than 10 s, and therefore, the dead chain formation is often dominated by chain transfer reaction, whose instantaneous MMD of dead polymer is shown by:

$$M_{inst.}(r) = r\tau_f^2 \exp[-\tau_f r] \quad (55)$$

Without using external CTA and by making $\bar{t}_{entry}$ large enough, it is not difficult to obtain an emulsion polymerization system in which the chain transfer to monomer dominates the dead polymer chain formation. Under such condition, the number fraction distribution of dead polymer chains is given by:

$$N(r) = C_m \exp(-C_m r) \quad (56)$$

where C_m the monomer transfer constant, $C_m = k_{fm}/k_p$.

Notably, when the monomer transfer reaction dominates the dead polymer chain formation, the chain length drift does not occur irrespective of the reactor types used, including the emulsion polymerization. Therefore, from the slope of $\ln[N(r)]$, one can experimentally determine the monomer transfer constant [165–169].

$$\frac{d\ln N(r)}{dr} = -C_m \quad (57)$$

When applying Equations (56) and (57) to emulsion polymerization, (1) the experimental condition must be set to satisfy the condition, $C_m \gg 1/(k_p[M]_p \bar{t}_{entry})$, and (2) the surfactant must not cause any reaction with radicals. In addition, (3) the total number of monomeric units polymerized in a particle, u must satisfy $uC_m \gg 1$ [170], because the polymer chain formed during Interval 1 and the final polymer chain that might be stopped abruptly may not follow the distribution given by Equation (56). This requirement implies that the polymerization must not be stopped at very low conversion region, which is a different requirement when using a bulk polymerization for the parameter estimation. Finally, careful attention must be paid to (4) the effect of chain transfer to polymer. The locus of polymerization in emulsion polymerization is the monomer-swollen polymer particle, and the polymer concentration in the polymerization locus is higher than the corresponding bulk polymerization from the beginning. Even for the reaction systems in which the polymer transfer reaction can be neglected in bulk polymerization, the effect of chain transfer to polymer may not be neglected in emulsion polymerization.

Figure 33 shows an example of the MC simulation results that involves the polymer transfer reactions [170]. The C_m value used was 5×10^{-5}, the polymer transfer constant, $C_{fp} = k_{fp}/k_p = 1 \times 10^{-5}$. The MMD shown is at the time when the total number of monomer units bound into polymer in a particle is $u = 1 \times 10^6$. More details on the simulation condition can be found therein [170].

Figure 33A shows the mass-fraction distribution when the independent variable is the logarithm of chain length, which is usually reported in the gel permeation chromatography

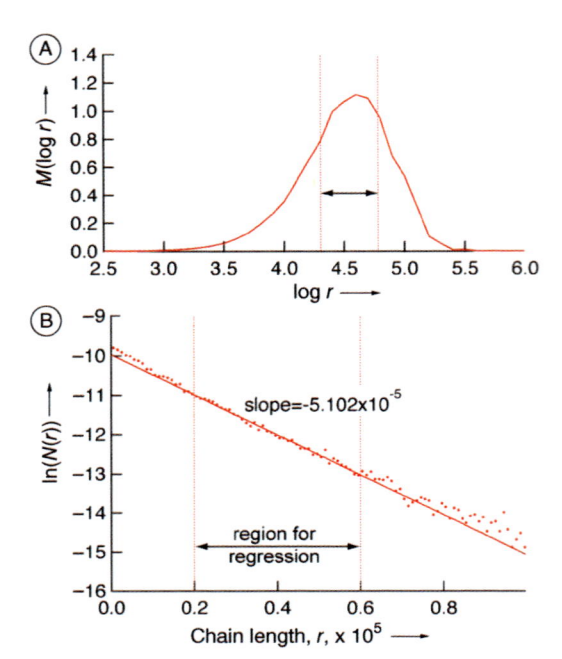

Figure 33. Monte Carlo simulation results for emulsion polymerization with polymer transfer reactions, with constant monomer/polymer ratio in the polymer particles [170] A) Mass-chain-length distribution as a function of the logarithm of the chain length r; B) Semi-log plot of the number fraction distribution.

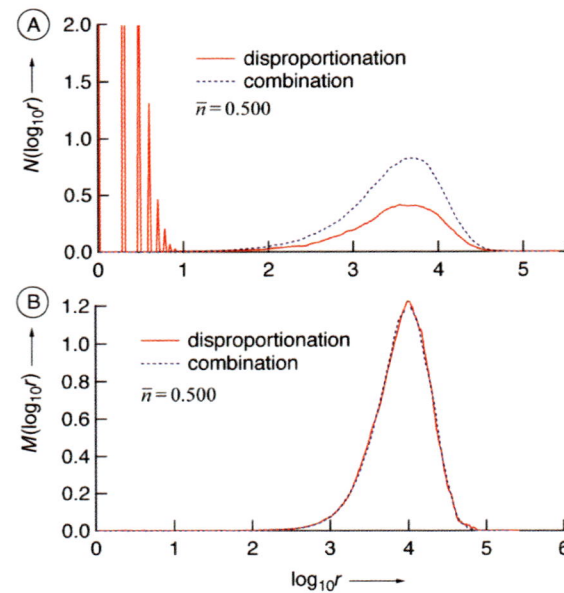

Figure 34. Calculated instantaneous chain length distribution formed in emulsion polymerization with $\bar{n} = 0.5$ and no radical exit [155]
A) Number fraction distribution; B) Mass fraction distribution

(GPC) measurements. Figure 33B shows the semi-log plot of the number fraction distribution $N(r)$, using the same MC simulation data. The distribution deviates from a straight line because of the polymer transfer reaction. However, by using the chain-length data around the peak region in $M(\log r)$, not the high MM tail, one can estimate the C_m value (5×10^{-5}) rather accurately, in spite of the existence of polymer-transfer reactions [170].

Because $\bar{t}_{entry}$ is large in usual emulsion polymerization, the chain-transfer reactions to monomer/CTA tend to be dominant to form dead polymer chains. Even when the polymerization rate is increased due to the gel effect (i.e., decrease in the bimolecular termination rate constant), the MMD still follows Equation (55), as long as most of dead polymer chains are formed by chain transfer reactions. Therefore, when the CTA is used effectively, the MMD is controlled by chain transfer to CTA whose instantaneous MMD is given by Equation (55) with $\tau_f = C_{fCTA}[CTA]_p/[M]_p$.

Theoretically, however, it is important to know the MMD formed by bimolecular termination reactions. Figure 34 shows the MC simulation results of the chain length distribution, assuming all polymer chains are formed by bimolecular termination with $\bar{n} = 0.5$. Figure 34A gives the number fraction distribution $N(\log_{10} r)$. When the bimolecular termination mode is by disproportionation, very large discrete peaks (spikes) appear for $r < 10$ (or $\log_{10} r < 1$), which correspond to the chains with length $r = 1, 2, 3$, and so on. With disproportionation termination, one half of dead polymer chains in number are oligomers. These oligomers are formed by the newly entered radicals that add a few monomeric units before almost instantaneous termination. Note that a radical can add $k_p[M]_p$ monomeric units in a second, and for example, assuming $k_p[M]_p = 1 \times 10^3$, 5 monomeric units are added in 5×10^{-3} s. As a limit of instantaneous termination, the PDI for the instantaneous MMD formed solely by disproportionation termination is $\bar{P}_{M,inst.}/\bar{P}_{n,inst.} = 4$. On the other hand, if the termination mode is by combination, the contribution of such oligomeric chain part is negligible in the dead polymer distribution, and the instantaneous MMD basically follows the most probable distribution given by Equation (47) with $\tau = 0$, whose PDI is $\overline{P}_{M,\text{inst.}}/\overline{P}_{n,\text{inst.}} = 2$.

Figure 34B shows the same distribution plotted on a mass basis, $M(\log_{10} r)$. In a GPC measurement, this type of distribution is obtained. The large spikes in the oligomeric range observed for the number fraction distribution disappear, because these spikes are large in number but negligibly small in weight. On the basis of Figure 34B, it is expected that the oligomeric spikes formed by disproportionation termination cannot be detected in the GPC measurement. In terms of the mass fraction distributions $M(\log_{10} r)$, the difference in the termination mode cannot be found, and both types of distribution agree completely. Note however, the agreement simply results from the inability of detecting oligomer chains formed in disproportionation termination.

With disproportionation termination, the oligomeric spikes move toward larger chain lengths as the $\bar{n}$-value increases, and the PDI decreases, as shown in Figure 35. The PDIs approach those for homogeneous polymerization, i.e. $\overline{P}_{M,\text{inst.}}/\overline{P}_{n,\text{inst.}} = 2$ for disproportionation and $\overline{P}_{M,\text{inst.}}/\overline{P}_{n,\text{inst.}} = 1.5$ for combination termination.

The MMD formed at $\bar{n} = 2$ is already very close to that formed in bulk polymerization [155]. As was shown in Figure 30, the M polymerization rate also reduces to that for bulk polymerization for $\bar{n} > 2$. The unique characteristics of emulsion polymerization kinetics are lost for $\bar{n} > 2$, and the pseudo-bulk behavior prevails, for which the simple bulk polymerization kinetics apply.

The deterministic calculation method for the MMD formed in linear emulsion polymerization, on the basis of the population balance equations, can be found in [171–177]. The complexity and uniqueness of emulsion polymerization kinetics result from the fact that the locus of polymerization is very small, often smaller than a few hundred nanometers. On the other hand, this characteristic is a great advantage for an MC simulation method [178]. Suppose the diameter of a particle consisting of polymer chains is $d = 100$ nm, the density of the polymer is 1 g/cm^3, and the molecular mass of monomeric unit in polymer is 100. In this case, the total number of monomeric units in the particle is 3×10^6. Assuming the number-average chain length of polymer molecules is 1×10^4, which is not unusually large in emulsion polymerization, the total number of polymer chains in a particle is only 300. For such a small number of polymer molecules, one can readily simulate the formation processes of all polymer molecules using the MC method with a well-designed simulation algorithm. In fact, some of the figures already shown in this and previous sections are based on the MC simulation results. In addition, it is straightforward to account for any kinetic event in the MC method. Some of the important results shown in the next sections are based on such a direct simulation method.

8.3.2. Nonlinear Polymerization

Nonlinear polymerization, such as with chain transfer to polymer and/or cross-linking reactions, involves reaction with dead polymer molecules. In emulsion polymerization, the polymer concentration is high from the beginning, and therefore, the branching and/or cross-linking density is higher in emulsion polymerization, compared with the corresponding bulk polymerization.

Figure 36 shows an example of the average cross-linking density development. The average cross-linking density is high from the early

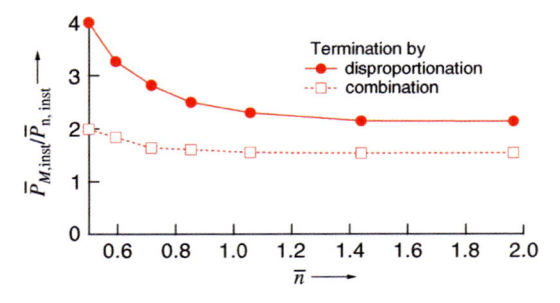

Figure 35. Calculated polydispersity index ($\overline{P}_{M,\text{inst.}}/\overline{P}_{n,\text{inst.}}$) as a function of $\bar{n}$, assuming all polymer chains are formed by bimolecular termination reactions in emulsion polymerization [155]

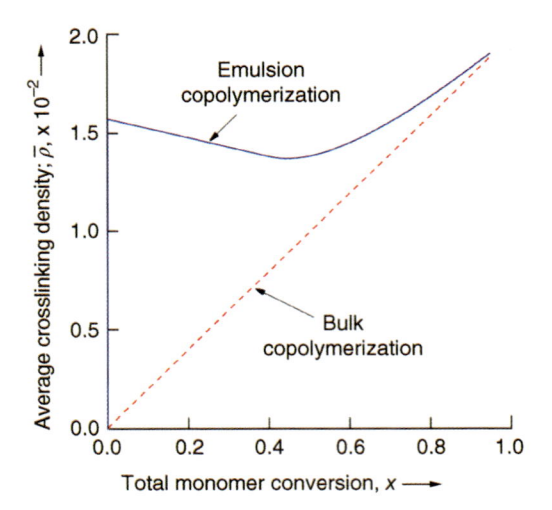

Figure 36. Calculated average cross-linking density development during bulk and emulsion copolymerization of vinyl/divinyl monomer when the reactivities of all double bonds are equal [178]
Initial mole fraction of divinyl monomer: $f_2^0 = 0.01$; monomer conversion at which monomer droplets disappear: $x_{c2} = 0.4$

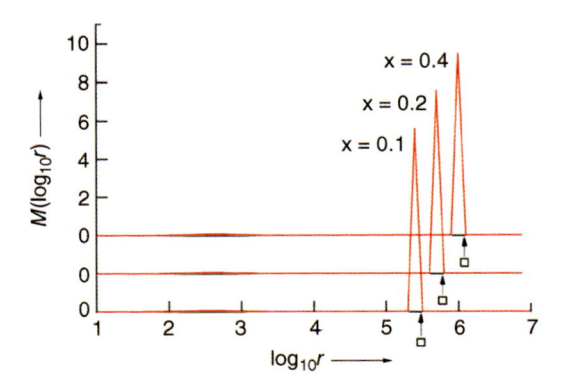

Figure 37. Example of the calculated mass fraction distribution $M(\log r)$ formed in emulsion cross-linking copolymerization, with microgels formed from the beginning [180]

stage of emulsion polymerization. A higher cross-linking density may result in unique behavior in emulsion cross-linking polymerization. A microgel may exist from very early stages of emulsion polymerization, and one polymer particle may essentially consist of a single cross-linked polymer molecule. In such a case, a sharp high molecular mass peak that proceeds to larger molecular mass with the progress of polymerization might be found, as the MC simulation shown in Figure 37 predicts. The arrows show the total number of monomeric units bound in polymer, which give the size of a dried polymer particle. For highly cross-linked cases, the MMD is essentially equivalent to the particle size distribution [180, 181]. For more mildly cross-linked systems, bimodal MMD may be formed due to the limitation of the maximum polymer size attainable in a polymer particle [180, 182].

Larger cross-linking density during Interval 1 leads to lower swelling ratio, resulting in a lower monomer concentration. Equation (44) predicts that a lower volumetric growth rate per particle μ results in a larger number of polymer particles N_T, which was reported in the experimental study [183].

For the cases with the polymer-transfer reaction, the branching reaction is enhanced by a larger polymer concentration in emulsion polymerization, similarly as in the case of a CSTR, discussed in Section (4.4). A notable difference of emulsion polymerization is that the size of a reaction locus is very small and there exists the molecular mass limit containable in a particle [178, 184, 185].

To highlight the most fundamental aspects of the limited space effect, consider the case where dead polymer chain formation is dominated by chain transfer to monomer and to polymer. In this case, the primary polymer chain length distribution follows the most probable distribution whose number fraction distribution is given by:

$$N_p(r) = (C_m + C_P)\exp[-(C_m + C_P)r] \qquad (58)$$

where $C_P = R_{fp}/R_p$.

Assuming a constant polymer/monomer ratio until the end of Interval 2, C_P is a constant. Therefore, the primary polymer chain length distribution is unchanged until the end of Interval 2. The probability that the chain end of a primary chain is connected, P_b is given by:

$$P_b = \frac{C_P}{C_m + C_P} \qquad (59)$$

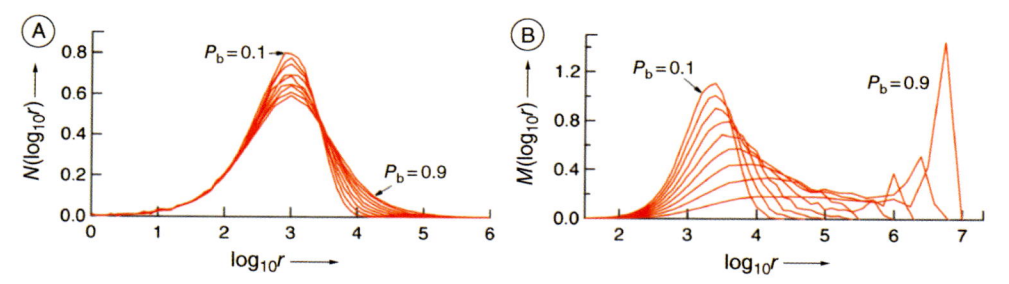

Figure 38. Calculated chain-length distribution for the branched polymer molecules formed in emulsion polymerization with $C_m + C_P = 1 \times 10^{-3}$, which means the number average chain length of the primary chain is $\bar{P}_{np} = 1000$ [185] A) Number fraction distribution; B) Mass fraction distribution

which is also a constant until the end of Interval 2.

Figure 38 shows the MC simulation results assuming constant values of C_p and P_b. In the figure, the value of P_b is changed from 0.1 to 0.9, with 0.1 increments. These distributions are at the time when the total number of monomeric units bound into polymer in a particle is $n = 1 \times 10^7$. Assuming the molecular mass of the monomeric unit 100 and the density of polymer 1 g/cm^3, $n = 1 \times 10^7$ corresponds to the dried polymer particle with diameter, $d = 147$ nm. The number fraction distribution $N(\log_{10} r)$ shows a smooth, single peak curve, irrespective of the P_b value. On the other hand, the mass fraction distribution $M(\log_{10} r)$, which is normally reported in the GPC analysis, becomes bimodal because of the limitation of the particle size, which corresponds to the maximum chain length, $\log_{10} n = 7$. A bimodal $M(\log_{10} r)$ is formed for $P_b > 0.5$.

Similarly with the case for a CSTR discussed in Section 4.4, the power low distribution, $M(r) \sim r^{-\alpha}$ is formed for a constant P_b, with a relationship, $\alpha = 1/P_b$ [185]. Even for the cases with the bimodal distribution in $M(\log_{10} r)$, the power law holds for the chain length region in between two peaks [185]. This feature could be used to determine the polymer transfer constant C_{fp}, on the basis of a well designed emulsion polymerization experiment.

With the MC simulation, one can observe the structure of each polymer molecule directly on a computer screen. An example of 2D structure is shown in Figure 39.

8.4. Effect of Small Reaction Loci in Reversible-Deactivation Radical Polymerization

As discussed in → Polymerization Processes, 1. Fundamentals and also in [164, 186], the reversible-deactivation radical polymerization (RDRP) in the IUPAC recommended name [96], often referred to as controlled/living radical polymerization, the pseudo-livingness is endowed by adding one of the reversible reactions depicted in Figure 40. In SRMP and ATRP, the equilibrium lies to the dormant (P$_i$X) side and [RGS] ≫ [Trap]. In RAFT, not necessarily but often, [RGS] ≪ [Trap].

RDRP is a free-radical polymerization, and therefore, the polymerization rate is

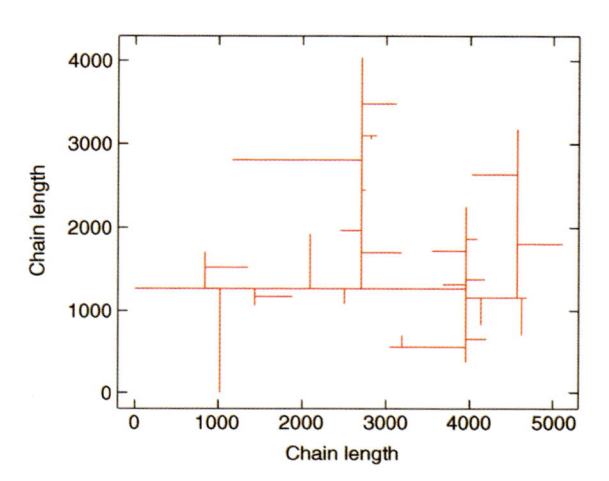

Figure 39. Example of a branched polymer structure formed in a model emulsion polymerization, with $P_b = 0.7$ and $C_m + C_P = 1 \times 10^{-3}$

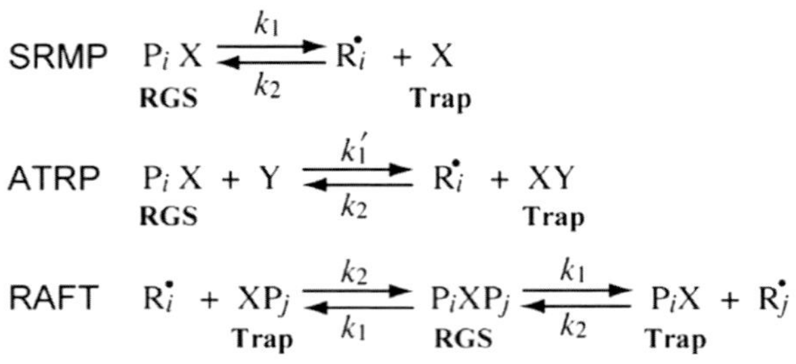

```
SRMP   P_iX  ⇌(k1/k2)  R_i• + X
        RGS            Trap

ATRP   P_iX + Y  ⇌(k1'/k2)  R_i• + XY
        RGS                  Trap

RAFT   R_i• + XP_j  ⇌(k2/k1)  P_iXP_j  ⇌(k1/k2)  P_iX + R_j•
         Trap                  RGS              Trap
```

Figure 40. Reversible-deactivation scheme in each type of RDRP
P_iX, XP_i = dormant polymer with chain length i; $R_i^\bullet$ = active polymer radical with chain length i; SRMP = stable-radical-mediated polymerization; ATRP = atom-transfer radical polymerization; RAFT = reversible-addition-fragmentation chain-transfer polymerization; RGS = radical generating species; Trap = trapping agent for the active radical

represented by $R_p = k_p[M][R^\bullet]$. In this section, however, the effect of small reaction locus (d smaller than a few hundred nanometers) is rationalized by using the following characteristic polymerization rate expression unique to RDRP (→ Polymerization Processes, 1. Fundamentals).

$$R_p = k_p[M]K\frac{[RGS]}{[Trap]} \quad (60)$$

where $K = k_1/k_2$, with $k_1 = k_1'[Y]$ for ATRP.

Here, we consider an ideal miniemulsion polymerization system with a uniform droplet diameter d_p, and each droplet can be considered as an independent batch reactor. Let us overview an important characteristics for each type of RDRP, based on Equation (60).

8.4.1. Stable Radical-Mediated Polymerization (SRMP) and Atom-Transfer Radical Polymerization (ATRP)

In SRMP and ATRP, $[RGS] \gg [Trap]$, and therefore, by making the droplet size smaller, the number of trapping agent (Trap) molecules becomes very small, while there still exists a large number of RGS molecules. When the number of Trap molecules in a particle becomes as small as unity, this single molecule concentration could be larger than that in the corresponding bulk polymerization, as was shown in Table 3. Because [Trap] is in the denominator in the polymerization rate expression given by Equation (60), the polymerization rate will become smaller than that for a bulk polymerization. The threshold droplet diameter below which the miniemulsion polymerization rate becomes smaller than the corresponding bulk polymerization, due to a high single-molecule concentration effect, can be estimated as follows.

The concentration of a single molecule in a particle with diameter d_p is given by $1/\{N_A(\pi/6)d_p^3\}$. When this concentration is equal to the trapping agent concentration in the corresponding bulk polymerization, the polymerization rate would be equal to that in bulk polymerization, given all the other conditions are the same. The particle size below which the polymerization rate becomes smaller than in the corresponding bulk polymerization, $d_{p,Trap}^{(1)}$ is given by:

$$d_{p,Trap}^{(1)} = \left(\frac{6}{\pi N_A[Trap]_{bulk}}\right)^{1/3} \quad (61)$$

Note that the trapping agent concentration in the corresponding bulk polymerization $[Trap]_{bulk}$ can be calculated in a straightforward manner by solving the material balance equations. For example, for SRMP,

$$\frac{d[R^\bullet]}{dt} = R_I - k_t[R^\bullet]^2 + k_1[PX] - k_2[R^\bullet][X] \quad (62)$$

$$\frac{d[X]}{dt} = k_1[PX] - k_2[R^\bullet][X] \quad (63)$$

are solved with the relationship, $[PX]_0 + [X]_0 = [PX] + [X]$, to determine $[Trap]_{bulk} = [X]$.

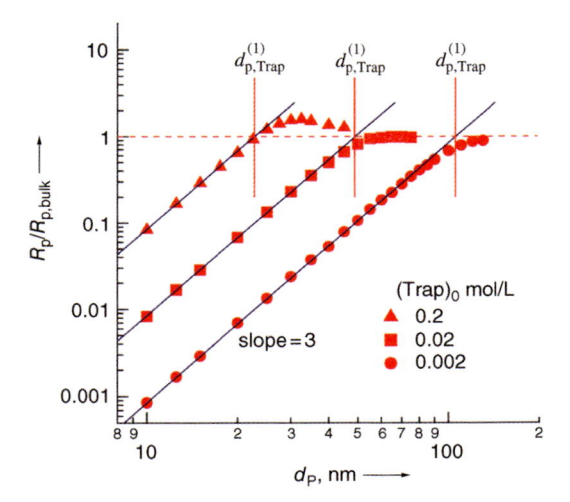

Figure 41. Calculated polymerization rates for the TEMPO-mediated styrene polymerization at 10% conversion with $[PX]_0 = 0.2$ mol/L [187].

The symbols in Figure 41 show the calculated polymerization rate for the 2,2,6,6-tetramethylpiperidyl 1-oxyl (TEMPO)-mediated styrene polymerization at 10 % conversion with $[PX]_0 = 0.2$ mol/L, reported in [187]. The $d_{p,\text{Trap}}^{(1)}$ values shown in the figure are calculated by using Equation (61), which predict the onset of significant polymerization rate decrease reasonably well [188].

Equation (60) shows that R_p is proportional to $1/[\text{Trap}]$. For the particle size region where the high single-molecule concentration effect is significant, [Trap] is given by $[\text{Trap}]_p = 1/(N_A v_p)$ and the particle volume $v_p \propto d_p^3$. Therefore, the relationship $R_p \propto 1/[\text{Trap}]_p \propto d_p^3$ holds for $d_p < d_{p,\text{Trap}}^{(1)}$. This is the reason for the lines with slope = 3 for $d_p < d_{p,\text{Trap}}^{(1)}$ in Figure 41. Lower polymerization rates in smaller polymer particles make it difficult to conduct the ab initio emulsion polymerization in SRMP and ATRP.

The decrease in R_p for smaller particles was reported experimentally [189]. Note however, desorption of trapping agents is neglected, which may cause deviation from the fundamental theory discussed herein.

The particle size region for which a slightly larger R_p, compared with $R_{p,\text{bulk}}$, is predicted to be formed for $d_p > d_{p,\text{Trap}}^{(1)}$. This is due to the statistical variation in the number of trapping agents in a particle [188], although the decreased frequency of termination for a radical due to the segregation of radicals may contribute the rate increase for bad living conditions in which the bimolecular termination is not suppressed well in bulk polymerization [190, 191]. The statistical variation effect is rationalized as follows. Because the number of monomer and RGS in a particle are large enough, the statistical variations of these components are negligible. On the other hand, the statistical variation effects may not be neglected when the average number of the trapping agents becomes smaller than about $\bar{n}_{\text{Trap}} = 10$. As an illustrative example, suppose that we have three particles containing 1, 2 and 3 trapping agents, respectively. In this case, the average number of trapping agents is $\bar{n}_{\text{Trap}} = 2$, and $[\text{Trap}]_{\text{bulk}} = \bar{n}_{\text{Trap}}/(N_A v_p)$ in the corresponding bulk polymerization, i.e., $R_{p,\text{bulk}} \propto 1/[\text{Trap}]_{\text{bulk}} \propto 1/\bar{n}_{\text{Trap}}$. On the other hand, in the miniemulsion polymerization, a strict representation of polymerization rate expression should be $R_p \propto \langle 1/[\text{Trap}]_i \rangle \propto \langle 1/n_{\text{Trap},i} \rangle$, where the subscript i indicate the ith particle and $\langle \rangle$ is the appropriate average for all particles. Assuming uniform particles, the above hypothetical example leads to $\langle 1/n_{\text{Trap},i} \rangle = (1 + 1/2 + 1/3)/3 = 0.611 > 0.5 = 1/\bar{n}_{\text{Trap}}$. Therefore, the miniemulsion polymerization rate is expected to be 1.22 times faster than the corresponding bulk polymerization. Because the average of the inverse numbers is always larger than the inverse of the average, the miniemulsion polymerization shows a larger polymerization rate than the corresponding bulk polymerization, when the statistical variation of n_{Trap} cannot be neglected. The effect of statistical variation becomes significant when $\bar{n}_{\text{Trap}} < 10$ [188], and the threshold particle diameter, below which R_p becomes noticeably larger than the corresponding bulk polymerization, $d_{p,\text{Trap}}^{(\text{sv})}$ could be estimated from:

$$d_{p,\text{Trap}}^{(\text{sv})} = \left(\frac{60}{\pi N_A [\text{Trap}]_{\text{bulk}}}\right)^{1/3} \quad (64)$$

Therefore, an acceleration window, $d_{p,\text{Trap}}^{(1)} < d_p < d_{p,\text{Trap}}^{(\text{sv})}$ in which $R_p/R_{p,\text{bulk}} > 1$ is predicted for SRMP and ATRP. On the other hand, the effect of the particle size distribution as well as the exit of trapping agent, which is not accounted for in the present fundamental theory, may need to be considered in the actual miniemulsion polymerization.

8.4.2. Reversible Addition–Fragmentation Chain Transfer (RAFT) Polymerization

In RAFT polymerization, the concentration of RGS (PXP) is smaller than that of Trap (XP or RAFT agent), and therefore, RGS is the candidate that may cause the high single-molecule concentration effect. Because [RGS] is the numerator term in Equation (60), a significant polymerization rate increase is expected for $d_p < d_{p,RGS}^{(1)}$. The threshold particle diameter, $d_{p,RGS}^{(1)}$ can be formulated as follows.

Practically, a significant polymerization rate increase due to the high single-molecule concentration effect occurs only for the systems with zero-one behavior. The zero–one behavior shown in Figure 28 is now modified for RAFT miniemulsion polymerization, as shown in Figure 42. During the active period, the number of propagating radical $n_{R^\bullet}$, or equivalently, the number of PXP molecules, n_{PXP} changes between zero and one repeatedly in a complementary manner.

The frequency of radical deactivation is $k_2[R^\bullet][XP]/[R^\bullet] = k_2[XP]\ s^{-1}$. Therefore, the average time for being a radical is $\bar{t}_{R^\bullet} = 1/(k_2[XP])$. Similarly, the average time for being an intermediate is $\bar{t}_{PXP} = 1/k_1$. The time fraction of the active radical period during $n = 1$ is given by:

$$\phi_A = \frac{\bar{t}_{R^\bullet}}{\bar{t}_{R^\bullet} + \bar{t}_{PXP}} = \frac{K}{K + [XP]} \tag{65}$$

where $K = k_1/k_2$.

Therefore, the effective PXP concentration in a zero-one particle is $\bar{n}(1-\phi_A)/(N_A v_p) = 6\bar{n}(1-\phi_A)/\left(\pi N_A d_p^3\right)$. By equating this concentration with [PXP]$_{bulk}$, the particle size below which the polymerization rate starts to become much larger than in the corresponding bulk polymerization, $d_{p,PXP}^{(1)}$ is given by:

$$d_{p,PXP}^{(1)} = \left(\frac{6\bar{n}(1-\phi_A)}{\pi N_A [PXP]_{bulk}}\right)^{1/3} \tag{66}$$

It is known that the RAFT polymerization rate shows retardation behavior by increasing the concentration of the RAFT agent. To rationalize the retardation, two conflicting models were proposed. One model [192] assumes that the intermediate radical, PXP terminates with the propagation radical $R^\bullet$, which is sometimes referred to as the intermediate termination (IT) model. On the other hand, a slower fragmentation of PXP can also cause retardation [193], which is called the slow fragmentation (SF) model. Both models fit with bulk polymerization data reasonably well, but the estimated k_1 value for the same reaction system could be more than 10^5 times larger for the IT model than the SF model [194]. The large difference in k_1 leads to a significant difference in the PXP concentration, i.e., [PXP]$_{bulk,IT}$ ≪ [PXP]$_{bulk,SF}$.

Equation (66) shows that larger [PXP]$_{bulk}$ results in a smaller threshold diameter. Figure 43 shows an example of the calculated threshold diameter change during RAFT polymerization of styrene [186]. In the IT model, a significant polymerization rate increase is predicted for $d_p < 212$ nm. On the other hand, the $d_{p,PXP}^{(1)}$ value is too small to actually conduct the miniemulsion polymerization.

Figure 44 shows the MC simulation results for the conversion development during

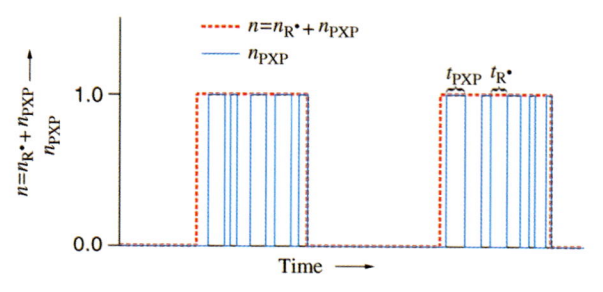

Figure 42. Schematic representation of the zero–one behavior in RAFT miniemulsion polymerization

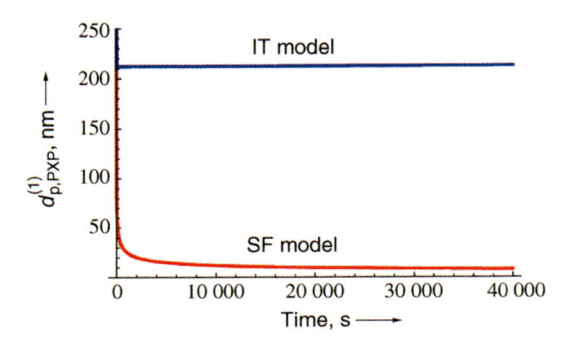

Figure 43. Calculated threshold diameter, using representative parameters for the styrene RAFT polymerization [186]

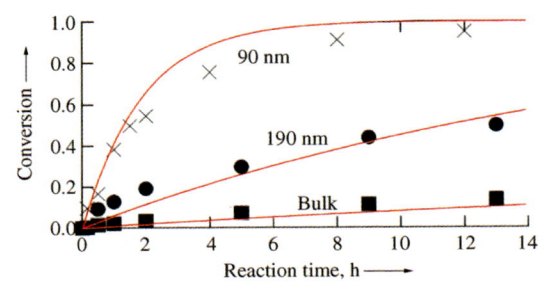

Figure 45. Conversion development during polystyryl dithiobenzoate mediated styrene miniemulsion polymerization at different particle sizes at 60 °C [195]
Solid lines: MC simulation results using the IT model; symbols: experimental data

miniemulsion polymerization, by using the same set of parameters as in Figure 43. In the IT model (Figure 44A), the miniemulsion polymerization shows significant polymerization rate increase for $d_p < 212$ nm, as predicted by Equation (66) and shown in Figure 43. In the case of SF model (Figure 44B), the $d_{p,PXP}^{(1)}$ value is too small, and the polymerization rate is not changed by the particle size.

This characteristic can be used for the model discrimination. The symbols in Figure 45 show the experimental results of the polystyryl dithiobenzoate mediated styrene miniemulsion polymerization at 60 °C, conducted by changing the particle size. Because a significant polymerization rate increase is observed in miniemulsion polymerization, it can be considered that the IT model applies to this reaction system [195]. The curves are the MC simulation results, using the IT model parameters. This conclusion agrees with the electron paramagnetic resonance (EPR) measurement results [196].

Various aspects of RDRP in emulsified systems are discussed in [120, 197–199].

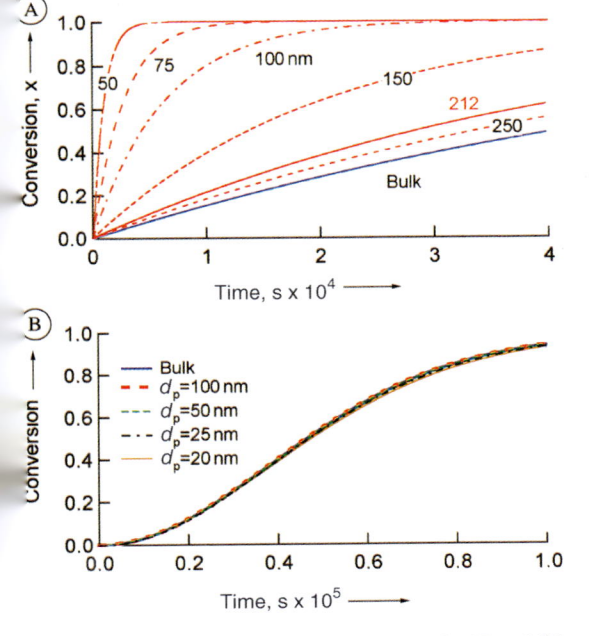

Figure 44. Monte Carlo simulation results for IT and SF model, with the same parameters as in Figure 43, showing the particle-size (d_p) effect on the miniemulsion polymerization rate A) IT model; B) SF model

List of Symbols

a	parameter for the calculation of the $\bar{n}$ value, $a = (8\alpha)^{0.5}$
a_s	surface area occupied by a unit amount of surfactant
ABS	acrylonitrile–butadiene–styrene rubber modified copolymer
ACA	aminocaproic acid
ATRP	atom-transfer radical polymerization
BHET	bis(hydroxyethyl)terephthalate
BR	batch reactor
CCD	chemical composition distribution
C_{fp}	polymer transfer constant ($=k_{fp}/k_p$)
C_m	monomer transfer constant ($=k_m/k_p$)
C_M	monomer concentration

CMC	critical micelle concentration	F_1	mole fraction of monomer 1 bound in polymer produced instantaneously in copolymerization
C_P	rate ratio between chain transfer to polymer and propagation, $C_P = R_{fp}/R_p$		
		$\overline{F}_1$	mole fraction of monomer 1 bound in accumulated copolymer
CSTR	continuous stirred tank reactor		
CTA	chain-transfer agent	F_M	molar flow rate of monomer
d	diameter	GPC	gel permeation chromatography
d_c	threshold particle diameter below which the miniemulsion polymerization rate starts to become larger than the corresponding bulk polymerization	HCSTR	homogeneous continuous stirred tank reactor
		HDPE	high-density polyethylene
		I	initiator
		$[I]_0$	Initial initiator concentration
d_p	diameter of polymer particle	$I_m(x)$	modified Bessel function of the first kind
$d_{p,Trap}^{(1)}$	threshold particle diameter below which the miniemulsion polymerization rate of RDRP starts to become *smaller* than the corresponding bulk polymerization due to a high *single-molecule concentration effect* of Trap		
		IT	intermediate termination in RAFT polymerization
		k_1	activation rate constant in RDRP
		k_2	deactivation rate constant in RDRP
		k_A	entry rate coefficient for the radicals from the water phase to the polymer particles
$d_{p,Trap}^{(sv)}$	threshold particle diameter below which the miniemulsion polymerization rate of RDRP starts to become *larger* than the corresponding bulk polymerization due to the *statistical variation effect* of Trap		
		k_f	desorption rate constant of a radical
		k_{fCTA}	rate constant for chain transfer to CTA
		k_{fm}	rate constant for chain transfer to monomer
$d_{p,PXP}^{(1)}$	threshold particle diameter below which the miniemulsion polymerization rate of RAFT starts to become *larger* than the corresponding bulk polymerization due to a high *single-molecule concentration effect* of PXP	k_{fp}	rate constant for chain transfer to polymer
		k_{fX}	rate constant for chain transfer to small molecule, X
		k_p	propagation rate constant
		k_t	rate constant for bimolecular termination ($=k_{tc}+k_{td}$)
$d_{p,RGS}^{(1)}$	threshold particle diameter below which the miniemulsion polymerization rate of RDRP starts to become *larger* than the corresponding bulk polymerization due to a high *single-molecule concentration effect* of RGS	k_{tc}	rate constant for bimolecular termination by combination
		k_{td}	rate constant for bimolecular termination by disproportionation
		k_{tw}	rate constant for bimolecular termination in the water phase
		L	reactor length
DMT	dimethyl terephthalate	L	segment length
D_w	diffusion coefficient of oligomer radicals in water phase	LDPE	low-density polyethylene
		LLDPE	linear low-density polyethylene
EG	ethylene glycol	m	dimensionless rate ratio, between radical exit and termination, $m = 2k_f v_p N_A/k_t$
$E(\theta)$	residence time distribution function		
f_1	mole fraction of monomer 1 in copolymerization		
		M	monomer
f_1^0	initial mole fraction of monomer 1 in copolymerization	[M]	monomer concentration
		$[M]_0$	initial monomer concentration

MC	Monte Carlo	P_b	probability that the head of a primary chain is connected to a backbone chain
$[M]_d$	monomer concentration in monomer droplets		
$\overline{M}_n$	number-average molecular mass	PEK	polyetherketone
M_m	molecular mass of monomer	PES	polyethersulfone
$[M]_p$	monomer concentration in polymer particles	PET	poly(ethylene telephtalate)
		PFR	plug flow reactor
MMD	molecular-mass distribution	P_i	number of moles of monomer i chemically bound in the polymer
$\overline{M}_M$	mass-average molecular mass		
$M(r)$	mass chain length distribution	$\overline{P}_n$	number-average chain length (degree of polymerization)
$M_{accum.}(r)$	mass chain length distribution of accumulated polymer		
$M_{inst.}(r)$	"instantaneous" mass chain length distribution	$\overline{P}_{n,inst.}$	number-average chain length of dead polymer chains formed instantaneously
$M^\bullet(r)$	mass chain length distribution of polymer radicals	PPS	poly(phenylene sulfide)
		P_r	polymer with chain length, r
$M(\rho_b)$	mass fraction of monomeric units in polymer whose expected branching density is ρ_b	$\overline{P}_M$	mass-average chain length (degree of polymerization)
		$\overline{P}_{M,inst.}$	mass-average chain length of dead polymer chains formed instantaneously
n	number of monomer types in multicomponent polymerization		
n	number of radicals in a particle, or the sum of R$^\bullet$ and PXP in the case of RAFT miniemulsion polymerization	PXP	intermediate adduct radical formed in RAFT polymerization
		$[PXP]_{bulk}$	PXP concentration in bulk RAFT polymerization
$\bar{n}$	average number of radicals in a polymer particle	Q_i	i-th moment of polymer distribution
n_{PXP}	number of PXP molecules in a particle for RAFT miniemulsion polymerization	r	polymer chain length (degree of polymerization)
		r_0	radius of unswollen polymer particle
$n_{R^\bullet}$	number of active radicals in a particle for RAFT miniemulsion polymerization	r_1, r_2	reactivity ratios
		R	gas constant
N	number of tanks in the tanks-in-series model	R$^\bullet$	active polymer radical
		$[R^\bullet]$	radical concentration
N	number of particles per liter of latex	$[R^\bullet]_{bulk}$	radical concentration in the case of bulk polymerization
N_A	Avogadro constant		
N_M	number of moles of monomer molecules	$[R^\bullet_w]$	radical concentration in the water phase
N_n	number of particles containing n radicals per unit volume of water-phase	RAFT	reversible-addition-fragmentation chain-transfer
		RDRP	reversible-deactivation radical polymerization
N_p	number of polymer particles per liter of latex	R_f	rate of chain transfer
$N(r)$	number chain length distribution (number-fraction of polymer molecules of length r)	R_{fX}	rate of chain transfer to small molecule X
		RGS	radical generating species in RDRP
N_T	number of polymer particles per liter of water-phase	R_I	initiation rate
		RIM	reaction injection molding
p	conversion of functional group in step polymerization	R_p	polymerization rate

R_t	termination rate (=$R_{tc}+R_{td}$)	α	dimensionless rate ratio, between radical entry and termination in a particle, $\alpha = 2\rho_A v_p N_A/(k_t N_T)$
R_{tc}	rate of termination by combination		
R_{td}	rate of termination by disproportionation	α_w	dimensionless rate ratio, between initiation in the water phase and termination in a particle, $\alpha_w = 2\rho_w v_p N_A/(k_t N_T)$
RTD	residence time distribution		
$<s^2>$	mean-square radius of gyration		
$[S]_0$	charged surfactant concentration		
SCSTR	segregated continuous stirred tank reactor	β	kinetic dimensionless parameter (=R_{tc}/R_p)
SF	slow fragmentation in RAFT polymerization	θ	dimensionless residence time = $t/\bar{t}$
		μ	volumetric growth rate per particle
$[Single]_p$	concentration of a single molecule in a polymer particle	ξ	conversion at which given primary chain was born
SRMP	stable radical-mediated polymerization	ξ	dimensionless rate ratio, between termination and propagation in a particle, $\xi = k_t/(k_p[M]_p v_p N_A)$
t	time		
$\bar{t}$	mean residence time	ρ_A	total radical entry rate from the aqueous phase to the polymer particles
$\bar{t}_{entry}$	average time interval between radical entry		
T	temperature	ρ_b	branching density
TPA	terephthalic acid	$\rho_b(\xi, \psi)$	expected branching density of the primary chain formed at $x = \xi$, when the conversion at the present time is ψ
Trap	trapping agent in RDRP		
$[Trap]_{bulk}$	concentration of trapping agent in bulk RDRP polymerization		
v	volumetric flow rate	$\bar{\rho}_b$	average branching density
v_p	volume of a polymer particle	ρ_m	density of monomer
u	total number of monomeric units polymerized in a particle	ρ_p	density of polymer
		$\bar{\rho}_s$	average scission density
U	number of monomeric units in a random walk segment	ρ_w	initiation rate in the water phase
		σ	interfacial tension
V	reactor volume	τ	kinetic dimensionless parameter [=$(R_{td}+R_f)/R_p$]
V_d	total volume of monomer droplets in the reactor		
		τ_f	kinetic dimensionless parameter [=R_f/R_p]
V_m	molar volume of monomer in polymer particle		
		ϕ_A	time fraction of the active radical period in RAFT polymerization
V_p	total volume of polymer particles in the reactor		
		ϕ_p	volume fraction of polymer in polymer particle
x	monomer conversion		
x_{c2}	monomer conversion at which Interval 2 ends in ab initio emulsion polymerization	$\Phi_p(n)$	number fraction of particles containing n radicals
		χ	Flory–Huggins polymer-solvent interaction parameter
X	small molecule that causes chain-transfer reaction	ψ	conversion at the present time
Y	parameter indicating the significance of termination with respect to radical entry, defined by $Y = \frac{k_{tw}k_t}{2k_A^2 v_p N_A N_T}$		
z	spatial coodinate		
α	power exponent for the power-law distribution		

General References

J.A. Biesenberger, D.H. Sebastian: *Principles of Polymerization Engineering*, John Wiley & Sons, New York 1983.

T. Meyer, J. Keurentjes (eds.): *Handbook of Polymer Reaction Engineering*, Wiley-VCH, Weinheim 2005.

J.M. Asua (ed.): *Polymer Reaction Engineering*, Blackwell, Oxford 2007.

K. Matyjaszewski, Y. Gnanou, L. Leibler (eds.): *Macromolecular Engineering*, Wiley-VCH, Weinheim 2007.

V. Mittal (ed.): *Miniemulsion Polymeriziation*, John Wiley & Sons, Hoboken 2010.

A.M. van Herk (ed.): *Chemistry and Technology of Emulsion Polymerisation*, 2nd ed., John Wiley & Sons, Chichester 2013.

E. Saldivar-Guerra, E. Vivaldo-Lima (eds.): *Handbook of Polymer Synthesis, Characterization, and Processing*, John Wiley & Sons, Hoboken 2013.

W.F. Reed, A.M. Alb (eds.): *Monitoring Polymerization Reactions*, John Wiley & Sons, Hoboken 2014.

Specific References

1 O. Levenspiel: *Chemical Reaction Engineering*, 3rd ed., Wiley, New York 1998.
2 E.B. Naumann, *J. Macromol. Sci. Rev. Macromol. Chem.* **10** (1974) 75.
3 J.A. Biesenberger, Z. Tadmor, *Polym. Eng. Sci.* **6** (1966) 299.
4 Z. Tadmor, J.A. Biesenberger, *Ind. Eng. Chem. Fundam.* **5** (1966) 336.
5 H. Gerrens, *Chem. Tech.* **12** (1982) June, 380.
6 H. Gerrens, *Chem. Tech.* **12** (1982) July, 434.
7 G.F. Froment, K.B. Bischoff: *Chemical Reactor Analysis and Design*, John Wiley & Sons, New York 1979.
8 J.M. Smith: *Chemical Engineering Kinetics*, McGraw-Hill, New York 1981.
9 J.A. Biesenberger, D.H. Sebastian: *Principles of Polymerization Engineering*, Wiley-Interscience, New York 1983.
10 S.K. Gupta, A. Kumar: *Reaction Engineering of Step Growth Polymerization*, Plenum Press, New York 1987.
11 H. Tobita, Y. Ohtani, *Polymer* **33** (1992) 801.
12 J.G. Watterson, J.W. Stafford, *J. Macromol. Sci., Chem.* **A 5** (1971) 679.
13 J.W. Stafford, *J. Polym. Sci., Polym. Chem. Ed.* **19** (1981) 3219.
14 H.-G. Elias, *J. Macromol. Sci., Chem.* **A 12** (1978) 183.
15 A. Kumar, R.K. Agarwal, S. Gupta, *J. Appl. Polym. Sci.* **27** (1982) 1759.
16 O. Saito, *J. Phys. Soc. Jpn.* **13** (1958) 198.
17 O. Saito in M. Dole (ed.): *The Radiation Chemistry of Macromolecules* vol. **1**, Academic Press, New York 1972, p. 223.
18 H. Tobita, *Macromol. React. Eng.* **4** (2010) 333.
19 L.J. Lee, *Rubber Chem. Technol.* **53** (1980) 542.
20 L.T. Manzione: "Reaction Injection Molding", *Encyclopedia of Polymer Science and Engineering*, vol. **14**, Wiley-Interscience, New York 1988, p. 72.
21 M. Szycher: "Reaction Injection Molding", *Szycher's Handbook of Polyurethanes*, 2nd ed., CRC Press, Boca Raton 2013, chap. 12.
22 M. Thakur: "Solid-State Polymerization", *Encyclopedia of Polymer Science and Technology*, vol. 15, Wiley-Interscience, New York 1989, p. 362.
23 F.K. Mallon, W.H. Ray, *J. Appl. Polym. Sci.* **69** (1998) 1233.
24 K.Z. Yao, K.B. McAuley, D. Berg, E.K. Marchildon, *Chem. Eng. Sci.* **56** (2001) 4801.
25 B. Gantillon, R. Spitz, T.F. McKenna, *Macromol. Mater. Eng.* **289** (2004) 88.
26 S.N. Vouyiouka, E.K. Karakatsani, C.D. Papaspyrides, *Prog. Polym. Sci.* **30** (2005) 10.
27 H.K. Reimschessel, *Macromol. Rev.* **12** (1977) 65.
28 J. Sebenda, *Prog. Polym. Sci.* **6** (1978) 123.
29 K. Tai, T. Tagawa, *Ind. Eng. Chem., Prod. Res. Dev.* **22** (1983) 192.
30 S.K. Gupta, A. Kumar, *J. Macromol. Sci., Rev. Macromol. Chem. Phys.* **C 26** (1986) 183.
31 S.K. Gupta, A. Kumar: *Reaction Engineering of Step-Growth Polymerization*, Plenum Press, New York 1987, chap. 7.
32 R.J. Welgos: Polyamide, Plastics, *Encyclopedia of Polymer Science and Engineering*, vol. 11, Wiley-Interscience, New York 1988, p. 445.
33 X. Wenhua, N. Huang, Z. Tang, R. Filippini-Fantoni, *Macromol. Mater. Eng.* **288** (2003) 235.
34 G. Odian: *Principles of Polymerization*, 4th ed., Wiley-Interscience, New York 2004, p. 569.
35 K. Udipi, R.S. Dave, R.L. Kruse, L.R. Stebbins, *Polymer* **38** (1997) 927.
36 J.M. Andrews, F.R. Jones, J.A. Semlyen, *Polymer* **15** (1974) 420.
37 J.A. Semlyen, *Adv. Polym. Sci.* **21** (1976) 41.
38 K. Tai, Y. Arai, T. Tagawa, *J. Appl. Polym. Sci.* **27** (1982) 731.
39 D.B. Jacobs, J. Zimmerman in C.E. Schildknecht, I. Skeist (eds.): *Polymerization Processes*, Wiley-Interscience, New York 1977, p. 424.
40 A. Gupta, K.S. Gandhi, *Ind. Eng. Chem. Prod. Res. Dev.* **24** (1985) 327.
41 S.A. Russell, D.G. Robertson, J.H. Lee, B.A. Ogunnaike, *Chem. Eng. Sci.* **53** (1998) 3685.
42 A Nisoli, M.F. Doherty, M.F. Malone, *Ind. Eng. Chem. Res.* **43** (2004) 428.
43 R.O. Pimentel, R. Giudici, *Ind. Eng. Chem. Res.* **45** (2006) 4558.
44 K.Y. Choi, K.B. McAuley: Step-Growth Polymerization, in J. M. Asua (ed.): *Polymer Reaction Engineering*, Blackwell, Oxford 2007, chap. 7.
45 M. Katz in C.E. Schildknecht, I. Skeist (eds.): *Polymerization Processes*, Wiley-Interscience, New York 1977, p. 468.
46 S.K. Gupta, A. Kumar: *Reaction Engineering of Step-Growth Polymerization*, Plenum Press, New York 1987, chap. 8.
47 M.J. Barandiaran, J.M. Asua, *Polymer* **31** (1990) 1347.
48 M.J. Barandiaran, J.M. Asua, *Polymer* **31** (1990) 1352.
49 K. Ravindranath, R.A. Mashelker, *J. Appl. Polym. Sci.* **26** (1981) 3179.
50 A. Kumar, V.K. Sukthankar, C.P. Vaz, S.K. Gupta, *Polym. Eng. Sci.* **24** (1984) 185.
51 K. Ravindranath, R.A. Mashelker, *Chem. Eng. Sci.* **41** (1986) 2969.
52 G.D. Lei, K.Y. Choi, *Ind. Eng. Chem. Res.* **31** (1992) 769.
53 I. Goodman: Polyesters, *Encyclopedia of Polymer Science and Technology*, vol. 12, Wiley-Interscience, New York 1988, p. 1.
54 K. Ravindranath, R.A. Mashelker, *Polym. Eng. Sci.* **22** (1982) 610.
55 T. Yamada, Y. Imamura, O. Makimura, *Polym. Eng. Sci.* **25** (1985) 788.
56 T. Yamada, Y. Imamura, O. Makimura, *Polym. Eng. Sci.* **26** (1986) 708.
57 M.R.P.F.N. Costa, R. Bachmann: "Polycondensation", in T. Meyer, J. Keurentjes (eds.): *Handbook of Polymer Reaction Engineering*, vol. 1, Wiley-VCH, Weinheim 2005, chap. 3.

58 A.E. Hamielec, J.F. MacGregor, A. Penlidis, *Makromol. Chem., Macromol. Symp.* **10/11** (1987) 521.
59 W.H. Stockmayer, *J. Chem. Phys.* **13** (1945) 199.
60 K.F. O'Driscoll, R. Knorr, *Macromolecules* **2** (1969) 507.
61 P.J. Flory: *Principles of Polymer Chemistry*, Cornell University Press, Ithaca, New York 1953, p. 385.
62 A.L. Barabasi: *Linked: The New Science of Networks*, Perseus Press, Cambridge, MA, 2002.
63 H. Tobita, *ePolymers* **4** (2004) 878.
64 H. Tobita, *Macromol. Mater. Eng.* **290** (2005) 363.
65 H. Tobita, *ePolymers* **5** (2005) 684.
66 D. Stauffer, *J. Chem. Soc. Faraday Trans.* **2** (1976) 1354.
67 D. Stauffer, A. Conoglio, M. Adam, *Adv. Polym. Sci.* **44** (1982) 103.
68 H. Tobita, *ePolymers* **4** (2004) 335.
69 K. Nagasubramanian, W.W. Graessley, *Chem. Eng. Sci.* **25** (1970) 1549.
70 H. Tobita, *Polym. React. Eng.* **1** (1993) 357.
71 H. Tobita, *Macromol. Theory Simul.* **23** (2014) 182.
72 H. Tobita, *J. Polym. Sci.: Part B: Polym. Phys.* **31** (1993) 1363.
73 H. Tobita, *Macromol. Theory Simul.* **5** (1996) 1167.
74 H. Tobita, *Macromol. React. Eng.* **7** (2013) 181.
75 K. Nitta, *J. Chem. Phys.* **101** (1994) 4222.
76 H. Tobita, *Macromol. React. Eng.* **9** (2015) 245.
77 B.H. Zimm, W.H. Stockmayer: *J. Chem. Phys.* **17** (1949) 1301.
78 H. Tobita, *Macromol. Theory Simul.* **9** (2000) 453.
79 H. Tobita, *J. Polym. Sci.: Part B: Polym. Phys.* **39** (2001) 2960.
80 D.J. Read, D. Auhl, C. Das, J. den Doelder, M. Kapnistos, I. Vittorias, T.C.B. McLeish, *Science* **333** (2011) 1871.
81 S. Beuermann, M. Buback, *Prog. Polym. Sci.* **27** (2002) 191.
82 C. Barner-Kowollik, M. Buback, M. Egorov, T. Fukuda, A. Goto, O.F. Olaj, G.T. Russell, P. Vana, B. Yamada, P.B. Zetterlund, *Prog. Polym. Sci.* **30** (2005) 605.
83 A.W. Hui, A.E. Hamielec, *J. Applied Polym. Sci.* **16** (1972) 749.
84 A. Husain, A.E. Hamielec, *J. Applied Polym. Sci.* **22** (1978) 1207.
85 A.E. Hamielec, J.F. MacGregor, S. Webb, T. Spychaj in K.H. Reichert, W. Geiseler (eds.): *Polymer Reaction Engineering*, Hüthig & Wepf Verlag, New York 1986, p. 185.
86 G.Z. Wu, L.A. Denton, R.L. Laurence, *Polym. Eng. Sci.* **22** (1982) 1.
87 N. Khac Tien, E. Flaschel, A. Renken in K.H. Reichert, W. Geiseler (eds.): *Polymer Reaction Engineering*, Hanser Publishers, New York 1983, p. 175.
88 T. Rintelen, K. Riederle, K. Kirchner in K.H. Reichert, W. Geiseler (eds.): *Polymer Reaction Engineering*, Hanser Publishers, New York 1983, p. 269.
89 H.K. Fauske, J.C. Leung, *Chem. Eng. Progr.* **81** (1985) 39.
90 J.C. Leung, H.K. Fauske, *Thermochim. Acta* **104** (1986) 13.
91 W.A. Pryor, *Free Radicals*, McGraw-Hill, New York 1966.
92 K. Kirchner, K. Riederle, *Angew. Makromol. Chem.* **111** (1983) 1.
93 C. Decker, *Prog. Polym. Sci.* **21** (1996) 593.
94 M.J. Barandiaran, J.M. Asua: "Removal of Monomers and VOCs from Polymers", in T. Meyer, J. Keurentjes (eds.): *Handbook of Polymer Reaction Engineering*, Vol. 1, Wiley-VCH, Weinheim 2005, chap. 18.
95 R.J. Albalak (ed.): *Polymer Devolatilization*, Marcel Dekker, New York 1996.
96 A.D. Jenkins, R.G. Jones, G. Moad, *Pure Appl. Chem.* **82** (2010) 483.
97 P.J. Flory: *Principles of Polymer Chemistry*, Cornell University Press, Ithaca 1953, p. 554.
98 J. Scheirs, D. Priddy (eds.): *Modern Styrenic Polymers*, John Wiley & Sons, Chichester 2003.
99 M. Fischer, G.P. Hellmann, *Macromolecules* **29** (1996) 2498.
100 E.R. Wagner, L.M. Robeson, *Rubber Chem. Technol.* **43** (1970) 1129–1137.
101 T.O. Craig, *J. Polym. Sci. Polym. Chem. Ed.* **12** (1974) 2105–2109.
102 N. Casis, D. Estenoz, L. Gugliotta, H. Oliva, G. Meira, *J. Appl. Polym. Sci.* **99** (2006) 3023.
103 G.R. Meira, C. Kiparissides: Free-Radical Polymerization: Heterogeneous Systems, in J.M. Asua (ed.) *Polymer Reaction Engineering*, Blackwell, Oxford 2007, chap. 4.
104 S. Slomkowski, J.V. Aleman, R.G. Gilbert, M. Hess, K. Horie, R.G. Jones, P. Kubisa, I. Meisel, W. Mormann, S. Penczek, R.F.T. Stepto, *Pure Appl. Chem.* **83** (2011) 2229.
105 R. Pelton, *Adv. Colloid Interface Sci.* **85** (2000) 1.
106 G. Talamini, *J. Polym. Sci. Polym. Phys. Ed.* **4** (1966) 535.
107 O.F. Olaj, *J. Makromol. Sci. C A* **11** (1977) 1307.
108 T.Y. Xie, A.E. Hamielec, P.E. Wood, E.R. Woods, *Polymer* **32** (1991) 537.
109 A. Krallis, C. Kotoulas, S. Papadopoulos, C. Kiparissides, J. Bousquet, C. Bonardi, *Ind. Eng. Chem. Res.* **43** (2004) 6382.
110 D.M. Rasmussen, *Chem. Eng.* **79** (1972) no. 21, 104.
111 J.B.P. Soares, T. McKenna, C.P. Cheng: Coordination Polymerization, in J.M. Asua (ed.): *Polymer Reaction Engineering*, Blackwell, Oxford 2007, chap. 2.
112 P. Galli, G. Vecellio, *Prog. Polym. Sci.* **26** (2001) 1287.
113 M. Covezzi, G. Mei, *Chem. Eng. Sci.* **56** (2001) 4059.
114 K.E.J. Barrett (ed.): *Dispersion Polymerization in Organic Media*, Wiley, London 1975.
115 S. Kawaguchi, K. Ito, *Adv. Polym. Sci.* **175** (2005) 299.
116 A.P. Richez, H.N. Yow, S. Biggs, O.J. Cayre, *Prog. Polym. Sci.* **38** (2013) 897.
117 D.H. Napper: *Polymeric Stabilization of Colloidal Dispersions*, Academic Press, London 1983.
118 D.J. Walbridge in K.E.J. Barrett (ed.): *Dispersion Polymerization in Organic Media*, Wiley, London 1975, chap 3.
119 S.M. Klein, V.N. Manoharan, D.J. Pine, F.F. Lange, *Colloid Polym. Sci.* **282** (2003) 7.
120 J.K. Oh, *J. Polym. Sci. Part A: Polym. Chem.* **46** (2008) 6983.
121 H. Zhang, *Eur. Polym. J.* **49** (2013) 579.
122 T.S. Ahmed, J.M. DeSimone, G.W. Roberts, *Chem. Eng. Sci.* **65** (2010) 651.
123 J. Jennings, M. Beija, J.T. Kennon, H. Willcock, R.K. O'Reilly, S. Rimmer, S.M. Howdle, *Macromolecules* **46** (2013) 6843.
124 T. Hales, *et al.*, arXiv:1501.02155v1.
125 T. Ngai, S.A.F. Bon (eds.): *Particle-Stabilized Emulsions and Colloids*, Royal Society of Chemistry, London 2015.
126 K.Y. Choi: "Continuous Processes for Radical Vinyl Polymerization", in M.K. Mishra, Y. Yagci (eds.): *Handbook of Vinyl Polymers*, CRC Press, Boca Raton 2009, chap. 12.
127 E. Lobry, T. Lasuye, C. Gourdon, C. Xuereb, *Chem. Eng. J.* **259** (2015) 505.
128 H.G. Yuan, G. Kalfas, W.H. Ray, *J. Macromol. Sci., Part C: Polym. Rev.* **31** (1991) 215.
129 E. Vivaldo-Lima, P.E. Wood, A.E. Hamielec, *Ind. Eng. Chem. Res.* **36** (1997) 939.
130 C. Kotoulas, C. Kiparissides, *Chem. Eng. Sci.* **61** (2006) 332.
131 C. Kotoulas, C. Kiparissides: "Suspension Polymerization", in J.M. Asua (ed.): *Polymer Reaction Engineering*, Blackwell, Oxford 2007, chap. 5.
132 T.Y. Xie, A.E. Hamielec, P.E. Wood, D.R. Woods, *J. Vinyl Technol.* **13** (1991) 2.

133 A.H. Alexopoulos, C. Kiparissides, *Chem. Eng. Sci.* **62** (2007) 3970.
134 M. Zerfa, B.W. Brooks, *Chem. Eng. Sci.* **52** (1997) 2421.
135 W.D. Harkins, *J. Am. Chem. Soc.* **69** (1947) 1428.
136 W.V. Smith, R.H. Ewart, *J. Chem. Phys.* **16** (1948) 592.
137 A.M. van Herk, R.G. Gilbert, in A.M. van Herk (ed.): *Chemistry and Technology of Emulsion Polymerization*, John Wiley & Sons, Chichester 2013, chap 3.
138 W.J. Priest, *J. Phys. Chem.* **56** (1952) 1077.
139 F.K. Hansen, J. Ugelstad, in I. Piirma (ed.): *Emulsion Polymerization*, Academic Press, New York 1982, p.51.
140 R.M. Fitch: *Polymer Colloids, A Comprehensive Introduction*, Academic Press, San Diego 1997, chap 2.
141 P.J. Feeney, D.H. Napper, R.G. Gilbert, *Macromolecules* **20** (1987) 2922.
142 M. Morton, S. Kaizerman, M.W. Altier, *J. Colloid Sci.* **9** (1954) 300.
143 J. Gardon, *J. Polym. Sci. Polym. Chem. Ed.* **6** (1968) 2859.
144 B.M.E. van der Hoff, *Adv. Chem. Ser.* **34** (1962) 6.
145 R.G. Gilbert: *Emulsion Polymerization, A Mechanistic Approach*, Academic Press, London 1995.
146 J. Gao, A. Penlidis, *Prog. Polym. Sci.* **27** (2002) 403.
147 M. Nomura, H. Tobita, K. Suzuki, *Adv. Polym. Sci.* **175** (2005) 1.
148 C.S. Chern, *Prog. Polym. Sci.* **31** (2006) 443.
149 S.C. Thickett, R.G. Gilbert, *Polymer* **48** (2007) 6965.
150 A.M. van Herk (ed.): *Chemistry and Technology of Emulsion Polymerization*, John Wiley & Sons, Chichester 2013.
151 M. Nomura, Y. Minamino, K. Fujita, M. Harada, *J. Polym. Sci., Polym. Chem. Ed.*, **20** (1982) 1261.
152 W.H. Stockmayer, *J. Polym. Sci.* **24** (1957) 314.
153 W.V. O'Toole, *J. Appl. Polym. Sci.* **9** (1965) 1291.
154 J.L. Garton, *J. Polym. Sci., A-1*, **6** (1968) 2859.
155 H. Tobita, Y. Takada, M. Nomura, *J. Polym. Sci., Polym. Chem.* **33** (1995) 441.
156 M. Nomura, K. Fujita, *Makromol. Chem., Suppl.* **10/11** (1985) 25.
157 N. Friis, A.E. Hamielec, *J. Polym. Sci., Polym. Chem. Ed.* **11** (1973) 3321.
158 M. Nomura, M. Harada, W. Eguchi, S. Nagata, *ACS Symp. Ser.* **24** (1976) 102.
159 F.K. Hansen, J. Ugelstad, *Makromol. Chem.* **180** (1979) 2423.
160 M. Nomura, in I. Piirma (ed.): *Emulsion Polymerization*, Academic Press, New York 1982, p. 191.
161 N. Bechthold, K. Landfester, *Macromolecules* **33** (2000) 4682.
162 F.J. Schork, Y. Luo, W. Smulders, J.P. Russum, A. Butte, K. Fontenot, *Adv. Polym. Sci.* **175** (2005) 129.
163 V. Mittal (ed.): *Miniemulsion Polymerization Technology*, John Wiley & Sons, Hoboken 2010.
164 H. Tobita, *Polymers* **3** (2011) 1944.
165 D.I. Christie, R.G. Gilbert, *Macromol. Chem. Phys.* **197** (1996) 197.
166 H.A.S. Schoonbrood, S.C.J. Pierik, B. Van den Reijen, J.P.A. Heuts, A.L. German, *Macromolecules* **29** (1996) 6717.
167 G. Moad, C.L. Moad, *Macromolecules* **29** (1996) 7727.
168 D. Kukulj, T.P. Davis, R.G. Gilbert, *Macromolecules* **31** (1998) 994.
169 S. Maeder, R.G. Gilbert, *Macromolecules* **31** (1998) 4410.
170 H. Tobita, *Macromol. Theory Simul.* **10** (2001) 676.
171 S. Katz, R. Shinnar, G.M. Saidal, *Adv. Chem. Ser.* **91** (1969) 145.
172 K.W. Min, H. Ray, *J. Macromol. Sci., Rev. Macromol. Chem.* **C11** (1974).
173 C.C. Lin, W.Y. Ciu, *J. Appl. Polym. Sci.* **23** (1979) 2049.
174 G. Lichti, R.G. Gilbert, D.H. Napper, *J. Polym. Sci., Polym. Chem.* **18** (1980) 1297.
175 G. Lichti, R.G. Gilbert, D.H. Napper, in I. Piirma (ed.): *Emulsion Polymerization*, Academic Press, New York 1982, p. 93.
176 E. Giannetti, G. Storti, M. Morbidelli, *J. Polym. Sci., Polym. Chem.* **26** (1988) 1835.
177 G. Storti, G. Polotti, M. Cociani, M. Morbidelli, *J. Polym. Sci., Polym. Chem.* **30** (1992) 731.
178 H. Tobita, *Acta Polym.* **46** (1995) 185.
179 H. Tobita, *Macromolecules* **25** (1992) 2671.
180 H. Tobita, K. Yamamoto, *Macromolecules* **27** (1994) 3389.
181 H. Tobita, M. Kumagai, N. Aoyagi, *Polymer* **41** (2000) 481.
182 H. Tobita, N. Aoyagi, S. Takamura, *Polymer* **42** (2001) 7583.
183 M. Nomura, K. Fujita, *Polym. Int.* **30** (1993) 483.
184 H. Tobita, *Macromolecules* **27** (2004) 585.
185 H. Tobita, *e-Polymers* **5** (2005) 684.
186 H. Tobita, *Macromol. React. Eng.* **4** (2010) 643.
187 P.B. Zetterlund, M. Okubo, *Macromol. Theory Simul.* **16** (2007) 221.
188 H. Tobita, *Macromol. Theory Simul.* **16** (2007) 810.
189 H. Maehata, C. Buragina, M. Cunnungham, *Macromolecules* **40** (2007) 7126.
190 H. Tobita, *Macromol. Theory Simul.* **20** (2011) 179.
191 P.B. Zetterlund, Y. Kagawa, M. Okubo, *Macromolecules* **42** (2009) 2488.
192 M.J. Monteiro, H. de Brouwer, *Macromolecules* **34** (2001) 349.
193 C. Barner-Kowollik, J.F. Quinn, T.L. Uyen Nguyen, J.P.A. Heuts, T.P. Davis, *Macromolecules* **34** (2001) 7849.
194 A.R. Wang, S. Zhu, Y. Kwak, A. Goto, T. Fukuda, M.S. Monteiro, *J. Polym. Sci., Polym. Chem.* **41** (2003) 2833.
195 K. Suzuki, Y. Kanematsu, T. Miura, M. Minami, S. Sato, H. Tobita, *Macromol. Theory Simul.* **23** (2014) 136.
196 W. Meiser, M. Buback, O. Ries, C. Ducho, A. Sidoruk, *Macromol. Chem. Phys.* **214** (2013) 924.
197 M.F. Cunningham, *Prog. Polym. Sci.* **33** (2008) 365.
198 P.B. Zetterlund, Y. Kagawa, M. Okubo, *Chem. Rev.* **108** (2008) 3747.
199 P.B. Zetterlund, *Polym. Chem.* **2** (2011) 534.

Plastics Processing, 1. Processing of Thermoplastics

Gert Burkhardt, Institut für Kunststoffverarbeitung, Aachen, Germany

Ulrich Hüsgen, Institut für Kunststoffverarbeitung, Aachen, Germany

Matthias Kalwa, Institut für Kunststoffverarbeitung, Aachen, Germany

Gerhard Pötsch, Institut für Kunststoffverarbeitung, Aachen, Germany

Claus Schwenzer, Institut für Kunststoffverarbeitung, Aachen, Germany

1.	Introduction	367
2.	Compounding Processes	368
2.1.	Mixing	369
2.2.	Rolling and Kneading	370
2.3.	Pelletizing	370
2.4.	Shredding and Grinding	371
2.5.	Storage and Transportation	372
3.	Pressureless Processing Techniques	373
3.1.	Casting	373
3.2.	Dipping	374
3.3.	Coating	375
3.4.	Foaming	377
4.	Processing under Pressure	379
4.1.	Compression Molding	379
4.2.	Rolling and Calendering	379
4.3.	Extrusion	381
4.4.	Blow Molding	389
4.5.	Injection Molding	393
5.	Forming Processes	401
5.1.	Drawing	401
5.2.	Thermoforming	402
	References	404

This article considers the processing of thermoplastics to semifinished and finished products, as well as the processing stages that occur prior to the final manufacturing step. The complexity and diversity of the subject limit the depth of the discussion here to the more fundamental aspects.

1. Introduction

Thermoplastics consist of macromolecules that are formed by addition polymerization [e.g., poly (vinyl)chloride], polycondensation (e.g., polycarbonate), polyaddition (e.g., polyurethane), or polymer transformation [e.g., transformation of poly(vinyl acetate) to poly(vinyl alcohol)]. The macromolecules can be linear or branched and are entangled. Additionally, they are held together by van der Waals forces. These can be classified according to their origin as dispersive, dipole, or induction forces and hydrogen bonds. Depending on their molecular structure, thermoplastics can be amorphous or semicrystalline.

The thermal and rheological behavior is decisive for processing and application characteristics. Below the softening temperature (glass transition temperature), thermoplastics are rigid and brittle. The intramolecular forces decrease with increasing temperature, which results in an increase in the mobility of the molecular chains. Above the glass transition temperature, amorphous polymers exhibit elastic behavior caused by chain entanglements. With semicrystalline materials, dimensional stability is maintained by the unmelted crystalline regions; the simultaneously existing "melted" amorphous regions give high impact strength. With increasing temperature, the chain entanglements loosen even more due to increased thermal motion; with semicrystalline materials this occurs even in crystalline zones at temperatures above the crystalline melting point. The result is the plastic melt required for processing. The transition from solid to melt is reversible; on cooling, the processes described above occur in the reverse order. A characteristic of the melt is that the material still possesses elastic properties due to the residual chain entanglements. Additionally, flowing of the melt leads to alignment of the molecules. The degree of orientation in a preferential direction depends on the temperature and the length of the molecular chains. When the material is cooled, this orientation is to some extent frozen in, a situation that may be desirable. For example, this leads to an increase in the mechanical properties in the direction of orientation. However, internal stresses are also frozen in, which relax when the temperature is increased and can lead to distortion of the finished product. To suppress this phenomenon, during molding, the melt must be held at the maximum temperature possible for a long residence time, to allow relaxation to occur. This is not always possible, because thermal damage can result.

The rheological properties of polymer melts are of considerable importance for processing. The strong dependence of viscosity on both temperature and shear velocity plays a significant role in the selection and design of processing machines and molds, as well as the choice of operating parameters. During processing a rapid input of energy is desirable to give a processible melt with sufficiently low viscosity as quickly as possible. Because of the poor thermal conductivity of polymers, conductive heating is unsuitable since it is too slow, and large temperature gradients could result. Convection also plays a minor role because high-molecular melts exhibit laminar flow at extremely low Reynolds numbers. Hence in most cases, the polymer is melted by internal friction (dissipation), generated by the introduction of mechanical energy and the resulting friction and shearing forces to which the molecules are exposed. Polar polymers can also be heated in a high-frequency electromagnetic field. Radiation heating, generally employing IR radiation, is typically used to warm surfaces and planar articles. However, the most important method is melting by dissipation in screw machines, combined with conductive heating via the contact surfaces.

Removal of heat from the melt is more difficult because the only mechanism available is thermal conduction. This can lead to major problems, especially with thick-walled moldings. Hence, the low thermal conductivity of polymers, and the resulting problems regarding heating and cooling, are decisive in the selection of processing technology and have a significant influence on process economics.

2. Compounding Processes

Compounding includes all the processing stages between the manufacture of the crude polymer and the molding step. The transition from compounding to processing can be a gradual change. Thus, storage, transportation, metering, mixing, plastication and granulation are regarded as compounding.

Compounding is necessary because either as-polymerized plastics cannot be processed directly or their properties are inadequate. Consequently, additives must be introduced to the crude polymer to allow the production of articles on an industrial scale.

Heat stabilizing additives enable the polymer to withstand processing temperatures without thermal degradation. Additionally, processing aids are employed which, for example, reduce melting time, influence melt viscosity, or decrease the material's residence time in the processing machine by preventing adhesion of the polymer to metallic machine components.

By using plasticizers, rigid and brittle polymers can be rendered flexible and ductile, opening up new applications.

Similarly, optical properties can also be varied by employing additives. By using pigments, polymer products can be manufactured in all colors imaginable, and gloss can be specifically influenced.

Fillers are increasing in importance not only because of their volume-enlarging effects but also because of their influence on processing and properties.

Another significant aspect of compounding is that many polymers are produced in a form unsuitable for further processing (e.g., as a fine powder). In such cases the material must be reshaped in an intermediate processing step known as pelletizing, which, for example, improves the feeding behavior in processing equipment.

Generally, polymer compounding is carried out by the polymer manufacturer, so the processor can purchase a compound that is ready for use. On an industrial scale, the only thermoplastic compounded by the processor is poly(vinyl chloride).

2.1. Mixing

The production of a homogeneous mixture of polymer and additives places high demands on machine and process. The objective of mixing is to distribute the additive throughout the crude polymer as evenly as possible without excessively stressing the polymer itself. The two basic processes are cold mixing at room temperature, in which the components are simply mixed, and hot mixing, during which absorption and diffusion occur.

Examples of cold mixers are the tumble mixer, in which mixing occurs simply under the action of gravitational forces, and the ribbon mixer, which mixes the materials with a spiral in a horizontal barrel.

Hot mixers are classified according to the method of introducing heat. The polymer can be heated by contact with heated metal components, by irradiation, by convection (e.g., in a plowshare mixer), or by dissipation of mechanical work in the product in the form of heat of friction. Because this form of energy introduction requires high mixer revolutions, these mixers are known as high-speed or turbo mixers. This type of mixing allows rapid heating of the materials and is therefore widely used throughout the plastics processing industry. It is described in more detail below, by using poly(vinyl chloride) (PVC) compounding as an example.

PVC is compounded in mixing vessels equipped with ring- or propeller-shaped mixing elements (Fig. 1). The vessel is usually double walled to allow better temperature control. The peripheral velocity of the mixing elements is generally 20 – 50 m/s. These simultaneously mix and heat the components by throwing them against the container walls and causing them to rise. This results in a whirling motion, with the material being circulated between the center of the container and the base of the vessel. The powder is heated by friction between particles and by particle collisions.

During the initial phase of mixing, the components are merely distributed homogeneously. Actual hot mixing begins above 50 °C. On further heating, during which the plastication temperature of PVC (ca. 150 °C) must not be exceeded, the additives are ground together, and meltable additives are melted. The PVC powder, which is not ground in this process, absorb

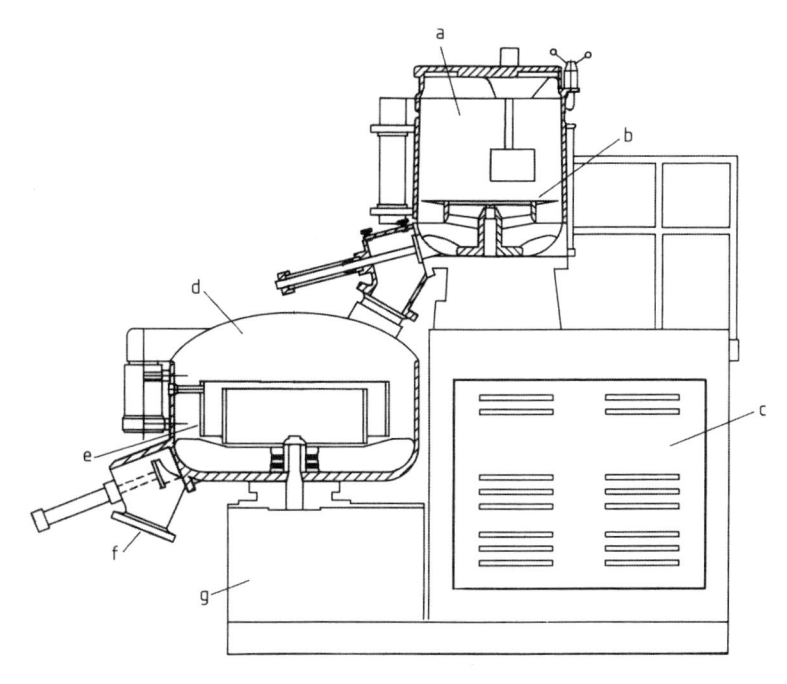

Figure 1. Heating/cooling mixer combination for compounding PVC
a) Heating mixer; b) Mixing element (propeller); c) High-speed drive; d) Cooling mixer; e) Stirrer; f) Outlet; g) Low-speed drive

the molten additives so that at the end of hot mixing a dry powder, the so-called dry blend, results. This material has a final temperature between 110 and 130 °C, and must be cooled in a cool mixer before it can be stored. It is allowed to fall into a water-cooled mixer situated beneath the hot mixer, where it is cooled to ca. 40 °C with constant, slow stirring to prevent fusion of the pellets.

The dry blend produced in this way can be processed without further compounding steps, as is typical for rigid PVC by use of twin screw extruders, or it can be pelletized, as is normal for plasticized PVC.

2.2. Rolling and Kneading

Thermoplastics are often plastified prior to processing. The material is compacted, melted, degassed (if necessary), homogenized, and modified with additives. The oldest process is *rolling*, which involves passing the plastic between two counterrotating, heated rolls where it is plastified and pressed to a thin layer. The plastic is heated by contact with the rolls and by friction in the buildup of material before the roll gap and during passage through the narrow gap between rolls. The major advantage of this system is that it is an open process which, in addition to facilitating addition of additives, also allows scrap material such as edge trimmings to be reintroduced without any difficulty. Because of the high surface-to-volume ratio, the material temperature can be controlled precisely. This is of considerable significance with thermally sensitive materials such as PVC. Furthermore, cleaning the machines is straightforward, and the process can be monitored visually.

A disadvantage is that rolling is a batch process and difficult to automate. More recently, a roll unit has been developed which can be operated in a continuous mode thanks to spiral-shaped grooves that are worked into the roll surfaces. The polymer is fed to one side of the rolls and, passing through the roll gap several times, is driven by the spiral grooves to the other side where it is taken off.

Internal mixers also process polymers batchwise. The material is kneaded between blades rotating in a double barrel and discharged when plastication is complete.

In continuous kneaders, the kneading blades are in the form of rotors with multiple screw channels on the feed side that transport the polymer to the actual kneading elements. The kneading zone is followed by a pump zone, which ejects the plasticated material.

The planetary roller extruder is based on the principle of a single-screw extruder, where the screw is designed as a spindle around which several planetary spindles rotate (Fig. 2 A). This design enables intensive shearing and mixing of the material due to repeated renewal of the material surface. Because of the thin layers of material and the large surfaces of the central and planetary spindles, precise temperature control is possible.

In the *plasticator* a feed screw transports the polymer to a conical rotor in a conical cylinder housing (Fig. 2 B). The conical gap can be adjusted by moving the cone relative to the housing. The excellent homogeneity of the plasticated product can be attributed to the fact that the powdered material is initially sintered and subsequently plastified. The material is spread out by the cone and repeatedly separated and relayered by spiral grooves or threads. Excellent temperature control is possible due to the large surface area. The plasticated material finally passes to further processing units via a discharge screw.

Additionally, a range of other compounding equipment is available, including cokneaders and co- or counterrotating twin-screw extruders. All possess various advantages and drawbacks; hence their selection is not necessarily a question of material-specific or process aspects alone. Plant conditions and economics also play major roles.

2.3. Pelletizing

Pelletizing is the manufacture of pellets (granules) of equal shape and size, to give optimum feed behavior on the processing equipment. The material to be granulated is melted in a single- or twin-screw extruder and fed to a pelletizer. Both hot and cold pelletizing processes are used. Depending on the manufacturing procedure, bead, cylinder, or cube granulate is obtained.

In cold pelletizing, strands, ribbons, or sheets are formed by a pelletizing die; after

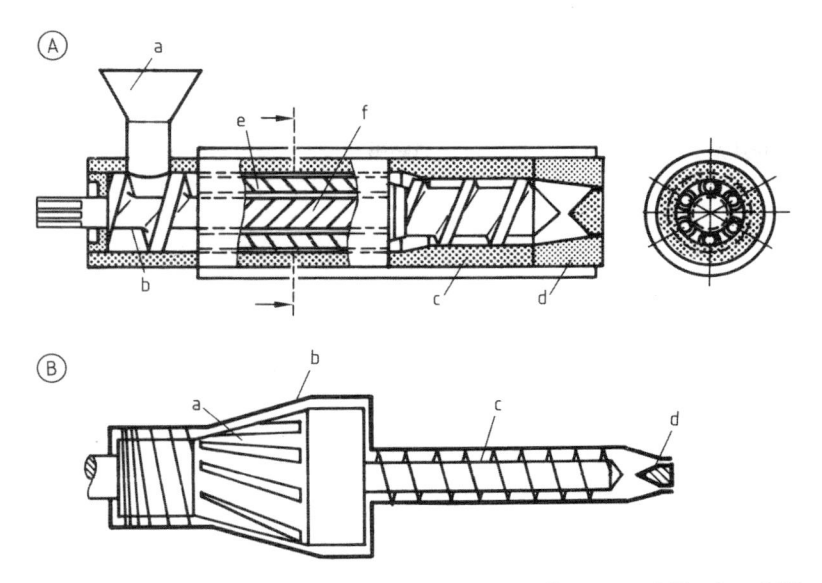

Figure 2. A) Planetary screw extruder: a) Hopper; b) Feeding screw; c) Housing; d) Die; e) Planetary spindle; f) Central spindle
B) Plasticator: a) Rotor; b) Housing; c) Discharge screw; d) Die

solidification, they are chopped to the desired shape by using rotating knives.

In hot pelletizing, the plasticated material is pressed through a die, and the exiting strand is chopped by a blade that rotates on the surface of the die plate. Cooling of the melt occurs after pelletizing, either with air (dry pelletizing) or water (wet pelletizing), although in the latter case, the water must be subsequently removed. With materials of very low viscosity and materials that tend to adhere, the melt passes through the die plate directly into a water bath (underwater pelletizing) where it is cut by blades rotating under the water surface. The resulting pellets must then be dried.

2.4. Shredding and Grinding

During the processing of thermoplastics, a variety of off-specification semifinished and finished products are generated, such as pipes, profiles, injection moldings, and hollow articles, along with scrap such as flash, die leakages, cutting die residues, and film and sheet edge trimmings. For reasons of economy and ecology, these materials are fed to reclaim systems (recycling of single-product manufacturing scrap).

These materials must be cut to produce a pellet form suitable for processing (chip granulate). In doing so however, a range of particle sizes is always obtained.

Cutting mills are generally employed for regranulating. Scrap material is added through a feed system: original-length pipes and profiles are fed horizontally; grinders for film edge trimmings can be incorporated directly in suction pipes. The rotor revolves with high momentum and is equipped with multiple cutting heads that operate against one or more stationary cutting edges mounted in the housing.

In some models, to increase efficiency the rotating cutting heads are divided and displaced with respect to each other. The feed material is cut in the cross direction and — with repeated reorientation — further chopped in subsequent cutting operations. When the desired particle size is reached, the material is ejected through a sieve whose pores determine the particle size. Heat generated in the cutting process can be removed by water cooling of the housing and rotor.

For processing larger plastic articles, a two-stage grinding process has proved effective. The first stage involves coarse chopping with a guillotine (for film rolls and start-up chunks, i.e., material formed before melt calibration commences) or by a preliminary cutting rotor (for bulky materials). The second stage normally involves the cutting mills described above.

To process thermoplastics in pressureless processes (see Chapter 3), they must generally be available as a flowable, finely divided powder. In many cases, they are not produced in this form during polymer syntheses, or compounding with additives in a mixing and granulation process may be necessary. Sinter powders are therefore often produced by grinding granular materials.

In the impact disk mill, the product passes from the feed hopper over a vibration-metering rim. A rotating wheel catapults the material with high velocity against a ribbed conical disk rotating in the opposite direction. The adjustable gap between the rotating rear disk and the conical front cover determines the fineness of the ground material. Powerful suction equipment then removes the powder from below. Heating of the plastic is prevented by the high air throughput and by water cooling the housing. The discharged powder is classified in a cyclone and then sieved. Coarse material is returned to the grinding mill.

2.5. Storage and Transportation

Automatic equipment has become well established in the polymer processing industry for storing and transporting base polymers and intermediate products. Compared to manual operations, these provide improved working conditions, reduction in material losses, lower workplace and machine contamination levels, and a reduction of accident risks.

For the storage of polymers, generally in powder or pellet form, large external silos are suitable. These can be freighted by road or rail. The outlet cone of the silo is generally fitted with a discharge mechanism. The formation of granule bridges is prevented by ventilation of the silo base or by using vibrating frames, bunker cushions, slotted shelves, or similar techniques.

Reserve materials or smaller charges and additives are shipped in bulk containers, drums, or sacks. These are emptied by special discharge equipment, tipper mechanisms, or sack shakers. Within a plant, small intermediate silos, (homogenizing) vessels, and containers are used to store raw materials and prepare polymer mixtures. Containers are preferred when frequent material changes are involved because they are portable and can be stacked. When fitted with a sliding floor, they can also be used as feeding vessels for the hoppers on processing equipment.

Powders and pellets are generally conveyed with pneumatic equipment. High-pressure conveyors (operating pressure $0.2 - 0.4$ MPa) are used for high outputs when filling storage silos from pressure-resistant rail or truck tanks. Medium-pressure conveyors operating at ca. 0.15 MPa are generally used for transport from the storage silos to day or intermediate containers. Low-pressure conveyors with radial blowers are used for short transport distances. Suction conveyors, operating at ca. 50 KPa are most suitable for feeding machine hoppers and gravimetric equipment. The transport interval can be controlled by using fill-level sensors, and air filters are automatically cleaned by back flushing. Mechanical conveyors include screw conveyors (also available as flexible versions) or oscillation conveyors in the form of tubes, channels, and spirals with electromagnetic vibrators or motor drives.

Oscillating displacement pumps, particularly piston, membrane, and bellows pumps, are used for metering liquids such as plasticizers and other additives for compounding. Solids are volumetrically metered by using single- or twin-screw equipment, vibration channels, rotating plates, or cellular wheel sluices. Gravimetric metering equipment can be used for all types of solids, regardless of particle size, homogeneity, flow properties, or adhesion. Weighing vessels, as used in batch processes, are also applicable for continuous metering by discharging with a screw mechanism whose rate of revolution is adjusted to match the weight of material required in a given time. With metering belts, either the belt speed or the rate of revolution of the feeding mechanism is adjusted.

The material flow in a processing plant, including compounding equipment, is shown schematically in Figure 3.

Some thermoplastics, such as polycarbonate (PC), polyamide (PA), and poly(methyl methacrylate) (PMMA), must be dried prior to processing; otherwise the mechanical or optical properties of the resulting products would be considerably impaired. Drying is carried out by heating for several hours in an oven at 80 to 120 °C, depending upon the material concerned.

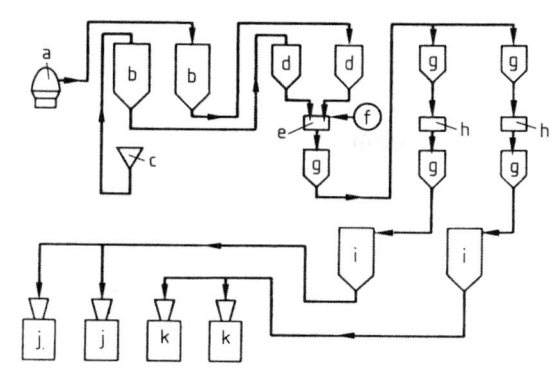

Figure 3. Schematic of a processing plant
a) Rail or truck container; b) Storage silo; c) Sack chute; d) Intermediate silo; e) Weighing; f) Additives; g) Interim storage; h) Mixer; i) Silo for one day's supply; j) Granulator; k) Processing machine

3. Pressureless Processing Techniques

Pressureless processes are used for starting materials that flow and can be formed under the action of gravity without application of external forces. These include monomers, polymer solutions and dispersions, heat-gelable PVC – plasticizer mixture (pastes), low-viscosity melts, and finely divided, meltable powders. The solid end product is formed by reaction, evaporation of the solvent or dispersant, or cooling.

3.1. Casting

Monomer Casting. Thick-walled articles, embeddings, sheets, blocks, and rods, preferentially of polystyrene, polyamides, and poly(methyl methacrylate), can be produced by polymerization of monomers in molds. Control of the heat of reaction necessitates prolonged reaction times. Adequate temperature regulation is required to avoid thermal stresses. The choice of mold material depends on the extent of use and the desired surface quality of the molding. Apart from metal molds with cooling channels, it is often possible to use simpler molds made of sheet metal, wood, ceramic, or glass. To guarantee ease of release and ensure smooth molding surfaces, even with porous materials, those components of the mold that come in contact with the monomers must be sealed by coating or impregnating.

Film Casting. Filtered polymer solutions, dispersions, or melts are cast onto a substrate (e.g., paper, textiles), directly onto a metal drum or rotating belt, or in the case of cellulose films, from a slot die into a precipitation bath.

The solvents are recovered by evaporation in heating and separation stages. The only major application of this process today is manufacture of cellulose acetate films, mainly for the photographic industry.

Slush Molding. Hollow articles made from PVC pastes are manufactured in a pressureless process using two-part metal molds. The molds are heated to the gelation temperature and filled with paste. The paste gels at the outermost edges to form a skin, whose thickness depends on the temperature and the composition of the paste. After excess paste is poured off, the layer remaining in the mold is completely gelled in an oven. After cooling, the hollow molding is removed from the mold.

Rotational Casting. A further development of slush molding is the manufacture of hollow articles from plasticized PVC by rotational casting of PVC pastes or powders. The method is also applicable to other thermoplastics or liquid monomers containing activators and polymerization catalysts.

In this case, only the amount of paste required for the manufacture of the article is metered to the mold, so that removal of excess paste and postgelation are not necessary. During the heating stage, the mold is rotated about two perpendicular axes (Fig. 4), causing uniform wetting of the cavity surface. To achieve optimum distribution of mold contents, the two axes

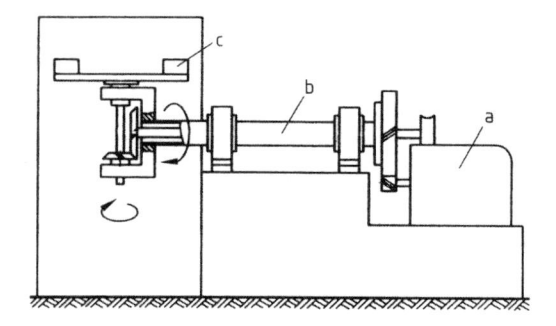

Figure 4. Rotational casting machine
a) Motor; b) Drive shaft; c) Molds

are rotated at (usually adjustable) different speeds (max. 30 rpm). The individual process steps (filling and closing the mold, rotating in an oven, cooling, and removal of the molding) are performed at separate stations. Either the molding is transported to the appropriate station on rails, or the heating oven or cooling air system can be brought to the rotational casting machine.

Numerous modifications of the process that permit shorter cycle times are known. In modern equipment, an oven is no longer used; instead the double-walled aluminum molds are heated with a circulating heat-transfer fluid. The direct-contact heating allows higher production rates. The gelation temperatures are 180 – 220 °C. Filling levels and cycle times are determined in pretests.

Advantages of this process include the low investment costs and the ability to manufacture moldings of various dimensions without pinch or weld seams, virtually free of stresses and with constant wall thicknesses.

Centrifugal casting is used in the manufacture of thick-walled, rotationally symmetrical moldings (e.g., pipes, sockets). In a variation of monomer casting — in this case, mainly limited to polyamide — polymerization occurs in the partially filled mold during rapid rotation about the central axis. The material in the mold is pressed outward and compacted by centrifugal forces. High-quality internal coatings (e.g., for pipes) can be applied directly by centrifugal casting, preferably using meltable powders (see Section 3.4).

3.2. Dipping

A mold or article may be dipped into solutions, dispersions, PVC pastes, melts, or meltable powders; after its removal, the material adhering to the surface is brought to its end-use state by heating or solidification.

Paste Dipping. For the manufacture of articles in which one side remains open (gloves, boots, etc.), a positive model is immersed and the coating is gelled. The process was adapted from the rubber industry; procedures and equipment are practically the same. The positive models are made of metal, glass, or porcelain.

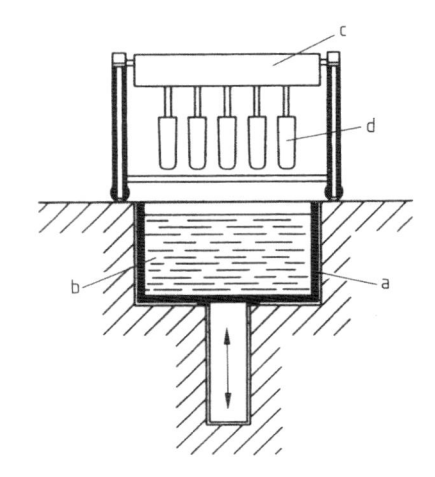

Figure 5. Dipping unit
a) Vessel; b) PVC paste; c) Dipping frame; d) Molds

For mass production, series of identical molds mounted on dipping frames are used (Fig. 5). Degassed PVC paste is placed in a vessel that is raised and lowered during the dipping process. In some cases, the vessel is stationary and the dipping frame is moved. The immersion and removal stages must be carried out relatively slowly to prevent the introduction of air into the paste and to allow it to flow homogeneously from the mold. The vessel or dipping frame must be moved smoothly without vibration to avoid imperfections or marks on the molded article. This is normally accomplished by using hydraulic drives.

During gelation by warming, the pastes temporarily become fluid. To prevent loss of material due to increased flow, it is necessary to heat by using a thermal shock, with rotation, or in an oil bath. Gelation is normally completed in a circulating air oven. The molds carrying the gelled coating are subsequently cooled to ca. 50 °C by immersion in water. At this temperature, even harder moldings can be removed easily from the molds, which are generally treated with talc before use.

Other items (usually made of metal, e.g., grids or handles for tools) can similarly be coated by using a dipping process. Cables and wires can be continually coated in this way. Thicker coatings can be obtained by repeatedly coating a form or object after gelation of the previous layer.

Powder Immersion (Fluidized-Bed Sintering). The dipping of articles directly into polymer melts has not gained major significance. In contrast, immersion of preheated and degreased (metallic) objects in plastic powders is widely used in the manufacture of plastic coatings. In the simplest case, the heated articles are immersed directly in the powder; a continuous coating is formed as a result of melting on the surface.

A more elegant and economical process, which is also applicable for large, irregularly shaped articles, is powder immersion using the fluidized-bed sinter technique. In this case, an immersion bath with fluidized powder is employed (Fig. 6). The upper floor of the immersion vessel has many small holes; a porous ceramic plate is usually employed.

The layer of powder at the base of the vessel is agitated by introducing a gas stream, usually an inert gas (e.g., nitrogen) to avoid oxidation reactions, filling the vessel with a fluidized powder. The preheated articles are immersed in the open bath. Thicker coatings can be obtained by repeating the heating and immersion stages. The powder must have good flow properties and give melts of low viscosity. To ensure pore-free coatings, the articles must usually be subjected to posttreatment, either in an oven or with a naked flame. This process is employed in the production of anticorrosion polyolefin or polyamide coatings. Today, fluidized-bed sintering is carried out both manually on individual articles or small production runs and on a fully automated, mass production scale.

3.3. Coating

The coating process is used to coat sheet materials such as paper and textiles and, in combination with release papers, in the production of self-supporting films. All products that can be cast maybe used for coating, such as polymer solutions, dispersions, and melts. PVC pastes, which are used in the production of coatings, floor tiles, and imitation leathers, are the most important.

Equipment. A coating machine consists of an unwinding station, an expander, coating head, gelling channel, cooling system, and a rewind station (Fig. 7). The unwind station is slowed by using disk brakes, whereas the rewind station is driven so that the product is held under tension but stretched as little as possible when running through the system. The expanders provide a smooth crease-free sheet. The PVC paste is applied to the moving sheet manually or mechanically by the coating head and then spread to the required thickness by a doctor blade.

The doctor blades are housed in a revolving holder so that they can easily be changed and adjusted in height and inclination. For thick coatings, doctor blades with a wide base (5 – 20 mm) are employed. These are designed with a sharp back edge and an undercut; any rising paste gathers here and can be removed periodically so that it does not drip onto the coating (Fig. 8). For thinner coatings, doctor blades with rounded edges 1 – 2 mm wide are used so that the paste is not forced too deeply into the fabric.

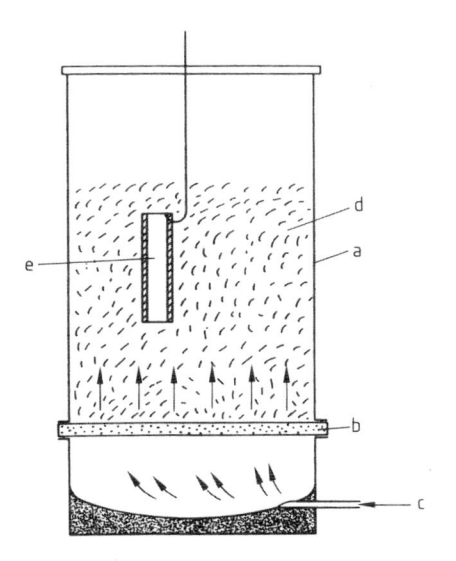

Figure 6. Fluidized-bed sintering unit
a) Immersion bath; b) Filter plate; c) Air inlet; d) Fluidized powder; e) Heated article

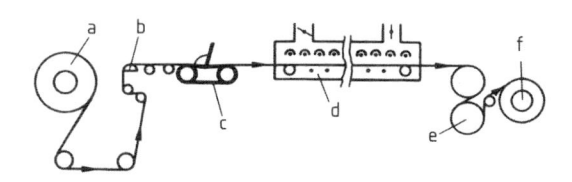

Figure 7. Coating unit
a) Unwinding station; b) Expander; c) Coating head; d) Gelling tunnel; e) Cooling roll; f) Rewind station

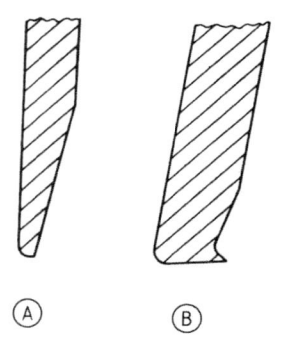

Figure 8. Doctor blades
A) Narrow base with rounded edges; B) Wide base with undercut

Various doctor blade systems are employed, depending on the composition of the paste, the coating material, and the intended use of the product (Fig. 9). With an air knife, the web runs freely between two rollers or metal bars; air knives are generally used with low-viscosity pastes. For many fabrics the knife-on-blanket doctor, in which the fabric lies on an endless rubber belt driven by two rollers, is suitable. This system requires intensive maintenance because of the ease of contamination and the high likelihood of swelling of the rubber belt caused by the plasticizers in the coating. With knife roll coaters, the sheet is supported by a revolving roller, which results in an exactly defined contact area. With steel rolls, complications caused by loose and irregular fabrics can occur, particularly with thin coatings (punctures, disturbances in the coating process); this can be prevented by coating the rollers with soft rubber. The reverse roll coater is employed for coated articles for technical applications in which a defined coating thickness and high operating speeds are important. Two rolls are positioned above a single lower roll (typically steel) over which the web is transported. These two rolls distribute the escaping paste and form it into a film. The separation between the rolls is adjustable and the rotation speeds can be regulated.

The length of the gelling tunnel depends on the required operating speed; it can be as long as 20 m. The heat required for gelation of the paste is supplied by either hot air or infrared dryers. With circulating air heat, the air is warmed by steam or oil heating or in heat exchangers using oil burners. Alternatively, the tunnel can be heated by the combustion gases of an oil heater. In modern equipment, air is directed at an oblique angle toward the web through nozzles arranged in the direction counter to the machine direction. With infrared heating the elements are combined in groups that can be slid into position and individually activated. In this way, heating — one of the more expensive aspects in a large-scale operation — can be adapted to suit the process in question. In some cases, circulating air heating is combined with infrared heaters.

The cooling system generally consists of two water-cooled, corrosion-resistant hollow drums. The cooling effect must be high enough that no adhesion between individual layers occurs during winding, nor is any subsequent material deformation (impression from the substrate web) possible. Continuous winding is performed with multistation winders.

Process. In a coating process, depending on the article being manufactured, various thicknesses can be applied in a single- or multiple-step system. Loosely woven fabrics are first smoothed on a textile calender (steel against paper rolls) at 80 – 120 °C. The coating weight of the paste depends on its composition, the water content of the filler system, and the heating system used in the gelling tunnel. With coating weights >500 g/m^2, the danger of blister formation during gelation exists; hence, the desired coating thickness must be applied in several layers.

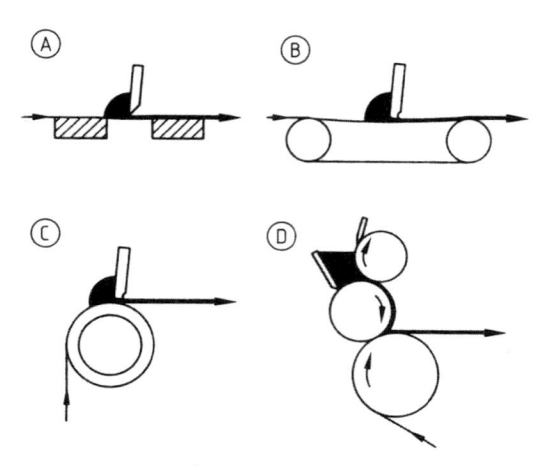

Figure 9. Doctor blade systems
A) Air knife; B) Knife-on-blanket doctor; C) Knife roll coater; D) Reverse roll coater

The primer layer must have excellent adhesion to the fabric; in general, compositions with high plasticizer content and low filler content are used. After gelation of the primer at ca. 130 °C, the filler or intermediate coating can be applied, in one or more stages. Pastes with high filter content or expanded pastes are often used. Each coating stage is followed by a preliminary gelation stage; if insufficient heating capacity is available in the gelling tunnel, thicker coatings must be exposed to an intermediate full gelation stage. The topcoat has the function of providing the product with an appropriately dry and durable surface. For particularly dry surfaces (e.g., necessary for upholstery) a finishing methacrylic coating is applied to seal the surface. This is usually carried out in a roll coating system with an immersion roll (Fig. 10). A mat-steel roll is immersed in a bath, and excess coating is removed by using a doctor blade. The sheet takes up the coating from the steel roll under pressure exerted by the rubber roll mounted above.

When required, a structured surface (e.g., grained) can also be generated in an embossing stage. The product is transported from an unwinding station and passed over a teflon-coated heating drum. After additional heating with infrared radiation, the sheet is embossed at 150 – 160 °C by a patterned rubber roll. The rubber roll is cooled constantly by spraying with water, which is subsequently removed by a nip roll. Before rewinding, the web is cooled by passing it over a series of cooling drums. Because the coating process normally involves multiple layers and generally higher working speeds (30 m/min), the surface is usually embossed in a separate process (embossing speed 1 – 5 m/min), particularly when frequent design changes are commonplace.

Foamed products can be produced by adding blowing agents or by employing mechanically foamed pastes (whipped foam; air is emulsified by rapid mixing). To attain specific foaming characteristics, precise temperature control and residence time in the gelling tunnel are required. For the manufacture of unsupported foamed sheets, or when light fabrics or nonwoven materials are being coated, indirect coating methods (transfer or countercoat techniques) are used. Either the coatings are applied in reverse order (topcoat first) onto endless metal or sieve belts, or carrier fabrics or release papers are employed as transport substrates; after being passed over the cooling drums, they are removed and wound individually.

3.4. Foaming (see also → Foamed Plastics)

In principle, all polymers can be foamed. To date however, only a limited number of materials have been employed in the manufacture of foams. These include the thermoplastics polystyrene, poly(vinyl chloride), and low-density polyethylene (LDPE), as well as phenolic and polyurethane resins.

Foams are generated by adding physical (gases and low-boiling compounds) or chemical blowing agents (gas-releasing compounds) to the polymer.

A basic requirement for foaming plastics is that the foaming process begin while the polymer is in a flowable form that allows bubbles to develop. Once the bubbles have reached the optimum size for the desired material properties, this condition must be fixed.

The flowability of the polymer to be foamed can be achieved in several ways. With thermoplastics, the polymer can be melted or used as a solution. With thermosets, flowability is obtained by using an incompletely cross-linked polymerization mixture or prepolymer.

The required bubbles can also be generated in various ways. With chemical blowing agents, the gas is evolved either as a byproduct of the polymerization reaction (polycondensation), as

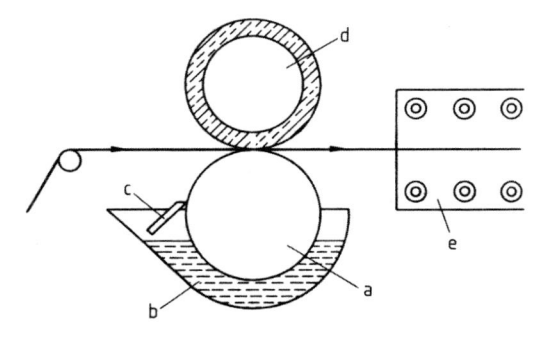

Figure 10. Immersion roll coater
a) Steel roll; b) Trough; c) Doctor blade; d) Rubber roll; e) Drying oven

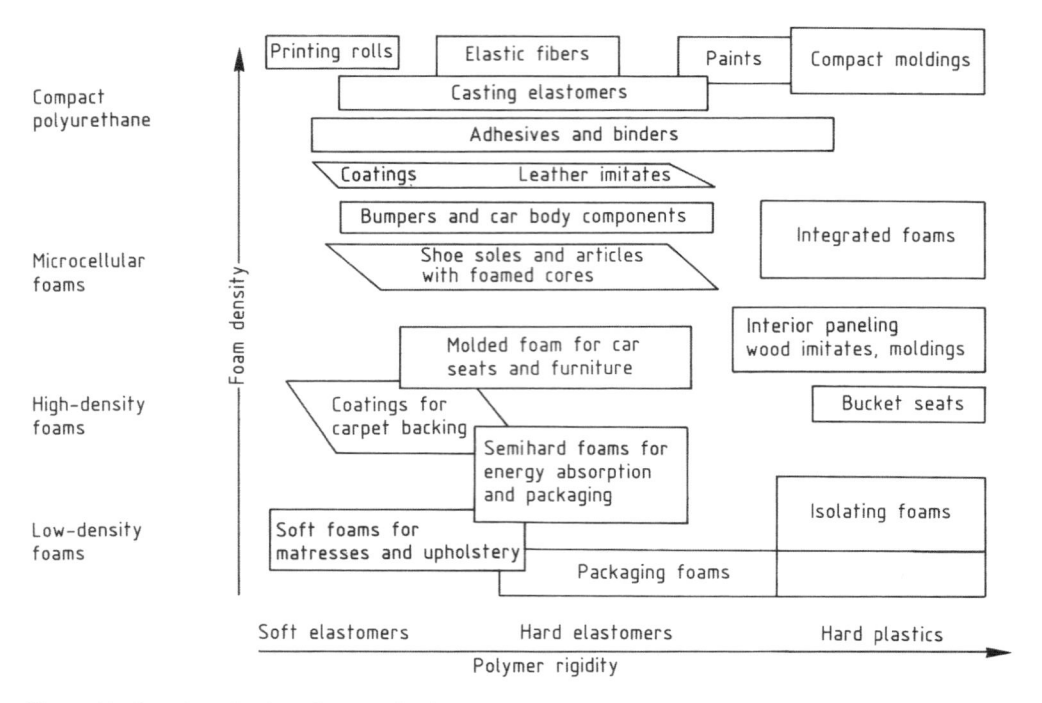

Figure 11. Overview of polyurethane applications

a product of a reaction between the polymer and the blowing agent, or as a result of chemical decomposition of the blowing agent.

Physical blowing agents include the mechanical whipping in of air or other gases and the evaporation of low-boiling liquids.

Polyurethane Foam (see also → Polyurethanes). Polyurethanes manufactured by reaction injection molding (RIM) techniques are processed exclusively by using the so-called one-shot process, which involves direct reaction of the starting materials to form the polymer; (i.e., the reaction begins immediately after mixing of individual components). The starting materials permit wide variations in density and hardness to be achieved. By modifying the chemical composition, soft, flexible, or hard articles can be produced. Accordingly, the range of applications for products made from polyurethane is broad and includes, for example,

1. Soft foams for matresses and upholstery
2. Hard foams for thermal insulation (construction industry, refrigerators)
3. Integral foams for automobile body components, where a variation in density over the cross section is required
4. Microcellular (compact) RIM components (e.g., computer housings)

Figure 11 illustrates various applications as a function of material density and stiffness. The incorporation of fabrics, long and short fibers, and fillers opens up additional areas of application.

Thermoplastic Foams . To foam a thermoplastic melt, a component must be added that is gaseous at predetermined temperature and pressure. Nitrogen is added directly to the melt by special gassing units; low-boiling hydrocarbon compounds can be added to the melt or during polymer synthesis. Chemical propellants are usually powdered substances that decompose at a certain temperature, generally releasing nitrogen.

The advantage of chemical blowing agents over direct gassing is that structural foam components can be manufactured on conventional injection molding equipment, and also that the raw material can be prepared simply by mixing the blowing agent with the granulate.

To guarantee optimum foaming of the melt, the gas formation reaction must go to completion within the prescrew zone. The reaction must not begin in the area around the feed opening of the plastication unit, however, because this would allow part of the gas to escape. To set the barrel temperature and obtain conditions that meet these requirements, the decomposition temperature must be known.

In conventional thermoplastic foam casting processes, the mold cavity is only partially filled, and complete filling is obtained as a result of the foaming reaction.

4. Processing under Pressure

This section considers processes in which solid thermoplastic materials are melted, molded to shape, and subsequently cooled. Because of the extremely high melt viscosities that can be involved, the molding process may require high pressure; for example, several tens of megapascals for extrusion and >100 MPa for injection molding.

4.1. Compression Molding

Compression molding of polymers to semifinished or finished products dates back to the onset of polymer processing but has meanwhile been almost totally displaced by injection molding thermoplastics. Compression molding is now used only in the manufacture of thick-walled components and components made of materials that can not be plastified in extruders [e.g., ultra-high molecular mass polyethylene (PE)]. However, compression molding still plays a significant role in the processing of elastomers and thermosets.

In its simplest form, a press consists of fixed and moving platens. The plunger and cavity components of the compression mold are mounted on these plates, with the parting plane horizontal. Closing and exertion of pressure are usually carried out hydraulically. In compression molding, the cold plastic is placed in the cavity and melted by heating the mold under slight pressure (compression). After melting, the mold is closed and pressure is applied. The molding cools via the mold surface, during which time the pressure is increased further to prevent void formation. Once the ejection temperature has been reached the press is released and the molding removed. The drawback of this process is the long cycle time, which is due to the material being heated and cooled by conduction via the mold surface. For this reason, compression molding is highly uneconomical.

One method of improving the economics is to employ multiplaten presses. In this case, multiple (up to 20) plane-parallel heatable and coolable press tables are loaded simultaneously with plastic, lying on press platens, by using special handling equipment and then pressed simultaneously.

In transfer molding, the starting materials are melted in a cylinder prior to injection into the mold cavity. The mold need only be cooled in this technique. Transfer molding is thus an intermediate between injection molding and compression molding.

4.2. Rolling and Calendering

Rolling mills consisting of only two rolls are employed exclusively in the compounding of thermoplastics. These so-called roll mills are used for additive compounding, plastication, and homogenization (see Section 2.2). The resulting product is a sheet that can be either granulated after cooling or fed in plasticated form directly to processing equipment. The main application is processing of PVC and rubber, typically in a batch process.

If a polymer is to be shaped by using rolls, then more than two rolls are necessary. Such machines are known as calenders and can be equipped with up to seven rolls, but more commonly come with four. Depending on the arrangement of the rolls, they are referred to as I, F, L, or Z calenders (Fig. 12). The rolls are generally made of highly polished cast iron and are individually driven and heated. Oil or water can be used as heating medium. A calender is generally operated with increasing temperature and speed from one roll to the next. The roll nips can be individually adjusted. Because of the high pressure in the nips, as well as the weight of the roll itself, large rolls tend to sag and hence

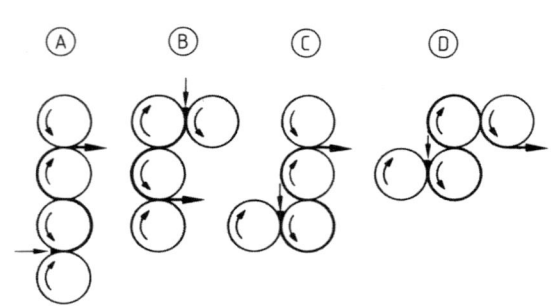

Figure 12. Configuration of calender rolls in four-roll calenders
A) I type; B) F type; C) L type; D) Z type

give an irregular thickness distribution across the width of the final product. This is counteracted by using convex-ground rolls, by roll bending using additional bearings, or by offsetting the roll axes with respect to each other. The roll mill itself must be extremely rigid and stable to withstand the forces to which it is exposed. For this reason, calenders are among the most expensive types of polymer processing equipment. Since they can be economically operated only under conditions of extremely high throughput, they are used for a few special products such as floor coverings and films made of plasticized PVC. Typical calenders have roll lengths (i.e., machine widths) of ca. 1.5 m, but calenders with widths exceeding 3 m are also used. Sheet thicknesses that can be manufactured range from 0.1 to 1 mm. Thinner gauges can be obtained by drawing down using faster speeds for the takeoff rolls. Takeoff speeds are typically 40 – 100 m/min, with throughputs up to several tonnes per hour.

Figure 13 illustrates a calendering line with a four-roll F calender. The roll arrangement shown is employed for processing plasticized PVC, since in this configuration plasticizer vapor that escapes can be removed easily by suction and thus prevented from condensing on the other rolls. The first roll nip is fed continuously from an extruder and conveyor belt. In the first nip a kneading stock is built up from material that is forced back by pressure in the nip. Figure 14

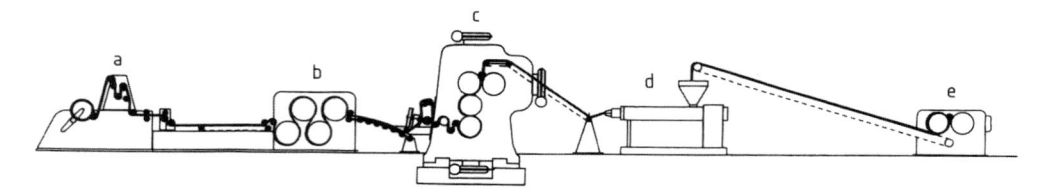

Figure 13. Calendering line
a) Winder and edge cutter; b) Cooling rolls; c) Four-roll calender, F type; d) Extruder; e) Mixing roll mill

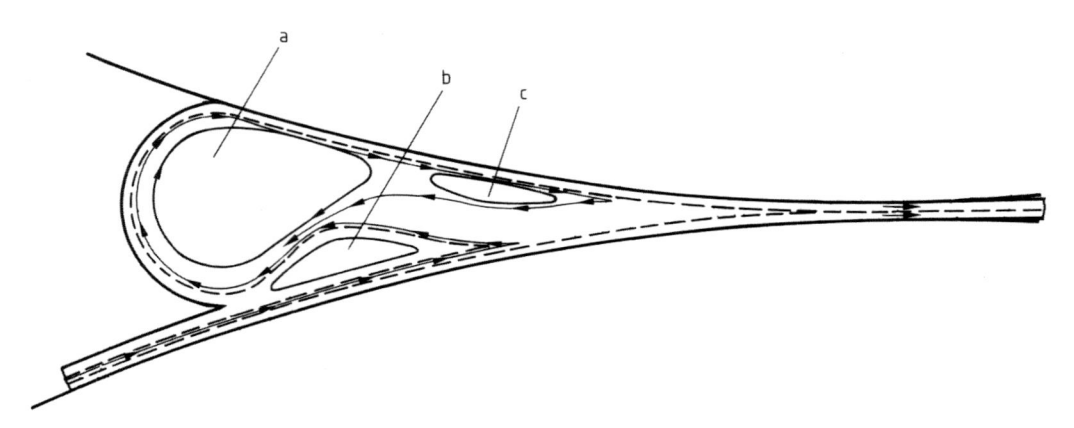

Figure 14. Flow relations in the roll gap
a) Kneading current; b) Feeding current; c) Outlet current

```
Single-screw extruders      Twin-screw extruders           Special types
        |                   (closely intermeshing)         Ram extruder
        |                         /        \               Planetary screw extruder
        |                        /          \              Cascade extruder
        |                  Corotating   Counterrotating    Weissenberg extruder
        |                        |       /      |          etc.
        |                        |      /       |
        |                    Cylindrical      Tapered
        |                        |
        |                        |
  Plasticating extruder      Melt extruder
        /     \
       /       \
Conventional  Forced-feed
```

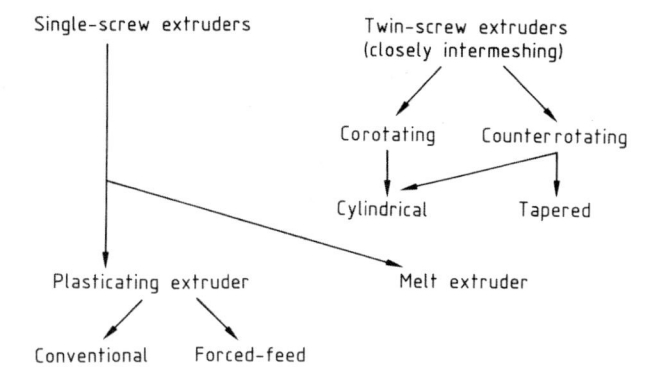

Figure 15. Classification of extruders

shows a cross section through this stock, illustrating the various currents that are formed. Whereas the lower part of the sheet can pass through the nip essentially unaffected, the remaining material is subject to the currents, which lead to continuous creation of new surfaces and increasing homogenization of the melt. Thus the formation of this kneading stock is of considerable significance for product quality. After preforming in the first nip, final shaping takes place at the next two nips, and the surface quality is established.

A process related to calendering is the roll melting process. In this case, cold polymer is fed to the initial roll nip from a hopper, or is preplasticated in an additional pair of rolls or an extruder. In this case the first roll nip must complete melting of the polymer. The material is then rolled and drawn to the desired thickness in subsequent roll nips. In the manufacture of coatings and laminates, preheated substrate is fed to the last roll nip so that the bond is formed under pressure. After this, both materials are removed simultaneously. Surface effects can be included by embossing with additional rolls.

4.3. Extrusion

Extrusion is the term used to describe the continuous manufacture of a semifinished plastic product (film, sheet, pipe, profile). An extrusion line consists of the following:

1. Extruder
2. Mold or die
3. Calibration and cooling
4. Takeoff
5. Finishing

Extruders. The extruder is the heart of every extrusion line; its task is to feed a homogeneous melt of predetermined temperature and pressure into the mold or die. Figure 15 provides an overview of the various types of extruder. An extruder normally consists of:

1. Machine frame
2. Drive (motor, gears, back pressure bearing)
3. Plasticating unit [screw(s), barrel, heating system]
4. Control panel (regulation, control mechanisms, electrical supplies)

Three types of plasticating extruder are most widely used:

1. Single-screw extruders with smooth feed bushes and three-zone screws
2. Single-screw extruders with grooved feed bushes and screws with mixing and sometimes shear zones
3. Twin-screw extruders with closely intermeshing, co- or counterrotating screws

Single-Screw Extruders. A distinction is drawn between melt and plasticating extruders. The basic concepts of both types are essentially identical. The different feed materials — melt extruders are fed with molten plastics, and plasticating extruders with free-flowing solid particles — require specific screw geometries.

Function		Characteristics of the process technology	Engineering solution
Transport	Solid	Low bulk density Low drag forces	High flight depth Grooved feed zone
	Melt	Higher melt density, usually barrel wall adhering	Lower flight depth
Plastication		Solid–melt mixture with low thermal conductivity and increasing density	Decreasing flight depth Shear elements
Homogenization		Inhomogeneous material and temperature distribution	Shallow-flighted mixing zone Mixing elements

Figure 16. The functions of a plasticating extruder

The functions of a melt extruder are limited to melt transportation and homogenization. For this reason, additional details of this type of extruder are not considered here. All further discussions refer to plasticating extruders. An extruder is normally classified by its screw diameter and effective screw length, which is given as a multiple of the screw diameter D (e.g., 50 mm/25 D).

The functions of a plasticating extruder are summarized in Figure 16. Transportation of the material in the extruder results from adhesion of the melt to the barrel and the screw, which move relative to one another. This mechanism is known as drag transportation. The resulting drag flow is superimposed by the pressure flow, whose magnitude and direction depend on the pressure drop along the length of the screw.

A third flow component is leakage flow, which is the flow that exists between the screw flight and the barrel wall. Because of the narrow gap width, the material is exposed to particularly high shear forces in this region.

Extruder Screws. The screw is the most important element of an extruder. Because of the multitude of demands (feeding, transportation, plastication, homogenization), the geometry varies along the screw length. Figure 17 shows typical screw geometries. The three-zone screw is the most widely used; it is made up of the following sections:

1. The feed zone, which draws the solid plastic from the hopper into the extruder and transports it
2. The compression zone, which compresses and plasticates material from the feed zone
3. The metering zone, which homogenizes the melt and heats it to the required temperature

Most thermoplastics can be processed with three-zone screws, although some polymers require additional shear and mixing zones.

Nevertheless, other screws have become established for special applications. The short-compression-zone is used for semicrystalline thermoplastics. The short-compression-zone results from the decrease in channel depth from the deep-flighted feed zone to the shallow-flighted metering zone.

The long-compression-zone screw, which is used for PVC, has a constantly increasing root diameter toward the tip (see Fig. 17 C). Extruders or screws designed for specific applications can be regarded as technically suitable when they fulfill the following conditions:

1. Constant, pulsation-free transport
2. Thermally and materially homogeneous melt
3. Ability to remain within the thermal, mechanical, and chemical material degradation limits

These demands must be met in all cases; additionally, for economic reasons, they must be

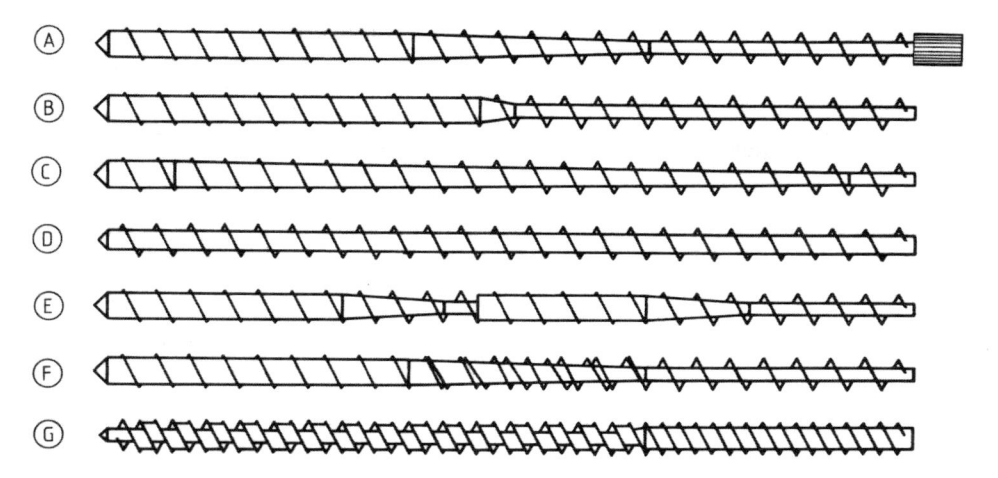

Figure 17. Plasticating screws in common use
A) Three-zone screw; B) Short-compression-zone screw; C) Long-compression-zone screw; D) Conveying screw (only in combination with grooved bushes); E) Degassing screw; F) Maillefer screw; G) Double-flighted wave screw (rarely used)

combined with high mass throughputs and low specific operating costs.

Modern extruders have screw diameters D ranging from 19 mm (laboratory extruder) to special designs of 300 mm. The length has increased from the 6 D of melt extruders for rubber processing to ca. 25 D for typical industrial plastication extruders. Plastication extruders with lengths up to 30 D are also available, and degassing extruders have lengths >35 D.

In addition to diameter and length, screws are classified according to the distribution of the functional zones and the compression ratio. The latter is defined as the ratio of the channel depth of the metering zone h_2 to that of the feed zone h_1. To ensure complete filling of the screw, the compression ratio must at least equal the ratio of the bulk density of the solid in the feed zone to the density of the melt in the metering zone.

The screws described here are employed in two different extruder concepts, which differ in the shape of the extruder barrel in the feed zone, the so-called feed bushes.

Conventional single-screw extruders have a smooth barrel surface in the feed zone; the metering zone is regarded as decisive for output. Molten material from the previous zones is homogenized in the metering zone and subsequently forced through the attached extrusion die under pressure.

In this design, the metering zone acts as a pressure generator. The pressure along the length of the screw depends not only on screw geometry, but also on the die back pressure.

Transport of the solid plastic is governed by coulombic friction between the barrel walls and the granules and between the granules themselves.

The result is that a strong correlation exists between the throughput behavior of the extruder and other operational parameters (temperature, back pressure, etc.). In particular, the temperature of the feed zone plays an important role. For this reason, the feed zone is generally cooled.

Fixed Transport Single-Screw Extruder. Feed bushes with axial grooves were developed to attain large, back-pressure-independent throughputs. The so-called forced-feed bush is part of the cylinder which forms the feeding zone; it is usually removable. The interruptions in the otherwise smooth barrel wall caused by the regularly spaced axial grooves prevent the cold feed polymer from rotating with the screw. Thus they increase the coefficient of friction of the barrel wall. This leads to increased throughput and generation in the feed zone of the pressure required for flow through the following zones.

The feed bush zone must be cooled to prevent melting of the plastic. Figure 18 shows a typical grooved bushing with conically formed

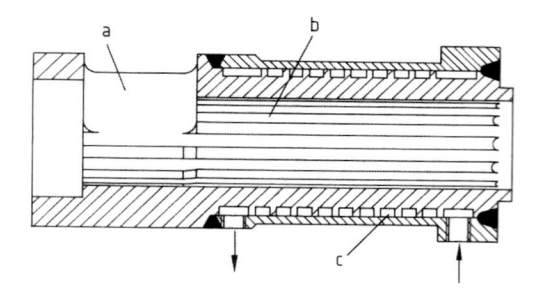

Figure 18. Forced-feed bush
a) Feed opening; b) Axial grooves; c) Cooling channels

grooves. This eliminates the possibility of mixing due to back flow caused by the back pressure. Pressure flow, drag flow, and leakage flow contributions are oriented in the same direction. For this reason, mixing zones must be used to obtain satisfactory melt homogeneity.

Plastication Process . The heat required to melt the plastic is introduced by dissipation and by heat-transfer mechanisms. This includes heating by conduction from the heated barrel to the cold material and the generation of heat by internal and external friction (dissipation) in and between the material.

Twin-Screw Extruders. Twin-screw extruders exist in co- and counterrotating versions; the latter are also available with conically tapered screws. The screws are non-intermeshing or closely intermeshing.

Corotating extruders are used in compounding, whereas counterrotating, intermeshing systems with cylindrical or tapered screws are employed for processing rigid PVC (dry blend).

Corotating Twin-Screw Extruders. With corotating twin-screw extruders, the melt contained in one screw channel is transferred to the other channel with each rotation. The transport mechanism (drag forces) is comparable to that of a single-screw extruder. The melt however is exposed to a greater shear stress due to the increased path length through the extruder.

This type of construction enables integration of closely intermeshing kneading and shear elements into the screw without any problem. Such extruders are almost exclusively employed for compounding.

Counterrotating Twin-Screw Extruders. The most widespread equipment for the extrusion of rigid PVC profiles and pipes consists of closely intermeshing, counterrotating twin-screw extruders. This type of extruder has a different transport mechanism. Each screw segment forms a sealed chamber, which means that during transport, virtually no exchange of material between neighboring chambers occurs over the full length of the screw from hopper to screw tip. Drag forces are not required for this forced conveying; hence, dissipation heating occurs to a considerably lesser extent.

Material heating occurs essentially by barrel heating, which can be controlled precisely and allows a mild treatment of sensitive products. The closely intermeshing screws form C-shaped chambers that are sealed by the screw flights. The C-shaped chamber is formed because the flight of one screw runs in the channel of the other and thus seals it (Fig. 19).

Generally, the counterrotating twin-screw extruder is operated with partially filled screw chambers. This is necessary to ensure a constant flux of material and an optimum balance of extruder, material, and mold. If the extruder were operated with full chambers, the volume of the individual chambers would have to be matched to the volume of polymer, which varies during heating and melting. Otherwise, very high pressures would be generated which lead to increased wear of the cylinder and screws.

The material being transported melts extremely slowly. Only in the final screw channels before the die does sintering occur, caused by the pressure buildup, and material is forced through the chamber gaps (leakage flow). This portion of material melts spontaneously under the short-term shear stress.

Although this transportation and melt mechanism guarantees well-controlled, mild treatment of the plastic, the product is often nonhomogeneous and insufficiently plastified.

The twin-screw extruder is a relatively expensive piece equipment due to the following factors:

1. Expensive bearings that are difficult to house

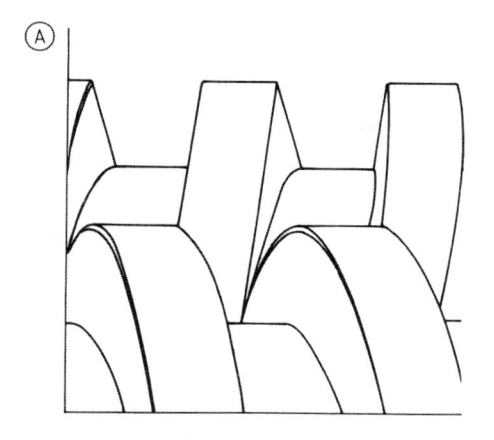

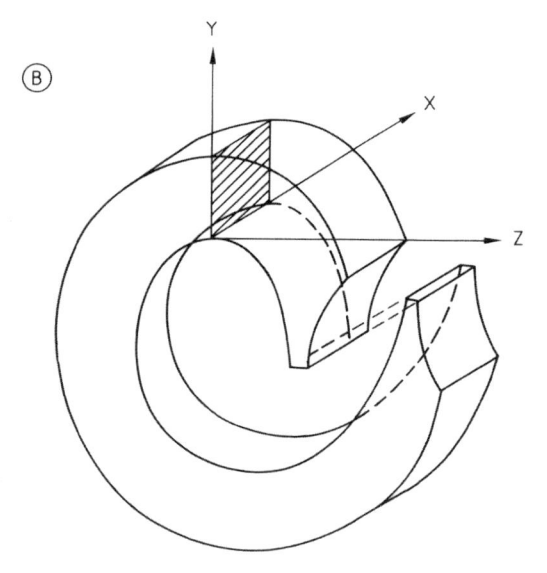

Figure 19. A) Intermeshing counterrotating twin screw; B) Strip of melt removed from one of the C-shaped chambers

2. Complex drives
3. Barrel with a figure-eight bore

The intermeshing of the screws and the pressure buildup in the last screw channels lead to extensive wear of screw and barrel. Extruders with diameters up to 160 mm are commonly employed.

Degassing Extruders. During the melting process, volatile components, such as moisture, decomposition products, and residual monomers, may degas. Since this can lead to defects in the extrudate, these gases must be removed from the system. This is carried out with degassing screws in which, after the compression zone, the channel depth is increased (decompression) so that the pressure drops to ambient levels. A degassing opening in the barrel is incorporated at this position, through which the material is degassed by vacuum. Ideally, at this point in the extruder the material should not be fully melted and should have as large a surface area as possible. The degassing zone is followed by further compression and metering zones (Fig. 17 E). A degassing screw can thus be regarded as a series connection for two screws, whereby the balance between the two is of importance in preventing the melt from leaking out of the degassing orifice.

Degassing extruders can be of both single- and twin-screw types. With counterrotating twin-screw extruders for PVC processing purposes, a degassing zone is standard.

Special Extruders. Apart from the extruders mentioned above, a range of extruders (Fig. 15) exist for special applications, all of which have individual advantages and drawbacks. The most significant of these are the cascade extruders in which two separate extruders are operated in series so as to divide the feeding and melting stages of the extrusion process from the homogenization and output stages, allowing better control and regulation of the process.

Ram extruders are piston plastication units employed in the processing of materials that are not suitable for processing with screws (ultra-high-molecular weight polyethylene, polytetrafluoroethylene).

Extruder Temperature Control. An extruder is heated by barrel heaters. Several heating zones are used to establish a temperature gradient along the barrel.

Commonly employed heating systems include the following:

1. Electrical heating with band heaters
2. Aluminum jackets with cast in heating wires and cooling pipes or ribs
3. Heating and cooling with a heat-transfer medium (oil, steam, pressurized water) that is in direct contact with the barrel walls (double-walled barrel with spiral ribs in the cavity)

The first two methods are sometimes combined with electric cooling fans. The third method is seldom used.

A thermocouple is located in the barrel wall of each heating zone, from which a temperature-dependent electrical signal is input to a control unit.

Additionally, during the processing of thermally sensitive materials such as PVC, the temperature of the screw is regulated with a heat-transfer oil.

Extrusion dies mounted on the extruder give the melt, the desired cross-sectional form. These forms include

1. Solid strand profiles (ribbon, quadratic, etc.)
2. Hollow profiles (shutters, window profiles, etc.)
3. Open profiles (U-shaped, etc.)
4. Pipes
5. Films
6. Sheets
7. Fibers

Furthermore, sheathing dies exist for cables, as well as a range of other constructions. The dies are heated, usually electrically.

All the dies contain a flow channel — the so-called runner system (melt distributor) through which the cylindrical melt strand leaving the extruder must flow. The following melt distributors are of importance.

The *spider* or *displacer* (*torpedo*) is the simplest runner system. It has a streamlined torpedo-shaped mandrel connected to the outer wall of the flow channel by so-called spider legs. The side facing the extruder is conical, but this changes to the desired outlet cross-sectional form as it approaches the die outlet (Fig. 20).

Applications: pipes, profiles, etc.

The *coat hanger distributor* (Fig. 21) spreads the melt over a required width. Melt enters the feed manifold, where the cross section tapers according to the decrease in mass flow, and flows over the so-called island region. This area contains choking elements (restrictor bars, adjustable lips), which can be used to correct irregular melt outputs across the die width.

A variation of the coat hanger distributor is the fishtail distributor in which the feed manifold has a geometrically simplified form.

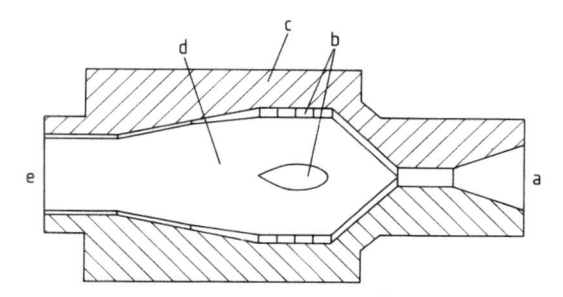

Figure 20. Spider-type manifold
a) Inlet; b) Spider legs; c) Die ring; d) Torpedo; e) Outlet

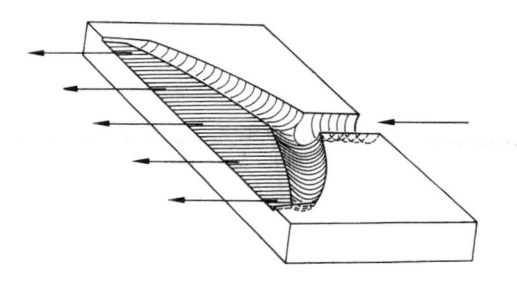

Figure 21. Coat hanger distributor

Applications: films, sheets, etc.

The *pinole* (side-fed die) is a cylindrical or conical coat hanger distributor (Fig. 22), which forms a tubular melt. What is characteristic,

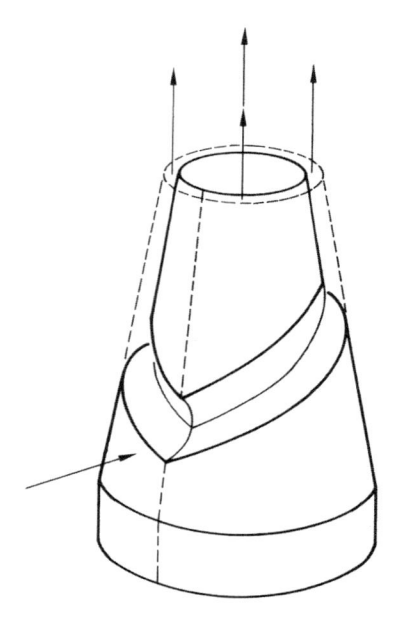

Figure 22. Pinole

however, is that the incoming flow direction is not parallel to the outgoing cylinder, but perpendicular to it.

Applications: pipes, tubular film, cable sheathing, etc.

The *spiral mandrel distributor* is a core with multiple spiral-flow channels of progressively decreasing depth in the direction of the die outlet (Fig. 23). This causes separation of the melt flow into axial and circumferential components. The result is a melt that is homogeneously distributed around the complete circumference and is free from the flow lines and seams that can occur, for example, at flow junctions following the spider legs and with pinole dies.

Application: blown films (tubes).

Subsequent Equipment. The melt leaving the extrusion die must be fixed in shape and dimension; this occurs in the calibration stage. The extrudate is pressed against the walls of the calibrator by compressed air or a vacuum and is cooled. After leaving the calibrator, it must be sufficiently solidified that during subsequent cooling stages no further distortion can occur and the dimensions are fully retained. Thus the length of the calibration zone must be adapted to the throughput and the geometry of the extrudate.

Transport of the extrudate through the calibrator requires high takeoff tensions due to the friction between the extrudate and the surface of the calibrator. Doing this without deformation by using only the cooling capacity of the calibrator is rarely possible, and it is usual to include a subsequent cooling zone.

For profiles, pipes, cables, and similar products, water baths through which the respective extrudate is passed are also employed. Flat extrudates are cooled with rolls. It is also possible to employ water sprays and air cooling mechanisms.

The cooling zone is followed by the takeoff unit whose objective is to remove the extrudate from the die (via calibrator and cooling zone) at a constant speed. Various takeoff systems are available such as band, roll, or caterpillar takeoffs. Elastic extrudates such as cables, fibers, films, and tubes are wound on reels following the takeoff. Rigid extrudates (pipes, sheets, profiles, etc.) are cut to length by saws or guillotines and then stacked.

Examples of Equipment. *Profile Extrusion Line.* Figure 24 shows a typical profile extrusion line. It consists of an extruder and a die followed by calibration, cooling, caterpillar takeoff, and cutting.

Flat film (sheet) extrusion lines (Fig. 25) consist of an extruder and a coat hanger die. These are followed by a roll mill, normally with three rolls, which calibrate and cool the film (sheet). Subsequent to this is a roller conveyor, which allows air cooling of the sheet and the takeoff roll. Finally, the film is wound, or the sheet is cut and stacked.

Blown film lines are employed to manufacture very wide films (Fig. 26). They consist of an extruder with a blown film head normally equipped with a spiral distributor. The vertically extruded tubular film is passed through a cooling ring where it is cooled by air. The tube is inflated to the desired circumference by blowing air into it. At the upper part of the tower, the tube is folded together on a collapsing board and the air is removed by nip rolls. Finally, the film is wound

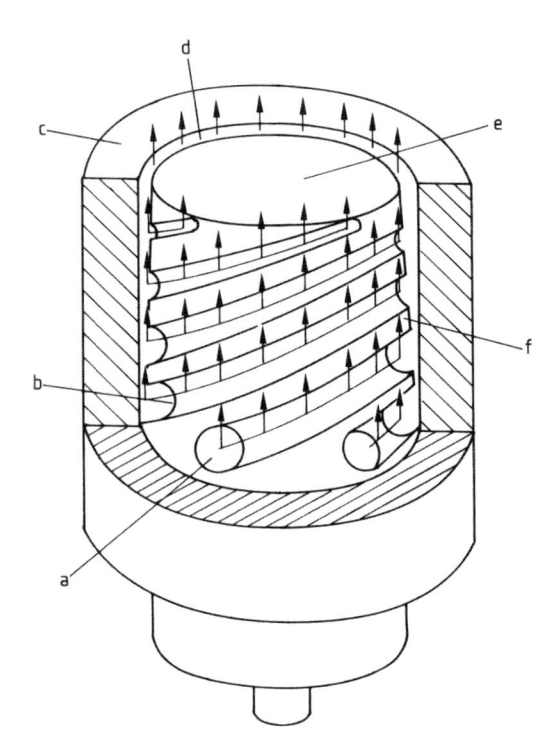

Figure 23. Spiral mandrel distributor
a) Feeding; b) Spiral channel; c) Casing; d) Annular gap; e) Mandrel; f) Steps

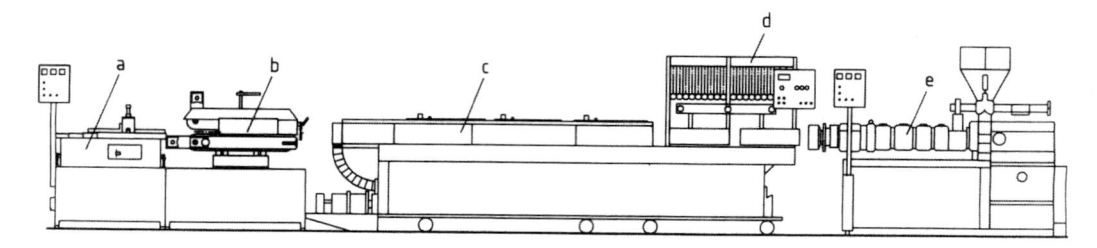

Figure 24. Profile extrusion line
a) Saw; b) Takeoff unit; c) Cooling system; d) Calibrator; e) Extruder

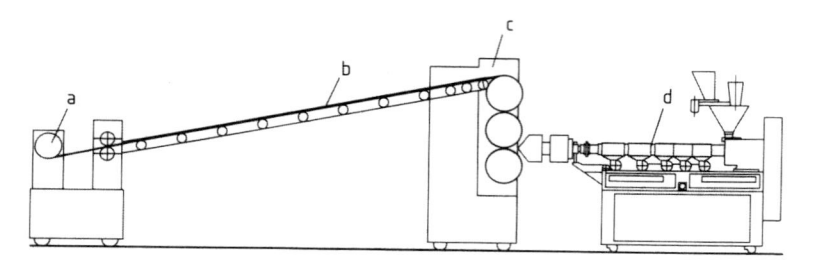

Figure 25. Flat film extrusion line
a) Winder; b) Roller belt; c) Roll mill; d) Extruder

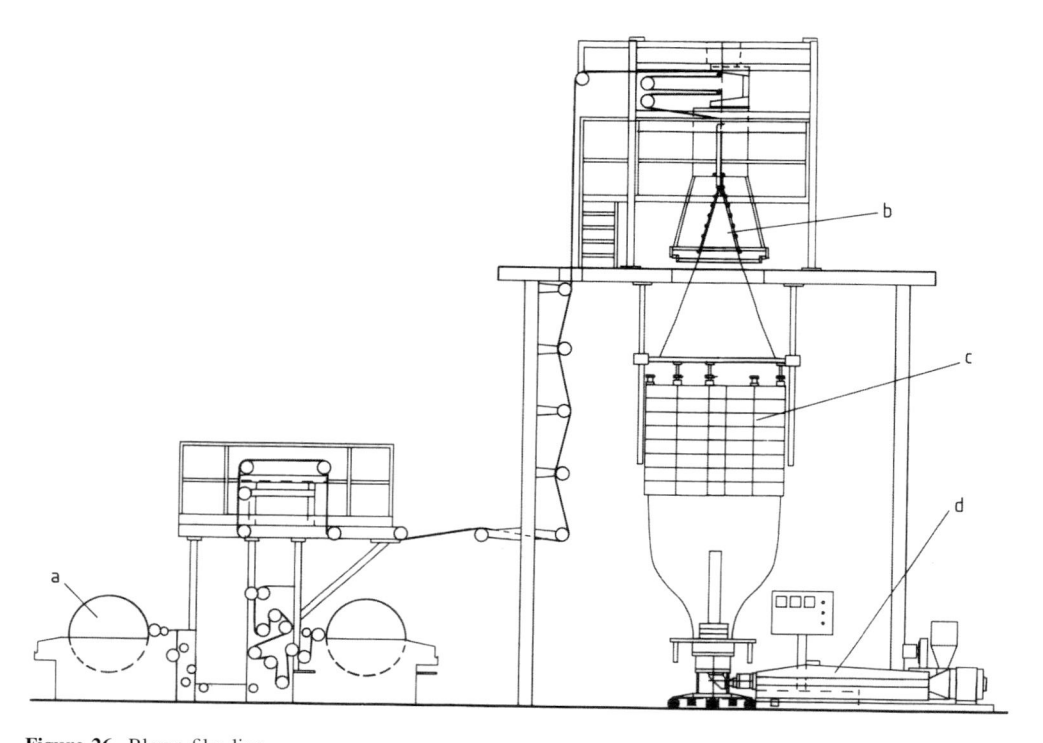

Figure 26. Blown film line
a) Twin winder; b) Guide rolls and collapsing boards; c) Calibrating basket; d) Extruder

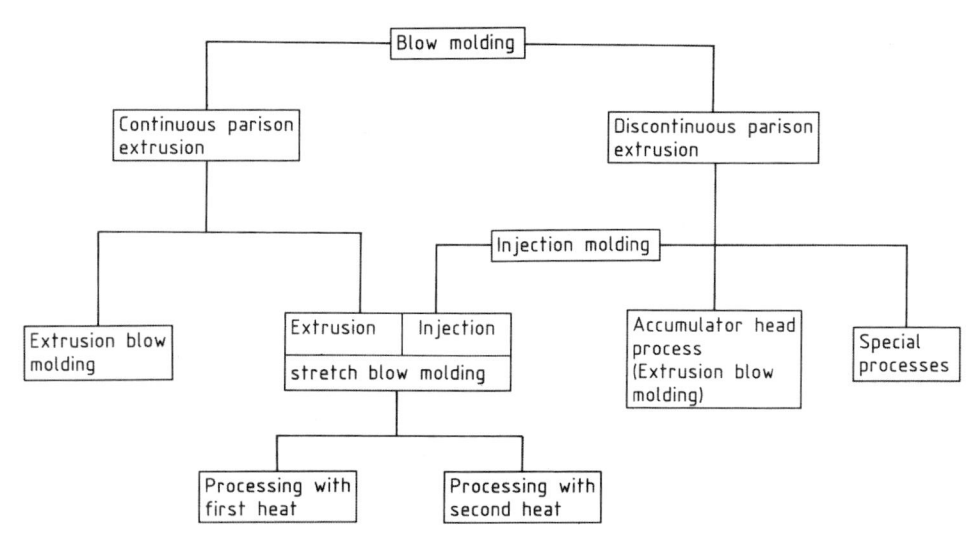

Figure 27. Blow molding processes

on reels. A calibration unit to maintain constant film bubble width can be mounted between the cooling ring and the collapsing board.

Coextrusion. The multitude of demands placed on modern extruded products can sometimes not be met by a single material. A solution to this problem is coextrusion, in which two or more molten materials are combined in a die under pressure. In adaptor coextrusion dies, the individual melt streams are combined in a so-called adaptor and subsequently shaped together in the melt distributor. In multiple manifold dies, each melt stream is individually shaped prior to being combined. Possible adhesion problems between various materials are addressed by incorporating coupling agents.

Well-known examples of the use of coextrusion include cable insulation and packaging films, where composite films with up to seven layers are used.

With coextrusion lines, the individual extruders and subsequent units are essentially identical to the corresponding units in the conventional lines.

4.4. Blow Molding

The most commonly employed techniques for manufacturing hollow articles from thermoplastic materials are extrusion blow molding and related processes such as stretch blow molding. Figure 27 gives an overview.

Blow-molded articles include simple products such as bottles and more complex components such as ventilation ducts, surfboards, suit cases, roof racks, or fuel tanks for automobiles. The volume of such articles can vary from a few milliliters (packaging for medical products) to ca. 13 000 L (tanks for heating oil).

Extrusion blow molding is made up of two parallel processes:

1. Continuous extrusion of a preform (parison)
2. Batchwise shaping of the preform in a mold by using compressed air

The individual steps of the process are shown in Figure 28. The first step begins when the extruded tube (parison) has reached the required length. The mold closes around the parison, which is then cut by a blade. The mold is then conveyed to the blowing station, the blow mandrel is inserted into the mold, and the actual blow process is carried out. After cooling, the mold is opened and the finished molding removed. The open mold then returns to the position below the extrusion die so that it can accept a new parison. Large and heavy molds are not moved; instead, the parison is cut by a gripper that transports it to the mold.

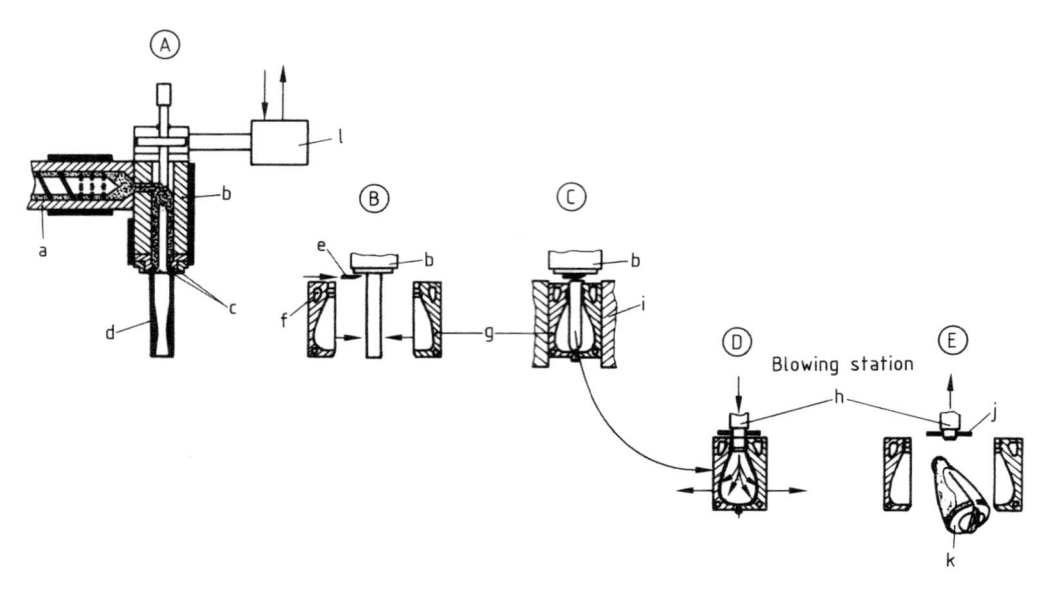

Figure 28. Extrusion blow molding process
A) Parison extrusion; B) Positioning of blow mold; C) Collecting and cutting of parison; D) Shaping and cooling; E) Demolding and flash trimming
a) Extruder; b) Crosshead die; c) Nozzle and mandrel; d) Parison; e) Cutter; f) Cooling channel; g) Blow mold; h) Blow mandrel; i) Closing unit; j) Stripper ring; k) Molding; l) Servohydraulic

An extrusion blow molding machine has four major components (Fig. 29):

1. Machine frame with closing mechanism and blowing station
2. Mold
3. Extruder
4. Crosshead and die

The machine frame is the basic framework of the machine and contains the closing mechanism as an integral component. The latter has the task of opening and closing the mold halves (similar to an injection molding machine), as well as transporting the mold between the die and the blowing station. The closing unit is normally of lighter construction than the counterparts

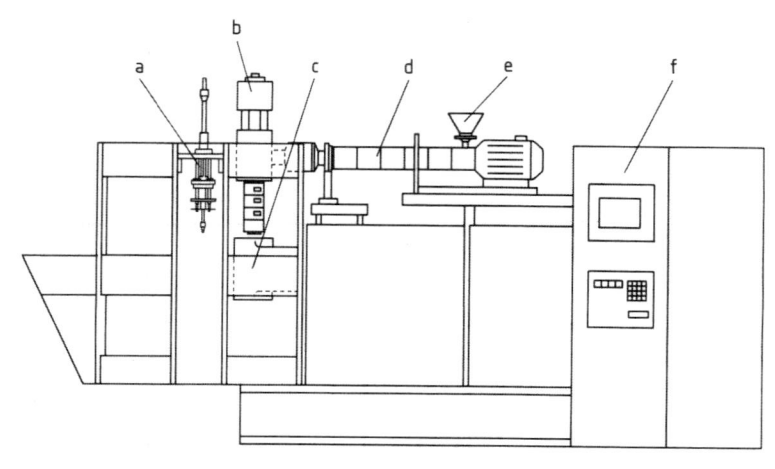

Figure 29. Extrusion blow molding machine (Krupp – Kautex type)
a) Blowing station; b) Blowing head; c) Mold; d) Extruder; e) Metering station; f) Control unit

involved in injection molding since the mold closing forces are relatively low because of the low mold internal pressure (ca. 0.8 MPa). The closing mechanism is driven hydraulically.

The blowing station is that part of the equipment in which the mold is located when the blowing mandrel is inserted and the actual blow process occurs.

The blowing mandrel introduces inflating air and shapes and calibrates the neck of the finished article. Figure 30 shows a closed mold containing a parison and the inserted blow mandrel.

The mold consists in general of two halves in which the negative contours of the molding have been incorporated. This has the function of shaping the parison, cooling the molding, and forming the welds. The latter are necessary because the parison is a portion of an extruded tube, which can only be blow molded when the ends have been sealed, preventing the escape of compressed air. In the case shown in Figure 30, one of the ends is sealed by the blow mandrel. At the other end, the tube is compressed during the closing motion by the pinch-off edges of the mold, and the molten material is welded together. These weld seams are a telltale characteristic of extrusion blow molded products. Thus, materials for blow molding must be weldable.

Cooling is possible on only one surface of the molding — the outer surface — by a water-cooled mold. Hence, cooling times are longer than those associated with injection molding, in which heat can also be removed through the core of the mold. This problem is counteracted by incorporating an exhaust air bore in the blowing mandrel. Compressed air flows through this and over a throttle (to maintain the required pressure); hence a circulating exchange of air takes place within the blow section, which also has some heat-transport effects (purging air technique).

Cooling times can be reduced further by using highly cooled nitrogen (N_2 process) as a blowing medium or adding liquid carbon dioxide (CO_2 process) to the compressed air.

Because compressed air seldom has a pressure greater than 0.4 – 0.6 MPa, no great demands are placed on the mechanical design of the mold. Due to its good thermal conductivity and workability, aluminum is widely used.

Like other processing techniques, the extruder must provide a thermally homogeneous melt stream of predetermined flow rate and pressure. For blow molding, single-screw extruders with a screw length of 20 – 25 D are generally used. A grooved and cooled feed zone is also used for some plastics (see Section 4.3).

In general, the screws are equipped with mixing zones to attain adequate melt homogeneity.

The so-called crosshead or parison die is flanged onto the extruder. This has the task of changing the melt flow direction from horizontal to vertical and forming a tube.

Near the outlet region, the flow channel is generally conical, and either the inner or the outer cone is movable, allowing continuous adjustment of parison wall thickness during extrusion. In this way, uneven blow ratios, which can occur with articles of complex geometry, can be accommodated and products with almost constant wall thicknesses can be manufactured (Fig. 31). This technique is referred to as wall thickness programming.

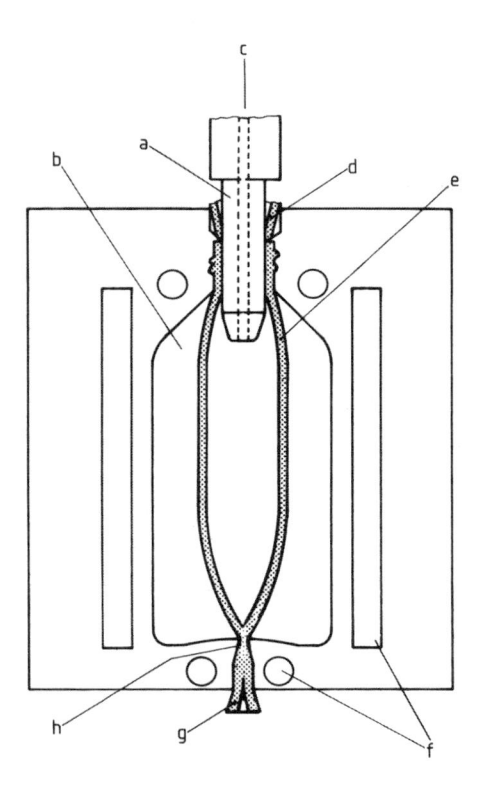

Figure 30. Blow mold
a) Blow mandrel; b) Mold cavity; c) Blowing air inlet; d) Top flash; e) Parison; f) Cooling channels; g) Bottom flash; h) Pinch-off blade

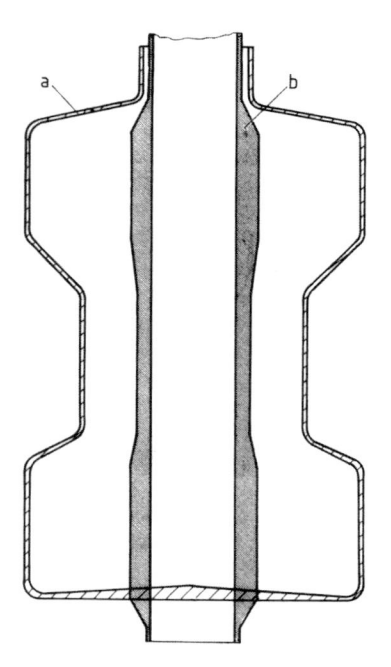

Figure 31. Wall-thickness profiled parison
a) Blown product; b) Parison

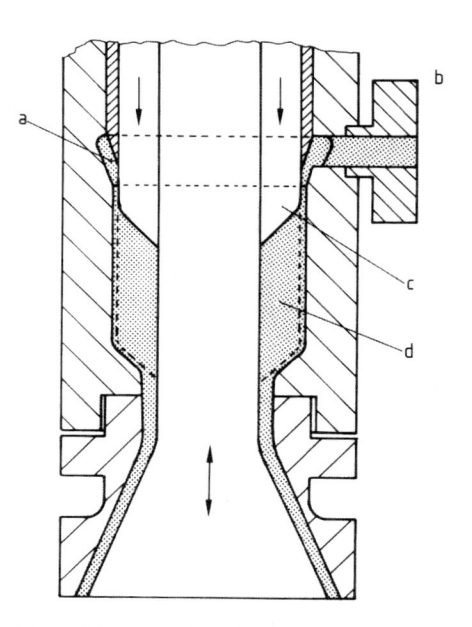

Figure 32. Accumulator head
a) Ring channel; b) Extruder; c) Annular piston; d) Accumulator chamber

A distinction is made between blow heads that function in a continuous mode and so-called accumulator heads. In the former, a continuously extruded tube is cut and passed to the blow mold in a cyclic operation. With large articles (i.e., molding volume greater than ca. 30 L), the parison itself is so heavy that it may deform under its own weight during extrusion. In such cases, accumulator heads are employed in which the melt is collected in a storage chamber and rapidly ejected by a piston to form the parison (Fig. 32). Modern storage heads are equipped with ring-shaped storage chambers, which are continuously fed with melt by one or more extruders according to the FIFO (first-in, first-out) principle.

Use of parison wall thickness regulators is also possible with storage heads.

Stretch blow molding is a special form of blow molding. The process exploits the fact that high degrees of orientation can be introduced in plastics by drawing at temperatures near the glass transition temperature or crystalline melting point. Mechanical properties are considerably improved in this way. In this process, the preform is drawn not only in the circumferential direction (as with extrusion blow molding) but also in the machine direction. The longitudinal drawing is brought about mechanically by using a ram. Because of the relatively low temperature at which this process is carried out, high deformation forces are required (air pressure up to 2 MPa).

Figure 33 illustrates the principal steps involved in stretch blow molding, beginning with an injection molded preform.

Preferred materials for use with this process are PVC, polypropylene (PP), and poly(ethylene terephthalate) (PETP).

The preforms for stretch blown moldings are prepared mostly by injection molding and are subsequently either oriented immediately in an injection stretch blow molding machine or first cooled and then fed to a separate stretch blow molding machine.

Preforms made of PVC for example, which are not suited for manufacture by injection molding, are prepared in an extrusion blow molding process and then processed on extrusion blow molding equipment.

Injection-molded preforms have the advantage of exact wall thickness distributions and a

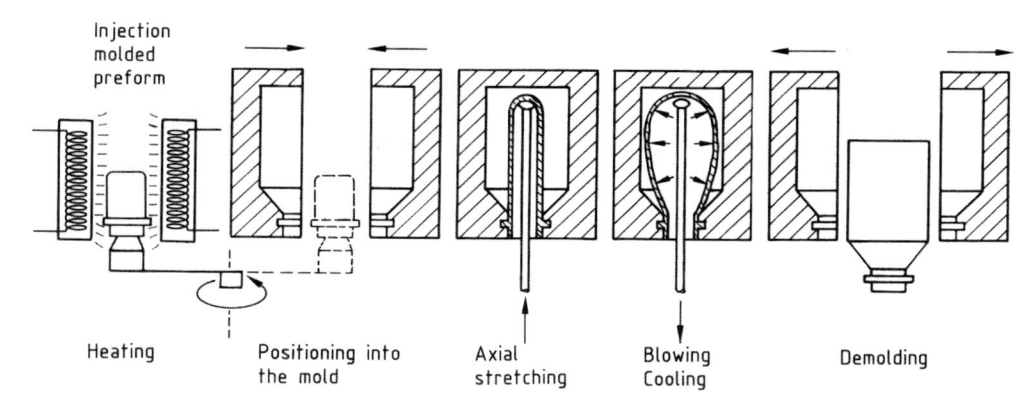

Figure 33. Stretch blow molding process

precisely calibrated neck. Additionally, they are free of weld seams.

4.5. Injection Molding

Injection molding is the most important process for producing moldings from thermoplastics, elastomers, and thermosets. The significance of injection molding is due to its ability to manufacture complex molding geometries in a single stage with high levels of reproducibility. A finishing operation (removing the runner, deflashing) is seldom necessary. Because of the high degree of automation possible, this molding process is also suitable for mass production operations.

Apart from its use for commodity articles, this technique is increasingly being used to produce technical moldings (e.g., compact discs, automobile bumpers).

The size of the molding can vary from a few tenths of a gram to several kilograms.

Injection molding machines consist of several individual units (Fig. 34). The most important components are the plastication unit for melting the plastic pellets, the clamping unit for holding and closing the injection mold, the control unit for coordinating the process, and a temperature control unit for cooling or heating the mold.

Injection molding is a discontinuous process in which plastic pellets, granules or powder is melted and injected under pressure into the cavity of a mold, where it is solidified by cooling or thermally cross-linked.

Process Stages. Production of an injection molding involves the following steps, which make up the injection molding cycle (Fig. 35):

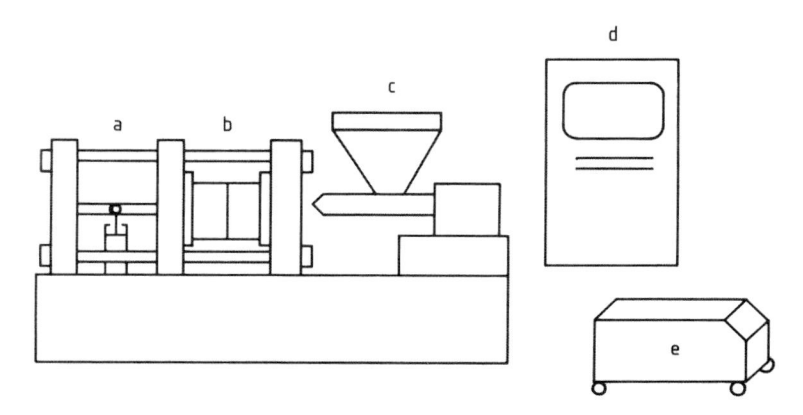

Figure 34. Schematic of an injection molding machine
a) Clamping unit; b) Mold; c) Plasticating unit; d) Control unit; e) Temperature control unit

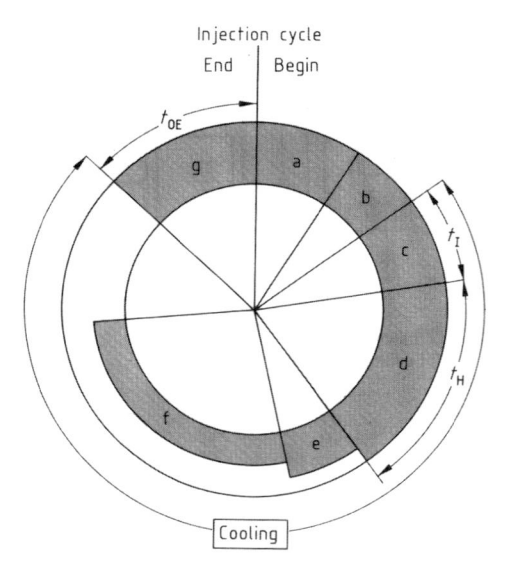

Figure 35. Injection molding cycle
a) Closing the mold; b) Moving the plasticating unit forwards; c) Injection; d) Holding pressure phase; e) Moving the plasticating unit backward; f) Metering; g) Opening the mold and ejection of molding

1. Closing the mold
2. Injection
3. Holding pressure stage
4. Cooling stage
5. Metering
6. Opening the mold
7. Ejection/removal of the molding

This cycle is repeated for each molding.

The cycle begins by closing the mold after ejection of the molding from the previous cycle (Fig. 36). The hot melt prepared in the front part of the injection cylinder is injected into the mold cavity through the runner system. Because of the increasing length of the flow path and the cooling of the melt in the mold, the pressure required to move the cylinder and polymer melt increases. Toward the end of the filling stage, the maximum injection pressure must be exerted. Depending on the material, molding, and process control, this can vary from several bar to 2000 bar. Injection times vary from several hundredths of a second to minutes.

The holding pressure phase begins at the end of the injection phase. Its objective is to counteract volume losses caused by cooling of the melt in

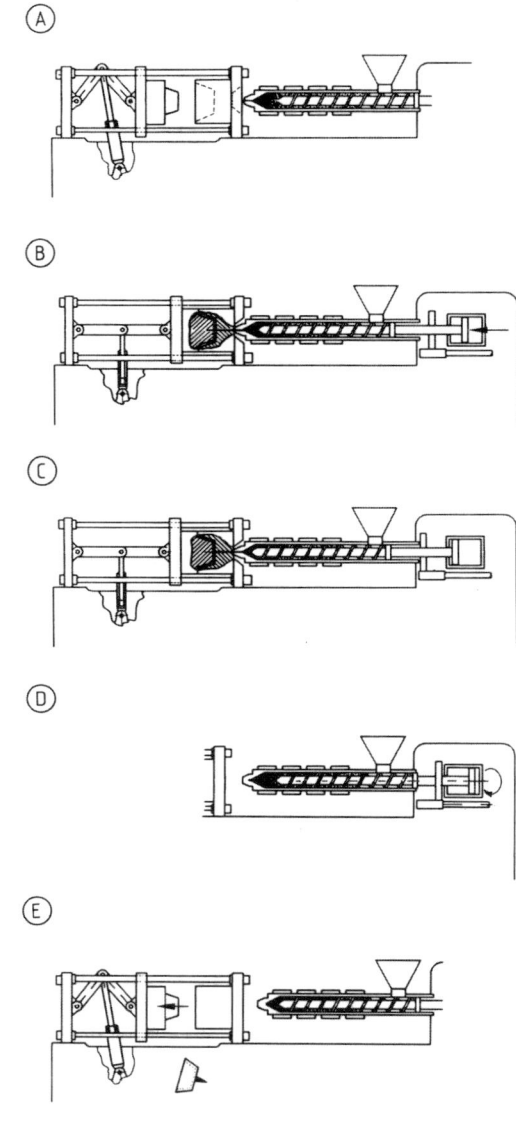

Figure 36. Machine status in various stages of injection molding cycle
A) Closing the mold; B) Injection; C) Holding pressure and cooling stage; D) Metering; E) Mold opening and ejection of molding

the mold and hence avoid sink marks and voids in the molding. At the beginning of the injection process, the melt, which is at 200–300 °C, begins to cool in the mold. After the holding pressure stage, the molding must be cooled until its dimensional stability is sufficient for ejection. The cooling time thus extends from the beginning of injection until the mold is opened.

At the end of the holding pressure phase, the plastication unit begins to prepare the plastic for the next cycle by rotating the screw. The screw moves helically backward in the plastication cylinder against a back pressure and thus transports melt to the area in front of the screw.

After the cooling stage, the mold is opened and the molding is removed mechanically, pneumatically, or by handling equipment.

The total cycle time depends on the maximum wall thickness and the geometrical complexity of the molding. With simple, thin-walled moldings, cycle times of ca. 1 s are possible, whereas with bulky, thick-walled moldings, cycle times of several minutes may be necessary. The process itself is coordinated by a control unit and is fully automated.

Machine Components. Plastication Unit. The plastication unit of an injection molding machine plastifies, meters, stores, and injects the plastic.

To meet these requirements, general-purpose screws are used (Fig. 37).

With these so-called reciprocating screws, the polymer is plasticated by the rotary motion of the screw as in an extruder (see Section 4.3) (Fig. 38). Simultaneously with the turning movement, the screw moves backward from the forward end position against a back pressure. This translatory motion results from transport of the melt into the area in front of the screw.

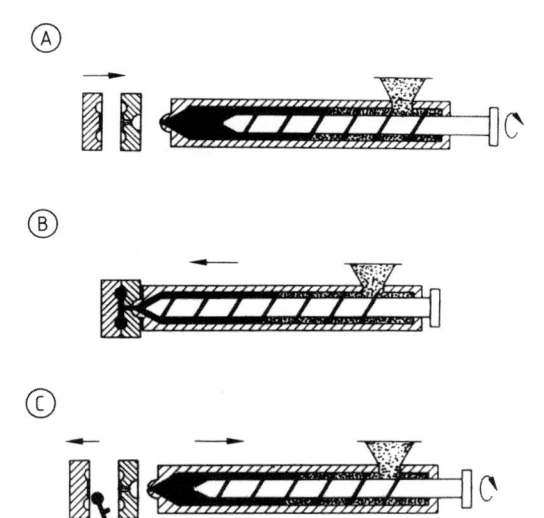

Figure 38. Reciprocating screw injection molding machine
A) Melt in front of the screw; B) Injection by screw stroke; C) Demolding and plasticating by screw rotation

At the end of the metering step, material is injected into the mold by a purely axial motion of the screw. Today, both the rotational and the translational movements of the screw are generally carried out hydraulically. However, using electric drives for the rotational motion is becoming more common, because they consume less energy than hydraulic systems.

The injection pressure for axial displacement of the screw is provided by a hydraulic cylinder.

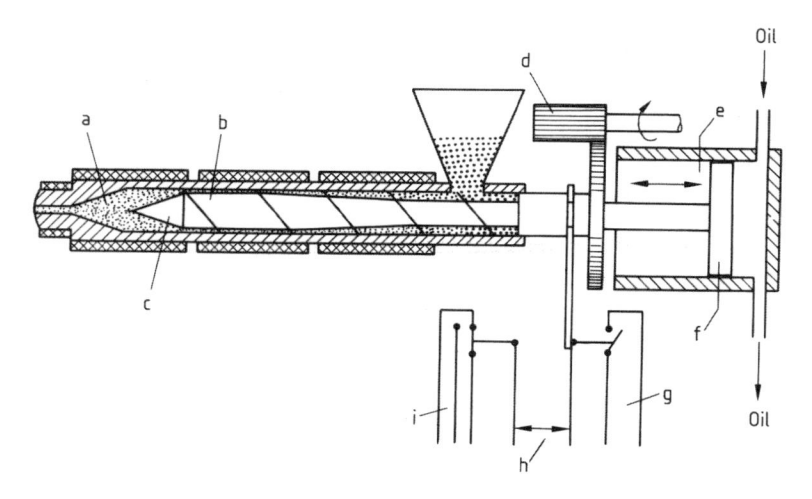

Figure 37. Conventional plasticating unit
a) Shot chamber; b) Screw; c) Screw tip; d) Screw drive; e) Hydraulic cylinder; f) Hydraulic piston; g) Back limit switch (stops rotation of screw); h) Injection stroke; i) Front limit switch (switching from injection pressure to holding pressure)

The general-purpose screw has three zones: (1) feed zone, (2) compression zone, and (3) metering zone.

The feed zone draws the polymer granules from a hopper into the screw. To achieve high mass throughputs, despite the low bulk density, the channel depth of the screw is relatively high.

In the compression zone, material transported from the feed zone is compressed.

In the output or metering zone, the melt is homogenized and heated to processing temperature.

Screw diameters range from 14 to 200 mm, and screw lengths, between 18 D and 24 D. To adapt the injection volume to the volume of the molding, usually the stroke of the screw (i.e., the distance between the front position after injection and the back position after plastication) is varied. The screw stroke cannot be increased to an unlimited extent by varying the back position, because when the screw is moved back the remaining screw length decreases. Therefore the effective length of the screw for melting and homogenizing the plastic is decreased, which results in a lower temperature for the plasticated material.

It is also possible to exchange the plastication unit and, therefore, the screw diameter and maximum stroke on a given machine.

During plastication, the machine nozzle that joins with the mold is sealed either by the runner of the molding or by a shutoff nozzle (Fig. 39) to prevent leakage of the melt caused by the pressure under which the plasticated material is held.

To prevent back flow of the melt during injection, a nonreturn valve is sometimes mounted at the tip of the screw (Fig. 40). The locking ring is pushed forward during plastication, which allows the melt to advance; during injection, the ring is pushed back, sealing the area in front of the screw.

Besides the general-purpose screw, screw geometries also exist that are specially adapted to suit the polymer to be processed. These include thermoset screws, rubber screws, and special screws. With degassing screws (Fig. 41), any moisture or residual monomer present in the melt is removed through an opening in the cylinder wall.

The most significant advantages of the reciprocating screw system include the mild plastication that allows the processing of thermally

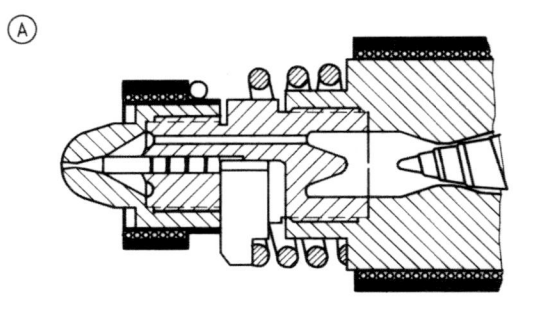

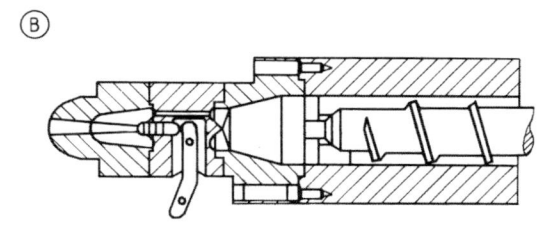

Figure 39. Shutoff nozzles
A) Needle valve nozzle (spring loaded); B) Needle valve nozzle (separately controlled)

sensitive products, the lower cycle time, and compact construction compared to ram plasticators. Apart from the horizontal construction illustrated in Figure 34, the plastication unit can also be mounted vertically.

Although such reciprocating screws are most frequently employed, specialty constructions also exist, for example ram extruders with screw preplastication units (Fig. 42). The polymer is plasticated in an extruder, and the resulting melt flows through a nonreturn valve into the reservoir chamber in front of the injection ram. Such systems are used only in special cases.

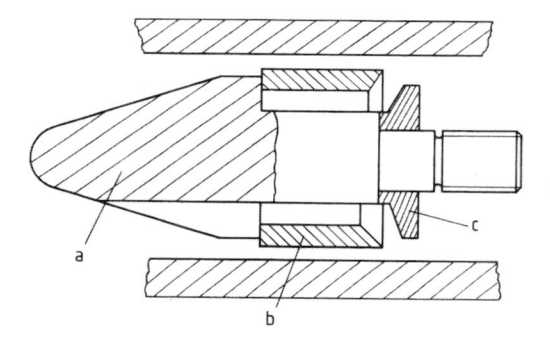

Figure 40. Nonreturn valve with a mobile locking ring
a) Screw tip; b) Sliding ring; c) Thrust ring

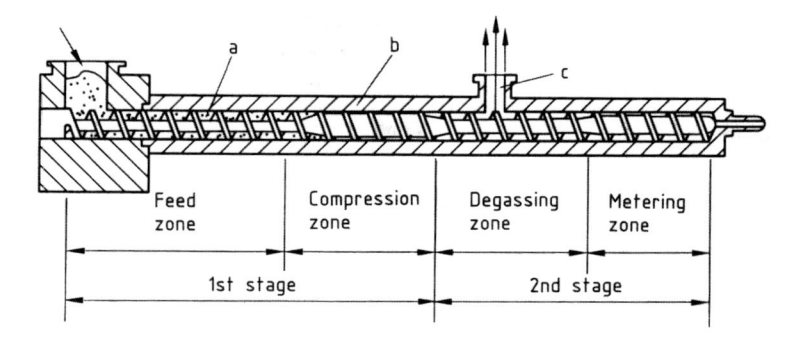

Figure 41. The degassing – plasticating unit of an injection molding machine
a) Melting; b) Cylinder; c) Degassing outlet

Clamping Unit. The clamping unit contains the halves of the mold, that are mounted on platens, and opens and closes the mold. Additionally, in most cases, mechanisms for ejecting the molding (e.g., cylinders for activating ejector pins) are also housed in the clamping unit.

During opening or closing of the mold, only one platen is moved. The platen on the side nearest the plastication unit is fixed to the machine frame and, hence, is immobile. The movement of the closing-side platen is guided by four tie rods (Fig. 43).

After closing, the full clamping force is exerted to prevent the mold from opening as a result of internal pressure during the injection and holding pressure phases. Otherwise, melt leaks in the parting surface (two mold halves)

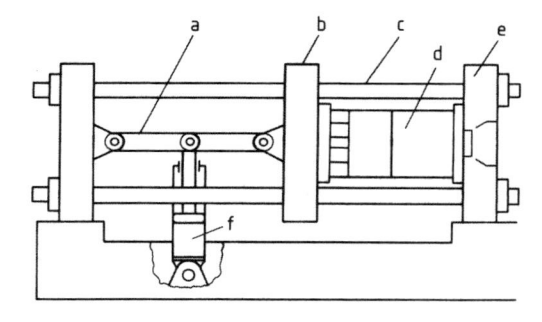

Figure 43. Toggle clamping unit
a) Toggle system; b) Moving platen; c) Tie rod; d) Mold; e) Fixed platen; f) Driving cylinder

and which leads to formation of flash and possibly damage to the mold.

Clamping forces of injection molding equipment range from 0.1 to 100 MN. Two main types of clamping mechanism are used:

1. Toggle clamping units (Fig. 43)
2. Fully hydraulic clamping units (Fig. 44)

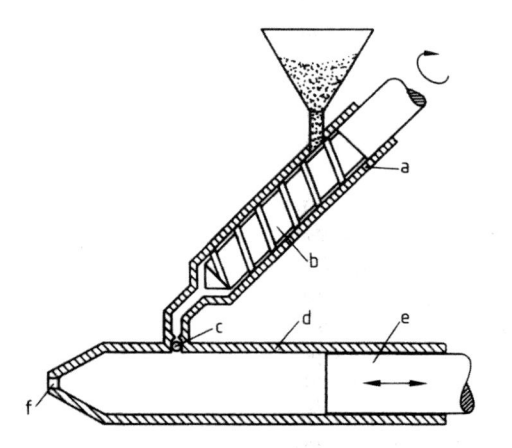

Figure 42. Injection molding with screw preplastication
a) Extruder barrel; b) Screw; c) Nonreturn valve; d) Injection cyclinder; e) Piston; f) Nozzle

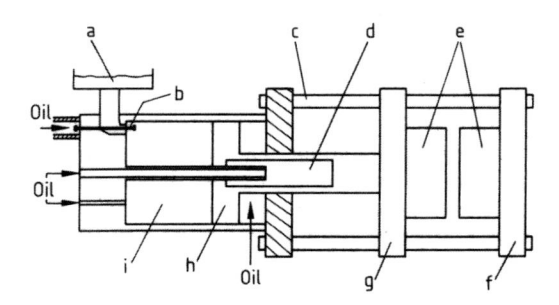

Figure 44. Hydraulic clamping unit
a) Oil tank; b) Suction valve; c) Tie rods; d) High-speed injection cylinder; e) Mold; f) Fixed platen; g) Moving platen; h) Main piston; i) Main cylinder

In the toggle clamping unit, the closing force is generated by extending a toggle driven by a hydraulic cylinder. The advantage of the toggle clamping unit is the favorable force – speed characteristic. High manufacturing costs are the main drawback.

With a fully hydraulic clamping unit, the clamping force is applied by a hydraulic cylinder alone. The advantage of this system is cheaper design; its disadvantage is the generally slower opening and closing motion.

Injection Molds. The central component in the injection molding process is the mold itself (Fig. 45). The injection mold does not form part of the machine, but must be constructed or modified for each molding. The number of design variations corresponds to the multitude of different injection-molded components. All molds, however, fulfill the following basic tasks. From a technological point of view, the mold must

1. Distribute the melt
2. Shape the melt to the desired molding geometry
3. Cool the melt (thermoplastics) or heat the contents of the mold (elastomers, thermosets)
4. Eject the molding

Additionally, the mold must meet design objectives, such as

1. Accomodate all forces
2. Transfer all motion
3. Guide mold components

To allow ejection of the molding, the mold consists of at least two parts. The front half, which contains the gate system and possibly also the flow channels, is mounted on the fixed, machine platen; the other half including the ejector mechanism is mounted on the moving platen of the clamping unit. Centering devices are used for precise adjustment.

In some cases, the molds have exchangeable inserts, which allow a basic unit to manufacture various moldings.

The mold is joined to the nozzle of the injection unit by the runner channel. The radius of the sprue bushing must always be greater than the radius of the nozzle, and similarly, the diameter of the sprue bushing bore must be larger than the nozzle bore. With voluminous moldings, a conical sprue gate is often used (Fig. 46). When the mold is opened, this is broken off at the

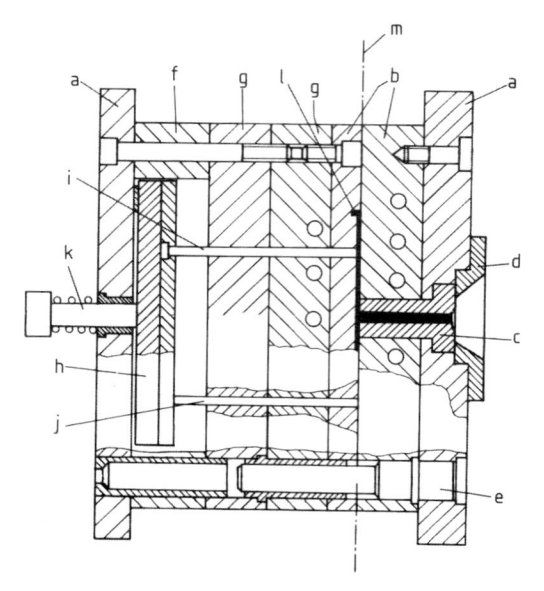

Figure 45. Injection mold
a) Base plate; b) Mold plate; c) Sprue bush; d) Locating ring; e) Guide pin; f) Supporting strip; g) Support plate; h) Ejector plate; i) Ejector pin; j) Ejector plate return pin; k) Ejector bolt; l) Molding; m) Parting line

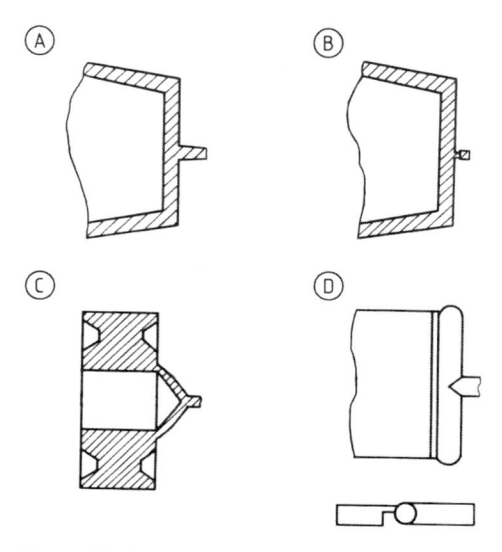

Figure 46. Gate types
A) Sprue gate; B) Pin gate; C) Disk gate; D) Film gate

nozzle. It remains as a stalk on the molding and must subsequently be removed. With a pin gate, the melt is injected through short, narrow holes into the mold cavity. During demolding, the gate breaks at the junction with the molding; no finishing process is required. Disk and film gates are used, for example, in the production of gear wheels or cartridges to avoid weld lines during filling. The film gate is advantageous with flat moldings because,in such cases,the linear filling avoids warpage and distortion of the molding.

Multicavity molds, which are used for simultaneous production of several moldings, are equipped with mold cavities connected to the gate channel by the manifold system. As with multiple gates on an individual mold, the gate and the runner system (hot runner mold) can be heated to reduce material requirements. Material in the hot runner system remains molten and is thus available for use in the next filling stage.

The runner system should be designed so that the melt fills each cavity uniformly and simultaneously at the same temperature and pressure. The position of the gate is chosen so as to avoid as much as possible the formation of weld lines that result when melt substreams combine. The position of the gate also determines the flow direction and hence the orientation on which mechanical strength, shrinkage, and distortion of the molding depend.

Special Processes. In addition to the classical injection molding process described above, several variations exist.

Injection – Compression Molding. With thin-walled, large-area moldings (i.e., when long flow lengths are involved), mold resistance is often so large that extremely high injection pressures must be exerted to fill the mold. With injection – compression molding, the mold is filled at only partial clamping pressure, causing the mold halves to open slightly (Fig. 47). To complete filling, the full clamping pressure is applied, whereby the clamping unit essentially operates as a press. The holding pressure exerted by the injection unit prevents the flow of the melt out of the mold cavity. Due to the homogeneous pressure over the complete injection area, moldings without warpage are obtained.

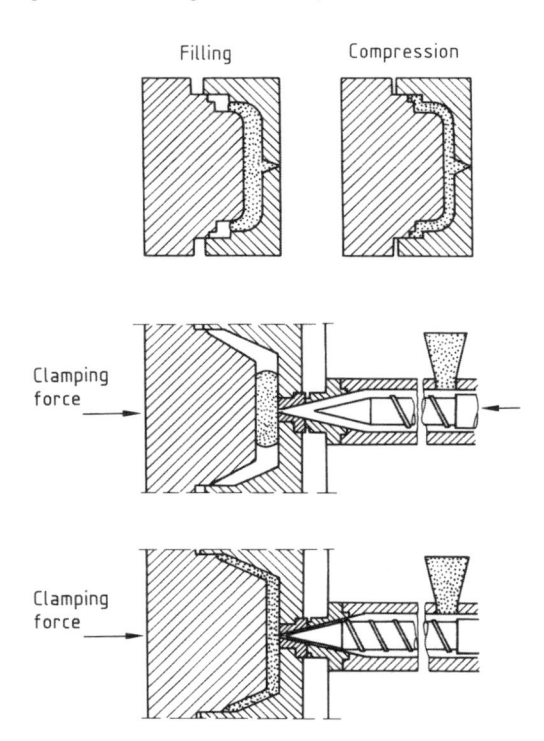

Figure 47. Injection – compression molding process

Intrusion Injection Molding. With thick-walled moldings, the injection pressure required for mold filling is usually so low that the pumping pressure from the rotating screw is sufficient. In the flow molding process (intrusion), the screw near the front end position plasticates the material and pumps the melt relatively slowly into the mold cavity under extrusion pressure alone. The holding pressure that counters shrinkage of the cooling molded component is applied by means of a short translational movement of the screw after the filling stage is completed. This process is used in the manufacture of thick-walled containers made of polyethylene and of PVC fittings whose volume exceeds the injection volume of the equipment by a factor of three to five.

Two-Component Injection Molding. The principal steps involved in two-component injection molding are shown in Figure 48.

The mold is first partially filled with material A. After the changeover point has been reached, complete filling takes place by injection of component B. During this stage, material A is

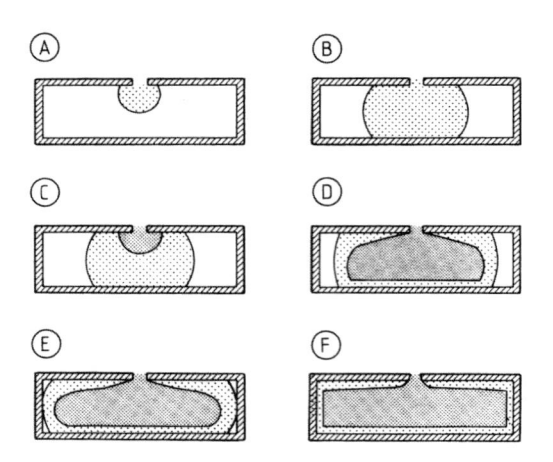

Figure 48. Two-component injection molding

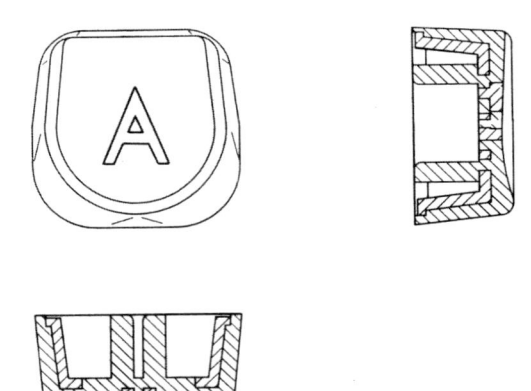

Figure 49. Two-color injection molding of a typewriter key

displaced by material B and forced against the walls of the mold. A multilayer molding is obtained in which the surface component A completely encloses the core component B. The external appearance of the molding and other surface properties are hence determined exclusively by material A.

The possible combinations of surface and core material are essentially unlimited, if adhesion between the two components is adequate and both materials exhibit similar shrinkage behavior. Two-component injection molding also offers economical and environmental advantages. The use of polymer waste and regranulate as the core material allows the molding to be manufactured at lower cost.

A second plastication cylinder is required for two-component injection molding.

Multicolor Injection Molding. In contrast to the two-component process, a sandwich-structured molding does not result with multicolor injection molding. In this case, all of the components form an integral part of the molding surface. This is illustrated in Figure 49 by using a typewriter key as an example. Another application is in the covers of automobile taillights; these are also injection molded with several colors.

Such combinations of different colors are achieved by rotating one-half of the mold (Fig. 50). First, one component is injected into a defined part of the mold cavity, while those regions of the cavity intended for components of another color are sealed. By rotating one-half of the mold, these regions are made available for filling with additional components.

Thermoplastic Foam Casting. Structured foam moldings are manufactured from thermoplastics (usually polystyrene, polystyrene copolymers,

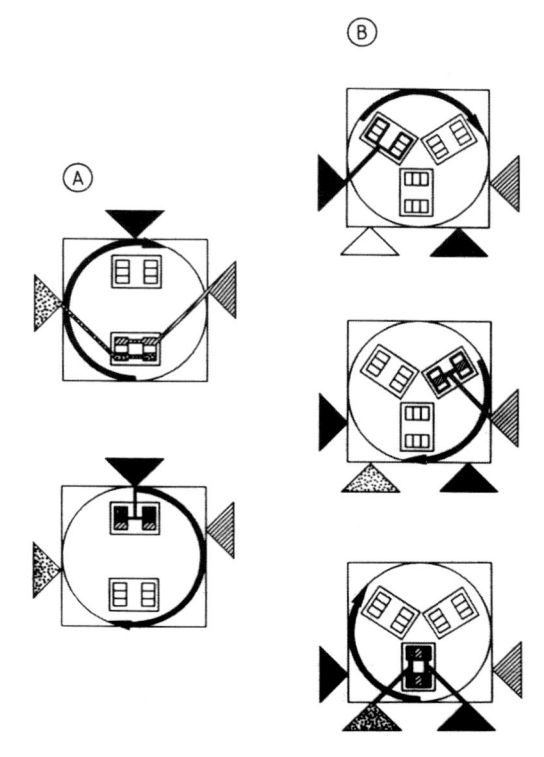

Figure 50. Multicolor injection molding
A) Three-color injection molding; B) Four-color injection molding

and polyolefins) containing a blowing agent. Moldings with a compact surface and a foamed core are obtained, which are employed mainly in the audio and furniture industries, as well as technical applications. Ram injection molding machines with screw preplastication units are generally used. The process requires lower injection pressures and closing forces than conventional injection molding techniques. The mold is only partially filled (50 – 80 %). Complete filling of the mold results subsequently from the foaming process. A holding pressure phase is hence unnecessary.

5. Forming Processes

The term forming refers to processes in which semifinished plastic products are heated and shaped under the influence of external forces. Apart from modifying the shape of an article, an objective of the process is often to increase material strength by introducing molecular orientation.

5.1. Drawing

To obtain an anisotropic material behavior in thermoplastics (usually an increase in mechanical stiffness and strength), the molecular chains must be oriented and the orientation frozen in. One way of achieving this is by stretching or drawing. The original semifinished product is heated to the drawing temperature and then mono- or biaxially drawn, followed by fixing the oriented state by cooling while the final shape is maintained. With amorphous thermoplastics, drawing is normally done at or around the glass transition temperature, and with semicrystalline materials, slightly below the crystalline melting temperature. When a drawn product is reheated, the forming process is reversed due to elastic deformations frozen into the article. If the draw ratio is not too high, the article may even revert to its original form. This ability to return to the original form can be desirable in some cases (e.g., shrink wrapping films). Generally, however, the product must be tempered (heat set) after drawing to relax any frozen-in elastic stresses.

The drawing of monofilaments and films is of major industrial importance.

Monofilaments are drawn in so-called godets (driven draw rolls). The fibers are fed through a preheating zone (oven or tempering bath) and taken off at high speed. The draw ratio is determined by the adjustable speed difference between individual godets. Typical draw ratios range from 1 : 3 to 1 : 12. Finally, drawn fibers pass through a tempering zone in which elastic deformations can relax while the orientation that has been introduced is retained. To compensate for shrinkage, the draw rolls run at slightly reduced speeds in this area. The whole process takes place simultaneously on many fibers, which are subsequently wound individually on spools. At this stage, a constant filament tension must be maintained. With multifilament machines, several hundred filaments are drawn simultaneously. Cold drawing is in principle a similar process, in which preheating is omitted. The fibers are heated by the internal shear to which the material is exposed on constriction.

Films are also drawn to increase their mechanical strength. Another reason, however, can be to reduce thickness. The film is heated by heated rolls and then passed over two pairs of rolls, with the second pair rotating at a higher speed than the first. This leads to monoaxial orientation in the machine direction. During this step, the film necks. If the film is to be biaxially oriented, it is held at both edges by a series of clips mounted on a chain that draw the web in the transverse direction (Fig. 51). This process can be carried out simultaneously with longitudinal drawing or separately in a second process step. Once the drawing process has been completed, the film is cooled and wound in a roll. As with calenders and fiber spinning equipment, biaxial drawing units

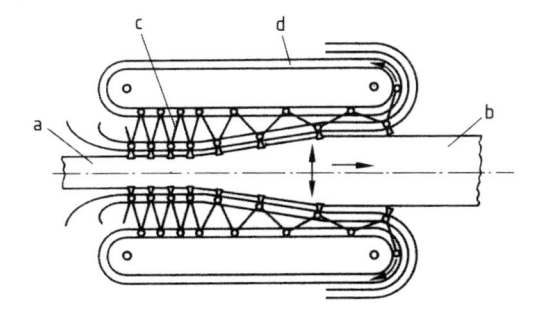

Figure 51. Biaxial simultaneous drawing
a) Initial sheet; b) Drawn sheet; c) Tenter chain; d) Guide frame

with integrated film extrusion are expensive. The most important oriented films include polyethylene, polypropylene, polystyrene, and polyester. Major products include films for magnetic tapes and for capacitors.

5.2. Thermoforming

The manufacture of three-dimensional moldings from flat plastic preforms such as films or sheets, under the influence of heat and pressure or vacuum, is a process referred to as thermoforming. In principle, the process can be carried out by a variety of techniques. Industrially, the most important method is the molding of a heated preform using a vacuum or compressed air, as well as mechanical stretching. Forming is carried out in the rubber-elastic temperature range, although in contrast to drawing, the introduction and freezing-in of orientation must be avoided as much as possible. For this reason, forming temperatures as high as possible are selected, which are essentially limited only by the onset of flow of the material, so that sufficient opportunity to relax is available. Depending on the material and temperature, geometric draw ratios of several hundred percent can be attained in thermoforming.

The range of products that can be manufactured by this process varies from packaging containers (up to 100 000 articles per hour) to bulky moldings such as swimming pools 8 m×4 m×1.5 m (1.5 articles per hour). Furthermore, this is also an important process for the production of automobile components such as dashboards and passenger compartment panels.

The preferred materials for processing include amorphous thermoplastics [PVC, polystyrene (PS), acrylonitrile – butadiene – styrene (ABS), styrene – acrylonitrile (SAN), cellulose acetate butyrate (CAB), PMMA, and PC], as well as semicrystalline materials (PP, PE, PETP) and special laminates that economically combine the favorable properties of several polymers in one film.

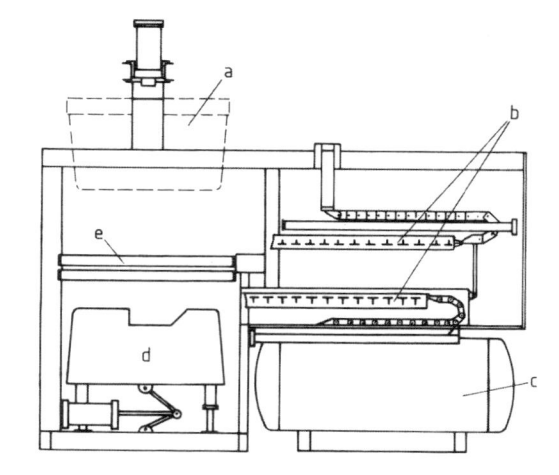

Figure 52. Single-station thermoforming machine
a) Top mold; b) Movable heaters; c) Vacuum tank; d) Bottom mold; e) Molding station

Thermoforming is a purely stretching technique that is carried out as a continuous process. The article to be formed is secured at the edges, and the geometric increase in area is thus accompanied by a reduction in thickness.

Two types of industrial thermoforming machines have become established:

1. Single-station machines
2. Multistation machines

With *single-station machines* (Fig. 52) the heating and forming steps take place in the same station. The heating system is generally conveyed to and from the article; occasionally, however, the secured preform is transported to the heating zone. With multistation machines (Fig. 53), heating and forming take place in separate stations.

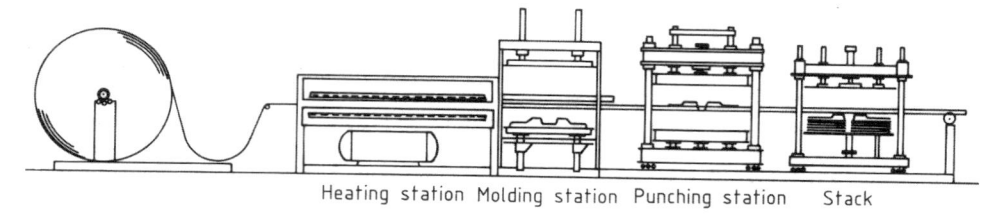

Figure 53. Multistation thermoforming machine

The technique of thermoforming polymers can be broken down into three basic process steps:

1. Heating
2. The transfer and forming process
3. Cooling

Heating, the first stage in all thermoforming techniques, is of decisive importance for the final quality of the molding. The process involves heating the sheet from its original temperature to the forming temperature.

Because films and sheets in thicknesses ranging from ca. 0.1 to 12 mm are thermoformed, a variety of preheating methods have evolved:

1. Convection heating
2. Contact heating
3. Heating with infrared radiation

In all the techniques, heating can be either single or double sided. With the exception of extremely thin films, two-sided heating is generally employed.

Heating with infrared radiation is used most widely today. Its major advantage is that because a portion of the energy penetrates directly into the bulk of material, high heat flow densities can be used to decrease the heating time without causing thermal damage to the polymer surface. Ceramic and quartz elements are the primary sources of radiation. The heating stations in modern thermoforming equipment consist of numerous, individually controlled radiator elements in a modular structure. In all thermoforming processes the heated preform is rapidly drawn biaxially. This generally takes place in half a second and thus can be assumed to be isothermal as long as it is a free draw (i.e., without contact to the mold). On contact with the walls of the mold, however, the material cools immediately. Particularly with semicrystalline thermoplastics, which are formed at or slightly below the crystalline melting point, this leads to a spontaneous increase in the resistance to deformation. To exert the necessary force, mechanical drawing mechanisms, compressed air, and vacuum are utilized.

With thermoforming, a distinction is drawn between male and female forming processes (Fig. 54). The decision to employ a male or a female process depends primarily on which side of the molding is to be replicated exactly. Only

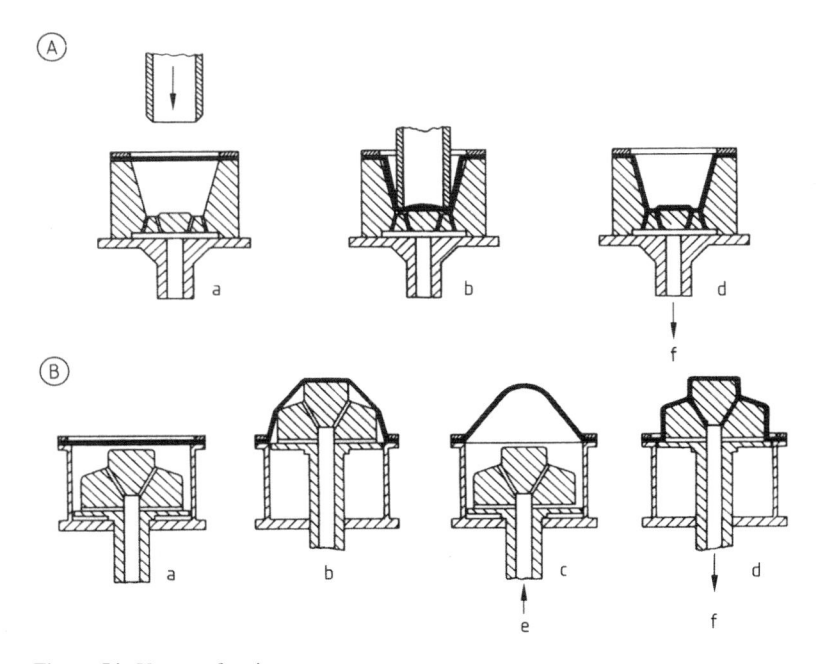

Figure 54. Vacuum forming
A) Female forming; B) Male forming
a) Original state; b) Mechanical prestretching; c) Pneumatic prestretching; d) End state; e) Compressed air; f) Vacuum

the surface in contact with the mold is dimensionally precise, whereas the other surface exhibits inherent inaccuracies, particularly in the unavoidable rounding of corners. The cooling process begins, by definition, once the film comes in contact with the mold and ends at the onset of the unmolding step.

For mass production purposes, the mold is equipped with temperature regulation. Mold temperatures vary from approximately 10 °C (PP packaging materials), to 70 °C [polystyrene (PS) inner panels for refrigerators], up to 100 °C [high-density polyethylene (HDPE) moldings]. Depending on the numbers of articles to be manufactured, the machines are fitted with single- or multicavity molds.

In most cases, one side of the material is cooled by contact with the mold, and the other surface is cooled by convection. Convective cooling occurs either via ambient air (free convection) or by use of a ventilator (forced convection). To intensify the cooling effect, atomized moisture is also added to the ventilated air (spray nozzle cooling).

The main advantages of thermoforming compared to alternative manufacturing techniques are the low mold costs, the short processing times, and when multicavity molds are used, the high throughputs. Additionally, reduced wall thicknesses are possible compared to injection molding. Thermoformed articles generally have about half the weight of equivalent injection-molded products.

In contrast to injection molding processes in which pressure acts only on the surface perpendicular to the flow direction during molding, in thermoforming the total mold surface is used. This is the reason thermoforming pressures are 20 – 30 times lower, and also the reason large moldings can be manufactured with lighter equipment, lower forming pressure, and lower energy consumption.

References

General Reference

1 O. Lauer: *Aufbereiten von Kunststoffen*, Hanser Verlag, München 1971. *Aufbereiten von PVC*, VDI-Gesellschaft Kunststofftechnik, Düsseldorf 1976. *PVC-Heißmischen (Dryblend)*, VDI-Gesellschaft Kunststofftechnik, Düsseldorf 1968. *Speichern, Fördern und Dosieren von Kunststoffen*, VDI-Gesellschaft Kunststofftechnik, Düsseldorf 1971. *Granulieren von thermoplastischen Kunststoffen*, VDI-Gesellschaft Kunststofftechnik, Düsseldorf 1974. W. Dalhoff: *Systematische Extruder-Konstruktion*, Krauskopf Verlag, Mainz 1974. H. Dominghaus: *Fortschrittliche Extrudertechnik*, VDI-Verlag, Düsseldorf 1970. *Der Extruder im Extrusionsprozeß*, VDI-Gesellschaft Kunststofftechnik, Düsseldorf 1989. H. Herrmann: *Schneckenmaschinen in der Verfahrenstechnik*, Springer Verlag, Berlin 1972. M. Jacobi: *Grundlagen der Extrudertechnik*, Hanser Verlag, München 1960. W. Mink: *Grundzüge der Extrudertechnik*, Zechner & Hüthig Verlag, Speyer 1973. H. Potente: *Modellgesetze für Ein- und Zweischneckenmaschinen*, Hanser Verlag, München 1981. C. Rauwendaal: *Polymer Extrusion*, Hanser Verlag, München 1986. G. Schenkel: *Kunststoff-Extrudertechnik*, Hanser Verlag, München 1963. E. G. Fischer: *Extrusion of Plastics*, Butterworths, London 1976. *Extrudieren von Schlauchfolien*, VDI-Gesellschaft Kunststofftechnik, Düsseldorf 1973. *Extrudieren von Rohren und Profilen*, VDI-Gesellschaft Kunststofftechnik, Düsseldorf 1974. *Extrudierte Feinfolien und Verbundfolien*, VDI-Gesellschaft Kunststofftechnik, Düsseldorf 1976. G. Menges, H. Recker: *Automatisierung in der Kunststoffverarbeitung*, Hanser Verlag, München 1986. *Rechnergesteuerte Extrusion*, VDI-Gesellschaft Kunststofftechnik, Düsseldorf 1976. Z. Tadmor: *Engineering Principles of Plasticating Extrusion I*, Klein Reinhold Book Corporation, New York 1970. W. Michaeli: *Extrusionswerkzeuge für Kunststoffe und Kautschuke*, Hanser Verlag, München 1990. F. Hensen, W. Knappe, H. Potente: *Handbuch der Kunststoffextrusionstechnik*, vols. I and II, Hanser Verlag, München 1989. W. Mink: *Grundzüge der Hohlkörperblastechnik*, Zechner & Hüthig Verlag, Speyer 1969. D. v. Rosato: *Blow Molding Handbook Dominick*, Hanser Verlag, München 1989. R. Holzmann: *Gestalten von Blasformteilen*, VDI-Verlag, Düsseldorf 1971. O. Plajer: *Werkzeuge für das Blasformen*, Zechner & Hüthig Verlag, Speyer 1969. *Spritzblasen*, VDI-Gesellschaft Kunststofftechnik, Düsseldorf 1976. J. L. Throne: *Thermoforming*, Hanser Verlag, München 1987. A. Thiel: *Grundzüge der Vakuumformung*, Zechner & Hüthig Verlag, Speyer 1969. K. Stoeckhert: *Kunststoff-Lexikon*, Hanser Verlag, München 1975. F. Johannaber, K. Stoeckhert: *Kunststoffmaschinenführer*, Hanser Verlag, München 1984. O. Scharz, F.-W. Ebeling, G. Lüpke, W. Schelter: *Kunststoffverarbeitung*, Vogel, Würzburg 1988. D. H. Morton-Jones: *Polymer Processing*, Chapman and Hall, New York 1989. G. Menges: *Einführung in die Kunststoffverarbeitung*, Hanser Verlag, München – Wien 1979. F. Johannaber: Spritzgießmaschinen, in: *Kunststoffmaschinenführer*, Hanser Verlag, München – Wien 1979. W. Elbe: "Untersuchungen zum Plastifizierverhalten von Schneckenspritzgießmaschinen," Dissertation, RWTH Aachen 1973. R. v. Hooven, A. J. Kaminski: "Entgasungsspritzgießen, ein Laborkonzept wird in die Praxis umgesetzt," *Plastverarbeiter* 31 (1990) no. 8, 441 – 446. Th. Copetti, R. Krebser: "Vollhydraulische und Kniehebelschließsysteme," *Kunststoffe* 70 (1980) no. 2, 821 – 825. G. Menges, P. Mohren: *Anleitung zum Bau von Spritzgießwerkzeugen*, Hanser Verlag, München – Wien 1983. S. Schäper: "Nukleierung thermodynamisch getriebener Polyurethan-Reaktionsschäume," Dissertation, RWTH Aachen 1977.

General Reference

2 W. Becker, D. Braun, B. Carlowitz: *Kunststoff-Handbuch*, vol. 1 Die Kunststoffe, Hanser Verlag, München 1990. W. Becker, D. Braun, W. Woebcken: *Kunststoff-Handbuch*, vol. 10 Duroplaste, Hanser Verlag, München 1988. H. Saechtling: *Kunststoff-Taschenbuch*, 24th. ed., Hanser Verlag, München 1989. K. Stoeckhert: *Kunststoff-Lexikon*, 6th. ed., Hanser Verlag, München 1975. DIN-Taschenbuch 21,Kunststoffnormen,8th. ed., Beuth-Verlag, Berlin 1979. W. Schönthaler: *Verarbeiten härtbarer Kunststoffe*, VDI-Verlag, Düsseldorf 1973. W. Bauer, W. Woebcken: *Verarbeitung duroplastischer Formmassen*, Hanser Verlag, München 1973. W. Bucksch, H. Briefs: *Preßwerkzeuge in der Kunststofftechnik*, 2nd. ed., Springer Verlag, Berlin 1962. K. Stoeckhert: *Formenbau für die Kunststoff-Verarbeitung*, Hanser Verlag, München 1969. Berufsgenossenschaft der chem. Industrie, *Merkblatt für das Verarbeiten von Polyester- und Epoxidharzen*, Verlag Chemie, Weinheim, Germany 1966. VDI-Richtlinie 2007, Epoxidharze im Fertigungsmittelbau, VDI-Verlag, Düsseldorf 1966. VDI-Richtlinie 2011, Faserstärkte Reaktionsharzformstoffe, VDI-Verlag, Düsseldorf 1973. H. Jahn: *Epoxidharze*, Verlag für die Grundstoffindustrie, Leipzig 1969. F. Breckner: *Glasfaserverstärkte Kunststoffe; Anwendungen, Eigenschaften*, 2nd ed., Verlag Fraunhofer-Gesellschaft, 1989. H. Domininghaus: *Die Kunststoffe und ihre Eigenschaften*, VDI-Verlag, Düsseldorf 1988. O. Schwarz: *Kunststoffkunde*, Vogel, Würzburg 1988. W. Michaeli, M. Wegener: *Einführung in die Technologie der Faserverbundwerkstoffe*, Hanser Verlag, München 1989. R. W. Meyer: *Handbook of Polyester Molding Compounds & Molding Technology*, Chapman and Hall, New York 1987. S. M. Lee: *International Encyclopedia of Composites*, VCH Publishers, New York 1990.

General Reference

3 W. Knig, L. Neder: "Schneiden mit ultraschallerregtem Messer," *Ind. Anz.* 107 (1985) 49. W. König, L. Neder: "Bessere Schnittflächen mit Ultraschall," *Plastverarbeiter* 37 (1986) no. 11. "Neue Werkstoffe und hohe Präzision — eine Herausforderung für die Fertigstellungstechnik" in AWK Aachener Werkzeugmaschinen-Kolloquium, (ed.): Produktionstechnik auf dem Weg zu integrierten Systemen, VDI-Verlag, Düsseldorf 1987. W. König, P. Graß, Ch. Wulf: "Machining of Fibre Reinforced Plastics," *CIRP Ann.* 34 (1985)no. 2, 537 – 548. F.-J. Trasser, M. Wehner,"Schnittflächen beim Laserstrahlschneiden von faserverstäkten Kunststoffen CO2-, Excimer-Laser — ein Verfahrensvergleich," *Ind. Anz.* 109 (1987) no. 43/44, 57 – 58. "Bearbeitungstechnik für Verbundwerkstoffe — eine vielschichtige Aufgabenstellung," *VDI-Fachtagung METAV*,June, 8 – 9, 1988. H. K. Tönshoff, V. Hohensee,"Bearbeitung faserverstärkter Kunststoffe," *ZWF, Z. Wirtsch. Fertigung* 81 (1986) no. 2, 106 – 111.

Further Reading

M. Chanda, S. K. Roy: *Plastics fabrication and recycling*, CRC Press, Boca Raton, Fla. 2009.

E. Cybulski: *Plastic conversion processes*, CRC Press/Taylor & Francis, Boca Raton, Fla. 2009.

E. Lokensgard: *Industrial plastics*, 5th ed., Delmar Cengage Learning, Clifton Park NY 2009.

P. A. Tres: *Designing plastic parts for assembly*, 6. ed., Hanser, Munich 2006.

M. Xanthos, D. B. Todd: *Plastics Processing*,Kirk Othmer Encyclopedia of Chemical Technology, 5th edition, vol. 19, p. 536–563, John Wiley & Sons, Hoboken, NJ, 2006, online: DOI: 10.1002/0471238961.1612011924011420.a01.pub2.

Plastics, Processing, 2. Processing of Thermosets

GERT BURKHARDT, Institut für Kunststoffverarbeitung, Aachen, Germany

ULRICH HÜSGEN, Institut für Kunststoffverarbeitung, Aachen, Germany

MATTHIAS KALWA, Institut für Kunststoffverarbeitung, Aachen, Germany

GERHARD PÖTSCH, Institut für Kunststoffverarbeitung, Aachen, Germany

CLAUS SCHWENZER, Institut für Kunststoffverarbeitung, Aachen, Germany

1.	Vinyl Ester Resins, and Phenol – Formaldehyde Resins..........	407
1.1.	Raw Materials...............	410
1.2.	Mold Materials..............	412
1.3.	Preparation of Resins.........	412
1.4.	Pressureless Processing without Fiber Reinforcement..........	413
1.5.	Pressureless Processing with Glass-Fiber Reinforcement..........	414
1.6.	Low-Pressure Processes........	416
1.7.	Press Molding...............	417
1.8.	Semicontinuous Processing Techniques..................	420
1.9.	Continuous Processes..........	422
2.	Curable Molding Compounds and Laminates................	423
2.1.	Processing Techniques.........	423
2.1.1.	Processing Curable Molding Compounds.................	423
2.1.2.	Manufacture of Laminated Plates..	428
2.2.	Processing Properties..........	429
2.3.	Processing Conditions.........	429
2.4.	Influence of Processing on Molding Properties....................	432
2.5.	Molds....................	433
2.6.	Machines..................	436
	References..................	437

1. Vinyl Ester Resins, and Phenol – Formaldehyde Resins

In curable polymers, the liquid or meltable resins as received from the supplier are rendered solid, insoluble, and unmeltable during processing by addition of a curing agent. This reaction can occur independently or be initiated by heat or ultraviolet radiation.

This two-phase formation of the resin has the following advantages: a multitude of processing and design possibilities are available when starting materials that are liquid or that can be melted are employed. The resins can be dyed and processed with or without fillers. By embedding reinforcing fibers, composite materials are obtained whose mechanical properties can be modified to suit the demands placed on the completed product simply by selecting the type, amount, and orientation of the fibers introduced, along with the processing technique employed. The properties of the finished article are governed not only by the choice of material, but also by the processing technology and the curing method.

Various processing techniques differ mainly in the complexity of the equipment used to provide the shape and to impregnate the reinforcing fibers. Customized moldings, small series, and bulky molding, as well as paneling and protective layers, can be manufactured manually with simple equipment. For large-scale operations, mechanized processes with a low manual cost contribution but correspondingly complex plants are available.

Table 1 gives an overview of the processing techniques for glass-fiber-reinforced composites. The use of more sophisticated processing machinery leads to improved reproducibility of properties, greater dimensional precision, and savings in terms of finishing operations. When selecting a process, one should consider the fact that the strength of the finished product depends strongly on fiber content. This is influenced, among other things, by the starting material and the processing pressure.

Table 2 lists a selection of mechanical characteristics of glass-fiber-reinforced unsaturated polyester (UP) resins with various types of

Table 1. Applications and characteristics of different processing methods for fiber–plastic composites*

	Manual processing	Fiber spray-up process	Vacuum bag process	Resin transfer molding	Cold liquid resin press molding	Hot liquid resin press molding	Press molding of molding compounds	Filament winding	Rotation molding	Pultrusion	Continuous lamination	Autoclave process
Application	single pieces, small series, large-area pieces	small to medium series, large-area pieces, linings	small series, large composite material	medium series	medium series	large series	large series	wound articles	rotationally symmetrical hollow moldings	solid and hollow profile	smooth and curved plates	high loaded components small series
Resins used	UP, EP	UP	UP, EP	UP	UP, EP	UP, EP	UP, EP, PF	UP, EP, VE	UP, EP, VE	UP, (EP)	UP	EP
Suitable reinforcing materials (glass content, wt%)	mat (20–30) fabric (35–50) rovings (50–70)	chopped strands (20–30)	mat (20–35) fabric (40–60)	mat (20–25) fabric (40–60)	mat (25–40) preform (25–40) fabric (50–60)	mat (25–45) preform (25–45) fabric (50–60)	mat (20–40) chopped strands (20–40)	rovings fabric (40–70) mat	mat (20–35) chopped strands (20–35) fabric (30–40)	rovings fabric (50–70)	mat (20–25)	fabric
Labor costs	high	high	high	medium	medium to high	medium	low	low medium to high	low	low	low	high
Investment	low	medium	low	medium	medium to high	high	high	high	high	high	very high	medium
Molds	open, one or multipart mold made of wood, metal, glass-fiber-reinforced plastic	open, one or multipart mold, made of wood, metal, glass-fiber-reinforced plastic	one part or with rigid counter-mold, made of wood, glass-fiber-reinforced plastic	closed, split mold made of glass-fiber-reinforced plastic	closed, split mold made of glass-fiber-reinforced plastic	closed, split mold made of steel	closed, split or multipart mold made of steel	one or multipart, steel mold core	one or multipart, rotor made of steel, aluminum	profile mold, sometimes with steel core		open, one part with vacuum bag
Economically feasible molding size, m²			0.5–30	0.5–20	0.3–5	0.3–5	0.05–3	0.1–100	1–100		width ≤5 m any length	
Cycle time	30 min to several days	30 min to several hours	30 min to several hours	30 min to several hours	5–30 min	2–10 min	2–5 min	depends on molding size	10 min to several hours			several hours
Processing temperature, °C	room temperature	room temperature	room temperature	room temperature	room temperature	80–120	120–160	room temperature	room temperature	≤160	room temperature and higher	80–200
Pressure, MPa			≤1	≤4	≤10	10–30	2–50	filament tension	centrifugal force			≤2

	1	2	3	4	5	6	7	8	9	10	11	12
Wall thickness, mm	2–10 usual	2–10 usual	1–5 usual	1–5	1.5–5	1–5	1–10	1–10	3–10	3–20	1–3	0.5–10
Wall thickness tolerances, %	mat ≤50, fabric ≤10	≤50 unavoidable	≤20	≤20	≤20	≤20	≤10	depends on reinforcement ≤20	≤20	≤10	≤10	≤5
Different wall thickness	possible	possible	limitedly possible	possible	barely possible	not possible	possible	possible	possible	usual in pulling direction	not possible	possible
Undercuts	possible with multi-part mold	possible with multi-part mold	limitedly possible	possible with multi-part mold	not possible	not possible	possible	not possible	not possible	perpendicular to the pulling direction possible	perpendicular to the pulling direction possible	possible with multipart mold
Stiffening layers	possible	possible	possible	possible	limitedly possible	possible	possible	possible	not possible	possible	not possible	not usual
Gel coat	usual one side	usual one side	one-sided; with rigid counter-mold both sides possible	both sides possible	limitedly possible	not possible	possible	possible	possible	not possible	possible	not usual
Surface	one side smooth	one side smooth	one side smooth; with rigid counter-mold both sides smooth	both sides smooth possible	both sides smooth, minor fiber structure	both sides smooth, minor fiber structure	both sides smooth	one side smooth	both sides smooth	all sides smooth	both sides smooth	one side smooth
Aftertreatment	edge trimming	edge trimming	edge trimming	deflashing, edge trimming	edge trimming	deflashing	deflashing	separate	separate	separate	separate, edge trimming	edge trimming

* UP = unsaturated polyester; EP = ethylene–propylene; PF = phenol–formaldehyde; VE = vinyl ester.

Table 2. Dependence of the properties of glass-fiber-reinforced UP moldings on reinforcing material and glass content

Property	Glass matss		Woven glass fabric		Glass rovings
Glass content					
wt %	20 – 30	40 – 50	40 – 50	50 – 60	70 – 80
vol %	11 – 17	25 – 32	25 – 32	32 – 42	53 – 66
Density, g/cm^3	1.3 – 1.5	1.5 – 1.75	1.5 – 1.75	1.6 – 1.85	1.9 – 2.1
Thermal expansion coefficient, 10^{-6} K^{-1}	40 – 30	24 – 20	22 – 18	18 – 16	14 – 12
Thermal conductivity, W K^{-1} m^{-1}	0.14 – 0.19	0.23 – 0.31	0.19 – 0.25	0.35 – 0.31	0.37 – 0.41
Tensile strength, N/mm^2	65 – 90	130 – 170	200 – 240	240 – 275	700 – 800
Breaking elongation, %	2	2	2	2	2
Modulus of elasticity (tensile), N/mm^2	5000 – 7000	9000 – 10 000	10 000 – 14 000	14 000 – 17 500	21 000 – 26 000
Bending strength, N/mm^2	115 – 145	180 – 220	220 – 260	260 – 300	400 – 500
Modulus of elasticity (bending), N/mm^2	5000 – 7000	9000 – 11 000	10 000 – 14 000	14 000 – 17 500	
Compressive strength, N/mm^2	110 – 135	165 – 200	150 – 180	180 – 200	
Shear modulus, N/mm^2	2000 – 3000	4000 – 5000	2000 – 3000	3000 – 4000	
Impact strength, kJ/m^2	25 – 55	75 – 105	130 – 160	160 – 190	

reinforcing material and fiber content. Figure 1 shows the market development for glass-fiber-reinforced polymers from 1986 to 1988.

Table 3 lists the main areas of application, and Table 4 the processing techniques that are employed.

1.1. Raw Materials

Selection. The most widely used reinforcing fibers are glass fibers (→ Fibers 5. Synthetic Inorganic, Section 2.1.) in the form of rovings, nonwoven fiber matting, and glass fabrics.

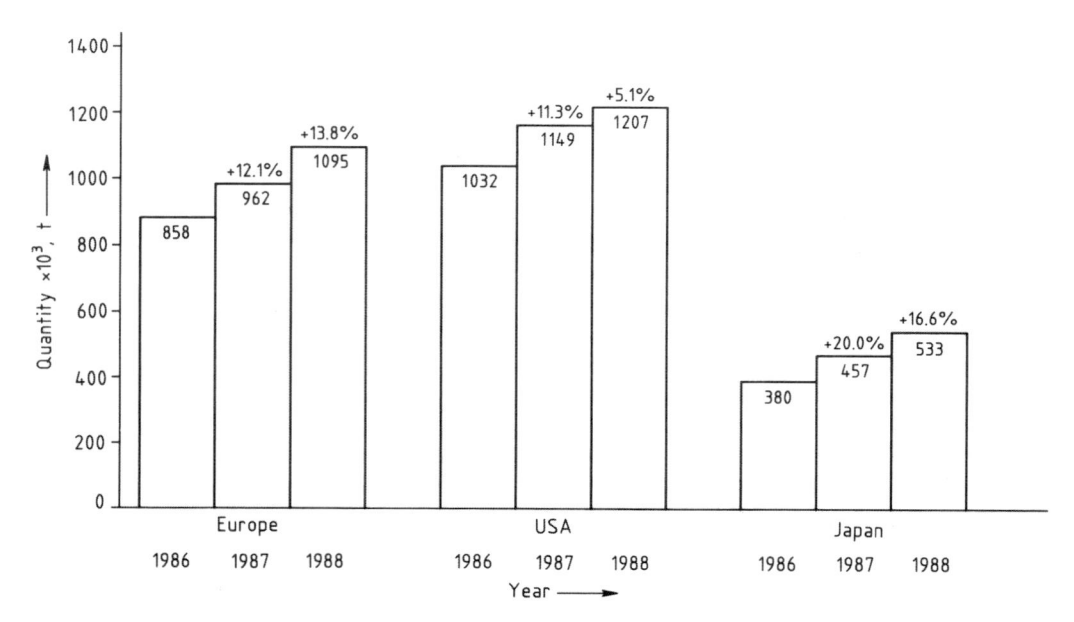

Figure 1. Development of glass-fiber-reinforced plastics market

Table 3. European glass-fiber-reinforced plastics market in 1988 according to end use

End use	Consumption, %
Construction	14.2
Sports and leisure	7.3
Industry and agriculture	18.0
Transportation	25.8
Electricity and electronics	18.1
Consumer goods	8.6
Military and other uses	8.0

Table 4. European glass-fiber-reinforced plastics market in 1988 divided according to process

Process	Use, %
Manual lamination	20.4
Fiber spraying	7.4
Press molding	29.2
Granulation (thermoplastic)	23.9
Injection molding (thermosets)	2.7
Continuous lamination	6.8
Pultrusion	1.5
Filament and rotational processes	4.5
Others	3.6

Carbon and aramid fibers, which were initially applied in aviation and space, are increasingly being used in industrial applications. Their significantly higher price compared with glass fibers is often justified by their much higher stiffness and strength and their more favorable thermal expansion behavior. All reinforcing fibers must be coated with a coupling agent to achieve satisfactory adhesion to the polymer matrix. Additionally, the coupling agents protect the fibers from abrasion during processing.

The resins are selected primarily for their final properties, corresponding to the demands placed on the finished product. With few exceptions, unsaturated polyester (UP), epoxy (EP), and vinyl ester (VE) resins can be processed by using all the techniques described below, although differing processing properties must be considered in selecting the resin. This applies particularly for phenol – formaldehyde (PF) resins, which cure via a condensation reaction and hence can be processed only under pressure.

The advantages of *EP resins* are air-drying cure; high filler levels possible; wide variation due to different combinations of resin and hardener; low volume shrinkage during curing (2 – 3 %), which begins before gelation and hence results in extremely good retention of dimensions and freedom from stress in the end product; no deformation during curing; high adhesive strength (i.e., good adhesion to the substrate and the fibers) and hence high mechanical strength of fiber-reinforced components, especially during dynamic loading.

In the latter case, the favorable mechanical properties of EP resins, including high fracture strain and tensile strength, also play an important role. High application temperatures may necessitate the use of EP resins. Certain types permit long-term service temperatures exceeding 200 °C.

Disadvantages of EP resins include high price; long curing times and hence higher processing costs; extreme care needed to meter resin and hardener; high viscosity, which makes degassing the resin and impregnating glass fibers difficult; sometimes difficulty in demolding as a result of low shrinkage.

The advantages of *UP resins* are low price; low curing times (i.e., rapid demolding is possible); easy handling due to low viscosity; low susceptibility to erroneous mixing of the final formulation.

Disadvantages of UP resins are higher volume shrinkage on curing (6 – 8 %), which makes it difficult to manufacture stress-free components and leads to a tendency to deform. In addition, during selection of the type of resin, attention must be paid to the reactivity. Low-reactivity resins cure with relatively low stresses, and the end products have good elasticity but generally exhibit poor resistance to corrosion and aging. High-reactivity products can be cured easily without heating, but removal of the heat of reaction must be effective to minimize stresses in the end product, shrinkage, and separation of the resin from the fibers, as well as possible thermal damage to the material itself. The long-term service temperature of components made from UP and VE resins is limited to 70 – 120 °C.

The advantages of *VE resins* compared to the chemically related UP resins include improved mechanical properties and, particularly, greater chemical stability, especially toward acidic media. Their disadvantage is the high price.

The PF resins represent one of the oldest and cheapest classes of polymer. Their advantages include very low manufacturing costs, excellent chemical and thermal stability, favorable combustion behavior, and low toxicity of combustion products. This has led to considerable importance in applications where safety is a prime consideration (aircraft inner paneling, transportation). The drawbacks of PF resins are poor mechanical properties (low fracture strain) along with problematic processing. Because the latter can be carried out only under pressure, techniques involving open molds cannot be employed. Only processes involving presses and autoclaves are of any importance.

Because of their high price, EP resins are employed where higher material costs are justified by improved adhesion to the substrate or improved mechanical properties (e.g., for surface protection and coatings).

Because they are easy to process, UP resins are employed preferentially for glass-fiber-reinforced products, with the exception of components that must withstand high dynamic loads, such as helicopter rotor blades; components for aircraft, rockets, and spacecraft; and skis.

Where corrosion problems are likely to be encountered, resins are selected according to their resistance to chemical attack.

Transportation, Storage, and Safety. The inflammability of resin systems depends on their chemical composition. During storage, transport, and processing, the regulations and safety information provided by the supplier must be followed.

Reactive resins must be stored below 20 °C, with the exclusion of atmospheric moisture and light. Resins supplied as a final mixture ready for direct reaction and including the curing agent — as is the case, for example, with pre-impregnated fiber materials known as prepregs — must be stored under refrigerated conditions at −18 °C.

During handling and processing of UP and VE resins, the skin and respiratory irritations that can result from the styrene component must be considered (MAK 20 ppm), along with the strong skin irritations that can be caused by several cross-linking agents. Particular care is necessary when dealing with peroxides, which are often used as hardeners (explosion hazard).

During the processing of EP and PF resins, safety gloves, clothes and glasses must be worn.

Several cross-linking agents for EP resins (e.g., amines) are strong skin irritants or can lead to allergic reactions. With PF resins, adequate ventilation must be ensured to remove reaction products.

1.2. Mold Materials

The choice of mold material depends on the processing technique and the number of articles to be manufactured. For casting, molds can be manufactured from flexible materials such as plasticized PVC or silicone rubber. These allow the production of moldings with undercuts. The number of castings is limited by the low solvent resistance.

Wood and plaster are used for making molding patterns; wood is also employed in manufacturing single-component molds for individual moldings or small production runs using pressureless processing techniques. The surface must be sealed with a paint that is resistant to aromatic compounds.

For the manufacture of glass-fiber-reinforced moldings in pressureless processes and the production of small series by cold press molding, a processor generally prepares his own molds from glass-fiber-reinforced UP or EP. The EP resins have the advantage of greater dimensional stability, but they must be coated to protect them from aromatic compounds. By casting a supporting substrate of polymer concrete or concrete on reverse surface, the molds can be made rigid.

For large production runs, metallic molds are most suitable. For hot press molding, heated molds made of tempered steel are employed. The sealing edges of molds for liquid resin press molding are formed as nip, shear, or pinch ridges. Molds for processing molding compounds and sheet molding compounds are described in Section 2.5.

To ensure faultless demolding, mold surfaces are treated with a release agent [wax, silicone grease, poly(vinyl alcohol)]. Prior to application of the release agent, the remains of any previous molding processes must be removed thoroughly.

1.3. Preparation of Resins

The resin components are either premixed by the supplier (e.g., prepregs) or metered and mixed by the processor. This is performed by automated mixers fitted directly to the processing machine (e.g., injection techniques such as resin transfer molding) or by hand mixing.

In all cases, the equipment (mixing vessels, stirrers, metering units, etc.) must be clean, dry, and free of residual deposits from previous batches. All components, including glass fibers must be thoroughly dried and, if necessary, warmed to room temperature before use. Mixing is best performed with a high-speed stirrer. The speed should be adjustable to avoid the inclusion of air bubbles. Highly viscous batches (e.g., EP resins) are prepared and held under vacuum until completely degassed.

The components are added in the following order: resin; styrene (with UP resins); plasticizer and diluent (with EP resins); hardener; additives such as stabilizers, lubricants, dyes, film formers, conductivity additives, thickeners, thixotropic agents, flame retardants, fillers, and accelerators. For properties and applications of additives, see → Plastics, Additives.

Materials that are difficult to disperse or additives that are not divided finely enough are made into a slurry before mixing by using small amounts of UP resin or EP thinner.

With UP resins, the curing reaction begins with addition of the second reactive component; initially the resin gels and then solidifies rapidly. Processing of the resin mixture must therefore be completed within the narrowly limited pot life dictated by formulation and processing temperature (i.e., before the onset of gelation). With resins that already contain an accelerator, the hardener is therefore always added as the final component. To avoid exceeding the pot life of UP resins prior to processing, the following procedure can be followed: all of the additives are mixed into the resin and the resulting batch is divided into two equal volumes. The hardener is added to one half, and the accelerator to the other. Immediately before processing, equal volumes of the two batches are mixed together.

With EP resins, the rate of hardening decreases with increasing viscosity. The pot life therefore does not end so sharply with the onset of gelation as with UP resins.

1.4. Pressureless Processing without Fiber Reinforcement

Casting and Embedding. Because of their light colors and low viscosity, UP resins are particularly suitable for casting figures and embedding specimens. The resin mixture can be adequately degassed by allowing it to stand for a short period prior to casting. For embedding two or more layers are usually cast. After the first layer has begun to cure, the item being embedded is laid on the surface and covered with further resin mixture. Because of shrinkage during curing, the casting must generally be finished by grinding and polishing.

EP resins are preferred for casting and embedding technical articles, where the slightly darker color is unimportant but high demands are placed on dimensional stability (e.g., for models) as well as embedding of electronic components. Figure 2 shows a schematic of a casting plant for EP resins. Resin and hardener are firstly melted or heated to lower the viscosity. Fillers and additives are then added. To degas the viscous resin mixture completely, mixing and casting must generally be done under vacuum.

When embedding electronic components, penetration of the resin mixture into joints and

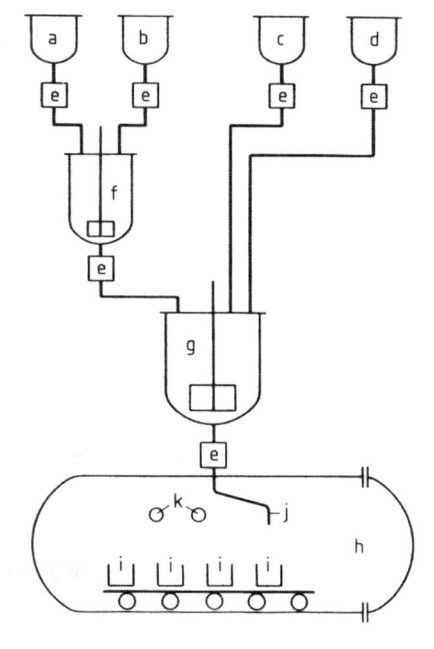

Figure 2. Schematic of a casting plant for epoxy resins
a) Heated storage tank for resin; b) Storage tank for fillers; c) Heated storage tank for hardener; d) Storage tank for other additives; e) Metering units; f) Heated, evacuable mixer; g) Heated and evacuated mixer for final mixing; h) Evacuable casting vessel; i) Casting mold; j) Swivel arm; k) Sight glasses

voids can be promoted by removing the vacuum once the casting stage has been completed.

In dipping processes, components are immersed in the resin mixture while under vacuum and exposed to atmospheric pressure after their removal.

In the drip process, the cold resin mixture drips onto a rotating, preheated substrate. During this process, the viscosity decreases and the resin penetrates any voids. This process is used among other things to bond axially parallel windings of wound articles (see Section 1.8).

Coating. Another important area for which EP resins are particularly well suited due to their excellent adhesion properties is in surface coatings. Coating is carried out with a liquid, low-viscosity resin – hardener system or with solid components dissolved in a solvent. Resin powders can be applied by fluidized-bed sintering, flame spraying, or electrostatic coating.

These processes and the continuous coating of printed paper or fabric webs with UP resins are described in → Plastics Processing, 1. Processing of Thermoplastics, Chapter 3.

Polymer Concrete. Both EP and UP resins are used in the manufacture of polymer plaster, mortar, and concrete. Because of their extremely low shrinkage and low viscosity, resins based on poly(methyl methacrylate) are also employed. These resins replace the conventional cement and water binder used for concrete. Gravel, pebbles, sand, and ground stone are used as additives. Polymer concretes have much shorter setting times and can be subjected to loading after about one day. Preparation and processing correspond with the techniques already in use with cement-based concrete.

The mechanical strength of polymer concrete is higher than that of cement concrete, and polymer concrete adheres well to set concrete. It is thus widely used to cast joints, for repair purposes, and above all as a coating for concrete surfaces when increased resistance toward acid, alkali, and salt solutions is required.

Another important application is the manufacture of bases for machines or machine housings. The advantage over the more common gray cast iron or cast alloy construction is the vibrational damping exhibited by polymer concrete.

The choice of resin depends mainly on the required thermal stability and chemical resistance, as well as the maximum allowable shrinkage.

For outdoor use, the sensitivity of the hardening reaction to water (soil moisture, rain) and temperature must be considered.

Foamed polymer concrete with low density, good mechanical properties, and low thermal conductivity can be produced by adding blowing agents to the resin mixture.

1.5. Pressureless Processing with Glass-Fiber Reinforcement

Gel Coating. The appearance of moldings made from glass-fiber-reinforced plastics (GRP) can be impaired by the presence of glass-fiber on the surface. Gel coatings are used to cover the glass fibers, giving a smoother surface and improving the resistance to weathering and chemicals. Such layers adhere better to the composite surface than subsequently applied paint coatings because they are anchored by chemical bonds. The gel coating process can be combined with most processing techniques.

To produce a gel coating, a layer of fiber-free resin is sprayed or painted onto the surface of the mold in a thickness of 300 – 600 µm. The coating can be dyed or filled to improve abrasion resistance. Before the following composite layer can be applied, the gel coating must be cured sufficiently so it is not swelled by styrene or solvents and cannot be penetrated by glass fibers during subsequent processing stages. The curing reaction must, however, not be so complete that bonding to the composite layer is reduced.

Manual Production of GRP Components. The manual process (Fig. 3) is the simplest method of producing GRP moldings and requires only

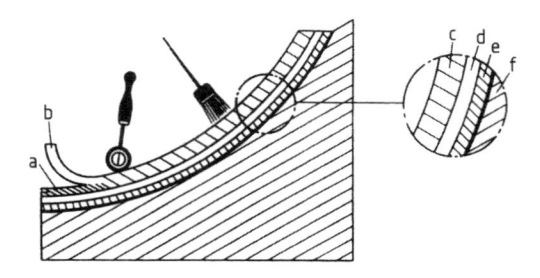

Figure 3. Manual production of GRP moldings
a) Resin; b) Fiber reinforcement; c) Laminate layer; d) Gel coat; e) Release agent; f) Mold

basic equipment. It is also the oldest processing technique. Production of GRP moldings began in the 1940s in the United States and, despite the development of machine processes, has retained a considerable portion of the market. The manual process is especially suitable for manufacturing large-area moldings; for small series using single-part, light-construction molds; and for reinforcing components made of thermoplastics. The moldings can be designed freely and are not subject to constraints regarding size.

A gel coat is first applied to the mold whose surfaces have been coated with a release agent. After initial curing, the first resin layer is applied and a layer of glass fibers, nonwoven glass-fiber matting, or glass fabric is applied by using a brush or fleece roll and a structured roll. Any air bubbles that form between the fine layer and the laminate must be removed. This primary composite layer is first cured slightly to prevent the ingress of air bubbles and movement of the layer during further processing. After this, additional composite layers are quickly applied wet-in-wet. The number of reinforcing layers depends on the strength required of the finished molding. The embedding of subcomponents made from other materials, such as stiffening ribs and screw thread sockets, is also possible. If required, a finishing surface coating, similar to the gel coating can be applied to cover the glass fibers on the surface. A film former (paraffin dissolved in styrene) is usually added to the topcoat when UP resins are used, to protect the surface of the composite from atmospheric oxygen. This also gives a tack-free surface even on curing at room temperature.

Before the molding is removed the curing reaction must be so far advanced that the article is not damaged by tearing or by separation of the resin and fibers (stress whitening). Subsequently, the product is stored in simple support equipment until curing is complete to avoid deformation, because upon hardening, all deformations become fixed. Moldings that are manufactured in single-part molds have a smooth surface only on the side that was in contact with the mold.

Attainable glass contents with glass matting range between 20 and 30 wt %, and with glass fabrics between 40 and 50 wt %. Suitable matrix materials include UP, VE, and EP resins.

Fiber Spray-up Process. The fiber spray-up process is a partially mechanized manual process. It is particularly useful for linings and in the manufacture of complex moldings. Only UP resins are used in this process.

After application of a gel coating, which for linings also serves to seal any pores, the resin mixture and chopped glass fibers are sprayed simultaneously onto the mold and then, in a manner similar to the fully manual process, are compressed with a roller and degassed with a structured roll. The fiber spray-up equipment illustrated in Figure 4 operates with two resin feeds, one of which contains the hardener, and the other the accelerator. The two components are fed to a double pistol and sprayed with compressed air. They are mixed immediately prior to contact with the mold. The glass rovings are fed to a chopping unit located between the two nozzles, where they are chopped to staple fibers and accelerated along with the two streams of resin onto the mold. With high-pressure spray equipment, accelerator-containing resin and hardener are pumped from storage tanks to the spray pistol via a mixing and metering unit.

With the fiber spray-up process, the layering of glass fibers and the glass content are comparable to those of manual processes. Maintaining constant wall thickness, however, requires skill and experience in spraying, compacting, and degassing the fibers, which initially lie very loosely. An advantage over the manual process is that large batches of resin with prolonged pot life can be processed, while very rapid curing times are also tolerated. Furthermore, the labor-intensive cutting of the glass matting to size is avoided, along with losses incurred during

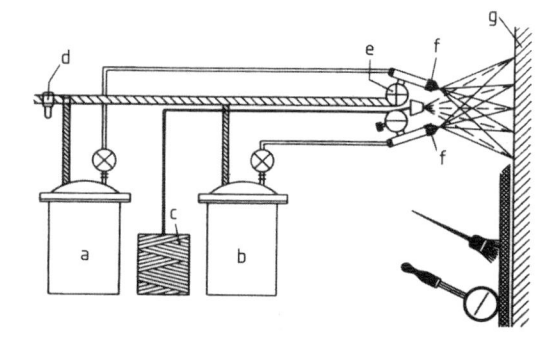

Figure 4. Fiber spray-up process
a) Resin with hardener; b) Resin with accelerator; c) Roving reel; d) Compressed air inlet; e) Fiber cutting device; f) Spray gun; g) Mold

trimming. Start-up and cleaning procedures for fiber spray equipment require considerable time. Usage is thus recommended only when extended production times without long interruptions can be expected.

1.6. Low-Pressure Processes

Low-pressure processes allow the production of articles of large surface area in small to medium production runs. Manufacturing times are usually shorter than with manual processes but longer than with compression molding. Low-pressure processes can be used to improve laminates manufactured with pressureless techniques (vacuum film process) and permit direct impregnation of fiber materials in a closed mold.

Vacuum bag process:

1. With single-component rigid mold and elastic sheet as countermold (Fig. 5): The resin mixture is applied to the mold; the glass-fiber reinforcement is added and, after an airtight seal is generated, impregnated with the resin by applying a vacuum.
2. Molding with two elastic sheets: A laminate that has been impregnated and degassed by hand is deep drawn by vacuum between two elastic films. This technique is suitable for the manufacture of domes.
3. Form and counter form are rigid: After the form has been coated with resin mixture, the fiber reinforcement and, when appropriate, the core material are applied, and the counterform, which is also coated with resin

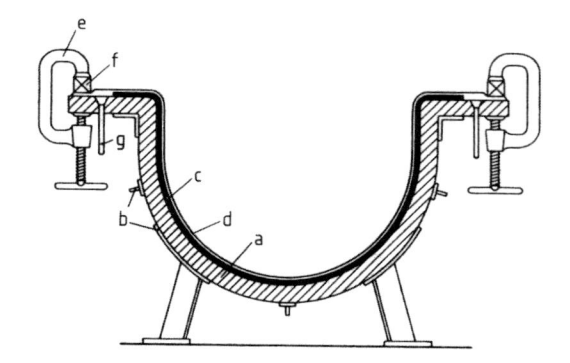

Figure 5. Vacuum bag molding
a) Mold; b) Reinforcing profiles; c) Laminate; d) Flexible sheet as countermold; e) Clamp; f) Sealing frame; g) Suction vents

mixture, is used to form an airtight seal. The necessary compression is generated by applying a vacuum. Two-sided topcoats are possible. This process is suitable for items of large surface area and is used primarily to coat plywood sheets and rigid foam cores for refrigerated trucks and containers.

Autoclave Processes. The low-pressure autoclave process is a variation of the vacuum bag technique. It is used mostly for the manufacture of high-quality components for the aviation and aerospace industries. The costs are high due to the use of a heated, pressurized autoclave and high-quality prepregs. Furthermore, the structure of the laminate is complex, containing release films and fabrics, suction, pressure distribution, and surface layers (Fig. 6). Cycle times can be several hours.

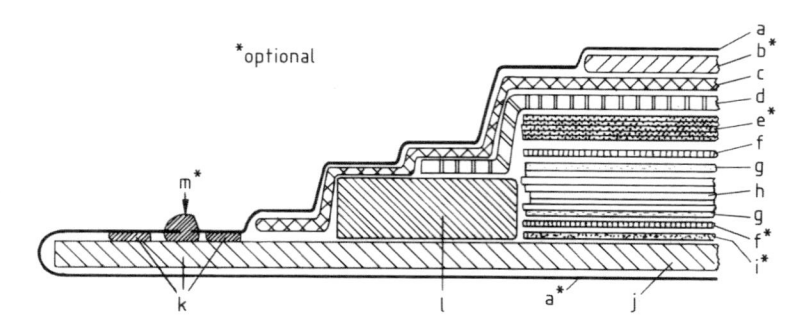

Figure 6. Structure of a laminate produced in the autoclave process
a) Vacuum bag; b) Compression plate; c) Pressure-distribution layer; d) Release film, perforated; e) Absorbent material; f) Release film or fabric; g) Peeling film; h) Fiber composite laminate; i) Release film; j) Mold; k) Sealing compound; l) Mold boundary; m) Double gasket

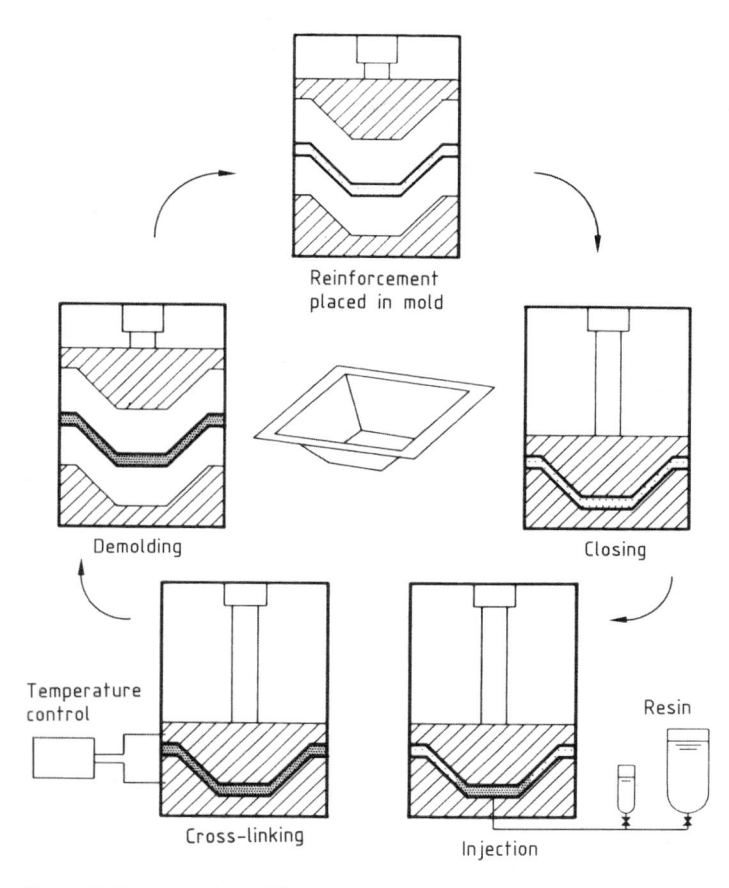

Figure 7. Resin transfer molding process

Resin Transfer Molding. In the resin transfer molding (RTM) process (Fig. 7), two-component or multicomponent rigid molds made from metal or GRP are employed, which are opened and closed by hydraulic mechanisms. The fiber reinforcement is cut to shape and placed in the mold. The mold is closed, and the resin mixture is injected from the bottom under pressure. Air release valves allow control of the resin flow. Application of a vacuum assists the impregnation stage. Large-surface moldings with topcoat matts on both surfaces can be manufactured with this technique.

Apart from moldings made from preshaped glass-fiber matting and UP resins (fiber content 20 – 40 vol %), which can withstand moderate loadings, the RTM process is equally suited for producing high-quality components from fiber fabrics and EP resins (fiber content up to 65 vol %). The injected resin must be of low viscosity and highly reactive. However, for modern resins, these requirements are associated with low softening temperatures and hence low use temperatures of the final molding.

1.7. Press Molding

Press molding allows the production of medium to large series of GRP moldings with smooth surfaces on both sides, good dimensional stability, and high glass content. The presses are generally hydraulic; two- or multipart molds are used. Labor costs are considerably lower than with pressureless processing.

Liquid Resin Press Molding. In liquid resin press molding the glass-fiber reinforcement is placed in the mold, and the resin mixture is poured into the center so that the flow paths to all sides of the mold are equally long. When the

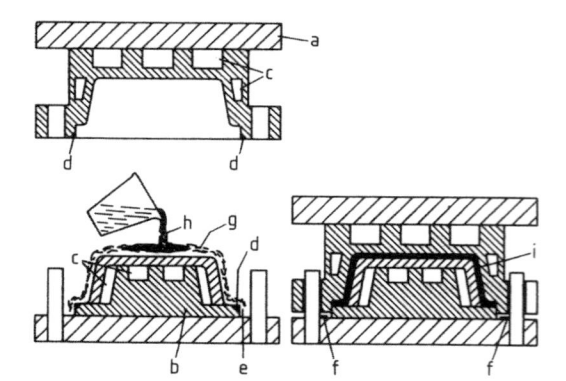

Figure 8. Liquid resin press molding
a) Female mold; b) Male mold; c) Heating channels; d) Cutting edges; e) Overflow; f) Excess; g) Fiber reinforcement; h) Resin; i) Molding

press is closed, the resin mixture is distributed evenly throughout the reinforcing material, and any air remaining in the mold is forced out of the cavity by the flow front (Fig. 8). The mold closing speed must be reduced significantly over the final 5 – 10 mm of the closing movement to avoid shifting or tearing of the fiber insert due to rapid flow of the resin and to prevent entrapment of air pockets.

Generally, glass matting is used as fiber reinforcement; glass fabrics or preforms are also used. In the production of preforms, glass fibers cut to a length of 20 – 50 mm are drawn onto a punched screen that exactly matches the shape of the item to be pressed. The fibers are bonded by spraying with a binder; after the binder has been dried, the preform can be processed. With moldings containing severe bends this process avoids the complex cutting and laying of glass matting, and overlap areas are avoided.

Topcoats can also be incorporated with liquid resin molding. However, this requires an extendable press table with two interchangeable lower mold halves to achieve acceptable cycle times. To obtain smooth surfaces a polymer or glass-fiber nonwoven fabric is best used as the first and last fiber layer. After being demolded, the pressed molding must be stored on support equipment to prevent distortions.

Generally UP resins are used; EP resins are less suited for liquid resin molding because of their long curing times. Filler contents up to 40 – 60 parts per 100 parts resin improve the surface quality by suppressing the fiber structure. Articles with varying wall thickness cannot be manufactured with wet pressing techniques because a constant glass content cannot be achieved at the junctions. The use of ejectors to assist in removing the molding is not recommended because of the possibility of malfunction caused by ingressing resin. They are usually not necessary because the mold must be slightly conical to avoid shifting or tearing of the glass-fiber reinforcement on closing.

Cold liquid resin press molding uses highly reactive UP resins containing hardener and accelerator. Pressing temperatures range from 20 to 60 °C. The amount of resin mixture is adjusted so that the resin begins to gel a few seconds after the mold closes. The mold surfaces are warmed by the heat of reaction. Additional heating of the mold is thus not necessary but does help to maintain a constant cycle time, particularly for the initial six to eight press cycles at the beginning of a shift or following a machine stoppage until the molds have reached their final temperature. Typical press pressures range from 0.1 to 1 MPa. When textile glass matting or preforms are used, glass contents of 25 – 35 wt % can be achieved.

Low pressures permit the use of molds made of GRP, supported by a polymer concrete or concrete backfilling for rigidity (Fig. 9). These molds are considerably cheaper than steel molds. Their lifetime is shorter; however, they are well suited for small- to medium-sized runs in which the required cycle times of 5 – 15 min are acceptable. The mold seal forms a nipping edge. By reducing the pressing layer thickness around this nipping edge to approximately one-third of the original thickness, the glass fibers are so compressed that they prevent leakage of the resin and permit the necessary pressure buildup.

Hot liquid resin press molding is suitable for the mass production of moldings. The presses and molds are described in → Plastics Processing, 1. Processing of Thermoplastics, Section 4.1. When processing UP resins, the cavity plate and core plate should be heated individually; temperature differences of 5 – 10 °C assist demolding, and the quality of the surface in contact with the warmer plate is improved.

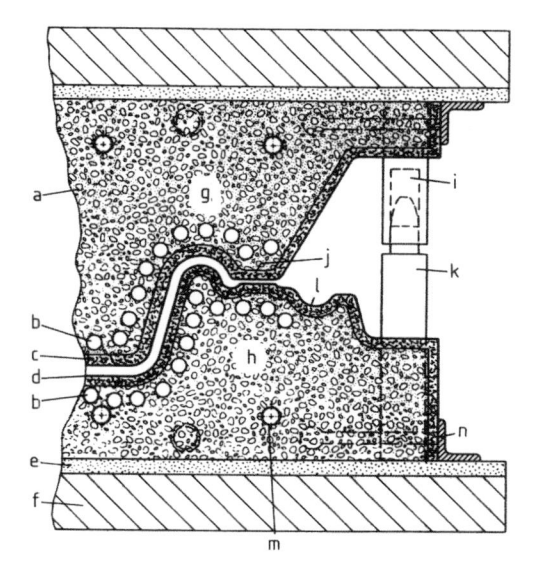

Figure 9. Compression mold made from glass-fiber-reinforced plastic with pinch-off edges for cold press molding
a) Backfilling of polymer concrete or cement concrete; b) Heating tube; c) Mold surface of glass-fiber-reinforced plastic; d) Gel coat; e) Rubber plate; f) Press table; g) Male mold; h) Female mold; i) Guide sleeve; j) Pinch-off edge; k) Guide pin; l) Overflow channel; m) Reinforcement of back filling; n) Frame for fixing mold

Liquid heating of the molds is preferable to electric heating because it allows removal of the heat of reaction. Press pressures of 0.5 – 5 MPa and temperatures of 80 – 120 °C are used.

In hot press molding, curing is initiated by external heat. UP resins can thus be used without the addition of accelerators. The pot life of the resin mixture ranges from several days to weeks at room temperature. The required press time is ca. 1 min per millimeter of wall thickness. Since the curing reaction begins shortly after the resin comes in contact with the mold surface, the filling and closing phases must be carried out rapidly. The mold seals form shear or immersion edges (Fig. 10).

Press Molding of Preimpregnated Fiber Articles. Preimpregnated textile fibers (prepregs) contain all the necessary components for manufacturing a pressed molding. The mixing of resins and the combining of resins with reinforcing material, which are necessary with wet pressing, are avoided during such processing.

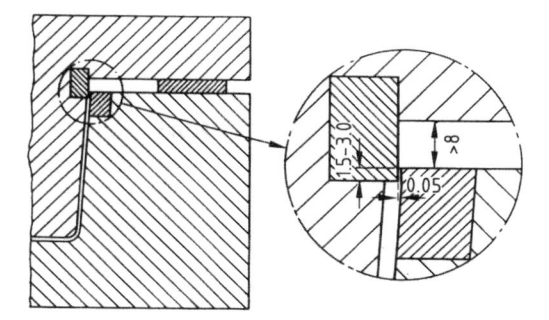

Figure 10. Shear edges on a steel compression mold

A distinction is made among preimpregnated fabrics, preimpregnated rovings, sheet molding compounds (DIN 16 913), and fiber-containing molding compounds (DIN 16 911 and 16 912). The processor can prepare these molding mixtures or purchase them ready for use. See → Plastics Processing, 1. Processing of Thermoplastics, Section 4.1 for processing molds and presses.

Prepregs in the form of fabrics or parallel glass, carbon, or aramid fibers are manufactured by impregnating the fibers with epoxy or phenolic resin. The resin is essentially solid, but still reactive and slightly tacky, and must be dissolved in a solvent or melted to penetrate the fibers completely. The resulting prepreg must be cut to fit the mold exactly because it cannot flow. Prepregs are employed mainly for highly loaded, thin-walled components (e.g., in aircraft). Besides press molding, processing with autoclave techniques is also common (see Section 1.6).

Sheet molding compounds are manufactured exclusively from UP resins. The resin mixture is composed of approximately equal weights of resin and filler (chalk, talc, clay), sometimes with addition of a lubricant. Curing agents with high initiation temperatures are used to give long pot lives at room temperature. To increase viscosity, thickening agents such as magnesium oxide are added, which cause a sudden thickening of the liquid to give a dry, leathery consistency. The thickening process takes from several hours to several days.

Impregnation is carried out immediately after the resin is mixed. Either glass is treated on equipment operating in a continuous process, or the resin mixture is applied on a film, cut glass

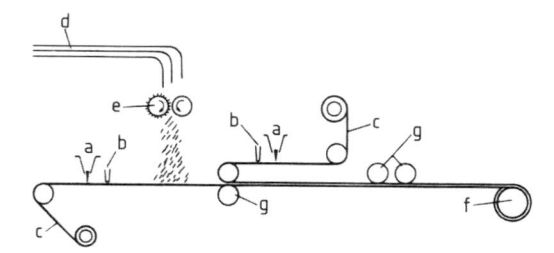

Figure 11. Production of sheet molding compounds
a) Feed tank for resin; b) Doctor blade; c) Cover sheet; d) Roving strands; e) Fiber cutting device; f) Winding of sheet molding compound; g) Impregnation of fibers with resin

fibers are added, and another film coated with resin is applied as a cover. After processing the fiber – resin mixture between several pairs of rolls, the sheet including the protective films is wound in a roll (Fig. 11). Because the thickening reaction continues during storage, the pot life is limited.

When working with sheet molding compounds preparation of the resin mixture is no longer necessary, and processing does not require the use of exactly cut sections. Because of the high viscosity, all components flow at the same rate under the molding pressure at the mold temperature. Thus, moldings with varying wall thickness can be manufactured, although in such cases, higher pressures are required than with liquid resin press molding.

To avoid the surface subsidence of areas with thick walls caused by the volume shrinkage of UP resins, low-shrinkage resin systems [low-profile (LP) resins] have been developed. These contain solutions of thermoplastics dissolved in styrene. On curing, part of the styrene contained in the thermoplastic particles evaporates prior to polymerization and causes foaming of the resin mixture, which in turn compensates for shrinkage.

Fiber-containing molding compounds are obtained by kneading chopped fibers into the resin mixture or by cutting continuous, impregnated roving strands to the desired length. These molding compounds can also be processed by injection press molding and injection molding (see → Plastics Processing, 1. Processing of Thermoplastics, Section 4.5). However, abrasion of metallic components by the hard glass fibers is a problem.

1.8. Semicontinuous Processing Techniques

Filament winding is a highly reproducible process for the manufacture of hollow articles with high mechanical strength. The equipment and the manufacturing technology are complex. Resins are selected according to their mechanical, thermal, and chemical loading; UP, EP, and VE resins are generally used. The fiber content of the product depends on the laminate structure, the process parameters, and the viscosity of the resin. Normal values range between 50 and 60 %.

The so-called lathe system is used for winding containers, container jackets, and pipes (Fig. 12). The winding equipment consists of a rotating winding mandrel and an oscillating fiber guide, the drives of which can both be regulated. Rovings are generally used for reinforcement, although woven tapes can also be employed. The rovings are fed through a resin impregnation bath, which is frequently combined with the fiber guide system to form a single unit. The impregnated fibers are then wound onto the core under constant tension and with an exact geometrical pattern governed by the ratio of the rotation speed of the core to the feeding speed of the fiber guide.

Modern winding equipment has up to six axes or degrees of freedom for the fiber guides and is usually computer controlled. This control system calculates data for the motion of individual axes, corresponding to the geometry of the winding mandrel and the desired winding pattern.

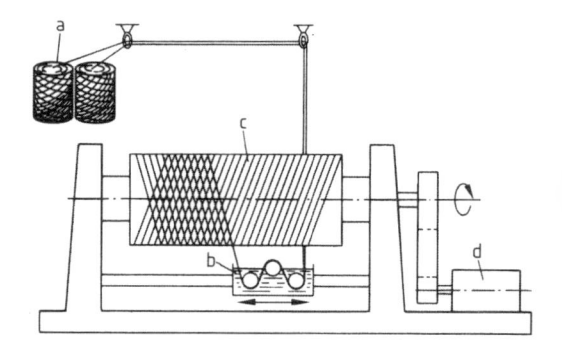

Figure 12. Filament winding (lathe system)
a) Roving reels; b) Support with impregnating bath and fiber guide; c) Winding mandrel with laminate; d) Drive

In parallel winding, individual roving strands are wound side by side without a gap, at a winding angle of almost 90°. To obtain sufficient strength in the axial direction, either warp-knitted fabrics are incorporated between the individually wound layers, or dry or impregnated glass fibers are sprayed between the winding layers.

With diagonal or cross winding, a winding angle of 30 – 70° is employed. The strength of the wound article can be adapted to the expected loading in the axial and radial directions by varying the winding angle.

Depending on the size of the molding, cold or hot curing resin systems are used for winding. Smaller moldings a few meters in length, which fit into hardening ovens or whose mandrels can be heated, are cured thermally to obtain superior material properties and extended resin pot lives. In contrast, large containers and pipes — diameters up to 12 m can be manufactured at present — must be cured in the cold or by using ultraviolet radiation.

Because the wound molding shrinks on the mandrel during curing, retractable winding mandrels are used to facilitate removal. Container bases, flanges, and supports are manufactured batchwise and are bonded to the container jacket or pipe by adhesion or lamination.

Feed silos; containers for drinks, chemicals, and heating oil; and pipes for the chemical industry (e.g., pipes for acid, wastewater, and exhaust air) are manufactured via winding processes. Mechanically loaded components including drive shafts and centrifuge drums can also be wound.

For the manufacture of highly stressed pressure vessels for the aerospace industry, equipment is used in which the winding mandrel rotates about two axes, or the fiber feed and the mandrel both rotate. Single fibers are used and wound in complex patterns.

A continuously operating winder has been designed for the mass production of pipes or container jackets in which an endless steel band is employed as the winding mandrel. This band runs spirally around a support using transport rolls, and is unwound at the outlet of machine and fed back through the inside of the core. The continually wound laminate comprising radially oriented rovings and axial fabric or matting reinforcement is cured by passage through a heated section after winding.

Rotation Molding. Cylindrical hollow articles with smooth internal and external surfaces can be manufactured by rotation molding. Glass matting or chopped glass fibers, sometimes in combination with textile fabrics, are employed as reinforcing materials. The choice of resin depends on the anticipated exposure to corrosive influences. The rotation equipment consists of a balanced cylindrical steel drum of adjustable rotation speed. With small-diameter pipes, the reinforcing matting or fabric is rolled in the stationary drum. On rotating, the strands are unrolled and pressed smoothly against the wall of the drum. The resin mixture is then distributed evenly across the complete length by use of a metering device. Due to the influence of centrifugal force, the reinforcing material is pressed against the wall and the resin mixture impregnates the fiber layers and displaces the air. Finishing coats can also be applied in this way. Pipes with smooth inner surfaces can be obtained by using a slight excess of resin. Generally, cold curing resin systems are employed. However, heating of the drum is advantageous, particularly with EP resins.

Larger vessels with diameters of several meters can be manufactured from UP resins by combining the rotation process with a modified fiber-spraying technique (Fig. 13). The mold is an aluminum rotor strengthened by external ribs and mounted on driven rolls. The feed vessel for the resin mixture and the fiber cutter are mounted in the head of a telescopic arm that is driven back and forth during coating in the rotating rotor. The glass fibers leaving the cutter are oriented by air blowing so

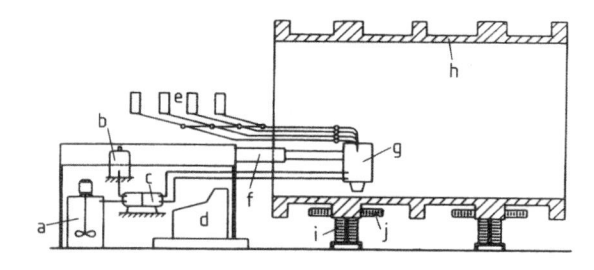

Figure 13. Centrifugal casting plant for large vessels
a) Vessel for resin and accelerator; b) Vessel for hardener; c) Mixing and metering pump; d) Control unit; e) Roving reels; f) Telescopic arm; g) Application unit with cutting device, compressed air inlet, and resin feed; h) Rotor; i) Driving rolls; j) Guide rolls

that the ratio of longitudinal to transverse orientation matches the expected mechanical loading of the container. Vertical containers, feed silos, and similar articles can be made with progressively increasing wall thickness from top to bottom. The roof of the container is placed in the rotor prior to rotation and is immediately bonded onto the container wall. The attainable glass content is 25 – 30 wt %.

1.9. Continuous Processes

Pultrusion. The pultrusion process is particularly suitable for mass production of uniaxially loaded solid and hollow profiles of any length. Labor costs are low, and equipment costs depend on the profile cross section. Glass rovings are used as reinforcement.

Figure 14 shows a vertical pultrusion machine. The rovings are wet with the resin compound in an immersion bath and then drawn into a heated profile mold. An inlet nozzle at the entrance of the mold retains any excess resin and entrapped air. The reactivity of the resin, curing temperature, temperature of the mold, and takeoff speed must be matched so that the resin mixture is dimensionally stable when it leaves the profile mold. Premature gelation causes adhesion to the mold, and delayed gelation can lead to deformation of the profile cross section. The solidified profile subsequently passes through a heating channel and is cured completely. The profile is then cut to length by a flying saw. Attainable glass contents vary between 50 and 75 wt %. The profiles can withstand high tensile forces in the fiber direction, but the strength in the transverse direction corresponds to that of the resin alone. This can be improved by employing textile tapes or looped rovings.

Continuous lamination is used for the manufacture of plane and longitudinally or laterally corrugated endless sheets of constant thickness. Glass matting or chopped glass fibers are used for reinforcement. Lightly colored, light-stabilized UP resins are employed. The production of corrugated sheets was the first large-scale processing technique for UP resins. It requires careful matching of resin – hardener system, curing temperature, and production speed to obtain an economical throughput via rapid curing times and to avoid a sudden curing reaction. This could lead to shrinkage of the resin from the fibers, which results in loss of transparency and embrittlement. Two processes have been developed for the production of corrugated sheets, which differ in the method of shaping. Figure 15 shows a machine for the production of

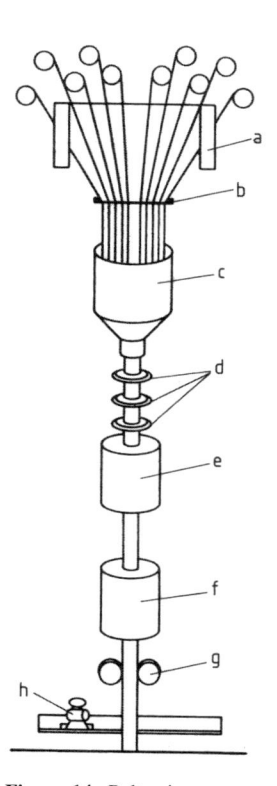

Figure 14. Pultrusion
a) Roving frame; b) Roving guide; c) Impregnating bath; d) Inlet nozzles; e) Profile mold; f) Heating channel; g) Takeoff unit; h) Cutting device

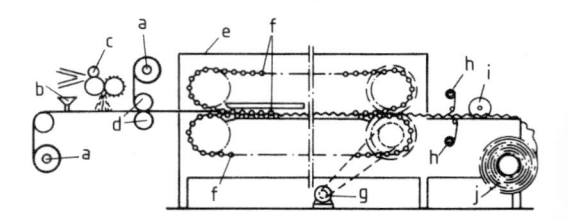

Figure 15. Continuous lamination of transversely corrugated sheets
a) Release film; b) Application of the resin; c) Fiber cutting device; d) Calibrating rolls; e) Hardening oven; f) Forming roller; g) Drive; h) Removal of release films; i) Trimming; j) Windup

transversely corrugated sheets. The resin is cast onto a cellulose or polyester release film and spread homogeneously by a doctor blade. The glass matting is then added, or glass fibers are distributed from a cutter unit fed with rovings positioned above the substrate. An open zone follows in which the resin is allowed to impregnate the fibers and air is forced out. Subsequently, a second release film is added from above, and a pair of calibrating rolls presses excess resin and air bubbles out of the laminate. The forming unit consists of a forming roller with displaced upper and lower rolls. The sheet passes between the shaping rolls in the curing oven, which consists of gelation and curing zones. The release films are then removed, the edges are trimmed, and the completed corrugated sheet is wound.

With longitudinal corrugating equipment, the impregnated laminate layer, calibrated between two films, is first shaped on a special forming element to a narrower-profile band and then passed through the curing oven on a longitudinally profiled slide system.

A significant improvement of the continuous process can be achieved by use of UV-curing UP resins because the hardening process is easier to control.

2. Curable Molding Compounds and Laminates

Moldings are manufactured from curable compounds under pressure and heat in closed molds; the molding compound is melted (or softened), shaped, and cured.

Terminology. Molding compounds are unshaped products that can be permanently molded to articles or semifinished products under pressure and temperature. Curable (thermoset) and uncurable (thermoplastic) molding compounds are available. Curable molding compounds always contain ca. 50 – 60 % filler and reinforcing materials. Press compounds are molding compounds that are suitable for press molding, injection press molding, or extrusion. Injection molding compounds are used for injection molding. The processing of curable molding compounds and the manufacture of laminates are described below; for low-pressure processes, see Section 1.6.

2.1. Processing Techniques

2.1.1. Processing Curable Molding Compounds

Curable molding compounds are formed and cured under the influences of temperature and pressure. On heating, the curable resin (which is present in the mass as a binder) softens, and the molding compound flows, filling the mold cavity. At the same time the cross-linking reaction begins, in which the binding agent and hence the molding compound harden. From this moment the material can no longer be melted; the curing (cross-linking) is irreversible. The molding is removed while still hot.

Flow – Cure Behavior. Figure 16 illustrates the flow – cure behavior of curable molding compounds. Figure 16 A shows how the molding

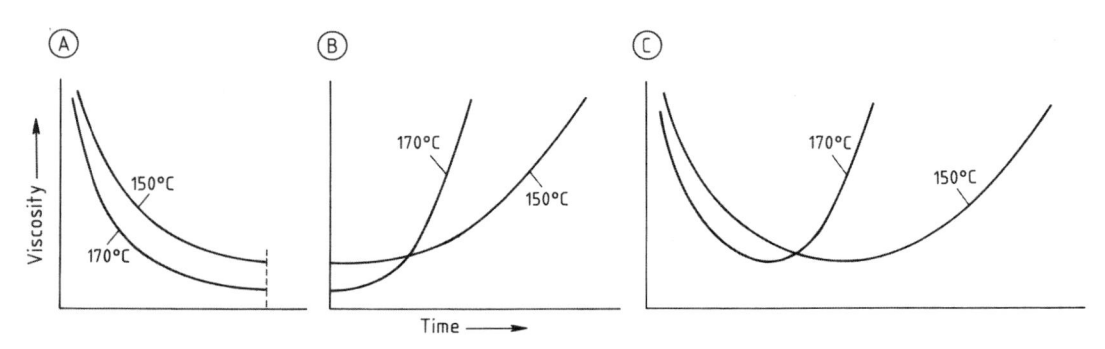

Figure 16. Flow – cure behavior of curable molding compounds
A) Molding compound without hardener, heated in the mold; B) Molding compound with hardener, heated immediately to mold temperature; C) Actual behavior of curable molding compound in press molding

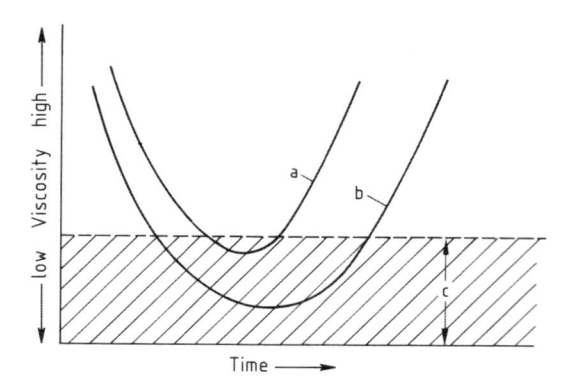

Figure 17. Flow – cure behavior for hard and soft flow
a) Hard flow; b) Soft flow; c) Possible processing range

compound is warmed and softens in the heated compression mold when no curing reaction takes place (this can be brought about by omitting the curing agent during preparation of the molding compound, which then behaves like a thermoplastic). Figure 16 B shows how the viscosity of the molding mass increases and the mass hardens if the curing agent is added suddenly and the mold is heated to 150 – 170 °C. Figure 16 C, the superposition of Figures 16 A and 70 B, shows the changes in viscosity observed when the mold is heated to 150 or 170 °C, filled with cold molding compound, and closed. Figure 17 shows the difference in behavior between hard and soft flow molding compounds. A hard-flowing compound generally cures faster than a similar compound with soft flow. The required viscosity of the molding compound depends on the design of the molding, the processing method, and the available forming pressure. In all cases, it must lie within the shaded zone in Figure 17; otherwise, the compound is too hard and the mold cavity will not be correctly filled.

Preparation Stages. Curable molding compounds are prepared by the manufacturers to be processed without further additives; the molding compound supplied to the processor contains all the necessary fillers, curing agents, stabilizers, lubricants, colorants, and flame retardants. The processor adds other materials only in exceptional cases. Occasionally, small amounts of lubricant can be added (e.g., 0.1 – 0.2 % zinc stearate) when particular processing conditions make this necessary. Molding compounds for screw injection molding machines can be colored by the processor, although it is difficult to manufacture uniformly colored molding in this way and the technique is seldomly used.

Release agents prevent adhesion of the cured resin to the surfaces of the mold that come in contact with the molding compound (mold cavity, gate, runner system, and all parts that could lead to the formation of flash). Waxes, metal soaps, fats, oils, and silicones, which are applied as solutions, sprays, or aqueous emulsions, or are melted and spread with a brush or compressed air, are available.

In the production of moldings that are to be painted, only release agents that do not impair adhesion of the paint should be used; otherwise, the molding would have to be degreased in an extra process stage.

Metering of the molding compound in a preliminary step is necessary for press molding, injection press molding, and pelletizing. Injection molding machines meter the material themselves (in the compressed state), as do plastication units of all types.

Automatic metering systems require satisfactory flow properties of the molding compound. Finely powdered compounds or those with high dust content require special accessories to permit automatic metering. Fibrous or doughy compounds can be metered only — if at all — by using special units.

Molding compounds are cut to the desired length and reweighed after the release film is removed. Exact cutting is not usually necessary.

Pelletizing of the molding mass is usually carried out for the following reasons:

1. Because the mold cavity is too small for the molding compound
2. To rationalize the feeding process
3. To permit high-frequency prewarming

Automated pelletizing machines operate with either constant pellet height or constant pelletizing force. Typical pelletizing pressures are 100 MPa and above. Circular pellets are generally used, with heights amounting to 20 – 60 % of the diameter. In special cases, square, rectangular, or ring-shaped pellets are produced.

Prewarming the molding compound simplifies processing in the case of press molding and injection press molding. Prewarmed compounds flow better and hence require lower press pressures and shorter closing times for press molding, as well as reduced injection pressure and shorter injection times for injection press molding; furthermore, the curing time is also reduced, particularly with press molding.

Phenolic and aminoplastic molding compounds evolve moisture and other volatile components during prewarming, which has a favorable effect on the properties of moldings. They exhibit reduced shrinkage, better electrical insulation, improved surface gloss, and reduced moisture absorption.

Polyester molding compounds incorporating styrene as a curing agent should be prewarmed only slightly or not at all.

Widely used prewarming processes are

1. High-frequency prewarming
2. Screw plastication
3. Oven warming (only applicable to phenolic and aminoplastic molding compounds, usually to remove moisture and other volatile components)
4. Infrared prewarming, where a thin layer of the compound is exposed to radiation

Molding compounds can be prewarmed quickly and homogeneously in a high-frequency field; pelletized materials are best suited. A prerequisite for a homogeneous temperature is constant thickness of the molding compound or constant pellet height and density. Typical prewarming times range between 0.5 and 2 min; temperatures of 100 – 140 °C are used.

Press Molding. A metered amount of the molding mass — either loose or in pellet form; as-received, prewarmed, or plasticated — is fed into the open mold, which has already been heated to curing temperature (Fig. 18). When the mold is closed, the mass softens as a result of heat flux from the mold walls and flows into the mold contours. Curing then takes place while the mold is closed. The hot molding is removed when sufficient curing has occurred and is then deflashed. Advantages of the process are the simplicity of the molds and presses; the low levels of filler orientation in the molding and, therefore, the almost isotropic properties; and the low molding shrinkage. Disadvantages include long curing times (especially when prewarming or preplastication is not employed), pronounced flash formation because the mold is closed only when the compound has begun to flow out of the mold (displacement of air), and the large deviations in mold-independent dimensions (see DIN 16 901).

Large, high-quality moldings (e.g., for automobile bodies) can be manufactured by press molding of molding compounds. Such moldings are processed by using computer-controlled, parallel-regulated hydraulic presses. The investment costs are high.

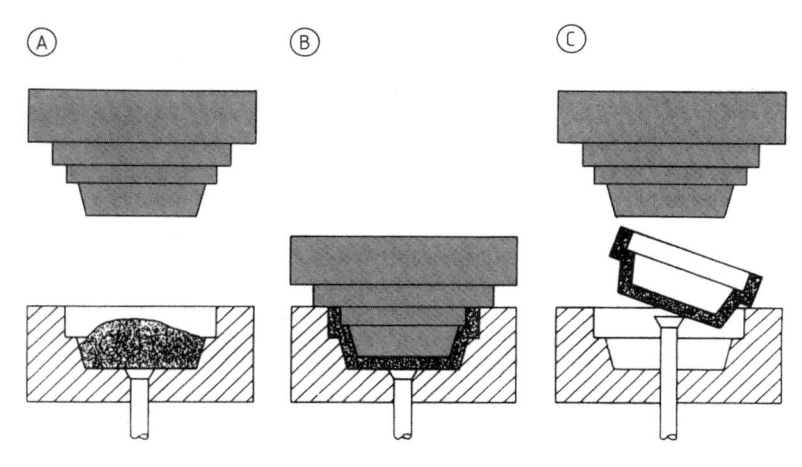

Figure 18. Press molding
A) Filling; B) Pressing; C) Ejection

The following stages are involved in press molding:

1. Filling the mold with the molding compound
2. Closing the press until the mass is squeezed out of the mold
3. Opening the press briefly ("breathing", optional)
4. Closing the press until curing is complete
5. Opening the press
6. Removal of the molding
7. Cleaning (blowing out) the mold

When metallic inserts are to be incorporated in the molding, they are placed in the mold prior to filling with the molding compound.

Injection Press Molding. In injection press molding (Fig. 19) a metered quantity of the molding compound, usually pelletized and prewarmed or preplasticated, is introduced into the injection cylinder of the mold. After the mold is closed, the compound is compressed by a piston, further plasticated, and injected through the gate into the mold cavities. The internal friction that results from this action leads to further heating, which accelerates the curing reaction. After curing, the molding and gate are either simultaneously or separately ejected from the open mold.

The advantage of injection press molding over press molding result from the fact that the mass is injected into the closed mold while hot and, hence, of relatively low viscosity:

1. Less flash is formed
2. The mold-independent dimensions are more exact
3. The manufacture of complicated contours and variable wall thicknesses is possible
4. Embedding of sensitive metallic components is possible
5. Curing times are reduced, particularly with thick-walled moldings

The disadvantages include the high consumption of molding compound since the material that remains in front of the injector piston and in the runner system also cures and cannot be reused. Additionally, depending on the gate geometry, variations in filler orientation can result, which in turn can lead to anisotropy of the strength and dimensional stability.

Injection Molding. In injection molding of curable molding compounds, the material is metered, plasticated, and injected into the heated mold by the screw of the injection cylinder (→ Plastics Processing, 1. Processing of Thermoplastics, Section 4.5). The cylinder is heated moderately, while the mold is heated to the final curing temperature. Due to shear heating of the compound (dissipation) during plastication in the cylinder and passage through the

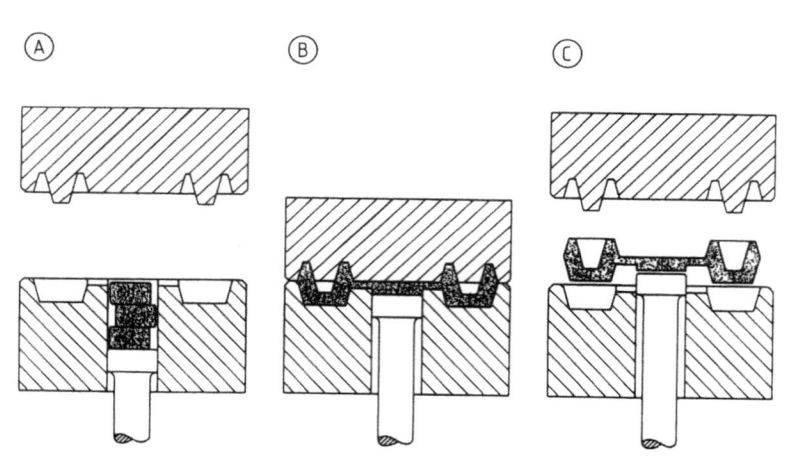

Figure 19. Injection press molding
A) Filling; B) Injection and pressing; C) Ejection

nozzle and gate, it is well plasticated and prewarmed. This leads to shorter curing times in the mold because only small amounts of heat need to be supplied conductively. The cycle times are short. More information on injection molding equipment is given in → Plastics Processing, 1. Processing of Thermoplastics, Section 4.5.

The advantages and disadvantages of injection molding compared to press molding are similar to those for injection press molding. The process cycle is generally shorter than with injection press molding, and pelletizing and high-frequency warming stages are not required. Additionally, a higher degree of automation is possible with injection molding.

Injection – compression molding (Fig. 20) is an adaptation of injection molding. With injection – compression molding, the mold is initially closed without pressure, and the plasticated molding mass is injected, whereby the mold opens up a few millimeters. The mold is then closed under pressure, and the compound is formed (see also → Plastics Processing, 1. Processing of Thermoplastics, Section 4.5).

Injection – compression molding offers the following advantages:

1. Filler orientation similar to that of press molding, which improves with increasing flow of the compound during pressing (the favorable filler orientation leads to more isotropic strength and improved dimensional precision)

2. Relatively low closing force required for pressing
3. Relatively low injection pressure and therefore lower back pressure at the screw and higher shot volume
4. Higher injection speeds, which can partially compensate for time losses caused by pressing
5. Disappearance of the gate during embossing (given appropriate mold construction)
6. Improved degassing
7. More uniform molding surface characteristics

Posttreatment. The molding is generally subjected to posttreatment after removal from the mold.

Setting. In certain cases, particularly when high dimensional precision is required, thermoset moldings are placed in a cooling template after being demolded.

Deflashing. Moldings produced by press, injection press, and injection molding have varying amounts of flash that must be removed. For reasons of economy, flash should be removed as quickly and simply as possible. For this, the flash must be as thin as feasible. Flash formation depends on the molding compound, processing technique, processing conditions, and above all on the quality of the mold. The finer the filler in the molding compound, the better the molding can be deflashed. Automatic deflashing is not possible with long-fiber fillers. Several of the more common deflashing processes are compared in Table 5.

Air-jet deflashing is the technically most developed and economical technique. With particle-jet deflashing, particles should be softer than the material from which the flash is to be removed. Organic materials are typically employed for jet particles (e.g., ground apricot stones or polyamide granules). The minimum particle diameter is ca. 0.5 mm; otherwise, the intensity of the collision is too low. Modern deflashing equipment has multiple nozzles, along with dust removal units and feed mechanisms for antistatic agents.

Polishing. If the mold has perfect surface quality, the resulting molding will have a smooth, glossy surface. This applies only to a limited

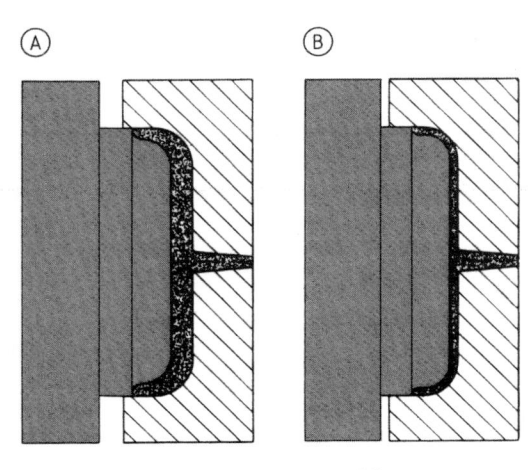

Figure 20. Injection – compression molding
A) Injection; B) Compression

Table 5. Deflashing of thermoset moldings

Deflashing	Application	Performance (number of articles per unit time)	Deflashing quality	Dust formation	Influence on surface quality
Mechanical treatment (drilling, filing)	flash thickness >0.4 mm or fiber-filled molding	low	high	low[b]	none
Abrasive belt treatment	flash in one or two planes (max.)	low	high	low[b]	damage possible
Drum deflashing	only with loose flash; no brittle parts and no sensitive metal inserts	high	moderate to good[a]	low[c]	low, not recommended for larger moldings with high surface quality
Air jet	flash thickness: 0.1 mm (max.)	high (several nozzles)	sufficient to good[a]	low[b]	none
Particle jet	flash thickness: 0.1 – 0.4 mm	high	good	low[b]	none

[a] Depending on the shape of the molding.
[b] With additional aids such as wet sawdust.
[c] Removal by suction is recommended.

extent to compounds with coarse fillers. The surface can be further improved by grinding and polishing.

Small moldings that are deflashed in drums can be simultaneously polished by addition of polishing agents.

Annealing. Postcuring is not usually necessary for thermoset moldings. However, heat treatment of the molding is often advantageous if it is to be thermally loaded during use. With polycondensation resins, annealing results in the slow discharge of volatile components, which in turn results in the advanced removal of postshrinkage, improved electrical properties, and increased application temperature.

General statements regarding duration and temperature of annealing are not possible; they depend on the case in question and must be determined experimentally. An extended treatment at lower temperature is always better for the material than a short period of annealing at higher temperature, even though the shrinkage is identical in both cases. This is particularly true for thermally sensitive aminoplastic materials.

2.1.2. Manufacture of Laminated Plates

Laminated plates are semifinished products manufactured from layered resin substrates with curable resins as binders. Phenolic, aminoplastic, unsaturated polyester, epoxy, and silicone resins are used as binders. Suitable resin substrates include paper (for paper-base laminate); cotton and glass fabrics (for fabric-based laminates); glass matting (for glass-mat-based laminates); and wood veneers (for synthetic resin laminated wood).

In the manufacture of laminated sheets, webs of the resin substrate are coated or impregnated with resin, dried (when necessary), cut, layered, and pressed between plane parallel plates. Phenolic and aminoplastic laminated papers and fabrics are usually pressed in multiplate presses at ca. 160 °C and 10 – 13 MPa (with synthetic resin laminated board at 25 MPa or higher).

Additionally, twin-belt presses are increasingly being employed because of their ability to function in a continuous process.

Laminates based on other resins are often pressed at considerably lower pressure (several bar).

Depending on binder, temperature, and laminate thickness, curing times vary from a few minutes to several hours. For some applications, laminated sheets are coated on one or both sides with special surface layers, for example:

1. With copper foil (laminated paper or fabric laminated to copper foil) for the production of printed circuits
2. Fine wood veneers for laminated board
3. With colored or printed, aminoplastic resin-impregnated papers for decorative laminated boards (e.g., for the furniture industry)

Aminoplastic-impregnated, decorative paper laminates can also be laminated directly to wood sheets (e.g., chip board).

Laminate moldings can also be pressed by using closed press molds once they have been appropriately cut and layered. Examples of this include pressed plywood moldings (for chairs) and pressed moldings made of polyester mat-based laminates.

2.2. Processing Properties

The processibility of a curable molding compound depends on structure, flow characteristics, and curing behavior.

The *structure* plays an important roll in automatic metering; consistent, rounded particles with good flow properties are preferred. The dust content should be low. Molding compounds containing fibrous reinforcing materials (e.g., cellulose, glass fibers) are granulated in many cases; glass-fiber compounds are also available in rod form. Impact-resistant molding compounds are employed in randomly oriented fiber structures; however, in such cases, the molding compound must be weighed and added by hand, and the mold must have a sufficiently large cavity.

Flow characteristics are adjusted by the manufacturer of the molding compound. With condensation resin molding compounds this is carried out by varying the extent of condensation of the binder during production. The flow level of a molding compound is selected according to the processing technique and the complexity of the molding. It also exerts an influence on the quality of the molding. Hard-flowing compounds generally give lower molding shrinkage, better surface gloss, and improved electrical insulation. However, they require higher processing pressure, often cannot be injected, and often cure more quickly.

The *curing behavior* of a molding compound can also be varied by the manufacture within certain limits. As a rule, a processor desires high curing speeds to attain high production throughputs. The higher the processing temperature, the faster is the curing reaction. However, the temperature can be increased only to a level at which the mold can still be filled without difficulty.

2.3. Processing Conditions

Processing guidelines can be established and described only for individual cases; they differ according to molding compound, processing technique, mold, and molding. Furthermore, processing difficulties or molding faults that may arise can be dealt with only as individual cases. The causes may lie in the molding compound, the processing equipment, or processing conditions so that a knowledge of all possible circumstances is necessary to make any evaluation or correction.

Table 6 gives an overview of processing conditions for the most commonly employed curable molding compounds.

Phenolic resin molding compounds are the thermosets that are easiest to process, regardless of whether press molding, injection press molding, or injection molding is used. This applies particularly to the standard compound of phenolic resin with sawdust (DIN 7708, type 3), but also to most other phenolic molding compounds.

Some fiber and macerate compounds, because of their structure, cannot be processed automatically (or only by using special equipment).

The most important operating conditions in press molding processes include temperature, pressing force, and curing time. When pressing phenolic molding compounds, the mold temperature is typically 160 – 170 °C. Higher temperature leads to quicker curing; hence the highest temperature possible is always used. The upper temperature limit is

1. For compounds containing sawdust, ca. 180 °C
2. For compounds with organic fibers or fabric cuttings, ca. 170 °C
3. For compounds with inorganic resin substrates, ca. 190 °C

In all cases, care must be taken to ensure that the temperature is not so high as to partially cure the material while the mold is closing and thus lead to faulty moldings.

The harder the flow characteristics, the larger the resin substrate, or the less intense the preheating, the higher is the pressing force required. Additionally, design, wall thickness, and height all play a role in determining the required pressure.

With *press molding*, curing time depends on the thickness of the pressed component, the mold temperature, the curing characteristics of the molding compound, and the prewarming.

Table 6. Processing conditions for curable molding compounds[*]

Process	Phenolic resin molding compound	Urethane resin molding compound	Melamine resin molding compound	Melamine–phenol resin molding compound	Polyester resin molding compound
Press molding					
Mold temperature, °C	160 – 190	130 – 160	150 – 170	150 – 170	130 – 190
Molding pressure, MPa	15 – 80	15 – 80	15 – 80	15 – 80	10 – 40
Approximate hardening time per millimeter of wall thickness, s	30 – 60	20 – 40	20 – 40	20 – 40	10 – 20
Injection press molding					
Mold temperature, °C	160 – 190	130 – 160	150 – 170	150 – 170	140 – 190
Outer injection pressure, MPa	50 – 200	50 – 200	50 – 200	50 – 200	30 – 80
Hardening time, s	40 – 120	30 – 120	30 – 120	30 – 120	10 – 60
Injection molding					
Temperature, °C					
Cylinder (feed zone)	65 – 85	70 – 80	65 – 80	65 – 80	40 – 60
Nozzle	85 – 120	95 – 125	90 – 120	90 – 110	60 – 90
Screw	65 – 85	70 – 80	65 – 120	65 – 80	40 – 60
Mold	110 – 140	120 – 140	130 – 140	125 – 135	80 – 110
Compound	170 – 190	140 – 160	160 – 180	160 – 180	150 – 190
Outer injection pressure, MPa	80 – 250	100 – 250	100 – 250	100 – 250	30 – 100
Hardening time[**], s	20 – 80	15 – 80	15 – 80	15 – 80	10 – 60

[*] Optimal conditions for individual cases must be determined experimentally. Some cases require conditions that do not lie in the range described here.
[**] Independent of wall thickness for thickness up to 20 mm.

For phenolic resins without prewarming, a mold temperature of 160 °C, and a molding thickness of 2 – 6 mm, a curing time of 30 – 60 s is estimated per millimeter of wall thickness, whereby the thickest part of the molding must be considered. Curing time can be decreased by increasing the mold temperature and particularly by pre-heating. In this way, the curing time even of thick-walled moldings can be reduced to 1 – 2 min.

Additionally, the curing time can be shortened by ventilation (i.e., temporary opening of the mold shortly after the initial closing stage, which allows any built-up gases to escape).

With *injection press molding*, the most important operating conditions are mold temperature, injection pressure, closing pressure, and curing time. The same applies in this case regarding mold temperature as reported above for pressing techniques.

In practice, injection pressures up to ca. 200 MPa, and in certain cases even higher, are employed. Often independent adjustment of the injection velocity and the injection force is desirable; too high an injection velocity can cause local scorching of the molding due to adiabatic overheating if air present in the mold cannot escape quickly enough.

The required injection pressure depends on the mold (molding design, gating), the molding compound (flow properties, resin substrate), and preheating. The curing time is generally only slightly dependent on wall thickness in injection press molding. When the gate is correctly dimensioned, the molding compound enters the cavity at the full mold temperature, and after the injection phase, only curing takes place. In practice, curing times in the injection press molding of phenolic compounds lie in the range of 0.7 – 2 min.

With *injection molding*, preparation stages (pelletizing, prewarming) are not required. However, several specific operating conditions in addition to those for injection press molding must be considered: the temperature of the plastication and injection units, screw rotation speed during plastication, back pressure during plastication, holding pressure, plastication time, and cycle time. In many cases, the mold temperature exceeds that used in press or injection press molding.

The injection cylinder is generally cooled in the feed zone. The next zone, the metering zone, is generally heated to 65 – 85 °C. The temperature of the front zone of the cylinder and the nozzle should be 85 – 120 °C. Up to a diameter of ca. 50 mm, control of the temperature of the screw is not normally necessary.

Fluid heating systems give good temperature consistency and permit removal of the excess heat of friction that can be generated under certain circumstances during plastication.

The temperature of the compound, measured on the plasticated material injected through the nozzle, is often considerably higher (20 – 40 K) than the temperature of the cylinder. Hotter molding compound can be injected more easily and hardens faster, but the risk of operational interruption is also greater.

Adjustment of the back pressure to counter the reverse movement of the screw during plastication allows control of the plastication stage. At low back pressure (3 – 5 MPa) plastication is still possible while the injection unit is withdrawn; high back pressure (8 – 15 MPa) leads to high compound temperature. Computer control of the back pressure opens up new possibilities for decreasing curing time.

The injection pressure in injection molding corresponds roughly with that of injection press molding with thoroughly preheated compounds. Depending on the flow rating and resin substrate of the molding compound, as well as the complexity of the mold, the injection pressure typically lies in the range 80 – 200 MPa, with an injection time of 3 – 10 s. After injection, a holding pressure is applied. The holding pressure should be 50 – 80 % of the injection pressure and be maintained until the material in the gate has hardened.

The curing time is generally lower with injection molding than with injection press molding. If the layout of the mold (gate) is correct, the curing times for phenolic molding compounds range from 0.3 to 1 min. They are independent of molding thickness over a wide range of values.

The cycle times are determined mainly by the curing time. Only in cases of very long plastication times or very short curing times does the cycle depend on plastication time rather than curing time. Whenever possible, the plastication time in such cases is reduced by processing with a faster screw.

Aminoplastic molding compounds (i.e., molding compounds with urea, melamine, or melamine – phenol resins) have similar processing characteristics to phenolic molding compounds. Hence, the processing information provided above for phenolic compounds is largely valid for aminoplastics. Deviations exist only for the processing temperature and curing time (see Table 6).

Urea resin molding compounds are particularly sensitive to overcuring, which can occur with standard curing times at 170 °C mold temperature, or with prolonged curing times at 160 °C. Melamine and melamine – phenolic resin molding compounds are less sensitive to overcuring, although excessive mold temperatures are not recommended.

With thin-walled moldings, the temperature can be increased to the upper limits of the regions mentioned above; with thick-walled components, however, lower temperatures are used. When press and injection press processes are employed, extensive high-frequency preheating is normal.

The curing time of aminoplastic molding compounds is usually 30 – 50 % shorter than that of corresponding phenolic compounds, given identical processing techniques, although mold temperatures with urea resin molding masses are as a rule 10 K lower.

Polyester resin molding compounds behave differently from phenolic and aminoplastic resins, the latter being condensation products that cure relatively mildly. In contrast, the polymerization reaction involved in curing polyester resins is usually strongly exothermic.

The mold temperature influences flow properties. If the mold is too hot, the resin cures before the mold cavity is completely filled; hence, the mold temperature has an upper limit. At the lower temperature limit the molding sticks in the mold. Between these two extremes, sufficient room for maneuverability exists to allow for malfunction-free operation. Additionally, the working temperature range is determined by the composition of the molding compound, particularly by the type of curing catalyst employed.

Except at low temperature, curing time is usually relatively short; in some cases, a few seconds suffice.

Air that is entrapped in the mold and unable to escape can easily lead to faults in the molding, particularly with injection press molding injection molding. This involves not only scorching but also insufficiently cured areas because atmospheric oxgen can disturb the curing reaction.

Contamination of the molding mass by traces of phenolic resins can impair curing; processing of phenoplastic and polyester resin molding masses should be done separately.

In mold construction, undercuts must be avoided for polyester resin molding masses, since moldings with undercuts cannot be demolded.

Ventilation of the mold is generally not recommended for processing polyester resin molding compounds, because the resulting decrease in the mold temperature is undesirable.

Polyester resin molding compounds are available as granules, long fibers (premix), and resin matting (sheet molding compounds). Granules are used for press, injection press, and injection molding.

Processing conditions are listed in Table 6.

Polyester resin molding compounds with randomly distributed long glass fibers — usually with styrene as cross-linking agent and moist doughy to semidry, strawlike consistency — can be processed at relatively low pressure (ca. 10 MPa for press molding and ≥ 30 MPa for injection press molding) and also injection molded by using special equipment. Additionally, polyester resin premix is frequently processed with low-pressure press molding techniques (see Section 1.6).

Styrene-free, dry polyester molding compounds are also available; they usually contain diallyl phthalate as cross-linking agent. These are also supplied as rods (chopped strands with parallel glass fibers) that require higher processing pressures.

Polyester resin sheet molding compounds are usually processed by press molding to give large-area moldings. Similarly, polyester sheet molding compounds are often processed by low-pressure press molding techniques (see Section 1.6).

2.4. Influence of Processing on Molding Properties

The properties of a plastic molding are determined not simply by material properties, but also determined by other factors, particularly the shape of the molding and the processing itself. For this reason, tabulated values determined on standard samples are valid only for those samples and cannot be transposed to any arbitrarily designed molding. The most important factors from a processing viewpoint are degree of curing and filler orientation.

The *degree of curing* depends on the original state of the molding compound and the processing procedure, in particular the variation of temperature with time and the length of the curing process. The degree of curing may not be constant throughout the molding. A thick-walled press molding, for instance, may be fully cured in the outer layers but still uncured in the center. With moldings made from urea resin compounds overcured, undercured, and well-cured regions may be found within one and the same article.

Overcuring can be recognized by the lack of surface gloss. In addition, bubbles can occur, and the molding may adhere to the steel walls of the mold. Overcuring of aminoplastics can lead to bubbles and, in some cases, cracks.

Variations in curing conditions can lead to internal stresses that might prove to be a drawback if the molding gives the appearance of being correctly cured. Undercuring of the complete molding is naturally a more significant problem; with aminoplastics this is also true of overcuring. Many properties depend on the degree of curing for phenolics and aminoplastics, these include surface gloss, stiffness, dimensions, water uptake, and electrical insulation properties.

With all processing techniques, *filler orientation* depends strongly on the flow process involved. Since, when processing thermosetting molding compounds, the tensile deformations exceed the shear deformations, the filler and reinforcing additives are oriented mainly perpendicular to the flow direction.

Filler orientation particularly determines the dimensional stability and strength of the molding. In injection press molding and injection molding, orientation can be influenced mainly by the gate position and less so by processing conditions. With press molding, the flow and hence the orientation can be influenced by the type of feeding, (e.g., concentrating the material in one position). Orientation has a greater effect with long-fiber reinforcement than with, for example, sawdust.

Molding shrinkage is defined as the difference between the dimensions of the mold and the corresponding dimensions of the molding. It is influenced significantly by the molding compound and the filler orientation, and additionally by the processing conditions, in particular the mold temperature and the pressure in the mold cavity. The influence of filler orientation on shrinkage arises because shrinkage is impeded in the direction parallel to the reinforcing fibers but not in the direction perpendicular to them.

The following applies to the relationship between shrinkage and processing conditions:

1. The higher the mold temperature, the greater is the molding shrinkage
2. The higher the pressure on the molding, the smaller is the process shrinkage
3. Low curing leads to increased molding shrinkage

Distortions arising from local differences in shrinkage are also influenced by filler orientation. The higher the content of inorganic components in a resin substrate, the less pronounced is the tendency to distort. In contrast, the longer the fibers in impact-resistant compounds, the greater is the distortion to be expected. Distortion can also result from local differences in the curing reaction (mold temperatures) or from inhomogeneous cooling of the molding.

The molding shrinkage can be reduced or even completely compensated by adding suitable fillers (e.g., thermoplastic powders), especially in the case of polyester resin molding compounds (low-profile systems).

Mechanical properties are often influenced strongly by filler orientation. With long-fiber reinforcement (glass fibers) and pronounced orientation, the strength in the direction of highest orientation can be several times greater than that in the least oriented direction. Materials containing short-fiber fillers such as sawdust exhibit similar differences in strength in the flow direction and perpendicular to it; however, these differences are small (typically 10 – 30 %). In contrast, curing conditions play only a secondary role in mechanical strength, if the molding has not been strongly over- or undercured.

The electrical insulation properties of phenolics and aminoplastics depend on the degree of curing. Undercuring has a negative effect, particularly on surface resistance.

To attain the best possible insulation characteristics:

1. As hard flowing a molding mass as possible should be used
2. The molding mass should be dried in an oven prior to processing
3. The molding must be perfectly cured

With aminoplastics, overcuring can be as detrimental as undercuring. Phenolic resins are less sensitive to overcuring.

For polyester molding masses, the relationship between degree of curing and electrical properties has minor practical significance. Undercured polyester moldings can be recognized visually and are not suitable for use even though the electrical characteristics are adequate.

2.5. Molds

The quality of the mold has a considerable influence on the manufacturing costs of the molding. With small production runs, the mold should be as simple as possible even if this necessitates use of a finishing process. For mass production, use of a more precise and possibly more complex mold is worthwhile if it avoids finishing.

Most molds for curable molding compounds are manufactured by machining or spark erosion of steel blocks. High-quality steels are employed, particularly for mold cavities. The forming surfaces are polished, hardened, and generally hard chromed to protect them against corrosion and wear. When cavities are manufactured by spark erosion, completely hardened steel is used. Nitriding is rarely employed in molds for processing thermosets. Mold components that are subject to the greatest wear should be easily replaceable or made from hard metals (e.g., gate inserts). With multicavity molds, the cavities themselves are often produced by hollowing with a stamp. Several systems of standard mold units are available with which mold manufacture can be rationalized. They include mold plates made of high-quality steel out of which the mold cavities can be cut, platens, guiding elements, and ejector systems that can be combined in a modular fashion.

Approximately 30 – 40 W/kg is sufficient to heat molds weighing 50 kg or more. Smaller molds and flat molds require up to ca. 80 W/kg. At least 20 mm of steel should separate the heating element from the mold cavity. The temperature sensor should be located between the heating element and the cavity. During operation, the mold temperature should not vary by more than 5 K. Ideally, each component of the mold should be equipped with its own heating and temperature control unit. Especially with small molds, insulation against heat loss is advantageous. Insulation by a thermal insulation sheet attached to the mounting platens is recommended in all cases to reduce heat losses.

In its simplest form, a *press mold* consists of a punch and a cavity (Fig. 18), generally electrically heated and mounted on the moving and fixed tables of a press. If required, an ejector system is built into the punch, cavity, or both, which releases the molding after the mold has been opened. The ejector pins are so distributed that the molding is ejected without indentation or deformation; the pressing surfaces are sufficiently large that they do not penetrate the molding even when it is still hot.

The formation of the flash surface plays a major role in a press mold (Fig. 21). One version or another is preferred, depending on whether a vertical or horizontal flash is desired (a function of the preferred deflashing method), the available molding pressure, how exact the metering is, whether much air or molding mass is driven out via the flash, and the degree of precision demanded for the thickness of the base.

For press moldings with undercuts, side openings, screw threads, metal implants, and other specialties, multicomponent molds, split molds, and molds with movable inserts, side slides, or supplements are employed. Multicavity molds have several identical or different cavities, with separate or common filling volumes.

Injection press molds usually consist of a top part, a bottom part with injection cylinder, and an injection plunger activated by the weaker plunger of the injection press (transfer press, two-plunger press) (Fig. 19). Channels lead from the free end of the injection cylinder to the mold cavities, which are located in the parting surface between the upper and lower parts of the mold. The narrowest part of the channel, the gate, is located at the junction with the mold cavity.

In another type of injection press mold (Fig. 22), the injection cylinder is incorporated in the top part of the mold, and the molding compound is injected through a gate in the top part into the mold cavity. These molds are used on single-plunger presses. Other mold constructions are also available that are suitable for injection pressing on single-plunger presses.

For *injection molding*, mold constructions typical for the injection molding of thermoplastics

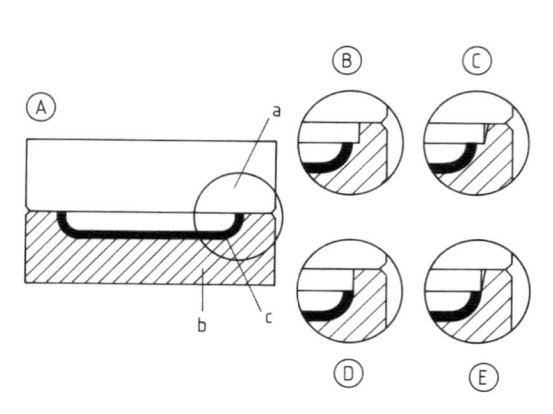

Figure 21. Shaping of the flash
A) Flash mold; B) Flash mold with charge cavity; C) Flash mold, charge cavity with inclined wall; D) Positive mold; E) Positive mold with inclined wall
a) Plunger; b) Punch; c) Mold

Figure 22. Injection press mold for single-plunger press
a) Top part of mold with injection cylinder; b) Bottom part of mold; c) Injection plunger; d) Ejector pin; e) Mold; f) Injection mass with gate

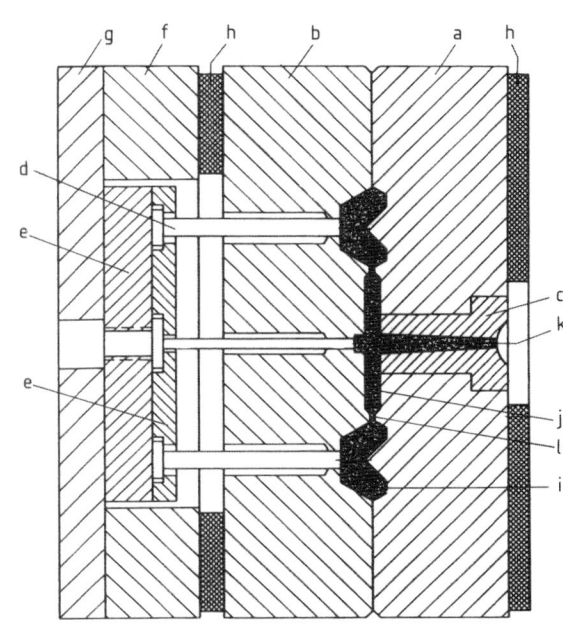

Figure 23. Injection mold
a), b) Mold plates; c) Sprue bush; d) Ejector pin; e) Ejector plates; f) Spacer; g) Adapter plate; h) Heat insulating plate; i) Mold; j) Runner; k) Sprue; l) Gate

are generally preferred; such molds (Fig. 23) have two mold plates, one containing the sprue bush and the other the ejector system.

The cavities and runner systems are incorporated in the parting surface between the two plates.

Molds for injection press molding and injection molding are also available as multi-component molds, split molds, molds with side slides, or screw thread inserts, etc. Molds for injection press and injection molding require venting channels if air contained in the cavity cannot escape quickly enough via leakages in the parting surface or ejector system during injection of the molding compound into the cavity. The channels are generally best positioned opposite the gate and should be approximately 0.1 – 0.2 mm deep and 2 – 8 mm wide. Alternatively, the closed mold can be evacuated prior to injection of the molding compound.

In molds for injection press and injection molding the gate is the narrowest point in the injection channel and lies directly in front of the junction with the mold cavity. The gate determines the flow of the molding compound during injection into the mold cavity and thus also the filler orientation and structure of the cured material.

With optimum dimensioning of the gate, the molding compound is subject to further heating as it passes through this final constriction so that it enters the mold cavity at a temperature just slightly below the mold temperature. If the gate is too narrow, the mold cavity is not completely filled; if it is too wide, the compound enters the cavity at too low a temperature and takes too long to cure. However, in injection press and injection molding, the conditions do not depend only on the gate, but also on the flow and curing properties of the molding compound along with the pressure and temperature.

As a rule of thumb for dimensioning the gate, with curable molding masses, for each cubic centimeter of cavity volume V, a gate cross section F of 1 mm² should be allowed: $F/\text{mm}^2 = V/\text{cm}^3 \pm 50\%$

The margin of $\pm 50\%$ permits special cases to be taken into account. Molding compounds with coarse and fibrous reinforcing materials require larger gates than those containing sawdust; furthermore, the flow properties of the compound, injection pressure, temperature, and design of the molding must also be considered.

The above rule is valid primarily for sheet or film gates (i.e., for rectangular cross sections with height-to-width ratios ranging from 1 : 10 to 1 : 200). Cylindrical pin gates can be made considerably narrower than estimated by the formula. Large amounts of material can generally pass through a pin gate without difficulty; cases are known in which the throughput is greater than 20 cm³/mm². The pin gate is also easier to machine; it can be removed more easily from the molding and normally does not require finishing. It does, however, cause pronounced filler orientation in the immediate vicinity of the gate.

The most homogeneous orientation is achieved with a film gate over the complete width or length of the molding. The film gate should not be thinner than 0.2 mm, and with larger moldings or with impact-resistant masses, no thinner than ca. 0.4 mm.

The runner system should have a much larger cross section, and thus offer less resistance to flow than the gate; depending on the gate type and the position of the runner system, the latter should be three to ten times larger in cross section than the gate. It should approach the mold cavity with as full a cross section as possible so that the actual gate is short.

2.6. Machines

Presses and Injection Presses. Motor-driven *toggle presses* permit pressing forces up to ca. 1 MN. The toggle is so designed that the press initially closes rapidly and with little force, but toward the end of the motion becomes slower and reaches full force only over the final few millimeters. Overloading is prevented by incorporating a slip clutch between the drive and the toggle. The movement and the development of force of the toggle press are in most cases well suited for the conventional processing of curable molding compounds by press molding, however they are unsuitable when the full molding pressure is required over a displacement of several centimeters (e.g., the motion of an injection plunger in injection press molding).

Hydraulic presses dominate for press forces >1 MN; they are widely employed, however, in laboratory presses down to 100 kN and are available up to ca. 100 MN. Depending on construction, distinctions are made between frame presses and column presses; downstroke presses (lower press table fixed, upper table moved by a piston) and upstroke presses; and presses with single drives (usually oil) or with group drives (usually water).

For high-quality moldings, electronically parallel controlled presses are often used. In addition to the main cylinders that activate the closing and pressing mechanism, additional cylinders are present to ensure the parallelism of the press table even under asymmetrical loading.

Two-piston presses, which are often employed in injection press molding, are usually equipped with a powerful upper piston to close the mold and a weaker lower piston to operate an injection cylinder or an ejector. Figure 24 is a schematic of a two-piston press (column form). The force ratio between the pistons of a two-piston press is generally between 2 : 1 and 8 : 1. Apart from those mentioned above, alternative constructions are available (e.g., with injection cylinders positioned above).

Large-surface presses, with pressing forces that are relatively small compared to the area of the plate, are used primarily for processing

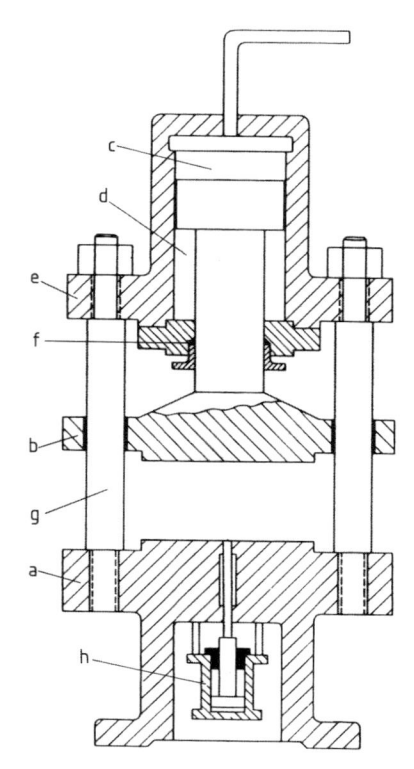

Figure 24. Hydraulic downstroke column press with hydraulic ejector mechanism
a) Fixed lower table; b) Movable upper table with differential piston for drawback; c) Pressure chamber for molding; d) Pressure chamber for drawback; e) Main cylinder; f) Concentric gasket; g) Pillars; h) Hydraulic ejector

resins with glass-fiber reinforcement, and also for low-viscosity polyester resin molding compounds with long glass fibers [e.g., DIN 16 911, type 801 (premix) or DIN 16 913, type 830].

Multiplaten presses with 2 – 20 parallel press platens are used to simultaneously press several identical items. Multiplaten presses, usually with steam-heated and, when required, water-cooled platens are used to press laminates (see Section 2.1.2). The platen size is up to 10 m^2, and the press force up to ca. 100 MN.

Semi- or fully automatic operation of all the press types outlined above is possible. In semi-automatic operation the process stages decisive for processing consistency — closing the press, controlling the curing time, opening the press, the injection stage in injection press molding, and ejection — are automated. The remaining steps — feeding the mold, removing the molding after completion of the program, and cleaning the mold — are carried out manually. With fully automatic plants, manual operations are limited to filling the molding compound hopper and transporting the completed moldings.

Injection Molding Equipment. Injection molding equipment for curable molding compounds (see → Plastics Processing, 1. Processing of Thermoplastics, Section 4.5) is similar to the screw injection molding machines used for processing thermoplastics. The major components are the clamping unit and the injection unit.

The clamping unit is a toggle or hydraulic press that closes and opens the mold and often activates the ejectors.

The injector unit, which consists of a cylinder (with cylinder head and nozzle), preferably fluid-heated, and a screw, meters and plasticates the molding compound and injects it into the mold. For transportation, metering, and plastication, the screw rotates, usually driven by an electric or hydraulic motor. For injection, it moves longitudinally (normally without rotation); in this case the drive is almost always hydraulic. The injection unit can be moved in the longitudinal direction so that the nozzle can either contact the mold sprue bush or be removed from it.

With the majority of machines, the closing and injection units are positioned horizontally one after the other. However, other constructions also exist, for instance, where the closing and injection unit can be rotated into place when needed.

Injection molding of curable molding compounds requires higher injection pressure than for thermoplastics (at least 150 MPa, or better still, 200 – 250 MPa). Many thermoplastic injection molding machines can be modified for use with curable molding compounds.

Injection molding machines for curable molding compounds generally have a capacity of ca. 50 – 1000 cm^3 and a closing force of 5 MN; larger machines are, however, also in operation.

References

General Reference

1 O. Lauer: *Aufbereiten von Kunststoffen*, Hanser Verlag, München 1971. *Aufbereiten von PVC*, VDI-Gesellschaft Kunststofftechnik, Düsseldorf 1976. *PVC-Heißmischen (Dryblend)*, VDI-Gesellschaft Kunststofftechnik, Düsseldorf 1968. *Speichern, Fördern und Dosieren von Kunststoffen*, VDI-Gesellschaft Kunststofftechnik, Düsseldorf 1971. *Granulieren von thermoplastischen Kunststoffen*, VDI-Gesellschaft Kunststofftechnik, Düsseldorf 1974. W. Dalhoff: *Systematische Extruder-Konstruktion*, Krauskopf Verlag, Mainz 1974. H. Domininghaus: *Fortschrittliche Extrudertechnik*, VDI-Verlag, Düsseldorf 1970. *Der Extruder im Extrusionsprozeß*, VDI-Gesellschaft Kunststofftechnik, Düsseldorf 1989. H. Herrmann: *Schneckenmaschinen in der Verfahrenstechnik*, Springer Verlag, Berlin 1972. M. Jacobi: *Grundlagen der Extrudertechnik*, Hanser Verlag, München 1960. W. Mink: *Grundzüge der Extrudertechnik*, Zechner & Hüthig Verlag, Speyer 1973. H. Potente: *Modellgesetze für Ein- und Zweischneckenmaschinen*, Hanser Verlag, München 1981. C. Rauwendaal: *Polymer Extrusion*, Hanser Verlag, München 1986. G. Schenkel: *Kunststoff-Extrudertechnik*, Hanser Verlag, München 1963. E. G. Fischer: *Extrusion of Plastics*, Butterworths, London 1976. *Extrudieren von Schlauchfolien*, VDI-Gesellschaft Kunststofftechnik, Düsseldorf 1973. *Extrudieren von Rohren und Profilen*, VDI-Gesellschaft Kunststofftechnik, Düsseldorf 1974. *Extrudierte Feinfolien und Verbundfolien*, VDI-Gesellschaft Kunststofftechnik, Düsseldorf 1976. G. Menges, H. Recker: *Automatisierung in der Kunststoffverarbeitung*, Hanser Verlag, München 1986. *Rechnergesteuerte Extrusion*, VDI-Gesellschaft Kunststofftechnik, Düsseldorf 1976. Z. Tadmor: *Engineering Principles of Plasticating Extrusion I*, Klein Reinhold Book Corporation, New York 1970. W. Michaeli: *Extrusionswerkzeuge für Kunststoffe und Kautschuke*, Hanser Verlag, München 1990. F. Hensen, W. Knappe, H. Potente: *Handbuch der Kunststoffextrusionstechnik*, vols. **I and II**, Hanser Verlag, München 1989. W. Mink: *Grundzüge der Hohlkörperblastechnik*, Zechner & Hüthig Verlag, Speyer 1969. D. v. Rosato: *Blow Molding Handbook Dominick*, Hanser Verlag, München 1989. R. Holzmann: *Gestalten von Blasformteilen*, VDI-Verlag, Düsseldorf 1971. O. Plajer: *Werkzeuge für das Blasformen*, Zechner & Hüthig Verlag, Speyer 1969. *Spritzblasen*, VDI-Gesellschaft Kunststofftechnik, Düsseldorf 1976. J. L. Throne: *Thermoforming*, Hanser Verlag, München 1987. A. Thiel: *Grundzüge der Vakuumformung*, Zechner & Hüthig Verlag, Speyer 1969. K. Stoeckhert: *Kunststoff-Lexikon*, Hanser Verlag, München 1975. F. Johannaber, K. Stoeckhert: *Kunststoffmaschinenführer*, Hanser Verlag, München 1984. O. Scharz, F.-W. Ebeling, G. Lüpke, W. Schelter: *Kunststoffverarbeitung*, Vogel, Würzburg 1988. D. H. Morton-Jones: *Polymer Processing*, Chapman and Hall, New York 1989. G. Menges: *Einführung in die Kunststoffverarbeitung*, Hanser Verlag, München-Wien 1979. F. Johannaber: Spritzgießmaschinen, in: *Kunststoffmaschinenführer*, Hanser Verlag, München – Wien 1979. W. Elbe: "Untersuchungen zum Plastifizierverhalten von Schneckenspritzgießmaschinen," Dissertation, RWTH Aachen 1973. R. v. Hooven, A. J. Kaminski: "Entgasungsspritzgießen, ein Laborkonzept wird in die Praxis umgesetzt," *Plastverarbeiter* **31** (1990) no. 8, 441 – 446. Th. Copetti, R. Krebser: "Vollhydraulische und Kniehebelschließsysteme," *Kunststoffe* **70** (1980) no. 2, 821 – 825. G. Menges, P. Mohren: *Anleitung zum Bau von Spritzgießwerkzeugen*, Hanser Verlag, München – Wien 1983. S. Schäper: "Nukleierung thermodynamisch getriebener Polyurethan-Reaktionsschäume," Dissertation, RWTH Aachen 1977.

General Reference

2 W. Becker, D. Braun, B. Carlowitz: *Kunststoff-Handbuch*, vol. **1** Die Kunststoffe, Hanser Verlag, München 1990. W. Becker, D. Braun, W. Woebcken: *Kunststoff-Handbuch*, vol. **10**

Duroplaste, Hanser Verlag, München 1988. H. Saechtling: *Kunststoff-Taschenbuch*, 24th. ed., Hanser Verlag, München 1989. K. Stoeckhert: *Kunststoff-Lexikon*, 6th. ed., Hanser Verlag, München 1975. DIN-Taschenbuch 21,Kunststoffnormen,8th. ed., Beuth-Verlag, Berlin 1979. W. Schönthaler: *Verarbeiten härtbarer Kunststoffe*, VDI-Verlag, Düsseldorf 1973. W. Bauer, W. Woebcken: *Verarbeitung duroplastischer Formmassen*, Hanser Verlag, München 1973. W. Bucksch, H. Briefs: *Preßwerkzeuge in der Kunststofftechnik*, 2nd. ed., Springer Verlag, Berlin 1962. K. Stoeckhert: *Formenbau für die Kunststoff-Verarbeitung*, Hanser Verlag, München 1969. Berufsgenossenschaft der chem. Industrie, *Merkblatt für das Verarbeiten von Polyester- und Epoxidharzen*, Verlag Chemie, Weinheim, Germany 1966. VDI-Richtlinie 2007, Epoxidharze im Fertigungsmittelbau, VDI-Verlag, Düsseldorf 1966. VDI-Richtlinie 2011, Faserverstärkte Reaktionsharzformstoffe, VDI-Verlag, Düsseldorf 1973. H. Jahn: *Epoxidharze*, Verlag für die Grundstoffindustrie, Leipzig 1969. F. Breckner: *Glasfaserverstärkte Kunststoffe; Anwendungen, Eigenschaften*, 2nd ed., Verlag Fraunhofer-Gesellschaft, 1989. H. Dominighaus: *Die Kunststoffe und ihre Eigenschaften*, VDI-Verlag, Düsseldorf 1988. O. Schwarz: *Kunststoffkunde*, Vogel, Würzburg 1988. W. Michaeli, M. Wegener: *Einführung in die Technologie der Faserverbundwerkstoffe*, Hanser Verlag, München 1989. R. W. Meyer: *Handbook of Polyester Molding Compounds & Molding Technology*, Chapman and Hall, New York 1987. S. M. Lee: *International Encyclopedia of Composites*, VCH Publishers, New York 1990.

General Reference

3 W. Knig, L. Neder: "Schneiden mit ultraschallerregtem Messer," *Ind. Anz.* **107** (1985) 49. W. König, L. Neder: "Bessere Schnittflächen mit Ultraschall," *Plastverabeiter* **37** (1986) no. 11. "Neue Werkstoffe und hohe Präzision — eine Herausforderung für die Fertigstellungstechnik" in AWK Aachener Werkzeugmaschinen-Kolloquium, (ed.): Produktionstechnik auf dem Weg zu integrierten Systemen, VDI-Verlag, Düsseldorf 1987. W. König, P. Graß, Ch. Wulf: "Machining of Fibre Reinforced Plastics," *CIRP Ann.* **34** (1985)no. 2, 537 – 548. F.-J. Trasser, M. Wehner,"Schnittflächen beim Laserstrahlschneiden von faserverstäkten Kunststoffen CO2-, Excimer-Laser — ein Verfahrensvergleich," *Ind. Anz.* **109** (1987) no. 43/44, 57 – 58. "Bearbeitungstechnik für Verbundwerkstoffe — eine vielschichtige Aufgabenstellung," *VDI-Fachtagung METAV*,June, 8 – 9, 1988. H. K. Tönshoff, V. Hohensee,"Bearbeitung faserverstärkter Kunststoffe," *ZWF, Z. Wirtsch. Fertigung* **81** (1986)no. 2, 106 – 111.

Further Reading

M. Chanda, S. K. Roy: *Plastics fabrication and recycling*, CRC Press, Boca Raton, Fla. 2009.

E. Cybulski: *Plastic conversion processes*, CRC Press/Taylor & Francis, Boca Raton, Fla. 2009.

E. Lokensgard: *Industrial plastics*, 5th ed., Delmar Cengage Learning, Clifton Park NY 2009.

P. A. Tres: *Designing plastic parts for assembly*, 6. ed., Hanser, Munich 2006.

M. Xanthos, D. B. Todd: *Plastics Processing*, Kirk Othmer Encyclopedia of Chemical Technology, 5th edition, vol. 19, p. 536–563, John Wiley & Sons, Hoboken, NJ, 2006, online: DOI: 10.1002/0471238961.1612011924011420.a01.pub2.

Plastics Processing, 3. Machining, Bonding, Surface Treatment

GERT BURKHARDT, Institut für Kunststoffverarbeitung, Aachen, Federal Republic of Germany

ULRICH HÜSGEN, Institut für Kunststoffverarbeitung, Aachen, Federal Republic of Germany

MATTHIAS KALWA, Institut für Kunststoffverarbeitung, Aachen, Federal Republic of Germany

GERHARD PÖTSCH, Institut für Kunststoffverarbeitung, Aachen, Federal Republic of Germany

CLAUS SCHWENZER, Institut für Kunststoffverarbeitung, Aachen, Federal Republic of Germany

1.	Machining	439	2.2.	Adhesion	453
1.1.	Estimating Machinability	439	2.3.	Mechanical Bonding	454
1.2.	Shaving Formation	440	3.	Surface Treatment	455
1.3.	Surface Quality	440	3.1.	Mechanical Processes	455
1.4.	Cutting Forces	442	3.1.1.	Hot Embossing	455
1.5.	Cutting Temperature and Tool Wear	442	3.1.2.	Flocking	456
1.6.	Guidelines for Machining Polymers	442	3.2.	Painting and Printing	456
1.7.	Processing of Fiber-Reinforced Plastics	443	3.2.1.	Painting	456
			3.2.2.	Printing	460
2.	Bonding	445	3.3.	Wet Chemical Deposition Processes	460
2.1.	Welding	446	3.3.1.	Chemical Deposition	460
2.1.1.	Heated-Tool Welding	447	3.3.2.	Electroplating of Plastics	461
2.1.2.	Hot-Gas Welding	450	3.4.	Physical Deposition Processes	463
2.1.3.	Radiation Welding	450	3.5.	Dry Reactive Processes	464
2.1.4.	Friction Welding	451	3.5.1.	Surface-Modifying Processes	464
2.1.5.	Induction Welding	452	3.5.2.	Coating Processes	466
				References	467

1. Machining

1.1. Estimating Machinability

The special properties and low cost of thermoformed plastics have led to their widespread use in the construction of machines and other equipment. Machining of components from plastics is significant because of the increasing number of design variations of manufactured components; the demand for greater stength and for cheap production of samples, individual articles, and small series; and the fabrication of thermoformed components (e.g., holes in PVC window frames for hinges). An important prerequisite for economical machining is the selection of optimum cutting conditions. This requires knowledge of the machinability of polymers and the wear resistance of the tools. A prerequisite for the economical utilization of machining techniques is a knowledge of the parameters that influence the machinability of the material. Experience from the metalworking industry cannot be applied directly to polymers. Even a comparison between the physical properties of metals and those of polymers shows that certain factors exerting a large influence on the machinability of metals are only slightly or not at all significant for plastics. Furthermore other factors can arise in polymer processing, such as the generation of dust, evolution of gaseous compounds, and melting of the component or of the shavings removed from it, that do not occur with metals.

The following criteria, which are also used in the metal industry, are used for judging machinability:

1. Effective machining forces
2. Resulting machining temperatures
3. Attainable surface quality, shaving formation, and wear of machining tools

1.2. Shaving Formation

With thermoplastics, shaving formation influences the attainable surface quality and the static and dynamic process controlling the cutting action and advancing force. The tendency to form long, ribbonlike shavings often hinders machining, and the poor thermal conductivity causes welding of the shavings, as well as distortion, leveling, melting, and smearing of the surface.

In turning polyamides, depending on machine conditions and material properties, three basic types of shaving may result, which can generally be broken down into continuous and discontinuous forms (Fig. 1). This figure shows that the shape of the shaving is a result of the type of shaving, which depends on water content, cutting geometry, and cutting speed, and its cross section.

The shape of the shaving is assessed by using the so-called shaving volume number, which is the ratio of the volume required by an amount of disordered shavings to the volume of material from which they were produced. It is most favorable for discontinuous shavings.

Drilling low-density polyethylene (LDPE), polyacetal, and polyamides with spiral drills gives similar shaving shapes, with three characteristic types (folded shavings, helical shavings, lamellar shavings) [4]. Milling of polyamides gives shaving typical of those obtained from machining metals (ribbon shavings, snarl shavings, spiral shavings, helical shavings, crumble shavings, needle shavings, and melted shavings).

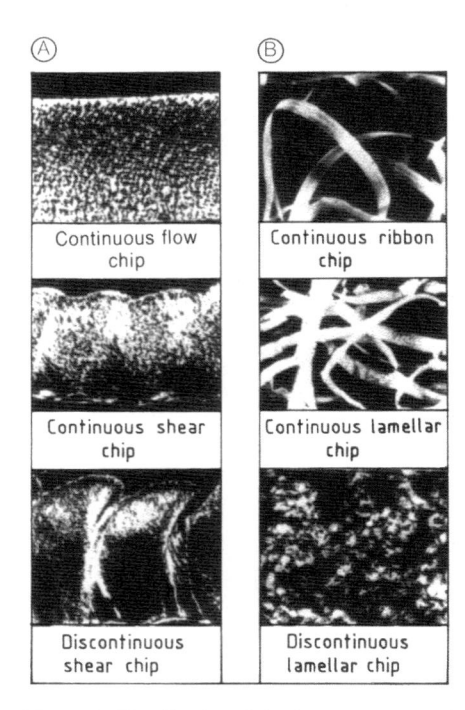

Figure 1. Classification of shaving types (A) and forms (B)

1.3. Surface Quality

In the qualitative evaluation of the turned surfaces of thermoplastic polymers, depending on shaving conditions, three different characteristic surface structures can arise (Fig. 2): (1) a groove profile, (2) a smeared groove profile, and (3) a groove profile with breakouts. The occurrence of surface breakouts under certain machining conditions during turning was first reported by SPUR and ZUG [6–8] for polyamides and polyacetal, and for poly(vinyl chloride) and polyacetal resins by FERICHMANN [9]. With several materials, the tendency to form surface breakouts is particularly pronounced (e.g., rigid PVC and bulk-polymerized polyamide [9], [10]).

As with metals, the surface roughness value T_t as defined by DIN 4760 can be used for a quantitative characterization of surface quality. It is the maximum vertical separation between the highest and the lowest points of a surface structure. Experimental determinations during turning of polyamide illustrate that the maximum roughness increases with increasing feed, whereby water-containing materials exhibit higher roughnesses than dry materials [6],

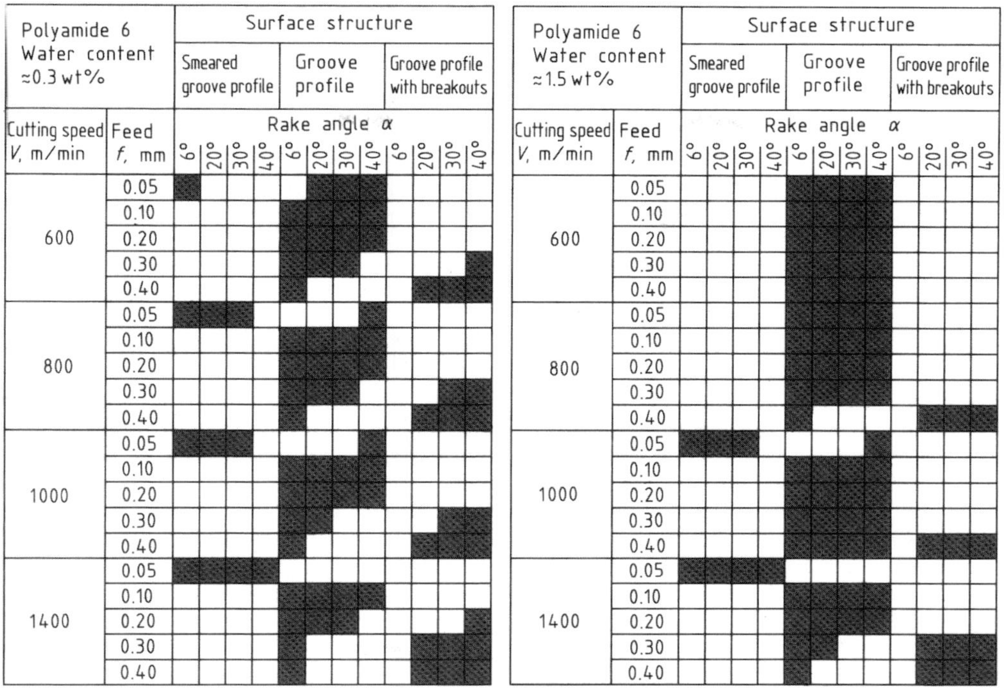

Figure 2. Occurrence of certain surface structures as a function of cutting speed, effective cutting angle, feed, and water content of Ultramid KR 1337

particularly under conditions of large feed. The surface roughness decreases with increasing cutting angle. For polyamide 6 containing 15 wt % glass fiber and polyamide 6 with 5 wt % graphite, increasing machining time causes an increase in surface roughness, whereas with the extruded materials polyamide 6 and polyamide 6 containing 1 wt % MoS_2, the surface roughness remains unchanged. The time dependence of the surface roughness with glass- and graphitefilled polyamides results from increased tool wear caused by the fillers.

In milling and grinding polyamides, surface melting generally occurs with tooth feeds $f < 0.1$ mm. For $f > 0.02$ mm, the surface roughness increases with increasing feed, whereby the best surface quality is attained with a rake angle of $\alpha = -6°$ to $+6°$. With short machining times of <30 s and cutting speeds $V<500$ m/min, the best surface quality is obtained with a small clearance angle ($\gamma<5°$). Longer cutting times and higher cutting speeds ($V>500$ m/min) require larger clearance angles due to increased warming of the cutting tool, particularly with small tooth feeds. Large stressed cross sections can lead to increased *breakouts* from the surface, proportional to the cutting speed. Surface *breakouts* occur more frequently with greater depth of cut and positive rake angle.

As a consequence of restricted shaving removal and the continuously changing cutting geometry of spiral drills, large surface roughness values result from drilling. The surface quality is influenced mainly by the heat generated during the machining process.

Drilling of polyamide and polyacetal showed that with high levels of feed, lamellar shavings occur that are accompanied by surface breakouts [10]. In contrast, in drilling LDPE, no lamellar shavings occur even at greater feeds, and the surface remains free of breakouts. With bulk-polymerized polyamide 6, polyamide 6 containing 0.5 wt % MoS_2, polyamide 6 containing 25 wt % glass fibers, polyamide 6 containing 35 wt % glass fibers, polyamide 66 containing 25 wt % glass fibers, polyamide 66 containing 35 wt % MoS_2 and polyacetal, the surface roughness increases once a threshold feed value has been exceeded ($f = 0.4$ mm; with polyacetal, $f = 0.8$ mm). An exception to this tendency is polyamide 6 containing 5 wt% MoS_2 at a cutting speed V of 40 m/min, and also LDPE. With polyamide the surface roughness decreases with increasing feed, whereas for LDPE the surface roughness is almost independent of feed. Since surface quality is reduced by melting of the bore wall, particularly at small feeds and high cutting speeds, it can be improved by cooling. Although external application of a drilling oil emulsion causes little or no improvement in surface quality, surface roughness can be reduced, especially at high cutting speeds, by cooling with compressed air via a series of internal cooling channels. The best surface quality is obtained at feeds of $f<0.8$ mm by circulating drilling oil through internal cooling channels.

1.4. Cutting Forces

The effective cutting and feed forces involved in machining thermoplastics are generally far lower than those required for metals. The cutting forces that occur on turning polyamides increase almost linearly with increasing feed and depth of cut, if shaving shape is constant. Another significant factor affecting cutting force is the machining angle. Thus, for the same shaving cross section, the cutting force during turning of polyamide at a rake angle of 50° can be 50% lower than at a rake angle of 0°. This applies only when the moisture content is constant [6]. The correlation is particularly significant for the widely used polyamides (PA 66, PA 6) because the high moisture absorption of these materials leads to considerable changes in mechanical properties. In turning polyamides, the cutting forces are approximately 30% lower for a material with a moisture content of 3.5 wt% than for a dry sample.

When drilling polycarbonate, polyethylene, poly(vinyl chloride), poly(methyl methacrylate), polystyrene, acrylonitrile–butadiene–styrene (ABS) polymers, polytetrafluoroethylene, polyacetal, polypropylene, and polyamide, the torque and feed force increase linearly or slightly degressively with increasing feed. The speed of the spiral drill has little influence on shaving force [11]. HOFFMANN investigated the influence of cooling lubricants on feed force and torque in drilling bulk-polymerized polyamide 6 and showed that when cooling lubricants are employed, the advancement force increases slightly because the strength of the polymer is extremely temperature dependent [4]. Whereas the torque increases significantly when internal compressed air cooling is employed, the lubricating effect that occurs when drilling oils are used causes it to increase to a considerably lesser extent due to the reduction of the fiber friction at the bore walls. The lowest values are obtained without cooling by using drills equipped with internal cooling channels.

1.5. Cutting Temperature and Tool Wear

The temperatures that arise during the machining of unfilled thermoplastics are so low that they do not influence the lifetime of the machining tools. Measurements carried out during drilling and turning have shown that the lowest temperatures occur in the machining of polyethylene, and the highest with polyamide. Under certain machining conditions, temperatures can exceed the melting point of the material being processed. However, due to the low thermal conductivity of polymers, this affects only the surface areas. The cutting temperature is particularly critical when gaseous decomposition products can form, which is especially applicable to nitrogen-containing polymers, phenol – formaldehyde resins, and fluoropolymers [12].

Tool wear is considerably lower in the machining of unfilled polymers than it is with metals [13]. In machining filled polymers, a significantly greater amount of wear occurs compared to unfilled plastics. This is especially true for graphite-filled materials, although glass-fiber-filled materials also cause increased tool wear [4, 6, 14, 15].

1.6. Guidelines for Machining Polymers

Tables 1–3 list guidelines for turning, milling, and drilling thermoplastics and thermosets with hard metal tools. Standardized tools are used for machining polymers, with cutting edges made

Table 1. Guidelines for turning plastics

Material	Type of turning	Depth of cut, mm	Cutting speed, m/min	Feed, mm	Cutting geometry, degrees		Attainable surface quality, µm
					Clearance angle γ	Rake angle α	
Thermoplastics							
Polyolefins	rough	3 – 5	250 – 400	0.5 – 1	5 – 15	≤ 0	6 – 10
	smooth	1 – 3	350 – 600	0.1 – 0.6			
Polystyrene	rough	1.5 – 4	80 – 120	0.1 – 0.2	8 – 12	≤ 0	3 – 6
	smooth	0.5 – 1.5	100 – 200	0.02 – 0.08			
PVC, rigid	rough	2 – 6	100 – 250	0.4 – 0.5	12 – 18	−5 to 0	3 – 10
	smooth	0.5 – 1	250 – 350	0.1 – 0.2			
Poly(methyl methacrylate)	rough	1.5 – 3	100 – 150	0.1 – 0.2	10 – 15	0 – 5	6 – 10
	smooth	0.5 – 1.5	200 – 350	0.05 – 0.1			
Polyamide	rough	2 – 4	100 – 200	0.2 – 0.5	5 – 15	35 – 45	3 – 6
	smooth	1 – 2	200 – 250	0.1 – 0.2			
Polyacetal	rough	2 – 5	120 – 200	0.2 – 0.3	10 – 15	−5 to 5	6 – 10
	smooth	0.4 – 2	180 – 300	0.1 – 0.25			
Thermosets							
Phenolics	rough	2 – 4	200 – 300	0.2 – 0.3	5 – 10	20 – 25	10 – 20
	smooth	0.5 – 1.5	350 – 450	0.08 – 0.15			
Aminoplastics	rough	1.5 – 3.0	200 – 350	0.1 – 0.15	5 – 10	20 – 25	10 – 20
	smooth	0.5 – 1.0	400 – 500	0.05 – 0.1			
Paper- and fabric-based laminate	rough	2	100 – 180	0.3 – 0.5	12 – 18	15 – 25	6 – 20
	smooth	0.5 – 2	200 – 250	0.1 – 0.2			
UP – GF*	rough	1.5 – 3	90 – 120	0.1 – 0.2	8 – 10	−5 to −2	10 – 20
	smooth	0.5 – 1.5	120 – 150	0.05 – 0.1			

*UP – GF = glass-fiber-reinforced unsaturated polyester resin.

of high-speed steel, hard metal, or ceramic. For the production of small series, bits with cutting edges made of high-speed steel can be used, whereas for mass production, hard metal is preferred because less wear occurs.

1.7. Processing of Fiber-Reinforced Plastics

The processing of fiber-reinforced plastics (see → Plastics, Processing, 2. Processing of

Table 2. Guidelines for milling plastics

Material	Depth of cut, mm	Cutting speed		Feed per tooth		Cutting geometry, degrees		Attainable surface quality, µm
		Rough, m/min	Smooth, m/min	Rough, mm	Smooth, mm	Clearance angle γ	Rake angle α	
Thermoplastics								
Polyolefins	1 – 5	250 – 350	300 – 500	0.2 – 0.3	0.05 – 0.2	15 – 20	−5 to 0	6 – 10
Polystyrene	1 – 5	150 – 200	250 – 400	0.3 – 0.5	0.04 – 0.1	20 – 25	10 – 20	6 – 20
PVC, rigid	3 – 8	250 – 350	350 – 550	0.4 – 0.6	0.05 – 0.3	25 – 30	20 – 25	6 – 20
Poly(methyl methacrylate)	1 – 5	130 – 150	180 – 300	0.1 – 0.2	0.03 – 0.1	25 – 30	20 – 25	4 – 8
Polyamides	1 – 5	100 – 140	150 – 180	0.2 – 0.5	0.05 – 0.1	25 – 30	20 – 25	6 – 10
Polyacetals	1 – 4	120 – 160	150 – 200	0.1 – 0.3	0.03 – 0.1	6 – 14	0 – 8	6 – 10
Thermosets								
Phenolics	3 – 7	150 – 250	200 – 300	0.3 – 0.6	0.08 – 0.4	20 – 30	20 – 25	6 – 10
Aminoplastics	3 – 7	150 – 250	200 – 300	0.4 – 0.8	0.08 – 0.3	20 – 30	20 – 25	6 – 10
Paper- and fabric-based laminate	1 – 4	80 – 150	150 – 250	0.2 – 0.4	0.08 – 0.1	20 – 25	20 – 30	6 – 10
UP – GF *	1 – 5	50 – 150	200 – 600	0.1 – 0.2	0.05 – 0.1	10 – 15	5 – 10	10 – 20

*UP – GF = glass-fiber-reinforced unsaturated polyester resin.

Table 3. Guidelines of drilling plastics [35]

Material	Average drill speed for a given drill diameter, min^{-1}			Feed, mm	Clearance angle γ, degrees	Rake angle α, degrees	Point angle σ, degrees	Recommended drill oversize, mm	Coolant
	4 mm	10 mm	20 mm						
Thermoplastics									
Polyolefins	4000	1800	1100	0.1 – 0.5	10 – 20	0 – 10	110 – 120	0.15 – 0.2	none
Polystyrene	800	550	180	0.1 – 0.3	6 – 10	5 – 8	49 – 90	0.05 – 0.08	water
PVC, hard	2000	950	630	0.2 – 0.6	10 – 30	0 – 5	110 – 130	0.06 – 0.09	compressed air
Poly(methyl methacrylate)	1000	600	300	0.2 – 0.3	8 – 12	10 – 15	60 – 90	0.05 – 0.08	water
Polyamides	7500	2500	1000	0.1 – 0.4	10 – 12	8 – 10	100 – 120	0.08 – 0.12	compressed air
Polyacetal	4800	2200	1300	0.1 – 0.25	10 – 15		60 – 100	0.08 – 0.10	none
Thermosets									
Phenolics	4000	1600	800	0.1 – 0.3	6 – 8	12 – 24	60 – 100	0.02 – 0.05	compressed air
Aminoplastics	4800	2000	850	0.05 – 0.15	5 – 10	10 – 15	60 – 120	0.06 – 0.08	compressed air
Paper- and fabric-based laminate	3500	1200	650	0.2 – 0.4	15 – 20	10 – 15	90 – 110	0.05 – 0.08	compressed air
UP – GF*	9000	5400	2800	0.05 – 0.1	10 – 12	14 – 16	140 – 150	0.03 – 0.05	water

*UP – GF = glass-fiber-reinforced unsaturated polyester resin.

Thermosets, Section 1.5) leads to more problems than the processing of unreinforced or filled polymers.

During the manufacture of a fiber composite component, several processing operations are generally required, in which the properties of the fiber material must be taken into account. Figure 3 gives an overview of the relevant properties and the fracture behavior of glass, carbon, and aramid fibers. The brittle fracture behavior of glass and carbon fibers contrasts with the ductile failure of aramid fibers, which tend to split in the fiber direction because of their molecular structure. In addition to the physical properties of individual fibers, the laminate structure, fiber orientation, fiber length, and fiber content of the product also influence processing characteristics. In general, machining operations are required at the beginning and the end of the manufacturing sequence: to cut the semifinished product to size and to produce the desired final contours. Continuous fiber sheet molding compounds are a widely used starting material. Trimming is generally performed manually with knives, scissors, or special equipment. Even this processing stage exhibits a range of problems and requires the use of tools adapted to suit the type of fiber employed. Difficulties also arise from adhesion of resin and fiber fragments to the equipment, which prevents a clean cut. Frequently, fibers are pulled out of the matrix, and ragged cut edges result. Prepreg trimming can be improved by using ultrasonically vibrating blades.

After shaping and curing, a finishing operation is generally necessary. This can involve edge trimming and deflashing, as well as generation of dimensionally precise functional surfaces. Finishing can be carried out by classical techniques such as milling, turning, and drilling, or by newer methods such as laser, water-jet, or abrasive water-jet cutting. None of these methods is suitable for all finishing operations. For example, laser and water-jet cutting are not well suited for generating high-quality functional surfaces, but have advantages in edge trimming and deflashing or in introducing *cutouts*.

Figure 4 shows the topography of cut surfaces generated by laser cutting, water-jet cutting, and milling of a sheet molding compound (SMC) material. The condensation of decomposed material and melt residues on surfaces cut by laser are typical of this method. The porous, relatively smooth surface is characteristic. Cracking of the surface indicates that the sample has suffered some thermal damage. Cracks are also apparent in the surface cut by a water jet. These are caused by local mechanical overloading of the fiber – matrix composite as a result of pressure buildup in the sample. Damage occurs slight unevenness is apparent, and the surface is macroscopically free from machining damage. Milling also permits the highest advancement speed. The brittle fracture characteristic of glass fibers is an advantage for mechanical methods such as water-jet cutting or milling, and the large difference between the melting points and decomposition temperatures of the fiber and matrix materials causes problems with processing by use of thermal energy.

Additionally, the high abrasiveness of glass and carbon fibers also causes problems. Milling thus necessitates the use of hard metal tools. Often polycrystalline diamond (PCD) cutting edges are preferred. Despite the considerably higher cost compared to hard metal tools, the use of PCD is often justified by improved processing quality and reduced machine down time due to longer lifetime of the tool.

In processing aramid-reinforced materials, the abrasive effects do not cause as many problems as the high toughness of the fibers and hence the ragged appearance of the cut edge. For this reason, opposing screw formed ground hard metal bits are preferentially employed to grind aramid composites in contrast to one- or two-bit heads to grind GRP and carbon-reinforced plastic (CRP) (Fig. 5).

A comparison of the various methods for finishing fiber-reinforced plastics is given in Table 4.

Apart from technological and economic aspects, environmetal aspects of the processes will become more important in the future, including the emission of pollutants and noise.

2. Bonding

Bonding processes for polymers can be categorized into three groups:

1. Welding
2. Adhesion
3. Mechanical bonding

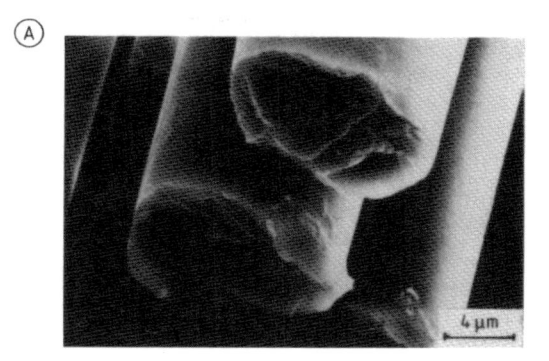

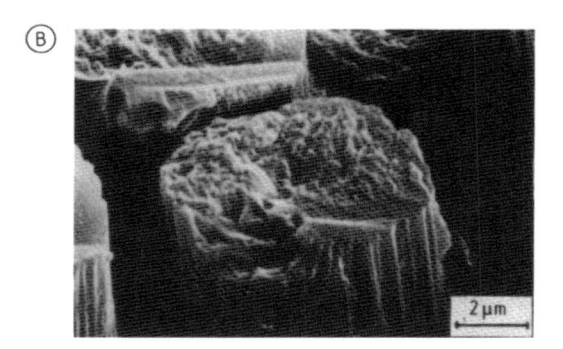

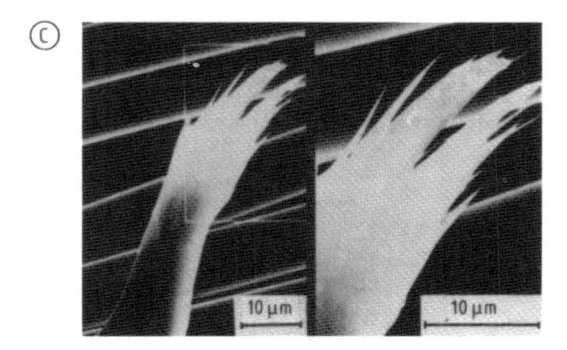

Figure 3. Properties and fracture characteristics of glass (A), carbon (B), and aramid (C) fibers

mainly to areas in the vicinity of the jet outlet because, in this position, support by underlying material is either partially or totally absent. Characteristic surface features include the almost parallel ridges oriented against the direction of advancement, whose course and size can be influenced over a wide range by the choice of process parameters. The best optical appearance is given by the milled surface. In this case, only

A designer employing polymers has a limited choice of bonding process because fully cross-linked polymers cannot be welded, and several non-cross-linked polymers can be adhered to only with extreme difficulty or to a certain degree. All polymers can be mechanically

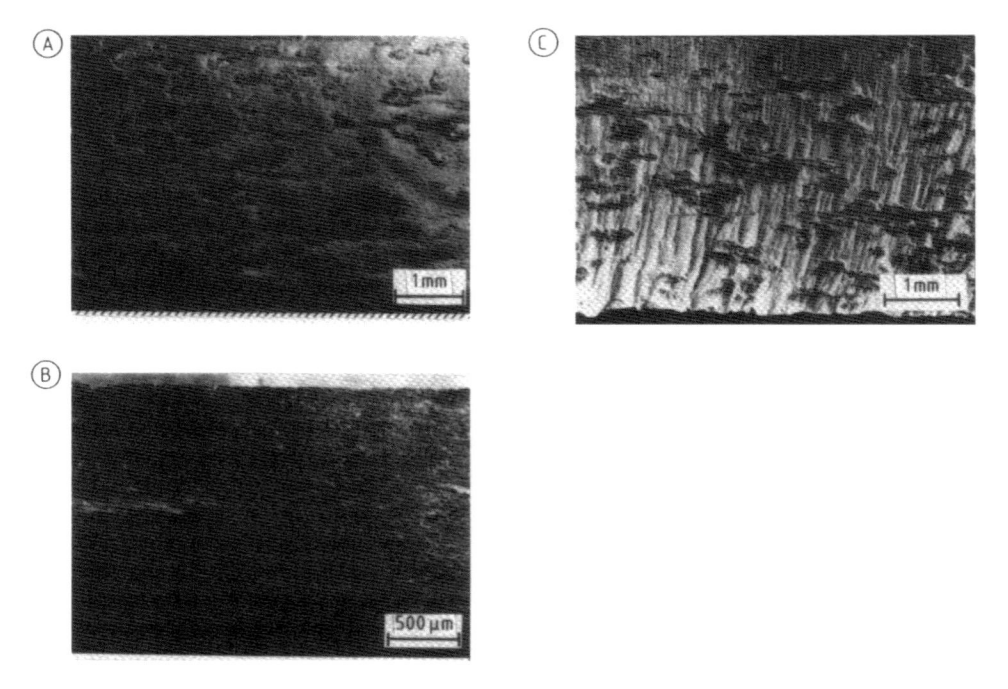

Figure 4. Cut surface of a polyester – glass fiber SMC component after laser cutting (A), milling (B), and water jet cutting (C)

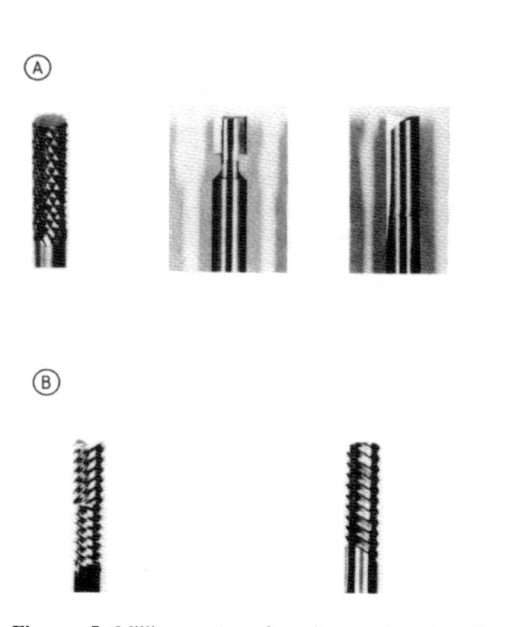

Figure 5. Milling cutters for glass and carbon-fiber-reinforced plastics (A) and aramid-fiber-reinforced plastics (B)

bonded to each other or to other materials, (e.g., by riveting, bolting, or clips). However, with fiber-reinforced polymers, because of anisotropy, clip and plug bonds are problematic, and the sensitivity to notch effects places constraints upon riveting and bolting.

Welding and adhesion yield permanent bonds; this also applies to riveting. All other mechanical bonds can be regarded as reversible. The choice of bonding technique is determined by processing factors, particularly bonding time. Welding and mechanical bonding require little time, whereas adhesion necessitates longer periods to allow for surface pretreatment and hardening of the adhesive.

2.1. Welding

Only polymers that can be softened or melted by heating can be welded (i.e., non-cross-linked polymers) [16, 17]. Exceptions to this include poly tetrafluoroethylene, heat-resistant polyaromatics such as polyimides and polybenzimidazoles, and ultrahigh-molecular-mass thermoplastics (e.g.,

Table 4. Criteria for evaluating the processing of fiber-reinforced plastics

	Process			
Criterion	Water-jet cutting	Abrasive water-jet cutting	Laser cutting	Contour milling
Mechanical stress	low	low	very low	medium
Thermal stress	very low	very low	very high	low – medium
Pollution	none	none	very high	high (dust)
Noise	high	high	medium	very high
Width of cut	low	medium	low	tool diameter
Shape of cut	slightly conical	conical	slightly conical	any
Minimum radius, mm	0.5 – 1.0	1.5 – 2.0	0.2 – 0.5	tool diameter
Symmetry of cut edge	parallel to jet direction	parallel to jet direction	parallel to jet direction	perpendicular to tool axis
Tool wear	low	medium – high	very low	medium – high
Acessibility of the workpiece	both sides	both sides	both sides	one side

UHMW-PE and cast PMMA), whose softening or melting temperature lies above their thermal decomposition temperature.

During welding, the polymers are heated in the region surrounding the weld seam until they pass into the plastic state. They can then be joined under pressure, whereby macromolecules of the two substrates undergo mutual diffusion processes.

Apart from possible thermal chain degradation, chemical reactions do not take place during polymer welding. To obtain an acceptable, materially intact joint, the material flow process in the plane of the seam must take place at a temperature comparable to that of the original forming process (injection molding, extrusion, etc.).

Various welding techniques are generally classified according to the way in which energy is introduced:

1. Thermal conduction
2. Convection
3. Radiation
4. Friction
5. Induction

The following welding techniques are commonly used today (the energy source involved is given in parentheses):

1. Heated-tool (HT) welding (thermal conduction)
    a. Direct heating of seam area
   HT butt welding
   Heated wedge welding
   Welding by bending
   HT socket welding
   Electric socket welding
    b. Indirect heating of seam area
   Thermal sealing
   Heat impulse welding
2. Hot-gas (HG) welding (convection)
   HG area welding *
   Draw welding *
   HG extrusion welding *
3. Radiation welding (thermal radiation)
   HT radiation welding
   Laser welding
   Laser extrusion welding *
4. Friction welding processes (friction)
    a. Heating via internal friction within the material
   High-frequency welding
   Ultrasonic welding
    b. Heating via external friction
   Rotation friction welding
   Vibration welding
5. Induction or electromagnetic welding (induction) *

(Asterisked processes require the use of welding rods)

2.1.1. Heated-Tool Welding

In HT techniques the welding zone is heated conductively by electrically heated metallic elements. In direct HT welding, heat flows directly from the heating element to the weld zone; with indirect HT welding, heat transport proceeds from an external source through the welding substrate to the welding zone. Because of the poor thermal conductivity of polymers, indirect

HT welding is used only for thin materials (films).

In *direct HT welding*, both welding partners are heated by the same heating element. For this, the components are mounted in their respective welding positions on two mechanisms that can be closed against each other. The heating element is introduced between the welding substrates, and the sections to be heated are pressed against the front and back surfaces.

Once the desired temperature has been reached, both welding partners are lifted briefly from the heating element, which is then removed, and the welding process is completed by contacting the welding partners under pressure. In all HT welding processes, the following parameters influence the quality of the weld seam:

1. Melt length (reduction in dimensions of the welding substrate during heating phase)
2. Heating time
3. Time required to remove heating element
4. Seam pressure
5. Holding time

Heated-tool butt welding is simple to automate, which is essential for reproducible welding seam quality. It requires short cycle times and is particularly suitable for welding partners with large weld seam areas (solid preforms, pipes, window frame profiles, transport pallets) and also for fine welds (e.g., taillights for automobiles). Due to melting of the bonding partners, a so-called weld bead is usually formed, which in some cases must be removed in a separate step. In the most favorable cases, the strength of the weld can reach the strength of the bonded materials.

Heated-wedge welding is a continuous bonding process for polymer films and plastic-coated fabrics (Fig. 6 A). In the manual process, the heated metal wedge and the pressure roll are operated by hand. In the mechanized version, the heated wedge and the transport and pressure roll are fixed, and sheets to be welded are drawn over the wedge. The thickness of overlapping films can reach 0.4 mm. Welding speeds up to 15 m/min can be reached.

Welding by bending is a manual technique used for sheets and plates (Fig. 6 B). The wedge-

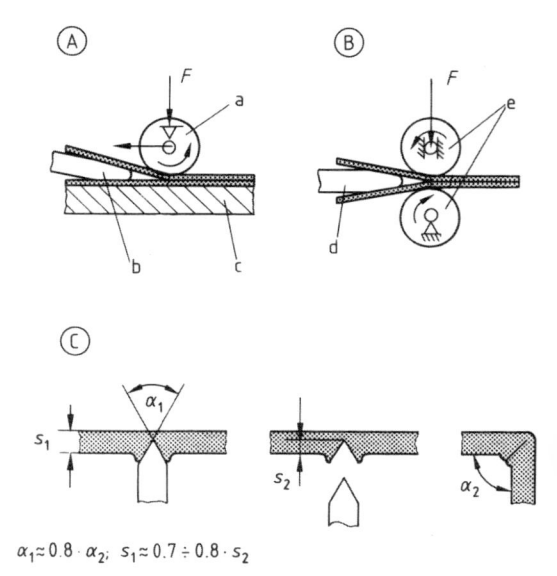

$\alpha_1 \approx 0.8 \cdot \alpha_2; \quad s_1 \approx 0.7 \div 0.8 \cdot s_2$

Figure 6. Manual (A) and mechanized (B) heated wedge welding, and welding by bending (C)
a) Pressure roll (manually operated); b) Heated wedge (movable); c) Support; d) Heated wedge (fixed); e) Transport and pressure rolls

shaped tip of the heated tool is inserted three-quarters of the way into the welding substrate and is removed after a period of heating (essentially pressureless). The sheets are welded together by bending. To ensure that sufficient welding pressure can be introduced, the angle at the tip of the heating tool should be approximately 20 % smaller than that at the opposing end. To reduce the heating period for thick sheets, a wedge-shaped groove can be milled into the sheet.

Heated-tool socket welding is a process for bonding pipes with sockets by means of an overlapping weld seam (Fig. 7 A). The heating element has the inner contours of the socket on one side and the outer contours of the pipe on the other. The dimensions of the heating element, pipe, and fitting are such that after heating, removal of the heating element, and repositioning, pressure is built up at the weld seam during welding. The technique is employed almost exclusively for pipes made of PE and PP.

Electric socket welding is another technique for bonding pipes. The injection-molded socket has electrical heating coils incorporated in the inner surface, which, after positioning at the end

An *indirect heated-tool* welding technique is thermal contact welding, which finds widespread application for moldings and films of low wall thickness (<1 mm) (Fig. 8). The surface of the continually heated element can be structured to permit combination of the welding process with embossing. Areas of application include packaging technology (heat sealing, welding on lids, closing plastic sacks, etc.). With heat impulse welding, the heating element is not heated continually (Fig. 8 B). At the onset of each welding phase, thin metal rods coated with an antiadhesive layer [generally polytetrafluoroethylene (PTFE)] are subjected to a short but powerful current impulse and are thus heated. The heat passes through the film to the weld zone by conduction. In the two-sided process, the temperature profile in the weld zone is symmetrical because of the arrangement of the heating bars, but it is asymmetrical in the one-side process. The main application is in packaging technology, although this technique is not suitable for welding large areas because of the heaters used (narrow bus bars). Because the welds cool via the ram and counterram, cycle times are shorter than in heat contact welding.

In all indirect HT welding techniques, a temperature profile across the thickness of the weld seam is formed during heating, with the highest temperature on the surface nearest the heating element. The welding parameters (HT temperature and heating time) must therefore be

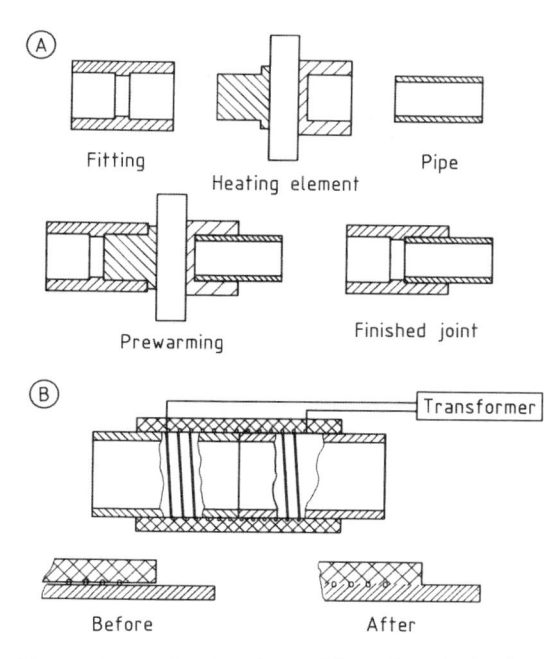

Figure 7. Heated-tool socket welding (A) and electric socket welding (B)

of the pipe, are connected to a welding transformer (Fig. 7 B). The inner surface of the sleeve and the outer surface of the pipe are plasticated. Because the sleeves are generally injection molded with intrinsic orientation, shrinkage occurs on heating, which generates the radial pressure required for welding. The coils remain within the bond.

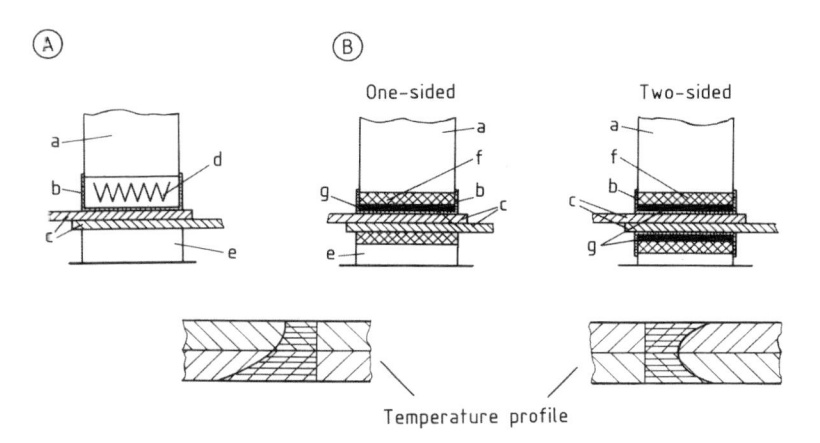

Figure 8. Thermal sealing (A) and heat impulse welding (B)
a) Ram; b) Antiadhesive coating; c) Films; d) Heating element, continuously heated; e) Counterram; f) Thermal insulation; g) Impulse heating bar, discontinuously heated

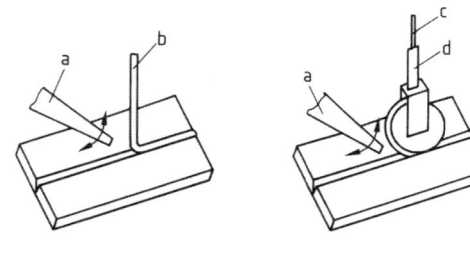

Figure 9. Hot-gas welding
a) Hot gas; b) Welding rod (manually guided); c) Welding rod; d) Pressure roll (manually guided)

adjusted so that a sufficiently high temperature is reached in the welding plane, without overheating the polymer on the surface.

2.1.2. Hot-Gas Welding

Hot-gas welding is usually carried out manually.

The heat-transfer medium is generally hot, clean compressed air; for oxidation-sensitive materials, inert gases such as nitrogen or carbon dioxide are also used. The surfaces to be joined are heated via the hot gas stream and welded under pressure, usually with welding rods (Fig. 9). This method is particularly suitable for assembly and repair, as well as machine and container construction. Use of hot-gas welding is recommended when access to the weld seams is difficult or when the materials concerned are difficult to weld.

The most important requirements for high weld quality are constant welding pressure, gas temperature, and gas flow rate, as well as precise guiding of the gas nozzle and welding rod.

Three variations of hot-gas welding exist.

In swirl welding, the welding substrate and welding rod are heated by a swirling motion of the circular hot-gas die. The welding rod (a few millimeters in diameter) is fed perpendicular to the surface to be welded (Fig. 9). With soft weld rods, a pressure roll is used to avoid folding of the welding rod.

In draw welding, the welding rod is warmed separately in a specially formed hot-gas nozzle. The metal housing removes heat from the gas stream and transfers it to the welding rod by conduction. The welding rod leaves the die in a

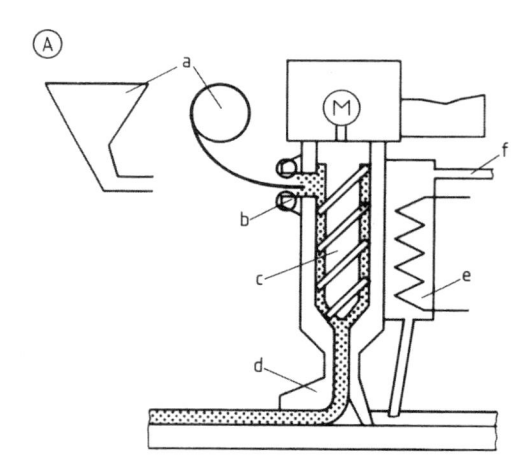

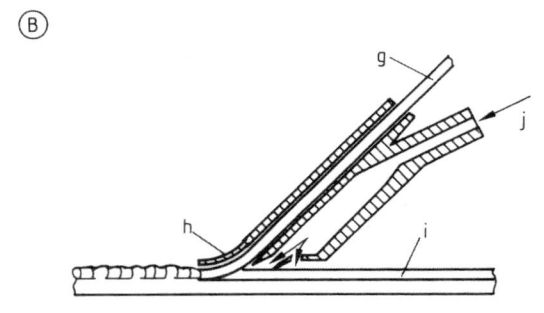

Figure 10. Extrusion (A) and draw (B) welding
a) Pellet hopper or wire reel; b) Wire feed; c) Extruder; d) Welding shoe; e) Heater; f) Air supply; g) Welding rod; h) Pressure lip; i) Part to be welded; j) Hot air

plasticated state and is pressed onto the basic material by a pressure lip (Fig. 10 B). Higher welding speeds are possible than with swirl welding.

Extrusion welding is mainly employed to bond thick-walled moldings because deep weld grooves can be filled in a single operation. The substrate is warmed with hot gas during which time the welding rod is fed as a plasticated ribbon directly from an extruder and pressed into the groove by a welding lip (Fig. 10 A).

2.1.3. Radiation Welding

Radiation welding techniques can generally be classified into two groups: those in which the thermal radiation from a heating element is utilized and those that employ lasers.

With HT radiation welding, the welding surfaces are not brought in contact with the heating element. Instead, heat transport from the heating element to the weld region takes place across a 0.5 – 1 mm air gap, mainly by radiation and partly by convection. The process is particularly advantageous for polymers that tend to adhere to the heating element and be drawn into fibers. To accelerate the heating process, radiation heating elements are generally operated at higher temperature than those employed in conductive methods. Care must be taken not to overheat the surfaces to be welded.

In laser welding, heating of nontransparent polymers is performed with a laser. This process is employed only in special cases for films and in winding fiber-reinforced polymer components with a thermoplastic matrix.

In laser extrusion welding, the substrate is heated by a laser instead of the hot gas used in conventional extrusion welding.

2.1.4. Friction Welding

Friction welding utilizes the heat of friction to plasticate the welding surface. A distinction is made between external and internal friction. Whereas the relative motions of the welding components cause heating via external friction, the electrical and mechanical damping characteristics of the material and the molecular vibrations lead to heating by internal friction.

Spin welding, a technique employing external friction, is suitable only for bonding components that are rotationally symmetrical. One of the welding substrates is rotated while the other is mounted on a bearing capable of rotating about the same axis, but initially prevented from doing so by a brake, and is lightly pressed against the rotating partner ($1 - 2$ N/mm^2). The relative motion ($0.8 - 2.5$ m/s) causes heating of the welding surfaces and finally melting. Once a large enough melt reservoir has been generated, the brake is released, and both components rotate at the same speed. If one component is particularly large or does not have rotational symmetry, rotation of the substrate can be stopped for welding. The weld must be cooled while still under compression. The technique can be carried out straightforwardly on a lathe. A disadvantage is the long setup time.

In vibration welding, direct heating of the weld plane is brought about by relative motion of the two partners while under pressure. An oscillating rotational (with rotationally symmetrical welding surfaces) or translational motion (frequencies of 100 – 300 Hz) is used. Important process parameters that must be adjusted to suit both the geometry of the articles to be welded and the material itself are

1. Friction speed or frequency and amplitude (up to 2.5 mm)
2. Friction time
3. Friction pressure (up to 4 N/mm^2)
4. Weld seam pressure
5. Holding time

Both processes are simple to automate and have short cycle times (ca. 5 s), so they are ideally suited for mass production operations. Additionally, vibration welding techniques can be used in the manufacture of hollow articles (e. g., brake fluid reservoirs and pressurized containers for automobile coolants) and also for bonding larger items (e.g., automobile bumpers) even when relatively soft thermoplastics are used.

A technique that utilizes the internal friction in a material is *ultrasonic welding* at a frequency of 20 – 60 kHz. The mechanical oscillation of a piezoelectric vibrator is introduced into one of the two components to be welded via a booster and a sonotrode. The components are mounted between the sonotrode and a stationary anvil. The sonotrode is shaped so that a standing mechanical wave is generated with its maximum amplitude at the contact surfaces of the two components. With appropriate shaping of the contact surfaces (energy directing), heating of the material by internal friction is rapid and concentrated in the contact area. Hence, short cycle times can be attained (0.2 – 1.5 s).

Many variations of this welding technique exist. The characteristic of near-field welding is the close proximity of the sonotrode contact surface and the welding surface (<6 mm). The method is employed to bond films and sheets of almost all thermoplastics. Only rigid polymers [such as styrene – acrylonitrile (SAN), ABS, PMMA, polyoxymethylene (POM), PC, and polystyrene (PS)] are suitable for distant field welding, where the energy losses are not too great across the longer ultrasonic pathway between the sonotrode contact and the welding surface. The main area of application of distant field welding involves the bonding of moldings (sometimes made from different polymers). The major process parameters that must be adjusted to suit the specific material and geometry of the welding components include

1. Sonotrode shape
2. Contact pressure
3. Amplitude of oscillation and ultrasonic power
4. Irradiation and holding time

A related process involves embedding metal inserts in thermoplastics (e.g., screw thread sockets) by mounting the metallic component on the sonotrode and pressing it into the polymer under the influence of ultrasonic irradiation. Ultrasonic equipment is also employed for riveting; a shaped sonotrode is used to form the head on a thermoplastic rivet.

Ultrasonic welding is suitable for use in mass production bonding operations and gives highly reproducible weld seam quality. By appropriate design of the welding surface, weld seam bead can be concealed behind undercuts and hence need not be removed in a finishing stage.

In high-frequency welding a polar material is heated by the molecular dipole oscillations generated by the alternating electric field of a capacitor (frequency ca. 27 MHz). By suitable shaping of the capacitor electrodes, heating can be limited to the immediate vicinity of the welding surface. Cycle times are then short. Electrodes are also employed to apply the bonding force to the welding partners. Specially shaped electrodes can be used to emboss the weld seam or to generate a weld seam with a cutting edge.

Suitable materials are polymers with dielectric loss factors greater than 0.01 [e.g., cellulose acetate (CA), PA, PMMA, polyurethane (PUR), and PVC]. This technique is used mainly for large-area welding of sheet materials. Product examples include PVC rainwear and inflatable life rafts.

High-frequency welding can be automated and used in mass production.

2.1.5. Induction Welding

In induction welding, weld fillers that contain electrically conducting particles are heated by eddy currents induced in the particles by an alternating electromagnetic field (frequency range 3 – 10 MHz) (Fig. 11). The weld filler is a (profiled) preform placed in a groove situated in the bonding surface. Welding takes place in an induction field with the components held under slight pressure. The composition and conducting particle content of the filler must be selected to suit the substrate to be welded. The filler can also be applied to the substrate during the manufacturing process by using two-component injection molding (see → Plastics Processing, 1. Processing of Thermoplastics, Section 4.5). Because heating of the substrate takes place by conduction, cycle times are longer than with high-frequency or ultrasonic welding.

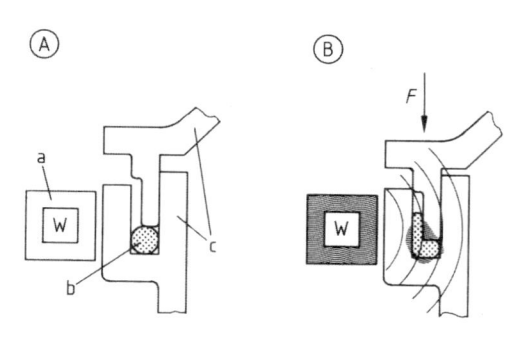

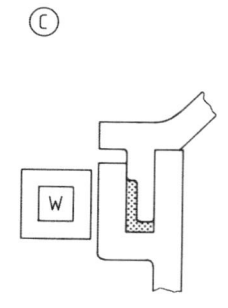

Figure 11. Induction welding
A) Before welding; B) During welding; C) After welding
a) Induction coil; b) Weld filler; c) Parts to be joined

The induction process is also used to open welded seams by placing the welded article in an induction field (e.g., opening containers for transporting corrosive materials).

2.2. Adhesion

In contrast to welding, all polymers can be bonded to themselves, each other, and other materials, such as wood, textiles, glass, and metals, by using adhesives [18–20], with the exception of solvent adhesion, which strictly speaking should be regarded as a welding process. During adhesion the substrates are not brought directly in contact with each other. In this case, a layer of similar or dissimilar polymer that can partially diffuse into the substrate surface is responsible for the bond. No or very little heat is necessary for adhesion, so that the structure and properties of the bonding substrates remain unchanged. In contrast to welding, adhesion can also be used for large-area bonding, a fact that is of considerable importance in the packaging industry. Adhesive bonds can be optically superior to welded bonds, which is important with transparent polymers. However, the manufacture of strong adhesive bonds often requires multistage processes and can take several days. The strength of polymer adhesive bonds can be nondestructively tested only to a very limited extent.

A wide range of adhesives with various adhesion and curing mechanisms are used for bonding plastics (see → Adhesives).

Some polymers can be softened by applying solvents and then fused together. Strictly speaking, this technique should be regarded as diffusion welding. Solvent-based adhesives that harden by evaporation of the solvents are also frequently employed. The solvents can also cause the surface of the bonding substrates to swell and hence can facilitate mutual diffusion of adhesive molecules and the molecules of the substrate. This is possible only with thermoplastics. Although adhesion with solvents or solvent-based adhesives is simple and economical, problems often arise with this process due to aging of the bond because the solvent cannot be removed completely from the joint after bonding.

Heat sealing and melt adhesives have also gained importance in bonding plastics. The highest adhesive bond strengths with polymers are generally obtained with chemically hardening one- or two-component adhesives. In most cases, single-component adhesives require heating for the curing reaction, whereas two-component systems also cure at room temperature.

Adhesion processes can now be extensively automated. Continuous adhesion process and short cycle times can be realized, particularly when heat sealing and melt adhesives are used.

Polymers can be classified by their adhesive bonding behavior as

1. Easily bondable polymers
2. Limitedly bondable polymers
3. Difficulty bondable polymers

The differences in bondability are due to variations in the solubility of the polymers and the polarity of the macromolecules (Table 5).

Further influences are the surface tension and hence wettability. Table 6 lists surface tensions for some polymers and adhesives. A prerequisite for adhesion is that the surface tension of the adhesive be lower than or equal to that of the polymer. Only when this applies does the binder adapt itself to the contours of the solid surface during the wetting stage so that the physical and chemical interfacial reactions required for good adhesion can occur. Only in solvent-based adhesives containing strongly

Table 5. Polarity and solubility of plastics

Plastic	Polarity	Solubility	Adhesive capacity
Polyethylene	nonpolar	very slightly soluble	poor
Polypropylene	nonpolar	slightly soluble	medium
Polytetrafluoroethylene	nonpolar	insoluble	very poor
Polyisobutylene	nonpolar	readily soluble	good
Polystyrene	nonpolar	soluble	good
Poly(vinyl chloride)	polar	soluble	good
Polyterephthalates	very polar	insoluble	medium
Poly(methyl methacrylate)	polar	soluble	good
Polyamide 66 – 610	polar	slightly soluble	medium
Polyamide 6 – 11	polar	slightly soluble to insoluble	medium to poor

Table 6. Surface tension of plastics and adhesives

	Surface tension, 10^{-3} N/m
Plastics	
Polytetrafluoroethylene	18
Polyethylene	31
Polystyrene	33
Poly(vinyl chloride)	40
Polyester	43
Cellulose film	45
Polyamide 66	46
Adhesives	
Acid-curing phenolic resins	78
Urea – formaldehyde	71
Phenol – resorcinol resins	48
Casein resins	47
Epoxy resins	30 – 47
Poly(vinyl acetate) rubber	38
Nitrocellulose	26

dissolving components can the wetting criterion be neglected to a certain extent. If the surface tension of the plastic is lower than that of the adhesive, the surface must be mechanically prepared or, in the case of difficultly bondable polymers, chemically pretreated.

Easily bondable polymers are: poly(vinyl chloride) with low plasticizer content, polystyrene, polyacrylates, polycarbonates, and polyurethanes, as well as polyester, epoxy, and phenolic resins in unreinforced and reinforced state. Whereas polystyrene can be bonded most easily with solvent-based adhesives, the remaining members of the group can be bonded reliably by a variety of adhesives without pretreatment or after mechanical roughening. The surfaces of the bonding substrates should be cleaned with solvents or mechanically even when easily bondable polymers are involved, to remove contaminants and residual release agents.

The limited bonding polymers include PVC with high plasticizer content, which must be thoroughly cleaned with organic solvents prior to application of the adhesive. The pronounced shrinkage and swelling behavior of PVC can lead to difficulties in using solvent-based adhesives. Plasticized PVC can be bonded quickly and reliably by using solvents alone (e.g., tetrahydrofuran). Greater problems are encountered with adhesion to synthetic rubbers and polyamide. The latter can be diffusion bonded by using formic acid; otherwise, two-component adhesives are required, which however, give strong bonds only if the surface of the polyamide substrate has been mechanically roughened. Furthermore, the adhesion to polyamide is strongly dependent on the chemical structure of the polymer.

Difficult-to-bond polymers include all polyolefins, fluoropolymers, polyacetals, and silicone resins. These polymers cannot be bonded in the untreated state. Polyolefins can be chemically etched and treated electrostatically by using corona discharge equipment or treated with an oxidizing flame (see Section 3.5.1). Care must be taken that the basic structure of the plastic is not altered by surface pretreatment. The greatest difficulties are encountered in adhesive bonding fluoroethylene – propylene copolymers and polytetrafluoroethylene, which can be bonded with adhesives only after etching with sodium in liquid ammonia or sodium naphthalenide. Although both processes are complicated, they give bonds of high mechanical strength and good aging resistance.

Current developments in adhesives are concerned with improving bond strength, resistance and stability, and simplifying processibility. An example involves single-component reactive adhesives based on cyanoacrylates, which are ideally suited for bonding rubber and whose hardening takes only seconds or minutes. High manufacturing speeds can also be attained with heat sealing and melt adhesives, whose major advantage is their freedom from solvents. Melt adhesives based on polyamides, linear polyesters, or aromatic polysulfones exhibit good thermal stability and have processing times of a fraction of a second. Of the two-component reaction adhesives, flexible polyurethanes have a broad range of uses for bonding polymers, including construction bonding and the manufacture of laminates such as packaging materials, light core sheets, film-laminated sheet metals, and metal-coated polymers.

2.3. Mechanical Bonding

Mechanical bonding processes include the conventional methods of bolting and riveting. The bolts and rivets can be made of the same or different materials with respect to the bonding substrates [21, 22]. The use of self-tapping screws avoids cutting threads in a previous process stage. Screw bonds that are loosened frequently and have a diameter of >4 mm should be equipped with a sunken or embedded metallic thread socket. Rivet bonds can be prepared by using the techniques employed in the metalworking industry. The rivet heads can be formed either by hot forming or cold pressing. Cycle times for riveting processes are short, and the mechanical strength of riveted bonds is satisfactory. Snap-in rivets are inserted in rivet holes and fixed to form a solid bond by spreading of the rivet shaft.

A particularly elegant bonding variation for polymers involves the so-called clamp or snap-in connectors in which protrusions or hooks become interlocked in correspondingly shaped undercuts when the two components are brought in contact. Bonds of this type can be rendered air- or watertight by use of gaskets or sealants. Mechanical bonds of this form can also be formed between polymers and other materials.

3. Surface Treatment

The surfaces of plastic components are treated to meet functional or decorative requirements that cannot be met by the base material alone [23, 24]. Surface treatment processes can be classified as follows:

1. Mechanical processes such as hot embossing and flocking
2. Painting and printing with solvent-based or chemically binding coatings
3. Wet chemical deposition processes such as electroplating or electroless metal deposition
4. Physical deposition of inorganic materials
5. Dry chemical processes such as fluorination, sulfonation, and plasma polymerization

3.1. Mechanical Processes

3.1.1. Hot Embossing

In hot embossing, a multilayer hot embossing sheet is applied to a substrate under pressure by a heated stamp. The term "embossing" is misleading in that it suggests the remolding of the substrate surface, which is not the case. Descriptions such as (color) film printing, (color) hot film printing, or dry printing are more apt, since, because of the way in which the (multilayer) decoration is applied, the initial surface topology remains unchanged, with the only exception being the gloss. Hot embossing can be used to coat the entire surface of an article (equivalent to painting), along with applying letters or a fine pattern of a different color to the substrate (e.g., scales and labeling on dials).

The hot stamp itself has the pattern to be embossed in relief on its surface (Fig. 12). The raised areas activate the hotmelt adhesive layer, which is located nearest the substrate during the stamping process; hence adhesion of the stamping laminate results in these areas. The deeper-cut areas of the stamp do not come in contact with the film, so no heat transfer and hence no activation of the adhesive occur, and the decorative and adhesive layers of the film are removed when the stamp is removed from the substrate.

A relief structure already present in the substrate surface can be highlighted by color by stamping with a flat stamp (relief stamping). With counterstamping, transparent materials are embossed on the reserve surface.

In the displacement stamping process, the stamp is generally planar or can be adapted to the slightly convex or concave shape of the article. In roll stamping, which is especially suited for large-area hot embossing and relief stamping, the stamp has the form of a roll that proceeds across the surface of the substrate with

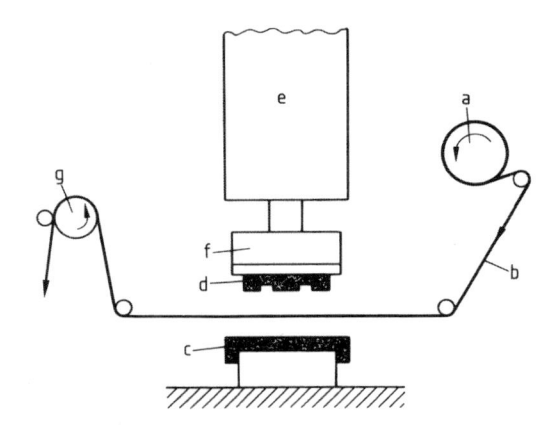

Figure 12. Hot embossing
a) Film roll; b) Film; c) Workpiece; d) Embossing stamp; e) Press; f) Heater; g) Feed roll

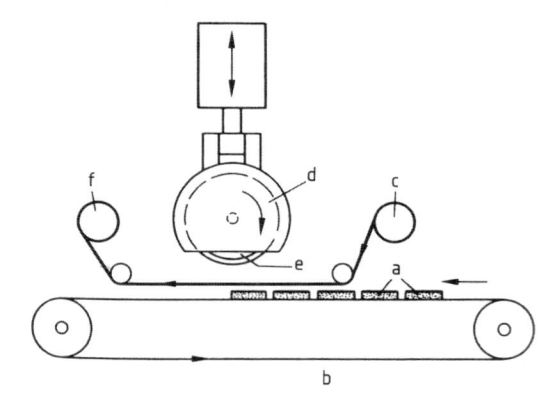

Figure 13. Roll stamping process
a) Workpieces; b) Conveyor belt; c) Roll of film; d) Heater; e) Embossing roll; f) Film windup

the stamping film positioned between them (Fig. 13).

As a rule, a thermoplastic substrate surface is necessary to obtain a satisfactory bond with the embossing film. The required thermoplastic surface can be produced on thermosets, metals, ceramics, or wood by applying paint coating. Embossing films are available in a multitude of decors, as mat or gloss films, in wood colors, single or multicolored, metallized (Fig. 14), etc.

3.1.2. Flocking

Flocking is used to produce a velvetlike surface. Colored fibers 0.3 – 0.5 mm in length are used, made from polyamide or viscose, for example. In mechanical flocking processes, which use vibration or compressed air, the flocks form an unoriented and nonuniform layer on the surface. For this reason, electrostatic processes, sometimes supplemented by mechanical techniques, are almost exclusively employed (Fig. 15). To ensure alignment in the electrostatic field, the

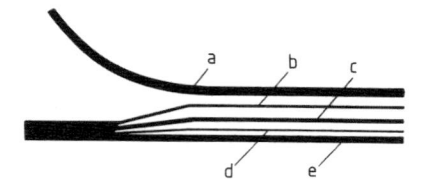

Figure 14. Construction of a metallized film
Layers c – e make up the embossing film.
a) Substrate sheet; b) Release film; c) Protective coating; d) Metallic layer; e) Adhesive layer

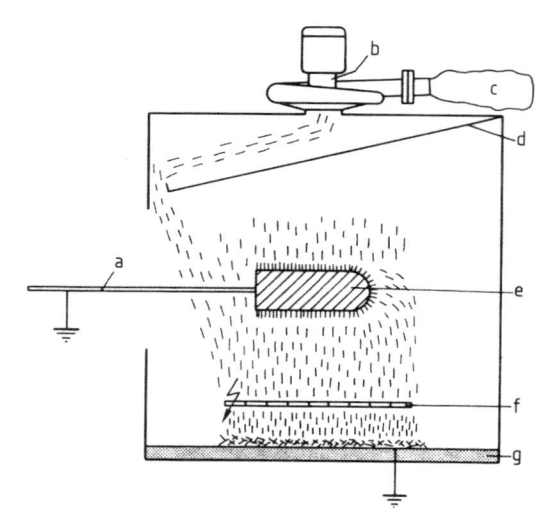

Figure 15. Electrostatic flocking
a) Rod; b) Extractor fan; c) Filter bag; d) Air-guiding plate; e) Molding; f) Electrode; g) Bottom tray

flocks are produced in a free-flowing, conducting form. Prior to flocking, the substrate surface is coated with an adhesive in which the flocks become permanently anchored. Sometimes, surface pretreatment is necessary (see Section 3.5.1). Equipment is available for flocking films, sheets, moldings, and hollow articles employed in the automobile industry, the packaging sector, toys, fashion jewelry, etc.

3.2. Painting and Printing

Whereas printing of plastic articles is carried out almost exclusively for decorative purposes, the reasons for painting polymers can be divided into three categories:

1. To meet functional requirements
2. For decorative purposes
3. As an intermediate step prior to further surface modification techniques, such as printing, film embossing, flocking, and metallizing

3.2.1. Painting

Previously, it was believed that, in contrast to metals, polymers did not have to be painted because they were not prone to corrosion.

However, this situation has changed considerably. Many plastic components can meet their respective specifications regarding optics, light, and weather resistance only when painted.

The thickness of the paint coatings varies between 5 and several hundred micrometers, depending on intended application.

When painting properties are considered, distinctions are drawn between thermosets and thermoplastics, between solvent-sensitive and solvent-resistant plastics, between solid and foamed polymers, and between foams with smooth and those with open-pored surfaces. The polymer and the coating must be matched with regard to their chemical compatibility, adhesion, and elasticity.

Surface State, Pretreatment. Critical demands are placed on the surface of the substrate to be painted and also on the design of the molding itself. The molding must be designed to suit the painting process (i.e., without sharp corners, edges, angles, or deep recesses and undercuts). Due to surface tension effects, paints recede from sharp edges (edge recession) and form a thin coating at these critical wear positions [26]. Undercuts and deep recesses are insufficiently coated in most spray painting processes. With larger planar areas, problems are caused by dust particles, especially with high-gloss coatings and metallization.

The surface to be coated must be clean (i.e., free from grease, dirt, sweat, and residual release agents, particularly those containing silicon). Contamination decreases adhesion and causes surface defects such as spots, dull or glossy areas, and flow disturbances. When necessary, cleaning procedures must be applied: cleaning with trichloroethylene vapor with compact thermosets, mixtures of organic solvents with degreasing and antistatic action, possibly supplemented by ultrasonic techniques or aqueous baths containing surfactants. Electrostatic charging, which causes wetting and flow problems and dust contamination, can be neutralized by blowing with ionized air.

With injection-molded articles, release agents and lubricants may concentrate in and around the weld lines and thus lead to surface defects in the coating. Stresses are often frozen into the molding due to poor selection of injection conditions and inappropriate choice of gate type or gate position [27]. In particular, in solvent-sensitive materials regions of increased internal tensile stress can lead to stress corrosion, deformation, and even loss of mechanical strength. Similarly, the type and amount of pigment in a polymer as well as choice of filler can influence results. Successful surface coating therefore requires knowledge of the type of polymer material, the process parameters for producing the molding, the quality of the coating material, and its processing conditions.

Besides these generally applicable aspects, several polymers require further specific measures prior to coating [26, 27]. Polyethylene, polypropylene, and epoxy resins are nonpolar and have very low surface tensions. To render them wettable and achieve satisfactory adhesion of a coating, they must be oxidized (see also Section 3.5.1) immediately before the coating is applied by

1. Low-pressure methods
2. Plasma treatment
3. Electrical methods at ambient pressure, such as corona treatment
4. Thermal processes with a strongly oxidizing flame (Kreidl – Traver process)
5. Immersion in chromic acid – sulfuric acid

However, antioxidants and stabilizers can reduce the effectiveness of such measures. With polypropylene and ethylene – propylene elastomers, the required paint adhesion can be attained by using special primers. Plasticized PVC and cellulose esters often contain plasticizers that can migrate into coatings and soften them. They must therefore be coated with paints that are resistant to migration of plasticizer.

Application Methods and Paint Systems. All nonelectrostatic coating methods (see → Plastics Processing, 1. Processing of Thermoplastics, Chapter 3) are in principle applicable for use with polymers: high-pressure spraying at 200 – 600 kPa with air atomization, high-pressure airless spraying at 10 – 20 MPa, dipping, flow coating, roller application, pouring, centrifugation, drum application, and brushing (see also → Paints and Coatings, 7. Paint Application, Chapter 3). Electrostatic spraying can be used with nonconducting polymers by coating them with a conducting solution [26, 28].

Apart from economic aspects, the form of the molding and the requirements placed on the surface, as well as the temperature and solvent sensitivity of various polymers, must be considered in selection of the coating method. Curing of the coating is generally accelerated by heating. Manufacturers' information on the heat resistance of the plastics is of limited value here. Heat resistance is determined on standard test specimens, whereas commercially manufactured moldings are seldomly produced under optimal conditions and have internal stresses and variations in wall thickness, and the coating and curing process often takes place under conditions of mechanical loading. This can cause deformation, losses in dimensional stability, and surface defects due to excessive penetration of solvents. Table 7 lists empirical heat resistances for fabricated plastic components. Bubble formation in the coating is possible at elevated temperature as a result of the evolution of formaldehyde from solid thermosets and from residual blowing agents in foamed thermosets.

Paint systems are selected according to the demands placed on the surface of the finished article and the coating characteristics of the substrate itself. These include physically drying paints (i.e., those that dry simply by evaporation of the solvent, e.g., nitrocellulose – resin lacquers and acrylic and vinyl paints); oxidative drying fatty-acid-containing resin paints; chemically drying multicomponent paints based on epoxy resins, polyurethanes, or unsaturated polyester resins; and resin stoving paints (see Section 3.2). For further information on various paint systems, see → Paints and Coatings, 1. Introduction.

The solvent sensitivity of many thermoplastics must be taken into account in formulating the paint (Table 8). Generally, solvent sensitivity increases with increasing temperature in the painting and drying process, as well as with increasing contact with the paint (e.g., in dipping). In particular, involatile solvents can diffuse deeper into the substrate surface and thus cause changes in molecular structure that lead to surface roughness and loss of mechanical strength. Highly solvent-sensitive polymers (polystyrene, polystyrene – butadiene, polycarbonate) allow only a limited choice of paints [29].

Polyurethane integral foams, thermosets, and elastomers have a closed, smooth surface. Differences in flexibility and foam weight influence the surface integrity and hence the coating characteristics. Single-layer coatings with smooth surfaces can be achieved if the surface quality is high. The best results are obtained with two-pack polyurethane paints based on aliphatic isocyanates. Apart from self-releasing polyurethane foams, the choice of release agent and its subsequent removal are particularly important. Thermoplastic cast foams exhibit an open-pore surface, which gives good adhesion to coatings and is less sensitive to solvent effects than unfoamed plastics. Smooth surfaces can be obtained only by multilayer coating with intermediate polishing. For this reason, paints with structured surfaces are generally employed.

Areas of Application. A precise distinction between functional and decorative coatings cannot generally be made [26]. Varnishes and paints that absorb UV radiation prevent photochemical degradation of ABS, PC, POM, and poly(propylene oxide) (PPO), which would otherwise result in discoloration, spots, loss of gloss, and embrittlement. Coatings increase the resistance to many aggressive media such as solvents, water, chemical cleaning baths, atmospheric influences, adhesives, and plasticizers. Furthermore, by using

Table 7. Empirical short-term heat resistance of fabricated plastic components

Plastic	Heat resistance, °C
Polystyrene – butadiene	50
Polystyrene	60
Poly(vinyl chloride)	60
Cellulose acetate and others	60
Styrene – acrylonitrile	70
Polyethylene	70
Acrylonitrile – butadiene – styrene	80
Poly(methyl methacrylate)	80
Poly(phenylene oxide)	110
Polyamide	110 – 120
Polypropylene	110
Polycarbonate	120
Polysulfone	150
Polyoxymethylene	150
Poly(ethylene terephthalate)	150
Polyurethane (foamed)	60
Urea – formaldehyde resin, melamine – formaldehyde resin	110
Phenol – formaldehyde resin	120 – 150
Unsaturated polyester	150

Table 8. Short-term resistance of plastics to solvents during painting*

Plastic	Alcohol and glycol ether	Esters	Ketones	Gasoline	Benzene	Chlorinated hydrocarbons
Polystyrene	+	−	−	+	−	−
Polystyrene – butadiene	+	−	−	+	−	−
Styrene – acrylonitrile	+	−	−	+	−	−
Acrylonitrile – butadiene – styrene	+	○	−	+	○	○
Poly(phenylene oxide)	+	○	−	+	○	−
Poly(methyl methacrylate)	−	○	○	+	+	+
Polycarbonate	+	−	−	+	○	−
Poly(vinyl chloride)	+	+	○	+	+	○
Cellulose acetate and others	+	○	○	+	+	+
Polyamide	+	+	+	+	+	+
Polysulfone	+	○	○	+	○	−
Polyoxymethylene	+	+	+	+	+	+
Polyethylene, polypropylene, epoxy resins	+	+	+	+	+	+
Poly(ethylene terephthalate)	+	+	+	+	+	+
Thermosets	+	+	+	+	+	+

*+ resistant; ○ partially resistant; − nonresistant.

colored paints, direct coloration of the polymer, which requires expensive high-quality pigments, can be avoided. Matt metallic coatings are obtained by using aluminum or bronze paints, of either the leafing or the nonleafing type. In leafing products the metal flakes float on the surface, are lighter in color, and are contact sensitive, whereas in nonleafing products they remain embedded in the coating and exhibit greater scratch and corrosion resistance.

The fire resistance of plastics containing flame retardants can be improved further by application of flame-retarding paints.

Antistatic coatings prevent electrostatic charging and consequent collection of dust on nonconducting polymer surfaces. Conductive coatings filled with graphite or metal powders (copper, silver, or their alloys) are used as internal coatings (e.g., in television sets and computers). Nonconducting metallic paints, which are used primarily for decorative purposes, prevent creep currents in electronic equipment.

During the electroplating of polymers (see Section 3.3.2), cover coatings are used for partial chromium plating. The coatings protect the substrate from damage caused by chemical baths; no metal is deposited on the coatings themselves. A sharp interface with the metal-plated areas is, however, practically impossible to achieve. If this is required, the entire molding should be chromium plated, and selected areas should be lacquered (e.g., automobile radiator grills).

Abrasion-resistant coatings consisting of polysiloxane resin are used to protect plastic lenses and mechanically loaded polycarbonate components from scratching.

Molds for polyurethane components can be sprayed with mold-release paints instead of release agents so that a coated molding is formed in the mold (in-mold coating).

Paints also serve as a base layer and adhesion promoter for printing and film embossing with simultaneous coloration.

In the manufacture of consistently colored or mat surfaces, and also to cover faults such as joints, colorless, mat, or pigmented scratch-resistant paints are utilized. Thermoplastic foam castings and glass fiber reinforced plastic components are coated with strongly filling paints to achieve optically satisfactory surfaces.

Solvents and other volatile components released during drying pass into the exhaust air in gaseous form; the exhaust must therefore be treated to comply with emission limits. Because vast amounts of air are involved in paint drying, treatment equipment is large and expensive. A trend toward the use of fewer solvent-based paints in favor of waterborne systems is apparent, particularly in the automobile industry.

3.2.2. Printing

In principle, all printing techniques are applicable to polymers [23, 30, 31]. The most suitable process for a given task depends on the material thickness and the substrate form (film, sheet, molding). Prior to printing, the material may have to be pretreated (see Section 3.5.1). For several applications (e.g., wood grain printing), a primer layer is applied before the actual print layer, and a topcoat protects the print from damage. Because the substrate is generally hard and nonabsorbent, indirect or transfer printing techniques are used, which employ an elastic intermediate substrate, generally made of rubber (e.g., gravure printing with horizontal or vertical rolls). The printed pattern is transferred, via a rubber roll, from a deeply etched, ink-filled steel applicator roll to the planar printing substrate.

Films are generally printed by using rotary techniques (flexo or gravure printing). Sheets that are to be vacuum remolded are printed in an offset process. The distorted printed pattern is straightened during remolding.

The most common technique — silk screen printing — is used for both sheets and moldings. The tightly woven fabric (usually made of polyester or polyamide) is mounted tautly in a frame and partly coated with a UV-cured solvent-resistant layer [30]. The sieve is positioned close to the article to be printed, and the ink is pressed through the screen onto the substrate by using an elastic doctor blade. The tension in the sieve ensures its withdrawal from the surface once the process is complete.

Moldings with complex shapes can be printed by indirect silk screen printing with an intermediate rubber substrate. With the tampo-print technique, also an indirect printing method, a negatively etched printing block is automatically filled with ink and doctored by a steel blade. The ink remaining in the etched areas is then transferred to the substrate by a tampon. The highly flexible printing tampon made of silicone rubber can be manufactured in a variety of forms so that irregularities and curvatures in the surface of the article to be printed do not play a significant role. Even multicolor wet-in-wet printing is possible.

3.3. Wet Chemical Deposition Processes

Wet chemical processes are used to deposit metallic layers on polymers. Because polymers are generally electrical insulators, components that are to be electroplated must first be coated with a metallic, conducting layer in an electroless process.

3.3.1. Chemical Deposition

Chemical deposition processes are based on the techniques used to manufacture mirrors: The metal (in the case of mirrors, usually silver) is deposited from an aqueous salt solution by a chemical reducing agent such as formaldehyde, glyoxal, or hydrazine.

Similar methods have been developed for depositing copper and nickel on polymer surfaces [23, 32]; these are generally immersion processes. The clean, wettable polymer surface is first "activated" or "seeded" with silver or palladium. For example, to obtain silver nucleation sites, the surface is immersed in a solution of tin (II) chloride, rinsed, and dipped in a basic silver nitrate solution. For palladium seeding a palladium salt solution is employed, followed by a reducing agent. A comparable effect can be achieved by immersion in a solution of colloidal palladium. Nucleation with $0.1 - 1.0$ mg/dm^2 of noble metal is a prerequisite for specific metal deposition by chemical metallizing techniques. The most common chemical metallizing baths currently employed contain the metal (Cu or Ni) as an aqueous solution of a complex salt along with the reducing agent (e.g., sodium hypophosphite or diethylaminoborane for nickel baths and formaldehyde for copper baths). The composition of the bath is adjusted so that metal deposition begins when the nucleated article is immersed and ceases when it has been removed. A coherent metal layer ca. 0.3 μm thick is deposited on the polymer typically within 10 min. This thin metallic layer can subsequently be thickened by electroplating.

With small production runs (e.g., jewelry), a spray process can also be employed for chemical metallization. A solution of a silver salt and a solution of a reducing agent are sprayed simultaneously onto an activated polymer surface by using a two-component spray gun. The components react to form a continuous layer of silver.

3.3.2. Electroplating of Plastics

The thin metal layers deposited by electroless methods are extremely sensitive to abrasion and corrosion. When demands other than decoration arise, such as hardness, abrasion resistance, corrosion resistance, and electrical conductivity, the thin layers must be reinforced electrolytically [23, 32]. Total metallic behavior of the layer is reached only with layer thicknesses of ca. 20 µm. However, because the Young's modulus of metals is approximately two orders of magnitude greater than that of the polymer, and the coefficient of thermal expansion is approximately an order of magnitude lower, with thick metal deposits, forces occur at the interface that cannot be withstood by the adsorptive forces arising between polymer and metal as a result of even the smallest mechanical or thermal loading.

Although electrolytic reinforcement normally leads merely to sheathing of the molding with metal, it is suitable for making small spherical articles such as typewriter heads. The poor adhesion of the metal is in fact desirable for electroforming [33] (e.g., for the manufacture of press matrices or injection molds from plastic models). The range of applications of metallizing techniques with polymeric moldings is limited by the poor adhesion of the metal, which cannot be improved by mechanical treatment of the plastic surface.

The term polymer electroplating is currently used to describe a technique that differs from the methods discussed above in that metal adhesion approximately two orders of magnitude greater can be obtained. Development of this process began in 1965 when it first became possible to produce a strong bond between a metal and a specially treated ABS graft polymer [34, 35]. Within a few years, a fully developed technology resulted that was adapted to suit ABS polymers [32] and that is still used to electroplate polymer moldings on a mass production scale with automated equipment.

The special significance of ABS graft polymers is a result of their structure. They have a

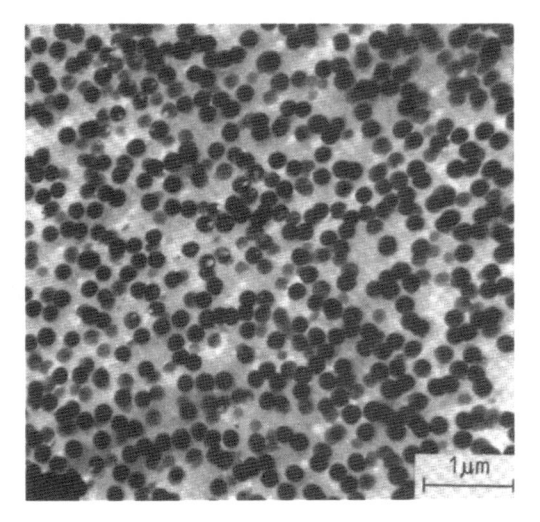

Figure 16. ABS graft polymer

styrene – acrylonitrile copolymer matrix in which fine spheres of an elastomer phase are dispersed (Fig. 16). This elastomer phase can be removed to a depth of ca. 1 µm by oxidation in a chromic acid – sulfuric acid etching bath [32, 36], without damaging the matrix material, in 5 – 10 min at 60 °C. A submicroscopic pore system with dimensions corresponding to the diameter of the elastomer particles is generated in the surface (Fig. 17).

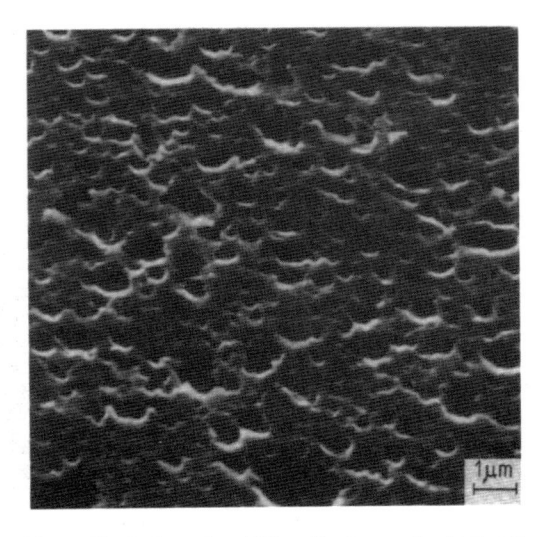

Figure 17. Surface of an ABS graft polymer after treatment in an etching bath

Figure 18. Underside of conducting metal layer (plastic removed by solvent)

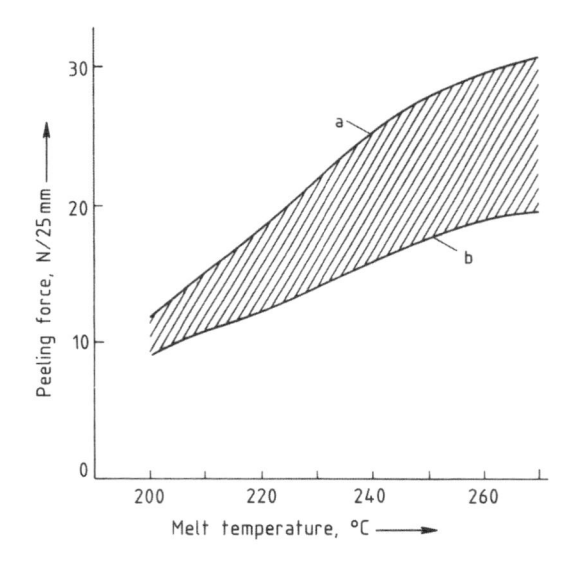

Figure 19. Peeling force as a function of melt temperature in injection molding for a plastic disk of 50 mm diameter and 2 mm thickness
a) Slow injection; b) Fast injection

After etching, the surface is activated, and nickel or copper can be deposited chemically (see Section 3.3.1). Initially the pores are filled, and subsequently a continuous, strongly anchored conductive metal layer is then formed that can be electrolytically reinforced, for example, with 20 µm of copper, 10 µm of nickel, or 0.2 – 0.8 µm of chromium by standard processes (see → Electrochemical and Chemical Deposition). The lower surface of the metal layer (Fig. 18) forms a positive replica of the pore system generated by etching. A polymer – metal composite is thus formed that is bonded so strongly that removal by force of the metal leads to cohesive fracture of the polymer in the vicinity of the metal layer. The adhesion of the metal layer is hence determined by the tensile strength of the outermost polymer layers [32, 37].

During the forming of thermoplastics by injection molding, orientation is frozen into the outermost layers, in particular, which can lead to a significant anisotropy in the mechanical properties. The processing conditions for injection molding (melt temperature, injection speed, mold temperature, wall thickness, position and design of the gate, etc.) often exert a decisive influence on the adhesion and the quality of electroplated moldings [32, 37]. Figure 19 shows the influence of melt temperature on peeling force (see below). Whereas with thick-walled, orientation-free (pressed) moldings, a peeeling force of >100 N per 25 mm can be reached, under less favorable processing conditions (high orientation) this value can fall below 10 N per 25 mm. Peeling forces >15 N per 25 mm are adequate to meet the majority of technical requirements.

Electroplated ABS is a composite material with a specific property profile [32, 37, 38]. The surface properties (gloss, hardness, abrasion resistance, corrosion resistance) are determined by the thickness and the structure of the metal layer.

Due to the reinforcing influence of the metal, the molding stiffness and strength increase proportionally to the ratio of the thickness of the metal to that of the polymer, up to approximately three to five times that of the unplated material. The creep behavior typical of polymers under loading is suppressed. The heat resistance of the plated molding is considerably superior to that of unplated materials.

The ABS electroplating process can also be applied to other polymers with only minor modifications. These include ABS blends with other polymers (e.g., PC) as well as polystyrene and poly(phenylene oxide) modified with butadiene rubber particles.

Polypropylene can be rendered electroplatable by using specific additives. In these PP products, the amorphous and crystalline regions are attacked by the etching bath to various extents. This leads to extensive fissured roughening of the polymer surface, which gives good anchorage and adhesion of the metal layer [23].

Most polymers can now be electroplated by employing other etching or pretreatment techniques [23, 39].

Quality Test. For planar surfaces, the peel strength (DIN 53 494) is used as a measure of adhesion. The quality of moldings is evaluated by means of alternating temperature tests (DIN 53 496).

Applications. Electroplated ABS moldings are employed where their low manufacturing costs offer advantages over metal components, or where their advantages relative to unplated material due to increased hardness, stiffness, dimensional stability, and reduced weight are apparent [23]. Particularly in the automobile, radio broadcasting, audio, electronic, lining, writing instrument, and watchmaking industries, electroplated polymers find application as fittings, covers, handles, buttons, levers, keys, frames, and housings. Uses range from radiator grills to dials for wristwatches. The emphasis has shifted from decorative effects (e.g., for fashion articles such as buckles, buttons, fasteners, and chains) to functional components such as chassis, shields, waveguides, and switches in the electrical industry.

3.4. Physical Deposition Processes

Various techniques are employed to produce transparent (layer thickness <0.2 µm) or high-gloss (layer thickness >0.2 µm) metallic coatings and abrasion- or scratch-resistant ceramic protective coatings.

Atmospheric Pressure Techniques. Metal spraying permits the rapid buildup of a relatively thick layer.

Metal powder (powder process) or wire (wire process) is melted (e.g., in a plasma burner) and sprayed onto the polymer surface by a gas jet. Rough metal layers with pores are formed rapidly, which exhibit low adhesion to the base material. For more details on thermal spraying of metals, see → Metals, Surface Treatment, Chapter 5.

Vacuum Techniques. The deposition of metals in a high vacuum is of greater importance than atmospheric pressure methods. The metal is evaporated and condenses on the cold surface of the substrate in a vacuum chamber ($10^{-2} - 10^{-3}$ Pa). It forms a dense film with high reflectivity. The following methods are used:

1. Thermal evaporation, whereby heat is generated by electrical heating, induction, or current flowing directly through the molten metal to be evaporated.
2. Electron beam evaporation, in which a high-energy electron beam is focused on the metal to be evaporated.
3. Cathodic sputtering in which a gas discharge (generally in an argon atmosphere) is ignited in the chamber: the chamber walls form the anode, while the metal is located on the bar or plate cathode. The energetic gas ions that bombard the metal surface dislodge particles, some of atomic dimensions, that deposit on the substrate surface. Magnetic fields are frequently employed to increase the efficiency of evaporation and reduce the thermal stress to which the substrate is exposed.

Thermal evaporation techniques do not permit deposition of alloys of defined composition because fractionation occurs as a result of differences in the boiling points of the alloy components. Deposition of alloys is possible by using electron beam evaporation and cathodic sputtering.

All evaporators operate with a directed jet of metal vapor so the vacuum chamber is not completely filled with an atmosphere of metal vapor, which explains why only the substrate surface that is in direct line with the evaporator is coated. For this reason, the molding surface should not be too rough and should have no undercuts.

Because of the time required to generate a high vacuum, substrates are introduced into the chamber in batches and attached to a turntable so that they can be coated successively.

The substrate surfaces are often activated prior to evaporation. A corona discharge is ignited in the chamber at ca. 10 Pa. The ions impinging on the substrate surface cause cleavage of chemical bonds near the surface and thus a roughening of the surface and the formation of polar structures. In this way, nucleation sites are generated for the condensing metal, and good adhesion of the coating is attained.

Apart from aluminum, which is used in most cases because of the high reflectivity that can be attained, silver, gold, copper, chromium, zinc, nickel, cobalt, tin, titanium, and indium are employed, as well as several nonmetallic materials such as borosilicate glass and silicon monoxide.

The thin metal films are generally extremely sensitive to mechanical damage and are prone to rapid attack by moisture and environmental chemicals (exceptions include aluminum, which on exposure to air immediately forms a hard protective oxide layer, and the noble metals). Therefore, in most cases the metal layers are covered with a protective coating.

All polymers are suitable as substrates providing they do not evolve any materials that could lead to disturbances in the high vacuum (e.g., plasticizer from PVC or moisture from PA 6, PA 66, and cellulose hydrate). Such substrates must be precoated with a primer that hinders degassing.

The metallic deposit reproduces the finest structures and defects of the substrate, and adhesion is seldom adequate. Hence, for these reasons as well, use of a primer can be beneficial.

This primer layer must form a smooth, defect-free, high-gloss surface and exhibit excellent adhesion to the substrate. After drying — almost always at elevated temperature to drive out any residual solvents — no disturbing substances may be released into the vacuum. Furthermore, the primer coating should exhibit good adhesion to the metal film and should not be colored. It should be unaffected by the metal film and resistant to any solvents in the protective coating.

Similarly, the protective coating must exhibit excellent adhesion to the metal and must not have any detrimental effects on the substrate, the primer coat, or the metallic layer.

The primer and protective coatings are colorless or colored with translucent, soluble, lightfast dyes. They can be used to imitate other metals such as chromium, copper, gold, and brass. Highly homogeneous but less lightfast colors are obtained by immersing the colorless protective coatings in dye solutions.

Transparent polymers such as poly(methyl methacrylate) and polycarbonate (Fig. 20) can

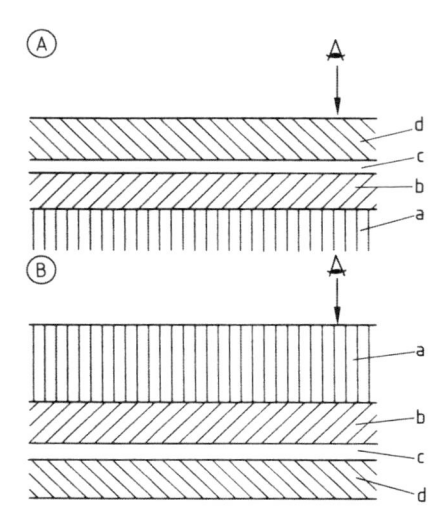

Figure 20. First-surface (A) and second-surface (B) metallization
a) Transparent substrate; b) Primer; c) Metal coating; d) Protective coating

be first-surface or second-surface metallized. With second-surface metallizing, the primer coating is often unnecessary, and the protective coating is generally highly pigmented and opaque.

Typical areas of application for polymer metallization include components for automobile interiors, taillight and headlight reflectors, sanitary fittings, signs, flashlight reflectors, lamp components, Christmas tree decorations, toys, fashion jewelry, caps for cosmetic bottles, frames for pictures and mirrors, and wristwatch dials.

3.5. Dry Reactive Processes

Dry reactive processes are employed to functionally modify the surface of a polymer or as a pretreatment for further coating processes.

Differences are drawn between:

1. Processes during which the substrate is itself chemically altered
2. Application of additional layers to the surface

3.5.1. Surface-Modifying Processes

Corona treatment is an electrical pretreatment process. The substrate is located at ambient pressure in a discharge field between two electrodes to which an a.c. voltage of 10 – 20 kV with a frequency of 20 – 40 kHz is applied. This technique is used mostly to activate the surface of sheets and films by passing them between two roll electrodes with a gap of ca. 1 mm. The attainable surface energies are high despite the relatively short treatment times. The treatment times depend on the electrical performance of the equipment although this is limited by possible electrical breakdown of the substrate. Whereas regularly shaped hollow moldings are relatively amenable to corona treatment by placing them on appropriately formed counterelectrodes, the technique is difficult to apply to irregularly shaped articles. In such cases, counterelectrode-free techniques are used in which, by applying very high voltages, the charge carriers can be sprayed from wire bundles onto the plastic component.

With substrates containing low molecular mass additives (e.g., lubricants, plasticizers), a decrease in surface energy with storage time following treatment can be observed. This is caused by migration of the low molecular mass components.

The toxic substances, such as nitrogen oxides and ozone, generated in large amounts in the discharge field must be removed.

Fluorination is employed not only to activate polymer surfaces for successive coating processes (adhesive bonding, painting, printing, etc.) but also to permanently modify surface properties such as wettability and permeability [42, 44]. The substrate is exposed to a nitrogen atmosphere containing up to 10 % fluorine at slightly reduced pressures and at temperatures up to 110 °C. In *oxyfluorination* [45], oxygen is added to the gas mixture. Apart from the increased surface energy, fluorinated substrate surfaces are characterized by a pronounced diffusion barrier effect toward organic solvents, for example. For this reason, the technique is employed to treat the inner surface of plastic fuel tanks to reduce permeation of the fuel. With oxyfluorination, good barrier effects are achieved even after short treatment times (ca. 10 min). With fluorination, better barrier effects can be achieved but only after extended treatment times. As a process for activating the substrate surface, fluorination has an advantage over corona treatment because it gives higher surface energies. Furthermore, the activating effect remains longer with substrates containing low molecular mass components because the rate of diffusion of these components in the fluorinated substrate layers near the surface (1 – 2 µm) is also reduced. However, longer treatment times are required, and equipment for the continuous fluorination of sheets is more expensive than corona machines. Fluorination has proved relatively problem free for the surface activation of irregularly shaped components.

Because the gaseous fluorinated products that are formed (e.g., hydrogen fluoride) and fluorine gas are severe pollutants, measures to guarantee compliance with permitted workplace and exhaust gas concentrations are necessary (operation at slightly reduced pressure, exhaust gas cleaning).

Flame treatment of plastics is a surface activation process that is operated at atmospheric pressure. The substrate surface is briefly exposed to the flame from a gas burner (propane, butane, town gas, or natural gas) adjusted to give strongly oxidizing conditions (excess oxygen). The substrate is oxidized at the surface, and the polar groups that are formed lead to improved wettability. Significant process parameters include the gas – air ratio, flame temperature, and duration of treatment, all of which must be adjusted so that the substrate is not heated too strongly and the surface is not visibly altered. The process is applicable to films, sheets, and moldings with large smooth surfaces. Flame treatment is employed primarily with polyolefins (e.g., PP automobile bumper linings). Because the activating effect is lost rapidly, flame-treated articles should be coated immediately after treatment.

Plasma treatment of plastics is also a surface activation process. It is carried out in vacuum chambers at 10 – 100 Pa with a nonthermal plasma (gas discharge) in a gas atmosphere consisting of an inert gas or reactive gases (e.g., oxygen). In the plasma, gas atoms and molecules are ionized and accelerated toward the substrate surface, which they impact with high energy, causing cleavage of chemical bonds and formation of polar groups. In *glow discharge* processes the plasma is fed with a d.c. or an a.c. voltage field (frequency range 50 Hz up to several tens of megahertz). The plasma burns between two electrodes located in the chamber (resistive coupling), whereby one of the electrodes can be an electrically conducting substrate. For treating sheets and films, the separation between electrodes at 200 – 400 V is considerably larger than with corona treatment and amounts to several tens of centimeters. When a glow discharge is used as a pretreatment (e.g., for metallizing) for moldings, the separation between the electrodes can be considerably larger and voltages up to 5 kV are applied.

When energy is introduced into the plasma by using *microwaves* (gigahertz range), significantly shorter treatment times (<1 s) are possible due to the increased charge density in the plasma.

These pretreatment techniques are used, for example, in the physical metallization or coating of films and moldings (automobile bumper covers [46]), and also in the treatment of carbon, aramid, and high-strength PE fibers to improve fiber-to-matrix adhesion when they are employed in fiber – polymer composites [47, 48].

Because plasma techniques require considerable expenditures for equipment due to the low pressures required, they are usually applied only when a subsequent vacuum coating process is intended.

3.5.2. Coating Processes

Gasplating is normally carried out at atmospheric or slightly elevated pressure. It is used in metallizing, where volatile metal compounds (e.g., nickel tetracarbonyl) are employed, which decompose on the preheated substrate surface and form a pore-free, homogeneous metal film.

Plasma polymerization is a vacuum coating process that is suitable for metallic substrates and also for thermally sensitive polymers due to the low thermal loading of the substrate [47, 49–56]. Monomer gases (e.g., ethylene, vinyl chloride, hexamethyldisiloxane) are vented into the vacuum chamber containing the substrate at 1 – 1000 Pa. A nonthermal plasma is ignited in which the monomer gas is activated. The plasma can be fed by an electrical d.c. or a.c. voltage field with frequencies of 50 Hz to several megahertz between two electrodes located within the chamber (resistive coupling) or outside it (capacitive coupling — not applicable to d.c. plasmas). Energy input using microwaves (generally 2.45 GHz) gives the highest attainable coating rate due to the high charge carrier density in the plasma. A distinction is made between processes in which the plasma burns directly on the surface to be coated and those in which it is remote from the surface (so-called downstream processes). In the latter, the monomers and (optional) propellant gas particles that are activated in the plasma (plasma reactor) must flow through a plasma-free zone to the substrate surface, whereby the first recombination reactions occur in the gas phase. Indirect ionization is employed with particularly sensitive monomers. Here, a propellant gas (e.g., a noble gas) is activated in the plasma, and the monomer gas is activated by ionized propellant ejected from the plasma. This prevents fragmentation of the monomer by direct contact with the plasma.

The activated particles and recombined intermediate products meet at the substrate surface and, in the course of further chemical reactions, form layers of highly cross-linked polymer. Low molecular mass, gaseous products that are not incorporated in the layers are removed by the vacuum pump. Apart from the composition of the feed gas, which can contain other inert or reactive gases in addition to the monomers, process parameters that influence the rate of deposition and the properties of the layer include reactor geometry, separation between plasma and substrate in downstream processes, substrate material and temperature, flow of feed gas, process pressure, excitation energy, excitation frequency, and coating time.

Because of the high degree of cross-linking, plasma-polymerized layers exhibit high density, mechanical stability, and chemical resistance. They also have electrical breakdown voltages and good heat resistance. Furthermore, even at very low thicknesses (10 nm) they are free of micropores, so that they are ideally suited as diffusion barrier layers. Due to the high energy of the particles formed during the process, the layers generally adhere well to the substrate.

As in metallizing in high vacuum, the properties of the layer can be varied across its thickness by varying the operating parameters or the composition of the process gas during coating.

The process is used in the coating of reflectors for headlights [46, 57]. These are mounted in charges on a rotating carousel in a high vacuum, where they are pretreated (glow discharge), coated with aluminum, and subsequently coated with a protective layer 20 – 30 nm thick by plasma polymerization with a process gas containing hexamethyldisiloxane and oxygen. Because of their extremely low thickness, the protective layers have no optical influence but are effective diffusion barriers (condensed water resistance is 1000 times greater than without the coating) and also have a high surface energy so that condensing materials (moisture or degassed products from other headlight components) spread across the surface to form thin films and hence have no optical effects. Other applications include the hydrophilic coating of contact lenses and the coating of medical materials to improve biocompatibility [58].

On a laboratory scale, a variety of layer properties are currently being investigated. For instance, with ethylene as a monomer, it has been possible to apply a layer to the inside of PE screw top bottles (1.2 L) that reduces the permeation of gasoline to < 2 % of that of uncoated bottles [59, 60]. The highly cross-linked polyethylene layer was coated onto the substrate within 40 s by using a microwave plasma; it had a high density (>1.5 kg/dm^3) and a thickness of 100 –200 nm. Because the coating also exhibits excellent adhesion and does not suffer appreciable loss of barrier properties even after severe deformation (drop test), it represents a viable alternative to fluorination in the manufacture of plastic fuel tanks.

An advantage of plasma polymerization is that a very high proportion of the monomers employed is incorporated into the film, and only small amounts are present in the evacuated residual gases. The material costs of the process are thus very low. Additionally, environmentally harmless materials are used in many cases so that the exhaust gases need not be cleaned. Because vacuum processes are involved, however, considerable equipment expenditure is necessary. Especially with microwave plasma polymerization, the process cycle time is influenced primarily by the capacity of the vacuum pumps (evacuation time), whereas the actual coating time is low (generally <1 min). For this reason, chambers for rough evacuation of the components to be coated through which the substrate passes to and from the actual process chamber, are useful.

References

General Reference

1 O. Lauer: *Aufbereiten von Kunststoffen*, Hanser Verlag, München 1971. *Aufbereiten von PVC*, VDI-Gesellschaft Kunststofftechnik, Düsseldorf 1976. *PVC-Heißmischen (Dryblend)*, VDI-Gesellschaft Kunststofftechnik, Düsseldorf 1968. *Speichern, Fördern und Dosieren von Kunststoffen*, VDI-Gesellschaft Kunststofftechnik, Düsseldorf 1971. *Granulieren von thermoplastischen Kunststoffen*, VDI-Gesellschaft Kunststofftechnik, Düsseldorf 1974. W. Dalhoff: *Systematische Extruder-Konstruktion*, Krauskopf Verlag, Mainz 1974. H. Dominghaus: *Fortschrittliche Extrudertechnik*, VDI-Verlag, Düsseldorf 1970. *Der Extruder im Extrusionsprozeß*, VDI-Gesellschaft Kunststofftechnik, Düsseldorf 1989. H. Herrmann: *Schneckenmaschinen in der Verfahrenstechnik*, Springer Verlag, Berlin 1972. M. Jacobi: *Grundlagen der Extrudertechnik*, Hanser Verlag, München 1960. W. Mink: *Grundzüge der Extrudertechnik*, Zechner & Hüthig Verlag, Speyer 1973. H. Potente: *Modellgesetze für Ein- und Zweischneckenmaschinen*, Hanser Verlag, München 1981. C. Rauwendaal: *Polymer Extrusion*, Hanser Verlag, München 1986. G. Schenkel: *Kunststoff-Extrudertechnik*, Hanser Verlag, München 1963. E. G. Fischer: *Extrusion of Plastics*, Butterworths, London 1976. *Extrudieren von Schlauchfolien*, VDI-Gesellschaft Kunststofftechnik, Düsseldorf 1973. *Extrudieren von Rohren und Profilen*, VDI-Gesellschaft Kunststofftechnik, Düsseldorf 1974. *Extrudierte Feinfolien und Verbundfolien*, VDI-Gesellschaft Kunststofftechnik, Düsseldorf 1976. G. Menges, H. Recker: *Automatisierung in der Kunststoffverarbeitung*, Hanser Verlag, München 1986. *Rechnergesteuerte Extrusion*, VDI-Gesellschaft Kunststofftechnik, Düsseldorf 1976. Z. Tadmor: *Engineering Principles of Plasticating Extrusion I*, Klein Reinhold Book Corporation, New York 1970. W. Michaeli: *Extrusionswerkzeuge für Kunststoffe und Kautschuke*, Hanser Verlag, München 1990. F. Hensen, W. Knappe, H. Potente: *Handbuch der Kunststoffextrusionstechnik*, vols. **I and II**, Hanser Verlag, München 1989. W. Mink: *Grundzüge der Hohlkörperblastechnik*, Zechner & Hüthig Verlag, Speyer 1969. D. v. Rosato: *Blow Molding Handbook Dominick*, Hanser Verlag, München 1989. R. Holzmann: *Gestalten von Blasformteilen*, VDI-Verlag, Düsseldorf 1971. O. Plajer: *Werkzeuge für das Blasformen*, Zechner & Hüthig Verlag, Speyer 1969. *Spritzblasen*, VDI-Gesellschaft Kunststofftechnik, Düsseldorf 1976. J. L. Throne: *Thermoforming*, Hanser Verlag, München 1987. A. Thiel: *Grundzüge der Vakuumformung*, Zechner & Hüthig Verlag, Speyer 1969. K. Stoeckhert: *Kunststoff-Lexikon*, Hanser Verlag, München 1975. F. Johannaber, K. Stoeckhert: *Kunststoffmaschinenführer*, Hanser Verlag, München 1984. O. Scharz, F.-W. Ebeling, G. Lüpke, W. Schelter: *Kunststoffverarbeitung*, Vogel, Würzburg 1988. D. H. Morton-Jones: *Polymer Processing*, Chapman and Hall, New York 1989. G. Menges: *Einführung in die Kunststoffverarbeitung*, Hanser Verlag, München-Wien 1979. F. Johannaber: Spritzgießmaschinen, in: *Kunststoffmaschinenführer*, Hanser Verlag, München – Wien 1979. W. Elbe:

"Untersuchungen zum Plastifizierverhalten von Schneckenspritzgießmaschinen," Dissertation, RWTH Aachen 1973. R. v. Hooven, A. J. Kaminski: "Entgasungsspritzgießen, ein Laborkonzept wird in die Praxis umgesetzt," *Plastverarbeiter* **31** (1990) no. 8, 441 – 446. Th. Copetti, R. Krebser: "Vollhydraulische und Kniehebelschließsysteme," *Kunststoffe* **70** (1980) no. 2, 821 – 825. G. Menges, P. Mohren: *Anleitung zum Bau von Spritzgießwerkzeugen*, Hanser Verlag, München – Wien 1983. S. Schäper: "Nukleierung thermodynamisch getriebener Polyurethan-Reaktionsschäume," Dissertation, RWTH Aachen 1977.

General Reference

2 W. Becker, D. Braun, B. Carlowitz: *Kunststoff-Handbuch*, vol. **1** Die Kunststoffe, Hanser Verlag, München 1990. W. Becker, D. Braun, W. Woebcken: *Kunststoff-Handbuch*, vol. **10** Duroplaste, Hanser Verlag, München 1988. H. Saechtling: *Kunststoff-Taschenbuch*, 24th. ed., Hanser Verlag, München 1989. K. Stoeckhert: *Kunststoff-Lexikon*, 6th. ed., Hanser Verlag, München 1975. DIN-Taschenbuch 21,Kunststoffnormen,8th. ed., Beuth-Verlag, Berlin 1979. W. Schönthaler: *Verarbeiten härtbarer Kunststoffe*, VDI-Verlag, Düsseldorf 1973. W. Bauer, W. Woebcken: *Verarbeitung duroplastischer Formmassen*, Hanser Verlag, München 1973. W. Bucksch, H. Briefs: *Preßwerkzeuge in der Kunststofftechnik*, 2nd. ed., Springer Verlag, Berlin 1962. K. Stoeckhert: *Formenbau für die Kunststoff-Verarbeitung*, Hanser Verlag, München 1969. Berufsgenossenschaft der chem. Industrie, *Merkblatt für das Verarbeiten von Polyester- und Epoxidharzen*, Verlag Chemie, Weinheim, Germany 1966. VDI-Richtlinie 2007, Epoxidharze im Fertigungsmittelbau, VDI-Verlag, Düsseldorf 1966. VDI-Richtlinie 2011, Faserverstärkte Reaktionsharzformstoffe, VDI-Verlag, Düsseldorf 1973. H. Jahn: *Epoxidharze*, Verlag für die Grundstoffindustrie, Leipzig 1969. F. Breckner: *Glasfaserverstärkte Kunststoffe; Anwendungen, Eigenschaften*, 2nd ed., Verlag Fraunhofer-Gesellschaft, 1989. H. Domininghaus: *Die Kunststoffe und ihre Eigenschaften*, VDI-Verlag, Düsseldorf 1988. O. Schwarz: *Kunststoffkunde*, Vogel, Würzburg 1988. W. Michaeli, M. Wegener: *Einführung in die Technologie der Faserverbundwerkstoffe*, Hanser Verlag, München 1989. R. W. Meyer: *Handbook of Polyester Molding Compounds & Molding Technology*, Chapman and Hall, New York 1987. S. M. Lee: *International Encyclopedia of Composites*, VCH Publishers, New York 1990.

General Reference

3 W. Knig, L. Neder: "Schneiden mit ultraschallerregtem Messer," *Ind. Anz.* **107** (1985) 49. W. König, L. Neder: "Bessere Schnittflächen mit Ultraschall," *Plastverabeiter* **37** (1986) no. 11. "Neue Werkstoffe und hohe Präzision — eine Herausforderung für die Fertigstellungstechnik" in AWK Aachener Werkzeugmaschinen-Kolloquium, (ed.): Produktionstechnik auf dem Weg zu integrierten Systemen, VDI-Verlag, Düsseldorf 1987. W. König, P. Graß, Ch. Wulf: "Machining of Fibre Reinforced Plastics," *CIRP Ann.* **34** (1985) no. 2, 537 – 548. F.-J. Trasser, M. Wehner,"Schnittflächen beim Laserstrahlschneiden von faserverstäkten Kunststoffen CO2-, Excimer-Laser — ein Verfahrensvergleich," *Ind. Anz.* **109** (1987) no. 43/44, 57 – 58. "Bearbeitungstechnik für Verbundwerkstoffe — eine vielschichtige Aufgabenstellung," *VDI-Fachtagung METAV*,June, 8 – 9, 1988. H. K. Tönshoff, V. Hohensee,"Bearbeitung faserverstärkter Kunststoffe," *ZWF, Z. Wirtsch. Fertigung* **81** (1986) no. 2, 106 – 111.

Specific Reference

4 V. Hoffmann: "Beitrag zur Untersuchung der Zerspanbarkeit thermoplastischer Kunststoffe beim Bohren," Dissertation, Techn. Univ. Berlin 1971.

5 E. Lemke: "Beitrag zur Untersuchung der Zerspanbarkeit von Plastomeren," Dissertation, Techn. Univ. Berlin 1972.

6 G. Zug: "Beitrag zur Untersuchung des Zerspanverhaltens von thermoplastischen Kunststoffen beim Drehvorgang," Dissertation, Techn. Univ. Berlin 1970.

7 G. Spur, G. Zug, *WT-Z. Ind. Fertigung* **59** (1969) no. 10, 485 – 490.

8 G. Spur, G. Zug, *ZWF Wirtsch. Fertigung* **66** (1971) no. 9, 441 – 445.

9 B. Ferichmann: "Ermittlung fertigungsgerechter Arbeitsbedingungen und Untersuchung des Zerspanungsverhaltens beim Drehen thermoplastischer Kunststoffe," Dissertation, RWTH Aachen 1966.

10 G. Spur, G. Zug, *Werkstatt Betr.* **101** (1968) no. 6, 325 – 328.

11 A. Kobayashi, *Machining of Plastics*, McGraw-Hill, New York 1967.

12 H. G. Schmidt, *ZWF Wirtsch. Fertigung* **60** (1965) no. 6, 286 – 287.

13 M. Masuko, S. Ammi, *CIRP Ann.* **13** (1966) no. 4, 389 – 397.

14 G. Spur, G. Zug, *ZWF Wirtsch. Fertigung* **66** (1971) no. 2, 47 – 51.

15 G. Spur, V. Hoffmann, *ZWF Wirtsch. Fertigung* **67** (1972) no. 2, 61 – 64.

16 M. N. Watson, *Joining Plastics in Production*, The Welding Institute, Cambridge 1989.

17 A. Neumann et al.: *Grundlagen der Schweißtechnik — Plastfügetechnik*, VEB Verlag Technik, Berlin 1985.

18 G. Habenicht: *Kleben*, Springer Verlag, Berlin 1986.

19 W. Brockmann, L. Dorn, H. Käufer: *Kleben von Kunststoff mit Metall*, Springer Verlag, Berlin 1989.

20 G. Menges, G. Stockhausen, M. Reinke: *Kleben — ein Nachschlagewerk über die Verwendung von Klebstoffen in der Kunststoffverarbeitung*, Herausgeb. IKV, Aachen 1981, Verlag TüV Rheinland, Köln 1991.

21 D. Wimmer: *Kunststoffgerecht konstruieren*, Hoppenstedt Verlag, Darmstadt 1989.

22 U. Delpy: *Schnappverbindungen aus Kunststoff*, Expert-Verlag, Ehningen 1989.

23 U. Zorll, E.-C. Schütze: *Kunststoffe in der Oberflächentechnik*, Kohlhammer, Stuttgart 1986.

24 H. Saechtling, W. Zebrowski: *Kunststoff-Taschenbuch*, 24th ed., Hanser Verlag, München–Wien 1989.

25 W. A. Zisman: *J. Paint Technol.* **44** (1972) 564.

26 P. Berns: *Ind. Lackierber.* **44** (1976) 125.

27 P. O. Damm: *Plastverarbeiter* **25** (1974) 683 – 687.

28 F. Gudehus: "Lackieren von Kunststoffteilen," *Kunststoffe* **78** (1988) 6.

29 P. Berns: *Kunststoffhandbuch*, vol. **5**, Polystyrol, Hanser Verlag, München 1969, pp. 369 – 376.

30 V. Bartelmäs: *Kunststoffhandbuch*, vol. **5**, Polystyrol, Hanser Verlag, München 1969, pp. 814 – 849.

31 W. Domininghaus: *Dekorieren von Kunststoffformteilen*, VDI-Verlag, Düsseldorf 1971.

32 *Kunststoffgalvanisierung — Handbuch für Theorie und Praxis*, Leuze Verl., Saalgau/Württ. 1973.
33 P. Spiro: *Electroforming*, R. Draper Ltd., Teddington, England, 1968.
34 K. Wiebusch, H. Hendus, E. Zahn: *Metallische überzüge auf Kunststoffen*, Hanser Verlag, München 1966, pp. 9 – 28.
35 K. Heymann, W. Riedel, G. Woldt: *Metallische überzüge auf Kunststoffen*, Hanser Verlag, München 1966, pp. 48 – 63.
36 K. Wiebusch: *Galvanotechnik* **59** (1968) no. 8, 640 –652.
37 K. Wiebusch: *Galvanotechnik* **61** (1970) no. 12, 984 –993.
38 K. Wiebusch: *Konstruktion, Fertigung und Anwendung feinwerktechnischer Kunststoffteile*, VDI-Verlag, Düsseldorf 1976, pp. 109 – 123.
39 W. Goldie: *Metallic Coating of Plastics Electrochem.*, Publ. Ltd., Ayr, UK, 1968.
40 K. Wiebusch: *Galvanotechnik* **61** (1970) no. 9, 704 –714.
41 R. P. Tison: *Plating* **1** (1973) 47 – 52.
42 C. Bliefert: "Fluorierung von Kunststoff-Kraftstoffbehältern," *Kunststoffberater* **6** (1986) 14 – 16.
43 Verein Dt. Ingenieure (ed.): *Sperrschichtbildung bei Kunststoff-Hohlkörpern*, VDI-Verlag, Düsseldorf 1986.
44 T. Volkmann, H. Widdecke: "Fluorination of Polyethylene Films," *Macromol. Chem. Macromol. Symp.* **25** (1989) 243 – 248.
45 T. Volkmann, H. Widdecke: "Oxifluorierung von Polyethylen," *Kunststoffe* **79** (1989) no. 8, 743 – 744.
46 J. Kieser: "Plasmapolymerisation als industriell einsetzbare Technik," VDI-Kolloquium *Dünnschichttechnologien*, 6./7. Nov., Hagen 1986.
47 R. H. Ludwig: "Plasmapolymerisation — ein Verfahren zur Erzeugung dünner Schichten," Dissertation, RWTH Aachen 1989.
48 H. Yasuda: "Tandem Plasma-Polymerization Apparatus for Continuous Coating of Fibers and Films," *ACS Symp. Ser.* **108**, (1979) 277 – 286.
49 G. Menges, P. Plein: "Plasmapolymerisation — maßgeschneiderte Beschichtungen für Kunststoffteile," *Kunststoffe* **78** (1988) no. 10, 1015 – 1018.
50 M. R. Wertheimer, M. Moisan: "Comparison of Microwave and Lower Frequency Plasmas for Film Deposition and Etching," *J. Vac. Sci. Technol. A* **3** (1985) no. 6, 2643 – 2649.
51 G. Menges *et al.*: "Plasmapolymerisation — ein zukunftsträchtiges Beschichtungsverfahren," *Coating* **1** (1989) 2 – 7.
52 R. Ludwig: "Plasmapolymerization — A New Technology for Surface Modification," *Annu. Tech. Conf. Soc. Plast. Eng.* 1989, 915 – 917.
53 H. V. Boening: *Plasma Science and Technology*, Hanser Verlag, München 1982.
54 G. Menges, P. Plein: "Die besonderen Schichten — Plasmapolymerisation steht vor breiter industrieller Nutzung," *Ind. Anz.* **109** (1987) no. 16, 23 – 24.
55 J. Kieser, M. Sellschopp: "Hochvernetze Polymerschichten mit einstellbaren Eigenschaften," *VDI-Nachr.* **41** (1984) 28.
56 J. Musil: "Microwave Plasma: Its Characteristics and Applications in Thin Film Technology," *Vacuum* **36** (1986) no. 1 – 3, 161 – 169.
57 G. Benz: "Schutzschichten durch Plasmapolymerisation," *Bosch Tech. Ber.* **8** (1986/1987).
58 Deutsche auf dünnem Eis, Industriemagazin, Jan., 1987, 124 – 129.
59 W. Michaeli *et al*: Plasmabehandelte PE-Fasern — eine Verstärkungsfaser ohne Haftungsprobleme, *J. Coating* (1991) 1.
60 W. Michaeli, M. Londschien, R. Ludwig: "Plasmapolymerisierte Sperrschichten," *Kunststoffe-Plastics* **8** (1990), 13 – 16.

Further Reading

M. Chanda, S. K. Roy: *Plastics fabrication and recycling*, CRC Press, Boca Raton, Fla. 2009.
E. Cybulski: *Plastic conversion processes*, CRC Press/Taylor & Francis, Boca Raton, Fla. 2009.
E. Lokensgard: *Industrial plastics*, 5th ed., Delmar Cengage Learning, Clifton Park NY 2009.
P. A. Tres: *Designing plastic parts for assembly*, 6. ed., Hanser, Munich 2006.
M. Xanthos, D. B. Todd: *Plastics Processing*,Kirk Othmer Encyclopedia of Chemical Technology, 5th edition, vol. 19, p. 536–563, John Wiley & Sons, Hoboken, NJ, 2006, online: DOI: 10.1002/0471238961.1612011924011420.a01.pub2.

Plastics, Properties and Testing

MANFRED STAMM, Max-Planck-Institut für Polymerforschung, Mainz, Federal Republic of Germany

BODO CARLOWITZ, Königstein im Taunus, Federal Republic of Germany

1.	States of Order of Polymers	471
1.1.	Glassy State and Melt	472
1.2.	Crystalline State	475
1.3.	Liquid Crystals	476
1.4.	Blends and Block Copolymers	477
2.	Properties and Testing	477
2.1.	Mechanical Properties and Their Testing	478
2.1.1.	Classification of Plastics Based on Mechanical Properties	478
2.1.2.	Viscoelastic Behavior of Polymers	480
2.1.3.	Mechanics of Deformation	483
2.1.4.	Friction and Wear	490
2.1.5.	Fracture Behavior	491
2.1.6.	Control of Mechanical Properties through Orientation, Internal Stresses, and Morphology	495
2.1.7.	Reinforced and Filled Plastics	497
2.2.	Acoustic Properties and Their Testing	497
2.3.	Thermal Properties and Their Testing	498
2.3.1.	Specific Heat	499
2.3.2.	Thermal Conductivity	499
2.3.3.	Thermal Expansion	500
2.4.	Electrical Properties and Their Testing	502
2.4.1.	Insulating Properties	502
2.4.2.	Dielectric Properties	503
2.4.3.	Dielectric Strength	504
2.4.4.	Electrets; Semiconducting and Conducting Polymers	506
2.5.	Optical Properties and Their Testing	506
2.5.1.	Refraction and Birefringence	506
2.5.2.	Transmittance and Gloss	507
2.5.3.	Color and Colorimetry	509
2.6.	Density and Physical-Chemical Properties and Their Testing	510
2.6.1.	Density	510
2.6.2.	Liquid Absorption and Swelling; Desorption of Plasticizers	510
2.6.3.	Permeation of Gases and Vapors	511
2.7.	Stability Properties and Their Testing	512
2.7.1.	Water Absorption	512
2.7.2.	Effect of Chemicals	512
2.7.3.	Effects of Heat, Light, and Weathering	513
2.7.4.	Effect of High-Energy Radiation	514
2.7.5.	Stability against Fire	515
2.7.6.	Effect of Microorganisms	516
2.8.	Processing Properties and Their Testing	516
2.8.1.	Molding Compounds	517
2.8.2.	Conventional Methods for Characterizing Flow Behavior	517
2.8.3.	Rheology of Plastics	518
	Abbreviations	521
	References	522

1. States of Order of Polymers

The state of order in plastics is, despite some peculiarities, very similar to that of low molecular mass compounds. Amorphous, crystalline, and liquid crystalline phases are observed, which exhibit short- and long-range order in one, two, or three dimensions. Therefore, the *characterization techniques* used to determine the state of order are also very similar. The most common techniques are summarized in Table 1, which also lists the typical dimensions and the principal information that can be obtained. One generally distinguishes among the following typical scales for the state of order:

1. The submolecular scale (typically 0.1 – 10 nm) where the state of order of chain segments and monomer units is considered

Table 1. Most common techniques for the determination of order in polymers

Technique	Information	Typical dimensions
Wide-angle X-ray scattering (WAXS)	crystallinity, orientation, crystal structure, near range order	0.1 – 1 nm
Small-angle X-ray scattering (SAXS)	morphology, long spacing, density fluctuations	1 – 100 nm
Small-angle neutron scattering (SANS)	chain conformation	1 – 100 nm
Electron microscopy (EM)	morphology,	
Electron diffraction (ED)	long spacing,	10 – 1000 nm
Electron energy loss spectroscopy (EELS)	crystal structure, element distribution	
Light scattering (LS)	larger structures,	
Microscopy	different phases,	1000 nm – 1 mm
Birefringence	orientation	

2. The molecular scale (typically 5 – 20 nm) where the spatial arrangement of one chain is investigated
3. The morphological scale (10 nm to several micrometers) where the superstructure formed by many polymer molecules is envisaged

The crystal structure, crystallinity, and orientation of polymers are determined by X-ray methods [1], whereas the crystal morphology or phase structure is obtained from electron microscopy [2] and light scattering [3]. The use of neutron scattering [4] offers the unique possibility of "seeing" the spatial arrangement or conformation of single molecules in the bulk by selective deuteration (Fig. 1).

The two *principal states of order* in polymers are the amorphous and the semicrystalline states. A peculiarity of polymers is that, in general, amorphous regions are also present in the crystalline state and the crystallinity can be varied over a wide range by means of the crystallization and annealing conditions. It is evident from Figure 1 that mechanical properties, such as the elasticity and the toughness of plastics, are determined largely by the connectivity and entanglement of the molecules. In particular, in the crystalline state the crystallites are interconnected by tie molecules, giving the material its mechanical strength. Hence the physical properties of polymers are determined not only by their chemical nature and composition but also by the state of order—the chain conformation, morphology, phase structure, and orientation of molecules. Therefore, the control and determination of the state of order are important areas of polymer science. They can determine the final use of products and also offer possibilities for modification and new areas of application. Examples include the stress-induced crystallization of rubber, which is undesirable because it makes the material brittle and unelastic, and the production of ultra-high-modulus polyethylene fibers by extreme orientation from a gel state to obtain a material that can be much tougher than steel in terms of the specific modulus.

Special morphologies and superstructures are observed in polymer blends and block copolymers, while the state of order of liquid crystalline polymers is between that of the crystalline and the amorphous materials.

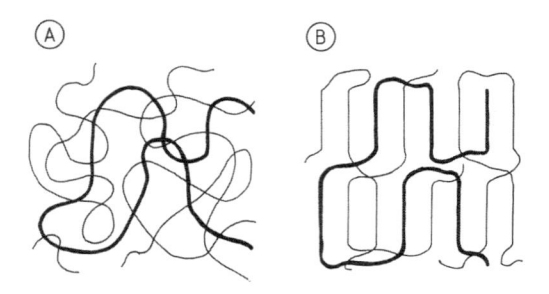

Figure 1. Schematic of chain conformations in the (A) amorphous and (B) semicrystalline state
A single chain is marked by the solid line and may represent a labeled (deuterated) chain among otherwise unlabeled molecules. In the amorphous state (melt or glass), molecules adopt a Gaussian coil conformation, whereas in the semicrystalline state, both amorphous and crystalline regions are present. During rapid crystallization from the melt, the overall chain conformation does not change significantly and chain segments are incorporated into crystalline lamellae.

1.1. Glassy State and Melt

Polymers in the glassy state and in the isotropic melt generally have a low degree of order: they are *amorphous*. Only short-range order is present, similar to that observed in simple liquids. Because polymers are large molecules, a distinction can be made between the intra- and intermolecular order of monomer units. Very stiff macromolecules have a high orientation correlation along the chain, while the center-of-mass positions of molecules with respect to one another are distributed statistically in space (i.e., a large intramolecular and no intermolecular order). However, in many cases, stiff molecules tend to form liquid crystalline phases (see Section 1.3).

Polymer Melt. Flexible chains form a Gaussian coil-like conformation [5], [6] with little intramolecular order. The *random-coil conformation* of the melt (Fig. 1 A) is also observed in dilute solution under theta conditions. In the ideal situation, no effective interaction occurs between a chain and its surroundings. The center-of-mass positions of the molecules are random, and the long-range order observed in the crystalline state is completely absent. This picture can be tested by small-angle neutron scattering in the melt [7], in which the radius of gyration R_g – a measure of the mean dimension of the chain and the chain conformation — can be obtained. Contrast for neutrons is generated by the deuteration of one chain in a matrix of otherwise undeuterated chains (Fig. 1). The deuterated chains are then "visible" against the others for neutrons. The assumption is that deuteration does not change the thermodynamics, and particularly the chain conformation, which to a good approximation is true in most cases [8], [9]. Investigations indicate that the chain conformation in the melt and under theta conditions is identical and, in many cases, can be described by an unperturbed coil. In the random-coil model with a random distribution of chain segments, the radius of gyration R_g should be proportional to the square root of the molecular mass M_w ($R_g = \alpha M_w^{0.5}$) [22], which to a good approximation is true for many polymers (Table 2).

The pair correlation function $h(r)$ of monomer units, irrespective of their connectivity to different chains, is a measure of the *near range order* of the monomer units (Fig. 2 A) [23]. In starting from one monomer unit in the melt, at a certain distance r_0 a peak in $h(r)$ is observed, corresponding to other monomer units at the nearest neighbor position, which form a first shell around the reference monomer. At larger distances, no spatial correlations exist and $h(r)$

Table 2. Molecular dimensions of some polymers in different states of order as determined by small-angle neutron scattering

Polymer	$\alpha = R_g/M_w^{0.5}$, nm $g^{-0.5}$	Reference
Melt		
Polyethylene	0.045	[7]
Polystyrene, atactic	0.028	[10]
Polybutadiene	0.035	[11]
Poly(ethylene oxide)	0.042	[12]
Polypropylene, isotactic	0.035	[13]
Glass		
Poly(methyl methacrylate), atactic	0.027	[14–16]
Polystyrene, atactic	0.028	[17]
Poly(vinyl chloride)	0.040	[18]
Poly(ethylene terephthalate)	0.039	[19]
Semicrystalline state		
Polyethylene	0.045	[20]
Polypropylene, isotactic	0.034 – 0.038	[13]
Polystyrene, isotactic	0.024 – 0.029	[21]
Poly(ethylene oxide)	0.052	[12]

becomes constant. It goes to zero at $r = 0$ since a double occupancy is excluded.

Because of interactions of various kinds this simple picture must be modified. Even in the case of largely noninteracting particles such as in a melt of *n*-alkanes or polyethylene indications can be found of additional near range order effects because of steric reasons [24], [25]. There is a slight tendency for neighboring *trans* sequences to align themselves parallel to one another giving rise to a small intermolecular interaction which can be observed by magnetic birefringence or depolarized light scattering [24]. These interactions do not, however, influence the conformations of the chains significantly as can be seen from small-angle neutron scattering, which is consistent with the unperturbed-coil picture of the single chains. Scattering data of isotactic, syndiotactic, and atactic poly(methyl methacrylate) [16], [26–28] and polycarbonate [29] over a wide angular range have been compared with model calculations [30], [31] based on the rotational isomeric state model [22] in which the rotational configurations (isomers) are taken into account with their proper statistical weight. The agreement is very good in those regimes where first at small angles the overall chain conformation and second at intermediate angles the segmental configuration dominate the scattering. The unperturbed-coil model thus seems to be a good approximation for the spatial arrangement of flexible polymer molecules in the amorphous state.

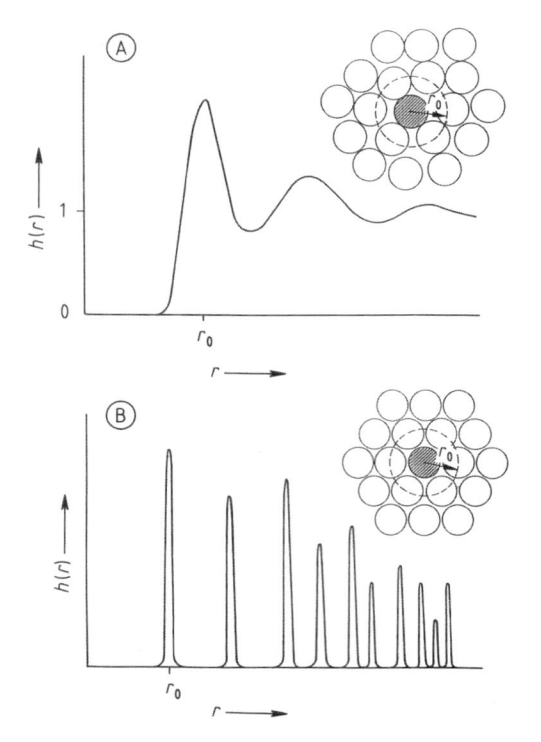

Figure 2. Distribution of monomer units $h(r)$ as a function of distance r in the (A) amorphous and (B) crystalline state for a simplified model in which monomer units are replaced by spheres

Connecting bonds to neighboring monomer units are not shown. The distance distribution $h(r)$ between centers of the spheres is continuous in the amorphous state and discrete in the crystalline state. Figure 2 A exhibits only a short-range, and Figure 2 B a long-range, positional order. The first coordination shell of radius r_0 is also indicated in the insert (dashed line).

Glassy State. The glassy state is of great importance for practical applications of plastics. Many atactic polymers do not crystallize when they are cooled down from the melt but form an amorphous glassy state. This can be described as a frozen melt in which the mobility of the chains is frozen in and the state of order of the melt at the glass transition temperature T_g is retained. The detailed nature of this transition is still not clear, but various aspects have been investigated and analyzed [32]. Many properties, such as mechanical modulus, dielectric constant, and specific heat, change at T_g (Fig. 3). A change in slope occurs at T_g for both volume V and enthalpy H, whereas a jump is observed for the specific heat c_p and the coefficient of thermal expansion α. The smooth transition in V and H and the jump in the second derivatives c_p and α suggests a second-order phase transition, which, however, cannot be demonstrated rigorously.

The glass transition temperature depends on the rate of cooling or heating, the frequency of the measuring technique (e.g., in dielectric or dynamic mechanical investigations), and the annealing conditions [32], [33]. The glassy state is thus not an equilibrium state. Since an equilibrium thermodynamic description of the glassy state [34] is questionable [35], a non-equilibrium free-volume theory [36] is mostly used, which assumes that the volume of the melt consists of the van der Waals volume of the molecules V_0 and an additional free volume V_f. For the movement of chains in the melt a certain number of voids or empty sites are required and

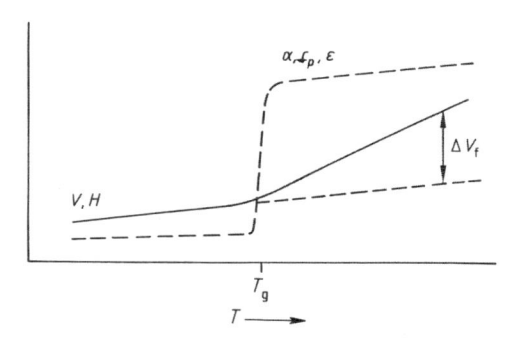

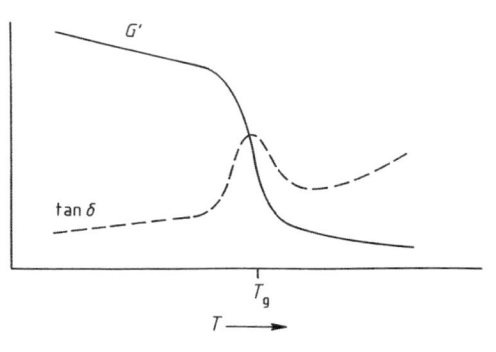

Figure 3. Schematic of the change of various properties of plastics at the glass transition temperature T_g: volume V, enthalpy H, linear coefficient of thermal expansion α, specific heat c_p, dielectric constant ε, shear modulus G', and mechanical loss factor $\tan \delta$

Also indicated in the upper figure is the increase of free volume ΔV_f with temperature according to the free-volume theory of the glass transition.

are taken into account by the free volume. The free volume of the melt decreases with decreasing temperature (Fig. 3) and is frozen in at T_g. At T_g the free volume is estimated to be 2.5 % of the total volume.

The empirical Williams – Landel – Ferry (WLF) approach [37] can be used to calculate the time – frequency – temperature behavior of the viscosity, diffusion constant, dielectric or mechanical relaxation, etc., and is an expression of the dynamic nature of the glassy state. Glass transition temperatures can be estimated from the chemical structure [38]. A list of the glass transition temperatures of many polymers is found in [39].

Blends; Polymer Orientation. In *blends* of glassy materials a single glass transition is observed when miscibility on a molecular level is achieved [40]. For blends of immiscible plastics the glass transition temperatures of the pure components are measured, which indicate phase segregation on the microscopic level. The addition of a low molecular mass component to plastics may lower T_g (plasticizer effect) [32], but can also cause crystallization or phase separation.

The order in amorphous polymers can be increased by *orientation* of the molecules, which is achieved, for instance, by mechanical deformation [41]. With some materials, stress-induced crystallization is observed (e.g., rubber) [42], which strongly affects mechanical properties. Similarly, on annealing above T_g some plastics [e.g., poly(vinyl chloride), polycarbonate, poly-(ethylene terephthalate] undergo partial crystallization [32]. In most crystalline plastics a large portion of the material is amorphous, and crystalline lamellae are formed between the amorphous regions.

1.2. Crystalline State

Some plastics can crystallize and form a crystalline state that depends largely on sample history and preparation. Only polymers with well-defined tacticity can crystallize, whereas atactic materials are generally amorphous. In contrast to low molecular mass materials, plastics form only a *semicrystalline state*, and generally both crystalline and amorphous regions are present in the polymeric crystalline state (Fig. 1 B). In the crystalline regions a long-range spatial order is present (Fig. 2), and monomer units occupy well-defined locations within the crystal. Crystal structures of polymers have been determined by X-ray and electron diffraction and are known for most crystalline polymers [39], [43]. The structural parameters depend, however, on sample treatment, and a different crystalline order may result from heating, quenching, annealing, orientation, or the application of pressure. The degree of crystallinity and the morphology (i.e., the superstructure of the crystallites) also depend on sample treatment.

Melt-Crystallized Material. During *crystallization from the melt* a lamellar morphology (Fig. 4) develops in which amorphous regions and crystalline lamellae form alternating layers with a well-defined periodicity L, typically in the range 5 – 30 nm. L is thus significantly smaller than the contour length (total length of the chain along its contour) of a polymer chain, and the molecules fold several times within the lamellar morphology (Fig. 1 B). The detailed chain conformation has long been under discussion [44], and various regular and statistical folding models have been introduced. This question has been investigated [4] in small-angle neutron scattering (SANS) experiments. One of the striking results from melt-quenched polyethylene is that the radius of gyration R_g does not change during crystallization (see Table 2), and the chain conformations in melt

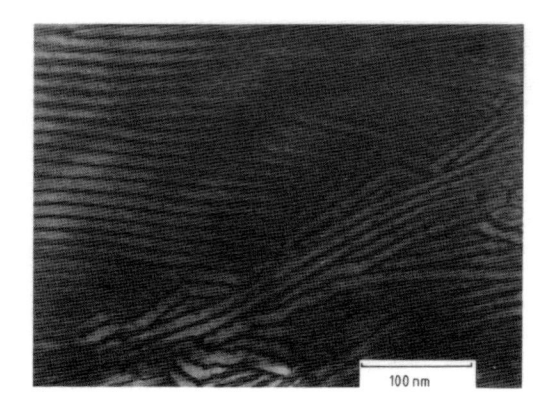

Figure 4. Transmission electron micrograph of polyethylene crystallized from the melt [2]
A thin cut of material has been stained to achieve a contrast between amorphous and crystalline regions. A side view through the crystalline lamellae is shown.

and quenched crystal are quite similar [7], [20]. Due to entanglements of the chains in the melt, long-range diffusion of segments is not possible, and only segments of the chain straighten during crystallization [45]. The overall chain conformation remains, however, essentially constant. The statistical folding model of FLORY [22], [46] for melt-crystallized polymers has been modified to include the true density of chains in the amorphous regions [47] and tie molecules between lamellae [48].

Solution-Crystallized Material. In polyethylene samples crystallized from solution a much more regular folding structure and regular morphology is observed (Fig. 5) [44]. Conformational models include "stacked-sheet" [45] or "superfolding" [49] structures in which crystalline stems of one molecule fold into stacks of an ordered arrangement along the (110) crystallographic direction, which is the main crystallization plane during crystallization from solution. Allowing the molecules during crystallization more time for rearrangement with respect to typical segmental relaxation times results

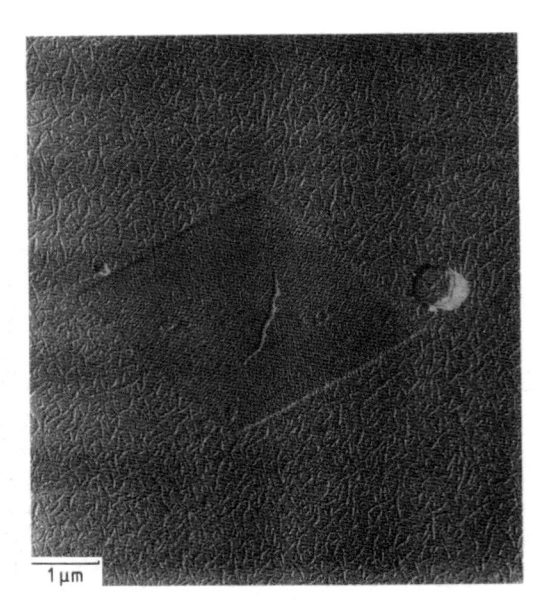

Figure 5. Surface replica [54] of a polyethylene single crystal crystallized from dilute solution
A top view of a crystal lamella is shown. As a result of decoration with low molecular mass polyethylene (evaporation of PE and epitaxial growth on the surface), four quadrants and pleating due to pyramidal shape can be seen.

generally in more regular chain folding structures; however, few experiments are performed on slowly crystallized or annealed samples [50]. Good agreement is obtained between SANS and IR experiments [51], which are both sensitive to chain conformation.

Degree of Crystallinity and Morphology. The *degree of crystallinity* depends on the material and on sample preparation. For rapidly melt-quenched materials it is near zero, and for some highly tactic or symmetric materials it can reach values of ca. 90%. Crystallinities of polyethylene and polytetrafluoroethylene for instance may be 80 – 90%, of highly isotactic polystyrene and isotactic polypropylene 60 – 70%, and for most other semicrystalline polymers < 50%.

The crystallinity is determined by X-ray diffraction, whereby the intensity of the amorphous halo is compared to the Bragg peak intensity [52]. The unstructured broad amorphous halo reflects the near range order in the amorphous regions (Fig. 2 A), while the crystalline long-range order gives rise to sharp Bragg peaks. To transform this ratio into absolute values a fully amorphous and fully crystalline sample must be investigated or other techniques (e.g., density measurements) have to be used for comparison. Depending on preparation conditions, different crystalline structures may be observed [43], [52]. Isotactic polypropylene for instance exhibits monoclinic, pseudohexagonal, and triclinic modifications [53].

The *morphology* of crystalline materials is determined by optical or electron microscopy [2], [52]. Whereas single crystals are observed in solution-crystallized material (Fig. 5), various morphologies occur in melt-crystallized materials. Crystalline lamellae (Fig. 4) may be arranged in spherulitic, dendritic, or other supramolecular structures [55]. Specific morphologies (e.g., fibrillar and "shish-kebab" structures) are observed in stretched plastic materials [56–58].

1.3. Liquid Crystals

Plastics can also form liquid crystalline ordered phases in which, like in low molecular mass compounds, smectic, nematic, and cholesteric phases are formed, depending on the composition and temperature [59], [60]. The state of order is between crystals and amorphous materials. In a nematic phase, for example, the center-of-mass position of the rigid mesogenic groups is random (like in an amorphous liquid), whereas mesogenic groups are preferentially oriented in one direction (as in a crystal). Long-range orientational order is present, but only short-range positional order.

Depending on the location of the mesogenic groups, liquid crystalline polymers are classified as side- chain and main- chain polymers (Fig. 6). A potential advantage of polymeric liquid crystals over the corresponding low molecular mass compounds is the possibility of freezing the ordered state by quenching the polymer below its crystallization or glass transition temperature. A major area of application is thus the production of ordered, high-modulus fibers from liquid crystalline phases [61].

Many main-chain liquid crystalline polymers are believed to be rigid, but they become more difficult to process with increasing rigidity. Various methods have been discussed [60], [62] for making them soluble and processable, including the grafting of flexible side chains (Fig. 6 C) or the introduction of flexible spacers between the mesogenic groups (Fig. 6 B). A potential application is the blending of rigid chains with flexible ones to form reinforced plastics on a molecular level (molecular composites) [63]. The field of liquid crystalline polymers is still growing, and quite different developments and applications may be envisaged.

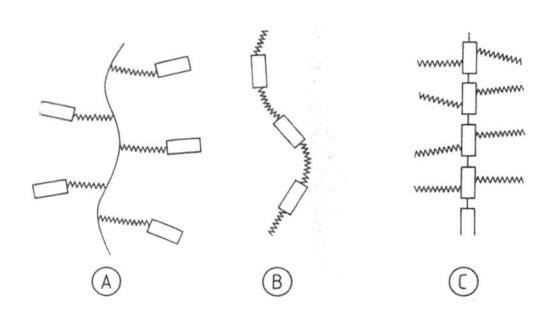

Figure 6. Different types of liquid crystalline polymers
Rigid mesogenic groups (▭), separated by flexible spacers (wiggly line), are incorporated into the side chain (A) and the main chain (B). For better processability, flexible groups may be attached to a rigid backbone (C).

1.4. Blends and Block Copolymers

Phase-separated structures are observed in blends and block copolymers in which the components are incompatible [64], [65]. The components are mostly amorphous but can also be crystalline. Observed morphologies can be very regular if, for instance, butadiene – styrene diblock copolymers with narrow molecular mass distributions are used. Depending on the molecular composition, cylindrical or lamellar ordered structures are formed [66]. In some cases [67], ordered crystalline (hexagonal) arrangements of phase-separated domains with macroscopic dimensions are observed. Within the domains, polymers may be amorphous or crystalline, while the superstructure of the domains forms a lattice.

The state of order of blends and block copolymers is very important for their applications since many plastics are blended to improve their properties. The structures can be quite complex, and the compatibility effect of additives acting at the interface between phases is not always easy to understand [68].

2. Properties and Testing

The term high polymers or plastics refers to compounds whose molecules contain more than 1000 – 1500 atoms. These substances display a number of special properties that are unknown in low molecular mass compounds. Such features occur in nearly all realms of physical and engineering properties, as described below.

Typically, the properties of plastics depend strongly on the structure and morphology, which result from the manufacturing process, as well as on loading parameters.

Over the course of plastics development, this specific behavior has dictated the methods used to investigate their properties. Plastics testing today is therefore an independent branch of engineering. Its procedures are discussed in parallel with the description of properties.

The standardization of testing methods (ISO, DIN, and other national standards) is also important for plastic materials. Standardization of plastics types based on their profile of properties is no longer customary except for a few groups of thermosets.

For thermoplastics a classification and nomenclature system has been developed that is consistent for ISO, DIN, and several other standards. For compatibility with modern data bank systems, as well as rationalization, a selection has been made from the large number of test methods available, and the profile of properties has been published as a "table of reference values" [69]. The table contains information on specimen preparation and data on properties related to processing, as well as mechanical, thermal, and optical properties. It is highly suitable for data collections because the properties listed can be compared readily.

2.1. Mechanical Properties and Their Testing

2.1.1. Classification of Plastics Based on Mechanical Properties

The mechanical properties of all groups of plastics cannot be discussed within a single context. Rather, three major classes of polymers exist that differ in their behavior under mechanical loading over a wide temperature range: thermoplastics, elastomers, and thermosets. Their differences in behavior result essentially from differences in physical as well as chemical structure.

Thermoplastics. Figures 7 A and 7 B plot the shear modulus versus temperature and the stress – strain curve for thermoplastics. This class includes un-cross-linked amorphous or partially crystalline plastics. At low temperature they are energy elastic, but above a certain temperature they exhibit viscous flow. Softening takes place in the glass transition range for amorphous thermoplastics (see Vibration. Test) and in the crystallite melting range for partially crystalline thermoplastics. Between the elastic and viscous-flow ranges there may be temperature range in which the material behaves in a rubberlike (entropy-elastic) manner.

In un-cross-linked substances such as thermoplastics, or in only partially cross-linked substances, entropy-elastic and viscous properties are superimposed. These substances are said to exhibit viscoelastic behavior (see Section 2.1.2).

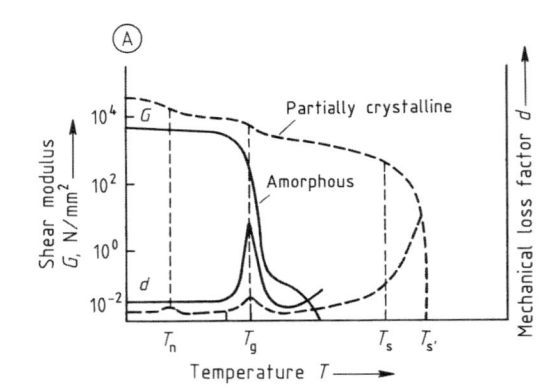

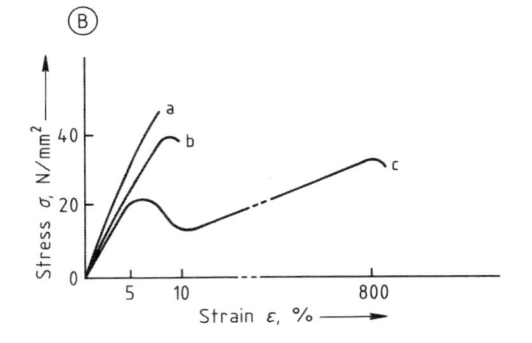

Figure 7. A) Temperature dependence of shear modulus G and mechanical loss factor d for amorphous and partially crystalline thermoplastics (schematic)
T_g = glass-transition temperature; T_n = temperature of a secondary relaxation mechanism; T_s = onset of melting of crystalline regions; $T_{s'}$ = end of melting of crystalline regions B) Stress – strain curve for thermoplastics (schematic) a) Amorphous material at moderate temperature; amorphous and partially crystalline material at low temperature or high strain rate; b) Amorphous and partially crystalline material at moderate temperature and strain rate; c) Partially crystalline material at moderate temperature and strain rate

Thermosets. Figures 8 A and 8 B present typical curves of shear modulus and loss factor versus temperature and of stress versus strain for thermosets (based on the curves for phenolic resins differing in degree of cross-linking). Thermosets are closely cross-linked polymers that exhibit energy-elastic or energy/entropy-elastic behavior at low temperature. At or above 60 °C, they show entropy-elastic behavior with very limited deformability. Figure 8 shows how, as cross-linking increases, the glass transition temperature increases and the main transition step becomes smaller, while the curve becomes flatter and the damping maximum becomes lower and broader. The steps disappear stress – strain curve. Elastomers are loosely cross-linked polymers that exhibit energy-elastic behavior at low temperature but do not exhibit viscous flow even at high temperature.

Above the glass transition temperature (20 °C or lower), elastomers exhibit rubberlike behavior. In this range, most have shear moduli between ca. 0.1 and 100 N/mm² and a high reversible deformability. The stress–strain curve has a characteristic S shape.

The glass transition temperature T_g is the mean temperature of the freezing range of a polymer, in which the microscopic Brownian motion of molecular chain segments becomes frozen and the polymer is converted to the glassy state. The temperature dependences of many physical properties change at the glass transition temperature (see Fig. 3). Some properties, such as the modulus, undergo a step-function change.

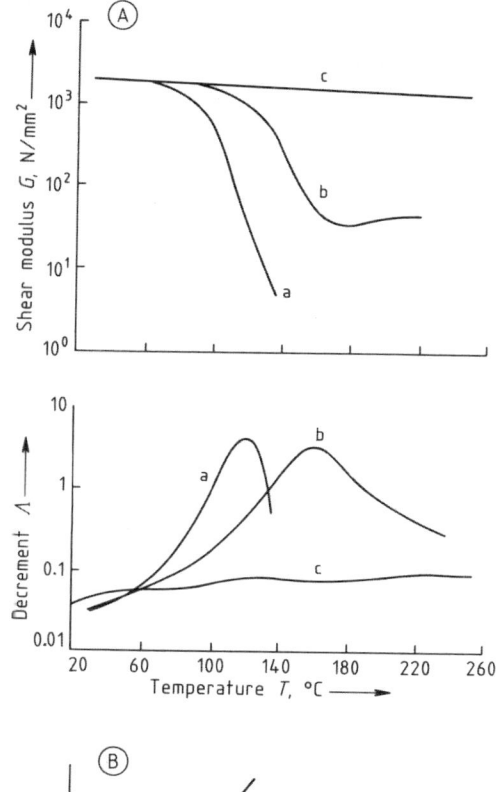

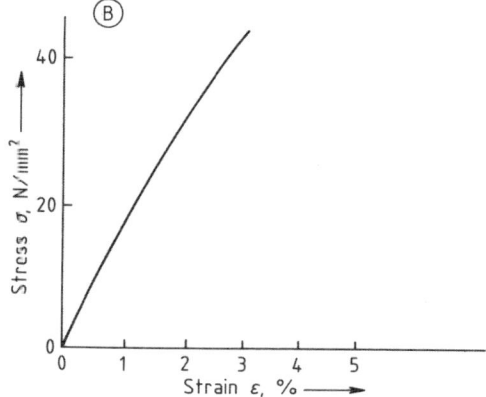

Figure 8. A) Temperature dependences of shear modulus G and logarithmic decrement Λ (as a measure of the loss factor d) for thermosets (phenol – formaldehyde resin) Cross-linking increases in the order a<b<c. B) Stress – strain curve for thermosets (schematic)

altogether upon curing and no main softening occurs in the closely-cross-linked network [70].

Elastomers. Figures 9 A and 9 B show curves of shear modulus and loss factor versus temperature for natural rubber, along with a schematic

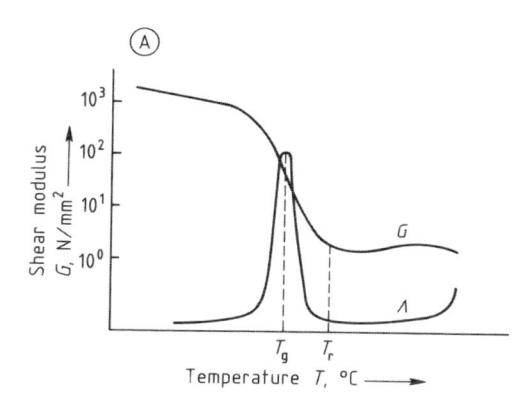

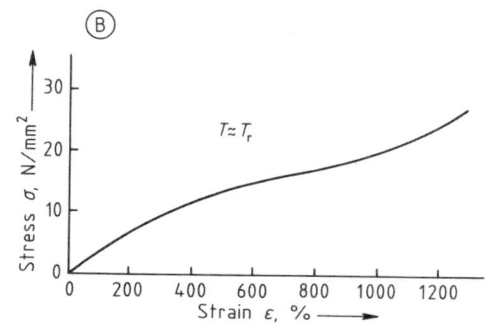

Figure 9. A) Temperature dependences of shear modulus G and logarithmic decrement Λ (as a measure of the loss factor d) for elastomers T_g = glass transition temperature; T_r = room temperature B) Stress – strain curve for elastomers (schematic)

Energy Elasticity, Entropy Elasticity, and Viscous Behavior. Strictly speaking, *energy-elastic* (perfectly elastic), mechanically and thermodynamically reversible behavior, in the sense of classical elasticity theory, does not occur in any material, particularly not in plastics. The main process in deformation is a distortion of the bond angles of the valence bonds, or else work is done against intra- or intermolecular forces (potential forces).

In *entropy-elastic* (rubberlike) substances, elasticity does not result from attractive forces but is a kinetic phenomenon like the volume elasticity of a gas. In the ideal case, the internal energy does not change when an amorphous, loosely cross-linked polymeric material network is deformed. Instead, the entropy decreases as the coiled chains are brought into a more highly ordered state. The hypothesis of constant internal energy implies that the work of deformation must be entirely transformed to heat; that is, a rubberlike substance (elastomer) must become heated when deformed. The assumption of decreasing entropy leads to the conclusion that the stress in a specimen must increase proportionally to the absolute temperature if the strain is held constant, or that if the tensile stress is held constant, the specimen must contract when heated.

Both effects have been the subject of extensive experimental and theoretical studies, initially in thermodynamic terms. However, complete understanding was achieved only with the application of the kinetic theory. This theory cannot be described in detail here [71], but chemical cross-linking of the chains is not absolutely necessary for rubberlike behavior. In partially crystalline polymers, for example, the crystallites act like cross-linking sites. In amorphous polymers under brief loads, entanglements of the chains can act as cross-links. When phase separation occurs in block copolymers, the regions composed of one of the two components play the role of cross-links (physical cross-linking).

Viscous Behavior can manifest itself in mechanically reversible creep (i.e., when stress is removed, the specimen returns to the starting position, not spontaneously but after a delay) or in flow (i.e., slipping of molecular chains past one another). Pure flow occurs (even if only as an approximation) in the melt. It is idealized as Newtonian flow, in which the deformation is mechanically and thermodynamically irreversible. However, this idealized state does not exist in polymers (see Section 2.8.3).

2.1.2. Viscoelastic Behavior of Polymers

In ideally elastic solids, the relationship between stresses and the corresponding strains is described by Hooke's law. The stresses and strains can be represented by diadics (second-order tensors), between whose coefficients linear relationships exist. This statement is the general form of Hooke's law. The mathematical expression is $\sigma_{kl} = c_{klmn} e_{mn}$, where σ_{kl} is the stress and e_{mn} is the strain. The 21 constants c_{klmn} required to describe this relationship in the linear range are known as elastic constants. They can be time dependent. In isotropic materials, the number of constants is reduced to two (Lamé constants), both of which can also be time dependent. In practice, other elastic constants or moduli, which themselves are functions of the Lamé constants, are also used. In particular, they are used as parameters in the attempts to mathematically describe the mechanical properties of elastic materials (see below).

In the nonlinear range, the number of elastic constants increases greatly, especially for anisotropic materials. Those of the 21 constants which vanish in the linear range for symmetry reasons have a finite, time-dependent value in the nonlinear range. The number of constants is so large that the measurement of all required constants and their time dependencies, although theoretically possible, is impractical. Thus, in this respect, there are limits to the exact description of the mechanical behavior. The problem is then to make an appropriate selection of constants so that the values measured adequately describe the mechanical behavior in practice.

For more than a century ago, many materials have been known that do not fit into this scheme because their behavior depends on their history. Such effects, not predicted by classical theory, were called "memory effects" (elastic aftereffects) [72], [73]. For many years, the study of this phenomenon was of only academic interest, but the situation changed with the development of plastics.

Plastics were found to have properties exhibited by both elastic and viscous substances; hence, the term viscoelasticity.

Mechanical behavior was studied by means of static and dynamic measurements involving static and fluctuating loads and deformations. The resulting coefficients, however, applied to only a limited experiment. The experimental results were systematized by using a heuristic principle, the superposition principle, which has since been recognized as the mathematical formulation of the fundamental assumption in the theory – assumption of the linearity of viscoelastic effects. Within the limits of this hypothesis, a coherent phenomenological theory could be constructed that left nothing unexplained and provided a complete picture of viscoelastic behavior. The theory cannot be derived and explained here; for a comprehensive account, see [74].

The effects and functions of the theory can be classed in two groups:

Group I, which relates to creep experiments or measurements with predetermined stress.

Group II, which relates to relaxation effects or measurements with predetermined strain.

With the superposition principle a linear viscoelastic theory could be constructed but stresses and strains can also be interrelated by linear differential equations of the general form $\mathbf{P}\sigma = \mathbf{Q}\,\varepsilon$, where $\mathbf{P}$ and $\mathbf{Q}$ are differential operators with respect to time.

This representation permits an especially advantageous mathematical description of linear viscoelastic behavior in relation to deformation problems with viscoelastic solids. The general explicit form of such differential equations is

$$a_0\sigma + a_1\frac{d\sigma}{dt} + a_2\frac{d^2\sigma}{dt^2} + \cdots = b_0\varepsilon + b_1\frac{d\varepsilon}{dt} + b_2\frac{d^2\varepsilon}{dt^2} + \cdots \quad (1)$$

Equations of this type arise when viscoelastic behavior is described in terms of mechanical models consisting of elastic springs subject to Hooke's law, combined with viscous dampers obeying Newton's law of viscosity. Such models are wholly fictitious, except that electrical networks described in a similar way are realistic (see Section 2.4.2).

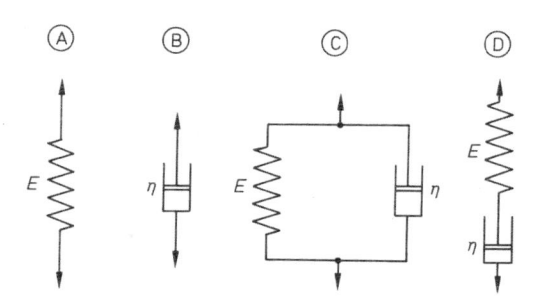

Figure 10. Simple models for describing linear mechanical behavior A) Spring; B) Damper; C) Voigt – Kelvin model [72], [73]; D) Maxwell model E = modulus of elasticity; η = viscosity

Mechanical Models and Relaxation – Time Spectra. The basic models are the Maxwell and Voigt – Kelvin models (see Fig. 10).

Maxwell model. The total strain consists of two parts, $\varepsilon = \varepsilon_{spring} + \varepsilon_{damper}$, where the stress is the same in damper and spring: $\sigma = E\,\varepsilon_{spring}$; $\sigma = E\,\tau\cdot\dot{\varepsilon}_{damper}$ where $E\,\tau = \eta$. Appropriate manipulations lead to the differential equation $\dot{\sigma} = E\,\dot{\varepsilon} - \sigma/\tau$ and hence to the following expressions: for the stress-relaxation test ($\varepsilon = \varepsilon_0$),

$$\sigma = \sigma_0 e^{-t/\tau} \quad (2)$$

and for the deformation (creep) test ($\sigma = \sigma_0$),

$$\varepsilon = (\sigma_0/E\tau)t + \sigma_0/E \quad (3)$$

Similar arguments for the *Voigt model* lead to the differential equation

$$\varepsilon = (\sigma/E\tau) - \varepsilon/\tau \quad (4)$$

and to $\sigma = E\,\varepsilon_0$ for the relaxation test and

$$\varepsilon = (\sigma_0/E)(1 - e^{-t/\tau}) \quad (5)$$

for the creep test.

These simple models are not, however, suitable for "correctly" describing creep or relaxation behavior, because the Maxwell model represents relaxation alone (stress-relaxation test) and the Voigt model, retardation behavior alone (creep test). The behavior of real polymers requires a combination of at least three or four basic models. A given viscoelastic behavior can be characterized by two arrangements of basic models: (1) a number of Voigt models in series plus a single spring and a single damper, also in series; and (2) a number of Maxwell elements in parallel.

The constants of the two arrangements are linked to one another, and either model can be calculated from the other. The mathematical proof of equivalence is laborious. The model selected depends on convenience. The Voigt type is preferable when the stress is given and the strain must be calculated, and the converse is true for the Maxwell type. Expanded types are referred to as "generalized" Voigt or Maxwell models.

The final step in the analysis is the transition from a generalized to a continuous mechanical model. Such infinite networks are characterized not by a finite number of constants but by a continuous function of an independent variable. These functions, which define continuous models, are called distribution functions. The extension of the theory is not simply a mathematically elegant solution. The restriction to a finite number of models, each having a well-defined characteristic time (retardation or relaxation time), would imply that processes in real viscoelastic polymers take place with very well-defined characteristic times, but this is true for only a few crystalline materials, not for polymers.

To clarify the situation, consider a generalized Maxwell model with relaxation times $\tau_1 < \tau_2 < \tau_3 < \cdots$. The model is determined if the corresponding spring constants M_i are given. The subdivision into different τ_i is made finer and finer, so that in the limit, the variable τ becomes continuous, taking on all values between 0 and ∞.

The "spring constant" associated with relaxation times between τ and $\tau + d\tau$ has a value $h(\tau) d\tau$, $0 < \tau < \infty$, where $h(\tau)$ is a continuous function of the relaxation time called the "distribution function of relaxation times" or, for brevity, the relaxation spectrum. If the relaxation spectrum $h(\tau)$ is known for all times, the mechanical behavior of the model is completely determined. In particular, the stress-relaxation modulus $M(t)$ is found by integrating over relaxation time (here with the use of a logarithmic time scale):

$$M(t) = \int_0^\infty H(\ln \tau) e^{-t/\tau} d\ln\tau \qquad (6)$$

The relaxation spectrum cannot, however, be measured directly; it must be calculated from measured phenomenological functions by means of special equations. The calculation is not simple in every case and can be done to only a rough approximation; the spectra are therefore of little importance for purposes of phenomenological description. Their value lies, instead, in the thermodynamic and structural interpretation of the relaxation process. Figure 11 shows relaxation spectra $H(\ln \tau)$ of polyethylene and polypropylene at various temperatures [75]. Inclusion of the temperature dependence leads to a threedimensional representation of the relaxation spectrum. Special reduction methods can be used to extend the spectra to shorter or longer times, after which the plots for polypropylene and polyethylene, for example, exhibit distinct maxima, whose interpretation is discussed in [75]. For example, in polypropylene, the principal maximum at short times is a criterion for mechanisms in amorphous domains, whereas the maximum at long times is a criterion for mechanisms in crystalline domains. Figure 12 shows the relaxation spectrum of polypropylene at 23 °C (after extension to short times) and the temperature traces of the two maxima [76].

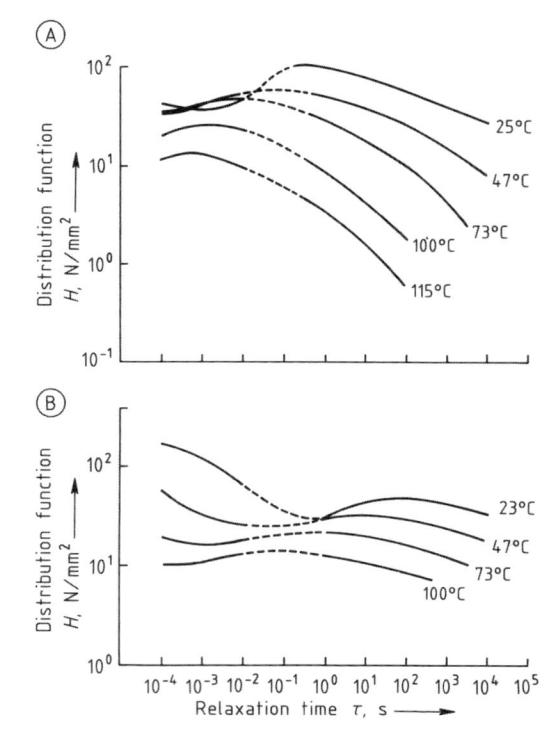

Figure 11. Relaxation spectra at various temperatures
A) Polyethylene; B) Polypropylene

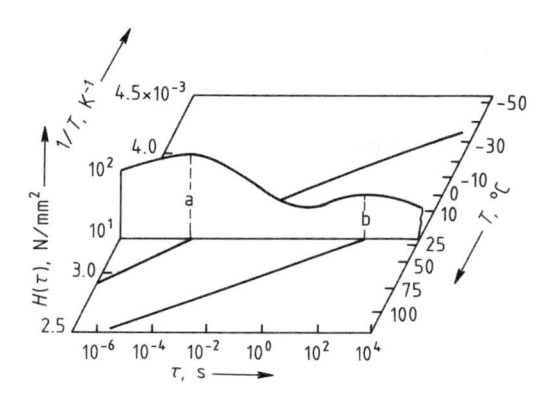

Figure 12. Relaxation spectra of polypropylene as a function of time and temperature [75], [76] a) Principal maximum; b) Crystalline maximum

As has been mentioned, relaxation spectra can be obtained from creep and relaxation tests only by indirect means; for immediate practical purposes, the results of these "phenomenological" experiments are therefore used directly. They cover loading times $t \geq 1$ s, with no upper bound on the time. These experiments are also referred to as static tests. The model theory of viscoelasticity is thus supplemented by a phenomenological treatment of viscoelasticity.

2.1.3. Mechanics of Deformation

Creep Modulus and Relaxation Modulus. The strongly time-dependent behavior of plastics (especially thermoplastics) under mechanical deformation, resulting from viscoelastic properties, is characterized by the time-dependent creep modulus (retardation modulus) and the time-dependent relaxation modulus. General laws for these can be derived from the viscoelasticity theory (see Section 2.1.2). For the relaxation test, this principle leads to an integral representation in the form

$$\sigma(t) = \int_0^t M(t-\xi)\dot{\varepsilon}(\xi)d\xi \qquad (7)$$

with variable of integration ξ [77].

In Equation (7), $M(t)$ is the relaxation modulus, which replaces the modulus of elasticity E, the shear modulus G, and the bulk modulus K; $\dot{\varepsilon}$ is the strain rate. The more specific notation $E_r(t)$ should be used for the modulus $E(t)$ derived from the relaxation test, to distinguish it from the reciprocal compliance $1/C(t) = E_c(t)$ discussed below, which is determined from the creep test. The superposition principle applied to the creep test leads to an integral representation

$$\varepsilon(t) = \int_0^t C(t-\xi)\sigma(\xi)d\xi \qquad (8)$$

which is analogous to Equation (7). As in the representation of $\sigma(t)$, $C(t-\xi)$ is the time-dependent elastic compliance, which is referred to as the creep compliance and takes the place of the elastic compliance $C_E(t)$, the shear compliance $C_G(t)$, and the bulk compliance $C_K(t)$. In practice, $C_E(t)$ is also written as $1/E_c(t)$.

The relaxation modulus $M(t)$ and the compliance $C(t)$ are linked by the general relation

$$\int M(t-\xi)C(\xi)d\xi = t$$

which implies the condition $M(t)\,C(t) \leq 1$. The equality holds if the time dependence is weak, so that $Er(t) = Ec(t)$; otherwise, $Er(t) < Ec(t)$.

If the time dependence is marked, the elastic modulus measured in the relaxation test is lower than that from the creep test. However, this holds only with no stress dependence in the creep test and no strain dependence in the relaxation test. This is not, however, the case in general, so in practice, $Er(t)$ can also be greater than $Ec(t)$.

The results of creep tests play an important role in construction with plastics. In the creep test, a stress σ is applied to the specimen at time $t = 0$ and held constant thereafter (relative to the initial cross section). The increase in strain with time (creep) $\varepsilon(t)$ is measured. Figure 13 shows how σ and ε vary with time in the creep test. The ε versus t curve is commonly referred to as the creep curve. For sufficiently low stress, Hooke's

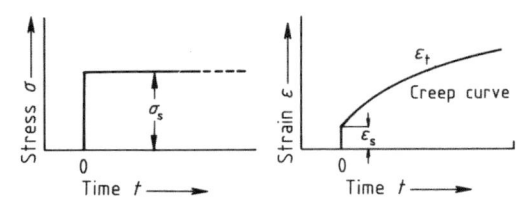

Figure 13. Variation of σ and ε with time in creep test
εs = instantaneous strain (immediately after the application of stress)

law in the form $Ec(t) = \sigma/\varepsilon(t)$ gives the time-dependent modulus $E(t)$ from the creep and the constant (initial) stress.

The creep modulus $Ec(t)$ is independent of the stress only if the stress lies in the Hooke's law range, where the stress and the strain measured after a given time are proportional to one another. Outside the linear range, the creep modulus decreases with increasing stress. When values are cited for the creep modulus, the stresses at which they were determined must be specified. In practice, uniaxial tensile loads are usually applied in creep testing, chiefly because this loading mode gives an identical state of stress at all points in the specimen. The performance of tensile creep tests is governed by DIN 53 444 [78]. The specimens used are the same as for the tensile test of DIN 53 455 [79]. Tensile specimens are loaded by the impactless application of weights, and strain is measured at fixed time intervals. A plot of the results gives the above-mentioned creep curve. The test involves three variables: stress, determined by load and (initial) specimen cross section; strain; and loading time. Two-dimensional plots, each with two explicit variables and one as parameter, yield the family of curves shown in Figure 14. For clarity's sake, only one stress – time curve (yield limit curve) and one isochron are plotted.

The creep modulus $Ec(t)$ is derived from a family of creep curves with stress as parameter; the result can be plotted, for example, versus time, as shown in Figure 15. Plotting points for the respective stress – strain values from the creep curve, with loading time as parameter, gives so-called isochronous stress – strain curves, which show the time and stress

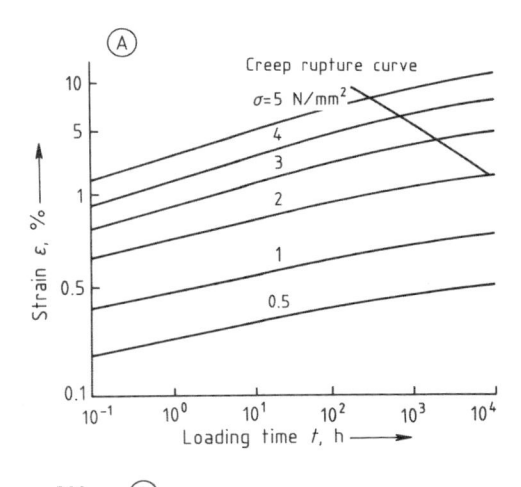

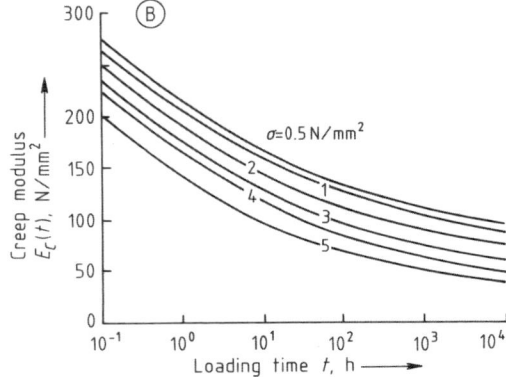

Figure 15. Creep curves and creep-modulus curves
A) Creep curves at 65 °C for high-density polyethylene with creep rupture curve from a long-term pressure test on high-density polyethylene pipes B) Creep-modulus curves derived from creep curves

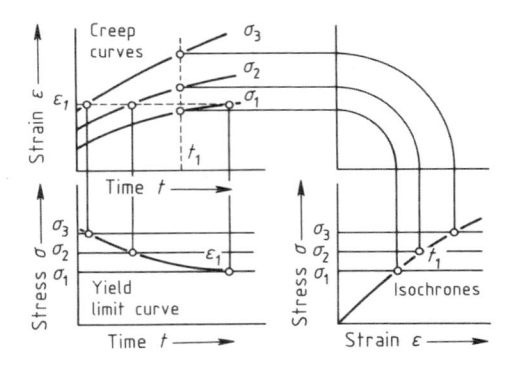

Figure 14. Derivation of yield limit curves and isochrons from creep curves

dependence clearly. Fields of isochrones for a number of temperatures convey much information to the designer (see Fig. 16).

The stress σ that leads to a total strain ε after time t is called the time yield limit $\sigma_{\varepsilon/t}$.

The curve connecting creep stresses for equal strains is now known as a "stress versus time curve" (formerly termed yield limit curve; see Fig. 14). The set of creep curves forms the creep diagram, which may also contain the test stress σ and the time t to fracture (or to a certain sign of damage). These give the creep rupture curve or a certain "damage curve." The term creep strength curve is also commonly used for the creep rupture curve, because ultimately it is the curve joining the creep strengths. The creep strength $\sigma_{B/t}$ is the

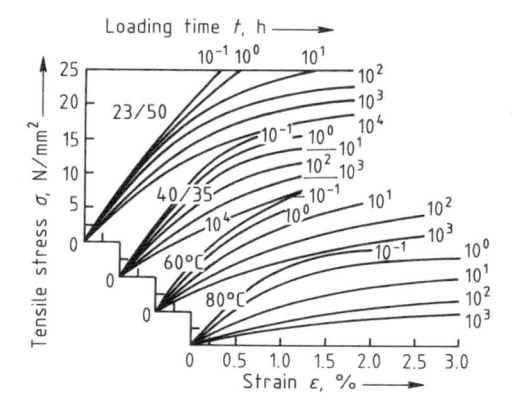

Figure 16. Isochronous stress – strain curves for acrylonitrile – butadiene – styrene; $ak = 12$ kJ/m2 [80] 23/50 = 23 °C/50 % relative humidity; 40/35 = 40 °C/35 % relative humidity

static load B, relative to the initial specimen cross section, that causes fracture of the specimen after a certain loading time t. The creep diagram can be used to estimate the time-dependent deformation occurring under uniaxial tensile loading and is a useful design aid for simple loading modes (Fig. 17). The time-dependent modulus of elasticity, however, is more universal in its application because, in many cases, it can also be used in the equations for materials engineering.

The relaxation test is experimentally far more difficult to perform than the creep test. The deformation must be held constant over time, and the mechanical stress must be measured as a function of time. For details, see [81].

Vibration Tests. Static tests involve loading times longer than ca. 1 s. If information about the millisecond range is needed, vibration tests must be performed. These usually cover two orders of magnitude in frequency; the ranges in conventional vibration tests are roughly between 10^{-1} and 10^4 Hz, corresponding to time t (if $t = 1/\omega$) in the range of 10^{-5} to 1 s.

As is customary in treatment of vibration, complex quantities are employed to describe the phenomena. The theory leads to the definition of a complex modulus (see Fig. 18):

$$M = |M|e^{i\delta} = M' + iM''$$

where M', the real part of M, is called the dynamic modulus (as distinct from the time-dependent modulus $M(t)$; see Creep Modulas and Ralaxation Modulus) or the storage modulus. It is a measure of the recoverable energy transformed in the deformation occurring during one vibration. The parameter M'', the imaginary part of M, is called the loss modulus. It is a measure of the vibrational energy unrecoverably transformed to heat (dissipated). The quotient of the real and imaginary parts of M, which is the tangent of the phase angle δ of the complex modulus, is known as the mechanical loss factor $d = M''/M' = \tan \delta$. It is a relative measure of vibrational energy losses compared to recoverable energy.

The chief methods used in plastics testing are the torsional vibration test and the bending vibration test. The various techniques for exciting torsional vibrations have in common that a specimen acting as a torsion spring is connected

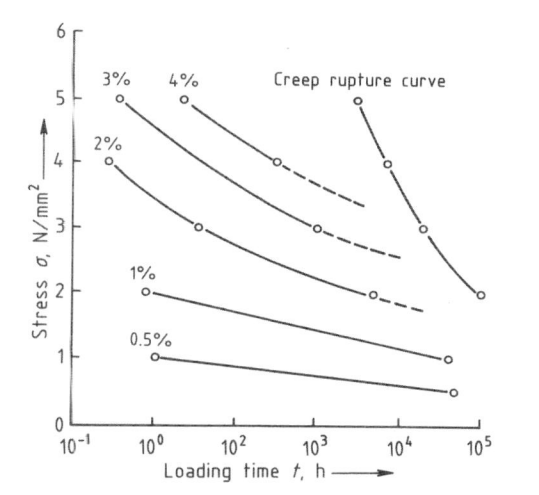

Figure 17. Creep diagram for high-density polyethylene with 0.5, 1, 2, 3, and 4 % yield limit curves and the creep-rupture curve

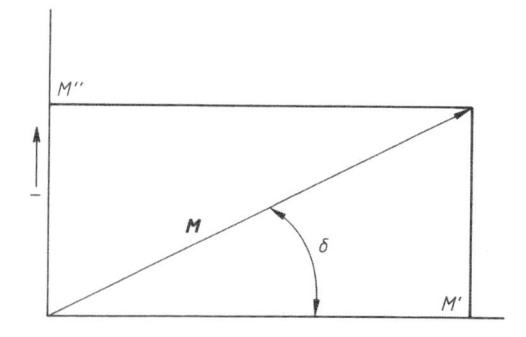

Figure 18. Representation of M in the complex plane

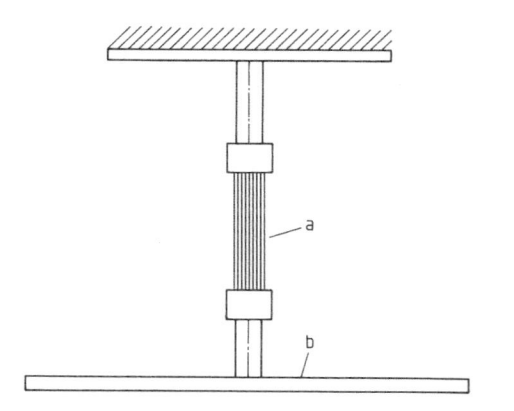

Figure 19. Schmieder – Wolf torsional pendulum a) Specimen; b) Rotating mass

to an oscillating mass. Figure 19 shows one possible arrangement.

If the upper specimen clamp is moved slightly, this torsional pendulum is set into damped natural oscillation. The logarithmic decrement $\Lambda = \ln(A_n/A_{n+1})$ and the mechanical loss factor $d \approx \Lambda/\pi$ (for $\Lambda \leq 2$) are calculated from the amplitudes A of successive oscillations. The shear modulus G is determined from the moment of intertia I of the oscillating mass, the square of the oscillation frequency f, and factors F_g and F_d relating to the dimensions and the damping effect, respectively:

$$G = If^2 F_g F_d$$

The test method is described in detail in DIN 53 445 [82]. Figure 20 shows how the curve of shear modulus G versus temperature T can be used to separate the various states of high polymers. The glassy range is characterized by high modulus and low damping. The diagram also shows clearly the temperature below which the substance described by this G versus T curve behaves as a solid. Temperature ranges in which the modulus is strongly temperature dependent, while the damping has a maximum, represent softening ranges. The measured results also show whether flow sets in as soon as softening begins or whether cohesive forces are still present. Figure 21 gives examples of shear modulus curves over a wide temperature range.

Because of the amount of information it provides, the torsional vibration test has also been incorporated into the table of reference values mentioned earlier (see Thermoplastics).

In the bending vibration test [83], the top end of a rod-shaped specimen is clamped in a fixed position and the rod is excited to steady-state resonance bending vibrations by an electromagnetic transducer at the bottom. The amplitude of the vibration is measured by a transducer near the clamping point. The modulus of elasticity is calculated from the resonant frequency, and the loss factor from the width of the resonance curve. The wide frequency range employed in this method (1 – 10 000 Hz) is an advantage of the bending vibration test. Because frequencies lie in the audible range, the test is particularly important for studies relating to acoustic applications of plastics for example for noise suppression [84]. The torsional vibration test gives the shear modulus G, whereas the bending vibration test yields the elastic modulus E. To a good approximation, the conversion formulas are

$$E = 2G(1+\mu) \text{ or}$$
$$E = \frac{3G}{1+G/(3K)}$$

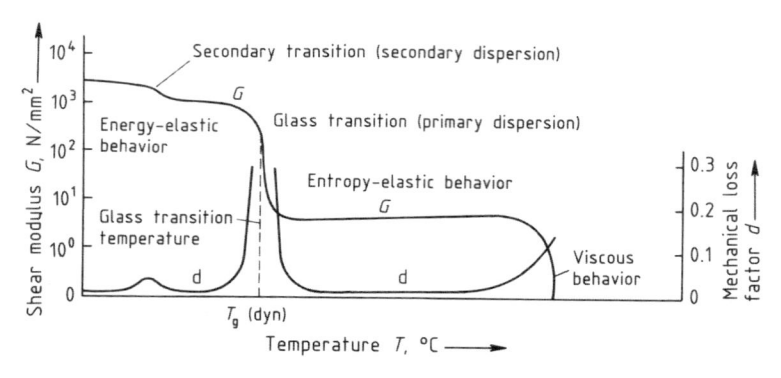

Figure 20. Definition of state regions based on the curve of shear modulus G versus temperature T (DIN 7724)

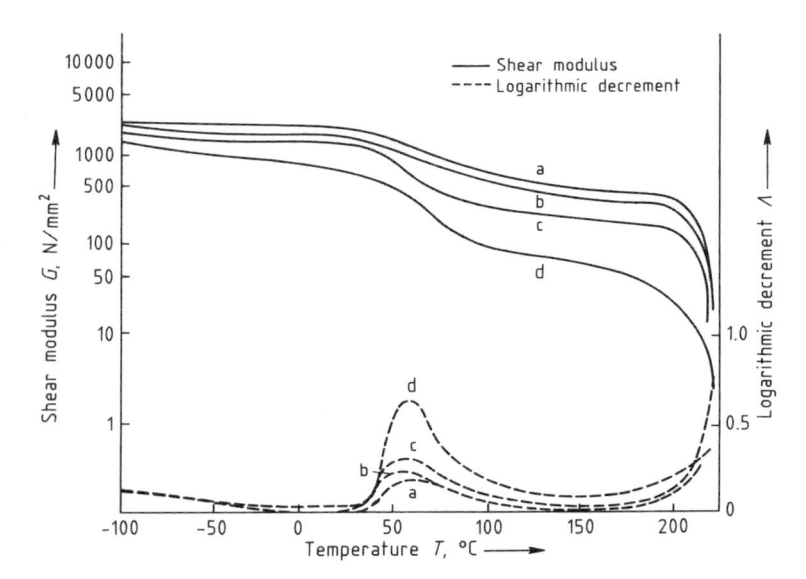

Figure 21. Temperature dependences of shear modulus and logarithmic decrement for poly(butylene terephthalate) a) 30 % Glass fiber; b) 20 % Glass fiber; c) 10 % Glass fiber; d) Unreinforced

where μ is Poisson's ratio and K is the bulk modulus; see also [85].

Behavior under Steady and Fluctuating Loads. To characterize the time-dependent load-deformation behavior, the laborious long-term and vibration tests are supplemented by steady-load and fluctuating-load tests. These techniques are important for product development, inspection, and preliminary materials selection. Tests with steady load include tensile, bending, and compressive tests, as well as hardness measurement; those with fluctuating load include flexural fatigue tests. Studies of friction and wear properties complete the series of mechanical measurements. For the designer, however, by far the most informative results are the temperature and time functions describing the long-term behavior under tensile loading together with the curves of shear modulus versus temperature. The predictive power of all other methods is limited by comparison [86]. The tensile test [87] is, however, largely exempt from the limitations mentioned. The elongation and strength values from this test, given uniform strain and a homogeneous stress distribution, allow a good assessment of the load deformation behavior. Figure 22 A presents three typical curve shapes [86], along with the stress and strain values listed in the table of reference values for plastics.

Outside the linear range of the stress – strain curve, the bending test [88] gives unrealistically high values. This test is therefore no longer included in the table of fundamental values [89].

Compression tests [90] are important in the evaluation of some thermoset molding compounds.

Static Modulus of Elasticity. The elastic modulus derived from stress – strain curves (or force –deformation curves) recorded in steady-load tests, like the values from tensile, bending, and compressive tests, is a single-point value with respect to strain rate. It has adequate predictive power for product development, inspection, and materials preselection in most cases. Derivation of the modulus for a tensile load is illustrated in Figure 16 B.

The applicable standard [91] specifies a uniform rate of extension (crosshead speed) of 1 mm/min and restriction of the evaluation to the stress – strain curve between 0.05 and 0.25 % strain. Because the stress-strain relationship for rigid plastics is quasi linear in this range, what is obtained is an approximate tangent modulus. A secant modulus is found for soft plastics with a nonlinear stress-strain relationship.

The standard does not, however, distinguish between tangent and secant moduli in the listing of values. Thus, the elastic modulus in this standard is given by

$$E = \frac{\sigma_{0.25} - \sigma_{0.05}}{0.002} \text{ (units of } E \text{ and } \sigma : \text{N/mm}^2\text{)}$$

where $\sigma_{0.25}$ is the stress at 0.25 % strain and $\sigma_{0.05}$ is the stress at 0.05 % strain. Table 3 lists tensile moduli of elasticity for several reinforced and unreinforced plastics.

Table 3. Static modulus of elasticity for unreinforced and reinforced plastics (maximum value in each case)

Product (abbreviations from DIN 7728)[*]	Elastic modulus E, N/mm^2 [**]
ABS, moderate impact strength	2 500 (T)
	2 300 (B)
ABS + 30 % GF	6 000 (T)
	4 500 (B)
PS	3 500 (T)
SB	3 000 (T)
SAN	3 800 (T)
ASA	2 900 (T)
PTFE	750 (T)
FEP	700 (T)
PFA	700 (T)
PE-HD (HDPE)	1 400 (T)
PP	1 800 (T)
PP + 40 % T	4 500 (T)
POM (Copolymer)	3 200 (T)
POM + 30 % GF	10 000 (T)
PBT	2 700 (T)
PBT + 30 % GS	4 000 (T)
PBT + 30 % GF	10 000 (T)
PBT + 30 % CF	16 000 (B)
PA 6 (d)	2 500 (T)
PA 6 + 30 % GF (d)	8 300 (B)
PA 6 + 30 % CF (d)	16 000 (B)
PA 66 (d)	3 000 (T)
PA 66 + 30 % GF (d)	10 000 (B)
PC	2 200 (T)
PC + 30 % GF	7 000 (T)
PPO	2 000 (T)
PPO + 30 % GF	9 000 (T)
PEEK	3 800 (B)
PEEK + 30 % CF	13 500 (B)
PPS	4 200 (B)
PPS + 40 % CF	19 600 (B)
PMMA	3 300 (T)
PVC	3 300 (T)
PVC-HI	2 600 (T)
PES	2 600 (B)
PES + 30 % CF	14 300 (B)

[*] GF = glass fiber; T = talc; GS = glass spheres; CF = carbon fibers; d = dry;
[**] T = tensile modulus; B = bending modulus.

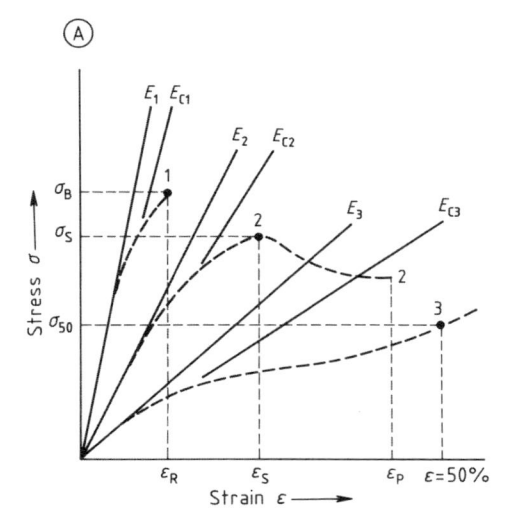

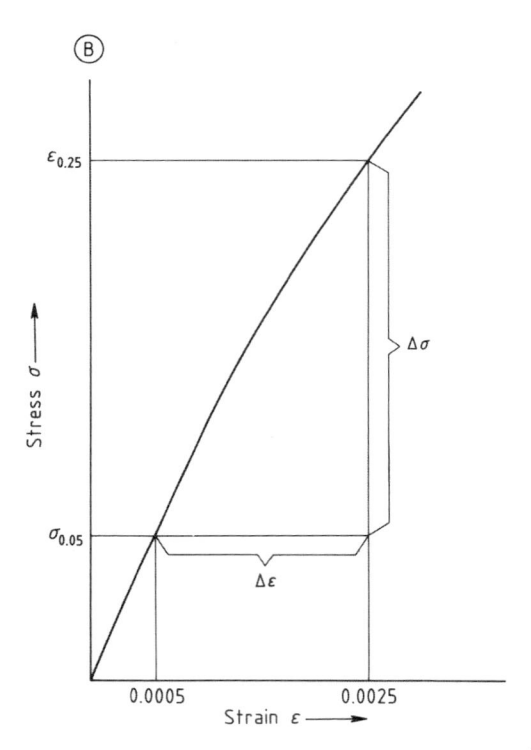

Figure 22. Evaluation of tests with steady load A) Values from tensile test, for three typical shapes of stress – strain curves [86] E = elastic modulus; E_c = creep modulus after 10^3 h; ε_R = elongation at rupture, ε_S = extension; σ_P = tensile strength; σ_S = yield stress; σ_{50} = stress at 50 % strain B) Determination of elastic modulus (DIN 53 457/1987) $\sigma_{0.25}$ = stress at 0.25 % strain; $\sigma_{0.05}$ = stress at 0.05 % strain; $\Delta\sigma = \sigma_{0.25} - \sigma_{0.05}$, $\Delta\varepsilon = 0.002$ (= 0.2 %)

Multiaxial Loading. The strengths measured under steady load, such as the tensile strength, apply to a one-dimensional load system. Application of these values to multiaxial stresses requires that the multiaxial states be linked with the individual strengths through fracture criteria that reduce the multiaxial state to a single reference stress. Definition of this reference stress depends on the fracture hypothesis selected. Finally, the reference stress is compared with the individual (usually tensile) strength. The fracture criteria used for metals cannot as a rule be applied to plastics, because isotropic fracture behavior – the precondition for most of the classical criteria – is generally not exhibited by plastics [92].

Classical failure criteria must therefore be modified to describe more accurately the failure behavior of polymers. A detailed discussion of these modifications can be found in [93].

Hardness. Hardness is not a fundamental property of plastics; the concept is ultimately subjective. Probably the most widely acknowledged definition is resistance to the penetration of another, harder body.

As in methods for metals, hardness testing on plastics employs a standard penetrator, often a hardened steel ball or steel pin, that is pressed by a defined force into a plane surface of the material being tested, with the depth of penetration being measured [94–97].

The viscoelastic behavior of plastics results in two basic difficulties: (1) the depth of penetration depends on the rate of loading, and (2) measuring the diameter of the impression is difficult. Because the impression springs back after the load is removed, it would have to be measured immediately. Thus, either the depth of penetration is measured after a given time under load, or a transparent penetrator is employed.

Not only does the hardness depend on time, temperature, and the nonlinear stress – strain relationship, it also depends on the magnitude of the forces applied, the modulus, and the shape of the penetrator. Accordingly, hardness measured by a given method cannot be compared with that obtained by another method. The hardness of elastomers and other "soft" materials is characterized by the Shore hardness value [97], while that of rigid thermoplastics and all thermosets is expressed as the ball indentation hardness [95].

Behavior under Fluctuating Load (Fatigue). A large fluctuating load applied many times can lead to material damage that results in failure (fatigue failure) at a load value far below the strengths measured in steady-load tests (tensile strength, tensile creep strength). Test methods devised for metals have been extended to plastics, with allowance for the fact that metals, in contrast to plastics, have low damping and high thermal conductivity. The loading frequencies for plastics are therefore markedly lower than for metals, lying between ca. 600 and 3000 min^{-1}.

Fluctuating loads are classified as loads with constant stress and those with constant strain. The many possible forms of loading cannot be discussed individually here. As Figure 23 shows, three loading ranges are distinguished: fluctuating compressive, alternating, and fluctuating tensile. In the repeated flexural fatigue test (Wöhler method), several (e.g., six to ten) equivalent specimens are subjected to appropriately graduated oscillating loads, and the number of cycles to failure is recorded for each. The same mean stress σ_m is chosen for all specimens, and the initial stress amplitudes are graded. The mean stresses and stress amplitudes, which vary within a test because of relaxation and heating, are then plotted against the common logarithm of the number of cycles. The curves obtained are known as run curves, and the complete set of curves obtained in the test is the family of run curves. In addition, the Wöhler curve proper ($\sigma_a - N$ or stress-life curve) is plotted; its points

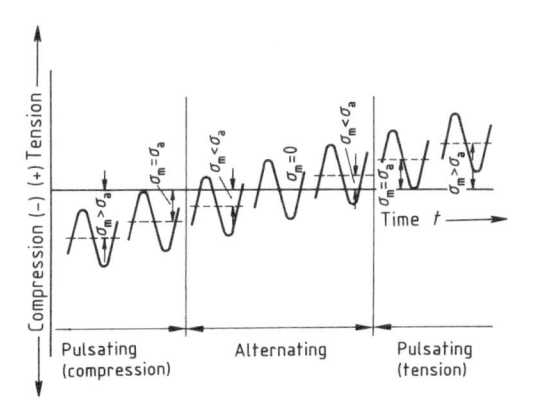

Figure 23. Load ranges in vibrational test σ_0 = maximum stress; σ_u = minimum stress; σ_m = mean stress [$\sigma_m = 0.5(\sigma_0 + \sigma_u)$]; σ_a = stress excursion [$\sigma_a = \pm 0.5(\sigma_0 - \sigma_u)$]

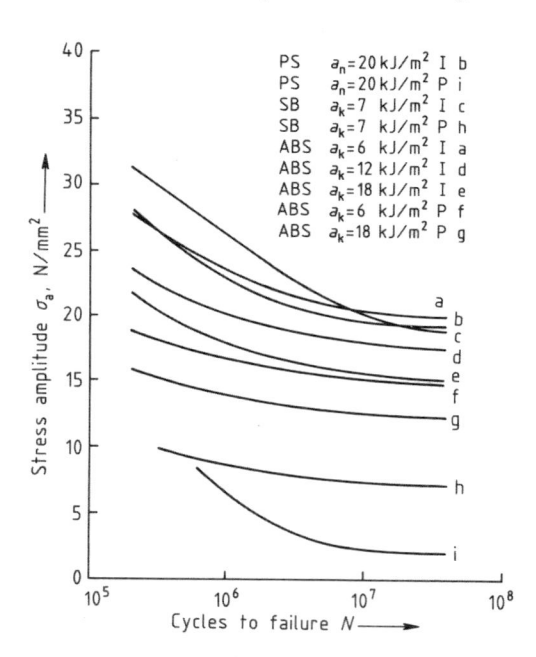

Figure 24. Wöhler curves for polystyrene, styrene – butadiene, and ABS, after Oberbach [71] a_n = impact toughness; a_k = notch impact toughness, S = injection-molded specimen; P = molded specimen

PS	$a_n = 20 \text{ kJ/m}^2$ I b
PS	$a_n = 20 \text{ kJ/m}^2$ P i
SB	$a_k = 7 \text{ kJ/m}^2$ I c
SB	$a_k = 7 \text{ kJ/m}^2$ P h
ABS	$a_k = 6 \text{ kJ/m}^2$ I a
ABS	$a_k = 12 \text{ kJ/m}^2$ I d
ABS	$a_k = 18 \text{ kJ/m}^2$ I e
ABS	$a_k = 6 \text{ kJ/m}^2$ P f
ABS	$a_k = 18 \text{ kJ/m}^2$ P g

represent pairs of initial stress amplitude σ_a and cycles to failure N, plotted with a logarithmic horizontal (N) axis and a linear vertical (σ_a) axis. Figure 24 presents Wöhler curves for polystyrene, styrene – butadiene, and acrylonitrile – butadiene – styrene (ABS) [80].

Service Temperature Range. The best way of assessing the service temperature range is to use curves showing the properties of interest as a function of temperature. In many cases, functional tests on a finished part can also be employed. Nonetheless, evaluating the service range – in relation to, say, mechanical behavior – by performing standard tests on specimens is often useful. This is successful, with some reservations, for determining the upper limit service temperature (heat stability) but not for determining brittleness temperatures. Attempts to devise a satisfactory technique for measuring these have not been successful.

The DIN heat stability tests are the Martens test [98] and the heat deflection temperature (HDT) test among [99]. The Vicat softening temperature (VST) indicates the softening behavior at the upper limit of the service temperature range [100].

Figure 25 shows a schematic of the HDT test. The heat deflection temperature is the temperature at which the bending test specimen, loaded in the middle, deforms by a given amount as its temperature is increased. Various test methods differ in the magnitude of the bending stresses applied. In the Martens test, the bending moment acts in a four-point loading scheme. The VST is the temperature at which a 1-mm^2 circular steel pin under a given force penetrates 1 mm vertically into the specimen as the temperature is increased.

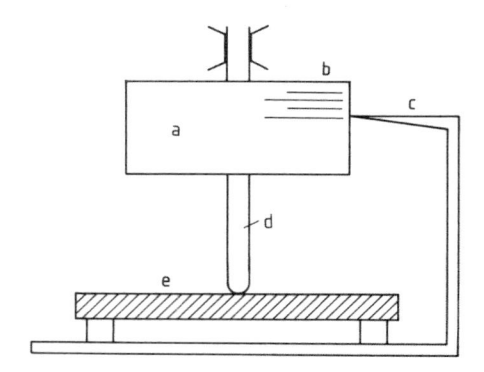

Figure 25. Setup for the HDT method (schematic)
a) Weight; b) Scale; c) Indicator; d) Bending mandrel; e) Specimen

2.1.4. Friction and Wear

Friction is defined in general as the resistance that must be overcome to set or maintain two contacting bodies in relative motion. The coefficient of friction μ is calculated from the normal force N between the contacting surfaces and the frictional force F opposing their motion: $\mu = F/N$. In other words, the frictional force is proportional to the normal force (first law of friction). This formula holds approximately for metals; the coefficient of friction for plastics decreases with increasing normal force. Friction measurements on plastics should therefore employ a graded series of normal forces.

The concept of friction is often broken down into static and dynamic friction, even though perfectly static friction is impossible. The term static friction normally refers to the amount of force required to induce a relative motion between two surfaces in contact; that is, what is measured is the frictional force with very slow movement.

Dynamic friction means the force that will maintain a constant, perceptible, relative motion between contacting surfaces. Velocity thus plays a role in determining the frictional force, even though for many materials the frictional force changes almost negligibly as relative velocity changes by several powers of ten. This finding corresponds to the third law of friction. For plastics, however, it does not hold perfectly. The frictional force may change by a factor of three or even four as the velocity varies over several powers of ten. Unfortunately, the velocity effect cannot be wholly separated from the temperature effect, but the velocity dependence must still be determined for plastics. The heat produced in a sliding contact is approximately proportional to the coefficient of friction, the sliding velocity, and the normal force.

Figure 26 shows how the coefficient of dynamic friction depends on the sliding velocity for ultrahigh molecular mass polyethylene under various pressures.

In general terms, wear refers to all abrasive processes involving mechanical energy. The mechanisms governing wear are complex and intimately interconnected (see also → Abrasion and Erosion). For plastics, relatively little experience has been gathered with regard to the effects of various parameters. The following methods are considered proven: The wear in dry sliding against a granular material is determined by the friction wheel technique [101]; a similar method is described in [102]. Another method measures the wear of plastics in dry sliding against a rotating steel shaft. This technique is used to test plastics for sliding bearings, especially those designed to run dry [102]. Table 4 lists values measured in a setup of this type.

Table 4. Wear (reduction in specimen diameter) of some plastics in dry sliding against 16 MnCr 5 steel

Product (abbreviations from DIN 7728)	Wear, µm per km of sliding distance
PA 66	0.09
PA 6	0.23
PA 610	0.32
PA 11	0.8
PET (PETP)	0.5
POM (copolymer)	8.9
PE-HD (HDPE)	1.0
PE-LD (LDPE)	7.4
PP	11
PTFE	21
PS	115
SAN	23
SB	51
ABS	8.4
PMMA	4.8
PVC	5.6
PC	22

2.1.5. Fracture Behavior

Plastics exhibit three types of deformation and fracture behavior:

1. Quasi-brittle fracture
2. Fracture after partially or completely irreversible shear deformation
3. Fracture by crazing after irreversible deformation of the specimen

This classification applies mainly to thermoplastics; fracture in thermosets is nearly always brittle. The fracture behavior of elastomers is described by the theory of BUECHE and HALPIN [103].

Brittle Fracture. The term brittle fracture denotes fracture with virtually no deformation, characterized by a high propagation velocity. If the latter is placed in the foreground, brittle fracture can be said to take place even if some plastic deformation occurs first. Brittleness and toughness as material properties are influenced by test conditions and depend on the temperature, the rate of loading, and the stress

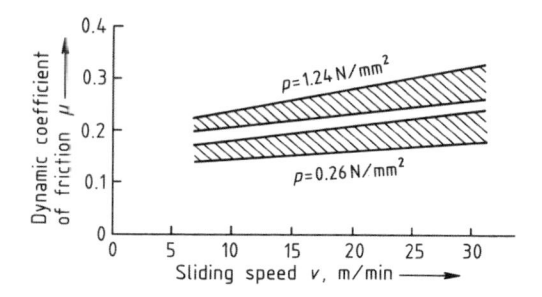

Figure 26. Dynamic coefficient of friction μ as a function of sliding velocity v for ultrahigh molecular mass polyethylene at various surface pressures p = mean surface pressure

state. In this sense, any plastic experiences low-deformation fracture in some range or ranges.

Despite their brittle behavior, plastics have a relatively high specific fracture energy (energy to produce 1 cm^2 of fracture surface), roughly a hundred times that of inorganic glasses. The reasons for this are that (1) the edges of an advancing fracture undergo substantial irreversible deformation, even in the glassy state; and, (2) the numerous crazes formed in the material act not as energy absorbers but as starting points for new fractures (see Fracture Mechaincs).

Fracture Mechanics. A few words should be said about the applicability of fracture mechanics to plastics. The fundamental idea of fracture mechanics is to analyze the mechanical behavior of an individual crack in a linearly elastic, homogeneous, and isotropic continuum. This basic concept is, however, valid only if the solid shows linearly elastic behavior even in its smallest regions, that is, at the tip of the crack. In other words, the material must be perfectly brittle. The energy transformation is then simply the conversion of supplied elastic energy to pure surface energy. Even for perfectly brittle glasses, however, the energy balance is not this simple; the "supply side" includes elastic energy derived from internal stresses, whereas on the "demand side" the liberation of plastic deformation energy plays a dominant part, especially for crack propagation in plastics. The specific surface energy is orders of magnitude lower than this deformation energy.

To describe this ductile behavior, the concept of crack resistance has been introduced, which describes ductile behavior without the necessity of knowing or postulating details about the shape, size, and time behavior of the plastic zone itself. Fracture mechanics becomes less and less valid as the crack-tip plasticity increases. A number of models of this plastic behavior exist, including that of DUGDALE.

In the area of *quasi-brittle fracture*, however, linearly elastic fracture mechanics (LEFM) offers a fairly good description even of viscoelastic materials. If certain fitting parameters are introduced, the framework of LEFM can also be extended to materials with small plastic deformation components. As the deformation behavior becomes more strongly plastic to completely plastic, the corrections are no longer sufficient and the linearly elastic solution ceases to be meaningful. For such cases, however, viscous fracture mechanics is based on a different set of principles and may provide a fracturemechanical description. The application of fracture mechanics to polymers is reviewed in [104].

Fracture after Partial or Complete Shear Deformation. A ductile fracture, in contrast to a brittle one, is characterized by large amounts of macroscopic deformation. The deformation may arise through uniform elongation or, in true ductile fracture, by yielding or necking. In metals and ceramics, direct inspection reveals lattice defects (dislocations) in this yielding region. Movements of the dislocations are responsible for observed plastic or permanent deformation; such motion does not occur below the tensile yield stress. This behavior provides a basis for clearly distinguishing between elastic and plastic (irreversible) strain in metals and ceramics. The curves plotted for plastics are very similar to those for metals (Fig. 27). The analogous behavior of metals and some plastics is underlined further by the occurrence of necking. In thermoplastics, for example, one or several such necking zones migrate over the specimen until the whole has experienced yielding (unless it has fractured first). The behavior can be described by the Eyring yield equation; detailed discussion and calculations based on a kink model are presented in [105].

In metals, the planes of maximal shear stress (e.g., a plane at 45° to the direction of uniaxial loading) exhibit so-called Lüders lines [106]. These are tracks left by slip lines along which the parts of the specimen are displaced. The lines

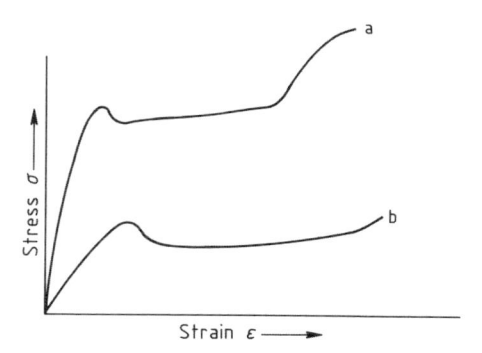

Figure 27. Stress – strain behavior of materials with yielding a) Metals; b) Thermoplastics

are regarded as evidence of a dislocation-dependent deformation process. Similar lines are observed in both crystalline and amorphous plastics. Their occurrence in crystalline plastics is interpreted in terms of a dislocation-dependent slip mechanism, but this interpretation does not apply to amorphous materials. ROSEN advanced a theory based on the observed birefringence [107]. According to this hypothesis, birefringence results from a new, noncrystalline state of aggregation. The lines are a domain of oriented material having a refractive index different from that of the surrounding material.

Fracture after Deformation with the Formation of Crazes (Flow Zones). Crazes or flow zones are planar inhomogeneities in a high polymer, produced by plastic deformation in the direction of an applied tensile stress [108–110]. The planar surface of the craze forms perpendicular to the maximum principal tensile stress (Fig. 28). A craze can be as large as 1 cm^2 in area but no more than 4 μm thick.

Surface crazes display arched crack edges or dislocations. In contrast to a crack, a craze contains material in a highly stretched, unthinned state and behaves as an optically homogeneous medium. The craze stands out sharply against the surrounding normal polymer material.

STERNSTEIN et al. have investigated the conditions under which deformation through craze formation (normal yielding) occurs [111], [112]. The results show that crazes can form only in the vicinity of a positive isotropic stress component (stress invariant I_1): $I_1 = \sigma_1 + \sigma_2 + \sigma_3 > 0$. However, even when this condition is fulfilled, crazes only occur when the maximum principal normal stress exceeds a time- and temperature-dependent critical value σ_{crit}.

The formation of flow zones in impact-resistant two-phase or multiphase systems is of great engineering importance. When flow zones develop in these systems, they act as energy absorbers. All analyses are based on the assumption that elastomer particles in such a system act as stress concentrators under mechanical load, because their elastic modulus is low compared to that of the matrix. BOHN has further shown that the soft phase must have a sufficiently uniform distribution and must be sufficiently cross-linked and grafted in order to produce and maintain the special dilatation stress state in the hard matrix on cooling [113].

Fracture Morphology of Plastics. So many factors affect fracture morphology in plastics that only a few examples can be discussed here. Brittle plastics exhibit a behavior similar to that studied in detail by SMEKAL for inorganic glasses. According to those findings, the fracture front advances radially in all directions from a primary notch site. The primary notch is surrounded by a smooth surface, the "mirror," which is bounded by a semicircle. SMEKAL states that the transport of heat in the initial phase (i.e., in the "mirror region" of the fracture) is an essential feature of the fracture phenomenon. This region of the fracture, which is traversed at a low velocity, is therefore referred to as the thermal phase of the fracture. The remainder of the process occurs at a very high fracture velocity (i.e., athermally), and this phase is called the athermal phase of the fracture. The fracture mirror and the athermal phase are also observed in brittle plastics, especially in fracturing after high strain rates (Fig. 29) [114]. The fracture mirror often displays periodic fine structures that can yield information about the rate of loading. In ductile materials, fissured morphologies are observed; these depend on, among other factors, orientation, stress distribution, and rate of loading.

A point of special interest is the fracture morphology of fiber-reinforced plastics. Fracture surfaces are observed by optical microscopy or by scanning electron microscopy at a magnification of 60 – 24 000.

Measurement of Fracture Behavior. Instrumental investigation of fracture behavior takes place in the context of steady-load and

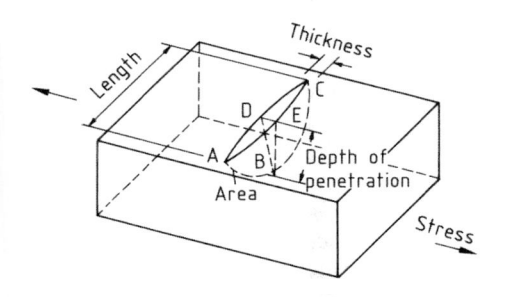

Figure 28. Schematic illustration of a craze AC = length; DE = thickness; EB = depth of penetration

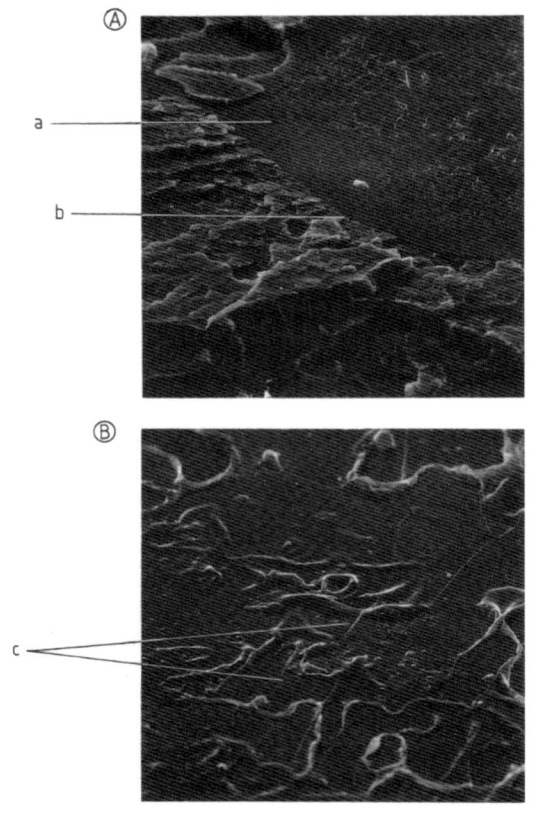

Figure 29. A) Boundary between thermal (mirror) and athermal fracture phases after impact tensile loading of polystyrene, [114] (scanning electron micrograph, ×180); B) Spickle structure in the mirror, with typical "thermal threads" (×1800) a) Mirror; b) Thermal – athermal phase boundary; c) Thermal threads

fluctuating-load mechanical tests, both short- and long-term, with tensile, compressive and bending loads. These tests are performed at moderate to low strain rates.

A special test procedure has been developed to study deformation and fracture behavior at high strain rates (impact behavior). To a large extent, plastics owe their great versatility to high impact strength combined with other qualities. Hence, many studies have dealt with the property of impact strength. The translation from values measured on test specimens to in-service performance (the behavior of moldings and semifinished products in a variety of applications) is more difficult for impact strength than for the mechanical properties already discussed.

For example, the influence of viscoelastic behavior can be estimated only in the quasi-linear range. At the extreme loads, required for evaluating performance under impact loading, which usually far exceed the linear range (especially for unreinforced materials), other tests must be used. These include impact flexural and impact tensile tests, both performed with notched-bar specimens. Studies are also carried out with multiaxial loading modes.

As in low-strain-rate tests, the stress – strain diagram or the deformation energy absorbed up to a certain damage level (usually to fracture) is determined.

Numerous standardized tests have been developed for this purpose. In general, however, they cannot be assessed in terms of how well they correlate with in-service performance under impact and shock loads.

The methods used to obtain impact values for inclusion in the table of reference values were chosen to satisfy a number of criteria, one being that the test should show a distinct difference in energy absorption between ductile fracture (in which the specimen is merely cracked) and brittle fracture.

This condition is met, for example, by the Izod notched-bar toughness test or the Charpy impact bending test; see Figure 30 [115–117]. Notched and unnotched specimens can be tested by either method. Both belong to the class of flexural impact tests in which the impact energy absorbed up to failure is measured. For reasons of measurement technique, no determination of the stress – strain relationship is included in this test. The V-shaped notches used have a 45° angle between sides and various notch root radii. Impact tensile tests are performed on plastics along with the flexural impact test,

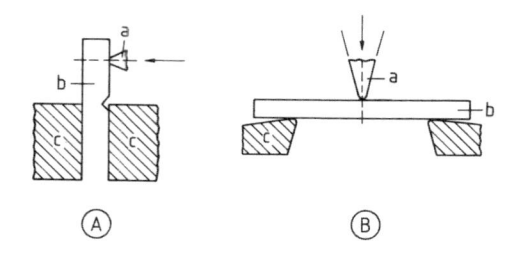

Figure 30. Setups for impact bending test A) Izod test, notched specimen; B) Charpy test, unnotched specimen a) Striker; b) Specimen; c) Clamp or support

especially when the specimens in the flexural impact or notched-bar flexural impact test do not fracture. Notched and unnotched specimens are used in the impact tensile test as well.

The notch action at the root of the notch is described by the theoretical stress concentration factor $a_k = \sigma_{max}/\sigma_n$, where σ_n is the nominal stress with no notch and σ_{max} is the maximum stress in the direction of loading at the root. For a notched flexural impact specimen (0.25-mm radius at notch root), a_k is ca. 2.4, and for a double-notched impact tensile specimen (DIN 53 448, 0.25-mm radius at notch roots), ca. 2.9 (values determined by photoelastic examination [114]). Thus the notch action is roughly comparable in both specimens even though the notch radii are very different. Stress distribution outside the notch region, however, are quite dissimilar.

In specimens with marked partly plastic behavior, the decrease in a_k from blunter to sharper notches is not very great. This is due to the fact that despite the large stress concentration factors, the partly plastic behavior prevents the maximum stress at the notch root reaching the high value predicted by elasticity theory [118].

Flexural impact and impact tensile tests are supplemented by a technique employing biaxial impact loading: the falling-weight impact test. A circular or square flat specimen is laid on a ring-shaped support and loaded in the center by the impact of a falling weight. Experimental setups of this kind are described in [119–121]; they are also used for investigating the impact behavior of films and finished or semifinished parts. Given a well-defined criterion of failure, the damaging energy at which a given percentage of specimens fail can be determined. Results are evaluated either statistically (probit method [122]) or by a staircase technique (up-and-down method [123]).

Methods that allow the plot of applied force versus deformation to be continued up to failure have greater predictive power than the impact and falling-weight tests described here, which yield the damaging energy or the proportion of damaged specimens.

In addition to the damaging force and deformation, these tests also measure the damaging energy for tensile loading or for biaxial loading by impact on the centrally loaded plate [124]. The test apparatus, which features automated performance and evaluation, is complicated.

Table 5. Damaging work W_s per unit thickness d, damaging force F_s, and damaging deformation l_s for several plastics, after DIN 53 443 Sheet 2

Product (abbreviations from DIN 7728)	W_s/d, Nm/mm	F_s, N	l_s, mm
PC	5500	9000	23
ABS	1450	3600	16
PS	30	300	3
PE	1550	2600	20
PVC-HI	3300	5400	18
PF Type 31	30	440	3

The devices for tensile loading are called high-speed tensile testers. Combined machines for high-speed tensile tests and impact penetration tests at high rates of deformation are also available. These devices have a maximum rated tensile force of 100 kN, with an impact velocity up to 20 m/s. The force is measured with piezoelectric pickups, and the strain is measured with inductive strain gauges or high-resolution precision laser systems [125].

Interpretation of results from the biaxial plate-loading test is difficult, chiefly because the state of orientation in the specimen varies from one plane layer to another. Apart from taking extreme orientation distributions into account, apparent anomalies in impact penetration test results can be explained only by analyses based on elasticity theory (e.g., the theory of sheets) [126].

Because of these difficulties, the results of impact tests are not generally regarded as material values.

If specimen fabrication and testing conditions are controlled very carefully, however, the tests are suitable for preliminary materials selection and especially for development in the field of impact-resistant plastics. Table 5 lists impact penetration test results for several plastics.

2.1.6. Control of Mechanical Properties through Orientation, Internal Stresses, and Morphology

The term orientation refers to the alignment of structural elements (i.e., polymer chains or chain segments, crystalline regions, or domains). Orientation takes place in the initial production process or results from subsequent drawing and is frozen in on transformation to the solid state, with simultaneous occurrence of disorientation processes. A material usually exhibits greater strength and less elongation in the orientation direction. Multiaxial states of orientation in molded parts can often influence mechanical behavior in an unforeseen way. Orientation is used in the production of high-strength monofilaments (fibers and wires) and biaxially drawn (and thus biaxially oriented) films with high impact penetration strength.

Orientation is of particular importance in the fabrication of film ribbons, where drawing ratios up to 1:15 are used. The anisotropy of the strength in the longitudinal and transverse directions, resulting from the orientation, can be so great that the ribbons split when loaded transversely. This splitting tendency is desirable in string and yarn. If splitting is to be avoided, the molecules must be oriented slightly transverse to the drawing direction during processing [127].

Molecular orientation in polymers can be determined by wide-angle X-ray diffraction (limited to polymers that are more than 10 % crystalline); birefringence (the second moment of the distribution is measured); linear dichroism (suitable for determining the amorphous and crystalline orientations separately); acoustic velocity (simple method for fibers); or nuclear resonance. An approximate measure of orientation is the shrinkage at high temperature, which results mainly from the prior molecular alignments and is thus an entropy-elastic effect. In contrast, energy-elastic effects play little part in shrinkage. Figure 31 shows how orientation (characterized by birefringence) influences the notched-bar impact bending strength.

Molded plastic parts can exhibit internal stresses, which may significantly affect mechanical properties. According to WÜBKEN [128], the three main causes of internal stresses in amorphous moldings are cooling stresses, stresses due to the expansion of parts ejected from the mold under pressure, and flow-related Hooke's law stresses. Deformations resulting from internal stresses are energy elastic and therefore very small in comparison to those associated with orientation. The internal stresses in a molding are at equilibrium. If, however, the stresses are not at equilibrium (e.g., on demolding), the injection-molded part deforms until equilibrium is reached. Internal stresses in components subject to compressive loads, such as pressure piping and pressure vessels, require special

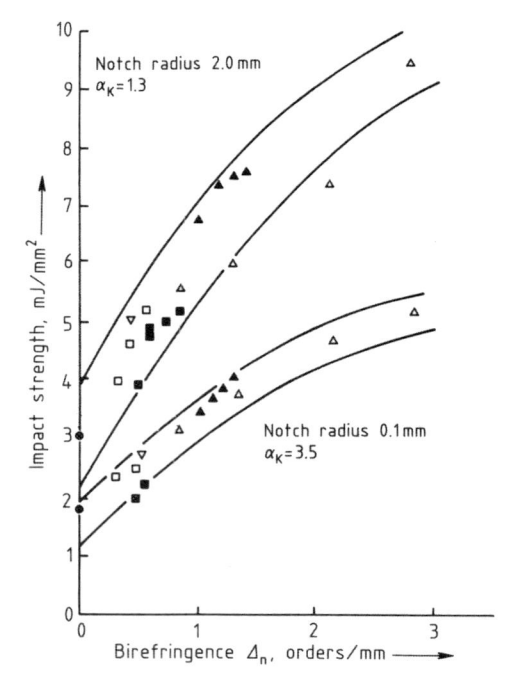

Figure 31. Notched-bar bending strength of polystyrene (specimen with double V notches) as a function of birefringence and notch geometry

attention. The addition of internal stresses to the multiaxial stresses produced by external forces can, under unfavorable conditions, lead to a reduction in the design creep strength. Internal stresses can be determined by deformation measurements on components [128]. A quasi-nondestructive test based on hardness measurements has been described [129] (see Section 2.7.2 for stress cracking).

Of the morphological states that influence mechanical properties, the only ones discussed here are crystallinity and, for amorphous plastics, particle structure. The partially crystalline state and the associated superlattice structures are discussed above (see Section 1.2). Higher crystallinity means higher rigidity and yield stress. Because crystallinity is correlated with density, measuring the density provides a simple criterion for evaluating the crystalline fraction.

Partially crystalline superlattice structures called spherulites (for their spherical shape) are formed when crystallization takes place from solution and especially from the melt. The spherulite diameter depends on the chemical structure of the macromolecules and the thermal prehistory. Polyethylene, for example, crystallizes mainly in fine spherulites having diameters of 5 – 200 μm (depending on the rate of cooling), whereas polypropylene and poly-1-butylene cooled at similar rates may form spherulites up to several millimeters in diameter.

In polymerization processes, spherulite size is often controlled by addition of nucleating agents in order to achieve desired properties, such as high rigidity or impact strength.

Rigidity (modulus, yield stress) generally increases with increasing spherulite radius, whereas toughness decreases, although exceptions can be found to this rule. Domain structures, which are especially marked in polymer blends and graft and block copolymers, also exhibit surprising engineering properties. The flow behavior, rigidity, and impact strength are improved.

Impact-resistant thermoplastics, which include, among others, grades of polystyrene and poly(vinyl chloride) (PVC), consist essentially of a continuous hard component (polystyrene, PVC) in which a soft component [e.g., polybutadiene, poly(vinyl acetate)] is dispersed in the form of domains. The number, shape, and distribution of the domains, as well as the structure of the matrix – domain phase boundaries, govern the properties of the blend. Other domain structures of engineering interest are found in block copolymers, such as those of butadiene and styrene [130].

If only a reduction in rigidity is required, and no special demands are imposed on the energy absorption in impact loading, internal or external plasticization can be used. Internal plasticization is the copolymerization of two compatible monomers whose glass transition temperatures differ greatly.

External plasticization is a solvation process in which low molecular mass plasticizers are bound to polymer macromolecules.

2.1.7. Reinforced and Filled Plastics

A reinforced plastic is a composite material with a plastic matrix. Such a composite has an elastic modulus and strength, in at least one direction, that far exceed those of the unreinforced matrix. Particle-reinforced composites (e.g., with talc, powdered chalk, powdered minerals, short fibers, or glass spheres) are distinguished from fiber-reinforced composites (e.g., with glass, aramid, or carbon fibers). Particle-reinforced thermoplastics have the advantage that they can be injection molded. The composites exhibit greatly improved rigidity, strength, and heat deflection temperature. Moldings exhibit reduced shrinkage and thermal expansion. Curable molding compounds also include particle-reinforced composites, some of which are free-flowing materials that can be injection molded.

Fiber-reinforced materials most commonly have thermoset matrices of unsaturated polyester (UP) or epoxy (EP) resin. An important class is glass-fiber-reinforced plastics. The various reinforcements (rovings, filaments, mats) have been standardized [131]. For further information and properties, see [132].

Reinforced thermosets tested in the longitudinal direction have moduli of ca. 30 000 N/mm^2 and tensile strengths of 700 N/mm^2; thermoplastics with carbon-fiber reinforcement (50 – 60 wt % carbon fiber) exhibit tensile strengths of 270 – 280 N/mm^2. The high carbon fiber content, however, lowers the volume resistivity to 10 – 100 $\Omega \cdot$ cm.

Additives and fillers are also used to modify plastics properties. Only a few examples can be mentioned here. Carbon black lowers resistivity (affording safety against electrostatic charging) and also reduces the coefficient of friction. Graphite, molybdenum disulfide, and carbon fibers lower the dynamic coefficient of friction. Metal powders (bronze, chromium – nickel steel) are used in bearing materials, valves, piston rings, and gaskets. Aluminum flakes provide electromagnetic shielding.

2.2. Acoustic Properties and Their Testing

Because most plastics have relatively high mechanical loss factors (at least two orders of magnitude higher than most metals), they can be used to attenuate solid-borne sound. This application is undoubtedly the most interesting in the field of acoustics, especially for noise suppression. Dimensionally stable plastics with high internal damping (i.e., high mechanical loss factor) include polypropylene and polyamide. Examples of acoustic damping applications are covers, gears, and rollers made of these materials. Unfortunately, glass-fiber reinforcement greatly lowers the loss factor, so that glass-fiber-reinforced plastics are not well suited to sound attenuation. The hum generated by metal plates can be suppressed by viscoelastic coatings. These are special amorphous thermoplastics, in particular vinyl acetate copolymers, or two-component systems based on polyurethane or epoxy resins. These coatings are improved further by active fillers such as vermiculite, a micalike clay material, or graphite.

To assess the damping properties of such coatings, the dynamic flexural rigidity and the bending wave rate must be known. These quantities are derived from the dynamic modulus of elasticity, which in turn is determined in flexural vibration tests performed at audio frequencies [83], [84].

Figure 32 shows curves of loss factor versus temperature for a 1-mm steel sheet with a 2-mm sound-deadening material and also for a steel – plastic – steel laminate (i.e., a symmetrical composite). Undamped steel sheet has a loss factor $d \approx 0.0001$; the maximum with heavy edge damping is 0.001. The laminates exhibit d values up to 1.0.

Homogeneous solid plastics do not attenuate airborne sound because even the lightest solids do not satisfy the requirement of wave impedance matching between air and absorbing medium. Plastics have a wave impedance several orders of magnitude greater than that of air. To suppress airborne sound, therefore, open-pored foams or fibrous materials are employed. Absorption takes place by friction between air particles and components of the skeleton. Open-pored foamed plastics used in this absorbing role include foamed urea – formaldehyde resin and nonrigid PVC foam. Foamed polystyrene slabs, even though they have closed pores, can be used for acoustic purposes if the surface is perforated.

If solid-borne sound is to be blocked (as opposed to attenuated), soft materials of low dynamic flexural rigidity are required. Low-frequency vibrations from machinery cannot, however, be blocked with plastics; instead, rubberized metal elements, steel springs, or air springs are employed.

Another application represents a far more important use of plastics in blocking solid-borne sound: the suppression of footfall noise by so-called floating floors. Acoustic requirements for buildings [133] are met by insulating slabs based on expanded-bead polystyrene foam, whose resilience is enhanced by postprocessing.

Plastics, even filled materials (e.g., with expanded glass or clay), are not very suitable for blocking airborne sound. They are not considered for this purpose by builders, because this type of soundproofing always requires heavy masses with the lowest possible flexural rigidity. Plastics with heavy fillers (PbO or $BaSO_4$) do meet the mass requirement and also have the high coincidence cutoff frequency needed for soundproofing [84]. Heavy, nonrigid plastic films are used in automobiles and machinery.

2.3. Thermal Properties and Their Testing

A knowledge of the thermodynamic properties of plastics – and, especially, of their temperature dependence – is important for the economics of production and for the quality and functioning of products. Relevant data include specific heat, specific enthalpy, thermal conductivity, thermal diffusivity, specific volume, compressibility, and coefficients of thermal volume and linear expansion. With this information, the processor and the user can calculate the heats needed for processing, mixing temperatures, volume changes on cooling, shrinkages during and after processing, and temperature-related dimensional changes in semifinished products and moldings. Furthermore, several of these quantities must be known if the reactions and internal changes taking place in plastics are to be understood.

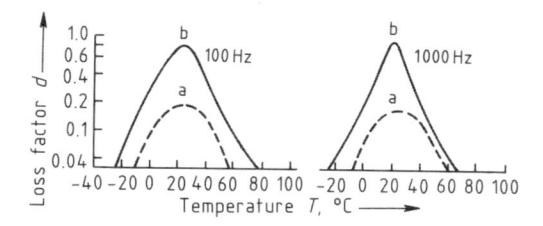

Figure 32. Mechanical loss factor d as a function of temperature T at 100 Hz and 1000 Hz for steel – plastic composite sheets a) 1-mm steel sheet with 2-mm sound-deadening material; b) Steel – plastic – steel composite (0.5 – 0.2 – 0.5 mm)

2.3.1. Specific Heat

The specific heat of a plastic is not always an unequivocal function of temperature. Crystalline and amorphous polymers do have comparable specific heats at very low temperature, but sizable differences occur in the vicinity of transformations (e.g., glass transition, crystallization, and melting). The specific heat in these ranges depends on the state (e.g., the degree of crystallization) of the polymer and thus on its thermal prehistory [134], [135]. If the specific heat at constant pressure c_p is known, the enthalpy differences ΔH between the initial T_0 and final temperatures T can be calculated by integration:

$$\Delta H (T)_{T_0} = H_T - H_{T_0} = \int_{T_0}^{T} c_p(T)\,dT$$

Enthalpy depends on pressure, and because of the high pressures that frequently occur in plastics processing this dependence must be taken into consideration. Figure 33 A shows schematically the specific heat as a function of temperature for amorphous thermoplastics. The steplike rise in the glass transition range is characteristic. Partially crystalline thermoplastics are distinguished by the fact that an additional enthalpy, the heat of fusion, must be supplied to melt crystalline regions. The heat of fusion results in a maximum in the curve of specific heat versus temperature for partially crystalline plastics, as shown schematically in Figure 33 B. With regard to the temperature dependence of specific heat and enthalpy, see [136], [137].

The enthalpy of the uncured starting product is of interest for the energy balance in the processing of thermosets.

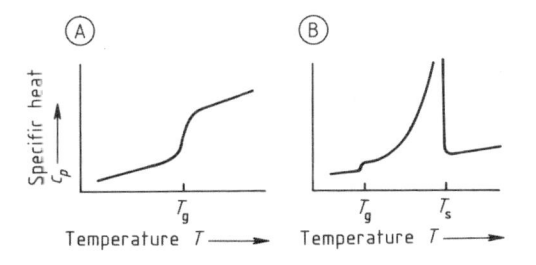

Figure 33. Temperature dependence of specific heat
A) Amorphous thermoplastics; B) Partially crystalline plastics T_s = crystallite melting point

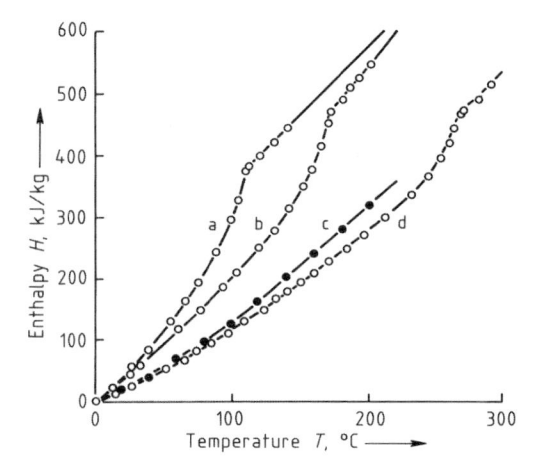

Figure 34. Enthalpy curves for some thermoplastics
a) Low-density polyethylene; b) Polypropylene; c) Polystyrene; d) Poly(ethylene terephthalate)

Measurements of the specific heat of plastics from nearly 0 K to ca. 600 K can be performed only with adiabatic calorimeters. A very simple, quick, and sufficiently accurate measurement can, however, be carried out with dynamic differential calorimeters [138]; the technique is called DSC (differential scanning calorimetry). A review of differential scanning calorimeters can be found in [139].

Not only does DSC yield the same results as adiabatic calorimetry, it also makes new types of measurements possible. Examples are the detection of glass-transition steps, exothermic crystallization, melting, and determination of stabilizer content [140].

Differential scanning calorimetry also permits indirect determination of, for example, crystalline content, degree of cross-linking, degree of curing, thermal stability, and plasticizer content. Studies of reaction kinetics are possible as well [141–143]. Figure 34 shows enthalpy curves for several commercially important thermoplastics.

2.3.2. Thermal Conductivity

Plastics are poor heat conductors, because they have virtually no free electrons available for conduction mechanisms such as those in metals.

For amorphous thermoplastics at 0 – 200 °C, the thermal conductivity lies between 0.125 and 0.2 W m^{-1} K^{-1}. Partially crystalline

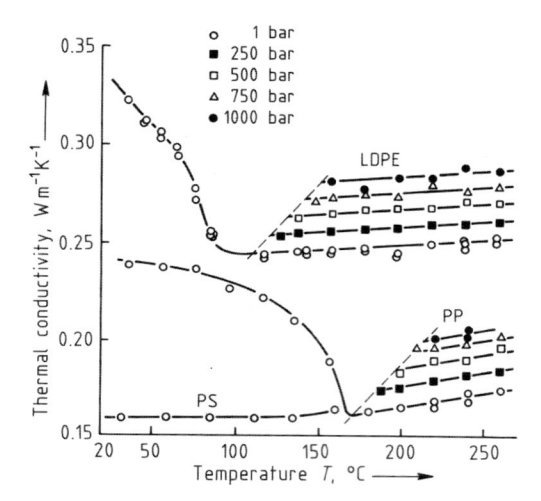

Figure 35. Thermal conductivity of low-density polyethylene, polypropylene, and polystyrene as a function of temperature [144]

thermoplastics, in contrast, have a higher thermal conductivity, because the ordered crystalline regions conduct better than the amorphous regions. As melting of the crystalline regions progresses, the thermal conductivity approaches that of the amorphous regions. Figure 35 plots the thermal conductivities of branched polyethylene, polypropylene, and polystyrene as functions of temperature and pressure [144]. Branching has a marked effect on thermal conductivity; molecular mass and molecular-mass distribution have no influence.

As cross-linking increases (thermosets, elastomers), thermal conductivity rises, because van der Waals bonds are progressively replaced by valence bonds with their greater molecular thermal conductivity. Moreover, the changes and transformations characteristic of polymers are less strongly marked in curves of thermal conductivity (and other thermal values) at higher degrees of cross-linking.

At content below 5 %, additives (air or even metals) do not greatly influence thermal conductivity. For higher filler contents, numerous mixture rules have been developed. They are, however, not universally applicable. The most reliable is still the geometric mean model of PRINGELHOF and THRONE, which was rediscovered by AGARI and UNO [145] and confirmed for many mixtures. According to this model, the logarithmus of the thermal conductivity increases linearly with increasing volumetric filler content:

$$\log \lambda = V \cdot \log \lambda_f + (1 - V) \cdot \log \lambda_m$$

where λ is the thermal conductivity of the mixture, λ_f the thermal conductivity of the filler, λ_m the thermal conductivity of the matrix, and V the filler content in vol %. Oriented plastics and plastics with oriented fillers, whose thermal conductivities differ from that of the matrix, exhibit anisotropic thermal conductivity.

For commercial closed-cell foamed plastics [polystyrene, poly(vinyl chloride), and phenolic resin], the thermal conductivity lies between approximately 0.035 and 0.042 W m^{-1} K^{-1}.

The thermal conductivities of solids are usually measured under steady-state conditions with the Poensgen two-plate apparatus [146]. A setup described by FRITZ and KÜSTER is suitable for the investigation of foamed materials [147]. A very simple device for determining the thermal conductivity of foamed materials with an accuracy sufficient for practical purposes is described in [148].

Two quantities derived from thermal conductivity are important in plastics processing. Cooling takes up a large part of the cycle time in injection molding, and a quantity called the thermal diffusivity a is used in calculating this process: $a = \lambda/c\varrho$, where λ is thermal conductivity, c is specific heat, and ϱ is density. Thermal diffusivity has the units m^2/s or m^2/h.

The second important quantity is the heat penetration coefficient $b = (\lambda c \varrho)^{1/2}$, expressed in J m^{-2} s$^{1/2}$ K. This coefficient determines the initial contact temperature T_c (in Kelvin) when two bodies A and B come in contact: $T_c = (b_A T_A + b_B T_B)/(b_A + b_B)$. The formula indicates, for example, the temperature that prevails at the mold wall when a hot plastic mass flows into a cooler mold and before heat removal by the cooling system becomes effective [149].

2.3.3. Thermal Expansion

The linear coefficient of thermal expansion lies between ca. 0.6×10^{-4} and 2.3×10^{-4} K^{-1} for thermoplastics or 0.2×10^{-4} and 0.6×10^{-4} K^{-1} for thermosets (in the service temperature range for each case).

The linear coefficient α can be measured directly with a dilatometer (e.g., a quartz-tube dilatometer as specified in ASTM D 696) or, more accurately, against a reference material with a differential dilatometer.

This coefficient can also be derived from measurements of the specific volume or of the volume coefficient of thermal expansion β. The approximate formula is $\alpha \approx \beta/3$.

The specific volume V is also measured in dilatometers [150–153].

If the specific volume is plotted against temperature for amorphous thermoplastics, the curve exhibits a kink near the glass-transition temperature. The expansion coefficient, which is the first derivative with respect to temperature, thus has a discontinuity (a step). This phenomenon can be interpreted with Eyring and Hirai's vacancy theory of liquids: As the temperature rises, the equilibrium vacancy concentration begins to increase at the glass transition. This increase manifests itself as an increase either in the free specific volume or in the slope of the specific volume–temperature curve. These effects are more marked for amorphous polymers than for partially crystalline ones, because the thermal behavior of partially crystalline materials is governed chiefly by the melting of crystallites.

At the onset of melting (i.e., as the structure begins to break down), the curve of specific volume versus temperature rises faster than linearly. After melting is complete, the curve becomes linear again but has a steeper slope than before melting; see Figure 36 [151].

On cooling from the melt, however, these discontinuities in the curves of specific volume versus temperature show a dependence on cooling rate. Figure 37 illustrates schematically what happens with amorphous materials. The vacancy concentration is somewhat lower in slow cooling than in fast cooling. The various branches correspond to states differing in vacancy concentration. Among the consequences of this behavior is that V versus T curves are measured only with increasing temperature because the heating curve cannot be reproduced on cooling.

p – V – T Diagrams. Because plastics are processed over a wide range of temperatures and pressures, the variation of the specific volume not only with temperature but also with pressure must be known. The literature describes several methods for investigating the relationship of p, V, and T [154–159].

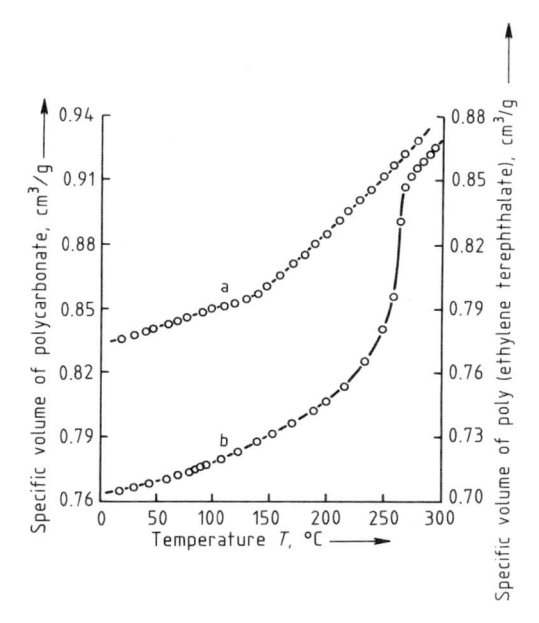

Figure 36. Specific volume as a function of temperature a) Polycarbonate, $\varrho_{20} = 1.2$ g/cm^3 (example of an amorphous product); b) Poly(ethylene terephthalate), $\varrho_{20} = 1.4$ g/cm^3 (example of a partially crystalline product) [154]

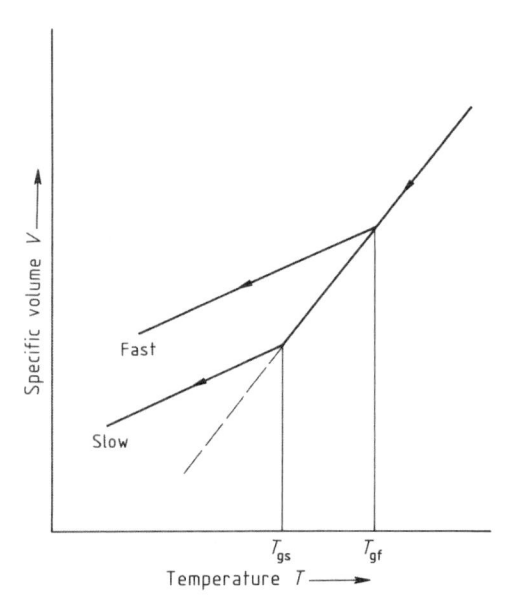

Figure 37. Temperature curves of specific volume for fast and slow cooling of an amorphous plastic T_{gs} = glass transition temperature for slow cooling; T_{gf} = glass transition temperature for fast cooling

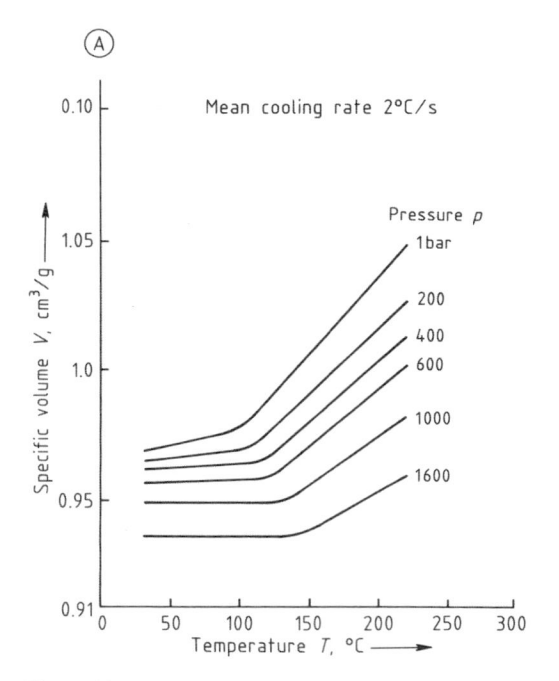

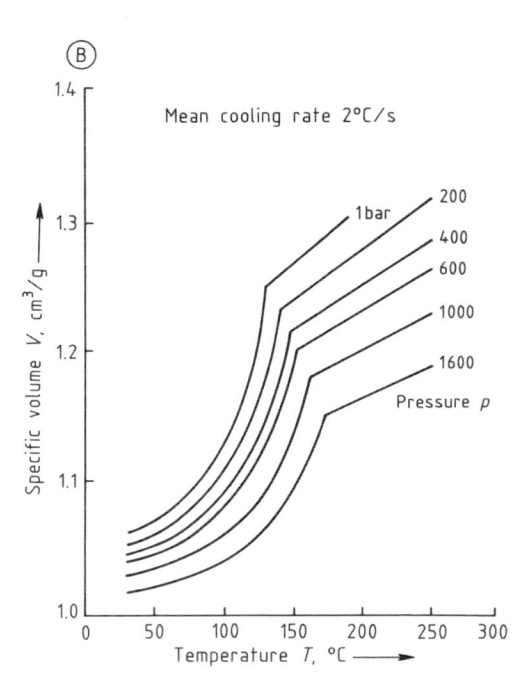

Figure 38. p–V–T diagrams A) Polystyrene; B) Polyethylene

In some methods, the specimen is placed in a cylinder and compressed from one or both ends by pistons. In another, the specimen is under hydrostatic pressure in a mercury-filled piezometer located in a pressure vessel filled with silicone oil. Figure 38 shows the p–V–T diagram of an amorphous plastic (polystyrene) and of a partially crystalline plastic (high-density polyethylene). As in the V versus T plots, the shape of the curve depends on the cooling rate.

2.4. Electrical Properties and Their Testing

Plastics owe their great variety of uses in electrical and electronic devices to their combination of electrical properties, rational fabrication methods, and widely variable engineering properties.

2.4.1. Insulating Properties

Volume Resistivity. Plastics used as electrical insulators have volume resistivities of 10^{10}–10^{19} Ω cm. Substantially lower values can be achieved with additives such as carbon black, graphite, carbon fibers, and metal powders or flakes, making possible a reduction in static charging due to high resistance, which can lead to safety problems in some instances (e.g., when flammable liquids are transferred from plastic vessels, when plastic flooring is employed in spaces where an explosion hazard exists, and in mining). In general, no hazardous charging occurs at resistivities of less than 10^{10} Ω cm.

Charge transport in plastics occurs mainly by ions and to only a minor extent by electrons. Whereas Ohm's law holds rigorously for metals, insulators not only have a charge carrier current but also an internal charge current resulting from polarization of the plastic. This internal charge current builds up a space charge, which decays on a scale of hours. The volume resistivity ϱ_D is a function of temperature, having the form $\varrho_D = \varrho_0 \exp(b/T)$, where ϱ_0 and b are constants and T is absolute temperature. Figure 39 shows how the volume resistivity of polypropylene films depends on temperature.

Surface Resistivity and Resistivity between Plugs. Because the insulation surface usually makes a large contribution to leakage, surface resistivity is also defined. This quantity does

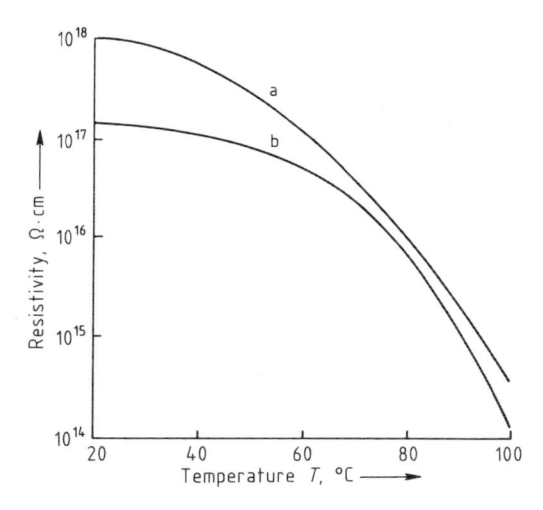

Figure 39. Resistivity of polypropylene films as a function of temperature a) Polypropylene; b) Biaxially drawn polypropylene

Figure 40. Schematic curves of ε'_r and $\varepsilon''_r = \varepsilon'_r \tan \delta$ versus frequency for high-polymer electrical insulators

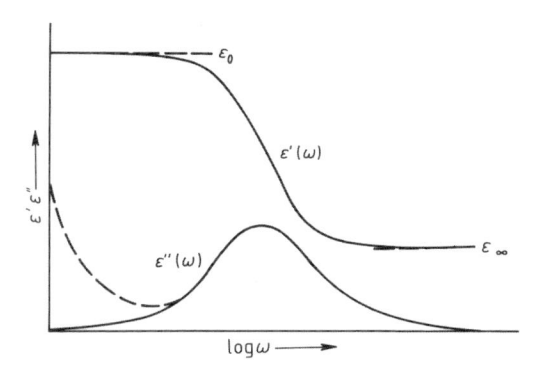

not, however, have the predictive power of volume resistivity, which is a material-specific value.

In the case of inhomogeneous insulators such as laminated composites or plastics containing electrically conducting fillers such as carbon black, the insulator resistance is often determined between two metal pins inserted into the material 25 mm apart (resistivity between plugs). This measurement includes an unknown contribution from surface resistivity.

Resistivity measurements are current – voltage measurements, preferably carried out at 100 or 1000 V d.c. with standard electrodes [158].

2.4.2. Dielectric Properties

Certain formal analogies exist between the relaxation behavior of viscoelastic bodies under mechanical loading, which can be described in terms of a complex modulus, and the dielectric behavior of plastics, which is characterized by a complex dielectric constant (permittivity) $\varepsilon^* = \varepsilon'(\omega) - i\,\varepsilon''(\omega)$, where ω is angular frequency, $\varepsilon'(\omega)$ the dielectric constant (permittivity), and $\varepsilon''(\omega)$ the dielectric loss factor. The ratio $\varepsilon''/\varepsilon'$ is equal to the tangent of the phase angle between the voltage and the current: $\tan \delta = \varepsilon''/\varepsilon'$.

Just as viscoelastic behavior in the mechanical case can be simulated by spring – damper models, dielectric behavior can be described in terms of electrical networks. The analogy is, as already mentioned, merely formal; it has to do with the fact that both phenomena come under a common theory, the theory of the elastic aftereffect (memory effect), advanced as early as 1893 by WIECHERT and extended by WAGNER in 1913 to the theory of the dielectric aftereffect.

The mathematics of viscoelastic and dielectric behavior is the same as that of the magnetic aftereffect, linear network theory, and linear integral equations. Figure 40 illustrates schematically how the dielectric values for high-polymer electrical insulators depend on the frequency of the applied alternating field. Here ε' is the relative permittivity ε'_r, which is a measure of the polarization attainable at any frequency: At low frequencies, the molecules have enough time to orient themselves in the field, so that ε' takes on its largest value. At higher frequencies, the field changes so rapidly that the dipoles can no longer follow and ε' declines to smaller values. The value of ε' falls off substantially when the period of the field is of the order of the dipole relaxation time. The value ε'' is a measure of the energy transformed in one oscillation period; this energy is extracted from the field and converted to heat as the dipoles orient themselves. This quantity passes through a maximum when the period of the field is equal to the relaxation time; it approaches zero at very low and very high frequencies.

Because $\varepsilon''/\varepsilon'_r = \tan \delta$, $\varepsilon''_r = \varepsilon'_r \tan \delta$. The product $\varepsilon'_r \tan \delta$ is also called the dielectric loss factor.

The measured ε''_r at low frequencies can be distorted by a metrological effect not related to polarization: the d.c. conductance, whose contribution to the total loss decreases with frequency (dashed curve in Fig. 40).

If plastics are to be heated in a high-frequency (HF) field, the loss factor must be made as large as possible (HF welding of polar plastics). In insulation, small tan δ values are desirable to minimize the evolution of heat. When polar plastics (e.g., PVC blends) are used in cable manufacturing, the loss factor maximum must be shifted to the highest possible temperature so that thermal breakdown will not be promoted at service voltages.

In nonpolar substances, an external a.c. field brings about polarization only by deforming the electron distribution. This displacement of the charge carriers is, however, independent of frequency, so that ε_r remains virtually constant over a wide frequency range. At very high frequencies (in the infrared), ε_r approaches the square of the optical refractive index.

The Clausius – Mosotti equation holds for nonpolar substances under certain conditions. This equation

$$\frac{\varepsilon - 1}{\varepsilon + 2} \frac{M_r}{\varrho} = \frac{4\pi}{3} N \alpha$$

describes the relationship between the dielectric constant and the density, where ε is the dielectric constant, N is Avogadro's number, M_r is molecular mass, ϱ is density, and α is molecular polarizability.

The temperature curve of tan δ for nonpolar straight-chain polymers generally shows two maxima of the loss factor: the α absorption (e.g., due to motion of the CH_2 dipoles in polyethylene) and the γ absorption (e.g., due to relaxation in crystalline regions in PE). Branching gives rise to a third absorption region, β absorption, which lies between the α and γ absorptions in the tan δ curves; see Figures 41 A and B.

Often, the loss factor and dielectric constant must be measured over a wide range of temperature and frequency (from a few hertz to more than 10 GHz, although most measurements are made between 50 and 100 MHz). The dielectric constant ranges from somewhat over 1 for foamed plastics up to 20 or more. Loss factors range from 3×10^{-5} for polyethylene with no residual catalyst to 0.1 or higher.

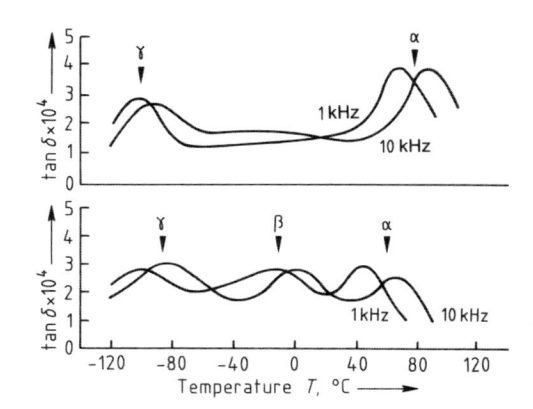

Figure 41. Plot of tan δ versus temperature at 1 and 10 kHz for polyethylene A) High-density polyethylene; B) Low-density (branched) polyethylene

Materials with anisotropic structure may display anisotropic dielectric properties. Measurement is complicated by the variety of methods needed to cover the wide frequency range. Up to about 10^8 Hz, however, only two measuring principles are needed. In general, the circuit used must contain coils, capacitors, and resistors. From the audio frequency range up to 100 kHz, bridge circuits are used without exception. In a.c. measurements, at least one branch of the bridge, not containing the specimen, must have a blind component, either an inductance or a capacitance. Examples of such bridges are the Schering type and the transformer measuring bridge. Automatic, self-balancing transformer bridges, which are unexcelled for quick routine measurements, are now available.

The range from 10 kHz to 100 MHz corresponds roughly to radio frequencies. Measurements in this range are performed by tuned-circuit methods (resonance techniques with tank circuits and a modified capacitance standard).

Above 10^8 Hz, different principles are employed. They include measuring-line methods and techniques with capacitively loaded coaxial resonators.

For further information on measuring methods see [159–162]. Table 6 lists dielectric values at 50 Hz and 20 °C for a number of plastics.

2.4.3. Dielectric Strength

Current theories of breakdown in a solid dielectric start with the assumption that

Table 6. Dielectric properties of some plastics at 20 °C and 50 Hz (mean values)

Product (abbreviations from DIN 7728)	Dielectric constant ε	Loss factor $\tan \delta$
PF 31.5	6 – 8	0.4
UP resins	3 – 3.8	0.003 – 0.01
EP resins	3.2 – 4.8	0.002 – 0.006
PE	2.28 – 2.34	0.0005
PP	2.27	0.0005
PVC, rigid (E)	3.8 – 4.3	0.03
PVC, rigid (S)	3.4 – 3.7	0.02
PVC/DOP 60:40	7.5 – 8.0	0.08
PA 6	4.3	0.03
PS	2.5	0.0003
PTFE	2.1	0.0001

breakdown is linked with the free electrons that are always present, which are set in motion by the electric field. Part of the energy obtained from the field is given up to the dielectric, and this process is decisive for the dielectric: If the field is strong enough, the energy absorbed from the field by the electrons exceeds the energy lost to the dielectric, and the electron energy increases without limit. This energy is transformed to heat and ultimately destroys the dielectric.

The theory of collisional ionization assumes that the electrons, after exceeding the maximum loss to the dielectric, are accelerated further until they have sufficient energy to liberate more electrons by collisional ionization. The result is an avalanche-like increase in the number of electrons.

In the theory of collective breakdown, in contrast, the energy excess is assumed to increase the thermal energy of the collective of free electrons. The resulting temperature rise is then transferred to the dielectric [163].

The breakdown voltage is measured with a 100- or 200-kV a.c., 50-Hz power supply, usually on plate or film specimens. The breakdown voltage per unit specimen thickness is the dielectric strength E_d, in kV/mm. This value is not, however, a specific material property; it depends on a number of factors, especially specimen thickness [164], [165].

A quantity of special engineering interest is the long-term dielectric strength – the breakdown strength as a function of the time over which the voltage is applied, often under service conditions [166]. The service life of an insulator is influenced by a variety of aging processes, including the buildup of space charge due to electron enrichment, electrical aging due to microcorona discharges, electrochemical aging due to mobile contaminants (e.g., water), and thermal aging [167].

Tracking Resistance. If contaminants and moisture on the surface of an insulating material form a conducting bridge between two contacts between which a voltage is imposed, surface leakage currents can damage the surface. This process is simulated in testing to assess tracking. Two standard methods are available, one with voltages up to 600 V (low-voltage tracking) [168] and the other up to 6 kV (high-voltage tracking) [169].

In the low-voltage method, contamination is simulated by a test solution dripped onto the surface. At a predetermined voltage selected from a series, five specimens must survive after 50 drops; failure is signaled by the presence of a given current. Variants of the test are described in [168].

The high-voltage method is intended for insulating materials for open-air service at high voltages. The specimen is tilted 45° from the horizontal, and a stream of a test solution is allowed to flow down the bottom of the specimen between two stainless steel electrodes. Two failure criteria can be applied: the current flowing over the specimen surface or the presence of tracks [169].

Other, application-oriented tests on insulating materials should be mentioned briefly.

The ability of insulating materials to withstand surface glow discharges is important for the service life of electrical equipment. The glow discharge is intimately related to breakdown behavior. Such discharges attack the surface both mechanically and chemically. In the test of this behavior, graduated voltages are imposed on a specimen with a point – plate electrode arrangement, and the time to breakdown is measured. The service life curve obtained in this way for the insulating material makes it possible to rank insulators with respect to glow discharge behavior [170], [171].

If insulators are in contact with conductors to which d.c. voltage is applied, corrosion of the conductors can result. This corrosion depends on the insulator – metal combination and is affected by temperature, atmospheric humidity, and exposure time. Electrolytic corrosion can take place on any base metal. This phenomenon is tested by imposing a d.c. voltage in a humid atmosphere; two brass electrodes are employed. Results are assessed by comparing the colors of the brass foils with a standard color scale [172], [173].

Plastics filled with conductive carbon black, aluminum flakes, stainless steel fibers, or metallized carbon fibers are suitable for shielding electromagnetic fields. These compounds, used for the housings of electronic devices, are supposed to prevent electromagnetic interaction of the devices.

Measurements of dielectric strength and surface resistance are not adequate for the evaluation of shielding qualities, and special methods must be used. Shielding effectiveness is measured in dB. A value of 30 – 40 dB is considered sufficient for plastics. For details, see [174], [175], and literature cited therein.

2.4.4. Electrets; Semiconducting and Conducting Polymers

Electrets are dielectrics bearing a quasi-permanent charge, in which positive and negative charges, either isolated or created by polarization, are kept apart for an extended time. The best-known electrets are fluorinated ethylene–propylene copolymers (EP-F), polycarbonate (PC), polytetrafluoroethylene (PTFE), polypropylene (PP), poly(ethylene terephthalate) (PETP), and polyphenylene ether (PPE).

Electrets are used in electroacoustic transducers (e.g., microphones), radiation dosimeters, and air filters (electrostatic dust collection).

Several polymers now under development exhibit piezoelectric and pyroelectric effects. The most important product in this area is poly(vinylidene fluoride) (PVDF). To achieve the strongest possible piezoelectric effect, the material is drawn and given a permanent dipole orientation in a strong electric field.

The true semiconductive and conductive polymers include polyacetylene, polypyrrole, poly(*p*-phenylene), and poly(*p*-phenylene sulfide). Detailed reviews of this area have been published by ROTH [176] and MAIR and ROTH [177]. Numerous literature references can be found in [178]. See also → Polymers, Electrically Conducting.

2.5. Optical Properties and Their Testing

Optical elements made of plastics include lenses, prisms, rear reflectors, glazings, lighting enclosures, optical-fiber light guides, compact discs, and many others. The most important optical properties for these applications are refractive index and lighting qualities (as defined, for example, in lighting standards), especially transmittance. In many other components for household and industry, aesthetic qualities such as color, gloss, and opacity are more important, but these properties are difficult to correlate with visual impressions.

2.5.1. Refraction and Birefringence

When a ray of light passes from one isotropic medium to another, the sine of the angle of incidence is in a constant ratio to the sine of the angle of refraction (both measured from the normal to the interface). If the light passes from a vacuum (or, less precisely, from air) into a denser medium, this ratio is called the refractive index n. The refractive index is dimensionless and depends on the wavelength of the light and on the temperature.

General methods for measuring refractive indices (i.e., not necessarily restricted to plastics) can be found in optics texts; standards also exist for measurements on plastics [179], [180]. Table 7 lists refractive indices for a number of plastics. Wavelength dependence is characterized by the dispersion. In practical optics, either the "dispersive power" $D_r = (n_F - n_C)/(n_D - 1)$ or its reciprocal, the Abbé number $\nu = 1/D_r$, is employed. Subscripts refer to wavelengths at which the indices are measured.

Birefringence is a measure of optical anisotropy. It is defined as the maximum difference between two refractive indices determined in two perpendicular directions. Some materials show isotropic behavior until they are deformed while in the glassy state. The deformation, which manifests itself in small changes in bond angles and interatomic distances, leads to the phenomenon *stress birefringence*, which is proportional to the applied stress. This behavior is utilized for, among other purposes, photoelastic examination, which enables the

Table 7. Refractive indices n_D^{20} of plastics

Product (abbreviations from DIN 7728)	Refractive index n_D^{20}
CA	1.49 – 1.51
CP	1.47 – 1.48
CAB	1.45 – 1.47
PTFE	1.35
FEP	1.338
PFA	1.35
ETFE	1.4028
POM	1.48
PA 6	1.53
PC	1.586
PE-(LDPE)	1.51
PE-(HDPE)	1.545
PMMA	1.4893 – 1.491
PP	1.5030
PAN	1.5187
PS	1.590 – 1.592
SAN	1.565 – 1.569
PVDC	1.60 – 1.63
PSU	1.633
PVC	1.52 – 1.55
PU	1.5 – 1.6
PVA	1.4665
PVDF	1.42
SB (30 % PS)	1.53
EP	1.55 – 1.60
UP (50 % styrene)	1.523 – 1.540
PVC + DOP 60/40	1.52
PF	1.7
UF	1.54 – 1.56

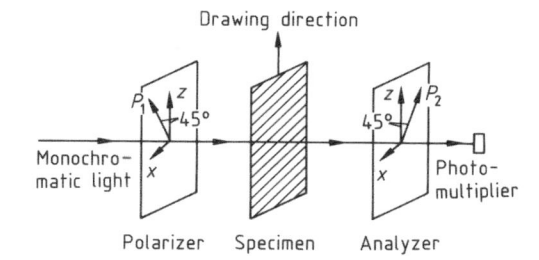

Figure 42. Setup for birefringence measurement [181]

specimen does not change, the transmitted light has zero intensity; if the specimen is birefringent, the intensity will be

$$I_b = I_0 \sin^2\left[\pi \frac{d}{\lambda}(n_x - n_z)\right]$$

where I_0 is the incident light intensity. From the intensity ratio I_t/I_0, the specimen thickness along the path of the light, and the wavelength of light employed, the birefringence $n_x - n_z$ can be calculated as

$$n_x - n_z = \frac{\lambda}{\pi \cdot d} \arcsin\sqrt{I/I_0}$$

This method is used when the degree of drawing and thus the degree of orientation are high.

A second method for measuring birefringence uses the Babinet compensator, which replaces the analyzer and the photomultiplier. Essentially, it consists of two quartz wedges that are adjusted so as to offset the phase shift δ between the x and z components of the emerging light vector.

2.5.2. Transmittance and Gloss

The light-transmitting qualities of plastics are described in terms of the visual impression that the specimen gives under precisely defined illumination. In the usual order, these terms are clear, almost clear, transparent, translucent, semiopaque, and opaque.

The customary definitions in lighting technology are as follows [182]:

Transmittance is the percentage of light transmitted without deflection. Without deflection means that only the components propagated in the direction of the incident ray are considered. Plastics with particularly high transmittances

investigation of plane and even triaxial states of stress in metal structures by using plastic models. The preferred material for this application is phenol – formaldehyde resin because it has a relatively high stress birefringence.

The other cause of birefringence in plastics is orientation of the polymer chains brought about in processing or by drawing. This *orientation birefringence* is greater than the stress birefringence. Birefringence measurement is one of the techniques used to determine the state of orientation in polymers.

Birefringence is measured by the transmission method (Fig. 42) [181]. A ray of monochromatic light is polarized so that the oscillation direction of the electric vector lies in the $x - z$ plane and makes a 45° angle with the z direction. The ray next passes through the specimen, in which the polymer molecules should be aligned preferably with the z axis, and then through an analyzer whose polarization direction P_z is perpendicular to that of the polarizer. The intensity of light transmitted by the analyzer is measured with a photomultiplier. If the polarization in the include poly(methyl methacrylate) (92 % at 380 – 780 nm), polycarbonate (88 %), and cellulose esters (86 % at 420 nm, 90 % at 660 nm for cellulose acetate). For some applications, the ultraviolet transmittance of these clear products can be reduced by addition of UV absorbers. Modified and filled polymers are clear or transparent only if their components have similar refractive indices or if the particle size of the incoherent phase is smaller than the wavelength of the light.

Haze refers to the portion (intensity) of light transmitted by a specimen that deviates from the direction of the incident light because of forward scattering in the material. Materials with high haze components are said to be semiopaque or translucent.

Both transmittance τ and haze vary with the geometry of the optical measuring system, which must therefore be precisely defined. The same holds for the type of light employed. Examples are Colorimetric Standard Illuminants A (incandescent lamp), C (same as A with Davis – Gibson filter), and D 65 (daylight with a color temperature of ca. 6500 K).

Once the spectral characteristics of the light and the spectral sensitivity of the human eye (characterized by "standard spectral curves") are known, mathematical formulas are applied to relate the wavelength dependence to the measured energy spectrum of the transmitted light. Finally, transmittance is determined.

The most common way of determining absorption spectra is by use of spectrophotometers. If the monochromator is placed in front of the specimen, the light, whose wavelength is continually varied, passes through the specimen, and the transmitted intensity of each wavelength is measured successively by a photocell. Alternatively, the light is dispersed only after it has passed through the specimen, and the intensities of the individual wavelength components are then led to the photocell. With this scanning arrangement, however, the long measuring time can be shortened considerably if all wavelength components are led simultaneously to a series of photocells. The result (spectral curve) is then determined in a few microseconds. Some applications of transmittance measurements in the visible spectrum include determination of the spectral or mean transmittances of colored and colorless plastics; color characterization (tristimulus values or total transmittance) for signal lights in air and marine navigation, railroads, and road transport; color matching; aging tests; damping measurements in light guides; and testing the lightfastness and weather resistance of colored plastics.

The haze of thin specimens with low scattering is measured in a standardized setup [183]. For reasons having to do with experimental technique and to bring the results into accord with subjectively assessed scattering (haze) properties, the scattered light measured is the component lying between two coaxial cones (about the axis of incidence) with opening angles of 12° and 80°.

For specimens with higher scattering (up to 30 %), the measuring system of ASTM 1003–61 is employed. The setup features an integrating sphere and can be reconfigured to measure the incident flux as well as the directly transmitted, scattered, or total transmitted flux. The parameters total transmittance, diffuse transmittance, and haze are then calculated with allowance for the scattering of the integrating sphere. High-quality transparent acrylates have haze values of the order of 0.15 %.

Gloss. HUNTER defines gloss as the ratio of the regularly (specularly) reflected light intensity to that of a high-quality optical mirror. Gloss can be defined in a number of other ways. Most glossmeters measure the specular gloss defined by HUNTER. Most instruments employ fixed incidence and measurement angles (e.g., 45° in the Lange glossmeter).

Highly polished, optically flat black glass plates are often used as reference standards. Their reflectivity can be calculated from the refractive index.

Measurements of the spectral reflectance curve (i.e., the dependence of reflectivity on incidence angle) give a more meaningful description of gloss. The instrument used for these measurements is a goniophotometer [184–187].

The values obtained by gloss measurement techniques, however, do not correlate perfectly with the visual impression of gloss. This becomes obvious when an inspector assessing gloss turns an object one way and another, thus scanning through many incidence and reflection angles. Polarization also plays a role in the impression of gloss.

2.5.3. Color and Colorimetry

Color is not an absolute property but depends on the surface condition of the specimen as well as the conditions under which it is examined. Nor does the term "artificial light" or "daylight" provide enough information; for example, the color of sunlight varies with the position of the sun in the sky. Two colors that the observer subjectively judges to be the same under given illumination may not match under other conditions. These problems are, of course, not specific to the plastics industry. The concepts of colorimetry, measuring methods, and test conditions are specified in DIN 5033 [188]. American and British standards also exist for this field [189], [190]. The CIE international vocabulary for lighting technology defines terms and nomenclature; problems involved in color differentiation are discussed in ISO 2573 [191], [192].

Very simple methods for assessing small color changes or departures from colorlessness in the direction of yellow (yellowing) are described in [191]. The objective determination of color with scales and formulas is a difficult undertaking but is still important in plastics production and applications.

Transparent, translucent, and opaque colored plastics must generally exhibit narrowly defined hues, whether they are to be used in signaling, advertising, or illumination. Consumer products and industrial articles must meet similar requirements, especially when colored plastics have to match other components in perceived color.

Problems of colorimetry can be mentioned only briefly here. The human eye interprets incident radiation with three distinct sensitivity functions at once. These are called the standard spectral curves and relate to the three primary colors red, green, and blue. One of these (green) is taken as the basis for the spectral curve describing the human perception of brightness. The standard spectral curves are now mathematically related to the energy distribution of the radiation giving rise to a color sensation and to the reflectance curve of the specimen in the visible spectrum (see also Section 2.5.2). The relations between the incident energy distribution (e.g., the Colorimetric Standard Illuminants) and the standard spectral curves are tabulated, so all that is necessary is to link

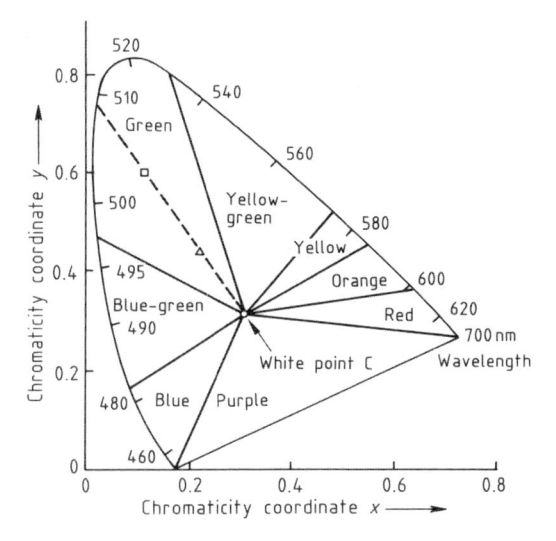

Figure 43. The color triangle

them with the reflectances. The resulting tristimulus values X, Y, and Z are normalized and can be used to generate a two-dimensional representation. The normalized values x, y, and z satisfy the condition $x+y+z=1$; thus the color location in the $x - y$ system, which formally characterizes the color, can be unambiguously determined. The colors of the spectrum, plotted in this way, lie on a horseshoe-shaped curve (Fig. 43). Together with the brightness reference value Y (which cannot, however, be presented in the color triangle), the point (x, y) completely describes a color. All natural colors lie inside the curve. The location of the "white point" depends on the type of illumination. The saturation decreases toward this point. Colors lying on a straight line radiating from the white point have the same hue. In Figure 37, for example, the square and the triangle are color locations with the same hue, and the square is more saturated than the triangle. It is important here that colors with the same color location are the same only for the illumination conditions illustrated, not necessarily under other illumination conditions. Such colors are said to be conditionally the same, or metameric colors. Colors that must appear the same under all illuminations must have identical reflectance curves.

In principle, any spectrophotometer that permits measurements in the visible spectrum and has a reflector attachment is suitable for colorimetry.

An important class of instruments, especially for routine color measurement, is the three-filter (tristimulus) type, which does not yield complete reflectance curves but facilitates the determination of X, Y, and Z values. These devices use three calibrated filters (tristimulus filters) to evaluate the fraction of incident light reflected from a specimen. The filters, in combination with the light source and the system of photoelectric cells, give the three standard spectral curves. Because the curves are not reproduced perfectly, the measurements do not rank as absolute values. However, these instruments are suitable for relative measurements (e.g., comparing the color of a plastic product with that of a color standard). However, only metameric color matching can be realized. As mentioned above, the geometry of the setup is also important for reflectance measurement. Geometries for colorimetry include "direct" illumination at 45° with observation at 0° and "diffuse" illumination at 0° with the integrating sphere used for reflected-light measurement. The appropriate geometry depends on the surface structure of the specimen. The most commonly used Colorimetric Standard Illuminant is D 65, which resembles the formerly customary type C, because the behavior of nonluminous colors in daylight is of particular interest. For color-matching purposes, fluorescent lamps or, even better, xenon arc lamps are used to provide artificial daylight.

The locations (in the color triangle) of standard and specimen must lie within a certain tolerance if they are to appear "the same." A number of workers have attempted to establish these tolerance ranges. Formulations to match standard colors can be found by computer calculations based on colorimetric measurements (Kubelka – Munk and other methods).

2.6. Density and Physical-Chemical Properties and Their Testing

2.6.1. Density

High-precision density measurements can be performed with relatively simple apparatus. Because some physical properties of plastics are single-valued functions of density, changes in these properties can be characterized indirectly on the basis of density measurements. An example of such a property is crystallinity, which in turn affects certain mechanical properties.

Density often occupies an important place in specifications; in the case of polyethylene, for example, it is essential for classification and identification [192].

The determination of density for solid plastics (except foamed plastics) is described in ISO 1183 [193]. The standard covers four methods, corresponding to the type of plastic in question. These methods are also contained in DIN 53 479 [194]; DIN 53 420 describes density measurements on foamed plastics. Compact, unfilled, unreinforced plastics have densities between ca. 900 kg/m^3 (polypropylene) and 2300 kg/m^3 (polytetrafluoroethylene).

2.6.2. Liquid Absorption and Swelling; Desorption of Plasticizers

Many polymers can absorb liquids (often up to an equilibrium concentration). The processes involved are adsorption, diffusion, and dissolution or swelling. Un-cross-linked amorphous polymers swell without limit in some liquid media and ultimately dissolve entirely (e.g., polystyrene in benzene). A table identifying good and poor solvents for some 60 important polymers appears in [195].

If the polymer is cross-linked or partially crystalline (with the crystallites acting like cross-link sites), swelling may be limited because the entropy-elastic restraint of the network prevents further liquid absorption.

Thermodynamic equilibrium is reached when the free swelling energy $\Delta F = \Delta H - T \Delta S$ vanishes or when the heat of swelling ΔH is equal to the product of the temperature T and the entropy of swelling ΔS. The time to equilibrium depends not only on the swelling agent and its concentration, but also on the geometry of the sample and, above all, the diffusion coefficient of the swelling agent in the polymer. Figure 44 shows the equilibrium water uptake of several plastics as a function of relative humidity.

The opposite process, liquid loss, occurs when low molecular mass plasticizers diffuse out of plastics and become desorbed.

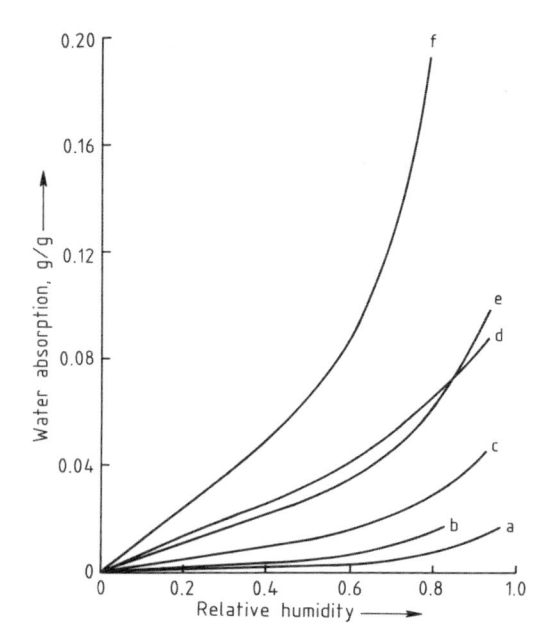

Figure 44. Equilibrium water absorption of some plastics as a function of relative humidity a) Poly(vinylidene chloride); b) Chlorinated rubber; c) Poly(vinyl butyral); d) Polyamide 66; e) Cellulose acetate; f) Poly(vinyl alcohol)

2.6.3. Permeation of Gases and Vapors

The permeation of gases and vapors in plastics can be described as a two-stage process. First, the permeating substance dissolves in the surface of the plastic; second, the molecules diffuse through the material, forming a concentration gradient, and then evaporate again. The concentration c of the permeating gas at the surface depends on the gas pressure p or the volumetric concentration in the gas space. For permanent gases (H_2, O_2, N_2, CO_2, noble gases, etc.), the concentration is directly proportional to the partial pressure p (Henry's law: $c = kp$). This simple relationship does not hold for the diffusion of vapors.

The steady-state diffusion of a gas or vapor is described by Fick's first law:

$$\frac{dQ}{dt} \cdot \frac{1}{A} = -D \frac{dc}{dx}$$

where dQ/dt is the flow velocity of the medium, A the flow area, D the diffusion constant, and dc/dx the concentration gradient. Transient diffusion is described by Fick's second law.

Consider a layer of finite thickness d and assume that D is independent of concentration. Integrating the above equation over the thickness gives

$$\frac{dQ}{dt} \cdot \frac{1}{A} = \frac{D(c_1 - c_2)}{d}$$

where c_1 and c_2 are the concentrations at the two surfaces. If concentration is expressed in terms of pressure, the result is

$$\frac{dQ}{dt} = D \cdot S \cdot A \frac{p_1 - p_2}{d}$$

where $D \cdot S$ is the permeability P; dQ/dt is constant in the steady state. In experiments, $p_1 - p_2 = \Delta p$ is held constant, so that

$$P = \frac{Qd}{\Delta t A \Delta p}$$

in units of $cm^3 \, cm^{-1} \, s^{-} \, mbar^{-1}$.

The practical quantity gas permeability is easier to work with. It is the volume of gas in cubic centimeters that permeates a 1-m^2 area of the product being tested in 24 h at a given temperature and a given pressure difference (e.g., 1 bar). The corresponding unit is $cm^3 \, m^{-2} \, d^{-1} \, bar^{-1}$. The unit employed for vapor permeability, especially water-vapor permeability, is $g \, m^{-2} \, d^{-1}$. The thickness of the test specimen must be stated, along with the gas and water-vapor permeabilities, because the units do not allow for reduction to unit thickness. Furthermore, the pressure difference must be known in the case of water-vapor permeability; in practice, this is accomplished by establishing a given atmospheric humidity difference.

The steady-state transport equation holds only for gases and vapors that are not very soluble in the polymer and do not react chemically with it. Vapors can often act as plasticizers, thus increasing diffusion.

The permeability is temperature dependent in all permeant – polymer systems. The diffusion coefficient for both gases and vapors increases with increasing temperature. With increasing temperature, the solubility factor S increases for gases but decreases for vapors.

Permeation is also influenced by the molecular size and polarity of the permeating substance. The morphology of the polymer (crystallinity, branching, orientation, etc.) plays an important role in the process.

Multilayer films (laminates) are commonly used in packaging. The P coefficient of such a composite can be calculated from the P values of the individual plies:

$$\frac{1}{P_L} = \frac{1}{L}\sum_{i=1}^{n}\frac{d_i}{P_i}$$

where L is the thickness of the composite, P_L is its permeability, and d_i and P_i are the thickness and the permeability of the ith ply.

This equation, however, applies only when permeability does not depend on pressure. The order of the plies has no effect in this case; otherwise, the overall permeability depends on the order of the plies.

Methods of measuring gas permeability include those employing pressure gauges and those involving carrier gases. In the first group, the quantity of gas permeating the specimen in a given time is found by measuring pressure and volume changes [196], [197]. In carrier-gas methods, gas flows through the specimen while the pressure remains the same on both sides. The test gas flows at a constant rate through a chamber on one side of the specimen, while a second (carrier) gas flows, also at a constant rate, through the chamber on the other side. The carrier gas delivers the permeating test gas to a detector, which can operate on the absorption or thermal conductivity principle [198]. A development in the carrier-gas technique was the insertion of a gas chromatograph column between the measuring cell and the detector [199].

To determine water-vapor permeability, an aluminum test cell filled with calcium chloride, for example, is employed as absorption chamber. The cell is closed by the specimen and sealed with wax at the edges. This apparatus is placed in an atmosphere having a well-defined temperature and humidity, and the weight gain of the absorption cell is measured at regular intervals [200], [201].

2.7. Stability Properties and Their Testing

The usefulness of plastics in many fields depends on their resistance to environmental factors. These complicated effects cannot always be simulated, and testing must often be accelerated. The loads applied are thus greater than would occur in the normal loading cycle. Accordingly, for short-term tests in particular, the correlation with in-service performance must be examined carefully and interpreted critically.

2.7.1. Water Absorption

Like other swelling agents (such as alcohols), water has a plasticizing effect on some plastics (polyamides, polyurethane). This action is based on the cleavage of hydrogen bonds between polar groups of adjacent macromolecules [202]. The result is a decrease in the glass-transition temperature of the amorphous regions in the plastic. The material, which is brittle and fragile at normal service temperatures and completely dry, becomes tough and flexible. Apart from the desired plasticization at room temperature, the dissolution of hydrogen bonds also leads to an increase in density, and thus modulus, at low temperature, although this effect generally has no practical importance.

Polymers whose main chain contains hydrolyzable groups (e.g., polyesters, polyamides) can be irreversibly damaged by water. In addition, water that has diffused into the material acts as a radical donor, thus promoting photochemical aging (see below) and, through the loosening of the polymer structure (hydrogen bond cleavage), facilitates the diffusion of oxygen and free radicals.

A number of standard test methods have been devised to measure water absorption quantitatively: water absorption in cold water [203], [204]; water absorption in boiling water [205], [206]; and behavior in humid air [207]. In the standards, water absorption is measured relative to a well-defined dry condition, to the dry condition with allowance for constituents lost to the water, or to the as-delivered condition of a specimen. Water absorption is stated as a percent weight change or as the absolute weight change in milligrams.

2.7.2. Effect of Chemicals

Water uptake and solubility in water are just one aspect of the more general concept of chemical resistance. The reason for treating the action of water on plastics separately is the ubiquity of this medium.

Plastics are resistant to the action of many chemicals, with the exception of certain classes of compounds. Organic solvents are among the most corrosive compounds towards plastics, since the plastics are themselves organic materials. Generally, nonpolar plastics are sensitive to nonpolar solvents, and polar plastics to polar solvents.

The standards for determining chemical resistance are ISO 175 and DIN 53 476 [208], [209]. For unstressed specimens, all that is necessary is to investigate how factors such as absorption, extraction, and chemical attack alter physical properties. Detailed tables of chemical resistance can be found in [210], [211].

Plastics in service are often subject to external mechanical stresses or residual stresses from production. A plastic may be stable against a given reagent when unstressed but fail in contact with the same reagent under stress. This phenomenon, called stress cracking, results from purely physical processes, in contrast to the stress corrosion cracking of metals (see also → Corrosion, 1. Electrochemical). The three most important processes are wetting, diffusion, and swelling. Wetting and swelling can increase the notch stress at weak points and microcracks that are always present in the material; diffusion is crucial for the transport of the medium to such weak points.

Methods for studying stress cracking are not completely analogous to practical experience. For example, the Bell test, specially conceived for polyethylene (PE), is very controversial. The Plax test described in ASTM 2561 has been devised for the evaluation of hollow articles made of PE. The setup enables hollow articles to be tested under constant internal pressure. The crack-opening medium is a 10 % aqueous solution of polyoxyethylnonylphenol. Cracking is detected with a simple electrical instrument.

2.7.3. Effects of Heat, Light, and Weathering

In DIN 50 035, Sheet 1, aging is defined as the totality of chemical and physical processes occurring irreversibly in a material over time. In this sense, the consequences of heat, light, and weathering are aging processes. The definition also implies that time is a significant variable in testing and, hence, that aging takes place in not spontaneous processes but relatively slow ones whose effects are seen only over a time span comparable to the service life. Thus the above definition excludes, for example, irreversible processes that are associated with processing operations [212].

Oxidation always accompanies aging due to heat and light. The most important apparatus for tests of thermal aging is the furnace. It can be operated with air recirculation and no fresh air, with fresh air circulation, or with fresh air metered into the recirculating air. Specimens can also be thermally aged by heating in a metal block, heating bath, or thermal analysis apparatus.

Light and Weathering. Weathering includes the action of many components: heat, water vapor, visible light, ultraviolet radiation, and the effects of oxygen and ozone. The synergism among these processes is often difficult to predict. In addition, the varying intensities of the components can have a more severe effect than a steady high level. The variability of the weather at any location and the difference between climatic regions render reliable correlations between an accelerated weathering test and natural weathering doubtful. Only when specimens are subjected to natural processes do weathering tests give reliable results. The possibility of accelerating the test consists in selecting a suitable weathering location. The three most important climatic regions for open-air weathering are temperate (e.g., Germany, with a mean global radiation of 400 MJ m^{-2} a^{-1}), warm-dry (Phoenix, Arizona, ca. 7900 MJ m^{-2} a^{-1}), and warm-humid (Bombay, 6300 MJ m^{-2} a^{-1}). The part of the spectrum responsible for photochemical aging of plastics ($\lambda \leq 380$ nm) accounts for only 4 – 6 % of global radiant energy; the remainder is transformed to heat when it is absorbed by the specimen. Open-air weathering tests are described in [213–215]. Resistance to natural weathering is expressed as the change in properties tested as a function of irradiation. Weatherfastness in the natural weathering test is expressed as the color change.

In many cases, open-air testing is done to determine not the weathering resistance or weatherfastness, but merely the stability to light or lightfastness (change in color only). These tests are governed by ISO 877 [216]. Color changes can be evaluated with the Woll lightfastness scale described in DIN 5003 or ISO 105 [217] or by the (more laborious) spectral method of DIN 5033 Sheet 4 or Sheet 6 (see Section 2.5.3).

Artificial weathering and artificial illumination allow a fairly constant radiation flux to be obtained, so that irradiation is roughly proportional to time. Various types of lamps are employed. High-pressure mercury-vapor lamps are suitable for testing if short wavelengths are filtered out. Carbon arc lamps are still used, even though their spectrum bears no resemblance to sunlight. Xenon arc lamps have recently found wide use in the plastics industry. With suitable filters, sunlight can be modeled well by this light source, although the acceleration over sunlight is not very great. To take one example, the action of xenon light on ABS is about fives times that of sunlight.

Although accelerated tests (i.e., simulated or artificial weathering and illumination) do not yield absolute results as to weather- or lightfastness, a number of standard methods have been in use for years [218–220]. Figure 45 shows how the impact strength of impact-resistant polystyrene decreases with time under xenon arc light. A comprehensive survey of the measurement of aging criteria, the extrapolation to determine service life, and the problem of reference standards is given in [212].

2.7.4. Effect of High-Energy Radiation

The service properties of all organic polymers are impaired when they are exposed to sufficient doses of high-energy radiation. The type of radiation (e.g., electrons, gamma rays, or reactor radiation) has no influence; only the absorbed dose of energy is important.

The dose rate may, however, be significant in the presence of atmospheric oxygen. In polymers that mainly undergo radiation cross-linking in the absence of oxygen, chain degradation occurs by radiative oxidation, leading to a loss of stability. Oxygen has little effect on polymers that undergo primarily radiative degradation. Degradation by radiative oxidation depends on time, and thus on dose rate, because it is determined on the one hand by oxygen diffusion into the material and on the other by decay of the peroxide radicals formed [221], [222]. The density of the polymer also plays a role in diffusion.

Plastics can be ranked in terms of their stability to high-energy radiation by means of half-value doses. Measurement of the half-value dose for the elongation at rupture has proved useful. Thus, the half-value dose is the dosage, in grays (1 Gy = 1 J/kg = 100 rad), at which the ultimate elongation falls to half the starting value (before irradiation). Figure 46 illustrates the situation with respect to tensile strength and elongation at rupture. At a certain dose rate (e.g., 2 kGy/h), the elongation reaches half of its starting value at a dose of 32 kGy. The tensile strength declines in a similar way, but drops to half the starting value only at 60 kGy.

The stability to high-energy radiation often depends strongly on the stabilizing system of the polymer; see Table 8 [221], which lists half-value doses (elongation) for polypropylene with various stabilizers. WILSKI points out that all the values in this table are "correct" in themselves, but only the value for the product with no stabilizer represents the true value applicable to this product type.

WÜNDRICH [222] and WILSKI [224] have provided a good introduction to this field, presenting detailed summaries of results with extensive references to other publications and values.

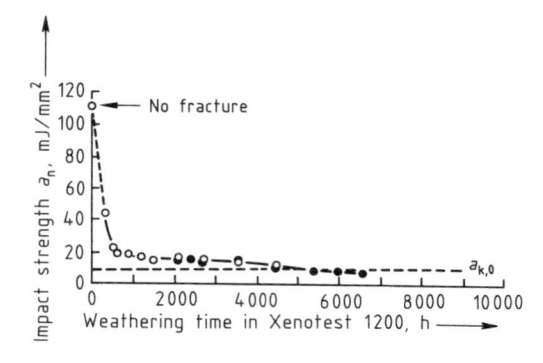

Figure 45. Impact toughness of high-impact-strength polystyrene (SB) as a function of illumination time in the Xenotest 1200 apparatus [212] $a_{k,0}$ is the notched impact strength of the unweathered probe

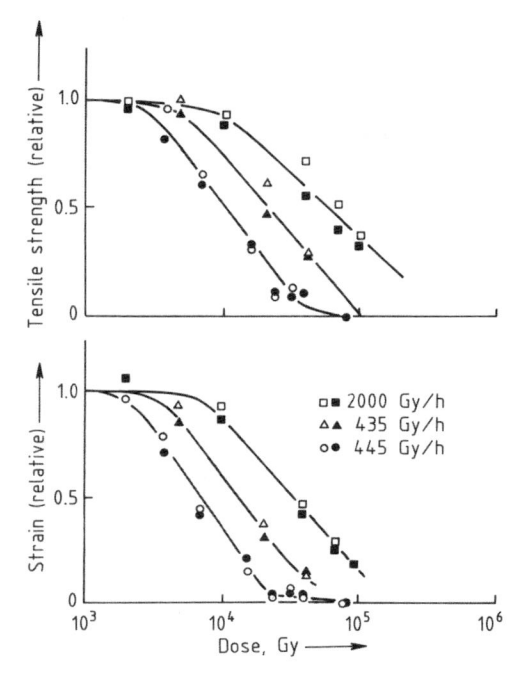

Figure 46. Tensile strength and elongation at rupture of polypropylene monofilaments (0.4-mm diameter) irradiated in air at various dose rates Initial tensile strength: 507 N/mm² (solid circles); 397 N/mm² (open circles) [223]

2.7.5. Stability against Fire

The combustion of plastics cannot be described quantitatively. It includes a number of steps, some of which have not yet been studied. For combustion to begin, three events are necessary: heating, decomposition, and

Table 8. Half value dose (extension) of Hostalen PPN polypropylene [223] *

Stabilizer system	Half-value dose, 10^3 Gy
No stabilizer	5
N,N'-Di-β-naphthyl-p-phenylenediamine (DPPD)	6
Trilauryl phosphite	6
Irganox 1076	14
Irganox 1010	15
Ionox 330	17
Phenothiazine	22
Ionol	26

* Dose rate 24 Gy/h in air, ^{60}Co radiation, molded sheets 0.5 mm thick. All specimens contain 0.2 % calcium stearate, stabilizer concentration 0.5 %.

Table 9. Decomposition temperature ranges T_d, flame initiation temperatures (FIT), self ignition temperatures (SIT), and heats of combustion ΔH for some plastics [225]

Product (abbreviations from DIN 7728)	T_d, °C	FIT, °C	SIT, °C	ΔH, kJ/kg
PE	340 – 440	340	350	46 500
PP	320 – 400	320	350	46 000
PS	300 – 400	350	490	42 000
PVC	200 – 300	390	450	20 000
ABS		390	480	36 000
PMMA	180 – 280	300	430	26 000
PA 6	300 – 350	420	450	32 000
PA 66	320 – 400	490	530	32 000
PTFE	500 – 550	560	580	
PAN	250 – 300	480	560	
Cellulose	280 – 380			17 500
Cotton		210	400	17 000

ignition of the polymer. Table 9 lists the decomposition, flame initiation, and self ignition temperatures, along with the heats of combustion, for several plastics. Thermoplastics tend to soften and melt in the presence of an ignition source. The intensity of burning, smoke evolution, and the formation of residues depend on the chemical structure.

Because of their three-dimensional cross-linked structure, thermosets burn without first softening and flowing. They also generate a smaller amount of decomposition products than thermoplastics. The action of heat on their surfaces results in carbonization. In some cases, ignition does not occur.

The requirements imposed on the fire performance of plastics, especially when they are employed in construction, electrical equipment, transportation, furniture, and appliances, mean that flame retardants must be used in many cases. Flame retardants fall into two classes, reactive and additive. Reactive flame retardants are chemically incorporated in the polymer molecule and are used chiefly in polyester and epoxy resins and in polyurethanes. They have a permanent action. Additive flame retardants are usually added only after polymerization. The use of high molecular mass flame retardants also has a permanent effect.

The most common flame retardants include chlorinated and brominated compounds, as well as phosphorus compounds. The literature contains discussions of synergistic effects between individual additives.

Fire-protection tests, also called test methods for evaluating fire performance, are so numerous – and the associated problems so complex – that only a few general points can be discussed here.

All authors who have dealt with this topic state that fire performance is not a material property but depends on flammability, flame propagation, heat evolution, and several associated phenomena: fume density, toxicity, and corrosiveness of the combustion gases [225]. A particularly important point is that general statements regarding flammability and combustion – such as flammable, flame resistant, combustible, self-extinguishing, difficultly flammable, and the like – without reference to a specific test method, can lead to serious misinterpretations. A numerical value referred to a specific test method or to an unambiguous code representing flammability test results must always be given.

Tests are performed on bench, laboratory, and full scales. Most of the standards relate to specific applications such as mining, electrotechnology, transportation, furniture and appliances, and construction.

Some important testing standards are listed below.

On a bench scale, the processes in incipient combustion are studied in *reaction to fire* tests. One such is described in ISO 1210.2 [226]: The flame from a Bunsen burner is tilted at 45° and applied to horizontal specimen. The plastic is assigned to one of three categories, depending on the flammability and the duration of burning, if any. The tests DIN 53 438 [227] and ASTM D 635–77 [228] are similar.

A somewhat controversial method is measurement of the *oxygen index*. The oxygen concentration at which the specimen just burns for 3 min is determined [229]. However, the oxygen content in the combustion mixture is unrealistic.

In electrotechnology, flammability measurements are material tests and product tests. The VDE standards include, in the first category, the Schramm–Zebrowski glow-bar test [230] and, in the second, the glowing-wire test [231].

Material-specific tests devised by Underwriters Laboratories (UL) in the United States have achieved worldwide recognition. Especially important test specifications include UL 94, which includes several fire safety tests on materials [232]. The UL 94 test is used in all plastics application areas except building construction.

Tests leading to the ratings 94 HB, 94 V-0, 94 V-1, and 94 V-2 are important. The letter "H" denotes horizontal clamping of the specimen, whereas "V" stands for vertical clamping. In the HB test, the flame from a Bunsen burner inclined at 45° is applied to the free end of the specimen. The evaluation criteria are burning rate or extinction of the specimen before a standard mark is reached.

In the V test, the specimen is suspended vertically and the flame from the upright burner is applied to its bottom end. The response is evaluated from continued burning, afterglow, and ignition of cotton wool by burning droplets. The best rating is 94 V-0. The test result depends on specimen thickness, which must therefore be stated.

The fire performance of plastics used in construction is evaluated in accordance with DIN 4102 [233]. This standard provides for rating as nonflammable (class A) or flammable (class B); flammable materials are further classified as difficultly flammable, normally flammable, and easily flammable. Most plastics, because of their organic nature, achieve only class B. In some cases, such as composites with inorganic materials, they are rated in class A.

2.7.6. Effect of Microorganisms

Synthetic polymers are generally not affected by microorganisms (bacteria, algae, and fungi). The picture is complicated by the presence of additives, reinforcements, plasticizers, and so forth. General instructions for the investigation of this problem in ISO 846 are set forth [234]. The result of any such action is characterized by visual description of the growth of the organisms. Two methods of study are available. In method A, determination is made of the extent to which plastics can serve as a nutrient substrate for microorganisms; in method B, the fungicidal properties of plastics are examined. See also [235].

2.8. Processing Properties and Their Testing

The processing of plastics requires a knowledge of the as-delivered properties of molding compounds, the flow behavior of thermoplastics, and the simultaneous flow and curing behavior during forming for thermosets. Molding compounds take the form of powders, granules, preformed pellets, liquids, or masses of doughy consistency; chips, sheets, and strips are sometimes encountered.

Important processing qualities of thermoplastics, curable molding compounds, and casting resins are identified and briefly described in the following material.

2.8.1. Molding Compounds

An essential condition for trouble-free automatic processing of thermoplastic and curable molding compounds is that the material flows freely. This quality is determined by measuring the time required for the compound to run through a funnel of a given size [236]. The discharge time of a powder can be influenced, for example, by the addition of aerosols.

The particle size and particle-size distribution govern the packing density and thus the filling of a completely positive mold or, in the case of a polymer in paste form [e.g., poly(vinyl chloride) in plasticizers], the formation of a gel. Screen analyses and visual counting methods are generally employed; microscopic examination is also used. Plastics in powder form are commonly analyzed on screens having openings of 63 – 2000 μm; the series usually has six screens with mesh sizes in a geometric series [237]. Often only two fractions are specified; in the case of vinyl chloride homopolymers and copolymers, for example, the oversize fractions on a 250-μm and a 63-μm screen are stated as the R (250) and R (63) values, respectively.

Particle-counting techniques based on optical measurements are important in the determination of particle-size distribution. In the laser granulometer, the beam from a helium–neon laser is guided through a cuvette that contains the specimen in dispersion form. The particles diffract the coherent beam, forming diffraction rings whose diameters are inversely proportional to the grain size. The size of the diffraction ring determines whether a particle is counted or not, depending on its spherical volume. The number of particles counted in a fraction (a given volume range) is converted to their volume and then to mass. In this way, the mass fraction is obtained as a function of the mean particle size (similar to screen analysis) [238].

The bulk density is informative in relation to metering. This value is the ratio of mass to volume when the molding compound is treated in a particular way (compacted, if apparent density is being measured). For long-strand fibers and chips, an apparent density is measured. The bulk density is easily determined with a filling funnel and a graduated flask.

The essential processing criterion for reaction resins is the viscosity as delivered; the viscosity after warm storage is also used for epoxy resins.

For reaction molding compounds, the time required to reach a given viscosity limit or the gelling time (open time, pot life) is measured under standard conditions [239], and the curing behavior (curing time) is determined.

The curing time is the time after which certain properties (e.g., heat deflection temperature, acetone-soluble matter, or styrene monomer content in the case of unsaturated polyester resins) no longer vary within certain limits.

2.8.2. Conventional Methods for Characterizing Flow Behavior

The only source of adequate information about the flow behavior of polymers is the flow curve (see Section 2.8.3). However, experimentally plotting a family of flow curves at a sufficient number of temperatures is laborious, and only in exceptional cases can this be done in the context of process monitoring. Attempts have therefore been made to characterize the flow behavior by simple, conventional measuring methods. This is especially appropriate for thermosetting molding compounds because curing takes place during processing and makes impossible the analysis of rheological phenomena in a simple way [240].

One of the most frequently used conventional test methods for the flow behavior of thermoplastics is determination of the melt flow index (MFI) or the melt volume index (MVI) [241]. For MFI, the quantity measured is the amount of polymer melt extruded from a nozzle in 10 min when a specified force is applied with a plunger (units of g/10 min). In automatic instruments, the plunger motion is measured as a function of time to give an MVI; the specimen volume expelled through the nozzle in a specified time under the stated conditions is determined (units of cm^3/10 min).

When the melt index values do not correlate with in-service behavior (chiefly for injection molding), the flow behavior is assessed on the basis of the spiral flow test. A specimen in the form of a spiral is made in an injection-molding machine; the flow qualities are evaluated in terms of the length of the spiral.

In production monitoring and control, the melt viscosity or melt index is measured continuously with bypass rheometers (Fig. 47) [242], [243].

The flow behavior of curable molding compounds is far more difficult to characterize than that of thermoplastics, because in thermosets a chemical curing reaction proceeds while flow continues during the production process. The names of the test methods for these resins reflect the shapes of the specimens molded in the test [244].

If information beyond that offered by these simple methods is required, the measurements must conform to well-defined and easily understood techniques, as in the case of thermoplastics. Such measurements have been described by KANAVEC and others [245], [246]. The variation of viscosity with time is determined in a rotational viscosimeter. Use of this method means that processing conditions can be varied to an extent comparable with plant practice. The Bauer and Eichler kneading-chamber method [247] and the Philips method [248] cannot be described in detail here.

2.8.3. Rheology of Plastics

Earlier (Section 2.8.2), the flow behavior of plastics melts was said to be describable only in terms of flow curves, that is, plots of shear stress versus shearing rate. The measurements are performed with simple shear deformation. Because deformations in processing nearly always involve extension, the extensional properties of the melt must often be investigated as well. For thermoplastics, the term melt refers to the condition above the glass-transition temperature for polymers that solidify in glassy form (e.g., polystyrene or polycarbonate) and to the condition above the maximum melting point of the crystalline regions for polymers that solidify in crystalline form.

The essential features in the rheology of thermoplastic melts are easily observable when a melt is extruded from a nozzle; see Figure 48 [249]. The behaviors illustrated at a, b, and c can be described as follows:

a. *Structural (non-Newtonian) viscosity.* The flow rate m increases faster than proportionally to the extrusion pressure p; that is, viscosity decreases as the mechanical load applied to the melt increases.
b. *Entropy elasticity* The molecules become oriented by the flow in the nozzle. This corresponds to a rubberlike elastic deformation, which is canceled as soon as the force is released outside the nozzle. The extrudate has a much larger diameter than the nozzle. This phenomenon is called extrudate swelling.
c. *Viscoelasticity* The coupling between viscous and elastic processes is time dependent.

This behavior implies that melts (and also solutions) of polymers are non-Newtonian, rubberlike, viscoelastic liquids.

Shear Viscosity. The flow behavior during shear deformation is described by the shear stress τ and the time derivative of the shear

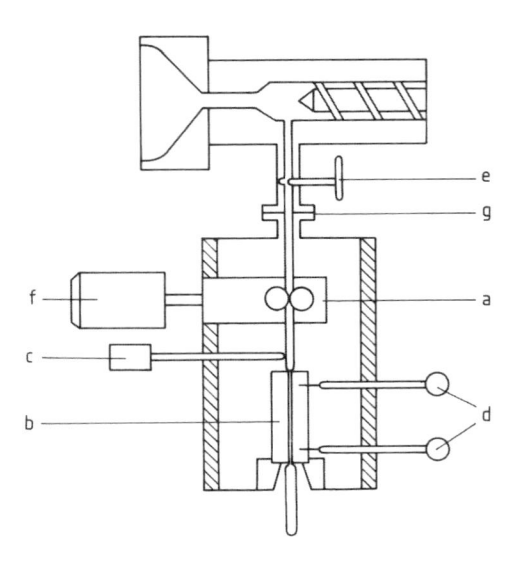

Figure 47. Göttfert bypass rheograph, open one-pump system with independent pressure and temperature controls
a) Melt pump; b) Nozzle; c) Pressure pickup; d) Temperature pickup; e) Shutoff valve; f) Pump drive motor; g) Flange

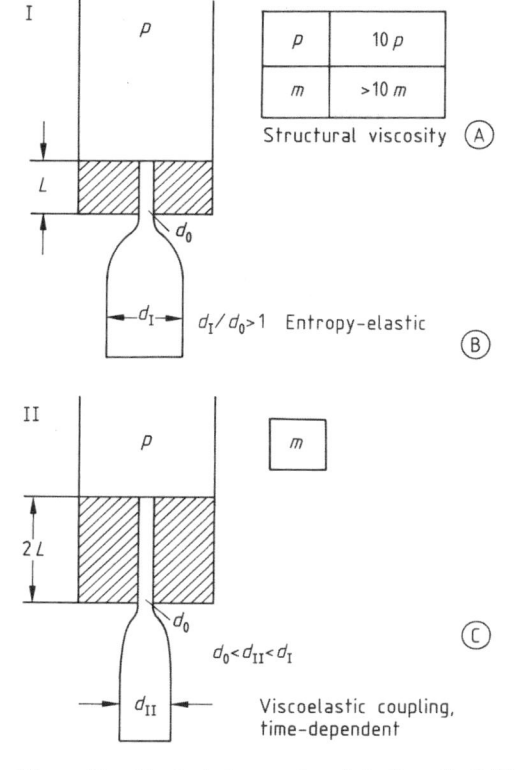

Figure 48. Rheological properties of plastic melts [249]

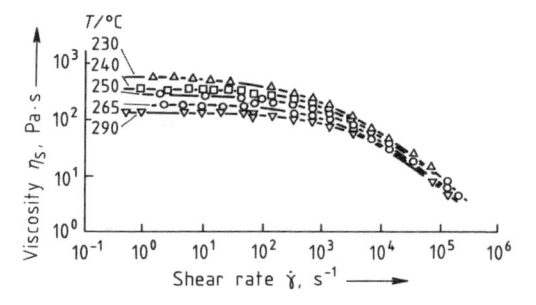

Figure 49. Viscosity of a polyamide 6 melt at various temperatures

γ, that is, the shear rate $\dot{\gamma}$. In addition to the nonlinear formulas $\tau = \mathbf{F}(\dot{\gamma})$ and $\dot{\gamma} = \mathbf{f}(\tau)$, a viscosity η is also defined for polymers: $\eta = \tau/\dot{\gamma}$.

This viscosity is independent of the shear rate only at very low values of the latter; that is, only in this region does a polymer melt behave like a Newtonian liquid (linear flow region). At the high shear rates used in plastics processing, viscosity decreases rapidly. The high rate of deformation obviously changes the tangled-chain structure in the melt. This concept gives rise to the term "structural" viscosity. Thus, quite generally, $\tau = \eta_s(\dot{\gamma})\,\dot{\gamma}$, where η_s is the structural viscosity.

The shear-rate-dependent viscosity at very low shear rates is called the zero-shear viscosity:

$$\eta_0 = \lim_{\dot{\gamma} \to 0} \eta_s(\dot{\gamma})$$

Figure 49 shows curves of viscosity versus shear rate for a Polyamide 6 melt at various temperatures.

In both the linear and the nonlinear regions, viscosity decreases with increasing temperature; see also [250].

The zero-shear viscosity η_0 is strongly dependent on the molecular mass above a critical value $M_r = M_c$. For commercial polymers, which always have $M_r > M_c$, this strong dependence of η_0 on M_r also manifests itself in the viscosity function $\eta(\dot{\gamma})$. The molecular-mass distribution, on the other hand, influences the elastic phenomena associated with viscous flow. Extrudate swelling becomes increasingly marked as the distribution becomes broader. The form of the viscosity function stays the same as molecular mass varies but changes with molecular-mass distribution.

Flow curves are usually determined with capillary viscosimeters. A disadvantage of the generally very convenient capillary viscosimeter is that the shear rate in the melt is not constant from point to point. Shear stress and shear rate increase from the center of the capillary toward the wall. The shear rate is zero at the capillary axis (see Fig. 50). In contrast, all particles of the liquid can be subjected to the same shear rate in rotational viscosimeters. This uniformity of shear rate cannot be achieved in a capillary.

Because it is easy to use, however, the capillary viscosimeter is employed widely in the roudetermination of flow curves. A standard method has been devised so that the results obtained can be compared [252]. The melt is forced through a capillary of specified size; either the test pressure at a specified volume flow rate or the volume flow rate at a specified test pressure is determined. In general, one charge in the cylinder yields several pairs of volume flow rate and test pressure values. The

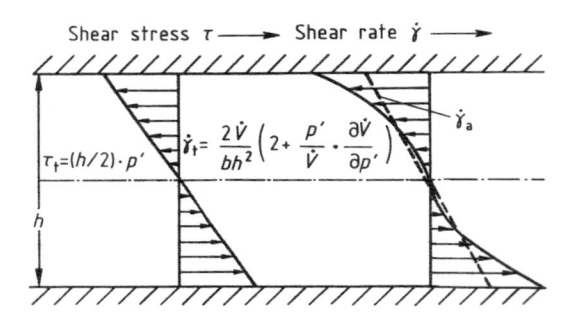

Figure 50. Steady-state shear flow in a flat-slot capillary [251] τ_t = true shear stress at wall; $\dot{\gamma}_t$ = true shear rate at wall; $\dot{\gamma}_a$ = apparent shear rate at wall; V = volume flow rate; p' = axial pressure gradient; b = capillary width Dashed curve: Newtonian liquid; solid curve: structurally viscous liquid

$$\tau_t = (h/2) \cdot p'$$

$$\dot{\gamma}_t = \frac{2\dot{V}}{bh^2}\left(2 + \frac{p'}{\dot{V}} \cdot \frac{\partial \dot{V}}{\partial p'}\right)$$

apparent shear rate $\dot{\gamma}_a$ is calculated from the flow rate and the capillary diameter, and the apparent shear stress τ_a is calculated from the capillary dimensions and the test pressure. The form $\dot{\gamma}_a = f(\tau_a)$ is recommended for flow curves used in comparisons (e.g., in quality control). The capillaries must have identical dimensions. The use of apparent flow curves and viscosity curves for the design of flow channels should be avoided. The Weissenberg and Rabinowitsch correction method [253], which cannot be discussed further here, gives the shear stress at the capillary wall (true shear stress τ_t) and the true shear rate $\dot{\gamma}_t$. The geometry-independent flow curve $\dot{\gamma}_t = f(\tau_t)$ and the viscosity function $\eta(\dot{\gamma}) = \tau_t/\dot{\gamma}_t$ are obtained from these values. The functions cited are indeed measured for the wall layer only, but they are valid in the bulk of the liquid as well, so that the subscript t can be omitted in the notation (units are s^{-1} for shear rate, Pa for shear stress, and Pa · s for viscosity). If the measured flow curves or viscosity functions are to be used for calculating steady-state shear flows in plastics-processing machinery, it is desirable to derive a mathematical approximation to the measured curves. Two approaches have proved useful: the Ostwald/de Waele power law approximation and the Carreau approximation [254], [255]. A large number of flow curves for commercial thermoplastics have been collected in [256].

Extensional Viscosity. During many processes employed in the plastics industry, such as melt spinning, film blowing, blow molding, coating at high takeoff speed, and injection molding, the melt is subjected not only to shearing but also to extension. Often the last deformation process performed on the melt before it hardens is an extension process that establishes the molecular orientation in the finished article. In contrast to low-viscosity Newtonian liquids with low tensile strength, plastic melts can be extended greatly without rupturing. The total extension consists of an entropy-elastic flow component and a viscous flow component; the ratio of the two components depends strongly on the external conditions and the product. The extensional deformation ε is characterized by the length change dl relative to the instantaneous length l: dε = dl/l (Hencky's measure).

Thus the Hencky extension ε of a rod with initial length l_0 is given by $\varepsilon = \int dl/l = \ln(l/l_0) = \ln\lambda$. The Hencky extension is thus equal to the natural logarithm of the degree of stretching λ, and the rate of extension is $\dot{\varepsilon} = \dot{l}/l_0$.

The measurement of stress – strain or tensile creep curves of melts is described in [256], [257]. An arrangement described by Münstedt is shown in Figure 51, whereas Figure 52 shows

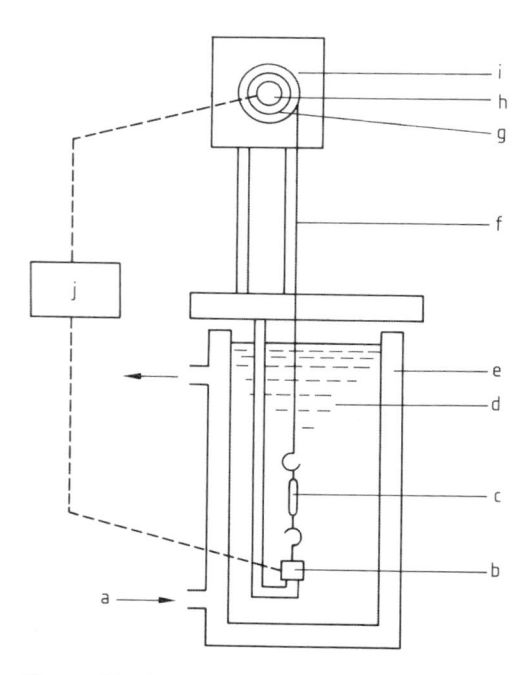

Figure 51. Münstedt extensional viscosimeter [256] a) Heating liquid; b) Load-measuring cell; c) Specimen; d) Oil bath; e) Temperature-controlled vessel; f) Belt; g) D.c. motor; h) Strain gauge; i) Takeup spool; k) Monitoring and control unit

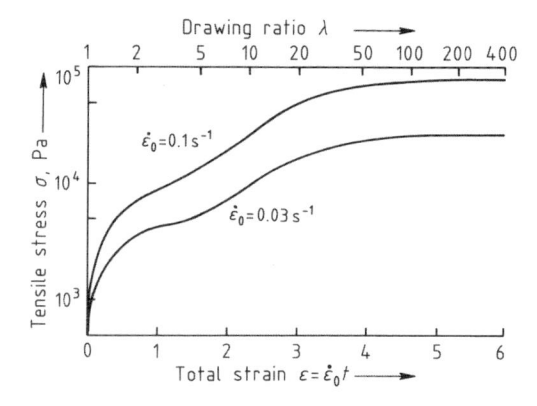

Figure 52. Stress – strain curves for low-density polyethylene melt at 150 °C for various constant rates of extension

stress – strain curves of a low-density polyethylene specimen at various constant extension rates [258].

The extensional viscosities derived from stress-strain curves are plotted as a function of either extension at constant tensile stress σ_0 or extension at constant extension rate $\dot{\varepsilon}$. The extensional viscosity $\eta_F(t)$ is given by $\eta_F(t) = \sigma(t)/\dot{\varepsilon}_0$, where $\sigma(t)$ is the measured tensile stress relative to the true cross-sectional area at time t, and $\dot{\varepsilon}_0$ is the extension rate determined in an extensional rheometer, which is constant over time; see also [259].

Finally, some flow anomalies of importance in plastics processing should be mentioned [259]. When a critical flow rate or wall shear stress value is exceeded, rough areas, termed melt fractures, are seen on the extrudate surface. These phenomena cannot be attributed to the critical turbulence of a flow, determined by the Reynolds number; they are due to the viscoelasticity of the melt (elastic turbulence). The onset of elastic turbulence is associated with a critical state of stress. This critical stress decreases as the molecular-mass distribution of the melt broadens. Modifications in the nozzle inlet reduce the occurrence of melt fracture [260].

If the melt is subjected to extensional loading at the nozzle outlet, periodic rupturing of the extrudate surface can be induced. The scaly surface is referred to as sharkskin. A periodic alternation of wall adhesion and wall slipping is observed at a critical wall stress in the nozzle, particular in high-density polyethylene melts. This phenomenon can also lead to a surface structure resembling melt fracture.

Abbreviations

a	constant
a	thermal diffusivity
a_k	impact strength (notched)
a_k	stress concentration factor
A	amplitude
A	flow area
b	constant
b	heat penetration coefficient
B	static load
c_p	specific heat
C	elastic constants
$C(t)$	compliance
d	mechanical loss factor
d	thickness of layer
D	diffusion constant
D_r	"dispersive power"
e	strain
$\dot{e}$	strain rate
E	modulus of elasticity
$E_c(t)$	creep modulus
E_d	dielectric strength
$E(t), E_r(t)$	time-dependent modulus of elasticity
f	oscillating frequency
F	friction force
F_d	factor for damping effects
F_g	factor for dimensional effects
G', G	shear modulus
$h(\tau)$	distribution function of the relaxation time or relaxation spectrum
$H(\ln \tau)$	relaxation spectra
I	isotropic stress invariant
I	moment of inertia
I_0	incident light intensity
K	bulk modulus
l	length
L	thickness of composite
M	complex modulus
M'	dynamic modulus
M''	loss modulus
$M(t)$	relaxation modulus
M_i	spring constant
n_C	refractive index at C-line
n_F	refractive index at F-line
N	cycles of failure
N	normal force
P	permeability
P	differential operator
P_z	polarization direction

Q	differential operator	τ	relaxation time
t	loading time	τ	shear stress
T_c	contact temperature	τ_a	apparent shear stress
v	Abbé number	τ_t	true shear stress
α	linear coefficient of thermal expansion	ω	angular frequency
α	molecular polarizability		
β	volume coefficient of thermal expansion		
γ	shear		
$\dot{\gamma}$	shear rate		
$\dot{\gamma}_a$	apparent shear rate		
$\dot{\gamma}_t$	true shear rate		
δ	phase shift		
$\tan \delta$	loss factor		
ε	total strain		
$\dot{\varepsilon}$	strain rate		
ε_0	constant strain during relaxation test		
ε	dielectric constant		
ε	extensional deformation		
$\dot{\varepsilon}$	extension rate		
$\varepsilon'(\omega)$	dielectric constant (permittivity)		
$\varepsilon''(\omega)$	dielectric loss factor		
ε^*	complex dielectric constant		
$\varepsilon', \varepsilon'_r$	relative permittivity		
ε''	dielectric loss factor		
η	viscosity		
η_0	zero shear viscosity		
η_F	extensional viscosity		
η_S	structure viscosity		
λ	thermal conductivity		
λ	logarithm of the degree of stretching		
Λ	logarithmic decrement		
μ	friction coefficient		
ξ	integration variable		
ϱ	density		
ϱ_0	constant		
ϱ_D	volume resistivity		
σ	stress		
$\dot{\sigma}$	stress rate		
σ_0	constant stress during creep test		
σ_a	initial stress; stress amplitude		
$\sigma_{B/t}$	creep strength		
σ_{crit}	stress exeding a critical value		
$\sigma_{\varepsilon/t}$	"time yield limit"		
σ_m	mean stress		
σ_{max}	maximal stress		
σ_n	nominal stress		

References

1. H. P. Klug, L. E. Alexander: *X-ray Diffraction Procedures*, J. Wiley, New York 1974.
2. I. Voigt-Martin, *Adv. Polym. Sci.* **67** (1985) 194.
3. B. J. Berne, R. Pecora: *Dynamic Light Scattering*, J. Wiley, New York 1976.
4. G. D. Wignall: Neutron Scattering,*Encyclopedia of Polymer Science and Engineering*, 2nd ed., vol. 10, J. Wiley, New York 1987.
5. P. J. Flory, *J. Chem. Phys.* **17** (1949) 303.
6. P. J. Flory: *Principles of Polymer Chemistry*, Cornell University Press, Ithaca 1953.
7. G. Lieser, E. W. Fischer, K. Ibel, *J. Pol. Sci.* **13** (1975) 29.
8. F. S. Bates, S. B. Dierker, G. D. Wignall, *Macromolecules* **19** (1986) 1938.
9. F. S. Bates, G. D. Wignall, *Phys. Rev. Let.* **57** (1986) 1429.
10. G. D. Wignall, D. G. H. Ballard, J. Schelten, *Eur. Polym. J.* **10** (1974) 861.
11. A. M. Fernandez, L. H. Sperling, G. D. Wignall, *Macromolecules* **19** (1986) 2572.
12. E. W. Fischer, M. Stamm, M. Dettenmaier, P. Herschenröder, *ACS Polymer Preprints* **20** (1979) (1) 219.
13. D. G. H. Ballard, P. Cheshire, G. W. Longman, J. Schelten, *Polymer* **19** (1978) 379.
14. R. G. Kirste, W. A. Kruse, J. Schelten, *Makromol. Chem.* **162** (1972) 299.
15. R. G. Kirste, W. A. Kruse, J. Schelten, *Kolloid Z. Z. Polym.* **251** (1973) 919.
16. R. G. Kirste, W. A. Kruse, K. Ibel, *Polymer* **16** (1975) 120.
17. D. G. H. Ballard, G. D. Wignall, J. Schelten, *Eur. Polym. J.* **9** (1973) 965.
18. P. Herschenröder, Thesis, University of Mainz 1978.
19. K. P. McAlea, J. M. Schultz, K. H. Gardner, G. D. Wignall, *Macromolecules* **18** (1985) 477.
20. J. Schelten et al.,*Polymer* **17** (1976) 751.
21. J. M. Guenet, *Polymer* **22** (1981) 313.
22. P. J. Flory: *Statistical Mechanics of Polymer Molecules*, J. Wiley, New York 1969.
23. A. Guinier: *X-ray Diffraction*, W. H. Freeman, San Francisco 1963.
24. E. W. Fischer et al., *Faraday Discuss. Chem. Soc.* **68** (1979) 26.
25. J. T. Bendler, *Macromolecules* **10** (1977) 162.
26. M. Dettenmaier et al., *Macromolecules* **19** (1986) 773.
27. H. Ito, T. P. Russell, G. D. Wignall, *Macromolecules* **20** (1987) 2213.
28. J. M. O'Reilley, D. M. Teegarden, G. D. Wignall, *Macromolecules* **18** (1985) 2747.
29. W. Gawrisch, M. G. Brereton, E. W. Fischer, *Polym. Bull.* **4** (1981) 687.
30. M. Vacatello, D. Y. Yoon, P. J. Flory, *Macromolecules* **23** (1990) 1993.
31. B. C. Laskowski, D. Y. Yoon, R. McLeand, R. J. Jaffe, *Macromolecules* **21** (1988) 1629.
32. E. R. Haward: *The Physics of Glassy Polymers, Materials Science Series*, Appl. Sci. Publ., London 1973. S. E. Keinath,

R. L. Miller, J. K. Rieke (eds.): *Order in the Amorphous "State" of Polymers*, Plenum Press, New York 1987.
33. L. C. E. Struik: *Physical Aging in Amorphous Polymers and Other Materials*, Elsevier Publ., Amsterdam 1978.
34. E. A. Di Marzio, J. H. Gibbs, *J. Polym. Sci.* **A 1** (1963) 1417. E. A. Di Marzio, J. H. Gibbs, *Macromolecules* **9** (1976) 763.
35. J. Jäckle, *Rep. Prog. Phys.* **49** (1986) 171.
36. A. J. Kovacs, *Fortschr. Hochpolym. Forsch.* **3** (1966) 394.
37. J. D. Ferry: *Viscoelastic Properties of Polymers*, 3rd ed., J. Wiley, New York 1980.
38. D. W. van Krevelen: *Properties of Polymers, Their Estimation and Correlation with Chemical Structure*, 2nd ed., Elsevier, Amsterdam 1976.
39. J. Bandrup, E. H. Immergut: *Polymer Handbook*, 3rd ed., J. Wiley, New York 1989.
40. D. R. Paul, S. Newman: *Polymer Blends*, vols. 1 and 2, Academic Press, New York 1978.
41. I. M. Ward: *Mechanical Properties of Solid Polymers*, 2nd ed., J. Wiley, Chichester 1983.
42. U. Eisele, *Progr. Colloid Polym. Sci.* **66** (1979) 59.
43. H. Tadokaro: *Structure of Crystalline Polymers*, J. Wiley, New York 1979.
44. *Faraday Discuss. Chem. Soc.* **68** (1979), various papers and references therein.
45. M. Stamm, E. W. Fischer, M. Dettenmaier, P. Convert, *Faraday Discuss. Chem. Soc.* **68** (1979) 263.
46. P. J. Flory, *J. Am. Chem. Soc.* **84** (1962) 2857.
47. D. Y. Yoon, P. J. Flory, *Macromolecules* **17** (1984) 868.
48. E. W. Fischer et al., *J. Polym. Sci. Polym. Phys. Ed.* **22** (1984) 1491.
49. S. J. Spell, D. M. Sadler, *Polymer* **25** (1984) 739.
50. J. M. Guenet, *Polymer* **22** (1981) 313. J. P. Guenet, C. Picot, *Macromolecules* **16** (1983) 205.
51. S. J. Spells, A. Keller, D. M. Sadler, *Polymer* **25** (1984) 749.
52. B. Wunderlich: "Crystal Structure, Morphology," Defects, *Macromolecular Physics*, vol. **1**, Academic Press, New York 1973.
53. A. Turner Jones, A. J. Cobbold, *J. Polym. Sci. Polym. Lett. Ed.* **6** (1968) 539. D. R. Morrow, *J. Macromol. Sci. Phys.* **3** (1969) 53. A. Turner Jones, *Polymer* **12** (1971) 487.
54. I. G. Voigt-Martin, private communication.
55. D. C. Bassett: *Principles of Polymer Morphology*, Cambridge Univ. Press 1981.
56. I. M. Ward: "The Preparation, Structure and Properties of Ultra-high Molecular Flexible Polymers," *Adv. Polym. Sci.* **70** (1985) 1.
57. H. H. Kausch: *Polymer Fracture*, 2nd ed., Springer Verlag, Berlin 1987.
58. U. Eisele: *Introduction to Polymer Physics*, Springer Verlag, Berlin 1990.
59. A. Blumstein (ed.): *Polymeric Liquid Crystals*, Plenum Press, New York 1985.
60. R. A. Weiss, C. K. Ober (eds.): "Liquid-crystalline Polymers," *ACS Symp*, Ser. **435**, Washington 1990.
61. J. M. G. Dobb, J. E. McIntyre (eds.): *Adv. Polym. Sci.* **60/61** (1984), various papers in this volume.
62. M. Ballauff, *Angew. Chem. Int. Ed. Engl.* **101** (1989) 261.
63. D. Dutta, H. Fruitwale, A. Kohli, R. A. Weiss, *Polym. Eng. Sci.* **30** (1990) 1005.
64. O. Olabisi, L. M. Roberson, M. T. Shaw: *Polymer-Polymer Miscibility*, Academic Press, New York 1979.
65. L. A. Utracki, R. A. Weiss (eds.): *Multiphase Polymers: Blends and Ionomers*," *ACS Symp*, Ser.**395**, Washington 1989.
66. G. E. Molan (ed.): *Colloidal and Morphological Behavior of Blockcopolymers*, Plenum Press, New York 1971.
67. F. S. Bates, J. H. Rosedale, G. H. Frederickson, *J. Chem. Phys.* **92** (1990) 6255.
68. R. Fayt, R. Jerome, P. Teyssie, in [65] Chap. 2, pp. 38 – 66. References for Chapter 2.
69. J. Schmitz, E. Bornschlegl, G. Dupp, G. Erhard, *Plastverarbeiter* **39** (1988) 50.
70. H. Oberst in G. Schreyer (ed.): *Konstruieren mit Kunststoffen*, Hanser Verlag, München 1972, p. 444.
71. L. R. G. Treloar in H. A. Stuart (ed.): *Die Physik der Hochpolymeren*, vol. 4, Springer Verlag, Berlin – Göttingen – Heidelberg 1956, p. 294 and p. 311.
72. Lord Kelvin: *Encyclopedia Britannica*, London 1875
73. W. Voigt, *Abh. Ges. Wiss. Göttingen, Math. Phys. Kl.* **36** (1890).
74. A. J. Staverman, F. Schwarzl in [71] p. 1 – 95.
75. H. Oberst, L. Bohn, *Rheol. Acta* **1** (1961) 608 – 617.
76. W. Retting, *J. Polym. Sci. Polym. Symp.* **42** (1973) 605 – 615.
77. A. J. Staverman, F. Schwarzl in [71] p. 11 – 29.
78. DIN 53 444, Prüfung von Kunststoffen, Zeitstandzugversuch, Entwurf 1987
79. DIN 53 455, Prüfung von Kunststoffen, Zugversuch, 1988.
80. K. Oberbach: *Kunststoff-Kennwerte für Konstrukteure*, Hanser Verlag, München 1975.
81. W. Kaufmann, D. Hoffmann, *Kunststoffe* **59** (1969) 173.
82. DIN 53 445, Prüfung von Kunststoffen, Torsionsschwingungsversuch, 1986.
83. DIN 53 440, Biegeschwingungsversuch, Blatt 1 – 3, 1973.
84. B. Carlowitz, *Plastverarbeiter* **31** (1980) 84.
85. H. Oberst in G. W. Becker,J. Meißner,H. Oberst,H. Thurn (eds.): *Elastische und viskose Eigenschaften von Werkstoffen*, Beuth-Vertrieb, Berlin –Köln – Frankfurt 1963, p. 19.
86. K. Oberbach, *Kunststoffe* **79** (1989) 713.
87. DIN 53 455, Prüfung von Kunststoffen, Zugversuch, 1988.
88. DIN 53 452, Prüfung von Kunststoffen, Biegeversuch, 1988.
89. K. Oberbach, L. Rupprecht: "Wege zur Reduzierung des Aufwandes bei der Kunststoffprüfung und zur Förderung der Vergleichbarkeit der ermittelten Kennwerte," *Vortrag auf der Fachtagung "Möglichkeiten der Rationalisierung bei Kunststoff-Prüfungen"*, Süddeutsches Kunststoff-Zentrum, Würzburg 2.–3. 12. 1986.
90. DIN 53 454, Prüfung von Kunststoffen, Druckversuch, Entwurf 1988.
91. DIN 53 457, Prüfung von Kunststoffen, Bestimmung des Elastizitätsmoduls im Zug-, Druck- und Biegeversuch, 1987.
92. W. Schneider, R. Bardenheier, *Z. Werkstofftech.* **6** (1975) 269 – 280.
93. R. Bardenheier, *Z. Werkstofftech.* **8** (1977) 379 –388.
94. ISO 2039, Plastics and Ebonit, Determination of Hardness by Ball Indentation Method, 1974.
95. DIN 53 456, Prüfung von Kunststoffen, Härteprüfung durch Eindruckversuch, 1973.
96. ISO 868, Plastics, Determination of Indentation Hardness by Means of a Durometer (Shore Hardness), 1978.
97. DIN 53 505, Härteprüfung nach Shore A und D, 1973.
98. DIN 53 462, Prüfung von Kunststoffen, Bestimmung der Formbeständigkeit in der Wärme nach Martens, 1987.
99. DIN 53 461, Prüfung von Kunststoffen, Bestimmung der Formbeständigkeitstemperatur, 1987.
100. DIN 53 460, Prüfung von Kunststoffen, Bestimmung der Vicat-Erweichungstemperatur.
101. DIN 53 754, Prüfung von Kunststoffen, Bestimmung des Abriebs nach dem Reibradverfahren.

102 H. Hachmann, E. Strickle, *Kunststoffe* **59** (1969) 47.
103 J. C. Halpin, *Rubber Chem. Technol.* **38** (1965) 1007 – 1038.
104 J. C. Williams: *Fracture Mechanics of Polymers*, Horwood, Chichester 1984.
105 A. S. Argon, *Philos. Mag.* **28** (1973) 839 – 865.
106 G. Jacoby, Ch. Cramer, *Rheol. Acta* 7 (1968) 23 –51.
107 B. Rosen: "Phenomenological Aspects of Progressive Weakening in the Highly Viscoelastic Glasses" in H. Rosen (ed.): *Fracture Processes in Polymeric Solids*, J. Wiley & Sons, New York 1964.
108 P. Beardmore, S. Rabinowitz, *J. Mater. Sci.* **10** (1975) 1763 – 1770.
109 R. P. Kambour, *J. Polym. Sci. Part A* **2** (1964) 4159 – 4163.
110 R. P. Kambour, *Macromol. Rev.* **7** (1973) 1 – 154.
111 S. S. Sternstein, L. Ongchin, A. Silverman, *Appl. Polym. Symp.* **7** (1968) 175 – 199.
112 S. S. Sternstein, F. A. Myers, *J. Macromol. Sci. Phys.* **B 8** (1973)no. 3 and 4, 539 – 571.
113 L. Bohn, *Angew. Makromol. Chem.* **20** (1971) 129.
114 B. Carlowitz: *Untersuchungen zum Bruchverhalten ausgewählter thermoplastischer Kunststoffe bei Schlagbeanspruchungen*, Technisch-wissenschaftlicher Bericht des IKV Aachen (ed.), Aachen 1979.
115 ISO 180, Determination of the IZOD-Impact Resistance of Rigid Plastics, 1982.
116 ISO 179, Determination of the CHARPY-Impact Resistance of Rigid Plastics, 1982.
117 DIN 53 453, Prüfung von Kunststoffen, Schlagbiegeversuch, 1982.
118 H. Dietmann, *Arch. Eisenhüttenwes.* **40** (1969) 1011 – 1022.
119 H. Grimminger, U. Troltenier, *Materialprüfung* **4** (1962) 4.
120 W. Grobusch, H. Jesse, *Kunststoffe* **58** (1968) 648.
121 DIN 53 443 Blatt 1, Prüfung von Kunststoffen, Stoßversuch Fallbolzenversuch, 1975.
122 J. Cornfield, N. Mantel, *J. Am. Stat. Assoc.* **45** (1950) 181.
123 M. S. Bartlett, *J. Statistical Society (Supplement)* **VIII** (1946) 113.
124 DIN 53 443 Blatt 2, Prüfung von Kunststoffen, Stoßversuch Durchstoßversuch mit elektronischer Meßwerterfassung.
125 Schenk & Co., Darmstadt, personal communication.
126 A. S. Wolmir: *Biegsame Platten und Schalen*, VEB-Verlag für Bauwesen, Berlin 1962.
127 E. Hensen in E. Hensen,W. Knappe,H. Potente (eds.): *Handbuch der Kunststoff-Extrusionsanlagen*, vol. **2**, Hanser Verlag, München – Wien 1986, p. 277.
128 G. Wübken, *Plastverarbeiter* **26** (1975) 17 – 23.
129 H. H. Racke, Th. Fett, *Instn. Mech. Engnrs., Experimental Stress Analysis, Paper 39*, London
130 G. Kämpf, *Angew. Makromol. Chem.* **60/61** (1977) 292 – 346.
131 DIN 61 853, 61 854, 61 855, Textilglas für die Kunststoffverstärkung, 1975.
132 R. Taprogge, R. Scharwächter, P. Hähnel: *Faserverstärkte Hochleistungs-Verbundwerkstoffe*, Vogel-Verlag, Würzburg 1975.
133 DIN 4109 Teil 1 – 5, Schallschutz im Hochbau,1962 – 1979.
134 Th. Grewer, H. Wilski, *Kolloid Z. Z. Polym.* **226** (1968) 46.
135 H. Wilski, *Kolloid Z. Z. Polym.* **238** (1970) 426.
136 G. Nachtrab, *Kunststoffe* **60** (1970) 261.
137 W. Knappe, G. Nachtrab, G. Weber, *Kunststoffe* **62** (1972) 455.
138 DIN 51 005, Thermische Analyse, Begriffe.
139 W. Hemminger, H. J. Seifert, *Nachr. Chem. Tech. Lab.* **34** (1986) M 1.
140 H. Wilski, D. Strobl, Unveröffentlichte Ergebnisse.
141 B. Wunderlich, H. Baur: "Heat Capacities of Linear High Polymers," *Adv. Polymer Sci.* **7** (1970) 151.
142 K.-H. Illers, *Makromol. Chem.* **127** (1969) 1.
143 B. Wunderlich in E. A. Turi (ed.): *Thermal Characterization of Polymeric Materials*, "The Basis of Thermal Analysis," Academic Press, New York 1981.
144 W. Dietz, *Tech. Mess. ATM* V 9213–6(1976) 123,153, 195.
145 Y. Agari, T. Uno, *J. Appl. Polym. Sci.* **32** (1986) 5705.
146 DIN 52 612, Teil 1, Bestimmung der Wärmeleitfähigkeit mit dem Plattengerät.
147 W. Fritz, W. Küster in *Wärme- und Stoffübertragung*, vol. **3**, Springer Verlag, Heidelberg 1979, p. 156.
148 S. Goldfein, J. Calderon, *J. Appl. Polym. Sci.* **9** (1965) 2985.
149 I. Catie: Dissertation, Aachen 1972.
150 K. Ueberreiter, W. Klein, *Chem. Tech.* **15** (1942) 5.
151 K. H. Hellwege, H. Hennig, W. Knappe, *Kolloid Z. Z. Polym.* **183** (1962) 110.
152 K. Ueberreiter, G. Kanig, *Glas Hochvak. Tech.* **2** (1953) 218.
153 H. Wilski, *Kunststoffe* **54** (1964) 10.
154 V.-H. Karl, F. Asmussen, K. Ueberreiter, *Angew. Makromol. Chem.* **62** (1977) 145.
155 V.-H. Karl, F. Asmussen, K. Ueberreiter, *Makromol. Chem.* **178** (1977) 2649.
156 V.-H. Karl, F. Asmussen, K. Ueberreiter, *Progr. Colloid Polym. Sci.* **64** (1978) 97.
157 G. Menges, P. Thienel, *Kunststoffe* **65** (1975) 696.
158 DIN 53 482/VDE 0303, Teil 3/05, Prüfung von Werkstoffen für die Elektrotechnik; Messung des elektrischen Widerstandes von nichtmetallenen Werkstoffen, 1983.
159 DIN VDE 0303, Teil 4, Bestimmungen für elektrische Prüfungen von Isolierstoffen; Bestimmung der dielektrischen Eigenschaften, 1969.
160 DIN VDE 0303, Teil 13, Prüfung von Isolierstoffen, Dielektrische Eigenschaften fester Isolierstoffe im Frequenzbereich von 8,2 bis 12,5 GHz, 1986.
161 Rohde & Schwarz-Mitteilung,*Meßgeräte*, *Meß-systeme*, München1988.
162 S. Roberts, A. v. Hippel, *J. Appl. Phys.* **17** (1946) 610.
163 D. Kind, H. Kärner: *Hochspannungs-Isoliertechnik (für Elektrotechniker)*, Vieweg & Sohn, Braunschweig 1982.
164 DIN 53 481/VDE 0303, Teil 2, VDE-Bestimmungen für elektrische Prüfungen von Isolierstoffen; Durchschlagspannung, Durchschlagfestigkeit, 1974.
165 E DIN VDE 0303, Teil 2/IEC 15 A (CO) 52, Prüfung von Isolierstoffen, Durchschlagspannung und Durchschlagfestigkeit bei technischen Frequenzen, 1987.
166 H. J. Mair: *Isolier- und Mantelwerkstoffe aus thermoplastischen Kunststoffen in Kabel und isolierten Leitungen*, VDI-Verlag, Düsseldorf 1984.
167 H. J. Mair in G. Braun,G. W. Becker,B. Carlowitz (eds.): "Die Kunststoffe, Chemie, Physik, Technologie," *Kunststoff-Handbuch*, 2nd ed., vol. **1**, Hanser Verlag, München 1990, p. 760.
168 DIN VDE 0303, Teil 1/IEC 112, Verfahren zur Bestimmung der Vergleichszahl und Prüfzahl der Kriechwegbildung auf festen Isolierstoffen unter feuchten Bedingungen, 1984.
169 E DIN VDE 0303, Teil 10/IEC 587, Prüfung von Isolierstoffen; Hochspannungskriechwegbildung, 1987.
170 DIN 83 485, Teil 1/VDE 0303 Teil 7, VDE-Bestimmungen für elektrische Prüfungen von Isolierstoffen; Verhalten unter Einwirkung von Oberflächen-Glimmentladungen, 1974.
171 E DIN VDE 0303, Teil 7 IEC 15 B (CO) 65, Prüfung von Isolierstoffen; Oberflächen-Glimmentladungen, 1987.
172 DIN VDE 0303, Teil 6, Bestimmung der elektrolytischen Korrosion, 1968.

173 E DIN VDE 0303, Teil 6/IEC 426, Bestimmung der korrosionsbegünstigenden Wirkung durch isolierende Werkstoffe, 1987.
174 K.-H. Möbius, *Kunststoffe* **78** (1988) 53.
175 K.-H. Möbius, *Kunststoffe* **78** (1988) 345.
176 S. Roth, *Phys. Bl.* **40** (1984) 321.
177 H. J. Mair, S. Roth (eds.): *Elektrisch leitende Kunststoffe*, Hanser Verlag, München – Wien 1986.
178 H.-G. Elias, F. Vowinkel: *Neue polymere Werkstoffe für die industrielle Anwendung*, Kapitel 14, Elektrisch leitfähige Polymere, Hanser Verlag, München – Wien 1983.
179 ISO R 489, Plastics, Determination of the Refractive Index of Transparent Plastics, 1966.
180 DIN 53 491, Prüfung von Kunststoffen, Bestimmung der Brechungszahl und Dispersion, 1955.
181 G. L. Wilkes, *Adv. Polym. Sci.* **8** (1971) 91 – 136.
182 DIN 5036, Teil 1 bis Teil 4, Strahlungsphysikalische und lichttechnische Eigenschaften von Materialien,1977, 1978, 1979.
183 DIN 53 490, Prüfung von Kunststoffen, Bestimmung der Trübung von durchsichtigen Kunststoffschichten.
184 DIN 67 530, Reflektometer als Hilfsmittel zur Glanzbeurteilung an ebenen Anstrich- und Kunststoffoberflächen.
185 H. Haussühl, K. Hamann, *Farbe Lack* **64** (1958) 642.
186 Th. Kosbahn, *Farbe Lack* **70** (1964) 693.
187 J. S. Christie, R. S. Hunter, *SPE Coloring and Decorating of Plastics* **9** (1975) 33.
188 DIN 5033, Blatt 1 – 8, Farbmessung,1966 – 1980.
189 ASTM E 308–66, Recommended Practice for Spectrophotometry and Description of Color in CIE 1931. System.
190 BS 4727, Part 4, Terms Particular to Light and Color, 1971.
191 ISO 3558, Plastics, Assessment of the Color of Near White or Near Colorless Materials.
192 DIN 16 776, Kunststoff-Formmassen, Polyethylen (PE)-Formmassen Einteilung und Bezeichnung, 1984.
193 ISO 1183, Methods for Determining the Density of Plastics excluding Cellular Plastics.
194 DIN 53 479, Prüfung von Kunststoffen und Elastomeren, Bestimmung der Dichte, 1976.
195 H. Descheimer, O. Fuchs in R. Nitsche,K. A. Wolf (eds.): *Kunststoffe*, vol. **1**, Springer Verlag, Berlin – Göttingen – Heidelberg 1962, pp. 733 –740.
196 ISO 2056, Plastics, Determination of the Gas Transmission Rate of Films and Thin Sheets under Atmospheric Pressure, Manometric Method, 1974.
197 DIN 53 380, Bestimmung der Gasdurchlässigkeit, 1969.
198 R. A. Pasternak et al., *J. Polym. Sci. Polym. Phys. Ed.* **8** (1970) 467.
199 D. G. Pye et al., *J. Appl. Polym. Sci.* **20** (1976) 287.
200 ISO 1195, Plastics, Determination of the Water Vapour Transmission Rate of Plastics Films and Thin Sheets, Dish Method, 1970.
201 DIN 53 122, Blatt 1, Bestimmung der Wasserdampfdurchlässigkeit, Gravimetrisches Verfahren, 1974.
202 K.-H. Illers, *Makromol. Chem.* **38** (1968) 168.
203 ISO 62 and Appendix 1, Plastics, Determination of Water Absorption.
204 DIN 53 495, Prüfung von Kunststoffen, Bestimmung der Wasseraufnahme nach Lagerung in kaltem Wasser, 1973.
205 ISO 117, Plastics, Determination of Boiling Water Absorption, 1959.
206 DIN 53 471, Prüfung von Kunststoffen, Bestimmung der Wasseraufnahme nach Lagerung in kochendem Wasser, 1976.
207 DIN 53 473, Prüfung von Kunststoffen, Bestimmung der Wasseraufnahme in feuchter Luft, 1973.
208 ISO 175, Plastics, Determination of the Effecting of Liquid Chemicals, 1981.
209 DIN 53 476, Prüfung von Kunststoffen, Bestimmung des Verhaltens gegen Flüssigkeiten, 1979.
210 B. Dolezel: *Die Beständigkeit von Kunststoffen und Gummi*, Hanser Verlag, München – Wien 1978, pp. 373 – 625.
211 B. Carlowitz: *Kunststoff-Tabellen*, 3rd ed., Hanser Verlag, München – Wien 1986.
212 J. Voigt, Untersuchung der Wärme- und Lichtbeständigkeit in I 101 I, p. 937.
213 DIN 53 386, Prüfung der Wetterbeständigkeit im Naturversuch (Freibewitterung), 1982.
214 ISO 4607, Methods of Exposure to Natural Weathering, 1978.
215 ASTM D 1435–75, Outdoor Weathering of Plastics, 1979.
216 ISO 877, Determination of Resistance to Change upon Exposure under Glass to Daylight, 1976.
217 ISO 105, Textiles, Tests for Color Fastness, 1978.
218 ISO 4892, Plastics, Methods of Exposure to Laboratory Light Sources, 1981.
219 DIN 53 387, Prüfung von Kunststoffen, Kurzprü-fung der Wetterbeständigkeit, 1982.
220 DIN 53 389, Prüfung von Kunststoffen, Kurzprü-fung der Lichtbeständigkeit, 1974.
221 K. Wündrich, unpublished results, 1979.
222 K. Wündrich, *Radiat. Phys. Chem.* **24** (1985) 503.
223 H. Wilski, *Radiat. Phys. Chem.* **29** (1987) 1 – 14.
224 H. Wilski in [167] p. 954.
225 J. Troitzsch: *Brandverhalten von Kunststoffen*, Hanser Verlag, München – Wien 1981.
226 ISO 1210.2, Determination of Flammability of Plastics in the Form of Bars, 1970.
227 DIN 53 438, Teil 1 – 3, Verhalten beim Beflammen mit einem Kleinbrenner, 1977.
228 ASTM D 635–77, Rate of Burning and/or Extent and Time of Burning of Self-Supporting Plastics in a Horizontal Position, 1977.
229 C. P. Fenimore, F. J. Martin, *Mod. Plast.* **43** (1966) 144.
230 VDE 0304, Teil 3/5.70, Bestimmungen für Prüfverfahren zur Beurteilung des thermischen Verhaltens fester Isolierstoffe; Teil 3 Brennverhalten, 1970.
231 VDE 0471, Teil 2, VDE-Bestimmungen für die feuersicherheitliche Prüfung von elektrotechnischen Erzeugnissen, ihren Baugruppen und Teilen; Glühdrahtprüfung, 1975.
232 Underwriter Laboratories: UL 94, 2nd ed. 1973: Tests for flammability of plastics materials for parts in devices and appliances.
233 DIN 4102, Teile 1 – 8, Brandverhalten von Baustoffen und Bauteilen,1977, 1978.
234 ISO 846, Determination of Behaviour under the Action of Fungi and Bacteria, Evaluation by Visual Examination or Measurement of Change in Mass or Physical Properties, 1978.
235 H. M. Haldenwanger: *Biologische Zerstörung der makromolekularen Werkstoffe*, Springer Verlag, Berlin – Heidelberg – New York 1970.
236 DIN 53 492, Bestimmung der Rieselfähigkeit von körnigen Kunststoffen, 1977.
237 DIN 4188, Drahtsiebböden, 1977.
238 Deutsche Solvay-Werke GmbH, Solvay-Information, *Bestimmung der Korngrößenverteilung von Vinylchlorid (VC)-Polymerisaten*, Rheinberg 1987.
239 DIN 16 945, Reaktionsharze, Reaktionsmittel, Reaktionsharzmassen; Prüfverfahren, 1976.
240 E. P. Weißler in [169] p. 613.
241 DIN 53 735, Bestimmung des Schmelzindex von Thermoplasten, 1983.

242 K.-H. Moos, *Kunststoffe* **75** (1985) 3.
243 A. Göttfert, *Kunststoffe* **76** (1986) 1200.
244 W. Schönthaler: *Verarbeiten härtbarer Kunststoffe*, VDI-Verlag, Düsseldorf 1973.
245 J. F. Kanavec: Inst. f. techn.-ökon. Information, Thema 7 Nr. V 56 – 66 Moskau 1956.
246 P. Ehrentraut, W. Dalhoff, *Kunststoffe* **57** (1967) 439.
247 W. Bauer, K. Eichler, *Ind. Anz.* **90** (1968) 187.
248 G. J. P. Dujardin, *Kunststoffe* **61** (1971) 177.
249 J. Meißner: "Rheologie von Polymer-Schmelzen," vorgetragen beim *Rheologie-Fortbildungskursus* vom 12. – 16. 10. 1981 an der ETH Zürich.
250 G. V. Vinogradov, A. Y. Malkin, *J. Polym. Sci. Polym. Phys. Ed.* **4** (1966) 135.
251 VDMA (ed.): *Kenndaten für die Verarbeitung thermoplastischer Kunststoffe*, Part 2 Rheologie, Part 4 Rheologie 2, Hanser Verlag, München – Wien 1982, 1986.
252 DIN 54 811, Bestimmung des Fließverhaltens von Kunststoffschmelzen mit einem Kapillar-Rheometer, 1984.
253 Z. Rabinowitsch, *Phys. Chem.* **145 A** (1929) 1.
254 K. Geiger, H. Kühnle, *Rheol. Acta* **23** (1984) 355.
255 P. J. Carreau: Ph. D. Thesis, University of Wisconsin, 1968.
256 H. Münstedt, *J. Rheol. (N.Y.)* **23** (1979) 421.
257 J. Meißner, *Chem. Eng. Commun.* **33** (1985).
258 H. M. Laun, H. Münstedt, *Rheol. Acta* **15** (1976) 517.
259 H. M. Laun, in [169] p. 247.
260 F. Ramsteiner, *Kunststoffe* **62** (1972) 766.

Further Reading

M. Chanda, S. K. Roy: *Plastics fundamentals, properties, and testing*, CRC Press, Boca Raton, Fla. 2009.

B. J. Furches: *Plastics Testing*, "Kirk Othmer Encyclopedia of Chemical Technology," 5th edition, vol. 19, p. 563–596, John Wiley & Sons, Hoboken, NJ, 2006, online: DOI: 10.1002/0471238961.1612011906211803.a01.pub2.

C. A. Harper: *Handbook of Plastics Technologies*, McGraw-Hill, New York, NY 2006.

H. Zweifel, R. D. Maier, M. Schiller: *Plastics Additives Handbook*, 6th ed., Hanser, München 2009.

Plastics, Additives

RAINER WOLF, Sandoz Huningue S.A., Huningue, France

BANSI LAL KAUL, Sandoz Huningue S.A., Huningue, France

1.	Introduction	527
2.	Antioxidants	528
2.1.	Purpose and Requirements	528
2.2.	Chemical Classes	529
2.3.	Mechanism of Action	531
2.4.	Test Methods	532
2.5.	Uses	532
2.6.	Legal Aspects	534
2.7.	Economic Aspects	534
3.	Light Stabilizers	535
3.1.	Purpose and Requirements	535
3.2.	Chemical Classes	536
3.3.	Mechanism of Action	536
3.4.	Test Methods	545
3.5.	Uses	545
3.6.	Economic Aspects	546
4.	Heat Stabilizers for Poly (Vinyl Chloride)	546
4.1.	Purpose and Requirements	546
4.2.	Chemical Classes	547
4.2.1.	Metal-Containing Stabilizers	547
4.2.2.	Metal-Free Stabilizers	548
4.3.	Mechanism of Action	548
4.4.	Test Methods	549
4.5.	Uses	549
4.6.	Economic Aspects	550
5.	Lubricants, Slip, Antiblocking, and Mold-Release Agents	551
5.1.	Purpose and Requirements	551
5.2.	Chemical Classes	551
5.3.	Mechanism of Action	552
5.4.	Test Methods	552
5.5.	Uses	553
5.6.	Economic Aspects	554
6.	Flame Retardants	554
6.1.	Purpose and Requirements	554
6.2.	Chemical Classes	555
6.3.	Mechanism of Action	559
6.4.	Test Methods and Standards	562
6.5.	Uses	562
6.6.	Toxicology	565
6.7.	Economic Aspects	565
7.	Fillers and Coupling Agents	565
7.1.	Purpose and Requirements	565
7.2.	Forms and Chemical Structures	566
7.3.	Mechanism of Action	566
7.4.	Test Methods	567
7.5.	Uses	567
7.6.	Economic Aspects	569
8.	Dyes and Pigments	569
8.1.	Requirements	569
8.2.	Pigments	570
8.2.1.	Dyes	571
9.	Miscellaneous Additives	572
9.1.	Nucleating Agents	572
9.2.	Antistatic Agents	573
9.3.	Impact Modifiers	574
9.4.	Chemical Blowing Agents	575
9.5.	Optical Brighteners	577
	References	577

1. Introduction

As plastics find new applications, they face increasingly stringent requirements regarding their service life, durability, and many other properties. A variety of approaches are taken to meet these requirements. One approach is to modify the chemistry of the macromolecules from which plastics are made (e.g., by using new monomers or copolymerization). Current developments in the area of polymer blends also offer a great deal of scope for innovation. Another important way of improving the properties of plastics is to employ additives. These agents have made a decisive contribution to the widespread use of plastics and promise to be useful tools for solving future problems.

Ullmann's Polymers and Plastics: Products and Processes
© 2016 Wiley-VCH Verlag GmbH & Co. KGaA, Weinheim
ISBN: 978-3-527-33823-8 / DOI: 10.1002/14356007.a20_459

Plastics additives can be classified into three main groups:

1. Additives that stabilize plastics against degradation and aging during processing or in use. Degradation usually involves chain cleavage of the macromolecules and can proceed through the addition of energy (e.g., shear forces, heat, UV light) or chemical attack (e.g., oxidation, hydrolysis). These additives are called antioxidants, light stabilizers, or heat stabilizers.
2. Additives that facilitate or control processing (e.g., lubricants, mold-release agents, or blowing agents).
3. Additives that impart new, desirable qualities to plastics, such as resistance to burning, transparency or color, improved mechanical or electrical properties, dimensional stability, and degradability. Such additives include flame retardants, fillers, dyes, pigments, antistatic agents, nucleating agents, optical brighteners, impact modifiers, and plasticizers. Plasticizers are discussed elsewhere, → Plasticizers.

Surveys of additive groups are given in [1], [2]. New developments are published monthly in [3]. The total amount of additives (excluding fillers and pigments) used worldwide by the plastics industry in 1990 is estimated at ca. 1.7×10^6 t (Fig. 1).

The concentration of additives varies greatly, ranging from a few parts per million for some stabilizers to more than 50 % for certain flame retardants or fillers.

Although additives can be added to the monomer prior to polymerization, they are usually introduced immediately after polymerization, blended, and extruded to form granular (pelletized) products and compounds. Many additives are not introduced until the granules are processed into moldings, films, or fibers.

Additives with monomer functions can be incorporated chemically into the plastic if they are introduced before polymerization. In the case of agents added before processing, the use of "masterbatches" (i.e., granular products containing the concentrated additive in a polymer vehicle) is often advisable.

Plastics generally contain many additives. Preblended additive systems and combination masterbatches are commercially available; they contain optimal proportions of additives that are mutually compatible or have a synergistic action.

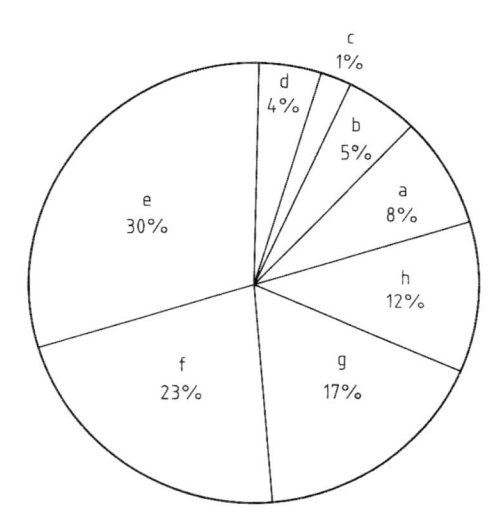

Figure 1. World consumption of plastics additives excluding fillers and pigments in 1990 (100 % = 1.7×10^6 t) a) Lubricants and related compounds; b) Antioxidants; c) Light stabilizers; d) Miscellaneous additives; e) Flame retardants; f) Plasticizers; g) Poly(vinyl chloride) heat stabilizers; h) Impact modifiers

2. Antioxidants

See also → Antioxidants.

2.1. Purpose and Requirements

Antioxidants prevent or retard the autoxidation of polymers and minimize associated damage (e.g., discoloration, reduction in gloss, cracking, embrittlement); i.e., they stabilize the physical properties of plastics. Oxidation reactions generally proceed via different mechanisms that depend on the structure of the polymer. They are often catalyzed by catalyst residues and contaminants. They are also accelerated by the addition of thermal or mechanical energy during plastics production and processing.

Three forms of stabilization are used: prestabilization, stabilization during processing, and long-term stabilization.

Most antioxidants are themselves oxidized and consumed in performing their function, so that the oxidation behavior of the additive in a given polymer is crucial for its effectiveness [4]. A number of other requirements apply to antioxidants:

1. They must be thermally stable and nonvolatile at processing temperatures
2. They must be soluble in polymers and no chalking should occur at service temperatures
3. They must not have an intrinsic color, and their oxidation products must have minimal color
4. Any acidic hydrolysis products must not corrode machinery
5. They must resist extraction
6. They must be odorless and tasteless
7. They must not create toxicity problems (many must be approved as indirect food additives)

2.2. Chemical Classes

Antioxidants are divided into two classes on the basis of their mode of action (see Section 2.3):

1. *Primary antioxidants*: sterically hindered phenols, secondary aromatic amines, and sterically hindered amines (HALS)
2. *Secondary antioxidants*: phosphites, phosphonites, thioethers (sulfides), and metal salts

Primary antioxidants can be used by themselves for prestabilization and long-term stabilization. Secondary antioxidants are used in combination with primary antioxidants, especially for stabilization during processing and for long-term stabilization under severe thermal conditions. Synergistic effects often occur in such systems (e.g., phenol and phosph(on)ites or phenol and thioethers). Antagonistic effects are also observed, however (e.g., between thioethers and HALS) [5].

The stabilizer system often includes other components that complex or neutralize degradation-promoting catalysts or acid traces. These may be metal deactivators (oxamides, hydrazones, and hydrazides) or acid scavengers [calcium or magnesium stearate, hydrotalcite (basic Mg/Al carbonate), and epoxy compounds].

Sterically Hindered Phenols. Industrial products are mostly derived from 2,6-di-*tert*-butylphenol (**1**; $n = 1$, R = H); or, less often, from 2-*tert*-butyl-6-methylphenol, 2-*tert*-butyl-5-methylphenol, or other hindered phenols. Tabulations are given in the literature [1], [6].

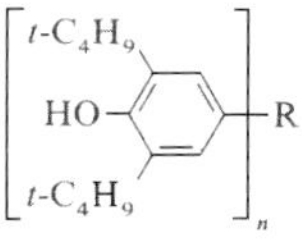

1

The simplest representative and oldest product of this class is 2,6-di-*tert*-butyl-4-methylphenol [*128-37-0*] (butylated hydroxytoluene, BHT) with $n = 1$ and R = CH_3. Variation of n from 1 to 4 and of R results in a range of hindered phenols with high molecular mass and thus low volatility, good solubility in certain polymers, and only slightly colored oxidation products. Products **2–4** are commonly used.

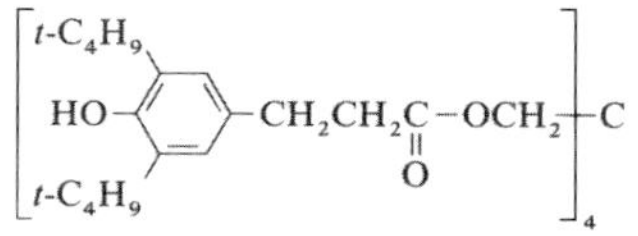

2 [*6683-19-8*]

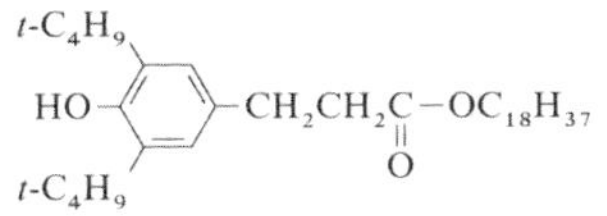

3 [*2082-79-3*]

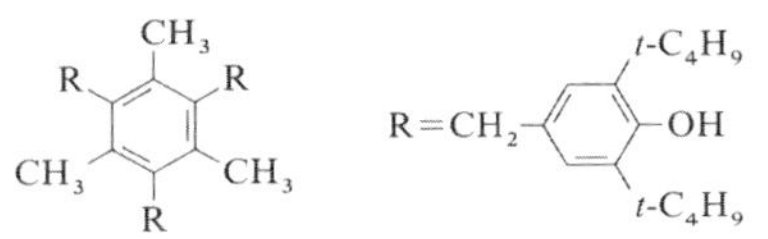

4 [*1709-70-2*]

The natural compound α-tocopherol [59-02-9] or vitamin E (**5**) is also used as an antioxidant [8], [8].

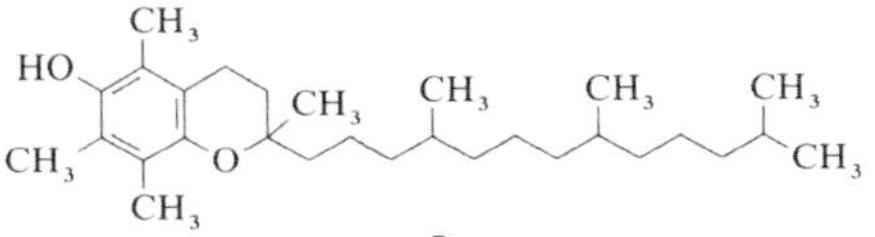

5

Secondary Aromatic Amines and Dihydroquinolines.
Industrial products are derived chiefly from diphenylamine or *p*-phenylenediamine. A widely used representative of this class is 4-isopropylaminodiphenylamine [101-72-4] (**6**):

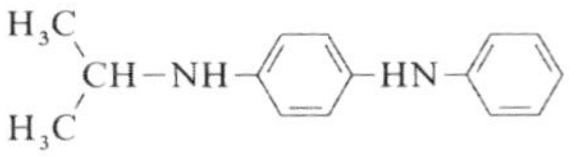

6

2,2,4-Trimethyl-1,2-dihydroquinoline [147-47-7] (**7**) should also be mentioned:

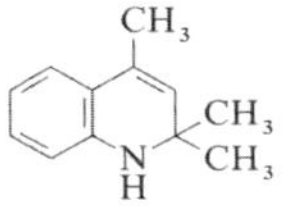

7

Sterically Hindered Amines (HALS).
The abbreviation HALS stands for "hindered amine light stabilizers." They are described in more detail in Chapter 3 and are based on the 2,2,6,6-tetramethylpiperidine structure.

Phosphites and Phosphonites.
The simplest industrial phosphite antioxidants are tris(nonylphenyl) phosphite [26523-78-4] (TNPP) (**8**) and tris(2,4-di-*tert*-butyl) phosphite [31570-04-4] (**9**).

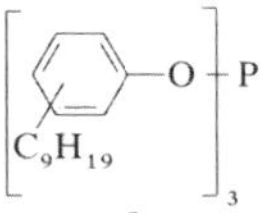

8

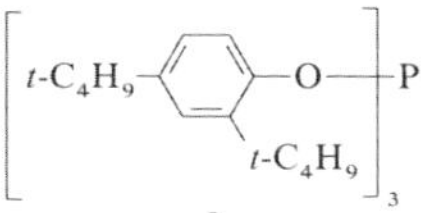

9

The cyclic phosphites **10**, **11** [9], and **12** [10] are also available:

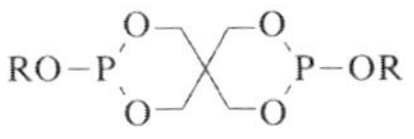

10a [26741-53-7], R = $\text{—}\!\!\!\bigcirc\!\!\!\text{—}$ *t*-C_4H_9
                             t-C_4H_9

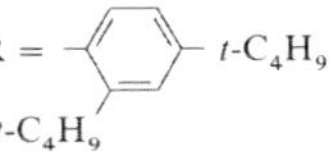

10b [3806-34-6], R = $C_{18}H_{37}$

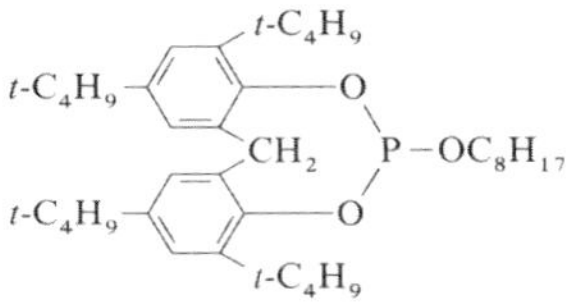

11 [126050-54-2]

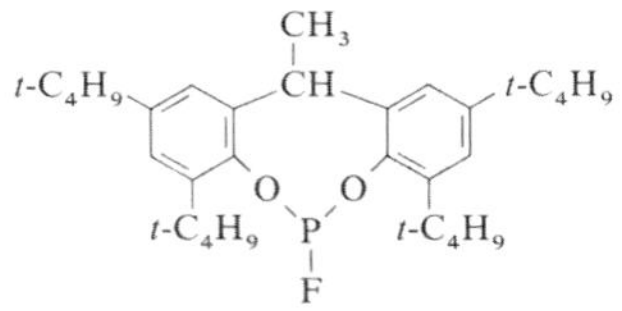

12 [118337-09-0]

Compound **13** is an industrially used phosphonite antioxidant [11]:

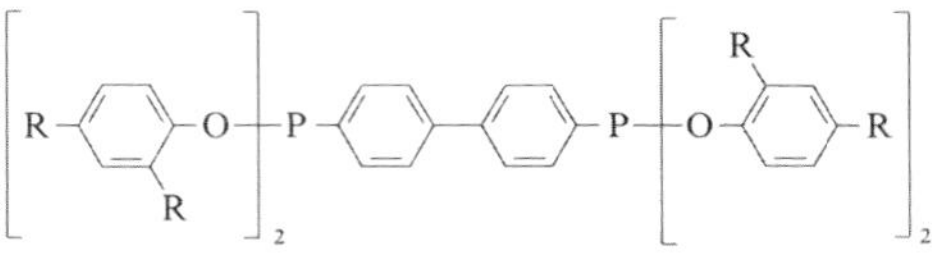

13 [38613-77-3], R = *t*-C_4H_9

The substituents influence not only the volatility and the solubility in polymers, but also the stability of the product to hydrolysis.

Organosulfur Compounds.
Organosulfur antioxidants in practical use are the dilauryl [123-28-4] (**14**, $n = 12$), dimyristyl [16545-54-3] (**14**, $n = 14$), and distearyl [693-36-7] (**14**, $n = 18$) esters of thiodipropionic acid (**14**) which are abbreviated as DLTDP, DMTDP, and DSTDP, respectively.

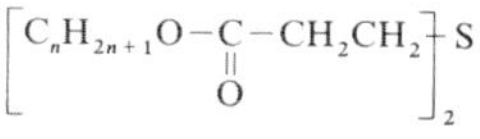

$$\left[C_nH_{2n+1}\text{—}O\text{—}\underset{\underset{O}{\|}}{C}\text{—}CH_2CH_2 \right]_2 S$$

14

Disulfides such as dioctadecanyl disulfide [*2500-88-1*] or higher molecular mass products such as **15** are used less often.

$$\left[C_{12}H_{25}-S-CH_2CH_2-\underset{\underset{O}{\|}}{C}-OCH_2 \right]_4 C$$

15 [*29598-76-3*]

Metal Compounds. Zinc and nickel dithiocarbamates or mercaptobenzimidazoles are sometimes used.

Metal Deactivators. The oxamides or hydrazine derivatives used as metal deactivators often have hindered phenol functions in the molecule as well, so that they also have a primary antioxidant action [1]. A typical structure is:

[Structure of compound **16** with t-C_4H_9 substituents on two phenol rings connected via $-CH_2CH_2-C(=O)-NHNH-C(=O)-CH_2CH_2-$ linker]

16 [*32687-78-8*]

2.3. Mechanism of Action

Thermooxidative degradation is an autocatalytic, free-radical chain reaction that involves initiation, propagation, branching, and termination stages. Primary radical sites R$^{\bullet}$ are initially formed by the abstraction of labile hydrogen atoms (e.g., tertiary or allylic hydrogens or those in the α-position to ether oxygen or amide nitrogen atoms). In the presence of oxygen, the primary radicals are oxidized to peroxy radicals ROO$^{\bullet}$. The peroxy radicals then take part in intermolecular (Eq. 1) and intramolecular (Eq. 2) chain propagation reactions to form new radical sites and hydroperoxides:

$$ROO^{\bullet} + RH \rightarrow ROOH + R^{\bullet} \qquad (1)$$

[Intramolecular chain propagation reaction showing rearrangement of peroxy radical to hydroperoxide] (2)

The chain propagation reactions are rate determining for thermooxidative degradation. Tertiary hydrogen atoms are about three times more reactive than secondary hydrogens. The hydroperoxides undergo thermal decomposition or participate in reactions catalyzed by trace metals to form oxy (RO$^{\bullet}$, HO$^{\bullet}$) or peroxy radicals (ROO$^{\bullet}$), which cleave additional bonds in the polymer and generate new radical sites in the chain or break the polymer chain to give carbonyl groups:

$$-R^1-\underset{\underset{O^{\bullet}}{|}}{\overset{\overset{R^2}{|}}{C}}-R^3- \longrightarrow -R^1-\overset{\overset{R^2}{|}}{C}=O + {}^{\bullet}R^3-$$

The extent of thermooxidative degradation in many polymers can be estimated via the concentration of peroxy radicals, hydroperoxide groups, and carbonyl groups. Macroradicals can also react with C=C bonds in the polymer, especially with vinyl groups [12], or can recombine and result in undesirable cross-linking (gel formation).

Antioxidants act by trapping free radicals formed in the polymer (primary antioxidants), reduce hydroperoxides to hydroxyl groups (secondary antioxidants), or form complexes with trace metals that would otherwise catalyze hydroperoxide decomposition (metal deactivators). Antioxidant mechanisms are described in [13–18]; see also → Antioxidants, Section 3.2.→ Antioxidants.

Hindered Phenols. Hindered phenols are radical-trapping antioxidants for oxy and especially peroxy radicals:

[Reaction of hindered phenol (OH with R^1 substituent) with RO$^{\bullet}$ + ROO$^{\bullet}$ giving phenoxy radical (O$^{\bullet}$) + ROH + ROOH]

The phenoxy radicals with their bulky substituents are stabilized by steric hindrance and cannot attack the polymer. These radicals are transformed to quinonoid structures that can recombine with other peroxy radicals:

[Reaction of quinonoid radical with ROO$^{\bullet}$ giving substituted quinonoid with OOR group]

The formation of yellow quinonoids is a drawback of this class. The intensity of the color depends on the structure of R^1; it is particularly high in the case of BHT due to the formation of stilbenequinones.

Secondary Aromatic Amines also act by trapping peroxy radicals. Since the products derived from secondary aromatic amines are generally dark in color, their use is restricted to colored elastomers. Their stabilizing action against ozone degradation should be noted.

HALS are used primarily as light stabilizers. They can trap both carbon radicals and peroxy radicals at ambient temperatures. The active species is a hindered nitroxyl radical, which is regenerated (see Chap. 3). HALS are effective long-term stabilizers [19], [20]; nonvolatile, high molecular mass derivatives are especially useful [21].

Phosphites and Phosphonites. The secondary phosphite and phosphonite antioxidants reduce hydroperoxides and are themselves oxidized to phosphates and phosphonates:

$$P^{III} + ROOH \rightarrow P^V + ROH$$

Peroxy radicals are analogously reduced to oxy radicals. Phosphites and phosphonites may also complex and thus deactivate trace metals (e.g., polymerization catalyst residues). Their oxidation products are colorless, so they are particularly successful as color stabilizers. Their action is especially useful in the processing of polymer melts, where they also protect phenolic antioxidants against premature oxidation.

Organosulfur Compounds decompose hydroperoxides. Sulfides, for example, are converted to sulfoxides, sulfenic acids, and sulfur dioxide, which in turn can reduce further hydroperoxides. Sulfur-containing stabilizers can also regenerate phenolic antioxidants from their quinonoid structures. Their action is most pronounced at higher temperatures in solid plastics. The odor of the reaction products can be a drawback.

Metal Deactivators act by chelating metal ions [22], particularly metals that have a number of oxidation states and catalyze hydroperoxide decomposition even at low temperatures. These metals include copper (from cablewires), iron, manganese (from fillers), and titanium (from polymerization catalysts).

For the effects of metals and metal deactivators, see [23].

2.4. Test Methods

The two most important test methods for antioxidants are the multiple extrusion test and the oven test (DIN 53 383, ASTM D 1870–68).

The *multiple extrusion test* examines degradation in the polymer melt (i.e., stabilization during processing). The antioxidant is blended with the polymer, and the resulting granules are regranulated up to ten times in succession in an extruder under defined temperature conditions. The yellowness index (ASTM D 1925–70) and the melt viscosity (melt flow index, MFI, ASTM D 1238–70) or solution viscosity are measured. The melt viscosity can also be determined on-line in an extruder equipped with a bypass [24].

In the *oven test*, polymer degradation is studied in the solid phase (i.e., long-term stabilization is determined). Test plates of the polymer are placed on rotating trays in special furnaces with air recirculation. Their appearance (browning, reduction in gloss and transparency), mechanical properties (brittleness, tensile strength, impact strength), and analytical data (carbonyl bands in the IR) are evaluated after various times. To shorten the test period, the oven test is often run at 100 – 150 °C, but correlation of the results with room temperature is problematic especially in the case of metal deactivators.

Other test methods include

1. Thermogravimetric analysis, especially for plastics such as polyoxymethylene that tend to undergo thermal depolymerization
2. Torque measurements on polymer melts in the Brabender, especially for plastics that tend to cross-link (e.g., polyethylene)
3. Transparency measurements on extruded films
4. Determination of the induction period in an oxygen absorption test at elevated temperature (OIT test)

2.5. Uses

Polyolefins. The optimal antioxidants for polyolefins depend on the type of resin, production process (catalyst; solution, suspension, or bulk polymerization), comonomer, filler, application (moldings, films, fibers), and other factors. Modern bulk polymerization processes, in which purification steps are omitted or polymers are produced directly as spherical granules, have led to reformulation of stabilization systems.

The primary antioxidant BHT, with its high volatility and yellowing tendency, is losing ground to higher molecular mass hindered phenols (e.g., **2-4**). For prestabilization up to 250 ppm of **3** is usually added at the polymerization stage; ca. 500 – 1000 ppm of a phenolic antioxidant is subsequently added for stabilization during processing and service. Higher concentrations are seldom used because of the risk of yellowing.

The main secondary antioxidants used during processing are the phosphites **8-10** and the phosphonite **13**, whereby especially **13** not only stabilizes the melt viscosity but also offers much better color stabilization than phosphites such as **8** and **9**. The phosph(on)ites are used at concentrations of 500 – 2000 ppm.

Figures 2 and 3 illustrate the stabilization obtained with phosphites and phosphonites during the processing of polypropylene and linear low-density polyethylene (LLDPE), respectively. Whereas polypropylene exhibits an increase in melt flow index (degradation by chain cleavage), LLDPE shows a decline (degradation dominated by cross-linking).

The selection of antioxidants also depends on the solubility in the polymer, which is especially critical for LLDPE. If the polymerization process already yields a granular product in the reactor, antioxidants can be applied to the reactor granules by spraying on the surface. Low-melting products such as **14** are especially suitable for this purpose.

Thioethers are preferentially used as synergists when severe thermal requirements apply to long-term stabilization. They are employed at concentrations of ca. 1000 – 3000 ppm.

Metal deactivators are used at the 500 – 5000 ppm level (higher concentrations in talc-filled polypropylene).

Other formulation guidelines are given in [1].

Styrene Polymers. Phenolic antioxidants (up to 2000 ppm) are added to high-impact polystyrene modified with polybutadiene. They are usually introduced before polymerization. Phosphites and phosphonites are sometimes employed as color stabilizers.

In acrylonitrile – butadiene – styrene copolymers (ABS), the antioxidants are usually introduced into the polybutadiene latex phase by

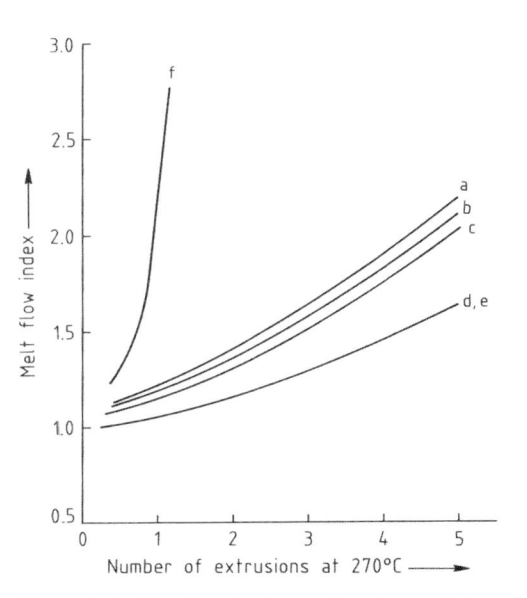

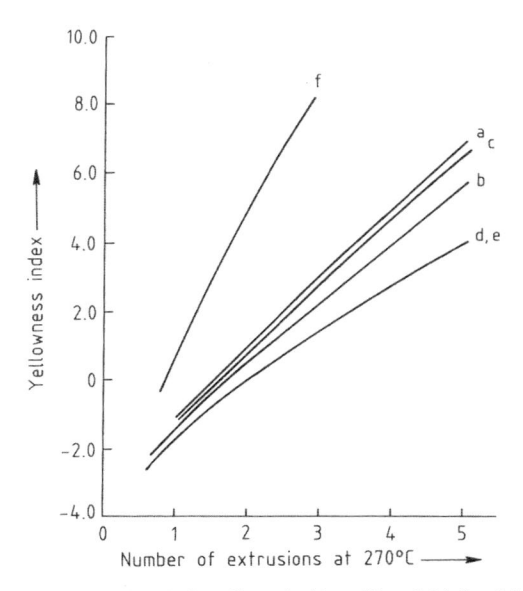

Figure 2. Stabilization of polypropylene in a multiple extrusion test with 0.1 % phenolic antioxidant (**2**) and 0.1 % calcium stearate as basic stabilization and 0.1 % phosph(on)ite process stabilizer a) Phosphite (**9**); b) Phosphite (**10a**); c) Phosphite (**12**); d) Phosphonite (**13**); e) Phosphite (**11**); f) No process stabilizer

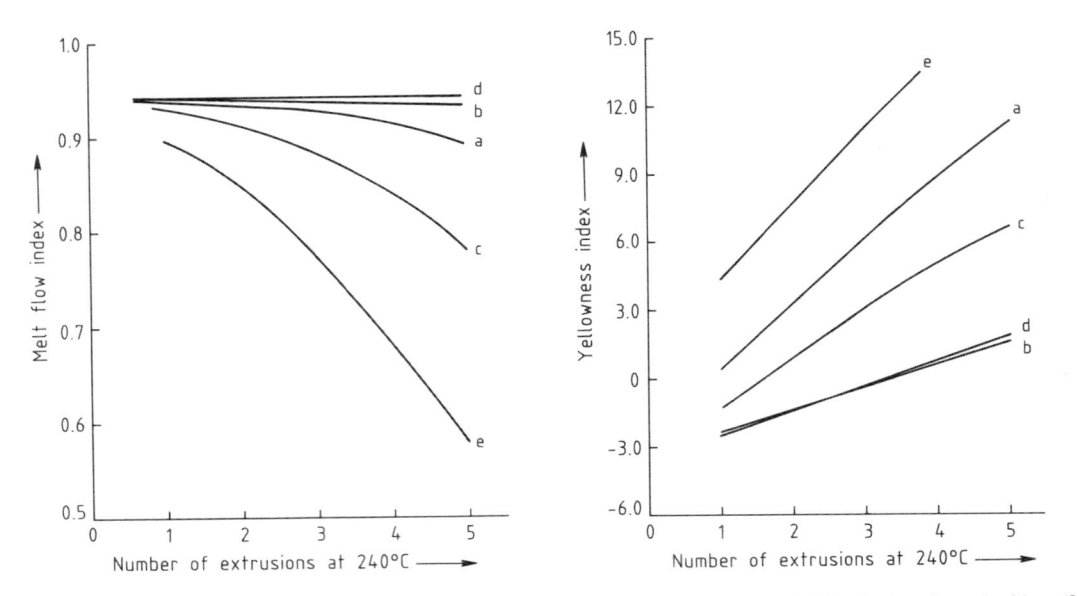

Figure 3. Stabilization of linear low-density polyethylene in a multiple extrusion test with 0.1 % phenolic antioxidant (**3**) and 0.1% calcium stearate as basic stabilization and 0.15% phosph(on)ite process stabilizer a) Phosphite (**9**); b) Phosphite (**10a**); c) Phosphite (**10b**); d) Phosphonite (**13**); e) No process stabilizer

emulsification before it is blended with polystyrene – polyacrylonitrile. Along with 1000 –2000 ppm of phenolic antioxidants, larger amounts of phosphite **8** or DLTDP are used.

Other Plastics [1]. Antioxidants play a less important role in plastics other than polyolefins and the styrene polymers described above. Secondary aromatic amines, BHT, TNPP, as well as high molecular mass phenols are used in elastomers and polyurethanes.

Although engineering thermoplastics (e.g., polycarbonates, polyesters, polyamides) are less sensitive to thermooxidative degradation, hindered phenols and phosph(on)ites are used for color stabilization or long-term stabilization.

2.6. Legal Aspects

Because plastics are widely used in food packaging and agriculture, antioxidants must be approved as indirect food additives by national agencies such as the FDA (USA) and the Bundesgesundheitsamt (Federal Health Office, Germany). Unified approvals are expected to apply throughout the EC as of 1993. Approval is based on toxicity and extraction tests (extraction of antioxidants from plastics by food simulants). Most well-established antioxidants have received broad approval for food use.

2.7. Economic Aspects

Table 1 presents 1990 market volumes of antioxidants. Polyolefins, ABS, and polystyrene account for ca. 80 – 90 % of the antioxidant market. Prices of antioxidants are given in [25].

Antioxidants are often marketed in the form of powder blends. The most important producers and trade names are listed in Table 2.

Table 1. Estimated market volume of antioxidants in 1990 (10^3 t)

Region	Phenols	Thioethers	Phosph (on)ites	Other	Total
United States	14	3	6	2	25
Western Europe	15	3	7	3	28
Japan	8	2	3	2	15
Other	10	2	3	2	17
Total	47	10	19	9	85

Table 2. Manufacturers and trade names of antioxidants

Manufacturer	Trade names	Phenols	Phosph(on)ites	Thioethers	Aromatic amines	Metal deactivators	Blends
Adeka Argus	Mark	X	X	X		X	X
Akzo Chemie	Perkanox	X					
American Cyanamid	Cyanox	X		X			X
BASF	Sicostab	X					X
Bayer	Vulkanox	X			X		
B. F. Goodrich Chem.	Goodrite	X			X		
Chem. Werke Lowi	Lowinox	X	X				
Ciba-Geigy	Irganox, Irgafos	X	X	X		X	X
Daiichi Kogyo Seiyaku	Lasumit			X			
Dover Chemicals	Doverphos		X				
Eastman Chemicals	Tenox	X				X	
Enichem Synthesis	Anox, Alkanox	X	X				X
Ethyl Corporation	Ethanox	X	X				
Ferro Corporation	Oxi-chek	X					
General Electric	Ultranox, Weston	X	X		X		X
W. R. Grace				X			
Hitachi	Cunox					X	
Hoechst	Hostanox	X	X	X		X	
ICI	Nonox, Topanol	X			X		X
Jokoku Chemicals			X				
Mitsubishi Petrochem.	Seenox	X		X			
Monsanto	Santonox, Santowhite Flectol	X			X		
Morton Thiokol	Carstab			X			
Olin Chemicals	Wytox	X	X				
Pennwalt	Anoxyn			X			
PMC Specialities Group	Prodox	X					
Raschig	Ralox	X					
Rhône-Poulenc	Rhodianox, Garbafix	X	X				
R. T. Vanderbilt	AgeRite, Vanox	X	X	X	X		
Sandoz Chemicals	Sandostab		X				
Schenectady Chemicals	Isonox	X					
Shell	Ionol, Ionox	X					
Shipro Kasei	Seenox	X		X			
Sumitomo	Sumilizer	X	X	X			
Ute Industries						X	
Uniroyal	Naugard, Aranox	X	X		X	X	X
Witco	Mark	X	X	X			X
Yoshitomo				X			

3. Light Stabilizers

3.1. Purpose and Requirements

Synthetic and natural polymers vary in their sensitivity to environmental influences. Ultraviolet radiation in sunlight plays a critical role because it has sufficient energy to break chemical bonds. The cleavage sites react with atmospheric oxygen and accelerate degradation (free-radical chain reactions; see Section 2.3). Many plastics therefore suffer yellowing, surface cracking, embrittlement, reduction of gloss, or chalking after a short time in outdoor service, and ultimately disintegrate. Light stabilizers are used to maximize protection against photodegradation.

Ultraviolet radiation is subdivided into UV-A ($\lambda = 320 - 400$ nm), UV-B ($\lambda = 280 - 320$ nm), and UV-C ($\lambda < 280$ nm). Only UV-A and UV-B reach the Earth's surface in sunlight, and UV-B is largely responsible for the degradation of plastics.

The difference in damage (yellowing) caused by UV-A and UV-B in poly(vinyl chloride) (PVC) is shown in Figure 4. Plastics are generally fairly stable in the long-wavelength UV-A (350 – 400 nm). Most plastics are sensitive to wavelengths of 290 – 320 nm.

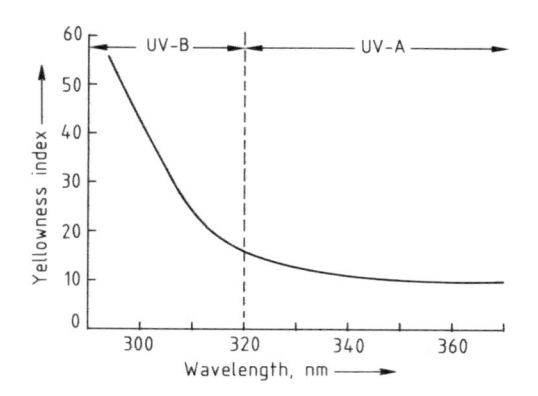

Figure 4. Damage in PVC as a function of the wavelength of UV light (300-h radiation)

Two methods are used to protect plastics against photodegradation: (1) incorporation of light stabilizers in the bulk plastic, and (2) coating with a light-stable or light-stabilized material that is largely opaque to the dangerous UV range.

Light stabilizers must have high light absorption; must resist extraction with water, hydrolysis, and thermal volatilization; and must themselves be stable toward UV radiation. Ideally, a light stabilizer should not be consumed while carrying out its function; it should operate in a closed cycle so that it still exists in active form even after a long period of weathering or use.

Other requirements include solubility in the polymer, stability under processing conditions, compatibility with other additives, and colorlessness. In certain applications, light stabilizers must be approved as indirect food additives.

The service life of a plastic is more than doubled by light stabilizers. Modern systems extend service life by a factor of 10–20 in many plastics, thus permitting their outdoor use.

3.2. Chemical Classes

Light stabilizers are classified in three groups according to their mode of action:

1. UV absorbers: 2-hydroxybenzophenones (BP-UVA), 2-hydroxyphenylbenzotriazoles (BT-UVA), oxalanilides (OA-UVA), 2-hydroxyphenyltriazines (TA-UVA), cinnamates (CA-UVA), salicylates (SA-UVA), and formamidines (FA-UVA)
2. Energy quenchers: nickel complexes (NIC)
3. Radical scavengers: hindered amines (HALS) and 4-hydroxybenzoates (HB)

Table 3 lists some of the most important light stabilizers now in use, along with their trade names and producers.

Stabilization can also be achieved by incorporation of carbon black, titanium dioxide, or zinc oxide. The effect of colored pigments varies: phthalocyanine blue pigments have a stabilizing action; organic red and yellow pigments sensitize photodegradation.

3.3. Mechanism of Action

Photodegradation of Polymers. Photodegradation is initiated when radiation is absorbed by chromophores in the polymer structure itself (e.g., C=C, C=O) or by impurities or defects. Energy can be dissipated from excited states by emission of radiation (fluorescence from short-lived singlet states or phosphorescence from long-lived triplet states), release of heat, or energy transfer to nearby chromophores. If none of these mechanisms is available, photodegradation takes place with bond cleavage and formation of radicals. These radical sites react with oxygen and initiate autoxidative chain reactions. Excited molecular segments can also transfer energy to oxygen, converting it to higher-energy singlet oxygen, which accelerates oxidative degradation.

Oxidation to carbonyl groups in the macromolecules results in chain cleavage via the Norrish reactions:

$$R-CH_2CH_2CH_2-\underset{\underset{O}{\|}}{C}-R' \xrightarrow{h\nu} \begin{array}{l} R-CH_2CH_2CH_2^{\cdot} + {}^{\cdot}\underset{\underset{O}{\|}}{C}-R' \\ \\ R-CH=CH_2 + CH_3-\underset{\underset{O}{\|}}{C}-R' \end{array}$$

Many plastics with aromatic residues (e.g., polycarbonates, polyesters, polyamides) tend to participate in "photo-Fries reactions", whose products may be colored or may be precursors of colored products:

Table 3. Light stabilizers

Structure, CAS registry no., code	Trade name (producer)
2-Hydroxybenzophenones	
HO–C(=O)–C₆H₄–OC₈H₁₇ (phenyl-CO-(2-hydroxy-4-octyloxyphenyl)) [1843-05-6] BP-UVA 1	Cyasorb UV 531 (American Cyanamid) Mark 1413 (Adeka Argus) Chimassorb 81 (Ciba-Geigy) UV-Chek AM 300 (Ferro) Hostavin ARO 8 (Hoechst) Rhodialux P (Rhône-Poulenc) Uvasorb 3 C (Sigma) Seesorb 102 (Shipro Kasei) Aduvex 248 (Shell) Lowilite 22 (Chem. Werke Lowi) Sumisorb 130 (Sumitomo) Viosorb 130 (Kyodo Yakuhin) Uvinul 408 (BASF)
phenyl-CO-(2-hydroxy-4-methoxyphenyl) [131-57-7] BP-UVA 2	Cyasorb UV 9 (American Cyanamid) Uvinul M-40 (BASF) UVA Bayer 325 (Bayer) Chimassorb 90 (Ciba-Geigy) Gafsorb 2H 4M (GAF) Rhodialux A (Rhône-Poulenc) Uvasorb MET (Sigma) Seesorb 101 (Shipro Kasei) Viosorb 110 (Kyodo Yakuhin) Sumisorb 110 (Sumitomo)
phenyl-CO-(2-hydroxy-4-hydroxyphenyl) [131-56-6] BP-UVA 3	Uvinul 400 (BASF) Aduvex 12 (Shell) Gafsorb 24 DH (GAF) DHB (Riedel de Haen) Rhodialux D (Rhône-Poulenc) Seesorb 100 (Shipro Kasei) Viosorb 100 (Kyodo Yakuhin)
phenyl-CO-(2-hydroxy-4-dodecyloxyphenyl) [2985-59-3] BP-UVA 4	Chimassorb 125 (Ciba-Geigy) Eastman Inhibitor DOBP (Eastman Chem.) Gafsorb 2H 4DD (GAF) Rhodialux 1200 (Rhône-Poulenc) Seesorb 103 (Shipro Kasei)
(2-hydroxyphenyl)-CO-(2-hydroxy-4-methoxyphenyl) [131-53-3] BP-UVA 5	Cyasorb UV 24 (American Cyanamid) Aduvex 24 (Shell) Sumisorb 140 (Sumitomo)
(2-hydroxy-4-methoxyphenyl)-CO-(2-hydroxy-4-methoxyphenyl) [131-54-4] BP-UVA 6	Univul D-49 (BASF) Aduvex 424 (Shell)

(continued)

Table 3. (*Continued*)

Structure, CAS registry no., code	Trade name (producer)
[131-55-5] BP-UVA 7 (2,2',4,4'-tetrahydroxybenzophenone)	Aduvex 412 (Shell) Uvinul D 50 (BASF)
[69119-80-8] BP-UVA 8	Mark LA 51 (Adeka Argus)

2-Hydroxyphenylbenzotriazoles

Structure, CAS registry no., code	Trade name (producer)
[2440-22-4] BT-UVA 1	Tinuvin P (Ciba-Geigy) Mark LA 32 (Adeka Argus) Uvasorb SV (Sigma) Seesorb 701 (Shipro Kasei) Lowilite 55 (Chem. Werke Lowi) Viosorb 520 (Kyodo Yakuhin) Sumisorb 200 (Sumitomo)
[3896-11-5] BT-UVA 2	Tinuvin 326 (Ciba-Geigy) Mark LA 36 (Adeka Argus) Seesorb 703 (Shipro Kasei) Viosorb 550 (Kyodo Yakuhin) Sumisorb 300 (Sumitomo)
[3147-75-9] BT-UVA 3	Cyasorb UV 5411 (American Cyanamid) Sumisorb 340 (Sumitomo) Viosorb 583 (Kyodo Yakuhin) Seesorb 709 (Shipro Kasei)
[3864-99-1] BT-UVA 4	Tinuvin 327 (Ciba-Geigy) Mark LA 34 (Adeka Argus) Seesorb 702 (Shipro Kasei) Viosorb 580 (Kyodo Yakuhin)
[3846-71-7] BT-UVA 5	Tinuvin 320 (Ciba-Geigy) Seesorb 705 (Shipro Kasei) Viosorb 582 (Kyodo Yakuhin) Sumisorb 320 (Sumitomo)
[25973-55-1] BT-UVA 6	Cyasorb 2337 (American Cyanamid) Tinuvin 328 (Ciba-Geigy) Seesorb 704 (Shipro Kasei) Viosorb 591 (Kyodo Yakuhin) Sumisorb 350 (Sumitomo)

Table 3. (*Continued*)

Structure, CAS registry no., code	Trade name (producer)
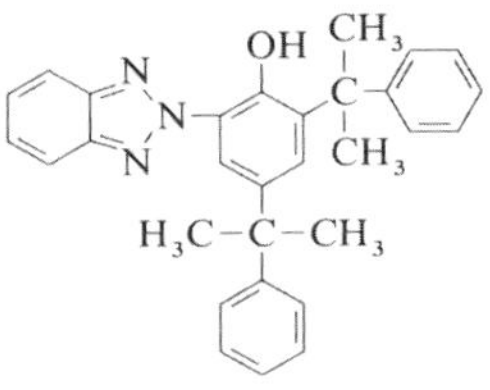 [70321-86-7] BT-UVA 7	Tinuvin 234 (Ciba-Geigy) Tinuvin 900 (Ciba-Geigy)
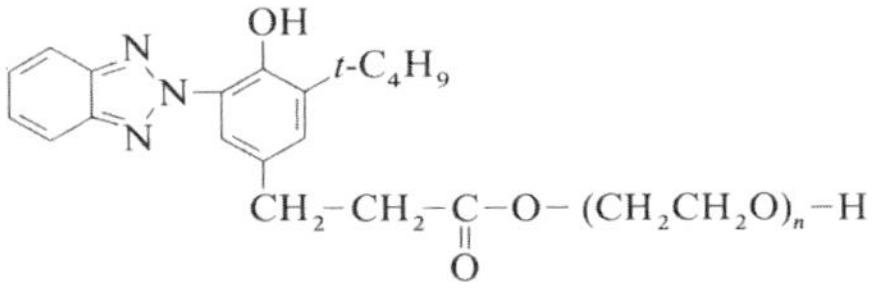 [104810-48-2] BT-UVA 8	Tinuvin 1130 (Ciba-Geigy)
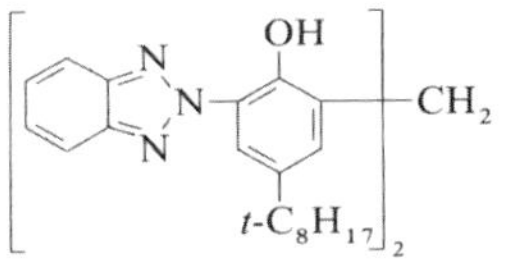 [103597-45-1] BT-UVA 9	Mark LA 31 (Adeka Argus)
2-Hydroxyphenyltriazines [2725-22-6] TA-UVA-1	Cyasorb 1164 (American Cyanamid)
Cinnamates [5232-99-5] CA-UVA-1	Uvinul N-35 (BASF) Seesorb 501 (Shipro Kasei) Viosorb 910 (Kyodo Yakuhin)
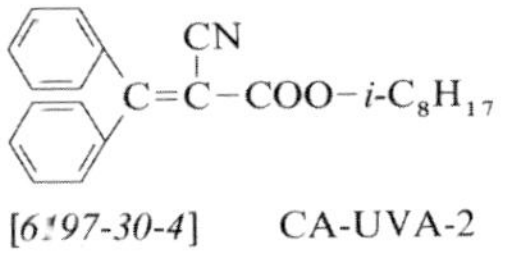 [6197-30-4] CA-UVA-2	Uvinul 539 (BASF)
Oxalanilides [23949-66-8] OA-UVA-1	Sanduvor VSU (Sandoz) Tinuvin 312 (Ciba-Geigy)

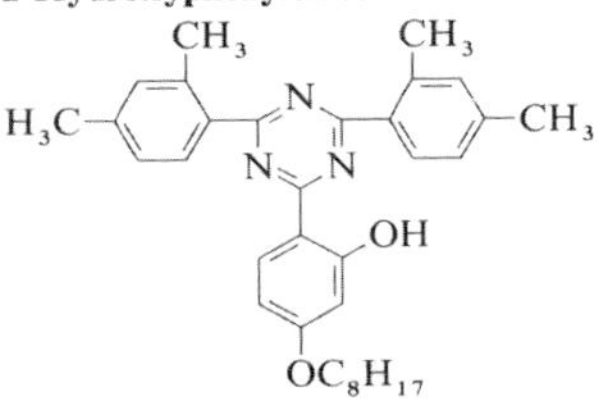

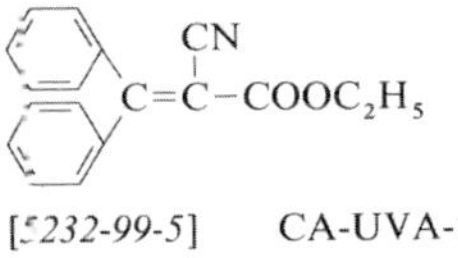

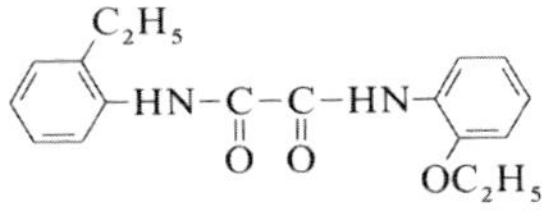

(*continued*)

Structure, CAS registry no., code	Trade name (producer)
$C_{12}H_{25}$—⟨⟩—HN—C—C—HN—⟨⟩ O O OC_2H_5 [82493-14-9] OA-UVA-2	Sanduvor 3206 (Sandoz)
Salicylates	
⟨⟩—C(=O)—O—⟨⟩ OH [118-55-8] SA-UVA-2	Eastman Inhibitor OPS (Eastman Chem.) Seesorb 201 (Shipro Kasei)
⟨⟩—C(=O)—O—⟨⟩—$t\text{-}C_4H_9$ OH [87-18-3] SA-UVA-2	Seesorb 202 (Shipro Kasei) Rhodialux K (Rhône-Poulenc) Viosorb 90 (Kyodo Yakuhin)
Formamidines	
C_2H_5O—C(=O)—⟨⟩—N=CH—N(CH_3)—⟨⟩ [57834-33-0] FA-UVA-1	Givsorb UV-1 (Givaudan)
C_2H_5O—C(=O)—⟨⟩—N=CH—N(C_2H_5)—⟨⟩ [65816-20-8] FA-UVA-2	Givsorb UV-2 (Givaudan)
4-Hydroxybenzoates	
$t\text{-}C_4H_9$ HO—⟨⟩—C(=O)—O—$C_{16}H_{33}$ $t\text{-}C_4H_9$ [67845-93-6] HB-1	Cyasorb UV 2908 (American Cyanamid)
$t\text{-}C_4H_9$ HO—⟨⟩—C(=O)—O—⟨⟩—$t\text{-}C_4H_9$ $t\text{-}C_4H_9$ $t\text{-}C_4H_9$ [4221-80-1] HB-2	Tinuvin 120 (Ciba-Geigy) Seesorb 712 (Shipro Kasei) UV-Chek AM 340 (Ferro) Viosorb 80 (Kyodo Yakuhin) Sumisorb 400 (Sumitomo)
Nickel complexes	
H_2N—R ⋮ O—Ni—O ⟨⟩—S—⟨⟩ $t\text{-}C_8H_{17}$ $t\text{-}C_8H_{17}$ R = C_4H_9 [14516-71-3] NIC-1	Cyasorb UV 1084 (American Cyanamid) Chimassorb N 705 (Ciba-Geigy) Rhodialux Q 84 (Rhône-Poulenc) Uvasorb Ni (Sigma)
same, R = C_8H_{17} [67668-65-9] NIC-2	Seesorb 612 NH (Shipro Kasei)

Table 3. (*Continued*)

Structure, CAS registry no., code	Trade name (producer)
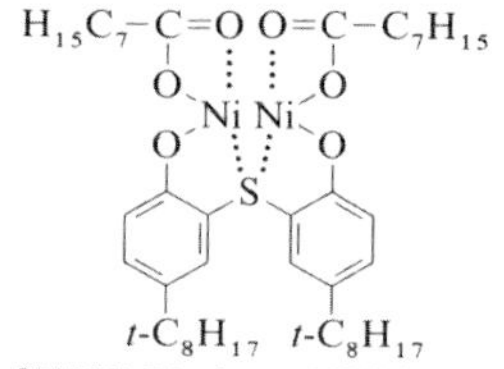 [Structure of bis(octylphenol sulfide) nickel complex with octanoate ligands, t-C_8H_{17} substituents] [*89073-38-1*] NIC-3	UV-Chek AM 205 (Ferro)
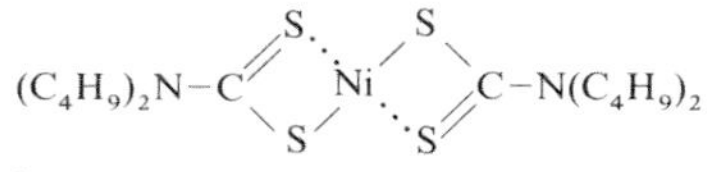 $(C_4H_9)_2N-C(S)(S)Ni(S)(S)C-N(C_4H_9)_2$ [*13927-77-0*] NIC-4	UV-Chek AM 104 (Ferro) Antigene NBC (Sumitomo) Vanox NBC (R. T. Vanderbilt)
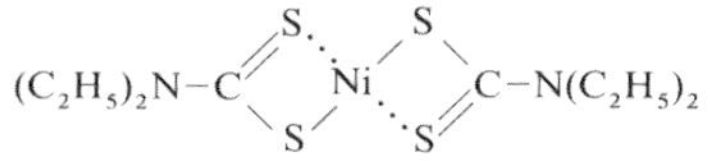 $(C_2H_5)_2N-C(S)(S)Ni(S)(S)C-N(C_2H_5)_2$ [*14267-17-5*] NIC-5	Robac Ni PP (Robinson Brothers)

Hindered amines

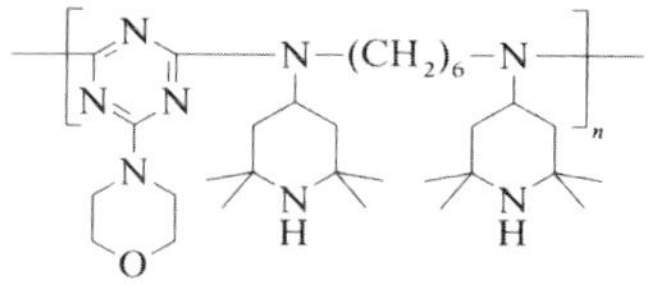 [Triazine-morpholine/piperidine polymer structure with N–(CH$_2$)$_6$–N linker] [*90751-07-8*] HALS 1	Cyasorb UV 3346 (American Cyanamid)
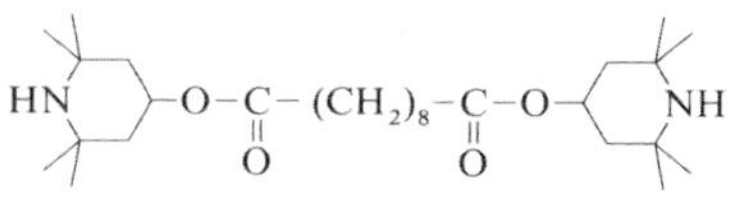 HN⟨piperidine⟩–O–C(=O)–(CH$_2$)$_8$–C(=O)–O–⟨piperidine⟩NH [*52829-07-9*] HALS 2	Tinuvin 770 (Ciba-Geigy) Mark LA 77 (Adeka Argus) Sanol LK 770 (Sankyo) Lowilite 77 (Chem. Werke Lowi)
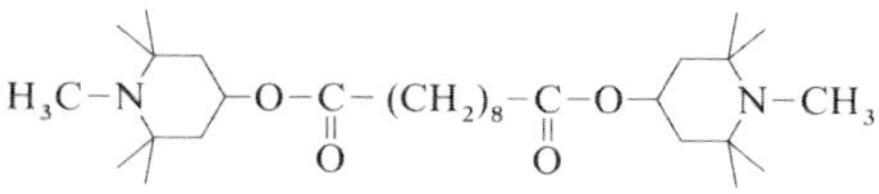 H_3C–N⟨piperidine⟩–O–C(=O)–(CH$_2$)$_8$–C(=O)–O–⟨piperidine⟩N–CH_3 [*41556-26-7*] HALS 3	Tinuvin 765 (Ciba-Geigy) Tinuvin 292 (Ciba-Geigy) Sanol 292 (Sankyo)
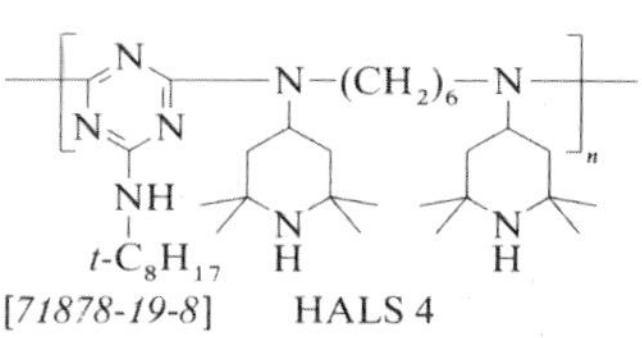 [Triazine-piperidine polymer with t-C_8H_{17}–NH substituent and N–(CH$_2$)$_6$–N linker] [*71878-19-8*] HALS 4	Chimassorb 944 (Ciba-Geigy)
R–NH–(CH$_2$)$_3$–N(R)–(CH$_2$)$_2$–N(R)–(CH$_2$)$_3$–NH–R [*106990-43-6*] HALS 5 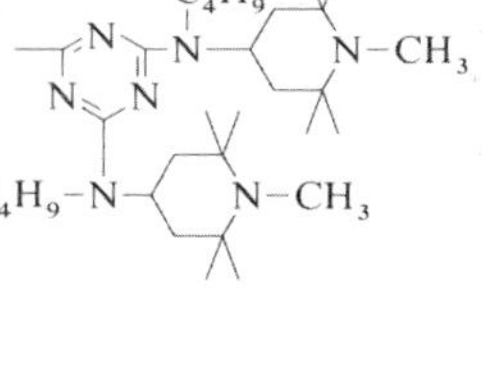	Chimassorb 119 (Ciba-Geigy)

R = [triazine substituted with C_4H_9-N⟨piperidine⟩N–CH_3 and C_4H_9–N⟨piperidine⟩N–CH_3 groups]

(*continued*)

Table 3. (*Continued*)

Structure, CAS registry no., code	Trade name (producer)
[62782-03-0] HALS 6 — piperidyl–O–C(O)–(CH$_2$)$_2$–C(O)–O–piperidyl	Tinuvin 780 (Ciba-Geigy)
[*70198-29-7*] HALS 7 — H–[O–piperidyl–N–(CH$_2$)$_2$–O–C(O)–(CH$_2$)$_2$–C(O)]$_n$–OCH$_3$	Tinuvin 622 (Ciba-Geigy)
[*64338-16-5*] HALS 8 — piperidyl with O–C(CH$_2$)$_{11}$ and C(O)–NH spiro	Hostavin N-20 (Hoechst)
[63843-89-0] HALS 10 — H$_3$C–N-piperidyl–O–C(O)–C(C$_4$H$_9$)(CH$_2$-Ar)–C(O)–O–piperidyl–N–CH$_3$; Ar = 3,5-di-*t*-C$_4$H$_9$-4-OH-phenyl	Tinuvin 144 (Ciba-Geigy)
[82537-67-5] HALS 10 — H$_3$C–C(O)–N-piperidyl with spiro C(O)–N–C$_{12}$H$_{25}$ / N–C(O)–H	Tinuvin 440 (Ciba-Geigy)
[*122586-52-1*] HALS 11 — H$_{17}$C$_8$O–N-piperidyl–O–C(O)–(CH$_2$)$_8$–C(O)–O–piperidyl–N–OC$_8$H$_{17}$	Tinuvin 123 (Ciba-Geigy)
M_r 1100–2500 [*102089-33-8*] HALS 12 — ⟵Si(CH$_3$)(CH$_2$–CH$_2$–CH$_2$O–piperidyl–NH)–O⟶$_n$	Uvasil 299 (Enichem)
R = C$_{12}$H$_{25}$ [*85099-51-0*] HALS 13 R = C$_{14}$H$_{29}$ [*85099-50-9*] piperidyl (HN) with O–C(CH$_2$)$_{11}$ and C(O)–N–CH$_2$–CH$_2$–C(O)–O–R spiro	Sanduvor 3050 (Sandoz)

Table 3. (*Continued*)

Structure, CAS registry no., code	Trade name (producer)
HALS 14 [116214-18-7] — piperidine derivative with -O-C(O)-O-O-C(CH$_3$)$_2$-C$_2$H$_5$ group	Lupersol HA 505 (Atochem)
HALS 15 [122035-71-6] — piperidine-HN-C(O)-C(O)-NHNH$_2$	Luchem HA-R 100 (Atochem)
HALS 16 R = C$_{12}$H$_{25}$ [119530-69-7]; R = C$_{14}$H$_{29}$ [119530-70-0] — piperidine-HN-CH$_2$CH$_2$-C(O)-O-R	Sanduvor 3052 (Sandoz)
HALS 17 [109423-00-9] — bis-piperidine glycoluril derivative	Uvinul 4049 H (BASF)
HALS 18 [79720-19-7] — piperidine-N-succinimide with -CH-C$_{12}$H$_{25}$ substituent	Cyasorb UV 3581 (American Cyanamid); Sanduvor 3055 (Sandoz)
HALS 19 [106917-30-0] — N-methyl piperidine-N-succinimide with -CH-C$_{12}$H$_{25}$ substituent	Cyasorb UV 3604 (American Cyanamid); Sanduvor 3056 (Sandoz)
HALS 20 [106917-31-1] — N-acetyl piperidine-N-succinimide with -CH-C$_{12}$H$_{25}$ substituent	Cyasorb UV 3668 (American Cyanamid); Sanduvor 3058 (Sandoz)

In semicrystalline plastics, UV radiation is scattered by the crystallites and increases the quantum efficiency of polymer degradation. Photodegradation occurs mainly in the amorphous, oxygen-permeable regions.

Ultraviolet Absorbers. The UV absorbers listed in Table 3 have intense absorption bands in the UV-B and some of them in the UV-A as well. They therefore filter out harmful radiation before it reaches the chromophores in the plastic.

The UV absorbers must dissipate the absorbed energy quickly as heat and not as radiation. Mechanisms proposed for this action include proton transitions in the excited singlet state [26–28]. Other quenching mechanisms have also been postulated [29].

The position and intensity of absorption can be modified by substituents on the aromatic rings. For example, alkoxy substituents in the 4- and 4'-positions of 2-hydroxybenzophenones bring about bathochromic shifts in the absorption maxima, as do alkyl substituents in the phenyl ring of 2-hydroxyphenylbenzotriazoles. The steepness of the UV absorption curve on the long-wavelength side is important with respect to intrinsic color—even slight residual absorption around 400 nm causes yellow coloration.

The UV absorbers hinder the photolysis of polymers by competing for UV absorption. They are thus particularly effective when degradation is due primarily to photolysis and not to oxidation. Since the intensity of radiation depends on the thickness of the absorbing layer (Lambert – Beer law), these substances are more effective within the plastic than at its surface. They are therefore only marginally effective in thin films and fibers.

Energy Quenchers. Light stabilizers based on nickel complexes, like some UV absorbers, are energy quenchers. They absorb energy from excited polymer regions or from singlet oxygen, and dissipate it as heat. The action of quenchers depends on concentration, not on layer thickness.

Radical Scavengers. Sterically hindered 4-hydroxybenzoates and amines act chiefly as radical scavengers. 4-Hydroxybenzoates can be regarded as photostable phenolic antioxidants [30]. The hindered amines (HALS) are

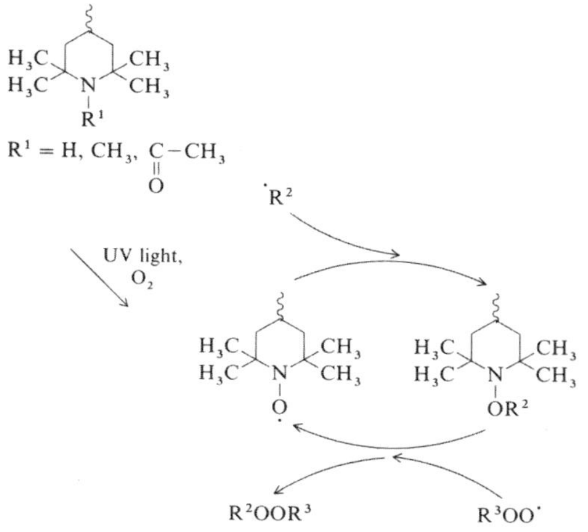

Figure 5. Stabilization mechanism of HALS

much more important, however; they all have a 2,2,6,6-tetramethylpiperidine ring [31]. The cyclic mode of action of HALS is shown in Figure 5 [32–43]. The stabilizing action is due not to the HALS compound itself $(N-R^1)$ but to the nitroxyl radical $(N-O^{\cdot})$ formed by photooxidation during weathering. The nitroxyl radical reacts with the polymer radicals formed by photolysis to give ethers $(N-O-R)$, which then react with peroxy radicals to regenerate the nitroxyl radical. Other mechanisms are based on the trapping of acyl peroxide radicals [44] or the decomposition of hydroperoxides [45].

The action of HALS depends on their concentration and not on layer thickness. They are therefore equally as effective at the surfaces of plastics in films, fibers, coatings, and thick layers. They are, however, sensitive to certain chemicals, such as organic thioether costabilizers [5], halogen flame retardants [46], kaolin fillers, pesticides [47], and acids. Synergistic effects between HALS and UV absorbers are widely used in industry, and synergies have also been observed with nickel quenchers [48]. A combination of low molecular mass HALS (which migrate to the surface) and high molecular mass compounds (which are more strongly anchored in the bulk polymer) is also recommended [21]. Recent developments include 2,2,6,6-tetramethylpiperazinones [49] and HALS that are chemically bound to polyolefins [50].

3.4. Test Methods

Light stabilizers are incorporated into polymers; processed into specimens in plate, film, fiber, or coating form; and then exposed to radiation in the presence or absence of moisture to determine their weather resistance and lightfastness, respectively.

Weathering may be natural or artificial (i.e., accelerated). Changes in color (DIN 53 236), yellowness index (ASTM D 1925–70), gloss (DIN 67 530), cracking (ISO 4628 Part 4), tensile strength, elongation at break (ASTM D 638, ISO IR 527), or impact resistance (ASTM D 256, ISO R 179) are used to assess lightfastness and weather resistance. The intensity of the IR carbonyl bands is also very often used as a criterion of photodegradation. See also → Paints and Coatings, 1. Introduction; → Plastics, Properties and Testing, Section 2.7.3..

Open-Air Weathering Tests are performed on a commercial basis in all climatic zones. Important sites are located in Central Europe (temperate climate), southern Florida (subtropical), Arizona (desert climate), Okinawa, Thailand, Australia, South Africa, Israel, and Singapore.

In *accelerated sunlight tests*, Fresnel reflectors are used to focus sunlight on the specimens (ASTM E 838–81). Examples of such tests include the Equatorial Mount with Mirrors for Acceleration (EMMA) test and the Equatorial Mount with Mirrors for Acceleration plus Water Spray (EMMAQUA) test, which are performed in Arizona.

In *artificial weathering* the specimens are subjected to precisely defined climatic conditions (temperature, irradiation, rainfall) in specially designed apparatus. Xenon arcs, carbon arcs, and fluorescent lamps are used as illumination sources (see Table 4); filtered xenon light best simulates global irradiation [51], [52].

Standards for artificial weathering tests include DIN 53 384, DIN 53 231, and ANSI/ASTM G 53–57.

3.5. Uses

Light stabilizers are employed mainly in plastics for outdoor service (e.g., garden furniture, stadium seating, window frames, floor coverings) films for protective coverage and agriculture; light fixtures and illuminated signs; automotive parts such as bumpers and body paints; and sports equipment and clothing. Light stabilizers extend the service lives of these articles by years.

The approximate breakdown of light stabilizer uses according to type of plastic is as follows:

Polyolefins (polyethylene, polypropylene)	55 %
Acrylic and polyurethane coatings	20 %
PVC	10 %
ABS and polystyrene	5 %
Unsaturated polyester resins	4 %
Miscellaneous substrates	6 %

The stabilizers are generally used in concentrations of 0.05 – 0.5 %, reaching 3 % in especially critical applications. Recommended formulations can be found in [1].

Polypropylene is particularly sensitive to photodegradation. The photostabilization of polypropylene is reviewed in [53]. Thick moldings are stabilized with 0.2 – 0.6 % of UV absorbers (e.g., BP-UVA 1, BT-UVA 2) and/or radical scavengers (e.g., HB-1 and 2, HALS 1 – 8). HALS (chiefly of high molecular mass, such as HALS 1, 4, or 5) are employed to stabilize films and fibers.

Polyethylene. Although polyethylene is more stable to light than polypropylene, it requires good stabilization for outdoor service. The

Table 4. Devices for accelerated weathering tests

Producer	Device	Light source
W.C. Heraeus GmbH, Hanau, Germany	Xenontest	xenon lamps
	Suntest	
Atlas Electric Devices Company, Chicago, USA	Weather-Ometer	xenon lamps
	Fade-Ometer	carbon-arc lamps
	Sunchex fluorescent tubes	UV
	UV-CON	
Q-Panel Company, Cleveland, USA	Q-U-V fluorescent tubes	UV
Weiss Technik AG, Reiskirchen, Germany	Global-UV-Testgerät Bauart BAM	special UV lamps
Dainippon Plastics, Tokyo, Japan	Eye Super UV Tester	special UV lamps

selection of light stabilizers is especially critical because of the danger of surface chalking due to migration of incompatible additives. Low-density and linear low-density polyethylene (LDPE and LLDPE) films contain up to 0.4 % of a nickel quencher and UV absorber combination (e.g., NIC-1 and BP-UVA 1) or of a polymeric HALS (e.g., HALS 1, 4, or 7). In filled grades, higher concentrations (ca. 1.5 %) can also produce more than a tenfold increase in service life in open-air weathering. In high-density polyethylene (HDPE) the best results are obtained with HALS (e.g., HALS 2, 4, or 7); 0.1 % is often sufficient.

Styrene Polymers. The UV absorbers and UV absorbers – HALS combinations (e.g., BT-UVA 1 with HALS 2) are recommended for polystyrene, ABS, and styrene – acrylonitrile copolymers (SAN) at a total concentration of ca. 1 % [54].

Poly(Vinyl Chloride). The lightfastness of PVC depends heavily on its heat stabilization or fillers. Barium and cadmium stabilizers, tin carboxylates, and titanium dioxide substantially improve lightfastness. Tin mercaptides are unsuitable for outdoor service or necessitate high levels of titanium dioxide. HALS are not very effective. For transparent PVC in particular, the use of 0.1 – 0.3 % UV absorbers (e.g., BP-UVA 1 or 2, BT-UVA 1 or 5, OA-UVA 1) is recommended.

Miscellaneous Plastics. Polyurethanes, especially those derived from aromatic isocyanates and polyether polyols, tend to yellow and show a deterioration of mechanical properties when exposed to light. Ternary combinations of HALS, UV absorbers, and phenolic antioxidants have a good stabilizing action, especially in foamed plastics [55].

The UV absorbers (e.g., benzotriazoles such as BT-UVA 1 and 2) and HALS are used to stabilize poly(methyl methacrylate) [56]; when colorlessness is important, oxalanilides (OA-UVA 1) are employed. Oxalanilides are also used in unsaturated polyesters and polycarbonates, but a satisfactory solution still has to be found to the problem of light-induced yellowing in polycarbonates.

Table 5. Estimated market volume of light stabilizers in 1990 (10^3 t)

Region	BP-UVA	BT-UVA	HALS	Others	Total
United States	1.1	1.7	2.4	0.6	5.8
Western Europe	1.0	1.4	2.2	0.6	5.2
Japan	0.4	1.3	1.3	0.3	3.3
Other	0.4	0.9	1.1	0.3	2.7
Total	2.9	5.3	7.0	1.8	17.0

3.6. Economic Aspects

Market volumes for light stabilizers are given in Table 5. The most important producers and trade names are listed in Table 3.

Roughly 50 % of the light stabilizers are sold to plastics processors in the form of masterbatches (concentrates containing ≤ 20 % light stabilizers in appropriate polymer vehicles). Combination masterbatches supplemented with pigments or flame retardants are being developed.

4. Heat Stabilizers for Poly (Vinyl Chloride)

4.1. Purpose and Requirements

Poly(vinyl chloride) (PVC) is thermally unstable. Heating results in elimination of HCl, the formation of polyene sequences, and rapid discoloration. This autocatalytic reaction begins at ca. 100 °C; at 180 °C a marked brown color occurs after a few minutes. Special heat stabilizers are therefore required for the processing and use of this high-tonnage plastic. These additives are rated according to several criteria besides their efficiency as heat stabilizers:

1. Versatility (use with mass, suspension, and emulsion PVC)
2. Effect on melt rheology of PVC
3. Lubricant action
4. Migration, plate-out
5. Compatibility with other additives and pigments
6. Effect on transparency
7. Effect on lightfastness
8. Effect on electrical insulation
9. Fogging (especially for automotive interior parts)

10. Occupational hygiene and approval as indirect food additives
11. Easy handling and costs

Combinations of stabilizers and costabilizers are normally used to obtain optimal effects. Many systems are employed in industry; PVC stabilization is discussed in detail in [57], [58].

4.2. Chemical Classes

Heat stabilizers for PVC are divided into metal-containing and metal-free stabilizers (costabilizers).

4.2.1. Metal-Containing Stabilizers

The strengths and weaknesses of the main metal-containing heat stabilizer classes are summarized in Table 6.

Lead Stabilizers include basic lead carbonates (e.g., $x\,\text{PbO}\cdot\text{PbCO}_3$), sulfates, phosphites, stearates, maleates, phthalates, salicylates, and silicates, as well as neutral lead stearate $\text{Pb}(\text{OOC}-\text{C}_{17}\text{H}_{35})_2$. They have an excellent cost – performance ratio.

Mixed Metal Stabilizers include systems of barium – cadmium, barium – zinc, and calcium – zinc compounds, mainly carboxylates (e.g., stearates, laurates, oleates, or alkylbenzoates) naphthenates, or phenolates. The Ba – Cd types are most effective, followed by Ba – Zn, and Ca – Zn. They must always be reinforced with costabilizers. Recent developments include a zinc glycinate $\text{H}_2\text{NCH}_2\text{COOZnOOCCH}_3$ [59], which is also used in combination with barium or calcium carboxylates.

Organotin Stabilizers are characterized by their tin content (4 – 25 %). They are classified as those containing sulfur (tin mercaptides, i.e., thiolates) and those containing none (tin carboxylates). The first group are the most efficient heat stabilizers. They can be employed without costabilizers and include mainly mono- and dialkyltin mercaptides, mercaptoacetates, mercaptopropionates, and thioglycolates with the following formula:

$$(R^1)_n Sn(SR^2)_{4-n}$$

where $n = 1$ or 2; R^1 = methyl, n-butyl, n-octyl, or n-dodecyl; and R^2 is, for example, $-\text{CH}_2\text{COO}$-alkyl, $-\text{CH}_2\text{CH}_2\text{COO}$-alkyl, or $-\text{CH}_2\text{CH}_2\text{OOC}$-alkyl. The alkyl group is often isooctyl.

"Methyltin stabilizers" (R^1 = methyl) offer better thermal stability than products containing higher alkyl groups, because the latter suffer thermal β-elimination. Cyclic dialkyltin sulfides are also employed, especially in multicomponent additive systems.

Most of the sulfur-free organotin stabilizers are dibutyltin or dioctyltin carboxylates,

Table 6. Strengths and weaknesses of metal-containing heat stabilizers for PVC *

| Property | Lead stabilizers | Mixed metal stabilizers | | Tin mercaptides | Tin carboxylates |
		Ba – Cd	Ba – Zn, Ca – Zn		
Versatility	–	–	–	+	+
Lubricant action	– (+ for stearates)	+	+	–	–
Migration/plate-out	+	–	+/–	+	+
Transparency of PVC	–	–	–	+	+
Light stability	+	+	–	–	+
Sulfur staining	–	–	+	–	+
Odor	+	–	+	–	+
Electrical insulating properties, water absorption	+	–	–	–	–
Toxicology, food approval in some cases	–	–	+	+	–

* + No problems, often beneficial effect; – Problematic

especially cyclic maleates, semiesters of maleic acid, or laurates.

Antimony Stabilizers include antimony thiolates [e.g., $Sb(SCH_2COOR)_3$].

4.2.2. Metal-Free Stabilizers

These costabilizers always have to be used in conjunction with metal-containing stabilizers. The most important costabilizers follow.

Phosphites. Tris(nonylphenyl) phosphite and mixed aromatic – aliphatic phosphites, such as diphenyldecyl phosphite, are often used.

Epoxy Compounds. The principal epoxy stabilizers are epoxidized plant oils, chiefly epoxidized soybean oil.

Polyols include pentaerythritol, trismethylolpropane, and tris(hydroxyethyl) cyanurate.

Other Costabilizers include β-aminocrotonates, dihydropyridines, and 1,3-diketones.

4.3. Mechanism of Action

The primary step in the thermal degradation of PVC is free-radical formation by the elimination of activated chlorine atoms, such as those on allyl groups or in tertiary positions. The chlorine atoms attack the polymer, resulting in hydrogen abstraction (formation of HCl) and double-bond formation [60].

$$-CH=CH-\underset{Cl}{CH}- \longrightarrow -CH=CH-\overset{\cdot}{C}H- + Cl^{\cdot}$$

$$Cl^{\cdot} + -CH_2-\underset{Cl}{CH}-CH_2-\underset{Cl}{CH}- \xrightarrow{-HCl}$$

$$-\overset{\cdot}{C}H-\underset{Cl}{CH}-CH_2-\underset{Cl}{CH}- \longrightarrow$$

$$-CH=CH-CH_2-\underset{Cl}{CH}- + Cl^{\cdot} \xrightarrow{-HCl}$$

$$-CH=CH-\underset{Cl}{\overset{\cdot}{C}H}-CH- \longrightarrow$$

$$-CH=CH-CH=CH- + Cl^{\cdot}, \text{ etc.}$$

Double-bond formation occurs in preference to oxidation of the radical sites to peroxy radicals in the presence of oxygen. Elimination of HCl with the formation of polyene structures continues. Products containing $\geq 6 - 7$ conjugated double bonds are colored.

Ionic mechanisms have also been proposed, but these are less prominent at higher temperature. The formation of carbenium chlorides in the polymer backbone results in deepening of the color ("halochromy"). Thermooxidative processes accelerate degradation.

Stabilizers neutralize labile chlorine atoms by reacting with them to form more stable compounds:

$$-CH=CH-\underset{Cl}{CH}- + (R^1)_2Sn(SR^2)_2 \longrightarrow$$
$$\text{Tin mercaptide}$$

$$-CH=CH-\underset{SR^2}{CH}- + (R^1)_2Sn-\underset{Cl}{SR^2} \quad (3)$$

$$-CH=CH-\underset{Cl}{CH}- + M(O-\underset{O}{\overset{\|}{C}}-R)_2 \longrightarrow$$
$$\text{Metal carboxylate}$$

$$-CH=CH-\underset{O-\overset{\|}{C}-R}{CH}- + M-O-\underset{O}{\overset{\|}{C}}-R \quad (4)$$

$$-CH=CH-\underset{Cl}{CH}- + -HC\underset{O}{-}CH- \longrightarrow$$
$$\text{Epoxy compound}$$

$$\begin{array}{c}-CH=CH-CH-\\ \underset{|}{O} \quad Cl \\ -CH-CH-\end{array} \quad (5)$$

$$-CH=CH-\underset{Cl}{CH}- + R_1-\overset{\|}{\underset{O}{C}}-CH_2-\overset{\|}{\underset{O}{C}}-R_2 \xrightarrow{-HCl}$$
$$\text{1,3-Diketone}$$

$$\begin{array}{c}-CH=CH-CH-\\ R_1-\overset{\|}{\underset{O}{C}}-CH-\overset{\|}{\underset{O}{C}}-R_2\end{array} \quad (6)$$

The exchange reactions (Eq. 3 and 4) (Frye – Horst theory [61]) form the basis for many mechanistic analyses. The metal stabilizers are finally converted to chlorides: R_2SnCl_2, $ZnCl_2$, $CdCl_2$, and $PbCl_2$. The metal stabilizers also react with and neutralize hydrogen chloride, which would catalyze further PVC degradation.

The metal chlorides $CdCl_2$ and especially $ZnCl_2$ are, however, Lewis acids and have a destabilizing influence because they catalyze further elimination of HCl. Consequently, although initial stabilization is excellent, further heating leads to a sudden deep coloration of the PVC ("zinc burning"). The remedy is to exploit the synergism of mixed metal stabilizers. Cadmium and zinc carboxylates are always used in combination with barium or calcium carboxylates, which exchange anions with zinc chloride and cadmium chloride. The zinc and cadmium stabilizers are therefore regenerated, with the formation of barium and calcium chlorides which do not enhance PVC degradation. Barium and calcium carboxylates also react preferentially with HCl.

No catalytic effect on degradation is observed with lead chloride and (di)alkyltin chlorides; the alkyl groups in the latter compounds have a deactivating action. For the effects of metal chlorides, see [62], [63]; for the mechanism of organotin stabilizers, see [64].

Costabilizers, such as polyols [63] and epoxy compounds, also neutralize harmful metal chlorides and improve long-term stabilization with calcium – zinc stabilizers. β-Diketones [65], [66] and dihydropyridines, in contrast, improve the initial color stabilization; the dihydropyridines hydrogenate polyene sequences and thereby shorten their length. Tin mercaptides and the thiols formed by their reaction with HCl can also react with polyene double bonds. The double bond in tin maleates can react with polyenes (Diels – Alder addition):

$$\begin{array}{c}CH-CH\\ \parallel \\ CH \quad CH\end{array} + R_2Sn\begin{array}{c}O\\ \parallel \\ O-C-CH\\ \parallel \\ O-C-CH\\ \parallel \\ O\end{array} \longrightarrow O=C\begin{array}{c}CH=CH\\ /\ \ \backslash\\ CH \quad CH\\ \backslash \ \ /\\ CH-CH\\ \end{array}C=O\begin{array}{c}\\ \\ \backslash \ / \\ O \ \ O\\ \backslash \ /\\ Sn\\ /\ \backslash\\ R \ \ R\end{array}$$

Finally, alkyltin mercaptides and phosphite costabilizers act as hydroperoxide decomposers, thus protecting against oxidative degradation.

4.4. Test Methods

The most important test methods are based on static and dynamic heating. Analytical methods such as the determination of dehydrochlorination at 180 – 200 °C are also common.

In the *static heating test*, films produced in a two-roll mill or plastisols applied to a substrate are heated in a furnace (e.g., to 180 – 200 °C), and the specimen color is observed as a function of time. In the *dynamic test*, PVC is treated in a two-roll mill, Brabender plastograph, or extruder at specified temperatures. Samples are withdrawn after various times or numbers of extrusion passes and pressed into films, whose color is then determined (usually via the yellowness index).

Because color development in PVC is fairly rapid, these tests (i.e., heat treatment) seldom require more than a few hours. The profile of the yellowness index curves reflects the effects of stabilizers, particularly with regard to short-term and long-term stabilization, and improvements due to costabilizers.

The effect of heat stabilizers on the processing and service qualities of PVC is the concern of extensive testing programs [67].

4.5. Uses

The choice of stabilizer system depends on the processing methods used (extrusion, injection molding, calendering, plastisol application), the intended uses of the PVC, and compatibility with other additives. Technologies and regulations vary greatly. The following survey can therefore give only general indications and trends. For further details, see the producers' literature.

Rigid PVC. Solid stabilizers or stabilizer blends are generally preferred for rigid PVC. Organotin stabilizers have found wide use because processing conditions are more severe than those for flexible PVC.

Window Frames and Sidings. Tin mercaptides are used in the United States; in Europe a shift from Ba–Cd stabilizers has been observed into Ca–Zn costabilizer systems via Ba–Cd–Pd and Pb formulas. Tin carboxylates are also used, especially for transparent or colored frames, because of their good weather resistance.

Pipes, Including Water-Supply Pipes. Tin mercaptides are used in the United States. Lead stabilizers predominate in Europe (except France) and Japan.

Beverage Bottles and Bottle Caps. The PVC is stabilized with food-approved tin mercaptides or Ca–Zn costabilizer systems containing epoxidized soybean oil, 1,3-diketones, or polyols as costabilizers.

Flexible PVC. Most of the stabilizers used for flexible (plasticized) PVC are in liquid form; Ba–Cd and Ba–Zn systems occupy the leading position, with a trend toward cadmium-free systems including Ca–Zn products, e.g., in foils, films, leather imitates (shoes, automotive interiors), hoses, and floor coverings. Lead stabilizers are used in electrical cables.

Stabilizer Content. Heat stabilizer contents of 0.3 – 5 % are common. Levels of organotin stabilizers are at the low end of the range ($\leq$ 2.5 %), mixed metal stabilizers in the middle, and lead stabilizers for cable insulation at the high end. Phosphite or 1,3-diketone costabilizers are generally used below 0.5 %, whereas epoxidized soybean oil is added in concentrations of 5 % or higher.

Legal Aspects. Some countries place strict limits on the use of certain stabilizer classes; for example Sweden and Japan restrict the use of cadmium stabilizers. Similar provisions are under discussion for the EC. Other broad limits apply to lead stabilizers in drinking-water pipes in France and the United States.

Stabilizers for products that come in contact with food (e.g., beverage bottles, packaging films) must be officially approved as indirect food additives. Many octyltin and dodecyltin mercaptides [e.g., di-*n*-octyltin bis(isooctylthioglycolate)], most of the Ca–Zn stabilizers, and many costabilizers (especially stearylbenzoylmethane, dihydropyridine and epoxidized oils) have such approval.

4.6. Economic Aspects

Table 7 lists estimated 1990 market volumes for metal-based heat stabilizers that are used as solids, pastes, or liquids. Dust-free solid forms such as granules, flakes, or "small bags" are preferred. "One-pack" systems, which contain primary and secondary stabilizers preblended with lubricants and other additives, are widely used. Most producers offer broad assortments, with individual products that are continually being modified and tailored to specific applications. Among the nonmetallic costabilizers, epoxidized soybean oil holds a dominant position with more than 100 000 t/a used worldwide in PVC.

Table 8 lists the most important producers and trade names of PVC heat stabilizers.

Table 7. Estimated market volume of PVC heat stabilizers in 1990 (10^3 t)

Region	Tin	Lead	Mixed metal	Total
United States	20	10	25	55
Western Europe	10	50	40	100
Japan	10	30	15	55
Other	10	30	15	55
Total	50	120	95	265

Table 8. Producers of PVC heat stabilizers

Producer	Trade name	Type [*]
Acima	Metastab	Sn, Me
Adeka Argus	Mark	Me, Pb, C, Sn
Akishima Chemicals		Me, Sn
Akzo	Interstab, Stanclere, Estabex	Me, Sn, Pb, C
BASF	Sicostab	Me
Bleiberger Bergwerks Union	Austrostab	Pb, Sn, Me, C
Cardinal Chemicals	CC	Sn
Chemson	Naftovin, Strandex	Pb, Me, Sn
Ciba-Geigy	Irgastab, Reoplast	Sn, Me, C
Commer	Prosper	Sn, Me, Pb, C
CWM O. Bärlocher	Bärostab, Okstan	Me, Sn, Pb, C
Ferro Corporation	Therm Chek	Me, Sn, C
Hammond Lead Products	Halstab	Pb
Hebron	Vinstab	Sn, Me, Pb, C
Henkel	Edenol, Stabilox, Stabiol	C
Hoechst	Hostastab	Sn, C
Katsuta Kako	Advastab	Me, C, Sn
Kemichrom	Kemistab	Sn, Me, Pb, C
Kyodo Chemicals		Sn, Me
Lagor	Lastab, Synesol	Me, Sn, C
Meister	Meister	Sn, Me, C
Morton Thiokol	Advastab	Sn
M & T Chemicals	Thermolite	Sn, Me, Pb, C
Nissan Ferro Organic Chem.	Therm Chek	Me
Nitto Kasei		Sn
Plastics Specialities & Technologies		Me
Reagens	Reatinor, Reablend	Sn, Me, Pb, C
Rhône-Poulenc	Rhodiastab	C
Sakai Kagaku		Me, C, Sn
Sankyo Organic Gosei	Stann	Sn
Swedstab	Swedstab	Sn, Me, C
Synthetic Products	Synpron	Sn, Me
Tinstab SA	Tinstab	Sn, Me, C
Tokyo Fine Chemicals	Embilizer	Sn
R. T. Vanderbilt	Vanstray, Barostab	Me
Witco Corporation	Mark	Me, Sn

[*] C=costabilizer; Me=mixed metal; Pb=lead; Sn=organotin.

5. Lubricants, Slip, Antiblocking, and Mold-Release Agents

5.1. Purpose and Requirements [68], [69]

Lubricants (→ Lubricants and Lubrication), slip, antiblocking, and mold-release agents (→ Release Agents) control the frictional and adhesive properties of plastics during processing and in service.

Lubricants also improve the dispersion of pigments and fillers in plastics. Uniform colors can be obtained by breaking up agglomerates in pigment preparations. Better dispersion of fillers (e.g., talc) improves flow limits and material properties.

Internal Lubricants reduce friction between polymer particles and molecules during the melting (plastication) of plastics and transport of the melt. They thus reduce energy consumption on plastication, lower melt viscosity, improve flow properties, increase output of processing machinery, and allow processing under less stringent conditions.

External Lubricants reduce the friction and adhesion of polymer melts on hot metal surfaces of processing machinery (e.g., extruder screws and cylinders, rolls). This reduces abrasion between the polymer melt and the metal, and improves melt flow. It also improves the gloss, smoothness, and regularity of the surface of the plastic.

Mold-Release Agents prevent the adhesion of molded plastics to metal cooling rolls and molds. *Slip agents* improve the frictional qualities of finished plastic surfaces (e.g., bearings and gears) and prevent adhesion between stacked films. *Antiblocking agents* prevent adhesion between film layers on rolls or in stacks when pressure and heat are applied.

Antislip Agents (adhesion promoters) impart limited tack to finished articles to reduce slip (e.g., between filled bags made of plastic films).

It is not always possible to assign a particular product to one of the above classes. For example, the boundary between internal and external lubricants depends on the type of polymer, the additive concentration, the presence of other lubricants, and the processing temperature. Many products also perform multiple functions: external lubricants are often effective as mold-release or slip agents. Carefully matched combinations of additives are often required. This group of additives must also comply with standards of odor and taste, migration behavior, product stabilization [70], and thermal stability.

Lubricants and related compounds may have disadvantageous effects on some product properties. Internal lubricants can act as plasticizers and thus reduce the heat distortion temperature. External lubricants can reduce transparency and impair the printability, cementability, and weldability of plastics. These additives also affect mechanical properties, especially in case of overlubrication, which causes inhomogeneities due to inadequate fusion of the material.

5.2. Chemical Classes

Many lubricants are amphoteric compounds containing both a long hydrocarbon chain and polar groups (e.g., hydroxyl, ester, or acid groups). Lubricants are characterized by their molecular mass, acid or hydroxyl number, saponification number, metal content, melting or dropping point, cold-setting temperature, thermal stability, volatility, density, viscosity, and (for solid agents) particle size. The action of a lubricant depends on its polarity. Polar molecules act as internal lubricants in polar polymers such as PVC and as external lubricants in nonpolar polymers such as polyolefins, and vice versa. The polarity is strongly influenced by the length of the hydrocarbon chain. Lubricants can be classified as fatty acids and their derivatives ($C_{12} - C_{22}$ chains), montanic acids and their derivatives ($C_{28} - C_{32}$), paraffins ($C_{20} - C_{70}$), and polyolefin waxes (molecular mass 2000 – 10 000).

The substances of greatest practical importance are listed below; for further details, see → Lubricants and Lubrication.

Fatty Acids, Fatty Alcohols, and Their Derivatives comprise the following groups:

1. Stearic, 12-hydroxystearic, palmitic, and behenic acids
2. Cetyl and stearyl alcohols

3. Metal stearates (Zn, Ca, Pb, Al, Mg, and Na)
4. Butyl stearate, tridecyl stearate, glycerol monostearate (GMS), glycerol monoricinoleate, glycerol monooleate, glycerol tristearate, glycerol tri-12-hydroxystearate (hydrogenated castor oil), trimethylolpropane tristearate, and pentaerythritol tetrastearate
5. Distearyl adipate, distearyl phthalate, cetyl palmitate, cetyl stearate, stearyl stearate, and behenyl behenate
6. Isostearamide, oleamide, erucamide, ethylene bissstearylamide (EDS or "amide wax")
7. Oligomeric fatty acid esters ("complex fatty acid esters") are also used. These are obtained by polycondensation of dicarboxylic acids (e.g., adipic acid), polyols (e.g., pentaerythritol), and monocarboxylic acids (e.g., stearic acid). They may be partially saponified or may contain calcium.

Montanic Acids and Their Derivatives (Montan Waxes). Montanic acids are obtained from lignite and consist of mixtures of $C_{28} - C_{32}$ acids. Derivatives include calcium montanates, sodium montanates, and esters with ethylene glycol, butanediol, glycerol, or pentaerythritol.

Paraffins may be derived from petroleum (straight-chain members are solids, branched-chain members are liquids) or are synthetic (Fischer – Tropsch) paraffins.

Polyolefin Waxes. The most common polyolefin waxes are polyethylene waxes with branched and linear structures; partially crystalline polypropylene waxes are less common. Polar polyethylene waxes are obtained by oxidation; they have acid numbers of ca. 20 – 70.

Other Chemical Classes. Fluorinated polymers, chlorinated paraffins, silicones, graphites, and other substances are used as lubricants and related compounds. Silica gel, chalk, and other minerals are employed as antiblocking agents and slip aids.

This article does not cover externally applied mold-release agents (e.g., waxes, fluorinated polymers, silicones, and metallic soaps), which are sprayed or brushed onto mold surfaces. These methods are used chiefly with thermosetting resins (e.g., polyurethane, unsaturated polyester, epoxy).

5.3. Mechanism of Action

The friction-lowering property of *lubricants* is due to their long hydrocarbon chains ($> C_{12}$). The lubricating effects depend on interactions between the lubricant and the polymer matrix (solubility of the lubricant in the polymer) and on the association between the lubricant and the metal surfaces. Unlike plasticizers, which are completely soluble in the polymer, lubricants should have only limited solubility; this is ensured by their amphoteric character.

Internal Lubricants [71] have a relatively high affinity for polymers. Polar lubricants in, for example, PVC are either dissolved in or strongly bound to the surface of the associations of polymer molecules that occur in the plastic state. Typical effects are a lowering of the softening temperature due to swelling, a reduction in melt viscosity, and frictional damping between polymer particles. Good internal lubricants have high saturation limits and haze points.

External Lubricants are strongly associated to the metal surfaces of processing equipment or form discrete phases within the polymer melt. They only have a low affinity for the polymers. Typical effects are a reduction in adherence to the metal and in friction between polymer particles, which leads to slower plastication. External lubricants often cause haze. Overlubrication must be avoided because some adhesion to the metal is necessary for conveyance of the melt in the extruder and also in calendering.

Mold-Release Agents form a film between the metal surface and the cooled plastic. *Slip agents* migrate to the plastic surface, where their hydrocarbon chains act as a lubricant. *Antiblocking agents* slightly roughen the film surface so that air is trapped between stacked layers. *Antislip agents* artificially create surface roughness where nonlubricating particles incorporated in the melt protrude from the film.

5.4. Test Methods

Methods based on practical applications are applied to lubricant testing. One of the most

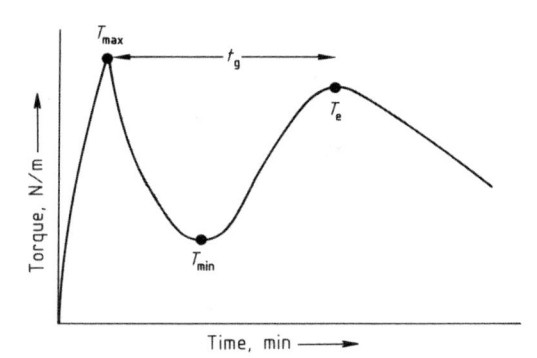

Figure 6. Schematic plastogram of PVC with lubricant (Brabender Plasticorder)

important is the plastograph (e.g., the Brabender Plasticorder). The formulated plastic powder is placed in a kneading chamber at a preset temperature (e.g., 140 – 160 °C) and plasticated by a twin-shaft device running at constant speed. The torque is plotted as a function of time (Fig. 6). The curve passes through a maximum torque value T_{max}, marking the onset of plastication in the powder mixture. It then declines to a minimum T_{min}, followed by an increase to a second maximum T_e as a result of growing friction. Plastication is complete at T_e. The time between T_{max} and T_e is called the plastication time t_g. The test is used to establish suitable ratios of internal and external lubricants. For example, a shorter t_g or an increase in the ratio $(T_{max} - T_{min})/t_g$ indicates internal lubrication. As the hydrocarbon chain in the lubricant becomes longer, the maximum kneading resistance declines.

Other tests employ extrusiometers (measurement of polymer melt temperature and pressure in the extruder cylinder and at the die), capillary viscometers (quantity of material discharged per unit time at constant pressure and temperature) [72], or roller mills (time that elapses between charging and adhesion of the polymer sheet on the rolls). In the spiral test, material is injected into a spiral mold and the distance the material travels in the runner is measured.

Slip and blocking effects are studied in skid and adhesion tests with films (e.g., ASTM D 1893 or D 3354). Haze and surface gloss measurements are an important component of all lubricant tests.

5.5. Uses

The main application of lubricants is in PVC (ca. 50 %), followed by styrene polymers, polyolefins, engineering thermoplastics, and thermoplastic elastomers.

Rigid PVC always requires lubricants for processing (usual content 1 – 4 %), and lubricant combinations (preformulated systems) are often employed. Selection of lubricants is dictated by the type of heat stabilizer used: stearic acid with lead stabilizers; 12-hydroxystearic acid with barium – cadmium stabilizers (to reduce plate-out); glycerol monostearate, montan waxes, or oxidized polyethylene waxes with tin stabilizers; and complex fatty acid esters with calcium – zinc stabilizers.

Fatty acid esters, calcium stearate, and fatty alcohols are typical internal lubricants. Montan waxes, amide waxes, paraffins, oxidized polyethylene waxes, and fatty acids act as external lubricants. Heavy-metal soaps and complex fatty acid esters fall between the two types. Lubricants allow processing under gentler conditions and thereby contribute to stabilization. A combination of calcium stearate as internal lubricant, paraffin wax as external lubricant, and oxidized polyethylene wax as mold-release agent has been used with success for PVC pipes.

Flexible PVC requires less lubricant (ca. 0.5 %). Liquid types, such as glycerol monooleate or paraffin oils, are used. Amide waxes serve as antiblocking agents.

The flow behavior of *ABS, SAN, and polystyrene* can be improved with butyl stearate, glycerol monostearate, fatty acid esters, paraffin oils, oxidized polyethylene waxes, and montan waxes. Zinc stearate and amide wax (0.1 – 2 %) are used as mold-release agents.

The calcium stearate used as an acid radical scavenger is often sufficient for *polyolefins*. Neutral polyethylene and polypropylene waxes are added to pigment preparations for polyethylene and polypropylene. Talc-filled polypropylene can contain ca. 0.5 % of partially saponified montan wax. A suitable mold-release and slip agent for films is 0.1 – 0.3 % of a fatty acid amide in polyethylene or of erucamide in polypropylene. Finally, 0.05 % of fluorinated polymers can be added to LLDPE films to combat melt fracture (a surface defect occurring at high viscosity and resulting from the alternating wall adhesion and slip of the extrudate). Highly effective slip aids for polyolefin films include micronized synthetic silica gels.

For *engineering thermoplastics* [polycarbonates, polyamides, poly(ethylene terephthalate) (PETP), poly(butylene terephthalate) (PBT)], thermally stable pentaerythritol esters of fatty acids are preferred; they have a better transparency than montanate esters. Amide wax can also serve as a mold-release agent. Filled thermoplastics represent a broad field for lubricants.

Because most lubricants are derived from natural sources, they are often approved for food packaging products.

5.6. Economic Aspects

Current worldwide consumption of lubricants and related compounds is ca. 130 000 – 140 000 t, with Western Europe, the United States, and the rest of the world accounting for a third each. Consumption is equally divided between the five following groups: fatty acids and esters, metallic soaps, amide waxes, paraffin and polyethylene waxes, and miscellaneous types.

Table 9 lists some important producers and trade names of lubricants and related compounds. The number of products on the market is relatively large because new formulations and easy-to-handle forms are continually appearing. Most products for PVC are sold to converters; those for other polymers are often sold to the polymer producers.

6. Flame Retardants [73–78] See also → Flame Retardants

6.1. Purpose and Requirements

The replacement of classical materials by combustible plastics in buildings, equipment, and vehicles creates a fire risk. Flame retardants are used to combat this danger. A useful measure of flammability for plastics is the limiting oxygen index (LOI), which denotes the volume fraction of oxygen in an oxygen – nitrogen mixture that just supports combustion of a well-defined

Table 9. Producers and trade names of lubricants and related compounds

Manufacturer	Trade name
Abril Industrial Waxes	Abril Wax
Akzo Chemicals	Dopral, Interstab, Armoslip, Aramid
Allied Signal	AC Polyethylene
Ashland Chem.	
Astor Wax	Comboloob
Axel Plastics Res. Labs.	Mold Wiz
CWM O. Bärlocher	Bärolub
BASF	Sicolub, Wachs BASF
Dr. T. Böhme	Tebestat
Chemson	Naftolube, Naftozin, Listab
Ciba-Geigy	Irgawax
Corriel	Lubriol, Synthewax, Glycmonos
Commer	Vinlub
Croda Universal	Crodamid, Crodacid
Deer Polymer	
Du Pont	Zonyl, Viton
Durham Chemicals	Durolube
Eastman	Epolene, Myverol, Myvaplex
Th. Goldschmidt	Tego, Tegin
Grace	Sylobloc
C. P. Hall	
Harcros Chemicals	Harochem, Harowax, Harogel, Lankroplast
Harwick Chemicals	Lubrox, Lubrex
Henkel	Loxiol, Loxamid
Hoechst	Hostalub, Hostamont, Hoechst Wachse
Hüls	Vestowax
Humko Chemicals (division of Witco)	Kemamide, Kemester, Mold Pro, Micro-ken
ICI	Atmer
Kao Corporation	
Kemichrom	Kemfluid
Lion Akzo	
Lonza	Glycolube
3 M	Dynamar
Marabushi Oil Chemicals	
Mathe Co.	Coad
Mazer Chemicals (division of PPG)	
Morton Thiokol	Advawax
M & T Chemicals	Stavinor
New Japan Chemical Co.	
Nippon Oil & Fats	
Nitto Kasei	
Reagens	Realube
Rohm & Haas	
Rousselot	
Sakai	
Speciality Products	Zn-Stearate
Struktol Co.	Polydis TR
Swedstab	Swedlub, Swedstab
Tonnan Kagaku Kogyo	
Unichema Chemie	Unislip, Uniwax, Estol, Priolube, Prifrac, Pristerene

specimen under standard conditions. A high LOI indicates a low flammability. The LOI values for the most important plastics follow:

Polyoxymethylene	15.5
Polyethylene	17.3
Poly(methyl methacrylate)	17.5
Polypropylene	17.6
Polyurethane	17–19
Polystyrene	18.0
Unsaturated polyester	18.5
ABS	18–19
Epoxy resin	19.0
PBT	21.5
Polyamide 12	21.6
PETP	22.0
Polycarbonate	23–25
Polyamide 6	23–26
PVC, flexible	24–25
Poly(phenylene oxide) – polystyrene	24–25
Polyamide 66	24–26
PVC, rigid	40–45
Polytetrafluoroethylene	95.0

Flame retardants in plastics affect ignitability, combustion rate, heat release, fuel contribution, smoke evolution, and the formation of toxic or corrosive gases. The significance of these parameters depends on the situation. Although flame retardants can be especially effective in the early stages of a fire and can reduce flammability, plastics modified with these additives are still combustible.

In many cases and especially at high concentrations, flame retardants lower the quality and increase the price of a plastic. They are therefore used only when mandated. Selection is determined by the effectiveness and price of the flame retardant, as well as the processibility of the plastic, and influences on its mechanical, optical, and electrical properties. Interactions with other additives (e.g., light stabilizers) must also be considered.

6.2. Chemical Classes

Flame retardants may be inorganic substances, halogenated organic compounds, organophosphorus compounds, or other organic substances. They are further classified as additive or reactive. Reactive flame retardants are incorporated into the polymer molecule by copolymerization, cross-linking, or grafting.

Inorganic Flame Retardants. Producers and trade names are summarized in Table 10.

Aluminum Trihydroxide, $Al(OH)_3$, and *magnesium dihydroxide,* $Mg(OH)_2$, [79] are the largest group of flame retardants in terms of

Table 10. Producers and trade names of inorganic flame retardants

Compound	CAS registry no.	Producer (trade name)
Aluminum trihydroxide, $Al(OH)_3$	[21645-51-2]	Aluchem; BA Chemicals, Alcan Aluminium (Baco Superfine, Ultrafine); Alcoa (Hydral, Flameguard); Atochem; Kaiser Aluminium; Martinswerke, Lonza (Martinal); Reynolds Metals; Solem Industries, J. M. Huber (Micral, Zerogen); Vereinigte Aluminium Werke (Apyral); Showa Aluminium Industries; Sumitomo Aluminium Smelting
Magnesium dihydroxide, $Mg(OH)_2$	[1309-48-4]	Dead Sea Bromine (FR-20); Kyowa Chemicals; Lonza (Magnifin); Morton Thiokol (Versamag); Solem Industries, J. M. Huber (Zerogen); Veitscher Magnesitwerke
Antimony trioxide, Sb_2O_3 (some as concentrate or in wetted form)	[1309-64-4]	Anzon (Firebloc, Envirostrand, Oncor); Asarco; Laurel Industries (Fire Shield); M & T Chemicals (Thermoguard); Amspec Chemical; BA Chemicals (Timonox)
Antimony pentoxide, Sb_2O_5	[1314-60-9]	Nyacol Products (Nyacol)
Sodium antimonates, e.g., $Na_2Sb_2O_6 \cdot x\,H_2O$		M & T Chemicals (Thermoguard)
Zinc borates, e.g., $4\,ZnO \cdot 6\,B_2O_3 \cdot 7\,H_2O$	[12536-65-1]	Borax Holdings; Climax Performance Materials (ZB-223); US Borax & Chemicals (Firebrake ZB)
Barium metaborate, e.g., $BaB_2O_4 \cdot x\,H_2O$		Buckman Laboratories (Busan)
Ammonium polyphosphate, $(NH_4PO_3)_n$ (some coated or encapsulated, some in systems with flame retardants containing N)	[68333-79-9]	Albright & Wilson (Amgard MC); Hoechst (Exolit); Monsanto (Phos-Chek); Great Lakes (Char Guard 329, CN 1197)
Red phosphorus	[7723-14-0]	Albright & Wilson (Amgard CRP); Hoechst (Exolit 405)

consumption. Particle size varies from 1 to 30 µm. High concentrations (40 – 60 wt %) of these additives are required and are associated with a severe impairment of mechanical properties. To overcome this, products coated with coupling agents (e.g., silanes, fatty acid esters) are recommended. The coating may, for example, improve the impact strength of the modified plastic or reduce the viscosity of prepolymers. Aluminum and magnesium hydroxides often suppress smoke evolution. Use is limited by their thermal stability (dehydration); $Al(OH)_3$ decomposes above ca. 200 °C, and $Mg(OH)_2$ above 300 – 340 °C.

Antimony Compounds. Antimony trioxide, Sb_2O_3; antimony pentoxide, Sb_2O_5; and sodium antimonates (e.g., $Na_2O \cdot Sb_2O_5 \cdot x\, H_2O$) are largely used as synergists with halogenated organic flame retardants or in PVC. In the absence of halogen compounds they have little effect. Antimony trioxide is sold in various purities, whitenesses, and grain sizes (mainly 1 – 3 µm, down to <1 µm for special grades). Antimony pentoxide is available in colloidal form (0.03 µm). Because of occupational health issues relating to antimony trioxide dust, nondusting preparations are preferred (i.e., concentrated masterbatches containing ca. 80 % of active agent, granules, or wetted grades).

Zinc Borates [80] often serve as a partial replacement for antimony trioxide, giving some advantages in suppression of smoke evolution, afterglow, weaker pigmentation, and lower cost. Combinations with aluminum trihydroxide are also employed. Negative effects on the stability of halogen-containing plastics must be considered.

Ammonium Polyphosphate, $(NH_4PO_3)_n$, is included in intumescent systems [81], [82] that contain polyols and melamine derivatives (→ Flame Retardants). The products are also marketed in coated or microencapsulated form to reduce water absorption, for example.

Microencapsulated or inertized *red phosphorus* is a very effective flame retardant that can be processed at up to 280 °C [82],[83]. Its drawbacks are a dark color and potential handling hazards.

Other Inorganic Flame Retardants include iron oxides (Fe_2O_3 and Fe_3O_4); hydrated calcium and magnesium carbonates; bismuth oxides and carbonates [84]; boric acid and borates; zinc hydroxystannate [$ZnSn(OH)_6$], zinc stannate ($ZnSnO_3$), and other zinc compounds [85]; and mineral substances that decompose endothermically at certain temperatures to release water or noncombustible gases. Smoke-inhibiting additives for PVC, unsaturated polyesters, or polyurethanes include calcium carbonate, tin oxides, copper oxide, molybdenum trioxide, and ammonium molybdate.

Halogenated Organic Flame Retardants. Halogenated compounds are effective flame retardants, especially in conjunction with antimony trioxide. They can therefore be used in low concentration so that they do not greatly affect product properties. A drawback is the fact that corrosive gases (HBr, HCl) are always produced in the event of fire and can cause secondary damage. Light stability of the plastic is also reduced, especially by brominated compounds.

The most important members of this group are brominated aromatics, above all decabromodiphenyl oxide and polymeric compounds. Some important producers and trade names are listed in Table 11. Aliphatic bromine compounds are often more effective but less thermally stable. This disadvantage is partially offset in brominated neopentyl structures that do not have hydrogen atoms in the α-position to the bromine. Hexabromocyclododecane is an example of a brominated cycloaliphatic compound; tetrabromobisphenol A is the most important reactive type.

Chlorinated flame retardants (Table 12) are only about half as effective as brominated ones because of the higher bond energy of chlorine. Chlorinated paraffins and chlorinated cycloaliphatics are mainly used [86]. Fluorine and iodine compounds are not employed as flame retardants.

Organophosphorus Flame Retardants. Important products and trade names are listed in Table 13.

The most important organophosphorus flame retardants are halogen-free triaryl phosphates. Phosphonates (e.g., dimethyl methylphosphonate), phosphine oxides, and halogenated

Table 11. Producers and trade names of brominated organic flame retardants

Structure	Name, CAS registry no.	Producer (trade name)
⟨Br$_x$⟩–O–⟨Br$_y$⟩ ; $x + y = 10$	decabromodiphenyl oxide [1163-19-5]	Dead Sea Bromine (FR 1210) Ethyl Corporation (Saytex 102) Great Lakes Chem. (DE-83)
⟨Br$_x$⟩–O–⟨Br$_y$⟩ ; $x + y = 8$	octabromodiphenyl oxide [32536-52-0]	Dead Sea Bromine (FR 1208) Ethyl Corporation (Saytex 111) Great Lakes Chem. (DE-79)
⟨Br$_x$⟩–O–⟨Br$_y$⟩ ; $x + y = 5$	pentabromodiphenyl oxide	Dead Sea Bromine (FR 1205) Ethyl Corporation (Saytex 115) Great Lakes Chem. (DE-71)
(tetradecabromodiphenoxybenzene structure)	tetradecabromodiphenoxybenzene [58965-66-5]	Ethyl Corporation (Saytex 120)
(bis(tribromophenoxy)ethane structure)	bis(tribromophenoxy)ethane [37853-59-1]	Great Lakes Chem. (FF 680)
(ethylene-bis(tetrabromophthalimide) structure)	ethylene-bis(tetrabromophthalimide) [32588-76-4]	Ethyl Corporation (Saytex BT 93)
(N-(2,4,6-tribromophenyl)maleinimide structure)	N-(2,4,6-tribromophenyl)maleinimide [59789-51-4]	Dead Sea Bromine (FR 1033)
(pentabromotoluene structure)	pentabromotoluene [87-83-2]	Dead Sea Bromine (FR 705)
(tetrabromobisphenol A structure)	tetrabromobisphenol A [79-94-7]	Dead Sea Bromine (FR 1524) Ethyl Corporation (Saytex RB 100) Great Lakes Chem. (BA 59)
HOCH$_2$–CH$_2$–O–⟨Br,Br⟩–C(CH$_3$)$_2$–⟨Br,Br⟩–O–CH$_2$–CH$_2$OH	tetrabromobisphenol A-bis(2-hydroxyethyl ether) [4162-45-2]	Dead Sea Bromine (FR 1525) Great Lakes Chem. (BA 50)
H$_2$C=CH–CH$_2$–O–⟨Br,Br⟩–C(CH$_3$)$_2$–⟨Br,Br⟩–O–CH$_2$–CH=CH$_2$	tetrabromobisphenol A-bis-(allyl ether) [25327-89-3]	Dead Sea Bromine (FR 2124) Great Lakes Chem. (BE 51)

(continued)

Table 11. (*Continued*)

Structure	Name, CAS registry no.	Producer (trade name)
	tetrabromophthalic anhydride [*632-79-1*]	Ethyl Corporation (Saytex RB 49) Great Lakes Chem. (PHT 4)
	mixed esters of tetrabromophthalic acid [*77098-07-8*] with diethylene glycol and propylene glycol [*75790-69-1*]	Ethyl Corporation (Saytex RB 79) Great Lakes Chem. (PHT 4-diol)
	poly(2,4,6-tribromostyrene) [*57137-10-7*]	Ferro Corporation (Pyro-chek 68, Pyro-chek LM)
	poly(tetrabromobisphenol-A)-polycarbonate [*28906-13-0*], [*71342-77-3*]	Great Lakes Chem. (BC 52, 58)
	poly(dibromophenylene oxide) [*26023-27-8*]	Great Lakes Chem. (PO 64)
	brominated epoxy resins	Makhteshim Chemical Works (F 2000 Series) M & T Chemicals (Thermoguard)
	poly(pentabromobenzyl acrylate) [*59447-57-3*]	Dead Sea Bromine (FR 1025)
	tetrabromobisphenol A-bis-(2,3-dibromopropyl ether) [*21850-44-2*]	Great Lakes Chem. (PE 68) Riedel de Haen (OBPE)
	hexabromocyclododecane [*3194-55-6*]	Dead Sea Bromine (FR 1206) Ethyl Corporation (Saytex HBCD) Great Lakes Chem. (CD 75-P) Riedel de Haen (HBCD)

Table 11. (Continued)

Structure	Name, CAS registry no.	Producer (trade name)
[Br-substituted ethylene-bis(dibromonorbornane dicarboximide) structure]	ethylene-bis(dibromonorbornane dicarboximide) [41291-34-3]	Ethyl Corporation (Saytex BN 451)
[triazine structure with R = OCH$_2$CHBrCH$_2$Br]	1,3,5-tris(2,3-dibromopropoxy)-2,4,6-triazine [52434-59-0]	Akzo Chem. (Interstab FR 930)
(HOCH$_2$)$_2$C(CH$_2$Br)$_2$	2,2-bis(bromomethyl)-1,3-propanediol [3296-90-0]	Dead Sea Bromine (FR 522) Ethyl Corporation (Saytex FR 1138)
HOCH$_2$–C(CH$_2$Br)$_3$	2,2-bis(bromomethyl)-3-bromo-1-propanol [1522-92-5]	Dead Sea Bromine (FR 513)
BrCH$_2$CHBrCH$_2$O–[Br$_2$-phenyl]–SO$_2$–[Br$_2$-phenyl]–OCH$_2$CHBrCH$_2$Br	3,5,3',5'-tetrabromo-4,4'-dihydroxydiphenylsulfone-bis-(2,3-dibromopropyl ether) [42757-55-1]	Marubeni (Non Nen 52)
Brominated polyols, some also containing chlorine		Dead Sea Bromine Solvay (IXOL B 251)

phosphate esters and phosphonates are also used. Advantages include low corrosivity of the combustion gases, lack of effect on polymer transparency, and suppression of afterglow. Drawbacks include volatility, sensitivity to hydrolysis, and negative effects on the heat distortion temperatures of plastics. Cyclic organophosphorus compounds (phosphorinanes, phosphazenes) overcome some of these problems.

Miscellaneous Organic Compounds. Organic additives occasionally used in flame retardant systems or for special types of plastics include the following:

1. Polyols (e.g., pentaerythritol) for intumescent systems [81]
2. Nitrogen-containing compounds (e.g., melamine or melamine cyanurate), often combined with (NH$_4$PO$_3$)$_n$
3. Silicones [87] as smoke inhibitors and impact modifiers when high concentrations of Al(OH)$_3$, Mg(OH)$_2$, or (NH$_4$PO$_3$)$_n$ are used as flame retardants
4. Free-radical formers (e.g., dicumyl peroxide or 2,3-dimethyl-2,3-diphenylbutane) together with brominated compounds, especially in expandable polystyrene [88]
5. Alkali-metal salts of organic sulfonates for polycarbonates [89]
6. Ferrocene as a smoke inhibitor in PVC

6.3. Mechanism of Action

Three elements are essential for a fire: heat, fuel, and air (oxygen) (Fig. 7). When heated, plastics undergo pyrolysis, forming characteristic degradation products, some of which appear as smoke whereas others serve as fuels that can be ignited once they are mixed with air. Oxidative combustion reactions in the gas phase (the "fire") are strongly exothermic. Thus, large amounts of thermal energy are returned to the plastic and to the surroundings. Pyrolysis and the associated fuel generation are thereby accelerated, and the fire propagates. Flame retardants interfere with these processes by one or more chemical or physical mechanisms [90].

Fuel Dilution. Noncombustible products (CO$_2$, H$_2$O) are generated by thermal decomposition of the flame retardant and released into the gas phase. Examples of such flame retardants are Al(OH)$_3$, Mg(OH)$_2$, hydrated carbonates, boric acid, and borates. The decomposition temperature of the flame retardant must be compatible with the processing temperature and pyrolysis temperature of the plastic.

Endothermic Action. If the pyrolytic decomposition of the flame retardant is endothermic, it can absorb part of the energy released during combustion. This mechanism often accompanies fuel dilution.

Protective Coating. The flame retardant forms a protective coating on the surface of the pyrolyzing plastic. It acts as thermal insulation and a barrier to oxygen, and also prevents combustible pyrolysis products and smoke particles from escaping into the gas phase. Most of the inorganic and organophosphorus flame retardants function by this mechanism. The protective layer can be glassy or have a porous carbon structure (intumescent char). In the latter case, the flame retardant system includes components that are responsible for char formation (e.g., polyols and phosphates) and expansion (e.g., melamine derivatives) [81], [91].

Pyrolysis Modification. Pyrolysis of the plastic can be chemically modified, for example, by dehydration in cellulosic polymers and polyurethanes or by inhibition of thermal depolymerization [e.g., in poly(methyl methacrylate)] to reduce the amount of combustible products generated. The proportion of solid carbon residue (char) is increased, and a noncombustible product (e.g., water) is formed. Organophosphorus compounds employ this mechanism; it is particularly effective in plastics containing oxygen. This mechanism may, however, increase smoke density.

Gas-Phase Quenching. Combustion processes in the gas phase are highly exothermic, free-radical chain reactions in which the fuel is oxidized by oxygen or hydroxyl radicals, for example:

$CO + HO^{\bullet} \rightarrow CO_2 + H^{\bullet} + \text{heat}$

$H^{\bullet} + O_2 \rightarrow HO^{\bullet} + O^{\bullet} + \text{heat}$

$RH + O^{\bullet} \rightarrow HO^{\bullet} + R^{\bullet} + \text{heat}$

Flame retardants that are active in the gas phase must neutralize the highest-energy radicals,

Table 12. Producers and trade names of chlorinated organic flame retardants

Structure	Name	CAS registry no.	Producer (trade name)
	Chlorinated paraffins, 30–70 % Cl, some also containing bromine		Atochem (Electrofine), Dover Chemical (Chlorez), Ferro, Hoechst, Hüls, ICI (Cerechlor), Occidental Chemical (Chlorowax)
	1,2,3,4,7,8,9,10,13,13,14,14-dodecachloro-1,4,4a,5,6,6a,7,10,10a,11,12,12a-dodecahydro-1,4:7,10-dimethanodibenzo(a,e)cyclooctene	[13560-89-9]	Occidental Chemical (Dechlorane Plus)
	1,4,5,6,7,7-hexachlorobicyclo[2.2.1]-5-heptene-2,3-dicarboxylic acid corresponding anhydride	[115-28-6] [115-27-5]	Occidental Chemical, Velsicol

Table 13. Organophosphorus flame retardants

Type or structure	Name, CAS registry no.	Producer (trade name)
Triaryl phosphates		
Triphenyl phosphate	[115-86-6]	Akzo (Phosflex)
Tricresyl phosphate	[1330-78-5]	Ciba-Geigy (Reofos)
Isopropylated triphenyl phosphate		FMC (Kronitex)
		Monsanto (Santicizer)
Other phosphates, phosphonates, or phosphinates		
$[C_6H_5-O]_2P(=O)-OC_{10}H_{21}$	diphenyl decyl phosphate [14167-87-4]	Monsanto (Santicizer)
$CH_3-P(=O)-(OCH_3)_2$	dimethyl methylphosphonate [756-79-6]	Akzo (Fyrol DMMP)
		Albright & Wilson (Amgard DMMP)
		Ciba-Geigy (Reoflam DMMP)
$\left[\begin{array}{c} H_3C \\ H_3C \end{array}\!\!>\!\!C\!\!<\!\!\begin{array}{c} CH_2O \\ CH_2O \end{array}\!\!>\!\!P(=\!S)\!-\!O\right]_2$	1,3,2-dioxaphosphorinane-2,2'-oxybis-(5,5-dimethyl-2,2'-disulfide) [4090-51-1]	Sandoz (Sandoflam 5060)
$\begin{array}{c} H_3C \\ \;\;\;\;\,\vert\,O\!-\!C\!=\!O \\ O\!=\!P \\ \;\;\;\;\;H_2C\!-\!CH_2 \end{array}$	2-methyl-2,5-dioxo-1,2-oxaphospholane [15171-48-9]	Hoechst
$(C_2H_5O)_2P(=O)-CH_2-N(CH_2CH_2OH)_2$	diethyl N,N-bis(2-hydroxyethyl)amino-methylphosphonate [2781-11-5]	Akzo (Fyrol 6)
Polyols containing phosphorus		Akzo (Fyrol 35)
		Hoechst (Exolit 413)
$O=P(OCH_2CH_2Cl)_3$	tris(2-chloroethyl) phosphate [115-96-8]	Akzo (Fyrol CEF)
		Albright & Wilson (Amgard TCEP)
		Bayer (Disflamoll)
		Hoechst (Genomoll P)
$O=P\!\!\left[\!OCH\!\!<\!\!\begin{array}{c}CH_3 \\ CH_2Cl\end{array}\right]_3$	tris(2-chloroisopropyl) phosphate [13674-84-5]	Akzo (Fyrol PCF)
		Albright & Wilson (Amgard TMPC)
$O=P\!\!\left[\!OCH\!\!<\!\!\begin{array}{c}CH_2Cl \\ CH_2Cl\end{array}\right]_3$	tris(2,2'-dichloroisopropyl) phosphate [13674-87-8]	Akzo (Fyrol FR-2)
		Albright & Wilson (Amgard TDCP)
$(ClCH_2CH_2O)_2\underset{O}{\overset{\|}{P}}\!-\!\underset{CH_3}{\overset{\vert}{CH}}\!-\!O\!\left[\!\underset{O}{\overset{OCH_2CH_2Cl}{\overset{\vert}{\underset{\|}{P}}}}\!\!-\!\!\underset{CH_3}{\overset{CH_2CH_2Cl}{\overset{\vert}{CH}}}\!-\!O\!\right]_n\!\!\underset{O}{\overset{\|}{P}}\!-\!OCH_2CH_2Cl$	polyphosphonate [59036-37-2]	Sandoz (Sandoflam 5087)
Halogenated phosphate esters and other proprietary phosphorus compounds		Akzo (Fyrol)
		Albright & Wilson (Amgard)
		Ciba-Geigy (Reoflam)
		FMC (PB-460, PB-528)
		Great Lakes Chem.
		Sandoz (Sandoflam)
$\left(\!\!\begin{array}{c} H_3C \\ H_3C \end{array}\!\!>\!\!C\!\!<\!\!\begin{array}{c} CH_2O \\ CH_2O \end{array}\!\!>\!\!P(=\!O)\!-\!OCH_2\!\right)_2\!\!C(CH_2Cl)_2$	1,3,2-dioxaphosphorinane-2,2'{[2,2-bis(chloromethyl)-1,3-propanediyl]bis(oxy)}-bis(5,5-dimethyl-2,2'-dioxide) [83044-97-7]	Sandoz (Sandoflam 5085)

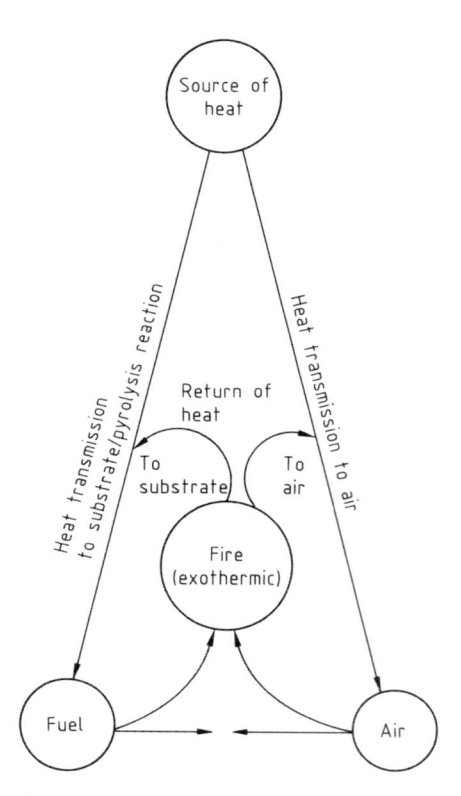

Figure 7. The Emmons fire triangle

such as HO· or H·. Halogenated compounds are especially effective in this respect. Pyrolysis produces hydrogen halides or, if antimony trioxide is present [92], antimony halides, which react as follows:

HBr + HO· → H_2O + Br·

HBr + H· → H_2 + Br·

$SbBr_3$ + H· → HBr + Br_2Sb·

Reactions of this type are less exothermic and block free-radical chain reactions. A drawback of this mechanism is that the gas phase becomes more corrosive. The smoke density may also increase due to inhibition of the combustion reactions.

6.4. Test Methods and Standards

The use of flame-retardant materials (especially in building materials, furnishings, electrical or electronics equipment, mining, and transportation) is covered by application-specific regulations. Many test methods and standards have been created to classify the burning behavior of materials but there is little correlation between them. The small-scale tests performed in materials development can only simulate actual burning behavior within limits; full-scale tests must therefore be performed.

Test methods and standards are specified by (inter)national standards organizations, industry groups or organizations, and government agencies. Numerous test methods are described in [78]. See also, → Flame Retardants, Chap. 3.. The methods encountered most frequently in development and testing laboratories are listed in Table 14. The relatively simple UL 94 test is most widely used and recognized. A standard test rod is clamped in a vertical position, and an igniting flame is applied to the lower end twice for 10 s. The second ignition takes place as soon as the flame from the first ignition is extinguished. The rating criteria are the afterburning time and the ignition of cotton wadding placed under the specimen caused by molten drops of material. The test gives a rating of V-0, V-1, or V-2:

V-0: = afterburning time of single specimen <10 s, of five specimens with two ignitions each <50 s; no ignition of wadding
V-1: = afterburning time of single specimen <30 s, of five specimens with two ignitions each <250 s; no ignition of wadding
V-2: = same as V-1, but with ignition of wadding

The specimens must not be completely burned, otherwise no such classification is given.

6.5. Uses

About 10 % of all plastics (chiefly PVC, ABS, polystyrene, unsaturated polyesters, polypropylene, polyethylene, and polyurethanes) contain flame retardants. The main applications are in building materials and furnishings (structural elements, roofing films, pipes, foamed plastics for insulation and furniture, wall and floor coverings); transportation (equipment and fittings for aircraft, ships, automobiles, and railroad cars); and in the electrical industry (cables; housings and components for television sets, office

Table 14. Test methods and standards (selected)

Test	Standard	Criteria	Applications
LOI test	ASTM D 2863	flammability	development laboratory
Vertical test	UL 94, UL 94–5 V	ignitability dripping	development laboratory electrical industry
Vertical test	DIN 54 332	ignitability	development laboratory floor coverings
Horizontal test	ASTM D 635	burning rate	development laboratory
Miscellaneous tests	FAR 25	ignitability burning rate smoke density	aircraft interiors
Horizontal test	FMVSS 302	burning rate	automotive interiors
Cone colorimeter	ISO DP 5660	ignitability heat release smoke toxic gases	development laboratory materials in buildings
Epiradiator	NF P 92–501	ignitability burning rate	materials in buildings
Surface spread of flame test	BS 476, Parts 5–7	ignitability burning rate	materials in buildings
NBS smoke chamber	ASTM E 662	smoke density	development laboratory aircraft interiors
Toxicity test	DIN 53 436	toxicity of pyrolysis gases	development laboratory materials in buildings
Brandschachttest	DIN 4102	flammability heat release	materials in buildings
Room corner test	NT Fire 025	flame spread smoke toxic gases	materials in buildings
Steiner tunnel test	ASTM E 84	flame spread smoke	materials in buildings

machines, and household appliances; lamination of printed circuits).

More than 90 % of the flame retardants used in thermoplastics are of the additive type. They are added before, during, or after polymerization, but usually when the polymers are processed into compounds or finished products. In the latter case, the retardant is often used as a highly concentrated masterbatch containing 50 – 80 % of the agent.

Reactive flame retardants are added to thermosetting resins before the cross-linking reaction.

Poly(Vinyl Chloride). Up to half of the dioctyl phthalate plasticizers in *flexible PVC* can be replaced by chlorinated paraffins or phosphate esters (e.g., triphenyl or tricresyl phosphate) [93]. Other common additives are Sb_2O_3 (2 – 4 %) and $Al(OH)_3$ (40 %).

In *rigid PVC*, Sb_2O_3 (1 – 4 %) and $Al(OH)_3$ (10 – 20 %) are employed and in some cases zinc borate and zinc stannate. MoO_3, ferrocene, or $CaCO_3$ can be used as smoke inhibitor.

Acrylonitrile – Butadiene – Styrene. A variety of halogenated flame retardants are used for ABS in conjunction with Sb_2O_3; most common for V-0 formulations is a combination of octabromodiphenyl oxide (ca. 20 %) and Sb_2O_3 (ca. 6 %). Halogen-free formulations, for example with $Mg(OH)_2$, are still under development.

Expandable Polystyrene. Hexabromocyclododecane is used at the 3 – 5 % level to comply with building codes; 0.5 % dicumyl peroxide or another free-radical former is added as synergist [88].

High-Impact Polystyrene. The V-0 grades contain ca. 12 % decabromodiphenyl oxide or tetrabromobisphenol A together with 3 – 4 % Sb_2O_3. Ethylene-bis(tetrabromophthalimide) and low molecular mass brominated polystyrenes are also suitable. For V-2 grades, brominated aliphatics are used. Red phosphorus (15 %) has been proposed for halogen-free V-0 grades.

Polyester Resins. The use of 50 – 60 % Al(OH)$_3$, preferably coated, is widespread. Halogenated reactive retardants, such as tetrabromophthalic anhydride and (for more light-stable grades) 2,2-bis(bromomethyl)-1,3-propanediol are also employed.

Low molecular mass brominated epoxy polymers of the F – 2000 series, together with Sb$_2$O$_3$, offer another alternative. Zinc borate or dimethyl methylphosphonate is also used.

Polyolefins. In polypropylene, 5 – 20 % tetrabromobisphenol A-bis(2,3-dibromopropyl ether) is used, along with 3 – 8 % Sb$_2$O$_3$ [94]. Other halogenated compounds are also used in combination with Sb$_2$O$_3$: ethylene-bis(tetrabromophthalimide), ethylene-bis(dibromonorbornane dicarboximide) or Dechlorane Plus. Halogen-free V-0 grades are obtained by using 25 – 35 % of an intumescent system based on ammonium polyphosphate – pentaerythritol – melamine derivatives or 50 – 60 % Al(OH)$_3$ or Mg(OH)$_2$ [95].

Chlorinated paraffins combined with Sb$_2$O$_3$ are used in LDPE films. Brominated aromatics or Dechlorane Plus with Sb$_2$O$_3$ are used for HDPE and cross-linked polyethylene. Levels of 30 % halogenated organics and 10 % Sb$_2$O$_3$ are required for V-0 grades. Halogen-free V-0 grades (e.g., cable sheathing compounds) must contain at least 60 % Al(OH)$_3$ or Mg(OH)$_2$; addition of zinc borate, zinc stannate, or silicones is recommended. Intumescent systems have also been proposed.

Polyurethanes. Flexible polyurethane foams (e.g., for upholstery) contain 20 – 30 % melamine together with a few percent of halogenated phosphate [tris(chloroethyl) phosphate, tris(2-chloroisopropyl) phosphate] or chlorinated paraffins. A combination of graphite and tris(2-chloroisopropyl) phosphate has also been successful. Other formulations include Al(OH)$_3$ or phosphorus-containing products such as dimethyl methylphosphonate, tris(2-chloroethyl) phosphate, tris(2-chloroisopropyl) phosphate, and ammonium polyphosphate.

Rigid foams employ phosphorus-containing or (preferably) halogenated diols and polyols [e.g., diethyl-*N,N*-bis(2-hydroxyethyl)aminomethyl phosphonate, 2,2-bis(bromomethyl)-1,3-propanediol or IXOL B 251] in combination with dimethyl methylphosphonate or halogenated phosphates.

Thermoplastic polyurethane can be rendered flame resistant with brominated compounds and Sb$_2$O$_3$ [96] or intumescent systems.

Polyamides. A variety of formulations are used for V-0 grades of polyamide 6 and polyamide 66 [97]. In systems based on brominated aromatics (15 – 20 %) and Sb$_2$O$_3$ (4 – 5 %), ethylene-bis(tetrabromophthalimide) and polymeric flame retardants such as poly(tribromostyrene), poly(dibromophenylene ether), and F – 2000 products dominate. Dechlorane Plus is also employed with synergists (e.g., Sb$_2$O$_3$, zinc borate, or Fe$_2$O$_3$ and Fe$_3$O$_4$).

In halogen-free systems, polyamides contain 7 – 10 % red phosphorus or, in the absence of glass-fiber fillers, melamine cyanurate. Tests with Mg(OH)$_2$ are under way.

Polyesters (PBT and PETP). The products used for polyesters are similar to those for polyamides, but combinations of brominated compounds (10 – 20 %) with Sb$_2$O$_3$ (3 – 5 %) clearly dominate. A cyclic methylphosphinic acid derivative 2-methyl-2,5-dioxo-1,2-oxaphospholane is chemically incorporated into polyester fibers at the condensation step.

Polycarbonates. Tetrabromobisphenol A is incorporated into polycarbonates as a comonomer; very small amounts of the alkali salts of aromatic sulfonic acids improve flame retardance [89]. Opaque V-0 polycarbonate grades are obtained by using 2 – 4 % brominated aromatics and 1 – 2 % Sb$_2$O$_3$ as additive retardants.

Poly(Methyl Methacrylate). The cast plastic can be treated with 13 – 20 % of a halogenated polyphosphonate (Sandoflam 5087) [98].

Poly(Phenylene Oxide) – Polystyrene. V-0 grades of blends of polystyrene and poly(phenylene oxide) can be obtained with ca. 15 % halogen-free triaryl phosphates.

Epoxy Resins. Reactive flame retardants (tetrabromobisphenol A and derivatives, glycidyl ethers of brominated aromatics) are employed for epoxy resins (minimum bromine content

20 %). For printed-circuit laminates, the use of Sb_2O_3 must be avoided. Tests on epoxy resins are being carried out with red phosphorus, Al(OH)$_3$, and zinc borate.

6.6. Toxicology

The question arises whether flame retardants, especially those with chemical action, increase the toxicity of combustion gases. Animal studies have shown that pyrolysis gases emitted by flame-retarded plastics are often less toxic than those from untreated plastics [99]. Standards governing toxicity tests on combustion gases include DIN 53 436 [100] and ISO/IEC/TR 9122–1. Carbon monoxide is the most significant factor in combustion-gas toxicity.

The inherent toxicity of some flame retardants and potential hazards during their processing and use in plastics have been discussed. Of special concern are the polybrominated diphenyl ethers because polybrominated dibenzodioxins and dibenzofurans have been detected in their decomposition products. The Verband der Chemischen Industrie (Chemical Industry Association) in Germany has decided to gradually phase out these retardants. In the United States, an initiative of the Brominated Fire Retardant Industry Panels (BFRIP) has led to in-depth studies on the relevance of the findings [101].

Chlorinated paraffins must be labeled in the United States if their residual carbon tetrachloride content exceeds 0.1 %.

6.7. Economic Aspects

The estimated current world market (1990) of flame retardants is 500 000 t, which is subdivided as follows:

Inorganic	290 000 t
Brominated organic	80 000 t
Chlorinated organic	50 000 t
Organophosphorus	60 000 t

The dominant inorganic product is Al(OH)$_3$, which is also the cheapest flame retardant available.

Producers and trade names of flame retardants are listed in Tables 10–13.

7. Fillers and Coupling Agents

7.1. Purpose and Requirements

Filled (or reinforced) plastics contain large amounts of fillers (20 – 50 wt % referred to the polymer or even occasionally higher). Fillers are divided into *extenders*, which merely provide bulk and lower product cost, and *reinforcements*, which improve several of the polymer properties. *Coupling agents* improve the adhesion between filler and polymer, preferably via chemical bonds; their use confers reinforcing properties on inexpensive extenders, improves the performance of reinforcements, and allows the filler content to be increased. The reinforcing effect of fillers depends on their chemistry, shape (fibers, flakes, spheres), and size (fiber length, particle size). The most important reinforcement effects are

1. Increased strength, ultimate elongation, rigidity (modulus of elasticity), and in some cases impact strength
2. Improved heat distortion temperature and dimensional stability, reduced shrinkage, and improved stability of mechanical values at high temperatures and over extended times (fatigue)
3. Modification of density (usually increased)
4. Improved chemical resistance and lower water absorption
5. Better surface quality and surface hardness

Special fillers are used to obtain compounds with the following properties:

1. Low flammability
2. Electrical conductivity and electromagnetic shielding
3. Radiation and UV shielding
4. Biodegradability
5. Noise suppression

Fillers are characterized by a number of parameters that include geographic origin (minerals); particle size and size distribution or fiber length; purity, especially heavy-metal content; water content; whiteness; oil adsorption value or plasticizer adsorption value (ASTM D 281–31, for PVC fillers); density; specific heat; and thermal conductivity. Consistency, especially with respect to purity and particle size, is an important criterion of quality.

Occupational health aspects (e.g., dust) and processing qualities (e.g., abrasive behavior toward machinery, effects on viscosity, and wettability – dispersability in the melt) are also important factors influencing the selection of fillers.

7.2. Forms and Chemical Structures

Spherical Fillers and Powders. Calcium carbonate is quantitatively the most important filler. Natural forms (e.g., chalk, marble, or limestone) are ground to powder in dry or wet form. Finer synthetic or precipitated forms [often abbreviated as CCPs (calcium carbonicum praecipitatum)] are obtained by passing carbon dioxide through milk of lime or by reacting calcium chloride and sodium carbonate.

Other mineral fillers include ground dolomite [$CaMg(CO_3)_2$], calcium sulfate (gypsum), and barium sulfate (ground barite or precipitated forms, which have a very high density of 4.5 g/cm^3).

Silica is used as a natural crystalline product (sand, quartz powder) or amorphous synthetic product. The latter is obtained by precipitation from a waterglass solution with acid or by flame hydrolysis from silicon tetrachloride (pyrogenic silica gel).

Solid and hollow glass microspheres of various sizes are employed. Hollow microspheres have the lowest density of all fillers (0.2 – 0.4 g/cm^3).

Other powders include wood flour, cork flour, starch, metal oxides (e.g., Al_2O_3, MgO, ZnO, TiO_2, iron oxides), and metal powders (e.g., Al, Cu, Ni, bronze, Zn, Ag-coated silicate spheres).

Carbon black is obtained by incomplete combustion of gaseous and liquid hydrocarbons. The quality and potential applications (black pigment, conductivity additive, or reinforcement) depend on the production process, purity, and particle size.

Flake fillers include talc, kaolin (China clay), mica, silicon carbide, aluminum trihydroxide, graphite, and metals (e.g., aluminum).

Fibrous Fillers and Reinforcements. A wide variety of inorganic and organic fibrous fillers and reinforcements are used. New ceramic and mineral fibers are being developed.

Carbon fibers are made primarily by oxidative thermal cyclization of polyacrylonitrile fibers. They are classified by their strength. Short fibers (e.g., 3 or 6 mm) are cut from filament yarns.

The starting material for the production of *glass fibers* is glass filament, which is cut into textile glass a few millimeters long. This product is milled to short fibers (e.g., 0.2 – 0.6 mm) when the reinforcement is blended with the polymer in the extruder.

Aramid fibers are obtained, for example, by the polycondensation of terephthaloyl chloride and *p*-phenylenediamine. They are cut to various lengths and used as reinforcement.

Single-crystal fibers (i.e., *whiskers*) can be based on metals, metal oxides, or metal carbides.

Coupling Agents. Typical silane finishes have the general structure $(X)_3Si(CH_2)_3$-Y, in which X is a short-chain alkoxy group (OCH_3, OC_2H_5), an acyloxy group, OH, or Cl; Y is a functional group such as mercapto, amino, azido, methacrylate, epoxy, an anhydride function, or a longer hydrocarbon chain.

Functional groups (e.g., vinyl) can also be attached directly to the silicon. Polymeric silanes [102] containing alkoxy groups and functional groups, as well as bis(trimethoxy) silyl, are also used. New developments include blends of silanes having different functions [103].

The *titanate and zirconate systems* have the general formula $X_mM(O\text{-}X\text{-}R\text{-}Y)_n$, where M denotes Ti or Zr; X denotes an alkoxy group; Z may be a carboxyl, sulfonyl, phosphate, or pyrophosphate group; R can be a long hydrocarbon chain; and Y is a functional group such as amino or methacryloxy. Not all the groups Z, R, and Y need be present. A survey of recommended types is given in [104].

Miscellaneous Coupling Agents. These include aluminates [104], zirconium aluminates [105], stearic acid and stearates, chlorinated paraffins, polyolefin waxes, and carboxylated polybutadienes, or polypropylenes modified with acrylic or maleic acid.

7.3. Mechanism of Action

Reinforcement depends on the transmission of mechanical energy (impact energy, tensile stress, etc.) from the polymer matrix to the filler. It increases with the strength of (chemical) bonding between the fillers and the matrix. Bonding also limits the mobility of the polymer chain and stiffens the material. Fibrous and flake fillers and reinforcements give rise to anisotropic properties.

Chemical bonding seldom occurs directly between the matrix and the filler; it is usually mediated by coupling agents. In silanes, zirconates, and titanates, for example, the X groups react directly, or after hydrolysis, with hydroxyl or carboxyl groups on the filler surface. The finish materials, particularly silanes, also undergo condensation to form higher molecular mass units:

$$\text{Filler surface} \begin{cases} -\!\!\!-\!\!\!-\text{O}-\text{Si}-(\text{CH}_2)_3\text{Y} \\ \quad\quad\quad\;\text{H} \quad \text{O} \\ -\!\!\!-\!\!\!-\text{O}\cdots\text{O}-\text{Si}-(\text{CH}_2)_3\text{Y} \\ \quad\quad\quad\text{H} \quad \text{O} \end{cases}$$

The Y functional groups react with the polymer (especially in thermosets) or are anchored to the polymer by physical mechanisms, for example, by penetration into the coiled structure in thermoplastics.

7.4. Test Methods

Fillers are tested on receipt, particularly for particle size (screen or sedimentation analysis), acidity, whiteness, density, oil adsorption, and plasticizer adsorption value [dioctyl phthalate (DOP) number]. Tests can be performed by standard methods [106]. Testing the adhesion between the filler and the polymer matrix and correlating it with important physical properties is especially difficult. Publications in this area deal chiefly with fiber-reinforced polymers [107].

7.5. Uses

Nearly all thermosetting and thermoplastic resins are filled or reinforced for special applications. The main applications are in PVC, followed by polyesters, polypropylene, and polyamides. The following examples give a brief summary of the types of fillers and reinforcements used and their effects. Further information can be found in the literature (e.g., [108]).

Calcium Carbonate is important in PVC. Natural forms of calcium carbonate are preferred for flexible PVC because they have lower plasticizer values than precipitated forms. Filler contents of 30 – 60 % or higher are common for cables, film, and floor covering, leading to substantial reductions in cost and additional savings of lubricants, plasticizers, white pigments, or stabilizers.

Precipitated calcium carbonates with surface treatments (coated CCPs) are preferred for rigid PVC. Contents of 10 – 20 %, occasionally up to 50 %, are used for window frames and other profiles, foam, and pipes. Important considerations, in addition to lower cost, are improved impact strength and stiffness, higher production rate, and improved surface appearance. $CaCO_3$ is valued for its low abrasivity in processing.

Other areas for $CaCO_3$ use include polyolefins. Coated types are used for polypropylene in industrial applications (increased heat distortion temperature, dimensional stability, rigidity, and impact strength). Films of LDPE and LLDPE can also be filled with 30 – 40 % $CaCO_3$, which provides slip and antiblocking effects (e.g., trash bags). Calcium carbonate is also used in polyamides, PETP, PBT, and semiopaque grades of other thermoplastics, such as poly(methyl methacrylate).

Barium Sulfate is employed in thermoplastics and provides acoustic damping, X-ray absorption, and increased density.

Kaolin is used in elastomers and flexible PVC for electric cables because of its high water resistance and excellent insulating properties. Kaolin has a high whiteness but also a higher plasticizer adsorption value than, for example, $CaCO_3$.

Mica is employed, for example, in automotive parts made of polypropylene.

Talc is a widely used filler for PVC and polyolefins. Polypropylene components gain in heat distortion temperature and rigidity but lose impact strength and thermooxidative stability.

Aluminum Trihydroxide and Magnesium Dihydroxide are used as flame retardants in polymers such as polypropylene, polyethylene (cable), PVC (carpet backing), and unsaturated polyesters. Filler contents are high (>50 %).

Zinc Oxide serves as an UV absorber in polyolefins.

Wood Flour and Starch. Wood flour is employed in acoustical components made of, for example, HDPE or PVC. The addition of starch, preferably in conjunction with oxidation promoters, is reported to render polyethylene films biodegradable.

Metal powders and carbon black are added to many polymers to improve thermal or electric conductivity.

Silica gel has many applications in thermoplastics (PVC, polyamides) and thermosets (polyester resins, epoxy resins). It is also used as antiblocking and slip agent in LDPE.

Glass spheres are used in polyolefins (including foams) and engineering thermoplastics (polyamides, PETP, PBT). Hollow glass spheres lower the density and result in high transparency.

Glass fibers are by far the most important reinforcement for plastics. For further details, see → Fibers, 5. Synthetic Inorganic, Section 2.1.. They are generally used at levels of 30 – 40 %, and improve tensile and flexural strength, elastic modulus, and heat distortion temperature. The reinforcing effect increases with fiber length and coupling agent. Fiber length must be matched to the elastic modulus of the matrix. Engineering thermoplastics [e.g., polyamides, poly(phenylene oxide), PETP, PBT] and ABS are usually supplied with glass-fiber reinforcement (typical length 0.2 mm) and used for housings or injection moldings in electrical and electronics equipment.

Polypropylene with glass-fiber reinforcement (typical length 0.66 mm) is also important, especially for automotive parts in the United States. Reinforcement results in polypropylene with properties similar to those of engineering thermoplastics. In the thermoset area, long fibers are used for reinforcement. Glass-fiber reinforcement is less common in polyethylene, PVC, and polystyrene.

Carbon Fibers are used mainly as short fibers processed into granular compounds. They provide dimensional stability, electrical conductivity, and (especially metal-coated grades) electromagnetic shielding. They are used in engineering thermoplastics, and epoxy resins. For further details, see → Fibers, 5. Synthetic Inorganic, Chap. 5..

Aramid Fibers (see also, → Fibers, 2. Structure, Section 4.2.) are used chiefly as long fibers (in thermoplastics, their length corresponds to that of the compound granules). They lend rigidity with a low density. Preferred substrates are polyamides, PETP, and epoxy resins.

Coupling Agents are employed at levels of 0.1 – 4 % referred to the filler. They are usually applied to the filler (coated fillers). Aqueous solutions, emulsions, and liquid products are applied by spraying. Coupling agents can, however, be also incorporated directly into the polymer or supplied as powdered concentrates in inert fillers and blended in at the compounding stage.

Silane Coupling Agents are applied to glass fibers, glass beads, and silicates such as mica, talc, wollastonite, and kaolin. They are also used in polymers treated with the flame retardants Al(OH)$_3$ and Mg(OH)$_2$. Silanes (except for special grades) are not, however, suitable for CaCO$_3$, carbon fibers, or metal oxides. They are useful for service in a humid atmosphere because they afford hydrolysis-resistant adhesion.

Titanates are especially good for CaCO$_3$, BaSO$_4$, carbon fibers, aramid fibers, metals, and metal oxides. They tend to become discolored in the presence of phenolic antioxidants. *Zirconates* are used for both CaCO$_3$ and silicates; they do not discolor with phenolic antioxidants.

Stearic Acid and Stearates are mainly used in CaCO$_3$, especially CCP grades. *Chemically modified polypropylene and polybutadiene* are used on glass fibers, mica, CaCO$_3$, and other fillers, chiefly in polypropylene.

Functional groups in the finish should be selected in accordance with the type of polymer.

Table 15. Producers and trade names of coupling agents

Producer	Trade name	Type
Akzo		titanates
BP Performance Polymers	Polybond	functionalized polymers
Cavedon	Cavco	zirconates, zirconium aluminates
Degussa		silanes
Dow Corning		silanes
Du Pont		titanates
Eastman	Epolene	functionalized polymers
Hercules	Az-cup	silanes
Hüls	Dynasilan, Polyvest	silanes, titanates, functionalized polymers
Kenrich Petrochemicals	Ken-React	titanates, zirconates, zirconium aluminates
Nippon Soda		titanates
PRC	Prosil	silanes
Titanium Intermediates		titanates
Union Carbide	Ucarsil PC	silanes
Wacker Chemie		silanes

Amino groups are suitable for polyamides, polycarbonates, PETP, PBT, PVC, and many thermosets. Methacrylate groups are used with polystyrene, ABS, SAN, and thermosets, as are epoxy and mercapto groups. Vinyl groups are employed for polystyrene, polyolefins, and PVC. Azido groups are recommended for polyolefins; inert, long-chain hydrocarbon groups can also be used instead of functional groups.

7.6. Economic Aspects

The market volume for fillers and fiber reinforcements is difficult to estimate but is probably on the order of at least 5×10^6 t/a. Mineral fillers are most important, followed by synthetic fillers and glass fibers. Coupling agents have reached a market volume of ca. 10 000 t/a. Because of the large number of suppliers, reference must be made to the literature [108], [109]. Table 15 lists some important producers and trade names of coupling agents.

8. Dyes and Pigments [110–112]

Although the main function of a colorant (i.e., a dye or a pigment) is to impart decorative character and sales appeal, it may also influence the functional properties of the substrate. The selection of colorants from a complex group of minerals and chemical substances has been cited as one of the higher forms of the "black art" of plastics compounding. The problem of colorant selection has become more acute because of the growing complexity of legislation dealing with product safety, occupational safety, and environmental protection. Thus the coloration of plastics is becoming an increasingly precise science.

8.1. Requirements [113], [114]

Modern color-delivery systems (particularly masterbatches) provide overall economy and ease of application. The main requirements for a colorant used in plastics include

1. Inertness under processing conditions (visible and invisible thermal fatigue)
2. Physical and chemical inertness toward the polymer
3. Compatibility with other additives (antagonism/synergism)
4. Required performance during the intended service life of the plastic product
5. Nontoxicity
6. Ability to be disposed of in discarded plastic items without causing extensive environmental nuisance (ecological safety)

Colorants are chosen from a wide range of organic and inorganic compounds. Pigments are particulate organic and inorganic solids that are virtually insoluble in the medium into which they are incorporated. The pigment particles must be dispersed in this medium. Dyes are organic compounds that are usually dissolved in the substrate. Incompatibility is manifested by "crocking", the migration of colorant to the free surface of the plastic from which it can be rubbed off.

Colorants are introduced into plastics by several methods [115]. The dry colorant may be blended with the polymer before processing (dry blending); dispersion or dissolution is less efficient than with color concentrates (masterbatches). Masterbatches are predispersed colorants in a physical form that is easy to handle and use (usually pellets, but also liquid or paste). They offer numerous advantages over powdered colorants, particularly more hygienic working conditions.

8.2. Pigments

8.2.1. Inorganic Pigments [116], [117] See also → Pigments, Inorganic, 1. General

Inorganic pigments are insoluble in plastics and hence do not bleed or migrate; they are therefore the colorants of choice for plastics. They generally exhibit good opacity and hiding power. Colors range from dull earth tones (iron oxides) to bright yellows and reds (pigments based on lead and cadmium).

Titanium Dioxide is the most widely used pigment; it imparts opacity and whiteness to all plastics. Rutile grades are preferred. They are produced by the chloride process, which gives a brighter product, or the sulfate process, which gives a product with a slight "bone" undertone but a lower abrasiveness. To enhance dispersibility and lower the photo- and antioxidant reactivity toward polymer resins, the pigments are usually coated with inorganic materials (e.g., silica or alumina). They are also aftertreated with organic compounds to render them hydrophobic and make them easier to disperse. Plastic-grade titanium dioxide pigments of high hiding power usually have a much finer particle size, in the 0.17 – 0.24 µm range.

Zinc Pigments. Although not as important as titanium dioxide, white zinc-based pigments do have technical advantages, particularly where wear of processing equipment and damage to the polymer during processing or in service are crucial.

Iron Oxide Pigments constitute a very important class of plastics colorants, especially because of their relatively low cost and because of toxicology and environmental considerations. The major shortcomings of iron oxide pigments are their rather dirty shades, ranging from buff yellow to brick red, brown, and black; their abrasiveness with respect to processing equipment; and the fact that they promote the degradation of sensitive polymers (e.g., PVC, polyolefins, and polyamides) by light and weathering. Moreover, the yellow, brown, and black iron oxides are stable only up to 180 °C.

Chromium Oxide Green. The dull green chromium oxide pigments have a poor tinting strength but outstanding heat, light, and chemical stability, making them suitable for all polymers. They are, however, difficult to disperse.

Lead Chromate Molybdate Pigments. The versatile group of lead chromate molybdate pigments ranges in color from primrose to lemon yellow, orange, scarlet, and even green. Silica-coated chromate and molybdate pigments have been developed to provide heat and weather resistance in plastics. Although they offer a combination of bright shades, high hiding power, excellent lightfastness, and low cost, the trend now is toward pigments that do not contain heavy metals.

Cadmium Pigments are ideal for coloring plastics and are the only bright colors available for many industrial polymers. They have excellent heat stability, even in highly aggressive industrial polymers such as nylons; a distinct hue ranging from yellow to orange, deep red, and maroon; and very good lightfastness. Stability under wet conditions is rather limited, however. Cadmium pigments do not cause dimensional instability in colored plastics. As with lead chromate pigments, however, stringent regulations and greater environmental consciousness are forcing plastics colorant users to phase out pigments based on cadmium.

Mixed Metal Oxide Pigments. In the search for lead- and cadmium-free formulations, metal oxide pigments (especially nickel antimony titanate and chromium antimony titanate) have recently attained prominence in both coating and plastics applications. Although their color intensity is low, they are usually combined with color-intense organic colorants to produce the required shades. Titanate pigments have exceptional heat, weather, and chemical stability. Cobalt aluminate pigments are used mainly as shading components.

Ultramarine Blue. Apart from their poor acid resistance, ultramarine blues exhibit excellent all-round fastness. Unlike the blue phthalocyanine pigments, ultramarine blues do not induce distortion in polyolefins. With their excellent heat stability, ultramarines are ideally suited for color correction of white and clear polymers.

Special-Effect Pigments [118]. Pearlescent (nacreous), iridescent, and metallic colorants are increasingly used in plastics. Synthetic nacreous pigments consist of TiO_2-coated mica. Colored pearlescent pigments (e.g., Iriodin, Merck) have an additional coat of iron oxide or chromium oxide. These pigments provide a pearly luster and are used, for example, in cosmetic packaging and fashion articles.

A metallic appearance can be imparted by metal flakes (e.g., aluminum).

8.2.2. Organic Pigments [116], [117], [119], [120] See also → Phthalocyanines, → Pigments, Organic

Organic pigments have cleaner and brighter shades and a much higher tinting strength than inorganic pigments. They tend to dissolve in polymers, however, particularly at very low concentration. The dissolved pigment can undergo thermal degradation [121]. Organic pigments generally produce translucent or semitransparent coloring. Many color formulations are derived from a combination of inorganic and organic pigments whose properties complement one another.

In contrast to inorganic pigments, it is difficult to indicate a particular class of organic pigments for a particular application.

The newer types of monoazo yellows and oranges (e.g., C.I. Pigment Yellow 180, 181, and 182; Orange 64) and monoazo yellow toners (C.I. Pigment Yellow 183, 190, and 191) possess improved heat stability up to 280 °C in polyolefins and even ABS, similar to those of disazo condensation pigments. Among monoazo red toners, C.I. Pigment Red 151 and 247 show good heat stability but rather poor lightfastness.

Diarylide yellows, oranges, and reds tend to undergo thermal decomposition in polyolefins above 200 °C, particularly at low concentration, generating potentially toxic 3,3'-dichlorobenzidine. Their use has recently been limited to PVC.

Azo condensation pigments have adequate heat stability for use in polyolefins and other resins processed at moderate temperature. They generally tend to be dirty in mass tones, however, especially reds. An exception is C.I. Pigment Red 242, a brilliant scarlet, which is of great interest as a heavy-metal replacement.

Except for their tendency to induce distortion in polyolefins, the phthalocyanine pigments, especially the blues, represent the nearest approaches to ideal pigments on the market. Not only are they thermally and photochemically stable, but they also exhibit excellent chemical stability (unlike many inorganic pigments).

Metal-complex pigments are generally rather dirty in shade, particularly in white reductions. C.I. Pigment Orange 68, a heterocyclic metal-complex pigment, is one of the substitutes for heavy-metal formulations in nylons and polycarbonates.

Polycyclic and some heterocyclic pigments (including diketopyrrolopyrrole pigments [122]) generally have very good heat stability in nonpolar polymers. In polar resins, especially nylons, they tend to poor heat and light stability. Of particular importance is the novel heterocyclic perinone pigment C.I. Pigment Yellow 192, which is an ideal cadmium substitute in nylons and polyesters.

Other organic pigments used to color plastics include isoindolines, isoindolinones, dioxazines, anthraquinonoids including indanthones, perylenes, perinones, thioindigoids, quinacridones, and quinophthalones.

8.2.3. Dyes [123]

Because of economy and ease of application, polymer-soluble dyes have been used to color polymers in brilliant and transparent shades. They show reasonably good lightfastness in transparent colorations. In combination with opacifying inorganic pigments, especially titanium dioxide, their lightfastness is rather poor. Polymer-soluble dyes are usually suitable for transparent plastics but not for polyolefins.

Dyes for plastics have a variety of chemical structures. Azo dyes are generally heat- and light-sensitive. Metal-complex azo dyes, especially the chromium complexes, exhibit good stability in nylons. Anthraquinone, quinophthalone, methine, naphthazine, perinone, cumarin, thioindigo, and thioxanthene dyes are usually employed.

Fluorescent pigments are used to produce eye-catching colorations, but tend to plate out in plastics [118]. They consist of finely divided resin particles that contain fluorescent dyes; see → Fluorescent Dyes, Chap. 12.

9. Miscellaneous Additives

9.1. Nucleating Agents

The properties of thermoplastics depend on their degree of crystallization and the morphology of their crystalline phases. Plastics may be amorphous (e.g., PVC, polystyrene, LDPE), of moderate crystallinity (e.g., LLDPE, polypropylene, polyamide 6, PETP), or highly crystalline (e.g., HDPE or polyamide 66). Chain mobility, tacticity, and side-chain volume determine the degree of crystallinity, as do the temperature profile, pressure, and orientation during polymer processing.

During cooling of polymer melts the disordered coiled structure is transformed into an ordered state with the formation of lamellar crystallites. The crystal nuclei can be residual, unmelted polymer crystallites already present in the melt (self-nucleation) or added agents (nucleating agents). The latter make crystallization energetically more favorable and increase the number of nuclei [124–127]. The resulting crystallites form a morphological superlattice structure (e.g., spherulites), which have a central primary nucleus and a radially symmetric structure with amorphous domains occurring between the spherulites. Scattering of light by coarse spherulites (> 0.5 µm) reduces the transparency of plastics.

Nucleating agents control the crystallization behavior of moderately crystalline plastics in the following ways:

1. Increasing the crystallization temperature by ca. 10 – 20 °C
2. Accelerating crystallization rate
3. Decreasing the spherulite diameter and narrowing the size distribution

Classification. Nucleating agents can be divided into

1. Inorganic fillers and pigments: talc, kaolin, silica gel, TiO_2
2. Salts of carboxylic acids [128]: sodium benzoate, aluminum 4-*tert*-butyl benzoate, sodium montanate, sodium β-naphthenate
3. Sodium organophosphates [129]
4. Dibenzylidene sorbitols [130], [131]
5. Waxes [132] and ionomers [133]
6. Systems of carboxylic acids and amines [134]

Some producers and trade names are listed in Table 16.

Properties. Nucleating agents offer some or all of the following advantages. Cycle times in injection molding are reduced by as much as 30 %. They improve optical properties (particularly transparency and gloss), mechanical properties, dimensional stability, and oxidation resistance(less oxygen diffusion in crystalline phases). They also decrease water absorption (polyamides).

Preferred nucleating agents have a high melting point, insolubility in polymer melts, good wettability and dispersability in melts, small particle size (e.g., 1 µm), adequate thermal stability and no odor. They are also approved as indirect food additives (desirable).

Test Methods. Several methods are employed to measure crystallinity.

FT – IR spectroscopy is the simplest way of determining the degree of crystallinity, (i.e., the mass fraction of the crystalline phase) [135].

Differential scanning calorimetry is used to measure the crystallization rate and the temperature at which the maximum crystallization rate occurs [136].

Table 16. Producers and trade names of nucleating agents

Manufacturer	Trade name	Type
Adeka Argus	Mark NA	organic sodium phosphates
Allied Signal	Aclyn	ionomer
Du Pont	Surlyn	ionomer
EC Chemical Co.	EC-1	dibenzylidene sorbitol
Hoechst	Hostalub, Hostamont	sodium montanate
ICI Speciality Chemicals	SCS, Clarifex	proprietary
Milliken Chemical Co.	Millad	dibenzylidene sorbitol
Mitsui Toatsu	NC	dibenzylidene sorbitol
New Japan Chem.	Gellal	dibenzylidene sorbitol
Sandoz Chemicals	Sandostab	aluminum *tert*-butyl benzoate
Schering	Geniset MD	dibenzylidene sorbitol
Shell	AL-PTBBA	aluminum *tert*-butyl benzoate
Witco Corp.	Mark	organic sodium phosphates

Polarization microscopy is useful for measuring spherulite size. Wide-angle X-ray diffraction is used for more detailed studies, for example, of allomorphic forms.

Uses. Nucleating agents are used mainly in polyamide 6, PETP, polyethylene, and polypropylene. Product applications include packaging films, beverage containers, video cassettes, and automotive parts. They are added at a level of 1000 – 5000 ppm and incorporated by dry blending, at the polycondensation stage, or in masterbatch form.

Shortening the injection-molding cycle time is a primary consideration for PETP; inorganic additives, waxes, or sodium salts are used. The situation is similar for polyamide 6, in which many mechanical properties can be improved. Sodium benzoate, aluminum benzoate, phenyl phosphates, and dibenzylidene sorbitols (clarifiers) are employed with polyolefins.

9.2. Antistatic Agents [137]

Plastic surfaces can become electrostatically charged through friction. Charge can build up during processing or service, for example, by the friction of air against the plastic surface. This has several disadvantages:

1. Dirty surface due to attracted dust (a problem with furniture, records, packaging, bottles, and cassettes)
2. Electric shock from walking on plastic floor coverings
3. Adhesion problems in processing films, fibers, foams, and powders
4. Functional disorders in electronic devices and computers
5. Spark generation and dust explosions

Surface-Active Antistatic Agents form a conductive surface layer, in the simplest case a thin film of water, that allows the charge to dissipate quickly or prevents it from building up. In some cases, other additives must be used to impart volume conductivity throughout the bulk of the plastic; these are referred to as *conductivity additives*.

Most plastics have surface resistances of $10^{14} - 10^{16}$ Ω, which can be lowered to $10^8 - 10^{10}$ Ω by antistatic agents utilizing the water-film mechanism. Certain applications, such as explosion hazard areas or electromagnetic shielding, call for even lower surface resistance or lower volume resistivity; conductivity additives can be used to obtain volume resistivities of $10^1 - 10^{-2}$ $\Omega \cdot$ cm.

Quality criteria for antistatic agents include the permanence of their effect, thermal stability, nonvolatility, color, product form, interaction with stabilizers (especially in PVC), and surface effects (tack, weldability, printability). Approval for use as an indirect food additive is often required.

Surface-Active Antistatic Agents are applied by solution spraying (external antistatic agents, nonpermanent) or incorporated into the plastic mass (internal antistatic agents). They are used at levels of 0.1 – 2.5 %, reaching 5 – 7 % in flexible PVC. Internal antistatic agents are amphoteric compounds with limited polymer compatibility (i.e., they tend to migrate to the surface, where their polar moieties "attract" and build up a water-film). Their rate of action depends on their diffusion rate, which is determined by their solubility, molecular mass, and structure, as well as on polymer type and morphology and the presence of other additives and fillers. Internal antistatic agents have a degree of permanence because when they are removed from the surface they are replaced by migration from the reservoir in the bulk plastic.

Surface-active antistatic agents may be cationic, anionic, and nonionic, or organometallic.

Cationic Antistatic Agents include quaternary ammonium salts and sulfonium or phosphonium salts, mainly chlorides, nitrates, hydrogen phosphates, or 4-toluenesulfonates with long hydrocarbon chains. Products derived from imidazoline and pyrazoline are also used.

These agents are employed in rigid PVC (although they may impair heat stabilization) and in polystyrene. Cationic antistatic agents are generally not approved for food contact and are relatively expensive.

Anionic Antistatic Agents include sodium alkyl sulfonates, alkyl phosphonates, and alkyl dithiocarbamates. They are used in rigid PVC (with less detrimental effect on heat stabilization).

Nonionic Antistatic Agents include ethoxylated (or propoxylated) fatty alcohols, fatty amines, or fatty acid amides: polyethylene glycol esters of fatty acids and alkylphenols; glyceryl esters of fatty acids; and sorbitol esters. They are used for many polymers, in particular polyolefins, flexible PVC, polystyrene, and ABS. Most of these agents have a liquid or waxy consistency.

Organometalic Compounds include neoalkyl titanates and zirconates [138] that have a high thermal stability and low migration; their effectiveness is less influenced by humidity. They are especially recommended for polyolefins, but also for PVC and polystyrene.

Conductivity Additives reduce the resistivity in a threshold fashion: when their concentration reaches a certain level (the percolation point), conductivity increases abruptly [139]. They are used at levels of 5 – 10 % and occasionally up to 20 %. Examples are carbon black, metal powders or fibers (e.g., Al or Cu), silver-coated silicates, nickel-coated chopped carbon or glass fibers, or highly conjugated organic polymers (e.g., polyacetylenes, polypyrroles, polythiophenes, and polyanilines), which are oxidized or reduced (doped) to form semiconducting polymers. A survey of conductive additives is given in [140].

Test Methods. The most important test methods measure the surface or volume resistivity (e. g., ASTM D 257 or DIN 53 482, respectively). The frictional charging tendency and the half-life of the accumulated charge (in seconds) can also be measured (DIN 53 486 E).

Empirical soiling tests (e.g., with cigarette ash) should also be mentioned. All testing should be performed under a controlled atmosphere (e.g., 22 °C and 50 % R.H.).

Economic Aspects. The annual worldwide consumption of surface-active antistatic agents is ca. 10 000 t; ethoxylated fatty amines are the leading group in terms of volume. Polyolefins, PVC, and polystyrene use more than 90 % of these products.

Important producers (and trade names) include Aceto Chem. (Catafor), Akzo (Armostat, Interstat), American Cyanamid (Cyastat), CWM Bärlocher (Bärostat), Bayer (Statexan), Dr. Th. Böhme (Tebestat), Chemax Inc. (Chemstat), Dai-ichi Kogyo Seiyaku (Resistat), Dow Chemical (Stature), Eastman (Myvaplex) B. F. Goodrich (StatRite), Henkel (Dehydat), Hoechst (Hostastat), ICI (Atmer), Kao Chem., Kenrich Petrochem. (Ken-Stat), Lonza (Glycolube), Marubishi Oil, Mazer Chemical Division of PPG (Larostat), Nippon Oil & Fats, Sandoz (Sandin), Sherex (Varstat), Swedstab (Swedstat), and Witco (Markstat, Kenamine).

Many producers are in the market with antistatic masterbatches that contain 10 – 50 % of the active ingredient in polymer vehicles, making the usually hygroscopic compounds easier to process.

9.3. Impact Modifiers

High-tonnage plastics such as PVC, polyolefins, or polystyrene have high rigidity but are brittle. Additives (impact modifiers) must therefore often be used to improve impact strength, especially at low temperature [141], [142]. In contrast to plasticizers, impact modifiers must not reduce but increase the heat distortion temperature. Other important selection criteria are their effect on weather resistance and transparency.

Impact modifiers are elastomeric copolymers with low glass transition temperatures. They are dispersed as discrete soft phases in the thermoplastic. Transfer of impact energy to the elastomer phases requires not only a good distribution but also adhesion at the interface between the two phases by chemical bonding or physical cross-linking. Elastomeric properties and adhesion dictate the structural principle of impact modifiers, which are naturally very substrate specific. Most such agents become active above a certain concentration (transition point), between 5 and 15 %.

Classification. Impact modifiers are frequently incorporated into the polymer during compounding; in PVC, they are blended in by the converter. The most important impact modifier classes follow.

Methyl Methacrylate – Butadiene – Styrene Copolymers (MBS). The MBS agents are prepared by graft polymerization of styrene and methyl methacrylate on polybutadiene. They are used chiefly in PVC, where they offer advantages in clear products (bottles) and, to a lesser extent, in polycarbonates (automotive parts). Because of their sensitivity to light, they are less suitable for outdoor service.

Acrylates include weatherfast polyacrylate esters based on poly(butyl acrylate) and poly (2-ethylhexyl acrylate), as well as skin – core graft copolymers of butyl acrylate and methyl methacrylate. In addition to pure polyacrylates ("all acrylics"), butyl acrylate – styrene – methyl methacrylate terpolymers are used which have good transparency and weather resistance. Acrylate impact modifiers are employed mainly in PVC (e.g., pipes, window frames, sidings). Polyacrylates functionalized with carboxyl groups are recommended for polyamides.

Graft Copolymers of Acrylonitrile and Styrene on Polybutadiene (ABS). ABS is a classical impact modifier for PVC (profiles, sheets, pipes, housings, bottles). The transparency of articles produced depends on the composition. ABS is less suitable for outdoor applications.

Chlorinated Polyethylene. Polyethylene containing 30 – 40 % chlorine is employed in PVC, lending high impact strength especially at low temperature.

Ethylene – Propylene – Ciene Copolymers (EPDM) are used in polypropylene, for example, in automotive parts (bumpers, instrument panels). Functionalized derivatives have been developed for engineering thermoplastics (polyamides, PBT, PETP), for example, by grafting on acrylic acid, maleic acid, or maleic anhydride. The carboxyl groups react with the end groups of the polymers.

Styrene – Butadiene Rubber. Finely divided styrene – butadiene rubber in polystyrene gives high-impact products with a better transparency than high-impact polystyrene (HIPS) produced by graft polymerization of styrene on polybutadiene.

Other Impact Modifiers include ethylene – vinyl acetate, acrylonitrile – styrene – acrylate, and styrene – maleic anhydride copolymers. New functionalized products are being developed for special applications, including those in the thermoset area.

Test Methods. Impact strength tests are specified in the following standards: ASTM D 256, ISO R 179, DIN 53 453. In PVC, for example, modifiers increase the impact value by a factor between 10 and 30.

Economic Aspects. Annual demand for impact modifiers is ca. 150 000 t; ca. 60 % of this is used in PVC. The most important class is MBS, followed by ethylene – propylene – diene copolymers, acrylates, and ABS.

Important producers (and trade names) include Arco Chem. (Arvyl), BASF (Vinuran), Bayer (Baymod, Novadur, Levapren), B. F. Goodrich, Dow Chemical, DSM (Keltan), Du Pont (Elvaloy), Exxon Chemical (Vistalon), General Electric Specialty Chemicals (Blendex), Hoechst (Hostapren), Hüls (Vestolit), Kanegafuchi Chem./Kaneka Texas/Kaneka Belgium (Kane Ace), Kureha Chem., M.A. Industries (MA),Metco America/Metablen b. V.(Metablen, Durastrength), Mitsubishi Rayon, Mobay (Baymod), Monsanto (Elix, Cadon), Nippon Gohsei (Soarblen), Nova Polymers (Novalar, Novalene), Polysar (Taktene), Rohm & Haas (Acryloid, Paraloid), Shell (Kraton), and Wacker Chemie (Vinnol).

9.4. Chemical Blowing Agents

Plastics are often converted into foamed products with physical or chemical blowing agents to obtain substantial savings in weight and improve insulating properties (thermal and acoustic). See also → Foamed Plastics. Physical agents include permanent gases (N_2, CO_2, air) and substances that are gases at the processing temperature (e.g., chlorofluorocarbons, dichloromethane, C_5 – C_7 hydrocarbons). Chemical blowing agents are usually insoluble in the polymer and undergo thermal decomposition or chemical reactions during processing to give gaseous products [143]. While there are technological limits to the use of permanent gases, chlorofluorocarbons have come under severe criticism because of their potential for depleting the ozone layer (Montreal Protocol, October 1987).

The most important criteria for chemical blowing agents are

1. Their decomposition temperature must be matched to the processing temperature of the plastic (150 – 250 °C). Decomposition should not take place spontaneously, should be only slightly exo- or endothermic, and should occur within a relatively narrow temperature window (5 – 15 °C).
2. The gas should be produced with a high yield (usually 100 – 225 mL/g) and should not be toxic, flammable, or corrosive.
3. The solid decomposition products should not interfere with processing (plate-out, migration, or discoloration).
4. They should have a small particle size, a narrow particle-size distribution, and be easily dispersed in the polymer.
5. Approval as an indirect food additive is often required.

Classification. The following are the most important classes of chemical blowing agents.

Azodicarbonamide [123-77-3], thermal decomposition 205 – 215 °C, gas yield 220 mL/g, gases produced N_2, CO (ca. 2 : 1), some NH_3, CO_2. The decomposition temperature can be lowered to 155 °C with "kickers", which are either metal compounds (e.g., ZnO, zinc stearate, Ba – Zn and K – Zn systems, and lead salts) or organic substances (e.g., acids, bases, urea). Azodicarbonamide is used in PVC, polyolefins, polystyrene, ABS, polyamides, poly(phenylene oxide), acrylates, and other resins.

4,4'-Oxybis(benzenesulfohydrazide) [80-51-3], thermal decomposition 150 – 160 °C, gas yield 125 mL/g, gases produced N_2, H_2O. This compound is suitable for plastics that are foamed at relatively low temperature: PVC, LDPE, ethylene – vinyl acetate copolymer, and polystyrene. In comparison with azodicarbonamide it has the advantage of colorlessness.

Cyanuric Trihydrazide [10105-42-7], trihydrazinotriazine, thermal decomposition 275 °C, gas yield 225 mL/g, gases produced N_2, NH_3. This compound is used in ABS, polypropylene, and polyamides.

p-Toluenesulfonyl Semicarbazide [10396-10-8], thermal decomposition 220 – 245 °C, gas yield 140 mL/g, gases produced N_2, CO_2. This compound is used in ABS, polypropylene, HDPE, polyamides, polystyrene, rigid PVC, and PBT.

Isatoic Anhydride [118-48-9], thermal decomposition 210 – 225 °C, gas yield 115 mL/g, gas produced CO_2. This compound is used in polystyrene, ABS, poly(phenylene oxide), PBT, polycarbonates, and polyamides.

5-Phenyltetrazole [18039-42-4], thermal decomposition 240 – 280 °C, gas yield 200 mL/g, gas produced N_2. This compound is used in high-temperature processing, especially in polycarbonates, PBT, poly(phenylene oxide), and polyamides.

Citric Acid and Sodium Bicarbonate, thermal decomposition 150 – 200 °C, gas produced CO_2. In contrast to the compounds listed above, this system generates blowing gases by an endothermic reaction. It can be used in polystyrene, polypropylene, ABS, PVC, and other plastics.

Inorganic Carbonates decompose to form CO_2 and are used in PVC, polystyrene, and polyolefins.

Most chemical blowing agents (ca. 60 %) are used in PVC, followed by polyolefins, including cross-linked polyethylene (ca. 30 %), polystyrene, and engineering thermoplastics. Polyurethanes and polystyrene are foamed mainly with physical blowing agents.

Chemical blowing agents are used at a concentration of 1 – 3 %, higher in cross-linked polyethylene. They are employed as powders or in other forms [144], including nondusting granules, masterbatches, or pastes in combination with plasticizers, kickers, and other additives. The blowing agent can be injected into polymer melts or preblended with the polymer granules and the compound fed to extrusion or injection molding.

Economic Aspects. The annual consumption of chemical blowing agents is ca. 15 000 t; azodicarbonamide is by far the most important product.

Important producers (and trade names) include Bayer/Mobay (Porofor), Boehringer Ingelheim (Hydrocerol), Dong Jim Chem. (Unicell), Eiwa Chem. (Neocellborn, Vinyfor), Fairmount Chem. (Azocel), Hebron (Vinstab, Hebron), Hoechst (Hostatron), J. M. Huber (Activex), Kum Yang (Cellcom), Lucidol/Pennwalt (Luperfoam), M & T (Azubol), Otsuka Chem. (Unifoam), Sankyo Kasei (Cellmic), Schering/Sherex (Ficel, Genitron), Toyo Hydrazine Ind. (Azobis), and Uniroyal/Olin (Celogen, Expandex, Kempore).

9.5. Optical Brighteners

Optical brighteners [145] enhance the whiteness and brilliance of plastics that have a slightly yellow intrinsic color (→ Optical Brighteners). They absorb UV light and emit part of it as fluoresence in the blue – violet region of the spectrum after $10^{-7} - 10^{-9}$ s.

Important criteria for optical brighteners are the hue of the emitted light (bluish, greenish, or reddish) and their lightfastness. They must also be soluble in the polymer substrate, thermally stable during processing, and resistant to migration.

Optical brighteners have conjugated systems based on stilbene, benzoxazole, coumarin, thiophene, oxadiazole, naphthalene, or triazine building blocks.

They are used in many plastics, at levels between 0.005 and 0.1 %; formulations with TiO_2, especially rutile, require the higher levels. Overdosing produces a greenish tinge. Optical brighteners are also used as concentrates in fillers or plasticizers, or as masterbatches.

Producers (and trade names) of optical brighteners suitable for use as plastics additives include BASF (Ultraphor), Bayer (Blankophor), Ciba-Geigy (Uvitex, Tinopal), Eastman Chemicals (Eastobrite, Kodel), Hoechst (Hostalux), ICI (Fluolite), Sandoz (Leucopur), Sigma (Optiblanc), and Sumitomo (Whitex, Whitefluor).

ABS	acrylonitrile – butadiene – styrene polymer
HIPS	high-impact polystyrene
PBT	poly(butylene terephthalate)
HDPE	high-density polyethylene
LDPE	low-density polyethylene
LLDPE	linear low-density polyethylene
PETP	poly(ethylene terephthalate)
PVC	poly(vinyl chloride)
SAN	styrene – acrylonitrile copolymer

References

1. R. Gächter, H. Müller: *Kunststoffadditive*, 3rd ed., Carl Hanser Verlag, München – Wien 1989.
2. J. T. Lutz: *Thermoplastic Polymer Additives*, Marcel Dekker, New York 1989.
3. *Additives for Polymers*, Elsevier, Oxford, published monthly.
4. E. L. Shanina, V. A. Belyakov, G. E. Zaikov, *Polym. Degr. Stab.* **27** (1990) 309.
5. K. Kikkawa, Y. Nakahara, Y. Ohkatsu, *Polym. Degr. Stab.* **18** (1987) 237.
6. T. J. Henman: *World Index of Polyolefin Stabilizers*, Kogan Page, London 1982.
7. BASF, US 4 806 580, 1987 (G. Bock, H. Trauth, W. Weber, P. Lechtken).
8. F. Nabholz, M. D. Castle: *International Conference on Advances in the Stabilization and Controlled Degradation of Polymers*, Luzern 1989.
9. Adeka Argus Chemical Co., EP 336 606, 1989 (K. Nishikawa, T. Haruna, M. Hamajima).
10. Ethyl Corp., EP 280 938, 1988 (L. P. J. Burton).
11. Sandoz AG, US 4 075 163, 1976 (K. Hofer, G. Tscheulin).
12. S. Moss, H. Zweifel, *Polym. Degr. Stab.* **25** (1989) 217.
13. G. Scott (ed.): *Developments in Polymer Stabilisation*, vols. **1 – 8**, Elsevier Applied Science, London – New York 1979 – 1987.
14. N. S. Allen (ed.): *Degradation and Stabilisation of Polyolefins*, Elsevier Applied Science Publishers, London – New York 1983.
15. N. Grassie, G. Scott: *Polymer Degradation and Stabilisation*, Cambridge University Press, Cambridge 1985.
16. P. P. Klemchuk, P. L. Horng, *Polym. Degr. Stab.* **7** (1984) 131.
17. G. Scott, *Chem. Ind. (London)* 1987, 841.
18. J. Pospisil, *Polym. Degr. Stab.* **20** (1988) 181.
19. N. S. Allen, J. T. Kotecha, J. L. Gardette, J. Lemaire, *Polym. Degr. Stab.* **11** (1985) 181.
20. D. Vyprachticky, J. Pospisil, J. Sedlar, *Polym. Degr. Stab.* **27** (1990) 227.
21. F. Gugumus, *Polym. Degr. Stab.* **24** (1989) 289.
22. E. Blatt, H. J. Griesser, J. H. Hodgkin, A. W. H. Mau, *Polym. Degr. Stab.* **25** (1989) 19.
23. Z. Osawa, *Polym. Degr. Stab.* **20** (1988) 203.
24. G. Menges, W. Michaeli, C. Schwenzer, L. Czyborra, *Plastverarbeiter* **40** (1989) 207.
25. E. M. Lewis, *Mod. Plast. Encyclopedia* **67** (1990) 165.
26. J. G. Calvert, J. N. Pitts: *Photochemistry*, Wiley, New York 1967, p. 534.
27. H. E. A. Kramer, *Farbe+Lack* **92** (1986) 919.
28. M. Allan, et al., *Polym. Degr. Stab.* **15** (1986) 311.
29. N. S. Allen, *Chem. Soc. Rev.* **15** (1986) 373.
30. N. S. Allen, A. Parkinson, F. F. Loffelmann, P. V. Susi, *Angew. Makromol. Chem.* **116** (1983) 203.
31. Sankyo KK, JP 7 1 31 734, 1968 (K. Murayama).
32. D. J. Carlsson, D. W. Grattan, T. Suprunchuk, D. M. Wiles, *J. Appl. Polym. Sci.* **22** (1978) 2217.
33. E. T. Denisov in [13] vol. 3 (1980) p. 1; vol. 5 (1982) p. 37.
34. D. M. Wiles, D. J. Carlsson, *Polym. Degr. Stab.* **3** (1980) 61; **6** (1984) 1.
35. N. S. Allen, *Polym. Degr. Stab.* **2** (1980) 129.
36. N. S. Allen, J. L. Gardette, J. Lemaire, *Polym. Degr. Stab.* **3** (1981) 199.

37. R. Bagheri, K. B. Chakraboty, G. Scott, *Polym. Degr. Stab.* **4** (1982) 1.
38. S. Al-Malaika, G. Scott in [14] p. 283.
39. N. S. Allen, J. L. Gardette, J. Lemaire, *Polym. Degr. Stab.* **8** (1984) 133.
40. G. Scott, *Polym. Degr. Stab.* **10** (1985) 97.
41. P. P. Klemchuk, M. E. Gande, *Polym. Degr. Stab.* **22** (1988) 241.
42. E. T. Denisov, *Polym. Degr. Stab.* **25** (1989) 209.
43. A. J. Chirinos Padron, *Rev. Macromol. Chem. Phys.* C **30** (1990) 107.
44. P. P. Klemchuk, M. E. Gande, *Makromol. Chem. Makromol. Symp.* **28** (1989) 117.
45. K. H. Chan, D. J. Carlsson, M. D. Wiles, *J. Polym. Sci. Polym. Lett. Ed.* **18** (1980) 607.
46. F. K. Meyer, F. Gugumus, E. Pedrazzetti, *Chemiefasern/Textilind.* **35** (1985) 840.
47. F. J. Barahona, J. M. G. Vasquez, *Plasticulture* **65** (1985) 3.
48. A. J. Chirinos-Padron, *Polym. Degr. Stab.* **25** (1989) 101.
49. B. F. Goodrich, EP 258 598, 1987 (J. T.-Y. Lai, P. N. Son).
50. G. Scott, *Makromol. Chem. Makromol. Symp.* **28** (1989) 59.
51. F. Gugumus in [13] vol. 8(1987).
52. D. Kockott, *Polym. Degr. Stab.* **25** (1989) 181.
53. N. S. Allen, A. Chirinos-Padron, T. J. Henman, *Polym. Degr. Stab.* **13** (1985) 31.
54. T. Kurumada, *Polym. Degr. Stab.* **19** (1987) 263.
55. F. Stohler, K. Berger, *Angew. Makromol. Chem.* **158/159** (1988) 233.
56. Röhm GmbH, EP 16 870, 1979 (L. Hosch).
57. K. S. Minsker, S. V. Kolesov, G. E. Zaikov: *Degradation and Stabilisation of Vinyl Chloride Based Polymers*, Pergamon Press, Oxford 1988.
58. E. D. Owen: *Degradation and Stabilisation of PVC*, Elsevier Applied Science Publishers, London 1984.
59. B. Y. K. Ho, *J. Vinyl Technol.* **6** (1984) 162.
60. R. Schlimper, *Plaste Kautsch.* **14** (1967) 657.
61. A. F. Frye, R. W. Horst, *J. Polym. Sci.* **40** (1959) 419; **45** (1960) 1; *J. Polym. Sci.*, **Part A 2** (1964) 1765, 1785, 1801.
62. H. O. Wirth, H. A. Müller, W. Wehner, *J. Vinyl Technol.* **1** (1979) 51.
63. D. Braun, D. Hepp, *Angew. Makromol. Chem.* **66** (1976) 23.
64. T. van Hoang, A. Michel, A. Guyot, *Polym. Degr. Stab.* **9** (1984) 73, 89.
65. A. Michel, T. van Hoang, B. Perrin, M. France-Llauro, *Polym. Degr. Stab.* **3** (1981) 107.
66. K. S. Minsker, S. V. Kolesov, V. M. Yanborisov, G. E. Zaikov, *Polym. Degr. Stab.* **15** (1986) 305.
67. J. Büssing in: *Kunststoff Handbuch 2/1, PVC,* 2nd ed., Carl Hanser Verlag, München – Wien 1986, p. 551.
68. T. Riedel in [1] p. 443.
69. K. Worschech in *Kunststoff Handbuch 2/1, PVC,* Carl Hanser Verlag, München 1986.
70. T. A. Skowronski, J. F. Rabek, B. Ranby, *Polym. Degr. Stab.* **12** (1985) 229.
71. G. Pfahler, T. Riedel, *Kunststoffe* **66** (1976) 694.
72. G. Plato, G. Schröter, *Kunststoffe* **50** (1960) 163.
73. J. W. Lyons: *The Chemistry and Uses of Fire Retardants*, Wiley-Interscience, New York 1970.
74. W. C. Kuryla, A. J. Papa: *Flame Retardancy of Polymeric Materials*, vols. **1 – 5**, Marcel Dekker, New York 1973 – 1979.
75. M. Lewin, S. M. Atlas, E. M. Pearce: *Flame Retardant Polymeric Materials*, vols. **1 – 2**, Plenum Press, New York – London 1975 – 1978.
76. A. H. Landrock: *Handbook of Plastics Flammability and Combustion Toxicology*, Noyes Publications, Park Ridge 1983.
77. C. F. Cullis, M. M. Hirschler: *The Combustion of Organic Materials*, Clarendon Press, Oxford 1981.
78. J. Troitzsch: *International Plastics Flammability Handbook*, Hanser Publishers, Munich – Vienna – New York 1990.
79. G. Kirschbaum, *Kunststoffe* **79** (1989) 1205.
80. W. J. Kennelly in: *Flame Retardants 90,* Elsevier Applied Science, London – New York 1990, p. 9.
81. G. Camino, L. Costa, G. Martinasso, *Polym. Degr. Stab.* **23** (1989) 359; **28** (1990) 17.
82. H. Staendeke, D. J. Scharf, *Kunststoffe* **79** (1989) 1200.
83. E. N. Peters in [74] vol. 5.
84. L. Costa, G. Camino, M. P. L. di Cortemiglia, *Polym. Degr. Stab.* **14** (1986) 113,115, 165.
85. P. A. Cusack, A. W. Mark, J. A. Pearce, S. J. Reynolds, *Fire Mater.* **14** (1989) 23.
86. R. L. Markezich, C. S. Ilardo, R. F. Mundhenke in: *Flame Retardants 90*, Elsevier Applied Science, London – New York 1990, p. 88.
87. R. B. Bush, *Plast. Eng.* **42** (1986) April, 29.
88. J. Eichhorn, *J. Appl. Polym. Sci.* **8** (1964) 2497.
89. V. Mark, *Org. Coat. Plast. Chem.* **43** (1980) 71.
90. L. Costa, G. Camino, *Polym. Degr. Stab.* **20** (1988) 271.
91. R. Delobel, M. LeBras, N. Ouassou, F. Alistiqsa, *J. Fire Sci.* **8** (1990) 85.
92. J. J. Pitts, P. H. Scott, D. G. Powell, *J. Cell. Plast.* **6** (1970) 35.
93. D. L. Buzzard in: *Fire Retardants 83,* The Plastics and Rubber Institute, London 1983, p. 21.
94. N. K. Jha, A. C. Misra, P. Bajaj, *Rev. Macromol. Chem. Phys.* **24** (1984) 69.
95. S. Miyata, T. Imahashi, H. Anabuki, *J. Appl. Polym. Sci.* **25** (1980) 415.
96. J. Sutker, *J. Fire Sci.* **1** (1983) 66.
97. J. G. Williams in: *Flame Retardants 83*, The Plastics and Rubber Institute, London 1983, p. 91.
98. R. Wolf, *Kunststoffe* **76** (1986) 943.
99. C. J. Hilado, *Mod. Plast.* **54** (1977) July, 64; **55** (1978) April, 92.
100. F. H. Prager et al., *J. Fire Sci.* **5** (1987) 308.
101. R. C. Kidder in *Flame Retardants 90*, Elsevier Applied Science, London – New York 1990, p. 9.
102. B. Akles, J. Steinmetz, J. Hogan, *Mod. Plast. Int.* **17** (1987) 49.
103. Union Carbide, US 4 481 322, 1984 (R. E. Godlewski, F. D. Osterholtz).
104. S. J. Monte, G. Sugerman; *Ken-React Reference Manual*, Kenrich Petrochemicals, Bayonne, N.J., 1985.
105. L. B. Cohen, *Plast. Eng.* **39** (1983) 29.
106. E. E. Lang in: *Kunststoff Handbuch 2/1, PVC*, 2nd ed., Carl Hanser Verlag, München – Wien 1986, p. 678.
107. D. L. Caldwell, F. M. Cortez, *Mod. Plast.* **65** (1988) Sept., 132.
108. H. P. Schlumpf in *Kunststoffadditive*, 3rd ed., Carl Hanser Verlag, München – Wien 1989, p. 549; R. Kleinholz, G. Heyn, R. Stolze in *Kunststoffadditive*, 3rd ed., Carl Hanser Verlag, München – Wien 1989, p. 617.
109. H. S. Katz, J. V. Milewski: *Handbook of Fillers and Reinforcements of Plastics*, Van Nostrand Reinhold, New York 1978.
110. B. L. Kaul, *J. Oil Colour Chem. Assoc.* **70** (1987) 349.
111. B. F. Greek, *Chem. Eng. News* **66** (1988) 35.
112. D. Spencer, *Mod. Plast.* **67** (1990) April, 19 (suppl. Waste Solution).
113. B. G. Murray in C. T. Patton (ed.): *Pigment Handbook*, vol. **2**, Wiley-Interscience, New York 1973, p. 277.
114. G. Sonn, *Am. Ink Maker* **65** (1987) June, 54.
115. W. Damm in R. Gächter, H. Müller (eds.): *Plastics Additives Handbook*, 2nd ed., Hanser Publishers, Wien 1987, p. 506.

116 A. P. Hopmeier: *Encyclopedia of Polymer Science and Technology*, vol. **10**, Wiley-Interscience, New York 1969, p. 157.
117 P. E. Lewis (ed.): *Pigment Handbook,* 2nd ed., vol. 2, Wiley-Interscience, New York 1987.
118 A. S. Wood, *Mod. Plast.* **66** (1989) Sept., 58.
119 W. Herbst, K. Hunger: *Industrial Organic Pigments*, VCH Verlagsgesellschaft, Weinheim, Germany 1992.
120 J. D. Sanders: *Pigments for Inkmakers*, SITA Technology, London 1989.
121 *Eur. Plast. News* **17** (1990) June, 44.
122 F. Baebler, Proc. of RETEC Meeting in Huron (USA), Sept. 1989, 205.
123 *Colour Index*, 3rd ed. and suppl., Society of Dyers and Colourists, Bradford 1982.
124 J. Jansen in [1] p. 893.
125 F. L. Binsbergen, *Polymer* **11** (1970) 253, 309.
126 F. L. Binsbergen, *J. Polym. Sci. Polym. Phys. Ed.* **11** (1973) 117.
127 A. M. Chatterjee, F. P. Price, S. Newman, *J. Polym. Sci. Polym. Phys. Ed.* **13** (1975) 2369, 2385, 2391.
128 J. M. Dekoninck, R. Legras, J. P. Mercier, *Polymer* **30** (1989) 910.
129 Adeka Argus, US 4 258 142, 1981 (T. Ohzeki, M. Ahutsu, J. Kawai); US 4 463 113, 1984 (Y. Nakahara, M. Ahutsu, T. Haruna, M. Takahashi).
130 C. C. Carroll, *Mod. Plast.* **61** (1984) Sept., 108.
131 New Japan Chem. Co., JP 53 117 044, 1977 (T. Kobayashi, T. Nobe).
132 L. Bourland, *Plast. Eng.* **43** (1987) July, 39.
133 V. Caldas, G. R. Brown, J. M. Willis, *Macromolecules* **23** (1990) 338.
134 ICI, EP 267 695, 1987 (J. P. Trotoir, C. Bath).
135 S. L. Hsu in: *Comprehensive Polymer Science,* vol. 1, Pergamon Press, Oxford 1989, p. 429.
136 J. Kaiser, *Kunststoffe* **80** (1990) 330.
137 G. Pfahler in [1] p. 779.
138 S. J. Monte, G. Sugerman in [104] suppl., p. 35.
139 B. Wessling, *Kunstst. J.* **21** (1987) 50.
140 P. Mapleston, *Mod. Plast. Int.* **19** (1989) no. 10, 66.
141 D. Hepp in [1] p. 525.
142 G. Menzel in: *Kunststoff Handbuch 2/1, PVC,* Carl Hanser Verlag, München – Wien 1986, p. 712.
143 H. Hurnik in [1] p. 843.
144 R. R. Puri in Abstracts 1st European Conference on High Performance Additives, British Plastics Federation, London 1988, p. 83.
145 K. Berger in [1] p. 807.

Further Reading

R. F. Grossman: *Handbook of Vinyl Formulating* Wiley, New York, 2008.
D. W. van Krevelen, K. te Nijenhuis: *Properties of Polymers, Fourth Edition: Their Correlation with Chemical Structure; their Numerical Estimation and Prediction from Additive Group Contributions*, Elesevier, Amsterdam, 2009.
M. Tolinski: *Additives for Polyolefins: Getting the Most out of Polypropylene*, Polyethylene and TPO, Elsevier, Amsterdam, 2009.
M. Xanthos: *Functional Fillers for Plastics*, Wiley-VCH, Weinheim 2010.
H. Zweifel: *Plastics Additives Handbook*, Carl Hanser Verlag, Munich, 2009.

Plasticizers

DAVID F. CADOGAN, ICI Chemicals and Polymers, Runcorn, Cheshire, United Kingdom

CHRISTOPHER J. HOWICK, ICI Chemicals and Polymers, Runcorn, Cheshire, United Kingdom

1.	Introduction	581
1.1.	Definition	581
1.2.	Types of Plasticization	581
1.3.	Types of Plasticizer	582
2.	Plasticizers in Common Use	582
2.1.	Market Overview	582
2.2.	Phthalate Esters	583
2.3.	Adipate Esters	583
2.4.	Trimellitate Esters	584
2.5.	Phosphate Esters	584
2.6.	Sebacate and Azelate Esters	584
2.7.	Polyester Plasticizers	585
2.8.	Sulfonate Esters	585
2.9.	Specialty Phthalate Esters	585
2.10.	Secondary Plasticizers	585
3.	Mechanism of Plasticizer Action	586
3.1.	The Lubricity Theory	586
3.2.	The Gel Theory	586
3.3.	The Free-Volume Theory	586
3.4.	Solvation – Desolvation Equilibrium	587
3.5.	Generalized Structure Theories	587
3.6.	Specific Interactions and Interaction Parameters	587
4.	Plasticized PVC	589
4.1.	Applications Technology	589
4.1.1.	Suspension PVC	589
4.1.2.	Emulsion PVC	589
4.2.	Effect of Plasticizer Choice on the Properties of Flexible PVC	590
4.2.1.	Plasticizer Efficiency	590
4.2.2.	High-Temperature Performance	590
4.2.3.	Low-Temperature Performance	591
4.2.4.	Gelation Properties	591
4.2.5.	Migration and Extraction	592
4.2.6.	Plastisol Viscosity and Viscosity Ageing	592
4.2.7.	Windscreen Fogging	593
4.2.8.	Summary	594
5.	Plasticization of Polymers Other Than PVC	595
5.1.	Requirements for Plasticization	595
5.2.	Plasticization of Other Polymers	595
6.	Toxicology and Environmental Aspects	596
6.1.	Toxicity Studies on Plasticizer Esters	596
6.2.	Plasticizers in the Environment	597
6.3.	Storage and Handling	599
	References	599

1. Introduction

1.1. Definition

A plasticizer is a substance incorporated into a material to increase its flexibility, workability, or distensibility. A plasticizer may reduce the melt viscosity, lower the temperature of the second-order transition, or lower the elastic modulus of the product. Plasticizers are inert, organic substances with low vapor pressures, predominantly esters, which react physically with high polymers to form a homogeneous physical unit, whether it be by means of swelling or dissolving or any other [1].

At present some 300 plasticizers are manufactured, of which at least 100 are of commercial importance. A list of some of the most widely used plasticizers is given in Table 1.

1.2. Types of Plasticization

Two principle ways of obtaining the plasticization effects described above are commonly utilized. Firstly, a rigid polymer may be *internally plasticized* by chemically modifying the polymer or the monomer so that the flexibility of the polymer is increased. A rigid polymer, however, can also be *externally plasticized* by addition of a

Table 1. Plasticizers in Common Use

Plasticizer	Abbreviation	Alcohol carbon number	CAS registry no.	M_r	Density at 20 °C, g/cm^3
Phthalates					
Diisobutyl phthalate	DIBP	4	[84-69-5]	278.3	1.039
Dibutyl phthalate	DBP	4	[84-74-2]	278.3	1.046
Diisoheptyl phthalate	DIHP	7	[41451-28-9]	362	0.991
L79 phthalate	L79P, 79P	7 – 9	[71888-89-6]	380	0.985
L711 phthalate	L711P, 711P	7 – 11	[68648-91-9]		0.971
Dioctyl phthalate	DOP	8	[117-81-7]	390	0.984
Diisooctyl phthalate	DIOP	8	[27554-26-3]	390	0.983
Dinonyl phthalate	DNP	9	[84-76-4]	419	0.97
Diisononyl phthalate	DINP	9	[28553-12-0] [68515-48-0]	419	0.975
Diisodecyl phthalate	DIDP	10	[68515-49-1]	447	0.967
L911 phthalate	L911P, 911P	9 – 11	[68515-43-5]	454	0.96
Diundecyl phthalate	DUP	11	[3648-20-2]	474	0.953
Diisoundecyl phthalate	DIUP	11	[85507-79-5]	474	0.962
Undecyl dodecyl phthalate	UDP	11 – 12	[68515-43-5]	488	0.957
Diisotridecyl phthalate	DTDP	13	[27253-26-5]	531	0.952
Butyl benzyl phthalate	BBP	4 – 7	[68515-47-9]	312.3	1.119
Adipates					
Dioctyl adipate	DOA	8	[103-23-1]	370.6	0.929
Diisononyl adipate	DINA	9	[33703-08-1]	398	0.929
Diisodecyl adipate	DIDA	10	[27178-16-1]	427.1	0.915
Trimellitates					
Trioctyl trimellitate	TOTM	8	[3319-31-1]	530	0.991
L79 trimellitate	L79TM	7 – 9	[68515-60-6]	530	0.996
L810 trimellitate	L810TM	8 – 10	[67989-23-5]	592	0.973
Phosphate esters					
Tri-2-ethylhexyl phosphate	TOF	8	[78-42-2]	434	0.926
2-Ethylhexyl diphenyl phosphate	Santicizer 141	6 and 8	[1241-94-7]	362.4	1.091
Tricresyl phosphate	TCP	6	[1330-78-5]	368.2	1.165

plasticizer, which imparts the desired flexibility but is not chemically changed by reaction with the polymer. The latter process has the advantage of lower cost and permits the fabricator freedom in devising formulations to manufacture items of differing flexibility simply by adding different amounts of plasticizer, within the constraints of polymer – plasticizer compatibility.

1.3. Types of Plasticizer

Plasticizers can be divided into two principal groups: primary and secondary. *Primary plasticizers* lower the glass transition temperature T_g and increase the elongation and softness of the polymer. *Secondary plasticizers*, when added to the polymer alone, do not bring about such changes and may also have limited compatibility with the polymer. However, when added to the polymer in the presence of a primary plasticizer, secondary plasticizers enhance the plasticizing performance of the primary plasticizer.

2. Plasticizers in Common Use

2.1. Market Overview

The list of commercially available plasticizers, covering applications throughout polymer science, is extensive. This reflects the development of plasticizers to meet the increasing demand for polymers, especially PVC, for new and higher quality applications. This article discusses the main groups of plasticizers; for further details, see [1–4] and references therein.

The Western European plasticizer market had a capacity of 969×10^3 t/a in 1989 and at that time was growing at a rate of 3 % per annum [5]. In 1996 production of plasticizer in Western Europe amounted to 1253×10^3 t/a and in the United States to 636×10^3 t/a [6]. In terms of plasticizer types, the majority of this tonnage (>85 %) is standard phthalate (esters of phthalic anhydride with $C_8 - C_{10}$ alcohols). Reasons for this are the relatively low price and ready availability of feedstocks. The remainder of the market is taken up by phthalate esters of other alcohols, speciality phthalates, adipates, trimellitates, and other esters.

2.2. Phthalate Esters (see also → Phthalic Acid and Derivatives)

Diesters of phthalic anhydride are produced commercially from $C_1 - C_{13}$ alcohols, although the $C_1 - C_3$ esters are generally too volatile to find widespread use. Di-n-butyl phthalate (DBP) and diisobutyl phthalate (DIBP) (linear and branched C_4 esters) are used in many PVC formulations, principally for ease of gelation. Owing to their relatively high volatility, in comparison with other phthalates, they are often used in conjunction with higher molecular mass esters. Diisopentyl phthalate (DIPP) is generally used in a similar manner. Diisoheptyl phthalate (DIHP), an ester based on a C_7 iso-alcohol from the oxo alcohol process, shows good plasticizing action with PVC as well as with a number of other polymers [4]. Its principle use in PVC, when compared with higher molecular mass plasticizers, lies in its gelation characteristics and balance of volatility and viscosity, the volatility of DIHP being only slightly higher than the corresponding C_8 phthalate (DIOP).

Di-2-ethylhexyl phthalate (DEHP) (also known as dioctyl phthalate, DOP) has long been accepted as the industry standard general purpose primary plasticizer.

Di-2-ethylhexyl phtalate, DEHP
(dioctyl phthalate, DOP)

This is because its properties are adequate for many applications in the flexible vinyl industry. DEHP shows good gelation characteristics, good softening action, and adequate viscosity properties in emulsion PVC pastes. Diisooctyl phthalate, produced from C_8 oxo alcohols, shows performance in PVC which is generally similar to DEHP. Di-n-octyl phthalate (DnOP), a C_8 phthalate ester derived from a linear alcohol, is also produced, but in far smaller quantities than the branched phthalates. The linear structure imparts improved low-temperature performance and lower viscosity.

Diisononyl phthalate (DINP) is produced from C_9 oxo alcohols and has similar plasticizing properties to DEHP. Its slightly higher molecular mass results in a somewhat lower plasticizer efficiency which requires a greater addition of plasticizer to impart an equivalent softness (these effects are discussed in detail in Section 4.2.1). The higher molecular mass alcohol also imparts improved high-temperature performance and resistance to extraction. Other C_9 phthalates are also produced [e.g., DNP, DINP(S)] and vary in the linearity of the alcohol component in the ester, the more linear varieties giving improved properties.

Diisodecyl phthalate (DIDP), the phthalate ester of C_{10} oxo alcohols, exhibits superior high-temperature performance to DINP and DEHP due to its higher molecular mass (lower volatility).

Diisodecyl phthalate, DIDP
(one of the possible isomers)

Its lower plasticizing efficiency compared to the lower phthalates requires greater plasticizer addition to impart the same softness, although this can, in some cases, be advantageous. Diundecyl phthalate (DUP), diisoundecyl phthalate (DIUP) and undecyl dodecyl phthalate (UDP) continue this trend. Ditridecyl phthalate (DTDP) is the highest molecular mass phthalate ester generally available in commercial quantities. Its high molecular mass results in low plasticizing efficiency, and some compatibility problems may occur at high plasticizer loadings. Its low volatility, however, also results in superior high-temperature performance and allows some strict electrical cable insulation and sheathing specifications to be met.

2.3. Adipate Esters

Alkyl esters of adipic acid have two principle advantages in the plasticization of PVC resins in comparison with the corresponding phthalate ester: (1) their lower intrinsic viscosity confers lower plastisol viscosity and good plastisol storage stability, enabling use in demanding applications and (2) they give superior low-temperature flexibility properties.

Of adipate esters in use in the flexible vinyl industry, di-2-ethylhexyl adipate, DEHA (also known as dioctyl adipate, DOA) is the most common.

Di-2-ethyl hexyl adipate, DEHA (dioctyl adipate, DOA)

It is widely used in flexible PVC food film (cling film). DEHA has good plasticizing efficiency and compatibility with PVC, and its very low viscosity enables it to be used in PVC plastisol applications where low viscosity is required, often in conjunction with a phthalate. Diisononyl adipate (DINA) is the ester of adipic acid with isononanol produced from oxo alcohols. It shows similar properties to DEHA. Its higher molecular mass does, however, increase its viscosity compared with DEHA, but this is accompanied by a reduction in volatility. The lower efficiency in comparison with DEHA, which requires a greater plasticizer addition, can result in lower plastisol viscosities. These effects are extended in moving to diisodecyl adipate (DIDA). Higher adipates (e.g., ditridecyl adipate, DTDA) are of limited use due to their lower compatibility with PVC.

2.4. Trimellitate Esters

Esters of trimellitic anhydride (TMA) have grown in popularity due to their excellent thermal properties and resistance to extraction. Tris-2-ethylhexyl trimellitate (TOTM), produced from trimellitic anhydride and 2-ethylhexanol, is the ester most widely used.

Tris-2-ethylhexyl trimellitate (trioctyl trimellitate, TOTM)

Although of a lower plasticizing efficiency than the corresponding phthalate, the thermal stability and extraction resistance of TOTM are superior. These properties can be improved still further by using trimellitate esters produced from mixed linear alcohols, e.g., L79 trimellitate and L810 trimellitate. The improved properties brought about by these esters are discussed further in Chapter 4.

2.5. Phosphate Esters

The principle advantages of phosphate esters such as tricresyl phosphate as plasticizers for PVC is their low volatility and their ability to impart fire-retardant properties to a PVC formulation.

Tricresyl phospate

Tris(2-ethylhexyl) phosphate shows good compatibility with PVC and also imparts good low-temperature performance in addition to good fire retardancy.

2-Ethylhexyl diphenyl phosphate has widespread use due to its combination of plasticizing efficiency, low-temperature properties, migration resistance, and fire retardancy.

2.6. Sebacate and Azelate Esters

Esters from the linear sebacic and azelatic acids are used in PVC applications for the same reasons as the corresponding adipates, i.e., low viscosity and good low-temperature performance. Di-2-ethylhexyl sebacate (DOS) and di-2-ethylhexyl azelate (DOZ) are the most widely used.

Di-2-ethylhexyl sebacate, DOS

In comparison to adipates, however, these esters tend to have a significantly higher cost and consequently only find use in extreme low-temperature applications.

2.7. Polyester Plasticizers

Polyester plasticizers, or polymeric plasticizers as they are more generally known, have found use due to their low volatility and high extraction resistance.

Polymeric plasticisers

They are typically based on condensation products of propane- or butanediols with adipic acid or phthalic anhydride. The growing polymer chain may then be end-capped with an alcohol or a monobasic acid, although non-end-capped polyesters can be produced by strict control of the reaction stoichiometry. Due to their higher molecular mass compared to other plasticizers they have much higher viscosities (3 – 10 Pa · s), which in some cases can make processing with PVC difficult.

2.8. Sulfonate Esters

Sufonate ester plasticizers for PVC are typically aryl esters of a $C_{13} - C_{15}$ alkane sulfonic acid. They are relatively efficient and easily processable plasticizers with good extraction properties.

2.9. Specialty Phthalate Esters

In addition to the phthalate plasticizers described in Section 2.2, other phthalate plasticizers are also commercially available. These are speciality products which find use as high-quality plasticizers in specific applications. Their quality tends to be reflected in their price since they require either a further manufacturing stage (e.g., transesterification) or the use of alcohol feedstocks of greater linearity than the standard phthalates.

Benzyl butyl phthalate (BBP) shows excellent compatibility and fusion characteristics with PVC and is widely used in vinyl flooring for these reasons and also due to its superior stain resistance.

Benzyl butyl phthalate, BBP

Linear and semilinear phthalates (e.g., DnOP, 911P, DUP) have lower viscosities than the corresponding branched ester. This results in lower plastisol viscosity and also lower plasticizer volatility. 911P, a phthalate ester based on linear and semilinear $C_9 - C_{11}$ alcohols, is the premium plasticizer for flexible PVC in automotive applications due to its low fogging (the condensation of volatile material on a car windscreen, causing impaired visibility); see Section 4.2.7.

2.10. Secondary Plasticizers

Secondary plasticizers (also referred to as extenders) continue to play a major role in flexible PVC technology. They do not have the same mode of softening action as a primary plasticizer but do impart softening behavior when used in combination with a primary plasticizer. In some very hard PVC formulations, however, a secondary plasticizer alone may be of use. The major secondary plasticizers are chlorinated hydrocarbons. These are hydrocarbons chlorinated to varying amounts, typically 30 – 70 %. For a given hydrocarbon, plasticizer viscosity increases with increasing chlorine content. Chlorinated hydrocarbons of the same chlorine content may, however, have different volatilities and viscosities if they are based on different hydrocarbons. The principal advantages of such materials are: (1) they improve fire retardancy due to their chlorine content and (2) formulations containing a secondary plasticizer are often cheaper. Precise knowledge of compatibility between standard plasticizers and chlorinated paraffins is required since some mixtures become incompatible with each other and the PVC resins at certain concentrations. Phthalate – chlorinated hydrocarbon compatibility decreases as the molecular mass of the phthalate and the plasticizer loading in the PVC increases; for specific examples, see [1].

3. Mechanism of Plasticizer Action

Several theories of plasticizer action have been proposed, ranging in detail and complexity. Whilst each theory is not exhaustive, an understanding of the plasticization process can be gained by combining ideas from each theory. The theories are discussed in detail in [1].

3.1. The Lubricity Theory

The lubricity theory is based on the assumption that the rigidity of the resin arises from intermolecular friction binding the chains together in a rigid network. On heating, these frictional forces are weakened so as to allow the plasticizer molecules between the chains. In the resulting mixture at room temperature the plasticizer molecules act as lubricants for the polymer chains.

3.2. The Gel Theory

The gel theory extends the lubricity theory in that it deals with the idea of the plasticizer acting by breaking the resin – resin attachments and interactions and by masking these centers of attachment from each other, preventing their reformation. Such a process may be regarded as being necessary but by itself insufficient to explain a completely plasticized system since while a certain concentration of plasticizer molecules will provide plasticization by this process the remainder will act more in accordance with the lubricity theory, with unattached plasticizer molecules swelling the gel and facilitating the movement of plasticizer molecules, thus imparting flexibility. Molecules acting by this latter action may in fact constitute the bulk of plasticizer molecules.

3.3. The Free-Volume Theory

The free-volume theory extends the above ideas and also allows a quantitative assessment of the plasticization process. The free volume of a polymer is described by Equation 1 and Figure 1

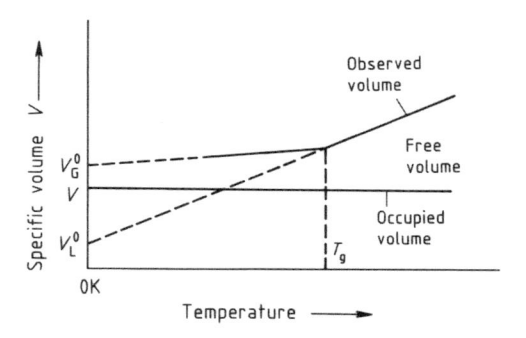

$$V_f = V_t - V^0 \tag{1}$$

Figure 1. The free-volume concept [4]

where

V_f = the free volume of the polymer
V_t = specific volume at temperature t
V^0 = specific volume at arbitrary reference point, usually 0K

and V_G^0 and V_L^0 are the specific volumes of the glass and liquid extrapolated to 0K.

Free volume is a measure of the internal space available in a polymer for the movement of the polymer chain, which imparts flexibility to the resin. Figure 1 shows how the amount of free volume in the polymer increases sharply at the glass transition temperature T_g. Consequently, the study of plasticization is a study of how to lower T_g, thus creating a polymer that is flexible at ambient temperature. Plasticizers therefore increase the free volume of the resin and also ensure that free volume is maintained as the resin – plasticizer mixture is cooled from the melt. Combining these ideas with the gel and lubricity theories implies that plasticizer molecules that do not interact with the polymer chain must simply fill free volume created by those molecules that do. These molecules may also be envisaged as providing a screening effect that prevents interactions between neighboring polymer chains, thus preventing reformation of the rigid polymer network on cooling.

For the plasticized resin, free volume can arise from:

1. Motion of the chain ends
2. Motion of the side chains
3. Motion of the main chain

These motions can be increased in a variety of ways, including:

1. Increasing the number of end groups
2. Increasing the length of the side chains
3. Increasing the possibility of main chain movement by the inclusion of segments of low steric hindrance and low intermolecular attraction
4. Introduction of a lower molecular mass compound which imparts the above properties
5. Raising the temperature

The introduction of a plasticizer, which has a lower molecular mass than the resin, can impart a greater free volume per volume of material since (1) there is an increase in the proportion of end groups and (2) it has a lower glass transition temperature T_g than the resin. A detailed mathematical treatment of this [7] can be carried out to explain the success of some plasticizers and the failure of others. Clearly, the use of a given plasticizer in a certain application will be a compromise between the above ideas and physical properties such as volatility, compatibility, high- and low-temperature performance, and viscosity. This choice will be application dependent, i.e., there is no ideal plasticizer for every application.

3.4. Solvation – Desolvation Equilibrium

Since plasticizer migrates from plasticized polymers, it is clear that plasticizer molecules are not bound permanently to the polymer as in an internally plasticized resin, but that an exchange or equilibrium mechanism is present. This implies that there is no stoichiometric relationship between polymer and plasticizer, although some quasistoichiometric relationships appear to exist [8], [9]. This idea is extended in Section 3.6.

3.5. Generalized Structure Theories

In their simplest form generalized structure theories attempt to produce a visual representation of the mechanism of plasticizer action [1]. The theories are based on the concept that if a small amount of plasticizer is incorporated into the polymer mass it imparts slightly more free volume and gives more opportunity for the movement of macromolecules. Many resins tend to become more ordered and compact as existing crystallites grow or new crystallites form at the expense of the more fluid amorphous material. For small additions of plasticizer, the plasticizer molecules may be totally immobilised by attachment to the resin by various forces. These tend to restrict the freedom of motion of small portions of the polymer molecule necessary for the absorption of mechanical energy. Therefore it results in a more rigid resin with a higher tensile strength and modulus than the base polymer itself. This phenomenon is therefore termed antiplasticization. A study by HORSLEY [10] used X-ray diffraction to show that small amounts of dioctyl phthalate (DOP) progressively increase the order in the PVC. Above these concentrations the order decreases and the polymer becomes plasticized (see also Section 3.6).

3.6. Specific Interactions and Interaction Parameters

Early attempts to describe PVC – plasticizer compatibility were based on the same principles as used to describe solvation, i.e., "like dissolves like" [11]. To obtain a quantitative measure of PVC – plasticizer compatibility a number of different parameters have been used. These are briefly described below.

The Hildebrand Solubility Parameter δ gives a measure of compatibility between plasticizer and a resin based on the solvating power of the plasticizer for the resin under investigation. For complex molecules this can be estimated [12] by using a set of additive constants for the more common groups in plasticizer molecules. These constants are designated F and the overall value for the plasticizer is given by the equation

$$\delta = \sum F/V \qquad (2)$$

i.e., a summation of each solubility constant for each molecular portion of the plasicizer molecule divided by the molecular volume. Data so calculated for different plasticizers can be compared, with the higher values suggesting greater compatibility.

The Polarity Parameters Φ were evaluated by VAN VEERSEN and MEULENBERG [13] and despite their apparent simplicity they show a good correlation with plasticizer activity for nonpolymeric plasticizers. The parameter is defined as:

$$\Phi = \frac{M(A_p/P_o)}{1000} \qquad (3)$$

where
M = molecular mass of plasticizer
A_p = number of carbon atoms in the plasticizer, excluding aromatic and carboxylic acid carbon atoms
P_o = number of polar (e.g., carbonyl) groups present

The factor of 1000 is used to produce values of a convenient size.

The Solid-Gel Transition Temperature T_m is a measure of plasticizer activity. It is the temperature at which a single grain of PVC dissolves in excess plasticizer. More efficient plasticizers have lower values of T_m due to their higher solvating power.

The Flory – Huggins Interaction Parameter χ. These ideas, based on a study of polymer miscibility, were applied to plasticizers by ANAGNOSTOPOULOS [14] according to the equation:

$$1/T_m = 0.002226 + 0.1351(1-\chi)/V_1 \qquad (4)$$

where V_1 is the molar volume of the plasticizer, calculated from the molecular mass and the density at T_m.

The Activity Parameter α is based on work by BIGG [15]. It is another measure of plasticizer activity and is defined as

$$\alpha = 1000\frac{(1-\chi)}{M} \qquad (5)$$

Summaries. Over the past few years these methods have been assessed and extended by many workers, in particular the Loughborough group [16], [17]. It was shown that solubility parameters are capable of classifying plasticizers of a given family in terms of their compatibility with PVC but that they are of limited use for comparing plasticizers of different families (e.g., phthalates with adipates). Polarity parameters provided useful predictions of the activity of monomeric plasticizers but again were not able to compare activity of plasticizers from different families. In all cases it was not possible to predict the behavior of polymeric plasticizers.

Specific Interactions. Ideas on the subject of specific interactions between PVC and a plasticizer molecule, as a basis of plasticization, can be considered a more detailed form of some of the ideas discussed above. Clearly some mechanism of attraction and interaction between PVC and plasticizer must exist for plasticizer to be retained in the polymer after processing.

The role of specific interactions in the plasticization of PVC has been proposed from work on specific interactions of esters in solvents (e.g., hydrogenated chlorocarbons) [18], work on blends of polyesters with PVC [19–25], and work on plasticized PVC itself [26–28]. Modes of interaction between the carbonyl functionality of the plasticizer ester or polyester were proposed, mostly on the basis of results from Fourier transform IR spectroscopy (FTIR). Shifts in the absorption frequency of the carbonyl group of the plasticizer ester to lower wavenumber, indicative of a reduction in polarity (i.e., some interaction between this functionality and the polymer) were reported [25–27]. Work performed with dibutyl phthalate [27] suggested an optimum concentration at which such interactions were maximized. Spectral shifts were in the range 3–8 cm^{-1}. Similar shifts were also reported in blends of PVC with polyesters [19–25], again showing a concentration dependence of the shift to lower wavenumber of the ester carbonyl absorption frequency.

Recent Studies. Some recent work has extended these ideas by using new analytical techniques, in particular molecular modeling and solid-state NMR spectroscopy.

Molecular Modeling. The computer modeling of molecules is a rapidly growing branch of chemistry (→ Molecular Modeling) [29–31]. High-resolution graphics and fast computers allow the operator to build molecules in minimum energy configurations and view them in real time. This model can be constructed from crystallographic coordinates available from data bases or by simple intervention from the operator. Molecular mechanics or quantum mechanics programs are then used to arrive at a likely structure.

A range of plasticizer molecule models and a model for PVC were generated and energy minimized to observe their most stable conformations. Such models highlighted the free volume increase caused by the mobility of the plasticizer alkyl chains. More detailed models were also produced to concentrate on the polar region of the plasticizer and its possible mode of interaction with the polymer. These showed the expected repulsion between areas on the polymer and plasticizer of like charge as well as attraction between the negatively charged portions of the plasticizer and positively charged portions of the PVC.

Solid-State NMR Spectroscopy. Recent advances in technology have made the study of solids by NMR techniques of considerably greater ease than in previous years. For the accumulation of solid state ^{13}C NMR spectra, cross polarisation magic angle spinning (CP-MAS) can be utilized to significantly reduce signal broadening effects present in the solid state but not in the liquid state. The technique was used to study the molecular effects of plasticization by comparing spectral shifts of PVC and plasticizer under various degrees of processing. For PVC plasticized with DIDP two processing temperatures, 130 °C and 170 °C, were used, representing a low degree and high degree of plasticization, respectively. The comparison of the spectra showed no shift in the resonance frequency of the carbonyl group with processing temperature. The most significant difference in the two spectra was in the aliphatic carbon resonances. The spectra of the more plasticized sample showed resonance shifts and increased resolution for these carbon atoms. This again shows a strong dependence of successful plasticization on the conformation of the alkyl chains of the plasticizer ester (linked to the increased free volume).

4. Plasticized PVC

4.1. Applications Technology

Over 90 % of plasticizer sales by volume are into the PVC industry. The reason is that the benefits imparted by the plasticization of PVC are far greater than those imparted to other polymers. PVC stands alone among polymers in its ability both to accept and retain large concentrations of plasticizer. This is due to a morphological form consisting of highly amorphous, semicrystalline, and highly crystalline regions. Without the wide range of additives available (e.g., plasticizers, stabilizers, fillers, lubricants, pigments) PVC would be of little use. The development of PVC as a commodity polymer is fundamentally linked to the development of its additives.

The two principle types of PVC are suspension and emulsion PVC. The plasticizer applications technology associated with these two forms are different and are discussed separately here. Details of the polymerization techniques giving rise to these two distinct polymer types are discussed in [32], [32]; see also → Poly(Vinyl Chloride)).

4.1.1. Suspension PVC

PVCs produced by suspension polymerization have a relatively large particle size (typically 100–150 µm). Additionally, these particles are highly porous and are therefore able to absorb large amounts of liquid plasticizer during a formulation mixing cycle. Flexible PVC formulations based on a suspension polymer are typically processed by a dry-blend process, during which all formulation ingredients are heated (typically 70–110 °C) and intimately mixed to form a dry powder (PVC dry-blend) which contains all formulation ingredients. This dry-blend can be either stored or processed immediately. Processing of suspension resin formulations is performed by a variety of techniques such as extrusion, injection molding, and calendering to totally fuse the formulation ingredients and form the desired product.

4.1.2. Emulsion PVC

Emulsion PVC resins differ from those manufactured by the suspension process principally in terms of particle size but also in the presence of emulsifiers and surfactants not used in suspension polymerization. The particle size of an emulsion polymer (typically 1–2 µm) is much lower than that of a suspension resin. This low particle size imparts a lack of porosity to the resin and thus the mixing of formulation ingredients in a dry-blending process is not possible.

A typical flexible PVC formulation prepared with an emulsion polymer will be a liquid or paste (plastisol). This consequently leads to a different applications technology area. PVC plastisols can be fused by techniques such as spreading, rotational molding, and slush molding. Plastisols may also be semi-gelled (i.e., enough heat is imparted to convert the plastisol into a solid but without the full development of tensile properties brought about by complete fusion) for storage.

As the formulation ingredients in a plastisol are in the liquid form, the viscosity of the plastisol is of major importance, and the intrinsic viscosity of the plasticizer has a major contribution to the plastisol viscosity, as do the polymerization conditions of the resin. The desired plastisol viscosity can be obtained by careful selection of polymer, plasticizer, and other formulation ingredients but also by shear rate. PVC plastisols are either (1) pseudoplastic or shear-thinning (viscosity decreases with shear), (2) near-Newtonian (viscosity remains nearly constant with shear), or (3) dilatant (viscosity increases with shear). Choice of plastisol ingredients to give desired viscosity effects will depend on the application requirements of the plastisol.

4.2. Effect of Plasticizer Choice on the Properties of Flexible PVC

This section discusses the effect of plasticizer choice on a range of properties of flexible PVC articles. Certain properties are more important for some applications than others, and hence some plasticizers find more extensive use in some application areas than others. The PVC technologist must ascertain the most important properties for the application in question and then make the correct choice of plasticizer.

4.2.1. Plasticizer Efficiency

Plasticizer efficiency is a measure of the concentration of plasticizer required to impart a specified softness to PVC. The softness may be measured as a British Standard Softness (BSS) or a Shore hardness (BSS 35 is equivalent to a Shore A hardness of 80 — test method BS 2782). For a given acid constituent of plasticizer ester, plasticizer efficiency decreases as the

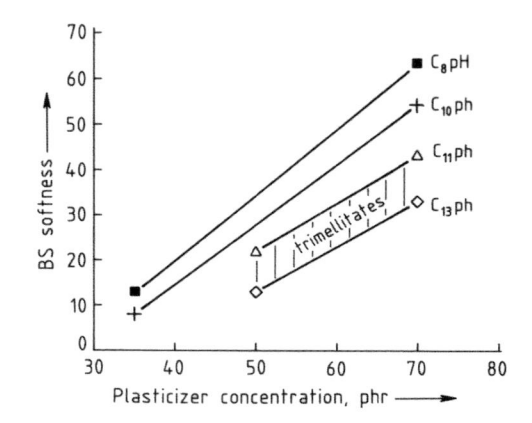

Figure 2. Relative efficiency of plasticizers
ph = phthalate

carbon number of the alcohol chain increases; e.g., for phthalate esters efficiency decreases in the order DBP > DIHP > DOP > DINP > DIDP > DTDP. For example, an additional six parts per hundred of PVC (phr) of DIDP than DOP is required to give a hardness of Shore 80 when all other formulation ingredients remain constant. In addition to the size of the alcohol chain, the amount of branching is also significant, with the more linear isomers being of greater efficiency. Choice of the acid constituent can also be significant. For the same alcohol constituent, phthalate and adipate esters are approximately equivalent but both are considerably more efficient than the corresponding trimellitate.

Figure 2 compares various plasticizers with respect to their ability to impart flexibility to a PVC resin. This shows the lower plasticizing efficiency of a C_{10} phthalate relative to a C_8 phthalate and also the lower plasticizing efficiency of the trimellitate esters compared to the phthalates.

4.2.2. High-Temperature Performance

High-temperature performance in flexible PVC is related to (1) plasticizer volatilization and (2) plasticizer degradation. Plasticizer volatilization, both from the finished article during use at elevated temperatures (e.g., electrical cable insulation) and also during processing (i.e., release of plasticizer fume) is directly related to the volatility of the plasticizer in use. Hence the higher molecular mass plasticizers give superior performance in this area. Figure 3

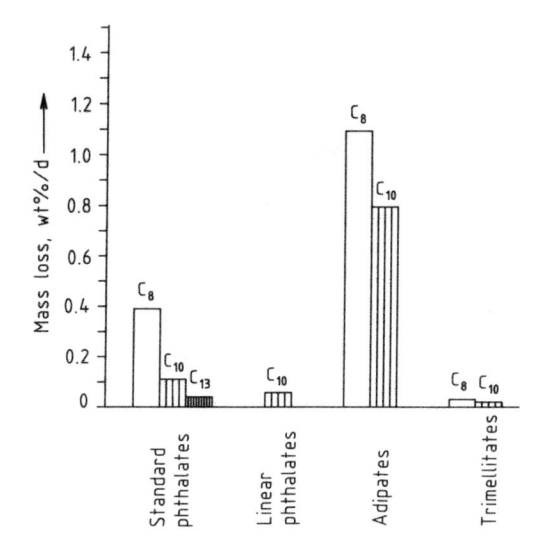

Figure 3. Volatile loss in use from plasticized PVC (50 phr plasticizer) for various plasticizers

shows the percentage mass loss from plasticized PVC (50 phr of plasticizer) for various plasticizers at 100 °C. The superior performance of the trimellitate esters and inferior performance of adipate esters, relative to the phthalate, is illustrated.

For phthalate plasticizers thermal stability decreases in the order DTDP > DIDP > DINP > DOP > DIHP > DBP. Higher molecular mass esters such as trimellitates improve this still further and are used in demanding cable specifications which have strict mass loss requirements. Polyester plasticizers give the best performance in this area, with performance improving with increasing molecular mass. Additionally, branched esters have somewhat higher volatilities than their linear equivalents.

High-temperature performance also includes the generation of fume in the workplace, for which the same structure relationships apply. Not only does excessive plasticizer volatilization have environmental consequences but inaccuracies in formulation can be incurred since not all the plasticizer enters the PVC resin, resulting in a harder material than calculated.

Plasticizer molecules can undergo thermal degradation at high temperature. Esters based on the more branched alcohol isomers are more susceptible to such degradation. This can, however, be offset by the incorporation of an antioxidant, and plasticizer esters for cable applications frequently contain a small amount of antioxidant.

4.2.3. Low-Temperature Performance

The ability of plasticized PVC to remain flexible at low temperature is of great importance in certain applications (e.g., external tarpaulins, underground cables). For this property the choice of the acid constituent of the plasticizer ester is of major importance, with the linear aliphatic adipic, sebacic, and azeleic acids giving excellent low-temperature flexibility compared to the corresponding phthalates and trimellitates.

There is also a major contribution to low-temperature performance from the alcohol portion of the ester; the greater the linearity of the plasticizer the greater the low-temperature flexibility. Figure 4 shows the cold flex temperature (the lowest temperature at which a sample of flexible PVC of standard dimensions can be twisted by 200° in the Clash and Berg test) for various PVC – plasticizer combinations. The increasing low-temperature performance with increasing alkyl chain length is illustrated, as is the superior low-temperature performance of the adipate esters.

4.2.4. Gelation Properties

The gelation characteristics of a plasticizer are related to its efficiency, and both properties are

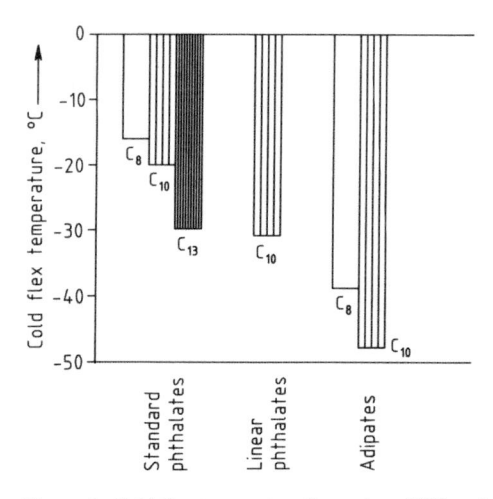

Figure 4. Cold flex temperature for various PVC – plasticizer combinations (50 phr plasticizer)

often discussed together. The gelation characteristics are a measure of the ability of a plasticizer to fuse with the polymer to give a product of maximum elongation and softness (i.e., maximum plasticization). Gelation properties are often measured either (1) as a processing temperature — the temperature to which the plasticizer and polymer must be heated in order to obtain these properties or (2) as a solution temperature, the temperature at which a grain of polymer dissolves in excess plasticizer, giving a measure of the solvating power of the plasticizer. Ease of gelation is related to plasticizer polarity and molecular size. The greater the polarity of a plasticizer molecule the greater the attraction it has for the PVC polymer chain and the less additional energy, in the form of heat, is required to cause maximum plasticizer – PVC interaction. The most active plasticizers are able to bring about these effects soon after the T_g of the polymer (70–80 °C) is reached, whereas the less active plasticizers require temperatures of ca. 180 °C for maximum elongation properties to be obtained. The polarity of the plasticizer is determined by both acid type and alcohol chain length. Aromatic acids, being of greater polarity, tend to show greater ease of gelation than aliphatic acid esters.

Molecular size also has a key contribution and explains why molecules of similar polarity can show different gelation properties. The smaller the plasticizer molecule the easier it is for it to enter the PVC matrix; larger molecules require more thermal energy to establish the desired interaction with the polymer. Since branching influences molecular size, this too has a contribution to gelation properties, with the more branched isomers showing greater activity. Thus for the phthalate esters ease of gelation decreases in the order BBP > DBP > DIHP > DOP > DINP > DIDP > DTDP. Figure 5 shows the greater ability of the more molecularly compact plasticizers to impart full gelation properties (i.e., maximum tensile strength).

4.2.5. Migration and Extraction

When plasticized PVC comes into contact with other materials, plasticizer may migrate from the plasticized PVC into the other material. The rate of migration depends not only on the plasticizer employed but also on the nature of the contact material.

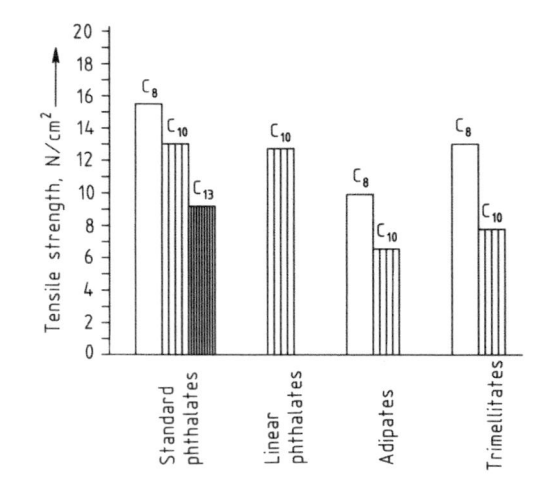

Figure 5. Tensile strength of various PVC – plasticizer combinations at 160 °C (50 phr plasticizer)

Plasticizer can also be extracted from PVC by a range of solvents, including water. How aggressive a particular solvent is depends on its molecular size and its compatibility with both the plasticizer and PVC. Water extracts plasticizer very slowly, oils are slightly more aggressive, and low molecular mass solvents are the most aggressive.

The key characteristics for migration and extraction resistance is molecular size. The larger the plasticizer the less it tends to migrate or be extracted. The extreme case is seen by the use of polymeric plasticizers in applications where excellent migration and extraction resistance is required. There is also a contribution from the linearity of the alcohol component of the plasticizer ester. The greater the linearity of the ester the greater its migration and extraction rate in comparison to the more branched isomers. Figure 6 shows the results of a standard test for the migration of plasticizer into rigid (unplasticized PVC). Equivalent masses of plaques containing different plasticizers were sandwiched between two rigid PVC discs and the plasticizer migration into the rigid PVC after the test was measured. The improved performance of the longer-chain esters and the trimellitate esters is illustrated.

4.2.6. Plastisol Viscosity and Viscosity Ageing

Plastisol viscosity and viscosity ageing are of great importance in emulsion PVC applications.

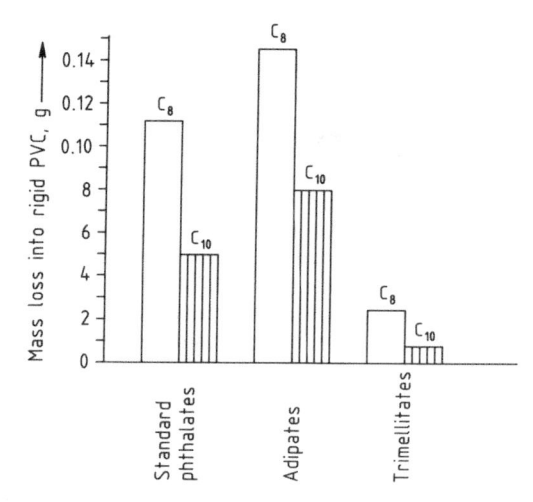

Figure 6. Mass loss into rigid (unplasticized) PVC (11d at 70 °C)

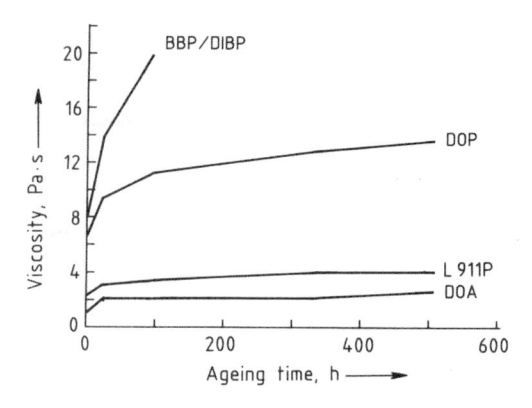

Figure 7. Viscosity ageing of PVC plastisols at 23 °C

Plastisol viscosity, for a given emulsion PVC resin, has a strong dependence on plasticizer viscosity, with the lower molecular mass and more linear esters showing the lowest viscosity and hence the lowest plastisol viscosity (i.e., plastisol viscosity, for a common set of other formulation ingredients, increases in the sequence DBP < DIHP < DOP < DINP < DIDP). In spite of these viscosity differences, however, if plastisols are formulated to equal softness, i.e., taking into account the efficiency of the plasticizers involved, more of the less efficient plasticizer must be employed to impart the same softness to the product being manufactured. The addition of this extra liquid to the plastisol may produce an equivalent viscosity to that of the plastisol with the less viscous plasticizer. Esters based on aliphatic acids, being of lower viscosity than the corresponding aromatic acids, show lower plastisol viscosities, and adipate esters have found widespread use in plastisol applications although to meet other requirements (e.g., volatility, gelation characteristics) they are often employed in a blend with other esters.

Plastisols are often mixed and then stored rather than processed immediately. It is of great importance in this case for the plasticizer to show little or no paste thickening action at the storage temperature, and clearly in this instance it is not advisable to use a plasticizer of too great an activity, since grain swelling, leading to plastisol viscosity increase, can occur at low temperatures for some active plasticizer systems. Figure 7 compares the ageing of plastisols based on various plasticizers. The molecularly compact plasticizers such as BBP and DIBP can enter the PVC matrix easily and cause a rapid increase in viscosity with time. The ageing characteristics improve as the polarity of the plasticizer is reduced. Greater linearity of the plasticizer (e.g., 911 phthalates) also improve this further, with adipates also showing very good viscosity stability.

4.2.7. Windscreen Fogging

The phenomenon of car windscreen fogging has been known for some time. Fogging is the condensation of volatile material on the car windscreen causing a decrease in visibility to the driver. Although this volatile material may arise from a variety of sources (e.g., exhaust fume sucked in through the ventilation system), material from inside the car (e.g., crash pads, rear shelves) may also contribute to windscreen fogging on account of the high temperatures which can arise in a car standing in sunlight. In the case of flexible PVC such a contribution may arise from emulsifiers in the polymer, stabilizers, and plasticizers. In each case manufacturers have studied their products in detail and now recommend low-fogging polymers, stabilizers, and plasticizers. Tests have been designed (e.g., DIN 75 201) to assess the fogging performance of both the PVC sheet and the raw materials used in its production. These tests involve heating a sheet of raw material for a specified period at a set temperature in an enclosed apparatus with a cooled glass plate

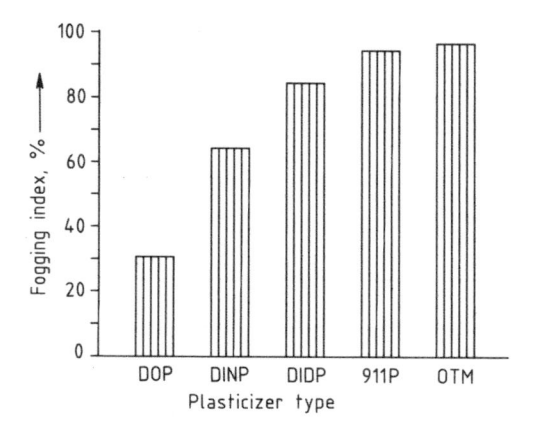

Figure 8. Fogging index of various plasticizers (6 h at 90 °C)

above the sheet or raw material. The reflectance of the glass plate before and after the test is then compared to ascertain the degree of fogging. Figure 8 shows the relationship between plasticizer molecular mass and linearity with fogging performance. The test was carried out at 90 °C for a 6 h duration (a variety of tests, each with their own specified temperature and time, are in use). In such a test, the fogging performance of a plasticizer is related to its (1) volatility, (2) refractive index, and (3) surface tension. The precise nature of (2) and (3) is somewhat complex [33] and attempts to improve the test are currently in progress. In the case of plasticizer volatility it is clear that a higher concentration of plasticizer leads to increased fogging in the test. The higher molecular mass and more linear plasticizers give superior performance, and trimellitate esters (in suspension PVC applications) and L911 phthalates, with their high degree of linearity, and consequently low viscosity for plastisol applications, have achieved widespread use as low-fogging plasticizers for these applications.

4.2.8. Summary

Table 2 gives a summary of the properties described above. For each criterion the effect of (1) increasing plasticizer concentration, (2) increasing the size of the plasticizer molecule, (3) increasing the linearity of the plasticizer molecule, and (4) changing the acid constituent of the ester is discussed. A tick indicates improved performance, a cross indicates poorer performance, and a dash indicate no effect; ticks and crosses in parentheses indicate that any changes tend to be marginal.

Table 2. Summary of the effect of plasticizer choice on the properties of flexible PVC

Property (at equal softness)	Increased concentration	Increased size	Increased linearity	Acid used*
Efficiency (higher)	—	✗	✓	ph = ad > tr
High-temperature properties	✗	✓	✓	tr > ph > ad
Low-temperature properties	✓	✓	✓	ad > ph = tr
Gelation temperature	(✓)	✗	(✗)	ph > ad ≥ tr
Migration/extraction	✗	✓	✗	tr > ph > ad
Plastisol viscosity	✓	(✓)	✓	ad > ph
Plastisol ageing	—	✓	(✓)	ad > ph
Fogging	✗	✓	✓	tr > ph > ad
Price	✓	✓	✗	ph > ad > tr

*ph = phthalate; ad = adipate; tr = trimellitate.

It can be seen from this section that there is clearly no perfect plasticizer for every application: choice will depend on the performance requirements of the article being manufactured and additionally price.

5. Plasticization of Polymers Other Than PVC

5.1. Requirements for Plasticization

The plasticization of PVC accounts for the vast majority of plasticizer sales. However, significant amounts of plasticizers are used in non-PVC polymers and this may become increasingly important in the future. Although PVC is unique in its ability to accept and retain large quantities of plasticizer, effective plasticization of other resins with slightly modified plasticizers may be possible if certain conditions, specific to the polymer, are met.

The factors which must be considered for the plasticization of a polymer are

1. The need — even though some polymers may be compatible with large concentrations of plasticizer the resultant softening may be of little use
2. Short- and long-term compatibility — the ability of a polymer to accept and retain the plasticizer

For a plasticizer to enter a polymer structure the polymer should be highly amorphous (e.g., crystalline nylon will retain only a small quantity of plasticizer if it retains its crystallinity). Once it has penetrated the polymer the plasticizer fills free volume and provides polymer chain lubrication, increasing rotation and movement.

The plasticizer content of a polymer may be increased by suppressing crystallization in the polymer, but if crystallization subsequently occurs the plasticizer is exuded. For highly crystalline resins, the small amounts of plasticizer allowable can change the nature of the small amorphous regions with a consequent change in properties.

Plasticizer compatibility has been studied in many resins and a summary of these studies is given in [4].

5.2. Plasticization of Other Polymers

Acrylic Resins. Although considerable information of the plasticization of acrylic resins is available, the subject is complicated by the fact that acrylic resins constitute a large family of polymers rather than a single polymeric species. An infinite variation in physical properties may be obtained through copolymerization of two or more acrylic monomers selected from the available esters of acrylic and methacrylic acid [34].

Plasticizers are used in the acrylics industry to produce tough, flexible coatings. Compatabilities with common plasticizers are up to 10 wt %, although in some cases (for low molecular mass plasticizers) it can be higher (e.g., a formulation of 100 phr PMMA, 150 phr DBP, 225 phr chalk, 25 phr resorcinol has been used [4]). PMMA is used in small amounts with PVC. Cast acrylics, however, require a high glass transition temperature and high rigidity, hence no plasticizer is required.

Listed plasticizers include all common phthalates and adipates.

There has been interest in the development of acrylic plastisols similar to those encountered with PVC. Clearly the same aspects of both plastisol viscosity and viscosity stability are important. Patents appear in the literature [35] indicating that the number of available plasticizers which show both good compatibility with arcrylic resins and satisfactory long term plastisol stability may be fewer than those showing equivalent properties with emulsion PVC resins.

Polyamides. The high degree of crystallinity of polyamides means that plasticization can occur only at very low levels. Plasticizers are used in nylon but are usually sulfonamide based since these are generally more compatible than phthalates (DEHP is 25 phr compatible, other phthalates less so; sulfonamides are compatible up to 50 phr).

Poly(ethylene terephthalate) (PETP) is crystalline and hence difficult to plasticize. Additionally, since PETP is used as high-strength film and textile fiber, plasticization is not usually required, although esters showing plasticizing properties with PVC may be used in small amounts as processing aids and external lubricants. Plasticizers have also been used to aid the injection molding of PETP, but only at low concentrations.

The main interest in plasticizers for PETP is in dyeing. Due to its lack of hydrogen bonds PETP is relatively difficult to dye. Plasticizers can increase the speed and intensity of dyeing. The compounds used, however, tend to be of low molecular mass since high volatility is required to enable rapid removal of plasticizer from the product.

Polyolefins. Interest has been shown in the plasticization of polyolefins [4], but their use generally results in a reduction of physical properties [16], and compatibility may be only 2 wt %. Most polyolefins give adequate physical properties without plasticization.

Plasticizers have been used with polypropylene to improve its elongation at break [15], although the addition of plasticizer can lower the glass transition temperature, room-temperature strength, and flow temperature. This can be overcome by simultaneous plasticization (ca. 15 wt %) and cross-linking. Plasticizers used include DOA.

Polystyrene is compatible with common plasticizers, but the modification of properties produced is of little value. Small amounts of plasticizer (e.g., DBP) are used as a processing aid.

Fluoroplastics. Conventional plasticizers are used as processing aids for fluoroplastics up to a concentration of 25 %. However, certain grades of Kel-F (polychlorotrifluoroethylene) contain up to 25 wt % plasticizer to improve elongation and increase softness. The plasticizers used are usually low molecular mass oily chloroethylene polymers [4].

Rubbers. Plasticizers have been used in rubber processing and formulations for many years [36], although phthalic and adipic esters have found little use since cheaper alternatives, e.g., heavy petroleum oils, coal tars, and other predominantly hydrocarbon products, are available for many types of rubber. Esters (e.g., DOA, DOP, DOS) can be used with latex rubber to produce large reductions in glass transition temperature. More polar elastomers such as nitrile rubber and chloroprene rubber are insufficiently compatible with hydrocarbons and require a more specialized type of plasticizer, e.g., a phthalate or adipate ester. Approximately 50 % of nitrile rubber used in Western Europe is plasticized at 10–15 phr (a total of $5-6 \times 10^3$ t/a), and 25 % of chloroprene at ca. 10 phr (ca. 2×10^3 t/a). Usage in other elastomers is very low but may increase due to toxicological concerns over polynuclear aromatic compounds.

Work on the use of high molecular mass esters in nitrile rubber has compared DINP with DBP and showed that at the 10 phr the nitrile rubber was effectively plasticized with DINP [37].

6. Toxicology and Environmental Aspects

6.1. Toxicity Studies on Plasticizer Esters

Acute Toxicity. The acute toxicity (LD_{50}) of plasticizers is extremely low (>2000 mg/kg). For example, DOP has an oral LD_{50} of >30 000 mg/kg.

In addition to their low acute toxicity many years of practical use coupled with animal tests show that plasticizers do not irritate the skin or mucous membranes.

Numerous long-term feeding trials on plasticizers have been carried out. The majority of these studies have been on di-2-ethylhexyl phthalate (DEHP, commonly called DOP) because it accounts for around 50 % of the plasticizer usage in Europe and has been considered as a model for the other phthalates. See also → Phthalic Acid and Derivatives, Section 6.2.5.).

Prior to 1980 such studies showed no adverse effects. However, in 1980 the results of a two-year feeding study carried out as part of the National Toxicology Program/National Cancer Institute (NTP/NCI) Bioassay Program in the United States indicated that DOP causes increased incidence of liver tumors in rats and mice and that di-2-ethylhexyl adipate (DEHA, commonly called DOA) had a similar effect in mice but not rats [38], [39]. In these studies the levels of plasticizers fed were very high, this being only possible because of their low acute toxicity. Their diets contained up to 12 000 ppm of DOP and 25 000 ppm of DOA (this is comparable with a human consuming 0.5 L of DOA every day).

A large number of more recent investigations [40–43] carried out in both Europe and the United States on a variety of plasticizers and different types of animals clearly show:

1. Plasticizers are not genotoxic.
2. Oral administration of plasticizers, fats, and other chemicals (such as hypolipidaemic drugs) to rodents causes a large increase of microbodies in the liver called peroxisomes. This peroxisome proliferation is considered by some authors to be linked to the formation of liver tumors.
3. The administration of plasticizers, fats, or hypolipidaemic drugs to nonrodent species such as the marmoset (a primate considered to be metabolically closer to humans) does not lead to peroxisome proliferation and liver damage. Some commonly used hypolipidaemic drugs (e.g., clofibrate), which cause peroxisome proliferation in rodents, have been used by humans for many years with no ill effects.
4. These species differences have also been observed in in vitro studies on the liver cells of rats, mice, guinea pigs, marmosets, and humans. Peroxisome proliferation was observed in the rat and mouse cells but not in those of humans, marmosets, or guinea pigs.
5. On the basis of these differences in species response (especially if the extremely high exposure in the animal tests is also taken into consideration) it is concluded that plasticizers pose no significant hazard to humans.

Human Exposure to Plasticizers. Experimental investigations and assessments in the United States and Europe have shown that the average human intake of plasticizers amounts to 2 g per person per year. This is mainly due to traces of DOA migrating from food packaging. The level of exposure is considerably below the dose at which toxic effects are to be expected. The no observed effect level (NOEL) for DOA and DOP is ca. 40 mg per kilogram body weight per day, or higher depending on the kind of effect, i.e., at least 1000 g per year for an average adult. There is thus a 500 fold safety factor between the estimated plasticizer intake of 2 g per person per year and the NOEL. Taking into consideration the difference in response between rodents and primates, the safety factor is in fact even greater.

The Opinion of Official Experts and Legislative Authorities. The IARC of the WHO has classified DOP as "an agent possibly carcinogenic to humans," largely because of the NTP/NCI rodent studies. However, IARC has not classified DOP as a carcinogen for regulatory purposes [44].

The EC Working Group "Classification and Labeling of Dangerous Substances" has examined all the toxicological data and has concluded that DOP should not be labeled as a human carcinogen [45].

In Germany the Advisory Body for Environmentally Relevant Existing Substances (BUA) has summed up the situation in their report No. 4 (1986) as follows [46]:

"Chronic damages have been noticed in rodents exposed to high DOP concentrations (rat: over 40–70 mg per kg bodyweight per day). The rat because of a characteristic feature in its DOP metabolism — different from the human metabolism — is especially sensitive to this substance. Thus, data obtained via tests with these animal species have no relevance for man. Tests with primates — having a metabolism comparable to that of humans — showed that DOP was rather ineffective. According to these findings there is no basis to suspect chronic damages due to DOP exposure at environmentally relevant concentrations."

Similarly the German Health Authority (BGA) has concluded that the carcinogenic effect of DOP in rats and mice is specific to this species and is not relevant to humans.

In the United States the FDA has not introduced any new regulations following the results of the NTP/NCI studies.

6.2. Plasticizers in the Environment

The release of plasticizers to the environment may occur during their production and distribution, during incorporation into PVC, and by loss from the finished article during use or after disposal.

The controlled nature of modern manufacturing processes makes it unlikely that significant loss of plasticizer to the environment occurs during its production (measurements of total emission from a modern plasticizer plant have given values of less than 0.01 %).

The extremely low solubility of plasticizers in water means that little is lost by leaching from flexible PVC either during use or after disposal. The increasing use of modern incineration techniques to dispose of domestic waste results in complete combustion of plasticizers to carbon dioxide and water.

The main way in which plasticizers enter the environment is therefore by evaporation during processing with PVC. The extent of such losses depends on the process and plasticizer used and has been estimated to vary from 0.02 % for injection molding up to 1 % for coating processes. However, these levels are being continually reduced by the installation of incineration, scrubbing, and filtration systems in processing plants.

Plasticizers can be detected in the environment in the ultratrace (parts per billion) range.

There are no indications of an accumulation of plasticizers in water, soil, or air because they are biologically and photochemically degraded. Degradation is particularly rapid under aerobic conditions to produce carbon dioxide and water as for other organic substances.

The occurrence and effects of phthalates in the environment has been reviewed in detail by ECETOC (European Chemical Industry Ecology and Toxicology Centre) [47]. It is clear from this report that in laboratory tests microorganisms and fish can take up dialkyl phthalates; however, they are rapidly excreted and no toxic effects have been seen. The report concludes that at the levels corresponding to current use and disposal practices, phthalates do not represent a hazard to the environment.

Health Aspects of Flexible PVC.

Small quantities of plasticizers may be lost from PVC by evaporation and extraction or by migration into other materials.

Plasticizer Evaporation From Flexible PVC. The use of plasticizers in the production of vinyl wall and floor covering can lead to plasticizer vapor being present in room air. However, the plasticizers used have extremely low vapor pressures and thus at 25 °C the maximum concentration of DOP is only ca. 0.01 mg/m^3. The concentration of the higher molecular mass plasticizers are even lower. Also, given the comparatively slow plasticizer release rate and normal ventilation, then the plasticizer concentration in reality is extremely low. The high concentrations sometimes reported must therefore be attributed to analytical errors. This is demonstrated by the fact that recent measurements of the air in an emission chamber containing 1 m^2 of PVC flooring for 96 h showed no detectable levels of plasticizer; the detection limit was 4 parts per billion.

Flexible PVC Toys. In the United States the Consumer Products Safety Commission (CPSC) has expressed concern that children might ingest plasticizers by sucking teething rings and pacifiers. The estimated intake of plasticizers by infants in this way varies over several orders of magnitude up to a maximum of 0.07 mg kg^{-1} d^{-1}. On the basis of the toxicological evidence this does not represent a health risk. For other toys, such as balls and dolls, the intake of plasticizers is negligible. The use of plasticizers in PVC toys for infants is regulated in some countries.

Medical Uses of Plasticized PVC. Flexible PVC has been established for many years as one of the most important materials used in the manufacture of a wide variety of medical products. The main advantages of PVC are its low toxicity and its flexibility, clarity, and sealing properties. These permit the manufacture of lightweight, breakage-resistant medical devices that can be supplied ready for use in sterile packages.

It is known that low levels of DOP migrate from flexible PVC blood bags into blood. There is published evidence [48] showing that this has a beneficial effect in that it significantly reduces the destruction of red cell walls compared to that which occurs on storage in other plastic or glass systems. DOP is the only plasticizer authorized by the European Pharmacopoeia for use in the manufacture of PVC blood bags.

Plasticized PVC in Food Packaging. Flexible PVC is used to produce food wrap film, cap seals for bottles and jars, and tubing for beverages. DOP, DINP, and DIDP are used in caps and seals, whereas DOA and polymeric adipates are used in the production of food wrap film.

DOA is used in this application because it gives a film with a high degree of cling which remains flexible at low temperatures. It also provides the high permeability to oxygen and water vapor that is necessary to ensure good food storage.

The migration of plasticizer from these packaging materials into foodstuffs has been measured in the UK by the Ministry of Agriculture Fisheries and Food (MAFF) both for domestic and retail packaging of food. These measurements together with a knowledge of the typical diet has enabled MAFF to assess the maximum possible intake of plasticizers as 0.02 mg per person per day of phthalates and 8 mg per person per day of DOA [49]. These estimates are based on a worst case situation, which assumes that all solid foodstuffs are purchased wrapped in cling film and that 50 % of these are subsequently rewrapped in the home.

The UK Committee on Toxicity of Chemicals in Food, Consumer Products, and the Environment (COT) has examined these data together with the relevant toxicological information and has concluded [50]:

"Toxic effects have been demonstrated for some plasticizers, but the safety margins between toxic doses and human intakes are large and we consider it very unlikely that there are any adverse health effects from the current or past use of plasticizers in food packaging materials."

Recent CEFIC studies [51], [52] involving the administration to humans of deuterium-labeled DOA have established the major human urinary metabolites and their excretion rates. This, together with analysis of 24 h urine samples from a limited survey of the UK population, has shown the average DOA intake to be 3 mg per person per day. This is well below the tolerable daily intake of 18 mg per person proposed by the EC Scientific Committee for Food and hence again confirms the suitability of plasticized PVC film for all food contact applications.

Regulations and Provisions. In terms of handling and storage, plasticizers are in general classified as nonhazardous.

In different countries the recommended 8 h TWA limit for occupational exposure varies between 2 and 10 mg/m^3.

These recommended exposure limits have been reviewed by the Health and Safety Authorities in the United Kingdom and Germany, and in the light of recent toxicological findings they have confirmed that the current limits of 5 and 10 mg/m^3, respectively, are satisfactory.

The regulations controlling the use of plasticizers in food contact applications differ from country to country. However, a common feature of the regulations is that plasticizer that migrates into foodstuffs should not have any adverse effects on their taste or smell and in general must not exceed 60 ppm. Because of the large volume of toxicological data indicating that phthalates and adipates pose no health risk to humans, regulatory authorities in most countries allow their use in caps, seals, tubing, and wrapping films.

6.3. Storage and Handling

Plasticizer esters are relatively inert, thermally stable liquids with high flash points and low volatility. Consequently they can be stored safely in mild steel storage tanks or drums for extended periods of time. Exposure to high temperatures for extended periods, as encountered in drums in hot climates, is not recommended since it may lead to a deterioration in product quality with respect to color, odor, and electrical resistance.

References

1 J. K. Sears, J. R. Darby: *The Technology of Plasticizers*, Wiley and Sons, New York 1982.
2 D. L. Buszard in W. V. Titow (ed.): *PVC Technology*, 4th ed., Elsevier, Amsterdam 1984.
3 *Ullmann*, 4th ed., **24**, 349 – 380.
4 *Kirk-Othmer*, 3rd ed., **18**, 111.
5 M. Kampmann, *Kunststoffe* **79** (1989) no. 10, 897.
6 K. Weissermel, H.-J. Arpe, *Industrielle Organische Chemie*, Wiley-VCH, Weinheim, Germany 1998, p. 428.
7 A. K. Doolittle, *J. Poly. Sci.* **2** (1947) no. 1, 121.
8 A. Hartmann, *Kolloid Z.* **142** (1955) 123.
9 R. S. Barshtein, G. A. Kotylarevski, *Sov. Plast. (Engl. Transl.)* **7** (1966) 18.
10 R. A. Horsley in P. Morgan (ed.): *Progress in Plastics*, Iliffe and Sons, New York 1957, pp. 77–88.
11 A. Kirkpatrick, *J. Appl. Phys.* **11** (1940) 255.
12 P. A. Small, *J. Appl. Chem.* **3** (1953) 76.
13 G. J. Van Veersen, A. J. Meulenberg, *SPE Tech. Pap.* **18** (1972) 314.
14 C. E. Anagnostopoulos, A. Y. Coran, W. R. Gamrath, *J. Appl. Polym. Sci.* **4** (1960) 181.
15 D. C. H. Bigg, *J. Appl. Polym. Sci.* **19** (1975) 3119.

16. S. V. Patel, M. Gilbert, *Plast. Rubber Processing Appl.* **6** (1986) 321.
17. L. Ramos de Valle, M. Gilbert, *Plast Rubber Processing Appl.* **13** (1990) 151.
18. A. Garton, P. Cousin, R. E. Prud'Homme, *J. Polym. Sci., Polym. Phys. Ed.* **21** (1983) 2275.
19. M. M. Coleman, J. Zarian, *J. Polym. Sci. Polym. Phys. Ed.* **12** (1979) 837.
20. M. M. Coleman, D. F. Varnell, *J. Polym. Sci. Polym. Phys. Ed.* **18** (1980) 1403.
21. J. J. Schmidt, J. A. Gardella, L. Salvati, *Macromolecules* **22** (1989) 4489.
22. M. Aubin, Y. Bedard, M. F. Morrissette, R. E. Prud'Homme, *J. Polym. Sci. Polym. Phys. Ed.* **21** (1983) 233.
23. M. B. Clark, C. A. Burkhardt, J. A. Gardella, *Macromolecules* **22** (1989) 4495.
24. D. J. Walsh, S. Rostami, *Adv. Polym. Sci. Rev.* **70** (1985) 119.
25. D. F. Varnell, M. M. Coleman, *Polymer* **22** (1981) 1324.
26. D. L. Tabb, J. L. Koenig, *Macromolecules* **8** (1975) no. 6, 929.
27. M. Theodorou, B. Jasse, *J. Polym. Sci. Polym. Phys. Ed.* **21** (1983) 2263.
28. E. Benedetti et al., *J. Polym. Sci. Polym. Phys. Ed.* **23** (1985) 1187.
29. *Chemistry in Britain*, Special Issue on Computational Chemistry, Nov. 1990.
30. *Chemistry and Industry*, Dec. 1990.
31. J. P. Sibilia: *A Guide to Materials Characterisation and Chemical Analysis*.VCH, Weinheim, Germany 1988.
32. M. J. Bunten, M. W. Newman, P. V. Smallwood, R. C. Stephenson: *Encyclopedia of Polymer Science and Engineering*, vol. **17 and suppl.**, John Wiley and Sons, New York 1989, pp. 241–392.
33. Fogging: A Problem that won't Go Away, *Proceedings of PRI Miniconference*, Oct. 1991, Plastics and Rubber Institute, London.
34. General Motors, US 3 178 386 (R. J. Hickman).
35. Teroson, GB 1 516 510; US 4 210 567.
36. S. S. Kurtz, J. S. Sweely, W. J. Stout in P. F. Bruins (ed.): *Plasticizer Technology*, Reinhold Publishing, New York 1965.
37. N. J. Clatden, C. J. Howick, unpublished work.
38. National Toxicology Programme, publication no. NIH-1768 (1980).
39. National Toxicology Programme, publication no. 82–1773 (1982).
40. P. Reddy et al., *CRC Crit. Rev. Toxicol.* **13** (1983).
41. P. Reddy et al., *Nature* **283** (1983) 397.
42. P. Schmezer et al., *Carcinogenesis* **9** (1988) 37.
43. *Di-2-Ethylhexyl Phthalate: A Critical Review of the Available Toxicological Information*, CEFIC, Brussels 1985.
44. *IARC Monogr.* **29** (1982) 281.
45. *Official Journal of the European Communities* **L 222** (17/8/1990) 49.
46. *Di-2-Ethylhexyl Phthalate*, BUA substance report no. 4, 1986.
47. *An Assessment of the Occurrence and Effects of Dialkyl Ortho-Phthalates in the Environment*, ECETOC technical report no. 19, Brussels 1985.
48. T. N. Estep et al., *Blood* **64** (1964) 1270.
49. Ministry of Agriculture, Fisheries and Food (UK), Plasticisers: Continuing Surveillance, *Food Surveillance Paper* no. 30, HMSO, London 1990.
50. Ministry of Agriculture, Fisheries and Food (UK), *Survey of Plasticizer Levels in Food Contact Materials in Food*, Surveillance Paper no. 21, HMSO, London 1987.
51. N. J. Loftus, W. J. Laird, J. E. Leeser, G. T. Stool, B. H. Woollen, *Hum. Exp. Toxicol.* **9** (1990) 326.
52. N. J. Loftus, W. J. Laird, M. F. Wilks, G. T. Steel, B. H. Woollen, British Toxicology Society Meeting, Edinburgh, September 1991.

Further Reading

D. F. Cadogan, C. J. Howick: *Plasticizers*, Kirk Othmer Encyclopedia of Chemical Technology, 5th edition, John Wiley & Sons, Hoboken, NJ, online DOI: 10.1002/0471238961.1612011903010415.a01.

C. A. Harper: *Handbook of Plastics Technologies*, McGraw-Hill, New York, NY 2006.

Y. Shashoua: *Conservation of Plastics*, Elsevier, Amsterdam 2008.